T0142363

Lecture Notes in Computer Science 13675

More information about this series at https://link.springer.com/bookseries/558

Shai Avidan · Gabriel Brostow ·
Moustapha Cissé · Giovanni Maria Farinella ·
Tal Hassner (Eds.)

Computer Vision – ECCV 2022

17th European Conference
Tel Aviv, Israel, October 23–27, 2022
Proceedings, Part XV

Editors
Shai Avidan
Tel Aviv University
Tel Aviv, Israel

Gabriel Brostow
University College London
London, UK

Moustapha Cissé
Google AI
Accra, Ghana

Giovanni Maria Farinella
University of Catania
Catania, Italy

Tal Hassner
Facebook (United States)
Menlo Park, CA, USA

ISSN 0302-9743 ISSN 1611-3349 (electronic)
Lecture Notes in Computer Science
ISBN 978-3-031-19783-3 ISBN 978-3-031-19784-0 (eBook)
https://doi.org/10.1007/978-3-031-19784-0

This Springer imprint is published by the registered company Springer Nature Switzerland AG
The registered company address is: Gewerbestrasse 11, 6330 Cham, Switzerland

Foreword

Organizing the European Conference on Computer Vision (ECCV 2022) in Tel-Aviv during a global pandemic was no easy feat. The uncertainty level was extremely high, and decisions had to be postponed to the last minute. Still, we managed to plan things just in time for ECCV 2022 to be held in person. Participation in physical events is crucial to stimulating collaborations and nurturing the culture of the Computer Vision community.

There were many people who worked hard to ensure attendees enjoyed the best science at the 16th edition of ECCV. We are grateful to the Program Chairs Gabriel Brostow and Tal Hassner, who went above and beyond to ensure the ECCV reviewing process ran smoothly. The scientific program includes dozens of workshops and tutorials in addition to the main conference and we would like to thank Leonid Karlinsky and Tomer Michaeli for their hard work. Finally, special thanks to the web chairs Lorenzo Baraldi and Kosta Derpanis, who put in extra hours to transfer information fast and efficiently to the ECCV community.

We would like to express gratitude to our generous sponsors and the Industry Chairs, Dimosthenis Karatzas and Chen Sagiv, who oversaw industry relations and proposed new ways for academia-industry collaboration and technology transfer. It's great to see so much industrial interest in what we're doing!

Authors' draft versions of the papers appeared online with open access on both the Computer Vision Foundation (CVF) and the European Computer Vision Association (ECVA) websites as with previous ECCVs. Springer, the publisher of the proceedings, has arranged for archival publication. The final version of the papers is hosted by SpringerLink, with active references and supplementary materials. It benefits all potential readers that we offer both a free and citeable version for all researchers, as well as an authoritative, citeable version for SpringerLink readers. Our thanks go to Ronan Nugent from Springer, who helped us negotiate this agreement. Last but not least, we wish to thank Eric Mortensen, our publication chair, whose expertise made the process smooth.

October 2022

Rita Cucchiara
Jiří Matas
Amnon Shashua
Lihi Zelnik-Manor

Foreword

Preface

Welcome to the proceedings of the European Conference on Computer Vision (ECCV 2022). This was a hybrid edition of ECCV as we made our way out of the COVID-19 pandemic. The conference received 5804 valid paper submissions, compared to 5150 submissions to ECCV 2020 (a 12.7% increase) and 2439 in ECCV 2018. 1645 submissions were accepted for publication (28%) and, of those, 157 (2.7% overall) as orals.

846 of the submissions were desk-rejected for various reasons. Many of them because they revealed author identity, thus violating the double-blind policy. This violation came in many forms: some had author names with the title, others added acknowledgments to specific grants, yet others had links to their github account where their name was visible. Tampering with the LaTeX template was another reason for automatic desk rejection.

ECCV 2022 used the traditional CMT system to manage the entire double-blind reviewing process. Authors did not know the names of the reviewers and vice versa. Each paper received at least 3 reviews (except 6 papers that received only 2 reviews), totalling more than 15,000 reviews.

Handling the review process at this scale was a significant challenge. To ensure that each submission received as fair and high-quality reviews as possible, we recruited more than 4719 reviewers (in the end, 4719 reviewers did at least one review). Similarly we recruited more than 276 area chairs (eventually, only 276 area chairs handled a batch of papers). The area chairs were selected based on their technical expertise and reputation, largely among people who served as area chairs in previous top computer vision and machine learning conferences (ECCV, ICCV, CVPR, NeurIPS, etc.).

Reviewers were similarly invited from previous conferences, and also from the pool of authors. We also encouraged experienced area chairs to suggest additional chairs and reviewers in the initial phase of recruiting. The median reviewer load was five papers per reviewer, while the average load was about four papers, because of the emergency reviewers. The area chair load was 35 papers, on average.

Conflicts of interest between authors, area chairs, and reviewers were handled largely automatically by the CMT platform, with some manual help from the Program Chairs. Reviewers were allowed to describe themselves as senior reviewer (load of 8 papers to review) or junior reviewers (load of 4 papers). Papers were matched to area chairs based on a subject-area affinity score computed in CMT and an affinity score computed by the Toronto Paper Matching System (TPMS). TPMS is based on the paper's full text. An area chair handling each submission would bid for preferred expert reviewers, and we balanced load and prevented conflicts.

The assignment of submissions to area chairs was relatively smooth, as was the assignment of submissions to reviewers. A small percentage of reviewers were not happy with their assignments in terms of subjects and self-reported expertise. This is an area for improvement, although it's interesting that many of these cases were reviewers hand-picked by AC's. We made a later round of reviewer recruiting, targeted at the list of authors of papers submitted to the conference, and had an excellent response which helped provide enough emergency reviewers. In the end, all but six papers received at least 3 reviews.

The challenges of the reviewing process are in line with past experiences at ECCV 2020. As the community grows, and the number of submissions increases, it becomes ever more challenging to recruit enough reviewers and ensure a high enough quality of reviews. Enlisting authors by default as reviewers might be one step to address this challenge.

Authors were given a week to rebut the initial reviews, and address reviewers' concerns. Each rebuttal was limited to a single pdf page with a fixed template.

The Area Chairs then led discussions with the reviewers on the merits of each submission. The goal was to reach consensus, but, ultimately, it was up to the Area Chair to make a decision. The decision was then discussed with a buddy Area Chair to make sure decisions were fair and informative. The entire process was conducted virtually with no in-person meetings taking place.

The Program Chairs were informed in cases where the Area Chairs overturned a decisive consensus reached by the reviewers, and pushed for the meta-reviews to contain details that explained the reasoning for such decisions. Obviously these were the most contentious cases, where reviewer inexperience was the most common reported factor.

Once the list of accepted papers was finalized and released, we went through the laborious process of plagiarism (including self-plagiarism) detection. A total of 4 accepted papers were rejected because of that.

Finally, we would like to thank our Technical Program Chair, Pavel Lifshits, who did tremendous work behind the scenes, and we thank the tireless CMT team.

October 2022

Gabriel Brostow
Giovanni Maria Farinella
Moustapha Cissé
Shai Avidan
Tal Hassner

Organization

Social and Student Activities Chairs

Tatiana Tommasi	Italian Institute of Technology, Italy
Sagie Benaim	University of Copenhagen, Denmark

Diversity and Inclusion Chairs

Xi Yin	Facebook AI Research, USA
Bryan Russell	Adobe, USA

Communications Chairs

Lorenzo Baraldi	University of Modena and Reggio Emilia, Italy
Kosta Derpanis	York University & Samsung AI Centre Toronto, Canada

Industrial Liaison Chairs

Dimosthenis Karatzas	Universitat Autònoma de Barcelona, Spain
Chen Sagiv	SagivTech, Israel

Finance Chair

Gerard Medioni	University of Southern California & Amazon, USA

Publication Chair

Eric Mortensen	MiCROTEC, USA

Area Chairs

Lourdes Agapito	University College London, UK
Zeynep Akata	University of Tübingen, Germany
Naveed Akhtar	University of Western Australia, Australia
Karteek Alahari	Inria Grenoble Rhône-Alpes, France
Alexandre Alahi	École polytechnique fédérale de Lausanne, Switzerland
Pablo Arbelaez	Universidad de Los Andes, Columbia
Antonis A. Argyros	University of Crete & Foundation for Research and Technology-Hellas, Crete
Yuki M. Asano	University of Amsterdam, The Netherlands
Kalle Åström	Lund University, Sweden
Hadar Averbuch-Elor	Cornell University, USA

Matthijs Douze	Facebook AI Research, USA
Mohamed Elhoseiny	King Abdullah University of Science and Technology, Saudi Arabia
Sergio Escalera	University of Barcelona, Spain
Yi Fang	New York University, USA
Ryan Farrell	Brigham Young University, USA
Alireza Fathi	Google, USA
Christoph Feichtenhofer	Facebook AI Research, USA
Basura Fernando	Agency for Science, Technology and Research (A*STAR), Singapore
Vittorio Ferrari	Google Research, Switzerland
Andrew W. Fitzgibbon	Graphcore, UK
David J. Fleet	University of Toronto, Canada
David Forsyth	University of Illinois at Urbana-Champaign, USA
David Fouhey	University of Michigan, USA
Katerina Fragkiadaki	Carnegie Mellon University, USA
Friedrich Fraundorfer	Graz University of Technology, Austria
Oren Freifeld	Ben-Gurion University, Israel
Thomas Funkhouser	Google Research & Princeton University, USA
Yasutaka Furukawa	Simon Fraser University, Canada
Fabio Galasso	Sapienza University of Rome, Italy
Jürgen Gall	University of Bonn, Germany
Chuang Gan	Massachusetts Institute of Technology, USA
Zhe Gan	Microsoft, USA
Animesh Garg	University of Toronto, Vector Institute, Nvidia, Canada
Efstratios Gavves	University of Amsterdam, The Netherlands
Peter Gehler	Amazon, Germany
Theo Gevers	University of Amsterdam, The Netherlands
Bernard Ghanem	King Abdullah University of Science and Technology, Saudi Arabia
Ross B. Girshick	Facebook AI Research, USA
Georgia Gkioxari	Facebook AI Research, USA
Albert Gordo	Facebook, USA
Stephen Gould	Australian National University, Australia
Venu Madhav Govindu	Indian Institute of Science, India
Kristen Grauman	Facebook AI Research & UT Austin, USA
Abhinav Gupta	Carnegie Mellon University & Facebook AI Research, USA
Mohit Gupta	University of Wisconsin-Madison, USA
Hu Han	Institute of Computing Technology, Chinese Academy of Sciences, China

Ivan Laptev	Inria Paris, France
Laura Leal-Taixé	Technical University of Munich, Germany
Erik Learned-Miller	University of Massachusetts, Amherst, USA
Gim Hee Lee	National University of Singapore, Singapore
Seungyong Lee	Pohang University of Science and Technology, Korea
Zhen Lei	Institute of Automation, Chinese Academy of Sciences, China
Bastian Leibe	RWTH Aachen University, Germany
Hongdong Li	Australian National University, Australia
Fuxin Li	Oregon State University, USA
Bo Li	University of Illinois at Urbana-Champaign, USA
Yin Li	University of Wisconsin-Madison, USA
Ser-Nam Lim	Meta AI Research, USA
Joseph Lim	University of Southern California, USA
Stephen Lin	Microsoft Research Asia, China
Dahua Lin	The Chinese University of Hong Kong, Hong Kong, China
Si Liu	Beihang University, China
Xiaoming Liu	Michigan State University, USA
Ce Liu	Microsoft, USA
Zicheng Liu	Microsoft, USA
Yanxi Liu	Pennsylvania State University, USA
Feng Liu	Portland State University, USA
Yebin Liu	Tsinghua University, China
Chen Change Loy	Nanyang Technological University, Singapore
Huchuan Lu	Dalian University of Technology, China
Cewu Lu	Shanghai Jiao Tong University, China
Oisin Mac Aodha	University of Edinburgh, UK
Dhruv Mahajan	Facebook, USA
Subhransu Maji	University of Massachusetts, Amherst, USA
Atsuto Maki	KTH Royal Institute of Technology, Sweden
Arun Mallya	NVIDIA, USA
R. Manmatha	Amazon, USA
Iacopo Masi	Sapienza University of Rome, Italy
Dimitris N. Metaxas	Rutgers University, USA
Ajmal Mian	University of Western Australia, Australia
Christian Micheloni	University of Udine, Italy
Krystian Mikolajczyk	Imperial College London, UK
Anurag Mittal	Indian Institute of Technology, Madras, India
Philippos Mordohai	Stevens Institute of Technology, USA
Greg Mori	Simon Fraser University & Borealis AI, Canada

Vittorio Murino	Istituto Italiano di Tecnologia, Italy
P. J. Narayanan	International Institute of Information Technology, Hyderabad, India
Ram Nevatia	University of Southern California, USA
Natalia Neverova	Facebook AI Research, UK
Richard Newcombe	Facebook, USA
Cuong V. Nguyen	Florida International University, USA
Bingbing Ni	Shanghai Jiao Tong University, China
Juan Carlos Niebles	Salesforce & Stanford University, USA
Ko Nishino	Kyoto University, Japan
Jean-Marc Odobez	Idiap Research Institute, École polytechnique fédérale de Lausanne, Switzerland
Francesca Odone	University of Genova, Italy
Takayuki Okatani	Tohoku University & RIKEN Center for Advanced Intelligence Project, Japan
Manohar Paluri	Facebook, USA
Guan Pang	Facebook, USA
Maja Pantic	Imperial College London, UK
Sylvain Paris	Adobe Research, USA
Jaesik Park	Pohang University of Science and Technology, Korea
Hyun Soo Park	The University of Minnesota, USA
Omkar M. Parkhi	Facebook, USA
Deepak Pathak	Carnegie Mellon University, USA
Georgios Pavlakos	University of California, Berkeley, USA
Marcello Pelillo	University of Venice, Italy
Marc Pollefeys	ETH Zurich & Microsoft, Switzerland
Jean Ponce	Inria, France
Gerard Pons-Moll	University of Tübingen, Germany
Fatih Porikli	Qualcomm, USA
Victor Adrian Prisacariu	University of Oxford, UK
Petia Radeva	University of Barcelona, Spain
Ravi Ramamoorthi	University of California, San Diego, USA
Deva Ramanan	Carnegie Mellon University, USA
Vignesh Ramanathan	Facebook, USA
Nalini Ratha	State University of New York at Buffalo, USA
Tammy Riklin Raviv	Ben-Gurion University, Israel
Tobias Ritschel	University College London, UK
Emanuele Rodola	Sapienza University of Rome, Italy
Amit K. Roy-Chowdhury	University of California, Riverside, USA
Michael Rubinstein	Google, USA
Olga Russakovsky	Princeton University, USA

Mathieu Salzmann	École polytechnique fédérale de Lausanne, Switzerland
Dimitris Samaras	Stony Brook University, USA
Aswin Sankaranarayanan	Carnegie Mellon University, USA
Imari Sato	National Institute of Informatics, Japan
Yoichi Sato	University of Tokyo, Japan
Shin'ichi Satoh	National Institute of Informatics, Japan
Walter Scheirer	University of Notre Dame, USA
Bernt Schiele	Max Planck Institute for Informatics, Germany
Konrad Schindler	ETH Zurich, Switzerland
Cordelia Schmid	Inria & Google, France
Alexander Schwing	University of Illinois at Urbana-Champaign, USA
Nicu Sebe	University of Trento, Italy
Greg Shakhnarovich	Toyota Technological Institute at Chicago, USA
Eli Shechtman	Adobe Research, USA
Humphrey Shi	University of Oregon & University of Illinois at Urbana-Champaign & Picsart AI Research, USA
Jianbo Shi	University of Pennsylvania, USA
Roy Shilkrot	Massachusetts Institute of Technology, USA
Mike Zheng Shou	National University of Singapore, Singapore
Kaleem Siddiqi	McGill University, Canada
Richa Singh	Indian Institute of Technology Jodhpur, India
Greg Slabaugh	Queen Mary University of London, UK
Cees Snoek	University of Amsterdam, The Netherlands
Yale Song	Facebook AI Research, USA
Yi-Zhe Song	University of Surrey, UK
Bjorn Stenger	Rakuten Institute of Technology
Abby Stylianou	Saint Louis University, USA
Akihiro Sugimoto	National Institute of Informatics, Japan
Chen Sun	Brown University, USA
Deqing Sun	Google, USA
Kalyan Sunkavalli	Adobe Research, USA
Ying Tai	Tencent YouTu Lab, China
Ayellet Tal	Technion – Israel Institute of Technology, Israel
Ping Tan	Simon Fraser University, Canada
Siyu Tang	ETH Zurich, Switzerland
Chi-Keung Tang	Hong Kong University of Science and Technology, Hong Kong, China
Radu Timofte	University of Würzburg, Germany & ETH Zurich, Switzerland
Federico Tombari	Google, Switzerland & Technical University of Munich, Germany

James Tompkin	Brown University, USA
Lorenzo Torresani	Dartmouth College, USA
Alexander Toshev	Apple, USA
Du Tran	Facebook AI Research, USA
Anh T. Tran	VinAI, Vietnam
Zhuowen Tu	University of California, San Diego, USA
Georgios Tzimiropoulos	Queen Mary University of London, UK
Jasper Uijlings	Google Research, Switzerland
Jan C. van Gemert	Delft University of Technology, The Netherlands
Gul Varol	Ecole des Ponts ParisTech, France
Nuno Vasconcelos	University of California, San Diego, USA
Mayank Vatsa	Indian Institute of Technology Jodhpur, India
Ashok Veeraraghavan	Rice University, USA
Jakob Verbeek	Facebook AI Research, France
Carl Vondrick	Columbia University, USA
Ruiping Wang	Institute of Computing Technology, Chinese Academy of Sciences, China
Xinchao Wang	National University of Singapore, Singapore
Liwei Wang	The Chinese University of Hong Kong, Hong Kong, China
Chaohui Wang	Université Paris-Est, France
Xiaolong Wang	University of California, San Diego, USA
Christian Wolf	NAVER LABS Europe, France
Tao Xiang	University of Surrey, UK
Saining Xie	Facebook AI Research, USA
Cihang Xie	University of California, Santa Cruz, USA
Zeki Yalniz	Facebook, USA
Ming-Hsuan Yang	University of California, Merced, USA
Angela Yao	National University of Singapore, Singapore
Shaodi You	University of Amsterdam, The Netherlands
Stella X. Yu	University of California, Berkeley, USA
Junsong Yuan	State University of New York at Buffalo, USA
Stefanos Zafeiriou	Imperial College London, UK
Amir Zamir	École polytechnique fédérale de Lausanne, Switzerland
Lei Zhang	Alibaba & Hong Kong Polytechnic University, Hong Kong, China
Lei Zhang	International Digital Economy Academy (IDEA), China
Pengchuan Zhang	Meta AI, USA
Bolei Zhou	University of California, Los Angeles, USA
Yuke Zhu	University of Texas at Austin, USA

Todd Zickler	Harvard University, USA
Wangmeng Zuo	Harbin Institute of Technology, China

Technical Program Committee

Davide Abati
Soroush Abbasi Koohpayegani
Amos L. Abbott
Rameen Abdal
Rabab Abdelfattah
Sahar Abdelnabi
Hassan Abu Alhaija
Abulikemu Abuduweili
Ron Abutbul
Hanno Ackermann
Aikaterini Adam
Kamil Adamczewski
Ehsan Adeli
Vida Adeli
Donald Adjeroh
Arman Afrasiyabi
Akshay Agarwal
Sameer Agarwal
Abhinav Agarwalla
Vaibhav Aggarwal
Sara Aghajanzadeh
Susmit Agrawal
Antonio Agudo
Touqeer Ahmad
Sk Miraj Ahmed
Chaitanya Ahuja
Nilesh A. Ahuja
Abhishek Aich
Shubhra Aich
Noam Aigerman
Arash Akbarinia
Peri Akiva
Derya Akkaynak
Emre Aksan
Arjun R. Akula
Yuval Alaluf
Stephan Alaniz
Paul Albert
Cenek Albl
Filippo Aleotti
Konstantinos P. Alexandridis
Motasem Alfarra
Mohsen Ali
Thiemo Alldieck
Hadi Alzayer
Liang An
Shan An
Yi An
Zhulin An
Dongsheng An
Jie An
Xiang An
Saket Anand
Cosmin Ancuti
Juan Andrade-Cetto
Alexander Andreopoulos
Bjoern Andres
Jerone T. A. Andrews
Shivangi Aneja
Anelia Angelova
Dragomir Anguelov
Rushil Anirudh
Oron Anschel
Rao Muhammad Anwer
Djamila Aouada
Evlampios Apostolidis
Srikar Appalaraju
Nikita Araslanov
Andre Araujo
Eric Arazo
Dawit Mureja Argaw
Anurag Arnab
Aditya Arora
Chetan Arora
Sunpreet S. Arora
Alexey Artemov
Muhammad Asad
Kumar Ashutosh
Sinem Aslan
Vishal Asnani
Mahmoud Assran
Amir Atapour-Abarghouei
Nikos Athanasiou
Ali Athar
ShahRukh Athar
Sara Atito
Souhaib Attaiki
Matan Atzmon
Mathieu Aubry
Nicolas Audebert
Tristan T. Aumentado-Armstrong
Melinos Averkiou
Yannis Avrithis
Stephane Ayache
Mehmet Aygün
Seyed Mehdi Ayyoubzadeh
Hossein Azizpour
George Azzopardi
Mallikarjun B. R.
Yunhao Ba
Abhishek Badki
Seung-Hwan Bae
Seung-Hwan Baek
Seungryul Baek
Piyush Nitin Bagad
Shai Bagon
Gaetan Bahl
Shikhar Bahl
Sherwin Bahmani
Haoran Bai
Lei Bai
Jiawang Bai
Haoyue Bai
Jinbin Bai
Xiang Bai
Xuyang Bai

Yang Bai
Yuanchao Bai
Ziqian Bai
Sungyong Baik
Kevin Bailly
Max Bain
Federico Baldassarre
Wele Gedara Chaminda Bandara
Biplab Banerjee
Pratyay Banerjee
Sandipan Banerjee
Jihwan Bang
Antyanta Bangunharcana
Aayush Bansal
Ankan Bansal
Siddhant Bansal
Wentao Bao
Zhipeng Bao
Amir Bar
Manel Baradad Jurjo
Lorenzo Baraldi
Danny Barash
Daniel Barath
Connelly Barnes
Ioan Andrei Bârsan
Steven Basart
Dina Bashkirova
Chaim Baskin
Peyman Bateni
Anil Batra
Sebastiano Battiato
Ardhendu Behera
Harkirat Behl
Jens Behley
Vasileios Belagiannis
Boulbaba Ben Amor
Emanuel Ben Baruch
Abdessamad Ben Hamza
Gil Ben-Artzi
Assia Benbihi
Fabian Benitez-Quiroz
Guy Ben-Yosef
Philipp Benz
Alexander W. Bergman
Urs Bergmann
Jesus Bermudez-Cameo
Stefano Berretti
Gedas Bertasius
Zachary Bessinger
Petra Bevandić
Matthew Beveridge
Lucas Beyer
Yash Bhalgat
Suvaansh Bhambri
Samarth Bharadwaj
Gaurav Bharaj
Aparna Bharati
Bharat Lal Bhatnagar
Uttaran Bhattacharya
Apratim Bhattacharyya
Brojeshwar Bhowmick
Ankan Kumar Bhunia
Ayan Kumar Bhunia
Qi Bi
Sai Bi
Michael Bi Mi
Gui-Bin Bian
Jia-Wang Bian
Shaojun Bian
Pia Bideau
Mario Bijelic
Hakan Bilen
Guillaume-Alexandre Bilodeau
Alexander Binder
Tolga Birdal
Vighnesh N. Birodkar
Sandika Biswas
Andreas Blattmann
Janusz Bobulski
Giuseppe Boccignone
Vishnu Boddeti
Navaneeth Bodla
Moritz Böhle
Aleksei Bokhovkin
Sam Bond-Taylor
Vivek Boominathan
Shubhankar Borse
Mark Boss
Andrea Bottino
Adnane Boukhayma
Fadi Boutros
Nicolas C. Boutry
Richard S. Bowen
Ivaylo Boyadzhiev
Aidan Boyd
Yuri Boykov
Aljaz Bozic
Behzad Bozorgtabar
Eric Brachmann
Samarth Brahmbhatt
Gustav Bredell
Francois Bremond
Joel Brogan
Andrew Brown
Thomas Brox
Marcus A. Brubaker
Robert-Jan Bruintjes
Yuqi Bu
Anders G. Buch
Himanshu Buckchash
Mateusz Buda
Ignas Budvytis
José M. Buenaposada
Marcel C. Bühler
Tu Bui
Adrian Bulat
Hannah Bull
Evgeny Burnaev
Andrei Bursuc
Benjamin Busam
Sergey N. Buzykanov
Wonmin Byeon
Fabian Caba
Martin Cadik
Guanyu Cai
Minjie Cai
Qing Cai
Zhongang Cai
Qi Cai
Yancheng Cai
Shen Cai
Han Cai
Jiarui Cai

Bowen Cai
Mu Cai
Qin Cai
Ruojin Cai
Weidong Cai
Weiwei Cai
Yi Cai
Yujun Cai
Zhiping Cai
Akin Caliskan
Lilian Calvet
Baris Can Cam
Necati Cihan Camgoz
Tommaso Campari
Dylan Campbell
Ziang Cao
Ang Cao
Xu Cao
Zhiwen Cao
Shengcao Cao
Song Cao
Weipeng Cao
Xiangyong Cao
Xiaochun Cao
Yue Cao
Yunhao Cao
Zhangjie Cao
Jiale Cao
Yang Cao
Jiajiong Cao
Jie Cao
Jinkun Cao
Lele Cao
Yulong Cao
Zhiguo Cao
Chen Cao
Razvan Caramalau
Marlène Careil
Gustavo Carneiro
Joao Carreira
Dan Casas
Paola Cascante-Bonilla
Angela Castillo
Francisco M. Castro
Pedro Castro
Luca Cavalli
George J. Cazenavette
Oya Celiktutan
Hakan Cevikalp
Sri Harsha C. H.
Sungmin Cha
Geonho Cha
Menglei Chai
Lucy Chai
Yuning Chai
Zenghao Chai
Anirban Chakraborty
Deep Chakraborty
Rudrasis Chakraborty
Souradeep Chakraborty
Kelvin C. K. Chan
Chee Seng Chan
Paramanand Chandramouli
Arjun Chandrasekaran
Kenneth Chaney
Dongliang Chang
Huiwen Chang
Peng Chang
Xiaojun Chang
Jia-Ren Chang
Hyung Jin Chang
Hyun Sung Chang
Ju Yong Chang
Li-Jen Chang
Qi Chang
Wei-Yi Chang
Yi Chang
Nadine Chang
Hanqing Chao
Pradyumna Chari
Dibyadip Chatterjee
Chiranjoy Chattopadhyay
Siddhartha Chaudhuri
Zhengping Che
Gal Chechik
Lianggangxu Chen
Qi Alfred Chen
Brian Chen
Bor-Chun Chen
Bo-Hao Chen
Bohong Chen
Bin Chen
Ziliang Chen
Cheng Chen
Chen Chen
Chaofeng Chen
Xi Chen
Haoyu Chen
Xuanhong Chen
Wei Chen
Qiang Chen
Shi Chen
Xianyu Chen
Chang Chen
Changhuai Chen
Hao Chen
Jie Chen
Jianbo Chen
Jingjing Chen
Jun Chen
Kejiang Chen
Mingcai Chen
Nenglun Chen
Qifeng Chen
Ruoyu Chen
Shu-Yu Chen
Weidong Chen
Weijie Chen
Weikai Chen
Xiang Chen
Xiuyi Chen
Xingyu Chen
Yaofo Chen
Yueting Chen
Yu Chen
Yunjin Chen
Yuntao Chen
Yun Chen
Zhenfang Chen
Zhuangzhuang Chen
Chu-Song Chen
Xiangyu Chen
Zhuo Chen
Chaoqi Chen
Shizhe Chen

Xiaotong Chen
Xiaozhi Chen
Dian Chen
Defang Chen
Dingfan Chen
Ding-Jie Chen
Ee Heng Chen
Tao Chen
Yixin Chen
Wei-Ting Chen
Lin Chen
Guang Chen
Guangyi Chen
Guanying Chen
Guangyao Chen
Hwann-Tzong Chen
Junwen Chen
Jiacheng Chen
Jianxu Chen
Hui Chen
Kai Chen
Kan Chen
Kevin Chen
Kuan-Wen Chen
Weihua Chen
Zhang Chen
Liang-Chieh Chen
Lele Chen
Liang Chen
Fanglin Chen
Zehui Chen
Minghui Chen
Minghao Chen
Xiaokang Chen
Qian Chen
Jun-Cheng Chen
Qi Chen
Qingcai Chen
Richard J. Chen
Runnan Chen
Rui Chen
Shuo Chen
Sentao Chen
Shaoyu Chen
Shixing Chen
Shuai Chen
Shuya Chen
Sizhe Chen
Simin Chen
Shaoxiang Chen
Zitian Chen
Tianlong Chen
Tianshui Chen
Min-Hung Chen
Xiangning Chen
Xin Chen
Xinghao Chen
Xuejin Chen
Xu Chen
Xuxi Chen
Yunlu Chen
Yanbei Chen
Yuxiao Chen
Yun-Chun Chen
Yi-Ting Chen
Yi-Wen Chen
Yinbo Chen
Yiran Chen
Yuanhong Chen
Yubei Chen
Yuefeng Chen
Yuhua Chen
Yukang Chen
Zerui Chen
Zhaoyu Chen
Zhen Chen
Zhenyu Chen
Zhi Chen
Zhiwei Chen
Zhixiang Chen
Long Chen
Bowen Cheng
Jun Cheng
Yi Cheng
Jingchun Cheng
Lechao Cheng
Xi Cheng
Yuan Cheng
Ho Kei Cheng
Kevin Ho Man Cheng
Jiacheng Cheng
Kelvin B. Cheng
Li Cheng
Mengjun Cheng
Zhen Cheng
Qingrong Cheng
Tianheng Cheng
Harry Cheng
Yihua Cheng
Yu Cheng
Ziheng Cheng
Soon Yau Cheong
Anoop Cherian
Manuela Chessa
Zhixiang Chi
Naoki Chiba
Julian Chibane
Kashyap Chitta
Tai-Yin Chiu
Hsu-kuang Chiu
Wei-Chen Chiu
Sungmin Cho
Donghyeon Cho
Hyeon Cho
Yooshin Cho
Gyusang Cho
Jang Hyun Cho
Seungju Cho
Nam Ik Cho
Sunghyun Cho
Hanbyel Cho
Jaesung Choe
Jooyoung Choi
Chiho Choi
Changwoon Choi
Jongwon Choi
Myungsub Choi
Dooseop Choi
Jonghyun Choi
Jinwoo Choi
Jun Won Choi
Min-Kook Choi
Hongsuk Choi
Janghoon Choi
Yoon-Ho Choi

Yukyung Choi
Jaegul Choo
Ayush Chopra
Siddharth Choudhary
Subhabrata Choudhury
Vasileios Choutas
Ka-Ho Chow
Pinaki Nath Chowdhury
Sammy Christen
Anders Christensen
Grigorios Chrysos
Hang Chu
Wen-Hsuan Chu
Peng Chu
Qi Chu
Ruihang Chu
Wei-Ta Chu
Yung-Yu Chuang
Sanghyuk Chun
Se Young Chun
Antonio Cinà
Ramazan Gokberk Cinbis
Javier Civera
Albert Clapés
Ronald Clark
Brian S. Clipp
Felipe Codevilla
Daniel Coelho de Castro
Niv Cohen
Forrester Cole
Maxwell D. Collins
Robert T. Collins
Marc Comino Trinidad
Runmin Cong
Wenyan Cong
Maxime Cordy
Marcella Cornia
Enric Corona
Huseyin Coskun
Luca Cosmo
Dragos Costea
Davide Cozzolino
Arun C. S. Kumar
Aiyu Cui
Qiongjie Cui
Quan Cui
Shuhao Cui
Yiming Cui
Ying Cui
Zijun Cui
Jiali Cui
Jiequan Cui
Yawen Cui
Zhen Cui
Zhaopeng Cui
Jack Culpepper
Xiaodong Cun
Ross Cutler
Adam Czajka
Ali Dabouei
Konstantinos M. Dafnis
Manuel Dahnert
Tao Dai
Yuchao Dai
Bo Dai
Mengyu Dai
Hang Dai
Haixing Dai
Peng Dai
Pingyang Dai
Qi Dai
Qiyu Dai
Yutong Dai
Naser Damer
Zhiyuan Dang
Mohamed Daoudi
Ayan Das
Abir Das
Debasmit Das
Deepayan Das
Partha Das
Sagnik Das
Soumi Das
Srijan Das
Swagatam Das
Avijit Dasgupta
Jim Davis
Adrian K. Davison
Homa Davoudi
Laura Daza
Matthias De Lange
Shalini De Mello
Marco De Nadai
Christophe De Vleeschouwer
Alp Dener
Boyang Deng
Congyue Deng
Bailin Deng
Yong Deng
Ye Deng
Zhuo Deng
Zhijie Deng
Xiaoming Deng
Jiankang Deng
Jinhong Deng
Jingjing Deng
Liang-Jian Deng
Siqi Deng
Xiang Deng
Xueqing Deng
Zhongying Deng
Karan Desai
Jean-Emmanuel Deschaud
Aniket Anand Deshmukh
Neel Dey
Helisa Dhamo
Prithviraj Dhar
Amaya Dharmasiri
Yan Di
Xing Di
Ousmane A. Dia
Haiwen Diao
Xiaolei Diao
Gonçalo José Dias Pais
Abdallah Dib
Anastasios Dimou
Changxing Ding
Henghui Ding
Guodong Ding
Yaqing Ding
Shuangrui Ding
Yuhang Ding
Yikang Ding
Shouhong Ding

Haisong Ding
Hui Ding
Jiahao Ding
Jian Ding
Jian-Jiun Ding
Shuxiao Ding
Tianyu Ding
Wenhao Ding
Yuqi Ding
Yi Ding
Yuzhen Ding
Zhengming Ding
Tan Minh Dinh
Vu Dinh
Christos Diou
Mandar Dixit
Bao Gia Doan
Khoa D. Doan
Dzung Anh Doan
Debi Prosad Dogra
Nehal Doiphode
Chengdong Dong
Bowen Dong
Zhenxing Dong
Hang Dong
Xiaoyi Dong
Haoye Dong
Jiangxin Dong
Shichao Dong
Xuan Dong
Zhen Dong
Shuting Dong
Jing Dong
Li Dong
Ming Dong
Nanqing Dong
Qiulei Dong
Runpei Dong
Siyan Dong
Tian Dong
Wei Dong
Xiaomeng Dong
Xin Dong
Xingbo Dong
Yuan Dong
Samuel Dooley
Gianfranco Doretto
Michael Dorkenwald
Keval Doshi
Zhaopeng Dou
Xiaotian Dou
Hazel Doughty
Ahmad Droby
Iddo Drori
Jie Du
Yong Du
Dawei Du
Dong Du
Ruoyi Du
Yuntao Du
Xuefeng Du
Yilun Du
Yuming Du
Radhika Dua
Haodong Duan
Jiafei Duan
Kaiwen Duan
Peiqi Duan
Ye Duan
Haoran Duan
Jiali Duan
Amanda Duarte
Abhimanyu Dubey
Shiv Ram Dubey
Florian Dubost
Lukasz Dudziak
Shivam Duggal
Justin M. Dulay
Matteo Dunnhofer
Chi Nhan Duong
Thibaut Durand
Mihai Dusmanu
Ujjal Kr Dutta
Debidatta Dwibedi
Isht Dwivedi
Sai Kumar Dwivedi
Takeharu Eda
Mark Edmonds
Alexei A. Efros
Thibaud Ehret
Max Ehrlich
Mahsa Ehsanpour
Iván Eichhardt
Farshad Einabadi
Marvin Eisenberger
Hazim Kemal Ekenel
Mohamed El Banani
Ismail Elezi
Moshe Eliasof
Alaa El-Nouby
Ian Endres
Francis Engelmann
Deniz Engin
Chanho Eom
Dave Epstein
Maria C. Escobar
Victor A. Escorcia
Carlos Esteves
Sungmin Eum
Bernard J. E. Evans
Ivan Evtimov
Fevziye Irem Eyiokur Yaman
Matteo Fabbri
Sébastien Fabbro
Gabriele Facciolo
Masud Fahim
Bin Fan
Hehe Fan
Deng-Ping Fan
Aoxiang Fan
Chen-Chen Fan
Qi Fan
Zhaoxin Fan
Haoqi Fan
Heng Fan
Hongyi Fan
Linxi Fan
Baojie Fan
Jiayuan Fan
Lei Fan
Quanfu Fan
Yonghui Fan
Yingruo Fan
Zhiwen Fan

Zicong Fan
Sean Fanello
Jiansheng Fang
Chaowei Fang
Yuming Fang
Jianwu Fang
Jin Fang
Qi Fang
Shancheng Fang
Tian Fang
Xianyong Fang
Gongfan Fang
Zhen Fang
Hui Fang
Jiemin Fang
Le Fang
Pengfei Fang
Xiaolin Fang
Yuxin Fang
Zhaoyuan Fang
Ammarah Farooq
Azade Farshad
Zhengcong Fei
Michael Felsberg
Wei Feng
Chen Feng
Fan Feng
Andrew Feng
Xin Feng
Zheyun Feng
Ruicheng Feng
Mingtao Feng
Qianyu Feng
Shangbin Feng
Chun-Mei Feng
Zunlei Feng
Zhiyong Feng
Martin Fergie
Mustansar Fiaz
Marco Fiorucci
Michael Firman
Hamed Firooz
Volker Fischer
Corneliu O. Florea
Georgios Floros
Wolfgang Foerstner
Gianni Franchi
Jean-Sebastien Franco
Simone Frintrop
Anna Fruehstueck
Changhong Fu
Chaoyou Fu
Cheng-Yang Fu
Chi-Wing Fu
Deqing Fu
Huan Fu
Jun Fu
Kexue Fu
Ying Fu
Jianlong Fu
Jingjing Fu
Qichen Fu
Tsu-Jui Fu
Xueyang Fu
Yang Fu
Yanwei Fu
Yonggan Fu
Wolfgang Fuhl
Yasuhisa Fujii
Kent Fujiwara
Marco Fumero
Takuya Funatomi
Isabel Funke
Dario Fuoli
Antonino Furnari
Matheus A. Gadelha
Akshay Gadi Patil
Adrian Galdran
Guillermo Gallego
Silvano Galliani
Orazio Gallo
Leonardo Galteri
Matteo Gamba
Yiming Gan
Sujoy Ganguly
Harald Ganster
Boyan Gao
Changxin Gao
Daiheng Gao
Difei Gao
Chen Gao
Fei Gao
Lin Gao
Wei Gao
Yiming Gao
Junyu Gao
Guangyu Ryan Gao
Haichang Gao
Hongchang Gao
Jialin Gao
Jin Gao
Jun Gao
Katelyn Gao
Mingchen Gao
Mingfei Gao
Pan Gao
Shangqian Gao
Shanghua Gao
Xitong Gao
Yunhe Gao
Zhanning Gao
Elena Garces
Nuno Cruz Garcia
Noa Garcia
Guillermo Garcia-Hernando
Isha Garg
Rahul Garg
Sourav Garg
Quentin Garrido
Stefano Gasperini
Kent Gauen
Chandan Gautam
Shivam Gautam
Paul Gay
Chunjiang Ge
Shiming Ge
Wenhang Ge
Yanhao Ge
Zheng Ge
Songwei Ge
Weifeng Ge
Yixiao Ge
Yuying Ge
Shijie Geng

Akshita Gupta
Ankush Gupta
Kamal Gupta
Kartik Gupta
Ritwik Gupta
Rohit Gupta
Siddharth Gururani
Fredrik K. Gustafsson
Abner Guzman Rivera
Vladimir Guzov
Matthew A. Gwilliam
Jung-Woo Ha
Marc Habermann
Isma Hadji
Christian Haene
Martin Hahner
Levente Hajder
Alexandros Haliassos
Emanuela Haller
Bumsub Ham
Abdullah J. Hamdi
Shreyas Hampali
Dongyoon Han
Chunrui Han
Dong-Jun Han
Dong-Sig Han
Guangxing Han
Zhizhong Han
Ruize Han
Jiaming Han
Jin Han
Ligong Han
Xian-Hua Han
Xiaoguang Han
Yizeng Han
Zhi Han
Zhenjun Han
Zhongyi Han
Jungong Han
Junlin Han
Kai Han
Kun Han
Sungwon Han
Songfang Han
Wei Han
Xiao Han
Xintong Han
Xinzhe Han
Yahong Han
Yan Han
Zongbo Han
Nicolai Hani
Rana Hanocka
Niklas Hanselmann
Nicklas A. Hansen
Hong Hanyu
Fusheng Hao
Yanbin Hao
Shijie Hao
Udith Haputhanthri
Mehrtash Harandi
Josh Harguess
Adam Harley
David M. Hart
Atsushi Hashimoto
Ali Hassani
Mohammed Hassanin
Yana Hasson
Joakim Bruslund Haurum
Bo He
Kun He
Chen He
Xin He
Fazhi He
Gaoqi He
Hao He
Haoyu He
Jiangpeng He
Hongliang He
Qian He
Xiangteng He
Xuming He
Yannan He
Yuhang He
Yang He
Xiangyu He
Nanjun He
Pan He
Sen He
Shengfeng He
Songtao He
Tao He
Tong He
Wei He
Xuehai He
Xiaoxiao He
Ying He
Yisheng He
Ziwen He
Peter Hedman
Felix Heide
Yacov Hel-Or
Paul Henderson
Philipp Henzler
Byeongho Heo
Jae-Pil Heo
Miran Heo
Sachini A. Herath
Stephane Herbin
Pedro Hermosilla Casajus
Monica Hernandez
Charles Herrmann
Roei Herzig
Mauricio Hess-Flores
Carlos Hinojosa
Tobias Hinz
Tsubasa Hirakawa
Chih-Hui Ho
Lam Si Tung Ho
Jennifer Hobbs
Derek Hoiem
Yannick Hold-Geoffroy
Aleksander Holynski
Cheeun Hong
Fa-Ting Hong
Hanbin Hong
Guan Zhe Hong
Danfeng Hong
Lanqing Hong
Xiaopeng Hong
Xin Hong
Jie Hong
Seungbum Hong
Cheng-Yao Hong
Seunghoon Hong

Yi Hong
Yuan Hong
Yuchen Hong
Anthony Hoogs
Maxwell C. Horton
Kazuhiro Hotta
Qibin Hou
Tingbo Hou
Junhui Hou
Ji Hou
Qiqi Hou
Rui Hou
Ruibing Hou
Zhi Hou
Henry Howard-Jenkins
Lukas Hoyer
Wei-Lin Hsiao
Chiou-Ting Hsu
Anthony Hu
Brian Hu
Yusong Hu
Hexiang Hu
Haoji Hu
Di Hu
Hengtong Hu
Haigen Hu
Lianyu Hu
Hanzhe Hu
Jie Hu
Junlin Hu
Shizhe Hu
Jian Hu
Zhiming Hu
Juhua Hu
Peng Hu
Ping Hu
Ronghang Hu
MengShun Hu
Tao Hu
Vincent Tao Hu
Xiaoling Hu
Xinting Hu
Xiaolin Hu
Xuefeng Hu
Xiaowei Hu
Yang Hu
Yueyu Hu
Zeyu Hu
Zhongyun Hu
Binh-Son Hua
Guoliang Hua
Yi Hua
Linzhi Huang
Qiusheng Huang
Bo Huang
Chen Huang
Hsin-Ping Huang
Ye Huang
Shuangping Huang
Zeng Huang
Buzhen Huang
Cong Huang
Heng Huang
Hao Huang
Qidong Huang
Huaibo Huang
Chaoqin Huang
Feihu Huang
Jiahui Huang
Jingjia Huang
Kun Huang
Lei Huang
Sheng Huang
Shuaiyi Huang
Siyu Huang
Xiaoshui Huang
Xiaoyang Huang
Yan Huang
Yihao Huang
Ying Huang
Ziling Huang
Xiaoke Huang
Yifei Huang
Haiyang Huang
Zhewei Huang
Jin Huang
Haibin Huang
Jiaxing Huang
Junjie Huang
Keli Huang
Lang Huang
Lin Huang
Luojie Huang
Mingzhen Huang
Shijia Huang
Shengyu Huang
Siyuan Huang
He Huang
Xiuyu Huang
Lianghua Huang
Yue Huang
Yaping Huang
Yuge Huang
Zehao Huang
Zeyi Huang
Zhiqi Huang
Zhongzhan Huang
Zilong Huang
Ziyuan Huang
Tianrui Hui
Zhuo Hui
Le Hui
Jing Huo
Junhwa Hur
Shehzeen S. Hussain
Chuong Minh Huynh
Seunghyun Hwang
Jaehui Hwang
Jyh-Jing Hwang
Sukjun Hwang
Soonmin Hwang
Wonjun Hwang
Rakib Hyder
Sangeek Hyun
Sarah Ibrahimi
Tomoki Ichikawa
Yerlan Idelbayev
A. S. M. Iftekhar
Masaaki Iiyama
Satoshi Ikehata
Sunghoon Im
Atul N. Ingle
Eldar Insafutdinov
Yani A. Ioannou
Radu Tudor Ionescu

Umar Iqbal
Go Irie
Muhammad Zubair Irshad
Ahmet Iscen
Berivan Isik
Ashraful Islam
Md Amirul Islam
Syed Islam
Mariko Isogawa
Vamsi Krishna K. Ithapu
Boris Ivanovic
Darshan Iyer
Sarah Jabbour
Ayush Jain
Nishant Jain
Samyak Jain
Vidit Jain
Vineet Jain
Priyank Jaini
Tomas Jakab
Mohammad A. A. K. Jalwana
Muhammad Abdullah Jamal
Hadi Jamali-Rad
Stuart James
Varun Jampani
Young Kyun Jang
YeongJun Jang
Yunseok Jang
Ronnachai Jaroensri
Bhavan Jasani
Krishna Murthy Jatavallabhula
Mojan Javaheripi
Syed A. Javed
Guillaume Jeanneret
Pranav Jeevan
Herve Jegou
Rohit Jena
Tomas Jenicek
Porter Jenkins
Simon Jenni
Hae-Gon Jeon
Sangryul Jeon
Boseung Jeong
Yoonwoo Jeong
Seong-Gyun Jeong
Jisoo Jeong
Allan D. Jepson
Ankit Jha
Sumit K. Jha
I-Hong Jhuo
Ge-Peng Ji
Chaonan Ji
Deyi Ji
Jingwei Ji
Wei Ji
Zhong Ji
Jiayi Ji
Pengliang Ji
Hui Ji
Mingi Ji
Xiaopeng Ji
Yuzhu Ji
Baoxiong Jia
Songhao Jia
Dan Jia
Shan Jia
Xiaojun Jia
Xiuyi Jia
Xu Jia
Menglin Jia
Wenqi Jia
Boyuan Jiang
Wenhao Jiang
Huaizu Jiang
Hanwen Jiang
Haiyong Jiang
Hao Jiang
Huajie Jiang
Huiqin Jiang
Haojun Jiang
Haobo Jiang
Junjun Jiang
Xingyu Jiang
Yangbangyan Jiang
Yu Jiang
Jianmin Jiang
Jiaxi Jiang
Jing Jiang
Kui Jiang
Li Jiang
Liming Jiang
Chiyu Jiang
Meirui Jiang
Chen Jiang
Peng Jiang
Tai-Xiang Jiang
Wen Jiang
Xinyang Jiang
Yifan Jiang
Yuming Jiang
Yingying Jiang
Zeren Jiang
ZhengKai Jiang
Zhenyu Jiang
Shuming Jiao
Jianbo Jiao
Licheng Jiao
Dongkwon Jin
Yeying Jin
Cheng Jin
Linyi Jin
Qing Jin
Taisong Jin
Xiao Jin
Xin Jin
Sheng Jin
Kyong Hwan Jin
Ruibing Jin
SouYoung Jin
Yueming Jin
Chenchen Jing
Longlong Jing
Taotao Jing
Yongcheng Jing
Younghyun Jo
Joakim Johnander
Jeff Johnson
Michael J. Jones
R. Kenny Jones
Rico Jonschkowski
Ameya Joshi
Sunghun Joung

Felix Juefei-Xu
Claudio R. Jung
Steffen Jung
Hari Chandana K.
Rahul Vigneswaran K.
Prajwal K. R.
Abhishek Kadian
Jhony Kaesemodel Pontes
Kumara Kahatapitiya
Anmol Kalia
Sinan Kalkan
Tarun Kalluri
Jaewon Kam
Sandesh Kamath
Meina Kan
Menelaos Kanakis
Takuhiro Kaneko
Di Kang
Guoliang Kang
Hao Kang
Jaeyeon Kang
Kyoungkook Kang
Li-Wei Kang
MinGuk Kang
Suk-Ju Kang
Zhao Kang
Yash Mukund Kant
Yueying Kao
Aupendu Kar
Konstantinos Karantzalos
Sezer Karaoglu
Navid Kardan
Sanjay Kariyappa
Leonid Karlinsky
Animesh Karnewar
Shyamgopal Karthik
Hirak J. Kashyap
Marc A. Kastner
Hirokatsu Kataoka
Angelos Katharopoulos
Hiroharu Kato
Kai Katsumata
Manuel Kaufmann
Chaitanya Kaul
Prakhar Kaushik
Yuki Kawana
Lei Ke
Lipeng Ke
Tsung-Wei Ke
Wei Ke
Petr Kellnhofer
Aniruddha Kembhavi
John Kender
Corentin Kervadec
Leonid Keselman
Daniel Keysers
Nima Khademi Kalantari
Taras Khakhulin
Samir Khaki
Muhammad Haris Khan
Qadeer Khan
Salman Khan
Subash Khanal
Vaishnavi M. Khindkar
Rawal Khirodkar
Saeed Khorram
Pirazh Khorramshahi
Kourosh Khoshelham
Ansh Khurana
Benjamin Kiefer
Jae Myung Kim
Junho Kim
Boah Kim
Hyeonseong Kim
Dong-Jin Kim
Dongwan Kim
Donghyun Kim
Doyeon Kim
Yonghyun Kim
Hyung-Il Kim
Hyunwoo Kim
Hyeongwoo Kim
Hyo Jin Kim
Hyunwoo J. Kim
Taehoon Kim
Jaeha Kim
Jiwon Kim
Jung Uk Kim
Kangyeol Kim
Eunji Kim
Daeha Kim
Dongwon Kim
Kunhee Kim
Kyungmin Kim
Junsik Kim
Min H. Kim
Namil Kim
Kookhoi Kim
Sanghyun Kim
Seongyeop Kim
Seungryong Kim
Saehoon Kim
Euyoung Kim
Guisik Kim
Sungyeon Kim
Sunnie S. Y. Kim
Taehun Kim
Tae Oh Kim
Won Hwa Kim
Seungwook Kim
YoungBin Kim
Youngeun Kim
Akisato Kimura
Furkan Osman Kınlı
Zsolt Kira
Hedvig Kjellström
Florian Kleber
Jan P. Klopp
Florian Kluger
Laurent Kneip
Byungsoo Ko
Muhammed Kocabas
A. Sophia Koepke
Kevin Koeser
Nick Kolkin
Nikos Kolotouros
Wai-Kin Adams Kong
Deying Kong
Caihua Kong
Youyong Kong
Shuyu Kong
Shu Kong
Tao Kong
Yajing Kong
Yu Kong

Zishang Kong
Theodora Kontogianni
Anton S. Konushin
Julian F. P. Kooij
Bruno Korbar
Giorgos Kordopatis-Zilos
Jari Korhonen
Adam Kortylewski
Denis Korzhenkov
Divya Kothandaraman
Suraj Kothawade
Iuliia Kotseruba
Satwik Kottur
Shashank Kotyan
Alexandros Kouris
Petros Koutras
Anna Kreshuk
Ranjay Krishna
Dilip Krishnan
Andrey Kuehlkamp
Hilde Kuehne
Jason Kuen
David Kügler
Arjan Kuijper
Anna Kukleva
Sumith Kulal
Viveka Kulharia
Akshay R. Kulkarni
Nilesh Kulkarni
Dominik Kulon
Abhinav Kumar
Akash Kumar
Suryansh Kumar
B. V. K. Vijaya Kumar
Pulkit Kumar
Ratnesh Kumar
Sateesh Kumar
Satish Kumar
Vijay Kumar B. G.
Nupur Kumari
Sudhakar Kumawat
Jogendra Nath Kundu
Hsien-Kai Kuo
Meng-Yu Jennifer Kuo
Vinod Kumar Kurmi
Yusuke Kurose
Keerthy Kusumam
Alina Kuznetsova
Henry Kvinge
Ho Man Kwan
Hyeokjun Kweon
Heeseung Kwon
Gihyun Kwon
Myung-Joon Kwon
Taesung Kwon
YoungJoong Kwon
Christos Kyrkou
Jorma Laaksonen
Yann Labbe
Zorah Laehner
Florent Lafarge
Hamid Laga
Manuel Lagunas
Shenqi Lai
Jian-Huang Lai
Zihang Lai
Mohamed I. Lakhal
Mohit Lamba
Meng Lan
Loic Landrieu
Zhiqiang Lang
Natalie Lang
Dong Lao
Yizhen Lao
Yingjie Lao
Issam Hadj Laradji
Gustav Larsson
Viktor Larsson
Zakaria Laskar
Stéphane Lathuilière
Chun Pong Lau
Rynson W. H. Lau
Hei Law
Justin Lazarow
Verica Lazova
Eric-Tuan Le
Hieu Le
Trung-Nghia Le
Mathias Lechner
Byeong-Uk Lee
Chen-Yu Lee
Che-Rung Lee
Chul Lee
Hong Joo Lee
Dongsoo Lee
Jiyoung Lee
Eugene Eu Tzuan Lee
Daeun Lee
Saehyung Lee
Jewook Lee
Hyungtae Lee
Hyunmin Lee
Jungbeom Lee
Joon-Young Lee
Jong-Seok Lee
Joonseok Lee
Junha Lee
Kibok Lee
Byung-Kwan Lee
Jangwon Lee
Jinho Lee
Jongmin Lee
Seunghyun Lee
Sohyun Lee
Minsik Lee
Dogyoon Lee
Seungmin Lee
Min Jun Lee
Sangho Lee
Sangmin Lee
Seungeun Lee
Seon-Ho Lee
Sungmin Lee
Sungho Lee
Sangyoun Lee
Vincent C. S. S. Lee
Jaeseong Lee
Yong Jae Lee
Chenyang Lei
Chenyi Lei
Jiahui Lei
Xinyu Lei
Yinjie Lei
Jiaxu Leng
Luziwei Leng

Jan E. Lenssen
Vincent Lepetit
Thomas Leung
María Leyva-Vallina
Xin Li
Yikang Li
Baoxin Li
Bin Li
Bing Li
Bowen Li
Changlin Li
Chao Li
Chongyi Li
Guanyue Li
Shuai Li
Jin Li
Dingquan Li
Dongxu Li
Yiting Li
Gang Li
Dian Li
Guohao Li
Haoang Li
Haoliang Li
Haoran Li
Hengduo Li
Huafeng Li
Xiaoming Li
Hanao Li
Hongwei Li
Ziqiang Li
Jisheng Li
Jiacheng Li
Jia Li
Jiachen Li
Jiahao Li
Jianwei Li
Jiazhi Li
Jie Li
Jing Li
Jingjing Li
Jingtao Li
Jun Li
Junxuan Li
Kai Li
Kailin Li
Kenneth Li
Kun Li
Kunpeng Li
Aoxue Li
Chenglong Li
Chenglin Li
Changsheng Li
Zhichao Li
Qiang Li
Yanyu Li
Zuoyue Li
Xiang Li
Xuelong Li
Fangda Li
Ailin Li
Liang Li
Chun-Guang Li
Daiqing Li
Dong Li
Guanbin Li
Guorong Li
Haifeng Li
Jianan Li
Jianing Li
Jiaxin Li
Ke Li
Lei Li
Lincheng Li
Liulei Li
Lujun Li
Linjie Li
Lin Li
Pengyu Li
Ping Li
Qiufu Li
Qingyong Li
Rui Li
Siyuan Li
Wei Li
Wenbin Li
Xiangyang Li
Xinyu Li
Xiujun Li
Xiu Li
Xu Li
Ya-Li Li
Yao Li
Yongjie Li
Yijun Li
Yiming Li
Yuezun Li
Yu Li
Yunheng Li
Yuqi Li
Zhe Li
Zeming Li
Zhen Li
Zhengqin Li
Zhimin Li
Jiefeng Li
Jinpeng Li
Chengze Li
Jianwu Li
Lerenhan Li
Shan Li
Suichan Li
Xiangtai Li
Yanjie Li
Yandong Li
Zhuoling Li
Zhenqiang Li
Manyi Li
Maosen Li
Ji Li
Minjun Li
Mingrui Li
Mengtian Li
Junyi Li
Nianyi Li
Bo Li
Xiao Li
Peihua Li
Peike Li
Peizhao Li
Peiliang Li
Qi Li
Ren Li
Runze Li
Shile Li

Sheng Li
Shigang Li
Shiyu Li
Shuang Li
Shasha Li
Shichao Li
Tianye Li
Yuexiang Li
Wei-Hong Li
Wanhua Li
Weihao Li
Weiming Li
Weixin Li
Wenbo Li
Wenshuo Li
Weijian Li
Yunan Li
Xirong Li
Xianhang Li
Xiaoyu Li
Xueqian Li
Xuanlin Li
Xianzhi Li
Yunqiang Li
Yanjing Li
Yansheng Li
Yawei Li
Yi Li
Yong Li
Yong-Lu Li
Yuhang Li
Yu-Jhe Li
Yuxi Li
Yunsheng Li
Yanwei Li
Zechao Li
Zejian Li
Zeju Li
Zekun Li
Zhaowen Li
Zheng Li
Zhenyu Li
Zhiheng Li
Zhi Li
Zhong Li
Zhuowei Li
Zhuowan Li
Zhuohang Li
Zizhang Li
Chen Li
Yuan-Fang Li
Dongze Lian
Xiaochen Lian
Zhouhui Lian
Long Lian
Qing Lian
Jin Lianbao
Jinxiu S. Liang
Dingkang Liang
Jiahao Liang
Jianming Liang
Jingyun Liang
Kevin J. Liang
Kaizhao Liang
Chen Liang
Jie Liang
Senwei Liang
Ding Liang
Jiajun Liang
Jian Liang
Kongming Liang
Siyuan Liang
Yuanzhi Liang
Zhengfa Liang
Mingfu Liang
Xiaodan Liang
Xuefeng Liang
Yuxuan Liang
Kang Liao
Liang Liao
Hong-Yuan Mark Liao
Wentong Liao
Haofu Liao
Yue Liao
Minghui Liao
Shengcai Liao
Ting-Hsuan Liao
Xin Liao
Yinghong Liao
Teck Yian Lim
Che-Tsung Lin
Chung-Ching Lin
Chen-Hsuan Lin
Cheng Lin
Chuming Lin
Chunyu Lin
Dahua Lin
Wei Lin
Zheng Lin
Huaijia Lin
Jason Lin
Jierui Lin
Jiaying Lin
Jie Lin
Kai-En Lin
Kevin Lin
Guangfeng Lin
Jiehong Lin
Feng Lin
Hang Lin
Kwan-Yee Lin
Ke Lin
Luojun Lin
Qinghong Lin
Xiangbo Lin
Yi Lin
Zudi Lin
Shijie Lin
Yiqun Lin
Tzu-Heng Lin
Ming Lin
Shaohui Lin
SongNan Lin
Ji Lin
Tsung-Yu Lin
Xudong Lin
Yancong Lin
Yen-Chen Lin
Yiming Lin
Yuewei Lin
Zhiqiu Lin
Zinan Lin
Zhe Lin
David B. Lindell
Zhixin Ling

Zhan Ling
Alexander Liniger
Venice Erin B. Liong
Joey Litalien
Or Litany
Roee Litman
Ron Litman
Jim Little
Dor Litvak
Shaoteng Liu
Shuaicheng Liu
Andrew Liu
Xian Liu
Shaohui Liu
Bei Liu
Bo Liu
Yong Liu
Ming Liu
Yanbin Liu
Chenxi Liu
Daqi Liu
Di Liu
Difan Liu
Dong Liu
Dongfang Liu
Daizong Liu
Xiao Liu
Fangyi Liu
Fengbei Liu
Fenglin Liu
Bin Liu
Yuang Liu
Ao Liu
Hong Liu
Hongfu Liu
Huidong Liu
Ziyi Liu
Feng Liu
Hao Liu
Jie Liu
Jialun Liu
Jiang Liu
Jing Liu
Jingya Liu
Jiaming Liu
Jun Liu
Juncheng Liu
Jiawei Liu
Hongyu Liu
Chuanbin Liu
Haotian Liu
Lingqiao Liu
Chang Liu
Han Liu
Liu Liu
Min Liu
Yingqi Liu
Aishan Liu
Bingyu Liu
Benlin Liu
Boxiao Liu
Chenchen Liu
Chuanjian Liu
Daqing Liu
Huan Liu
Haozhe Liu
Jiaheng Liu
Wei Liu
Jingzhou Liu
Jiyuan Liu
Lingbo Liu
Nian Liu
Peiye Liu
Qiankun Liu
Shenglan Liu
Shilong Liu
Wen Liu
Wenyu Liu
Weifeng Liu
Wu Liu
Xiaolong Liu
Yang Liu
Yanwei Liu
Yingcheng Liu
Yongfei Liu
Yihao Liu
Yu Liu
Yunze Liu
Ze Liu
Zhenhua Liu
Zhenguang Liu
Lin Liu
Lihao Liu
Pengju Liu
Xinhai Liu
Yunfei Liu
Meng Liu
Minghua Liu
Mingyuan Liu
Miao Liu
Peirong Liu
Ping Liu
Qingjie Liu
Ruoshi Liu
Risheng Liu
Songtao Liu
Xing Liu
Shikun Liu
Shuming Liu
Sheng Liu
Songhua Liu
Tongliang Liu
Weibo Liu
Weide Liu
Weizhe Liu
Wenxi Liu
Weiyang Liu
Xin Liu
Xiaobin Liu
Xudong Liu
Xiaoyi Liu
Xihui Liu
Xinchen Liu
Xingtong Liu
Xinpeng Liu
Xinyu Liu
Xianpeng Liu
Xu Liu
Xingyu Liu
Yongtuo Liu
Yahui Liu
Yangxin Liu
Yaoyao Liu
Yaojie Liu
Yuliang Liu

Gautam B. Machiraju
Spandan Madan
Mathew Magimai-Doss
Luca Magri
Behrooz Mahasseni
Upal Mahbub
Siddharth Mahendran
Paridhi Maheshwari
Rishabh Maheshwary
Mohammed Mahmoud
Shishira R. R. Maiya
Sylwia Majchrowska
Arjun Majumdar
Puspita Majumdar
Orchid Majumder
Sagnik Majumder
Ilya Makarov
Farkhod F. Makhmudkhujaev
Yasushi Makihara
Ankur Mali
Mateusz Malinowski
Utkarsh Mall
Srikanth Malla
Clement Mallet
Dimitrios Mallis
Yunze Man
Dipu Manandhar
Massimiliano Mancini
Murari Mandal
Raunak Manekar
Karttikeya Mangalam
Puneet Mangla
Fabian Manhardt
Sivabalan Manivasagam
Fahim Mannan
Chengzhi Mao
Hanzi Mao
Jiayuan Mao
Junhua Mao
Zhiyuan Mao
Jiageng Mao
Yunyao Mao
Zhendong Mao
Alberto Marchisio
Diego Marcos
Riccardo Marin
Aram Markosyan
Renaud Marlet
Ricardo Marques
Miquel Martí i Rabadán
Diego Martin Arroyo
Niki Martinel
Brais Martinez
Julieta Martinez
Marc Masana
Tomohiro Mashita
Timothée Masquelier
Minesh Mathew
Tetsu Matsukawa
Marwan Mattar
Bruce A. Maxwell
Christoph Mayer
Mantas Mazeika
Pratik Mazumder
Scott McCloskey
Steven McDonagh
Ishit Mehta
Jie Mei
Kangfu Mei
Jieru Mei
Xiaoguang Mei
Givi Meishvili
Luke Melas-Kyriazi
Iaroslav Melekhov
Andres Mendez-Vazquez
Heydi Mendez-Vazquez
Matias Mendieta
Ricardo A. Mendoza-León
Chenlin Meng
Depu Meng
Rang Meng
Zibo Meng
Qingjie Meng
Qier Meng
Yanda Meng
Zihang Meng
Thomas Mensink
Fabian Mentzer
Christopher Metzler
Gregory P. Meyer
Vasileios Mezaris
Liang Mi
Lu Mi
Bo Miao
Changtao Miao
Zichen Miao
Qiguang Miao
Xin Miao
Zhongqi Miao
Frank Michel
Simone Milani
Ben Mildenhall
Roy V. Miles
Juhong Min
Kyle Min
Hyun-Seok Min
Weiqing Min
Yuecong Min
Zhixiang Min
Qi Ming
David Minnen
Aymen Mir
Deepak Mishra
Anand Mishra
Shlok K. Mishra
Niluthpol Mithun
Gaurav Mittal
Trisha Mittal
Daisuke Miyazaki
Kaichun Mo
Hong Mo
Zhipeng Mo
Davide Modolo
Abduallah A. Mohamed
Mohamed Afham Mohamed Aflal
Ron Mokady
Pavlo Molchanov
Davide Moltisanti
Liliane Momeni
Gianluca Monaci
Pascal Monasse
Ajoy Mondal
Tom Monnier

Aron Monszpart
Gyeongsik Moon
Suhong Moon
Taesup Moon
Sean Moran
Daniel Moreira
Pietro Morerio
Alexandre Morgand
Lia Morra
Ali Mosleh
Inbar Mosseri
Sayed Mohammad Mostafavi Isfahani
Saman Motamed
Ramy A. Mounir
Fangzhou Mu
Jiteng Mu
Norman Mu
Yasuhiro Mukaigawa
Ryan Mukherjee
Tanmoy Mukherjee
Yusuke Mukuta
Ravi Teja Mullapudi
Lea Müller
Matthias Müller
Martin Mundt
Nils Murrugarra-Llerena
Damien Muselet
Armin Mustafa
Muhammad Ferjad Naeem
Sauradip Nag
Hajime Nagahara
Pravin Nagar
Rajendra Nagar
Naveen Shankar Nagaraja
Varun Nagaraja
Tushar Nagarajan
Seungjun Nah
Gaku Nakano
Yuta Nakashima
Giljoo Nam
Seonghyeon Nam
Liangliang Nan
Yuesong Nan
Yeshwanth Napolean
Dinesh Reddy Narapureddy
Medhini Narasimhan
Supreeth Narasimhaswamy
Sriram Narayanan
Erickson R. Nascimento
Varun Nasery
K. L. Navaneet
Pablo Navarrete Michelini
Shant Navasardyan
Shah Nawaz
Nihal Nayak
Farhood Negin
Lukáš Neumann
Alejandro Newell
Evonne Ng
Kam Woh Ng
Tony Ng
Anh Nguyen
Tuan Anh Nguyen
Cuong Cao Nguyen
Ngoc Cuong Nguyen
Thanh Nguyen
Khoi Nguyen
Phi Le Nguyen
Phong Ha Nguyen
Tam Nguyen
Truong Nguyen
Anh Tuan Nguyen
Rang Nguyen
Thao Thi Phuong Nguyen
Van Nguyen Nguyen
Zhen-Liang Ni
Yao Ni
Shijie Nie
Xuecheng Nie
Yongwei Nie
Weizhi Nie
Ying Nie
Yinyu Nie
Kshitij N. Nikhal
Simon Niklaus
Xuefei Ning
Jifeng Ning
Yotam Nitzan
Di Niu
Shuaicheng Niu
Li Niu
Wei Niu
Yulei Niu
Zhenxing Niu
Albert No
Shohei Nobuhara
Nicoletta Noceti
Junhyug Noh
Sotiris Nousias
Slawomir Nowaczyk
Ewa M. Nowara
Valsamis Ntouskos
Gilberto Ochoa-Ruiz
Ferda Ofli
Jihyong Oh
Sangyun Oh
Youngtaek Oh
Hiroki Ohashi
Takahiro Okabe
Kemal Oksuz
Fumio Okura
Daniel Olmeda Reino
Matthew Olson
Carl Olsson
Roy Or-El
Alessandro Ortis
Guillermo Ortiz-Jimenez
Magnus Oskarsson
Ahmed A. A. Osman
Martin R. Oswald
Mayu Otani
Naima Otberdout
Cheng Ouyang
Jiahong Ouyang
Wanli Ouyang
Andrew Owens
Poojan B. Oza
Mete Ozay
A. Cengiz Oztireli
Gautam Pai
Tomas Pajdla
Umapada Pal

Simone Palazzo
Luca Palmieri
Bowen Pan
Hao Pan
Lili Pan
Tai-Yu Pan
Liang Pan
Chengwei Pan
Yingwei Pan
Xuran Pan
Jinshan Pan
Xinyu Pan
Liyuan Pan
Xingang Pan
Xingjia Pan
Zhihong Pan
Zizheng Pan
Priyadarshini Panda
Rameswar Panda
Rohit Pandey
Kaiyue Pang
Bo Pang
Guansong Pang
Jiangmiao Pang
Meng Pang
Tianyu Pang
Ziqi Pang
Omiros Pantazis
Andreas Panteli
Maja Pantic
Marina Paolanti
Joao P. Papa
Samuele Papa
Mike Papadakis
Dim P. Papadopoulos
George Papandreou
Constantin Pape
Toufiq Parag
Chethan Parameshwara
Shaifali Parashar
Alejandro Pardo
Rishubh Parihar
Sarah Parisot
JaeYoo Park
Gyeong-Moon Park
Hyojin Park
Hyoungseob Park
Jongchan Park
Jae Sung Park
Kiru Park
Chunghyun Park
Kwanyong Park
Sunghyun Park
Sungrae Park
Seongsik Park
Sanghyun Park
Sungjune Park
Taesung Park
Gaurav Parmar
Paritosh Parmar
Alvaro Parra
Despoina Paschalidou
Or Patashnik
Shivansh Patel
Pushpak Pati
Prashant W. Patil
Vaishakh Patil
Suvam Patra
Jay Patravali
Badri Narayana Patro
Angshuman Paul
Sudipta Paul
Rémi Pautrat
Nick E. Pears
Adithya Pediredla
Wenjie Pei
Shmuel Peleg
Latha Pemula
Bo Peng
Houwen Peng
Yue Peng
Liangzu Peng
Baoyun Peng
Jun Peng
Pai Peng
Sida Peng
Xi Peng
Yuxin Peng
Songyou Peng
Wei Peng
Weiqi Peng
Wen-Hsiao Peng
Pramuditha Perera
Juan C. Perez
Eduardo Pérez Pellitero
Juan-Manuel Perez-Rua
Federico Pernici
Marco Pesavento
Stavros Petridis
Ilya A. Petrov
Vladan Petrovic
Mathis Petrovich
Suzanne Petryk
Hieu Pham
Quang Pham
Khoi Pham
Tung Pham
Huy Phan
Stephen Phillips
Cheng Perng Phoo
David Picard
Marco Piccirilli
Georg Pichler
A. J. Piergiovanni
Vipin Pillai
Silvia L. Pintea
Giovanni Pintore
Robinson Piramuthu
Fiora Pirri
Theodoros Pissas
Fabio Pizzati
Benjamin Planche
Bryan Plummer
Matteo Poggi
Ashwini Pokle
Georgy E. Ponimatkin
Adrian Popescu
Stefan Popov
Nikola Popović
Ronald Poppe
Angelo Porrello
Michael Potter
Charalambos Poullis
Hadi Pouransari
Omid Poursaeed

Shraman Pramanick
Mantini Pranav
Dilip K. Prasad
Meghshyam Prasad
B. H. Pawan Prasad
Shitala Prasad
Prateek Prasanna
Ekta Prashnani
Derek S. Prijatelj
Luke Y. Prince
Véronique Prinet
Victor Adrian Prisacariu
James Pritts
Thomas Probst
Sergey Prokudin
Rita Pucci
Chi-Man Pun
Matthew Purri
Haozhi Qi
Lu Qi
Lei Qi
Xianbiao Qi
Yonggang Qi
Yuankai Qi
Siyuan Qi
Guocheng Qian
Hangwei Qian
Qi Qian
Deheng Qian
Shengsheng Qian
Wen Qian
Rui Qian
Yiming Qian
Shengju Qian
Shengyi Qian
Xuelin Qian
Zhenxing Qian
Nan Qiao
Xiaotian Qiao
Jing Qin
Can Qin
Siyang Qin
Hongwei Qin
Jie Qin
Minghai Qin
Yipeng Qin
Yongqiang Qin
Wenda Qin
Xuebin Qin
Yuzhe Qin
Yao Qin
Zhenyue Qin
Zhiwu Qing
Heqian Qiu
Jiayan Qiu
Jielin Qiu
Yue Qiu
Jiaxiong Qiu
Zhongxi Qiu
Shi Qiu
Zhaofan Qiu
Zhongnan Qu
Yanyun Qu
Kha Gia Quach
Yuhui Quan
Ruijie Quan
Mike Rabbat
Rahul Shekhar Rade
Filip Radenovic
Gorjan Radevski
Bogdan Raducanu
Francesco Ragusa
Shafin Rahman
Md Mahfuzur Rahman Siddiquee
Hossein Rahmani
Kiran Raja
Sivaramakrishnan Rajaraman
Jathushan Rajasegaran
Adnan Siraj Rakin
Michaël Ramamonjisoa
Chirag A. Raman
Shanmuganathan Raman
Vignesh Ramanathan
Vasili Ramanishka
Vikram V. Ramaswamy
Merey Ramazanova
Jason Rambach
Sai Saketh Rambhatla
Clément Rambour
Ashwin Ramesh Babu
Adín Ramírez Rivera
Arianna Rampini
Haoxi Ran
Aakanksha Rana
Aayush Jung Bahadur Rana
Kanchana N. Ranasinghe
Aneesh Rangnekar
Samrudhdhi B. Rangrej
Harsh Rangwani
Viresh Ranjan
Anyi Rao
Yongming Rao
Carolina Raposo
Michalis Raptis
Amir Rasouli
Vivek Rathod
Adepu Ravi Sankar
Avinash Ravichandran
Bharadwaj Ravichandran
Dripta S. Raychaudhuri
Adria Recasens
Simon Reiß
Davis Rempe
Daxuan Ren
Jiawei Ren
Jimmy Ren
Sucheng Ren
Dayong Ren
Zhile Ren
Dongwei Ren
Qibing Ren
Pengfei Ren
Zhenwen Ren
Xuqian Ren
Yixuan Ren
Zhongzheng Ren
Ambareesh Revanur
Hamed Rezazadegan Tavakoli
Rafael S. Rezende
Wonjong Rhee
Alexander Richard

Christian Richardt
Stephan R. Richter
Benjamin Riggan
Dominik Rivoir
Mamshad Nayeem Rizve
Joshua D. Robinson
Joseph Robinson
Chris Rockwell
Ranga Rodrigo
Andres C. Rodriguez
Carlos Rodriguez-Pardo
Marcus Rohrbach
Gemma Roig
Yu Rong
David A. Ross
Mohammad Rostami
Edward Rosten
Karsten Roth
Anirban Roy
Debaditya Roy
Shuvendu Roy
Ahana Roy Choudhury
Aruni Roy Chowdhury
Denys Rozumnyi
Shulan Ruan
Wenjie Ruan
Patrick Ruhkamp
Danila Rukhovich
Anian Ruoss
Chris Russell
Dan Ruta
Dawid Damian Rymarczyk
DongHun Ryu
Hyeonggon Ryu
Kwonyoung Ryu
Balasubramanian S.
Alexandre Sablayrolles
Mohammad Sabokrou
Arka Sadhu
Aniruddha Saha
Oindrila Saha
Pritish Sahu
Aneeshan Sain
Nirat Saini
Saurabh Saini
Takeshi Saitoh
Christos Sakaridis
Fumihiko Sakaue
Dimitrios Sakkos
Ken Sakurada
Parikshit V. Sakurikar
Rohit Saluja
Nermin Samet
Leo Sampaio Ferraz Ribeiro
Jorge Sanchez
Enrique Sanchez
Shengtian Sang
Anush Sankaran
Soubhik Sanyal
Nikolaos Sarafianos
Vishwanath Saragadam
István Sárándi
Saquib Sarfraz
Mert Bulent Sariyildiz
Anindya Sarkar
Pritam Sarkar
Paul-Edouard Sarlin
Hiroshi Sasaki
Takami Sato
Torsten Sattler
Ravi Kumar Satzoda
Axel Sauer
Stefano Savian
Artem Savkin
Manolis Savva
Gerald Schaefer
Simone Schaub-Meyer
Yoni Schirris
Samuel Schulter
Katja Schwarz
Jesse Scott
Sinisa Segvic
Constantin Marc Seibold
Lorenzo Seidenari
Matan Sela
Fadime Sener
Paul Hongsuck Seo
Kwanggyoon Seo
Hongje Seong
Dario Serez
Francesco Setti
Bryan Seybold
Mohamad Shahbazi
Shima Shahfar
Xinxin Shan
Caifeng Shan
Dandan Shan
Shawn Shan
Wei Shang
Jinghuan Shang
Jiaxiang Shang
Lei Shang
Sukrit Shankar
Ken Shao
Rui Shao
Jie Shao
Mingwen Shao
Aashish Sharma
Gaurav Sharma
Vivek Sharma
Abhishek Sharma
Yoli Shavit
Shashank Shekhar
Sumit Shekhar
Zhijie Shen
Fengyi Shen
Furao Shen
Jialie Shen
Jingjing Shen
Ziyi Shen
Linlin Shen
Guangyu Shen
Biluo Shen
Falong Shen
Jiajun Shen
Qiu Shen
Qiuhong Shen
Shuai Shen
Wang Shen
Yiqing Shen
Yunhang Shen
Siqi Shen
Bin Shen
Tianwei Shen

Xi Shen
Yilin Shen
Yuming Shen
Yucong Shen
Zhiqiang Shen
Lu Sheng
Yichen Sheng
Shivanand Venkanna Sheshappanavar
Shelly Sheynin
Baifeng Shi
Ruoxi Shi
Botian Shi
Hailin Shi
Jia Shi
Jing Shi
Shaoshuai Shi
Baoguang Shi
Boxin Shi
Hengcan Shi
Tianyang Shi
Xiaodan Shi
Yongjie Shi
Zhensheng Shi
Yinghuan Shi
Weiqi Shi
Wu Shi
Xuepeng Shi
Xiaoshuang Shi
Yujiao Shi
Zenglin Shi
Zhenmei Shi
Takashi Shibata
Meng-Li Shih
Yichang Shih
Hyunjung Shim
Dongseok Shim
Soshi Shimada
Inkyu Shin
Jinwoo Shin
Seungjoo Shin
Seungjae Shin
Koichi Shinoda
Suprosanna Shit
Palaiahnakote Shivakumara
Eli Shlizerman
Gaurav Shrivastava
Xiao Shu
Xiangbo Shu
Xiujun Shu
Yang Shu
Tianmin Shu
Jun Shu
Zhixin Shu
Bing Shuai
Maria Shugrina
Ivan Shugurov
Satya Narayan Shukla
Pranjay Shyam
Jianlou Si
Yawar Siddiqui
Alberto Signoroni
Pedro Silva
Jae-Young Sim
Oriane Siméoni
Martin Simon
Andrea Simonelli
Abhishek Singh
Ashish Singh
Dinesh Singh
Gurkirt Singh
Krishna Kumar Singh
Mannat Singh
Pravendra Singh
Rajat Vikram Singh
Utkarsh Singhal
Dipika Singhania
Vasu Singla
Harsh Sinha
Sudipta Sinha
Josef Sivic
Elena Sizikova
Geri Skenderi
Ivan Skorokhodov
Dmitriy Smirnov
Cameron Y. Smith
James S. Smith
Patrick Snape
Mattia Soldan
Hyeongseok Son
Sanghyun Son
Chuanbiao Song
Chen Song
Chunfeng Song
Dan Song
Dongjin Song
Hwanjun Song
Guoxian Song
Jiaming Song
Jie Song
Liangchen Song
Ran Song
Luchuan Song
Xibin Song
Li Song
Fenglong Song
Guoli Song
Guanglu Song
Zhenbo Song
Lin Song
Xinhang Song
Yang Song
Yibing Song
Rajiv Soundararajan
Hossein Souri
Cristovao Sousa
Riccardo Spezialetti
Leonidas Spinoulas
Michael W. Spratling
Deepak Sridhar
Srinath Sridhar
Gaurang Sriramanan
Vinkle Kumar Srivastav
Themos Stafylakis
Serban Stan
Anastasis Stathopoulos
Markus Steinberger
Jan Steinbrener
Sinisa Stekovic
Alexandros Stergiou
Gleb Sterkin
Rainer Stiefelhagen
Pierre Stock

Ombretta Strafforello
Julian Straub
Yannick Strümpler
Joerg Stueckler
Hang Su
Weijie Su
Jong-Chyi Su
Bing Su
Haisheng Su
Jinming Su
Yiyang Su
Yukun Su
Yuxin Su
Zhuo Su
Zhaoqi Su
Xiu Su
Yu-Chuan Su
Zhixun Su
Arulkumar Subramaniam
Akshayvarun Subramanya
A. Subramanyam
Swathikiran Sudhakaran
Yusuke Sugano
Masanori Suganuma
Yumin Suh
Yang Sui
Baochen Sun
Cheng Sun
Long Sun
Guolei Sun
Haoliang Sun
Haomiao Sun
He Sun
Hanqing Sun
Hao Sun
Lichao Sun
Jiachen Sun
Jiaming Sun
Jian Sun
Jin Sun
Jennifer J. Sun
Tiancheng Sun
Libo Sun
Peize Sun
Qianru Sun
Shanlin Sun
Yu Sun
Zhun Sun
Che Sun
Lin Sun
Tao Sun
Yiyou Sun
Chunyi Sun
Chong Sun
Weiwei Sun
Weixuan Sun
Xiuyu Sun
Yanan Sun
Zeren Sun
Zhaodong Sun
Zhiqing Sun
Minhyuk Sung
Jinli Suo
Simon Suo
Abhijit Suprem
Anshuman Suri
Saksham Suri
Joshua M. Susskind
Roman Suvorov
Gurumurthy Swaminathan
Robin Swanson
Paul Swoboda
Tabish A. Syed
Richard Szeliski
Fariborz Taherkhani
Yu-Wing Tai
Keita Takahashi
Walter Talbott
Gary Tam
Masato Tamura
Feitong Tan
Fuwen Tan
Shuhan Tan
Andong Tan
Bin Tan
Cheng Tan
Jianchao Tan
Lei Tan
Mingxing Tan
Xin Tan
Zichang Tan
Zhentao Tan
Kenichiro Tanaka
Masayuki Tanaka
Yushun Tang
Hao Tang
Jingqun Tang
Jinhui Tang
Kaihua Tang
Luming Tang
Lv Tang
Sheyang Tang
Shitao Tang
Siliang Tang
Shixiang Tang
Yansong Tang
Keke Tang
Chang Tang
Chenwei Tang
Jie Tang
Junshu Tang
Ming Tang
Peng Tang
Xu Tang
Yao Tang
Chen Tang
Fan Tang
Haoran Tang
Shengeng Tang
Yehui Tang
Zhipeng Tang
Ugo Tanielian
Chaofan Tao
Jiale Tao
Junli Tao
Renshuai Tao
An Tao
Guanhong Tao
Zhiqiang Tao
Makarand Tapaswi
Jean-Philippe G. Tarel
Juan J. Tarrio
Enzo Tartaglione
Keisuke Tateno
Zachary Teed

Ajinkya B. Tejankar
Bugra Tekin
Purva Tendulkar
Damien Teney
Minggui Teng
Chris Tensmeyer
Andrew Beng Jin Teoh
Philipp Terhörst
Kartik Thakral
Nupur Thakur
Kevin Thandiackal
Spyridon Thermos
Diego Thomas
William Thong
Yuesong Tian
Guanzhong Tian
Lin Tian
Shiqi Tian
Kai Tian
Meng Tian
Tai-Peng Tian
Zhuotao Tian
Shangxuan Tian
Tian Tian
Yapeng Tian
Yu Tian
Yuxin Tian
Leslie Ching Ow Tiong
Praveen Tirupattur
Garvita Tiwari
George Toderici
Antoine Toisoul
Aysim Toker
Tatiana Tommasi
Zhan Tong
Alessio Tonioni
Alessandro Torcinovich
Fabio Tosi
Matteo Toso
Hugo Touvron
Quan Hung Tran
Son Tran
Hung Tran
Ngoc-Trung Tran
Vinh Tran
Phong Tran
Giovanni Trappolini
Edith Tretschk
Subarna Tripathi
Shubhendu Trivedi
Eduard Trulls
Prune Truong
Thanh-Dat Truong
Tomasz Trzcinski
Sam Tsai
Yi-Hsuan Tsai
Ethan Tseng
Yu-Chee Tseng
Shahar Tsiper
Stavros Tsogkas
Shikui Tu
Zhigang Tu
Zhengzhong Tu
Richard Tucker
Sergey Tulyakov
Cigdem Turan
Daniyar Turmukhambetov
Victor G. Turrisi da Costa
Bartlomiej Twardowski
Christopher D. Twigg
Radim Tylecek
Mostofa Rafid Uddin
Md. Zasim Uddin
Kohei Uehara
Nicolas Ugrinovic
Youngjung Uh
Norimichi Ukita
Anwaar Ulhaq
Devesh Upadhyay
Paul Upchurch
Yoshitaka Ushiku
Yuzuko Utsumi
Mikaela Angelina Uy
Mohit Vaishnav
Pratik Vaishnavi
Jeya Maria Jose Valanarasu
Matias A. Valdenegro Toro
Diego Valsesia
Wouter Van Gansbeke
Nanne van Noord
Simon Vandenhende
Farshid Varno
Cristina Vasconcelos
Francisco Vasconcelos
Alex Vasilescu
Subeesh Vasu
Arun Balajee Vasudevan
Kanav Vats
Vaibhav S. Vavilala
Sagar Vaze
Javier Vazquez-Corral
Andrea Vedaldi
Olga Veksler
Andreas Velten
Sai H. Vemprala
Raviteja Vemulapalli
Shashanka Venkataramanan
Dor Verbin
Luisa Verdoliva
Manisha Verma
Yashaswi Verma
Constantin Vertan
Eli Verwimp
Deepak Vijaykeerthy
Pablo Villanueva
Ruben Villegas
Markus Vincze
Vibhav Vineet
Minh P. Vo
Huy V. Vo
Duc Minh Vo
Tomas Vojir
Igor Vozniak
Nicholas Vretos
Vibashan VS
Tuan-Anh Vu
Thang Vu
Mårten Wadenbäck
Neal Wadhwa
Aaron T. Walsman
Steven Walton
Jin Wan
Alvin Wan
Jia Wan

Jun Wan
Xiaoyue Wan
Fang Wan
Guowei Wan
Renjie Wan
Zhiqiang Wan
Ziyu Wan
Bastian Wandt
Dongdong Wang
Limin Wang
Haiyang Wang
Xiaobing Wang
Angtian Wang
Angelina Wang
Bing Wang
Bo Wang
Boyu Wang
Binghui Wang
Chen Wang
Chien-Yi Wang
Congli Wang
Qi Wang
Chengrui Wang
Rui Wang
Yiqun Wang
Cong Wang
Wenjing Wang
Dongkai Wang
Di Wang
Xiaogang Wang
Kai Wang
Zhizhong Wang
Fangjinhua Wang
Feng Wang
Hang Wang
Gaoang Wang
Guoqing Wang
Guangcong Wang
Guangzhi Wang
Hanqing Wang
Hao Wang
Haohan Wang
Haoran Wang
Hong Wang
Haotao Wang
Hu Wang
Huan Wang
Hua Wang
Hui-Po Wang
Hengli Wang
Hanyu Wang
Hongxing Wang
Jingwen Wang
Jialiang Wang
Jian Wang
Jianyi Wang
Jiashun Wang
Jiahao Wang
Tsun-Hsuan Wang
Xiaoqian Wang
Jinqiao Wang
Jun Wang
Jianzong Wang
Kaihong Wang
Ke Wang
Lei Wang
Lingjing Wang
Linnan Wang
Lin Wang
Liansheng Wang
Mengjiao Wang
Manning Wang
Nannan Wang
Peihao Wang
Jiayun Wang
Pu Wang
Qiang Wang
Qiufeng Wang
Qilong Wang
Qiangchang Wang
Qin Wang
Qing Wang
Ruocheng Wang
Ruibin Wang
Ruisheng Wang
Ruizhe Wang
Runqi Wang
Runzhong Wang
Wenxuan Wang
Sen Wang
Shangfei Wang
Shaofei Wang
Shijie Wang
Shiqi Wang
Zhibo Wang
Song Wang
Xinjiang Wang
Tai Wang
Tao Wang
Teng Wang
Xiang Wang
Tianren Wang
Tiantian Wang
Tianyi Wang
Fengjiao Wang
Wei Wang
Miaohui Wang
Suchen Wang
Siyue Wang
Yaoming Wang
Xiao Wang
Ze Wang
Biao Wang
Chaofei Wang
Dong Wang
Gu Wang
Guangrun Wang
Guangming Wang
Guo-Hua Wang
Haoqing Wang
Hesheng Wang
Huafeng Wang
Jinghua Wang
Jingdong Wang
Jingjing Wang
Jingya Wang
Jingkang Wang
Jiakai Wang
Junke Wang
Kuo Wang
Lichen Wang
Lizhi Wang
Longguang Wang
Mang Wang
Mei Wang

Min Wang
Peng-Shuai Wang
Run Wang
Shaoru Wang
Shuhui Wang
Tan Wang
Tiancai Wang
Tianqi Wang
Wenhai Wang
Wenzhe Wang
Xiaobo Wang
Xiudong Wang
Xu Wang
Yajie Wang
Yan Wang
Yuan-Gen Wang
Yingqian Wang
Yizhi Wang
Yulin Wang
Yu Wang
Yujie Wang
Yunhe Wang
Yuxi Wang
Yaowei Wang
Yiwei Wang
Zezheng Wang
Hongzhi Wang
Zhiqiang Wang
Ziteng Wang
Ziwei Wang
Zheng Wang
Zhenyu Wang
Binglu Wang
Zhongdao Wang
Ce Wang
Weining Wang
Weiyao Wang
Wenbin Wang
Wenguan Wang
Guangting Wang
Haolin Wang
Haiyan Wang
Huiyu Wang
Naiyan Wang
Jingbo Wang
Jinpeng Wang
Jiaqi Wang
Liyuan Wang
Lizhen Wang
Ning Wang
Wenqian Wang
Sheng-Yu Wang
Weimin Wang
Xiaohan Wang
Yifan Wang
Yi Wang
Yongtao Wang
Yizhou Wang
Zhuo Wang
Zhe Wang
Xudong Wang
Xiaofang Wang
Xinggang Wang
Xiaosen Wang
Xiaosong Wang
Xiaoyang Wang
Lijun Wang
Xinlong Wang
Xuan Wang
Xue Wang
Yangang Wang
Yaohui Wang
Yu-Chiang Frank Wang
Yida Wang
Yilin Wang
Yi Ru Wang
Yali Wang
Yinglong Wang
Yufu Wang
Yujiang Wang
Yuwang Wang
Yuting Wang
Yang Wang
Yu-Xiong Wang
Yixu Wang
Ziqi Wang
Zhicheng Wang
Zeyu Wang
Zhaowen Wang
Zhenyi Wang
Zhenzhi Wang
Zhijie Wang
Zhiyong Wang
Zhongling Wang
Zhuowei Wang
Zian Wang
Zifu Wang
Zihao Wang
Zirui Wang
Ziyan Wang
Wenxiao Wang
Zhen Wang
Zhepeng Wang
Zi Wang
Zihao W. Wang
Steven L. Waslander
Olivia Watkins
Daniel Watson
Silvan Weder
Dongyoon Wee
Dongming Wei
Tianyi Wei
Jia Wei
Dong Wei
Fangyun Wei
Longhui Wei
Mingqiang Wei
Xinyue Wei
Chen Wei
Donglai Wei
Pengxu Wei
Xing Wei
Xiu-Shen Wei
Wenqi Wei
Guoqiang Wei
Wei Wei
XingKui Wei
Xian Wei
Xingxing Wei
Yake Wei
Yuxiang Wei
Yi Wei
Luca Weihs
Michael Weinmann
Martin Weinmann

Congcong Wen
Chuan Wen
Jie Wen
Sijia Wen
Song Wen
Chao Wen
Xiang Wen
Zeyi Wen
Xin Wen
Yilin Wen
Yijia Weng
Shuchen Weng
Junwu Weng
Wenming Weng
Renliang Weng
Zhenyu Weng
Xinshuo Weng
Nicholas J. Westlake
Gordon Wetzstein
Lena M. Widin Klasén
Rick Wildes
Bryan M. Williams
Williem Williem
Ole Winther
Scott Wisdom
Alex Wong
Chau-Wai Wong
Kwan-Yee K. Wong
Yongkang Wong
Scott Workman
Marcel Worring
Michael Wray
Safwan Wshah
Xiang Wu
Aming Wu
Chongruo Wu
Cho-Ying Wu
Chunpeng Wu
Chenyan Wu
Ziyi Wu
Fuxiang Wu
Gang Wu
Haiping Wu
Huisi Wu
Jane Wu
Jialian Wu
Jing Wu
Jinjian Wu
Jianlong Wu
Xian Wu
Lifang Wu
Lifan Wu
Minye Wu
Qianyi Wu
Rongliang Wu
Rui Wu
Shiqian Wu
Shuzhe Wu
Shangzhe Wu
Tsung-Han Wu
Tz-Ying Wu
Ting-Wei Wu
Jiannan Wu
Zhiliang Wu
Yu Wu
Chenyun Wu
Dayan Wu
Dongxian Wu
Fei Wu
Hefeng Wu
Jianxin Wu
Weibin Wu
Wenxuan Wu
Wenhao Wu
Xiao Wu
Yicheng Wu
Yuanwei Wu
Yu-Huan Wu
Zhenxin Wu
Zhenyu Wu
Wei Wu
Peng Wu
Xiaohe Wu
Xindi Wu
Xinxing Wu
Xinyi Wu
Xingjiao Wu
Xiongwei Wu
Yangzheng Wu
Yanzhao Wu
Yawen Wu
Yong Wu
Yi Wu
Ying Nian Wu
Zhenyao Wu
Zhonghua Wu
Zongze Wu
Zuxuan Wu
Stefanie Wuhrer
Teng Xi
Jianing Xi
Fei Xia
Haifeng Xia
Menghan Xia
Yuanqing Xia
Zhihua Xia
Xiaobo Xia
Weihao Xia
Shihong Xia
Yan Xia
Yong Xia
Zhaoyang Xia
Zhihao Xia
Chuhua Xian
Yongqin Xian
Wangmeng Xiang
Fanbo Xiang
Tiange Xiang
Tao Xiang
Liuyu Xiang
Xiaoyu Xiang
Zhiyu Xiang
Aoran Xiao
Chunxia Xiao
Fanyi Xiao
Jimin Xiao
Jun Xiao
Taihong Xiao
Anqi Xiao
Junfei Xiao
Jing Xiao
Liang Xiao
Yang Xiao
Yuting Xiao
Yijun Xiao

Yao Xiao
Zeyu Xiao
Zhisheng Xiao
Zihao Xiao
Binhui Xie
Christopher Xie
Haozhe Xie
Jin Xie
Guo-Sen Xie
Hongtao Xie
Ming-Kun Xie
Tingting Xie
Chaohao Xie
Weicheng Xie
Xudong Xie
Jiyang Xie
Xiaohua Xie
Yuan Xie
Zhenyu Xie
Ning Xie
Xianghui Xie
Xiufeng Xie
You Xie
Yutong Xie
Fuyong Xing
Yifan Xing
Zhen Xing
Yuanjun Xiong
Jinhui Xiong
Weihua Xiong
Hongkai Xiong
Zhitong Xiong
Yuanhao Xiong
Yunyang Xiong
Yuwen Xiong
Zhiwei Xiong
Yuliang Xiu
An Xu
Chang Xu
Chenliang Xu
Chengming Xu
Chenshu Xu
Xiang Xu
Huijuan Xu
Zhe Xu
Jie Xu
Jingyi Xu
Jiarui Xu
Yinghao Xu
Kele Xu
Ke Xu
Li Xu
Linchuan Xu
Linning Xu
Mengde Xu
Mengmeng Frost Xu
Min Xu
Mingye Xu
Jun Xu
Ning Xu
Peng Xu
Runsheng Xu
Sheng Xu
Wenqiang Xu
Xiaogang Xu
Renzhe Xu
Kaidi Xu
Yi Xu
Chi Xu
Qiuling Xu
Baobei Xu
Feng Xu
Haohang Xu
Haofei Xu
Lan Xu
Mingze Xu
Songcen Xu
Weipeng Xu
Wenjia Xu
Wenju Xu
Xiangyu Xu
Xin Xu
Yinshuang Xu
Yixing Xu
Yuting Xu
Yanyu Xu
Zhenbo Xu
Zhiliang Xu
Zhiyuan Xu
Xiaohao Xu
Yanwu Xu
Yan Xu
Yiran Xu
Yifan Xu
Yufei Xu
Yong Xu
Zichuan Xu
Zenglin Xu
Zexiang Xu
Zhan Xu
Zheng Xu
Zhiwei Xu
Ziyue Xu
Shiyu Xuan
Hanyu Xuan
Fei Xue
Jianru Xue
Mingfu Xue
Qinghan Xue
Tianfan Xue
Chao Xue
Chuhui Xue
Nan Xue
Zhou Xue
Xiangyang Xue
Yuan Xue
Abhay Yadav
Ravindra Yadav
Kota Yamaguchi
Toshihiko Yamasaki
Kohei Yamashita
Chaochao Yan
Feng Yan
Kun Yan
Qingsen Yan
Qixin Yan
Rui Yan
Siming Yan
Xinchen Yan
Yaping Yan
Bin Yan
Qingan Yan
Shen Yan
Shipeng Yan
Xu Yan

Sheng Ye
Nanyang Ye
Yufei Ye
Xiaoqing Ye
Ruolin Ye
Yousef Yeganeh
Chun-Hsiao Yeh
Raymond A. Yeh
Yu-Ying Yeh
Kai Yi
Chang Yi
Renjiao Yi
Xinping Yi
Peng Yi
Alper Yilmaz
Junho Yim
Hui Yin
Bangjie Yin
Jia-Li Yin
Miao Yin
Wenzhe Yin
Xuwang Yin
Ming Yin
Yu Yin
Aoxiong Yin
Kangxue Yin
Tianwei Yin
Wei Yin
Xianghua Ying
Rio Yokota
Tatsuya Yokota
Naoto Yokoya
Ryo Yonetani
Ki Yoon Yoo
Jinsu Yoo
Sunjae Yoon
Jae Shin Yoon
Jihun Yoon
Sung-Hoon Yoon
Ryota Yoshihashi
Yusuke Yoshiyasu
Chenyu You
Haoran You
Haoxuan You
Yang You
Quanzeng You
Tackgeun You
Kaichao You
Shan You
Xinge You
Yurong You
Baosheng Yu
Bei Yu
Haichao Yu
Hao Yu
Chaohui Yu
Fisher Yu
Jin-Gang Yu
Jiyang Yu
Jason J. Yu
Jiashuo Yu
Hong-Xing Yu
Lei Yu
Mulin Yu
Ning Yu
Peilin Yu
Qi Yu
Qian Yu
Rui Yu
Shuzhi Yu
Gang Yu
Tan Yu
Weijiang Yu
Xin Yu
Bingyao Yu
Ye Yu
Hanchao Yu
Yingchen Yu
Tao Yu
Xiaotian Yu
Qing Yu
Houjian Yu
Changqian Yu
Jing Yu
Jun Yu
Shujian Yu
Xiang Yu
Zhaofei Yu
Zhenbo Yu
Yinfeng Yu
Zhuoran Yu
Zitong Yu
Bo Yuan
Jiangbo Yuan
Liangzhe Yuan
Weihao Yuan
Jianbo Yuan
Xiaoyun Yuan
Ye Yuan
Li Yuan
Geng Yuan
Jialin Yuan
Maoxun Yuan
Peng Yuan
Xin Yuan
Yuan Yuan
Yuhui Yuan
Yixuan Yuan
Zheng Yuan
Mehmet Kerim Yücel
Kaiyu Yue
Haixiao Yue
Heeseung Yun
Sangdoo Yun
Tian Yun
Mahmut Yurt
Ekim Yurtsever
Ahmet Yüzügüler
Edouard Yvinec
Eloi Zablocki
Christopher Zach
Muhammad Zaigham Zaheer
Pierluigi Zama Ramirez
Yuhang Zang
Pietro Zanuttigh
Alexey Zaytsev
Bernhard Zeisl
Haitian Zeng
Pengpeng Zeng
Jiabei Zeng
Runhao Zeng
Wei Zeng
Yawen Zeng
Yi Zeng

Yiming Zeng
Tieyong Zeng
Huanqiang Zeng
Dan Zeng
Yu Zeng
Wei Zhai
Yuanhao Zhai
Fangneng Zhan
Kun Zhan
Xiong Zhang
Jingdong Zhang
Jiangning Zhang
Zhilu Zhang
Gengwei Zhang
Dongsu Zhang
Hui Zhang
Binjie Zhang
Bo Zhang
Tianhao Zhang
Cecilia Zhang
Jing Zhang
Chaoning Zhang
Chenxu Zhang
Chi Zhang
Chris Zhang
Yabin Zhang
Zhao Zhang
Rufeng Zhang
Chaoyi Zhang
Zheng Zhang
Da Zhang
Yi Zhang
Edward Zhang
Xin Zhang
Feifei Zhang
Feilong Zhang
Yuqi Zhang
GuiXuan Zhang
Hanlin Zhang
Hanwang Zhang
Hanzhen Zhang
Haotian Zhang
He Zhang
Haokui Zhang
Hongyuan Zhang
Hengrui Zhang
Hongming Zhang
Mingfang Zhang
Jianpeng Zhang
Jiaming Zhang
Jichao Zhang
Jie Zhang
Jingfeng Zhang
Jingyi Zhang
Jinnian Zhang
David Junhao Zhang
Junjie Zhang
Junzhe Zhang
Jiawan Zhang
Jingyang Zhang
Kai Zhang
Lei Zhang
Lihua Zhang
Lu Zhang
Miao Zhang
Minjia Zhang
Mingjin Zhang
Qi Zhang
Qian Zhang
Qilong Zhang
Qiming Zhang
Qiang Zhang
Richard Zhang
Ruimao Zhang
Ruisi Zhang
Ruixin Zhang
Runze Zhang
Qilin Zhang
Shan Zhang
Shanshan Zhang
Xi Sheryl Zhang
Song-Hai Zhang
Chongyang Zhang
Kaihao Zhang
Songyang Zhang
Shu Zhang
Siwei Zhang
Shujian Zhang
Tianyun Zhang
Tong Zhang
Tao Zhang
Wenwei Zhang
Wenqiang Zhang
Wen Zhang
Xiaolin Zhang
Xingchen Zhang
Xingxuan Zhang
Xiuming Zhang
Xiaoshuai Zhang
Xuanmeng Zhang
Xuanyang Zhang
Xucong Zhang
Xingxing Zhang
Xikun Zhang
Xiaohan Zhang
Yahui Zhang
Yunhua Zhang
Yan Zhang
Yanghao Zhang
Yifei Zhang
Yifan Zhang
Yi-Fan Zhang
Yihao Zhang
Yingliang Zhang
Youshan Zhang
Yulun Zhang
Yushu Zhang
Yixiao Zhang
Yide Zhang
Zhongwen Zhang
Bowen Zhang
Chen-Lin Zhang
Zehua Zhang
Zekun Zhang
Zeyu Zhang
Xiaowei Zhang
Yifeng Zhang
Cheng Zhang
Hongguang Zhang
Yuexi Zhang
Fa Zhang
Guofeng Zhang
Hao Zhang
Haofeng Zhang
Hongwen Zhang

Hua Zhang
Jiaxin Zhang
Zhenyu Zhang
Jian Zhang
Jianfeng Zhang
Jiao Zhang
Jiakai Zhang
Lefei Zhang
Le Zhang
Mi Zhang
Min Zhang
Ning Zhang
Pan Zhang
Pu Zhang
Qing Zhang
Renrui Zhang
Shifeng Zhang
Shuo Zhang
Shaoxiong Zhang
Weizhong Zhang
Xi Zhang
Xiaomei Zhang
Xinyu Zhang
Yin Zhang
Zicheng Zhang
Zihao Zhang
Ziqi Zhang
Zhaoxiang Zhang
Zhen Zhang
Zhipeng Zhang
Zhixing Zhang
Zhizheng Zhang
Jiawei Zhang
Zhong Zhang
Pingping Zhang
Yixin Zhang
Kui Zhang
Lingzhi Zhang
Huaiwen Zhang
Quanshi Zhang
Zhoutong Zhang
Yuhang Zhang
Yuting Zhang
Zhang Zhang
Ziming Zhang
Zhizhong Zhang
Qilong Zhangli
Bingyin Zhao
Bin Zhao
Chenglong Zhao
Lei Zhao
Feng Zhao
Gangming Zhao
Haiyan Zhao
Hao Zhao
Handong Zhao
Hengshuang Zhao
Yinan Zhao
Jiaojiao Zhao
Jiaqi Zhao
Jing Zhao
Kaili Zhao
Haojie Zhao
Yucheng Zhao
Longjiao Zhao
Long Zhao
Qingsong Zhao
Qingyu Zhao
Rui Zhao
Rui-Wei Zhao
Sicheng Zhao
Shuang Zhao
Siyan Zhao
Zelin Zhao
Shiyu Zhao
Wang Zhao
Tiesong Zhao
Qian Zhao
Wangbo Zhao
Xi-Le Zhao
Xu Zhao
Yajie Zhao
Yang Zhao
Ying Zhao
Yin Zhao
Yizhou Zhao
Yunhan Zhao
Yuyang Zhao
Yue Zhao
Yuzhi Zhao
Bowen Zhao
Pu Zhao
Bingchen Zhao
Borui Zhao
Fuqiang Zhao
Hanbin Zhao
Jian Zhao
Mingyang Zhao
Na Zhao
Rongchang Zhao
Ruiqi Zhao
Shuai Zhao
Wenda Zhao
Wenliang Zhao
Xiangyun Zhao
Yifan Zhao
Yaping Zhao
Zhou Zhao
He Zhao
Jie Zhao
Xibin Zhao
Xiaoqi Zhao
Zhengyu Zhao
Jin Zhe
Chuanxia Zheng
Huan Zheng
Hao Zheng
Jia Zheng
Jian-Qing Zheng
Shuai Zheng
Meng Zheng
Mingkai Zheng
Qian Zheng
Qi Zheng
Wu Zheng
Yinqiang Zheng
Yufeng Zheng
Yutong Zheng
Yalin Zheng
Yu Zheng
Feng Zheng
Zhaoheng Zheng
Haitian Zheng
Kang Zheng
Bolun Zheng

Haiyong Zheng
Mingwu Zheng
Sipeng Zheng
Tu Zheng
Wenzhao Zheng
Xiawu Zheng
Yinglin Zheng
Zhuo Zheng
Zilong Zheng
Kecheng Zheng
Zerong Zheng
Shuaifeng Zhi
Tiancheng Zhi
Jia-Xing Zhong
Yiwu Zhong
Fangwei Zhong
Zhihang Zhong
Yaoyao Zhong
Yiran Zhong
Zhun Zhong
Zichun Zhong
Bo Zhou
Boyao Zhou
Brady Zhou
Mo Zhou
Chunluan Zhou
Dingfu Zhou
Fan Zhou
Jingkai Zhou
Honglu Zhou
Jiaming Zhou
Jiahuan Zhou
Jun Zhou
Kaiyang Zhou
Keyang Zhou
Kuangqi Zhou
Lei Zhou
Lihua Zhou
Man Zhou
Mingyi Zhou
Mingyuan Zhou
Ning Zhou
Peng Zhou
Penghao Zhou
Qianyi Zhou
Shuigeng Zhou
Shangchen Zhou
Huayi Zhou
Zhize Zhou
Sanping Zhou
Qin Zhou
Tao Zhou
Wenbo Zhou
Xiangdong Zhou
Xiao-Yun Zhou
Xiao Zhou
Yang Zhou
Yipin Zhou
Zhenyu Zhou
Hao Zhou
Chu Zhou
Daquan Zhou
Da-Wei Zhou
Hang Zhou
Kang Zhou
Qianyu Zhou
Sheng Zhou
Wenhui Zhou
Xingyi Zhou
Yan-Jie Zhou
Yiyi Zhou
Yu Zhou
Yuan Zhou
Yuqian Zhou
Yuxuan Zhou
Zixiang Zhou
Wengang Zhou
Shuchang Zhou
Tianfei Zhou
Yichao Zhou
Alex Zhu
Chenchen Zhu
Deyao Zhu
Xiatian Zhu
Guibo Zhu
Haidong Zhu
Hao Zhu
Hongzi Zhu
Rui Zhu
Jing Zhu
Jianke Zhu
Junchen Zhu
Lei Zhu
Lingyu Zhu
Luyang Zhu
Menglong Zhu
Peihao Zhu
Hui Zhu
Xiaofeng Zhu
Tyler (Lixuan) Zhu
Wentao Zhu
Xiangyu Zhu
Xinqi Zhu
Xinxin Zhu
Xinliang Zhu
Yangguang Zhu
Yichen Zhu
Yixin Zhu
Yanjun Zhu
Yousong Zhu
Yuhao Zhu
Ye Zhu
Feng Zhu
Zhen Zhu
Fangrui Zhu
Jinjing Zhu
Linchao Zhu
Pengfei Zhu
Sijie Zhu
Xiaobin Zhu
Xiaoguang Zhu
Zezhou Zhu
Zhenyao Zhu
Kai Zhu
Pengkai Zhu
Bingbing Zhuang
Chengyuan Zhuang
Liansheng Zhuang
Peiye Zhuang
Yixin Zhuang
Yihong Zhuang
Junbao Zhuo
Andrea Ziani
Bartosz Zieliński
Primo Zingaretti

Nikolaos Zioulis
Andrew Zisserman
Yael Ziv
Liu Ziyin
Xingxing Zou
Danping Zou
Qi Zou
Shihao Zou
Xueyan Zou
Yang Zou
Yuliang Zou
Zihang Zou
Chuhang Zou
Dongqing Zou
Xu Zou
Zhiming Zou
Maria A. Zuluaga
Xinxin Zuo
Zhiwen Zuo
Reyer Zwiggelaar

Contents – Part XV

WaveGAN: Frequency-Aware GAN for High-Fidelity Few-Shot Image Generation

Mengping Yang[1,2], Zhe Wang[1,2(✉)], Ziqiu Chi[1,2], and Wenyi Feng[1,2]

[1] Department of Computer Science and Engineering, East China University of Science and Technology, Shanghai, China
{mengpingyang,chiziqiu,Y10200096}@mail.ecust.edu.cn,
wangzhe@ecust.edu.cn

[2] Key Laboratory of Smart Manufacturing in Energy Chemical Process, East China University of Science and Technology, Shanghai, China

Abstract. Existing few-shot image generation approaches typically employ fusion-based strategies, either on the image or the feature level, to produce new images. However, previous approaches struggle to synthesize high-frequency signals with fine details, deteriorating the synthesis quality. To address this, we propose WaveGAN, a frequency-aware model for few-shot image generation. Concretely, we disentangle encoded features into multiple frequency components and perform low-frequency skip connections to preserve outline and structural information. Then we alleviate the generator's struggles of synthesizing fine details by employing high-frequency skip connections, thus providing informative frequency information to the generator. Moreover, we utilize a frequency L_1-loss on the generated and real images to further impede frequency information loss. Extensive experiments demonstrate the effectiveness and advancement of our method on three datasets. Noticeably, we achieve new state-of-the-art with FID 42.17, LPIPS 0.3868, FID 30.35, LPIPS 0.5076, and FID 4.96, LPIPS 0.3822 respectively on Flower, Animal Faces, and VGGFace. GitHub: https://github.com/kobeshegu/ECCV2022_WaveGAN.

Keywords: GANs · Few-shot learning · Image generation · Wavelet trasformation

1 Introduction

Recent years have witnessed remarkable developments in visual generative tasks with the rapid progress of generative models, especially Generative Adversarial Networks (GANs) [19,21–23]. Despite being applied to various domains, the success of GANs mainly comes from immense data, and GANs struggle to synthesize high-quality images given insufficient data. Few-shot learning [7], being proposed to improve the generalization ability in the limited-data scenarios,

Supplementary Information The online version contains supplementary material available at https://doi.org/10.1007/978-3-031-19784-0_1.

S. Avidan et al. (Eds.): ECCV 2022, LNCS 13675, pp. 1–17, 2022.
https://doi.org/10.1007/978-3-031-19784-0_1

has gained extensive attention and focused research. However, most of existing few-shot algorithms are designed for classification [27] and segmentation [25] problems, few studies address few-shot image generation. Therefore, exploring and facilitating the generation quality in the few-shot regime is necessary.

Few-shot image generation aims at generating novel images for a category when given a few images from the same category. Specifically, the model is first trained on an auxiliary dataset (seen classes) with sufficient data in an episode-based manner [37], *i.e.*, feeding a specific number of images (*e.g.*, 3, 5) into the model in each episode. The trained model is then expected to produce new images when given a few images of a category from a new dataset (unseen classes). There is no overlap between the auxiliary dataset and the new testing dataset. The model is encouraged to capture the transferable ability from seen classes to unseen classes to generate new images for unseen classes.

Previous methods try to 1) transform intra-class information [1], 2) design new optimization schemes by combining GANs with meta-learning [4,26], and 3) fuse the given images [12,14,15] to address few-shot image generation. Among these methods, LoFGAN [12] achieves current state-of-the-art performance by fusing local representations based on the semantic similarity of features. However, existing approaches ignore the enormous impact of the frequency information throughout the generating process. F-principal [39] proves that neural networks preferentially fit frequency signals from low to high. Consequently, the model tends to generate frequencies with higher priority and more superficial complexity, *i.e.*, only generating low-frequency signals.

Fig. 1. Visualization of the transformed frequency components. LL denotes low-frequency component, and LH, HL, HH denote high-frequency components.

We visualize different frequency components of real images in Fig. 1. The low-frequency component (*i.e.*, LL) contains general information like the overall surface, outline, and structure. While rich details and perceptible information like the leaves of flowers, the tongue of dogs, and the hair of human, lie in the high-frequency components (*i.e.*, LH, HL, HH). Rich details can be obtained by adding all high-frequency components (*i.e.*, $LH+HL+HH$) together. Since high frequency components contain meticulous information, losing them may lead the generator to synthesize blurry images with more aliasing artifacts. This issue highlights the necessity of considering frequency signals in generating images, especially high-frequency signals as the generator usually eschews them [18,39].

This paper proposes WaveGAN, an innovative and effective approach to ameliorate the few-shot synthesis quality from the perspective of frequency domains. We first perform wavelet decomposition to transform the encoded features from the spatial domain to multiple frequency domains, comprising low and high-frequency components. Then we feed the low-frequency component to the behind layers of the encoder via low-frequency skip connections, maintaining the overall outline and structural patterns. To mitigate the pressure of the generator to generate high-frequency signals and provide more details to the decoder, we directly feed the decomposed high-frequency signals to the decoder. Two strategies are designed to aggregate the high-frequency signals, namely WaveGAN-M and WaveGAN-B. Both of them are effective and can provide high-frequency information to the decoder. The high-frequency components are then precisely reconstructed back into the original features with our inverse frequency transformation operations, guaranteeing the minimal loss of high-frequency signals. In addition, we apply frequency L_1-loss to the generated images and the real images, which is complementary to spatial losses and impedes losing frequency information. Our primary contributions can be summarized as follows:

- We propose WaveGAN, the first few-shot image generation method that exploits frequency components to promote synthesis quality. Adding low and high-frequency skip connections to the generator, our WaveGAN alleviates the generator's struggles to encode high-frequency signals and provide more perceptible information, resulting in favorable generation quality.
- We design two techniques to aggregate the high-frequency information for reconstructing frequency signals back to the original features, *i.e.*, WaveGAN-M and WaveGAN-B, which preserve fine details and statistical properties. We also present frequency L_1-loss to avoid losing frequency information.
- We conduct comprehensive experiments on three datasets. Both qualitative and quantitative results demonstrate the superiority and effectiveness of our method. Notably, our model outperforms the state-of-the-art approach with significant FID improvements (*e.g.*, from **102.07** to **30.35** on Animal Face).

2 Related Work

Generative Adversarial Network. Generative Adversarial Networks (GANs) have made significant progress since the pioneering work in [10]. Benefit from remarkable ability of capturing the data distribution, GANs have been successfully applied in various visual domains, including image generation [21,23], video generation [38], image-to-image translation [32,35], etc. Typically, a GAN model consists of a generator and a discriminator, and the two networks are updated alternatively in an adversarial manner. Training a GAN is notoriously formidable as it requires massive data and computation resources, and the adversarial training may make the model diverge. The discriminator is tend to overfitting when given limited data, resulting in poor generation quality. Several works have been proposed to mitigate the discriminator overfitting. Different data augmentation techniques, including differentiable [43], non-leaking [20] and adaptive pseudo augmentation [17] are designed to expand the limited training data. Lecam [36] regularizes the output of the discriminator to avoid overfitting. Unlike these efforts made for unconditional image generation with limited data, in this paper, we seek to generate novel images for one specific category when given a few images from this category.

Wavelet Transformation in GANs. Decomposing given signals into different frequency components, wavelet transformation has made great success in various generative tasks such as style transfer [40], image reconstruction [18], image inpainting [41], image editing [9] and image super-resolution [6,16]. These approaches try to narrow the information gaps in the frequency domain to boost the model's performance. For example, Jiang *et al.* propose focal frequency loss to avoid the loss of important frequency information for image reconstruction tasks [18]. WaveFill [41] decomposes images into multiple frequency components and fills the corrupted image regions with decomposed signals, which achieves superior image inpainting. Different from these methods, we try to generate realistic and plausible images when given only a few data. We are interested in the influence of frequency information on the challenging few-shot image generation.

Few-Shot Image Generation. Inspired by the human's great generalization ability from a few observations, few-shot image generation models try to generate new images given a few images. Existing few-shot image generation approaches can be roughly divided into three categories: 1) Optimization-based, 2) Fusion-based, and 3) Transformation-base methods. DAGAN [1] transforms combined projected latent codes and encoded images to new images. The optimized-based methods FIGR [4] and DAWSON [26] combine generative models with optimization-based meta learning Reptile [29] and MAML [8], respectively. The fusion-based methods fuse the local feature [12] or the input images [14,15] to synthesis novel images. GMN [2] combines VAE [24] with Matching Networks [37] to capture the few-shot distribution. MatchingGAN [14] matches random vectors with given real images and mapping the fused features to novel images. F2GAN [15] further improves MatchingGAN with a fusing-and-filling paradigm. By fusing local representations with semantic similarity, LoFGAN [12] promotes the generation quality. Notably, zero-shot or few-shot text-to-image generation methods [11,33,34] have made great progress recently. Differently, this paper focuses on the problem of few-shot image generation for generating new images for a given class as defined in Sect. 3.1.

However, existing methods ignore the influence of frequency components on the quality of generated images, leading the generator to synthesize unfavorable images with more artifacts and fewer details. In this paper, we present a frequency-aware model that can generate appealing and photorealistic images by adding low and high-frequency skip connections to the generator. Such design mitigates the generator's pressure of synthesising high-frequency signals. Our work explores an effective solution for few-shot image generation from the frequency domain perspective, which complements previous fusion-based methods.

3 Methodology

3.1 Overview

Problem Definition. Given K images from a new class, our model's goal is to synthesis diverse and plausible images for the given class. The number of images K defines a K-shot image generation task. Generally, this task is accomplished in two phases, *i.e.*, training and testing. The datasets is first split into seen classes $\mathbb{C}_s$ and unseen classes $\mathbb{C}_u$, where $\mathbb{C}_s$ and $\mathbb{C}_u$ have no overlap. In the training phase, a substantial amount of K-shot image generation tasks sampled from $\mathbb{C}_s$ are fed into the model, expecting the model to transfer the knowledge of generating new images learned from $\mathbb{C}_s$ to $\mathbb{C}_u$. In the testing phase, the model takes images from $\mathbb{C}_u$ as input to synthesis new images.

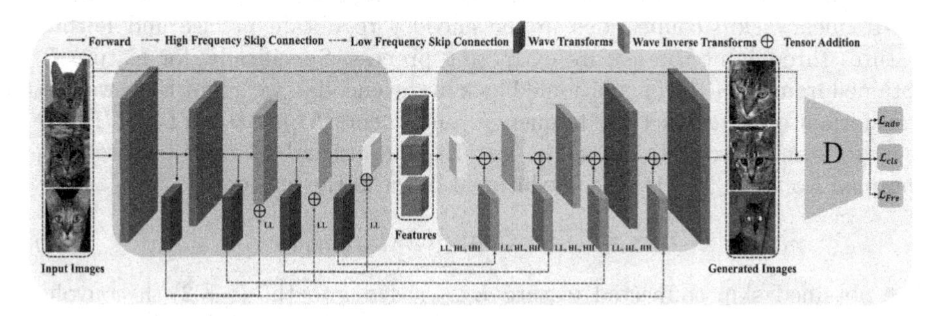

Fig. 2. The overall framework of our WaveGAN. We employ low frequency (LL) skip connections in the encoder and high frequency skip connection (LH, HL, HH) in the decoder to provide rich details to improve synthesis quality.

Overall Framework. As shown in Fig. 2, our model consists of a WaveEncoder, a WaveDecoder, and a Discriminator, the WaveEncoder and Wavedecoder constitute our generator. The WaveEncoder extracts feature representations of images, while the WaveDecoder maps the feature representation to new images. We perform wavelet transformation to the encoded features and obtain multiple frequency components. Then we employ low-frequency skip connections in the encoder to preserve the overall structure and outline. We exploit high-frequency skip connections to provide detailed information to the decoder. The wavelet inverse transformation module reconstructs these high-frequency signals to the original features. The high-frequency signals contain rich details and perceptible information, enabling the generator to synthesis high-quality images. The real and generated images are then fed into the discriminator to train the whole model. Next, we elaborate our WaveEncoder and WaveDecoder in detail.

3.2 WaveEncoder

Our WaveEncoder is composed of convolutional blocks and wavelet transformation blocks. The convolutional operations extract features for the decoder to produce new images. To disentangle the extracted features into multiple frequency components, we adopt a simple yet effective wavelet transformation, *i.e.*, Haar wavelet [5]. Haar wavelet contains two operations: wavelet transform and inverse wavelet transformation, and four kernels, namely LL^T, LH^T, HL^T, and HH^T.

$$L^\top = \frac{1}{\sqrt{2}}\left[1\,1\right], \quad H^\top = \frac{1}{\sqrt{2}}\left[-1\,1\right] \tag{1}$$

where L and H denote the low and high pass filters, respectively. The low pass filter focuses on low-frequency signals containing the outline and structural information. In contrast, the high pass filter emphasizes high-frequency signals that capture fine-grained details like subtle edges and contour (See Fig. 1 and 5).

The wavelet transformation decomposes features into frequency components LL, LH, HL, HH. Among these frequency signals, LL captures the overall appearance and basic object structures of images (See Fig. 1). Thus we employ low-frequency skip connections in the encoder to obtain precise and faithful features throughout the feature extracting process. Specifically, for feature E_i obtained from the i-th convolutional block in the encoder, we adopt Haar wavelet transformation to extract the frequency components LL_i, LH_i, HL_i, HH_i. We then perform tenor addition on the low frequency signal LL_i and feature E_{i+1} obtained from the $(i+1)$-th convolutional block in the encoder.

$$E_{i+2} = LL_i + ConvBlock_{i+1}(E_{i+1}) \tag{2}$$

The obtained skip connected feature E_{i+2} is fed into the $(i+2)$-th convolutional block. The low-frequency skip connections contribute to the fidelity of the generated images. The experimental proofs are given in Sect. 4.4.

3.3 WaveDecoder

High-frequency components contain rich details of images. However, deep networks usually fit frequency signals from low to high, making it difficult for the generator to produce high-frequency information since it generates frequencies with higher priority. To alleviate the encoder's pressure to synthesis rich details and provide fine-grained information to the decoder, we directly feed the decomposed high-frequency signals LH, HL, and HH into the WaveDecoder via high-frequency skip connections. Specifically, for the i-th layer of the encoder, we perform wavelet transformation on the features and obtain high frequency components LH_i, HL_i, HH_i, then we feed the inversed components to the $(n-i)$-th layer of the decoder as exhibited in Fig. 2, where n is the number of all layers. This operation encourages the decoder to synthesis images with more details and fewer artifacts. We employ wavelet inverse transformation to reconstruct high-frequency signals back to original features. Our wavelet inverse transformation can be categorized into Mean and Base-index inverse transformation based on how these high-frequency components are aggregated.

Mean Inverse Transformation. As presented in Fig. 3, we calculate the frequency element-wise average of all high-frequency components of K features from the same category, and take the averaged results as the input of our Mean inverse transformation module.

$$HF_M = \sum_{i=1}^{K} HF_i, HF_i \in \{LH_i, HL_i, HH_i\} \tag{3}$$

Although providing high-frequency signals to the decoder facilitates the generated images' quality, the Mean inverse transformation may not be suited for one specific image as the averaged frequency information may shift the frequency signals. The averaged frequency information becomes more neutral with the number of training images K increases, leading to a decrease in generalization. This conjecture may go against our common sense that the generalization ability should improve as the number of images increases. We analyze this is because different images, even from the same category, have different frequency signals in the frequency domain. The experiments in Sect. 4 confirm our analysis that the averaged frequency transformation may fail to generalize when K is bigger. To improve the generalization ability of our inverse transformation in the frequency domain, we design shots-agnostic Base-index inverse transformation.

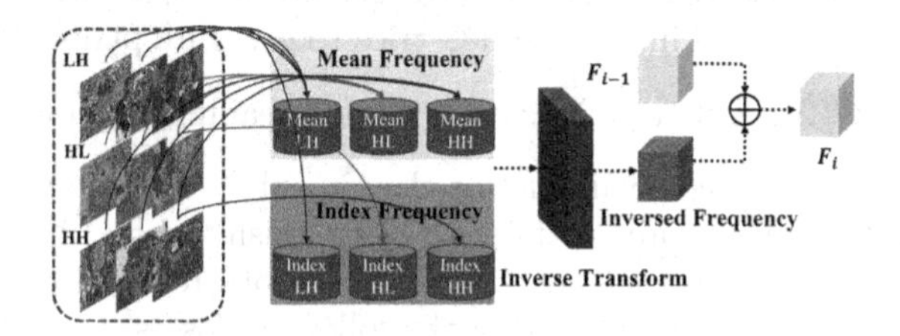

Fig. 3. Illustration of our Mean and Base-index inverse transformation.

Base-Index Inverse Transformation. Our Base-index inverse transformation is implemented based on the Local representation Fusion (LoF) strategy of LoFGAN [12]. We first give a brief introduction to LoFGAN. Given encoder features $\mathbb{F} = \mathbb{E}(X) \in \mathbb{R}^{k \times w \times h \times c}$, LoFGAN randomly selects one base feature $f_{base} \in \mathbb{R}^{w \times h \times c}$ and views the rest $(K-1)$ features as reference, *i.e.*, $f_{ref} \in \mathbb{R}^{(k-1) \times w \times h \times c}$. LoFGAN fuses the local features based on the calculated semantic similarity map and replaces the closest base feature with the fused feature.

To provide customized high-frequency signals to the generated images, we recode the index of the selected base feature of the LoF module. As illustrated in Fig. 3, instead of calculating the averaged frequency signals, we explicitly feed the high-frequency component corresponding to the recorded index i.

$$HF_B = HF_i, HF_i \in \{LH_i, HL_i, HH_i\} \tag{4}$$

The high-frequency components are exact frequency signals of the selected feature, thus providing customized rich details and perceptible information to the decoder. The generalization ability of our Base-index inverse transformation would not deteriorate with the training number increases.

After obtaining the aggregate high-frequency signals LH, LH, HH, we perform our inverse transformation to reconstruct these signals back to the original features. Concretely, we first perform transposed-convolution on each frequency component, then sum up all the output features. The summed result convert frequency signals back to original features precisely, theoretical analysis can be found in [40]. The inverse transformation can be formally expressed as:

$$F_{IF} = \sum TransConv(HF), HF \in \{HF_M, HF_B\} \tag{5}$$

HF is obtained from either the Mean or the Base-index inverse transformation. Element-wise tensor addition are employed to integrate the inversed high-frequency components with the former feature, *i.e.*, $F_i = F_{i-1} + F_{IF(n-i+1)}$. Such branches assuage the generator's dilemmas in producing limited high-frequency containing fine details.

3.4 Optimization Objective

Our model has two networks to optimize, the generator (G) and the discriminator (D). The input of G is real images, and G tries to generate plausible and diverse new images $\hat{x} = G(X)$. Let $X = \sum_{i=1}^{K} x_i$ denotes the real images, and $c(x_i)$ denotes the labels for the image x_i (available for the $\mathbb{C}_s$ only). The inputs of D are the real and generated images, and D tries to distinguish the real images from the generated ones. The generator and the discriminator are updated alternatively in an adversarial manner by optimizing the following losses.

Frequency L_1-Loss. We employ the frequency L_1-loss on the transformed frequency components of generated images and the real images, impeding losing the frequency information. Besides, the frequency loss complements existing spatial losses. We perform wavelet transformation on the generated and real images to compute the frequency L_1 loss.

$$\mathcal{L}_{Fre} = \sum \|Fre_x - Fre_{\hat{x}}\|_1, Fre \in \{LL, LH, HL, HH\} \tag{6}$$

Local Reconstruction Loss. We adopt local reconstruction loss to constrain the model to maintain the local features.

$$\mathcal{L}_{rec} = \|\hat{x} - \mathrm{LFM}(X, \boldsymbol{\alpha})\|_1 \tag{7}$$

where $\boldsymbol{\alpha}$ denotes the coefficient vector to fuse the features in the local fusion module (LoF).

Adversarial Loss. Following [12] and [31], we adopt the hinge version of adversarial loss to optimize the generator and the discriminator.

$$\begin{aligned} \mathcal{L}_{adv}^{D} &= \max(0, 1 - D(x)) + \max(0, 1 + D(\hat{x})) \\ \mathcal{L}_{adv}^{G} &= -D(\hat{x}) \end{aligned} \tag{8}$$

Classification Loss. Classification loss constrains the model to synthesis images that belong to one specific category. We add an auxiliary classifier to the generator and the discriminator following ACGAN [31]. The classification loss encourages the discriminator to identify which category an image belongs to while enabling the generator to synthesis images that belong to one specific category.

$$\begin{aligned} \mathcal{L}_{cls}^{D} &= -\log P(c(x) \mid x) \\ \mathcal{L}_{cls}^{G} &= -\log P(c(\hat{x}) \mid \hat{x}) \end{aligned} \tag{9}$$

Our model is optimized with the following objective function with the linear combination of the above losses.

$$\begin{aligned} \mathcal{L}_{G} &= \mathcal{L}_{adv}^{G} + \lambda_{cls}^{G}\mathcal{L}_{cls}^{G} + \lambda_{Fre}\mathcal{L}_{Fre} + \lambda_{rec}\mathcal{L}_{rec}^{G} \\ \mathcal{L}_{D} &= \mathcal{L}_{adv}^{D} + \lambda_{cls}^{D}\mathcal{L}_{cls}^{D} \end{aligned} \tag{10}$$

4 Experiments

Datasets. We use three popular datasets in the few-shot image generation community to evaluate the performance of our model, namely Flower [30], Animal Faces [28] and VGGFace [3]. These datasets are split into seen classes $\mathbb{C}_s$ and unseen classes $\mathbb{C}_u$. $\mathbb{C}_s$ is used in the training stage, while $\mathbb{C}_u$ is used in the testing stage. The adopted datasets are split in Table 1 following [12] and [15].

Table 1. The split of experimental datasets.

Datasets	#Total classes	#Seen classes	#Unseen classes	#Images/class
Flower	102	85	17	40
Animal Faces	149	119	30	100
VGGFace	2354	1802	552	100

Evaluation and Baselines. We evaluate the quality of the generated images with two commonly used metrics: Fréchet Inception Distance (FID) [13] and Learned Perceptual Image Patch Similarity (LPIPS) [42]. The two metrics are calculated in the same setting with [12]. We compare our model with several few-shot image generation approaches, namely FIGR [4], GMN [2], DAWSON [26], DAGAN [1], MathingGAN [14], F2GAN [15] and LoFGAN [12]. We re-implement the current state-of-the-art LoFGAN for fair comparisons (denoted as "LoFGAN‡"), and all methods are evaluated under the same conditions. Implementation details are given in the appendix.

4.1 Quantitative Evaluation

We first train the model with $\mathbb{C}_s$ and then use the data from $\mathbb{C}_u$ to synthesis novel images for quantitative evaluation. Following LoFGAN [12] and [15], we split each unseen class into two parts, $\mathbb{S}_{sup}$ and $\mathbb{S}_{que}$, the images in $\mathbb{S}_{sup}$ are fed into the model to generate images. We generate 128 images for each class (denoted as $\mathbb{S}_{gen}$), $\mathbb{S}_{gen}$ and $\mathbb{S}_{que}$ are used to compute FID (lower is better) and LPIPS (higher is better) scores to evaluate the synthesis quality. The quantitative results of our model and the baselines are given in Table 2. All results in the table are conducted under 3-shot setting for both training and testing stages.

Table 2. Quantitative comparison results of our model and the baselines on FID and LPIPS. † results are quoted from LoFGAN [12]. ‡ results are re-impelemented under the same condition with our model for fair comparison. The best and the second-ranked results are **bold** and underlined, respectively.

Method	Type	Flowers		Animal Faces		VGGFace	
		FID (↓)	LPIPS (↑)	FID (↓)	LPIPS (↑)	FID (↓)	LPIPS (↑)
FIGR† [4]	Optimization	190.12	0.0634	211.54	0.0756	139.83	0.0834
DAWSON† [26]	Optimization	188.96	0.0583	208.68	0.0642	137.82	0.0769
DAGAN† [1]	Transformation	151.21	0.0812	155.29	0.0892	128.34	0.0913
GMN† [2]	Fusion	200.11	0.0743	220.45	0.0868	136.21	0.0902
MatchingGAN† [14]	Fusion	143.35	0.1627	148.52	0.1514	118.62	0.1695
F2GAN† [15]	Fusion	120.48	0.2172	117.74	0.1831	109.16	0.2125
MatchingGAN+LoFGAN† [12]	Fusion	86.59	0.3704	112.99	0.5024	22.99	0.2687
LoFGAN† [12]	Fusion	79.33	0.3862	112.81	0.4964	20.31	0.2869
LoFGAN‡	Fusion	81.70	0.3768	102.07	0.5005	16.82	0.3041
WaveGAN-M (ours)	Fusion	63.79	0.3709	50.98	0.5014	8.62	**0.3822**
WaveGAN-B (ours)	Fusion	**42.17**	**0.3868**	**30.35**	**0.5076**	**4.96**	0.3255

As can be observed from Table 2, our waveGAN achieves the lowest FID and the highest LPIPS on all the datasets, and both WaveGAN-M and WaveGAN-B outperform the baseline models. Notably, Our model achieves much better FID results than the current state-of-the-art LoFGAN. Specifically, WaveGAN-B achieves the FID of less than 5 (**4.96**) on the challenging VGGFace dataset, while LoFGAN obtains 16.82. And WaveGAN-B lowers the FID from 102.07 (*resp.*, 81.70) to **30.35** (*resp.*, **42.17**) on Animal Faces (*resp.*, Flower). Such significant improvements on the quantitative metrics demonstrate that our model could generate plausible and vivid images. Since the upper bound of FID scores equals 0 and we achieve a single digit for the first time, demonstrating the efficacy of our method. As for the LPIPS metric, we calculate its upper bound by measuring the real images' LPIPS score and obtain 0.4393 for Flower, 0.5729 for Animal Faces, and 0.4389 for VGGFace. Our model yields favorable LPIPS scores that approach the upper bound, further substantiating the effectiveness of our model.

4.2 Qualitative Evaluation

We present the visualization results of LoFGAN [12] and our WaveGAN-B for qualitative comparison in Fig. 4. For each real image, we give two fake images generated by LoFGAN and our WaveGAN-B. As can be observed from the figure, images generated by our model are more plausible than that of LoFGAN.

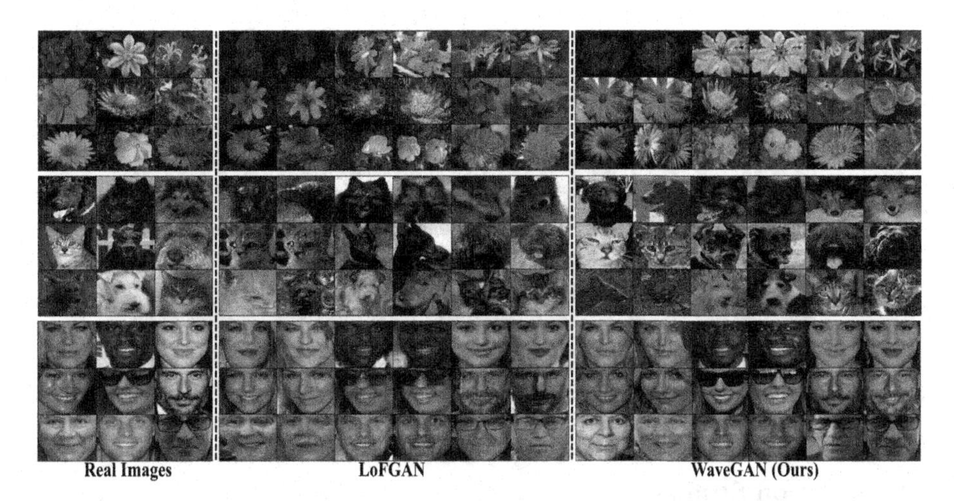

Fig. 4. Qualitative results of our WaveGAN and LoFGAN. The left-most three columns are real images, we give two generated images for each class of real image.

Moreover, images generated by our model contain rich details and perceptible information. Take the generated flowers as examples, the horizontal and vertical orientation of petals, the details of stamens, and the shapes of the leaves of images synthesised by our model are more reasonable and realistic. Moreover, animal and human face images generated by LoFGAN are distorted with blurry unfavorable artifacts, features like the eyes of cats, hair of dogs are even misplaced. By contrast, animal and human face images generated by our model have higher fidelity and even look indistinguishable from real images.

4.3 Visualization of the Frequency Components of Generated Images

We visualize the frequency components of images generated by our WaveGAN and LoFGAN [12] in Fig. 5. As observable in Fig. 5, in addition to the fact that our WaveGAN produces more realistic visual images, the decomposed high-frequency components of our WaveGAN contain more details and perceptible information than LoFGAN. Specifically, the frequency components of LoFGAN contain only the surface and texture information of images, indicating that the generator of LoFGAN fails to synthesize high-frequency information. In comparison, our WaveGAN is frequency-aware and can produce high-frequency signals that contain more fine details and statistical properties. Further, the frequency components of WaveGAN capture delicate information that is less noticeable (*e.g.*, the glasses frames and flower rhizomes in the second and fifth row of Fig. 5, respectively). Such observation further demonstrates the effectiveness and advancement of our method.

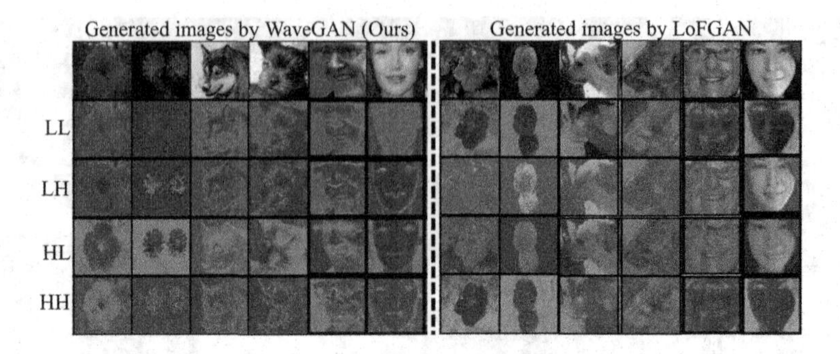

Fig. 5. Visualization results of the frequency components of images generated by our WaveGAN and LoFGAN [12].

4.4 Ablation Studies

We conduct ablation studies to evaluate the effectiveness of each component of the proposed WaveGAN. There are three main components of our WaveGAN, namely 1) the low-frequency skip connection, 2) the high-frequency skip connection, and 3) frequency L_1-loss. We remove each component and keep other settings unchanged to validate their contributions. Besides, we remove the LoF module to investigate the influence of local fusion on our model. We also test the contribution of each component for our two transformation techniques (*i.e.*, WaveGAN-M and WaveGAN-B) quantitatively in the appendix.

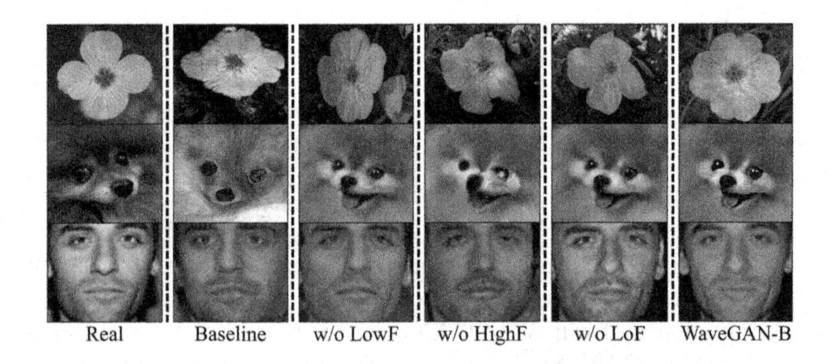

Fig. 6. Visualization comparison results of ablation studies.

We give visualization results of ablation studies in Fig. 6. The quantitative results of ablation studies are given in the appendix. Combining the qualitative and quantitative results, we can draw four conclusions as follows: 1) The skip connection of both low and high-frequency signals play an essential role in our model, and our high-frequency skip connections play a predominant role in our model. 2) High-frequency information provides detailed information to the generated images, and low-frequency information provides the overall outline of images (compare baseline with other generated images in Fig. 6). 3) Our full model achieves the best results and can generate satisfactory images. 4) Our method complements to the local feature fusion approach.

4.5 Augmentation for Classification

To further investigate the quality of generated images, we augment the datasets with images generated by our model for downstream image classification tasks. Specifically, we first pre-train a ResNet18 network with seen classes following [12] and [15], we train the ResNet18 model for 100 epochs with batch size of 4. Then we split the unseen datasets into $\mathbb{D}_{\text{train}}$, $\mathbb{D}_{\text{test}}$ and $\mathbb{D}_{\text{val}}$. For each category of the flower dataset, the number of train, test, and valid images are 10, 15, and 15, respectively. For each category of Animal Faces and VGGFace dataset, the number of train, test and valid images are 30, 35, and 35, respectively. We train a new classifier using the pre-trained model on seen classes with $\mathbb{D}_{\text{train}}$ without any augmentation, which is denoted as "Base". Then we generate images to augment $\mathbb{D}_{\text{train}}$ with LoFGAN and our WaveGAN. The number of augmented images is 30 for Flower dataset and 50 for Animal Face and VGGFace datasets.

Table 3. Classification results of augmentation.

Datasets	Base	LoFGAN	WaveGAN-M (ours)	WaveGAN-B (ours)
Flower	64.71	80.78	70.20	**84.71**
Animals	20.00	26.10	31.81	**32.19**
VGGFace	50.76	64.74	62.96	**77.36**

The classification results are given in Table 3. Compared with the results without any augmentation, our model achieves significant improvements, and our WaveGAN-B outperforms LoFGAN and WaveGAN-M obviously. The effectiveness of using the generated images to augment the training dataset substantiates that our model can produce high-quality images, and the improvement on the classification accuracy provides a new data augmentation strategy for solving few-shot image classification problems.

4.6 Influence of the Number of Shots

All the experiments conducted before are 3-shot image generation tasks. We wonder to know the influence of different shots on our model. We perform different shots of experiments with $K \in \{2, 3, 5, 7, 9\}$. The number of training and testing images for different shots of the experiment are the same. Figure 7 demonstrates the performance of our WaveGAN and LoFGAN. We can observe from the figure that when K is relatively small, WaveGAN-M is better than LoFGAN. However, the performance of WaveGAN-M degrades with the number of images increases,

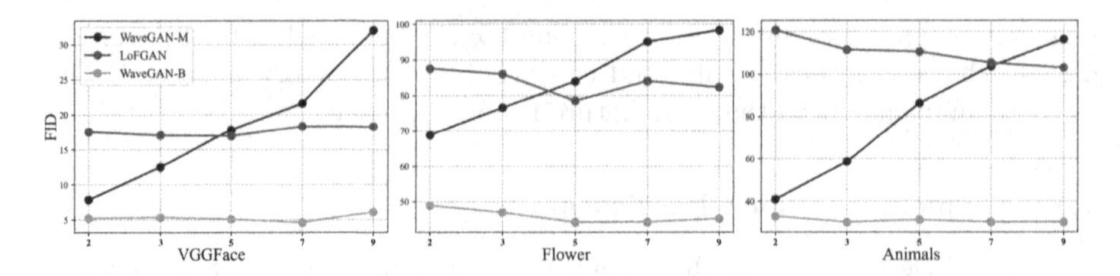

Fig. 7. Comparison results of our model under different shots of generation tasks. The ordinate denotes FID scores, and the abscissa denotes different shots K.

making WaveGAN-M inadequate for image generation tasks with larger K. Such phenomena corroborates our analysis in Sect. 3 that the averaged transformation may fail to generalize to one specific image. By contrast, our WaveGAN-B has relatively low sensitivity to K and is shots-agnostic, thus WaveGAN-B has stronger generalization ability for different shots generation tasks, which manifests the superiority of our WaveGAN-B.

5 Conclusion and Acknowledgments

Conclusion. In this paper, we propose WaveGAN, the first few-shot image generation model that ameliorates the synthesis quality from the frequency domain perspective. The key ingredients of our method are a WaveEncoder and a WaveDecoder. Our WaveEncoder performs wavelet transformation on the different levels of features to obtain frequency signals. We feed the decomposed low-frequency signals to the behind layers of the encoder, and we feed the high-frequency signals to our WaveDecoder. Our design mitigates the generator's struggles of generating rich details for images, especially when limited data are available. We further perform frequency L_1-loss to maintain the frequency information of real images, facilitating the fidelity of generated images. Experimentally, our WaveGAN yields significant improvements on three challenging datasets, and the visualization and downstream classification results demonstrate that our WaveGAN can produce realistic images. Besides, the ablation studies suggest the efficacy of each component of our method and substantiate that our approach complements the existing local fusion-based strategy. Hopefully, our WaveGAN may inspire researchers to explore few-shot image generation from the frequency domain perspective.

Acknowledgments. This work is supported by Shanghai Science and Technology Program "Distributed and generative few-shot algorithm and theory research" under Grant No. 20511100600 and "Federated based cross-domain and cross-task incremental learning" under Grant No. 21511100800, Natural Science Foundation of China under Grant No. 62076094, Chinese Defense Program of Science and Technology under Grant No. 2021-JCJQ-JJ-0041, China Aerospace Science and Technology Corporation Industry-University-Research Cooperation Foundation of the Eighth Research Institute under Grant No. SAST2021-007.

References

1. Antoniou, A., Storkey, A., Edwards, H.: Data augmentation generative adversarial networks. arXiv preprint arXiv:1711.04340 (2017)
2. Bartunov, S., Vetrov, D.: Few-shot generative modelling with generative matching networks. In: International Conference on Artificial Intelligence and Statistics, pp. 670–678. PMLR (2018)
3. Cao, Q., Shen, L., Xie, W., Parkhi, O.M., Zisserman, A.: VGGFace2: a dataset for recognising faces across pose and age. In: 2018 13th IEEE International Conference on Automatic Face & Gesture Recognition (FG 2018), pp. 67–74. IEEE (2018)
4. Clouâtre, L., Demers, M.: FIGR: few-shot image generation with reptile. arXiv preprint arXiv:1901.02199 (2019)
5. Daubechies, I.: The wavelet transform, time-frequency localization and signal analysis. IEEE Trans. Inf. Theory **36**(5), 961–1005 (1990)
6. Deng, X., Yang, R., Xu, M., Dragotti, P.L.: Wavelet domain style transfer for an effective perception-distortion tradeoff in single image super-resolution. In: Proceedings of the IEEE/CVF International Conference on Computer Vision, pp. 3076–3085 (2019)
7. Fei-Fei, L., Fergus, R., Perona, P.: One-shot learning of object categories. IEEE Trans. Pattern Anal. Mach. Intell. **28**(4), 594–611 (2006)
8. Finn, C., Abbeel, P., Levine, S.: Model-agnostic meta-learning for fast adaptation of deep networks. In: International Conference on Machine Learning, pp. 1126–1135. PMLR (2017)
9. Gao, Y., et al.: High-fidelity and arbitrary face editing. In: Proceedings of the IEEE/CVF Conference on Computer Vision and Pattern Recognition, pp. 16115–16124 (2021)
10. Goodfellow, I., et al.: Generative adversarial nets. Adv. Neural. Inf. Process. Syst. **27** (2014)
11. Gu, S., et al.: Vector quantized diffusion model for text-to-image synthesis. In: Proceedings of the IEEE/CVF Conference on Computer Vision and Pattern Recognition, pp. 10696–10706 (2022)
12. Gu, Z., Li, W., Huo, J., Wang, L., Gao, Y.: LoFGAN: fusing local representations for few-shot image generation. In: Proceedings of the IEEE/CVF International Conference on Computer Vision, pp. 8463–8471 (2021)
13. Heusel, M., Ramsauer, H., Unterthiner, T., Nessler, B., Hochreiter, S.: GANs trained by a two time-scale update rule converge to a local nash equilibrium. Adv. Neural. Inf. Process. Syst. **30** (2017)
14. Hong, Y., Niu, L., Zhang, J., Zhang, L.: MatchingGAN: matching-based few-shot image generation. In: 2020 IEEE International Conference on Multimedia and Expo (ICME), pp. 1–6. IEEE (2020)
15. Hong, Y., Niu, L., Zhang, J., Zhao, W., Fu, C., Zhang, L.: F2GAN: fusing-and-filling GAN for few-shot image generation. In: Proceedings of the 28th ACM International Conference on Multimedia, pp. 2535–2543 (2020)
16. Huang, H., He, R., Sun, Z., Tan, T.: Wavelet domain generative adversarial network for multi-scale face hallucination. Int. J. Comput. Vis. **127**(6), 763–784 (2019)
17. Jiang, L., Dai, B., Wu, W., Loy, C.C.: Deceive D: adaptive pseudo augmentation for GAN training with limited data. Adv. Neural. Inf. Process. Syst. **34** (2021)
18. Jiang, L., Dai, B., Wu, W., Loy, C.C.: Focal frequency loss for image reconstruction and synthesis. In: Proceedings of the IEEE/CVF International Conference on Computer Vision, pp. 13919–13929 (2021)

19. Karras, T., Aila, T., Laine, S., Lehtinen, J.: Progressive growing of GANs for improved quality, stability, and variation. In: International Conference on Learning Representations (2018)
20. Karras, T., Aittala, M., Hellsten, J., Laine, S., Lehtinen, J., Aila, T.: Training generative adversarial networks with limited data. Adv. Neural. Inf. Process. Syst. **33**, 12104–12114 (2020)
21. Karras, T., et al.: Alias-free generative adversarial networks. Adv. Neural. Inf. Process. Syst. **34** (2021)
22. Karras, T., Laine, S., Aila, T.: A style-based generator architecture for generative adversarial networks. In: Proceedings of the IEEE/CVF Conference on Computer Vision and Pattern Recognition, pp. 4401–4410 (2019)
23. Karras, T., Laine, S., Aittala, M., Hellsten, J., Lehtinen, J., Aila, T.: Analyzing and improving the image quality of StyleGAN. In: Proceedings of the IEEE/CVF Conference on Computer Vision and Pattern Recognition, pp. 8110–8119 (2020)
24. Kingma, D.P., Welling, M.: Auto-encoding variational bayes. arXiv preprint arXiv:1312.6114 (2013)
25. Li, G., Jampani, V., Sevilla-Lara, L., Sun, D., Kim, J., Kim, J.: Adaptive prototype learning and allocation for few-shot segmentation. In: Proceedings of the IEEE/CVF Conference on Computer Vision and Pattern Recognition, pp. 8334–8343 (2021)
26. Liang, W., Liu, Z., Liu, C.: Dawson: a domain adaptive few shot generation framework. arXiv preprint arXiv:2001.00576 (2020)
27. Liu, B., et al.: Negative margin matters: understanding margin in few-shot classification. In: Vedaldi, A., Bischof, H., Brox, T., Frahm, J.-M. (eds.) ECCV 2020. LNCS, vol. 12349, pp. 438–455. Springer, Cham (2020). https://doi.org/10.1007/978-3-030-58548-8_26
28. Liu, M.Y., Huang, X., Mallya, A., Karras, T., Aila, T., Lehtinen, J., Kautz, J.: Few-shot unsupervised image-to-image translation. In: Proceedings of the IEEE/CVF International Conference on Computer Vision, pp. 10551–10560 (2019)
29. Nichol, A., Schulman, J.: Reptile: a scalable metalearning algorithm. arXiv preprint arXiv:1803.02999 **2**(3), 4 (2018)
30. Nilsback, M.E., Zisserman, A.: Automated flower classification over a large number of classes. In: 2008 Sixth Indian Conference on Computer Vision, Graphics & Image Processing, pp. 722–729. IEEE (2008)
31. Odena, A., Olah, C., Shlens, J.: Conditional image synthesis with auxiliary classifier GANs. In: International Conference on Machine Learning, pp. 2642–2651. PMLR (2017)
32. Park, T., Efros, A.A., Zhang, R., Zhu, J.-Y.: Contrastive learning for unpaired image-to-image translation. In: Vedaldi, A., Bischof, H., Brox, T., Frahm, J.-M. (eds.) ECCV 2020. LNCS, vol. 12354, pp. 319–345. Springer, Cham (2020). https://doi.org/10.1007/978-3-030-58545-7_19
33. Ramesh, A., Dhariwal, P., Nichol, A., Chu, C., Chen, M.: Hierarchical text-conditional image generation with clip latents. arXiv preprint arXiv:2204.06125 (2022)
34. Ramesh, A., et al.: Zero-shot text-to-image generation. In: International Conference on Machine Learning, pp. 8821–8831. PMLR (2021)
35. Richardson, E., et al.: Encoding in style: a StyleGAN encoder for image-to-image translation. In: Proceedings of the IEEE/CVF Conference on Computer Vision and Pattern Recognition, pp. 2287–2296 (2021)

36. Tseng, H.Y., Jiang, L., Liu, C., Yang, M.H., Yang, W.: Regularizing generative adversarial networks under limited data. In: Proceedings of the IEEE/CVF Conference on Computer Vision and Pattern Recognition, pp. 7921–7931 (2021)
37. Vinyals, O., Blundell, C., Lillicrap, T., Kavukcuoglu, K., Wierstra, D.: Matching networks for one shot learning. In: Proceedings of the 30th International Conference on Neural Information Processing Systems, pp. 3637–3645 (2016)
38. Wang, L., Ho, Y.S., Yoon, K.J., et al.: Event-based high dynamic range image and very high frame rate video generation using conditional generative adversarial networks. In: Proceedings of the IEEE/CVF Conference on Computer Vision and Pattern Recognition, pp. 10081–10090 (2019)
39. Xu, Z.Q.J., Zhang, Y., Luo, T., Xiao, Y., Ma, Z.: Frequency principle: Fourier analysis sheds light on deep neural networks. arXiv preprint arXiv:1901.06523 (2019)
40. Yoo, J., Uh, Y., Chun, S., Kang, B., Ha, J.W.: Photorealistic style transfer via wavelet transforms. In: Proceedings of the IEEE/CVF International Conference on Computer Vision, pp. 9036–9045 (2019)
41. Yu, Y., et al.: WaveFill: a wavelet-based generation network for image inpainting. In: Proceedings of the IEEE/CVF International Conference on Computer Vision, pp. 14114–14123 (2021)
42. Zhang, R., Isola, P., Efros, A.A., Shechtman, E., Wang, O.: The unreasonable effectiveness of deep features as a perceptual metric. In: Proceedings of the IEEE Conference on Computer Vision and Pattern Recognition, pp. 586–595 (2018)
43. Zhao, S., Liu, Z., Lin, J., Zhu, J.Y., Han, S.: Differentiable augmentation for data-efficient GAN training. Adv. Neural. Inf. Process. Syst. **33**, 7559–7570 (2020)

End-to-End Visual Editing with a Generatively Pre-trained Artist

Andrew Brown[1,2](✉), Cheng-Yang Fu[2], Omkar Parkhi[2], Tamara L. Berg[2], and Andrea Vedaldi[1,2]

[1] Visual Geometry Group, University of Oxford, Oxford, UK
abrown@robots.ox.ac.uk
[2] Meta AI Research, New York, US
{chengyangfu,omkar,tlberg,vedaldi}@fb.com
https://www.robots.ox.ac.uk/abrown/E2EVE/

Abstract. We consider the targeted image editing problem, namely blending a region in a source image with a driveg that specifiesthe desired change. Differently from prior works, we solve this problem by learning a conditional probability distribution of the edits, *end-to-end* in code space. Training such a model requires addressing the lack of example edits for training. To this end, we propose a self-supervised approach that simulates edits by augmenting off-the-shelf images in a target domain. The benefits are remarkable: implemented as a state-of-the-art auto-regressive transformer, our approach is simple, sidesteps difficulties with previous methods based on GAN-like priors, obtains significantly better edits, and is efficient. Furthermore, we show that different blending effects can be learned by an intuitive control of the augmentation process, with no other changes required to the model architecture. We demonstrate the superiority of this approach across several datasets in extensive quantitative and qualitative experiments, including human studies, significantly outperforming prior work.

1 Introduction

A key part of the creative process is the ability to combine known factors in novel ways. For instance, we can imagine how a dress would look like with a different v-neck, or our bedroom would look like with the large windows we have seen in a magazine. In this paper, we thus consider the problem of generating new variants of a source image, guided by another image containing a feature, such as a component of a dress or window style, that we wish to change in the source. For additional control, we wish the edit operation to focus on a particular target region of the source, leaving the context as unchanged as possible while maintaining the realism of the edit (see Fig. 1).

Supplementary Information The online version contains supplementary material available at https://doi.org/10.1007/978-3-031-19784-0_2.

S. Avidan et al. (Eds.): ECCV 2022, LNCS 13675, pp. 18–35, 2022.
https://doi.org/10.1007/978-3-031-19784-0_2

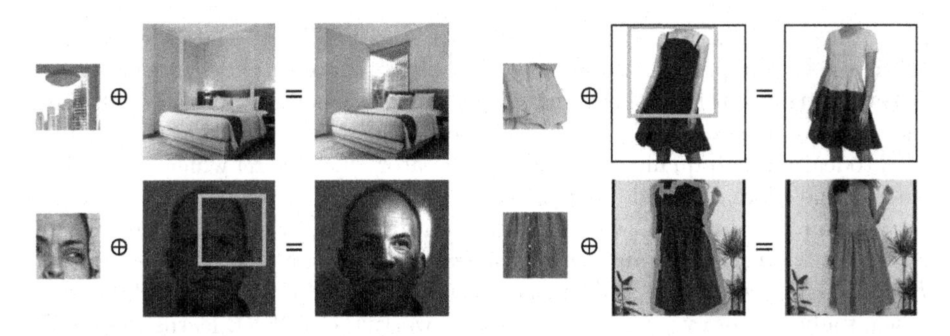

Fig. 1. E2EVE combines a driver and source image (resp. to the left and right of the $\oplus$ symbol), generating a new version of the source that resembles the driver in the edit region (marked in blue). The generated output looks realistic while faithfully resembling the driver. Our method can be trained to work well on different types of images, including bedrooms, dresses, and faces, and can use regions of arbitrary shape, from rough rectangles to pixel-accurate segmentations (bottom-right). (Color figure online)

Prior works consider image editing tasks, but often guided by a *textual* description of the desired change [21,43,51]. We argue that specifying edits visually rather than textually offers a far more fine grained and explicit level of control, ultimately resulting in a more useful editor[1]. Formally, we can describe the editing process as drawing a sample from a conditional image distribution $P(\hat{x}|x, y, R)$, where x is the source image, y is the driver image, R the edit region and $\hat{x}$ is an edited version of the source x. The goal of the edit $\hat{x}$ is to look natural while being close to the source x everywhere except for the region R, where it should resemble the driver image y.

A main challenge in learning the model $P(\hat{x}|x, y, R)$ is the lack of suitable training data, namely quadruplets $(\hat{x}, x, y, R)$, that exemplify the desired mapping. Most authors have thus proposed to focus on learning an unconditional prior distribution $P(x)$ of images, for which abundant training data is usually available, and then seeking an edit $\hat{x}$ that is both likely according to the prior and close in some sense to the source x and the driver y. This can be achieved in a pre-processing [8,76] or post-processing stage [4], and often uses a Generative Adversarial Network (GAN) to model the prior $P(x)$. Although demonstrating some impressive results, such approaches offer limited control on the edit $\hat{x}$, which either shows only a weak dependency on the driver image y, or does not stay on the image manifold $P(x)$, resulting in undesirable artifacts.

In this work, we overcome these challenges by considering a different approach where we learn the conditional distribution $P(\hat{x}|x, y, R)$ directly, *end-to-end* in code space. For this, we propose new ways of *synthesising* suitable training quadruplets $(\hat{x}, x, y, R)$ on a large scale and without requiring manual intervention. We do this in a self-supervised manner: given an image $\hat{x}$, we select an edit region R at random and use it to decompose the image into source x and driver y images, so that the edit can be written as a (random) function $(x, y, R) = f(\hat{x})$

[1] After all, a picture is worth a thousand words!.

of $\hat{x}$. A shortcoming is that such x and y are statistically correlated, whereas in a "creative" edit process a user must be able to choose y independently of x. A key contribution is to show that, if the process f is carefully designed, then the resulting images x and y are independent *enough*, meaning that they can be used to learn a high-quality conditional generator $P(\hat{x}|x, y, R)$ which works even when x and y are sampled independently.

We pair this intuition with the adoption of state-of-the-art auto-regressive image modelling using transformers for learning and sampling the distribution $P(\hat{x}|x, y, R)$. Overall, our End-to-End Visual Editor (E2EVE) approach has significant advantages over prior image editing works: (1) E2EVE learns to *directly* map the input to the output representations; (2) hence, based on extensive qualitative, quantitative and human-analysis experiments on several datasets, it results in higher quality edits that are simultaneously more dependent on the driver image and more natural looking than prior works based on GANs and attention which train task-agnostic priors not necessarily optimal for editing; (3) it is generally easier to implement and tune than GAN-based alternatives; and (4) it is efficient because it allows the direct sampling of edits without involving the expensive pre- or post-processing steps required by some prior methods.

2 Related Work

Targeted Generative Image Editing. Approaches for editing images in targeted locations include spatial manipulation of objects [77], adding or removing a closed-set of objects [5,6], or text-driven manipulation using CLIP [4,58]. These GAN-based approaches have some downsides: First, they require inverting GANs to represent the input image—a difficult problem [1,2,7,25,38,77] which can limit the *editability* of the images [2,61]. Second, they are not trained end-to-end. Third, text-driven approaches offer limited fine-grained control of shape and texture [4]. We address such shortcomings by training a sequence-to-sequence model end-to-end for the task of targeted image-based visual editing.

Note that the targeted image manipulation task which we address, differs from spatially-conditioned generation [31,37,45,64,68,75], or global image manipulation, where an entire image is generated/manipulated [35,49]. In global manipulation works, interpolations in the latent space are located which correspond to edits over the entire image, either via visual attribute classifiers [10,33,57,71], unsupervised disentanglement [26,47,56,63,67], or via image-text similarity [3,11,22,40,46,55,68]. Different to in-painting [29,39], in our task, generation depends on the driver image as well as image context.

Image Composition. Image composition combines possibly inconsistent images into a single cohesive output. Previous approaches include collaging nearest neighbors [30], composing foregrounds and backgrounds [62], composing a closed set of visual attributes [42,48], or using semantic pyramids [59]. Others fuse images by projecting their composition on the manifold generated by a GAN via inversion [2,8,23,32,70,76]. These methods work well if the driver image is sufficiently aligned to the source (which usually requires manual intervention), but worse than our end-to-end model when this is not the case.

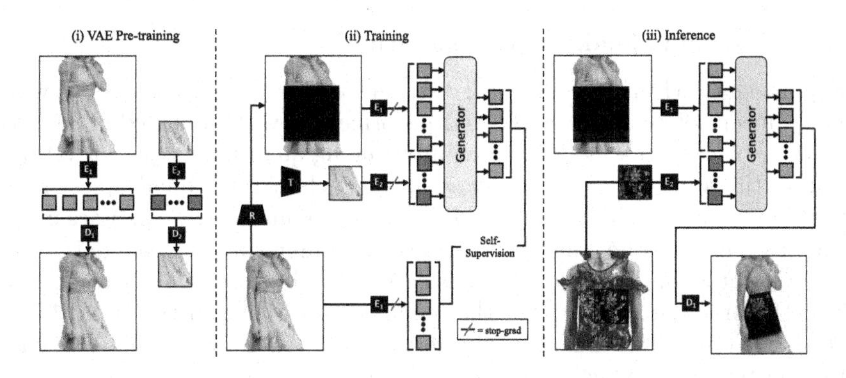

Fig. 2. The E2EVE approach: (i) **VAE Pre-Training**: we train two quantized VAEs (one for whole images, and one for image patches), each consisting of an Encoder, E, and decoder, D. (ii) **Training**: Each data sample produces a masked source image (via the operation R), and driver image (via random transformation T on the masked region). Given these conditioning inputs, the model is self-supervised to predict the data sample. Following prior work, the VAEs are kept frozen while training the generator. (iii) **Inference**: E2EVE generates edited images when the source and driver are sampled independently from different images.

Two Stage Image Synthesis. We adopt state-of-the-art two stage auto-regressive models [9,12,15,18,19,52,69] for the image generator. These models scale better than sequence-to-sequence [14,50] models applied directly to pixels by reducing first the dimensionality of images via a discrete autoencoder [19,20,44,53]. Others have recently built on this work for text- or class-driven image manipulation [11,17,66], whereas we consider image-driven editing [74]. Shows a (single) qualitative result for image-based out-painting using BERT [14], where a small set of tokens in the output sequence is fixed (termed *preservation controls*), and paired training data is sourced from the same image. We do not solve the out-painting problem, but the targeted editing problem, with the added challenge of preserving context while mixing images with very different statistics.

3 Method

We wish to learn a model that can "naturally" blend a given source image x with a user-provided driver image y. Formally, we denote with $x, \hat{x} \in \mathbb{R}^{3\times H\times W}$ the source and output (RGB) images and with $y \in \mathbb{R}^{3\times H'\times W'}$ the driver image (where, usually, $H > H'$ and $W > W'$). Furthermore, we target the edit operation on a region $R \in \{0,1\}^{H\times W}$, expressed as a binary mask. We cast the problem as one of learning a conditional probability distribution $P(\hat{x}|x, y, R)$ and then sample the output image $\hat{x}$ conditioned on the source image x, the driver image y, and the edit region R (Fig. 1).

Next, we discuss the advantages and requirements of this approach (Sect. 3.1), propose a self-supervised learning formulation that does not require any manually-provided labels to train the model (Sect. 3.2), and give the technical details of the neural network that learns the conditional distribution (Sect. 3.3). An overview of our training and inference settings is shown in Fig. 2.

3.1 End-to-End Conditional Generation

Our approach is to learn a conditional distribution $P(\hat{x}|x, y, R)$ *end-to-end*, which we do by means of an auto-regressive transformer network discussed in Sect. 3.3. In order to train such a model, we require training quadruplets $(\hat{x}, x, y, R)$ sampled from the joint distribution $P(\hat{x}, x, y, R)$. Each of these quadruplets represents the outcome of a "creative" process, where a human artist combines images x and y to generate a new image $\hat{x}$. Because obtaining such training data would require the intervention of human artists, it would be very difficult to obtain a sufficiently large dataset to learn the required conditional distribution. Hence, much of the research in image editing focuses on how to avoid this bottleneck and use instead data which is readily available.

A popular approach is to consider an *indirect* formulation and learn instead an unconditional image distribution $P(x)$, for instance expressed as a GAN generator $x = G(z)$. Then, the output image $\hat{x} = G(z^*)$ is "sampled" via an optimization process such as $z^* = \mathrm{argmin}_z\, d(G(z)|_R, y)$ where $d(\hat{x}|_R, y)$ measures compatibility between the region R of the generated image $\hat{x}$ and the driver image y. The advantage is that the model G can be learned from a collection $\mathcal{X}$ of unedited images $x \sim P(x)$, which is often easy to obtain at scale. The disadvantage is that this model is not optimized for the final task of image editing.

By contrast, in our approach we learn directly the model $P_\theta(\hat{x}|x, y, R)$ minimizing the standard negative log-likelihood loss:

$$\theta^* = \underset{\theta}{\mathrm{argmin}} \left(-\frac{1}{|\mathcal{T}|} \sum_{(\hat{x},x,y,R)\in\mathcal{T}} \log P_\theta(\hat{x}|x, y, R) \right) \tag{1}$$

where $\mathcal{T}$ is a large collection of training quadruplets. Once learned, we can directly draw samples $\hat{x} \sim P_{\theta^*}(\hat{x}|x, y, R)$. The main challenge is how to obtain the training set $\mathcal{T}$. The key to our method is a way of constructing $\mathcal{T}$ from $\mathcal{X}$ in an automated fashion, at no extra cost. This is explained in the next section.

3.2 Synthesizing a Dataset of Meaningful Edits

Given a training set $\mathcal{X}$ of unedited images x, the goal is to create a dataset $\mathcal{T}$ of "edits" $(\hat{x}, x, y, R)$ consisting of the generated image $\hat{x}$, the source image x, the diver image y, and the edit region R. The difficulty is that these quadruplets should be representative of a "creative" process where the generated image $\hat{x}$ is a meaningful blend of the source and driver images, x and y, according to a human artist. Specifically, $\hat{x}$ should resemble x as much as possible except in the region R, where it should take the character of y, but without introducing unnatural artifacts (e.g., simply pasting y on top of x would not do).

We propose to build such quadruplets as follows (see also Fig. 4). We sample an output image $\hat{x}$ from the unedited collection $\mathcal{X}$, thus pretending that the latter is, in fact, the result of an edit operation. Then, we define the source and driver images for this virtual edit as follows:

$$x = (1 - R) \odot \hat{x}, \qquad y = T(R \odot \hat{x}), \tag{2}$$

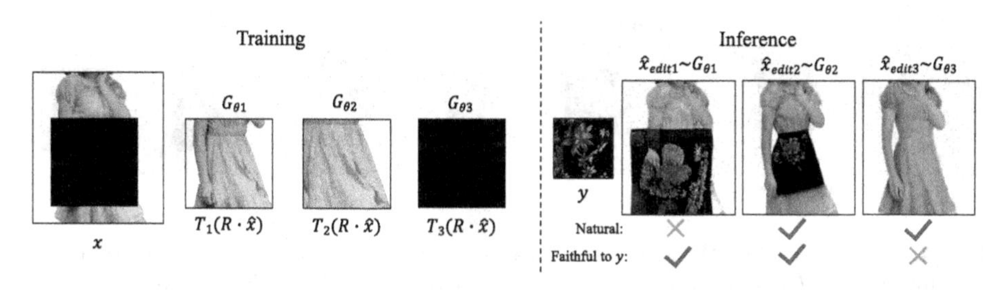

Fig. 3. Intuition for augmenting training inputs. During training, the driver image y is computed by applying a random transformation T to the masked region of the source image x. Left: Three options for T that are used to train three different generators, $G_{\theta n}$. Right: Samples from the three trained models when conditioned on x and an independently sampled y. An optimal choice of T removes just enough information that generated images are both natural-looking, and are faithful to the driver image y.

where R is the mask of a random image region, $\odot$ is the element-wise product (where broadcasting is used as required), and $T : \mathbb{R}^{3\times H\times W} \rightarrow \mathbb{R}^{3\times H'\times W'}$ is a *random image transformation*, also known as an "augmentation".

By optimizing the log-likelihood loss in eq. (1), the model $P_\theta(\hat{x}|x, y, R)$ learns to predict the full image $\hat{x}$ from x (which misses the region R), and y, which preserves some information about the missing region. Because $\hat{x}$ is originally an unedited image, the model learns to predict a natural-looking output.

The key design choice here is the random transformation T. For example, if we set $T = 1$ to be the identity function, then the output image can be reconstructed exactly as $\hat{x} = x + y$; in this case, the model $P_\theta(\hat{x}|x, y, R)$ learns to paste y onto x, which is uninteresting (see model G_{θ_1} in Fig. 3). Inversely, if we set $T = 0$ to be the null function, then y does not provide any information; in this case, the model $P_\theta(\hat{x}|x, y, R)$ learns to inpaint $\hat{x}$, filling in the missing region in a non-trivial manner, but ignoring the driver image y altogether (see model G_{θ_3} in Fig. 3). The augmentations T should find a *sweet spot* and remove just the right amount of information from y (see model G_{θ_2} in Fig. 3). While finding the optimal choice for the random transformations T is ultimately an empirical process, we describe next some important design criteria that were crucial for our results.

Decorrelating Source and Driver Images. A difficulty with our approach is that, because both source image x and driver image y are derived from the same image $\hat{x}$, they are *not* independent but *paired*. This is a problem because the user should be free to choose almost *any* driver image y for editing, so the generator must work well for *unpaired* inputs x and y too. We can approximate this condition by making x and y as uncorrelated as possible during training.

While we cannot achieve this result exactly, we propose two methods to approximate it, *block* and *free-form*, both shown in Fig. 4. **Block edits** are simple: we let the transformation $T(R \odot \hat{x})$ take a further sub-crop R_T of the crop R it is given as input, thus removing the most direct source of correlation between

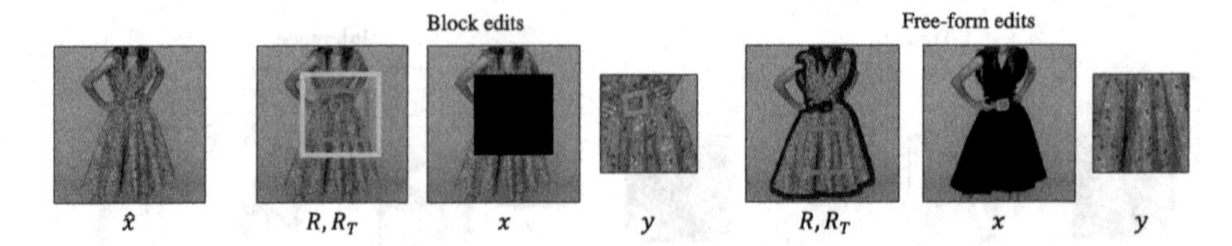

Fig. 4. *Left: block edits.* We sample an output image $\hat{x}$ and generate the corresponding edit input (x, y, R) by sampling a square region R, a transformation T cutting a sub-region R_T from R, and extracting the source image x and a driver y from these two. *Right: free-form edits.* The same approach, but this time R is a pixel-accurate segmentation mask extracted manually or automatically (R_T remains a square sub-region).

x and y: spatial continuity. Furthermore, we found empirically that if the sub-crop is always centered in R and of a fixed relative size w.r.t. R, then the model learns, as one would expect, to paste this crop in the middle. Instead, we let T further randomize the position (`pos_augment`) and size (`size_augment`) of the sub-crop relative to the edit region during training. We parameterize the sub-cropping operation via α, which defines the ratio of the sub-crop width, to the edit region width. The `size_augment` operation allows α to vary during training. The model hence learns to find a meaningful placement for the patch y in the context of x, without assuming spatial continuity or a specific geometric arrangement. Optionally, we further decorrelate source and driver images via **free-form edits**. The difference is that we let R be the output of a semantic segmentation network, while T still takes a square sub-crop R_T from region R. Because y is fully contained in the edit region R and the latter separates a foreground object from the background, this significantly reduces the correlation between x and y. While this approach requires additional machinery (the segmentation network), empirically it can obtain impressive results (Fig. 5).

Controlling the Learned Editor. In order to favour generalization, the augmentations above should remove as much information as possible from the crop y *except* for the information that the editor should transfer from the driver image y to the generated image $\hat{x}$. For example, it would be possible to consider further augmentations such as color jitter, but this would cause the editor to learn to ignore the color, which we usually wish to transfer. In general, by choosing different augmentations we can *control* what information the editor learns to transfer from the driver image to the generated one (e.g., style and colour), and what to ignore (e.g., the specific spatial arrangement).

3.3 Two-Stage Conditional Auto-regressive Image Generation

In order to implement the conditional distribution $P_\theta(\hat{x}|x, y, R)$, we use an *auto-regressive (AR) model.* AR models have been shown to be highly expressive for image generation [19,52], they can be conditioned on multiple signals elegantly and without architectural changes, and, unlike GANs [54], are mode-covering.

In practice, this leads to more varied generation results and the ability to model datasets with more variation. We summarise next how this model is applied to our case and point the reader to the supp. mat. for additional details.

The goal is to model a conditional distribution $P(\hat{x}|c)$, where c lumps together all conditioning information. An AR model further decomposes $\hat{x} = (\hat{x}_1, \dots, \hat{x}_M)$ into M components and factorizes the distribution as the product $P(\hat{x}|c) = \prod_{m=1}^{M} P(\hat{x}_m|\hat{x}_1, \dots, \hat{x}_{m-1}, c)$. The model is trained by minimizing the negative log likelihood (1) (avoiding unstable adversarial techniques used in GANs). For modelling images x, a challenge lies in finding a suitable decomposition, such that the individual factors $P(\hat{x}_m|\hat{x}_1, \dots, \hat{x}_{m-1}, c)$ can be implemented effectively. To this end, we build upon the two stage process of Esser *et al.* [19] and use a transformer on top of a discrete autoencoder. Note that the focus of this work is on end-to-end targeted image editing and the training formulation; we describe the two-stage auto-regressive method for completeness and reproducibility.

Specifically, in the first stage we learn a compressed and discretized representation $z = \Phi(\hat{x}) \in \{1, \dots, K\}^M$ of the images, where here K is the size of the discrete encoding space and M the resulting number of discrete tokens. For this, we use the VQ-GAN [19], achieving 16-fold image compression (we use separate encoders for x and y). Naturally, the encoder comes with a paired decoder $\hat{x} = \Psi(z)$ that allows to reconstruct the image from the code. This achieves two important goals: it allows (1) to scale the generator model to higher resolution images, and (2) to predict discrete distributions for the second stage.

The second stage uses a transformer to model the factors $P(\hat{x}_m|\hat{x}_1, \dots, \hat{x}_{m-1}, c)$. Specifically, the conditioning information $c = (x, y, R)$ consists of source image x, driver image y and region R. The sequence of tokens $S_m = (z_1, \dots, z_m) \oplus \Phi(x) \oplus \Phi(y)$ (where $\oplus$ denotes concatenation), comprising the partially-predicted output tokens along with the conditioning tokens, is fed to the transformer to output a K-dimensional histogram $P(z_m = \cdot|S_m)$. Spatial encodings are added to the tokens, but there is no need to explicitly encode the region R as it can be inferred from x because $x = (1 - R) \odot \hat{x}$ has a R-shaped 'hole'. While this way of encoding for R may seem naïve, it is in fact simple and powerful: prior work such as EdiBERT [32] use "occlusion tokens", and hence lose the ability to express pixel-accurate edit regions, which we can do effortlessly.

In practice, we train a GPT-2 [50] style transformer. Because the factors $P(\hat{x}_m|\hat{x}_1, \dots, \hat{x}_{m-1}, c)$ require each predicted token to depend only on those prior to it in the sequence, GPT-2 uses causal masking allowing only unidirectional attention towards earlier tokens in the sequence. All factors are trained efficiently in parallel using *teacher forcing* [24,65]. During inference, only the conditioning information is provided so the target sequence is predicted iteratively, sampling one symbol z_m at a time from the corresponding histogram. In practice, inference is faster than for some GAN alternatives, as shown in the sup. mat.

4 Experiments

We compare our method to others that, given a source image x and a driver image y, produce one or more edits $\hat{x}$. Good edits have three properties: (1) *naturalness* (the edit $\hat{x}$ looks like a sample from the prior $P(x)$); (2) *locality* ($\hat{x}$ is close to x outside the edit region R—although a certain amount of slack is necessary to allow the edit to blend in naturally); and (3) *faithfulness* ($\hat{x}$ resembles y within the edit region R). Achieving only one of the three objectives is trivial (for example, setting $\hat{x} = x$ ignoring y is natural and local but unfaithful whereas copying y on top of x is local and faithful but unnatural) so a good model must seek for a trade-off between these properties.

Measuring these properties is not entirely trivial; for the quantitative analysis, we take the standard FID measure for naturalness [27], the L^1 distance $\|(1 - R) \odot (\hat{x} - x)\|_1$ to measure locality, and a retrieval approach to measure faithfulness. For the latter, we consider a set $\mathcal{Y}_{\text{dstr}}$ of 100 distractor images of the same size as the driver image y, use the edited region $\hat{x}|_R$ as a query, and find its nearest neighbour $y^* = \text{argmin}_{\hat{y} \in \{y\} \cup \mathcal{Y}_{\text{dstr}}} d(\hat{x}|_R, \hat{y})$, incurring the loss $\delta_{y^* \neq y}$. In this expression, $d(\cdot, \cdot)$ is the Inception v3 [60] feature distance (pre-trained on ImageNet [13]), which is the same encoder used for FID calculation.

Evaluation Data. Recall that our goal is to evaluate the quality of automated editing algorithms. To feed such algorithms, we need triplets (x, y, R) consisting of a source image x, a driver image y and an edit region R. In Sect. 3.2 we explained how to build such a dataset for the purpose of *training* our model—with the added complexity that, for training, we *also* need to know the result $\hat{x}$ of the edit process. We could use the same dataset for evaluation, but this would unfairly advantage our model. Instead, since knowing the output $\hat{x}$ is *not* required to measure naturalness, locality and faithfulness, we are free to choose *new* and less constrained triplets (x, y, R) for evaluation, resulting in more challenging edits and a fairer evaluation. However, we wish to avoid too many cases in which cohesive blending is impossible (*e.g.*, where y is a patch of sky and $x|_R$ is a face). Hence, we build evaluation triplets as follows: given a sample image x and an edit region R, we define $y = x'|_R$ to be a crop taken at the same spatial location from a *different* image x' in the dataset. The effect is to (very) weakly constrain $x|_R$ and y to be compatible (e.g., both sky regions, or face regions) by exploiting the photographer bias in the datasets we consider.

We conduct experiments on three datasets: (1) the private *Dresses-7M* dataset containing 7 million images mainly depicting a woman wearing a dress; (2) LSUN bedrooms [72] containing 3M images of bedrooms; and (3) FFHQ [35], containing 70K aligned faces. We sample x by considering 1024 images from the validation sets of *Dresses-7m* and FFHQ, and 256 for LSUN bedrooms (due to its small size). For each (x, y, R), we consider 10 edit samples and obtain naturalness, locality and faithfulness by averaging over all images thus generated (totalling 10,240 and 2,560 samples, respectively). All images in this paper are from UnSplash[2] (dresses and bedrooms), or DFDC [16] (faces).

Implementation Details. We use a transformer architecture with 24 layers, 16 head multi-head attention, embedding size 1024, and we train it using standard

[2] www.unsplash.com.

cross-entropy loss. The masked source image $(1-R)\odot x$ and the driver image y are encoded using VQ-GANs with 16× compression and 1024 codebook size. Source x and output $\hat{x}$ images have resolution 256×256 and y resolution 64×64. The token sequences S_m (comprising the coded x, y and partial $\hat{x}$) has maximum length 516. Our final model uses `pos_augment` and `size_augment`—the latter forms y by taking a sub-crop in the region R, with α varying from 0.4 to 0.7. We use a batch size of 512, and the AdamW [41] optimizer with learning rate 4.5e-6. During inference, for each input (x, y, R), we generate 20 samples $\hat{x}$ and keep the 10 with highest similarity to y (a method termed *Filter*, in Table 2). In order to focus the sampling on more realistic/likely outputs we use *nucleus sampling* [28] with a p-value of 0.9. Following prior work [19,52], the VQ-GANs are kept frozen when training the transformer.

Baselines. We compare our method against the following image composition baselines: (1) *Copy-paste* generates $\hat{x}$ by pasting y onto x at the specified location R; (2) *Inpaint* ablates our method by removing the tokens y from the input, thus generating $\hat{x}$ by inpainting the region R unconditionally while disregarding y; (3) *GAN inv*, inspired by [32,70,76], takes the copy-paste output and uses the StyleGANv2 [36] or StyleGANv2-ADA [34] networks to re-encode and thus denoise the resulting image via GAN inversion [2], "blending" the edit; (4) *EdiBERT* [32] is a related transformer-based approach, which iteratively refines the output of copy-paste output using BERT [14] (for fairness, we use the same VQ-GAN and sample filtering by similarity to driver as for our method); (5) *In-Domain GAN* [76] uses a regularised form of GAN inversion to blend source and driver images. Pre-trained models are available for all test datasets except *Dresses-7M*; unfortunately, we were unable to successfully train the GAN-based models on the latter (possibly due to the significant diversity of this data), so in this case we limit the other baselines. Some off-the-shelf models are trained on the validation sets that we use for testing, which disadvantages our approach in the comparison. For more details on experimental settings, please see supp. mat.

4.1 Quantitative Evaluation

Block Edits. Table 1 reports the evaluation metrics for all baselines and datasets. Our approach significantly outperforms others in **naturalness**: because our method is trained explicitly with the goal of blending source and driver images, it works even for cases where the images are poorly aligned, where prior works based on fitting priors on unedited images fail (see Fig. 5).

The copy-paste baseline outperforms other methods on the **faithfulness** metric but has very poor naturalness—this is expected as the edited image contains a 1-to-1 copy of the driver image. The opposite is true for the inpainting baselines, which attain good naturalness but very poor faithfulness as they ignore the driver image altogether. Our method is second only to copy-paste in faithfulness while also scoring best in naturalness. Other baselines sit somewhere in between, but generally do not fair very well in faithfulness because, by projecting the composite image to the prior manifold, they distort the cue too much.

Table 1. Results for *block-* and *free-form edits*. Naturalness is computed over the whole image (Image), and just the edit-region (Edit-R) using FID. Faithfulness is computed via retrieval, where R@K measures whether or not the sample is retrieved in the top-k instances. Locality is measured using the L^1 distance outside of the edit-region

		Naturalness (↓)		Faithfulness (↑)			Locality (↓)
		Image	Edit-R	R@1	R@5	R@20	(L1)
Dresses-7m (*block-edits*)	Baseline: Copy-Paste	21.457	35.924	1.000	1.000	1.000	0.000
	Baseline: Inpaint	15.797	25.769	0.071	0.214	0.515	0.095
	EdiBERT [32]	17.193	32.621	0.554	0.837	0.963	**0.052**
	(ours) E2EVE	**14.411**	**24.743**	**0.797**	**0.937**	**0.978**	0.056
FFHQ (*block-edits*)	Baseline: Copy-Paste	33.330	25.811	1.000	1.000	1.000	0.000
	Baseline: Inpaint	18.328	12.665	0.421	0.704	0.895	0.139
	GAN inv [2]: StyleGANv2	26.583	16.223	0.590	0.823	0.948	0.198
	GAN inv [2]: StyleGANv2-Ada	26.657	16.290	0.593	0.821	0.949	0.199
	In-domain [76]	19.880	14.270	0.539	0.800	0.938	0.178
	EdiBERT [32]	13.192	12.230	0.718	0.925	0.983	**0.093**
	(ours) E2EVE	**12.770**	**10.574**	**0.853**	**0.970**	**0.994**	0.106
LSUN Bedrooms (*block-edits*)	Baseline: Copy-Paste	24.402	28.828	1.000	1.000	1.000	0.000
	Baseline: Inpaint	15.080	21.493	0.113	0.297	0.596	0.161
	GAN inv [2]: StyleGANv2	23.735	33.530	0.405	0.689	0.866	0.259
	In-domain [76]	32.333	43.544	0.171	0.363	0.608	0.208
	EdiBERT [32]	16.518	27.528	0.537	0.816	0.946	**0.111**
	(ours) E2EVE	**14.107**	**22.187**	**0.789**	**0.923**	**0.981**	0.119
Dresses-7m (*free-form edits*)	Baseline: Copy-Paste	23.107	58.259	0.581	0.700	0.817	0.000
	Baseline: Inpaint	13.718	24.516	0.193	0.385	0.659	0.103
	EdiBERT [32]	15.277	27.359	0.650	0.843	0.937	0.079
	(ours) E2EVE	**14.000**	**25.973**	**0.814**	**0.920**	**0.951**	**0.072**

As for **locality**, copy-paste is also optimal, as it does not change the context region at all. Compared to non-trivial baselines, our method is first or second best in this metric, affecting the context region much less than GAN methods. However, EdiBERT is also very competitive as it is designed to leave the context nearly exactly unchanged (via *periodic collage* of the output with the context). However, relaxing locality is often necessary to obtain a more reasonable blending effect—an intuitive fact that we show qualitatively in Fig. 5f.

Finally, we also conduct a **human-study** on the *Dresses-7M* dataset using Amazon Mechanical Turk. We showed 256 edit samples using our method and EdiBERT to 3 human assessors each, asking two questions: which of the two outputs is more realistic, and which is more faithful to the driver image. The results show that by majority vote, human annotators think that our samples are more natural 83.2% of the time, and more faithful 80.5% of the time. The key to the performance improvements are that E2EVE is trained *end-to-end* for targeted visual editing, whereas prior work are not. Furthermore, we outperform EdiBERT, which uses the same VQ-GAN with a 50% larger transformer.

Free-Form Edits. Here, we use a semantic segmentation network to extract the edit region R from image x, while we define y as a crop taken from within a semantic region of a different image x' (see Sect. 3.2) We conduct experiments on *Dresses-7M* and compare to EdiBERT, with results in Table 1. Our approach again outperforms the previous methods in terms of **naturalness** and **faithfulness** and, this time, **locality** too, because we can better capture irregular

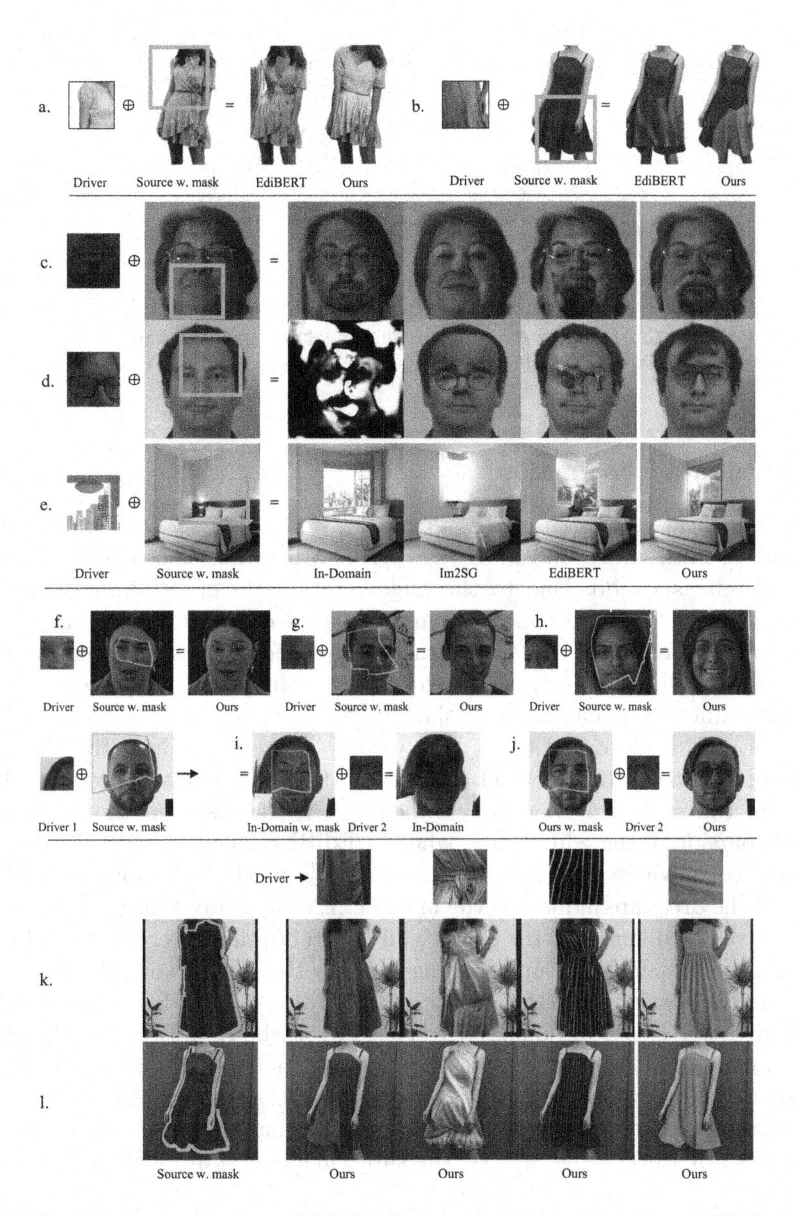

Fig. 5. Qualitative results from E2EVE. Rows a, b: Block edits from E2EVE trained on *Dresses-7M*. Row c, d: Comparisons to prior work trained on FFHQ. Row e: Comparisons to prior work trained on LSUN-Bedrooms. Rows f, g, h: Edits from E2EVE on FFHQ using random masks of increasing size. Rows i, j: one edit after another (sequential) on the same source image using prior work, and E2EVE, respectively. Rows k, l: *free-form* edits from E2EVE trained on *Dresses-7M*. Please zoom in for details. In each case, the masked region in the source image is that contained within the blue line. (Color figure online)

edit regions compared to EdiBERT which results in blocky artifacts. Additional experiments with random free-form masks can be found in the supp. mat.

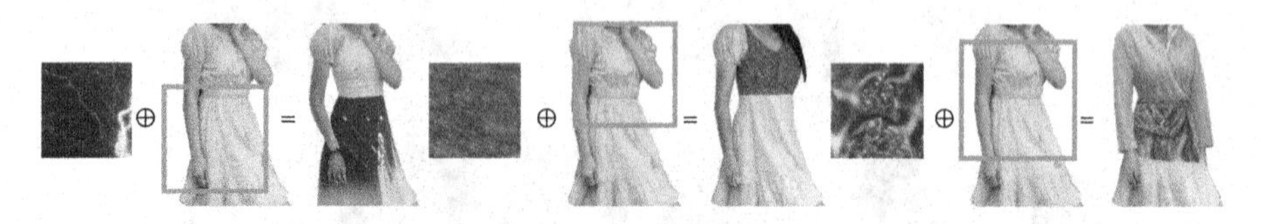

Fig. 6. E2EVE generalises surprisingly well to out-of-domain driver images (*e.g.* images of weather and nature). Three examples using the same source image from *block-edit Dresses-7M* model. As shown, E2EVE generates cohesive and varied samples.

4.2 Qualitative Evaluation

We show qualitative comparisons against prior work in Fig. 5. Our edits combine naturalness, faithfulness and locality, whereas others fail at achieving all three goals as well as we do. Due to the augmentations in our training edits, our method is better able to cope with uncorrelated driver images y than other approaches that only rely on a pre-learned unconditional prior distribution $P(x)$. For example, in Fig. 5c, d our approach can successfully mix images coming from faces with different gender or pose, showing better naturalness and faithfulness. As for locality, while EdiBERT is highly competitive in Table 1, this comes at a cost: in Fig. 5a, d our method achieves better naturalness by coloring both sleeves in the same way and by completing the glasses even though part of them lie outside of the edit region, whereas EdiBERT cannot. Fig. 5k, l shows free-form edits where the entire clothing item is masked. Although structural details of the dress are hidden by the mask, E2EVE generates natural and varied structure that is different to the source and faithful to the driver image. In Fig. 5e E2EVE generates more natural looking samples than prior work that edit with respect to the spatial geometry of the room. We also see that E2EVE generalises surprisingly well to out-of-domain driver images, as shown in Fig. 6 for the block-edit *Dresses-7M* model. In Fig. 5f, g, h, E2EVE generates impressive samples with small, medium or large random free-form masks. In Fig. 5j E2EVE makes sequential edits with random Rs while maintaining naturalness – empirically this fails with prior works (Fig. 5i). See the supp. mat. for additional results.

4.3 Ablations

In Table 2 we analyse and ablate design choices in E2EVE. We report additional metrics: negative log likelihood (NLL) on the validation set and sample diversity, computed pairwise between samples from the same inputs using LPIPS [73].

Table 2. Model ablations and sweeps for *block edits.* Key: NLL: Negative log likelihood. α, pos-aug, size-aug: the parameters used to define the sub-cropping transformation T. Filter: filtering E2EVE samples by visual similarity to the driver image. 2VQ: using two VQ-GANs rather than one. Datasets: D: *Dresses-7m*, B: Bedrooms, F-FFHQ.

							Naturalness (↓)		Faithfulness (↑)			Locality (↓)	NLL (↓)	Diversity (↑)	
	α	pos-aug	size-aug	Filter	2VQ	Data	Image	Edit-R	R@1	R@5	R@20	(L1)		Image	Edit-R
a.	0.8	✗	✗	✓	✓	D	17.241	30.076	0.882	0.980	0.996	0.056	2.181	0.135	0.309
b.	0.6	✗	✗	✓	✓	D	15.593	29.364	0.920	0.986	0.997	0.056	1.704	0.137	0.315
c.	0.4	✗	✗	✓	✓	D	13.967	26.975	0.811	0.954	0.988	0.056	1.594	0.139	0.327
d.	0.0	✗	✗	✓	✓	D	15.797	25.769	0.071	0.214	0.515	0.095	1.537	0.190	0.419
e.	0.6	✓	✗	✓	✓	D	15.605	26.513	0.887	0.980	0.997	0.056	1.518	0.142	0.338
f.	0.5-0.6	✓	✓	✓	✓	D	14.951	26.186	0.856	0.968	0.992	0.056	1.460	0.143	0.344
g.	0.4-0.7	✗	✓	✓	✓	D	14.589	27.824	0.864	0.970	0.992	0.056	1.494	0.139	0.328
h.	0.4-0.7	✓	✓	✓	✓	D	14.411	24.743	0.797	0.937	0.960	0.056	1.448	0.143	0.344
i.	0.4-0.7	✓	✓	✗	✓	D	13.913	24.583	0.611	0.817	0.929	0.056	1.448	0.145	0.351
j.	0.4-0.7	✓	✓	✗	✓	B	14.347	22.998	0.636	0.831	0.929	0.119	2.942	0.287	0.460
k.	0.4-0.7	✓	✓	✗	✓	F	12.636	10.699	0.723	0.899	0.976	0.106	2.392	0.203	0.321
l.		EdiBERT [32]		✗		B	16.643	29.775	0.356	0.627	0.823	0.111	-	0.291	0.575
m.		EdiBERT [32]		✗		F	13.036	12.891	0.536	0.778	0.925	0.093	-	0.181	0.423
n.	0.4-0.7	✓	✓	✓	✗	D	14.107	23.916	0.720	0.891	0.963	0.056	1.454	0.144	0.347

Starting from the construction of the training edits $\mathcal{T}$ (Sect. 3.2), reducing α (rows a–d) means removing more of the image $\hat{x}$ from the crop y. As predicted in Sect. 3.2, removing information from y increases naturalness (lower FID) at the expense of weaker faithfulness (lower R). $\alpha = 0.6$ provides a balance. `pos_augment` (row e vs. b) increases naturalness by preventing the model from simply pasting the driver image in the centre of the edit region. `size_augment` (row h) randomizes the choice of α in a range during training, so that the editor learns to automatically resize the driver image as needed. This significantly improves naturalness at the cost of a reduction of faithfulness (row h vs. e). In part, this is likely due to limitations of the retrieval model used to measure faithfulness, which struggles to cope with geometric deformations even when they preserve the style of the driver. α and augmentation have no effect on the locality (rows a–h). A benefit is that our final model generates more diverse samples (row h vs g) by learning to place the driver image at different positions and sizes. Interestingly, NLL is also minimised by the final model despite the fact that y is *less* correlated to $\hat{x}$ than in other cases: this is likely because additional augmentations reduce overfitting to the training data. For further discussion, please see supp. mat.

5 Conclusions, Limitations and Future Work

We present E2EVE, a new approach for targeted visual image editing. The key innovation is an effective method for self-supervising the model end-to-end, based on only an unlabelled collection of natural images. E2EVE learns a conditional image generator that responds well to diverse user inputs, significantly outperforming prior work qualitatively and quantitatively without any manual supervision. Limitations remain: our data generation technique might be difficult to extend to text-based edits and some proposed edits are unreasonable (see sup. mat.) because the model lacks a full understanding of the semantic content of images. Furthermore, as our model is unsupervised and data-driven, it may contain surprising unwanted biases. Next steps include extending E2EVE beyond images to spatial editing in 3D scenes and spatio-temporal video editing.

Acknowledgements. We are grateful to the advice and support of Yanping Xie, Antoine Toisoul, Thomas Hayes, and the EdiBERT authors.

References

1. Abdal, R., Qin, Y., Wonka, P.: Image2stylegan: How to embed images into the stylegan latent space? In: ICCV (2019)
2. Abdal, R., Qin, Y., Wonka, P.: Image2stylegan++: how to edit the embedded images? In: CVPR (2020)
3. Abdal, R., Zhu, P., Femiani, J., Mitra, N.J., Wonka, P.: Clip2stylegan: unsupervised extraction of stylegan edit directions. arXiv:2112.05219 [cs.CV] (2021)
4. Bau, D., et al.: Paint by word. arXiv:2103.10951 [cs.CV] (2021)
5. Bau, D., et al.: Semantic photo manipulation with a generative image prior. ACM Trans, Graph (2019)
6. Bau, D., Zhu, J.Y., Strobelt, H., Lapedriza, A., Zhou, B., Torralba, A.: Understanding the role of individual units in a deep neural network. In: Proceedings of the National Academy of Sciences (2020)
7. Bau, D., et al.: Inverting layers of a large generator. In: ICLR 2019 Debugging Machine Learning Models Workshop (2019)
8. Chai, L., Wulff, J., Isola, P.: Using latent space regression to analyze and leverage compositionality in gans. In: ICLR (2021)
9. Chen, M., et al.: Generative pretraining from pixels. In: ICML (2020)
10. Choi, Y., Choi, M., Kim, M., Ha, J.W., Kim, S., Choo, J.: Stargan: unified generative adversarial networks for multi-domain image-to-image translation. In: CVPR (2018)
11. Crowson, K.: VQGAN-CLIP (2021). https://github.com/nerdyrodent/VQGAN-CLIP
12. Dai, B., Wipf, D.: Diagnosing and enhancing VAE models. In: ICLR (2019)
13. Deng, J., Dong, W., Socher, R., Li, L.J., Li, K., Fei-Fei, L.: Imagenet: a large-scale hierarchical image database. In: CVPR (2009)
14. Devlin, J., Chang, M.W., Lee, K., Toutanova, K.: Bert: pre-training of deep bidirectional transformers for language understanding. In: NAACL (2019)
15. Ding, M., et al.: Cogview: mastering text-to-image generation via transformers. In: NeurIPS (2021)
16. Dolhansky, B., et al.: The deepfake detection challenge dataset (2020)
17. Esser, P., Rombach, R., Blattmann, A., Ommer, B.: Imagebart: bidirectional context with multinomial diffusion for autoregressive image synthesis. In: NeurIPS (2021)
18. Esser, P., Rombach, R., Ommer, B.: A disentangling invertible interpretation network for explaining latent representations. In: CVPR (2020)
19. Esser, P., Rombach, R., Ommer, B.: Taming transformers for high-resolution image synthesis. In: CVPR (2021)
20. Fauw, J.D., Dieleman, S., Simonyan, K.: Hierarchical autoregressive image models with auxiliary decoders. arXiv:1903.04933 [cs.CV] (2019)

21. Gafni, O., Polyak, A., Ashual, O., Sheynin, S., Parikh, D., Taigman, Y.: Make-a-scene: scene-based text-to-image generation with human priors (2022)
22. Galatolo., F., Cimino., M., Vaglini, G.: Generating images from caption and vice versa via clip-guided generative latent space search. In: Proceedings of the International Conference on Image Processing and Vision Engineering (2021)
23. Ghosh, P., Zietlow, D., Black, M.J., Davis, L.S., Hu, X.: Invgan: invertible gans. arXiv:2112.04598 [cs.CV] (2021)
24. Goyal, A., Lamb, A., Zhang, Y., Zhang, S., Courville, A., Bengio, Y.: Professor forcing: a new algorithm for training recurrent networks. In: NeurIPS (2016)
25. Guan, S., Tai, Y., Ni, B., Zhu, F., Huang, F., Yang, X.: Collaborative learning for faster stylegan embedding. arXiv:2007.01758 [cs.CV] (2020)
26. Härkönen, E., Hertzmann, A., Lehtinen, J., Paris, S.: Ganspace: discovering interpretable gan controls. arXiv:2004.02546 [cs.CV] (2020)
27. Heusel, M., Ramsauer, H., Unterthiner, T., Nessler, B., Hochreiter, S.: Gans trained by a two time-scale update rule converge to a local nash equilibrium. In: NeurIPS (2017)
28. Holtzman, A., Buys, J., Du, L., Forbes, M., Choi, Y.: The curious case of neural text degeneration. In: ICLR (2020)
29. Iizuka, S., Simo-Serra, E., Ishikawa, H.: Globally and locally consistent image completion. ACM Trans. Graph. **36**(4), 1–14 (2017)
30. Isola, P., Liu, C.: Scene collaging: analysis and synthesis of natural images with semantic layers. In: ICCV (2013)
31. Isola, P., Zhu, J.Y., Zhou, T., Efros, A.A.: Image-to-image translation with conditional adversarial networks. In: CVPR (2017)
32. Issenhuth, T., Tanielian, U., Mary, J., Picard, D.: Edibert, a generative model for image editing. arXiv:2111.15264 [cs.CV] (2021)
33. Jahanian, A., Chai, L., Isola, P.: On the "steerability" of generative adversarial networks. In: ICLR (2020)
34. Karras, T., Aittala, M., Hellsten, J., Laine, S., Lehtinen, J., Aila, T.: Training generative adversarial networks with limited data. In: NeurIPS (2020)
35. Karras, T., Laine, S., Aila, T.: A style-based generator architecture for generative adversarial networks. In: CVPR (2019)
36. Karras, T., Laine, S., Aittala, M., Hellsten, J., Lehtinen, J., Aila, T.: Analyzing and improving the image quality of StyleGAN. In: CVPR (2020)
37. Kim, H., Choi, Y., Kim, J., Yoo, S., Uh, Y.: Exploiting spatial dimensions of latent in gan for real-time image editing. In: CVPR (2021)
38. Lipton, Z.C., Tripathi, S.: Precise recovery of latent vectors from generative adversarial networks. arXiv:1702.04782 [cs.LG] (2017)
39. Liu, G., Reda, F.A., Shih, K.J., Wang, T.C., Tao, A., Catanzaro, B.: Image inpainting for irregular holes using partial convolutions. In: ECCV (2018)
40. Liu, X., et al.: More control for free! image synthesis with semantic diffusion guidance. arXiv:2112.05744 [cs.CV] (2021)
41. Loshchilov, I., Hutter, F.: Decoupled weight decay regularization. In: ICLR (2019)
42. Mokady, R., Benaim, S., Wolf, L., Bermano, A.: Mask based unsupervised content transfer. arXiv:1906.06558 [cs.CV] (2018)
43. Nichol, A., et al.: Glide: towards photorealistic image generation and editing with text-guided diffusion models (2021)
44. van den Oord, A., Vinyals, O., Kavukcuoglu, K.: Neural discrete representation learning. arXiv:1711.00937 [cs.LG] (2017)
45. Park, T., Liu, M.Y., Wang, T.C., Zhu, J.Y.: Semantic image synthesis with spatially-adaptive normalization. In: CVPR (2019)

46. Patashnik, O., Wu, Z., Shechtman, E., Cohen-Or, D., Lischinski, D.: Styleclip: Text-driven manipulation of stylegan imagery. arXiv:2103.17249 [cs.CV] (2021)
47. Peebles, W., Peebles, J., Zhu, J.-Y., Efros, A., Torralba, A.: The hessian penalty: a weak prior for unsupervised disentanglement. In: Vedaldi, A., Bischof, H., Brox, T., Frahm, J.-M. (eds.) ECCV 2020. LNCS, vol. 12351, pp. 581–597. Springer, Cham (2020). https://doi.org/10.1007/978-3-030-58539-6_35
48. Press, O., Galanti, T., Benaim, S., Wolf, L.: Emerging disentanglement in auto-encoder based unsupervised image content transfer. In: ICLR (2019)
49. Radford, A., Metz, L., Chintala, S.: Unsupervised representation learning with deep convolutional generative adversarial networks. arXiv:1511.06434 [cs.LG] (2016)
50. Radford, A., Wu, J., Child, R., Luan, D., Amodei, D., Sutskever, I.: Language models are unsupervised multitask learners. OpenAI blog **1**(8), 9 (2019)
51. Ramesh, A., Dhariwal, P., Nichol, A., Chu, C., Chen, M.: Hierarchical text-conditional image generation with clip latents (2022)
52. Ramesh, A., et al.: Zero-shot text-to-image generation. In: ICML (2021)
53. Razavi, A., van den Oord, A., Vinyals, O.: Generating diverse high-fidelity images with vq-vae-2. In: NeurIPS (2019)
54. Salimans, T., Goodfellow, I., Zaremba, W., Cheung, V., Radford, A., Chen, X.: Improved techniques for training gans. arXiv:1606.03498 [cs.LG] (2016)
55. Schaldenbrand, P., Liu, Z., Oh, J.: Styleclipdraw: Coupling content and style in text-to-drawing synthesis. arXiv:2111.03133 [cs.CV] (2021)
56. Schwettmann, S., Hernandez, E., Bau, D., Klein, S., Andreas, J., Torralba, A.: Toward a visual concept vocabulary for gan latent space. In: ICCV (2021)
57. Shen, Y., Gu, J., Tang, X., Zhou, B.: Interpreting the latent space of gans for semantic face editing. In: CVPR (2020)
58. Shi, J., Xu, N., Zheng, H., Smith, A., Luo, J., Xu, C.: Spaceedit: learning a unified editing space for open-domain image editing. arXiv:2112.00180 [cs.CV] (2021)
59. Shocher, A. et al.: Semantic pyramid for image generation. In: CVPR (2020)
60. Szegedy, C., Vanhoucke, V., Ioffe, S., Shlens, J., Wojna, Z.: Rethinking the inception architecture for computer vision. In: CVPR (2016)
61. Tov, O., Alaluf, Y., Nitzan, Y., Patashnik, O., Cohen-Or, D.: Designing an encoder for stylegan image manipulation. arXiv:2102.02766 [cs.CV] (2021)
62. Tsai, Y.H., Shen, X., Lin, Z.L., Sunkavalli, K., Lu, X., Yang, M.H.: Deep image harmonization. In: CVPR (2017)
63. Voynov, A., Babenko, A.: Unsupervised discovery of interpretable directions in the gan latent space. In: ICML (2020)
64. Wang, T.C., Liu, M.Y., Zhu, J.Y., Tao, A., Kautz, J., Catanzaro, B.: High-resolution image synthesis and semantic manipulation with conditional gans. In: CVPR (2018)
65. Williams, R.J., Zipser, D.: A learning algorithm for continually running fully recurrent neural networks. Neural Comput. **1**(2), 270–280 (1989)
66. Wu, C., et al.: N\" uwa: visual synthesis pre-training for neural visual world creation. arXiv:2111.12417 [cs.CV] (2021)
67. Wu, Z., Lischinski, D., Shechtman, E.: Stylespace analysis: Disentangled controls for stylegan image generation. arXiv:2011.12799 [cs.CV] (2020)
68. Xia, W., Yang, Y., Xue, J.H., Wu, B.: Tedigan: text-guided diverse face image generation and manipulation. In: CVPR (2021)
69. Xiao, Z., Yan, Q., Chen, Y.A., Amit, Y.: Generative latent flow. arXiv:1905.10485 [cs.CV] (2019)
70. Xu, Y., Shen, Y., Zhu, J., Yang, C., Zhou, B.: Generative hierarchical features from synthesizing images. In: CVPR (2021)

71. Yang, C., Shen, Y., Zhou, B.: Semantic hierarchy emerges in deep generative representations for scene synthesis. Int. J. Comput. Vis. **129**(5), 1451–1466 (2021). https://doi.org/10.1007/s11263-020-01429-5
72. Yu, F., Zhang, Y., Song, S., Seff, A., Xiao, J.: Lsun: construction of a large-scale image dataset using deep learning with humans in the loop. arXiv:1506.03365 [cs.CV] (2015)
73. Zhang, R., Isola, P., Efros, A.A., Shechtman, E., Wang, O.: The unreasonable effectiveness of deep features as a perceptual metric. In: CVPR (2018)
74. Zhang, Z., et al.: UFC-BERT: unifying multi-modal controls for conditional image synthesis. In: NeurIPS (2021)
75. Zhao, S., et al.: Large scale image completion via co-modulated generative adversarial networks. In: ICLR (2021)
76. Zhu, J., Shen, Y., Zhao, D., Zhou, B.: In-domain gan inversion for real image editing. In: Vedaldi, A., Bischof, H., Brox, T., Frahm, J.-M. (eds.) ECCV 2020. LNCS, vol. 12362, pp. 592–608. Springer, Cham (2020). https://doi.org/10.1007/978-3-030-58520-4_35
77. Zhu, J.-Y., Krähenbühl, P., Shechtman, E., Efros, A.A.: Generative visual manipulation on the natural image manifold. In: Leibe, B., Matas, J., Sebe, N., Welling, M. (eds.) ECCV 2016. LNCS, vol. 9909, pp. 597–613. Springer, Cham (2016). https://doi.org/10.1007/978-3-319-46454-1_36

High-Fidelity GAN Inversion with Padding Space

Qingyan Bai[1], Yinghao Xu[2], Jiapeng Zhu[3], Weihao Xia[4], Yujiu Yang[1(✉)], and Yujun Shen[5]

[1] Shenzhen International Graduate School, Tsinghua University, Shenzhen, China
bqy20@mails.tsinghua.edu.cn, yang.yujiu@sz.tsinghua.edu.cn
[2] CUHK, Shatin, Hong Kong
xy119@ie.cuhk.edu.hk
[3] HKUST, Sai Kung, Hong Kong
[4] University College London, London, UK
[5] ByteDance Inc., Shenzhen, China

Abstract. Inverting a Generative Adversarial Network (GAN) facilitates a wide range of image editing tasks using pre-trained generators. Existing methods typically employ the latent space of GANs as the inversion space yet observe the insufficient recovery of spatial details. In this work, we propose to involve the *padding space* of the generator to complement the latent space with spatial information. Concretely, we replace the constant padding (*e.g.*, usually zeros) used in convolution layers with some instance-aware coefficients. In this way, the inductive bias assumed in the pre-trained model can be appropriately adapted to fit each individual image. Through learning a carefully designed encoder, we manage to improve the inversion quality both qualitatively and quantitatively, outperforming existing alternatives. We then demonstrate that such a space extension barely affects the native GAN manifold, hence we can still reuse the prior knowledge learned by GANs for various downstream applications. Beyond the editing tasks explored in prior arts, our approach allows a more flexible image manipulation, such as the separate control of face contour and facial details, and enables a novel editing manner where users can *customize* their own manipulations highly efficiently. (Project page can be found here.)

1 Introduction

Generative Adversarial Network (GAN) [12] has received wide attention due to its capability of synthesizing photo-realistic images [7,24,26]. Recent studies have shown that GANs spontaneously learn rich knowledge in the training process, which can be faithfully used to control the generation [6,21,39]. However, the

Q. Bai and Y. Xu—Equal contribution.

Supplementary Information The online version contains supplementary material available at https://doi.org/10.1007/978-3-031-19784-0_3.

S. Avidan et al. (Eds.): ECCV 2022, LNCS 13675, pp. 36–53, 2022.
https://doi.org/10.1007/978-3-031-19784-0_3

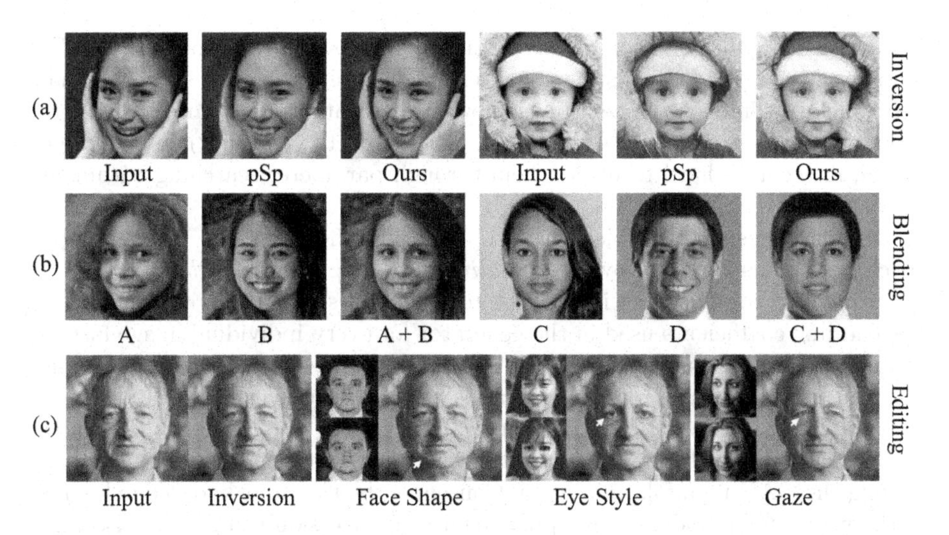

Fig. 1. Inversion and editing results obtained by PadInv. (a) Our method better reconstructs the out-of-distribution objects (*e.g.*, hands and hat) than pSp [36]. (b) Introducing the padding space to complement the latent space allows separate control of face contour and facial details, facilitating face blending. (c) Versatile manipulations can be *customized* with *only one* image pair (*i.e.*, on the left of each editing result)

emerging controllability is hard to apply to real-world scenarios. That is because the generator in a GAN typically learns to render an image from a randomly sampled latent code, and hence lacks the ability to make inferences on a target sample, limiting its practical usage.

The advent of GAN inversion techniques (*i.e.*, inverting the generation process of GANs) [35,53,55] appears to fill in this gap. The core thought is to convert a given image to some GAN-interpretable representations [46], which can be decoded by the generator to reconstruct the source. That way, real image editing can be simply achieved by manipulating the inverted representations, reusing the pre-trained generator as a learned renderer.

The crux of GAN inversion is to find the appropriate representations that can recover the input as much as possible. A common practice is to regularize the representations within the latent space of GANs [35,36,41,53], which best matches the generation mechanism (*i.e.*, the latent code uniquely determines the synthesis with the generator fixed). However, merely using the latent space suggests unsatisfactory reconstruction performance. The major reason causing such an issue is the inadequate recovery of some out-of-distribution objects, like the hands and hat in face images shown in Fig. 1a.

In this work, we re-examine the procedure of how an image is produced, oriented to GANs with convolution-based generators, and get a deeper understanding of why some spatial details cannot be well recovered. Recall that, to maintain the spatial dimensions of the input feature map, a convolution layer is asked to pad the feature map (*e.g.*, usually with zeros) before convolution. The padding is *not learned* in the training phase and hence introduces some inductive bias [45], which may cause the "texture-sticking" problem [24]. For instance, some padded constants may encode hair information such that hair gets stuck to some synthesized pixels however the latent code varies [24]. Hereafter, when those pixels are filled with other objects (*e.g.*, hat) in the target image, it becomes hard to invert them through parameter searching within the latent space.

To alleviate such a problem, we propose a high-fidelity GAN inversion approach, termed as **PadInv**, by *incorporating the padding space of the generator as an extended inversion space* in addition to the latent space. Concretely, we adapt the padding coefficients used in the generator for every individual image instead of inheriting the inductive bias (*e.g.*, zero padding) assumed in GAN training. Such an instance-aware reprogramming is able to complement the latent space with adequate spatial information, especially for those out-of-distribution cases. It is noteworthy that the convolutional weights are still shared across samples, leaving the GAN manifold preserved. Consequently, the prior knowledge learned in the pre-trained model is still applicable to the inversion results. We also carefully tailor an encoder that is compatible with the newly introduced padding space, such that an image can be inverted accurately and efficiently.

We evaluate our algorithm from the perspectives of both inversion quality and image editing. On the one hand, PadInv is capable of recovering the target image with far better spatial details than state-of-the-art methods, as exhibited in Fig. 1a. On the other hand, our approach enables two novel editing applications that have not been explored by previous GAN inversion methods. In particular, we manage to control the image generation more precisely, including the *separate manipulation of spatial contents and image style.* As shown in Fig. 1b, we achieve face blending by borrowing the contour from one image and facial details from another. Furthermore, we come up with an innovative editing manner, which *allows users to define their own editing attributes.* As shown in Fig. 1c, versatile manipulations are customized with *only one* image pair, which can be created super efficiently either by graphics editors like Photoshop (*e.g.*, face shape) or by the convenient copy-and-paste (*e.g.*, eye style).

2 Related Work

Generative Adversarial Networks (GANs). Recent years have witnessed the tremendous success of GANs in producing high-resolution, high-quality, and highly-diverse images [7,24–26]. Existing studies on GAN interpretation have affirmed that, pre-trained GAN models own great potential in a range of downstream applications, such as object classification [11,46], semantic segmentation [50], video generation [24], and image editing [5,13,15,21,29,31,38,39,43, 47,51,52].

GAN Inversion. GAN inversion [44,55] aims at finding the reverse mapping of the generator in GANs. Active attempts broadly fall into two categories, which are optimization-based [8,13,18,30,32,35] and learning-based [2,33,34,36,41,46, 53–55]. However, they typically employ the native latent space of the generator as the inversion space yet observe inadequately recovered spatial details.

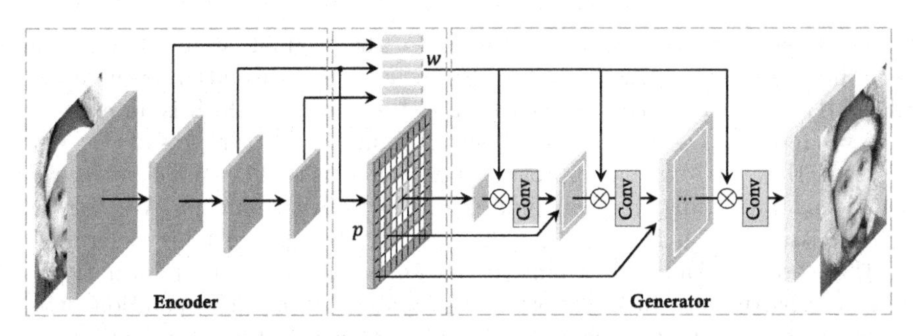

Fig. 2. Framework of the proposed PadInv, which learns an encoder to invert the generation of a pre-trained convolution-based generator, $G(\cdot)$. Given a target image, our encoder not only maps it to layer-wise latent codes (*i.e.*, $\mathbf{w}$), but also produces a coefficient tensor (*i.e.*, $\mathbf{p}$) to *replace the padding* used in the convolution layers of $G(\cdot)$. $\otimes$ denotes the AdaIN [17] operation. The convolutional kernels of $G(\cdot)$ are *fixed* in the training phase and *shared* across all samples

Some also [3,10,37] try to fine-tune the pre-trained generator, leading to better inversion but much slower speed. To make GAN inversion spatially aware, some attempts [1,26] incorporate the noise space of StyleGAN [25], while some concurrent studies [22,42,56] propose to invert an image to an intermediate feature map such that only half of the generator will be used as the decoder. By contrast, we involve the *padding space* of the generator to complement the latent space with spatial information. In doing so, the prior knowledge contained in the pre-trained model is barely affected, and hence our inversion results support various editing tasks well.

Padding in Convolutional Neural Networks (CNNs). The side effect of constant padding used in CNNs has recently been studied [4,19,20,24,27,46]. It is revealed that padding would implicitly offer the absolute spatial location as the inductive bias [19,20,27], which may bring blind spots for object detectors [4]. A similar phenomenon is also observed in generative models [45]. The unwanted spatial information would be utilized by the generator to learn fixed texture in some coordinates, also known as the "texture-sticking" problem [24]. In this work, we manage to *convert such a side effect to an advantage for GAN inversion* through replacing the constant padding assumed in the pre-trained generator with instance-aware coefficients.

3 Method

3.1 Preliminaries

The StyleGAN family [24–26] has significantly advanced image generation of convolution-based GANs. With the start-of-the-art synthesis performance, they are widely used for GAN inversion studies [1,2,22,35,36,41,42,46,53,56]. Like most GAN variants [12], the generator of StyleGAN takes a latent code, $\mathbf{z}$, as the input and outputs an image, $\mathbf{x} = G(\mathbf{z})$. Differently, StyleGAN [25] learns a more disentangled latent space, $\mathcal{W}$, on top of the original latent space, $\mathcal{Z}$, and feeds the disentangled latent code, $\mathbf{w}$, to every single convolution layer through AdaIN [17]. Prior arts have verified that $\mathcal{W}^+$ space [35], which is an extension of $\mathcal{W}$ via repeating $\mathbf{w}$ across layers, yields the most promising results for GAN inversion. However, existing approaches still suffer from the insufficient recovery of some spatial details, like hats and earrings within face images. We attribute such an issue to the fact that AdaIN acts on the feature map globally [17, 25], therefore, the latent code fails to provide adequate spatial information for image reconstruction. In this work, we propose an innovative solution, which complements the latent space with the padding space of the generator.

3.2 Padding Space for GAN Inversion

Recall that the StyleGAN generator [25,26] stacks a series of convolution layers (*i.e.*, with 3×3 kernel size) for image generation. For each layer, it pads the input feature map before convolution to keep its spatial dimensions unchanged. Zero padding is commonly adopted and frozen in GAN training. Recent studies [24,46] have pointed out that, the constant padding acts as an inductive bias through exposing the spatial location to the generator [46], and hence causes the "texture-sticking" problem [24]. In other words, some coordinates may *get stuck with fixed texture* no matter what latent code is given [24]. Such a side effect of padding complicates the inversion task, especially when the target image contains out-of-distribution objects (*e.g.*, a hat appears at the pixels that are supposed to be hair as encoded by padding). From this perspective, merely employing the latent space as the inversion space is highly insufficient for a fair reconstruction.

To tackle the obstacle, we propose to *adapt the padding coefficients* used in the generator for high-fidelity GAN inversion. That way, the inductive bias offered in the generator can be appropriately adjusted to fit every individual target. We call the parameter space, which is constructed by those instance-aware convolutional paddings, as the padding space, $\mathcal{P}$. Intuitively, $\mathcal{P}$ is incorporated as an extended inversion space to complement the latent space (*i.e.*, $\mathcal{W}^+$) with spatial information. Besides, the StyleGAN generator learns image synthesis starting from a constant tensor [25]. The constant input can be viewed as a special padding (*i.e.*, from a tensor with shape 0×0) without performing convolution. In our approach, the initial constant is also adapted for each target sample.

3.3 Encoder Architecture

To accelerate the inversion speed, we learn an encoder, $E(\cdot)$, following prior works [2,36,41,53]. The architecture of the encoder is carefully tailored to get compatible with the newly introduced padding space, $\mathcal{P}$. In particular, in addition to the layer-wise latent codes, $\{\mathbf{w}_\ell\}_{\ell=1}^L$, our encoder also predicts a coefficient map from the input, as shown in Fig. 2. Here, L stands for the total number of layers in the generator. Values from the coefficient map will be used to replace the paddings used in $G(\cdot)$, denoted as $\{\mathbf{p}_\ell\}_{\ell=0}^L$, at the proper resolution. $\mathbf{p}_0$ represents the initial constant input. For example, the coefficients at the outer of the central 18×18 patch are borrowed as the padding for the convolution at 16×16 resolution. Note that, we only apply one set of coefficients for each resolution to avoid the two convolution layers of the same resolution from sharing the same padding. In practice, we replace $\mathbf{p}_0, \mathbf{p}_1, \mathbf{p}_3, \mathbf{p}_5, \ldots$, keeping the padding for other layers untouched.

Our encoder backbone is equipped with several residual blocks [16], and a Feature Pyramid Network (FPN) [28] is adopted to inspect the input at multiple levels. The outputs of the last three blocks (*i.e.*, with the smallest resolutions) are employed to learn the latent codes, $\{\mathbf{w}_\ell\}_{\ell=1}^{L}$, in a hierarchical manner. They correspond to the generator layers in reverse order. Taking an 18-layer StyleGAN generator as an example, the last block produces latent codes for layers 1–3, the second last for layers 4–7, while the third last for layers 8–18, respectively. Meanwhile, we also choose a max resolution[1] where padding replacement will be performed. The coefficient map is projected (*i.e.*, implemented with a convolution layer with 1×1 kernel size) from the FPN feature with target resolution, and then properly resized to allow padding in the generator. For instance, given the max resolution as 32, we convolve the 32×32 FPN feature and upsamples the convolution result to 34×34, which is the minimum size to pad 32×32 features. More details can be found in *Supplementary Material.*

3.4 Training Objectives

Encoder. We develop the following loss functions for encoder training.

1. *Reconstruction loss.* Image reconstruction is the primary goal of GAN inversion. Hence, we optimize the encoder with both pixel guidance and perceptual guidance, as

$$\mathcal{L}_{pix} = ||\mathbf{x} - G(E(\mathbf{x}))||_2, \tag{1}$$

$$\mathcal{L}_{per} = ||\phi(\mathbf{x}) - \phi(G(E(\mathbf{x})))||_2, \tag{2}$$

where $|| \cdot ||_2$ denotes the l_2 norm. $\phi(\cdot)$ is a pre-trained perceptual feature extractor [49], which is popularly used for GAN inversion [36,53].

2. *Identity loss.* This loss function is particularly designed for inverting face generation models, following [36]. It can help the encoder to focus more on the face identity, which is formulated as

$$\mathcal{L}_{id} = 1 - \cos(\psi(\mathbf{x}), \psi(G(E(\mathbf{x})))), \tag{3}$$

where $\cos(\cdot, \cdot)$ computes the cosine similarity. $\phi(\cdot)$ is a pre-trained identity feature extractor [9]. This loss will be *disabled* for non-face models.

3. *Adversarial loss.* To avoid blurry reconstruction [36], the encoder is also asked to compete with the discriminator, resulting in an adversarial training manner. The adversarial loss is defined as

$$\mathcal{L}_{adv} = - \mathop{\mathbb{E}}_{\mathbf{x} \sim \mathcal{X}} [D(G(E(\mathbf{x})))], \tag{4}$$

[1] We empirically verify that 32 is the best choice in Sect. 4.2.

where $\mathbb{E}[\cdot]$ and $\mathcal{X}$ are the expectation operation and the real data distribution, respectively. $D(\cdot)$ stands for the discriminator.

4. *Regularization loss.* Recall that StyleGAN generator maintains an averaged latent code, $\overline{\mathbf{w}}$, which can be viewed as the statistics of the latent space $\mathcal{W}$. We expect the inverted code to be subject to the native latent distribution as much as possible [41]. Hence, we regularize it with

$$\mathcal{L}_{reg} = ||E_{latent}(\mathbf{x}) - \overline{\mathbf{w}}||_2, \tag{5}$$

where $E_{latent}(\cdot)$ indicates the latent part of the encoder, excluding padding.

To summarize, the full objective for our encoder is:

$$\mathcal{L}_E = \lambda_{pix}\mathcal{L}_{pix} + \lambda_{per}\mathcal{L}_{per} + \lambda_{id}\mathcal{L}_{id} + \lambda_{adv}\mathcal{L}_{adv} + \lambda_{reg}\mathcal{L}_{reg}, \tag{6}$$

where λ_{pix}, λ_{per}, λ_{id}, λ_{adv}, and λ_{reg} are loss weights to balance different terms.

Discriminator. The discriminator is asked to compete with the encoder, as

$$\mathcal{L}_D = \mathop{\mathbb{E}}_{\mathbf{x}\sim\mathcal{X}}[D(G(E(\mathbf{x})))] - \mathop{\mathbb{E}}_{\mathbf{x}\sim\mathcal{X}}[D(\mathbf{x})] - \frac{\gamma}{2}\mathop{\mathbb{E}}_{\mathbf{x}\sim\mathcal{X}}[||\nabla_{\mathbf{x}}D(\mathbf{x})||_2], \tag{7}$$

where γ is the hyper-parameter for gradient penalty [14].

4 Experiments

In this part, we conduct extensive experiments to evaluate the effectiveness of the proposed PadInv. Sect. 4.1 introduces the experimental settings, such as datasets and implementation details. In Sect. 4.2, we experimentally show the superiority of the proposed method PadInv in terms of inversion quality and real image editing with off-the-shelf directions. We also conduct ablation studies. In Sect. 4.3, we explore the property of the proposed padding space, which enables separate control of spatial contents and image style. A face blending application is introduced to further present the superiority of the padding space. Sect. 4.4 shows a novel editing method that could be achieved by providing customized image pairs. We also quantitatively evaluate the editing effects of semantic directions in $\mathcal{W}+$ space and $\mathcal{P}$ space.

4.1 Experimental Settings

We conduct experiments on the high-quality face datasets FFHQ [25] and CelebA-HQ [23,31] for face inversion. We use the former 65k FFHQ faces as the training set (the rest 5k for further evaluation) and test set of CelebA-HQ for quantitative evaluation. For scene synthesis, we adopt LSUN Church and Bedroom [48] and follow the official data splitting strategy. The GANs to invert are pre-trained following StyleGAN [25]. As for the loss weights, we set $\lambda_{id} = 0.1$ and $\lambda_{adv} = 0$ for encoders trained on FFHQ, $\lambda_{adv} = 0.03$ and $\lambda_{id} = 0$ for encoders trained on LSUN. For all the experiments, we set $\lambda_{pix} = 1$, $\lambda_{per} = 0.8$, $\lambda_{reg} = 0.003$, and $\gamma = 10$. Our codebase is built on Hammer [40].

Table 1. Quantitative comparisons between different inversion methods on three datasets, including FFHQ (human face) [25], LSUN Church (outdoor scene) [48], and LSUN Bedroom (indoor scene) [48]

	Face			Church		Bedroom	
	MSE↓	LPIPS↓	Time ↓	MSE↓	LPIPS↓	MSE↓	LPIPS↓
StyleGAN2 [26]	0.020	0.09	122.39	0.220	0.39	0.170	0.42
ALAE [34]	0.15	0.32	**0.0208**	–	–	0.33	0.65
IDInvert [53]	0.061	0.22	0.0397	0.140	0.36	0.113	0.41
pSp [36]	0.034	0.16	0.0610	0.127	0.31	0.099	0.34
e4e [41]	0.052	0.20	0.0610	0.142	0.42	–	–
Restyle$_{pSp}$ [2]	0.030	0.13	0.2898	0.090	0.25	–	–
Restyle$_{e4e}$ [2]	0.041	0.19	0.2898	0.129	0.38	–	–
Ours	**0.021**	**0.10**	0.0629	**0.086**	**0.22**	**0.054**	**0.21**

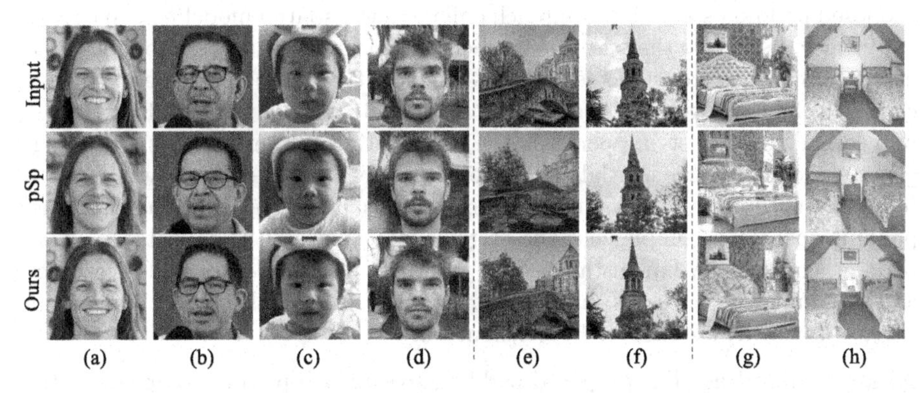

Fig. 3. Qualitative comparisons between our approach and the baseline, pSp [36], where we better recover the spatial details of the input images

4.2 GAN Inversion Performance

Quantitative Results. We quantitatively evaluate the quality of the reconstructed images in Table 1. Pixel-wise MSE and perceptual LPIPS [49] are adopted as metrics to evaluate the reconstruction performance. We also include the inference time of face inversion to measure model efficiency. As shown in Table 1, our approach leads to a significant improvement on face images compared with encoder-based baselines. It even achieves a comparable performance to the optimization-based StyleGAN2 with better efficiency. When applied to more challenging datasets *e.g.*, church and bedroom, StyleGAN2 fails to reconstruct images well because $\overline{\mathbf{w}}$ as the starting point for optimization cannot handle the unaligned data with large spatial variations. However, our method is robust and still outperforms all baselines. It demonstrates that the proposed padding space can provide fine-grained spatial details of images, resulting in better reconstruction quality.

Fig. 4. Real image editing with *off-the-shelf* semantics identified by InterFaceGAN [39] (smile and gender), HiGAN [47] (plant), and LowRankGAN [51] (sky)

Qualitative Results. In Fig. 3, we compare the reconstructed images of our method with the baseline method pSp [36] on the face, church, and bedroom. Even though pSp can restore the foreground of the given face images, it struggles to reconstruct spatial details, as in Fig. 3 (a), (b), (c), (d). Thanks to the proposed padding space, our approach enhances the spatial modeling capacity of the inversion space so that the fine-grained background texture, microphone, and headwear can be reconstructed well, and the identity can be preserved. When it comes to challenging church and bedroom datasets with large spatial variations, pSp can achieve satisfying performance on restoring the texture but still struggles to reconstruct the spatial structure of the given images, as shown in Fig. 3 (e). Our method can preserve structural details better because the proposed padding space can complement the original $\mathcal{W}+$ space with the instance-aware padding coefficients. For instance, the structural information of buildings, such as the number of spires, is preserved more. The bridge, sheet and painting in Fig. 3(e), (g), (h) are also well reconstructed.

Ablation Studies. The proposed padding space is adopted to complement the $\mathcal{W}+$ space, and thus its scale is critical to the quality of the reconstructed image. We adjust its scale by varying the padding layers. Table 2 shows the reconstruction performance on the test sets of CelebA-HQ and LSUN Church. We use pSp [36] as the baseline, which only utilizes $\mathcal{W}+$ space for inversion. We firstly extend the inversion space with $\mathbf{p}_0$, which represents the constant input of the generator. The performance is slightly improved due to the introduction of the case-specific information. Then we progressively enlarge the padding space by predicting padding coefficients for more layers as shown in Fig. 2. As the padding space increases, the reconstruction performance becomes better. This demonstrates that more padding coefficients can provide more precise information for spatial modeling, resulting in the improvement of image reconstruction. When padding layers come to 9 (resolution of 64×64), the inversion performance gets a drop. We assume that the parameter of the padding space is too large, leading to overfitting on the training set. Thus the largest index of padding layers is chosen as 7 (resolution of 32×32) to maintain the reconstruction performance. We also study the effects of the regularization loss $\mathcal{L}_{reg}$. This regularization is designed to encourage the inverted code in $\mathcal{W}+$ space to be subject to the native latent distribution. Equipping $\mathcal{L}_{reg}$ to the model leads to a slight drop in reconstruc-

Table 2. Ablation studies on the number of layers to replace padding, as well as the regularization loss, $\mathcal{L}_{reg}$, introduced in Eq. (5)

	Face		Church	
	MSE↓	LPIPS↓	MSE↓	LPIPS↓
Baseline	0.0344	0.161	0.1272	0.311
$+\mathbf{p}_0$	0.0343	0.156	0.1257	0.307
$+\mathbf{p}_{0,1}$	0.0341	0.154	0.1241	0.309
$+\mathbf{p}_{0,1,3}$	0.0252	0.115	0.1005	0.253
$+\mathbf{p}_{0,1,3,5}$	0.0190	0.091	0.0891	0.218
$+\mathbf{p}_{0,1,3,5,7}$	0.0186	0.090	0.0838	0.209
$+\mathbf{p}_{0,1,3,5,7,9}$	0.0216	0.108	0.0979	0.271
$+\mathbf{p}_{0,1,3,5,7}$ & $\mathcal{L}_{reg}$	0.0214	0.103	0.0866	0.222

Fig. 5. Analysis of the padding effect on LSUN Church [48]. We fix the latent code as the statistical average and interpolate the padding from the fixed constants in the generator to the coefficients specifically learned for inversion. It verifies that padding encodes the spatial information

tion performance, but the editing ability becomes better as in *Supplementary Material*.

Editability. Having verified our method in better reconstruction quality, here we explore its editability. We utilize off-the-shelf semantic directions from [39,47,51] to edit the inversion results. Figure 4 presents the results of manipulating faces and churches. Our method can preserve most other details when manipulating a particular facial attribute. For churches, the semantics changes smoothly on the high-fidelity reconstructed images. These editing results support that the extra padding space can not only invert the given images in high quality but also facilitate them with good properties on image manipulation.

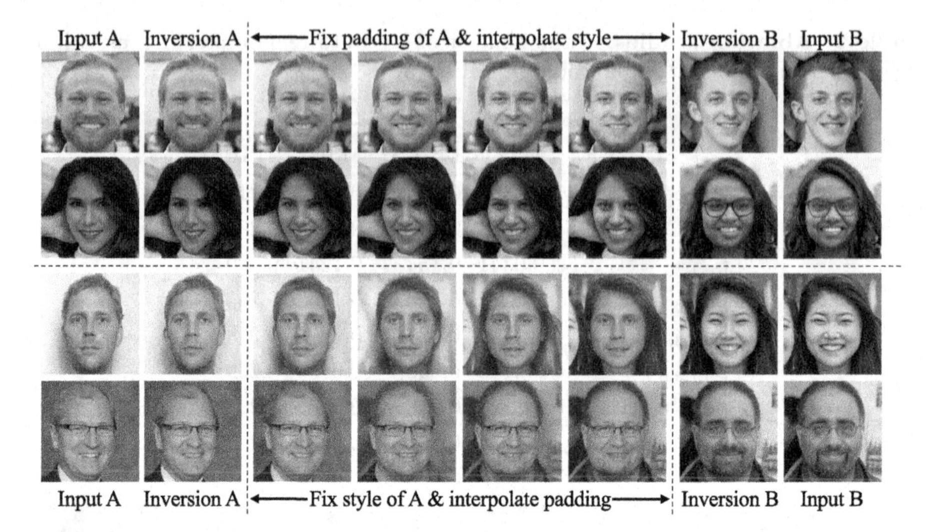

Fig. 6. Analysis of the extended inversion space on FFHQ [25]. We perform interpolation both in the latent space and in the padding space. It turns out that the latent and padding tends to encode style and spatial information, respectively

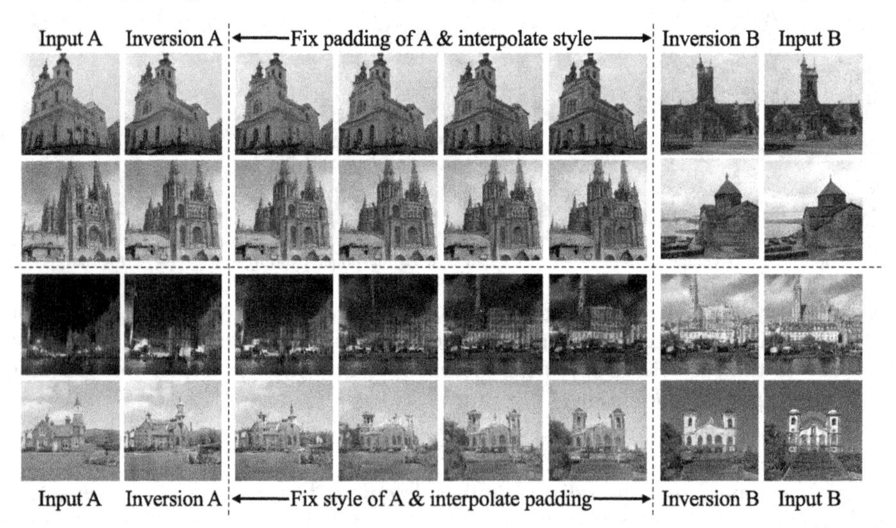

Fig. 7. Analysis of the extended inversion space on LSUN Church [48]. We perform interpolation both in the latent space and in the padding space

4.3 Separate Control of Spatial Contents and Image Style

Property of Padding Space. In this part, we investigate the proposed padding space's properties using interpolation experiments on the FFHQ and LSUN test sets. We first conduct interpolation experiments between a real image and the average image synthesized with the average latent code $\overline{\mathbf{w}}$ and original paddings. The real image is inverted with our encoder into the style latent codes and padding coefficients. The paddings between the average image and the inversion result are then interpolated, while the latent codes from $\mathcal{W}+$ space are kept

Fig. 8. Face blending results by borrowing the inverted latent of A and the inverted padding of B. Different from style mixing, our approach enables the separate control of the facial details (*e.g.*, the features of A), and the face contour (B)

as $\overline{\mathbf{w}}$. Figure 5 demonstrates the interpolation results on the test sets of LSUN Church. The results of interpolation on FFHQ and LSUN Bedroom are shown in *Supplementary Material*. The style of the interpolated images is nearly identical, while the spatial structure is gradually transformed from the average image to the given image. It suggests that the learned padding coefficients encode the structural information of the given images.

Interpolation experiments between two real images are further conducted to validate the separate controllability of the spatial contents and image style. We first invert two real images A and B to $\mathcal{P}$ and $\mathcal{W}+$ space. Then we fix one of the paddings or latent codes and interpolate the other. As illustrated in Fig. 6 and Fig. 7, when padding coefficients are kept to be fixed, the style varies smoothly when the latent code changes. Interpolating paddings with fixed style latent codes leads to the results with the same style and different spatial structures. The interpolation results on LSUN Bedroom can be found in *Supplementary Material*. Above interpolation results indicate that the paddings of $\mathcal{P}$ space encode structure information such as the pose and shape, which is complementary to style information encoded in latent codes of $\mathcal{W}+$ space. We can leverage the properties of two spaces to achieve the independent control of spatial contents and image style.

Application on Face Blending. As discussed in Sect. 4.3, paddings of $\mathcal{P}$ space control the spatial structure of the image. For face images, it controls the face shape and background. Based on this property, we can naturally achieve face blending which integrates the shape and background of one face and facial features of another. Specifically, we invert a real source face A and target portrait B by the encoder. Then the blended face can be synthesized by sending the latent codes of A and padding coefficients of B to the pre-trained generator. In Fig. 8, we compare our face blending results with naively mixing the latter 11 style latent codes as pSp [36]. Obviously, the naive style mixing in pSp introduces redundant information of the source image A and cannot keep the background and hair of the target image B. However, our method can transfer the identity information of face A well and inherit the structure of target face B perfectly. It

Fig. 9. Visual results on attribute editing within the latent space and the padding space

Table 3. Qualitative comparisons on latent editing and padding editing. "Factor" indicates the ratio of MSE_p to MSE_s

Attribute	MSE_p	MSE_s	Factor
Face shape	$116 \times 1e{-4}$	$9 \times 1e{-4}$	12.89:1
Bangs	$471 \times 1e{-4}$	$65 \times 1e{-4}$	7.25:1
Glasses	$245 \times 1e{-4}$	$42 \times 1e{-4}$	5.83:1
Lipstick	$0.437 \times 1e{-4}$	$6 \times 1e{-4}$	1:13.73
Eyebrow	$0.880 \times 1e{-4}$	$14 \times 1e{-4}$	1:15.90

is worth noting that, the whole face blending process does not require any pretrained segmentation or landmark models. It leverages the structure information encoded in $\mathcal{P}$ space, enabling the independent control of the face structure and details.

4.4 Manipulation with Customized Image Pair

Implementation. Based on the padding space, we propose a novel editing method. We find the semantic direction defined by one pair of customized image labels can be successfully applied to arbitrary samples. The paired images can be obtained by human retouching or naive copy-and-paste. Specifically, we firstly invert the paired label images A and A' into $\mathcal{P}$ and $\mathcal{W}+$ space. The difference of the padding coefficients $\mathbf{n_p}$ in $\mathcal{P}$ space and difference of style codes $\mathbf{n_s}$ in $\mathcal{W}+$ space can be viewed as the semantic directions defined by the given paired images. Then other real images can be edited using $\mathbf{n_p}$ and $\mathbf{n_s}$. Considering that the paddings and latent codes encode various information, we propose to apply the semantic direction of one (either $\mathcal{P}$ or $\mathcal{W}+$) space, to maintain the editing performance.

Evaluation and Analysis. We firstly explore the effects of $\mathbf{n_p}$ and $\mathbf{n_s}$ with respect to various attributes. As shown in Fig. 9, for a spatial attribute such as face shape, editing with $\mathbf{n_p}$ works fine while $\mathbf{n_s}$ rarely affects the image. However, $\mathbf{n_s}$ works well for the style-related attributes like lipstick while the editing effect brought by $\mathbf{n_p}$ can be ignored. To quantitatively evaluate the editing effects of $\mathbf{n_p}$ and $\mathbf{n_s}$, MSE_p and MSE_s between the inversion and editing results in the non-edited region like [51] are calculated and reported. Specifically, MSE_p is calculated over the 50 randomly-sampled faces and their editing results driven by $\mathbf{n_p}$. MSE_s is obtained in a similar manner with $\mathbf{n_s}$. Table 3 presents the quantitative results of semantic directions in different space. The factor between MSE_p and MSE_s is also reported for better comparison. We discover that $\mathbf{n_p}$ influences target faces far more than $\mathbf{n_s}$ for spatial attributes such as face shape and bangs. When it comes to style-related attributes like lipstick and brows, $\mathbf{n_s}$ appears to change the inversion images much more than $\mathbf{n_p}$. Given this property, we can only use the $\mathbf{n_p}$ or the $\mathbf{n_s}$ for spatial or style-related editing, respectively. The use of semantic directions separately is extremely beneficial in avoiding unwanted

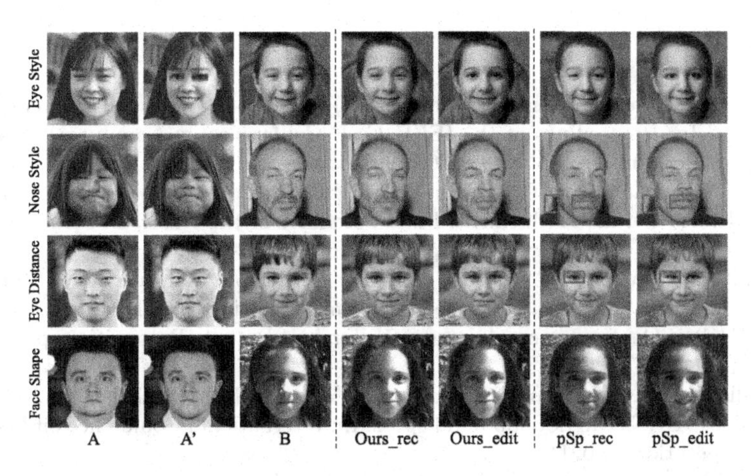

Fig. 10. Customized manipulation defined by *only one* image pair (A and A'). Compared to pSp [36], which introduces unwanted changes to the target image (B), our approach presents a more precise editing performance

changes to the inversion result. Figure 10 presents the editing results of $\mathbf{n_p}$ and $\mathbf{n_s}$. We also include the one-shot editing results of pSp, which can be achieved in $\mathcal{W}+$ space. Naively computing latent difference and applying these directions in $\mathcal{W}+$ space with pSp often interferes with other unconcerned attributes. For example, a change to eye or nose style can be accompanied by a change in the background. Besides, the face shape direction in $\mathcal{W}+$ space produces a larger smile in the portrait, whereas ours does not affect any attributes other than face shape. Additional customized manipulation results of other attributes are provided in *Supplementary Material*.

5 Conclusion

In this work, we attribute the unsatisfactory GAN inversion performance to inductive bias brought by zero padding. Thus we propose padding space ($\mathcal{P}$ space) to complement $\mathcal{W}+$ space for the reconstruction of spatial details, particularly for the out-of-distribution objects. A carefully designed encoder is further proposed to learn the latent code and the padding coefficients jointly. The experiments demonstrates that the padding coefficients can encode rich spatial information such as pose and contour and thus our method enables two novel applications, *i.e.*, face blending and one-shot customized editing. However, the distribution gap between the learned paddings and original features occasionally causes artifacts at the edge. This phenomenon can be largely relieved by adversarial training. We strongly oppose the abuse of our method in violating privacy and security. We hope it can be used to improve the fake detection systems.

Acknowledgement. This work was supported in part by the National Natural Science Foundation of China (Grant No. 61991450) and the Shenzhen Key Laboratory of Marine IntelliSense and Computation (ZDSYS20200811142605016). We thank Zhiyi Zhang for the technical support.

References

1. Abdal, R., Qin, Y., Wonka, P.: Image2StyleGAN++: how to edit the embedded images? In: Proceedings of the IEEE/CVF Conference on Computer Vision and Pattern Recognition, pp. 8296–8305 (2020)
2. Alaluf, Y., Patashnik, O., Cohen-Or, D.: ReStyle: a residual-based StyleGAN encoder via iterative refinement. In: Proceedings of the IEEE/CVF International Conference on Computer Vision, pp. 6711–6720 (2021)
3. Alaluf, Y., Tov, O., Mokady, R., Gal, R., Bermano, A.: HyperStyle: StyleGAN inversion with hypernetworks for real image editing. In: Proceedings of the IEEE/CVF Conference on Computer Vision and Pattern Recognition, pp. 18511–18521 (2022)
4. Alsallakh, B., Kokhlikyan, N., Miglani, V., Yuan, J., Reblitz-Richardson, O.: Mind the Pad – CNNs can develop blind spots. In: International Conference on Learning Representations (2021)
5. Bau, D., et al.: Paint by word. arXiv preprint arXiv:2103.10951 (2021)
6. Bau, D., et al.: GAN dissection: visualizing and understanding generative adversarial networks. In: International Conference on Learning Representations (2019)
7. Brock, A., Donahue, J., Simonyan, K.: Large scale GAN training for high fidelity natural image synthesis. In: International Conference on Learning Representations (2019)
8. Creswell, A., Bharath, A.A.: Inverting the generator of a generative adversarial network. IEEE Trans. Neural Netw. Learn. Syst. **30**(7), 1967–1974 (2018)
9. Deng, J., Guo, J., Xue, N., Zafeiriou, S.: ArcFace: additive angular margin loss for deep face recognition. In: Proceedings of the IEEE/CVF Conference on Computer Vision and Pattern Recognition, pp. 4690–4699 (2019)
10. Dinh, T.M., Tran, A.T., Nguyen, R., Hua, B.S.: HyperInverter: improving StyleGAN inversion via hypernetwork. In: Proceedings of the IEEE/CVF Conference on Computer Vision and Pattern Recognition, pp. 11389–11398 (2022)
11. Donahue, J., Simonyan, K.: Large scale adversarial representation learning. Adv. Neural Inform. Process. Syst. **32** (2019)
12. Goodfellow, I., et al.: Generative adversarial nets. Adv. Neural Inform. Process. Syst. (2014)
13. Gu, J., Shen, Y., Zhou, B.: Image processing using multi-code GAN prior. In: Proceedings of the IEEE/CVF Conference on Computer Vision and Pattern Recognition, pp. 3012–3021 (2020)
14. Gulrajani, I., Ahmed, F., Arjovsky, M., Dumoulin, V., Courville, A.C.: Improved training of wasserstein GANs. Adv. Neural Inform. Process. Syst. **30** (2017)
15. Härkönen, E., Hertzmann, A., Lehtinen, J., Paris, S.: GANSpace: discovering interpretable GAN controls. Adv. Neural Inform. Process. Syst. **33**, 9841–9850 (2020)
16. He, K., Zhang, X., Ren, S., Sun, J.: Deep residual learning for image recognition. In: Proceedings of the IEEE Conference on Computer Vision and Pattern Recognition, pp. 770–778 (2016)
17. Huang, X., Belongie, S.: Arbitrary style transfer in real-time with adaptive instance normalization. In: Proceedings of the IEEE International Conference on Computer Vision, pp. 1501–1510 (2017)
18. Huh, M., Zhang, R., Zhu, J.-Y., Paris, S., Hertzmann, A.: Transforming and projecting images into class-conditional generative networks. In: Vedaldi, A., Bischof, H., Brox, T., Frahm, J.-M. (eds.) ECCV 2020. LNCS, vol. 12347, pp. 17–34. Springer, Cham (2020). https://doi.org/10.1007/978-3-030-58536-5_2

19. Islam, M.A., Jia, S., Bruce, N.D.: How much position information do convolutional neural networks encode? arXiv preprint arXiv:2001.08248 (2020)
20. Islam, M.A., Kowal, M., Jia, S., Derpanis, K.G., Bruce, N.D.: Position, padding and predictions: a deeper look at position information in CNNs. arXiv preprint arXiv:2101.12322 (2021)
21. Jahanian, A., Chai, L., Isola, P.: On the "steerability" of generative adversarial networks. In: International Conference on Learning Representations (2020)
22. Kang, K., Kim, S., Cho, S.: GAN inversion for out-of-range images with geometric transformations. In: Proceedings of the IEEE/CVF International Conference on Computer Vision, pp. 13941–13949 (2021)
23. Karras, T., Aila, T., Laine, S., Lehtinen, J.: Progressive growing of GANs for improved quality, stability, and variation. In: International Conference on Learning Representations (2018)
24. Karras, T., et al.: Alias-free generative adversarial networks. Adv. Neural Inform. Process. Syst. **34**, 852–863 (2021)
25. Karras, T., Laine, S., Aila, T.: A style-based generator architecture for generative adversarial networks. In: Proceedings of the IEEE/CVF Conference on Computer Vision and Pattern Recognition, pp. 4401–4410 (2019)
26. Karras, T., Laine, S., Aittala, M., Hellsten, J., Lehtinen, J., Aila, T.: Analyzing and improving the image quality of StyleGAN. In: Proceedings of the IEEE/CVF Conference on Computer Vision and Pattern Recognition, pp. 8110–8119 (2020)
27. Kayhan, O.S., Gemert, J.C.V.: On translation invariance in CNNs: convolutional layers can exploit absolute spatial location. In: Proceedings of the IEEE/CVF Conference on Computer Vision and Pattern Recognition, pp. 14274–14285 (2020)
28. Lin, T.Y., Dollár, P., Girshick, R., He, K., Hariharan, B., Belongie, S.: Feature pyramid networks for object detection. In: Proceedings of the IEEE Conference on Computer Vision and Pattern Recognition, pp. 2117–2125 (2017)
29. Ling, H., Kreis, K., Li, D., Kim, S.W., Torralba, A., Fidler, S.: EditGAN: high-precision semantic image editing. Adv. Neural Inform. Process. Syst. **34**, 16331–16345 (2021)
30. Lipton, Z.C., Tripathi, S.: Precise recovery of latent vectors from generative adversarial networks. In: International Conference on Learning Representations (2017)
31. Liu, Z., Luo, P., Wang, X., Tang, X.: Deep learning face attributes in the wild. In: Proceedings of the IEEE International Conference on Computer Vision, pp. 3730–3738 (2015)
32. Pan, X., Zhan, X., Dai, B., Lin, D., Loy, C.C., Luo, P.: Exploiting deep generative prior for versatile image restoration and manipulation. In: IEEE Transactions on Pattern Analysis and Machine Intelligence (2020)
33. Perarnau, G., Van De Weijer, J., Raducanu, B., Álvarez, J.M.: Invertible conditional GANs for image editing. In: Advances in Neural Information Processing Systems (2016)
34. Pidhorskyi, S., Adjeroh, D.A., Doretto, G.: Adversarial latent autoencoders. In: Proceedings of the IEEE/CVF Conference on Computer Vision and Pattern Recognition, pp. 14104–14113 (2020)
35. Rameen, A., Yipeng, Q., Peter, W.: Image2StyleGAN: how to embed images into the StyleGAN latent space? In: Proceedings of the IEEE/CVF International Conference on Computer Vision, pp. 4432–4441 (2019)
36. Richardson, E., et al.: Encoding in style: a StyleGAN encoder for image-to-image translation. In: Proceedings of the IEEE/CVF Conference on Computer Vision and Pattern Recognition, pp. 2287–2296 (2021)

37. Roich, D., Mokady, R., Bermano, A.H., Cohen-Or, D.: Pivotal tuning for latent-based editing of real images. arXiv preprint arXiv:2106.05744 (2021)
38. Shen, Y., Gu, J., Tang, X., Zhou, B.: Interpreting the latent space of GANs for semantic face editing. In: Proceedings of the IEEE/CVF Conference on Computer Vision and Pattern Recognition, pp. 9243–9252 (2020)
39. Shen, Y., Yang, C., Tang, X., Zhou, B.: InterFaceGAN: interpreting the disentangled face representation learned by GANs. IEEE Trans. Pattern Anal. Mach. Intell. (2020)
40. Shen, Y., Zhang, Z., Yang, D., Xu, Y., Yang, C., Zhu, J.: Hammer: an efficient toolkit for training deep models (2022). https://github.com/bytedance/Hammer
41. Tov, O., Alaluf, Y., Nitzan, Y., Patashnik, O., Cohen-Or, D.: Designing an encoder for StyleGAN image manipulation. ACM Trans. Graph. **40**(4), 1–14 (2021)
42. Wang, T., Zhang, Y., Fan, Y., Wang, J., Chen, Q.: High-fidelity GAN inversion for image attribute editing. In: Proceedings of the IEEE/CVF Conference on Computer Vision and Pattern Recognition, pp. 11379–11388 (2022)
43. Wu, Z., Lischinski, D., Shechtman, E.: StyleSpace analysis: disentangled controls for StyleGAN image generation. In: Proceedings of the IEEE/CVF Conference on Computer Vision and Pattern Recognition, pp. 12863–12872 (2021)
44. Xia, W., Zhang, Y., Yang, Y., Xue, J.H., Zhou, B., Yang, M.H.: GAN inversion: a survey. IEEE Trans. Pattern Anal. Mach. Intell. (2022)
45. Xu, R., Wang, X., Chen, K., Zhou, B., Loy, C.C.: Positional encoding as spatial inductive bias in GANs. In: Proceedings of the IEEE/CVF Conference on Computer Vision and Pattern Recognition, pp. 13569–13578 (2021)
46. Xu, Y., Shen, Y., Zhu, J., Yang, C., Zhou, B.: Generative hierarchical features from synthesizing images. In: Proceedings of the IEEE/CVF Conference on Computer Vision and Pattern Recognition, pp. 4432–4442 (2021)
47. Yang, C., Shen, Y., Zhou, B.: Semantic hierarchy emerges in deep generative representations for scene synthesis. Int. J. Comput. Vis. **129**(5), 1451–1466 (2021). https://doi.org/10.1007/s11263-020-01429-5
48. Yu, F., Seff, A., Zhang, Y., Song, S., Funkhouser, T., Xiao, J.: LSUN: construction of a large-scale image dataset using deep learning with humans in the loop. arXiv preprint arXiv:1506.03365 (2015)
49. Zhang, R., Isola, P., Efros, A.A., Shechtman, E., Wang, O.: The unreasonable effectiveness of deep features as a perceptual metric. In: Proceedings of the IEEE Conference on Computer Vision and Pattern Recognition, pp. 586–595 (2018)
50. Zhang, Y., et al.: DatasetGAN: efficient labeled data factory with minimal human effort. In: Proceedings of the IEEE/CVF Conference on Computer Vision and Pattern Recognition, pp. 10145–10155 (2021)
51. Zhu, J., et al.: Low-rank subspaces in GANs. Adv. Neural Inform. Process. Syst. **34**, 16648–16658 (2021)
52. Zhu, J., Shen, Y., Xu, Y., Zhao, D., Chen, Q.: Region-based semantic factorization in GANs. In: International Conference on Machine Learning (2022)
53. Zhu, J., Shen, Y., Zhao, D., Zhou, B.: In-domain GAN inversion for real image editing. In: Vedaldi, A., Bischof, H., Brox, T., Frahm, J.-M. (eds.) ECCV 2020. LNCS, vol. 12362, pp. 592–608. Springer, Cham (2020). https://doi.org/10.1007/978-3-030-58520-4_35
54. Zhu, J., Zhao, D., Zhang, B., Zhou, B.: Disentangled inference for GANs with latently invertible autoencoder. Int. J. Comput. Vis. **130**, 1259–1276 (2022). https://doi.org/10.1007/s11263-022-01598-5

55. Zhu, J.-Y., Krähenbühl, P., Shechtman, E., Efros, A.A.: Generative visual manipulation on the natural image manifold. In: Leibe, B., Matas, J., Sebe, N., Welling, M. (eds.) ECCV 2016. LNCS, vol. 9909, pp. 597–613. Springer, Cham (2016). https://doi.org/10.1007/978-3-319-46454-1_36
56. Zhu, P., Abdal, R., Femiani, J., Wonka, P.: Barbershop: GAN-based image compositing using segmentation masks. arXiv preprint arXiv:2106.01505 (2021)

Designing One Unified Framework for High-Fidelity Face Reenactment and Swapping

Chao Xu[1], Jiangning Zhang[1,3], Yue Han[1], Guanzhong Tian[2], Xianfang Zeng[1], Ying Tai[3], Yabiao Wang[3], Chengjie Wang[3], and Yong Liu[1(✉)]

[1] APRIL Lab, Zhejiang University, Hangzhou, China
{21832066,186368,22132041,zzlongjuanfeng}@zju.edu.cn,
yongliu@iipc.zju.edu.cn

[2] Ningbo Research Institute, Zhejiang University, Hangzhou, China
gztian@zju.edu.cn

[3] YouTu Lab, Tencent, Shanghai, China
{yingtai,caseywang,jasoncjwang}@tencent.com

Abstract. Face reenactment and swapping share a similar identity and attribute manipulating pattern, but most methods treat them separately, which is redundant and practical-unfriendly. In this paper, we propose an effective end-to-end unified framework to achieve both tasks. Unlike existing methods that directly utilize pre-estimated structures and do not fully exploit their potential similarity, our model sufficiently transfers identity and attribute based on learned disentangled representations to generate high-fidelity faces. Specifically, *Feature Disentanglement* first disentangles identity and attribute unsupervisedly. Then the proposed *Attribute Transfer* (AttrT) employs learned *Feature Displacement Fields* to transfer the attribute granularly, and *Identity Transfer* (IdT) explicitly models identity-related feature interaction to adaptively control the identity fusion. We joint AttrT and IdT according to their intrinsic relationship to further facilitate each task, *i.e.*, help improve identity consistency in reenactment and attribute preservation in swapping. Extensive experiments demonstrate the superiority of our method. Code is available at https://github.com/xc-csc101/UniFace.

Keywords: Face swapping · Face reenactment · Unified framework

1 Introduction

Recent research has witnessed the development of face reenactment and swapping due to their extensive applications in the metaverse. Face reenactment aims to transfer the attributes (*e.g.*, pose and expression) from target to another

C. Xu and J. Zhang—Equal contributions.

Supplementary Information The online version contains supplementary material available at https://doi.org/10.1007/978-3-031-19784-0_4.

S. Avidan et al. (Eds.): ECCV 2022, LNCS 13675, pp. 54–71, 2022.
https://doi.org/10.1007/978-3-031-19784-0_4

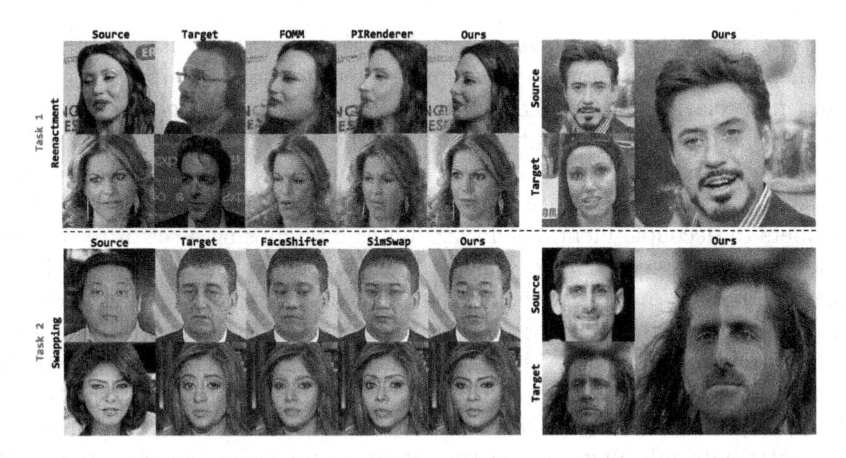

Fig. 1. Comparison with SOTA methods. Top part shows the results of challenging situations in face reenactment, *e.g.*, large pose, occlusion, and extreme lighting. Our method is significantly better than FOMM [34] and PIRenderer [31] with higher realism and better source identity preservation. **Bottom part** shows that our method achieves better performance both on the source identity integration and target attributes preservation (*i.e.*, especially on mouth and eyes regions) than SOTA FaceShifter [23] and SimSwap [5] in face swapping. Images are from official attached results or released codes for fair comparisons. We further present the results of *in-the-wild* situations in the right part. Our method could generalize to out-of-domain pairs and generate high-fidelity faces. Please zoom in for more details.

source face while keeping the identity of the source face unchanged. Similarly, face swapping aims to transfer the identity of the source face into the target face while keeping the attributes of the target face unchanged. Although these two tasks have the same pattern, *i.e.*, recombining the corresponding identity and attribute from source and target faces, current methods treat them independently and design specific networks for each task, which lack universal applicability. In this paper, we focus on exploiting a unified end-to-end framework for both tasks.

Some previous works have the same spirits as ours. Ngo *et al.* [26] learn isolated disentangled representations from input face and its corresponding 2D landmarks to guide the generation of the transformed face. Other works borrow help from the 3D information. Peng *et al.* [29] and Cao *et al.* [4] incorporate 3DMM [2,9] to decompose a face into pose, expression, and identity coefficients, and then recombined factors of the specific task are used to synthesize transformed face. However, **there are two continuously critical issues:** *1) How to get rid of pre-trained structure information.* Obtaining the landmarks and 3DMM coefficients requires excessive annotation effort and complicated preprocessing. Besides, as shown in the top part of Fig. 1, when under some challenging situations, *e.g.*, occlusion, extreme lighting, and large pose, the structure representations are ambiguous and would cause degradation problems. *2) How to sufficiently transfer the identity and attribute representations and facilitate* *corresponding tasks in a joint framework.* The above methods just recombine different facial parameters for face swapping or reenactment accordingly, ignoring the intrinsic relationship between two tasks and resulting in poor performance.

In this paper, we are dedicated to solving the above problems. First, instead of relying on fixed identity and attribute models to extract corresponding embeddings, we design a *Feature disentanglement* module that consists of two embedders to decouple identity and attributes by imposing a set of reconstruction losses in face reenactment. Our attribute embedder learns only attributes-related descriptors, and identity embedder encodes face to low-resolution semantic feature maps with sufficient identity information. Second, to produce more high-fidelity transformed faces based on disentangled representations, we devise *Attribute Transfer* (AttrT) and *Identity Transfer* (IdT) to process attributes and identity, respectively. For face reenactment, recent works [3,26,28,44] implicitly extract attribute information into the latent vector space, which significantly leads to spatial information loss. Consequently, other works [11,31,34] explicitly predict flow fields to transform the source face spatially, but they are still dependent on structure information. To align the attributes of the source face with the target more efficiently, our AttrT learns *Feature Displacement Fields* (FDF) from attribute representation and applies them to source identity features. A powerful learned decoder with rich facial prior is followed to eliminate unnatural texture introduced by the warping operation and generate contents that do not exist in the source faces end-to-end. For face swapping, to align the identity of the target face with the source face, mainstream methods [5,23,39] employ AdaIN [16] to transfer the identity information of the source face into the target. However, lack of semantic interaction makes these methods prone to aggregate inappropriate identity cues to the target, leading to low identity consistency. Inspired by self-attention [45], we design an efficient IdT module that performs identity interaction more granularly to control the identity-related feature fusion.

More importantly, we joint AttrT and IdT in a unified framework according to the intrinsic relationship of two tasks, termed *Feature Transfer*, which could further tackle the problems of identity inconsistency in isolated face reenactment and the dilemma of attributes preservation in isolated face swapping. Specifically, the IdT in face reenactment serves as an identity enhancement module, which allows the ambiguity warped feature to learn detailed texture information from the source face, resulting in higher identity consistency between the transformed and source faces. For face swapping, we apply AttrT on source features to align their attributes with the target face before sending to IdT. The aligned source face makes our method preserve the attributes of the target face excellently while not harming the identity modification performance, as depicted in Fig. 1.

In summary, we make the following three contributions:

- We propose a novel end-to-end unified framework to achieve both face reenactment and swapping. Our method does not rely on prior knowledge during inference stage to disentangle identity and attribute representations.
- We thoroughly exploit the intrinsic relationship between face reenactment and swapping, and design the novel joint-learned AttrT and IdT that transfer the

attributes and maintain identity consistency for reenactment, simultaneously facilitate attributes preservation and integrate identity for swapping.
- Abundant experiments qualitatively and quantitatively demonstrate the superiority of our method to generate high-fidelity faces, *i.e.*, more attributes-alike identity-preserving reenacted faces and identity-consistent attributes-preserving swapped faces than both SOTA unified and isolated methods.

2 Related Work

Unsupervised Disentanglement. Unsupervised separation of face attributes is very common in face manipulation. Zheng *et al.* [49] disentangle facial semantics from StyleGAN without external supervision. Liu *et al.* [24] adopt structure-texture independent network to achieve attribute disentanglement. Some methods [40,44] of face reenactment extract structure cues from drive face by imposing reconstruction loss. Similar to these works, we design a feature disentanglement module to decouple identity and attribute features unsupervisedly.

Face Reenactment. Face reenactment aims to animate the source face into pose and expression of the target face, which could be roughly divided into two categories: instruction-based and warping-based. Instruction-based methods animate the source face instructed by the target structure. Works [6,13,15,41, 46] adopt landmarks and segmentation maps to indicate the facial attribute. Recently, 3DMMs have been proven to be very effective for modeling faces. Some works [19,21] fit 3DMMs by pose and expression from the target face, and identity from the source face to achieve reenactment. Moreover, some AdaIN-based [16] methods [43,44] encode target attributes in the latent vector space, which is later injected into the source face.

However, the above methods could not explicitly indicate the transformations between the source and target faces. Subsequently, warping-based methods have grown to be popular. X2Face [40] learns flow fields from the target face, which are used to warp the embedded source face at the pixel level, but it suffers unnatural head deformations. In the follow-up work, MonkeyNet [33] uses sparse key points to predict flow fields for source appearance driving. FOMM [34] utilizes relative key-point locations to further improve identity preservation. Recently, several image-level warping-based works [11,31,48] separate motion estimation and warped source face refinement into two stages. However, when under extreme conditions, structural priors are not reliable, leading to identity degradation. Unlike the above, our method is independent of facial structure, and the identity transfer further boosts the model's robustness to extreme conditions.

Face Swapping. Face swapping aims to change the facial identity but keep other face attributes constant. Early efforts [36,37] rely on 3D geometrical operations to transfer identity. However, these approaches fail to produce high-fidelity images since their performance depends on the accuracy of the non-trainable external 3D models. Recently, with the development of GANs [12],

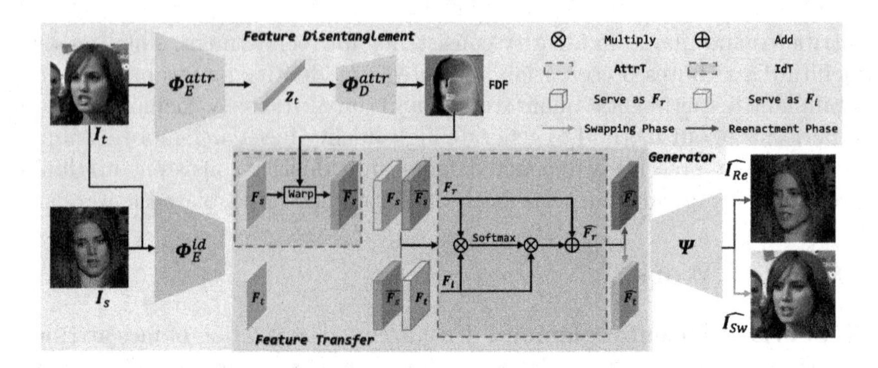

Fig. 2. Overview of the proposed method. Given a source face $\boldsymbol{I}_s$ and target face $\boldsymbol{I}_t$, *Feature Disentanglement* first extracts the disentangled representations, obtaining accurate *Feature Displacement Fields* (FDF) generated from attributes vector $\boldsymbol{Z}_t$, and identity features $\boldsymbol{F}_s$ and $\boldsymbol{F}_t$, which are then sent to *Feature Transfer*. Specifically, *Attribute Transfer* (AttrT) uses accurate FDF to puppeteer $\boldsymbol{F}_s$ into the aligned $\bar{\boldsymbol{F}}_s$ with the desired attributes. *Identity Transfer* (IdT) receives the $\boldsymbol{F}_t$ (serve as $\boldsymbol{F}_r$) and $\bar{\boldsymbol{F}}_s$ (serve as $\boldsymbol{F}_i$) to transfer the identity information from source to the target if in swapping task, otherwise fed with the $\bar{\boldsymbol{F}}_s$ (serve as $\boldsymbol{F}_r$) and $\boldsymbol{F}_s$ (serve as $\boldsymbol{F}_i$) to preserve the identity details after global transformation in reenactment task. Subsequently, the reenacted features $\hat{\boldsymbol{F}}_s$ and swapped features $\hat{\boldsymbol{F}}_t$ go through a powerful *Generator* $\boldsymbol{\Psi}$ to synthesize $\hat{\boldsymbol{I}}_{Re}$ and $\hat{\boldsymbol{I}}_{Sw}$, respectively.

learning-based methods have made significant progress. IPGAN [1] recombines learned disentangled identity and attribute embeddings to generate swapped faces. But vectorized representations inevitably bring information loss. Consequently, FaceShifter [23] integrates the identity and attributes by a carefully designed fusion module. SimSwap [5] proposes a weak feature matching loss to improve target attributes preservation. HifiFace [39] uses an extra 3D shape model to pay more attention to shape change. MegaFS [50] follows the GAN-Inversion [18] paradigm to generate high-resolution swapped faces. However, the above methods use direct concatenation, or AdaIN [16], which fails to model identity-related feature interaction and is prone to introduce identity-unrelated cues, resulting in poor identity consistency and attribute preservation. To alleviate these problems, we design identity transfer based on self-attention to finely integrate identity information and further plug with attribute transfer to help maintain attributes.

3 Method

In this paper, we propose a novel efficient paradigm to complete face reenactment and swapping in an end-to-end unified framework. As shown in Fig. 2, given a target face $\boldsymbol{I}_t$ that provides attribute cues and a source face $\boldsymbol{I}_s$ that provides identity cues, the model learns to animate the source $\boldsymbol{I}_s$ guided by the FDF extracted from $\boldsymbol{I}_t$ to generate reenacted face $\hat{\boldsymbol{I}}_{Re}$, while integrate the identity information from $\boldsymbol{I}_s$ into the target $\boldsymbol{I}_t$ to generate swapped face $\hat{\boldsymbol{I}}_{Sw}$.

3.1 Architecture

Feature Disentanglement. The previous methods [4,26,29] that use identity and structure priors to extract corresponding disentangled representations, which are fixed as the information guidance during training. However, such representations would be inaccurate under challenging conditions, leading to performance degradation. Here we introduce *Feature Disentanglement* module consisting of two embedders to decouple identity and attribute features. Specifically, Identity Embedder employs an encoder $\boldsymbol{\Phi}_E^{id}$ that embeds $\boldsymbol{I}_t$ and $\boldsymbol{I}_s \in \mathbb{R}^{3\times H\times W}$ to low-resolution semantic feature maps, obtaining $\boldsymbol{F}_t$ and $\boldsymbol{F}_s \in \mathbb{R}^{C\times H/8\times W/8}$, which contain more identity-related spatial details than that in vector space.

$$\boldsymbol{F}_t, \boldsymbol{F}_s = \boldsymbol{\Phi}_E^{id}(\boldsymbol{I}_t), \boldsymbol{\Phi}_E^{id}(\boldsymbol{I}_s). \tag{1}$$

Attribute Embedder is designed in an Encoder-Decoder architecture, the encoder $\boldsymbol{\Phi}_E^{attr}$ embeds $\boldsymbol{I}_t$ to disentangled attribute vector $\boldsymbol{Z}_t \in \mathbb{R}^{512}$, which is sent to the followed $\boldsymbol{\Phi}_D^{attr}$ to estimate *Feature Displacement Fields* $\boldsymbol{W}_{fdf} \in \mathbb{R}^{2\times H/4\times H/4}$.

$$\boldsymbol{W}_{fdf} = \boldsymbol{\Phi}_D^{attr}(\boldsymbol{\Phi}_E^{attr}(\boldsymbol{I}_t)). \tag{2}$$

To guide these embedders to learn the desired disentangled descriptors, we impose several reconstruction losses during the reenactment training phase to achieve disentanglement in a fully unsupervised manner.

Attribute Transfer. Current image-level warping-based methods usually cause severe artifacts, which require further refinement in the two-stage paradigm. To align the source face with the desired attributes and preserve visual textures more efficiently, we employ *Attribute Transfer* module that adopts the warping operation on source identity features. Specifically, we first downsample FDF from the Attribute Embedder to match the resolution of the source features. Such design is resolution-free for different identity features, and a higher resolution of FDF contains fine-grained attribute details. As shown in Fig. 2, FDF clearly displays the approximate location and movement of each facial region, *e.g.*, eyes, mouth, left face, and right face, indicating the coordinate offsets specifying which position in the source feature maps could be sampled to generate the targets. Then the warped source features $\bar{\boldsymbol{F}}_s$ could be calculated by the equation:

$$\bar{\boldsymbol{F}}_s = \mathrm{AttrT}(\boldsymbol{F}_s, \mathrm{RS}(\boldsymbol{W}_{fdf})), \tag{3}$$

where RS is resize operation, $\bar{\boldsymbol{F}}_s$ is the coarse feature maps with desired pose and expression like target face while keeping the same identity of the source face.

Identity Transfer. To effectively model the identity-related feature interaction and finely aggregate identity information between identity and reference faces, we propose a novel *Identity Transfer* module, which is well designed based on the self-attention mechanism. As shown in Fig. 2, given identity features $\boldsymbol{F}_i$ and reference features $\boldsymbol{F}_r$ with the same dimensions, the query is extracted by one convolution from $\boldsymbol{F}_r$, and the key and value are extracted from $\boldsymbol{F}_i$ in the same way, obtaining $\boldsymbol{Q}_r, \boldsymbol{K}_i, \boldsymbol{V}_i \in \mathbb{R}^{C/4 \times H/8 \times W/8}$ with reduced channel numbers. Then $\boldsymbol{Q}_r$ and $\boldsymbol{K}_i$ are employed to calculate the correlation matrix $\boldsymbol{M}$, which further multiplies $\boldsymbol{V}_i$ to obtain $\boldsymbol{F}_{i \to r}$. A zero-initialized learned scale parameter α is applied on $\boldsymbol{F}_{i \to r}$ to control the identity transfer flow when added to the $\boldsymbol{F}_r$:

$$\boldsymbol{F}_{i \to r} = \mathrm{softmax}(\boldsymbol{Q}_r(\boldsymbol{K}_i)^T)\boldsymbol{V}_i = \boldsymbol{M}\boldsymbol{V}_i, \quad (4)$$

$$\hat{\boldsymbol{F}}_r = \alpha \boldsymbol{F}_{i \to r} + \boldsymbol{F}_r. \quad (5)$$

Benefiting from representing identity with feature maps, IdT is allowed to learn the explicit correlation between identity and reference faces on identity-related regions and transfer the identity information to the reference face adaptively.

Generator. To generate authentic faces, we adopt the stylemap resizer along with the synthesis network of the StyleMapGAN [20] as our *Generator* $\boldsymbol{\Psi}$. Specifically, the stylemap resizer is a common decoder with the corresponding feature size to the encoder $\boldsymbol{\Phi}_E^{id}$. We further add a skip connection that brings the texture details from the $\boldsymbol{\Phi}_E^{id}$ to keep the facial contents and background. The synthesis network could learn more sufficient facial prior than the typical decoder, enabling us to generate more realistic faces. Such a powerful generator with skip connection successfully embeds the transformed features to the high-fidelity faces:

$$\hat{\boldsymbol{I}_{Re}}, \hat{\boldsymbol{I}_{Sw}} = \boldsymbol{\Psi}(\hat{\boldsymbol{F}}_s, \boldsymbol{F}_s^i), \boldsymbol{\Psi}(\hat{\boldsymbol{F}}_t, \boldsymbol{F}_t^i), \quad (6)$$

where i indicates the feature maps of i-th layer in $\boldsymbol{\Phi}_E^{id}$.

3.2 Face Reenactment and Swapping

The isolated AttrT is designed for face reenactment and IdT is for face swapping, we joint two modules in a unified framework, termed *Feature Transfer*, which enables each task to capture complementary information from the other. Specifically, as shown in Fig. 2, for face reenactment, we send $\boldsymbol{I}_t$ to attribute embedder and $\boldsymbol{I}_s$ to identity embedder, AttrT first spatially transforms source feature $\boldsymbol{F}_s$ by FDF. Subsequently, the warped source feature $\bar{\boldsymbol{F}}_s$ is in the same pose and expression as the target face. A followed IdT borrows the source identity cues from $\boldsymbol{F}_s$ (serve as $\boldsymbol{F}_i$ in Sect. 3.1) to $\bar{\boldsymbol{F}}_s$ (serve as $\boldsymbol{F}_r$) for better identity preservation. For face swapping, the Identity Embedder is given $\boldsymbol{I}_t$ and $\boldsymbol{I}_s$. AttrT first aligns the $\boldsymbol{F}_s$ with the attributes of $\boldsymbol{I}_t$, obtaining $\bar{\boldsymbol{F}}_s$ (serve as $\boldsymbol{F}_i$), which along with $\boldsymbol{F}_t$ (serve as $\boldsymbol{F}_r$) are sent into IdT for identity transfer. The aligned source features help preserve the attributes when performing identity-related feature interaction. By fully exploiting the intrinsic relationship of two tasks, the joint AttrT and IdT effectively achieve both tasks, while solving the challenges of identity inconsistency in reenactment and attribute preservation in swapping.

3.3 Objective Functions

We use several loss terms to train our unified framework: reconstruction loss $\mathcal{L}_{rec}$, perceptual loss $\mathcal{L}_p$, and adversarial loss $\mathcal{L}_{adv}$ for face reenactment. Based on these three losses, identity loss $\mathcal{L}_{id}$ and contextual loss $\mathcal{L}_{ctx}$ are adopted for face swapping. Thus, the total loss for two tasks are defined as follow:

$$\begin{aligned}\mathcal{L}_{all}^{Re} &= \lambda_{adv}\mathcal{L}_{adv} + \lambda_{rec}\mathcal{L}_{rec} + \lambda_p\mathcal{L}_p, \\ \mathcal{L}_{all}^{Sw} &= \lambda'_{adv}\mathcal{L}_{adv} + \lambda'_{rec}\mathcal{L}_{rec} + \lambda'_p\mathcal{L}_p + \lambda'_{id}\mathcal{L}_{id} + \lambda'_{ctx}\mathcal{L}_{ctx}.\end{aligned} \tag{7}$$

Reconstruction Loss. When the source and target faces are from the same identity, we define a reconstruction loss to calculate pixel-level errors:

$$\begin{aligned}&\mathcal{L}_{rec} = \left\|\hat{\boldsymbol{I}} - \boldsymbol{I}_t\right\|_2 \quad \textit{if } \boldsymbol{I}_s, \boldsymbol{I}_t \textit{ in same identity}, \\ &\text{where } \hat{\boldsymbol{I}} \in \{\hat{\boldsymbol{I}_{Re}}, \hat{\boldsymbol{I}_{Sw}}\}.\end{aligned} \tag{8}$$

Perceptual Loss. Besides measuring the difference between two faces at the pixel level, we adopt LPIPS [47] loss to calculate the semantic errors:

$$\mathcal{L}_p = \left\|\phi_{vgg}(\hat{\boldsymbol{I}}) - \phi_{vgg}(\boldsymbol{I}_t)\right\|_2, \tag{9}$$

where $\phi_{vgg}(\cdot)$ represents the pre-trained VGG16 [35] network.

Identity Loss. We calculate the cosine similarity to estimate the identity consistent between swapped and source faces:

$$\mathcal{L}_{id} = 1 - \cos(R(\boldsymbol{I}_s), R(\hat{\boldsymbol{I}_{Sw}})), \tag{10}$$

where $R(\cdot)$ is a pre-trained ArcFace [8] network.

Contextual Loss. We utilize contextual loss [25] to mitigate the effect of excessive information on the swapped faces:

$$\mathcal{L}_{ctx} = -\log(\mathrm{CX}(\phi_{vgg}(\hat{\boldsymbol{I}}_t), \phi_{vgg}(\boldsymbol{I}_{Sw}))). \tag{11}$$

Adversarial Loss. We adopt adversarial training to ensure the authenticity of the transformed faces:

$$\mathcal{L}_{adv} = \mathbb{E}_{x\sim p_x}\left[\log D\left(x\right)\right] + \mathbb{E}_{\tilde{x}\sim p_{\tilde{x}}}\left[\log\left(1 - D\left(\tilde{x}\right)\right)\right], \tag{12}$$

where p_x ans $p_{\tilde{x}}$ are real and generated image distributions.

4 Experiments

4.1 Datasets and Implementations Details

Datasets. For face reenactment, we leverage the VoxCeleb2 [7] dataset, which contains over 1 million utterances for over 6,000 celebrities. Following the preprocessing method in FOMM [34], we crop faces from the original videos and resize them to 256 × 256 for training and testing. For face swapping, we adopt the CelebA-HQ [17] dataset, which consists of 28,000 faces for training and 2,000 for testing. We use the FaceForensics++ [32] dataset for further evaluation.

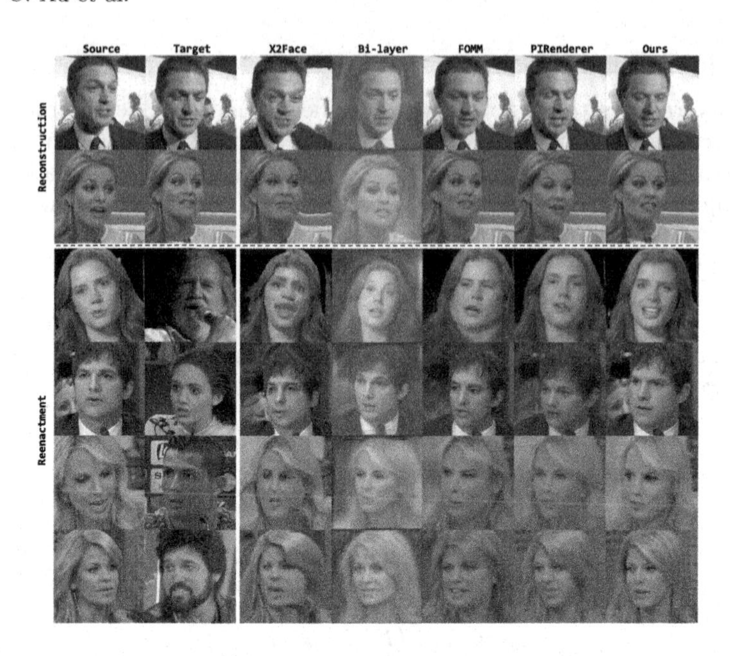

Fig. 3. Qualitative comparison with four SOTA methods on VoxCeleb2 [7] test set. The top shows the results of the reconstruction task and the bottom of the reenactment.

Evaluation Metric. For face reenactment, we use FID [14] to evaluate the realism of the generated images. The Average Pose Errors (APE) and Average Expression Errors (AEE) are used to estimate the motion accuracy, and Cosine SIMilarity (CSIM) is used to measure identity preservation. D3DFR [10] is used to extract the pose and expression coefficients. CosFace [38] is used to extract identity embedding. We additionally use LPIPS [47] to evaluate reconstruction quality. For face swapping, we evaluate the performance by IDentity Retrieval (IDRet), APE, and AEE. IDRet retrieves the closest face to evaluate identity modification while APE and AEE evaluate attributes preservation.

Implementation Details. For face reenactment, we perform reconstruction during training, *i.e.*, randomly sample the source and target faces from the same video. The values of the loss weights are set to $\lambda_{adv} = 1$, $\lambda_{rec} = 5$, $\lambda_p = 5$. For face swapping, we train on the CelebA-HQ dataset with resized 256×256 faces. The values of the loss weights are set to $\lambda^{'}_{adv} = 1$, $\lambda^{'}_{rec} = 1$, $\lambda^{'}_{p} = 1$, $\lambda^{'}_{id} = 2.5$, $\lambda^{'}_{ctx} = 0.5$. The ratio of the training data with $\boldsymbol{I}_t = \boldsymbol{I}_s$ and $\boldsymbol{I}_t \neq \boldsymbol{I}_s$ is set to $1:4$. We train two tasks in stages. First, face reenactment is trained for $500K$ iterations from scratch. Then we only load the trained attribute embedder weights for initialization, and then train the face swapping stage with $500K$ iterations. Both tasks adopt Adam optimizer with $\beta_1 = 0$, $\beta_2 = 0.99$, $lr = 1e-4$ for updating whole model weights, using 2 T V100 GPUs and 8 batch size.

Fig. 4. Qualitative comparison with three SOTA unified methods and two SOTA isolated methods. Images are from official attached results in the corresponding paper.

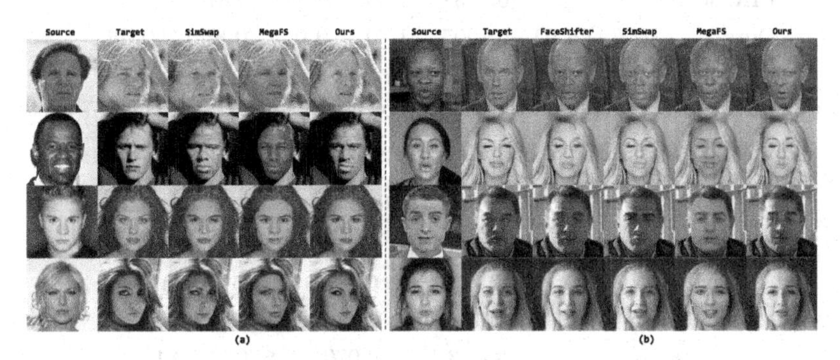

Fig. 5. Qualitative comparison with SOTA methods. (a) The results on CelebA-HQ [17] test set. We not compare with FaceShifter since its official codes are not available. (b) The results on FaceForensics++ [32]. FaceShifter attaches official results in this dataset.

4.2 Comparison with SOTAs

Qualitative Results. For face reenactment, we conduct two tasks: *Reconstruction* where the source and target faces are of the same identity, and *Reenactment* where the source face is driven to mimic the motions of another cross-identity individual. We perform qualitative comparisons with X2Face [40], Bi-layer [42], FOMM [34], and PIRenderer [31]. As shown in Fig. 3, we sample two pairs for reconstruction and four pairs for reenactment from VoxCeleb2 test set. It can be seen that X2Face suffers from severe warping artifacts. Bi-layer over-smooths the facial details and could not generate authentic background. Recent FOMM and PIRenderer could generate high-quality results, but they fail to handle extreme cases. As shown in row 3, they are sensitive to the lighting in the target face due to their dependence on structural information. In rows 4 and 6, the source identity could not be well-preserved when the source face shape is very different from the target. Besides, under the large-pose conditions in rows 5 and 6, they fail to maintain source identity and are less realistic. In contrast, our method successfully generates more realistic results with accurate pose and expression while still preserving the source identity in various conditions (Fig. 4).

For face swapping, we first compare our method with three SOTA unified works (Peng *et al.* [29], UniFaceGAN [4], and FSGAN [27]) and two SOTA isolated methods (FaceShifter [23] and FaceInpainter [22]). Obviously, recent

Table 1. Quantitative results on the tasks of reconstruction for VoxCeleb2 [7] test set. **Bold** and underline represent optimal and suboptimal results. The up arrow indicates that the larger the value, the better the model performance, and vice versa.

Method	FID↓	LPIPS↓	AEE↓	APE↓	CSIM↑	Auth.↑
X2Face [40]	93.99	0.3612	2.76	0.194	0.6122	0.12
Bi-layer [42]	149.14	0.5443	1.92	0.055	0.7002	-
FOMM [34]	48.96	**0.1817**	**1.55**	0.044	0.8462	0.30
PIRenderer [31]	55.57	0.2634	2.09	0.059	0.8186	0.25
Ours	**45.91**	0.1907	1.70	**0.042**	**0.8607**	**0.33**

Table 2. Quantitative results on the tasks of reenactment for VoxCeleb2 [7] test set.

Method	FID↓	AEE↓	APE↓	CSIM↑	Auth.↑
X2Face [40]	119.32	3.99	0.274	0.4329	0.09
Bi-layer [42]	195.67	3.18	0.073	0.5353	-
FOMM [34]	72.20	3.25	**0.067**	0.5365	0.26
PIRenderer [31]	73.73	**3.08**	0.079	0.4737	0.21
Ours	**54.34**	3.15	0.070	**0.5703**	**0.44**

Table 3. Quantitative results on the tasks of face swapping for FaceForensics++ [32].

Method	IDRet↑	AEE↓	APE↓	Id-Attr.↑	Auth.↑
DeepFakes [30]	79.84	3.75	0.260	0.04	0.02
FaceShifter [23]	92.59	3.47	0.206	0.11	0.13
SimSwap [5]	90.02	**3.13**	**0.039**	0.13	0.08
MegaFS [50]	93.87	3.45	0.202	0.13	0.10
Ours	**99.45**	3.21	0.052	**0.59**	**0.67**

unified works fail to preserve the attributes of the target face (*e.g.*, skin color and gaze) and maintain the low identity consistency. Our method exhibits a strong identity modification ability as SOTA FaceInpainter while achieving more high-fidelity swapped faces with authentic skin textures. We further conduct a series of qualitative experiments to compare with FaceShifter, SimSwap [5], and MegaFS [50] on CelebA-HQ and FaceForensics++ dataset. As shown in Fig. 5, we sample eight pairs of significant gaps between pose, expression, skin color, and lighting. Notably, FaceShifter could produce highly identifiable faces but unable to keep the details of the attributes, *e.g.*, closed eyes in row 2 of Fig. 5(b), while SimSwap improves the attributes preservation but in poor identity consistency, *e.g.*, pupil color in row 1 and small mouth in row 2 of Fig. 5(b). Comparing

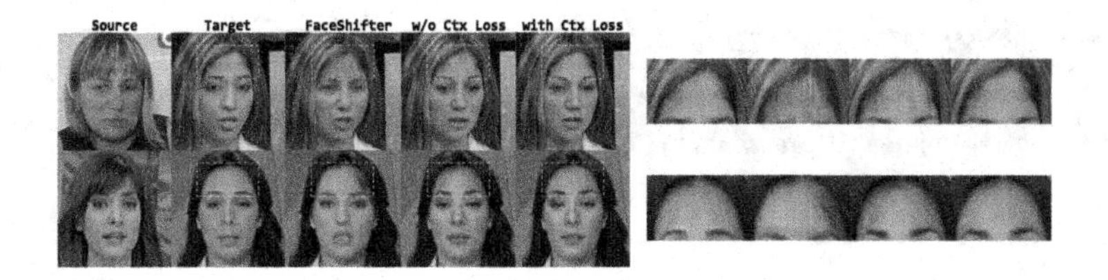

Fig. 6. Qualitative ablation study for contextual loss of face swapping. We zoom in the red dotted rectangles of the forehead area for more clear comparison. (Color figure online)

with SOTA unified and isolated methods shows the superiority of our method to generate both identity-consistent and attributes-preserving faces.

Quantitative Results. We first evaluate the effectiveness of the face reenactment with different SOTA methods on VoxCeleb2. We randomly sample 1,000 videos from the test set and set 6 random seeds to generate 6K pairs in total. The reconstruction results are summarized in Table 1 and the reenactment ones are in Table 2. It can be seen that FOMM and PIRenderer achieve impressive results on AEE and APE due to the accurate structure prior in most conditions. However, as facial structures involve identity information, they would inevitably cause poor identity preservation when guiding the reenactment, which can be inferred from the low CSIM. Besides, these methods struggle to handle challenging conditions and generate unrealistic facial textures, resulting in low FID. The above observations are consistent with the qualitative results in Fig. 3.

For face swapping, we conduct quantitative comparisons on FaceForensics++ and follow the settings in MegaFS, which carefully check the aligned faces and manually categorize all videos into 885 identities. As depicted in Table 3, SimSwap preserves attributes of the target face better but a relatively poor performance on IDRet. Our method achieves comparable results in attribute preservation and maintains the highest IDRet, outperforming MegaFS with a large margin. Results show that our method is better considering both identity consistency with the source and attribute preservation with the target. Our method is not compared with works [4,29] quantitatively due to unavailability of their codes.

Human Study. We conduct the human study to evaluate the performance of each method in two tasks. Concretely, we randomly sample 200 pairs from the corresponding test set. Each pair is compared 5 times by different volunteers. For face reenactment, the volunteers are asked to choose the most realistic image among the generated results of all methods. Similarly, for face swapping, the volunteers are invited to select the one that most resembles the source and shares similar attributes with the target, as well as the one that looks most realistic. The results are shown in Table 1, Table 2, and Table 3 of the Id-Attr. (abbreviated

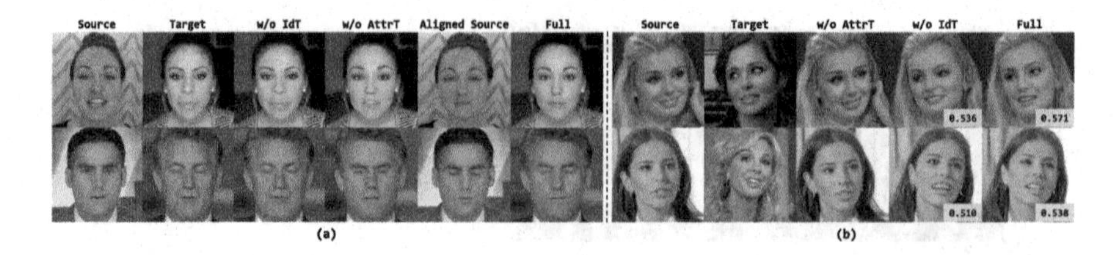

Fig. 7. Qualitative ablation study of our method with different components. (a) The results of face swapping, please attention to the mouth of row 1 and eyes of row 2. (b) The results of face reenactment, we tag CSIM on reenacted faces for clear comparison.

Table 4. Quantitative ablation study of our approach with different components. We mark the optimal and suboptimal results without considering meaningless values.

Method	Face Swapping			Face Reenactment			
	IDRet↑	AEE↓	APE↓	FID↓	AEE↓	APE↓	CSIM↑
w/o IdT	0.05	0.02	0.000	<u>56.28</u>	<u>3.16</u>	<u>0.071</u>	<u>0.5570</u>
w/o AttrT	<u>99.43</u>	<u>3.30</u>	<u>0.058</u>	44.45	4.34	0.424	0.9884
Ours	**99.45**	**3.21**	**0.052**	**54.34**	**3.15**	**0.070**	**0.5703**

from Identity-consistent and Attribute-preserving) and Auth. (abbreviated from Authentic), our generated faces are preferred by volunteers, meaning that we can generate higher fidelity images than SOTA methods on both tasks.

4.3 Ablation Study and Further Analysis

Loss Functions. To evaluate the effectiveness of contextual loss, we report the ablation study in Fig. 6. We sample two pairs of significant identity differences in the forehead area. FaceShifter tends to show excessive source identity and cause apparent artifacts. Comparing the results of columns 3 and 4, our method without contextual loss can already alleviate texture distortions due to the effectively semantic interactions. The last column exhibiting a better performance illustrates that contextual loss further helps preserve the target textures. We further evaluate the effectiveness of identity and adversarial losses on Face-Forensicss++. IDRet suffers a sharp decline without identity loss, from 99.45 to 0.08, while FID shows higher value without adversarial loss, from 3.29 to 4.01.

Network Components. We perform qualitative and quantitative experiments to evaluate the effectiveness of AttrT and IdT. As shown in Fig. 7(a), IdT is critical for identity transfer since our model without it fails to generate identity-consistent faces. From columns 1 and 5, AttrT aligns the source face with the target face in terms of attributes, *e.g.*, closed mouth and closed eyes. Comparing the results of columns 4 and 6, we can observe that the aligned source features help achieve more attributes-preserving results. Similarly, we can draw the conclusions from Fig. 7(b) that AttrT successfully transfers the accurate motions

Fig. 8. (a) Image retrieval using the disentangled attribute embeddings. The Top-4 retrieved images have similar poses and expressions but different identity from the query. (b) Attention visualization of IdT. The color bars indicate activation values. The points in the target face could correctly match similar semantic areas in the source. (Color figure online)

from the target, and IdT further boosts the identity consistency with the source face on the reenactment task, as depicted in columns 4 and 5. Moreover, the above observations could also be summarized from Table 4, the sharp decreasing of IDRet, and rapidly rising of AEE and APE illustrate that IdT and AttrT are indispensable to the corresponding tasks. In the meantime, AttrT significantly improves the attributes preservation for face swapping with a smaller AEE and APE, and IdT keeps better identity preservation for face reenactment with the improvement of CSIM.

Interpretability of AttrT. Since AttrT receives the FDF from the attribute embedder, we first perform a qualitative experiment to demonstrate the disentanglement of learned attribute representations. This experiment is based on the intuition that a well-disentangled attribute descriptor contains almost no identity information and can be used to purely measure the similarity of pose and expression between different identity individuals. Specifically, we randomly sample three images from the VoxCeleb2 test set and select their most consistent faces with all test images, according to the cosine similarity between two attribute embeddings. The results are shown in Fig. 8(a), where the retrieved images have similar poses and expressions to those of the query images. Besides, we visualize the feature maps and FDF in Fig. 9. As expected, FDF explicitly models the absolute pose and expression of the target face, which can adaptively animate the source to desired attributes.

Interpretability of IdT. To better understand the effect of the IdT, we visualize the attention maps in Fig. 8(b). Specifically, we select four points from different regions in the target face, *i.e.*, forehead, eyes, face, and background. The visualized attention maps indicate that each location pays more attention to semantically similar areas, allowing the explicitly identity-related semantic interaction to achieve the generation of highly identity-consistent.

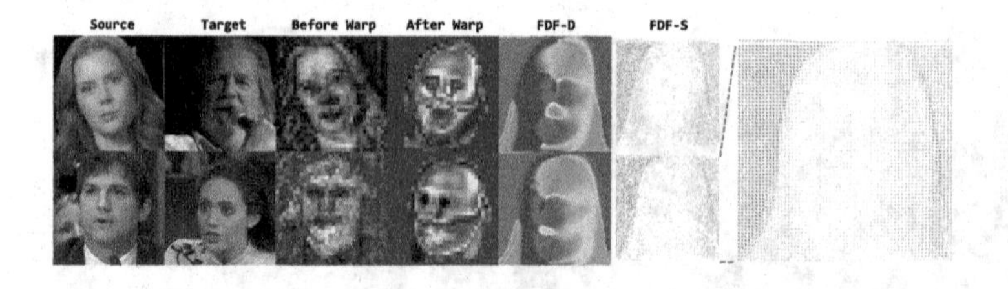

Fig. 9. The visualization of feature maps before and after warping, and FDF represented in dense flow (FDF-D) and sparse displacement (FDF-S). These two pairs are sampled from Fig. 3. We enlarge one FDF-S to show better motion details.

5 Conclusions

In this paper, we propose a novel end-to-end unified paradigm to complete face reenactment and swapping. Our method shows several appealing properties: 1) We extract identity and attributes during inference stage without any prior knowledge, which is more robust under some challenging conditions. 2) To allow sufficient feature interaction, we design a novel AttrT to transfer attributes for reenactment and IdT to integrate identity for swapping. 3) We effectively joint AttrT and IdT by exploiting their underlying similarity to achieve better identity consistency in reenactment and better attribute preservation in swapping.

Acknowledgments. This work is supported by the Key R&D Program Project of Zhejiang Province (2021C01035).

References

1. Bao, J., Chen, D., Wen, F., Li, H., Hua, G.: Towards open-set identity preserving face synthesis. In: Proceedings of the IEEE Conference on Computer Vision and Pattern Recognition, pp. 6713–6722 (2018)
2. Blanz, V., Vetter, T.: A morphable model for the synthesis of 3D faces. In: Proceedings of the 26th Annual Conference on Computer Graphics and Interactive Techniques, pp. 187–194 (1999)
3. Burkov, E., Pasechnik, I., Grigorev, A., Lempitsky, V.: Neural head reenactment with latent pose descriptors. In: Proceedings of the IEEE/CVF Conference on Computer Vision and Pattern Recognition, pp. 13786–13795 (2020)
4. Cao, M., et al.: UniFaceGAN: a unified framework for temporally consistent facial video editing. IEEE Trans. Image Process. **30**, 6107–6116 (2021)
5. Chen, R., Chen, X., Ni, B., Ge, Y.: SimSwap: an efficient framework for high fidelity face swapping. In: Proceedings of the 28th ACM International Conference on Multimedia, pp. 2003–2011 (2020)
6. Chen, Z., Wang, C., Yuan, B., Tao, D.: PuppeteerGAN: arbitrary portrait animation with semantic-aware appearance transformation. In: Proceedings of the IEEE/CVF Conference on Computer Vision and Pattern Recognition, pp. 13518–13527 (2020)
7. Chung, J.S., Nagrani, A., Zisserman, A.: VoxCeleb2: deep speaker recognition. arXiv preprint arXiv:1806.05622 (2018)

8. Deng, J., Guo, J., Xue, N., Zafeiriou, S.: ArcFace: additive angular margin loss for deep face recognition. In: Proceedings of the IEEE/CVF Conference on Computer Vision and Pattern Recognition, pp. 4690–4699 (2019)
9. Deng, Y., Yang, J., Chen, D., Wen, F., Tong, X.: Disentangled and controllable face image generation via 3D imitative-contrastive learning. In: Proceedings of the IEEE/CVF Conference on Computer Vision and Pattern Recognition, pp. 5154–5163 (2020)
10. Deng, Y., Yang, J., Xu, S., Chen, D., Jia, Y., Tong, X.: Accurate 3D face reconstruction with weakly-supervised learning: from single image to image set. In: Proceedings of the IEEE/CVF Conference on Computer Vision and Pattern Recognition Workshops (2019)
11. Doukas, M.C., Zafeiriou, S., Sharmanska, V.: HeadGAN: one-shot neural head synthesis and editing. In: Proceedings of the IEEE/CVF International Conference on Computer Vision, pp. 14398–14407 (2021)
12. Goodfellow, I., et al.: Generative adversarial nets. In: Advances in Neural Information Processing Systems, vol. 27 (2014)
13. Ha, S., Kersner, M., Kim, B., Seo, S., Kim, D.: MarioNETte: few-shot face reenactment preserving identity of unseen targets. In: Proceedings of the AAAI Conference on Artificial Intelligence, vol. 34, pp. 10893–10900 (2020)
14. Heusel, M., Ramsauer, H., Unterthiner, T., Nessler, B., Hochreiter, S.: GANs trained by a two time-scale update rule converge to a local Nash equilibrium. In: Advances in Neural Information Processing Systems, vol. 30 (2017)
15. Huang, P.H., Yang, F.E., Wang, Y.C.F.: Learning identity-invariant motion representations for cross-id face reenactment. In: Proceedings of the IEEE/CVF Conference on Computer Vision and Pattern Recognition, pp. 7084–7092 (2020)
16. Huang, X., Belongie, S.: Arbitrary style transfer in real-time with adaptive instance normalization. In: Proceedings of the IEEE International Conference on Computer Vision, pp. 1501–1510 (2017)
17. Karras, T., Aila, T., Laine, S., Lehtinen, J.: Progressive growing of GANs for improved quality, stability, and variation. arXiv preprint arXiv:1710.10196 (2017)
18. Karras, T., Laine, S., Aittala, M., Hellsten, J., Lehtinen, J., Aila, T.: Analyzing and improving the image quality of StyleGAN. In: Proceedings of the IEEE/CVF Conference on Computer Vision and Pattern Recognition, pp. 8110–8119 (2020)
19. Kim, H., et al.: Deep video portraits. ACM Trans. Graph. (TOG) **37**(4), 1–14 (2018)
20. Kim, H., Choi, Y., Kim, J., Yoo, S., Uh, Y.: Exploiting spatial dimensions of latent in GAN for real-time image editing. In: Proceedings of the IEEE/CVF Conference on Computer Vision and Pattern Recognition, pp. 852–861 (2021)
21. Koujan, M.R., Doukas, M.C., Roussos, A., Zafeiriou, S.: Head2Head: video-based neural head synthesis. In: 2020 15th IEEE International Conference on Automatic Face and Gesture Recognition, FG 2020, pp. 16–23. IEEE (2020)
22. Li, J., Li, Z., Cao, J., Song, X., He, R.: Faceinpainter: high fidelity face adaptation to heterogeneous domains. In: Proceedings of the IEEE/CVF Conference on Computer Vision and Pattern Recognition, pp. 5089–5098 (2021)
23. Li, L., Bao, J., Yang, H., Chen, D., Wen, F.: FaceShifter: towards high fidelity and occlusion aware face swapping. arXiv preprint arXiv:1912.13457 (2019)
24. Liu, K., Cao, G., Zhou, F., Liu, B., Duan, J., Qiu, G.: Towards disentangling latent space for unsupervised semantic face editing. IEEE Trans. Image Process. **31**, 1475–1489 (2022)

25. Mechrez, R., Talmi, I., Zelnik-Manor, L.: The contextual loss for image transformation with non-aligned data. In: Proceedings of the European Conference on Computer Vision (ECCV), pp. 768–783 (2018)
26. Ngo, L.M., Karaoglu, S., Gevers, T., et al.: Unified application of style transfer for face swapping and reenactment. In: Proceedings of the Asian Conference on Computer Vision (2020)
27. Nirkin, Y., Keller, Y., Hassner, T.: FSGAN: subject agnostic face swapping and reenactment. In: Proceedings of the IEEE/CVF International Conference on Computer Vision, pp. 7184–7193 (2019)
28. Nitzan, Y., Bermano, A., Li, Y., Cohen-Or, D.: Face identity disentanglement via latent space mapping. arXiv preprint arXiv:2005.07728 (2020)
29. Peng, B., Fan, H., Wang, W., Dong, J., Lyu, S.: A unified framework for high fidelity face swap and expression reenactment. IEEE Trans. Circuits Syst. Video Technol. **32**, 3673–3684 (2021)
30. Perov, I., et al.: DeepFaceLab: a simple, flexible and extensible face swapping framework. arXiv preprint arXiv:2005.05535 (2020)
31. Ren, Y., Li, G., Chen, Y., Li, T.H., Liu, S.: PIRenderer: controllable portrait image generation via semantic neural rendering. In: Proceedings of the IEEE/CVF International Conference on Computer Vision, pp. 13759–13768 (2021)
32. Rossler, A., Cozzolino, D., Verdoliva, L., Riess, C., Thies, J., Nießner, M.: Faceforensics++: learning to detect manipulated facial images. In: Proceedings of the IEEE/CVF International Conference on Computer Vision, pp. 1–11 (2019)
33. Siarohin, A., Lathuilière, S., Tulyakov, S., Ricci, E., Sebe, N.: Animating arbitrary objects via deep motion transfer. In: Proceedings of the IEEE/CVF Conference on Computer Vision and Pattern Recognition, pp. 2377–2386 (2019)
34. Siarohin, A., Lathuilière, S., Tulyakov, S., Ricci, E., Sebe, N.: First order motion model for image animation. In: Advances in Neural Information Processing Systems, vol. 32, pp. 7137–7147 (2019)
35. Simonyan, K., Zisserman, A.: Very deep convolutional networks for large-scale image recognition. arXiv preprint arXiv:1409.1556 (2014)
36. Thies, J., Zollhöfer, M., Nießner, M.: Deferred neural rendering: image synthesis using neural textures. ACM Trans. Graph. (TOG) **38**(4), 1–12 (2019)
37. Thies, J., Zollhofer, M., Stamminger, M., Theobalt, C., Nießner, M.: Face2Face: real-time face capture and reenactment of RGB videos. In: Proceedings of the IEEE Conference on Computer Vision and Pattern Recognition, pp. 2387–2395 (2016)
38. Wang, H., et al.: CosFace: large margin cosine loss for deep face recognition. In: Proceedings of the IEEE Conference on Computer Vision and Pattern Recognition, pp. 5265–5274 (2018)
39. Wang, Y., et al.: HifiFace: 3D shape and semantic prior guided high fidelity face swapping. arXiv preprint arXiv:2106.09965 (2021)
40. Wiles, O., Koepke, A., Zisserman, A.: X2Face: a network for controlling face generation using images, audio, and pose codes. In: Proceedings of the European Conference on Computer Vision (ECCV), pp. 670–686 (2018)
41. Wu, W., Zhang, Y., Li, C., Qian, C., Loy, C.C.: ReenactGAN: learning to reenact faces via boundary transfer. In: Proceedings of the European Conference on Computer Vision (ECCV), pp. 603–619 (2018)
42. Zakharov, E., Ivakhnenko, A., Shysheya, A., Lempitsky, V.: Fast bi-layer neural synthesis of one-shot realistic head avatars. In: Vedaldi, A., Bischof, H., Brox, T., Frahm, J.-M. (eds.) ECCV 2020. LNCS, vol. 12357, pp. 524–540. Springer, Cham (2020). https://doi.org/10.1007/978-3-030-58610-2_31

43. Zakharov, E., Shysheya, A., Burkov, E., Lempitsky, V.: Few-shot adversarial learning of realistic neural talking head models. In: Proceedings of the IEEE/CVF International Conference on Computer Vision, pp. 9459–9468 (2019)
44. Zeng, X., Pan, Y., Wang, M., Zhang, J., Liu, Y.: Realistic face reenactment via self-supervised disentangling of identity and pose. In: Proceedings of the AAAI Conference on Artificial Intelligence, vol. 34, pp. 12757–12764 (2020)
45. Zhang, H., Goodfellow, I., Metaxas, D., Odena, A.: Self-attention generative adversarial networks. In: International Conference on Machine Learning, pp. 7354–7363. PMLR (2019)
46. Zhang, J., et al.: FReeNet: multi-identity face reenactment. In: Proceedings of the IEEE/CVF Conference on Computer Vision and Pattern Recognition, pp. 5326–5335 (2020)
47. Zhang, R., Isola, P., Efros, A.A., Shechtman, E., Wang, O.: The unreasonable effectiveness of deep features as a perceptual metric. In: Proceedings of the IEEE Conference on Computer Vision and Pattern Recognition, pp. 586–595 (2018)
48. Zhang, Z., Li, L., Ding, Y., Fan, C.: Flow-guided one-shot talking face generation with a high-resolution audio-visual dataset. In: Proceedings of the IEEE/CVF Conference on Computer Vision and Pattern Recognition, pp. 3661–3670 (2021)
49. Zheng, Y., Huang, Y.K., Tao, R., Shen, Z., Savvides, M.: Unsupervised disentanglement of linear-encoded facial semantics. In: Proceedings of the IEEE/CVF Conference on Computer Vision and Pattern Recognition, pp. 3917–3926 (2021)
50. Zhu, Y., Li, Q., Wang, J., Xu, C.Z., Sun, Z.: One shot face swapping on megapixels. In: Proceedings of the IEEE/CVF Conference on Computer Vision and Pattern Recognition, pp. 4834–4844 (2021)

Sobolev Training for Implicit Neural Representations with Approximated Image Derivatives

Wentao Yuan[1,2], Qingtian Zhu[1], Xiangyue Liu[1], Yikang Ding[1], Haotian Zhang[1](✉), and Chi Zhang[1]

[1] Megvii Research, Beijing, China
zhanghaotian@megvii.com
[2] Peking University, Beijing, China

Abstract. Recently, Implicit Neural Representations (INRs) parameterized by neural networks have emerged as a powerful and promising tool to represent different kinds of signals due to its continuous, differentiable properties, showing superiorities to classical discretized representations. However, the training of neural networks for INRs only utilizes input-output pairs, and the derivatives of the target output with respect to the input, which can be accessed in some cases, are usually ignored. In this paper, we propose a training paradigm for INRs whose target output is image pixels, to encode image derivatives in addition to image values in the neural network. Specifically, we use finite differences to approximate image derivatives. We show how the training paradigm can be leveraged to solve typical INRs problems, i.e., image regression and inverse rendering, and demonstrate this training paradigm can improve the data-efficiency and generalization capabilities of INRs. The code of our method is available at https://github.com/megvii-research/Sobolev_INRs.

Keywords: Implicit neural representations · Finite differences · Sobolev training

1 Introduction

A promising recent direction in computer vision and computer graphics is encoding complex signals implicitly by aid of multi-layer perceptron (MLP) named Implicit Neural Representations (INRs) [16,19,20,23,26–28,30]. The MLPs (more specifically, coordinate-based MLPs) take low-dimensional coordinates as inputs and are trained to output a representation of shape, density, and/or colors at each input location.

Compared to classical alternatives, i.e., discrete grid-based representation, INRs offer two main benefits. (a) Due to the fact that MLPs are continuous

This work is done by the first four authors as interns at Megvii Research.

Supplementary Information The online version contains supplementary material available at https://doi.org/10.1007/978-3-031-19784-0_5.

S. Avidan et al. (Eds.): ECCV 2022, LNCS 13675, pp. 72–88, 2022.
https://doi.org/10.1007/978-3-031-19784-0_5

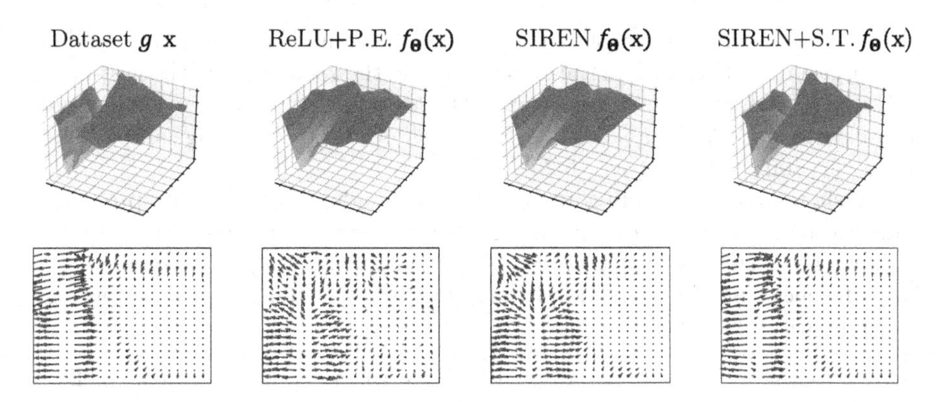

Fig. 1. Visualized approximation results of different training schemes and different activation functions. P.E. stands for positional encoding [19,31]; S.T. stands for Sobolev training. Top: dataset g (the target function to approximate) as well as approximated functions f_{Θ} (functions parameterized by neural networks) represented by MLPs within an image patch of 20×20 pixels. Bottom: vector fields of the first-order derivatives, where red points are training samples (in a total proportion of 6.25%) and remaining black points are unseen samples. When activated by sine functions and trained with additional supervision on derivatives, the network demonstrates the best approximation of both function values and derivatives.

functions, they are significantly more memory efficient than discrete grid-based representations and theoretically, signals parameterized by MLPs can be presented in arbitrary resolution. (b) INRs are naturally differentiable, so gradient descent (GD) can be applied to optimize MLPs using modern auto-differentiation tools, e.g., PyTorch [25], TensorFlow [1], JAX [5].

However, as will be discussed below, the training paradigm adopted by most of the INRs only utilizes input-output pairs, which brings about bad generalization capabilities on unseen coordinates. In other words, if trained with low-resolution training data, MLPs will yield heavily degraded predictions when the target outputs are high-resolution. Based on this observation, an intuitive idea is that we can take advantage of the differentiability of INRs to improve the generalization capabilities. In this paper, we propose a training paradigm for INRs whose training data and the desired results are both images, with both supervision on values and derivatives enforced. Considering that the ground truth representation (the "continuous" image) is agnostic, finite differences are applied to approximate numerical first-order derivatives on images.

Further, we study some important peripheral problems relevant to the proposed training paradigm, e.g., the choice of activation functions, the number of layers in MLPs and the choice of image filters.

We find that in practice, most of the recent INRs build on ReLU-based MLPs [8,19,31] fail to represent well the derivatives of target outputs, even if the corresponding supervision is provided. For this reason, we use periodic activation functions [22,26], i.e., sine function, to improve MLPs' convergence property of derivatives. As is shown in Fig. 1, INRs trained with traditional value-based schemes fail to explicitly constraint the behavior at unseen coordinates, leading to poor generalization both in the image field and the vector field of the first-order derivatives. Besides, with sine-activated MLPs trained with both value loss and derivative loss, INRs are able to generalize well even if the training samples only take a very limited proportion (6.25% in Fig. 1).

We show how the training paradigm can be adapted to specific problems, namely direct image regression [26,31] and indirect inverse rendering [19,31]. Experiments results demonstrate that our training paradigm can improve the data-efficiency and generalization capabilities of INRs. The main contributions are summarized below.

- We propose a training paradigm for INRs whose training data as well as target outputs are image pixels. With the paradigm, image values and image derivatives are encoded into the INRs. Concretely, we use finite differences to approximate the derivatives of the ground truth signal. To the best of our knowledge, supervising network derivatives in INRs is minimally explored.
- We study the factor of activation functions in INRs when applying the proposed training paradigm. By experiments, We use sine functions for better convergence property of derivatives.
- Experiment results on two typical INRs problems demonstrate that with the proposed paradigm we are able to train better INRs - representations that require fewer training samples and generalise better on unseen inputs.

2 Related Work

2.1 Implicit Neural Representations

Implicit Neural Representations (INRs), which usually represent an object as a multi-layer perceptron model that maps low-dimensional coordinates to signal values, have drawn a lot of attention recently. Since this representation is continuous and can capture fine details of signals, it has been widely applied to novel view synthesis [14,15,19,24,32,35,36], shape representation [3,6,9,11,17,23,34,38], and multi-view 3D reconstruction [7,16,21,39]. As a milestone, Mildenhall et al. [19] propose a novel view synthesis method that learns an implicit representation for a specific 3D scene by using a set of multi-view calibrated images. Recent work from Park et al. [23] puts forward to learn a Signed Distance Function (SDF) to represent the shape of objects, and different SDFs can be inferred with different input latent codes. However, the (approximated) ground truth derivative information of the target signal has rarely been utilized to train INRs.

2.2 Derivative Supervision

Derivative information for neural networks has also been exploited in some previous work. Hornik [12] proves the universal approximation theorems for neural networks in Sobolev spaces - a vector space of functions equipped with a

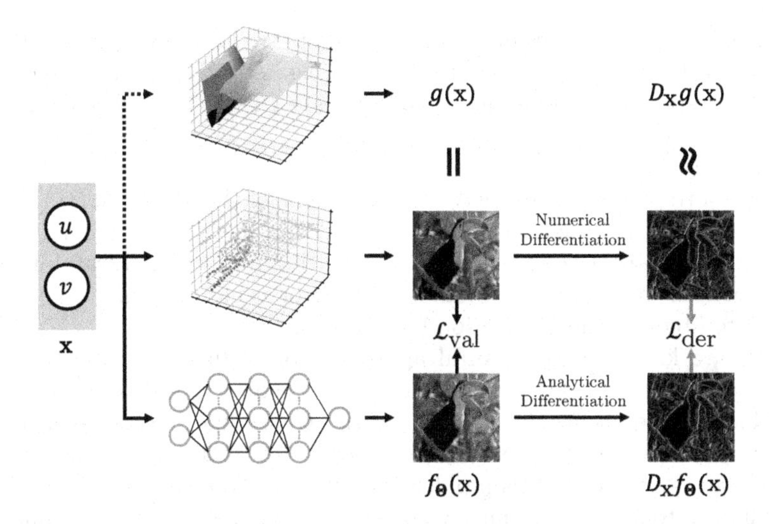

Fig. 2. An overview of the proposed training paradigm for INRs. Top: the underlying continuous signal g and its derivatives $D_{\mathbf{x}}g$. Middle: the pixels of image $\mathbf{I}$, which can be considered as discrete but exact sampling of g and its derivatives are obtained by numerical approximation at image space. Bottom: the MLP denoted as f_Θ used as a general approximator whose derivatives are analytical thanks to auto-differentiation tools. In the proposed training paradigm, we enforce the supervision on both values and derivatives.

norm that is a combination of ℓ_p-norms of the function together with its derivatives up to a given order. Hornik [12] shows that neural networks with non-constant, bounded, continuous activation functions, with continuous derivatives up to order K are universal approximators in the Sobolev spaces of order K. Further, Czarnecki et al. [10] prove ReLU-based MLPs are universal approximators for function values and first-order derivatives theoretically and propose Sobolev training for neural networks. However, only the scenarios with analytical derivatives are covered and the ReLU-based MLPs show relatively poor convergence properties in practice. Gropp et al. [11] propose an *Eikonal term* which is a first-order derivative related penalty term to regularize INRs. The Eikonal term is different from our explicit derivative supervision with ground truth derivatives approximation. Sitzmann et al. [26] verify that INRs with periodic activation functions can fit derivatives robustly, while the supervision on both values and derivatives and the property of Sobolev training in INRs remain unexplored.

3 Methodology

3.1 Formulation

Considering a continuous target signal $g : \mathbb{R}^D \to \mathbb{R}^C$, which is sampled discretely at $\{\mathbf{x}_i\}_{i=1}^N$, the goal of INRs is to approximate g with a neural network (typically a MLP) $f_\Theta : \mathbb{R}^D \to \mathbb{R}^C$, parameterized by weights Θ. The inputs to the MLP are typically low dimensional coordinates $\mathbf{x} \in \mathbb{R}^D$, e.g., $D = 2$ for image coordinates, and the corresponding outputs are signal values, e.g., RGB colors. Training pairs of most recent INRs are denoted as $\{(\mathbf{x}_i, g(\mathbf{x}_i))\}_{i=1}^N$, where N is the number of total training samples.

By measuring the distance between predictions of the MLP $\{f_\Theta(\mathbf{x}_i)\}_{i=1}^N$ and ground truth signal values $\{g(\mathbf{x}_i)\}_{i=1}^N$ with certain metrics, we actually obtain an optimization objective for tuning Θ. Benefiting from the natural differentiability of MLPs, we are able to minimize the error using gradient descent (GD) implemented by auto differentiation frameworks.

MLPs are known as a powerful approximator of functions and with their continuous property, we are able to interpolate the discrete signal up to an arbitrarily higher resolution. However, given the traditional training scheme, the resulting ability to interpolate at unseen coordinates is usually not satisfactory. In this paper, we alleviate this problem by leveraging the supervision on derivatives, which is to say, we optimize Θ by supervising not only signal values, but also derivatives at given coordinates.

The overall training paradigm is illustrated in Fig. 2. We therefore construct a training dataset $\{(\mathbf{x}_i, g(\mathbf{x}_i), D_\mathbf{x} g(\mathbf{x}_i))\}_{i=1}^N$ for INRs. Since the analytical expression of g is unknown, its derivatives require to be approximated with finite differences, which will be elaborated in Sect. 3.2. Considering that the task is a standard regression task, we apply ℓ-2 error as the loss function. With the training set, the optimization objective evolves to a combination of both supervision on signal values and first-order derivatives:

$$\begin{aligned} \Theta &= \arg\min_{\Theta} \frac{1}{N} \sum_{i=1}^{N} [\mathcal{L}_{\text{val}}(g(\mathbf{x}_i), f_\Theta(\mathbf{x}_i)) + \lambda \mathcal{L}_{\text{der}}(D_\mathbf{x} g(\mathbf{x}_i), D_\mathbf{x} f_\Theta(\mathbf{x}_i))] \\ &= \arg\min_{\Theta} \frac{1}{N} \sum_{i=1}^{N} (\|g(\mathbf{x}_i) - f_\Theta(\mathbf{x}_i)\|_2^2 + \lambda \|D_\mathbf{x} g(\mathbf{x}_i) - D_\mathbf{x} f_\Theta(\mathbf{x}_i)\|_2^2). \end{aligned} \tag{1}$$

We additionally extend this paradigm to enable both pre-processing of $\mathbf{x}$ and post-processing of $f_\Theta(\mathbf{x})$, for instances that INRs are used to be an intermediate representation where the inputs to the MLP are calculated from $\mathbf{x}$ and the outputs of the MLP require further processing to obtain the final results. To formulate, we parameterize the MLP as f_Θ, and denote the pre-processing and post-processing as p and q respectively. The target signal g is therefore approximated as $q(f_\Theta(p(\mathbf{x})))$. Assuming that both p and q are differentiable and non-parametric, we can integrate these two processes into f_Θ and follow the aforementioned training paradigm. We will carry out experiments on tasks satisfying this configuration in Sect. 4.2. Specially, if p and q are identical mappings, the case degenerates to the original form.

Though can be applied under more circumstances, in this paper, we set the scope of target signals to $g : \mathbb{R}^2 \to \mathbb{R}^3$, that $\{\mathbf{x}_i\}_{i=1}^N$ are image coordinates and $\{g(\mathbf{x}_i)\}_{i=1}^N$ are pixel values (colors). The loss supervision on derivatives $\mathcal{L}_{\text{der}}$ in Eq. (1) is composed of partial derivatives w.r.t. u and v respectively.

3.2 Approximate Derivatives with Finite Differences

To enable Sobolev training on image coordinates, the first-order derivative of the target signal g is supposed to be known as a precondition. However, the only information accessible at this stage is discrete sampled values $\{g(\mathbf{x}_i)\}_{i=1}^N$ so we need to approximate derivatives from these values with finite differences, or namely numerical differentiation (Fig. 2). Since we only consider cases that $\mathbf{x} \in \mathbb{R}^2$, the partial derivatives are

$$\begin{aligned} D_u g(u,v) &= \frac{g(u+h,v) - g(u-h,v)}{2h}, \\ D_v g(u,v) &= \frac{g(u,v+h) - g(u,v-h)}{2h}. \end{aligned} \tag{2}$$

In this paper, we mainly apply the Sobel operator to obtain first-order derivatives of images. It defines a template that convolves the image $\mathbf{I}$ to obtain the partial derivative w.r.t. u while the transposed template for the partial derivative w.r.t. v:

$$D_u = \begin{bmatrix} -1 & 0 & 1 \\ -2 & 0 & 2 \\ -1 & 0 & 1 \end{bmatrix} * \mathbf{I}, D_v = \begin{bmatrix} -1 & -2 & -1 \\ 0 & 0 & 0 \\ 1 & 2 & 1 \end{bmatrix} * \mathbf{I}, \tag{3}$$

where $*$ denotes the 2D convolution operation. We also study the choice of image filters for derivatives in Sect. 4.3, the conclusion of which is that enforcing the supervision of derivatives matters, regardless of the specific choice of filters.

3.3 Activation Functions

Activation functions are usually non-linear functions applied after each fully-connected layer. ReLU (Rectified Linear Units) [2] is a universally applied choice for its simple design, strong biological motivations and mathematical justifications. Due to the fact that the ReLU function are piecewise linear and its second derivative is zero everywhere, ReLU can not fitting derivatives well in practice even the positional encoding is applied, just as shown in Fig. 3a. On the other hand, [22,26] attempt to explore the potential of the sine function as activation functions due to its periodicity, boundedness, arbitrary-order differentiability. The superior convergence property of sine has been proved in [26] when supervising the derivative only. In our setting, we try to replace ReLU with sine for better convergence property in Sobolev space. As can be indicated from Fig. 3a, sine indeed converges to smaller derivative difference (close to 0) than ReLU when the training in Sobolev space is enabled.

However, studying the properties of activation functions in-depth is a complicated problem and in this paper, we mainly study the training paradigm applied to MLPs with activation functions of (a) ReLU, whose first-order derivative is piece-wise smooth, and (b) sine, whose arbitrary-order derivative is itself (with phase shift). We also study other activation functions in the Supplementary Material.

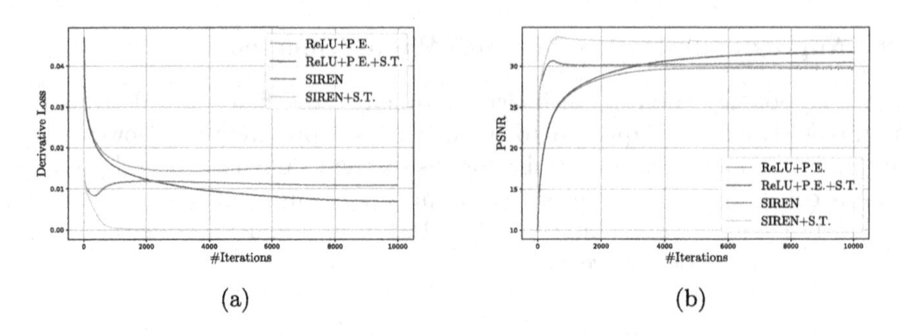

Fig. 3. Convergence situations of $\mathcal{L}_{\text{der}}$ (a) and PSNR (b) w.r.t. iterations.

4 Experiments

In this section, we study the effectiveness of the proposed training paradigm on specific tasks that satisfy the preconditions. We sort tasks involving INRs into two categories, namely direct and indirect ones. Direct tasks are those where the supervision labels are in the same space as the network outputs, where p and q defined in Sect. 3.1 are both identical mappings. While in indirect tasks, the network outputs are processed afterwards so that the observations in the same space as the supervision labels are produced. We give an example of direct and indirect tasks respectively in Sect. 4.1 and Sect. 4.2.

4.1 Direct: Image Regression

A typical direct task is image regression. Given a RGB image $\mathbf{I}$, a MLP is appointed to regress the mapping from image coordinates to RGB colors $g : \mathbb{R}^2 \rightarrow \mathbb{R}^3$. This task is fundamental regarding demonstrating the representation ability of INRs.

Implementation

Dataset. We construct a dataset based on the Set 5 dataset [4] to demonstrate the great interpolatability of MLPs when trained with additional supervision on derivatives. We split the pixels within an image into two groups, namely training samples and evaluation samples, by performing nearest-neighbor downscaling by a factor of 4. This is to say, for an image patch of 4×4, we have 1 training sample and the remaining 15 pixels as evaluation samples.

Network Architecture. For network architecture, we implement a fully-connected MLP with 4 activated linear layers with 256 hidden units each and 1 output linear layer. For the ReLU-based implementation, we follow the conclusion in [31,37] and apply positional encoding (P.E.) to the inputs (additional 20 channels).

Table 1. Quantitative results on the task of image regression on Set 5 dataset [4]. When trained with additional supervision on derivatives and activated by sine and ReLU functions, the network achieves the best and second-best performance respectively in terms of both PSNR and SSIM. We also provided interpolated results by bilinear and bicubic interpolation. best second-best (Best viewed online)

Method	Mean	*Baby*	*Bird*	*Butterfly*	*Head*	*Woman*
			PSNR ↑			
Bilinear	24.14	26.98	24.88	18.68	28.04	22.10
Bicubic	23.63	26.52	24.46	18.23	27.43	21.52
ReLU+P.E. [19]	26.59	29.80	28.18	20.54	29.47	24.95
ReLU+P.E.+S.T.	28.63	31.93	31.22	21.93	31.11	26.97
SIREN [26]	26.22	30.48	28.07	20.73	28.22	23.62
SIREN+S.T.	30.28	33.36	33.34	24.42	31.64	28.62
			SSIM ↑			
Bilinear	0.716	0.778	0.754	0.636	0.674	0.736
Bicubic	0.705	0.773	0.746	0.625	0.655	0.727
ReLU+P.E. [19]	0.740	0.792	0.808	0.636	0.690	0.775
ReLU+P.E.+S.T.	0.809	0.867	0.888	0.708	0.757	0.827
SIREN [26]	0.722	0.849	0.804	0.698	0.635	0.625
SIREN+S.T.	0.890	0.918	0.955	0.870	0.795	0.910

Training and Evaluation. We train the network with all training samples with Adam optimizer for 50k iterations, at a learning rate of 10^{-4}. Following the common practice in image super resolution [13,29,33], we measure PSNR and SSIM on the Y channel and ignore 4 pixels from the image border. For PSNR, we only take the evaluation samples into account.

Results. The quantitative scores are shown in Table 1, where we compare the proposed training paradigm with the previous value-based training paradigm, and with different activation functions. When trained with $\mathcal{L}_{\text{val}}$ alone, ReLU and sine (SIREN [26]) functions lead to comparable results; while with additional supervision on image derivatives, a huge performance improvement has been gained with both activations in both metrics. It is worth noting that in all experiments, INRs demonstrate greater interpolatability than rule-based interpolation methods, i.e., bilinear and bicubic interpolation. Figure 4 gives some example regions where our method produces clearer images and sharper edges and also approximates better the derivatives. It can be observed that sine-based MLPs outperform ReLU-based ones when Sobolev training is enabled. Even so, Sobolev training still improves the results of ReLU-based INRs by a large margin. As additional evidences, Fig. 3b shows the growth of PSNR w.r.t. the number

Table 2. Quantitative results on the task of inverse rendering on LLFF dataset [18]. When trained with additional supervision on derivatives and activated by sine functions, the network achieves the best performance in terms of both PSNR and SSIM. best second-best (Best viewed online)

Method	Mean	*Fern*	*Flower*	*Fortress*	*Horns*	*Leaves*	*Orchids*	*Room*	*T-Rex*
				PSNR ↑					
ReLU+P.E. [19]	23.42	21.82	25.03	27.89	24.75	18.14	18.92	27.22	23.62
ReLU+P.E.+S.T.	23.74	22.99	25.27	27.90	24.50	19.30	19.40	27.10	23.49
SIREN[26]+P.E.	23.38	21.61	25.30	27.60	24.73	18.22	18.99	26.89	23.69
SIREN+P.E.+S.T.	24.38	23.63	25.99	28.30	25.24	20.02	19.74	27.76	24.40
				SSIM ↑					
ReLU+P.E. [19]	0.706	0.651	0.755	0.769	0.729	0.535	0.544	0.877	0.791
ReLU+P.E.+S.T.	0.708	0.680	0.757	0.762	0.711	0.559	0.554	0.878	0.768
SIREN [26]+P.E.	0.682	0.596	0.739	0.763	0.720	0.495	0.525	0.851	0.768
SIREN+P.E.+S.T.	0.751	0.721	0.800	0.811	0.756	0.629	0.592	0.893	0.807

of iterations, where the combination of Sobolev training and sine leads to the fastest convergence and the best performance.

4.2 Indirect: Inverse Rendering

INRs are recently attention-drawing for their great power demonstrated when representing 3D scenes. NeRF [19], which is a typical example of the kind, leverages a MLP $f_\Theta : \mathbb{R}^5 \rightarrow \mathbb{R}^4$, to regress the mapping from (x, y, z, θ, ϕ) to (r, g, b, σ), where x, y, z are 3D coordinates and θ, ϕ are the view directions while the outputs are emitted colors $\mathbf{c} = (r, g, b)$ and volume density σ. In the pipeline of NeRF, the 3D sampling points (x, y, z), as well as view directions (θ, ϕ), of the MLP are determined from 2D image coordinates and corresponding calibrated camera parameters. After obtaining (r, g, b, σ) for each sampling point, a numerical approximation of volume rendering is applied to obtain the integral colors. Considering that both pre-processing and post-processing are differentiable, we can adapt the extended formulation of proposed training paradigm mentioned in Sect. 3.1 to perform inverse rendering of scenes.

Implementation

Dataset. We carry out experiments on LLFF dataset [18]. Similarly, we downsample the training images to 189×252 while the resolution of target images to render is 756×1008.

Network Architecture. We implement a simplified version of NeRF [19] $(x, y, z) \rightarrow (r, g, b, \sigma)$, removing the skip connection, the strategy of hierarchical sampling and view dependence. For network architecture, we implement a fully-connected MLP with 8 activated linear layers with 256 hidden units each and 1 output layer. We apply positional encoding (additional 60 channels) to both sine-activated and ReLU-activated MLPs.

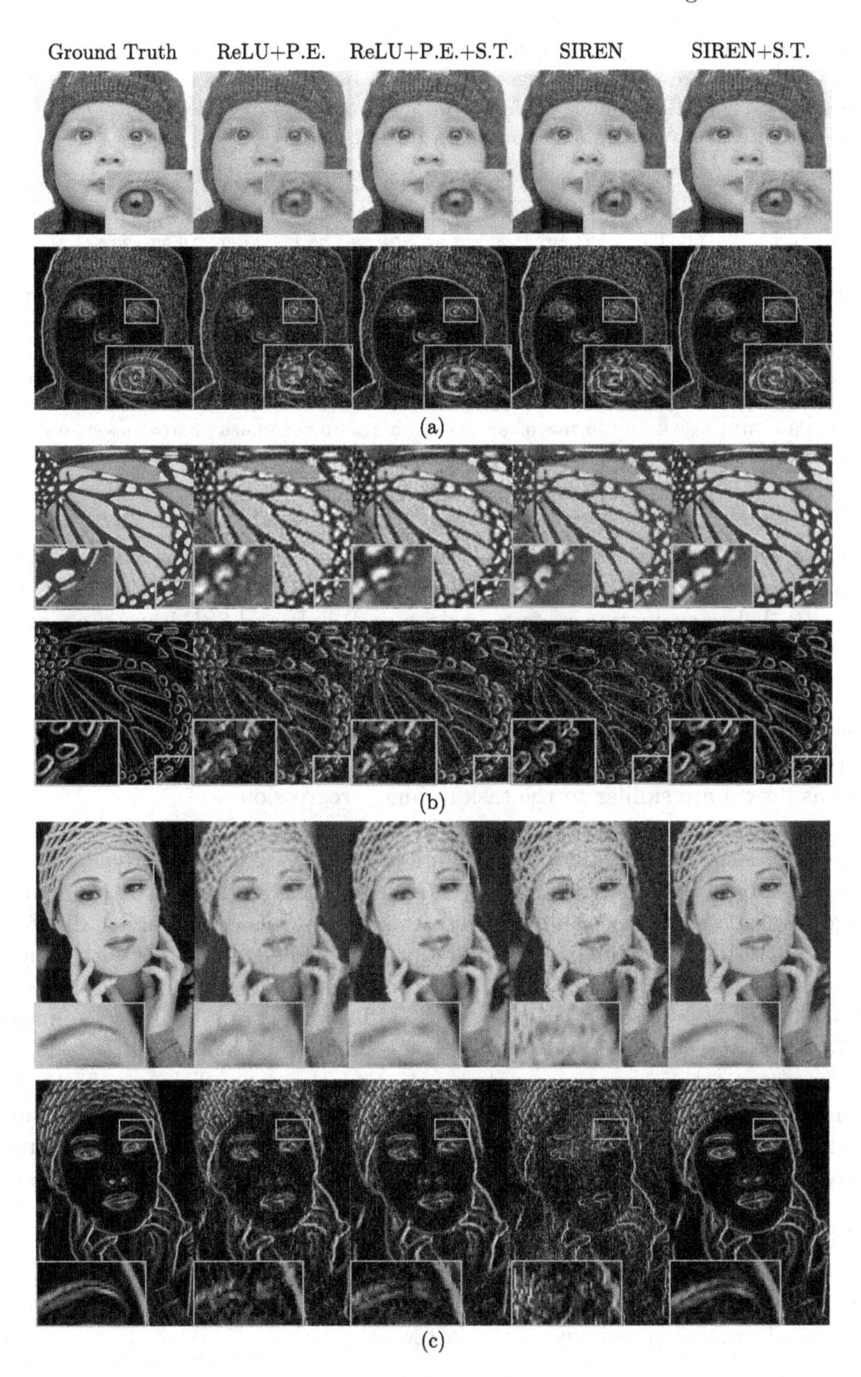

Fig. 4. Results of regressed images and derivatives of the set *Baby* (a), *Butterfly* (b) and *Woman* (c). When sine functions are adopted for activating linear layers and Sobolev training is enabled, the MLP yields the best visual effects, especially at high-frequency sharp edges, while ReLU-based MLPs fail to regress these details.

Table 3. Results for comparisons on the task of inverse rendering on LLFF dataset [18] when the training resolution is larger. It is worth noting that even only trained with 1/16 samples, the MLP trained with the proposed paradigm is able to outperform the ReLU-activated network on both metrics in the set of *Flower*.

$H \times W$	Method	Mean	*Fern*	*Flower*	*Fortress*	*Horns*	*Leaves*	*Orchids*	*Room*	*T-Rex*
				PSNR ↑						
756 × 1008	ReLU+P.E	**24.69**	24.26	25.92	**28.58**	25.32	20.39	**19.99**	**28.54**	24.52
378 × 504	ReLU+P.E	24.55	23.93	25.90	28.46	25.22	20.20	19.97	28.30	24.42
378 × 504	SIREN+P.E.+S.T	24.67	**24.27**	**26.16**	28.37	**25.37**	**20.64**	19.94	27.99	**24.61**
189 × 252	SIREN+P.E.+S.T	24.38	23.63	25.99	28.30	25.24	20.02	19.74	27.76	24.40
				SSIM ↑						
756 × 1008	ReLU+P.E	0.755	0.744	0.788	0.792	0.746	0.646	0.606	**0.900**	**0.816**
378 × 504	ReLU+P.E	0.750	0.734	0.785	0.790	0.744	0.637	0.603	0.897	0.813
378 × 504	SIREN+P.E.+S.T	**0.766**	**0.753**	**0.805**	**0.815**	**0.762**	**0.666**	**0.614**	0.899	0.815
189 × 252	SIREN+P.E.+S.T	0.751	0.721	0.800	0.811	0.756	0.629	0.596	0.893	0.807

Training and Evaluation. At the stage of training, we train the network with a batch size of 128 with Adam optimizer for 400k iterations, at a learning rate of 5×10^{-4}. For evaluation metrics, we report PSNR and SSIM on all channels (RGB) of the rendered images.

Results. Table 2 demonstrates the PSNR and SSIM on all scenes of LLFF dataset [18] and Fig. 5 provides some results of novel view synthesis. The conclusions drawn are similar to the task of image regression.

Data-Efficiency. To provide more insights of the data-efficiency brought about by the proposed training paradigm, we also conduct experiments with different image resolution, namely different amount of training samples under different training paradigms. For the valued-based paradigm, we train MLPs with images of resolution 378×504 and 756×1008 respectively and render novel views of 756×1008. For the proposed paradigm, we additionally train MLPs with images of 378×504 and render novel views of 756×1008.

The quantitative comparisons demonstrating the data-efficiency of Sobolev training are shown in Table 3. The following two points can be observed and indicated. (a) Even with only 1/16 training samples, the Sobolev trained MLP sometimes exceeds the ReLU-activated MLP with the value-based training paradigm, e.g., on *Flower* in terms of PSNR and SSIM, on *Fortress* and *Horns* in terms of SSIM. (b) The Sobolev trained MLP using images of 378×504 gets similar PSNR and even higher SSIM compared to the ReLU-activated MLP trained with the value-based paradigm using images of 756×1008. Note that the former case is trained only with 1/4 samples as the latter.

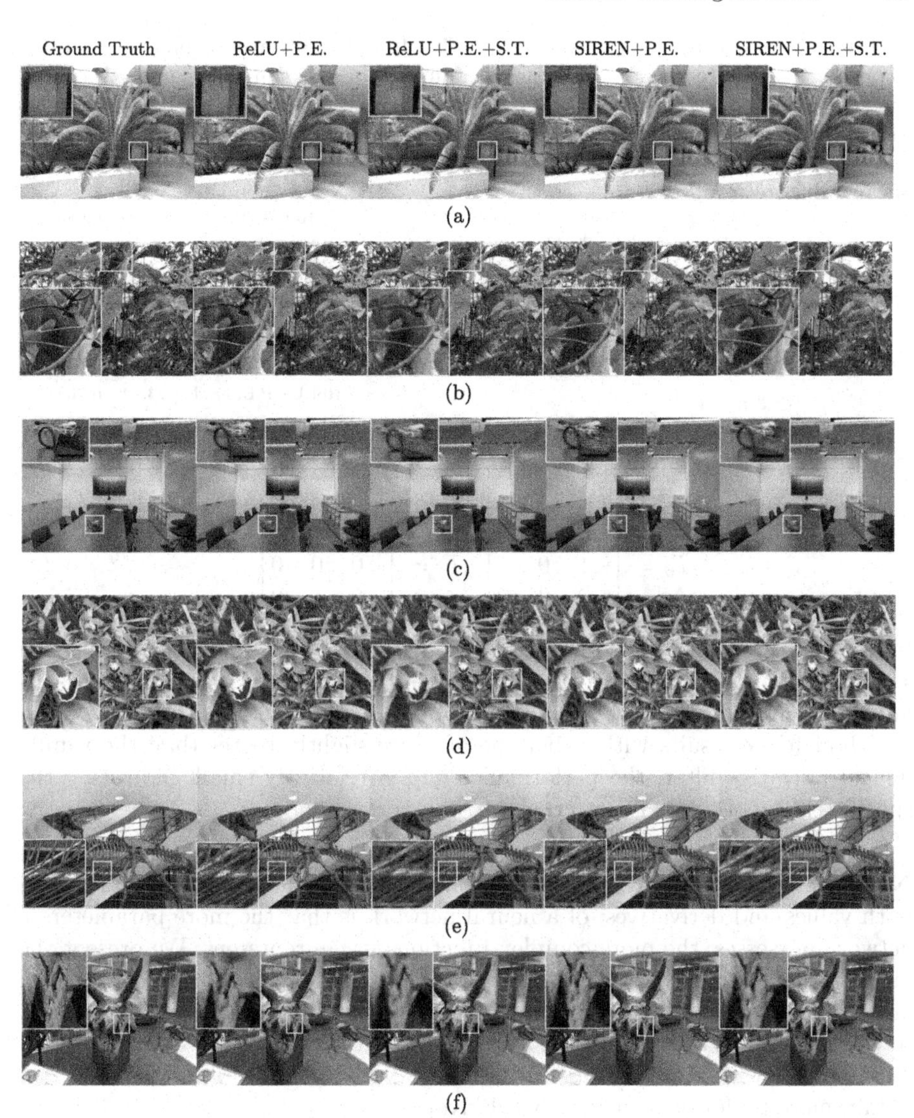

Fig. 5. Results of novel view synthesis by INRs in the task of inverse rendering. Top to bottom: *Fern* (a), *Leaves* (b), *Room* (c), *Orchids* (d), *T-Rex* (e), *Horns* (f). Supervision on derivatives, applied with sine-activated MLPs, helps to render clear images and sharp edges at unseen views.

4.3 Ablation Study

Image Derivative Filters. We compare the network performance with another image filter as finite differences approximating derivatives, namely vanilla image derivative filter whose templates are

Table 4. Quantitative comparison on the choice of image filters when approximating derivatives of g. Experiments are carried out with MLPs activated by sine functions. PSNR and SSIM stand for mean values over all scenes.

Task	Filter	PSNR	SSIM
Image Regression	Vanilla	30.24	0.886
	Sobel	**30.28**	**0.890**
Inverse Rendering	Vanilla	24.28	0.746
	Sobel	**24.38**	**0.751**

Table 5. Quantitative comparison on the number of layers in MLPs under the task of inverse rendering. PSNR and SSIM stand for mean values over all scenes.

#layers	Method	PSNR	SSIM
4	ReLU+P.E	23.29	0.685
	SIREN+P.E	23.17	0.654
	ReLU+P.E.+S.T	23.35	0.689
	SIREN+P.E.+S.T	23.66	0.697
8	ReLU+P.E	23.42	0.706
	SIREN+P.E	23.38	0.682
	ReLU+P.E.+S.T	23.74	0.708
	SIREN+P.E.+S.T	**24.38**	**0.751**

$$T_u = \begin{bmatrix} 0 & 0 & 0 \\ -1 & 0 & 1 \\ 0 & 0 & 0 \end{bmatrix}, T_v = \begin{bmatrix} 0 & -1 & 0 \\ 0 & 0 & 0 \\ 0 & 1 & 0 \end{bmatrix}. \tag{4}$$

It considers the same neighborhood as Sobel operator but removes the spatial smoothing operation included in Sobel operator. The quantitative comparisons are shown in Table 4.

Though the results with Sobel operator are slightly better than the vanilla image derivative filter, the performance gap is relatively small, indicating the specific choice of filters is not crucial to the performance.

Network Capacity. A simple intuition towards the fitting ability, including both values and derivatives, of a neural network is that the more parameters a network possesses, the more complex functions it can represent. We present the quantitative results obtained with networks with different number of layers (4 and 8 respectively) in the task of inverse rendering in Table 5. Other experiment settings are aligned with Sect. 4.2. It can be indicated that when the layer number increases to 8 from 4, the representation ability of INRs is stronger so it can obtain more performance gain from derivatives.

5 Discussions

5.1 Future Work

To restate, the preconditions of the proposed training paradigm are that (a) the raw dataset is in the form of $\{(\mathbf{x}_i, g(\mathbf{x}_i))\}_{i=1}^N$, where $\{\mathbf{x}_i\}_{i=1}^N$ are spatial coordinates; (b) the derivatives of g can be obtained or approximated with limited $\{g(\mathbf{x}_i)\}$; (c) $\{\mathbf{x}_i\}$ do not necessarily be the direct inputs of the MLP, and the target values $\{g(\mathbf{x}_i)\}$ do not necessarily be the direct outputs if and only if the pre-processing p is differentiable w.r.t. to $\mathbf{x}$ and the post-processing q is differentiable w.r.t. $f_{\Theta}(p(\mathbf{x}))$.

Theoretically, referring to the formulation in Sect. 3.1, the training paradigm can further generalize to more tasks, e.g., audio regression (please refer to the Supplementary Material) and video regression, as long as their partial derivatives can be obtained by numerical approximation. Take the task of video regression $(u, v, t) \rightarrow (r, g, b)$ as an example, whose partial derivatives w.r.t. u and v are the same on images. We can approximate the partial derivative w.r.t. t by subtracting consecutive frames.

5.2 Limitations

We summarize the limitations of the proposed training paradigm as below.

- The computation of analytical derivatives is computationally expensive and occupies a considerable amount of memory. For example, both PyTorch [25] and TensorFlow [1] cannot calculate derivatives for each sample in a batch separately, leading to redundant computation.
- Though the proposed training paradigm can be applied to various INRs-based tasks, the scope of application is still limited. Since the derivatives are approximated with finite differences, the distribution of $\{\mathbf{x}_i\}$ is supposed to be as uniform as possible at its domain, which is not always satisfied.

6 Conclusion

In this paper, we put forward a novel training paradigm for INRs in Sobolev space, where supervision on both signal values and first-order derivatives is enforced for network training. The formulation of the task is generalized to any circumstance that the inputs are 2D image coordinates and the outputs are image values (e.g., RGB) as long as the pre-processing is differentiable w.r.t. the coordinates and the post-processing is differentiable w.r.t. the MLP's outputs. We obtain approximated image derivatives by the Sobel operator.

Experiments are carried out on the task of image regression (direct) and inverse rendering (indirect) and results show that the training paradigm brings remarkable improvement to the quality of reconstructed images, especially when applied together with a MLP whose activation functions are sine.

In addition, we study some important and interesting peripheral problems relevant to the proposed training paradigm and give insights based on our observations.

Acknowledgements. The authors would like to thank Zhongtian Zheng from Peking University for his valuable advice.

References

1. Abadi, M., et al.: TensorFlow: a system for Large-Scale machine learning. In: 12th USENIX Symposium on Operating Systems Design and Implementation, OSDI 2016, pp. 265–283 (2016)

2. Agarap, A.F.: Deep learning using rectified linear units (ReLU). arXiv preprint arXiv:1803.08375 (2018)
3. Atzmon, M., Lipman, Y.: SAL: sign agnostic learning of shapes from raw data. In: Proceedings of the IEEE/CVF Conference on Computer Vision and Pattern Recognition, pp. 2565–2574 (2020)
4. Bevilacqua, M., Roumy, A., Guillemot, C., Alberi-Morel, M.L.: Low-complexity single-image super-resolution based on nonnegative neighbor embedding (2012)
5. Bradbury, J., et al.: JAX: composable transformations of Python+NumPy programs (2018). https://github.com/google/jax
6. Chabra, R., et al.: Deep local shapes: learning local SDF priors for detailed 3D reconstruction. In: Vedaldi, A., Bischof, H., Brox, T., Frahm, J.-M. (eds.) ECCV 2020. LNCS, vol. 12374, pp. 608–625. Springer, Cham (2020). https://doi.org/10.1007/978-3-030-58526-6_36
7. Chen, A., et al.: MVSNeRF: fast generalizable radiance field reconstruction from multi-view stereo. In: Proceedings of the IEEE/CVF International Conference on Computer Vision, pp. 14124–14133 (2021)
8. Chen, Y., Liu, S., Wang, X.: Learning continuous image representation with local implicit image function. In: Proceedings of the IEEE/CVF Conference on Computer Vision and Pattern Recognition, pp. 8628–8638 (2021)
9. Chen, Z., Zhang, H.: Learning implicit fields for generative shape modeling. In: Proceedings of the IEEE/CVF Conference on Computer Vision and Pattern Recognition, pp. 5939–5948 (2019)
10. Czarnecki, W.M., Osindero, S., Jaderberg, M., Swirszcz, G., Pascanu, R.: Sobolev training for neural networks. In: Advances in Neural Information Processing Systems, vol. 30 (2017)
11. Gropp, A., Yariv, L., Haim, N., Atzmon, M., Lipman, Y.: Implicit geometric regularization for learning shapes. In: International Conference on Machine Learning, pp. 3789–3799. PMLR (2020)
12. Hornik, K.: Approximation capabilities of multilayer feedforward networks. Neural Netw. **4**(2), 251–257 (1991)
13. Jo, Y., Kim, S.J.: Practical single-image super-resolution using look-up table. In: Proceedings of the IEEE/CVF Conference on Computer Vision and Pattern Recognition, pp. 691–700 (2021)
14. Lin, C.H., Ma, W.C., Torralba, A., Lucey, S.: BARF: bundle-adjusting neural radiance fields. In: Proceedings of the IEEE/CVF International Conference on Computer Vision, pp. 5741–5751 (2021)
15. Martin-Brualla, R., Radwan, N., Sajjadi, M.S., Barron, J.T., Dosovitskiy, A., Duckworth, D.: NeRF in the wild: neural radiance fields for unconstrained photo collections. In: 2021 IEEE/CVF Conference on Computer Vision and Pattern Recognition (CVPR), pp. 7206–7215. IEEE Computer Society (2021)
16. Mescheder, L., Oechsle, M., Niemeyer, M., Nowozin, S., Geiger, A.: Occupancy networks: learning 3D reconstruction in function space. In: Proceedings of the IEEE/CVF Conference on Computer Vision and Pattern Recognition, pp. 4460–4470 (2019)
17. Michalkiewicz, M., Pontes, J.K., Jack, D., Baktashmotlagh, M., Eriksson, A.: Implicit surface representations as layers in neural networks. In: 2019 IEEE/CVF International Conference on Computer Vision (ICCV), pp. 4742–4751. IEEE Computer Society (2019)
18. Mildenhall, B., et al.: Local light field fusion: practical view synthesis with prescriptive sampling guidelines. ACM Trans. Graph. (TOG) **38**(4), 1–14 (2019)

19. Mildenhall, B., Srinivasan, P.P., Tancik, M., Barron, J.T., Ramamoorthi, R., Ng, R.: NeRF: representing scenes as neural radiance fields for view synthesis. In: Vedaldi, A., Bischof, H., Brox, T., Frahm, J.-M. (eds.) ECCV 2020. LNCS, vol. 12346, pp. 405–421. Springer, Cham (2020). https://doi.org/10.1007/978-3-030-58452-8_24
20. Niemeyer, M., Mescheder, L., Oechsle, M., Geiger, A.: Differentiable volumetric rendering: learning implicit 3D representations without 3D supervision. In: Proceedings of the IEEE/CVF Conference on Computer Vision and Pattern Recognition, pp. 3504–3515 (2020)
21. Oechsle, M., Peng, S., Geiger, A.: UNISURF: unifying neural implicit surfaces and radiance fields for multi-view reconstruction. In: 2021 IEEE/CVF International Conference on Computer Vision (ICCV), pp. 5589–5599 (2021)
22. Parascandolo, G., Huttunen, H., Virtanen, T.: Taming the waves: sine as activation function in deep neural networks (2017)
23. Park, J.J., Florence, P., Straub, J., Newcombe, R., Lovegrove, S.: DeepSDF: learning continuous signed distance functions for shape representation. In: Proceedings of the IEEE/CVF Conference on Computer Vision and Pattern Recognition, pp. 165–174 (2019)
24. Park, K., et al.: Nerfies: deformable neural radiance fields. In: Proceedings of the IEEE/CVF International Conference on Computer Vision, pp. 5865–5874 (2021)
25. Paszke, A., et al.: Pytorch: an imperative style, high-performance deep learning library. In: Advances in Neural Information Processing Systems, vol. 32 (2019)
26. Sitzmann, V., Martel, J., Bergman, A., Lindell, D., Wetzstein, G.: Implicit neural representations with periodic activation functions. In: Advances in Neural Information Processing Systems, vol. 33, pp. 7462–7473 (2020)
27. Sitzmann, V., Rezchikov, S., Freeman, B., Tenenbaum, J., Durand, F.: Light field networks: neural scene representations with single-evaluation rendering. In: Advances in Neural Information Processing Systems, vol. 34, pp. 19313–19325 (2021)
28. Sitzmann, V., Zollhöfer, M., Wetzstein, G.: Scene representation networks: continuous 3D-structure-aware neural scene representations. In: Advances in Neural Information Processing Systems, vol. 32 (2019)
29. Soh, J.W., Park, G.Y., Jo, J., Cho, N.I.: Natural and realistic single image super-resolution with explicit natural manifold discrimination. In: Proceedings of the IEEE/CVF Conference on Computer Vision and Pattern Recognition, pp. 8122–8131 (2019)
30. Suhail, M., Esteves, C., Sigal, L., Makadia, A.: Light field neural rendering. In: Proceedings of the IEEE/CVF Conference on Computer Vision and Pattern Recognition, pp. 8269–8279 (2022)
31. Tancik, M., et al.: Fourier features let networks learn high frequency functions in low dimensional domains. In: Advances in Neural Information Processing Systems, vol. 33, pp. 7537–7547 (2020)
32. Tewari, A., et al.: Advances in neural rendering. In: ACM SIGGRAPH 2021, pp. 1–320. ACM (2021)
33. Yang, W., Zhang, X., Tian, Y., Wang, W., Xue, J.H., Liao, Q.: Deep learning for single image super-resolution: a brief review. IEEE Trans. Multimed. **21**(12), 3106–3121 (2019)
34. Yariv, L., et al.: Multiview neural surface reconstruction by disentangling geometry and appearance. In: Advances in Neural Information Processing Systems, vol. 33, pp. 2492–2502 (2020)

35. Yu, A., Li, R., Tancik, M., Li, H., Ng, R., Kanazawa, A.: PlenOctrees for real-time rendering of neural radiance fields. In: Proceedings of the IEEE/CVF International Conference on Computer Vision, pp. 5752–5761 (2021)
36. Yu, A., Ye, V., Tancik, M., Kanazawa, A.: pixeLNeRF: neural radiance fields from one or few images. In: Proceedings of the IEEE/CVF Conference on Computer Vision and Pattern Recognition, pp. 4578–4587 (2021)
37. Yüce, G., Ortiz-Jiménez, G., Besbinar, B., Frossard, P.: A structured dictionary perspective on implicit neural representations. arXiv preprint arXiv:2112.01917 (2021)
38. Zhang, J., Yang, G., Tulsiani, S., Ramanan, D.: NeRS: neural reflectance surfaces for sparse-view 3D reconstruction in the wild. In: Advances in Neural Information Processing Systems, vol. 34 (2021)
39. Zhang, J., Yao, Y., Quan, L.: Learning signed distance field for multi-view surface reconstruction. In: Proceedings of the IEEE/CVF International Conference on Computer Vision, pp. 6525–6534 (2021)

Make-A-Scene: Scene-Based Text-to-Image Generation with Human Priors

Oran Gafni(✉), Adam Polyak, Oron Ashual, Shelly Sheynin, Devi Parikh, and Yaniv Taigman

Meta AI Research, Ramat Gan, Israel
oran.gafni@gmail.com

Abstract. Recent text-to-image generation methods provide a simple yet exciting conversion capability between text and image domains. While these methods have incrementally improved the generated image fidelity and text relevancy, several pivotal gaps remain unanswered, limiting applicability and quality. We propose a novel text-to-image method that addresses these gaps by (i) enabling a simple control mechanism complementary to text in the form of a scene, (ii) introducing elements that substantially improve the tokenization process by employing domain-specific knowledge over key image regions (faces and salient objects), and (iii) adapting classifier-free guidance for the transformer use case. Our model achieves state-of-the-art FID and human evaluation results, unlocking the ability to generate high fidelity images in a resolution of 512×512 pixels, significantly improving visual quality. Through scene controllability, we introduce several new capabilities: **(i)** Scene editing, **(ii)** text editing with anchor scenes, **(iii)** overcoming out-of-distribution text prompts, and **(iv)** story illustration generation, as demonstrated in the story we wrote.

Keywords: Text-to-image · Image synthesis · Multi-modal

1 Introduction

"A poet would be overcome by sleep and hunger before being able to describe with words what a painter is able to depict in an instant."

Similar to this quote by Leonardo da Vinci [27], equivalents of the expression "A picture is worth a thousand words" have been iterated in different languages and eras [1,14,25], alluding to the heightened expressiveness of images over text, from the human perspective. There is no surprise then, that the task of text-to-image generation has been gaining increased attention with the recent success of text-to-image modeling via large-scale models and datasets. This new capability of effortlessly bridging between the text and image domains enables new forms of creativity to be accessible to the general public.

Supplementary Information The online version contains supplementary material available at https://doi.org/10.1007/978-3-031-19784-0_6.

S. Avidan et al. (Eds.): ECCV 2022, LNCS 13675, pp. 89–106, 2022.
https://doi.org/10.1007/978-3-031-19784-0_6

Fig. 1. Make-A-Scene: Samples of generated images from text inputs (a), and a text and scene input (b). Our method is able to both generate the scene (a, bottom left) and image, or generate the image from text and a simple sketch input (b, center).

While current methods provide a simple yet exciting conversion between the text and image domains, they still lack several pivotal aspects:

(i) Controllability. The sole input accepted by the majority of models is text, confining any output to be controlled by a text description only. While certain perspectives can be controlled with text, such as style or color, others such as structure, form, or arrangement can only be loosely described at best [46]. This lack of control conveys a notion of randomness and weak user-influence on the image content and context [33]. Controlling elements additional to text have been suggested by [68], yet their use is confined to restricted datasets such as fashion items or faces. An earlier work by [23] suggests coarse control in the form of bounding boxes resulting in low resolution images.
(ii) Human perception. While images are generated to match human perception and attention, the generation process does not include any relevant prior knowledge, resulting in little correlation between generation and human attention. A clear example of this gap can be observed in person and face generation, where a dissonance is present between the importance of face pixels from the human perspective and the loss applied over the whole image [28,65]. This gap is relevant to animals and other salient objects as well.
(iii) Quality and resolution. Although quality has gradually improved between consecutive methods, the previous state-of-the-art methods are still limited to an output image resolution of 256×256 pixels [40,45]. Alternative approaches propose a super-resolution network which results in less favorable visual and quantitative results [12]. Quality and resolution are strongly linked, as scaling up to a resolution of 512×512 requires a substantially higher quality with fewer artifacts than 256×256.

In this work, we introduce a novel method that successfully tackles these pivotal gaps, while attaining state-of-the-art results in the task of text-to-image generation. Our method provides a new type of control complementary to text, enabling new-generation capabilities while improving structural consistency and quality. Furthermore, we propose explicit losses correlated with human preferences, significantly improving image quality, breaking the common resolution barrier, and thus producing results in a resolution of 512×512 pixels.

Our method is comprised of an autoregressive transformer, where in addition to the conventional use of text and image tokens, we introduce implicit conditioning over optionally controlled scene tokens, derived from segmentation maps. During inference, the segmentation tokens are either generated independently by the transformer or extracted from an input image, providing freedom to impel additional constraints over the generated image. Contrary to the common use of segmentation for explicit conditioning as employed in many GAN-based methods [24,42,61], our segmentation tokens provide implicit conditioning in the sense that the generated image and image tokens are not constrained to use the segmentation information, as there is no loss tying them together. In practice, this contributes to the variety of samples generated by the model, producing diverse results constrained to the input segmentations.

We demonstrate the new capabilities this method provides in addition to controllability, such as (i) complex scene generation (Fig. 1), (ii) out-of-distribution generation (Fig. 3), (iii) scene editing (Fig. 4), and (iv) text editing with anchored scenes (Fig. 5). We additionally provide an example of harnessing controllability to assist with the creative process of storytelling in the supplementary materials.

While most approaches rely on losses agnostic to human perception, this approach differs in that respect. We use two modified Vector-Quantized Variational Autoencoders (VQ-VAE) to encode and decode the image and scene tokens with explicit losses targeted at specific image regions correlated with human perception and attention, such as faces and salient objects. The losses contribute to the generation process by emphasizing the specific regions of interest and integrating domain-specific perceptual knowledge in the form of network feature-matching.

While some methods rely on image re-ranking for post-generation image filtering (utilizing CLIP [44] for instance), we extend the use of classifier-free guidance suggested for diffusion models [20,53] by [22,40] to transformers, eliminating the need for post-generation filtering, thus producing faster and higher quality generation results, better adhering to input text prompts.

An extensive set of experiments is provided to establish the visual and numerical validity of the our different contributions.

2 Related Work

Image Generation. Recent advancements in deep generative models have enabled algorithms to generate high-quality and natural-looking images. Generative Adversarial Networks (GANs) [17] facilitate the generation of high fidelity images [3,29,30,56] in multiple domains by simultaneously training a generator network G and a discriminator network D, where G is trained to fool D, while D is trained to judge if a given image is real or fake. Concurrently to GANs, Variational Autoencoders (VAEs) [31,57] have introduced a likelihood-based approach to image generation. Other likelihood-based models include autoregressive models [8,13,41,43] and diffusion models [11,20,21]. While the former model image pixels as a sequence with autoregressive dependency between each pixel, the latter synthesizes images via a gradual denoising process. Specifically, sampling starts with a noisy image which is iteratively denoised until all denoising steps are performed. Applying both methods directly to the image pixel-space can be challenging. Consequently, recent approaches either compress the image to a discrete representation [13,58] via Vector Quantized Variational Autoencoders (VQ-VAEs) [58], or downsample the image resolution [11,21]. Our method is based on autoregressive modeling of the discrete image representation.

Fig. 2. Qualitative comparison with previous work. The text and generated images for [40,45,66] were taken from [40]. For CogView [12] we use the released 512×512 model weights, applying self-reranking of 60 for post-generation selection. When zoomed-in, our results can be seen to be of higher-quality, higher-resolution, with crisp details.

Fig. 3. Overcoming out-of-distribution text prompts with scene control. By introducing simple scene sketches (bottom right) as additional inputs, our method is able to overcome unusual objects and scenarios presented as failure cases in previous methods.

Image Tokenization. Image generation models based on discrete representation [12,13,45,47,58] follow a two-stage training scheme. First, an image tokenizer is trained to extract a discrete image representation. In the second stage, a generative model generates the image in the discrete latent space. Inspired by Vector Quantization (VQ) techniques, VQ-VAE [58] learns to extract a discrete latent representation by performing online clustering.

VQ-VAE-2 [47] presented a hierarchical architecture composed of VQ-VAE models operating at multiple scales, enabling faster generation compared with pixel space generation. The DALL-E [45] text-to-image model used dVAE, which uses gumbel-softmax [26,38], relaxing the VQ-VAE's online clustering. Recently, VQGAN [13] added adversarial and perceptual losses [67] on top of the VQ-VAE reconstruction task, producing reconstructed images with higher quality. In our work, we modify the VQGAN framework by adding perceptual losses to specific image regions, such as faces and salient objects, which further improve the fidelity of the generated images.

Image-to-Image Generation. Generating images from segmentation maps or scenes can be viewed as a conditional image synthesis task [24,37,42,60,61,70]. Specifically, this form of image synthesis permits more controllability over the desired output. CycleGAN [70] trained a mapping function from one domain to the other. UNIT [37] projected two different domains into a shared latent space and used a per-domain decoder to re-synthesize images in the desired domain. Both methods do not require supervision between domains. pix2pix [24] utilized conditional GANs together with a supervised reconstruction loss. pix2pixHD [61] improved the latter by increasing output image resolution thanks to improved network architecture. SPADE [42] introduced a spatially-adaptive normalization

Fig. 4. Generating images through edited scenes. For an input text (a) and the segmentations extracted from an input image (b), we can re-generate the image (c) or edit the segmentations (d) by replacing classes (top) or adding classes (bottom), generating images with new context or content (e).

Fig. 5. Generating new image interpretations through text editing and anchor scenes. For an input text (a) and image (b), we first extract the semantic segmentation (c), we can then re-generate new images (d) given the input segmentation and edited text. Purple denotes text added or replacing the original text.

layer which elevated information lost in normalization layers. [15] introduced face-refinement to SPADE through a pre-trained face-embedding network inspired by face-generation methods [16]. Unlike the aforementioned, our work conditions jointly on text and segmentation, enabling bi-domain controllability.

Text-to-Image Generation. Text-to-image generation [12,40,45,54,63,64,66,69,71] focuses on generating complex scenes from standalone text description. Preliminary work on text-to-image conditioned RNN-based DRAW [18] on text [39]. With the introduction of GANs, text conditioned image generation improved [48]. AttnGAN [63] introduced an attention component which enabled the generator network to attend to relevant words in the text. DM-GAN [71] introduced a dynamic memory component while DF-GAN [54] employed a novel fusion block to fuse text information into image features. Contrastive learning further improved the results of DM-GAN [64] while XMC-GAN [66] used contrastive learning to maximize the mutual information between image and text.

DALL-E [45] and CogView [12] trained an autoregressive transformer [59] on text and image tokens, demonstrating convincing zero-shot capabilities on the MS-COCO dataset. GLIDE [40] used diffusion models conditioned on images. Inspired by the high-quality unconditional images generation model, GLIDE employed guided inference with and without a classifier network to generate high-fidelity images. LAFITE [69] employed a pre-trained CLIP [44] model to project text and images to the same latent space, training text-to-image models without text data. Similarly to DALL-E and CogView, we train an autoregressive transformer model on text and image tokens. Our main contributions are introducing additional controlling elements in the form of a scene, improve the tokenization process, and adapt classifier-free guidance to transformers.

3 Method

Our model generates an image given a text input and an optional scene layout (segmentation map). As demonstrated in our experiments, by conditioning over the scene layout, our method provides a new form of implicit controllability, improves structural consistency and quality, and adheres to human preference (as assessed by our human evaluation study). In addition to our scene-based approach, we extended our aspiration of improving the general and perceived quality with a better representation of the token space. We introduce several modifications to the tokenization process, emphasizing awareness of aspects with increased importance in the human perspective, such as faces and salient objects. To refrain from post-generation filtering and further improve the generation quality and text alignment, we employ classifier-free guidance.

We follow next with a detailed overview of the proposed method, comprised of (i) scene representation and tokenization, (ii) attending human preference in the token space with explicit losses, (iii) the scene-based transformer, and (iv) transformer classifier-free guidance. Aspects commonly used prior to this method are not extensively detailed below, whereas specific settings for all elements can be found in the appendix.

3.1 Scene Representation and Tokenization

The scene is composed of a union of three complementary semantic segmentation groups - panoptic, human, and face. By combining the three extracted semantic segmentation groups, the network learns to both generate the semantic layout and condition on it while generating the final image. The semantic layout provides additional global context in an implicit form that correlates with human preference, as the choice of categories within the scene groups, and the choice of the groups themselves are a prior to human preference and awareness. We consider this form of conditioning to be implicit, as the network may disregard any scene information, and generate the image conditioned solely on text. Our experiments indicate that both the text and scene firmly control the image.

In order to create the scene token space, we employ VQ-SEG: a modified VQ-VAE for semantic segmentation, building on the VQ-VAE suggested for semantic segmentation in [13]. In our implementation the inputs and outputs of VQ-SEG are m channels, representing the number of classes for all semantic segmentation groups $m = m_p + m_h + m_f + 1$, where m_p, m_h, m_f are the number of categories for the panoptic segmentation [62], human segmentation [34], and face segmentation extracted with [5] respectively. The additional channel is a map of the edges separating the different classes and instances. The edge channel provides both separations for adjacent instances of the same class, and emphasis on scarce classes with high importance, as edges (perimeter) are less biased towards larger categories than pixels (area).

3.2 Adhering to Human Emphasis in the Token Space

We observe an inherent upper-bound on image quality when generating images with the transformer, stemming from the tokenization reconstruction method. In other words, quality limitations of the VQ image reconstruction method inherently transfer to quality limitations on images generated by the transformer. To that end, we introduce several modifications to both the segmentation and image reconstruction methods. These modifications are losses in the form of emphasis (specific region awareness) and perceptual knowledge (feature-matching over task-specific pre-trained networks).

Face-Aware Vector Quantization. While using a scene as an additional form of conditioning provides an implicit prior for human preference, we institute explicit emphasis in the form of additional losses, explicitly targeted at specific image regions.

We employ a feature-matching loss over the activations of a pre-trained face-embedding network, introducing "awareness" of face regions and additional perceptual information, motivating high-quality face reconstruction. Before training the face-aware VQ (denoted as VQ-IMG), faces are located using the semantic segmentation information extracted for VQ-SEG. The face locations are then used during the face-aware VQ training stage, running up to k_f faces per image from the ground-truth and reconstructed images through the face-embedding network. The face loss can then be formulated as following:

$$\mathcal{L}_{\text{Face}} = \sum_k \sum_l \alpha_f^l \| \, \text{FE}^l(\hat{c}_f^k) - \text{FE}^l(c_f^k) \|, \tag{1}$$

where the index l is used to denote the size of the spatial activation at specific layers of the face embedding network FE [6], while the summation runs over the last layers of each block of size 112×112, 56×56, 28×28, 7×7, 1×1 (1×1 being the size of the top most block), $\hat{c}_f^k$ and c_f^k are respectively the reconstructed and ground-truth face crops k out of k_f faces in an image, α_f^l is a per-layer normalizing hyperparameter, and $\mathcal{L}_{\text{Face}}$ is the face loss added to the VQGAN losses defined by [13].

Face Emphasis in the Scene Space. While training the VQ-SEG network, we observe a frequent reduction of the semantic segmentations representing the face parts (such as the eyes, nose, lips, eyebrows) in the reconstructed scene. This effect is not surprising due to the relatively small number of pixels that each face part accounts for in the scene space. A straightforward solution would be to employ a loss more suitable for class imbalance, such as focal loss [35]. However, we do not aspire to increase the importance of classes that are both scarce, and of less importance, such as fruit or a tooth-brush. Instead, we (1) employ a weighted binary cross-entropy face loss over the segmentation face parts classes, emphasizing higher importance for face parts, and (2) include the face parts edges as part of the semantic segmentation edge map mentioned above. The weighted binary cross-entropy loss can then be formulated as following:

$$\mathcal{L}_{\text{WBCE}} = \alpha_{cat}\, \text{BCE}(s, \hat{s}), \tag{2}$$

where s and $\hat{s}$ are the input and reconstructed segmentation maps respectively, α_{cat} is a per-category weight function, BCE is a binary cross-entropy loss, and $\mathcal{L}_{\text{WBCE}}$ is the weighted binary cross-entropy loss added to the conditional VQ-VAE losses defined by [13].

Object-Aware Vector Quantization. We generalized and extend the face-aware VQ method to increase awareness and perceptual knowledge of objects defined as "things" in the panoptic segmentation categories. Rather than a specialized face-embedding network, we employ a pre-trained VGG [52] network trained on ImageNet [32], and introduce a feature-matching loss representing the perceptual differences between the object crops of the reconstructed and ground-truth images. By running the feature-matching over image crops, we are able to increase the output image resolution from 256×256 by simply adding to VQ-IMG an additional down-sample and up-sample layer to the encoder and decoder respectively. Similarly to Eq. 1, the loss can be formulated as:

$$\mathcal{L}_{\text{Obj}} = \sum_k \sum_l \alpha_o^l \| \text{VGG}^l(\hat{c}_o^k) - \text{VGG}^l(c_o^k) \|, \tag{3}$$

where $\hat{c}_o^k$ and c_o^k are the reconstructed and input object crops respectively, VGG^l are the activations of the $l-th$ layer from the pre-trained VGG network, α_o^l is a per-layer normalizing hyperparameter, and $\mathcal{L}_{\text{Obj}}$ is the object-aware loss added to the VQ-IMG losses defined in Eq. 1.

3.3 Scene-Based Transformer

The method relies on an autoregressive transformer with three independent consecutive token spaces: text, scene, and image, as depicted in the appendix. The token sequence is comprised of n_x text tokens encoded by a BPE [50] encoder, followed by n_y scene tokens encoded by VQ-SEG, followed by n_z image tokens encoded and decoded by VQ-IMG.

Prior to training the scene-based transformer, each encoded token sequence corresponding to a [text, scene, image] triplet is extracted using the corresponding encoder, producing a sequence that consists of:

$$\begin{aligned} t_x, t_y, t_z &= \text{BPE}(i_x), \text{VQ-SEG}(i_y), \text{VQ-IMG}(i_z), \\ t &= [t_x, t_y, t_z], \end{aligned}$$

where i_x, i_y, i_z are the input text, scene and image respectively, $i_x \in \mathbb{N}^{d_x}$, d_x is the length of the input text sequence, $i_y \in \mathbb{R}^{h_y \times w_y \times m}$, $i_z \in \mathbb{R}^{h_z \times w_z \times 3}$, h_y, w_y, h_z, w_z are the height and width dimensions of the scene and image inputs respectively, BPE is the Byte Pair Encoding encoder, t_x, t_y, t_z are the text, scene and image input tokens respectively, and t is the complete token sequence.

3.4 Autoregressive Transformer Classifier-Free Guidance

Inspired by the high-fidelity of unconditional image generation models, we employ classifier-free guidance [9,22,44]. Classifier-free guidance is the process of guiding an unconditional sample in the direction of a conditional sample. To support unconditional sampling we fine-tune the transformer while randomly replacing the text prompt with padding tokens with a probability of p_{CF}. During inference, we generate two parallel token streams: a conditional token stream conditioned on text, and an unconditional token stream conditioned on an empty text stream initialized with padding tokens. For transformers, we apply classifier-free guidance on logit scores:

$$\begin{aligned} \text{logits}_{cond} &= T(t_y, t_z | t_x), \\ \text{logits}_{uncond} &= T(t_y, t_z | \emptyset), \\ \text{logits}_{cf} &= \text{logits}_{uncond} + \alpha_c \cdot (\text{logits}_{cond} - \text{logits}_{uncond}), \end{aligned}$$

where $\emptyset$ is the empty text stream, $logits_{cond}$ are logit scores outputted by the conditioned token stream, $logits_{uncond}$ are logit scores outputted by the unconditioned token stream, α_c is the guidance scale, and $logits_{cf}$ is the guided logit scores used to sample the next scene or image token, T is an autoregressive transformer based on the architecture of GPT-3 [4]. Note that since we use an autoregressive transformer, we use $logits_{cf}$ to sample once and feed the same token (image or scene) to the conditional and unconditional stream.

4 Experiments

Our model achieves state-of-the-art results in human-based and numerical metric comparisons. Samples supporting qualitative advantage are provided in Fig. 2. Additionally, we demonstrate new creative capabilities possible with this method's new form of controllability. Finally, to better assess the effect of each contribution, an ablation study is provided.

Experiments were performed with a 4 billion parameter transformer, generating a sequence of 256 text tokens, 256 scene tokens, and 1024 image tokens, that are then decoded into an image with a resolution of 256×256 or 512×512 pixels (depending on the model of choice).

Datasets. The scene-based transformer is trained on a union of CC12m [7], CC [51], and subsets of YFCC100m [55] and Redcaps [10], amounting to 35m text-image pairs. MS-COCO [36] is used as well unless otherwise specified. VQ-SEG and VQ-IMG are trained on CC12m, CC, and MS-COCO.

Metrics. The goal of text-to-image generation is to generate high-quality and text-aligned images from a human perspective. Different metrics have been suggested to mimic the human perspective, where some are considered more reliable than others. We consider human evaluation the highest authority when evaluating image quality and text-alignment, and rely on FID [19] to increase evaluation confidence and handle cases where human evaluation is not applicable. We do not use IS [49] as it has been noted to be insufficient for model evaluation [2].

4.1 Comparison with Previous Work

The task of text-to-image generation does not contain absolute ground-truths, as a specific text description could apply to multiple images and vice versa. This constrains evaluation metrics to evaluate distributions of images, rather than specific images, thus we employ FID [19] as our secondary metric.

Baselines. We compare our results with several state-of-the-art methods using the FID metric and human evaluators (AMT) when possible. **DALL-E** [45] provides strong zero-shot capabilities, similarly employing an autoregressive transformer with VQ-VAE tokenization. We train a re-implementation of DALL-E with 4B parameters to enable human evaluation and fairly compare both methods employing an identical VQ method (VQGAN). **GLIDE** [40] demonstrates vastly improved results over DALL-E, adopting a diffusion-based [53] approach with classifier-free guidance [22]. We additionally provide an FID comparison with **CogView** [12], **LAFITE** [69], **XMC-GAN** [66], **DM-GAN(+CL)** [64], **DF-GAN** [54], **DM-GAN** [71], **DF-GAN** [54] and, **AttnGAN** [63].

Human Evaluation Results. Human evaluation with previous methods is provided in Table 1. In each instance, human evaluators are required to choose between two images generated by the two models being compared. The two models are compared in three aspects: (i) image quality, (ii) photorealism (which image appears more real), and (iii) text alignment (which image best matches the text). Each question is surveyed using 500 image pairs, where 5 different evaluators answer each question, amounting to 2500 instances per question for a given comparison. We compare our 256×256 model with our re-implementation of DALL-E [45] and CogView's [12] 256×256 model. CogView's 512×512 model is compared with our corresponding model. Results are presented as a percentage of majority votes in favor of our method when comparing between a certain model and ours. Compared with the three methods, ours achieves significantly higher favorability in all aspects.

FID Comparison. FID is calculated over a subset of $30k$ images generated from the MS-COCO validation set text prompts with no re-ranking, and provided in Table 1. The evaluated models are divided into two groups: trained with and without (denoted as filtered) the MS-COCO training set. In both scenarios our model achieves the lowest FID. In addition, we provide a loose practical lower-bound (denoted as ground-truth), calculated between the training and validation subsets of MS-COCO. As FID results are approaching small numbers, it is interesting to get an idea of a possible practical lower-bound.

Generating Out of Distribution. Methods that generate from text inputs only are more confined to generate within the training distribution, as demonstrated by [40]. Unusual objects and scenarios can be challenging to generate, as certain objects are strongly correlated with specific structures, such as cats with four legs or cars with round wheels. The same is true for scenarios, "a mouse hunting a lion" is most likely not a scenario easily found within the dataset. Using scenes in the form of simple sketches as inputs, we are able to attend to these uncommon objects and scenarios, as demonstrated in Fig. 3, despite the fact that some objects do not exist as categories in our scene (such as the mouse and lion). We solve the category gap by using categories that may be close in certain aspects (elephant instead of mouse, cat instead of lion). In practice, for non-existent categories several categories could be used instead.

4.2 Scene Controllability

Samples are provided in Fig. 1, 3, 4, 5 and in the supplementary with both our 256×256 and 512×512 models. In addition to generating high fidelity images from text only, we demonstrate the applicability of scene-wise image control and maintaining consistency between generations.

Scene Editing and Anchoring. Rather than editing certain regions of images as demonstrated by [45], we introduce new capabilities of generating images from existing or edited scenes. In Fig. 4, two scenarios are considered. In both scenarios the semantic segmentation is extracted from an input image, and used

Table 1. Comparison with previous work (FID and human preference). FID is calculated over a subset of $30k$ images generated from the MS-COCO validation set text prompts. When possible, we include models trained with and without (filtered) the MS-COCO training set. In both scenarios our model achieves state of the art results, correlating with visual samples and human evaluation. We add a loose practical lower-bound (denoted as ground-truth), calculated between the training and validation subsets of MS-COCO. Human evaluation is shown as a percentage of majority votes in favor of our method when comparing between a certain model and ours.

Model	FID↓	FID↓ (filt.)	Image quality	Photo-realism	Text alignment
AttnGAN [63]	35.49	-	-	-	-
DM-GAN [71]	32.64	-	-	-	-
DF-GAN [54]	21.42	-	-	-	-
DM-GAN+CL [64]	20.79	-	-	-	-
XMC-GAN [66]	9.33	-	-	-	-
DALL-E [45]	-	34.60	81.8%	81.0%	65.9%
CogView$_{256}$ [12]	-	32.20	92.2%	94.2%	92.2%
CogView$_{512}$ [12]	-	36.53	91.1%	88.2%	87.8%
LAFITE [69]	8.12	26.94	-	-	-
GLIDE [40]	-	12.24	-	-	-
Ours$_{256}$	**7.55**	**11.84**			
Ground-truth	2.47	-	-	-	-

to re-generate an image conditioned on the input text. In the top row, the scene is edited, replacing the 'sky' and 'tree' categories with 'sea', and the 'grass' category with 'sand', resulting in a generated image adhering to the new scene. A simple sketch of a giant dog is added to the scene in the bottom row, resulting in a generated image corresponding to the new scene without any change in text.

Figure 5 demonstrates the ability to generate new interpretations of existing images and scenes. After extracting the semantic segmentation from a given image, we re-generate the image conditioned on the input scene and edited text.

Storytelling Through Controllability. To demonstrate the applicability of harnessing scene control for story illustrations, we wrote a children story, and illustrated it using our method. The main advantages of using simple sketches as additional inputs in this case, are (i) that authors can translate their ideas into paintings or realistic images, while being less susceptible to the "randomness" of text-to-image generation, and (ii) improved consistency between generation. A short video of the story and process can be found in this video.

4.3 Ablation Study

An ablation study of human preference and FID is provided in Table 2 to assess the effectiveness of our different contributions. Settings in both studies are similar to the comparison made with previous work (Sect. 4.1). Each row corresponds to a model trained with the additional element, compared with the model without that specific addition for human preference. We note that while the lowest FID is attained by the 256 × 256 model, human preference favors the 512 × 512 model with object-aware training, particularly in quality. Furthermore, we re-examine the FID of the best model, where the scene is given as an additional input, to gain a better notion of the gap from the lower-bound (Table 1).

Table 2. Ablation study (FID and human preference). FID is calculated over a subset of 30*k* images generated from the MS-COCO validation set text prompts. Human evaluation is shown as a percentage of majority votes in favor of the added element compared to the previous model. Classifier-free guidance is denoted as CF.

Model	FID↓	Image quality	Photo-realism	Text alignment
Base	18.01	-	-	-
+Scene tokens	19.16	57.3%	65.3%	58.3%
+Face-aware	14.45	63.6%	59.8%	57.4%
+CF	**7.55**	76.8%	66.8%	66.8%
+Obj-aware$_{512}$	8.70	62.0%	53.5%	52.2%
+CF with scene input	4.69	-	-	-

5 Conclusion

The text-to-image domain has witnessed a plethora of novel methods aimed at improving the general quality and adherence to text of generated images. While some methods propose image editing techniques, progress is not often directed towards enabling new forms of human creativity and experiences. We attempt to progress text-to-image generation towards a more interactive experience, where people can perceive more control over the generated outputs, thus enable real-world applications such as storytelling. In addition to improving the general image quality, we focus on improving key image aspects we deem significant in human perception, such as faces and salient objects, resulting in higher favorability of our method in human evaluations and objective metrics.

References

1. Advice, S.G.S.: Syracuse post standard, March 28, 18 (1911)
2. Barratt, S., Sharma, R.: A note on the inception score. arXiv preprint arXiv:1801.01973 (2018)

3. Brock, A., Donahue, J., Simonyan, K.: Large scale GAN training for high fidelity natural image synthesis. arXiv preprint arXiv:1809.11096 (2018)
4. Brown, T., et al.: Language models are few-shot learners. In: Advances in Neural Information Processing Systems, vol. 33, pp. 1877–1901 (2020)
5. Bulat, A., Tzimiropoulos, G.: How far are we from solving the 2D & 3D face alignment problem? (and a dataset of 230,000 3D facial landmarks). In: International Conference on Computer Vision (2017)
6. Cao, Q., Shen, L., Xie, W., Parkhi, O.M., Zisserman, A.: VGGFace2: a dataset for recognising faces across pose and age. arXiv preprint arXiv:1710.08092 (2017)
7. Changpinyo, S., Sharma, P., Ding, N., Soricut, R.: Conceptual 12M: pushing web-scale image-text pre-training to recognize long-tail visual concepts. In: Proceedings of the IEEE/CVF Conference on Computer Vision and Pattern Recognition, pp. 3558–3568 (2021)
8. Chen, M., et al.: Generative pretraining from pixels. In: III, H.D., Singh, A. (eds.) Proceedings of the 37th International Conference on Machine Learning. Proceedings of Machine Learning Research, vol. 119, pp. 1691–1703. PMLR (2020). https://proceedings.mlr.press/v119/chen20s.html
9. Crowson, K.: Classifier Free Guidance for Auto-regressive Transformers (2021). https://twitter.com/RiversHaveWings/status/1478093658716966912
10. Desai, K., Kaul, G., Aysola, Z., Johnson, J.: RedCaps: web-curated image-text data created by the people, for the people. arXiv preprint arXiv:2111.11431 (2021)
11. Dhariwal, P., Nichol, A.: Diffusion models beat GANs on image synthesis. In: Advances in Neural Information Processing Systems, vol. 34 (2021)
12. Ding, M., et al.: CogView: mastering text-to-image generation via transformers. In: Advances in Neural Information Processing Systems, vol. 34 (2021)
13. Esser, P., Rombach, R., Ommer, B.: Taming transformers for high-resolution image synthesis. In: Proceedings of the IEEE/CVF Conference on Computer Vision and Pattern Recognition, pp. 12873–12883 (2021)
14. Fourcade, M.M.: L'Arche de Noé: réseau Alliance, 1940–1945. Plon (1968)
15. Gafni, O., Ashual, O., Wolf, L.: Single-shot freestyle dance reenactment. In: Proceedings of the IEEE/CVF Conference on Computer Vision and Pattern Recognition, pp. 882–891 (2021)
16. Gafni, O., Wolf, L., Taigman, Y.: Live face de-identification in video. In: Proceedings of the IEEE/CVF International Conference on Computer Vision, pp. 9378–9387 (2019)
17. Goodfellow, I., et al.: Generative adversarial nets. In: Advances in Neural Information Processing Systems, vol. 27 (2014)
18. Gregor, K., Danihelka, I., Graves, A., Rezende, D., Wierstra, D.: DRAW: a recurrent neural network for image generation. In: International Conference on Machine Learning, pp. 1462–1471. PMLR (2015)
19. Heusel, M., Ramsauer, H., Unterthiner, T., Nessler, B., Hochreiter, S.: GANs trained by a two time-scale update rule converge to a local Nash equilibrium. In: Advances in Neural Information Processing Systems, vol. 30 (2017)
20. Ho, J., Jain, A., Abbeel, P.: Denoising diffusion probabilistic models. In: Advances in Neural Information Processing Systems, vol. 33, pp. 6840–6851 (2020)
21. Ho, J., Saharia, C., Chan, W., Fleet, D.J., Norouzi, M., Salimans, T.: Cascaded diffusion models for high fidelity image generation. J. Mach. Learn. Res. **23**(47), 1–33 (2022)
22. Ho, J., Salimans, T.: Classifier-free diffusion guidance. In: NeurIPS 2021 Workshop on Deep Generative Models and Downstream Applications (2021)

23. Hong, S., Yang, D., Choi, J., Lee, H.: Inferring semantic layout for hierarchical text-to-image synthesis. In: Proceedings of the IEEE Conference on Computer Vision and Pattern Recognition, pp. 7986–7994 (2018)
24. Isola, P., Zhu, J.Y., Zhou, T., Efros, A.A.: Image-to-image translation with conditional adversarial networks. In: Proceedings of the IEEE Conference on Computer Vision and Pattern Recognition, pp. 1125–1134 (2017)
25. Ivan, T.: Fathers and Sons. Pandora's Box (2017)
26. Jang, E., Gu, S., Poole, B.: Categorical reparameterization with Gumbel-Softmax. arXiv preprint arXiv:1611.01144 (2016)
27. Janson, H.W., Janson, A.F., Marmor, M.: History of Art. Thames and Hudson London (1991)
28. Judd, T., Durand, F., Torralba, A.: A benchmark of computational models of saliency to predict human fixations (2012)
29. Karras, T., et al.: Alias-free generative adversarial networks. In: Advances in Neural Information Processing Systems, vol. 34 (2021)
30. Karras, T., Laine, S., Aittala, M., Hellsten, J., Lehtinen, J., Aila, T.: Analyzing and improving the image quality of StyleGAN. In: Proceedings of the IEEE/CVF Conference on Computer Vision and Pattern Recognition, pp. 8110–8119 (2020)
31. Kingma, D.P., Welling, M.: Auto-encoding variational bayes. arXiv preprint arXiv:1312.6114 (2013)
32. Krizhevsky, A., Sutskever, I., Hinton, G.E.: ImageNet classification with deep convolutional neural networks. In: Advances in Neural Information Processing Systems, vol. 25 (2012)
33. Li, B., Qi, X., Lukasiewicz, T., Torr, P.: Controllable text-to-image generation. In: Advances in Neural Information Processing Systems, vol. 32 (2019)
34. Li, P., Xu, Y., Wei, Y., Yang, Y.: Self-correction for human parsing. IEEE Trans. Pattern Anal. Mach. Intell. (2020). https://doi.org/10.1109/TPAMI.2020.3048039
35. Lin, T.Y., Goyal, P., Girshick, R., He, K., Dollár, P.: Focal loss for dense object detection. In: Proceedings of the IEEE International Conference on Computer Vision, pp. 2980–2988 (2017)
36. Lin, T.-Y., et al.: Microsoft COCO: common objects in context. In: Fleet, D., Pajdla, T., Schiele, B., Tuytelaars, T. (eds.) ECCV 2014. LNCS, vol. 8693, pp. 740–755. Springer, Cham (2014). https://doi.org/10.1007/978-3-319-10602-1_48
37. Liu, M.Y., Breuel, T., Kautz, J.: Unsupervised image-to-image translation networks. In: Advances in Neural Information Processing Systems, vol. 30 (2017)
38. Maddison, C.J., Mnih, A., Teh, Y.W.: The concrete distribution: a continuous relaxation of discrete random variables. arXiv preprint arXiv:1611.00712 (2016)
39. Mansimov, E., Parisotto, E., Ba, J.L., Salakhutdinov, R.: Generating images from captions with attention. arXiv preprint arXiv:1511.02793 (2015)
40. Nichol, A., et al.: Glide: towards photorealistic image generation and editing with text-guided diffusion models. arXiv preprint arXiv:2112.10741 (2021)
41. Van den Oord, A., Kalchbrenner, N., Espeholt, L., Vinyals, O., Graves, A., et al.: Conditional image generation with PixelCNN decoders. In: Advances in Neural Information Processing Systems, vol. 29 (2016)
42. Park, T., Liu, M.Y., Wang, T.C., Zhu, J.Y.: Semantic image synthesis with spatially-adaptive normalization. In: Proceedings of the IEEE/CVF Conference on Computer Vision and Pattern Recognition, pp. 2337–2346 (2019)
43. Parmar, N., et al.: Image transformer. In: International Conference on Machine Learning, pp. 4055–4064. PMLR (2018)

44. Radford, A., et al.: Learning transferable visual models from natural language supervision. In: International Conference on Machine Learning, pp. 8748–8763. PMLR (2021)
45. Ramesh, A., e al.: Zero-shot text-to-image generation. In: International Conference on Machine Learning, pp. 8821–8831. PMLR (2021)
46. Ramesh, A., et al.: Zero-shot text-to-image generation (ICML spotlight) (2021). https://icml.cc/virtual/2021/spotlight/9430
47. Razavi, A., Van den Oord, A., Vinyals, O.: Generating diverse high-fidelity images with VQ-VAE-2. In: Advances in Neural Information Processing Systems, vol. 32 (2019)
48. Reed, S., Akata, Z., Yan, X., Logeswaran, L., Schiele, B., Lee, H.: Generative adversarial text to image synthesis. In: International Conference on Machine Learning, pp. 1060–1069. PMLR (2016)
49. Salimans, T., Goodfellow, I., Zaremba, W., Cheung, V., Radford, A., Chen, X.: Improved techniques for training GANs. In: Advances in Neural Information Processing Systems, vol. 29 (2016)
50. Sennrich, R., Haddow, B., Birch, A.: Neural machine translation of rare words with subword units. arXiv preprint arXiv:1508.07909 (2015)
51. Sharma, P., Ding, N., Goodman, S., Soricut, R.: Conceptual captions: a cleaned, hypernymed, image alt-text dataset for automatic image captioning. In: Proceedings of the 56th Annual Meeting of the Association for Computational Linguistics (Volume 1: Long Papers), pp. 2556–2565 (2018)
52. Simonyan, K., Zisserman, A.: Very deep convolutional networks for large-scale image recognition. arXiv preprint arXiv:1409.1556 (2014)
53. Sohl-Dickstein, J., Weiss, E., Maheswaranathan, N., Ganguli, S.: Deep unsupervised learning using nonequilibrium thermodynamics. In: International Conference on Machine Learning, pp. 2256–2265. PMLR (2015)
54. Tao, M., et al.: DF-GAN: deep fusion generative adversarial networks for text-to-image synthesis. arXiv preprint arXiv:2008.05865 (2020)
55. Thomee, B., et al.: YFCC100M: the new data in multimedia research. Commun. ACM **59**(2), 64–73 (2016)
56. Tseng, H.Y., Jiang, L., Liu, C., Yang, M.H., Yang, W.: Regularizing generative adversarial networks under limited data. In: Proceedings of the IEEE/CVF Conference on Computer Vision and Pattern Recognition, pp. 7921–7931 (2021)
57. Vahdat, A., Kautz, J.: NVAE: a deep hierarchical variational autoencoder. In: Advances in Neural Information Processing Systems, vol. 33, pp. 19667–19679 (2020)
58. Van Den Oord, A., Vinyals, O., et al.: Neural discrete representation learning. In: Advances in Neural Information Processing Systems, vol. 30 (2017)
59. Vaswani, A., et al.: Attention is all you need. In: Advances in Neural Information Processing Systems, vol. 30 (2017)
60. Wang, T.C., et al.: Video-to-video synthesis. arXiv preprint arXiv:1808.06601 (2018)
61. Wang, T.C., Liu, M.Y., Zhu, J.Y., Tao, A., Kautz, J., Catanzaro, B.: High-resolution image synthesis and semantic manipulation with conditional GANs. In: Proceedings of the IEEE Conference on Computer Vision and Pattern Recognition, pp. 8798–8807 (2018)
62. Wu, Y., Kirillov, A., Massa, F., Lo, W.Y., Girshick, R.: Detectron2 (2019). https://github.com/facebookresearch/detectron2

63. Xu, T., et al.: AttnGAN: fine-grained text to image generation with attentional generative adversarial networks. In: Proceedings of the IEEE Conference on Computer Vision and Pattern Recognition, pp. 1316–1324 (2018)
64. Ye, H., Yang, X., Takac, M., Sunderraman, R., Ji, S.: Improving text-to-image synthesis using contrastive learning. arXiv preprint arXiv:2107.02423 (2021)
65. Yun, K., Peng, Y., Samaras, D., Zelinsky, G.J., Berg, T.L.: Studying relationships between human gaze, description, and computer vision. In: Proceedings of the IEEE Conference on Computer Vision and Pattern Recognition, pp. 739–746 (2013)
66. Zhang, H., Koh, J.Y., Baldridge, J., Lee, H., Yang, Y.: Cross-modal contrastive learning for text-to-image generation. In: Proceedings of the IEEE/CVF Conference on Computer Vision and Pattern Recognition, pp. 833–842 (2021)
67. Zhang, R., Isola, P., Efros, A.A., Shechtman, E., Wang, O.: The unreasonable effectiveness of deep features as a perceptual metric. In: Proceedings of the IEEE Conference on Computer Vision and Pattern Recognition, pp. 586–595 (2018)
68. Zhang, Z., et al.: M6-UFC: unifying multi-modal controls for conditional image synthesis. arXiv preprint arXiv:2105.14211 (2021)
69. Zhou, Y., et al.: LAFITE: towards language-free training for text-to-image generation. arXiv preprint arXiv:2111.13792 (2021)
70. Zhu, J.Y., Park, T., Isola, P., Efros, A.A.: Unpaired image-to-image translation using cycle-consistent adversarial networks. In: Proceedings of the IEEE International Conference on Computer Vision, pp. 2223–2232 (2017)
71. Zhu, M., Pan, P., Chen, W., Yang, Y.: DM-GAN: dynamic memory generative adversarial networks for text-to-image synthesis. In: Proceedings of the IEEE/CVF Conference on Computer Vision and Pattern Recognition, pp. 5802–5810 (2019)

3D-FM GAN: Towards 3D-Controllable Face Manipulation

Yuchen Liu[1(✉)], Zhixin Shu[2], Yijun Li[2], Zhe Lin[2], Richard Zhang[2], and S.Y. Kung[1]

[1] Princeton University, Princeton, USA
{yl16,kung}@princeton.edu
[2] Adobe Research, San Jose, USA
{zshu,yijli,zlin,rizhang}@adobe.com

Abstract. 3D-controllable portrait synthesis has significantly advanced, thanks to breakthroughs in generative adversarial networks (GANs). However, it is still challenging to manipulate existing face images with precise 3D control. While concatenating GAN inversion and a 3D-aware, noise-to-image GAN is a straight-forward solution, it is inefficient and may lead to noticeable drop in editing quality. To fill this gap, we propose 3D-FM GAN, a novel conditional GAN framework designed specifically for **3D**-controllable **F**ace **M**anipulation, and does not require any tuning after the end-to-end learning phase. By carefully encoding both the input face image and a physically-based rendering of 3D edits into a StyleGAN's latent spaces, our image generator provides high-quality, identity-preserved, 3D-controllable face manipulation. To effectively learn such novel framework, we develop two essential training strategies and a novel multiplicative co-modulation architecture that improves significantly upon naive schemes. With extensive evaluations, we show that our method outperforms the prior arts on various tasks, with better editability, stronger identity preservation, and higher photo-realism. In addition, we demonstrate a better generalizability of our design on large pose editing and out-of-domain images.

1 Introduction

Face manipulation with precise control has long attracted attention from computer vision and computer graphics community for its application in face recognition, photo editing, visual effects, and AR/VR applications, etc. In the past, researchers developed 3D morphable face models (3DMMs) [4,26,29], which provide an explainable and disentangled parameter space to control face attributes of identity, pose, expression, and illumination. However, it is still challenging to render photo-realistic face manipulations with 3DMMs.

Y. Liu—Work done during an internship at Adobe Research.

Supplementary Information The online version contains supplementary material available at https://doi.org/10.1007/978-3-031-19784-0_7.

S. Avidan et al. (Eds.): ECCV 2022, LNCS 13675, pp. 107–125, 2022.
https://doi.org/10.1007/978-3-031-19784-0_7

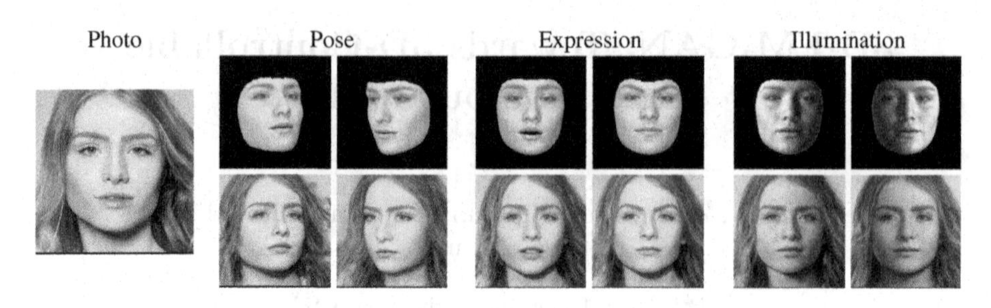

Fig. 1. With explicit 3D controls of pose, expression, and illumination, presented as identity-preserved rendered faces (top row), 3D-FM GAN provides controllable and disentangled face manipulations on real world images (bottom row) with strong identity preservation and high photo-realism.

In recent years, generative adversarial networks (GANs) [15] have demonstrated promising results in photo-realistic face synthesis [23,24] by mapping random noise to image domain. While latent space exploration has been attempted [1,17,39], it requires a lot of human labor to discover meaningful directions, and the editings could still be entangled. As such, the variants of conditional GANs are widely studied for identity-preserved face manipulations [3,8,47,56]. Nonetheless, they either only allow control on a single facial attribute or require reference images/human annotations for face editing.

More recently, several works introduced 3D priors into GANs [10,14,40,44] for controllable synthesis. However, most of them are learned for noise-to-image random face generation, which does not naturally fit with the image manipulation task. Hence, they require time-consuming optimization in the test time, and the inverted latent codes may not lie in the manifold for high-quality editing [45,50]. Moreover, photo-realistic synthesis remains challenging [25,40].

To this end, we propose **3D-FM GAN**, a novel framework particularly designed for high-quality **3D**-controllable **F**ace **M**anipulation. Specifically, we perform a learning process to solve the image-to-image translation/editing problem with a conditional StyleGAN [24]. Different from prior 3D GANs trained for random sampling, we train our model exactly for existing face manipulation and do not require optimization/manual tuning after the learning phase. As shown in Fig. 1, with a single input face image, our framework manages to produce photo-realistic disentangled editing on attributes of head pose, facial expression, and scene illumination, while faithfully preserving the face identity.

Our framework leverages face reconstruction networks and a physically-based renderer, where the former estimate the input 3D coefficients and the latter embeds the desired manipulations, e.g., pose rotation, into an identity-preserved rendered face. A StyleGAN [24] conditional generator then takes in both the original image and the manipulated face rendering to synthesize the edited face. The consistent identity information provided by the input and the rendered edit signals spontaneously creates a strong synergy for identity preservation in manipulation. Moreover, we develop two essential training strategies, reconstruction and disentangled training, to help our model gain abilities of identity preservation and 3D editability. As we find an interesting trade-off between identity and editability in the network structure and the simple encoding strategy is sub-optimal, we propose a novel multiplicative co-modulation architecture for our framework. This structure stems from a comprehensive study to understand how to encode different information in the generator's latent spaces, where it achieves the best performance. We conduct extensive qualitative and quantitative evaluations on our model and demonstrate good disentangled editing ability, strong identity preservation, and high photo-realism, outperforming the prior arts in various tasks. More interestingly, our model can manipulate artistic faces which are out of our training domain, indicating its strong generalization ability.

Our contributions can be summarized as follows.

(1) We propose 3D-FM GAN, a novel conditional GAN framework that is specifically designed for precise, explicit, high-quality, 3D-controllable face manipulation. Unlike prior works, our training objective is strongly consistent with the task of existing face editing, and our model does not require any optimization/manual tuning after the end-to-end learning process.
(2) We develop two essential training strategies, reconstruction and disentangled training to effectively learn our model. We also conduct a comprehensive study of StyleGAN's latent spaces for structural design, leading to a novel multiplicative co-modulation architecture with strong identity-editability trade-off.
(3) Extensive quantitative and qualitative evaluations demonstrate the advantage of our method over prior arts. Moreover, our model also shows a strong generalizability to edit artistic faces, which are out of the training domain.

2 Related Works

3D Face Modelling. 3D morphable models (3DMMs) [4,5,26,28,29,46] have long been used for face modelling. In 3DMMs, human faces are normally parametrized by texture, shape, expression, skin reflectance and scene illumination in a disentangled manner to enable 3D-controllable face synthesis. However, 3DMMs require expensive data of 3D [29] or even 4D [26] scans of human heads to build, and the rendered images often lack photo-realism due to the low-dimensional linear representation as well as the absence of modelling in hair, mouth cavity, and fine details like wrinkles. With 3DMMs, many methods attempt to estimate the 3D parameters of 2D images [11–13,33–37,42,49]. This is normally achieved by optimization or a neural networks to extract 3D parameters from a face image and/or landmarks [11,13,37]. In our work, we use 3DMMs for 3D face representation and adopt face reconstruction networks to provide the basis of 3D editing signals from 2D images. As such, our model can be trained solely with 2D images to gain 3D controllability.

GAN. Recently, unconditional GANs show promising results in synthesizing photo-realistic faces [22–24]. While latent space exploration [17,39,41] has proved to be effective, it requires extensive human labors to obtain meaningful control for generation. As such, a rich set of literatures propose to use conditional GANs [3, 8,21,30,31,38,47,48,51,56] for controllable identity-preserved image synthesis by disentangling identity and non-identity factors.

Noticeably, DR-GAN [47] and TP-GAN [21] disentangles identity and pose to allow frontal view synthesis, while Zhou et al. [56] extracts spherical harmonic lighting from source image for portrait relighting. However, these works can only manipulate one attribute of the faces, whilst we are able to conduct disentangled editing on pose, lighting, and expression under a unified framework. We even outperform some of them on the tasks that they are solely trained on.

To transfer multiple attributes, Bao et al. [3] and Xiao et al. [51] extract identity from one image and facial attributes from another one for reference-based generation. StarGAN [8] leverages multiple labeled datasets to learn attributes translation. In contrast, our model provide manipulations with just a single input, and it does not require any labeled information/datasets for training.

3D Controllable GAN. In line with our work, several prior methods [6,10, 14,25,27,40,43,44] introduce 3D priors into GANs to achieve 3D controllability over face attributes of expression, pose, and illumination.

Deng et al. [10] enforce the input space of its GAN to bear the same disentanglement as the parameter space of a 3DMM to achieve controllable face generation. GIF [14] conditions the space of StyleGAN's layer noise on render images from FLAME [26] to control pose, expression, and lighting. Tewari et al. [44] leverage a pretrained StyleGAN and learn a RigNet to manipulate latent vectors with respect to the target editing semantics. However, these approaches are all trained for random face generation, not for existing face manipulation. Although GAN inversion can well project existing images in their latent spaces for good reconstruction, these latent codes may not fall on the manifold with good editability, leading to noticeable quality drop after manipulation. On the contrary, our model is trained exactly for the task of real face editing, which demonstrates a clear improvement in manipulation quality upon these works.

While CONFIG [25] does not need GAN inversion for real image manipulation, its parametric editing space doesn't inherit identity information from the input images, resulting in a clear identity loss. Moreover, our novel generator architecture also provides us with a larger range of editability and higher photorealism upon them. While PIE [43] proposes a specialized GAN inversion process to be later combined with StyleRig for real image manipulation, we again find our approach provides better quality and higher efficiency. Compared to a more recent VariTex [6] approach which can not synthesize background and rigid body like glasses, our method produces much more realistic outputs.

3 Methodology

3.1 Overview and Notations

In Fig. 2, we show the workflow of 3D-FM GAN which consists of: the generator **G**, the face reconstruction network **FR**, and the renderer **Rd**. Given an input face image $P \in \mathbb{R}^{H \times W \times 3}$, it first estimates the lighting and 3DMM parameters

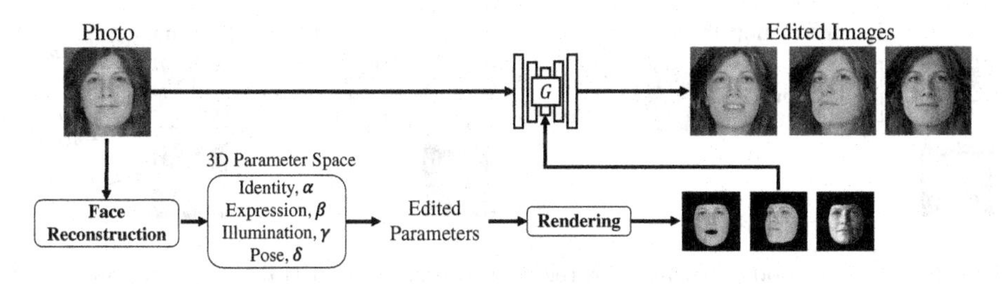

Fig. 2. Workflow of 3D-FM GAN. Given a photo input, our framework first extracts its 3D parameter by face reconstruction and then renders identity-preserved 3D-manipulated edit faces. The input photo and the edit signals are later jointly encoded into a generator to synthesize various photo editings.

of the face $p = \mathbf{FR}(P)$, $p = (\alpha, \beta, \gamma, \delta) \in \mathbb{R}^{254}$. Naturally, p has disentangled controllable components for identity $\alpha \in \mathbb{R}^{160}$, expression $\beta \in \mathbb{R}^{64}$, lighting $\gamma \in \mathbb{R}^{27}$, and pose $\delta \in \mathbb{R}^{3}$ [29]. The disentangled editing is then achieved in this parameter space, where we keep the identity factor α unchanged but adjust β, γ, δ to the desired semantics for the expression, lighting, and pose, which returns a manipulated parameter $\hat{p}$ to render an image $\hat{R} = \mathbf{Rd}(\hat{p})$ [4]. Finally, the manipulated photo is generated by feeding P and $\hat{R}$ through the generator as $\hat{P} = \mathbf{G}(P, \hat{R})$. In this way, the synthesized output $\hat{P}$ will preserve the identity from P, while its expression, illumination, and pose, follow the control from $\hat{p}$.

3.2 Dataset

The training data are in the form of photo and render image pairs $(P,\ R)$, where P and R share the same attributes of identity, expression, illumination, and pose. We construct our dataset with both the FFHQ data and synthetic data. We show examples of the data pairs in Fig. 3.

Fig. 3. Examples of photo and render image pairs $(P,\ R)$. **Left:** FFHQ data. Each identity just has one corresponding image. **Right:** Synthetic data from [10]. We generate multiple images for an identity with varied expression, pose, and illumination.

FFHQ. FFHQ [23] is a human face photo dataset, where most identities only have one corresponding image. For each of the training image P, we extract its render counterpart by $R = \mathbf{Rd}(\mathbf{FR}(P))$ to form the $(P,\ R)$ pair.

Synthetic Dataset. We also require a dataset where each identity has multiple images with various attributes of expression, pose, and illumination. Such a dataset is crucial for model to perform learning for editing. While this kind of high-quality dataset is not publicly available, we leverage DiscoFaceGAN [10], $\mathbf{G_d}$, to synthesize one as follows.

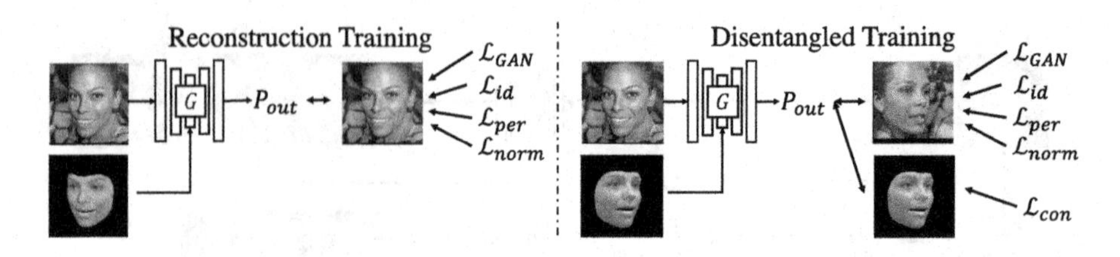

Fig. 4. Proposed model learning strategies of reconstruction training (**Left**) and disentangled training (**Right**).

Given a parameter p of our 3D parameter space and a noise vector $n \in \mathbb{R}^{32}$, $\mathbf{G_d}$ synthesizes a photo image $P = \mathbf{G_d}(p, n)$ that resembles the identity, expression, illumination, and pose of its render counterpart $R = \mathbf{Rd}(p)$. We can thus generate multiple images of the same identity with other attributes varied following the steps below: (1) randomly sample a 3D parameter $p_1 = (\alpha_1, \beta_1, \gamma_1, \delta_1)$ and a noise n; (2) keep α_1 unchanged and re-sample M - 1 tuples of (β, γ, δ) such that we have $p_2 = (\alpha_1, \beta_2, \gamma_2, \delta_2)$, ..., $p_M = (\alpha_1, \beta_M, \gamma_M, \delta_M)$; (3) Use $\mathbf{G_d}$ and $\mathbf{Rd}$ to generate photo-render pairs of (P_1, R_1), ... (P_M, R_M), where $P_i = \mathbf{G_d}(p_i, n)$ and $R_i = \mathbf{Rd}(p_i)$. Such process is iterated for N identities to form a dataset of $N \times M$ pairs. Examples of such image pairs are in Fig. 3 (**Right**).

3.3 Training Strategy

While **FR** and **Rd** do not require further tuning, we design two strategies, reconstruction and disentangled training (Fig. 4) to train **G**. We find the former helps for identity preservation while the latter ensures editability (Fig. 17). Formally, we denote the input pair (P_{in}, R_{in}) and its output $P_{out} = \mathbf{G}(P_{in}, R_{in})$.

Reconstruction Training. We first equip **G** with the ability to reconstruct P_{in} from (P_{in}, R_{in}). In this case, we want P_{out} to be as similar as P_{in}, and set the target output $P_{tg} = P_{in}$. We first define a face identity loss with a face recognition network $\mathbf{N_f}$ [9] :

$$\mathcal{L}_{id} = ||\mathbf{N_f}(P_{out}) - \mathbf{N_f}(P_{tg})||_2^2 \tag{1}$$

We also enforce P_{out} and P_{tg} to have similar low-level features and high-level perception by imposing an $\ell 1$ loss and a perceptual loss based on LPIPS [53]:

$$\mathcal{L}_{norm} = ||P_{out} - P_{tg}||_1 \tag{2}$$

$$\mathcal{L}_{per} = LPIPS(P_{out}, P_{tg}) \tag{3}$$

Finally, we adopt the GAN loss, $\mathcal{L}_{GAN}$ such that the generated images P_{out} shall match the distribution of P_{tg}. In this way, our loss is constructed as:

$$\mathcal{L}_{rec} = \mathcal{L}_{GAN} + \lambda_1 \mathcal{L}_{id} + \lambda_2 \mathcal{L}_{norm} + \lambda_3 \mathcal{L}_{per} \tag{4}$$

where $\lambda_1, \lambda_2, \lambda_3$ are the weights for different losses. We use both synthetic and FFHQ datasets for this procedure.

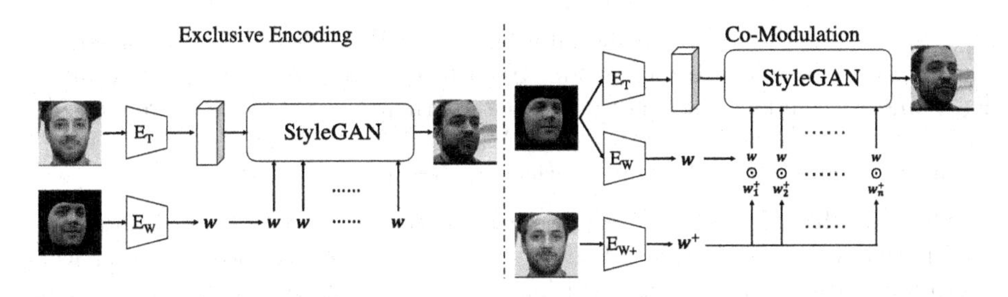

Fig. 5. Our generator **G** can take forms of both exclusive modulation (**Left**) and co-modulation (**Right**) architectures.

Disentangled Training. To achieve our goal, only teaching the model how to "reconstruct" is not sufficient. Thus, we propose a disentangled training strategy to enable editing, which can only be achieved by the synthetic dataset as it has multiple images of the same identity with varying attributes.

Specifically, we first sample two pairs, (P_{in}^1, R_{in}^1) and (P_{in}^2, R_{in}^2), from the same identity. Then, given P_{in}^1 and R_{in}^2, we want our model to produce P_{in}^2, which shares the same edit signal in R_{in}^2 while the same identity of P_{in}^1. In this case, we set $P_{out}^1 = \mathbf{G}(P_{in}^1, R_{in}^2)$ and $P_{tg}^1 = P_{in}^2$, and impose the prior defined loss of $\mathcal{L}_{GAN}$, $\mathcal{L}_{id}$, $\mathcal{L}_{norm}$, and $\mathcal{L}_{per}$ between P_{out}^1 and P_{tg}^1. Different from reconstruction, we also inject a content loss to better capture the target editing signals, where we set $R_{tg}^1 = R_{in}^2$ and define the loss as:

$$\mathcal{L}_{con} = ||M \odot (P_{out}^1 - R_{tg}^1)||_2^2 \tag{5}$$

M is the face region that R_{tg}^1 has non-zero pixels and $\odot$ is the element-wise multiplication. To sum up, the loss of our disentangled training is:

$$\mathcal{L}_{dis} = \mathcal{L}_{GAN} + \lambda_1 \mathcal{L}_{id} + \lambda_2 \mathcal{L}_{norm} + \lambda_3 \mathcal{L}_{per} + \lambda_4 \mathcal{L}_{con} \tag{6}$$

with the same weights of $\lambda_1, \lambda_2, \lambda_3$ as reconstruction training. To use the loaded data more efficiently, we repeat the same procedure for $P_{out}^2 = \mathbf{G}(P_{in}^2, R_{in}^1)$.

Learning Schedule. In practice, we alternate between reconstruction and disentangled training: for every S iterations, we do 1 step of disentangled training and S - 1 steps of reconstruction. Moreover, as reconstruction can be performed by both synthetic and FFHQ datasets, we carry out our learning in two phases. In phase-1, we takes synthetic data for both training strategies. In phase-2, we switch to FFHQ for reconstruction while still uses synthetic data for disentangled training. Figure 17 shows the advantages of this 2-phase learning.

3.4 Architecture

Our conditional generator **G** is composed of a set of encoders **E** and a StyleGAN [24] generator $\mathbf{G_s}$. We utilize three latent spaces of $\mathbf{G_s}$ for information encoding, namely, the input tensor space $\mathcal{T} \in \mathbb{R}^{512\times4\times4}$, the modulation space

$\mathcal{W} \in \mathbb{R}^{512}$, and the extended modulation space $\mathcal{W}^+ \in \mathbb{R}^{512 \times L}$ (L: number of layers in $\mathbf{G_s}$). We denote the encoder to each of these spaces as $\mathbf{E_T}$, $\mathbf{E_W}$, and $\mathbf{E_{W^+}}$ and we conduct the study of what "information" (photo P or render R) to be encoded into which "space" ($\mathcal{T}$, $\mathcal{W}$, and $\mathcal{W}^+$), where we experiment both exclusive modulation and co-modulation architectures.

Exclusive Modulation. Naively, we can exclusively encode P or R into the modulation space ($\mathcal{W}/\mathcal{W}^+$). While one is encoded into the modulation space, the other one can only be encoded into $\mathcal{T}$. Figure 5 (**Left**) shows an example of such architectures, where R is encoded into $\mathcal{W}$ and P is encoded into $\mathcal{T}$. This structure is denoted as Render-$\mathcal{W}$. Whether R or P is used for modulation and whether the space is $\mathcal{W}$ or $\mathcal{W}^+$ provides us with 4 variants of exclusive modulation in total, and we investigate all of them.

Co-Modulation. We further investigate to encode both P and R into $\mathcal{W}$ and $\mathcal{W}^+$ and combine their embeddings for final modulation. A representative of such a architecture is shown in Fig. 5 (**Right**), where R is encoded into $\mathcal{W}$, and P is encoded into $\mathcal{W}^+$. In particular, the modulation signal for layer l is obtained by $\mathcal{W}_l^+ \odot \mathcal{W}$ where $\mathcal{W}_l^+ \in \mathbb{R}^{512}$ is the l-th column of $\mathcal{W}^+$ and $\odot$ is the element-wise multiplication. Unlike prior works that use concatenation or a recent approach of tensor transform plus concatenation scheme [55] to combine $\mathcal{W}_l^+$ and $\mathcal{W}$, we find our multiplicative co-modulation owns the best effectiveness. Moreover, in Fig. 5, we also encode R into $\mathcal{T}$ to further improve the identity-editability trade-off.

4 Experiment

4.1 Experimental Setup

We adopt the face reconstruction network **FR** [11], the 3DMM [29], and the renderer **Rd** [4]. Our conditional generator **G** consists of the StyleGAN [24] generator $\mathbf{G_s}$ and ResNet [18] encoders **E**. Specifically, $\mathbf{E_T}$ and $\mathbf{E_W}$ are based on ResNet-18 structure, where $\mathbf{E_T}$ outputs the feature prior to final pooling and $\mathbf{E_W}$ outputs the layer after that. We use a 18-layer PSP encoder [32] as $\mathbf{E_{W^+}}$. The discriminator architecture is the same as [24]. The synthetic data are generated by DiscoFaceGAN [10], where we set N and M to 10000 and 7. We use the first 65k FFHQ images (sorted by file names) for training and the rest 5k images as held-out testing set. All images (render and photo input, model output) are of 256 px resolution. We set S to 2 and use a batch size of 16 for both reconstruction and disentangled training. We set λ_1, λ_2, λ_3, and λ_4 to be 3, 3, 30, and 20. The model is learned in 2-phase, where phase-1 takes 140k iterations followed by 280k updates of phase-2.

4.2 Evaluation Metrics

We develop several quantitative metrics with the held-out 5k FFHQ images (denoted as $\mathbb{P}$) to evaluate identity preservation (**Identity**), editing controllability (**Face Content Similarity and Landmark Similarity**), and photorealism (**FID**) of our model **G** for image manipulation.

Manipulated Images. For each $P \in \mathbb{P}$, we first get $p = \mathbf{FR}(P) = (\alpha, \beta, \gamma, \delta)$ and then re-sample (β, γ, δ) to form edited control parameters $\hat{p}$. The editing signals and the manipulated images are thus $\hat{R} = \mathbf{Rd}(\hat{p})$ and $\hat{P} = \mathbf{G}(P, \hat{R})$. We generate 4 $\hat{P}$ for each P.

Identity. For each $(P, \hat{P})$ pair, we measure the identity preservation by computing the cosine similarity of $< \mathbf{N_f}(P), \mathbf{N_f}(\hat{P}) >$.

Landmark Similarity. For each $(\hat{P}, \hat{R})$ pair, we use a landmark detection network $\mathbf{N_l}$ [7], to extracts both of their 68 2D landmarks. The similarity metric is defined as $||\mathbf{N_l}(\hat{P}) - \mathbf{N_l}(\hat{R})||_2^2$.

Face Content Similarity. For each $(\hat{P}, \hat{R})$ pair, we follow Eq. 5 to measure the face content similarity.

FID. We denote all edited images as $\hat{\mathbb{P}}$ and measure FID [19] between $\mathbb{P}$ and $\hat{\mathbb{P}}$ to evaluate $\mathbf{G}$'s photo-realism.

4.3 Architectures Evaluation

Exclusive Modulation. We first evaluate exclusive modulation architectures in Table 1, where we observe a trade-off between identity preservation and editability. For example, Photo-$\mathcal{W}^+$ shows the best identity preservation (**Id**), while its editability of landmark (**LM**) and face content similarity (**FC**) is the worst. On the other hand, Photo-$\mathcal{W}$ and Render-$\mathcal{W}^+$ owns strong **LM** and **FC**, yet their **Id** are much poorer and they even have issues on photo-realism. Moreover, we see that Render-$\mathcal{W}$ provides us with a decent editability, while improves a lot on **Id** compared to Render-$\mathcal{W}^+$ and Photo-$\mathcal{W}$. For good identity preservation, we design a co-modulation architecture based on Render-$\mathcal{W}$ and Photo-$\mathcal{W}^+$.

Co-Modulation. Based on the study above, we find that encoding P into $\mathcal{W}^+$ produces the best identity preservation, while encoding R into $\mathcal{W}$ provides good editability. Thus, we investigate three 2-encoder co-modulation architectures where $\mathbf{E_W}$ encodes R and $\mathbf{E_{W^+}}$ encodes P. Combining $\mathcal{W}^+$and $\mathcal{W}$ are achieved via multiplication, concatenation, and a variant of concatenation named tensor transform in [55]. From Tab. 1, we find the multiplicative co-modulation achieves the best results from all perspectives. This could be accounted by the fact that modulation itself is a multiplicative operation and thus merging signals together multiplicatively would provide the best synergy. We further propose a

Table 1. Quantitative measurement of identity preservation (**Id**), editing control (**LM** & **FC**), and photo-realism (**FID**) for different architectures. ↑ means the higher the better, and vice versa for ↓.

Model		Metrics			
Mod. scheme	Type	Id.↑	LM.↓	FC.↓	FID↓
Exclusive modulation	Render-$\mathcal{W}$	0.57	17.8	0.021	14.5
	Render-$\mathcal{W}^+$	0.46	16.2	0.019	25.8
	Photo-$\mathcal{W}$	0.50	15.6	0.018	13.7
	Photo-$\mathcal{W}^+$	0.66	27.3	0.033	12.3
2-Encoder Co-Mod	Concat.	0.64	59.3	0.031	17.8
	Tensor	0.62	24.9	0.028	18.6
	Multi.	0.66	22.2	0.025	12.4
3-E Co-Mod	Multi.	0.66	17.2	0.020	12.2

3-encoder multiplicative co-modulation architecture (bottom of Fig. 5) to boost the editability, which achieves the best trade-off from our observation.

Visualization. We show a visual comparison among Render-$\mathcal{W}$ (**Col. 3**), Photo-$\mathcal{W}^+$ (**Col. 4**), and the 3-encoder co-modulation scheme (**Col. 5**) with the same set of inputs (**Col. 1 & 2**) in Fig. 6. In the first row, we find that Render-$\mathcal{W}$ has a clear identity loss, while Photo-$\mathcal{W}^+$ can hardly manipulate the light intensity, showing inferior editability. Moreover, Render-$\mathcal{W}$ and Photo-$\mathcal{W}^+$ both generate artifacts in the second row. On the contrary, the co-modulation scheme improves the identity-editability by combining the merits from both schemes: good editability from Render-$\mathcal{W}$ and strong identity preservation from Photo-$\mathcal{W}^+$.

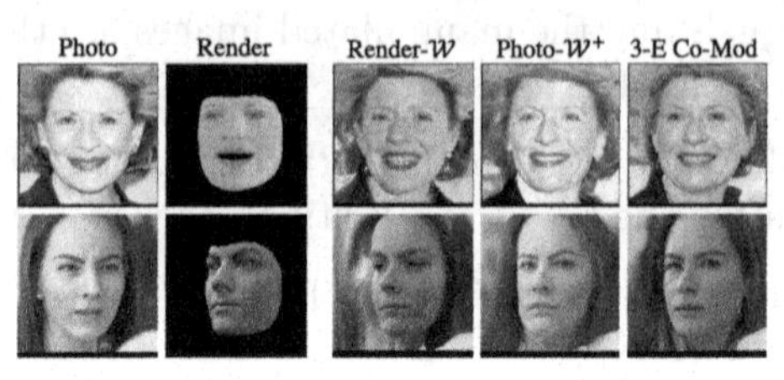

Fig. 6. Visual comparison among architectures. The co-modulation scheme takes advantages from both: good editability from Render-$\mathcal{W}$ and strong identity preservation from Photo-$\mathcal{W}^+$.

4.4 Controllable Image Synthesis

We apply our 3-encoder co-modulation architecture to several image manipulation tasks where it all shows good editing controllability, strong identity preservation, and high photo-realism. More samples are in Supplementary.

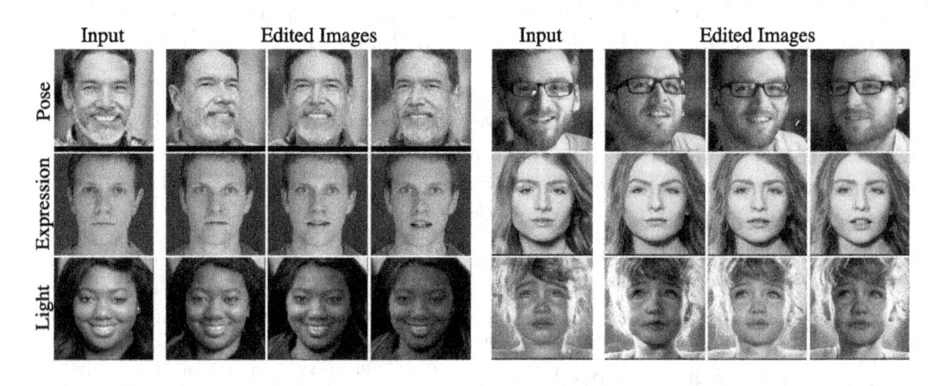

Fig. 7. Our model provides a variety of disentangled controls for pose (**Row 1**), expression (**Row 2**), and illumination (**Row 3**). It shows strong preservation across diverse identities and for facial details like glasses.

Disentangled Editing. Figure 1 and Fig. 7 show the results of single factor editing, where we only change one factor of pose, expression, and illumination at a time. Our model provides highly disentangled editing for the edited factors, while all others remain the same. Moreover, it shows strong preservation for the identity across people with diverse ages, genders, etc., and subtle facial details like the glasses and the teeth.

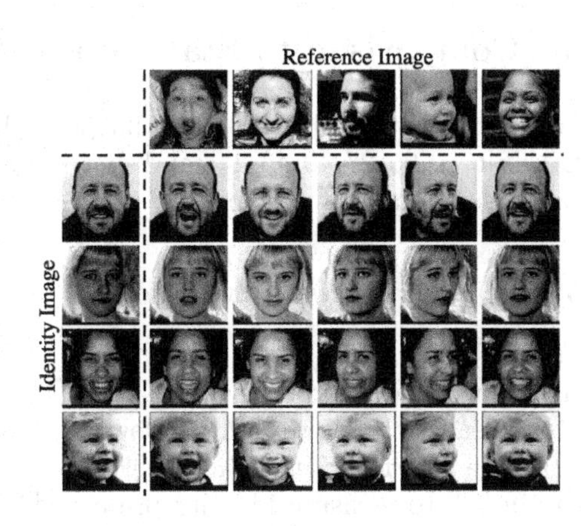

Fig. 8. Reference based face generation. The facial attributes of pose, expression, and illumination are extracted from the reference images to manipulate the identity images.

Reference-Based Synthesis. Our model can also perform image manipulation based on reference images shown in Fig. 8. With the pose, expression, and illumination extracted from the reference images, we re-synthesize our identity images to bear these editing facial attributes while the identities are still well preserved.

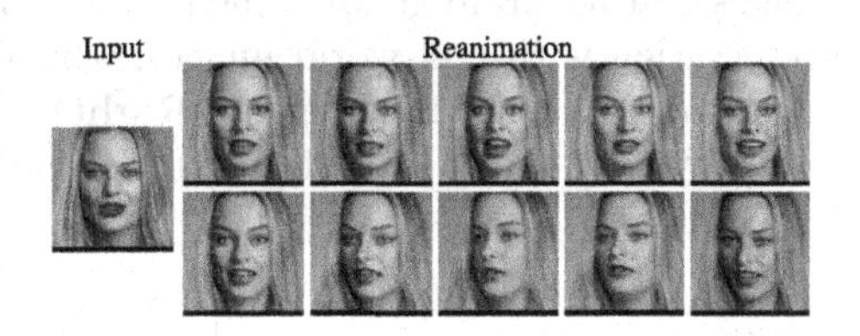

Fig. 9. Image reanimation. Our model again well preserves the identity and some subtle facial attributes like the dark lipstick.

Face Reanimation. Our model can also be applied for face reanimation, as shown in Fig. 9. With a single input photo image, we provide a series of editing render signals to make it animated, where the identity of the person is well preserved across frames. Moreover, our model again well preserves facial details like the dark lipstick.

Artistic Images Manipulation. We further perform manipulation on artistic faces [52] in Fig. 10. Surprisingly, although our model is only trained on photography faces, it can still provide controllable and identity-preserved editing on artistic images that are out of the training domain. This well indicates the strong generalizability of our model.

5 Comparison to State of the Arts

We compare with prior 3D-controllable GANs [6,10,25,27,40,43,44], and show more results in Supplementary.

Fig. 10. Although our model is solely trained on photo faces, it demonstrates a strong generalizability to manipulate artistic faces.

5.1 Quantitative Comparison

From Fig. 11 (**Left**), we clearly find that our model produces the most photo-realistic images with the lowest FID. We also follow the similar strategy in [40] to measure identity preservation, where we use all frontal images from the held-out FFHQ set and perform pose editing at different angles to compute the identity cosine similarity between the edited faces and the original ones. While prior methods evaluates the preservation between their generated images that naturally fit with their latent manifolds, we are assessing the identity preservation with real world images, which represents a more challenging task. Surprisingly, as shown in Fig. 11 (**Right**), our model still outperforms prior arts in all rotation angles on a harder task, and it can well preserve identity even at large angles.

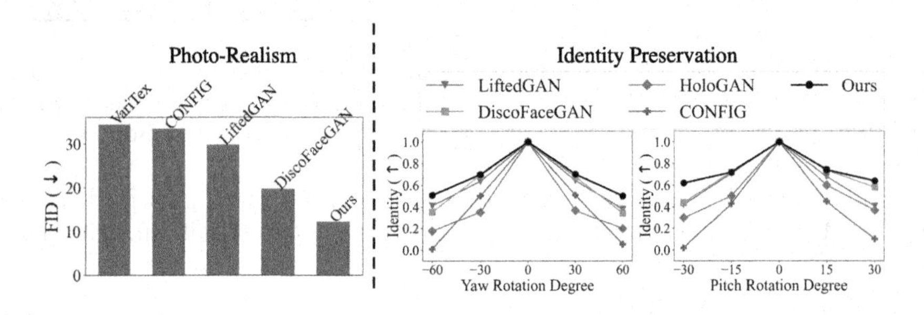

Fig. 11. Quantitative comparison with prior arts. Our method achieves the best photo realism (**Left**) and better identity preservation (**Right**) at different rotation angles.

5.2 Visual Comparison

DiscoFaceGAN. We first compare 3D-FM GAN with the direct combination of GAN inversion [2] + noise-to-image, 3D GAN, here DiscoFaceGAN (DFG) [10] for image manipulation in Fig. 12. Although GAN inversion successfully retrieves latent codes that well project the image in DFG, manipulating these codes for high-quality editing is still challenging. On the contrary, our approach provides both good image reconstruction and high-quality disentangled editing.

We further notice DFG is primarily trained to disentangle its λ space where 3DMM parameter p lies in, while its image embedding is conducted in $\mathcal{W}+$

Fig. 12. Comparing 3D-FM GAN with GAN inversion + DiscoFaceGAN (DFG) [10] for face editing. Although GAN inversion allows good projection with DFG, providing faithful manipulation with the inverted codes remains challenging. On the contrary, our method achieves good reconstruction and high-quality disentangled editing.

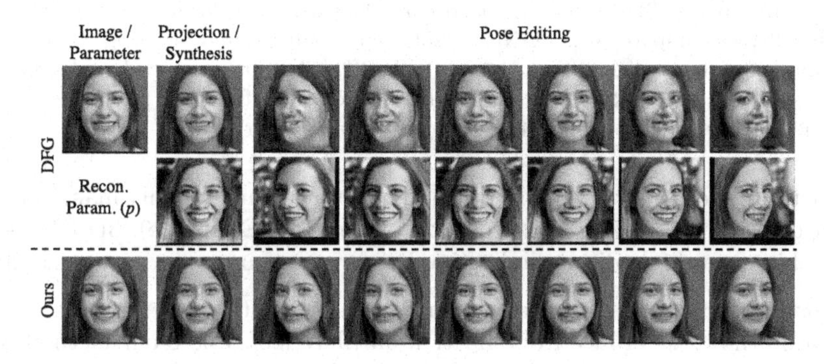

Fig. 13. Analysis of $\mathcal{W}+$ and λ space of DiscoFaceGAN (DFG) [10]. While DFG's image embedding is performed in $\mathcal{W}+$ space (**Row 1**) for editing, we also extract the input's 3DMM parameter p by **FR** and conduct the same editing in λ space (**Row 2**). While its λ space provides realistic synthesis, its inverted code in $\mathcal{W}+$ falls off the manifold of good editability trained in λ. In contrast, 3D-FM GAN (**Row 3**) uses the same editing spaces for training and testing, which easily leads to high-quality editing.

space[1]. This already creates an obvious disparity between training and test time tasks as different latent spaces are used. We thus analyze how DFG behaves in these two spaces in Fig. 13, where we retrieve p from photo input by **FR** and perform the same editing in both λ and $\mathcal{W}+$ space. We clearly see that DFG's λ space is well trained for realistic disentangled synthesis, yet its $\mathcal{W}+$ space is not. This suggests that despite the inverted code in $\mathcal{W}+$ can well embed the image, it may not lie on the manifold with good editability trained from λ. On the contrary, our method utilizes the same editing space for both training and testing, and this consistency guarantees the high-quality manipulation.

Other 3D GANs. We further compare 3D-FM GAN with CONFIG [25], StyleRig [44], and PIE [44] on disentangled image editing and reference-based synthesis tasks in Fig. 14. Our model clearly shows a larger range of pose editability and better identity preservation over CONFIG. Compared to GAN inversion [2] + StyleRig, 3D-FM GAN again provides more realistic synthesis with much less

[1] In [10], it claims that DFG's λ space is not feasible for image embedding.

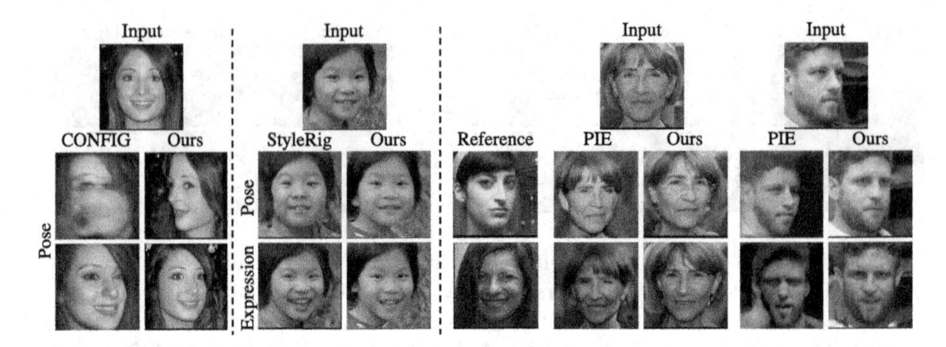

Fig. 14. Comparing 3D-FM GAN with other 3D-aware GANs for image manipulation. Specifically, we compare with CONFIG [25] on pose editing and StyleRig [44] on both pose and expression editing. We compare with PIE [43] on a reference-based synthesis task, where the pose, expression, and light are extracted from the reference images. Our method again shows the best editing results over all prior arts.

artifacts around the face. While PIE could not provide high-quality manipulation when the input is at large pose rotation angle (**2nd Example**), 3D-FM GAN still achieves faithful editing, indicating its advantage in better generalizability.

Frontalization. In Fig. 15, we compare our model with prior methods [21,31, 47,54] on the tasks of face frontalization on LFW [20], where our method best preserves the face identity and produce more photo-realistic images.

Relighting. We show portrait relighting comparison with [10,37] on Multi-PIE [16] in Fig. 16. While SfsNet and DFG do not synthesize realistic manipulation with artifacts in background and around the face, our method shows higher photo-realism and can preserve background pattern like the clothes around the neck. Moreover, DFG completely changes the skin tone of the person, whilst our method meshes the extreme indoor light with the skin tone more naturally.

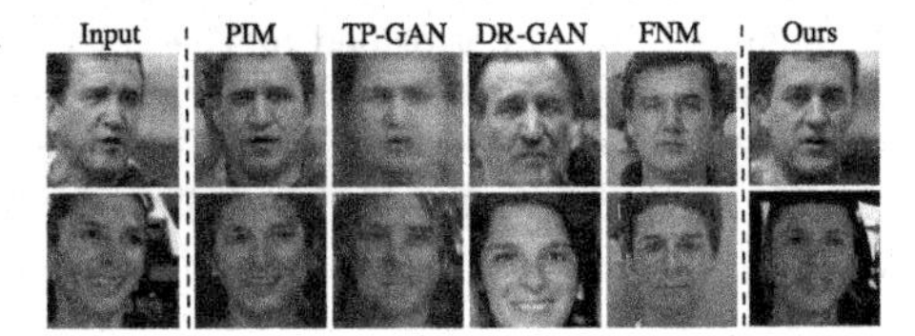

Fig. 15. Face frontalization on LFW [20] images. Our model preserves the best identity with a higher photo realism.

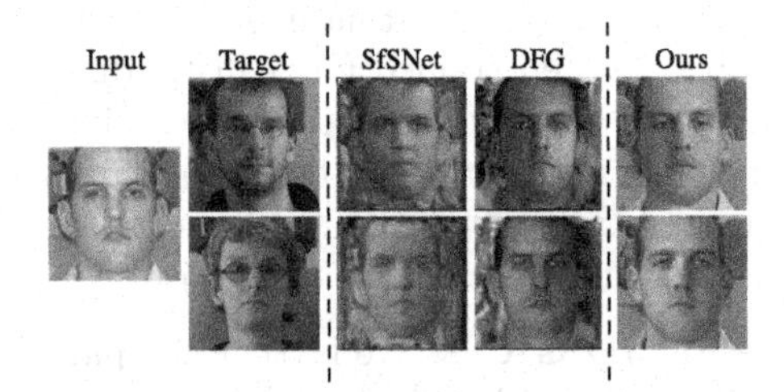

Fig. 16. Comparing prior arts on portrait relighting with Multi-PIE [16] images. Our method provides a higher photo-realism, and merges the indoor light with the person's skin tone more naturally.

6 Ablation Study

6.1 Training Strategy

In Sect. 3.3, we propose to do alternate training between reconstruction and disentanglement. To understand

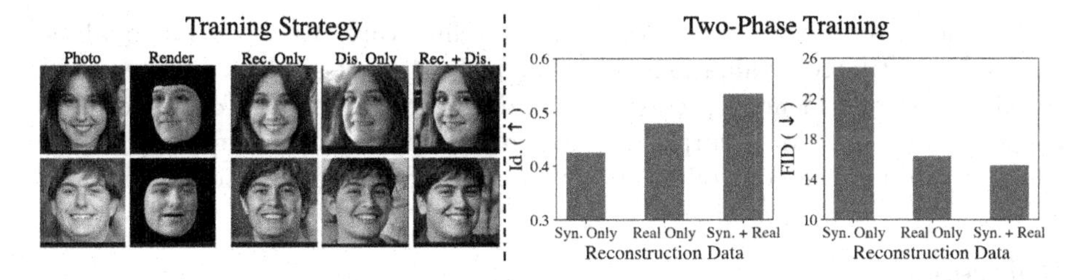

Fig. 17. Ablation study. **Left:** Effectiveness of alternate training. Compared to reconstruction or disentangled only training, alternate training scheme acquire information for both editing and identity preservation. **Right:** Effectiveness of two-phase training. Using FFHQ for reconstruction significantly improves the photo-realism. The two-phase scheme, fine-tuning with synthetic reconstruction first and then switch to FFHQ, further improves identity preservation.

its effectiveness, we conduct a study and find both of them are essential to learn high-quality identity-preserved editing with results shown in Fig. 17 (**Left**). Specifically, we perform 140k training iterations with synthetic data on a Render-$\mathcal{W}$ with the following variants: (1) reconstruction only training (**Col. 3**); (2) disentangled only training (**Col. 4**); (3) alternate training (70k iterations for each) between reconstruction and disetanglement (**Col. 5**). While reconstruction only training enables good identity preservation, it's hard for the model to respond to the editing signals. On the other hand, disentangled only training provides good editability, but fails to preserve the identity like face shapes, ages, etc. Different from them, alternating between these two strategies helps the model achieve a much better performance as it picks up information from both sides.

6.2 Two-Phase Training

To study our two-phase training scheme, where different data are used for reconstruction training, we adopt a Render-$\mathcal{W}$ architecture and train for 280k iterations with the following schedules: (1) synthetic data only reconstruction; (2) real data only reconstruction; (3) 140K iterations of synthetic reconstruction followed by 140K iterations of real reconstruction. In Fig. 17 (**Right**), we see that incorporating real data for reconstruction training is crucial for achieving high photo-realism. Moreover, the two-phase training scheme, (3), yields the best identity preservation.

7 Conclusion

In this work, we propose 3D-FM GAN, a novel framework for high-quality, 3D-controllable, existing face manipulation. Unlike prior works, our model is trained exactly for the task of face manipulation, and does not require any manual tuning after the learning phase. We design two training strategies that are both essential for the model to gain abilities of high-quality, identity-preserved editing. We also study the information encoding scheme on StyleGAN's latent spaces, which leads us to a novel multiplicative co-modulation architecture. We carry out qualitative and quantitative evaluations on our model, where it all demonstrates good editability, strong identity preservation and high photo-realism, outperforming the state of the arts. More surprisingly, our model shows a strong generalizability, where it can perform controllable editing on out-of-domain artistic faces.

References

1. Abdal, R., Qin, Y., Wonka, P.: Image2stylegan: how to embed images into the stylegan latent space? In: Proceedings of the IEEE/CVF International Conference on Computer Vision, pp. 4432–4441 (2019)
2. Abdal, R., Qin, Y., Wonka, P.: Image2stylegan++: how to edit the embedded images? In: Proceedings of the IEEE/CVF Conference on Computer Vision and Pattern Recognition, pp. 8296–8305 (2020)
3. Bao, J., Chen, D., Wen, F., Li, H., Hua, G.: Towards open-set identity preserving face synthesis. In: Proceedings of the IEEE Conference on Computer Vision and Pattern Recognition, pp. 6713–6722 (2018)
4. Blanz, V., Vetter, T.: A morphable model for the synthesis of 3D faces. In: Proceedings of the 26th Annual Conference on Computer Graphics And Interactive Techniques, pp. 187–194 (1999)
5. Booth, J., Roussos, A., Ponniah, A., Dunaway, D., Zafeiriou, S.: Large scale 3D morphable models. Int. J. Comput. Vis. **126**(2), 233–254 (2018)
6. Bühler, M.C., Meka, A., Li, G., Beeler, T., Hilliges, O.: Varitex: variational neural face textures. arXiv preprint arXiv:2104.05988 (2021)
7. Bulat, A., Tzimiropoulos, G.: How far are we from solving the 2D & 3D face alignment problem? (and a dataset of 230,000 3D facial landmarks). In: International Conference on Computer Vision (2017)
8. Choi, Y., Choi, M., Kim, M., Ha, J.W., Kim, S., Choo, J.: StarGAN: unified generative adversarial networks for multi-domain image-to-image translation, pp. 8789–8797 (2018)
9. Deng, J., Guo, J., Xue, N., Zafeiriou, S.: ArcFace: additive angular margin loss for deep face recognition. In: Proceedings of the IEEE/CVF Conference on Computer Vision and Pattern Recognition, pp. 4690–4699 (2019)
10. Deng, Y., Yang, J., Chen, D., Wen, F., Tong, X.: Disentangled and controllable face image generation via 3D imitative-contrastive learning. In: Proceedings of the IEEE/CVF Conference on Computer Vision and Pattern Recognition, pp. 5154–5163 (2020)
11. Deng, Y., Yang, J., Xu, S., Chen, D., Jia, Y., Tong, X.: Accurate 3D face reconstruction with weakly-supervised learning: from single image to image set. In: Proceedings of the IEEE/CVF Conference on Computer Vision and Pattern Recognition Workshops (2019)
12. Gecer, B., Ploumpis, S., Kotsia, I., Zafeiriou, S.: Ganfit: generative adversarial network fitting for high fidelity 3D face reconstruction. In: Proceedings of the IEEE/CVF Conference on Computer Vision and Pattern Recognition, pp. 1155–1164 (2019)
13. Genova, K., Cole, F., Maschinot, A., Sarna, A., Vlasic, D., Freeman, W.T.: Unsupervised training for 3D morphable model regression. In: Proceedings of the IEEE Conference on Computer Vision and Pattern Recognition, pp. 8377–8386 (2018)

14. Ghosh, P., Gupta, P.S., Uziel, R., Ranjan, A., Black, M.J., Bolkart, T.: Gif: generative interpretable faces. In: 2020 International Conference on 3D Vision (3DV), pp. 868–878. IEEE (2020)
15. Goodfellow, I.,et al.: Generative adversarial nets. In: Advances in Neural Information Processing Systems, vol. 27 (2014)
16. Gross, R., Matthews, I., Cohn, J., Kanade, T., Baker, S.: Multi-pie. Image Vis. Comput. **28**(5), 807–813 (2010)
17. Härkönen, E., Hertzmann, A., Lehtinen, J., Paris, S.: Ganspace: discovering interpretable GAN controls. arXiv preprint arXiv:2004.02546 (2020)
18. He, K., Zhang, X., Ren, S., Sun, J.: Deep residual learning for image recognition. In: Proceedings of the IEEE Conference on Computer Vision and Pattern Recognition, pp. 770–778 (2016)
19. Heusel, M., Ramsauer, H., Unterthiner, T., Nessler, B., Hochreiter, S.: Gans trained by a two time-scale update rule converge to a local nash equilibrium. In: Advances in neural information processing systems, vol. 30 (2017)
20. Huang, G.B., Ramesh, M., Berg, T., Learned-Miller, E.: Labeled faces in the wild: a database for studying face recognition in unconstrained environments. Technical report. 07–49, University of Massachusetts, Amherst (2007)
21. Huang, R., Zhang, S., Li, T., He, R.: Beyond face rotation: global and local perception GAN for photorealistic and identity preserving frontal view synthesis. In: Proceedings of the IEEE International Conference on Computer Vision, pp. 2439–2448 (2017)
22. Karras, T., Aila, T., Laine, S., Lehtinen, J.: Progressive growing of GANS for improved quality, stability, and variation. arXiv preprint arXiv:1710.10196 (2017)
23. Karras, T., Laine, S., Aila, T.: A style-based generator architecture for generative adversarial networks. In: Proceedings of the IEEE/CVF Conference on Computer Vision and Pattern Recognition, pp. 4401–4410 (2019)
24. Karras, T., Laine, S., Aittala, M., Hellsten, J., Lehtinen, J., Aila, T.: Analyzing and improving the image quality of StyleGAN. In: Proceedings of the IEEE/CVF Conference on Computer Vision and Pattern Recognition, pp. 8110–8119 (2020)
25. Kowalski, M., Garbin, S.J., Estellers, V., Baltrušaitis, T., Johnson, M., Shotton, J.: CONFIG: controllable neural face image generation. In: Vedaldi, A., Bischof, H., Brox, T., Frahm, J.-M. (eds.) ECCV 2020. LNCS, vol. 12356, pp. 299–315. Springer, Cham (2020). https://doi.org/10.1007/978-3-030-58621-8_18
26. Li, T., Bolkart, T., Black, M.J., Li, H., Romero, J.: Learning a model of facial shape and expression from 4d scans. ACM Trans. Graph. **36**(6), 194–1 (2017)
27. Nguyen-Phuoc, T., Li, C., Theis, L., Richardt, C., Yang, Y.L.: HoloGAN: unsupervised learning of 3D representations from natural images. In: Proceedings of the IEEE/CVF International Conference on Computer Vision, pp. 7588–7597 (2019)
28. Parke, F.I.: A parametric model for human faces. The University of Utah (1974)
29. Paysan, P., Knothe, R., Amberg, B., Romdhani, S., Vetter, T.: A 3D face model for pose and illumination invariant face recognition. In: 2009 sixth IEEE International Conference on Advanced Video and Signal Based Surveillance, pp. 296–301. IEEE (2009)
30. Pumarola, A., Agudo, A., Martinez, A.M., Sanfeliu, A., Moreno-Noguer, F.: GANimation: anatomically-aware facial animation from a single image. In: Proceedings of the European Conference on Computer Vision (ECCV), pp. 818–833 (2018)
31. Qian, Y., Deng, W., Hu, J.: Unsupervised face normalization with extreme pose and expression in the wild. In: Proceedings of the IEEE/CVF Conference on Computer Vision and Pattern Recognition, pp. 9851–9858 (2019)

32. Richardson, E., et al.: Encoding in style: a StyleGAN encoder for image-to-image translation. In: Proceedings of the IEEE/CVF Conference on Computer Vision and Pattern Recognition, pp. 2287–2296 (2021)
33. Richardson, E., Sela, M., Or-El, R., Kimmel, R.: Learning detailed face reconstruction from a single image. In: Proceedings of the IEEE Conference on Computer Vision and Pattern Recognition, pp. 1259–1268 (2017)
34. Saito, S., Wei, L., Hu, L., Nagano, K., Li, H.: Photorealistic facial texture inference using deep neural networks. In: Proceedings of the IEEE Conference on Computer Vision and Pattern Recognition, pp. 5144–5153 (2017)
35. Sanyal, S., Bolkart, T., Feng, H., Black, M.J.: Learning to regress 3D face shape and expression from an image without 3D supervision. In: Proceedings of the IEEE/CVF Conference on Computer Vision and Pattern Recognition, pp. 7763–7772 (2019)
36. Sela, M., Richardson, E., Kimmel, R.: Unrestricted facial geometry reconstruction using image-to-image translation. In: Proceedings of the IEEE International Conference on Computer Vision, pp. 1576–1585 (2017)
37. Sengupta, S., Kanazawa, A., Castillo, C.D., Jacobs, D.W.: SfSNeT: learning shape, reflectance and illuminance of faces in the wild'. In: Proceedings of the IEEE Conference on Computer Vision and Pattern Recognition, pp. 6296–6305 (2018)
38. Shen, Y., Luo, P., Yan, J., Wang, X., Tang, X.: FaceID-GAN: learning a symmetry three-player GAN for identity-preserving face synthesis. In: Proceedings of the IEEE Conference on Computer Vision and Pattern Recognition, pp. 821–830 (2018)
39. Shen, Y., Zhou, B.: Closed-form factorization of latent semantics in GANS. In: Proceedings of the IEEE/CVF Conference on Computer Vision and Pattern Recognition, pp. 1532–1540 (2021)
40. Shi, Y., Aggarwal, D., Jain, A.K.: Lifting 2D StyleGAN for 3D-aware face generation. In: Proceedings of the IEEE/CVF Conference on Computer Vision and Pattern Recognition, pp. 6258–6266 (2021)
41. Shoshan, A., Bhonker, N., Kviatkovsky, I., Medioni, G.: Gan-control: explicitly controllable GANs. arXiv preprint arXiv:2101.02477 (2021)
42. Slossberg, R., Shamai, G., Kimmel, R.: High quality facial surface and texture synthesis via generative adversarial networks. In: Proceedings of the European Conference on Computer Vision (ECCV) Workshops (2018)
43. Tewari, A., et al.: PIE: portrait image embedding for semantic control. ACM Trans. Graph. (TOG) **39**(6), 1–14 (2020)
44. Tewari, A., et al.: StyleRig: rigging StyleGAN for 3D control over portrait images. In: Proceedings of the IEEE/CVF Conference on Computer Vision and Pattern Recognition, pp. 6142–6151 (2020)
45. Tov, O., Alaluf, Y., Nitzan, Y., Patashnik, O., Cohen-Or, D.: Designing an encoder for StyleGAN image manipulation. ACM Trans. Graph. (TOG) **40**(4), 1–14 (2021)
46. Tran, L., Liu, F., Liu, X.: Towards high-fidelity nonlinear 3D face morphable model. In: Proceedings of the IEEE/CVF Conference on Computer Vision and Pattern Recognition, pp. 1126–1135 (2019)
47. Tran, L., Yin, X., Liu, X.: Disentangled representation learning GAN for pose-invariant face recognition. In: Proceedings of the IEEE Conference on Computer Vision and Pattern Recognition, pp. 1415–1424 (2017)
48. Usman, B., Dufour, N., Saenko, K., Bregler, C.: PuppetGAN: cross-domain image manipulation by demonstration. In: Proceedings of the IEEE/CVF International Conference on Computer Vision, pp. 9450–9458 (2019)
49. Ververas, E., Zafeiriou, S.: SliderGAN: synthesizing expressive face images by sliding 3D blendshape parameters. Int. J. Comput. Vis. **128**(10), 2629–2650 (2020)

50. Xia, W., Zhang, Y., Yang, Y., Xue, J.H., Zhou, B., Yang, M.H.: Gan inversion: a survey. arXiv preprint arXiv:2101.05278 (2021)
51. Xiao, T., Hong, J., Ma, J.: ELEGANT: exchanging latent encodings with GAN for transferring multiple face attributes. In: Proceedings of the European Conference on Computer Vision (ECCV), pp. 168–184 (2018)
52. Yaniv, J., Newman, Y., Shamir, A.: The face of art: landmark detection and geometric style in portraits. ACM Trans. Graph. (TOG) **38**(4), 1–15 (2019)
53. Zhang, R., Isola, P., Efros, A.A., Shechtman, E., Wang, O.: The unreasonable effectiveness of deep features as a perceptual metric. In: Proceedings of the IEEE Conference on Computer Vision and Pattern Recognition, pp. 586–595 (2018)
54. Zhao, J., et al.: Towards pose invariant face recognition in the wild. In: Proceedings of the IEEE Conference on Computer Vision and Pattern Recognition, pp. 2207–2216 (2018)
55. Zhao, S., et al.: Large scale image completion via co-modulated generative adversarial networks. arXiv preprint arXiv:2103.10428 (2021)
56. Zhou, H., Hadap, S., Sunkavalli, K., Jacobs, D.W.: Deep single-image portrait relighting. In: Proceedings of the IEEE/CVF International Conference on Computer Vision, pp. 7194–7202 (2019)

Multi-Curve Translator for High-Resolution Photorealistic Image Translation

Yuda Song, Hui Qian, and Xin Du(✉)

Zhejiang University, Hangzhou, China
{syd,qianhui,duxin}@zju.edu.cn

Abstract. The dominant image-to-image translation methods are based on fully convolutional networks, which extract and translate an image's features and then reconstruct the image. However, they have unacceptable computational costs when working with high-resolution images. To this end, we present the Multi-Curve Translator (MCT), which not only predicts the translated pixels for the corresponding input pixels but also for their neighboring pixels. And if a high-resolution image is downsampled to its low-resolution version, the lost pixels are the remaining pixels' neighboring pixels. So MCT makes it possible to feed the network only the downsampled image to perform the mapping for the full-resolution image, which can dramatically lower the computational cost. Besides, MCT is a plug-in approach that utilizes existing base models and requires only replacing their output layers. Experiments demonstrate that the MCT variants can process 4K images in real-time and achieve comparable or even better performance than the base models on various photorealistic image-to-image translation tasks.

Keywords: High-resolution · Real-time · Image-to-image translation

1 Introduction

Image-to-image (I2I) translation aims to translate images from a source domain to a target domain. Many computer vision tasks, such as image denoising [64], dehazing [4], colorization [65], attribute editing [8], and style transfer [11], can be posed as I2I translation problems. Some approaches [21,23,39,48,68] use a universal framework to handle various I2I translation problems. No matter the training scheme and the addressed problem, their network architectures are generally based on fully convolutional networks (FCNs) [40]. However, the computational cost of FCN is proportional to the input image pixels, making high-resolution (HR) images be a considerable obstacle to employing these methods. For example, CycleGAN [68] requires 56.8G multiply-accumulate operations (MACs) to

Supplementary Information The online version contains supplementary material available at https://doi.org/10.1007/978-3-031-19784-0_8.

S. Avidan et al. (Eds.): ECCV 2022, LNCS 13675, pp. 126–143, 2022.
https://doi.org/10.1007/978-3-031-19784-0_8

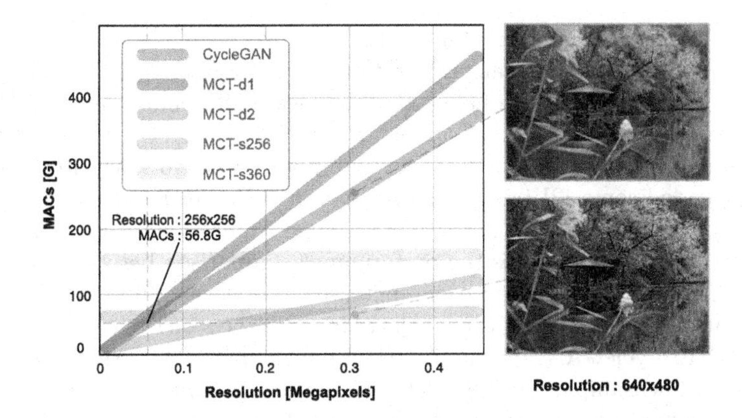

Fig. 1. An example comparing CycleGAN and its MCT variant on `autumn2summer`. The only difference between two models is the output layer. `d2` means the input image of the backbone network is downsampled by a factor of 2, and `s256` means the input image of the backbone network is downsampled to 256×256.

process a 256×256 image and requires 7.2T MACs when working with a 4K (3840×2160) image, which is unacceptable even for high-performance GPUs.

To this end, some researchers design lightweight networks [2,35] or employ model compression [32,52] to save computational cost. However, designing and training a new lightweight FCN is not easy since it involves a trade-off between efficiency and effectiveness. And repeating this procedure for every I2I translation task can be highly time-consuming and power-consuming. Therefore, we prefer to propose a more flexible approach to the problem. We found that some photorealistic I2I translation methods [28,34,41] apply post-processing techniques that constrain the mapping to be spatially smooth to preserve the image's structure information. So why not just predict a spatially smooth mapping to approximate this translation process in the image space? We can downsample HR images and use the downsampled images to predict the mappings for the original images. In this way, we can feed low-resolution (LR) images to the backbone networks, which exponentially reduces the computational cost. Besides, we can try to reuse the existing FCN architectures.

At this point, the key to the problem lies in designing a mapping that can provide sufficient I2I translation capability but has a much lower computational cost than the backbone network. To address the above challenges, we propose an I2I translator, dubbed **M**ulti-**C**urve **T**ranslator (MCT). Specifically, we take an existing FCN as the backbone network and find that it only predicts the output pixels for their corresponding input pixels. So we increase its last layer's output channels to make each output pixel indicate a set of mapping functions in the form of curves. We quantize these curves as look-up tables (LUTs) [24,37], then given the output pixel (*i.e.*, curves' parameters) responding to the input pixel of the downsampled image, we can derive the output pixels for all pixels in the full-resolution image's corresponding region. Besides reducing the computational cost, MCT has additional advantages. Firstly, an FCN's receptive field is limited, so it may not extract meaningful semantic information when processing HR images. But for MCT, we can adjust the downsampling ratio to change the backbone network's receptive field size dynamically. Secondly, since MCT only requires increasing the output channels of the FCN, it is easy to employ it on another I2I translation task without designing a new network architecture.

We extended some I2I translation models to their MCT variants and found that they have significant advantages in saving computational cost and preserving details. Figure 1 illustrates the performance comparison between CycleGAN and its MCT variant. Because we can increase the downsampling ratio to reduce the computational cost, the MCT-CycleGAN can always be less computationally intensive than CycleGAN. In practice, the input images of the MCT's backbone network are downsampled to 256×256 (MCT-s256 in Fig. 1), consistent with the training set's image size to minimize the gap between inference and training. In this case, the gap between CycleGAN and MCT-CycleGAN becomes increasingly large as the input image size grows. Specifically, when processing 4K images, the computational cost of MCT-CycleGAN is only 0.8% of that of CycleGAN, leading to the former being $40\times$ faster than the latter on GPUs. Finally, MCT enables the input image's high-frequency information to flow easily to the output image, making the trees sharper to improve image realism. While MCT looks appealing, it is to be noted that MCT focuses on photorealistic image-to-image translation and does not work well on more general image-to-image translation tasks, which is our primary future work.

2 Methodology

2.1 Problem Formulation and Prior Work

Let $x \in \mathcal{X}$ and $y \in \mathcal{Y}$, the goal of I2I translation is to learn a mapping $G\colon \mathcal{X} \rightarrow \mathcal{Y}$ such that the distribution $p(G(x))$ is as close as possible to the distribution $p(y)$. Although models for different I2I translation tasks are trained in different manners, they commonly use FCN-based models [23,43,48,68]. We assume that G is an FCN with weight θ, then the translated image $\tilde{y}$ can be formulated as:

$$\tilde{y} = G(x; \theta). \tag{1}$$

However, the FCN's computational cost is proportional to the image pixels [18], and our goal is to break it. We divide the mapping into two components: the translator G that translates the images from $\mathcal{X}$ to $\mathcal{Y}$ and the encoder E that predicts the translator's parameters. Assuming that the encoder E with weight θ encodes the parameters of G from the condition z, the translation is:

$$\tilde{y} = G(x; E(z; \theta)). \tag{2}$$

There have been several works based on similar ideas. Since conventional image processing methods commonly employ filters [15,56,61], KPN [44] predicts the convolutional filter G with parameter $\theta_i = E(x;\theta)_i$ for each pixel x_i and applies it to its spatial support $\Omega(x_i)$ to obtain $\tilde{y}_i$ for burst image denoising. It can be formulated as:

$$\tilde{y}_i = G(\Omega(x_i); E(x;\theta)_i). \tag{3}$$

However, it still requires to perform the FCN on the HR image, leading to no reduction in its computational cost. Besides, the convolutional filter is still a linear mapping, which has limited translation capability.

Another feasible solution is HyperNetwork [14], which uses a network to generate the weights of another network and is originally designed for neural network compression. Following some HyperNetwork-based works [10,27,47,51] on image processing tasks, we can use a encoder E to predict the weights of a lightweight FCN G on the downsampled image $x\downarrow$, which can be formulated as:

$$\tilde{y} = G(x; E(x\downarrow;\theta)). \tag{4}$$

If E is fed with only fixed-size images $x\downarrow$, the total computational cost of the model grows slowly with the size of the input images [55]. In our experiments, this plain idea works well for photo retouching but not other tasks.

We try to combine the two approaches above to overcome their respective shortcomings. We expect the encoder E to encode the parameter maps $F = E(x\downarrow;\theta)$, in which each cell F_j contains a set of translator's parameters:

$$\tilde{y}_i = G(x_i; F_j). \tag{5}$$

Since the parameter maps F cannot be aligned with x, we also need to define the relation between i and j. We will detail our MCT in the next subsection.

Similar works to MCT consist of bilateral learning [12,59,67], curve mapping [25,31,46] and 3D LUTs [60,63], as they are all based on slicing operation. In contrast to bilateral learning-based methods, MCT predicts pixel values rather than affine transformations, which allows us to directly constrain the output of MCT to prevent falling into poor solutions when training on unpaired datasets. Curve-based methods usually predict a global transformation, which prevents them from working on more challenging I2I tasks such as daytime translation. 3D LUTs predict global transformations like curve mappings, but they have a stronger translation capability. Ideally, we could introduce spatial coordinates to extend 3D LUTs to 5D LUTs, but this would lead to an unacceptable computational cost and memory consumption. From the implementation perspective, MCT extends the curve-based methods by introducing spatial coordinates and channel interactions to improve translation capability, which can be implemented using 3D LUTs. More importantly, MCT is a plug-in module that does not rely on fancy backbone networks and loss functions, and it can be trained directly on small-sized images, dramatically reducing the effort to modify the methods.

2.2 Multi-Curve Translator

Recalling Eq. (5), our goals are to 1) design the encoder E; 2) design the translator G; and 3) define the relation between i and j. We desire our approach to be plug-in for the existing I2I translation models. Since the models of I2I translation tasks are often based on FCNs, we directly use these networks as

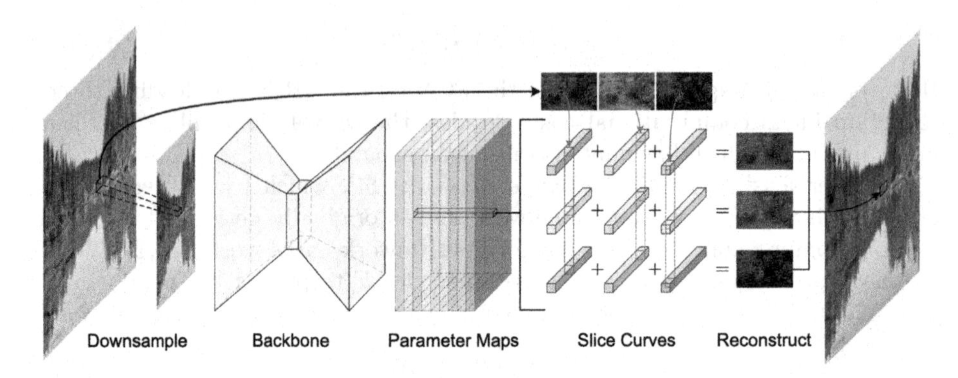

Fig. 2. Inference workflow of MCT. The backbone network receives the downsampled image and predicts the curve parameter maps with the same spatial size as the downsampled image. A cell of curve parameter maps consists of 9 sets of curves in the form of 1D LUTs, responsible for translating the corresponding region in the HR image.

our base models (*i.e.*, backbone networks) to eliminate the effort involved in designing the encoder E. Then the only modification needed is to increase their last layer's output channels to match the parameters of the translator G. Given that $x \in \mathcal{R}^{H \times W \times 3}$ and $x\downarrow \in \mathcal{R}^{H_d \times W_d \times 3}$, then $F \in \mathcal{R}^{H_d \times W_d \times C}$, where C is the number of parameters of G.

Reviewing the idea of HyperNetwork, a simple idea is to employ an FCN as the translator G. However, since convolutional layers lead to a large C, we seek another expressive nonlinear mapping with fewer parameters to replace the FCN. We found that some curve-based methods [25,31,46] achieved better performance than FCN-based methods [7,22,58] on photo retouching. These curve-based methods make the network regress the knot points of the curve to mimic the color adjustment curve tool. Although these methods implement knot points in different ways, they are all equivalent to 1D LUTs [24,37]. We illustrate the transformation function using the curve in the form of a 1D LUT for a grayscale image. Given a 1D LUT $\mathbf{T} = \{t_{(k)}\}_{k=0,\ldots,M-1}$ (*i.e.*, M knot points), pixel $x_{(\mathrm{i,j})}$ can find its location z in the LUT via a lookup operation:

$$\mathrm{z} = x_{(\mathrm{i,j})} \cdot \frac{M-1}{C_{max}}, \tag{6}$$

where C_{max} is the maximum pixel value. Since z may not be an integer, we should derive the output pixel value via interpolation. Let $d_{\mathrm{z}} = \mathrm{z} - \lfloor \mathrm{z} \rfloor$, where $\lfloor \cdot \rfloor$ is the floor function. Given that $\lceil \cdot \rceil$ is the ceil function, we derive output pixel value $y_{(\mathrm{i,j})}$ via linear interpolation:

$$\tilde{y}_{(\mathrm{i,j})} = (1 - d_{\mathrm{z}}) \cdot t_{(\lfloor \mathrm{z} \rfloor)} + d_{\mathrm{z}} \cdot t_{(\lceil \mathrm{z} \rceil)}. \tag{7}$$

Finally we need to define the relation between i and j. We can upsample the parameter maps to make their resolution the same as x (*i.e.*, $F \uparrow \in \mathcal{R}^{H \times W \times C}$). Unfortunately, while the computational cost of this operation is acceptable, it produces larger parameter maps, which may consume a lot of memory. Inspired

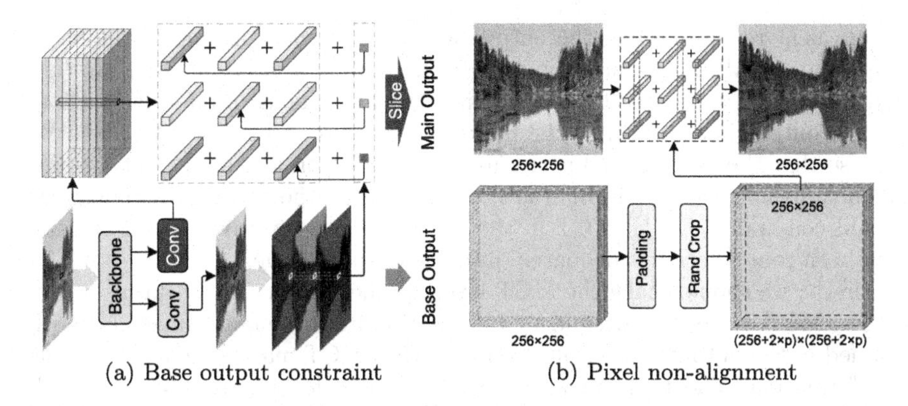

(a) Base output constraint (b) Pixel non-alignment

Fig. 3. Two strategies to train MCT variants in a stable manner. (a) Use the translated LR image as the base of the parameter maps. We can constrain the base output in the training phase to ensure that the backbone network does not degrade. (b) Train the MCT using only LR images. We use padding and random cropping to obtain parameter maps that are not pixel-aligned with the LR images.

by bilateral grid [6], we employ a 3D LUT $\mathbf{T} \in \mathcal{R}^{H_d \times W_d \times M}$. Given a grayscale pixel $x_{(\mathrm{i,j})}$, its location $(\mathrm{x,y,z})$ in the 3D LUT lattice is:

$$\mathrm{x} = \mathrm{i} \cdot \frac{H_d-1}{H-1},\ \mathrm{y} = \mathrm{j} \cdot \frac{W_d-1}{W-1},\ \mathrm{z} = x_{(\mathrm{i,j})} \cdot \frac{M-1}{C_{max}}. \tag{8}$$

Let $d_{\mathrm{x}} = \mathrm{x} - \lfloor \mathrm{x} \rfloor$ and $d_{\mathrm{y}} = \mathrm{y} - \lfloor \mathrm{y} \rfloor$, we extend Eq. (7) to trilinear interpolation to slice the output pixel:

$$\begin{aligned}
\tilde{y}_{(\mathrm{i,j})} = &(1-d_{\mathrm{x}})(1-d_{\mathrm{y}})(1-d_{\mathrm{z}})t_{(\lfloor \mathrm{x} \rfloor, \lfloor \mathrm{y} \rfloor, \lfloor \mathrm{z} \rfloor)} + d_{\mathrm{x}} d_{\mathrm{y}} d_{\mathrm{z}} t_{(\lceil \mathrm{x} \rceil, \lceil \mathrm{y} \rceil, \lceil \mathrm{z} \rceil)} \\
&+ (1-d_{\mathrm{x}}) d_{\mathrm{y}} (1-d_{\mathrm{z}}) t_{(\lfloor \mathrm{x} \rfloor, \lceil \mathrm{y} \rceil, \lfloor \mathrm{z} \rfloor)} + d_{\mathrm{x}}(1-d_{\mathrm{y}}) d_{\mathrm{z}} t_{(\lceil \mathrm{x} \rceil, \lfloor \mathrm{y} \rfloor, \lceil \mathrm{z} \rceil)} \\
&+ d_{\mathrm{x}} d_{\mathrm{y}} (1-d_{\mathrm{z}}) t_{(\lceil \mathrm{x} \rceil, \lceil \mathrm{y} \rceil, \lfloor \mathrm{z} \rfloor)} + (1-d_{\mathrm{x}})(1-d_{\mathrm{y}}) d_{\mathrm{z}} t_{(\lfloor \mathrm{x} \rfloor, \lfloor \mathrm{y} \rfloor, \lceil \mathrm{z} \rceil)} \\
&+ d_{\mathrm{x}}(1-d_{\mathrm{y}})(1-d_{\mathrm{z}}) t_{(\lceil \mathrm{x} \rceil, \lfloor \mathrm{y} \rfloor, \lfloor \mathrm{z} \rfloor)} + (1-d_{\mathrm{x}}) d_{\mathrm{y}} d_{\mathrm{z}} t_{(\lfloor \mathrm{x} \rfloor, \lceil \mathrm{y} \rceil, \lceil \mathrm{z} \rceil)}.
\end{aligned} \tag{9}$$

For RGB color images, we employ the channel-crossing strategy [54]. Specifically, 9 curves should be learned, corresponding to $\{\mathbf{T}^{p \to q}\}_{p,q \in \{R,G,B\}}$ respectively ($C = 9M$). Let $\mathbf{T}(\cdot)$ denote Eq. (8–9), we derive output pixel via:

$$\tilde{y}^q_{(\mathrm{i,j})} = \mathbf{T}^{R \to q}(x^R_{(\mathrm{i,j})}) + \mathbf{T}^{G \to q}(x^G_{(\mathrm{i,j})}) + \mathbf{T}^{B \to q}(x^B_{(\mathrm{i,j})}). \tag{10}$$

Since the translator consists of a large number of curves, we call it Multi-Curve Translator (MCT). Figure 2 shows how MCT processes a HR image.

2.3 Training Strategy

Although MCT appears to be more complex than reconstructing images using the convolutional layer, its last layer still outputs pixel values. As a special case, when only the LUTs $\{\mathbf{T}^{p \to p}\}_{p \in \{R,G,B\}}$ are included and $M = 1$, MCT is equivalent to upsampling the output image of the base model. However, we found that the MCT variant is more like to fall into poor solutions than the base model. We review the MCT and find that the input image's information flows into the output image through the backbone network and slicing operation. We suppose that the cause of the performance degradation is that the network has difficulty balancing the information flowing through the two routes. Therefore, we add constraints to the MCT in the training phase to drive the information from both routes to flow adequately into the output image.

Firstly, we should make the MCT leverage the information of downsampled images. MCT makes high-frequency information from the input image be easily retained in the output image, but we found that MCT may learn a simple color transformation. The problem arises because the information flows too easily from the input to the output, leading to a "short circuit" phenomenon that traps the network in a poor local optimum solution. Recalling the special case of MCT, we find that $\{\mathbf{T}^{p \to p}\}_{p \in \{R,G,B\}}$ can be decomposed into LUTs and biases, as shown in Fig. 3(a). Specifically, we use the last layer of the base model to predict the reconstructed image and add its pixel values as biases to the corresponding LUTs. So we can obtain the base output $\tilde{y}_b$ and the main output $\tilde{y}_m$ by $[\tilde{y}_b, \tilde{y}_m] = [G_b(x), G_m(x)]$ and constrain $\tilde{y}_b$ at training phase to ensure that the backbone network does not degrade. This strategy has a bonus that the pre-trained base model's weight can be fully utilized, including the output layer. Therefore, we employ this strategy even if we do not need to constrain the base output. In extreme cases, we can fine-tune only the added output convolutional layers to make the training faster.

Secondly, we should make the MCT leverage the information of HR images. MCT allows us to perform the backbone network on LR images, dramatically reducing the computational cost of translating HR images. However, we are still unable to use HR images as training data during training directly. The reasons are threefold: 1) loading and preprocessing HR images takes a lot of time, resulting in inefficient training; 2) The discriminator's computational cost remains proportional to the input image pixels. 3) Existing datasets often provide low-resolution images. For this reason, we still use LR images to train the MCT. As shown in Fig. 3(b), we first pad each side of the parameter maps by size p with duplication and randomly crop them to the size before padding, then the image and the parameter maps are not pixel-wise aligned, forcing MCT to extract high-frequency information from the image. We can also achieve more complex pixel misalignment by adding small random noise to x and y, but this does not visibly improve performance in our experiments.

3 Applications

We apply MCT to extend some representative I2I translation methods. Unless otherwise noted, we set $H_d = W_d = 256$ and $M = 8$, and employ pixel unaligned training strategy with $p = 1$ but do not constrain the base output.

3.1 Photorealistic I2I Translation

We refer here to the I2I translation tasks done with GANs [13]. We perform the daytime translation (`day2dusk`) and season translation (`summer2autumn`) for experiments. To extend to HR scenes, we collected new unpaired datasets from Flickr[1] with image resolutions ranging from 480p to 8K. Each domain of the datasets contains 2200 images, of which 2000 LR images are downsampled for training and the remaining 200 HR images for testing.

We employ CycleGAN [68] and UNIT [39] as the base models, which use different training procedures. When training the MCT variants, we add constraints to the base output y_b when updating the generator. Let the conventional generator's loss function be $\mathcal{L}_{base}$, then the loss function of the MCT variant is $\mathcal{L} = \mathcal{L}_{base} + \lambda\mathcal{L}_{reg}$, where $\lambda = 1$ for CycleGAN and $\lambda = 10$ for UNIT. $\mathcal{L}_{reg}$ is a cycle-consistency loss [26,68] constrainting the base output:

$$\mathcal{L}_{reg} = \left\| G_b^{y \to x}(G_m^{x \to y}(x)) - x \right\|_1 . \tag{11}$$

3.2 Style Transfer

Style transfer aims at transferring the style from a reference image to a content image and is divided into two types: artistic style transfer and photorealistic style transfer. We only study the photorealistic style transfer since it fits our motivation. We use the Microsoft COCO dataset [38] to train the base models and their MCT variants, and the test set consists of the examples provided by DPST [41] with image resolutions ranging from 720p to 4K.

We use AdaIN [20] and WCT^2 [62] as the base models since they employ different training schemes. AdaIN is designed for artistic style transfer, with few constraints on preserving high-frequency information. It employs a weighted combination of the content loss $\mathcal{L}_c$ and the style loss $\mathcal{L}_s$ with the weight λ, *i.e.* $\mathcal{L} = \mathcal{L}_c + \lambda\mathcal{L}_s$. Both $\mathcal{L}_c$ and $\mathcal{L}_s$ use pre-trained VGG-19 [53] to compute the loss function without constraining the pixels of the images. So we add a gradient loss $\mathcal{L}_g$ to prompt the preservation of the geometric structure:

$$\mathcal{L}_g = \left\| \nabla_h G(x) - \nabla_h x \right\|_2^2 + \left\| \nabla_v G(x) - \nabla_v x \right\|_2^2 , \tag{12}$$

where ∇_h (∇_v) denotes the gradient operator along the horizontal (vertical) direction. The modified AdaIN's full objective is $\mathcal{L} = \mathcal{L}_c + \lambda_1\mathcal{L}_s + \lambda_2\mathcal{L}_g$, where $\lambda_1 = 1$ and $\lambda_2 = 100$. The WCT^2's scheme is special because it only requires the output image to reconstruct the input during training and performs WCT [33] sequentially at each scale to achieve stylization during testing. Let the reconstruction loss function of WCT^2 be $\mathcal{L}_{rec}(G(x), x)$, and its MCT variant's loss function is:

$$\mathcal{L} = \mathcal{L}_{rec}(G_b(x), x) + \mathcal{L}_{rec}(G_m(x), x). \tag{13}$$

WCT^2's training scheme makes the HR image's low-frequency information flow easily to the output image, so we perform the grayscale operation on the HR

[1] https://www.flickr.com/.

image. When performing stylization, we further match the HR image's brightness with the reference image's brightness to prevent the brightness of the HR image from being retained in the output image.

3.3 Image Dehazing

Image dehazing aims to recover clean images from hazy images, which is essential for subsequent high-level tasks. We choose 6000 synthetic image pairs from RESIDE dataset [30] for training, 3000 from the ITS subset, and 3000 from the OTS subset. We use two datasets, named SOTS [30] and HazeRD [66], to evaluate the performance of the methods. Note that the SOTS has more image pairs while the HazeRD has 4K-resolution image pairs.

We take GCANet [5] and MSBDN [9] as the base models. GCANet expands the receptive field by dilated convolution [36], while MSBDN uses upsampling and downsampling operations. For simplicity, we use only the $\mathcal{L}_1$ loss function to train the models instead of using the original training scheme of the base models. Note that for supervised training, the base output constraint is optional.

3.4 Photo Retouching

Photo retouching aims to adjust an image's brightness, contrast, and so on to make the image fit people's aesthetics. We choose 4500 image pairs from the MIT-Adobe-5K dataset [3] for training and the remaining 500 image pairs for testing with image resolutions ranging from 2K to 6K. We also employ an unpaired training scheme, with the 2250 images as domain $\mathcal{X}$ and the remaining 2250 images as domain $\mathcal{Y}$ in the training set.

We use DPED [22] and DPE [7] as the base models. Specifically, DPED uses a residual network [16] without downsampling and upsampling, leading to a small receptive field. But large scale context is critical for photo retouching, which is used to sense the illumination and contrast of an image [7,12]. So we set $H_d = W_d = 32$ for DPED to enlarge the receptive field without modifying the network architecture. Since color mapping is critical for photo retouching, we set $p = 8$ for DPE. For paired training, we use the $\mathcal{L}_1$ loss function to train the models. We employ the CycleGAN's training scheme for unpaired training rather than the base models' training scheme for comparison purposes.

4 Experiments

4.1 Runtime

We have shown the advantages of the MCT in terms of computational cost in Fig. 1, but MACs are indirect metrics of speed [42], which is an unconvincing indicator. So we test the runtime of the base models and their MCT variants on multiple hardware platforms. Specifically, we use the PyTorch framework to test each method's frames per second (FPS) in float32 data format and set the mini-batch size to 1. Given the size of the input image, we randomly generate

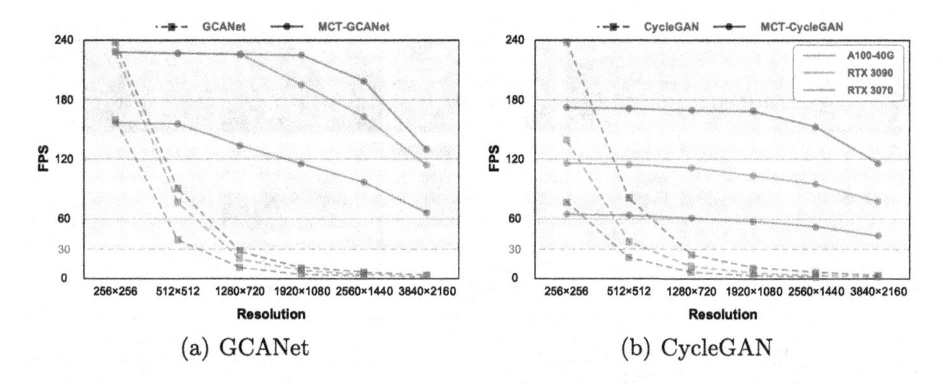

(a) GCANet (b) CycleGAN

Fig. 4. Runtime comparison of the base models and their MCT variants on GPUs. No data for base models means that they runs with an OOM at that resolution.

200 images and compute the FPS for a single experiment by recording the total time to process the 200 images. Then we repeated each experiment 10 times and took the median of the 10 results as the final result. Figure 4 illustrates the FPS of base models and their variants running on 3 models of GPUs.

If we feed a 256×256 image to the models, the MCT variants do not have any advantage over the base models since we set $H_d = W_d = 256$ in the experiment. On the other hand, the large 7×7 convolution kernels for the output layer introduce an additional 25% computational cost to the backbone network since the output channels of the MCT-CycleGAN's output layer increase. Finally, the curve slicing operation contains some operations with low computational cost but high memory access cost (*e.g.* indexing), further increasing the MCT variants' runtime. Fortunately, the computational cost of MCT's backbone network does not vary with the image size, which gives it a distinct advantage when working with HR images. When the input image size is 512×512, the MCT variants are significantly faster than the base models. Moreover, the gap between the MCT variants and the base models becomes increasingly large as the input image size grows. Taking 30 FPS as the cut-off for whether a model can run in real-time, the MCT variants can process 4K images in real-time on 3 models of GPUs. As a comparison, CycleGAN takes 40× longer to process a 4K image (116.0 FPS vs. 2.7 FPS on A100), even with an out-of-memory (OOM) on RTX 3070. The computational cost of the curve slicing operation is so low that it accounts for less than 1% of the overall computational cost for processing 4K images. Still, it introduces a high memory access cost, making the curve slicing operation limited by the GPU's memory bandwidth. Finally, MCT-GCANet processes 256×256 and 512×512 images at almost the same speed on the A100 and RTX 3090, probably due to the limitations of CPU performance and PyTorch runtime.

4.2 Qualitative Comparison

We qualitatively compare the base models with their MCT variants on four I2I translation tasks, and Fig. 5 shows some examples.

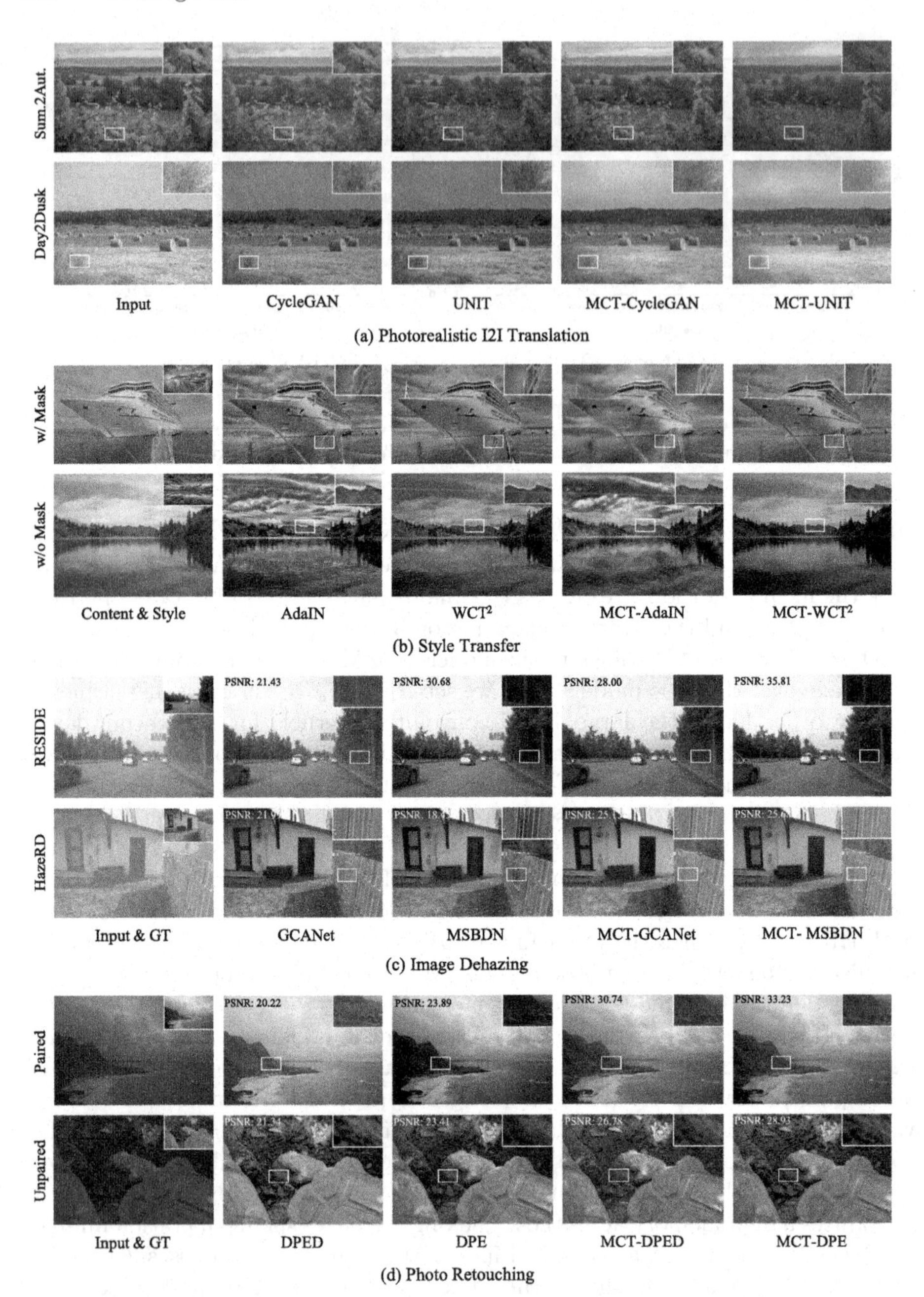

Fig. 5. Qualitative comparison of four I2I translation tasks, each consisting of examples in two experimental settings. From top to bottom are (a) photorealistic I2I translation, (b) style transfer, (c) image dehazing, and (d) photo retouching.

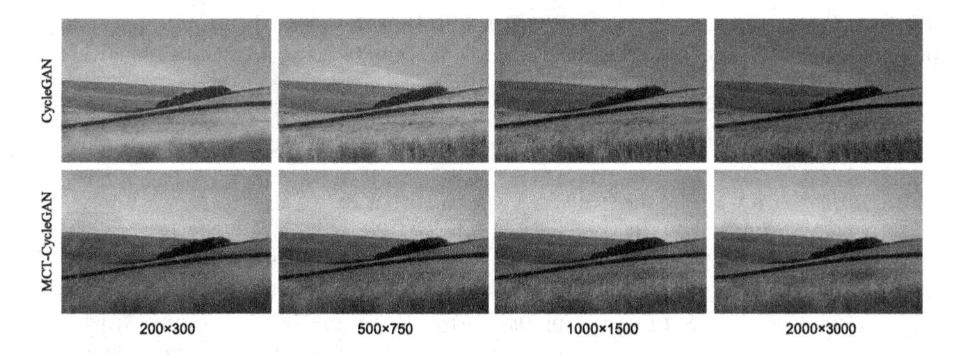

Fig. 6. Comparison of CycleGAN and MCT-CycleGAN when processing the same image in different resolution versions on day2dusk.

Almost all base models have a non-global and fixed-size receptive field. In contrast, the receptive field of MCT variants grows larger as the input image becomes larger. In the day2dusk example, the base models only change the colors of the ground and sky and do not generate sunset light. This is because the resolution of the input image is so high that the base models with limited receptive fields cannot determine the sky area. Figure 6 futher shows the results when CycleGAN and MCT-CycleGAN process the same image in different resolution versions. CycleGAN works fine on low-resolution images but cannot translate HR images well. The same problem occurs when MSBDN processes HR images in the HazeRD dataset, where significant black artifacts appear on the white railings because the receptive field of MSBDN is not large enough to capture the railings' semantic information. The DPED's receptive field is small, so it tends to adjust the image's brightness locally to normal brightness, but the image lacks contrast globally. By lowering the resolution of the downsampled images, the MCT variant of DPED has a large receptive field so that it can better capture global illumination, leading to visually more pleasing results. In short, the MCT's receptive field is dynamic, which helps to capture the HR images' semantics.

MCT's curve slicing operation allows the backbone network to focus more on region semantics than retaining the high-frequency information. This is evident in the comparison of AdaIN with its MCT variant. The original network architecture of AdaIN does not contain any skip connection, resulting in the high-frequency information that VGG-19 loses not being recovered. Therefore, even after adjusting the weights of content loss and style loss and introducing the gradient loss, AdaIN still cannot reconstruct the input image's high-frequency information. For example, the text and railing are blurred, and the texture of the mountain is lost. In contrast, its MCT variant can preserve the high-frequency information in the input image by curve slicing operation without being limited by the network architecture. However, for network architectures like GCANet, the high-frequency information flow is only shifted from skip connections to the curve slicing operation, which does not produce visible differences in the output image details. We consider that MCT is more likely to retain high-frequency information of the input image.

Table 1. Quantitative comparison (PSNR & SSIM) of the image dehazing (upper) and photo retouching (lower). FPS is measured on 4K images using a single A100-40G.

	Compared models					Base models		MCT Variants	
	MS [49]	DHN [4]	AOD [29]	GFN [50]	MGBL [67]	GCANet	MSBDN	GCANet	MSBDN
SOTS	20.31	21.02	20.27	23.52	24.50	25.09	28.56	25.71	**28.70**
	0.862	0.881	0.864	0.915	0.920	0.923	**0.966**	0.927	0.962
HazeRD	15.35	15.42	15.44	14.62	16.06	16.69	16.23	**17.19**	16.81
	0.634	0.622	0.660	0.580	0.794	0.825	0.805	0.810	**0.840**
FPS	13.6	14.8	41.6	3.0	120.8	3.1	2.2	**131.1**	35.9
	WB [19]	HDR [12]	UPE [57]	LPF [45]	LPTN [35]	DPED	DPE	DPED	DPE
Paired	–	23.15	23.24	24.48	23.86	24.11	24.14	24.73	**25.10**
	–	0.918	0.893	0.887	0.885	0.886	0.934	0.936	**0.941**
Unpaired	18.57	21.63	21.59	21.34	22.02	22.29	20.92	22.81	**23.09**
	0.701	0.885	0.884	0.866	0.879	0.884	0.854	0.902	**0.905**
FPS	0.13	14.3	15.9	2.3	37.9	2.5	10.8	**181.1**	162.4

4.3 Quantitative Comparison

We quantitatively compare the performance on image dehazing and photo retouching because the images from these two tasks have corresponding ground truth to compute PSNR and SSIM. In addition to the base models and their MCT variants, we trained some representative compared models. Table 1 shows the results. Note that HDRNet and DUPE use the open source Tensorflow [1] implementation, while the others use PyTorch framework.

For image dehazing, it can be seen that GCANet and MSBDN are powerful models, which are significantly better than the other compared models on the SOTS dataset. And the MCT variants can achieve comparable or even better performance than the base models. MSBDN performs overfitting and shows a significant performance degradation on the HazeRD dataset. In contrast, its MCT variant has a significantly higher SSIM, which indicates that more image detail is retained. For photo retouching, DPED and DPE are not state-of-the-art methods. But their MCT variants outperform the base models and compared models since the curve-based methods are in line with the image retouching process. The DPED's low SSIM is due to the small receptive field that cannot extract image contrast information effectively. Finally, DPE performs poorly in the unpaired training setting, which may be because the CycleGAN's training strategy is not suitable for it.

In terms of runtime, FCN-based methods are significantly slower than slice-based methods, even for lightweight networks such as AODNet and LPTN. Since we set $H_d = W_d = 32$ for MCT-DPED, it runs faster than MCT-DPE. Note that DUPE and HDRNet are both much slower than MCT-DPE, which is mainly due to the inefficient open-source implementations. In our experiments, they can all reach about 180 FPS using 3D LUTs.

Table 2. Quantitative comparison (FID/user study) of the photorealistic I2I translation. The percentage of user study results indicates the preferred model outputs out of 95 responses. Lower is better for FID, and higher is better for the user study.

	day2dusk	dusk2day	summer2autumn	autumn2summer
CycleGAN	89.00/32.6%	94.17/48.4%	**101.98**/47.4%	100.34/40.0%
MCT-CycleGAN	**81.67/67.4%**	**92.14/51.6%**	103.45/**52.6%**	**94.72/60.0%**
UNIT	92.14/29.5%	96.66/**55.8%**	105.15/42.1%	95.18/**50.5%**
MCT-UNIT	**84.22/70.5%**	**93.14**/44.2%	**103.43/57.9%**	**91.35**/49.5%

Table 2 shows the quantitative comparison results of the photorealistic I2I translation. We first compare the methods using Fréchet Inception Distance (FID) [17]. Unlike the usual experimental setup, all models translate HR images and then downsample the output images to 256 × 256 to compute the FID. Although these tasks are not difficult I2I translation problems, since the input images are usually high-resolution, the base models are often not performing as well as their MCT variants. We also released questionnaires to colleagues who were not involved in this work to conduct a small user study. Specifically, 19 users participated in this experiment, and we provided 5 randomly selected samples for each task. Overall, most users recognized the ability of MCT to retain high-frequency information. In particular, for the `day2dusk` task, most users felt that the output of the MCT variants was much better than the base models.

5 Discussion

Contributions. In this paper, we propose to modify the network's output layer for the I2I translation task. The network is extended to predict the output pixels for the input pixel's neighboring pixels. Since the pixels lost during downsampling are the neighboring pixels of the pixels that remain, the modified network can receive LR images to predict the mapping for process HR images. For the adversarial training, we introduced two additional training strategies to stabilize the training. Experimental results show that it can perform comparable or even better than the conventional models but translate 4K images in real-time on various photorealistic I2I translation tasks.

Limitations. MCT is a trade-off between translation capability and speed, and it cannot be applied to the more difficult I2I translation tasks. For tasks that the I2I translation process greatly changes the shape and texture of the objects in the image (*i.e.*, high-frequency information), such as `dog2cat`, MCT is helpless. In future research, we hope to improve its capabilities further to make it be applied to more I2I translation tasks.

References

1. Abadi, M., et al.: TensorFlow: a system for large-scale machine learning. In: OSDI (2016)
2. Anokhin, I., et al.: High-resolution daytime translation without domain labels. In: CVPR (2020)
3. Bychkovsky, V., Paris, S., Chan, E., Durand, F.: Learning photographic global tonal adjustment with a database of input/output image pairs. In: CVPR (2011)
4. Cai, B., Xu, X., Jia, K., Qing, C., Tao, D.: Dehazenet: an end-to-end system for single image haze removal. IEEE TIP (2016)
5. Chen, D., et al.: Gated context aggregation network for image dehazing and deraining. In: WACV (2019)
6. Chen, J., Paris, S., Durand, F.: Real-time edge-aware image processing with the bilateral grid. ACM TOG **26**(3), 103-es (2007)
7. Chen, Y.S., Wang, Y.C., Kao, M.H., Chuang, Y.Y.: Deep photo enhancer: unpaired learning for image enhancement from photographs with GANs. In: CVPR (2018)
8. Choi, Y., Choi, M., Kim, M., Ha, J.W., Kim, S., Choo, J.: StarGAN: unified generative adversarial networks for multi-domain image-to-image translation. In: CVPR (2018)
9. Dong, H., et al.: Multi-scale boosted dehazing network with dense feature fusion. In: CVPR (2020)
10. Fan, Q., Chen, D., Yuan, L., Hua, G., Yu, N., Chen, B.: Decouple learning for parameterized image operators. In: ECCV (2018)
11. Gatys, L.A., Ecker, A.S., Bethge, M.: Image style transfer using convolutional neural networks. In: CVPR (2016)
12. Gharbi, M., Chen, J., Barron, J.T., Hasinoff, S.W., Durand, F.: Deep bilateral learning for real-time image enhancement. ACM TOG **36**(4), 1–12 (2017)
13. Goodfellow, I., et al.: Generative adversarial nets. In: NeurIPS (2014)
14. Ha, D., Dai, A., Le, Q.V.: Hypernetworks. In: ICLR (2017)
15. He, K., Sun, J., Tang, X.: Guided image filtering. IEEE TPAMI **36**(6), 1397–1409 (2012)
16. He, K., Zhang, X., Ren, S., Sun, J.: Deep residual learning for image recognition. In: CVPR (2016)
17. Heusel, M., Ramsauer, H., Unterthiner, T., Nessler, B., Hochreiter, S.: GANs trained by a two time-scale update rule converge to a local Nash equilibrium. NeurIPS (2017)
18. Howard, A.G., et al.: MobileNets: efficient convolutional neural networks for mobile vision applications. arXiv preprint arXiv:1704.04861 (2017)
19. Hu, Y., He, H., Xu, C., Wang, B., Lin, S.: Exposure: a white-box photo post-processing framework. ACM TOG **37**(2), 1–17 (2018)
20. Huang, X., Belongie, S.: Arbitrary style transfer in real-time with adaptive instance normalization. In: ICCV (2017)
21. Huang, X., Liu, M.Y., Belongie, S., Kautz, J.: Multimodal unsupervised image-to-image translation. In: ECCV (2018)
22. Ignatov, A., Kobyshev, N., Timofte, R., Vanhoey, K., Van Gool, L.: DSLR-quality photos on mobile devices with deep convolutional networks. In: ICCV (2017)
23. Isola, P., Zhu, J.Y., Zhou, T., Efros, A.A.: Image-to-image translation with conditional adversarial networks. In: CVPR (2017)

24. Karaimer, H.C., Brown, M.S.: A software platform for manipulating the camera imaging pipeline. In: Leibe, B., Matas, J., Sebe, N., Welling, M. (eds.) ECCV 2016. LNCS, vol. 9905, pp. 429–444. Springer, Cham (2016). https://doi.org/10.1007/978-3-319-46448-0_26
25. Kim, H.-U., Koh, Y.J., Kim, C.-S.: Global and local enhancement networks for paired and unpaired image enhancement. In: Vedaldi, A., Bischof, H., Brox, T., Frahm, J.-M. (eds.) ECCV 2020. LNCS, vol. 12370, pp. 339–354. Springer, Cham (2020). https://doi.org/10.1007/978-3-030-58595-2_21
26. Kim, T., Cha, M., Kim, H., Lee, J.K., Kim, J.: Learning to discover cross-domain relations with generative adversarial networks. In: ICML (2017)
27. Klocek, S., Maziarka, Ł, Wołczyk, M., Tabor, J., Nowak, J., Śmieja, M.: Hypernetwork functional image representation. In: Tetko, I.V., Kůrková, V., Karpov, P., Theis, F. (eds.) ICANN 2019. LNCS, vol. 11731, pp. 496–510. Springer, Cham (2019). https://doi.org/10.1007/978-3-030-30493-5_48
28. Laffont, P.Y., Ren, Z., Tao, X., Qian, C., Hays, J.: Transient attributes for high-level understanding and editing of outdoor scenes. ACM TOG **33**(4), 1–11 (2014)
29. Li, B., Peng, X., Wang, Z., Xu, J., Feng, D.: AOD-Net: All-in-one dehazing network. In: ICCV (2017)
30. Li, B., et al.: Benchmarking single-image dehazing and beyond. IEEE TIP **28**(1), 492–505 (2018)
31. Li, C., Guo, C., Ai, Q., Zhou, S., Loy, C.C.: Flexible piecewise curves estimation for photo enhancement. arXiv preprint arXiv:2010.13412 (2020)
32. Li, M., Lin, J., Ding, Y., Liu, Z., Zhu, J.Y., Han, S.: GAN compression: Efficient architectures for interactive conditional GANs. In: CVPR (2020)
33. Li, Y., Fang, C., Yang, J., Wang, Z., Lu, X., Yang, M.H.: Universal style transfer via feature transforms. In: NeurIPS (2017)
34. Li, Y., Liu, M.Y., Li, X., Yang, M.H., Kautz, J.: A closed-form solution to photorealistic image stylization. In: ECCV (2018)
35. Liang, J., Zeng, H., Zhang, L.: High-resolution photorealistic image translation in real-time: a Laplacian pyramid translation network. In: CVPR (2021)
36. Liang-Chieh, C., Papandreou, G., Kokkinos, I., Murphy, K., Yuille, A.: Semantic image segmentation with deep convolutional nets and fully connected CRFs. In: ICLR (2015)
37. Lin, H.T., Lu, Z., Kim, S.J., Brown, M.S.: Nonuniform lattice regression for modeling the camera imaging pipeline. In: Fitzgibbon, A., Lazebnik, S., Perona, P., Sato, Y., Schmid, C. (eds.) ECCV 2012. LNCS, vol. 7572, pp. 556–568. Springer, Heidelberg (2012). https://doi.org/10.1007/978-3-642-33718-5_40
38. Lin, T.-Y., et al.: Microsoft COCO: common objects in context. In: Fleet, D., Pajdla, T., Schiele, B., Tuytelaars, T. (eds.) ECCV 2014. LNCS, vol. 8693, pp. 740–755. Springer, Cham (2014). https://doi.org/10.1007/978-3-319-10602-1_48
39. Liu, M.Y., Breuel, T., Kautz, J.: Unsupervised image-to-image translation networks. In: NeurIPS (2017)
40. Long, J., Shelhamer, E., Darrell, T.: Fully convolutional networks for semantic segmentation. In: CVPR (2015)
41. Luan, F., Paris, S., Shechtman, E., Bala, K.: Deep photo style transfer. In: CVPR (2017)
42. Ma, N., Zhang, X., Zheng, H.T., Sun, J.: Shufflenet v2: practical guidelines for efficient CNN architecture design. In: ECCV (2018)
43. Mechrez, R., Talmi, I., Zelnik-Manor, L.: The contextual loss for image transformation with non-aligned data. In: ECCV (2018)

44. Mildenhall, B., Barron, J.T., Chen, J., Sharlet, D., Ng, R., Carroll, R.: Burst denoising with kernel prediction networks. In: CVPR (2018)
45. Moran, S., Marza, P., McDonagh, S., Parisot, S., Slabaugh, G.: DeepLPF: deep local parametric filters for image enhancement. In: CVPR (2020)
46. Moran, S., McDonagh, S., Slabaugh, G.: Curl: neural curve layers for global image enhancement. In: ICPR (2021)
47. Muller, L.K.: Overparametrization of hypernetworks at fixed FLOP-count enables fast neural image enhancement. In: CVPRW (2021)
48. Park, T., Efros, A.A., Zhang, R., Zhu, J.-Y.: Contrastive learning for unpaired image-to-image translation. In: Vedaldi, A., Bischof, H., Brox, T., Frahm, J.-M. (eds.) ECCV 2020. LNCS, vol. 12354, pp. 319–345. Springer, Cham (2020). https://doi.org/10.1007/978-3-030-58545-7_19
49. Ren, W., Liu, S., Zhang, H., Pan, J., Cao, X., Yang, M.-H.: Single image dehazing via multi-scale convolutional neural networks. In: Leibe, B., Matas, J., Sebe, N., Welling, M. (eds.) ECCV 2016. LNCS, vol. 9906, pp. 154–169. Springer, Cham (2016). https://doi.org/10.1007/978-3-319-46475-6_10
50. Ren, W., et al.: Gated fusion network for single image dehazing. In: CVPR (2018)
51. Shaham, T.R., Gharbi, M., Zhang, R., Shechtman, E., Michaeli, T.: Spatially-adaptive pixelwise networks for fast image translation. In: CVPR (2021)
52. Shu, H., et al.: Co-evolutionary compression for unpaired image translation. In: ICCV (2019)
53. Simonyan, K., Zisserman, A.: Very deep convolutional networks for large-scale image recognition. In: ICLR (2015)
54. Song, Y., Qian, H., Du, X.: Starenhancer: learning real-time and style-aware image enhancement. In: ICCV (2021)
55. Song, Y., Zhu, Y., Du, X.: Model parameter learning for real-time high-resolution image enhancement. IEEE SPL **27**, 1844–1848 (2020)
56. Tomasi, C., Manduchi, R.: Bilateral filtering for gray and color images. In: ICCV (1998)
57. Wang, R., Zhang, Q., Fu, C.W., Shen, X., Zheng, W.S., Jia, J.: Underexposed photo enhancement using deep illumination estimation. In: CVPR (2019)
58. Wei, C., Wang, W., Yang, W., Liu, J.: Deep retinex decomposition for low-light enhancement. In: BMVC (2018)
59. Xia, X., et al.: Joint bilateral learning for real-time universal photorealistic style transfer. In: Vedaldi, A., Bischof, H., Brox, T., Frahm, J.-M. (eds.) ECCV 2020. LNCS, vol. 12353, pp. 327–342. Springer, Cham (2020). https://doi.org/10.1007/978-3-030-58598-3_20
60. Yang, C., Jin, M., Jia, X., Xu, Y., Chen, Y.: AdaInt: learning adaptive intervals for 3D lookup tables on real-time image enhancement. In: CVPR (2022)
61. Yin, H., Gong, Y., Qiu, G.: Side window filtering. In: CVPR (2019)
62. Yoo, J., Uh, Y., Chun, S., Kang, B., Ha, J.W.: Photorealistic style transfer via wavelet transforms. In: ICCV (2019)
63. Zeng, H., Cai, J., Li, L., Cao, Z., Zhang, L.: Learning image-adaptive 3D lookup tables for high performance photo enhancement in real-time. IEEE TPAMI (2020)
64. Zhang, K., Zuo, W., Chen, Y., Meng, D., Zhang, L.: Beyond a gaussian denoiser: Residual learning of deep CNN for image denoising. IEEE TIP (2017)
65. Zhang, R., Isola, P., Efros, A.A.: Colorful image colorization. In: Leibe, B., Matas, J., Sebe, N., Welling, M. (eds.) ECCV 2016. LNCS, vol. 9907, pp. 649–666. Springer, Cham (2016). https://doi.org/10.1007/978-3-319-46487-9_40

66. Zhang, Y., Ding, L., Sharma, G.: Hazerd: an outdoor scene dataset and benchmark for single image dehazing. In: ICIP (2017)
67. Zheng, Z., et al.: Ultra-high-definition image dehazing via multi-guided bilateral learning. In: CVPR (2021)
68. Zhu, J.Y., Park, T., Isola, P., Efros, A.A.: Unpaired image-to-image translation using cycle-consistent adversarial networks. In: ICCV (2017)

Deep Bayesian Video Frame Interpolation

Zhiyang Yu[1,2], Yu Zhang[2], Xujie Xiang[2,3], Dongqing Zou[2,4], Xijun Chen[1(✉)], and Jimmy S. Ren[2,4]

[1] Harbin Institute of Technology, Harbin, China
chenxijun@hit.edu.cn
[2] SenseTime Research and Tetras. AI, Beijing, China
zhangyulb@gmail.com
[3] Beihang University, Beijing, China
[4] Qing Yuan Research Institute, Shanghai Jiao Tong University, Shanghai, China

Abstract. We present deep Bayesian video frame interpolation, a novel approach for upsampling a low frame-rate video temporally to its higher frame-rate counterpart. Our approach learns posterior distributions of optical flows and frames to be interpolated, which is optimized via learned gradient descent for fast convergence. Each learned step is a lightweight network manipulating gradients of the log-likelihood of estimated frames and flows. Such gradients, parameterized either explicitly or implicitly, model the fidelity of current estimations when matching real image and flow distributions to explain the input observations. With this approach we show new records on 8 of 10 benchmarks, using an architecture with half the parameters of the state-of-the-art model. Code and models are publicly available at https://github.com/Oceanlib/DBVI.

Keywords: Video frame interpolation · Deep Bayesian · Image reconstruction

1 Introduction

To meet the requirement of power consumption and image quality, modern video sensors often work with limited frame aquisition rate. Video Frame Interpolation (VFI) [3,7,15,16,20,25,41] is an important computational approach that synthesizes the missing intermediate frames between consecutive frames of a low frame-rate video. When capable of performing interpolation at any continuous time step, VFI is able to temporally upsample a video to any desired resolution, creating smooth slow-motion effect at various degrees of natural movements [4,15].

Z. Yu and X. Xiang - The work is done during an internship at SenseTime Research and Tetras. AI.

Supplementary Information The online version contains supplementary material available at https://doi.org/10.1007/978-3-031-19784-0_9.

S. Avidan et al. (Eds.): ECCV 2022, LNCS 13675, pp. 144–160, 2022.
https://doi.org/10.1007/978-3-031-19784-0_9

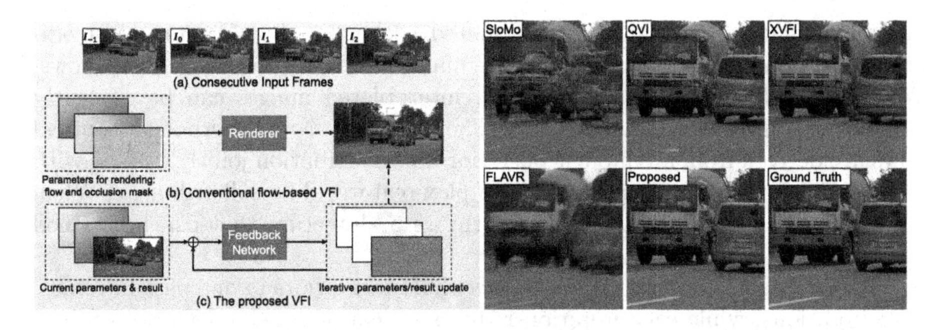

Fig. 1. Motivation of the proposed approach. Given input consecutive frames (a), most conventional approaches predict parameters for novel frame rendering in one pass (b) while ours generates learned updates in multiple iterations (c) to simplify prior modeling. This yields improved interpolation results in challenging scenarios compared with recent models like SloMo [15], QVI [41], XVFI [32], and FLAVR [16].

Solving continuous-time VFI requires representing pixel movements continuously. It was commonly achieved by fitting parametric pixel trajectory models to the optical flows extracted from sparse observed frames [6,15,41]. Intermediate pixel flows are resampled at sub-frame time steps, providing warping fields to reorganize the pixels from known input frames to synthesize the intermediate view, possibly with external geometric guidance (*e.g.* depth [3]). However, as inputs of VFI are sparse scene captures with low temporal frequency, inferring the underlying full frequency motion is highly ill-posed. Several works tackle the ill-poseness by learning a function mapping the low frame-rate input to its high frame-rate counterpart, fitted from raw data (*e.g.* [7,11,16]). Yet without continuous representation, they lose the flexibility for continuous-time frame interpolation.

If viewed from optimization perspective, continuous-time VFI methods predict the parameters (*e.g.* intermediate scene geometries like flows and occlusion) of an inference model (*e.g.* image renderer with scene geometries as rendering parameters) to best explain the measurements (*e.g.* input frames). Since there are far more predictive parameters than measurements, effective priors are required. In most contemporary works [3,15,27,41], parameters are predicted from sparse measurements in one pass, having few chance to feedback and improve rendering errors (as in Fig. 1(b)). Consequently, burden is left to the network to encode strong priors into the parameters (*e.g.* joint space of parameters of natural scene geometry), which is prohibitively difficult even for large architecture.

In this work we propose deep Bayesian video frame interpolation, a continuous-time VFI framework with iterative learned task priors. Our framework learns posterior distributions of the intermediate optical flows and frames to be interpolated, optimized via learned gradient descent [1,2] for fast convergence within a few iterations. Each iteration evaluates the gradients of the log-likelihood modelling fidelity of current estimations of image and flows. Such first-order gradients are altered by a network encoding learned prior of higher-order information, such that after updated with altered gradients, intermediate estimations fall effectively onto the manifold of natural images and flows. We show log-likelihood gradients *w.r.t.* interpolated images can be explicitly defined with intuitive interpretations. The gradients *w.r.t.* interpolated optical flows are derived from an implicit posterior flow distribution jointly trained with other components to account for the complex real-world flow distribution. When unfolded, the learned gradient steps form an architecture that can be trained end-to-end.

Our approach amortizes learning the whole task priors in one pass into multiple iterations, while each iteration reduces to learning a simpler prior on how to manipulate gradients towards smaller rendering errors in a principled optimization framework. Evaluated with extensive experiments, this method sets new records on 8 of 10 public benchmarks, outperforms state-of-the-art VFI model [16] with half parameter size.

Our work aims to introduce 3 folds of contributions: 1) a principled contiuous-VFI framework, which produces high-quality interpolation results with less parameters; 2) a deep parameterization of gradients of arbitary posterior distributions that can be end-to-end optimized as a network component; 3) a unified and fair benchmarking of VFI models on standard datasets.

2 Related Works

Video frame interpolation has became an active research topic in computer vision since more than one decade earlier [22,39,43]. Though, the performance of VFI methods are effectively boosted in recent year [15], with emergence of deep learning architectures and high-quality training data. Since then, various methods were proposed. Among them, a large body of works treat VFI as a novel view rendering problem by warping the observed video frames.

Flow-based rendering predicts pixel flows of intermediate time steps by analyzing the optical flows extracted from input video. As input frames are sparse, explicit flow priors are placed to yield continuous sampling of flows at arbitary time. Early work [15] imposes linear flow prior, which was further extended to quadratic [41] or even cubic [6] trajectory models to accurately represent local movements of pixels. A challenge here is that to get rid of occlusion and confliction in flow-based warping, flows starting from the intermediate frame to input frames need to be known, while the resampled flows are in reverse order. Heuristic flow reversing was thus proposed, by interpolating at occluded pixels with various local kernels [4,15,32,41] possibly guided by scene geometry like depth [3]. Instead of reversing flows, [24] addresses occlusion-ware warping with forward splatting, using color constancy as guidance. As an important component, learnable flow refinement modules were applied to improve the heuristically interpolated flows. To name a few, they include bilateral cost volumes [26,27], spatial pyramidal flow upsampling [32], temporal pyramidal refinement [6] and learnable median filtering [41].

Kernel-based rendering learns linear kernels to aggregate input pixels within a local spatiotemporal window to produce interpolation results. Such kernels were implemented as separable convolutions [25], deformable convolutions [5,20, 30,31] and trilinear sampling kernels [21]. However, most such methods cannot address VFI at arbitary time steps as learned kernels are time-specific.

Instead of explicit frame rendering, several works learn direct mapping from low frame-rate video to its higher frame-rate counterpart from data. It makes VFI benefit from the strong pattern fitting ability of deep networks, implemented with channel attention [7] or 3D convolution [16]. Again, they do not address arbitary-time VFI for the descrete representations of neural network, and have risk of over-fitting to the low-level statistics of a dataset.

Our method starts from a different perspective, treating directly the inverse problem nature of VFI and solving it with gradient-based Bayesian optimization [1,2]. Though actively studied for low-level tasks like denoising [19,36], demosaicking [18] and 3D view synthesis [9], it is by the first time explored for VFI to the best of our knowledge. The method can be viewed as combining existing directions but in different way — it learns frame interpolation by fitting the data, while at the same time evaluates current rendering errors on the observed frames and flows and feedback such errors to boost next iteration. This makes it robust to overfitting, as each of its step is to fix local estimation errors instead of encoding all the appearance and motion dynamics in a single pass.

3 The Proposed Method

3.1 Background and Notations

Given consecutive frames $\mathbf{I}_{-1}$, $\mathbf{I}_0$, $\mathbf{I}_1$, $\mathbf{I}_2$, where $\mathbf{I}_i \in [0,1]^{H\times W\times 3}$, $\forall i \in \{-1,0,1,2\}$, we aim at interpolating intermediate frame $\mathbf{I}_t$ at any arbitrary time $t \in (0,1)$. Following prior works [15,32,41], we start by extracting optical flows from input frames, then fit a continuous motion model that can be resampled at time t to get flows $\mathbf{F}_{0\to t}, \mathbf{F}_{1\to t} \in \mathbb{R}^{H\times W\times 2}$. To render the target frame $\mathbf{I}_t$, such forward flows were heuristically reversed and then refined in previous works, yielding backward-stage flows $\mathbf{F}_{t\to 0}$, $\mathbf{F}_{t\to 1}$ so as to make rendering possible via warping:

$$\mathbf{I}_{\mathrm{t}}^* = (1-t)\odot \mathbf{M}\odot \phi_{\mathcal{B}}(\mathbf{I}_0,\mathbf{F}_{t\to 0}) + t\odot (1-\mathbf{M})\odot \phi_{\mathcal{B}}(\mathbf{I}_1,\mathbf{F}_{t\to 1}), \tag{1}$$

where $\mathbf{M}$ is the blending mask, $\odot$ is the Hadamard product, and ϕ_B stands for the backward warping function [14]. However, reversing forward flows and inferring occlusions at unknown time step are both challenging, while the rendering-driven scheme (1) poses strong assumption on the quality of them.

Differing from previous works, we solve VFI with a deep Bayesian approach, optimizing for results that can best explain the occurrence of input data regularized with priors. Specfically we learn directly the posterior distributions of $\mathbf{I}_t$, conditioned on the observed frames and flows:

$$\mathbf{I}_{\mathrm{t}}^* = \arg\max_{\mathbf{I}_t} P(\mathbf{I}_t|\mathbf{I}_0,\mathbf{I}_1,\mathbf{F}_{0\to t},\mathbf{F}_{1\to t}). \tag{2}$$

3.2 Deep Bayesian Video Frame Interpolation

To derive a tractable solution of (2), we make the following simplification

$$P(\mathbf{I}_t|\mathbf{I}_0,\mathbf{I}_1,\mathbf{F}_{0\to t},\mathbf{F}_{1\to t}) = \prod_{i\in\{0,1\}} P(\mathbf{I}_t|\mathbf{I}_i,\mathbf{F}_{i\to t}), \tag{3}$$

where $\mathcal{I}_t$ is assumed that can be independently explained by image and flows at either frame 0 or 1. Since the forward flows $\mathbf{F}_{0\to t}$ and $\mathbf{F}_{1\to t}$ are estimated from low frame-rate video frames, errors inevitably exist. To account for it, we also incorporate the refined task-specific forward flows as latent variables:

$$\begin{aligned} P(\mathbf{I}_t|\mathbf{I}_i,\mathbf{F}_{i\to t}) &= \int_{\Delta\mathbf{F}_{i\to t}} P(\mathbf{I}_t|\mathbf{I}_i,\mathbf{F}_{i\to t},\Delta\mathbf{F}_{i\to t})P(\Delta\mathbf{F}_{i\to t}|\mathbf{I}_i,\mathbf{F}_{i\to t})d\Delta\mathbf{F}_{i\to t} \\ &\approx P(\mathbf{I}_t|\mathbf{I}_i,\mathbf{F}_{i\to t},\Delta\hat{\mathbf{F}}_{i\to t})P(\Delta\hat{\mathbf{F}}_{i\to t}|\mathbf{I}_i,\mathbf{F}_{i\to t}), \end{aligned} \tag{4}$$

where refined flows are modelled by the residual flows $\Delta\mathbf{F}_{i\to t}$. The approximation step replaces the integral, which enumerates all possible $\Delta\mathbf{F}_{i\to t}$s to be calculated, with the mode of the conditional $\Delta\hat{\mathbf{F}}_{i\to t} = \arg\max_{\Delta\mathbf{F}_{i\to t}} P(\Delta\mathbf{F}_{i\to t}|\mathbf{I}_i,\mathbf{F}_{i\to t})$, for tractability. Integrating (4) into (3) and taking the negative logarithm, we arrive at the following minimization problem

$$\min_{\mathbf{I}_t,\{\Delta\hat{\mathbf{F}}_{i\to t}\}} -\sum_{i\in\{0,1\}} \left(\log P(\mathbf{I}_t|\mathbf{I}_i,\mathbf{F}_{i\to t},\Delta\hat{\mathbf{F}}_{i\to t}) + \log P(\Delta\hat{\mathbf{F}}_{i\to t}|\mathbf{I}_i,\mathbf{F}_{i\to t})\right). \tag{5}$$

The non-linear problem (5) can be optimized via iterative gradient descent:

$$\mathbf{I}_t^{(k+1)} = \mathbf{I}_t^{(k)} - \lambda_{\mathbf{I}}\frac{\partial L}{\partial \mathbf{I}_t}, \quad \Delta\hat{\mathbf{F}}_{i\to t}^{(k+1)} = \Delta\hat{\mathbf{F}}_{i\to t}^{(k)} - \lambda_{\mathbf{F}}\frac{\partial L}{\partial \Delta\hat{\mathbf{F}}_{i\to t}}, \tag{6}$$

where $L(\mathbf{I}_t,\{\Delta\hat{\mathbf{F}}_{i\to t}\};\{\mathbf{I}_i\},\{\mathbf{F}_{i\to t}\})$ is the objective of (5) ($\{\cdot\}$ represents the collection of variables of different is for short), $\lambda_{\mathbf{I}}$ and $\lambda_{\mathbf{F}}$ are step sizes. For $\frac{\partial L}{\partial \mathbf{I}_t}$,

$$\begin{aligned} \frac{\partial L}{\partial \mathbf{I}_t} &= -\sum_{i\in\{0,1\}} \frac{\partial \log P(\mathbf{I}_t|\mathbf{I}_i,\mathbf{F}_{i\to t},\Delta\hat{\mathbf{F}}_{i\to t})}{\partial \mathbf{I}_t} \\ &= -\sum_{i\in\{0,1\}} \left(\frac{\partial \log P(\mathbf{I}_i|\mathbf{I}_t,\mathbf{F}_{i\to t},\Delta\hat{\mathbf{F}}_{i\to t})}{\partial \mathbf{I}_t} + \frac{\partial \log P(\mathbf{I}_t|\mathbf{F}_{i\to t},\Delta\hat{\mathbf{F}}_{i\to t})}{\partial \mathbf{I}_t}\right), \end{aligned} \tag{7}$$

where the first term represents gradients of the *log-likelihood* that input frame $\mathbf{I}_i$ occurs given the estimations $\mathbf{I}_t$ and $\Delta\hat{\mathbf{F}}_{i\to t}$, while the second term are gradients of the *conditional prior* of the interpolated frames. Iteratively evaluating gradients and performing updates can take large numbers of steps to converge. We leverage recent advances on learned gradient descent [1], performing updates with learned gradients conditioned on the classical ones:

$$\mathbf{I}_t^{(k+1)} = \mathbf{I}_t^{(k)} + \mathcal{G}_{\mathbf{I}}\left(\left\{\frac{\partial \log P(\mathbf{I}_i|\mathbf{I}_t^{(k)},\mathbf{F}_{i\to t},\Delta\hat{\mathbf{F}}_{i\to t}^{(k)})}{\partial \mathbf{I}_t}\right\},\mathbf{I}_t^{(k)},\{\mathbf{F}_{i\to t}\},\{\Delta\hat{\mathbf{F}}_{i\to t}^{(k)}\}\right). \tag{8}$$

Note that conditional priors are implicitly folded into the learned gradient network $\mathcal{G}_{\mathbf{I}}$. Conditioning on the likelihood gradients encodes the local loss landscape into iterative minimization, where gradient directions and magnitudes signify the gap of current estimation to explain the given observations.

The partial gradients $\frac{\partial L}{\partial \Delta\hat{\mathbf{F}}_{i\to t}}$ in (5) are comprised exactly of the conditional likelihood and prior terms. Therefore, its learned update rule can be defined:

$$\Delta\hat{\mathbf{F}}_{i\to t}^{(k+1)} = \Delta\hat{\mathbf{F}}_{i\to t}^{(k)} + \mathcal{G}_{\mathbf{F}}\left(\frac{\partial \log P(\mathbf{I}_t^{(k)}|\mathbf{I}_i, \mathbf{F}_{i\to t}, \Delta\hat{\mathbf{F}}_{i\to t}^{(k)})}{\partial \Delta\hat{\mathbf{F}}_{i\to t}}, \Delta\hat{\mathbf{F}}_{i\to t}^{(k)}, \mathbf{I}_i, \mathbf{F}_{i\to t}\right). \quad (9)$$

With learned gradients, we unroll a small number of K update steps. The overall optimization procedure can be unfolded as a specific architecture, trained on a dataset with paired low frame-rate and high frame-rate video frames.

3.3 Formulating the Gradients

Each iteration of (8) and (9) requires evaluating specific log-likelihood gradients. We show that gradients corresponding to the incremental image (8) has explicit and interpretable representations under simple distributions, resembling warped reconstruction errors with current flow estimations. For the flow log-likelihood gradients (9), we propose an implicit representation via deep parameterization to make them tractable and computationally efficient.

Explicit Image Log-Likelihood Gradients. Image likelihood in (8) models the occurrence of input video frame $\mathbf{I}_i$, given intermediate estimations $\mathbf{I}_t^{(k)}$ and $\Delta\hat{\mathbf{F}}_{i\to t}^{(k)}$. We define this conditional likelihood as a pixel-independent Gaussian distribution, centered at the warped version of $\mathbf{I}_t^{(k)}$:

$$\mathbf{I}_i \sim \mathcal{N}\left(\phi_{\mathcal{B}}(\mathbf{I}_t^{(k)}, \mathbf{F}_{i\to t} + \Delta\hat{\mathbf{F}}_{i\to t}^{(k)}), \sigma_{\mathbf{I}}^2\right), \quad (10)$$

where $\phi_{\mathcal{B}}$ is the bilinear warping function, and $\sigma_{\mathbf{I}}$ is set to a constant. If we write $\hat{\mathbf{W}}_{i\to t}^{(k)}$ as the matrix absorbing bilinear warping weights from flows $\mathbf{F}_{i\to t} + \Delta\hat{\mathbf{F}}_{i\to t}^{(k)}$, then the warping can be represented by matrix multiplication $\hat{\mathbf{W}}_{i\to t}^{(k)}\mathbf{I}_t^{(k)}$. Gradients of this Gaussian likelihood are then $(\hat{\mathbf{W}}_{i\to t}^{(k)})^{\mathrm{T}}(\hat{\mathbf{W}}_{i\to t}^{(k)}\mathbf{I}_t^{(k)} - \mathbf{I}_i)$. It can be intuitively interpreted as evaluating the errors of explaining $\mathbf{I}_i$ when $\mathbf{I}_t^{(k)}$ is warped to time i, while gathering such errors at time t by reverse splatting.

Implicit Flow Log-Likelihood Gradients. The flow-side likelihood from (9) aims to explain the occurrence of $\mathbf{I}_t^{(k)}$ given $\mathbf{I}_i$ and estimated flows $\hat{\mathbf{F}}_{i\to t}^{(k)}$. Unlike image likelihood, defining it with simple distribution by image warping is tricky, as forward flows $\hat{\mathbf{F}}_{i\to t}^{(k)}$ do not convey sufficient scene geometry (*e.g.* occlusion relaionship) for warping. Instead of heuristic flow reversal [3,15,32,41], we model such distribution implicitly with learnable parameterization. By the Bayesian formula relating likelihood, posterior and prior, the likelihood gradients can be implicitly represented with the gradients derived from posterior and prior:

$$\frac{\partial \log P(\mathbf{I}_t^{(k)}|\cdot, \Delta\hat{\mathbf{F}}_{i\to t}^{(k)})}{\partial \Delta\hat{\mathbf{F}}_{i\to t}} = \frac{\partial \log P(\Delta\hat{\mathbf{F}}_{i\to t}^{(k)}|\mathbf{I}_t^{(k)}, \cdot)}{\partial \Delta\hat{\mathbf{F}}_{i\to t}} - \frac{\partial \log P(\Delta\hat{\mathbf{F}}_{i\to t}^{(k)}|\cdot)}{\partial \Delta\hat{\mathbf{F}}_{i\to t}}, \tag{11}$$

where we omit $\mathbf{I}_i$ and $\mathbf{F}_{i\to t}$ in the conditions to ease presentation. As prior is folded into gradient network, we propose to evaluate the log-posterior gradients (the first term of the righthand of (11)) instead. This resembles flow estimation from a frame pair $\mathbf{I}_t^{(k)}$ and $\mathbf{I}_i$, where in the existing literature, the flow posterior distribution is mostly a Laplacian of fixed variance (with ℓ_1 loss as negative log-density) [8,35,37]. However, to ensure such simple distribution works, it requires presence of ground-truth optical flows, which are absent in our task.

To learn flow posterior gradients without explicitly specifying its underlying distribution $P(\Delta\hat{\mathbf{F}}_{i\to t}^{(k)}|\mathbf{I}_t^{(k)}, \cdot)$, we propose a deep reparameterization approach by leveraging *normalizing-flow-based invertible layers* [17]. We assume an invertible, injective and twice-differentiable function $f : \mathbb{R}^m \mapsto \mathbb{R}^m$ transforms the flow maps $\Delta\hat{\mathbf{F}}_{i\to t}^{(k)}$ to a multi-dimensional latent representation $\mathbf{X}_{i\to t}^{(k)} = f(\Delta\hat{\mathbf{F}}_{i\to t}^{(k)})$. By the change of variables and chain rule of gradients,

$$\frac{\partial \log P(\Delta\hat{\mathbf{F}}_{i\to t}^{(k)})|\mathbf{I}_t^{(k)}, \cdot)}{\partial \Delta\hat{\mathbf{F}}_{i\to t}} = \frac{\partial \log \left|\det\left(\frac{\partial \mathbf{X}_{i\to t}}{\partial \Delta\hat{\mathbf{F}}_{i\to t}}\right)\right|}{\partial \Delta\hat{\mathbf{F}}_{i\to t}} + \frac{\partial \mathbf{X}_{i\to t}}{\partial \Delta\hat{\mathbf{F}}_{i\to t}} \frac{\partial \log P(\mathbf{X}_{i\to t}^{(k)}|\mathbf{I}_t^{(k)}, \cdot)}{\partial \mathbf{X}_{i\to t}}. \tag{12}$$

As $\mathbf{X}_{i\to t}$ resides in deep latent space, we assume it follows multi-variate Gaussian $P(\mathbf{X}_{i\to t}^{(k)}|\mathbf{I}_t^{(k)}, \cdot) = \mathcal{N}\left(\mu(\mathbf{I}_t^{(k)}, \cdot), \mathbf{\Sigma}(\mathbf{I}_t^{(k)}, \cdot))\right)$, whose mean and covariance are predictive as intermediate network outputs. Gradients of Gaussian log-likelihood *w.r.t.* $\mathbf{X}_{i\to t}$ can be efficiently evaluated:

$$\frac{\partial \log P(\mathbf{X}_{i\to t}^{(k)}|\mathbf{I}_t^{(k)}, \cdot)}{\partial \mathbf{X}_{i\to t}} = \mathbf{\Sigma}^{-1}(\mathbf{I}_t^{(k)}, \cdot)\left(\mathbf{X}_{i\to t}^{(k)} - \mu(\mathbf{I}_t^{(k)}, \cdot)\right). \tag{13}$$

To ensure PSD of covariance matrix and avoid inversion, we let $\mathbf{\Sigma} = \mathbf{S}\mathbf{S}^{\mathrm{T}}$ and parameterize $\mathbf{S}^{-1}$ instead of $\mathbf{\Sigma}$. Other gradient terms of (12) can be computed by once and twice differentiating though the layers of the function f[1].

The modeling (12) can be thought of transforming the flow predictions $\Delta\hat{\mathbf{F}}_{i\to t}$ into a deep embedded space such that the fidelity of flows are measureable with learned metric defined by a simple distribution. The first and second gradient terms of (12) serve as volume-preserving scaling and biases of the deep gradients, ensuring real gradients *w.r.t.* $\Delta\hat{\mathbf{F}}_{i\to t}$ are computed no matter what the transformation is. The transformation is not explicitly defined, but implicitly optimized towards explaining the training data. By stacking many invertible layers to form f, we theoretically can model the gradients of arbitarily complex posterior distributions of optical flows to match the real distribution.

One thing to take care of in practice is that the forward flows $\Delta\hat{\mathbf{F}}_{i\to t}$ have only two channel dimensions, which is too restricted to learn expressive deep latent

[1] Please refer to the supplementary material for more discussions on implementation.

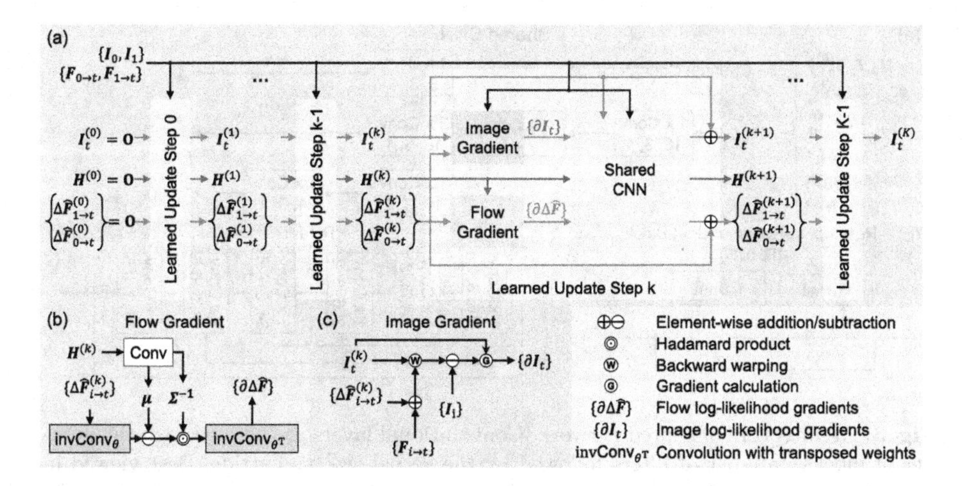

Fig. 2. Pipeline of our approach, best viewed in color. Please see text for details.

representations (since invertible layers preserve dimensionality). We therefore lift $\varDelta\hat{\mathbf{F}}_{i\rightarrow t}$ to an augmented M-dimensional space by adding more channels, while only the first two channels are used for flow modeling. Such additional channels are initialized with zeros at beginning, while learned during iterations.

3.4 Implementation

Architecture. The per-iteration networks $\mathcal{G}_{\mathbf{I}}$ and $\mathcal{G}_{\mathbf{F}}$ have many shared inputs (*e.g.* $\{\mathbf{F}_{i\rightarrow t}\}$, $\{\varDelta\hat{\mathbf{F}}^{(k)}_{i\rightarrow t}\}$ and $\{\mathbf{I}^{(k)}_t\}$). As shown in Fig. 2, we therefore use a single network to reduce redundant calculation and explore the complementary cues. At each iteration, this shared network generates learned updates of image and flows, which are element-wise added to previous estimations to get current results $\mathbf{I}^{(k)}_t$ and $\{\varDelta\hat{\mathbf{F}}^{(k)}_{i\rightarrow t}\}$. In addition to image and flow updates, the shared network also outputs additional intermediate features $\mathbf{H}^{(k)}$, which summarizes historical information and are passed across iterations, as suggested in [2,9].

In each iteration, $\mathbf{I}^{(k)}_t$, $\{\mathbf{I}_i\}$, $\{\varDelta\hat{\mathbf{F}}^{(k)}_{i\rightarrow t}\}$, and $\{\mathbf{F}_{i\rightarrow t}\}$ are processed by the Image Gradient Module to get the gradients $\partial\mathcal{I}_t$, as described in Sect. 3.3. For flow gradients, first note that flows are lifted to high-dimensional space ($\varDelta\hat{\mathbf{F}}_{i\rightarrow t}$ is set 16-dimensional in our implementation). The Flow Gradient Module then predicts the mean and inverse of co-variance (or precision) of 16-dimensional Gaussian from the last hidden features $\mathbf{H}^{(k)}$. For efficiency we assume the precision matrix is diagonal, using exponential layer on the top to ensure positiveness of precision. Remember that $\{\varDelta\hat{\mathbf{F}}^{(k)}_{i\rightarrow t}\}$ go through invertible transformation for gradient evaluation. In practice, we model it with a single invertible convolution layer [17]. Though stacking more layers is advantageous for performance in principle, we find it already effective in experiments (Sect. 4) while highly efficient.

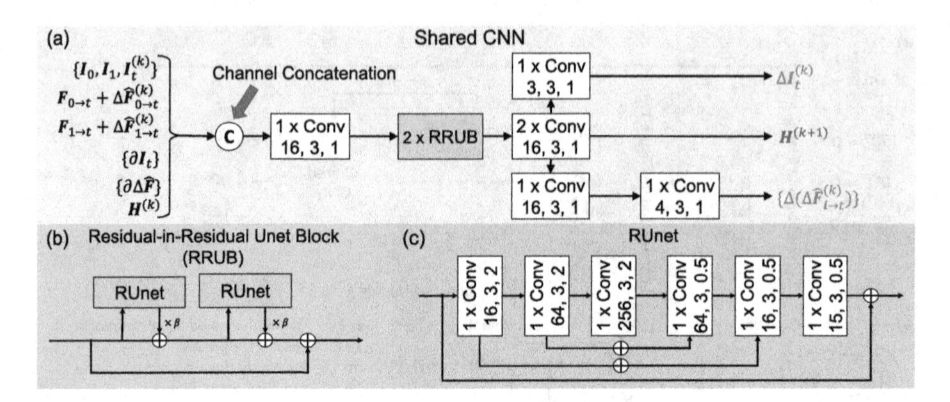

Fig. 3. Architecture of shared network. Convolutional layers are labeled with the number of blocks, number of filters followed by the kernel size and stride. Best viewed in color. Please see text for details.

As shown in Fig. 2(b), differentiation through this transformation is simply a matrix operation with the transpose of the convolution weights.

After calculated, image and flow gradients are concatenated with other conditional inputs and fed into the shared CNN, whose configuration is shown in Fig. 3. After an input-transform convolution, two Residual-in-Residual UNet Blocks (RRUB) are adopted. RRUB is a simplified version of the RRDB block [38], where its dense block [13] is replaced with a light-weight UNet [29]. The UNet has 3 scales, with skip connections as feature addition between encoder and decoder. Group normalization [40] is used with group size 4, accompanied with GELU activation [12], for all the convolutional layers in the shared network.

Initialization. At the beginning, $\mathbf{I}_t^{(0)}$, $\mathbf{H}^{(0)}$ and the lifted $\Delta\hat{\mathbf{F}}_{i\to t}$ are initialized with zeros. The forward flows $\{\mathbf{F}_{i\to t}\}$ are estimated based on the quadratic motion model proposed in [41], where optical flows are computed by RAFT [37].

Loss Function. We use ℓ_1 loss only to jointly train all the iterations:

$$\mathcal{L} = \sum_{k=1}^{K} \alpha_k \left\| \mathbf{I}_t^{gt} - \mathbf{I}_t^{(k)} \right\|_1, \tag{14}$$

where α_k is the weight for kth iteration. Weights are set empirically, increasing monotonously with the iterations (please see Sect. 4.1 for detailed parameters).

4 Experiments

4.1 Experimental Setup

Datasets. We evaluate extensively on 10 benchmarks: GoPro [23], Adobe240 [34], X4K1000FPS [32], Vimeo-90K [42], UCF101 [33], DAVIS [28] and the *easy*,

Table 1. Quantitative results on 8× interpolation in terms of PSNR/SSIM on GoPro, X4K1000FPS and Adobe240 datasets. Best results highlighted in red.

	GoPro	X4K1000FPS	Adobe240	Param (M)	TFLOPS
SloMo [15]	29.71/0.924	25.07/0.795	29.63/0.927	39.61	0.624
QVI [41]	30.52/0.941	28.06/0.855	31.41/0.955	29.23	1.075
DAIN [3]	29.53/0.920	27.28/0.835	30.53/0.939	24.03	5.785
EDSC [5]	29.20/0.916	25.30/0.811	29.87/0.931	8.95	0.260
FLAVR [16]	31.10/0.942	24.50/0.791	30.92/0.938	42.06	3.793
XVFI [32]	29.80/0.925	28.42/0.881	29.74/0.930	5.61	0.676
Ours	31.73/0.947	31.10/0.928	33.28/0.965	15.18	1.284

Table 2. Quantitative results following benchmarking protocol of the X4K1000FPS dataset [32] for 8× interpolation. Best results highlighted in red.

	AdaCoF [20]	FeFlow [10]	DAIN [3]	SloMo [15]	FLAVR [16]	XVFI [32]	QVI [41]	Ours
PSNR	25.81	25.16	27.52	27.77	27.92	30.12	29.96	32.89
SSIM	0.772	0.783	0.821	0.849	0.853	0.870	0.892	0.939

medium, *hard* and *extreme* subsets from SNU-FILM [7]. Among them, GoPro, X4K1000FPS and Adobe240 are used to evaluate 8× interpolation, while others are 2×.

Unified Evaluation Protocols with Guaranteed Reproducibility[2]. For 8× interpolation, we follow the same settings in [16] and use the official train/test split of GoPro to perform evaluations on this dataset. For X4K1000FPS, we use both the model pretrained on GoPro and retrain it on X4K1000FPS to fairly compare with existing methods evaluated with these two different protocols. For Adobe, there is no standard test dataset. We thus follow [32] and randomly extract non-overlapping clips containing complex motions. Note our sampled test set is 3× larger than that of [32]. Downsampling as in [41] is not applied.

For 2× interpolation, our model is trained on the train split of Vimeo-90K. As various methods including ours require 4 consecutive frames as input, we use the *septuplet* subset for evaluation. The pre-trained model is tested on the test dataset of Vimeo-90K, UCF101 and DAVIS, which is the same setting of [16]. We follow [5,7,27] to report the results on the 4 subsets from SUN-FILM. The Peak Signal-to-Noise Ratio (PSNR) and Structural Similarity Index Measure (SSIM) are used for quantitative evaluation, averaged across all the interpolated frames.

Note that evaluation protocols of previous VFI models for multiple interpolation are not totally unified. For example, some were trained with private training data [6,15,41], or tested on different subsets selected from a particular dataset [16,32]. As a contribution of this work, we present here a more unified benchmarking. Specifically, we retrained SloMo [15], QVI [41], DAIN [3],

[2] In the supplementary material we provide a more detailed description.

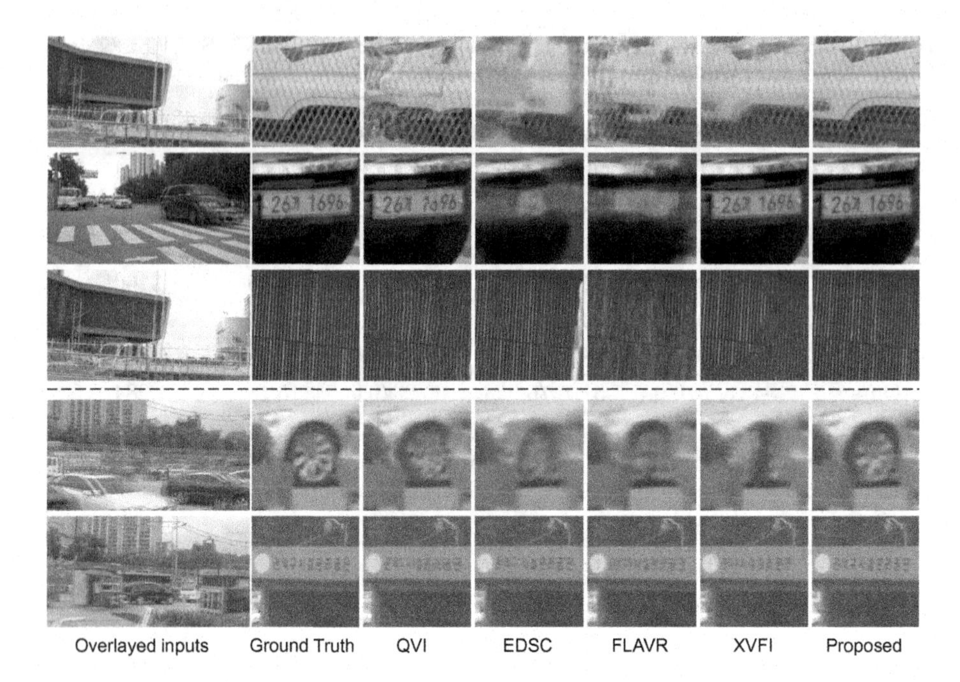

Fig. 4. Visual comparisons on X4K1000FPS (top 3 rows) and GoPro (bottom 2 rows) datasets, where each row shows an example. We overlay the nearest 2 input frames to illustrate the input motion. More results can be found in the supplementary material.

CAIN [7], FLAVR [16] and XVFI [32], which have open-source code, on different datasets with reproducibility guarantee. We reuse the results of several works (*e.g.* AdaCoF [20] and FeFlow [10]) in case that their evaluation protocols tightly follow the standard (and ours). However, for methods without training code or image results released, we either use their pretrained model (*e.g.* EDSC [38] and ABME [27]) or the reported results from related paper with the same evaluation protocol with ours (*e.g.* Softsplat [24] results are copied from [31]).

Training Details. For 8$\times$ MFI, we use $K = 4$ iterations, while the per-iteration weight α_k is $\{0.2, 0.4, 1.0, 1.0\}$. We train the network for 200 epochs with Adam optimizer and batch size 16. The learning rate is initialized as 5×10^{-4} and decreased by a factor of 0.4 at the 80th, 120th and 160th epochs. Data augmentations like random cropping, flipping, and color jittering are adopted. For 2$\times$ VFI, we use $K = 6$ iterations with all corresponding α_k set to 1.0. The batch size is 32, and learning rate is decreased at the 110th and 135th epochs, respectively.

4.2 Comparisons with State-of-the-Art Models

8$\times$ Interpolation. We report quantitative results on 8$\times$ VFI on GoPro, Adobe240 and X4K1000FPS in Table 1, as well as TFLOPS and the model size

Table 3. Quantitative results on 2× interpolation in terms of PSNR/SSIM with model size reported in number of parameters. Best results highlighted in red.

	Vimeo-90K (septulets)	UCF101	DAVIS	SNU-FILM				Param(M)
				Easy	Medium	Hard	Extreme	
SloMo [15]	34.43/0.969	32.45/0.967	26.10/0.862	36.12/0.984	33.44/0.972	29.17/0.928	24.14/0.843	39.61
QVI [41]	34.98/0.970	32.87/0.966	27.20/0.874	39.53/0.990	36.43/0.983	31.07/0.947	24.96/0.856	29.23
CAIN [7]	34.69/0.969	32.40/0.966	27.12/0.872	39.33/0.989	35.34/0.977	30.15/0.933	24.88/0.855	42.78
ABME [27]	35.67/0.972	32.81/0.969	27.00/0.868	39.59/0.990	35.77/0.977	30.58/0.936	25.42/0.864	18.1
EDSC [5]	34.52/0.967	32.67/0.968	26.28/0.849	40.01/0.990	35.37/0.978	29.59/0.926	24.39/0.843	8.95
XVFI [32]	35.21/0.970	32.68/0.968	26.89/0.868	39.21/0.989	34.96/0.977	29.43/0.928	24.02/0.841	5.61
DAIN [3]	33.57/0.964	31.65/0.963	26.61/0.867	38.53/0.988	34.34/0.974	29.50/0.930	24.54/0.851	24.03
BMBC [26]	34.76/0.965	32.61/0.955	26.42/0.868	39.90/0.991	35.34/0.978	29.34/0.927	23.65/0.837	11.01
Softsplat [24]	35.76/0.972	32.89/0.970	27.42/0.878	–	–	–	–	12.46
VFIT-B [31]	36.96/0.978	33.44/0.971	28.09/0.888	–	–	–	–	28.09
FLAVR [16]	36.30/0.975	33.33/0.971	27.44/0.873	40.44/0.991	36.37/0.981	30.87/0.942	25.18/0.862	42.06
Ours	36.17/0.976	33.01/0.970	28.61/0.905	40.46/0.991	36.95/0.985	31.68/0.953	25.90/0.876	21.69

in terms of number of parameters of different methods. Following [16], all the models for evaluation are trained on the GoPro dataset except EDSC, for which we use the officially pretrained model due to lacking the training code. The proposed approach achieves the best results across all the datasets, achieving at least 0.63 dB improvement with comparable FLOPS and greatly reduced size of model (15M parameters compared with 42M of FLAVR and 29M of QVI). For those with less parameters than ours, we achieve nearly 2 dB improvements at least.

8× Interpolation Trained on X4K1000FPS. We also make fair comparisons on the recently proposed X4K1000FPS dataset [32], by retraining our model on it and comparing with the benchmarking results from [32], which is summarized in Table 2. This dataset is more challenging due to the large motion universally appeared. Our approach achieves 2.7 dB improvement than previous leading result. Figure 4 shows visual comparisons of different methods. Our approach shows advantage in challenging cases such as occlusion and high frequency textures, where many existing methods would fail. Due to space limit, please refer to our supplementary material for more results.

Single-Frame (2×) Interpolation. Though our method is for continuous-time VFI, we evaluate it on additional 7 datasets from the single frame interpolation literature for completeness. Results are shown in Table 3. Following the convention [16], the models are trained on the *septuplet* subset of Vimeo-90K. For EDSC and BMBC the official pretrained models are used due to lacking the training code. Results of Softsplat and VFIT-B are taken from the literature [31] due to absence of both source code and models. Though not designed for single-frame interpolation, the proposed method achieves best results on 5 of 7 datasets. Our method performs worse than FLAVR and VFIT-B, which adopt larger architectures than ours, on Vimeo-90K and UCF101. These two datasets have relatively slow motion, so that advantage of our method is less significant. While, we show much better results than them on DAVIS.

Table 4. Ablation analysis on the GoPro dataset.

Component				PSNR	SSIM
Image gradient	Flow gradient	Flow update	Deep reparam		
✗	✗	✗	✗	30.56	0.935
✗	✗	✓	✗	30.72	0.937
✗	✓	✓	✓	30.78	0.937
✓	✗	✗	✗	31.28	0.942
✓	✗	✓	✗	31.52	0.944
✓	✓	✓	✗	31.53	0.944
✓	✓	✓	✓	31.73	0.947

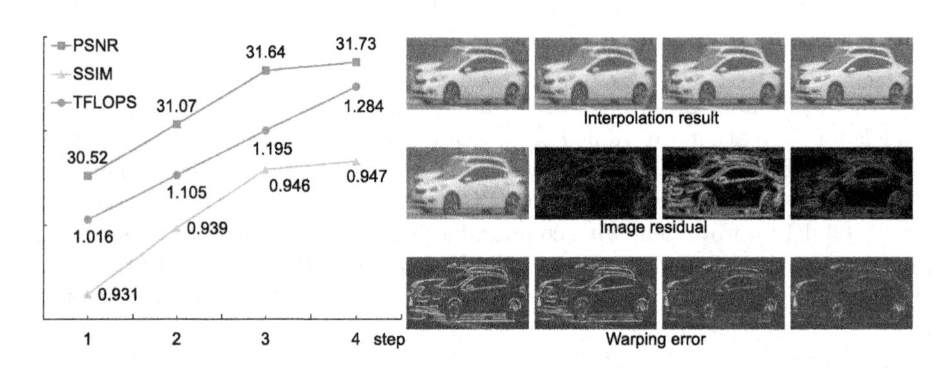

Fig. 5. Analyzing the performance of each iteration on the GoPro dataset. Left: quantitative performance in terms of PSNR, SSIM and TFLOPS as a function of iteration step. Right: the per-step interpolation result (top), learned image residual (middle) as well as the rendering error on one input image by warping the result (visualized as heatmap).

4.3 Performance Analysis

In the following, we conduct several experiments to analyze the performance of the proposed approach under several conditions. All the experiments are conducted on the GoPro dataset.

Ablation Analysis. The aim of this experiment is to show impact of each proposed module. Here, excluding the Image/Flow Gradient module means that image/flow gradient calculation is removed when forming the input of each iteration. By comparing the 3rd, 5th and 7th rows, we conclude that dropping either module would cause loss of performance. We also evaluate the case that flow updates are not learned at each iteration. By comparing 4th and 5th rows, we show there will be 0.3 dB PSNR loss. Finally, we evaluate the effectiveness of the proposed deep reparameterization of flow gradients. To this end, we place 2-variate Gaussian on the 2-dimensional optical flows directly, instead of the transformed 16-dimensional deep features. The model capacity is preserved by

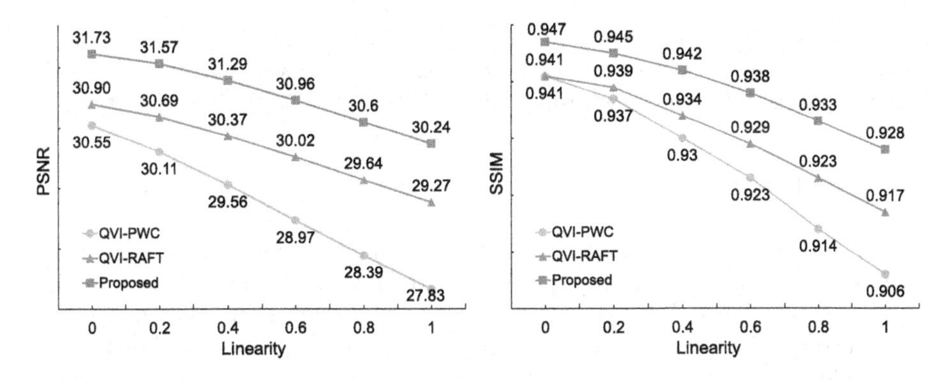

Fig. 6. Analyzing the robustness *w.r.t.* the quality of initial optical flows on different methods. PSNR and SSIM results are shown in the left and right, respectively.

adding extra convolutional layers when processing optical flows for fairer comparison. By comparing 6th and 7th rows, this variant results into 31.53 dB PSNR with a loss of 0.2 dB, demonstrating the effectiveness of deep gradient reparameterization (Table 4).

Analyzing Per-Iteration Performance. We evaluated the performance of per-iteration interpolation result in the left of Fig. 5. As expected, the result get consistently improved with more iterations. Two iterations of our model already surpasses all methods in Table 1 except FLAVR with comparable FLOPS and more iterations up to 4 can further outperform FLAVR and achieve the leading result. In the right of Fig. 5, we visualize the results on one example. As can be seen, the interpolation result gets consistently sharper, while the rendering error gets smaller, with iteration goes. The learned per-iteration image residuals identify pixels with high rendering errors in the previous step. It might be correlated with different pixels at different iterations, as a result of the quality of previously estimated image and flow results. However, at the last iteration, the error often converges and the residual becomes subtle.

Analyzing Robustness to Motion Estimation. Many existing approaches and our method assume the input of high-quality optical flows. In this experiment, we aim to see what happens if the initially estimated pixel motion are of worse quality. For comparision we choose QVI as baseline, which proposed the quadratic model for intermediate flow estimation, as also followed by our work. Since QVI computes optical flows with PWCNet [35] while we use RAFT [37], for fairness we have also trained a variant, QVI-RAFT, which improves QVI with RAFT optical flows. For this evaluation, we replace the quadratic model with the simpler linear one as proposed in [44]. The interpolated forward flows from quadratic and linear models are alpha blended with different weights in $[0, 1]$. In this way, we artifically deteriorate the quality of initial flow estimations by setting a higher weight for the linear motion model. Figure 6 shows the performance as a function of the linearity, measured as blending weight of forward flows generated by linear motion model. As expected, when motion estimation results get worse, so will be the final performance. However, our approach still achieves consistently the best results with initial motion of various quality.

5 Conclusion

In this work we present deep Bayesian video frame interpolation, a lightweight approach showing new records on 8 of 10 VFI benchmarks. Our approach by the first time formulates VFI with a posterior maximization framework optimized by learned gradient descent, whose gradient terms are principally defined.

In addition, we provide standard benchmarking results on the GoPro [23] and X4K1000FPS [32] datasets, and unified evaluation protocols on GoPro. We hope this facilitates future VFI research.

References

1. Adler, J., Öktem, O.: Solving ill-posed inverse problems using iterative deep neural networks. Inverse Prob. **33**(12), 124007 (2017)
2. Adler, J., Öktem, O.: Learned primal-dual reconstruction. IEEE Trans. Med. Imag. (TMI) **37**(6), 1322–1332 (2018)
3. Bao, W., Lai, W.S., Ma, C., Zhang, X., Gao, Z., Yang, M.H.: Depth-aware video frame interpolation. In: IEEE Conference on Computer Vision and Pattern Recognition (CVPR), pp. 3703–3712 (2019)
4. Bao, W., Lai, W.S., Zhang, X., Gao, Z., Yang, M.H.: Memc-net: motion estimation and motion compensation driven neural network for video interpolation and enhancement. IEEE Trans. Pattern Anal. Mach. Intell. (TPAMI) **43**(3), 933–948 (2019)
5. Cheng, X., Chen, Z.: Multiple video frame interpolation via enhanced deformable separable convolution. In: IEEE Transactions on Pattern Analysis and Machine Intelligence (TPAMI) (2021)
6. Chi, Z., Nasiri, R.M., Liu, Z., Lu, J., Tang, J., Plataniotis, K.N.: All at once: temporally adaptive multi-frame interpolation with advanced motion modeling. In: European Conference on Computer Vision (ECCV), vol. 12372, pp. 107–123 (2020)
7. Choi, M., Kim, H., Han, B., Xu, N., Lee, K.M.: Channel attention is all you need for video frame interpolation. In: AAAI Conference on Artificial Intelligence (AAAI), pp. 10663–10671 (2020)
8. Dosovitskiy, A., et al.: Flownet: learning optical flow with convolutional networks. In: IEEE International Conference on Computer Vision (ICCV), pp. 2758–2766 (2015)
9. Flynn, J., et al.: Deepview: view synthesis with learned gradient descent. In: IEEE Conference on Computer Vision and Pattern Recognition (CVPR), pp. 2367–2376 (2019)
10. Gui, S., Wang, C., Chen, Q., Tao, D.: Featureflow: robust video interpolation via structure-to-texture generation. In: IEEE Conference on Computer Vision and Pattern Recognition (CVPR), pp. 14004–14013 (2020)

11. Gupta, A., Aich, A., Roy-Chowdhury, A.K.: Alanet: adaptive latent attention network for joint video deblurring and interpolation. In: ACM International Conference on Multimedia (ACMMM), pp. 256–264 (2020)
12. Hendrycks, D., Gimpel, K.: Gaussian error linear units (gelus). Arxiv preprint, 1606.08415 [cs.CV] (2016)
13. Huang, G., Liu, Z., Van Der Maaten, L., Weinberger, K.Q.: Densely connected convolutional networks. In: IEEE Conference on Computer Vision and Pattern Recognition (CVPR), pp. 4700–4708 (2017)
14. Jaderberg, M., Simonyan, K., Zisserman, A., Kavukcuoglu, K.: Spatial transformer networks. In: Advances in Neural Information Processing Systems (NeurIPS), pp. 2017–2025 (2015)
15. Jiang, H., Sun, D., Jampani, V., Yang, M., Learned-Miller, E.G., Kautz, J.: Super slomo: high quality estimation of multiple intermediate frames for video interpolation. In: IEEE Conference on Computer Vision and Pattern Recognition (CVPR), pp. 9000–9008 (2018)
16. Kalluri, T., Pathak, D., Chandraker, M., Tran, D.: Flavr: flow-agnostic video representations for fast frame interpolation. In: IEEE Conference on Computer Vision and Pattern Recognition (CVPR) (2021)
17. Kingma, D.P., Dhariwal, P.: Glow: generative flow with invertible 1×1 convolutions. In: Advances in Neural Information Processing Systems (NeurIPS), pp. 10236–10245 (2018)
18. Kokkinos, F., Lefkimmiatis, S.: Iterative joint image demosaicking and denoising using a residual denoising network. IEEE Transa. Image Process. (TIP) **28**(8), 4177–4188 (2019)
19. Kokkinos, F., Lefkimmiatis, S.: Iterative residual cnns for burst photography applications. In: IEEE Conference on Computer Vision and Pattern Recognition (CVPR), pp. 5929–5938 (2019)
20. Lee, H., Kim, T., Chung, T.y., Pak, D., Ban, Y., Lee, S.: Adacof: adaptive collaboration of flows for video frame interpolation. In: IEEE Conference on Computer Vision and Pattern Recognition (CVPR), pp. 5316–5325 (2020)
21. Liu, Z., Yeh, R.A., Tang, X., Liu, Y., Agarwala, A.: Video frame synthesis using deep voxel flow. In: IEEE International Conference on Computer Vision (ICCV), pp. 4463–4471 (2017)
22. Mahajan, D., Huang, F.C., Matusik, W., Ramamoorthi, R., Belhumeur, P.: Moving gradients: a path-based method for plausible image interpolation. ACM Trans. Graph. (TOG) **28**(3), 1–11 (2009)
23. Nah, S., Hyun Kim, T., Mu Lee, K.: Deep multi-scale convolutional neural network for dynamic scene deblurring. In: IEEE Conference on Computer Vision and Pattern Recognition (CVPR), pp. 3883–3891 (2017)
24. Niklaus, S., Liu, F.: Softmax splatting for video frame interpolation. In: IEEE Conference on Computer Vision and Pattern Recognition (CVPR), pp. 5437–5446 (2020)
25. Niklaus, S., Mai, L., Liu, F.: Video frame interpolation via adaptive separable convolution. In: IEEE International Conference on Computer Vision (ICCV), pp. 261–270 (2017)
26. Park, J., Ko, K., Lee, C., Kim, C.: BMBC: bilateral motion estimation with bilateral cost volume for video interpolation. In: European Conference on Computer Vision (ECCV), vol. 12359, pp. 109–125 (2020). https://doi.org/10.1007/978-3-030-58568-6_7

27. Park, J., Lee, C., Kim, C.S.: Asymmetric bilateral motion estimation for video frame interpolation. In: IEEE International Conference on Computer Vision (ICCV), pp. 14539–14548 (2021)
28. Perazzi, F., Pont-Tuset, J., McWilliams, B., Van Gool, L., Gross, M., Sorkine-Hornung, A.: A benchmark dataset and evaluation methodology for video object segmentation. In: IEEE Conference on Computer Vision and Pattern Recognition (CVPR), pp. 724–732 (2016)
29. Ronneberger, O., Fischer, P., Brox, T.: U-net: Convolutional networks for biomedical image segmentation. In: International Conference on Medical Image Computing and Computer-assisted Intervention (MICCAI), pp. 234–241 (2015). https://doi.org/10.1007/978-3-319-24574-4_28
30. Shi, Z., Liu, X., Shi, K., Dai, L., Chen, J.: Video frame interpolation via generalized deformable convolution. IEEE Trans. Multimedia (TMM) **24**, 426–439 (2021)
31. Shi, Z., Xu, X., Liu, X., Chen, J., Yang, M.H.: Video frame interpolation transformer. Arxiv preprint, 2111.13817 [cs.CV] (2021)
32. Sim, H., Oh, J., Kim, M.: Xvfi: Extreme video frame interpolation. In: IEEE International Conference on Computer Vision (ICCV), pp. 14489–14498 (2021)
33. Soomro, K., Zamir, A.R., Shah, M.: Ucf101: A dataset of 101 human actions classes from videos in the wild. Arxiv preprint, 1212.0402 [cs.CV] (2012)
34. Su, S., Delbracio, M., Wang, J., Sapiro, G., Heidrich, W., Wang, O.: Deep video deblurring for hand-held cameras. In: IEEE Conference on Computer Vision and Pattern Recognition (CVPR), pp. 1279–1288 (2017)
35. Sun, D., Yang, X., Liu, M.Y., Kautz, J.: Pwc-net: Cnns for optical flow using pyramid, warping, and cost volume. In: IEEE Conference on Computer Vision and Pattern Recognition (CVPR), pp. 8934–8943 (2018)
36. Sun, L., Dong, W., Li, X., Wu, J., Li, L., Shi, G.: Deep maximum a posterior estimator for video denoising. Int. J. Comput. Vis. (IJCV) **129**(10), 2827–2845 (2021). https://doi.org/10.1007/s11263-021-01510-7
37. Teed, Z., Deng, J.: Raft: recurrent all-pairs field transforms for optical flow. In: European Conference on Computer Vision (ECCV), vol. 12347, pp. 402–419 (2020). https://doi.org/10.1007/978-3-030-58536-5_24
38. Wang, X., et al.: ESRGAN: enhanced super-resolution generative adversarial networks. In: European Conference on Computer Vision Workshops (ECCVW), vol. 11133, pp. 63–79 (2018)
39. Werlberger, M., Pock, T., Unger, M., Bischof, H.: Optical flow guided tv-l 1 video interpolation and restoration. In: Computer Vision and Pattern Recognition workshops (CVPRW), pp. 273–286 (2011)
40. Wu, Y., He, K.: Group normalization. In: European Conference on Computer Vision (ECCV), vol. 11217, pp. 3–19 (2018)
41. Xu, X., Siyao, L., Sun, W., Yin, Q., Yang, M.H.: Quadratic video interpolation. Adv. Neural Inf. Process. Syst. (NeurIPS) **32**, 1645–1654 (2019)
42. Xue, T., Chen, B., Wu, J., Wei, D., Freeman, W.T.: Video enhancement with task-oriented flow. Int. J. Comput. Vis. (IJCV) **127**(8), 1106–1125 (2019). https://doi.org/10.1007/s11263-018-01144-2
43. Yu, Z., Li, H., Wang, Z., Hu, Z., Chen, C.W.: Multi-level video frame interpolation: exploiting the interaction among different levels. IEEE Trans. Circuits Syst. Video Technol. **23**(7), 1235–1248 (2013)
44. Yu, Z., et al.: Training weakly supervised video frame interpolation with events. In: IEEE International Conference on Computer Vision, (ICCV), pp. 14589–14598 (2021)

Cross Attention Based Style Distribution for Controllable Person Image Synthesis

Xinyue Zhou[1], Mingyu Yin[1], Xinyuan Chen[3], Li Sun[1,2(✉)], Changxin Gao[4], and Qingli Li[1]

[1] Shanghai Key Laboratory of Multidimensional Information Processing, East China Normal University, Shanghai, China
sunli@ee.ecnu.edu.cn
[2] Key Laboratory of Advanced Theory and Application in Statistics and Data Science, East China Normal University, Shanghai, China
[3] Shanghai AI Laboratory, Shanghai, China
[4] Huazhong University of Science and Technology, Wuhan, China

Abstract. Controllable person image synthesis task enables a wide range of applications through explicit control over body pose and appearance. In this paper, we propose a cross attention based style distribution module that computes between the source semantic styles and target pose for pose transfer. The module intentionally selects the style represented by each semantic and distributes them according to the target pose. The attention matrix in cross attention expresses the dynamic similarities between the target pose and the source styles for all semantics. Therefore, it can be utilized to route the color and texture from the source image, and is further constrained by the target parsing map to achieve a clearer objective. At the same time, to encode the source appearance accurately, the self attention among different semantic styles is also added. The effectiveness of our model is validated quantitatively and qualitatively on pose transfer and virtual try-on tasks. Codes are available at https://github.com/xyzhouo/CASD.

Keywords: Person image synthesis · Pose transfer · Virtual try-on

1 Introduction

Synthesizing realistic person images under explicit control of the body pose and appearance has many potential applications, such as person reID [6,38,47], video generation [16,40] and virtual clothes try-on [3,7,9,34], *etc.* Recently, the conditional GAN is employed to transfer the source style into the specified target pose. The generator connects the intended style with the required pose in its different layers. *E.g.*, PATN [50], HPT [41], ADGAN [23] insert several repeated modules with the same structure to combine style and pose features. However,

Supplementary Information The online version contains supplementary material available at https://doi.org/10.1007/978-3-031-19784-0_10.

S. Avidan et al. (Eds.): ECCV 2022, LNCS 13675, pp. 161–178, 2022.
https://doi.org/10.1007/978-3-031-19784-0_10

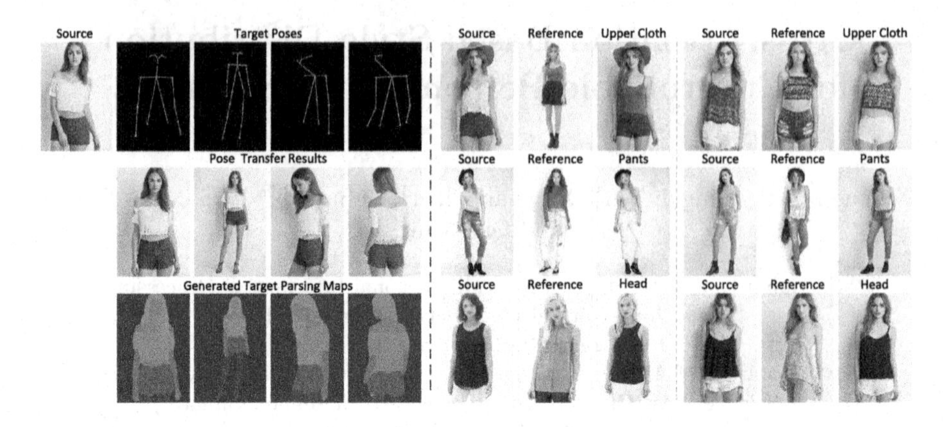

Fig. 1. Left: given the source image and target pose, our model is able to transfer the pose and generate the target parsing map as required. Note that we have only a single training stage without independent generation for the target parsing map. However, our model still synthesizes it precisely by cross attention based style distribution module. Right: Our model also enables virtual try-on and head (identity) swapping by explicitly controlling the poses and per-body-part appearance of source and reference images.

these modules are usually composed of common operations, such as Squeeze-and-Excitation (SE) [11] or Adaptive Instance Normalization (AdaIN) [12], which lacks the ability to align source style with target pose.

In contrast, the 2D or 3D deformation is applied in the task with a clearer motivation. DefGAN [28], GFLA [26] and Intr-Flow [15] estimate the correspondence between the source and target pose to guide the spread of appearance features. Although these methods generate realistic texture, they may produce noticeable artifacts when faced with large deformations. Besides, more than one training stages are often needed, and the unreliable flow from the first stage limits the quality of results.

This paper aims for the better fusion on features of both source image and target pose in a single training stage. Instead of directly estimating the geometry deformation and warping source features to fulfill target pose, we propose a simple cross attention based style distribution (CASD) module to calculate between the target pose and the source style represented by each semantic, and distribute the source semantic styles to the target pose. The basic idea is to employ the coarse fusion features under target pose as queries, requiring the source styles from different semantic components as keys and values to update and refine them. Following ADGAN [23], appearance within each semantic region is described by a style encoder, which extracts the color and texture within the corresponding region (such as head, arms, or legs, *etc.*). The style features are dynamically distributed by the CASD block for each query position under the target pose. Particularly, values from each semantic are softly weighted and summed together according to the attention matrix, so that they are matched with target pose. The aligned feature can be further utilized to affect the input to the decoder.

To further improve the synthesis quality, we have some special designs within CASD block. First, to tightly link styles from different semantics, the self attention is performed among them, making each style no longer independent with others. Second, another routing scheme, in the same size with attention matrix, is also employed for style routing. It is directly predicted from the target pose without exhaustive comparisons with keys of styles. Third, extra constraint from target parsing map is incorporated on the attention matrix, so that the attention head has a clearer motivation. In this way, our attention matrix represents the predicted target parsing map. Additionally, our model can also achieve virtual try-on and head (identity) swapping based on reference images by exchanging the specific semantic region in style features. Figure 1 shows some applications of our model. The contributions of the paper can be summarized into following aspects.

- We propose cross attention based style distribution (CASD) module for controllable person image synthesis, which softly selects the source style represented by each semantic and distributes them to the target pose.
- We intentionally add self attention to connect styles from different semantic components, and let the model predict the attention matrix based on the target pose. Moreover, the target parsing maps are used as the ground truths for the attention matrix, giving the model an evident object during training.
- We can achieve applications in image manipulation by explicit controlling over body pose and appearance, *e.g.*, pose transfer, parsing map generation, virtual try-on and head(identity) swapping.
- Extensive experiments on DeepFashion dataset validates the effectiveness of our proposed model. Particularly, the synthesis quality has been greatly improved, indicated by both quantitative metrics and user study.

2 Related Work

Human pose transfer is first proposed in [20], and becomes well developed in recent years due to the advancement in image synthesis. Most of the existing works need paired training data, which employ the ground truth under target pose during training. Though a few of them are fully unsupervised [5,21,25,27,30,44,44,48], they are not of our major concern. Previous research can be characterized into either two- (or multi-) stage or one-stage methods. The former first generates coarse images or foreground masks, and then gives them to the second stage generator as input for refinement. In [1], the model first segments the foreground from image into different body components, and then applies learnable spatial deformations on them to generate the foreground image. GFLA [26] pretrains a network to estimate the 2D flow and occlusion mask based on source image, source and target poses. Afterwards, it uses them to warp local patches of the source to match the required pose. Li *et.al.* [15] fit a 3D mesh human model onto the 2D image, and train the first stage model to predict the 3D flow, which is employed to warp the source appearance in the second stage. LiquidGAN [16] also adopts the 3D model to guide the geometry deformation within the foreground region. Although geometry-based methods generate realistic texture, they may fail to extract accurate motions, resulting in noticeable artifacts. On the other hand, without any deformation operation, PISE [43] and SPGnet [19] synthesize the target parsing maps, given the source masks, source poses and target poses as input in the first stage. Then it generates the image with the help of them in the second stage. These work show that the parsing maps under target pose have potential to be exploited for pose transfer.

Compared to two-stage methods, the single stage model has light training burden. Different from [1], DefGAN [28] explicitly computes the local 2D affine transformation between source and target patches, and applies the deformation to align source features to the target pose. PATN [50] proposes a repeated pose attention module, consisting of the computation like SE-Net, to combine features from the appearance and pose. ADGAN [23] uses a texture encoder to extract style vectors within each semantic, and gives them to several AdaIN residual blocks to synthesize the target image. XingGAN [31] proposes two types of cross attention blocks to fuse features from the target pose and source appearance in two directions, repeatedly. Although these models design the fusion block of pose and appearance style, they lack the operation to align source appearance with the target pose. CoCosNet [44,48] computes the dense correspondences between cross-domain images by attention-based operation. However, each target position is only related to a local patch of the source image, which implies that the correlation matrix should be a sparse matrix, and the dense correlation matrix leads to quadratic memory consumption. Our model deals with this problem by an efficient CASD block.

Attention and transformer modules first appear in NLP [33], which enlarge the receptive field in a dynamic way. Non-local network [36] is its first attempt in image domain. The scheme becomes increasingly popular in various tasks including image classification [4,17,32,35], object detection [2,49] and semantic segmentation [13,46] due to its effectiveness. There are basically two different ways for it which are self and cross attention. Self attention projects the queries, keys and values from the same token set, while cross attention usually obtains keys and values from one set, and queries from another one. The computation process then becomes the same, measuring the similarity between queries and keys to form an attention matrix, which is used to weight values to update query tokens. Based on the repeated attention module, multi-stage transformer can be built. Note that adding the MLP (FFN) and residual connection between stages are crucial and become a designing routine, for which we also follows.

3 Method

3.1 Overview Framework

Given a source image I_s under the pose P_s, our goal is to synthesize a high fidelity image $\hat{I}_t$ under a different target pose P_t. $\hat{I}_t$ should not only fulfil the pose requirement, but also have the same appearance with I_s. Figure 2 shows the overview of the proposed generation model. It consists of a pose encoder E_p, a semantic region style encoder E_s and a decoder Dec. Besides, there are

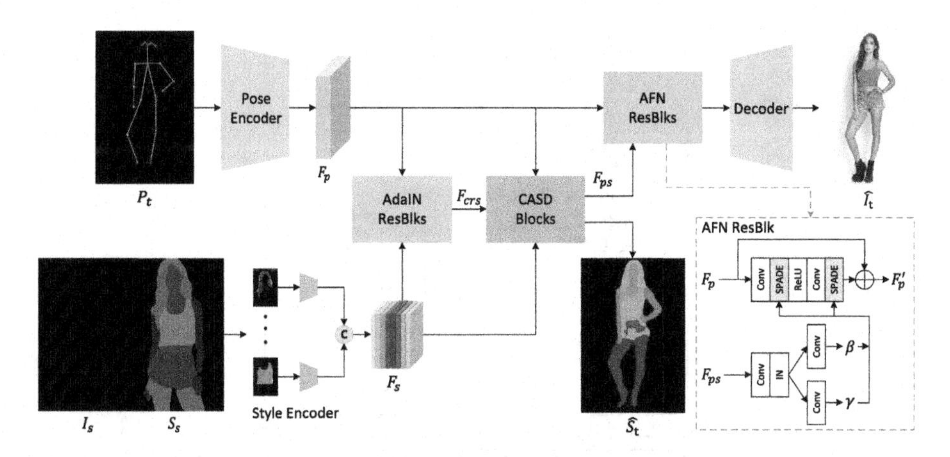

Fig. 2. Overview architecture of our proposed generator. There are separate pose and style encoders, with their outputs F_p and F_s being fused by AdaIN ResBlks and Cross Attention based Style Distribution (CASD) blocks, and giving the pose-aligned feature F_{ps} as the output. Then, the same F_{ps} is adapted to decoder through AFN ResBlks. The key component, CASD Block, consists of both self and cross attention, and can also output the predicted target parsing maps $\hat{S}_t$.

several cross attention based style distribution (CASD) blocks which are the key components in our generator. Before and after the attention, there are several AdaIN residual blocks and Aligned Feature Normalization (AFN) residual blocks with the similar design as [24,42]. The former coarsely adapts the source style to the target pose, while the latter incorporates the pose-aligned feature from the CASD blocks into the decoder. Both of them are learnable, which change the feature statistics in their affecting layers.

As is shown in Fig. 2, the desired P_t is directly used as the input by encoder E_p, which describes the key point positions of human body. For each point, we make a single channel heatmap with a predefined standard deviation to describe its location. Except the individual point, we additionally adopt straight lines between selected points to better model the pose structure. There are totally 18 points and 12 lines, so $P_t \in \mathbb{R}^{H \times W \times 30}$. To facilitate accurate style extraction from I_s, we follow the strategy in [23], which employs the source parsing map $S_s \in \mathbb{R}^{H \times W \times N_s}$ to separate the full image into regions, so that E_s independently encodes the styles in different semantics. N_s is the total number of semantics in the parsing map. During training, the ground truth image I_t and its corresponding parsing map S_t are exploited. Note that $F_p \in \mathbb{R}^{H \times W \times C}$ and $F_s \in \mathbb{R}^{N_s \times 1 \times 1 \times C}$ represent the pose and style features from E_p and E_s, respectively. They are utilized by the CASD blocks, whose details are introduced in the following section. Moreover, the CASD block is repeated by multiple times, *e.g.* twice, gradually completing the fusion between F_p and F_s, and forming a better aligned feature F_{ps}. Finally, F_{ps} is given to AFN from the side branch to take its effect. Besides, our CASD block can also output the predicted target parsing map $\hat{S}_t$ by constraining the cross attention matrix.

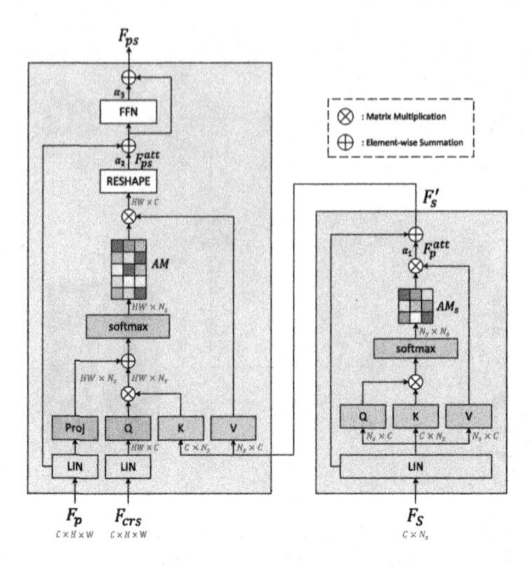

Fig. 3. Illustrations of the cross attention based style distribution (CASD) block. On the right, self attention is performed on the source style features F_s, giving the updated F_s' as the output. On the left, cross attention is computed between the coarsely aligned feature F_{crs} and the updated style feature F_s'. The pose F_p also joins the cross attention. And by constraining the cross attention matrix AM, predicted target parsing maps $\hat{S}_t$ is generated. Note that α_{1-3} indicates the learnable scaling factors.

3.2 Pre- and Post-attention Style Injection

Since the source style $F_s^i \in \mathbb{R}^{1\times1\times C}$ is independently encoded by a shared-weight encoder E_s, where $i = 1, 2, \cdots, N_s$ is the semantic index, they may not appropriate for style injection together. So we first combine F_s^i from different semantic regions through an MLP, and give results to AdaIN ResBlks to roughly combine F_s with F_p, specifying the coarse fusion F_{crs} reflecting the target pose P_t, which then participates cross attention as queries in the CASD blocks.

After the CASD blocks, we have F_{ps} which is obviously superior to F_{crs}, and is suitable for the utilization by *Dec*. Instead of giving F_{ps} directly to *Dec*, we design an AFN ResBlks to employ it as a conditional feature. Within the block, an offset β and a scaling factor γ are first predicted. Then, they take effect through AFN, which performs the conditional normalization according to β and γ. Note that F_{ps}, β and γ are in the same size.

3.3 Cross Attention Based Style Distribution Block

The computation in CASD block includes two stages, which are self attention and cross attention stages, as is depicted in Fig. 3. The two types of attention are carried out in sequence, and they together align the source style F_s according to the target pose feature F_p. We describe them in the following two sections.

Self Attention for Style Features. The self attention is performed among F_s^i, so that each F_s^i of a particular semantic is connected with others F_s^j where $j \neq i$. We simplify the classic design on self attention module in transformer [33]. Traditionally, there are three learnable projection heads W_Q, W_K and W_V, and they are in the same shape $W_Q, W_K, W_V \in \mathbb{R}^{C \times C}$. These heads are responsible for mapping the input tokens F_s into the query Q, key K and value V. In our application, to reduce learnable parameters, we omit W_Q and W_K, and directly use F_s as both query and key. However, we keep W_V and it yields $V \in \mathbb{R}^{N_s \times C}$ in the same dimension of F_s. The self attention, computing the update style feature F_s^{att}, can be summarized into Eq. (1).

$$\begin{aligned} F_s^{att} &= \text{Attention}(Q, K, V) = \text{Softmax}(QK^{\mathrm{T}}/\sqrt{C})V \\ Q &= F_s, \quad K = F_s, \quad V = F_s W_V \end{aligned} \tag{1}$$

Here the attention module essentially compares the similarities of different semantic styles, so that each F_s^i in F_s absorbs information from other style tokens F_s^j. Note that we also follow the common designs in transformer. Particularly, there is a residual connection between F_s and F_s^{att}, so they are added, and given to later layers. The final style is denoted by F_s', as is shown on the right of Fig. 3. Moreover, different from traditional transformer, we employ Layer Instance Normalization (SW-LIN) proposed in [39] to replace LN for better synthesis.

Cross Attention. The cross attention module further adapts the source style F_s' into the required pose, shown on the left of Fig. 3. Such attention is carried out across different domains, between the coarse fusion F_{crs} output from AdaIN resblks and the style feature F_s'. Therefore, different from the previous self attention, its result F_{ps} has spatial dimensions, and actually reflects how to distribute the source style under an intended pose. In this module, F_{crs} from AdaIN ResBlks is treated as queries Q. Specifically, there are totally $H \times W$ unique queries. Each of them is a C-dim vector. F_s' provides the keys K for comparison with Q and values V for soft selection.

Here we aggregate the common attention computation as is shown in Eq. (2). $F_{ps}^{att} \in \mathbb{R}^{H \times W \times C}$ is the updated amount on query after the attention, in the same size with F_{crs}.

$$\begin{aligned} F_{ps}^{att} &= \text{Attention}(Q, F_p, K, V) = AM \cdot V \\ &= \left(\text{Softmax}\left(\frac{QK^{\mathrm{T}}}{\sqrt{C}} + \text{Proj}(F_p)\right)\right) V \end{aligned} \tag{2}$$

Note that we set $Q = F_{crs} W_Q$, $K = F_s' W_K$ and $V = F_s' W_V$, so the attention actually combines the features F_{crs} with source styles F_s'. In Eq. (2), the first term in the bracket indicates the regular attention matrix which exhaustively computes the similarity between every Q-K pair. It is of the shape $H \times W \times N_s$ and determines the possible belonging semantic for each position. Since there are projection heads W_Q and W_K to adjust query and key, the attention matrix is fully dynamic, which is harmful to model convergence. Our solution is to add the second term with the same shape as the first one, forming the augmented attention matrix AM, and let it participate value routing. Proj$(\cdot)$ is a linear projection head which directly outputs a routing scheme based only on F_p. Therefore, it implies that the model is able to predict the target parsing map for each position, given the encoded pose feature F_p. Some recent works like SPGNet [19] or PISE [43] has a separate training stage to generate the target parsing map according to the required pose. Our model has a similar intention, but it is more convenient with only a single training stage. In the following section, we add a constraint on the attention matrix to generate the predicted target parsing map.

After the cross attention in Eq. (2), we follow the routine in transformer, which first makes the element-wise summation between F_{ps}^{att} and F_p, then gives the result to an FFN, leading to a better pose feature F_{ps} which combines the source style for the next stage.

3.4 Learning Objectives

Similar to the previous method [23,50], we employ the adversarial loss L_{adv}, reconstruction loss L_{rec}, perceptual loss L_{perc} and contextual loss L_{CX} as our learning objectives. Additionally, we also adopt an attention matrix cross-entropy loss L_{AMCE} and an LPIPS loss L_{LPIPS} to train our model. The full learning objectives are formulated in Eq. (3),

$$\begin{aligned} L_{full} = \lambda_{adv} L_{adv} + \lambda_{rec} L_{rec} + \lambda_{perc} L_{perc} + \lambda_{CX} L_{CX} \\ + \lambda_{AMCE} L_{AMCE} + \lambda_{LPIPS} L_{LPIPS} \end{aligned} \tag{3}$$

where λ_{adv}, λ_{rec}, λ_{perc}, λ_{CX}, λ_{AMCE} and λ_{LPIPS} are hyper-parameters controlling the relative importance of these objectives. They are detailed as follows.

Attention Matrix Cross-Entropy Loss. To train our model with an evident object, we adopt cross-entropy loss to constrain the attention matrix AM close to target parsing map S_t, which is defined as:

$$L_{AMCE} = -\sum_{i=1}^{H}\sum_{j=1}^{W}\sum_{c=1}^{N_s} S_t\left(i,j,c\right)\log\left(AM\left(i,j,c\right)\right). \tag{4}$$

where i, j denote the position of spatial dimension in the attention matrix AM, and c denotes the position of semantic dimension in attention matrix AM. By employing this loss in the training process, our model can generate the predicted target parsing map in a single stage.

Adversarial loss. We adopt a pose discriminator D_p and a style discriminator D_s to help G generate more realistic result in adversarial training. Specifically, real pose pairs (P_t, I_t) and fake pose pairs $(P_t, \hat{I}_t)$ are feed into D_p for pose consistency. Meanwhile, real image pairs (I_s, I_t) and fake image pairs $(I_s, \hat{I}_t)$ are feed into D_s for style consistency. Note that both discriminators are trained with G in an end-to-end way to promote each other.

$$L_{adv} = \mathbb{E}_{I_s,I_t,P_t}\left[\log\left(D_s\left(I_s,I_t\right)\cdot D_p\left(P_t,I_t\right)\right)\right] + \mathbb{E}_{I_s,P_t}\left[\log\left(1-D_s\left(I_s,G\left(I_s,P_t\right)\right)\right)\cdot\left(1-D_p\left(P_t,G\left(I_s,P_t\right)\right)\right)\right] \tag{5}$$

Reconstruction and Perceptual Loss. The reconstruction loss L_{rec} is used to encourage the generated image $\hat{I}_t$ to be similar with ground-truth I_t at the pixel level, which is computed as $L_{rec} = \| \hat{I}_t - I_t \|_1$. The perceptual loss calculates the L_1 distance between the features extracted from the pre-trained VGG-19 network [29]. It can be written as $L_{perc} = \sum_i \| \phi_i\left(\hat{I}_t\right) - \phi_i\left(I_t\right) \|_1$, where ϕ_i is the feature map of the i-th layer of the pre-trained VGG-19 network.

Contextual Loss. We also adopt contextual loss, which is first proposed in [22], aiming to measure the similarity between two non-aligned images for image transformation. It is computed in Eq. (6).

$$L_{CX} = -\log\left(CX\left(F^l\left(\hat{I}_t\right), F^l\left(I_t\right)\right)\right) \tag{6}$$

Here $F^l(\hat{I}_t)$ and $F^l(I_t)$ denotes the feature extracted from layer $l = relu\{3\_2, 4\_2\}$ of the pre-trained VGG-19 for images $\hat{I}_t$ and I_t, respectively, and CX denotes the cosine similarity metric between features.

LPIPS Loss. In order to reduce distortions and learn perceptual similarities, we integrate the LPIPS loss [45], which has been shown to better preserve image quality compared to the more standard perceptual loss:

$$L_{LPIPS} = \| F\left(\hat{I}_t\right) - F\left(I_t\right) \|_2 \tag{7}$$

where $F(\cdot)$ denotes the perceptual feature extracted from pre-trained VGG-16 network.

4 Experiments

4.1 Experimental Setup

Dataset. We carry out experiments on DeepFashion (In-shop Clothes Retrieval Benchmark) [18], which contains 52, 712 high-quality person images with the resolution of 256×256. Following the same data configuration in [50], we split this dataset into training and testing subsets with 101,966 and 8,570 pairs, respectively. Additionally, we use the segmentation masks obtained from the human parser [8]. Note that the person ID of the training and testing sets do not overlap.

Evaluation Metrics. We employ four metrics SSIM [37], FID [10], LPIPS [45] and PSNR for evaluation. Peak Signal to Noise Ratio (PSNR) and Structural Similarity Index Measure (SSIM) is the most commonly used in image generation task with known ground truths. The former utilizes the mean square error to give an overall evaluation, while the latter calculates the global variance and

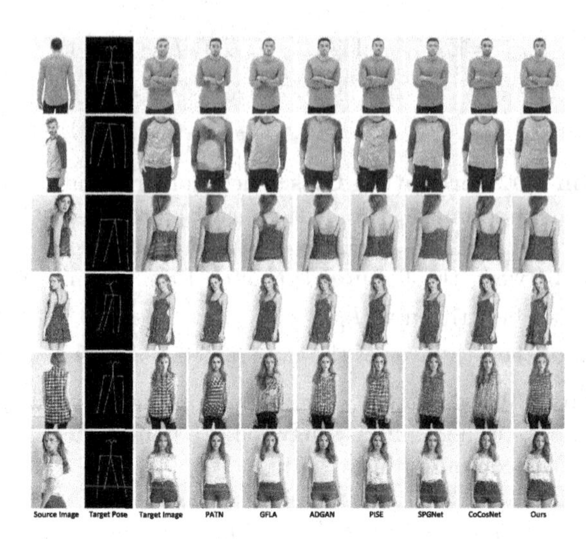

Fig. 4. Qualitative comparison between our method and other state-of-the-arts. The target ground truths and the synthesized results from each models are listed in rows.

mean to assess the structural similarity. Meanwhile, Learned Perceptual Image Patch Similarity (LPIPS) is another metric to compute the distance between the generations and ground truths in the perceptual domain. Besides, Fréchet Inception Distance (FID) is employed to measure the realism of the generated images. It calculates the Wasserstein-2 distance between the distributions of the generated and real data.

Implementation Details. Our method is implemented in PyTorch and trained 2 NVIDIA Tesla-A100 GPUs with the batch size being equal to 16. We adopt Adam optimizer [14] with $\beta_1 = 0.5$, $\beta_2 = 0.999$ to train our model for around 330k iterations, using the same epochs with other works [23]. The weights of the learning objectives are set as: $\lambda_{AMCE} = 0.1$, $\lambda_{LPIPS} = 1$, $\lambda_{rec} = 1$, $\lambda_{perc} = 1$, $\lambda_{adv} = 5$ and $\lambda_{CX} = 0.1$, without tuning. The number of semantic part is $N_s = 8$, which includes the ordinary semantics as background, pants, hair, glove, face, dress, arms and legs. Furthermore, the learning rate is initially set to 0.001, and linearly decayed to 0 after 115k iterations. Following above configuration, we alternatively optimize the generator and two discriminators. We train our model for pose transfer task as described in Sect. 3, after the convergence of training, we use the same trained model for all tasks, *e.g.*, pose transfer, virtual try-on, head(identity) swapping and parsing map generation.

4.2 Pose Transfer

In this section, we compare our method with several state-of-the-art methods, including PATN [50], GFLA [26], ADGAN [23], PISE [43], SPGNet [19] and CoCosNet [44]. Quantitative and qualitative results as well as user study are

Table 1. Comparisons on metrics for image quality and user study. SSIM, FID, LPIPS and PSNR are the quantitative metrics for synthesized images. R2G, G2R and Jab are metrics computed from users' feedback.

Models	SSIM↑	FID↓	LPIPS↓	PSNR↑	R2G↑	G2R↑	Jab↑
PATN [50]	0.6709	20.7509	0.2562	31.14	19.14	31.78	0.26%
GFLA [26]	0.7074	**10.5730**	0.2341	31.42	19.53	35.07	13.72%
ADGAN [23]	0.6721	14.4580	0.2283	31.28	23.49	38.67	11.17%
PISE [43]	0.6629	13.6100	0.2059	31.33	–	–	14.89%
SPGNet [19]	0.6770	12.2430	0.2105	31.22	19.47	36.80	17.26%
CoCosNet [44]	0.6746	14.6742	0.2437	31.07	–	–	13.73%
Ours	**0.7248**	11.3732	**0.1936**	**31.67**	**24.67**	**40.52**	**28.96%**

conducted to verify the effectiveness of our method. All the results are obtained by directly using the source code and well-trained models published by their authors. Since CoCosNet uses a different train/test split, we directly uses its well-trained model on our test set.

Quantitative Comparison. The quantitative results are listed in Table 1. Notably, our method achieves the best performance on most metrics compared with the other methods, which can be attributed to the proposed CASD block.

Qualitative Comparison. In Fig. 4, we compare the generated results from different methods. It can be observed that our method produces more realistic and reasonable results (*e.g.*, the second, third and penultimate rows). More importantly, our model can well retain the details from the source image (*e.g.*, the fourth and last rows). Moreover, even if target pose is complex (*e.g.*, the first row), our method can still generate it precisely.

User Study. While both quantitative and qualitative comparisons can evaluate the performance of the generated results in different aspects, human pose transfer tasks tend to be user-oriented. Therefore, we conduct a user study with 30 volunteers to evaluate the performance in terms of human perception. The user study consists of two parts. (*i*) Comparison with ground-truths. Following [19], we randomly select 30 real images and 30 generated images from test set and shuffle them. Volunteers are required to determine whether a given image is real or fake within a second. (*ii*) Comparison with the other methods, we present volunteers 30 random selected image pairs that include source image, target pose, ground-truth and images generated by our method and baselines. Volunteers are asked to select the most realistic and reasonable image with respect to the source image and ground truth. Note that we shuffle all the generated images for fairness. The results are shown in the right part of Table 1. Here we adopt three metrics, namely **R2G:** the percentage of the real images treated as the generated images; **G2R:** the percentage of the generated images treated as real images; **Jab:** the percentage of images judged to be the best among all models.

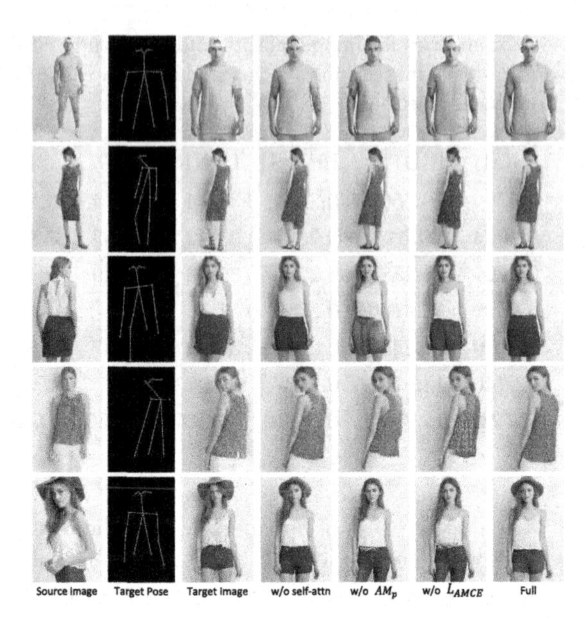

Fig. 5. The qualitative results of ablation study. The ground truths and the synthesize images from each ablation model are listed in columns.

Table 2. Quantitative ablations on each proposed component in the full model. The performances of the final model are given the last row. In the above three rows, we intentionally exclude one component from the full model. Details are given in Sect. 4.3.

Model	SSIM↑	FID↓	LPIPS↓	PSNR↑
w/o self-attn	0.7201	13.2462	0.2017	31.52
w/o AM_p	0.7213	12.4265	0.1985	31.48
w/o L_{AMCE}	0.7156	14.7513	0.2126	31.41
Ours-Full	**0.7248**	**11.3732**	**0.1936**	**31.67**

Higher values of these three metrics mean better performance. We can observe that our model achieves the best results, especially about 11% higher than the 2-nd best one on Jab.

4.3 Ablation Study

In this section, we perform ablation study to further verify our assumptions and evaluate the contribution of each component in our model. We implement 3 variants by alternatively removing a specific component from the full model (w/o self-attn, w/oAM_p, w/o L_{AMCE}).

W/o self-attn. This model removes self attention in CASD blocks, only uses cross attention, which directly feeds F_s into cross attention as Key and Value.

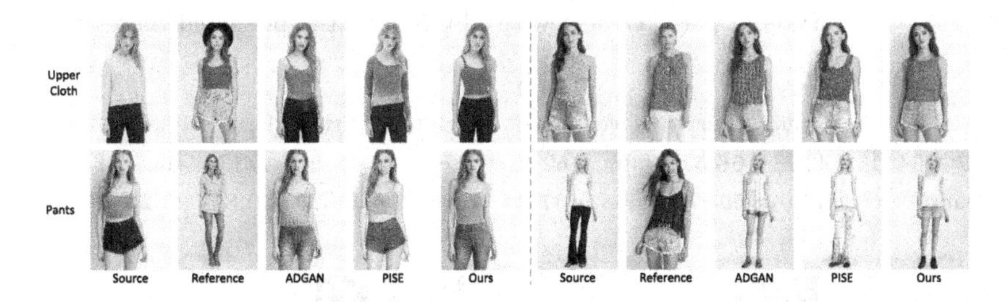

Fig. 6. The visual comparisons with other state-of-the-art methods on virtual try-on.

Table 3. Comparisons of the FID score and user study with other state-of-the-art methods on virtual try-on and head (identity) swapping tasks.

Method	Upper cloth		Pants		Head	
	FID↓	Jab↑	FID↓	Jab↑	FID↓	Jab↑
ADGAN	14.3720	24.67%	14.4446	29.67%	14.4596	23.58%
PISE	14.0537	22.92%	14.2874	22.25%	14.3647	30.75%
Ours	**12.5376**	**52.41%**	**12.5456**	**48.08%**	**12.6578**	**45.67%**

W/o AM_p. The model removes $AM_p = \text{Proj}(Q)$ in CASD blocks in Eq. (2), which will not let the model predict the attention matrix based on the target pose.

W/o L_{AMCE}. The model does not adopt L_{AMCE} loss defined in Eq. (4) for training, so it can not be explicitly guided by the target parsing map during cross attention.

Full model. It includes all components and achieves the best performance on all quantitative metrics, as is shown Table 2. Meanwhile, it also gives the best visual results as is shown in Fig. 5. It's shown that by removing any parts of our proposed model would lead to a performance drop.

4.4 Virtual Try-On and Head Swapping

Benefiting from the semantic region style encoder, our model can also achieve controllable person image synthesis based on reference images by exchanging the channel feature of specific semantic region in the style features (*e.g.*, upper-body transfer, lower-body transfer and head swapping) without further training. We compare our method with ADGAN [23] and PISE [43]. The visual comparisons are shown in Fig. 6. We observe that our model can reconstruct target part and retain other remaining parts more faithfully. In addition, when transferring the lower-body, PISE cannot transfer the target pants to the source person, it will retain the shape of the source person's pants and only transfer the texture.

For more comprehensive comparisons, quantitative comparison and user study are also conducted. The results are shown in Table 3. In the user study,

Table 4. Comparison of per-class IoU with SPGNet on the predicted target parsing maps.

Model	Pants	Hair	Gloves	Face	U-clothes	Arms	Legs	Bkg	Avg
SPGNet [19]	42.18	**66.51**	**9.36**	62.46	67.87	58.46	44.13	84.65	61.89
Ours	**49.02**	65.87	5.50	**67.24**	**76.72**	**59.90**	**50.81**	**90.28**	**66.34**

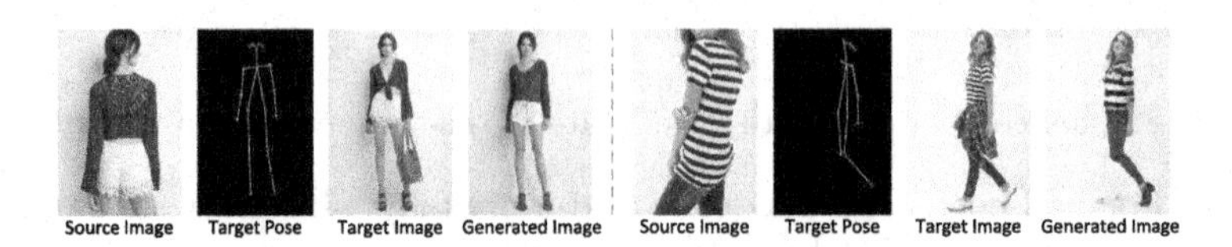

Fig. 7. Failure cases caused by incomprehensible garment (left) or pose (right).

we randomly select 40 results generated by our method and the other compared methods for each task, and then we invite 30 volunteers to select the most realistic results. **Jab** is the percentage of images judged to be the best among all methods.

4.5 Target Parsing Map Synthesis

Moreover, to intuitively understand of our CASD blocks, we further show the predicted target parsing maps in Fig. 1. It shows when given the source image and various target poses, our model can not only transfer the poses, but also synthesize the target parsing maps, though we do not separately build a model to do this. We list the Intersection over Union (IoU) metric between predictions from our method and [8] for all semantics in Table 4. For major semantics, we achieve higher IoU than SPGNet [19]. Note that our model gives the final synthesis images and target parsing map in one stage. The synthesized paired data can be used as training data for segmentation.

5 Limitations

Although our method produces impressive results in most cases, it still fails to generate incomprehensible garments and poses. As shown in Fig. 7, a specific knot on a blouse fails to generate and a person in a rare pose can not be synthesized seamlessly. We believe that training the model with more various images will alleviate this problem.

6 Conclusion

This paper presents a cross attention based style distribution block for a single-stage controllable person image synthesis task, which has strong ability to align the source semantic styles with the target poses. The cross attention based style distribution block mainly consists of self and cross attention, which not only captures the source semantic styles accurately, but also aligns them to the target pose precisely. To achieve a clearer objective, the AMCE loss is proposed to constrain the attention matrix in cross attention by target parsing map. Extensive experiments and ablation studies show the satisfactory performance of our model, and the effectiveness of its components. Finally, we show that our model can be easily applied to virtual try-on and head (identity) swapping tasks.

Acknowledgement. This work is supported by the Science and Technology Commission of Shanghai Municipality No. 19511120800, Natural Science Foundation of China No. 61302125 and No. 62102150, and ECNU Multifunctional Platform for Innovation (001).

References

1. Balakrishnan, G., Zhao, A., Dalca, A.V., Durand, F., Guttag, J.: Synthesizing images of humans in unseen poses. In: Proceedings of the IEEE Conference on Computer Vision and Pattern Recognition, pp. 8340–8348 (2018)
2. Carion, N., Massa, F., Synnaeve, G., Usunier, N., Kirillov, A., Zagoruyko, S.: End-to-end object detection with transformers. In: Vedaldi, A., Bischof, H., Brox, T., Frahm, J.-M. (eds.) ECCV 2020. LNCS, vol. 12346, pp. 213–229. Springer, Cham (2020). https://doi.org/10.1007/978-3-030-58452-8_13
3. Dong, H., Liang, X., Shen, X., Wu, B., Chen, B.C., Yin, J.: FW-GAN: flow-navigated warping GAN for video virtual try-on. In: Proceedings of the IEEE/CVF International Conference on Computer Vision, pp. 1161–1170 (2019)
4. Dosovitskiy, A., et al.: An image is worth 16×16 words: transformers for image recognition at scale. arXiv preprint arXiv:2010.11929 (2020)
5. Esser, P., Sutter, E., Ommer, B.: A variational U-net for conditional appearance and shape generation. In: Proceedings of the IEEE Conference on Computer Vision and Pattern Recognition, pp. 8857–8866 (2018)
6. Ge, Y., et al.: FD-GAN: pose-guided feature distilling GAN for robust person re-identification. arXiv preprint arXiv:1810.02936 (2018)
7. Ge, Y., Song, Y., Zhang, R., Ge, C., Liu, W., Luo, P.: Parser-free virtual try-on via distilling appearance flows. In: Proceedings of the IEEE/CVF Conference on Computer Vision and Pattern Recognition, pp. 8485–8493 (2021)
8. Gong, K., Liang, X., Zhang, D., Shen, X., Lin, L.: Look into person: self-supervised structure-sensitive learning and a new benchmark for human parsing. In: Proceedings of the IEEE Conference on Computer Vision and Pattern Recognition, pp. 932–940 (2017)
9. Han, X., Wu, Z., Wu, Z., Yu, R., Davis, L.S.: Viton: an image-based virtual try-on network. In: Proceedings of the IEEE Conference on Computer Vision and Pattern Recognition, pp. 7543–7552 (2018)
10. Heusel, M., Ramsauer, H., Unterthiner, T., Nessler, B., Hochreiter, S.: GANs trained by a two time-scale update rule converge to a local nash equilibrium. Adv. Neural. Inf. Process. Syst. **30**, 1–12 (2017)
11. Hu, J., Shen, L., Sun, G.: Squeeze-and-excitation networks. In: Proceedings of the IEEE Conference on Computer Vision and Pattern Recognition, pp. 7132–7141 (2018)

12. Huang, X., Belongie, S.: Arbitrary style transfer in real-time with adaptive instance normalization. In: Proceedings of the IEEE International Conference on Computer Vision, pp. 1501–1510 (2017)
13. Huang, Z., Wang, X., Huang, L., Huang, C., Wei, Y., Liu, W.: CCNet: criss-cross attention for semantic segmentation. In: Proceedings of the IEEE/CVF International Conference on Computer Vision, pp. 603–612 (2019)
14. Kingma, D.P., Ba, J.: Adam: a method for stochastic optimization. arXiv preprint arXiv:1412.6980 (2014)
15. Li, Y., Huang, C., Loy, C.C.: Dense intrinsic appearance flow for human pose transfer. In: Proceedings of the IEEE/CVF Conference on Computer Vision and Pattern Recognition, pp. 3693–3702 (2019)
16. Liu, W., Piao, Z., Min, J., Luo, W., Ma, L., Gao, S.: Liquid warping GAN: a unified framework for human motion imitation, appearance transfer and novel view synthesis. In: Proceedings of the IEEE/CVF International Conference on Computer Vision, pp. 5904–5913 (2019)
17. Liu, Z., et al.: Swin transformer: hierarchical vision transformer using shifted windows. arXiv preprint arXiv:2103.14030 (2021)
18. Liu, Z., Luo, P., Qiu, S., Wang, X., Tang, X.: DeepFashion: powering robust clothes recognition and retrieval with rich annotations. In: Proceedings of IEEE Conference on Computer Vision and Pattern Recognition (CVPR) (2016)
19. Lv, Z., et al.: Learning semantic person image generation by region-adaptive normalization. In: Proceedings of the IEEE/CVF Conference on Computer Vision and Pattern Recognition, pp. 10806–10815 (2021)
20. Ma, L., Jia, X., Sun, Q., Schiele, B., Tuytelaars, T., Van Gool, L.: Pose guided person image generation. arXiv preprint arXiv:1705.09368 (2017)
21. Ma, L., Sun, Q., Georgoulis, S., Van Gool, L., Schiele, B., Fritz, M.: Disentangled person image generation. In: Proceedings of the IEEE Conference on Computer Vision and Pattern Recognition, pp. 99–108 (2018)
22. Mechrez, R., Talmi, I., Zelnik-Manor, L.: The contextual loss for image transformation with non-aligned data. In: Proceedings of the European Conference on Computer Vision (ECCV), pp. 768–783 (2018)
23. Men, Y., Mao, Y., Jiang, Y., Ma, W.Y., Lian, Z.: Controllable person image synthesis with attribute-decomposed GAN. In: Proceedings of the IEEE/CVF Conference on Computer Vision and Pattern Recognition, pp. 5084–5093 (2020)
24. Park, T., Liu, M.Y., Wang, T.C., Zhu, J.Y.: Semantic image synthesis with spatially-adaptive normalization. In: Proceedings of the IEEE/CVF Conference on Computer Vision and Pattern Recognition, pp. 2337–2346 (2019)
25. Pumarola, A., Agudo, A., Sanfeliu, A., Moreno-Noguer, F.: Unsupervised person image synthesis in arbitrary poses. In: Proceedings of the IEEE Conference on Computer Vision and Pattern Recognition, pp. 8620–8628 (2018)
26. Ren, Y., Yu, X., Chen, J., Li, T.H., Li, G.: Deep image spatial transformation for person image generation. In: Proceedings of the IEEE/CVF Conference on Computer Vision and Pattern Recognition, pp. 7690–7699 (2020)
27. Sanyal, S., et al.: Learning realistic human reposing using cyclic self-supervision with 3D shape, pose, and appearance consistency. In: Proceedings of the IEEE/CVF International Conference on Computer Vision, pp. 11138–11147 (2021)
28. Siarohin, A., Sangineto, E., Lathuiliere, S., Sebe, N.: Deformable GANs for pose-based human image generation. In: Proceedings of the IEEE Conference on Computer Vision and Pattern Recognition, pp. 3408–3416 (2018)
29. Simonyan, K., Zisserman, A.: Very deep convolutional networks for large-scale image recognition. arXiv preprint arXiv:1409.1556 (2014)

30. Song, S., Zhang, W., Liu, J., Mei, T.: Unsupervised person image generation with semantic parsing transformation. In: Proceedings of the IEEE/CVF Conference on Computer Vision and Pattern Recognition, pp. 2357–2366 (2019)
31. Tang, H., Bai, S., Zhang, L., Torr, P.H.S., Sebe, N.: XingGAN for person image generation. In: Vedaldi, A., Bischof, H., Brox, T., Frahm, J.-M. (eds.) ECCV 2020. LNCS, vol. 12370, pp. 717–734. Springer, Cham (2020). https://doi.org/10.1007/978-3-030-58595-2_43
32. Touvron, H., Cord, M., Douze, M., Massa, F., Sablayrolles, A., Jégou, H.: Training data-efficient image transformers & distillation through attention. In: International Conference on Machine Learning, pp. 10347–10357. PMLR (2021)
33. Vaswani, A., et al.: Attention is all you need. In: Advances in Neural Information Processing Systems, pp. 5998–6008 (2017)
34. Wang, B., Zheng, H., Liang, X., Chen, Y., Lin, L., Yang, M.: Toward characteristic-preserving image-based virtual try-on network. In: Proceedings of the European Conference on Computer Vision (ECCV), pp. 589–604 (2018)
35. Wang, W., et al.: Pyramid vision transformer: a versatile backbone for dense prediction without convolutions. arXiv preprint arXiv:2102.12122 (2021)
36. Wang, X., Girshick, R., Gupta, A., He, K.: Non-local neural networks. In: Proceedings of the IEEE Conference on Computer Vision and Pattern Recognition, pp. 7794–7803 (2018)
37. Wang, Z., Bovik, A.C., Sheikh, H.R., Simoncelli, E.P.: Image quality assessment: from error visibility to structural similarity. IEEE Trans. Image Process. **13**(4), 600–612 (2004)
38. Wei, L., Zhang, S., Gao, W., Tian, Q.: Person transfer GAN to bridge domain gap for person re-identification. In: Proceedings of the IEEE Conference on Computer Vision and Pattern Recognition, pp. 79–88 (2018)
39. Xu, W., Long, C., Wang, R., Wang, G.: DRB-GAN: a dynamic ResBlock generative adversarial network for artistic style transfer. In: Proceedings of the IEEE/CVF International Conference on Computer Vision, pp. 6383–6392 (2021)
40. Yang, C., Wang, Z., Zhu, X., Huang, C., Shi, J., Lin, D.: Pose guided human video generation. In: Proceedings of the European Conference on Computer Vision (ECCV), pp. 201–216 (2018)
41. Yang, L., et al.: Towards fine-grained human pose transfer with detail replenishing network. IEEE Trans. Image Process. **30**, 2422–2435 (2021)
42. Yin, M., Sun, L., Li, Q.: Novel view synthesis on unpaired data by conditional deformable variational auto-encoder. In: Vedaldi, A., Bischof, H., Brox, T., Frahm, J.-M. (eds.) ECCV 2020. LNCS, vol. 12373, pp. 87–103. Springer, Cham (2020). https://doi.org/10.1007/978-3-030-58604-1_6
43. Zhang, J., Li, K., Lai, Y.K., Yang, J.: PISE: person image synthesis and editing with decoupled GAN. In: Proceedings of the IEEE/CVF Conference on Computer Vision and Pattern Recognition, pp. 7982–7990 (2021)
44. Zhang, P., Zhang, B., Chen, D., Yuan, L., Wen, F.: Cross-domain correspondence learning for exemplar-based image translation. In: Proceedings of the IEEE/CVF Conference on Computer Vision and Pattern Recognition, pp. 5143–5153 (2020)
45. Zhang, R., Isola, P., Efros, A.A., Shechtman, E., Wang, O.: The unreasonable effectiveness of deep features as a perceptual metric. In: Proceedings of the IEEE Conference on Computer Vision and Pattern Recognition, pp. 586–595 (2018)
46. Zheng, S., et al.: Rethinking semantic segmentation from a sequence-to-sequence perspective with transformers. In: Proceedings of the IEEE/CVF Conference on Computer Vision and Pattern Recognition, pp. 6881–6890 (2021)

47. Zheng, Z., Zheng, L., Yang, Y.: Unlabeled samples generated by GAN improve the person re-identification baseline in vitro. In: Proceedings of the IEEE International Conference on Computer Vision, pp. 3754–3762 (2017)
48. Zhou, X., et al.: CoCosNet v2: full-resolution correspondence learning for image translation. In: Proceedings of the IEEE/CVF Conference on Computer Vision and Pattern Recognition, pp. 11465–11475 (2021)
49. Zhu, X., Su, W., Lu, L., Li, B., Wang, X., Dai, J.: Deformable DETR: deformable transformers for end-to-end object detection. arXiv preprint arXiv:2010.04159 (2020)
50. Zhu, Z., Huang, T., Shi, B., Yu, M., Wang, B., Bai, X.: Progressive pose attention transfer for person image generation. In: Proceedings of the IEEE/CVF Conference on Computer Vision and Pattern Recognition, pp. 2347–2356 (2019)

KeypointNeRF: Generalizing Image-Based Volumetric Avatars Using Relative Spatial Encoding of Keypoints

Marko Mihajlovic[1,2(✉)], Aayush Bansal[2], Michael Zollhöfer[2], Siyu Tang[1], and Shunsuke Saito[2]

[1] ETH Zürich, Zürich, Switzerland
marko.mihajlovic@inf.ethz.ch
[2] Reality Labs Research, Pittsburgh, USA
https://markomih.github.io/KeypointNeRF

Abstract. Image-based volumetric humans using pixel-aligned features promise generalization to unseen poses and identities. Prior work leverages global spatial encodings and multi-view geometric consistency to reduce spatial ambiguity. However, global encodings often suffer from overfitting to the distribution of the training data, and it is difficult to learn multi-view consistent reconstruction from sparse views. In this work, we investigate common issues with existing spatial encodings and propose a simple yet highly effective approach to modeling high-fidelity volumetric humans from sparse views. One of the key ideas is to encode relative spatial 3D information via sparse 3D keypoints. This approach is robust to the sparsity of viewpoints and cross-dataset domain gap. Our approach outperforms state-of-the-art methods for head reconstruction. On human body reconstruction for unseen subjects, we also achieve performance comparable to prior work that uses a parametric human body model and temporal feature aggregation. Our experiments show that a majority of errors in prior work stem from an inappropriate choice of spatial encoding and thus we suggest a new direction for high-fidelity image-based human modeling.

Keywords: Neural radiance field · Pixel-aligned features

1 Introduction

3D renderable human representations are an important component for social telepresence, mixed reality, and a new generation of entertainment platforms. Classical mesh-based methods require dense multi-view stereo [34,57,58] or

M. Mihajlovic—The work was primarily done during an internship at Meta.

Supplementary Information The online version contains supplementary material available at https://doi.org/10.1007/978-3-031-19784-0_11.

S. Avidan et al. (Eds.): ECCV 2022, LNCS 13675, pp. 179–197, 2022.
https://doi.org/10.1007/978-3-031-19784-0_11

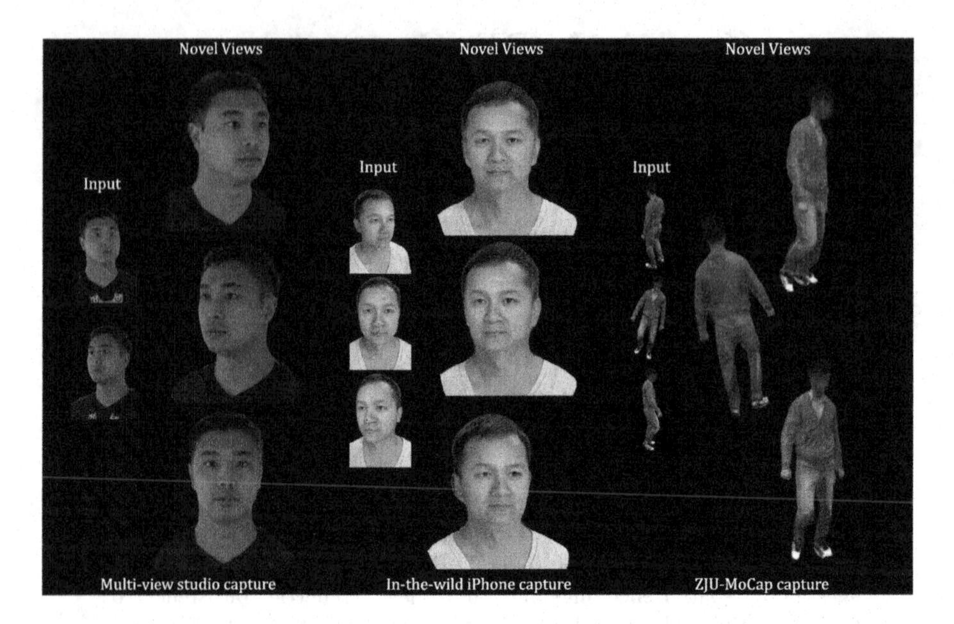

Fig. 1. **(a)** Our approach allows us to synthesize high-fidelity volumetric humans from two or three views of unseen subjects. **(b)** Model learned on studio captures can be used without modification on in-the-wild iPhone captures; and **(c)** finally, the same approach also allows us to synthesize novel views of unseen human subjects (faces are blurred). The figure is best viewed in electronic format.

depth sensor fusion [68]. The fidelity of these methods is limited due to the difficulty of accurate geometry reconstruction. Recently, neural volumetric representations [29,40] enable high-fidelity human reconstruction, especially where accurate geometry is difficult to obtain (e.g., hair). By injecting human-specific parametric shape models [7,32], extensive multi-view data capture can be reduced to sparse camera setups [15,43]. However, these learning-based approaches are subject-specific and require days of training for each individual subject, which severely limits their scalability. Democratizing digital volumetric humans requires an ability to instantly create a personalized reconstruction of a user from two or three snaps (from different views) taken from their phone. Towards this goal, we learn to generalize metrically accurate image-based volumetric humans from two or three views.

Fully convolutional pixel-aligned features utilizing multi-scale information have enabled better performance for various 2D computer vision tasks [6,11,31, 41]. Pixel-aligned representations applied to occupancy and texture fields [49], and neural radiance fields [27,46,67] have enabled generalization to unseen subjects. Pixel-aligned neural fields infer field quantities given a pixel location and spatial encoding function (to avoid ray-depth ambiguity). Different spatial encoding functions [16,19,21,46] have been proposed in the literature. However, the effect of spatial encoding is not fully understood. In this paper, we provide an extensive analysis of spatial encodings for modeling pixel-aligned neural radiance fields for human faces. Our experiments show that the choice of spatial encoding influences the reconstruction quality and generalization to novel identities and views. The models that use camera depth overfit to the training distribution, and multi-view stereo constraints are less robust to sparse views with large baselines.

We present a simple yet highly effective approach based on sparse 3D keypoints to address the limitations of existing approaches. 3D keypoints are easy to obtain using an off-the-shelf 2D keypoint detector [11] and a simple triangulation of multi-views [18]. We treat 3D keypoints as spatial anchors and encode relative 3D spatial information to those keypoints. Unlike global spatial encoding [49,50,67], the relative spatial information is agnostic to camera parameters. This property allows the proposed approach to be robust to changes in pose. 3D keypoints also allow us to use the same approach for both faces and human bodies. Our approach achieves state-of-the-art performance for generating volumetric humans for unseen subjects from sparse-and-wide two or three views, and we can also incorporate more views to further improve performance. We also achieve performance comparable to Neural Human Performer (NHP) [27] on full-body human reconstruction. NHP relies on accurate parametric body fitting and temporal feature aggregation, whereas our approach employs 3D keypoints alone. Our method is not biased [56] to the distribution of the training data. We can use the learned model (without modification) to never-before-seen iPhone captures. We attribute our ability to generalize image-based volumetric humans to an unseen data distribution to our choice of spatial encoding. Our key contributions include:

- A simple approach that leverages sparse 3D keypoints and allows us to create high-fidelity state-of-the-art volumetric humans for unseen subjects from widely spread out two or three views.
- Extensive analysis on the use of spatial encodings to understand their limitations and the efficacy of the proposed approach.
- We demonstrate generalization to never-before-seen iPhone captures by training with only a studio-captured dataset. To our knowledge, no prior work has shown these results.

2 Related Work

Our goal is to create high-fidelity volumetric humans for unseen subjects from as few as two views.

Classical Template-Based Techniques: Early work for human reconstruction [22] required dense 3D reconstruction from a large number of images of the subject and non-rigid registration to align a template mesh to 3D point clouds. Cao et al. [10] employ coarse geometry along with face blendshapes and a morphable hair model to address restrictions posed by dense 3D reconstruction. Hu et al. [20] retrieves hair templates from a database and carefully composes facial and hair details. Video Avatar [1] obtains a full-body avatar based on a monocular video captured using silhouette-based modeling. The dependence on geometry and meshes restricts the applicability of these methods to faithfully reconstruct regions such as the hair, mouth, teeth, etc., where it is non-trivial to obtain accurate geometry.

Neural Rendering: Neural rendering [54,55] has tackled some of the challenges classical template-based approaches struggle with by directly learning components of the image formation process from raw sensor measurements. 2D neural rendering approaches [25,33,35,37,44,45] employ surface rendering and a convolutional network to bridge the gap between rendered and real images. The downside of these 2D techniques is that they struggle to synthesize novel viewpoints in a temporally coherent manner. Deep Appearance Models [28] employ a coarse 3D proxy mesh in combination with view-dependent texture mapping to learn personalized face avatars from dense multi-view supervision. Using a 3D proxy mesh significantly helps with viewpoint generalization, but the approach faces challenges in synthesizing regions for which it is hard to obtain good 3D reconstruction, such as the hair and mouth interior. Current state-of-the-art methods such as NeuralVolumes [29] and NeRF [40] employ differentiable volumetric rendering instead of relying on meshes. Due to their volumetric nature, these methods enable high-quality results even for regions where estimating 3D geometry is challenging. Various extensions [5,30,63] have further improved quality. These methods require dense multi-view supervision for person-specific training and take 3–4 days to train a single model.

Sparse View Reconstruction: Large scale deployment requires approaches that allow a user to take two or three pictures of themselves from multi-views and generate a digital human from this data. The use of pixel-aligned features [49,50] further allows the use of large datasets for learning generalized models from sparse views. Different approaches [13,14,60,67] have combined multi-view constraints and pixel-aligned features alongside NeRF to learn generalizable view-synthesis. In this work, we observe that these approaches struggle to generate fine details given sparse views for unseen human faces and bodies.

Learning Face and Body Reconstruction: Generalizable parametric mesh [4,32,66] and implicit [2,36,38,61,65] body models can provide additional constraints for learning details from sparse views. Recent approaches have incorporated priors specific to human faces [8,9,15,17,48,59,70] and human bodies [42,43,62,64,65,69,71,72] to reduce the dependence on multi-view captures. Approaches such as H3DNet [47] and SIDER [12] use signed-distance functions (SDFs) for learning geometry priors from large datasets and perform test-time fine-tuning on the test subject. PaMIR [72] uses volumetric features guided by a human body model for better generalization. Neural Human Performer [27] employs SMPL with pixel-aligned features and temporal feature aggregation. In this work, we observe that the use of human 3D-keypoints provides necessary and sufficient constraints for learning from sparse-view inputs. Our approach has high flexibility since it only relies on 3D keypoints alone and thus enables us to work both on human faces and bodies. Prior methods have also employed various forms of spatial encoding for better learning. For example, PVA [46] and PortraitNeRF [16] use face-centric coordinates. ARCH/ARCH++ [19,21] use canonical body coordinates. In this work, we extensively study the role of spatial encoding, and found that the use of a relative depth encoding using 3D keypoints leads to the best results. Our findings enable us to learn a representation that generalizes to never-before-seen iPhone camera captures for unseen human faces. In addition to achieving state-of-the-art results on volumetric face reconstruction from as few as two images, our approach can also be used for synthesizing novel views of unseen human bodies and achieves competitive performance to prior work [27] in this setting.

3 Preliminaries: Neural Radiance Fields

NeRF [40] is a continuous function that represents the scene as a volumetric radiance field of color and density. Given a 3D point and a viewing direction $d \in \mathbb{R}^3$, NeRF estimates RGB color values and density (c, σ) that are then accumulated via quadrature to calculate the expected color of each camera ray $r(t) = o + td$:

$$C(r) = \int_{t_n}^{t_f} \exp\left(-\int_{t_n}^{t} \sigma(s)\,ds\right) \sigma(t) c(t, d)\,dt, \tag{1}$$

where t_n and t_f define near and far bounds.

Pixel-Aligned NeRF. One of the core limitations of NeRF is that the approach requires per-scene optimization and does not work well for extremely sparse input views (e.g., two images). To address these challenges, several recent methods [46, 60, 67] propose to condition NeRF on pixel-aligned image features and generalize to novel scenes without retraining.

Spatial Encoding. To avoid the ray-depth ambiguity, pixel-aligned neural fields [46,49,50,67] attach spatial encoding to the pixel-aligned feature. PIFu [49] and related methods [50,67] use depth value in the camera coordinate space as spatial encoding, while PVA [46] uses coordinates relative to the head position. However, we argue that such spatial encodings are global and sub-optimal for learning generalizable volumetric humans. In contrast, our proposed relative spatial encoding provides a localized context that enables better learning and is more robust to changes in human pose.

4 KeypointNeRF

Our method is based on a radiance field function:

$$f\left(X, d | \{(I_n, P_n)\}_{n=1}^{N}\right) = (c, \sigma) \tag{2}$$

that infers a color $c \in \mathbb{R}^3$ and a density $\sigma \in \mathbb{R}$ value for any point in 3D space given its position $X \in \mathbb{R}^3$ and its viewing direction $d \in \mathbb{R}^3$ as input. In addition,

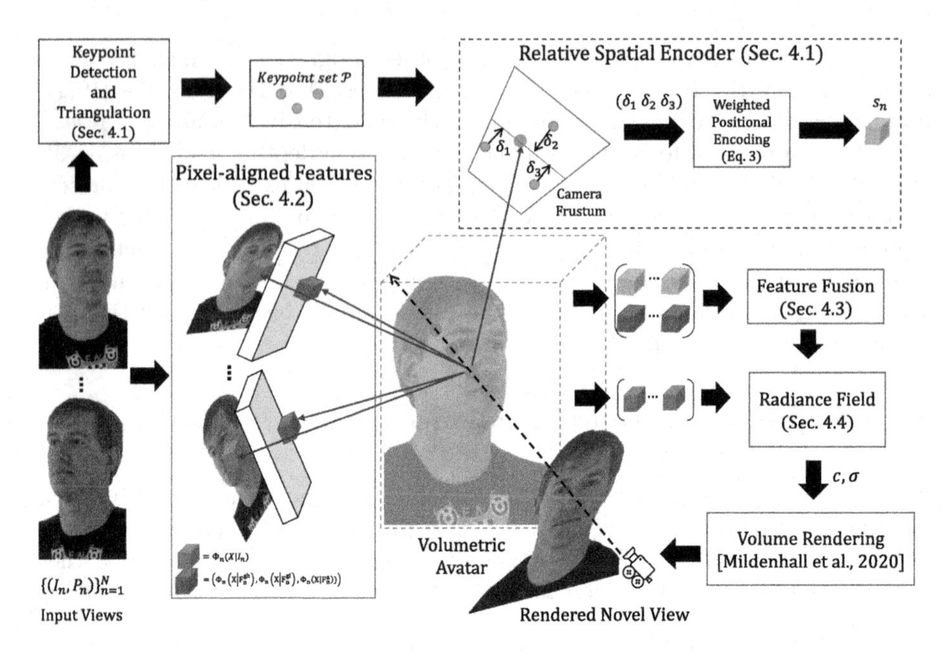

Fig. 2. Overview. We learn a generalizable image-based neural radiance representation for volumetric humans. Given a sparse set of input images $\{I_n\}_{n=1}^N$ and their respective camera parameters P_n, we first detect keypoints and lift them to 3D $\mathcal{P}$. The keypoints are used to provide the relative spatial encoding (Sect. 4.1). The input images are simultaneously encoded via convolutional encoders and provide image-guided pixel-aligned features (Sect. 4.2). These two types of encoding are fused (Sect. 4.3) and condition the radiance field (Sect. 4.4). The radiance field is decoupled by two MLPs, one that directly predicts view-independent density value σ, and the other one which predicts blending weights that are used to output the final color value c by blending image pixel values $\{\Phi(X|I_n)\}_{n=1}^N$. The predicted color and density values are accumulated along the ray via volume rendering [39] to render the volumetric representation from novel views. The rendered example in the figure is an actual output of our method for the displayed two input images of an unseen subject.

the function has access to the N input images I_n and their camera calibrations P_n. We model the function f as a neural network that consists of four main parts; a spatial keypoint encoder (Sect. 4.1), two convolutional image encoders that extract pixel-aligned features (Sect. 4.2), an MLP fusion network that aggregates multiple pixel-aligned features (Sect. 4.3), and two MLPs that predict density σ and color values c (Sect. 4.4). The high-level overview of our method is illustrated in Fig. 2 and in the following we further describe its components.

4.1 Relative Spatial Keypoint Encoding

Our method first leverages an off-the-shelf keypoint regressor [11] to extract K 2D keypoints from at least two input views. Then these points are triangulated and lifted to 3D $\mathcal{P} = \{p_k \in \mathbb{R}^3\}_{k=1}^K$ by using the direct linear transformation algorithm [18]. To spatially encode the query point X, we first compute the depth values of the query point and all keypoints w.r.t each input view $z(p_k|P_n)$ by the 2D projection and then estimate their relative depth difference $\delta_n(p_k, X) = z(p_k|P_n) - z(X|P_n)$. We further employ positional encoding $\gamma(\cdot)$ [40] and the Gaussian exponential kernel to create the final relative spatial encoding for the query point X as:

$$s_n(X|\mathcal{P}) = \Big[\exp(\frac{-l_2(p_k, X)^2}{2\alpha^2})\gamma(\delta_n(p_k, X))\Big]_{k=1}^K, \tag{3}$$

where α is a fixed hyper-parameter that controls the impact of each keypoint. We set this value to 5 cm for facial keypoints and to 10 cm for the human skeleton.

4.2 Convolutional Pixel-Aligned Features

In addition to the spatial encoding $s_n(X)$, we extract pixel-aligned features for the query point X by encoding the input images $I_n \in \mathbb{R}^{H \times W \times 3}$ using two convolutional encoders.

Image Encoders. The first image encoder uses a single HourGlass [41] network that generates both deep low-resolution $F_n^{gl} \in \mathbb{R}^{H/8 \times W/8 \times 64}$ and shallow high-resolution $F_n^{gh} \in \mathbb{R}^{H/2 \times W/2 \times 8}$ feature maps. This network learns a geometric prior of humans and its output is used to condition the density estimation network. The second encoder is a convolutional network with residual connections [23] that encodes input images $F_n^a \in \mathbb{R}^{H/4 \times W/4 \times 8}$ and provides an alternative pathway for the appearance information which is independent of the density prediction branch in the spirit of DoubleField [52]. Please see the supplemental material for further architectural details.

Pixel-Aligned Features. To compute the pixel-aligned features, we project the query point on each feature plane $x = \pi(X|P_n) \in \mathbb{R}^2$ and bi-linearly interpolate the grid values. We define this operation of computing the pixel-aligned features (2D projection and interpolation) by the operator $\Phi_n(X|F)$, where F can represent any grid of vectors for the nth camera: $F_n^{gl}, F_n^{gh}, F_n^a, I_n$.

4.3 Multi-view Feature Fusion

To model a multi-view consistent radiance field, we need to fuse the per-view spatial encodings s_n (Eq. 3) and the pixel-aligned features Φ_n.

The spatial encoding s_n is first translated into a feature vector via a single-layer perceptron. This feature is then jointly blended with the deep low-resolution pixel-aligned feature $\Phi_n(X|F_n^{gl})$ by a two-layer perceptron. The output is then concatenated with the shallow high-resolution feature $\Phi_n(X|F_n^{gh})$ and processed by an additional two-layer perceptron that outputs per-view 64-dimensional feature vector that jointly encodes the blended spatial encoding and the pixel-aligned information. These multi-view features are then fused into a single feature vector $G_X \in \mathbb{R}^{128}$ by the mean and variance pooling operators as in [60].

4.4 Modeling Radiance Fields

The radiance field is modeled via decoupled MLPs for density σ and color c prediction.

Density Fields. The density network is implemented as a four-layer MLP that takes as input the geometry feature vector G_X and predicts the density value σ.

View-Dependent Color Fields. We implement an additional MLP to output the consistent color value c for a given query point X and its viewing direction d by blending image pixel values $\{\Phi(X|P_n)\}_{n=1}^N$ similarly to IBRNet [60]. The input to this MLP is *1)* the extracted geometry feature vector G_X that ensures geometrically consistent renderings, *2)* the additional pixel-aligned features $\Phi_n(X|F_n^a)$, *3)* the corresponding pixel values $\Phi_n(X|I_n)$, and *4)* the view direction that is encoded as the difference between the view direction d and the camera views along with their dot product.

These inputs are concatenated and augmented with the mean and variance vectors computed over the multi-view pixel-aligned features, and jointly propagated through a nine-layer perceptron with residual connections which predicts blending weights for each input view $\{w_n\}_{n=1}^N$. These blending weights form the final color prediction by fusing the corresponding pixel-aligned color values:

$$c = \sum_{n=1}^{N} \frac{\exp(w_n)\Phi_n(X|I_n)}{\sum_{i=1}^{N} \exp(w_i)}. \tag{4}$$

5 Novel View Synthesis

Given our radiance field function $f(X, d) = (c, \sigma)$, we render novel views via the volume rendering Eq. (1), in which we define the near and far bound by analytically computing the intersection of the pixel ray and a geometric proxy that over-approximates the volumetric human and use the entrance and exit points as near and far bounds respectively. For the experiments on human heads, we use a sphere with a radius of 30 cm centered around the keypoints, while for the human bodies we follow the prior work [27,43] and use a 3D bounding box. Similar to NeRF [40], we employ a coarse-to-fine rendering strategy, but we employ the same network weights for both levels.

5.1 Training and Implementation Details

To train our network, we render $H' \times W'$ patches (as in [51]) by accumulating color and density values for 64 sampled points along the ray for the coarse and 64 more for the fine rendering. Our method is trained end-to-end by minimizing the mean ℓ_1-distance between the rendered and the ground truth pixel values and the VGG [53]-loss applied over rendered and ground truth image patches:

$$\mathcal{L} = \mathcal{L}_{\text{RGB}} + \mathcal{L}_{\text{VGG}}. \tag{5}$$

The use of the VGG loss for NeRF training was also leveraged by the concurrent methods [3,64] to better capture high-frequency details. The final loss $\mathcal{L}$ is minimized by the Adam optimizer [26] with a learning rate of $1e^{-4}$ and a batch size of one. For the other parameters, we use their defaults. The background from all training and test input images is removed via an off-the-shelf matting network [24]. Additionally for more temporally coherent novel-view synthesis at inference time, we clip the maximum of the dot product (introduced in Sect. 4.4) to 0.8 when the number of input images is two in the supplementary video.

6 Experiments

In this section, we validate our method on three different reconstruction tasks and datasets: *1)* reconstruction of human heads from images captured in a multi-camera studio, *2)* reconstruction of human heads from in-the-wild images taken with the iPhone's camera, and *3)* reconstruction of human bodies on the public ZJU-MoCap dataset [43]. As evaluation metrics, we follow prior work [27,46,60] and report the standard SSIM and PSNR metrics.

6.1 Reconstruction of Human Heads from Studio Data

Dataset and Experimental Setup. Our captured data consists of 29 1280×768-resolution cameras positioned in front of subjects. We use a total of 351 identities and 26 cameras for training and 38 novel identities for evaluation. At inference time, we reconstruct humans only from 2–3 input views.

Baselines. As baseline, we employ the current state-of-the-art model IBR-Net [60]. In addition, we add several other baselines by varying different types of encoding for the query points in our proposed reconstruction pipeline. Specifically, *1)* our pipeline without any encoding, *2)* with the camera z encoding used in [49,50], *3)* with the encoding of xyz coordinates in the canonical space of a human head that is used in [46], *4)* relative encoding of xyz as the distance between the query point and estimated keypoints, *5)* our relative spatial encoding without distance weighing ($\alpha \rightarrow \infty$ in Eq. 3), and *6)* the proposed weighted relative encoding as described in the method section (Sect. 4.3). The last three models use a total of 13 facial keypoints that are visualized in Fig. 1. All methods are trained with a batch size of one for 150k training steps, except IBRNet [60] which was trained for 200k iterations. For more comparisons and baselines we refer the reader to the supplementary material and video.

Table 1. Studio capture results. Quantitative comparison with IBRNet [60] and different types of spatial encoding. Visual results are provided in Fig. 3

		SSIM↑	PSNR↑
	PVA [46]	81.95	25.87
	IBRNet [60]	82.39	27.14
Our pipeline	1) no encoding	84.38	27.16
	2) camera z encoding [49]	77.86	22.66
	3) canonical xyz encoding [46]	83.11	26.33
	4) relative xyz encoding	83.66	27.05
	5) relative z encoding	84.72	27.66
	6) KeypointNeRF (distance weighted relative z encoding, Eq. 3)	85.19	27.64

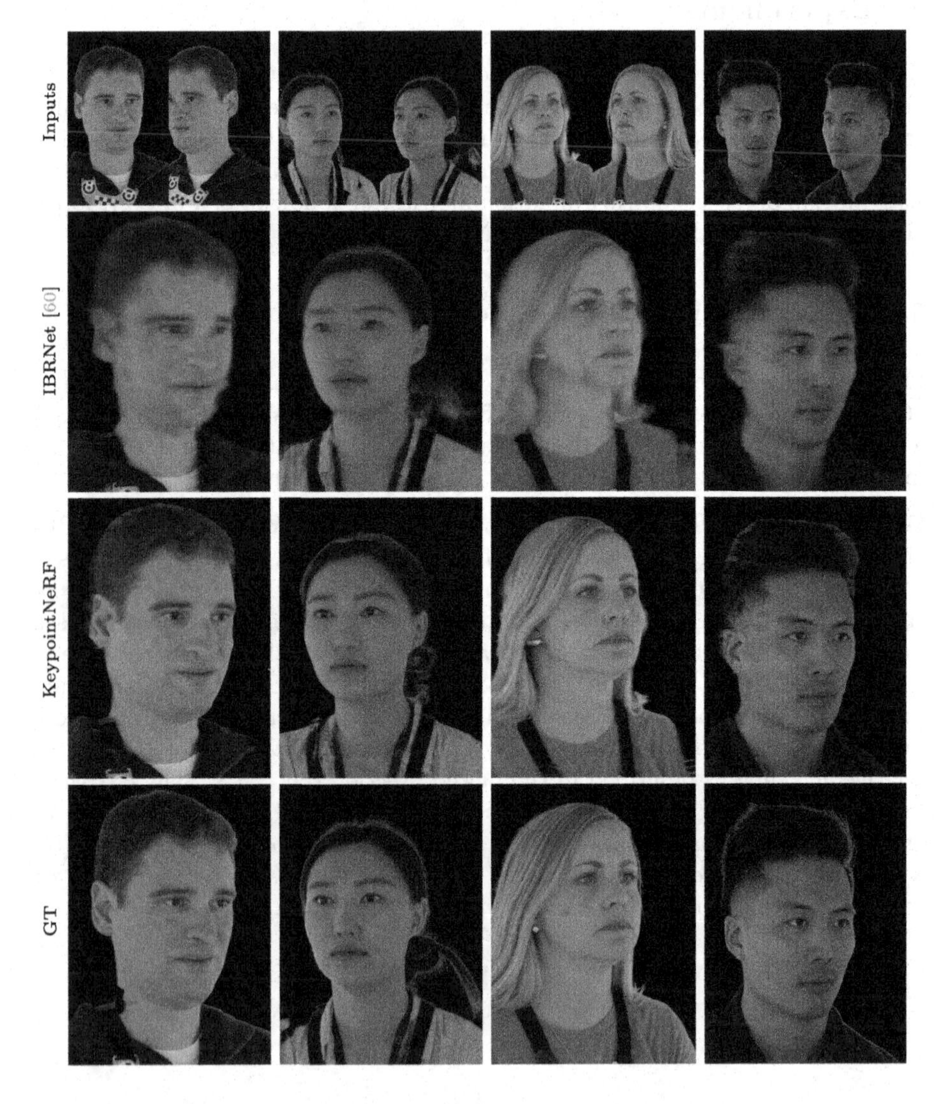

Fig. 3. Studio capture results. Reconstruction results on held-out subjects from only two input views. Best viewed in electronic format.

Results. We provide novel view synthesis (Fig. 3) results for unseen identities that have been reconstructed from only two input images. The results clearly demonstrate that the rendered images of our method are significantly sharper compared to the baselines and are of significantly higher quality. This improvement is confirmed by the quantitative evaluation (Table 1) which further indicates that the proposed distance weighting of the relative spatial encoding improves the reconstruction quality. The third-best performing method is our pipeline without any spatial encoding. However, such a method does not generalize well to other capture systems as we will demonstrate in the next section on in-the-wild captured data.

Table 2. Robustness to different noise levels. Perturbing our keypoints vs. perturbing head position in the canonical xyz encoding used in [46]. Our proposed encoding demonstrates significantly slower performance degradation as the noise increases

Noise level [*mm*]	Canonical xyz encoding		Our keypoint encoding	
	SSIM↑	PSNR↑	SSIM↑	PSNR↑
No noise	83.65	27.05	85.19	27.64
1	82.79	26.24	85.20	27.64
2	82.26	26.05	85.08	27.62
3	81.58	25.48	84.96	27.59
4	80.92	25.08	84.95	27.56
5	80.36	25.16	84.80	27.51
10	76.62	22.23	83.69	27.10
15	73.33	20.26	82.33	26.47
20	70.40	18.87	81.17	25.77

Robustness to Different Noise Levels. We evaluate the robustness of our relative spatial encoding and the encoding in canonical space proposed by Raj et al. [46] by adding different noise levels to the estimated keypoints and the head center respectively. The results reported in Table 2 show that our proposed encoding based on keypoints is significantly more robust compared to modeling in an object-specific canonical space. Note that the canonical encoding requires head template fitting for which we used the ground truth estimation from all views and which in practice is very erroneous or infeasible from two views.

Dynamic Scenes. Although our model is trained only with a neutral face, it generalizes well to dynamic expressions and outperforms the baseline methods. We evaluate the trained models of 38 test subjects performing 8 different expressions and report results in Table 3.

Table 3. Dynamic expressions. Our model is more accurate than the baselines

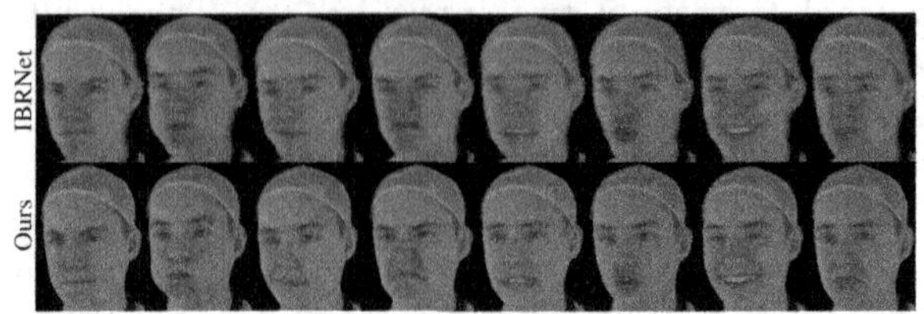

	SSIM↑	PSNR↑
IBRNet [60]	82.64	26.79
Ours (no keypoints)	84.97	27.14
Ours	**85.31**	**27.30**

6.2 Reconstruction from In-the-wild Captures

Setup. To tackle the problem of reconstructing humans in the wild, we acquire a small dataset of subjects by taking several photos with an iPhone and estimate camera parameters. We directly use intrinsic from manufacturing information of iPhone and extrinsic is computed by multi-view RGB-D fitting as in [9]. We evaluate the reconstruction methods trained on the studio captured data in Sect. 6.1 without any retraining.

As input, all methods take three 1920 × 1024-resolution images of a person and predict a radiance field that is then rendered from novel views. In Fig. 4, we display rendered novel views of IBRNet [60], our method without any spatial encoding, and our method with the proposed spatial encoding. The baseline methods produce significantly worse results with lots of blur and cloudy artifacts, whereas our method can reliably reconstruct the human heads. This improvement is quantitatively supported by computed SSIM and PSNR on novel held-out views of the visualized four subjects (Table 4). This experiment demonstrates that our relative spatial encoding is the crucial component for cross-dataset generalization. Please see the supplementary material for more visualizations.

Table 4. In-the-wild captures.

	SSIM↑	PSNR↑
IBRNet [60]	81.74	18.45
Ours (no keypoints)	79.50	19.79
KeypointNeRF	**86.73**	**25.29**

6.3 Reconstruction of Human Bodies

Additionally, we demonstrate that our method is suitable for reconstructing full volumetric human bodies without relying on template fitting of parametric human bodies [32]. We use the public ZJU [43] dataset in order to follow the experimental setup used in [27], so that we could closely compare our method's ability to reconstruct human bodies to the current state-of-the-art method without changing any experimental variables. We follow the standard training-test split of frames and use a total of seven subjects for training and three for validation. At inference time, all methods use three input views. We compare our method with the generalizable volumetric methods: pixelNeRF [67], PVA [46], the current state-of-the-art Neural Human Performer (NHP) [27], and our method without weighting the relative spatial encoding in Eq. 3. We report

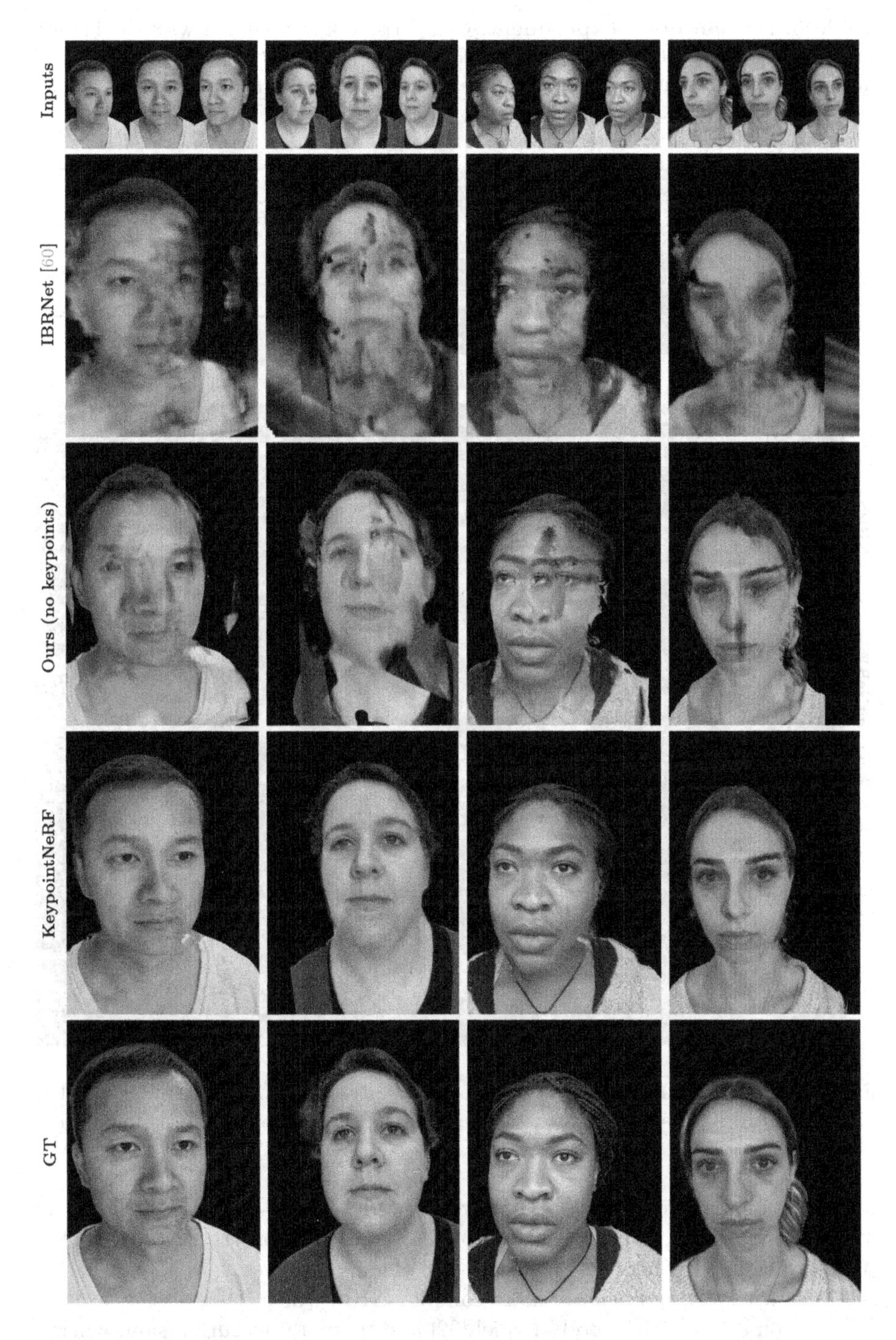

Fig. 4. In-the-wild captures. Our approach produces high-quality reconstructions from three iPhone images, while all baselines show significant artifacts.

Table 5. Human body experiment. Comparison of our method with the baseline methods pixelNeRF [67], PVA [46], and Neural Human Performer (NHP) [27]; 'no w.' in the table means our method without weighting the relative spatial encoding (Eq. 3)

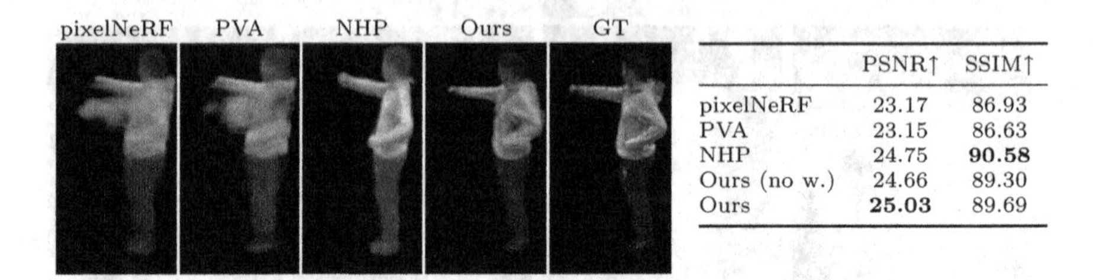

	PSNR↑	SSIM↑
pixelNeRF	23.17	86.93
PVA	23.15	86.63
NHP	24.75	**90.58**
Ours (no w.)	24.66	89.30
Ours	**25.03**	89.69

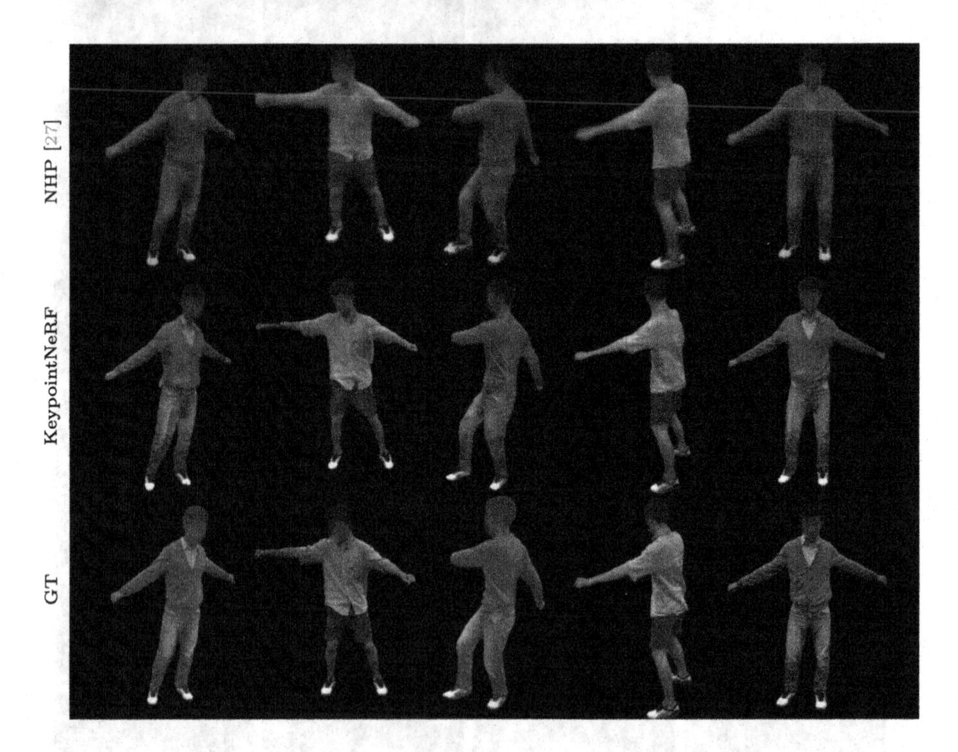

Fig. 5. Human body experiment. Comparison of NHP [27] and our method on unseen identities from the ZJU-MoCap dataset [43]. Best viewed in electronic format.

results on unseen identities for 438 novel views in Table 5 and side-by-side qualitative comparisons with NHP in Fig. 5. The results demonstrate that weighting the spatial encoding benefits reconstruction of human bodies as well. Our method is on par with the significantly more complex NHP, which relies on the accurate registration of the SMPL body model [32] and temporal feature fusion, whereas ours only requires skeleton keypoints.

7 Conclusion

We present a simple yet highly effective approach for generating high-fidelity volumetric humans from as few as two input images. The key to our approach is a novel spatial encoding based on relative information extracted from 3D keypoints. Our approach outperforms state-of-the-art methods for head reconstruction and better generalizes to challenging out-of-domain inputs, such as selfies captured in the wild by an iPhone. Since our approach does not require a parametric template mesh, it can be applied to the task of body reconstruction without modification, where it achieves performance comparable to more complicated prior work that has to rely on parametric human body models and temporal feature aggregation. We believe that our local spatial encoding based on keypoints might also be useful for many other category-specific neural rendering applications.

Acknowledgments. We thank Chen Cao for the help with the in-the-wild iPhone capture. M. M. and S. T. acknowledge the SNF grant 200021 204840.

References

1. Alldieck, T., Magnor, M., Xu, W., Theobalt, C., Pons-Moll, G.: Video based reconstruction of 3D people models. In: Proceedings of the IEEE/CVF Conference on Computer Vision and Pattern Recognition (2018)
2. Alldieck, T., Xu, H., Sminchisescu, C.: imGHUM: implicit generative models of 3D human shape and articulated pose. In: Proceedings of the IEEE/CVF Conference on Computer Vision and Pattern Recognition (CVPR) (2021)
3. Alldieck, T., Zanfir, M., Sminchisescu, C.: Photorealistic monocular 3D reconstruction of humans wearing clothing. In: Proceedings of the IEEE/CVF Conference on Computer Vision and Pattern Recognition, pp. 1506–1515 (2022)
4. Anguelov, D., Srinivasan, P., Koller, D., Thrun, S., Rodgers, J., Davis, J.: SCAPE: shape completion and animation of people. In: ACM SIGGRAPH 2005 Papers, pp. 408–416 (2005)
5. Athar, S., Shu, Z., Samaras, D.: Flame-in-NeRF: neural control of radiance fields for free view face animation. arXiv preprint arXiv:2108.04913 (2021)
6. Bansal, A., Chen, X., Russell, B., Gupta, A., Ramanan, D.: PixelNet: representation of the pixels, by the pixels, and for the pixels. arXiv preprint arXiv:1702.06506 (2017)
7. Blanz, V., Vetter, T.: A morphable model for the synthesis of 3D faces. In: Proceedings of the 26th Annual Conference on Computer Graphics and Interactive Techniques, pp. 187–194 (1999)
8. Buehler, M.C., Meka, A., Li, G., Beeler, T., Hilliges, O.: VariTex: variational neural face textures. In: Proceedings of the IEEE/CVF International Conference on Computer Vision (2021)
9. Cao, C., et al.: Authentic volumetric avatars from a phone scan. ACM Trans. Graph. (TOG) **41**, 1–19 (2022)
10. Cao, C., Wu, H., Weng, Y., Shao, T., Zhou, K.: Real-time facial animation with image-based dynamic avatars. ACM Trans. Graph. **35**(4) (2016)

11. Cao, Z., Simon, T., Wei, S.E., Sheikh, Y.: Realtime multi-person 2D pose estimation using part affinity fields. In: Proceedings of the IEEE Conference on Computer Vision and Pattern Recognition, pp. 7291–7299 (2017)
12. Chatziagapi, A., Athar, S., Moreno-Noguer, F., Samaras, D.: SIDER: single-image neural optimization for facial geometric detail recovery. arXiv preprint arXiv:2108.05465 (2021)
13. Chen, A., et al.: MVSNeRF: fast generalizable radiance field reconstruction from multi-view stereo. In: Proceedings of the IEEE International Conference on Computer Vision (ICCV) (2021)
14. Chibane, J., Bansal, A., Lazova, V., Pons-Moll, G.: Stereo radiance fields (SRF): learning view synthesis from sparse views of novel scenes. In: IEEE Conference on Computer Vision and Pattern Recognition (CVPR). IEEE (2021)
15. Gafni, G., Thies, J., Zollhofer, M., Niessner, M.: Dynamic neural radiance fields for monocular 4D facial avatar reconstruction. In: Proceedings of the IEEE/CVF Conference on Computer Vision and Pattern Recognition (CVPR) (2021)
16. Gao, C., Shih, Y., Lai, W.S., Liang, C.K., Huang, J.B.: Portrait neural radiance fields from a single image. arXiv preprint arXiv:2012.05903 (2020)
17. Grassal, P.W., Prinzler, M., Leistner, T., Rother, C., Nießner, M., Thies, J.: Neural head avatars from monocular RGB videos. In: Proceedings of the IEEE/CVF Conference on Computer Vision and Pattern Recognition, pp. 18653–18664 (2022)
18. Hartley, R., Zisserman, A.: Multiple View Geometry in Computer Vision. Cambridge University Press, Cambridge (2003)
19. He, T., Xu, Y., Saito, S., Soatto, S., Tung, T.: ARCH++: animation-ready clothed human reconstruction revisited. In: Proceedings of the IEEE International Conference on Computer Vision (ICCV) (2021)
20. Hu, L., et al.: Avatar digitization from a single image for real-time rendering. ACM Trans. Graph. (TOG) **36**(6), 1–14 (2017)
21. Huang, Z., Xu, Y., Lassner, C., Li, H., Tung, T.: ARCH: animatable reconstruction of clothed humans. In: Proceedings of the IEEE/CVF Conference on Computer Vision and Pattern Recognition (CVPR) (2020)
22. Ichim, A.E., Bouaziz, S., Pauly, M.: Dynamic 3D avatar creation from hand-held video input. ACM Trans. Graph. (TOG) **34**(4), 1–14 (2015)
23. Johnson, J., Alahi, A., Fei-Fei, L.: Perceptual losses for real-time style transfer and super-resolution. In: Leibe, B., Matas, J., Sebe, N., Welling, M. (eds.) ECCV 2016. LNCS, vol. 9906, pp. 694–711. Springer, Cham (2016). https://doi.org/10.1007/978-3-319-46475-6_43
24. Ke, Z., Sun, J., Li, K., Yan, Q., Lau, R.W.: MODNet: real-time trimap-free portrait matting via objective decomposition. In: AAAI (2022)
25. Kim, H., Garrido, P., Tewari, A., Xu, W., Thies, J., Nießner, M., Pérez, P., Richardt, C., Zollöfer, M., Theobalt, C.: Deep video portraits. ACM Trans. Graph. (TOG) **37**(4), 163 (2018)
26. Kingma, D.P., Ba, J.: Adam: a method for stochastic optimization. In: Proceedings of International Conference on Learning Representations (2015)
27. Kwon, Y., Kim, D., Ceylan, D., Fuchs, H.: Neural human performer: learning generalizable radiance fields for human performance rendering. In: Advances in Neural Information Processing Systems (NeurIPS). Curran Associates, Inc. (2021)
28. Lombardi, S., Saragih, J., Simon, T., Sheikh, Y.: Deep appearance models for face rendering. ACM Trans. Graph. (TOG) **37**(4), 1–13 (2018)
29. Lombardi, S., Simon, T., Saragih, J., Schwartz, G., Lehrmann, A., Sheikh, Y.: Neural volumes: learning dynamic renderable volumes from images. ACM Trans. Graph. (TOG) (2019)

30. Lombardi, S., Simon, T., Schwartz, G., Zollhoefer, M., Sheikh, Y., Saragih, J.: Mixture of volumetric primitives for efficient neural rendering. arXiv preprint arXiv:2103.01954 (2021)
31. Long, J., Shelhamer, E., Darrell, T.: Fully convolutional networks for semantic segmentation. In: Proceedings of the IEEE Conference on Computer Vision and Pattern Recognition, pp. 3431–3440 (2015)
32. Loper, M., Mahmood, N., Romero, J., Pons-Moll, G., Black, M.J.: SMPL: a skinned multi-person linear model. ACM Trans. Graph. (TOG) **34**(6), 1–16 (2015)
33. Martin-Brualla, R., et al.: LookinGood: enhancing performance capture with real-time neural re-rendering. arXiv preprint arXiv:1811.05029 (2018)
34. Matusik, W., Buehler, C., Raskar, R., Gortler, S.J., McMillan, L.: Image-based visual hulls. In: ACM SIGGRAPH, pp. 369–374 (2000)
35. Meka, A., et al.: Deep relightable textures: volumetric performance capture with neural rendering. ACM Trans. Graph. (TOG) **39**(6), 1–21 (2020)
36. Mihajlovic, M., Saito, S., Bansal, A., Zollhoefer, M., Tang, S.: COAP: compositional articulated occupancy of people. In: Proceedings of the IEEE/CVF Conference on Computer Vision and Pattern Recognition (CVPR) (2022)
37. Mihajlovic, M., Weder, S., Pollefeys, M., Oswald, M.R.: DeepSurfels: learning online appearance fusion. In: Proceedings of the IEEE/CVF Conference on Computer Vision and Pattern Recognition, pp. 14524–14535 (2021)
38. Mihajlovic, M., Zhang, Y., Black, M.J., Tang, S.: LEAP: learning articulated occupancy of people. In: Proceedings of the IEEE/CVF Conference on Computer Vision and Pattern Recognition (CVPR) (2021)
39. Mildenhall, B., et al.: Local light field fusion: practical view synthesis with prescriptive sampling guidelines. ACM Trans. Graph. (TOG) **38**, 1–14 (2019)
40. Mildenhall, B., Srinivasan, P.P., Tancik, M., Barron, J.T., Ramamoorthi, R., Ng, R.: NeRF: representing scenes as neural radiance fields for view synthesis. In: Vedaldi, A., Bischof, H., Brox, T., Frahm, J.-M. (eds.) ECCV 2020. LNCS, vol. 12346, pp. 405–421. Springer, Cham (2020). https://doi.org/10.1007/978-3-030-58452-8_24
41. Newell, A., Yang, K., Deng, J.: Stacked hourglass networks for human pose estimation. In: Leibe, B., Matas, J., Sebe, N., Welling, M. (eds.) ECCV 2016. LNCS, vol. 9912, pp. 483–499. Springer, Cham (2016). https://doi.org/10.1007/978-3-319-46484-8_29
42. Peng, S., et al.: Animatable neural radiance fields for modeling dynamic human bodies. In: Proceedings of the IEEE/CVF International Conference on Computer Vision, pp. 14314–14323 (2021)
43. Peng, S., et al.: Neural body: implicit neural representations with structured latent codes for novel view synthesis of dynamic humans. In: Proceedings of the IEEE/CVF Conference on Computer Vision and Pattern Recognition (CVPR) (2021)
44. Prokudin, S., Black, M.J., Romero, J.: SMPLpix: neural avatars from 3D human models. In: Proceedings of the IEEE/CVF Winter Conference on Applications of Computer Vision, pp. 1810–1819 (2021)
45. Raj, A., Tanke, J., Hays, J., Vo, M., Stoll, C., Lassner, C.: ANR: articulated neural rendering for virtual avatars. In: Proceedings of the IEEE/CVF Conference on Computer Vision and Pattern Recognition, pp. 3722–3731 (2021)
46. Raj, A., et al.: PVA: pixel-aligned volumetric avatars. In: Proceedings of the IEEE/CVF Conference on Computer Vision and Pattern Recognition (CVPR) (2021)

47. Ramon, E., et al.: H3D-net: few-shot high-fidelity 3D head reconstruction. arXiv preprint arXiv:2107.12512 (2021)
48. Rebain, D., Matthews, M., Yi, K.M., Lagun, D., Tagliasacchi, A.: LOLNeRF: learn from one look. In: Proceedings of the IEEE/CVF Conference on Computer Vision and Pattern Recognition, pp. 1558–1567 (2022)
49. Saito, S., Huang, Z., Natsume, R., Morishima, S., Kanazawa, A., Li, H.: PIFu: pixel-aligned implicit function for high-resolution clothed human digitization. In: Proceedings of the IEEE International Conference on Computer Vision (ICCV) (2019)
50. Saito, S., Simon, T., Saragih, J., Joo, H.: PIFuHD: multi-level pixel-aligned implicit function for high-resolution 3D human digitization. In: Proceedings of the IEEE/CVF Conference on Computer Vision and Pattern Recognition (CVPR) (2020)
51. Schwarz, K., Liao, Y., Niemeyer, M., Geiger, A.: GRAF: generative radiance fields for 3D-aware image synthesis. In: Advances in Neural Information Processing Systems (NeurIPS). Curran Associates, Inc. (2020)
52. Shao, R., Zhang, H., Zhang, H., Cao, Y., Yu, T., Liu, Y.: DoubleField: bridging the neural surface and radiance fields for high-fidelity human rendering. arXiv preprint arXiv:2106.03798 (2021)
53. Simonyan, K., Zisserman, A.: Very deep convolutional networks for large-scale image recognition. arXiv preprint arXiv:1409.1556 (2014)
54. Tewari, A., et al.: State of the art on neural rendering. In: Computer Graphics Forum, vol. 39, pp. 701–727. Wiley Online Library (2020)
55. Tewari, A., et al.: Advances in neural rendering. arXiv preprint arXiv:2111.05849 (2021)
56. Torralba, A., Efros, A.A.: Unbiased look at dataset bias. In: CVPR 2011, pp. 1521–1528. IEEE (2011)
57. Vlasic, D., Baran, I., Matusik, W., Popović, J.: Articulated mesh animation from multi-view silhouettes. ACM Trans. Graph. **27**(3), 97 (2008)
58. Vlasic, D., et al.: Dynamic shape capture using multi-view photometric stereo. ACM Trans. Graph. **28**(5), 174 (2009)
59. Wang, L., Chen, Z., Yu, T., Ma, C., Li, L., Liu, Y.: FaceVerse: a fine-grained and detail-controllable 3D face morphable model from a hybrid dataset. In: Proceedings of the IEEE/CVF Conference on Computer Vision and Pattern Recognition, pp. 20333–20342 (2022)
60. Wang, Q., et al.: IBRNet: learning multi-view image-based rendering. In: Proceedings of the IEEE/CVF Conference on Computer Vision and Pattern Recognition, pp. 4690–4699 (2021)
61. Wang, S., Mihajlovic, M., Ma, Q., Geiger, A., Tang, S.: Metaavatar: learning animatable clothed human models from few depth images. Adv. Neural Inf. Process. Syst. **34** (2021)
62. Wang, S., Schwartz, K., Geiger, A., Tang, S.: ARAH: animatable volume rendering of articulated human SDFs. In: European conference on computer vision (2022)
63. Wang, Z., et al.: Learning compositional radiance fields of dynamic human heads. In: Proceedings of the IEEE/CVF Conference on Computer Vision and Pattern Recognition (CVPR) (2021)
64. Weng, C.Y., Curless, B., Srinivasan, P.P., Barron, J.T., Kemelmacher-Shlizerman, I.: HumanNeRF: free-viewpoint rendering of moving people from monocular video. arXiv preprint arXiv:2201.04127 (2022)

65. Xu, H., Alldieck, T., Sminchisescu, C.: H-NeRF: Neural radiance fields for rendering and temporal reconstruction of humans in motion. Adv. Neural Inf. Process. Syst. **34** (2021)
66. Xu, H., Bazavan, E.G., Zanfir, A., Freeman, W.T., Sukthankar, R., Sminchisescu, C.: Ghum & ghuml: Generative 3D human shape and articulated pose models. In: Proceedings of the IEEE/CVF Conference on Computer Vision and Pattern Recognition, pp. 6184–6193 (2020)
67. Yu, A., Ye, V., Tancik, M., Kanazawa, A.: pixelNeRF: neural radiance fields from one or few images. In: Proceedings of the IEEE/CVF Conference on Computer Vision and Pattern Recognition, pp. 4578–4587 (2021)
68. Yu, T., et al.: DoubleFusion: real-time capture of human performances with inner body shapes from a single depth sensor. In: Proceedings of the IEEE Conference on Computer Vision and Pattern Recognition, pp. 7287–7296 (2018)
69. Zhao, H., et al.: High-fidelity human avatars from a single RGB camera. In: Proceedings of the IEEE/CVF Conference on Computer Vision and Pattern Recognition, pp. 15904–15913 (2022)
70. Zheng, Y., Abrevaya, V.F., Bühler, M.C., Chen, X., Black, M.J., Hilliges, O.: IM Avatar: Implicit morphable head avatars from videos. In: Proceedings of the IEEE/CVF Conference on Computer Vision and Pattern Recognition (CVPR) (2022)
71. Zheng, Z., Huang, H., Yu, T., Zhang, H., Guo, Y., Liu, Y.: Structured local radiance fields for human avatar modeling. In: Proceedings of the IEEE/CVF Conference on Computer Vision and Pattern Recognition, pp. 15893–15903 (2022)
72. Zheng, Z., Yu, T., Liu, Y., Dai, Q.: PaMIR: parametric model-conditioned implicit representation for image-based human reconstruction. IEEE Trans. Pattern Anal. Mach. Intell. (2021)

ViewFormer: NeRF-Free Neural Rendering from Few Images Using Transformers

Jonáš Kulhánek[1,2](✉), Erik Derner[1], Torsten Sattler[1], and Robert Babuška[1,3]

[1] Czech Institute of Informatics, Robotics and Cybernetics, Czech Technical University in Prague, Prague, Czech Republic
jonas.kulhanek@cvut.cz
[2] Faculty of Electrical Engineering, Czech Technical University in Prague, Prague, Czech Republic
[3] Cognitive Robotics, Faculty of 3mE, Delft University of Technology, Delft, The Netherlands

Abstract. Novel view synthesis is a long-standing problem. In this work, we consider a variant of the problem where we are given only a few context views sparsely covering a scene or an object. The goal is to predict novel viewpoints in the scene, which requires learning priors. The current state of the art is based on Neural Radiance Field (NeRF), and while achieving impressive results, the methods suffer from long training times as they require evaluating millions of 3D point samples via a neural network for each image. We propose a 2D-only method that maps multiple context views and a query pose to a new image in a single pass of a neural network. Our model uses a two-stage architecture consisting of a codebook and a transformer model. The codebook is used to embed individual images into a smaller latent space, and the transformer solves the view synthesis task in this more compact space. To train our model efficiently, we introduce a novel *branching attention* mechanism that allows us to use the same model not only for neural rendering but also for camera pose estimation. Experimental results on real-world scenes show that our approach is competitive compared to NeRF-based methods while not reasoning explicitly in 3D, and it is faster to train.

Keywords: Novel view synthesis · Neural rendering · Localization

1 Introduction

Image-based novel view synthesis, *i.e.*, rendering a 3D scene from a novel viewpoint given a set of context views (images and camera poses), is a long-standing

Supplementary Information The online version contains supplementary material available at https://doi.org/10.1007/978-3-031-19784-0_12.

S. Avidan et al. (Eds.): ECCV 2022, LNCS 13675, pp. 198–216, 2022.
https://doi.org/10.1007/978-3-031-19784-0_12

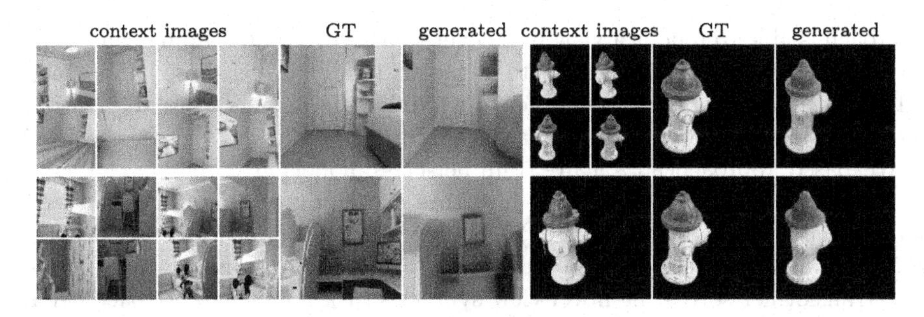

Fig. 1. Our novel view synthesis method renders images of previously unseen objects based on a few context images. It operates in 2D space without any explicit 3D reasoning (as opposed to NeRF-based approaches [51,72]). The results are shown on the CO3D [51] (**right**) and InteriorNet [32] (**left**) datasets rendered for unseen scenes

problem in computer graphics with applications ranging from robotics (*e.g.* planning to grasp objects) to augmented and virtual reality (*e.g.* interactive virtual meetings). Recently, the field has gained a lot of popularity thanks to Neural Radiance Field (NeRF) methods [2,40] that were successfully applied to the problem and outperformed prior approaches. We distinguish between two variants of the view synthesis problem. The first variant renders a novel view from multiple context images taken from similar viewpoints [40,69]. Only a (very) sparse set of context images is provided in the second variant [51,72], *i.e.*, larger viewpoint variations and missing observations need to be handled. The latter task is much more difficult as it is necessary to learn suitable priors that can be used to predict unseen scene parts. This paper focuses on the second variant.

Recently, generalizable NeRF-based approaches have been proposed to tackle this problem by learning priors for a class of objects and scenes [51,72]. Instead of learning a radiance field for each scene, they use context views captured from the target scene to construct the radiance field on the fly by projecting the image features from all context views into 3D. Highly optimized NeRF approaches [22,43,50,71] can be sped up by tuning or caching the radiance field representation [43], although often requiring lots of images per scene. To the best of our knowledge, these techniques do not apply to generalizable NeRF-based methods that do not learn a scene-specific radiance field, and take thousands of GPU-hours to train [51]. In contrast, 2D-only feed-forward networks can be highly efficient. However, explicitly encoding 3D geometric principles in them can be challenging. In our work, we thus pose the question: *Is reasoning in 3D necessary for high-quality novel view synthesis, or can a purely image-based method achieve a competitive performance?*

Recently, Rombach *et al.* [54] successfully tackled single-view novel view synthesis, where the model was able to predict novel views without explicit 3D reasoning. Inspired by these findings, we tackle the more complex problem of multi-view novel view synthesis. To answer the question, we propose a method with no explicit 3D reasoning able to predict novel views using multiple context images in a forward pass of a neural network. We train our model on a large collection of diverse scenes to enable the model to learn 3D priors implicitly. Our approach is able to render a view in a novel scene, unseen at training time, three orders of magnitude faster than state-of-the-art (SoTA) NeRF-based approaches [51], while also being ten times faster to train. Furthermore, we are able to train a single model to render multiple classes of scenes (see Fig. 1), whereas the SoTA NeRF-based approaches typically train per-class models [51].

Our model uses a two-stage architecture consisting of a Vector Quantized-Variational Autoencoder (VQ-VAE) codebook [45] and a transformer model. The codebook model is used to embed individual images into a smaller latent space. The transformer solves the novel view synthesis task in this latent space before the image is recovered via a decoder. This enables the codebook to focus on finer details in images while the transformer operates on shorter input sequences, reducing the quadratic memory complexity of its attention layer.

For training, we pass a sequence of views into the transformer and optimize it for all context sizes at the same time, effectively utilizing all images in the training batch, which is different from other methods [20,21,46,48] that train only one query view. Unlike autoregressive models [21,46,48], we do not decode images token-by-token but all tokens are decoded at once which is both faster and mathematically exact (while autoregressive models rely on greedy strategies). Our approach can be considered a combination of autoregressive [47,68] and masked [18] transformer models. With the standard attention mechanism, the complexity would be quadratic in the number of views, because we would have to stack different query views corresponding to different context sizes along the batch dimension. Therefore, we propose a novel attention mechanism called *branching attention* with constant overhead regardless of how many query views we optimize. Our attention mechanism also allows us to optimize the same model for the camera pose estimation task – predicting the query image's camera pose given a set of context views. Since this task can be considered an "inverse" of the novel view synthesis task [70], we consider the ability to perform both tasks via the same model to be an intriguing property. Even though the localization results are not yet competitive with state-of-the-art localization pipelines, we achieve a similar level of pose accuracy as comparable methods such as [1,60].

In summary, this paper makes the following contributions: **1**) We propose an efficient novel view synthesis approach that does not use explicit 3D reasoning. Our two-stage method consisting of a codebook model and a transformer is competitive with state-of-the-art NeRF-based approaches while being more efficient to train. Compared to similar methods that do not use explicit 3D reasoning [15,20,66], our approach is not only evaluated on synthetic data but performs well on real-world scenes. **2**) Our transformer model is a combination of an autoregressive and a masked transformer. We propose a novel attention mechanism called *branching attention* that allows us to optimize for multiple context sizes at once with a constant memory overhead. **3**) Thanks to the branching attention, our model can both render a novel view from a given pose and predict the pose for a given image. **4**) Our source code and pre-trained models are publicly available at https://github.com/jkulhanek/viewformer.

2 Related Work

Novel view synthesis has a long history [12,63]. Recently, deep learning techniques have been applied with great success, enabling higher realism [16,24,38,52,53]. Some approaches use explicit reconstructed geometry to warp context images into the target view [16,24,52,53,65]. In our approach, we do not require any proxy geometry and only operate on 2D images.

Neural Radiance Field methods [2,27,36,38,40,50,71] use neural networks to represent the continuous volumetric scene function. To render a view, for each pixel in the image plane, they project a ray into 3D space and query the radiance field in 3D points along each ray. The radiance field is trained for each scene separately. Some methods generalize to new scenes by conditioning the continuous volumetric function on the context images [55,64], which allows them to utilize trained priors and render views from scenes on which the model was not trained, much like our approach. Other approaches remove the trainable continuous volumetric scene function altogether. Instead, they reproject the context image's features into the 3D space and apply the NeRF-based rendering pipeline on top of this representation [25,51,67,69,72]. Similarly to these methods, our approach also utilizes few context views (less than 20), and it also generalizes to unseen objects. However, we do not use the continuous volumetric function nor the reprojection into the 3D space. A different approach, IBRNet [69], learns to copy existing colours from context views, effectively interpolating the context views. Unlike ours, it thus cannot be applied to the settings where the object is not covered enough by the context views [25,51,67,72].

A different line of work directly maps 2D context images to the 2D query image using an end-to-end neural network [15,20,66]. GQN-based methods [15,20,66] apply a CNN to context images and camera poses and combine the resulting features. While some GQN methods [15,20] do not use any explicit 3D reasoning (same as our approach), Tobin *et al.* [66] uses an epipolar attention to aggregate the features from the context views. We optimize our model on all context images and fully utilize the training sequences, whereas GQN methods optimize only a single query view.

A recent work by Rombach *et al.* [54] proposed an approach for novel view synthesis without explicit 3D modeling. They used a codebook and a transformer model to map a single context view to a novel view from a different pose. Their approach is limited in its scope to mostly forward-facing scenes where it is easier to render the novel view given a single context view and the poses have to be close to one another. It cannot be extended to more views due to the limit on the sequence size of the transformer model. In contrast, in our approach, we focus on using multiple context views, which we tackle through the proposed branching attention. Furthermore, we can jointly train the same model for both the novel view synthesis and camera pose estimation and our decoding is faster because we decode the output at once instead of autoregressive decoding.

Visual Localization. There is an enormous body of work tackling the problem of localization, where the goal is to output the camera pose given the camera image. *Structure-based* approaches use correspondences between 2D pixel positions and 3D scene coordinates for camera pose estimation [6,11,34,37,56,58,62]. Our method does not explicitly reason in 3D space, and the camera pose is instead predicted by the network. Simple *image retrieval* (IR) approaches store a database of all images with camera poses and for each query image they try to find the most similar images [9,10,17,26,59,74] and use them to estimate the pose of the query. IR methods can also be used to select relevant images for accurate pose estimation [4,26,56,74,75].

Pose regression methods train a convolutional neural network (CNN) to regress the camera pose of an input image. There are two categories: *absolute pose regression* (APR) methods [5,8,14,28,30,33,41,60] and *relative pose regression* (RPR) methods [1,19,31,33,39]. It was shown [59] that APR is often not (much) more accurate than IR. RPR methods do not train a CNN per scene or a set of scenes, but instead, condition the CNN on a set of context views. While our approach performs relative pose regression, the main focus of our method is on the novel view synthesis. Some pose regression methods use novel view synthesis methods [14,41,42,44], however, they assume there is a method that generates images, whereas our method performs both the novel view synthesis and camera pose regression in a single model. *Iterative refinement* pose regression methods [57,70] start with an initial camera pose estimate and refine it by an iterative process, however, our approach generates novel views and the camera pose estimates in a single forward pass.

3 Method

In this work, we tackle the problem of image-based novel view synthesis – given a set of *context* views, the algorithm has to generate the image it would most likely observe from a *query* camera pose. We focus on the case where the number of context views is small, and the views sparsely cover the 3D scene. Thus, the algorithm must hallucinate parts of the scene in a manner consistent with the context views. Therefore, it is necessary to learn a prior over a class of scenes (*e.g.*, all indoor environments) and use this prior for novel scenes. Besides rendering novel views, our model can also perform camera pose estimation, *i.e.*, the "inverse" of the view synthesis task: given a set of context views and a query image, the model outputs the camera pose from which the image was taken.

Our framework consists of two components: a codebook model and a transformer model. The codebook is used to map images to a smaller discrete latent space (*code space*), and back to the image space. In the *code space*, each image is represented by a sequence of *tokens*. For the novel view synthesis task, the transformer is given a set of context views in the code space and the query camera pose, and it generates an image in the *code space*. The codebook then maps the image tokens back to the image space. See Fig. 2 for an overview. For the camera pose estimation task, the transformer is given the set of context views and the query image in the code space, and it generates the camera pose using a regression head attached to the output of the transformer corresponding to the query image tokens.

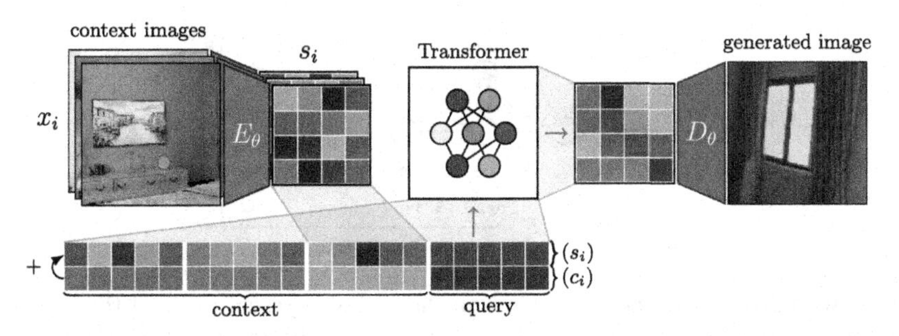

Fig. 2. Inference pipeline. The context images x_i are encoded by the codebook's encoder E_θ to the code representation s_i. We embed all tokens in s_i, and add the transformed camera pose c_i. The transformer generates the image tokens which are decoded by the codebook's decoder D_θ

Having the codebook and the transformer as separate components was inspired by the recent work on image generation [21,48,54]. The main motivation is to decrease its sequence size, because the required memory grows quadratically with it. It also allows us to separate image generation and view synthesis, enabling us to train the transformer more efficiently in a simpler space.

Codebook model is a VQ-VAE [45,49], which is a variational autoencoder with a categorical distribution over the latent space. The model consists of two parts: the encoder E_θ and decoder D_θ. The encoder first reduces the dimension of the input image from 128×128 pixels to 8×8 tokens by several strided convolution layers. The convolutional part is followed by a quantization layer, which maps the resulting feature map to a discrete space. The quantization layer stores n_{lat} embedding vectors of the same dimension as the feature vectors returned by the convolutional part of the encoder. It encodes each point of the feature map by returning the index of the closest embedding vector. The output of the encoder at position (i, j) for image x is:

$$\arg\min_k \|(f_\theta^{(enc)}(x))_{i,j} - W_k^{(emb)}\|_2, \tag{1}$$

where $W^{(emb)} \in \mathbb{R}^{n_{lat} \times d_{lat}}$ is the embedding matrix with rows $W_k^{(emb)}$ of length d_{lat} and $f_\theta^{(enc)}$ is the convolutional part of the encoder. The decoder then performs an inverse operation by first encoding the indices back to the embedding vectors by using $W^{(emb)}$ followed by several convolutional layers combined with upscaling to increase the spatial dimension back to the original image size.

Since the operation in Eq. (1) is not differentiable, we approximate the gradient with a straight-through estimator [3] and copy the gradients from the decoder input to the encoder output. The final loss for the codebook is a weighted sum of three parts: the pixel-wise mean absolute error (MAE) between the input image and the reconstructed image, the perceptual loss between the input and reconstructed image [21], and the commitment loss [45,49] $\mathcal{L}_c$, which encourages the output of the encoder to stay close to the chosen embedding vector to prevent it from fluctuating too frequently from one vector to another:

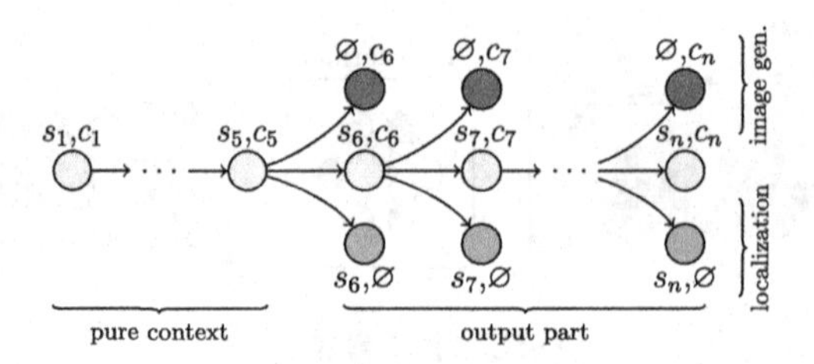

Fig. 3. Branching attention mechanism: the nodes represent parts of the processed sequence. Starting in any node and tracing the arrows backwards gives the sequence over which the attention is computed, *e.g.*, node $s_7, \varnothing$ attends to s_1, c_1, s_2, c_2, ..., $s_7, \varnothing$. Blue and red nodes in the last transformer block are used in the loss computation (Colour figure online)

$$\mathcal{L}_c = \min_k ||f_\theta^{(enc)}(x)_{i,j} - \text{sg}(W_k^{(emb)})||_2^2, \tag{2}$$

where sg is the stop-gradient operation [45]. We use the exponential moving average updates for the codebook [45]. See [45,49] for more details on the codebook, and the *supp. mat.* for the architecture details.

Transformer. We first describe the case of image generation and extend the approach to camera pose estimation later. We optimize the transformer for multiple context sizes and multiple query views in the batch at the same time. This has two benefits: it will allow the trained model to handle different context sizes, and the model will fully utilize the training batch (multiple images will be targets in the loss function). Each training batch consists of a set of n views. Let $(x_i)_{i=1}^n$ be the sequence of images under a random ordering and $(c_i)_{i=1}^n$ be the sequence of the associated camera poses. Let us also define the sequence of images transformed by the encoder E_θ parametrized by θ as $s_i = E_\theta(x_i)$, $i = 1, \ldots, n$. Note that each s_i is itself a sequence of tokens. With this formulation, we generate the next image in the sequence given all the previous views, effectively optimizing all different context sizes at once. Therefore, we model the probability $p(s_i|s_{<i}, c_{\leq i})$. Note that we do not optimize the first $n_{\min}$ views (called the *pure context*), because they usually do not provide enough information for the task.

In practice, we need to replace the tokens corresponding to each query view with mask tokens to allow the transformer to decode them in a single forward pass. For the image generation task, the tokens of the last image in the sequence are replaced with special mask tokens λ, and, for the localization task, the tokens of the last image do not include the camera pose (denoted as $\varnothing$). However, if we replaced the tokens in the training batch, the next query image would not be able to perceive the original tokens. Therefore, we have to process both the original and the masked tokens. For the i-th query image, we need the sequence of $i-1$ context views ending with masked tokens at the i-th position. We can represent the sequences as a tree (see Fig. 3) where different endings branch off the shared trunk. By following a leaf node back to the root of the tree, we recover the original sequence corresponding to the particular query view.

For localization, we train the model to output the camera pose c_i given $s_{\leq i}$ and $c_{<i}$. For image generation, this leads to $n - n_{\min}$ sequences. We attach a regression head to the hidden representation of all tokens of the last image in the sequence. The query image tokens form the input, and we mask the camera poses by replacing the camera pose representation with a single trainable vector.

Branching Attention. In this section, we introduce the *branching attention* which computes attention over the tree shown in Fig. 3, and allows us to optimize the transformer model for all context sizes and tasks very efficiently. Note that we have to forward all tree nodes through all layers of the transformer. Therefore, the memory and time complexity is proportional to the number of nodes in the tree and thus to the number of views and tasks.

The input to the branching attention is a sequence of triplets of keys, values, and queries: $\left((K^{(i)}, Q^{(i)}, V^{(i)})\right)_{i=0}^{p}$ for $p = 2$, because we train the model on two tasks. Each element in the sequence corresponds to a single row in Fig. 3 and $i = 0$ is the middle row. All $K^{(i)}$, $Q^{(i)}$, $V^{(i)}$ have the size $(nk^2) \times d_m$ where d_m is the dimensionality of the model and k is the size of the image in the latent space. The output of the branching attention is a sequence $\left(R^{(i)}\right)_{i=0}^{p}$. The case of $R^{(0)}$ is handled differently because it corresponds to the trunk shared for all tasks and context sizes. Let us define a lower triangular matrix $M \in \mathbb{R}^{n \times n}$, where $m_{i,j} = 1$ if $i \leq j$. We compute the causal block attention as:

$$R^{(0)} = (\mathrm{softmax}(Q^{(0)}(K^{(0)})^T) \odot M \otimes \mathbf{1}^{k^2 \times k^2})V^{(0)}, \tag{3}$$

where $\otimes$ and $\odot$ are the Kronecker and element-wise product, respectively, and $\mathbf{1}^{m \times n}$ is a matrix of ones. Equation (3) is similar to normal masked attention [68] with the only difference in the causal mask. In this case, we allow the model to attend to all previous images and all other vectors from the same image. For $i > 0$ we can compute $R^{(i)}$ as follows:

$$D = Q^{(i)}(K^{(0)})^T, \tag{4}$$

$$C = \begin{bmatrix} Q^{(i)}_{1:k^2}(K^{(i)}_{1:k^2})^T \\ \vdots \\ Q^{(i)}_{(n-1)\cdot k^2+1:n\cdot k^2}(K^{(i)}_{(n-1)\cdot k^2+1:n\cdot k^2})^T \end{bmatrix}, \tag{5}$$

$$S = \mathrm{softmax}([D, C]) \odot [(M - I) \otimes \mathbf{1}^{k^2 \times k^2}), \mathbf{1}^{nk^2 \times k^2}], \tag{6}$$

$$S' = S_{\cdot,1:n\cdot k^2}\,,\; S'' = S_{\cdot,n\cdot k^2+1:(n+1)\cdot k^2}, \tag{7}$$

$$R^{(i)} = S'V^{(0)} + \begin{bmatrix} S''_{1:k^2} V^{(i)}_{1:k^2} \\ \vdots \\ S''_{n\cdot k^2+1:(n+1)\cdot k^2} V^{(i)}_{n\cdot k^2+1:(n+1)\cdot k^2} \end{bmatrix}. \tag{8}$$

Matrix D represents the unmasked raw attention scores between i-th queries and keys from all previous images. Matrix C contains the raw pairwise attention scores between i-th queries and i-th keys (the ending of each sequence). Then, the softmax is computed to normalize the attention scores and the causal mask is applied to the result, yielding the attention matrix S, and the respective values are weighted by the computed scores. In particular, the scores contained in the last k^2 columns of the attention matrix are redistributed back to the associated i-th values. The result $R^{(0)}$ corresponds to the nodes in the middle row in Fig. 3, whereas $R^{(i)}, i > 0$ are the other nodes.

Transformer Input and Training. To build the input for the transformer, we first embed all image tokens into trainable vector embeddings of length d_m. Before passing camera poses to the network, we express all camera poses relative to the first context camera pose in the sequence. We represent camera poses by concatenating the 3D position with the normalized orientation quaternion (a unit quaternion with a positive real part). Finally, we transform the camera poses with a trainable feed-forward neural network in order to increase the dimension to the same size as image token embeddings d_m in order to be able to sum them.

Similarly to [47], we also add the positional embeddings by summing the input sequence with a sequence of trainable vectors. However, our positional embeddings are shared for all images in the sequence, *i.e.*, the i-th token of every image will share the same positional embedding.

The output of the last transformer block is passed to an affine layer followed by a softmax layer, and it is trained using the cross-entropy loss to recover the last k^2 tokens $(s_{j,1}, \ldots, s_{j,k^2})$. For the localization task, the output is passed through a two-layer feed-forward neural network, and it is trained using the mean square error to match the ground-truth camera pose of the last k^2 tokens. Note that we compute the losses over position and orientation separately and add them together without weighing.[1] Since we attach the pose prediction head to the hidden representation of all tokens of the query image, we obtain multiple pose estimates. During inference, we simply average them.

4 Experiments

To answer the question of whether explicit 3D reasoning is really needed for novel view synthesis, we designed a series of experiments evaluating the proposed approach. First, we evaluate the codebook, whose performance is the upper bound on what we can achieve with the full pipeline. We next compare our method to GQN-based methods [14,20,66] that also do not use continuous volumetric scene representations. We continue by evaluating our approach on other synthetic data. Then, we compare our approach to state-of-the-art NeRF-based approaches on a real-world dataset. Finally, we show our model's localization performance.

We evaluate our approach on both real and synthetic datasets: a) **Shepard-Metzler-7-Parts (SM7)** [20,61] is a synthetic dataset, where objects composed of 7 cubes of different colours are rotated in space. b) **ShapeNet** [13] is a synthetic dataset of simple objects. We use 128×128 pixel images rendered by [64] containing two categories: cars and chairs. c) **InteriorNet** [32] is a collection of interior environments designed by 1,100 professional designers. We used the publicly available part of the dataset (20k scenes with 20 images each). While

[1] We tried dynamic weighting as described in [29], but it performed worse.

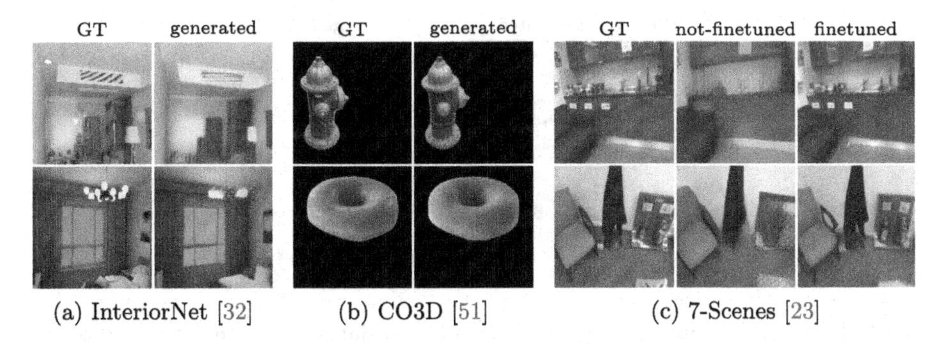

(a) InteriorNet [32] (b) CO3D [51] (c) 7-Scenes [23]

Fig. 4. Codebook evaluation on multiple datasets comparing the ground truth (**GT**) with the reconstructed image. For the 7-Scenes dataset, we compare the model fine-tuned and not-finetuned on the 7-Scenes dataset

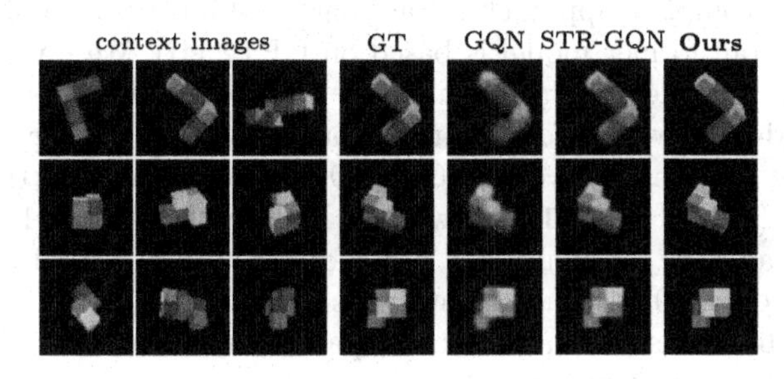

Fig. 5. Results on the SM7 dataset. We compare against GQN [20] and STR-GQN [15]

the dataset is synthetic, the renderings are similar to real-world environments. The first 600 environments serve as our test set. d) **Common Objects in 3D (CO3D)** [51] is a real-world dataset containing 1.5 million images showing almost 19k objects from 51 MS-COCO [35] categories (*e.g.*, apple, donut, vase, etc.). The capture of the dataset was crowd-sourced. e) **7-Scenes** [23] is a real-world dataset depicting 7 indoor scenes as captured by a Kinect RGB-D camera. The dataset consists of 44 sequences of 500–1,000 frames each and it is a standard benchmark for visual localization [1,8,30,31,39].

Codebook Evaluation. First, we evaluate the quality of our codebooks by measuring the quality of the images generated by the encoder-decoder architecture without the transformer. We trained codebooks of size 1,024 using the same hyperparameters for all experiments using an architecture very similar to [21]. The training took roughly 480 GPU-hours. A detailed description of the model and the hyperparameters is given in *supp. mat.* as well as in the published code.

Examples of reconstructed images are shown in Fig. 4. As can be seen, although losing some details and image sharpness, the codebooks can recover the overall shape well. The results show that using the codebook leads to good results, even though we use only 8×8 codes to represent an image. In some images, there are noticeable artifacts. In our analysis, we pinpointed the perceptual loss to be the cause, but removing the perceptual loss led to more blurry images. Further analysis of the codebooks is included in the *supp. mat.*

Fig. 6. Evaluation of our method on the InteriorNet dataset with the context size 19

Full Method Evaluation. The transformer is trained using only the tokens generated by the codebook. Having verified that our codebooks work as intended, we evaluate our complete approach in the context of image synthesis. The architecture of our transformer model is based on GPT2 [47]. We give more details on the architecture, the motivation, and the hyperparameters in the *supp. mat.*

The **SM7** dataset was used to compare our approach to other methods that only operate in 2D image space [15,20,66]. Our method achieved the best mean absolute error (MAE) of **1.61**, followed by E-GQN [66] with 2.14, STR-GQN [14] with 3.11 and the original GQN [20] method with MAE 3.13. The results were averaged over 1,000 scenes (context size was 3) and computed on images with size 64×64 pixels. A qualitative comparison is shown in Fig. 5.

We use the **InteriorNet** dataset because of its large size and realistic appearance. The models pre-trained on it are also used in other experiments. Since each scene provides 20 images, we use 19 context views. Figure 6 shows images generated by the model trained for both the localization and novel view synthesis tasks.

ShapeNet Evaluation. We used the InteriorNet pre-trained model and we fine-tuned it on the ShapeNet dataset. We trained a single model for both categories (cars and chairs) using 3 context views. The training details and additional results are given in *supp. mat.* We show the qualitative comparison with PixelNeRF [72] in Fig. 7. PixelNeRF trained a different model for each category.

The results show that our method achieves good visual quality overall, especially on the cars dataset. However, the geometry is slightly distorted on the chairs. Compared to PixelNeRF, it prefers to hallucinate a part of the scene instead of rendering a blurry image. This can cause some neighboring views to have a different colour or shape in places where the scene is less covered by context views. However, this problem can be reduced by simply adding the previously generated view to the set of context views. See the video in the *supp. mat.*

Common Objects in 3D. In order to show that we can transfer a model pre-trained on synthetic data to real-world scenes, we evaluate our method on the CO3D dataset [51]. We compare our approach with NeRF-based methods using the results reported in [51]. Unfortunately, we tried to train the PixelNeRF

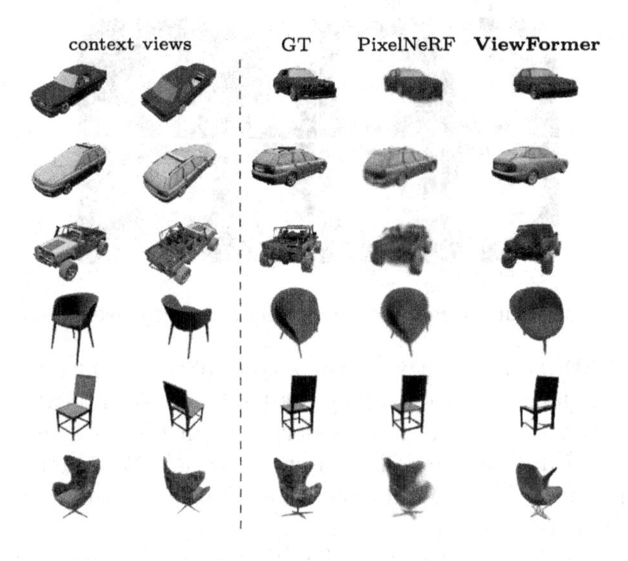

Fig. 7. ShapeNet qualitative comparison with PixelNeRF [72] using 2 context views

Table 1. Novel view synthesis results on the CO3D dataset [51] on all categories and 10 categories from [51]. We compare ViewFormer with and without localization ('no-loc') trained on all categories ('@ all cat.') and 10 selected categories ('@ 10 cat.'). We show the PSNR and LPIPS for seen and unseen scenes ('train' and 'test') and test PSNR with varying context size. The best value is **bold**; the second is underlined

EC	Method	3D	avg. test PSNR↑	avg. test LPIPS↓	avg. train PSNR↑	avg. train LPIPS↓	PSNR↑ @ # ctx. size 9	7	5	3	1
all categories	**ViewFormer @ all cat.**	✗	15.3	**0.23**	15.6	**0.22**	16.1	15.9	15.5	15.1	13.7
	ViewFormer no-loc @ all cat.	✗	15.4	**0.23**	15.8	**0.22**	16.2	16.0	15.6	15.2	13.8
	NerFormer [51]	✗	**15.7**	0.24	16.5	0.24	**16.7**	**16.4**	**16.1**	**15.5**	**13.9**
	SRN+WCE	✗	14.2	0.27	16.3	0.25	14.4	14.3	14.3	14.2	13.5
	SRN+WCE+γ	✗	13.7	0.28	**17.1**	0.25	14.0	13.8	13.9	13.7	13.2
	NeRF+WCE [25]	✗	11.6	0.27	12.6	0.27	11.9	11.8	11.8	11.6	10.8
10 categories	**ViewFormer @ 10 cat.**	✗	15.6	**0.25**	16.6	0.23	16.5	16.3	15.8	15.3	14.0
	ViewFormer no-loc @ 10 cat.	✗	15.6	**0.25**	17.1	**0.22**	16.5	16.2	15.8	15.3	14.0
	ViewFormer @ all cat.	✗	16.0	**0.25**	16.4	0.24	17.0	16.7	16.3	15.7	14.3
	ViewFormer no-loc @ all cat.	✗	16.1	**0.25**	16.6	0.23	17.0	16.8	16.3	15.8	14.3
	NerFormer [51]	✓	**17.6**	0.27	**17.9**	0.26	**18.9**	**18.6**	**18.1**	**17.1**	**15.1**
	SRN+WCE+γ	✓	14.4	0.27	17.6	0.24	14.6	14.5	14.6	14.5	13.9
	SRN+WCE	✓	14.6	0.27	16.6	0.26	14.9	14.8	14.8	14.6	13.9
	NeRF+WCE [25]	✓	13.8	0.27	14.3	0.27	12.6	14.5	14.4	14.2	13.8
	IPC+WCE	✓	13.5	0.37	14.1	0.36	13.8	13.8	13.7	13.6	12.6
	P3DMesh	✓	12.4	0.26	17.2	0.23	12.6	12.5	12.5	12.5	12.1
	NV+WCE	✓	11.6	0.35	12.3	0.34	11.7	11.6	11.6	11.6	11.3

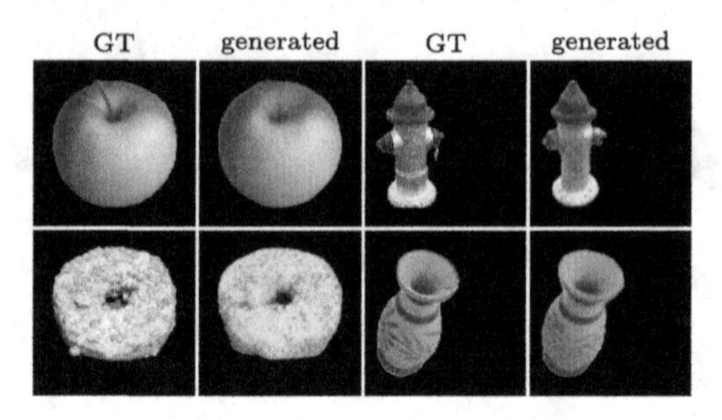

Fig. 8. Evaluation of our method on the CO3D dataset [51] with the context size 9

[72] on the CO3D dataset, but were not able to obtain good results. Therefore we omit it from the comparison. While the baselines are trained separately per category, we train two transformer models: one on the 10 categories used for evaluation in [51] and one for all dataset categories. We fine-tune the model trained on the InteriorNet dataset. The context size is 9. Additional details and hyperparameters are given in *supp. mat.*

The testing set of each category in the CO3D dataset is split into two subsets: 'train' and 'test' containing unseen images of objects seen and unseen during training respectively. We use the evaluation procedure provided by Reizenstein *et al.* [51]. It evaluates the model on 1,000 sequences from each category with context sizes 1, 3, 5, 7, 9. The PSNR) and the LPIPS distance [73] are reported. Note that the PSNR is calculated only on foreground pixels. For more details on the evaluation procedure and the details of compared methods, please see [51].

Table 1 shows results of the evaluation on all CO3D categories and on the 10 categories used for evaluation in [51]. Our method is competitive even though it does not explicitly reason in 3D as other baselines, does not utilize object masks, and even though we trained a single model for all categories while other baselines are trained per category. Note that on the whole dataset, the top-performing method, NerFormer [51], was trained for about 8400 GPU-hours while training our codebook took 480 GPU-hours, training the transformer on InteriorNet took 280 GPU-hours, and fine-tuning the transformer took 90 GPU-hours, giving a total of 850 GPU-hours. Also, note that rendering a single view takes 178 s for the NerFormer and only 93 ms for our approach.

The results show that our model has a large capacity (it is able to learn all categories while the baselines are only trained on a single category), and it benefits from more training data as can be seen when comparing models trained on 10 and all categories. We also observe that models achieve a higher performance on 10 categories than on all categories, suggesting that the categories selected by the authors of the dataset are easier to learn or of higher quality. All our models outperform all baselines in terms of LPIPS, which indicates that the images can look more realistic while possibly not matching the real images very precisely.

Figure 1 and 8 show qualitative results. Our method is able to generalize well to unseen object instances, although it tends to lose some details. To answer the original question if explicit 3D reasoning is needed for novel view synthesis, based on our results, we claim that even without explicit 3D reasoning, we can achieve similar results, especially when the data are noisy, *e.g.* a real-world dataset.

Evaluating Localization Accuracy on 7-Scenes. We compare the localization part of our approach to methods from the literature on the 7-Scenes dataset [23]. Due to space constraints, here we only summarize the results of the comparisons. Detailed results can be found in the *supp. mat.*

Our approach performs similar to existing APR and RPR techniques that also use only a single forward pass in a network [1,8,30,60], but worse than iterative approaches such as [19] or methods that use more densely spaced synthetic views as additional input [41]. Note that these approaches that do not use 3D scene geometry are less accurate than state-of-the-art methods based on 2D-3D correspondences [7,56,58]. Overall, the results show that our approach achieves a similar level of pose accuracy as comparable methods. Furthermore, our approach is able to perform both localization and novel view synthesis in a simple forward pass, while other methods can only be used for localization.

5 Conclusions and Future Work

This paper presents a two-stage approach to novel view synthesis from a few sparsely distributed context images. We train our model on classes of similar 3D scenes to be able to generalize to a novel scene with only a handful of images as opposed to NeRF and similar methods that are trained per scene. The model consists of a VQ-VAE codebook [45] and a transformer model. To efficiently train the transformer, we propose a novel branching attention module. Our approach, ViewFormer, can render a view from a previously unseen scene in 93 ms without any explicit 3D reasoning and we train a single model to render multiple categories of objects, whereas NeRF-based approaches train per-category models [51]. We show that our method is competitive with SoTA NeRF-based approaches especially on real-world data, even without any explicit 3D reasoning. This is an intriguing result because it implies that either current NeRF-based methods are not utilizing the 3D priors effectively or that a 2D-only model is able to learn it on its own without explicit 3D modeling. The experiments also show that ViewFormer outperforms other 2D-only multi-view methods.

One limitation of our approach is the large amount of data needed, which we tackle through pre-training on a large synthetic dataset. Also, we need to fine-tune both the codebook and the transformer to achieve high-quality results on new datasets, which could be resolved by utilizing a larger codebook trained on more data. Using more tokens to represent images should increase the rendering quality and pose accuracy. We also want to experiment with a simpler architecture with no codebook and larger scenes, possibly of outdoor environments.

Acknowledgement. This work was supported by the European Regional Development Fund under projects IMPACT (reg. no. CZ.02.1.01/0.0/0.0/15_003/0000468) and Robotics for Industry 4.0 (reg. no. CZ.02.1.01/0.0/0.0/15_003/0000470), the EU Horizon 2020 project RICAIP (grant agreement No 857306), the Grant Agency of the Czech Technical University in Prague (grant no. SGS22/112/OHK3/2T/13), and the Ministry of Education, Youth and Sports of the Czech Republic through the e-INFRA CZ (ID:90140).

References

1. Balntas, V., Li, S., Prisacariu, V.: RelocNet: continuous metric learning relocalisation using neural nets. In: Proceedings of the European Conference on Computer Vision (ECCV), pp. 751–767 (2018)
2. Barron, J.T., Mildenhall, B., Tancik, M., Hedman, P., Martin-Brualla, R., Srinivasan, P.P.: Mip-NeRF: a multiscale representation for anti-aliasing neural radiance fields. In: Proceedings of the IEEE/CVF International Conference on Computer Vision, pp. 5855–5864 (2021)
3. Bengio, Y., Léonard, N., Courville, A.: Estimating or propagating gradients through stochastic neurons for conditional computation. arXiv preprint arXiv:1308.3432 (2013)
4. Bhayani, S., Sattler, T., Barath, D., Beliansky, P., Heikkilä, J., Kukelova, Z.: Calibrated and partially calibrated semi-generalized homographies. In: Proceedings of the IEEE/CVF International Conference on Computer Vision (ICCV) (2021)
5. Blanton, H., Greenwell, C., Workman, S., Jacobs, N.: Extending absolute pose regression to multiple scenes. In: Proceedings of the IEEE/CVF Conference on Computer Vision and Pattern Recognition Workshops, pp. 38–39 (2020)
6. Brachmann, E., Rother, C.: Visual camera re-localization from RGB and RGB-D images using DSAC. IEEE Trans. Pattern Anal. Mach. Intell. **44**(9), 5847–5865 (2021)
7. Brachmann, E., Rother, C.: Visual camera re-localization from RGB and RGB-D images using DSAC. IEEE Trans. Pattern Anal. Mach. Intell. **44**, 5847–5865 (2021)
8. Brahmbhatt, S., Gu, J., Kim, K., Hays, J., Kautz, J.: Geometry-aware learning of maps for camera localization. In: Proceedings of the IEEE Conference on Computer Vision and Pattern Recognition, pp. 2616–2625 (2018)
9. Camposeco, F., Cohen, A., Pollefeys, M., Sattler, T.: Hybrid camera pose estimation. In: Proceedings of the IEEE Conference on Computer Vision and Pattern Recognition, pp. 136–144 (2018)
10. Cao, S., Snavely, N.: Graph-based discriminative learning for location recognition. In: Proceedings of the IEEE Conference on Computer Vision and Pattern Recognition, pp. 700–707 (2013)
11. Cavallari, T., et al.: Real-time RGB-D camera pose estimation in novel scenes using a relocalisation cascade. IEEE Trans. Pattern Anal. Mach. Intell. **42**(10), 2465–2477 (2019)
12. Chan, S., Shum, H.Y., Ng, K.T.: Image-based rendering and synthesis. IEEE Signal Process. Mag. **24**(6), 22–33 (2007)
13. Chang, A.X., et al.: ShapeNet: an information-rich 3D model repository. arXiv preprint arXiv:1512.03012 (2015)
14. Chen, S., Wang, Z., Prisacariu, V.: Direct-PoseNet: absolute pose regression with photometric consistency. arXiv preprint arXiv:2104.04073 (2021)
15. Chen, W.C., Hu, M.C., Chen, C.S.: STR-GQN: Scene representation and rendering for unknown cameras based on spatial transformation routing. In: Proceedings of the IEEE/CVF International Conference on Computer Vision, pp. 5966–5975 (2021)
16. Choi, I., Gallo, O., Troccoli, A., Kim, M.H., Kautz, J.: Extreme view synthesis. In: Proceedings of the IEEE/CVF International Conference on Computer Vision, pp. 7781–7790 (2019)

17. Derner, E., Gomez, C., Hernandez, A.C., Barber, R., Babuška, R.: Change detection using weighted features for image-based localization. Robot. Auton. Syst. **135**, 103676 (2021)
18. Devlin, J., Chang, M.W., Lee, K., Toutanova, K.: BERT: pre-training of deep bidirectional transformers for language understanding. In: Proceedings of the 2019 Conference of the North American Chapter of the Association for Computational Linguistics: Human Language Technologies, Volume 1 (Long and Short Papers), pp. 4171–4186. Association for Computational Linguistics, Minneapolis (2019)
19. Ding, M., Wang, Z., Sun, J., Shi, J., Luo, P.: CamNet: coarse-to-fine retrieval for camera re-localization. In: Proceedings of the IEEE/CVF International Conference on Computer Vision, pp. 2871–2880 (2019)
20. Eslami, S.A., et al.: Neural scene representation and rendering. Science **360**(6394), 1204–1210 (2018)
21. Esser, P., Rombach, R., Ommer, B.: Taming transformers for high-resolution image synthesis. In: Proceedings of the IEEE/CVF Conference on Computer Vision and Pattern Recognition, pp. 12873–12883 (2021)
22. Garbin, S.J., Kowalski, M., Johnson, M., Shotton, J., Valentin, J.: FastNeRF: high-fidelity neural rendering at 200FPS. In: Proceedings of the IEEE/CVF International Conference on Computer Vision, pp. 14346–14355 (2021)
23. Glocker, B., Izadi, S., Shotton, J., Criminisi, A.: Real-time RGB-D camera relocalization. In: 2013 IEEE International Symposium on Mixed and Augmented Reality (ISMAR), pp. 173–179. IEEE (2013)
24. Hedman, P., Philip, J., Price, T., Frahm, J.M., Drettakis, G., Brostow, G.: Deep blending for free-viewpoint image-based rendering. ACM Trans. Graph. (TOG) **37**(6), 1–15 (2018)
25. Henzler, P., et al.: Unsupervised learning of 3D object categories from videos in the wild. In: Proceedings of the IEEE/CVF Conference on Computer Vision and Pattern Recognition, pp. 4700–4709 (2021)
26. Irschara, A., Zach, C., Frahm, J.M., Bischof, H.: From structure-from-motion point clouds to fast location recognition. In: 2009 IEEE Conference on Computer Vision and Pattern Recognition, pp. 2599–2606. IEEE (2009)
27. Jain, A., Tancik, M., Abbeel, P.: Putting NeRF on a diet: Semantically consistent few-shot view synthesis. In: Proceedings of the IEEE/CVF International Conference on Computer Vision, pp. 5885–5894 (2021)
28. Kendall, A., Cipolla, R.: Modelling uncertainty in deep learning for camera relocalization. In: 2016 IEEE International Conference on Robotics and Automation (ICRA), pp. 4762–4769. IEEE (2016)
29. Kendall, A., Cipolla, R.: Geometric loss functions for camera pose regression with deep learning. In: Proceedings of the IEEE Conference on Computer Vision and Pattern Recognition, pp. 5974–5983 (2017)
30. Kendall, A., Grimes, M., Cipolla, R.: PoseNet: a convolutional network for real-time 6-DOF camera relocalization. In: Proceedings of the IEEE International Conference on Computer Vision, pp. 2938–2946 (2015)
31. Laskar, Z., Melekhov, I., Kalia, S., Kannala, J.: Camera relocalization by computing pairwise relative poses using convolutional neural network. In: Proceedings of the IEEE International Conference on Computer Vision Workshops, pp. 929–938 (2017)
32. Li, W., et al.: InteriorNet: mega-scale multi-sensor photo-realistic indoor scenes dataset. In: British Machine Vision Conference (BMVC) (2018)
33. Li, X., Ling, H.: TransCamP: graph transformer for 6-DoF camera pose estimation. ArXiv abs/2105.14065 (2021)

34. Li, Y., Snavely, N., Huttenlocher, D., Fua, P.: Worldwide pose estimation using 3D point clouds. In: Fitzgibbon, A., Lazebnik, S., Perona, P., Sato, Y., Schmid, C. (eds.) ECCV 2012. LNCS, vol. 7572, pp. 15–29. Springer, Heidelberg (2012). https://doi.org/10.1007/978-3-642-33718-5_2
35. Lin, T.-Y., et al.: Microsoft COCO: common objects in context. In: Fleet, D., Pajdla, T., Schiele, B., Tuytelaars, T. (eds.) ECCV 2014. LNCS, vol. 8693, pp. 740–755. Springer, Cham (2014). https://doi.org/10.1007/978-3-319-10602-1_48
36. Liu, L., Gu, J., Zaw Lin, K., Chua, T.S., Theobalt, C.: Neural sparse voxel fields. Adv. Neural Inf. Process. Syst. **33**, 15651–15663 (2020)
37. Lynen, S., et al.: Large-scale, real-time visual-inertial localization revisited. Int. J. Robot. Res. **39**(9), 1061–1084 (2020)
38. Martin-Brualla, R., Radwan, N., Sajjadi, M.S., Barron, J.T., Dosovitskiy, A., Duckworth, D.: NeRF in the wild: neural radiance fields for unconstrained photo collections. In: Proceedings of the IEEE Conference on Computer Vision and Pattern Recognition, pp. 7210–7219 (2021)
39. Melekhov, I., Ylioinas, J., Kannala, J., Rahtu, E.: Relative camera pose estimation using convolutional neural networks. In: Blanc-Talon, J., Penne, R., Philips, W., Popescu, D., Scheunders, P. (eds.) ACIVS 2017. LNCS, vol. 10617, pp. 675–687. Springer, Cham (2017). https://doi.org/10.1007/978-3-319-70353-4_57
40. Mildenhall, B., Srinivasan, P.P., Tancik, M., Barron, J.T., Ramamoorthi, R., Ng, R.: NeRF: representing scenes as neural radiance fields for view synthesis. In: Vedaldi, A., Bischof, H., Brox, T., Frahm, J.-M. (eds.) ECCV 2020. LNCS, vol. 12346, pp. 405–421. Springer, Cham (2020). https://doi.org/10.1007/978-3-030-58452-8_24
41. Moreau, A., Piasco, N., Tsishkou, D., Stanciulescu, B., de La Fortelle, A.: LENS: localization enhanced by NeRF synthesis. In: 5th Annual Conference on Robot Learning (2021)
42. Mueller, M.S., Sattler, T., Pollefeys, M., Jutzi, B.: Image-to-image translation for enhanced feature matching, image retrieval and visual localization. ISPRS Ann. Photogram. Remote Sens. Spat. Inf. Sci. **4**, 111–119 (2019)
43. Müller, T., Evans, A., Schied, C., Keller, A.: Instant neural graphics primitives with a multiresolution hash encoding. arXiv preprint arXiv:2201.05989 (2022)
44. Ng, T., Lopez-Rodriguez, A., Balntas, V., Mikolajczyk, K.: Reassessing the limitations of CNN methods for camera pose regression. arXiv:2108.07260 (2021)
45. van den Oord, A., Vinyals, O., kavukcuoglu, k.: Neural discrete representation learning. In: Guyon, I., et al. (eds.) Advances in Neural Information Processing Systems, vol. 30. Curran Associates, Inc. (2017)
46. Parmar, N., et al.: Image transformer. In: International Conference on Machine Learning, pp. 4055–4064. PMLR (2018)
47. Radford, A., Wu, J., Child, R., Luan, D., Amodei, D., Sutskever, I.: Language models are unsupervised multitask learners (2019)
48. Ramesh, A., et al.: Zero-shot text-to-image generation. arXiv preprint arXiv:2102.12092 (2021)
49. Razavi, A., van den Oord, A., Vinyals, O.: Generating diverse high-fidelity images with VQ-VAE-2. In: Wallach, H., Larochelle, H., Beygelzimer, A., d' Alché-Buc, F., Fox, E., Garnett, R. (eds.) Advances in Neural Information Processing Systems, vol. 32. Curran Associates, Inc. (2019)
50. Reiser, C., Peng, S., Liao, Y., Geiger, A.: KiloNeRF: speeding up neural radiance fields with thousands of tiny MLPs. In: Proceedings of the IEEE/CVF International Conference on Computer Vision, pp. 14335–14345 (2021)

51. Reizenstein, J., Shapovalov, R., Henzler, P., Sbordone, L., Labatut, P., Novotny, D.: Common objects in 3D: large-scale learning and evaluation of real-life 3D category reconstruction. In: Proceedings of the IEEE/CVF International Conference on Computer Vision, pp. 10901–10911 (2021)
52. Riegler, G., Koltun, V.: Free view synthesis. In: Vedaldi, A., Bischof, H., Brox, T., Frahm, J.-M. (eds.) ECCV 2020. LNCS, vol. 12364, pp. 623–640. Springer, Cham (2020). https://doi.org/10.1007/978-3-030-58529-7_37
53. Riegler, G., Koltun, V.: Stable view synthesis. In: Proceedings of the IEEE/CVF Conference on Computer Vision and Pattern Recognition, pp. 12216–12225 (2021)
54. Rombach, R., Esser, P., Ommer, B.: Geometry-free view synthesis: transformers and no 3D priors. In: Proceedings of the IEEE/CVF International Conference on Computer Vision, pp. 14356–14366 (2021)
55. Saito, S., Huang, Z., Natsume, R., Morishima, S., Kanazawa, A., Li, H.: PIFu: pixel-aligned implicit function for high-resolution clothed human digitization. In: Proceedings of the IEEE/CVF International Conference on Computer Vision, pp. 2304–2314 (2019)
56. Sarlin, P.E., Cadena, C., Siegwart, R., Dymczyk, M.: From coarse to fine: robust hierarchical localization at large scale. In: CVPR (2019)
57. Sarlin, P.E., et al.: Back to the feature: learning robust camera localization from pixels to pose. In: Proceedings of the IEEE/CVF Conference on Computer Vision and Pattern Recognition, pp. 3247–3257 (2021)
58. Sattler, T., Leibe, B., Kobbelt, L.: Efficient & effective prioritized matching for large-scale image-based localization. IEEE Trans. Pattern Anal. Mach. Intell. **39**(9), 1744–1756 (2016)
59. Sattler, T., Zhou, Q., Pollefeys, M., Leal-Taixe, L.: Understanding the limitations of CNN-based absolute camera pose regression. In: Proceedings of the IEEE/CVF Conference On computer Vision and Pattern Recognition, pp. 3302–3312 (2019)
60. Shavit, Y., Ferens, R., Keller, Y.: Learning multi-scene absolute pose regression with transformers. arXiv preprint arXiv:2103.11468 (2021)
61. Shepard, R.N., Metzler, J.: Mental rotation of three-dimensional objects. Science **171**(3972), 701–703 (1971)
62. Shotton, J., Glocker, B., Zach, C., Izadi, S., Criminisi, A., Fitzgibbon, A.: Scene coordinate regression forests for camera relocalization in RGB-D images. In: CVPR (2013)
63. Shum, H., Kang, S.B.: Review of image-based rendering techniques. In: Visual Communications and Image Processing 2000, vol. 4067, pp. 2–13. International Society for Optics and Photonics (2000)
64. Sitzmann, V., Zollhöfer, M., Wetzstein, G.: Scene representation networks: continuous 3D-structure-aware neural scene representations. Adv. Neural Inf. Process. Syst. **32**, 1121–1132 (2019)
65. Thies, J., Zollhöfer, M., Theobalt, C., Stamminger, M., Nießner, M.: Image-guided neural object rendering. In: 8th International Conference on Learning Representations. OpenReview. net (2020)
66. Tobin, J., Zaremba, W., Abbeel, P.: Geometry-aware neural rendering. Adv. Neural. Inf. Process. Syst. **32**, 11559–11569 (2019)
67. Trevithick, A., Yang, B.: GRF: learning a general radiance field for 3D representation and rendering. In: Proceedings of the IEEE/CVF International Conference on Computer Vision, pp. 15182–15192 (2021)
68. Vaswani, A., et al.: Attention is all you need. In: Advances in neural information processing systems, pp. 5998–6008 (2017)

69. Wang, Q., et al.: IBRNET: learning multi-view image-based rendering. In: Proceedings of the IEEE/CVF Conference on Computer Vision and Pattern Recognition, pp. 4690–4699 (2021)
70. Yen-Chen, L., Florence, P., Barron, J.T., Rodriguez, A., Isola, P., Lin, T.Y.: iNeRF: inverting neural radiance fields for pose estimation. In: IEEE/RSJ International Conference on Intelligent Robots and Systems (IROS) (2021)
71. Yu, A., Li, R., Tancik, M., Li, H., Ng, R., Kanazawa, A.: PlenOctrees for real-time rendering of neural radiance fields. In: Proceedings of the IEEE/CVF International Conference on Computer Vision, pp. 5752–5761 (2021)
72. Yu, A., Ye, V., Tancik, M., Kanazawa, A.: pixelNeRF: neural radiance fields from one or few images. In: Proceedings of the IEEE/CVF Conference on Computer Vision and Pattern Recognition, pp. 4578–4587 (2021)
73. Zhang, R., Isola, P., Efros, A.A., Shechtman, E., Wang, O.: The unreasonable effectiveness of deep features as a perceptual metric. In: Proceedings of the IEEE Conference on Computer Vision and Pattern Recognition, pp. 586–595 (2018)
74. Zhang, W., Kosecka, J.: Image based localization in urban environments. In: Third International Symposium on 3D Data Processing, Visualization, and Transmission (3DPVT 2006), pp. 33–40. IEEE (2006)
75. Zhou, Q., Sattler, T., Pollefeys, M., Leal-Taixé, L.: To learn or not to learn: visual localization from essential matrices. In: 2020 IEEE International Conference on Robotics and Automation (ICRA), pp. 3319–3326 (2020)

L-Tracing: Fast Light Visibility Estimation on Neural Surfaces by Sphere Tracing

Ziyu Chen[1], Chenjing Ding[2,3], Jianfei Guo[3], Dongliang Wang[2,3], Yikang Li[2,3], Xuan Xiao[4], Wei Wu[2,3], and Li Song[1(✉)]

[1] Department of Electronic Engineering, Shanghai Jiao Tong University, Shanghai, China
ziyu.sjtu@gmail.com, song_li@sjtu.edu.cn
[2] SenseTime Research, Shanghai, China
{dingchenjing,wangdongliang,wuwei}@senseauto.com
[3] Shanghai AI Laboratory, Shanghai, China
{guojianfei,liyikang}@pjlab.org.cn
[4] Department of Mechanical Engineering, Tsinghua University, Beijing, China
x-xiao20@mails.tsinghua.edu.cn

Abstract. We introduce a highly efficient light visibility estimation method, called L-Tracing, for reflectance factorization on neural implicit surfaces. Light visibility is indispensable for modeling shadows and specular of high quality on object's surface. For neural implicit representations, former methods of computing light visibility suffer from efficiency and quality drawbacks. L-Tracing leverages the distance meaning of the Signed Distance Function(SDF), and computes the light visibility of the solid object surface according to binary geometry occlusions. We prove the linear convergence of L-Tracing algorithm and give out the theoretical lower bound of tracing iteration. Based on L-Tracing, we propose a new surface reconstruction and reflectance factorization framework. Experiments show our framework performs nearly ~10x speedup on factorization, and achieves competitive albedo and relighting results with existing approaches.

Keywords: Surface reflectance factorization · Sphere tracing · Inverse rendering · 3D deep learning

1 Introduction

Reconstructing an object's surface, reflectance property, and environment illumination from multi-view images is a fundamental problem in computer vision

The work is done when Ziyu Chen is an intern at SenseTime.

Supplementary Information The online version contains supplementary material available at https://doi.org/10.1007/978-3-031-19784-0_13.

S. Avidan et al. (Eds.): ECCV 2022, LNCS 13675, pp. 217–233, 2022.
https://doi.org/10.1007/978-3-031-19784-0_13

and graphics. Deep learning methods for objects' reconstruction save human resources for producing high-quality 3D models. Recent works [24,29,33] adopt neural volume rendering to learn detailed surfaces of high quality from multi-view images. Researchers [2,3,5,22,26,28,32,36,39] further explore the field that decomposes the reflectance of neural surfaces and environment illumination, to produce renderable 3D models with de-illuminated textures.

In the area of surface reflectance and ambient illumination decomposition, estimating whether the surface point is visible to light sources is called light visibility estimation. Since the color of each observed pixel is aggregated by the reflected rays on surface from light sources, light visibility is important for modeling shadows and the specular effect.

Recently, there are several works studying reflectance factorization of neural volume representation [3,26,37,39]. The light visibility in these studies is regarded as the possibility the light passing to a certain point. Under this assumption, the differential probability distribution along each ray is needed to compute light visibility by volumetric integration. This is computationally expensive since hundreds of points on each ray are sampled to compute the discrete form of volumetric integration.

NeRFactor [39] and NeRV [26] reduce the computational cost by estimating light visibility via a Multi Layer Perceptron(MLP), which is trained by precomputed light visibility caches. These methods only reduce the computational cost of inference time, the caching process still depends on volumetric integration, consuming hundreds of hours which buries a heavy burden at the training time.

We note that previous work performing reflectance decomposition on neural implicit shapes suffer from the following drawbacks: 1)Some methods [3,12, 32,37] fail to explicitly model the light visibility, where the learning process entirely depends on neural networks to learn the geometry occlusion and shadows. 2)Some methods [26,39] show poor computation efficiency because of using the volumetric integration to compute the light visibility. In this way, computing the light visibility once usually queries hundreds of points along the ray for the density. 3)Under the differential probability assumption of neural volume representation, some methods [3,26,39] incorrectly model the objects as translucent volumes instead of solid spatial boundaries, producing low-quality surfaces.

Our goal is to present a highly efficient and accurate light visibility estimation method that explicitly respects solid objects and binary geometry occlusions. We make up the third drawback by using Signed Distance Function(SDF) as our shape representation. SDF is a continuous implicit surface representation that is easily optimized by gradient decent. The signed distance induces a highly efficient ray tracing algorithm: sphere tracing. We propose L-tracing based on sphere tracing, to solve the computational problem for light visibility estimation. Different from the former assumption, the light visibility in our work is estimated according to hard spatial occlusions instead of integrated visible probability, which improves the learning of sharp shadows and specular effects. Moreover, L-tracing achieves ~10x faster than volumetric integration. We also provide the convergence proof of L-tracing.

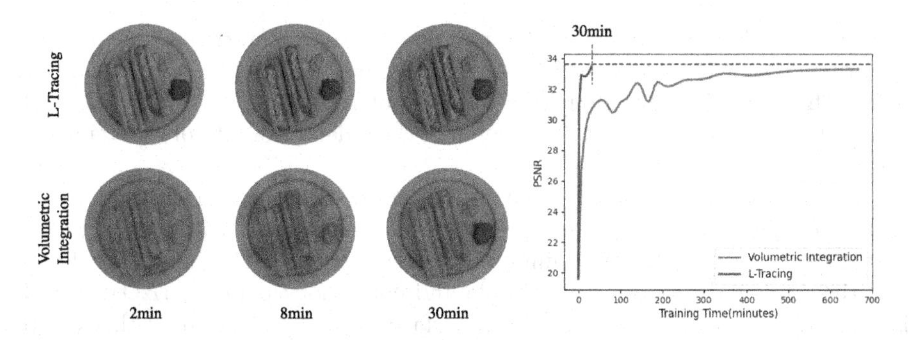

Fig. 1. L-Tracing is a highly efficient light visibility estimation method for reflectance factorization on neural implicit surfaces. We show that our L-Tracing based reflectance factorization framework adapted from NeRFactor [39] produces photo-realistic novel view images with nearly 10x speedup, compared with the same framework applying volumetric integration for light visibility estimation.

We propose a two-stage learning framework for reconstructing objects' surface and decomposing the reflectance and environment illumination. First, we adopt recent works [29] of reconstructing SDF from multi-view images without object masks via differentiable volume rendering. Second, we apply L-Tracing as our light visibility estimation method, and jointly optimize the albedo, reflectance and environment illumination in a similar manner to [39]. As shown in Fig. 1, we compare the convergence speed and novel views' quality of the reflectance factorization stage based on L-Tracing against those based on volumetric integration. The results show that L-Tracing is far more efficient than volumetric integration.

In summary, our work makes the following contributions:

- We propose L-tracing, a highly efficient light visibility estimation method on neural implicit surfaces, to accelerate the rendering process for reflectance decomposition tasks. We also prove the convergence of L-Tracing theoretically.
- We propose a two-stage framework for neural implicit surface reconstruction and reflectance factorization using the proposed L-Tracing as the light visibility estimation method.
- L-Tracing is ~10x faster than estimating light visibility based on volumetric integration. The rendered novel views and relighting images from the reconstructed object surface and reflectance achieve comparable quality with recent works across different metrics.

2 Related Works

2.1 Neural Shape Representation

Neural shape representations are widely used in 3D shape reconstruction and inverse rendering tasks. NeRF [21] and its variants [8–10,17,20,23,27,30,35,38] use volume differentiable rendering on volumetric representation achieving good quality on novel view synthesis. Although they have achieved great success in the field of free-view interpolation, they struggle at recovering solid object boundaries (object surfaces) due to the lack of surface definition of volume representation.

Recent works [6,19,25,34] studying 3D surface reconstruction tend to use SDF as learnable shape representation. SDF is a function of spatial coordinates, with its value indicating the minimum distance between the input coordinate and the surface. Because of the simple and continuous parameterization, SDF is suitable to learn well-shaped neural surfaces. [24,29,33] explored the way to reconstruct an implicit surface by combining the advantages of both volume rendering and surface rendering. In our work, we use SDF as our neural implicit shape representation, and adopt [29] as our first stage of reconstructing well-shaped surfaces from multi-view unmasked images. We then apply L-tracing on trained neural surfaces to decompose the reflectance and material from unknown ambient illumination.

2.2 Reflectance and Illumination Estimation

Decomposing the object surface into materials, reflectance, and ambient illumination from multi-view images is one of the main concerns of the inverse rendering tasks. Since this is a highly ill-posed problem, traditional works [1,4,14–16] often rely on priors.

Recent works tend to use the neural rendering method [2,3,5,12,13,16,22, 26,28,31,32,36,37,39] to learn the object BRDF and environment illumination from image collections. NeRD [3] decompose reflectance from images under different illumination. PhySG [37] models geometry by SDF and the environment illumination by the mixtures of spherical Gaussians. Neural Ray-Tracing [12] leverages sphere tracing to model multi-bounce reflection. NeRD [3], PhySG [37] and Neural Ray-Tracing [12] don't model the light visibility, the intensity of the incoming light is learned totally by deep networks. NeRV [26] and NeRFactor [39] model the light visibility by volumetric integration of the density along the ray and train a MLP for light visibility prediction. However, learning from the network loses details of high frequency topology, which will be discussed in the experiment section. Instead of doing reflectance factorization on neural representations, [22] jointly optimize the explicit geometry, material and illumination, the images are rendered by a differentiable rasterizer with deferred shading.

As demonstrated in Sect. 1, the works mentioned above either fail to model light visibility physically, or do modeling at an expensive computation cost by volumetric integration. Our proposed L-Tracing models the direct illumination of the environment, estimating the light visibility based on physical geometry occlusions, accelerates the learning process for reflectance factorization.

3 Light Visibility Estimation

3.1 Preliminaries

We first review the volume rendering process that is developing rapidly in the area of novel view synthesis. Consider rendering a ray starting from $\boldsymbol{o}$ and in the normalized direction of $\boldsymbol{d}$. The points on this ray is noted by $\{\boldsymbol{r}(t) = \boldsymbol{o} + t\boldsymbol{d} \mid t \geq 0\}$. The color of the pixel corresponding to this ray is:

$$L(\boldsymbol{o}, \boldsymbol{d}) = \int_0^{+\infty} w(t)c(\boldsymbol{r}(t), \boldsymbol{d})\mathrm{d}t \tag{1}$$

where $w(t)$ denotes the ratio of each sampled point's color $c(\boldsymbol{r}(t), \boldsymbol{d})$ contributing to the corresponding pixel color $L(\boldsymbol{o}, \boldsymbol{d})$. $w(t)$ is further expanded into

$$w(t) = T(t)\sigma(\boldsymbol{r}(t)), \quad \text{where } T(t) = \exp\left(-\int_0^t \sigma(\boldsymbol{r}(u))\mathrm{d}u\right) \tag{2}$$

$T(t)$ denotes the probability of the ray being transmitted to $\boldsymbol{r}(t)$, and $\sigma(\boldsymbol{r}(t))$ denotes the differential probability the ray being omitted at the position $\boldsymbol{r}(t)$.

The light visibility of spatial point is defined by the accumulated transmittance on the path. Figure 2(b) demonstrates the light visibility of the volumetric integration process. Consider a ray emitted from the light source noted by $\{\boldsymbol{r}_l(t) = \boldsymbol{x}_l + t\boldsymbol{d}_l \mid t \geq 0\}$, and the spatial point along the ray noted by $\boldsymbol{x}_\mathrm{s} = \boldsymbol{r}_l(t_s)$. The light visibility v based on volumetric integration is:

$$v(\boldsymbol{x}_l, \boldsymbol{x}_\mathrm{s}) = 1 - \int_0^{t_s} T(t)\sigma(\boldsymbol{r}_l(t))\mathrm{d}t \tag{3}$$

The light visibility computed by volumetric integration models the probability of the light being blocked along the way, which is a continuous value ranging from 0 to 1. Hence, the volumetric integration process is considered physically incorrect as for direct illumination and binary geometry occlusions, because in most of the real-world cases objects are typically solid and not translucent. Motivated by this, we introduce L-Tracing, a highly efficient light visibility estimation method for neural implicit surfaces that respect solid objects and hard geometry occlusions.

3.2 L-Tracing

To alleviate the physical incorrectness of volume rendering's light visibility and accelerate the process of computing light visibility, we propose L-Tracing, adopting sphere tracing for light visibility estimation on neural surfaces encoded by learned SDF. L-Tracing computes whether the light path is occluded by a surface leveraging the signed distance output.

Denoting the signed distance function as f, the neural implicit surface is defined as the zero-level set of f:

$$\mathcal{S} = \left\{\boldsymbol{x} \in \mathbb{R}^3 \mid f(\boldsymbol{x}) = 0\right\} \tag{4}$$

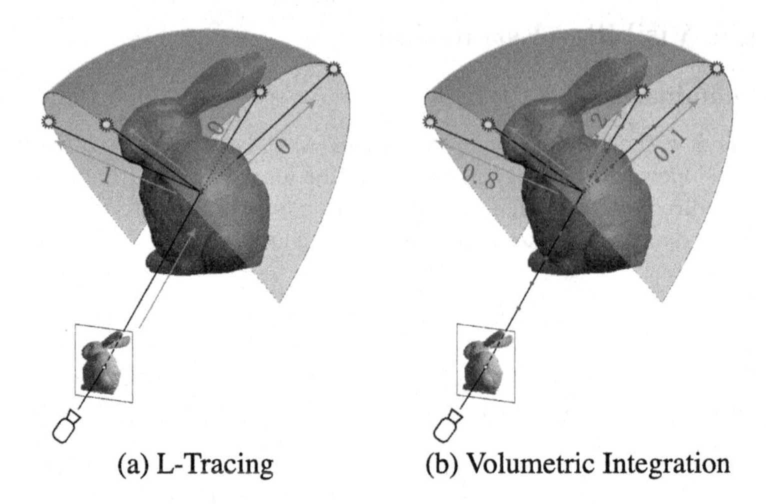

Fig. 2. Illustration of Light Visibility Computation. (a) shows binary visibility computed by L-Tracing, the surface point is invisible to the right lights because of the solid geometry occlusion, thus the light visibility is 0. (b) shows the continuous visibility computed by volumetric integration. Although the surface point is invisible to the right lights, volumetric integration still computes the number greater than 0.

The light visibility v from light source $\boldsymbol{x}_l$ to surface point $\boldsymbol{x}_s \in \mathcal{S}$ is computed following Algorithm 1. We first calculate the upper bound of the tracing range, that is the euclidean distance between $\boldsymbol{x}_l$ and $\boldsymbol{x}_s$. At the start of the iteration, a tracing point on light position is queried for its signed distance (sd for short), to decide the signed distance step for the next iteration. If the tracing point moves beyond the range, the returned light visibility v is 1, indicating there is no surface that occludes the light path. If the tracing point is trapped after T iterations, then return light visibility $v = 0$, which means the light is blocked by the surface.

According to the assumption of L-Tracing, light visibility models binary geometry occlusions on the solid object surface, which helps better produce photo-realistic shadows and specular effects. This will be discussed in Sect. 5.1. In our work under the direct-illumination-only setting, light visibility is represented by the binary value in $\{0, 1\}$ which represents the binary geometry occlusion, instead of the continuous value ranging from 0 to 1 that derived from volumetric integration. Besides, L-Tracing exhibits good performance on computational efficiency, because of the linear convergence, enabling computing light visibility in fewer iterations with very high accuracy.

Convergence Proof. We prove that the high computational efficiency of L-Tracing comes from the linear convergence of sphere tracing. In Fig. 3 we show two surface conditions of L-Tracing. Among the Figure, $\boldsymbol{O}_k = \boldsymbol{x}_l + t_k \boldsymbol{d}_l$ denotes the tracing point at k_{th} iteration, $\boldsymbol{Q}$ denotes the observed surface point, $\boldsymbol{B}_k$ is the nearest surface points of $\boldsymbol{O}_k$. The distance between $\boldsymbol{Q}$ and $\boldsymbol{O}_k$ is ϵ_k. f_k is the signed distance of $\boldsymbol{O}_k$.

Algorithm 1. L-Tracing

Input: $\boldsymbol{x}_l, \boldsymbol{x}_s, f$
Output: v

```
1: r = x_s − x_l
2: far = ||r||_2
3: d = r/||r||_2
4: t = 0
5: for i in range(T) do
6:     pts = x_l + d ∗ t
7:     sd = f(pts)
8:     t = t + sd
9:     if t > far or t < 0 then
10:        return 1
11:    end if
12: end for
13: return 0
```

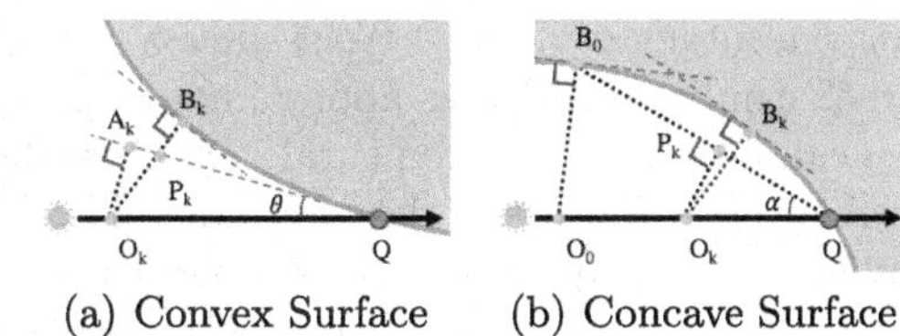

(a) Convex Surface (b) Concave Surface

Fig. 3. Illustration of L-Tracing. We consider the ray of the light source is encountered with (a) a convex surface, (b) a concave surface. Since the plane surface can be regarded as the special case of (b), we don't list it separately for discussion.

When the surface is convex, as shown in Fig. 3(a). We have $|\boldsymbol{O}_k\boldsymbol{B}_k| = f_k$, $|\boldsymbol{O}_k\boldsymbol{A}_k| = \epsilon_k sin\theta$, and $|\boldsymbol{O}_k\boldsymbol{A}_k| \leq |\boldsymbol{O}_k\boldsymbol{P}_k| \leq |\boldsymbol{O}_k\boldsymbol{B}_k|$ (the equal sign should be assigned when the surface can be represented as a local plane), from them we derive:

$$\epsilon_k sin\theta \leq f_k \tag{5}$$

As the tracing point in $(k+1)_{th}$ and k_{th} iteration satisfy $\epsilon_{k+1} = \epsilon_k - f_k$, from Eq. 5 we have $\epsilon_{k+1} \leq \epsilon_k - \epsilon_k sin\theta$. Since ϵ is monotonic and bounded during iterations(Please refer to the supplementary material for the Proof), $\epsilon_k > 0$ is always satisfied. Then we get the following inequality:

$$0 < \frac{\epsilon_{k+1}}{\epsilon_k} \leq 1 - sin\theta \tag{6}$$

As for convex surface, we calculate the upper bound of the ratio between ϵ_{T-1} and ϵ_0:

$$0 < \frac{\epsilon_{T-1}}{\epsilon_0} \leq (1 - sin\theta)^T \leq \frac{\epsilon}{\epsilon_0} \tag{7}$$

where ϵ is the accuracy of convergence. Since $\theta \in (0, 90°)$, after a transformation, we have:

$$T \geq \frac{log\frac{\epsilon}{\epsilon_0}}{log(1 - sin\theta)} \tag{8}$$

When the surface is concave, as shown in Fig. 3(b). We have $\angle \boldsymbol{B}_0\boldsymbol{Q}\boldsymbol{O}_0 = \alpha$; $|\boldsymbol{O}_k\boldsymbol{B}_k| = f_k$; $\boldsymbol{O}_k\boldsymbol{P}_k \perp \boldsymbol{B}_0\boldsymbol{Q}$; Since $|\boldsymbol{O}_k\boldsymbol{P}_k| < |\boldsymbol{O}_k\boldsymbol{B}_k|$ is always satisfied, we get $\epsilon_k sin\alpha = |\boldsymbol{O}_k\boldsymbol{P}_k| \leq f_k$, from Eq. 6, the following inequality is derived:

$$0 < \frac{\epsilon_{T-1}}{\epsilon_0} \leq (1 - sin\alpha)^T \leq \frac{\epsilon}{\epsilon_0} \tag{9}$$

According Eq. 9, we get the theoretical minimum iterations needed for achieving the accuracy $\epsilon = 0.001$. We assume a ray intersects surface with an angle less than 5° can be regarded as approximately parallel to the surface, thus, θ_{min} and α_{min} can be set to 5°. The neural implicit surface is trained within a unit sphere spherical bounding area, the max range is 2, thus $\epsilon_0 = 2$. We can compute the number of the theoretical minimum iteration is 83. However, this limit condition exists only in a very small probability, we set T to 20 through experimental comparison as discussed in Sect. 5.4.

4 Reflectance Factorization Based on L-Tracing

We design a two-stage training framework based on L-Tracing, as shown in Fig. 4, to decompose the material and reflectance of neural surface from posed multi-view images under unknown illumination, enabling novel view interpolation and relighting. Our framework starts from training an neural surface leveraging recent works applying volume rendering to SDF reconstruction. After which we perform reflectance factorization on the well trained shape, with L-Tracing as the light estimation method.

4.1 Shape Learning

Our goal is to learn a neural surface, from which we can extract the 3D mesh with de-illuminated texture and reflectance for rendering. Neural volumetric field shows the strong ability to handle complex geometry occlusions(e.g. the abrupt depth change), however, the free space artifacts prevent neural volume field from modeling well-shaped surfaces. Inspired by the recent works applying volume rendering on the neural implicit surface, we apply SDF as our shape representation, leveraging its simple parameterization, we refer to the [29] for further details. Figure 4 Stage one shows the process of implicit surface learning.

4.2 Reflectance Factorization

Our Bidirectional Reflectance Distribution Function(BRDF) model is designed similar to NeRFactor [39]. We model the Lambertian reflection fully with the albedo MLP $\boldsymbol{a}$, and the specular spatial-varying component with the BRDF MLP $\boldsymbol{f}_{\mathrm{r}}$. The formulation of our BRDF model is:

$$\boldsymbol{R}\left(\boldsymbol{x}_{\mathrm{surf}}, \boldsymbol{\omega}_{\mathrm{i}}, \boldsymbol{\omega}_{\mathrm{o}}\right) = \frac{\boldsymbol{a}\left(\boldsymbol{x}_{\mathrm{surf}}\right)}{\pi} + \boldsymbol{f}_{\mathrm{r}}\left(\boldsymbol{f}_{\mathrm{z}}(\boldsymbol{x}_{\mathrm{surf}}), \boldsymbol{n}(\boldsymbol{x}_{\mathrm{surf}}), \boldsymbol{\omega}_{\mathrm{i}}, \boldsymbol{\omega}_{\mathrm{o}}\right) \tag{10}$$

The BRDF MLP models the function of the surface point's normal $\boldsymbol{n}(\boldsymbol{x}_{\mathrm{surf}})$, the incoming light direction $\boldsymbol{\omega}_{\mathrm{i}}$, the out-coming light direction $\boldsymbol{\omega}_{\mathrm{o}}$, and the latent material code $\boldsymbol{z}$. According to NeRFactor, the BRDF MLP is pre-trained on the MERL dataset [18] to learn the priors from real-world reflectance. During training, we fix the parameters of BRDF MLP, then recover the surface BRDF by learning the material latent code of the corresponding surface point. We regard

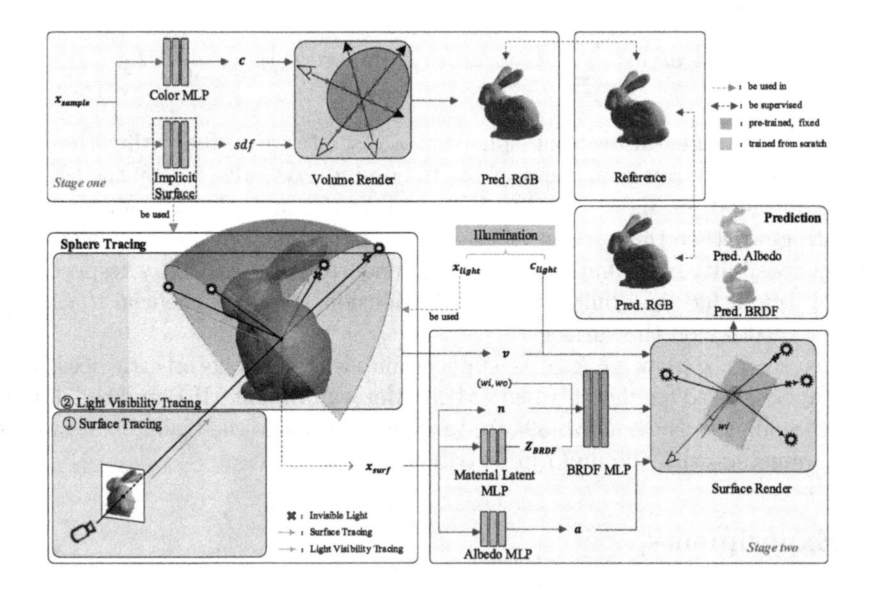

Fig. 4. Overview of Our framework. The training procedure is divided into two stages. **Stage One** is the training of neural implicit surface. The input is the point sets sampled on the rays, $\boldsymbol{c}$ is the space color of each point. The predicted ray color is rendered by volume rendering. **Stage Two** shows how the surface is decomposed. With $\boldsymbol{x}_{\text{surf}}$ obtained by sphere tracing, the light visibility v is computed by L-Tracing, $\boldsymbol{n}$ denotes the surface normal, $\boldsymbol{a}$ is the predicted albedo, BRDF MLP is pre-trained on MERL dataset to learn reflectance priors. The RGB images and Albedo images are rendered by the surface rendering module. In both stages, we use L2 RGB loss to compute the gradients of the networks.

the BRDF MLP as the material database and material latent code as the index. By querying the index to the BRDF MLP with related directions and normal, it outputs the final BRDF result.

Our illumination is an HDR light probe image [7] in the latitude-longitude format. The light sources are located at a sphere, m denotes their total number. $\boldsymbol{L}_{\text{j}}$ denote the color of the j_{th} light source. Obeying the physical energy law of rendering [11], we define A_{j} as the area of the j_{th} light source, indicating the proportion of its energy on the entire sphere, thus $\sum_{0}^{m-1} A_{\text{j}} = 4\pi$.

We directly use L-Tracing to estimate the light visibility while training. As shown in Fig. 4, we first compute the intersection surface points $\boldsymbol{x}_{\text{surf}}$ (in the following, we use $\boldsymbol{x}_{\text{s}}$ for short), then apply L-Tracing to estimate the light visibility $V = \{v_0, v_1, \ldots v_{m-1}\}$. Since we train our reflectance decomposition under unknown illumination, the predicted illumination greatly contributes to our final output, we only model one-bounce reflection in our work. After we obtain the surface point $\boldsymbol{x}_{\text{s}}$, the corresponding surface normal computed from SDF gradient, the predicted albedo, and the material latent code, we render an image with learned illumination by surface rendering. The render formulation is:

$$\boldsymbol{L}_{\mathrm{o}}\left(\boldsymbol{x}_{\mathrm{s}}, \boldsymbol{\omega}_{\mathrm{o}}\right)=\sum_{\mathrm{j}=0}^{m-1} \boldsymbol{R}\left(\boldsymbol{x}_{\mathrm{s}}, \boldsymbol{\omega}_{\mathrm{i}}, \boldsymbol{\omega}_{\mathrm{o}}\right)\left(\boldsymbol{\omega}_{\mathrm{i}} \cdot \boldsymbol{n}(\boldsymbol{x}_{\mathrm{s}})\right) \cdot v_{\mathrm{j}} \cdot \frac{A_{\mathrm{j}}}{4 \pi} \cdot \boldsymbol{L}_{\mathrm{j}} \tag{11}$$

This is the integral over the sphere in a discrete form, indicating the color of the out-coming ray is the sum of the in-coming rays reflected by the surface. In this formulation, light visibility is served as the mask for in-coming lights, indicating whether the surface is visible to the lights. We note the way that the light visibility contributes to the final rendered color explicitly respects the solid object surface and binary geometry occlusions, that is beneficial to casting sharper shadows on the surface.

We use multi-view images as supervision, with L2 photometric loss, and smoothness loss to regularize albedo MLP, Material Latent MLP and the illumination. Similar to NeRFactor [39], we randomly initialize the trainable networks and parameters with the uniform distribution.

5 Experiments

In this section, we test our framework based on L-Tracing quantitatively and qualitatively. We 1)compare L-Tracing with volumetric integration on light visibility estimation, 2)compare our factorization results with alternative approaches, 3)show the novel view interpolation results of ours and other works.

5.1 Light Visibility Estimation

We show the light visibility estimation results derived from I. **L-Tracing on Neural Surface(NS+LT):** our proposed framework based on L-Tracing; II. **Volume Integration on Neural Surface(NS+VI):** based on SDF shape representation, convert signed distance to density for volume integration; III. **Volume Integration on Neural Volume(NV+VI):** based on neural volume representation, directly apply volumetric integration on neural volumes.

Visual Quality. In order to show the limitation and robustness of L-Tracing, alongside "mean", the mean visibility across all lights, the light visibility of single point light with no ambient illumination(One-Light-at-A-Time, OLAT) is listed, too. The results in Fig. 5 show, NV+VI struggles with the noise, especially in OLAT2 because of the free space artifacts. NS+VI tends to produce soft shadows since the light visibility is a continuous number ranging from 0 to 1, modeling the direct illumination incorrectly. NS+LT makes highly noise-free results on different illuminations and recovers sharper shadows since it is based on solid surfaces and binary geometry occlusions.

Computational Efficiency. High computational efficiency for light visibility estimation is our heart contribution. In Table 1, we recorded the time cost on light visibility computation of L-Tracing and volumetric integration. The test is done on one NVIDIA RTX 3090 GPU, we input a batch of surface points with different batch size. For L-Tracing, we set the tracing iteration to 20(according

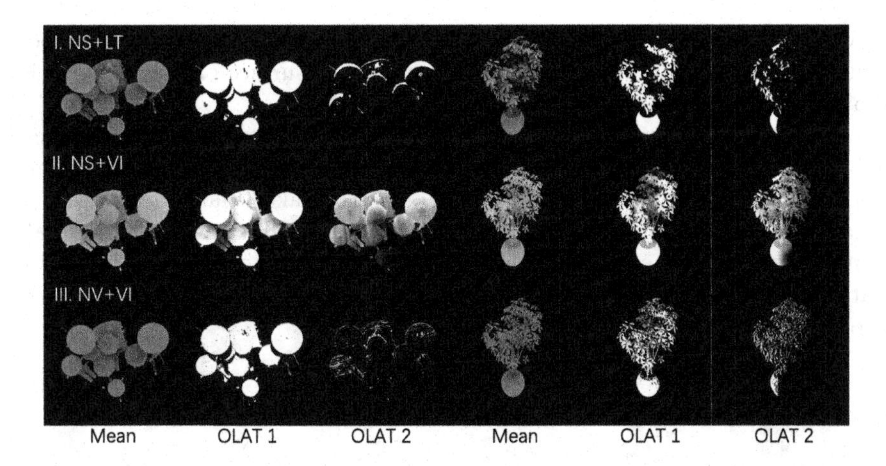

Fig. 5. Light Visibility Visualization. The light visibility is estimated on the NeRFactor synthetic dataset. The color denotes the lighting intensity on the surface. "Mean" denotes the mean light visibility across all light sources, while "OLAT1" and "OLAT2" denotes the light visibility under two different single light source points. Please refer to the supplementary material for more results.

Table 1. Efficiency Comparison. For each surface point, we compute the light visibility of 512 light sources.

Points	L-Tracing	Volumetric integration
1024	1.07 s	13.36 s
2048	2.10 s	26.86 s
4096	4.13 s	64.75 s

Table 2. Iteration Ablation. Novel view interpolation quality of our framework in different tracing iterations

Iterations	PSNR↑	SSIM↑	LPIPS↓
20	31.194	0.943	0.063
40	31.655	0.944	0.061
80	31.889	0.946	0.060

to Sect. 5.4, 20 iterations is sufficient for estimation), for volumetric integration, referring to [39], we set the sampling number of coarse points to 64 and 128 for fine sample. Results show that L-Tracing is about 13 times faster than volumetric integration.

5.2 Reflectance Factorization

We design a framework, as described in Sect. 4, for reflectance factorization. We compare it with Nvdiffrec [22], and NeRFactor [39] under two settings: 1) the original setting **NeRFactor**: precompute light visibility by volume integration, train a light visibility MLP using the precomputed caches. 2) **NeRFactor + VI**: directly use precomputed light visibility caches for factorization.

We visualize the predicted albedos of all methods in Fig. 6. Albedos predicted by L-Tracing based scheme shows similar predictions on details with those from NeRFactor, due to the similar factorization setting. Albedos from Nvdiffrec show uneven color blocks on the object surface, the reason is that the low-level smooth regularization leads the networks to distinguish the surface on the same region into different materials.

We record the metrics of albedo in Table 3. As there is an inherent scale ambiguity in inverse rendering problems: the ambiguity between the illuminant's average color and the albedo's average color. To make a fair comparison, we rescaled the albedos of all methods by scaling each RGB channel with a scalar to minimize the mean squared error with ground truth albedos. The result shows that our method performs on par with alternative approaches on albedo recovery across different metrics.

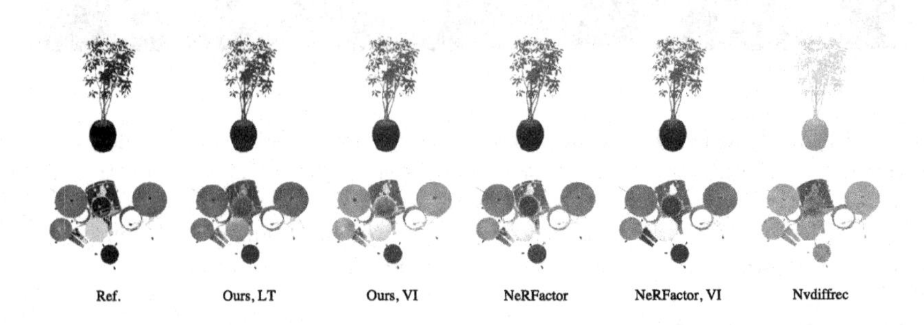

Fig. 6. Albedo Recovery. We show the recovered albedo on NeRFactor synthetic data. The albedos predicted by Ours and NeRFactor are all scaled to minimize the error with ground truth. The rightmost images were provided by Nvdiffrec's [22] authors. We show more recovered albedos in the supplementary material.

5.3 Novel View Interpolation and Relighting

Novel View Interpolation. We record the novel view synthesis quality in Table 3, it shows the works specializing in different areas all achieve good results, since the listed method commonly use the multi-view images as supervision, volume rendering based methods show a gift in learning multi-view consistency, which makes the results robust. We note that the L-Tracing based scheme shows little quality reduction on novel view interpolation.

Novel View Relighting. To evaluate the quality of light visibility and recovered material, we relight the learned surfaces and reflectance with different HDR probes. Table 3 shows that the L-Tracing based scheme produces comparable relighting quality with recent works. We surpassed all compared methods on SSIM.

We show the relighting results of Lego and Hotdog in Fig. 8. The relighting results are close to the reference intuitively from vision. We select one region on the images of Hotdog under "Sunrise" to compare the relighting details, as shown in Fig. 7. Ours+VI lost the shadows on the left side because the volume integration based light visibility estimation fails to model geometry occlusion

Table 3. Quantitative Evaluation. We calculate the metrics on four scenes of NeRFactor synthetic dataset. Each scene includes 8 novel validation views, for each view, 8 probe relighting images and 1 ground truth albedo are provided. The reported numbers are the arithmetic mean over 4 scenes. Following [22], we rescaled the albedos of all methods before measuring the errors, to eliminate the scale ambiguity. To make a fair comparison, the rendered albedo and relighting images are all blended with the same masks. The results of NeRF are token from the Table 4 of Nvdiffrec [22] paper.

Method	View interpolation			Albedo			HDR relighting		
	PSNR↑	SSIM↑	LPIPS↓	PSNR↑	SSIM↑	LPIPS↓	PSNR↑	SSIM↑	LPIPS↓
NeRF [21]	31.080	0.956	0.064	—	—	—	—	—	—
NeuS [29]	30.193	0.944	0.081	—	—	—	—	—	—
Nvdiffrec [22]	30.640	0.965	0.044	26.205	0.929	0.078	25.502	0.919	0.073
NeRFactor [39]	31.869	0.944	0.076	27.829	0.943	0.065	26.016	0.915	0.090
NeRFactor, VI [39]	30.288	0.916	0.097	27.686	0.941	0.070	24.913	0.870	0.122
Ours, VI	32.229	0.947	0.059	27.594	0.935	0.076	25.878	0.917	0.075
Ours, LT	31.194	0.943	0.063	27.296	0.933	0.080	25.586	0.920	0.080

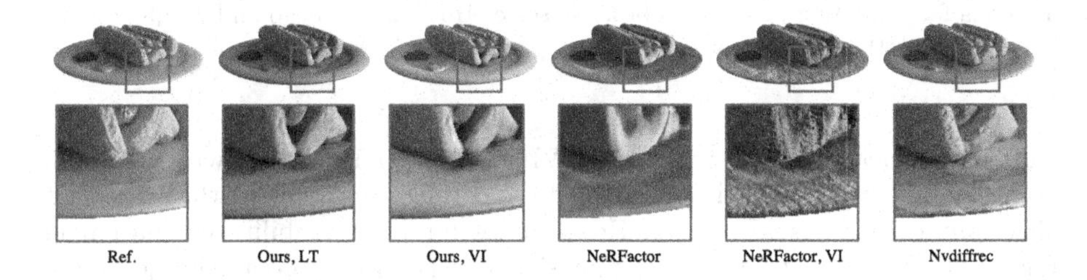

Fig. 7. Relighting Details. We select one area of Hotdog to compare the shape details and relighted shadow. The images are rendered with a new ambient illumination "Sunrise" in the HDR format.

correctly. NeRFactor produces smooth shape details and soft shadows because the learned normals and light visibility through MLPs omit the high-frequency component. NeRFactor+VI directly uses the light visibility integrated on the neural volumes, struggling with the free space artifacts. Nvdiffrec appears to have uneven surface color blocks due to the uneven albedo. L-Tracing based scheme not only learns the detailed surface with high smoothness leveraging the neural implicit surface representation, but also produces sharper shadows, since L-Tracing respects solid object surfaces and binary geometry occlusions.

5.4 Ablation Study

Light visibility $\{0, 1\}$ vs. $[0, 1]$. As L-Tracing estimates light visibility based on geometry occlusions, the binary numbers in $\{0, 1\}$ indicates whether the surface point is visible to the lights. Volume integration assumes the light visibility is the probability the light isn't blocked, thus the number is continuous between 0 and 1. To showcase the effectiveness of these two assumptions. We compare

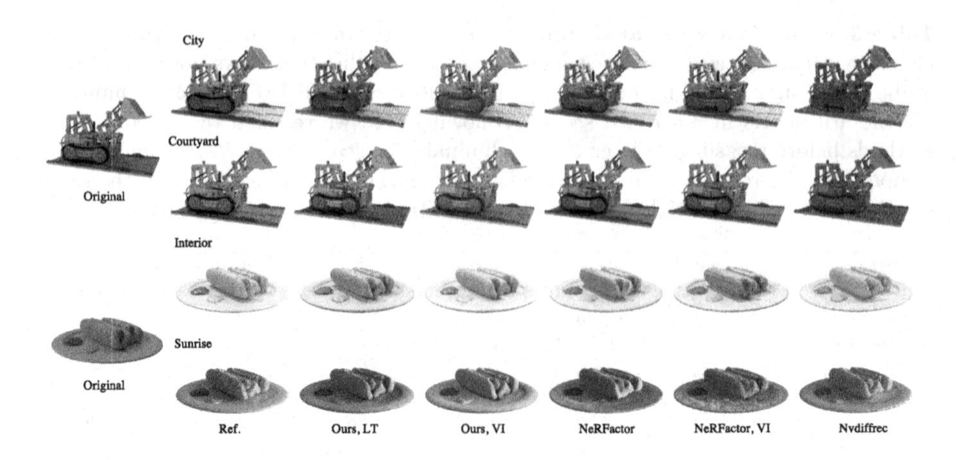

Fig. 8. Novel View Relighting. The factorization results produced by these methods enable us to perform novel view relighting. The illumination representation of listed methods for relighting is HDR probes. We select Hotdog and Lego and relighted with four different HDR probes for comparison, please see supplementary for more.

Ours+LT(which is $\{0, 1\}$) with Ours+VI(which is $[0, 1]$). As shown in Table 3, Ours+LT shows a small drop on albedo and relighting quality, that does not affect the overall visual quality. However, as for light visibility estimation in Fig. 5, NS+VI seems to destroy the physical consistency of observed objects. On drums OLAT2, light sheds on most surface that is invisible to the light source, while NS+LT shows sharp and clear shadows. These comparisons show that L-Tracing out-performs volume integration in terms of the definition of light visibility.

Tracing Iteration. The novel view synthesis results in Table 2 show that reducing the iteration from 80 to 20 doesn't significantly reduce the quality of factorization, indicating that it rarely needs 83 iterations(which is the theoretical minimum iterations) to compute the accurate light visibility. We also visualize the light visibility estimated in different tracing iterations in supplementary.

6 Conclusions

We have proposed L-Tracing, a highly efficient but also accurate enough light visibility estimation method on neural implicit surfaces. L-Tracing respects the solid object surface and the binary geometry occlusions, as a result, produces sharper shadows and specular. Moreover, L-Tracing is ~10x faster for light visibility computation than volumetric integration. Furthermore, since our proposed framework models the direct illumination efficiently, we believe that L-Tracing can be further applied to model multi-bounce reflection. All above, it is potential to apply L-Tracing for large scale scene factorization which is highly demanded for computation efficiency and modeling accuracy.

Acknowledgement. The authors Ziyu Chen and Li Song were supported by the Shanghai Key Laboratory of Digital Media Processing and Transmissions.

References

1. Barron, J.T., Malik, J.: Shape, illumination, and reflectance from shading. IEEE Trans. Pattern Anal. Mach. Intell. **37**(8), 1670–1687 (2014)
2. Bi, S., et al.: Neural reflectance fields for appearance acquisition. arXiv preprint arXiv:2008.03824 (2020)
3. Boss, M., Braun, R., Jampani, V., Barron, J.T., Liu, C., Lensch, H.: NeRD: neural reflectance decomposition from image collections. In: Proceedings of the IEEE/CVF International Conference on Computer Vision, pp. 12684–12694 (2021)
4. Boss, M., Groh, F., Herholz, S., Lensch, H.P.: Deep dual loss BRDF parameter estimation. In: MAM@ EGSR, pp. 41–44 (2018)
5. Boss, M., Jampani, V., Braun, R., Liu, C., Barron, J., Lensch, H.: Neural-PIL: neural pre-integrated lighting for reflectance decomposition. In: Advances in Neural Information Processing Systems, vol. 34 (2021)
6. Chen, Z., Zhang, H.: Learning implicit fields for generative shape modeling. In: Proceedings of the IEEE/CVF Conference on Computer Vision and Pattern Recognition, pp. 5939–5948 (2019)
7. Debevec, P.: Rendering synthetic objects into real scenes: bridging traditional and image-based graphics with global illumination and high dynamic range photography. In: ACM SIGGRAPH 2008 Classes, pp. 1–10 (2008)
8. Deng, K., Liu, A., Zhu, J.Y., Ramanan, D.: Depth-supervised NeRF: fewer views and faster training for free. In: Proceedings of the IEEE/CVF Conference on Computer Vision and Pattern Recognition, pp. 12882–12891 (2022)
9. Fridovich-Keil, S., Yu, A., Tancik, M., Chen, Q., Recht, B., Kanazawa, A.: Plenoxels: radiance fields without neural networks. In: Proceedings of the IEEE/CVF Conference on Computer Vision and Pattern Recognition, pp. 5501–5510 (2022)
10. Garbin, S.J., Kowalski, M., Johnson, M., Shotton, J., Valentin, J.: FastNeRF: high-fidelity neural rendering at 200FPS. In: Proceedings of the IEEE/CVF International Conference on Computer Vision, pp. 14346–14355 (2021)
11. Kajiya, J.T.: The rendering equation. In: Proceedings of the 13th Annual Conference on Computer Graphics and Interactive Techniques, pp. 143–150 (1986)
12. Knodt, J., Baek, S.H., Heide, F.: Neural ray-tracing: learning surfaces and reflectance for relighting and view synthesis. arXiv preprint arXiv:2104.13562 (2021)
13. Kuang, Z., Olszewski, K., Chai, M., Huang, Z., Achlioptas, P., Tulyakov, S.: NeROIC: neural rendering of objects from online image collections. arXiv preprint arXiv:2201.02533 (2022)
14. Lensch, H.P.A., Kautz, J., Goesele, M., Heidrich, W., Seidel, H.-P.: Image-based reconstruction of spatially varying materials. In: Gortler, S.J., Myszkowski, K. (eds.) EGSR 2001. E, pp. 103–114. Springer, Vienna (2001). https://doi.org/10.1007/978-3-7091-6242-2_10
15. Lensch, H.P., Lang, J., Sá, A.M., Seidel, H.P.: Planned sampling of spatially varying BRDFs. In: Computer Graphics Forum, vol. 22, pp. 473–482. Wiley Online Library (2003)

16. Li, Z., Shafiei, M., Ramamoorthi, R., Sunkavalli, K., Chandraker, M.: Inverse rendering for complex indoor scenes: shape, spatially-varying lighting and svbrdf from a single image. In: Proceedings of the IEEE/CVF Conference on Computer Vision and Pattern Recognition, pp. 2475–2484 (2020)
17. Martin-Brualla, R., Radwan, N., Sajjadi, M.S., Barron, J.T., Dosovitskiy, A., Duckworth, D.: NeRF in the wild: neural radiance fields for unconstrained photo collections. In: Proceedings of the IEEE/CVF Conference on Computer Vision and Pattern Recognition, pp. 7210–7219 (2021)
18. Matusik, W.: A data-driven reflectance model. Ph.D. thesis, Massachusetts Institute of Technology (2003)
19. Michalkiewicz, M., Pontes, J.K., Jack, D., Baktashmotlagh, M., Eriksson, A.: Implicit surface representations as layers in neural networks. In: Proceedings of the IEEE/CVF International Conference on Computer Vision, pp. 4743–4752 (2019)
20. Mildenhall, B., Hedman, P., Martin-Brualla, R., Srinivasan, P.P., Barron, J.T.: NeRF in the dark: high dynamic range view synthesis from noisy raw images. In: Proceedings of the IEEE/CVF Conference on Computer Vision and Pattern Recognition, pp. 16190–16199 (2022)
21. Mildenhall, B., Srinivasan, P.P., Tancik, M., Barron, J.T., Ramamoorthi, R., Ng, R.: NeRF: representing scenes as neural radiance fields for view synthesis. In: Vedaldi, A., Bischof, H., Brox, T., Frahm, J.-M. (eds.) ECCV 2020. LNCS, vol. 12346, pp. 405–421. Springer, Cham (2020). https://doi.org/10.1007/978-3-030-58452-8_24
22. Munkberg, J., et al.: Extracting triangular 3D models, materials, and lighting from images. In: Proceedings of the IEEE/CVF Conference on Computer Vision and Pattern Recognition, pp. 8280–8290 (2022)
23. Niemeyer, M., Geiger, A.: GIRAFFE: representing scenes as compositional generative neural feature fields. In: Proceedings of the IEEE/CVF Conference on Computer Vision and Pattern Recognition, pp. 11453–11464 (2021)
24. Oechsle, M., Peng, S., Geiger, A.: UNISURF: unifying neural implicit surfaces and radiance fields for multi-view reconstruction. In: Proceedings of the IEEE/CVF International Conference on Computer Vision, pp. 5589–5599 (2021)
25. Park, J.J., Florence, P., Straub, J., Newcombe, R., Lovegrove, S.: DeepSDF: learning continuous signed distance functions for shape representation. In: Proceedings of the IEEE/CVF Conference on Computer Vision and Pattern Recognition, pp. 165–174 (2019)
26. Srinivasan, P.P., Deng, B., Zhang, X., Tancik, M., Mildenhall, B., Barron, J.T.: NeRV: neural reflectance and visibility fields for relighting and view synthesis. In: Proceedings of the IEEE/CVF Conference on Computer Vision and Pattern Recognition, pp. 7495–7504 (2021)
27. Sun, C., Sun, M., Chen, H.T.: Direct voxel grid optimization: super-fast convergence for radiance fields reconstruction. In: Proceedings of the IEEE/CVF Conference on Computer Vision and Pattern Recognition, pp. 5459–5469 (2022)
28. Verbin, D., Hedman, P., Mildenhall, B., Zickler, T., Barron, J.T., Srinivasan, P.P.: Ref-NeRF: structured view-dependent appearance for neural radiance fields. In: Proceedings of the IEEE/CVF Conference on Computer Vision and Pattern Recognition, pp. 5491–5500 (2022)
29. Wang, P., Liu, L., Liu, Y., Theobalt, C., Komura, T., Wang, W.: NeuS: learning neural implicit surfaces by volume rendering for multi-view reconstruction. In: Advances in Neural Information Processing Systems, vol. 34, pp. 27171–27183 (2021)

30. Wang, Z., Wu, S., Xie, W., Chen, M., Prisacariu, V.A.: NeRF-: neural radiance fields without known camera parameters. arXiv preprint arXiv:2102.07064 (2021)
31. Wimbauer, F., Wu, S., Rupprecht, C.: De-rendering 3D objects in the wild. In: Proceedings of the IEEE/CVF Conference on Computer Vision and Pattern Recognition, pp. 18490–18499 (2022)
32. Wizadwongsa, S., Phongthawee, P., Yenphraphai, J., Suwajanakorn, S.: NeX: real-time view synthesis with neural basis expansion. In: Proceedings of the IEEE/CVF Conference on Computer Vision and Pattern Recognition, pp. 8534–8543 (2021)
33. Yariv, L., Gu, J., Kasten, Y., Lipman, Y.: Volume rendering of neural implicit surfaces. In: Advances in Neural Information Processing Systems, vol. 34 (2021)
34. Yifan, W., Wu, S., Oztireli, C., Sorkine-Hornung, O.: ISO-points: optimizing neural implicit surfaces with hybrid representations. In: Proceedings of the IEEE/CVF Conference on Computer Vision and Pattern Recognition, pp. 374–383 (2021)
35. Yu, A., Ye, V., Tancik, M., Kanazawa, A.: pixelNeRF: neural radiance fields from one or few images. In: Proceedings of the IEEE/CVF Conference on Computer Vision and Pattern Recognition, pp. 4578–4587 (2021)
36. Zhang, J., Yang, G., Tulsiani, S., Ramanan, D.: NeRS: neural reflectance surfaces for sparse-view 3D reconstruction in the wild. In: Advances in Neural Information Processing Systems, vol. 34 (2021)
37. Zhang, K., Luan, F., Wang, Q., Bala, K., Snavely, N.: PhySG: inverse rendering with spherical gaussians for physics-based material editing and relighting. In: Proceedings of the IEEE/CVF Conference on Computer Vision and Pattern Recognition, pp. 5453–5462 (2021)
38. Zhang, K., Riegler, G., Snavely, N., Koltun, V.: NeRF++: analyzing and improving neural radiance fields. arXiv preprint arXiv:2010.07492 (2020)
39. Zhang, X., Srinivasan, P.P., Deng, B., Debevec, P., Freeman, W.T., Barron, J.T.: NeRFactor: neural factorization of shape and reflectance under an unknown illumination. ACM Trans. Graph. (TOG) **40**(6), 1–18 (2021)

A Perceptual Quality Metric for Video Frame Interpolation

Qiqi Hou(✉), Abhijay Ghildyal, and Feng Liu

Portland State University, Portland, USA
{qiqi2,abhijay,fliu}@pdx.edu

Abstract. Research on video frame interpolation has made significant progress in recent years. However, existing methods mostly use off-the-shelf metrics to measure the quality of interpolation results with the exception of a few methods that employ user studies, which is time-consuming. As video frame interpolation results often exhibit unique artifacts, existing quality metrics sometimes are not consistent with human perception when measuring the interpolation results. Some recent deep learning-based perceptual quality metrics are shown more consistent with human judgments, but their performance on videos is compromised since they do not consider temporal information. In this paper, we present a dedicated perceptual quality metric for measuring video frame interpolation results. Our method learns perceptual features directly from videos instead of individual frames. It compares pyramid features extracted from video frames and employs Swin Transformer blocks-based spatio-temporal modules to extract spatio-temporal information. To train our metric, we collected a new video frame interpolation quality assessment dataset. Our experiments show that our dedicated quality metric outperforms state-of-the-art methods when measuring video frame interpolation results.

1 Introduction

Video frame interpolation aims to generate frames between consecutive frames. It has attracted a lot of attention because of its wide applications in video editing [53], video generation [38,39], optical flow estimation [49,89], and video compression [87]. Consequentially, great progress has been made and many video frame interpolation methods are available [4,15,30,31,40,42,57,58,64,72,73].

To evaluate the quality of video frame interpolation results, most interpolation methods rely on traditional metrics, such as Peak Signal-to-Noise Ratio (PSNR) and Structural Similarity Index Measure (SSIM) [79]. These metrics estimate perceptual similarity between images by comparing the difference between pixels or carefully designed low-level visual patterns. While such methods achieve promising results, adapting them to measure video frame interpolation results is challenging. First, these metrics have been designed for general tasks such as video compression and other common low-level computer vision tasks, like super resolution, deblurring, and denoising, not specifically for video

S. Avidan et al. (Eds.): ECCV 2022, LNCS 13675, pp. 234–253, 2022.
https://doi.org/10.1007/978-3-031-19784-0_14

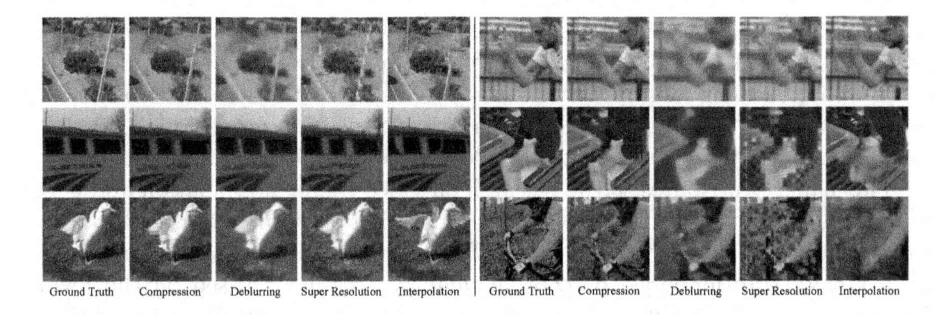

Fig. 1. Unique distortions in video frame interpolation results.

frame interpolation. Hence, these metrics are not optimal for measuring the quality of video frame interpolation results. As shown in Fig. 1, video frame interpolation methods exhibit unique distortions. For example, the telegraph pole is twisted out of shape and the wings of the duck have ghosting artifacts. Second, traditional metrics rely on low-level features, such as per-pixel distance, and structure similarity [79]. However, human perception is highly complex, and hence, low-level features are insufficient for assessing the perceptual quality [97]. These shortcomings motivate us to develop a dedicated quality metric for video frame interpolation.

Our research is inspired by the recent deep learning approaches to image quality assessment [6,17,20,21,32,33,35,36,41,76,91,97]. Deep learning approaches, such as LPIPS [97], take an image and its reference image as input and output the perceptual score. Instead of comparing low-level features, they learn perceptual similarity from the internal features of deep convolutional networks. Such methods have shown great success for image perceptual quality assessment. However, applying these metrics to video frame interpolation results frame by frame ignores the temporal information in the videos, which often make these metrics inconsistent with human perception.

This paper presents a learned perceptual video similarity metric dedicated to video frame interpolation results. It takes an interpolated video and its reference ground-truth video as input and outputs the perceptual similarity between them. It builds upon state-of-the-art image perceptual quality metrics but extends to videos, leading to a naturally spatio-temporal architecture. Specifically, our network extracts the features from each frame. Then, we design a dedicated spatio-temporal module to capture the spatio-temporal information from videos. This spatio-temporal module consists of the Swin Transformer blocks [47], which can explore the interaction among different regions, learn to attend the important regions, and extract spatio-temporal video information.

To train our network, we collected a large Video Frame Interpolation Perceptual Similarity (VFIPS) dataset that contains 25,887 video triplets. Each video triplet contains two interpolated videos, their reference video, and the corresponding perceptual preference. Our dataset consists of various artifacts generated by eleven state-of-the-art video frame interpolation methods [4,5,15,30,40,42,46,57–59,64]. With this dataset, we train our video perceptual metric using Binary Cross Entropy (BCE) as the loss function in an end-to-end manner. Our experiments show that our metric outperforms existing metrics significantly when measuring video frame interpolation results.

This paper contributes to the research on video frame interpolation and perceptual quality assessment by 1) providing the first video perceptual similarity metric dedicated to video frame interpolation, 2) designing a novel neural network architecture for video perceptual quality assessment based on the Swin Transformers, and 3) building a large video frame interpolation perceptual similarity dataset.

2 Related Work

This paper investigates quality assessment of video frame interpolation results. Below, we first briefly survey video frame interpolation methods. Then we discuss traditional video quality metrics and recent deep-learning-based approaches. We also discuss video quality assessment datasets and vision transformers.

Video Frame Interpolation. Video frame interpolation aims to estimate intermediate frames between two consecutive frames. They can be classified into three categories: kernel-based methods [59–62,92], phase-based methods [54,55], and flow-based methods [4,5,13,16,18,28–30,40,42,46,48,57,58,64,65,68,70,73, 94,96]. Kernel-based methods [59–61,92] estimate a kernel for each pixel in the frame indicating the weights for its neighboring pixels to synthesize the new pixel. They generate the synthesized frame in a local convolutional manner. Phase-based methods [54,55] generate intermediate frames by per-pixel phase modification. Currently, most of methods are flow based [4,5,30,40,42,46,48,57, 58,64,68,70,73,94]. Typically, their networks contain two modules: optical flow module and frame synthesized module. The optical flow module is used to estimate the optical flow between input frames. By warping the input frame based on the estimated optical flow, the frame synthesized module synthesizes the intermediate frames. Most of the video frame interpolation methods use PSNR and SSIM as their evaluation metrics. Some methods [40,58,93] also explore Interpolation Error (IE) [2], Normalized Interpolation Error (NIE) [2], and Learned Perceptual Image Patch Similarity (LPIPS) [97] to measure their results. However, these metrics are insufficient to evaluate the perceptual similarity. Traditional quality metrics, such as PSNR, only use low-level features and LPIPS lacks temporal information. In this paper, we propose a learned video quality metric dedicated to video frame interpolation.

Video Quality Metrics. Based on the availability of the reference video, traditional video quality metrics can be classified into No-Reference (NR), Reduced-Reference (RR), and Full Reference (FR). Due to the limitation of space, we refer readers to a comprehensive survey for traditional methods [85]. Our method is most related to the FR methods, which have access to the reference video. Traditional FR methods infer the perceptual similarity from the differences from pixels or low-level features [8,23,27,50,69,81–84,86,88,90]. For instance, Wange *et al.* observed that the human visual system (HVS) is adaptive to structural similarity [79]. They evaluated the perceptual similarity from the structural deformity between a reference and distortion image. In comparison to SSIM, Pinson *et al.* had a better correlation with HVS on video quality assessment [66]. They also considered multiple perceptual-based features, including blurring, unnatural motion, noise, color distortion, and block distortion. Li *et al.* proposed a multi-method metric to fuse multiple video perceptual quality metrics using a support vector machine [44]. It can preserve the strength of the individual metrics and achieve promising results. However, these metrics only use low-level features, which is insufficient for many nuances of human perception. Recently, with the success of deep learning, many research works introduce the deep learning method to video quality assessment [12,78]. However, these metrics are designed for the traditional artifacts like video compression, while our method focuses on artifacts that arise from video frame interpolation.

Deep Learning Based Quality Metrics. Our method is most related to the recent deep learning based metrics for image quality assessment [6,17,20,21,32, 33,35,36,41,76,91,97]. Instead of using low-level features, these methods learn features from deep convolutional networks and train a network in an end-to-end manner. For full-reference image quality assessment (FR-IQA), Kim *et al.* estimated the perceptual similarity using a deep network [32]. Their method takes a distorted images and an error map as input and estimates the sensitivity map. They get the subjective score by multiplying the sensitivity map with the error map. Zhang *et al.* explored the convolutional features from a classification network [97]. By comparing the features extracted using the convolutional network, their method is able to predict the perceptual similarity robustly. They also built a large-scale dataset for image perceptual quality assessment, which contains traditional distortions and CNN-based distortions. Their dataset also contains distortions from frame interpolation. However, it is designed for image quality assessment and only contains single images. In contrast, our dataset contains videos for frame interpolation, which enable us to learn the temporal information. Ding *et al.* proposed an image quality metric with tolerance to texture resampling [21]. Bhardwaj *et al.* trained their metric using the information theory-guided loss function [6]. Czolbe *et al.* trained an image quality metric for GAN using Watson's perceptual model [17]. While these methods have achieved great success on image quality assessment, directly adapting these methods to videos will be disadvantageous to these methods because they ignore the spatio-temporal consistency. In contrast, our work builds upon these deep learning methods and considers the spatio-temporal information in our spatio-temporal network module, thus providing a dedicated solution to video frame interpolation quality assessment.

Video Quality Datasets. Traditional video quality datasets focus on video compression and transmission [1,3,24,25,56,69,71]. Video Quality Experts Group (VQEG) dataset was proposed for the secondary distribution of television [1]. Most videos in the VQEG dataset are interlaced. To address this problem, Kalpana *et al.* built the Laboratory for Image and Video Engineering (LIVE) Video Quality Database [3,71]. It is designed for the H.264 compression, MPEG-2 compression, and video transmission. In their recent works, they extended the dataset for the distortion in the streaming [24,56]. Li *et al.* proposed a video quality dataset for video compressions, including H.264/AVC, HEVC, and VP9 [69]. Different from these datasets, our work focuses on distortions from video frame interpolation. Recently, Danier *et al.* proposed the BVI-VFI dataset for video frame interpolation [19]. It is a relatively small dataset and hence not suitable for training deep learning models. Therefore, we make use of it as a test dataset to compare our method against various traditional and recently proposed metrics.

Vision Transformer. There has been considerable interest in vision transformers due to their impressive performance in image classification [22,43,47], object detection [9], and image restoration [11]. Recently, the vision transformers have also been introduced for the image quality assessment [14,95]. Their networks combined vision transformer [22] and CNN, and achieved promising results for image quality assessment. Our work adapts the recent proposed Swin Transformer [47] for video quality assessment.

3 Video Frame Interpolation Quality Dataset

We collected a Video Frame Interpolation Perceptual Similarity (VFIPS) dataset. It consists of a wide variety of distortions from video frame interpolation. Each sample consists of two videos synthesized from different interpolation methods, its reference video, and its perceptual judgments $h \in \{0, 0.33, 0.66, 1\}$. We can denote it as $\{\mathbf{V}_A, \mathbf{V}_B, \mathbf{V}_R, h\}$. Below, we briefly describe how we prepare and annotate this dataset.

3.1 Data Preparation

Source Videos. From YouTube, we collected 96 videos under the Creative Commons Attribution license (reuse allowed), which enables us to share this dataset with the community. Among them, 45 videos are 120 fps, and 51 videos are 60 fps. Most videos last over 10 min. The resolutions of the source videos are either 1080p or 4K. We downsampled these videos to 540p with the bicubic downsampling function from OpenCV. From these videos, we extracted video clips for quality assessment. After eliminating the video clips containing cut shots or interlaced frames, we curate, in total, 23,856 12-frame video clips. These 12-frame sequences serve as a reference in our dataset.

Videos Synthesized by Frame Interpolation Methods. We generated the synthesized videos for each reference video with 11 state-of-the-art video frame interpolation methods, including SepConv [59], Super-SloMo [30], CtxSyn [57], CyclicGen [46], MEMC-Net [5], DAIN [4], CAIN [15], RRIN [42], AdaCoF [40],

BMBC [64], and SoftSplat [58]. We set ×4 scale for 120-fps videos and ×2 scale for 60-fps videos.

For each reference sequence, we need to compare two video frame interpolation results. To make it easy to view and find differences between videos, we keep the size of the cropped video patches as 256 × 256. Specifically, we calculated the mean ℓ_1 error map between two synthesized videos and selected the patch location with the highest error by sliding patch windows.

3.2 Annotation

Compared to annotating images, annotating videos often requires much greater effort from users. For instance, a video contains many frames, but the distortions might only show in one or just a few frames. In such cases, a user needs to play the video multiple times to analyze differences and make a final judgment. In our case, a user often takes more than 1 min to annotate 1 sample. Hence, can we make use of some existing perceptual quality metrics to help us annotate more samples?

Recent developments in perceptual quality metrics show that considerable progress has been made with regard to correlating these metrics with human judgment [6,17,21,32,76,97]. The result of a single image quality assessment can provide helpful prior knowledge for the quality assessment of videos. For instance, by visual inspection, we found that if all frames of one video have significantly better quality than the frames in another video, this video often has a better quality. Based on this observation, we make use of the widely adopted state-of-the-art image quality metric LPIPS [97] to help annotate samples where two videos have significantly different LPIPS scores. These LPIPS annotated examples are only used for training and not used for testing. While they are not perfect, they can be used to train a good VFI quality metric. A similar strategy was also adopted in the PIEAPP metric [67]. For samples which is hard to make a judgment via LPIPS, we collect human annotations.

Automatic Annotation. We select 500 video triplets for our study to test the hypothesis that *when the difference in LPIPS scores of two videos is significant, the judgment via LPIPS is consistent with the human judgment.* We collect human annotations for these 500 triplets. Each video triplet contains one reference video and two synthesized videos. For each synthesized video, we calculate its LPIPS score as the mean of the LPIPS scores of the frames in that video. We tested whether the judgments via LPIPS are consistent with the human judgments. Evidence from this study suggests that the judgment via LPIPS can be reliably used for annotating triplets with large LPIPS difference.

Figure 2 shows the LPIPS difference versus the human judgment, where the LPIPS difference is the absolute difference between the LPIPS scores of a pair of videos. The judgment via LPIPS is consistent with the human judgment with the increase in the LPIPS difference. We can find that 97.6% judgments via LPIPS are consistent with the human judgments when the LPIPS difference is greater than 0.15. Based on this observation, we select 0.15 as the LPIPS-difference

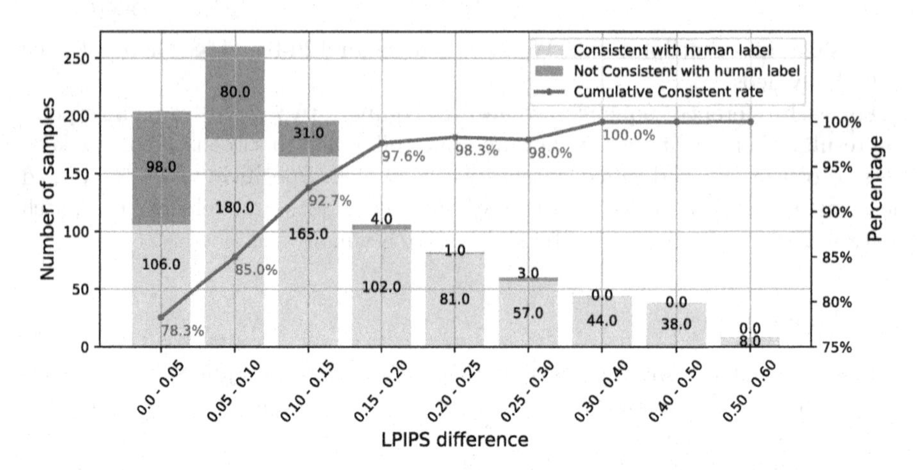

Fig. 2. LPIPS difference vs. human labeling result.

threshold. If a new video triplet's LPIPS difference is greater than 0.15, we take the LPIPS judgment as its ground truth label. For the rest of the video triplets where LPIPS difference is less than 0.15, we collect human annotations. It helps save time and effort involved in the recruitment and collection of high-quality human annotations. In the end, for our dataset, we have 18,939 video triplets with ground truth labels annotated via LPIPS.

Manual Annotation. For the samples where LPIPS-difference is less than 0.15, we collect judgments on perceptual quality by humans.

User Interface. For the annotation, we presented participants with a reference video and two corresponding video frame interpolation results. The videos were played simultaneously to help participants compare the two synthesized videos. For judgment, participants were asked which of the two synthesized videos has a higher perceptual quality. We played the videos in a loop without interruption or human intervention. Figure 3 shows the user interface. It has four options {"A(Sure)", "A(Maybe)", "B(Maybe)", "B(Sure)"}. Please note that the middle point in Fig. 3 is not an option.

Playback Setting. To find the best playback setting, we tested video clips of length from 4 to 16 frames. If a video is too short, human annotators would not get enough temporal information to make a decision. But a too-long video often leads to ambiguity. For example, the quality of a video can be high in the first half but be poor in the last half. In such cases, it is difficult for users to make a decision. We empirically found that a video clip with 12 frames works well for users to annotate the quality. We also explored the playback speed in our study. At a high fps, such as 8 fps, users will find it extremely hard to compare the distortions, as each frame only shows for a very short time. However, a low fps will make it harder to judge the temporal consistency. We empirically found that 2 fps works well.

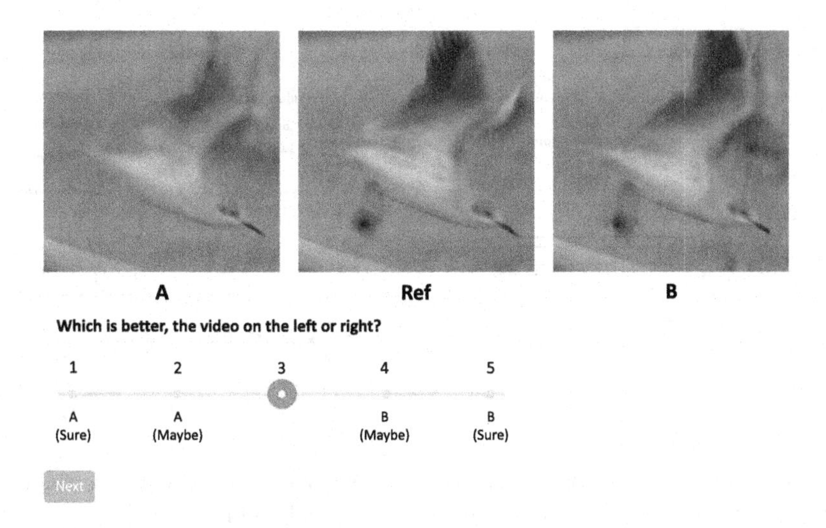

Fig. 3. User interface. Two frame interpolation results and one reference are played simultaneously in a loop.

We collected annotations on 5,948 video triplet samples from 41 participants. Following LPIPS [97], each sample was labeled by 3 participants, and the mean of the judgments was taken to decide which video had a higher quality. All participants that we recruited were volunteers from diverse backgrounds. For our training dataset, we randomly select 5,353 samples out of the 5,948 human-annotated video triplets and all 18,939 sample triplets with judgment via LPIPS. The rest 595 human-annotated samples were used for validation.

4 Video Frame Interpolation Quality Metric

Our method takes a video $\mathbf{V} = \{\mathbf{I}_0, \mathbf{I}_1, \cdots, \mathbf{I}_N\}$ and its reference video $\mathbf{V}_R = \{\mathbf{I}_{R,0}, \mathbf{I}_{R,1}, \cdots, \mathbf{I}_{R,N}\}$ as inputs and estimates the perceptual similarity d between these two videos, where N is the number of frames.

Figure 4 shows the architecture of our network. Our network first extracts the feature maps for each image in the videos. For each feature map, we design a spatio-temporal (ST) module to capture the spatio-temporal information and estimate the perceptual similarity of features. Finally, we predict the perceptual similarity between the input video $\mathbf{V}$ and its reference $\mathbf{V}_R$ by averaging the similarity across all features. Our network is trained in a Siamese manner. Below we describe our network in more detail.

Feature Extraction. As shown in Fig. 4, we extract features from the video $\mathbf{V}$ and its reference video $\mathbf{V}_R$. Specifically, we build a pyramid network with five levels. Each level contains two 3×3 `Conv2D` layers. The stride of the second `Conv2D` layer is set to 2 to have a large valid receptive field. From the first to

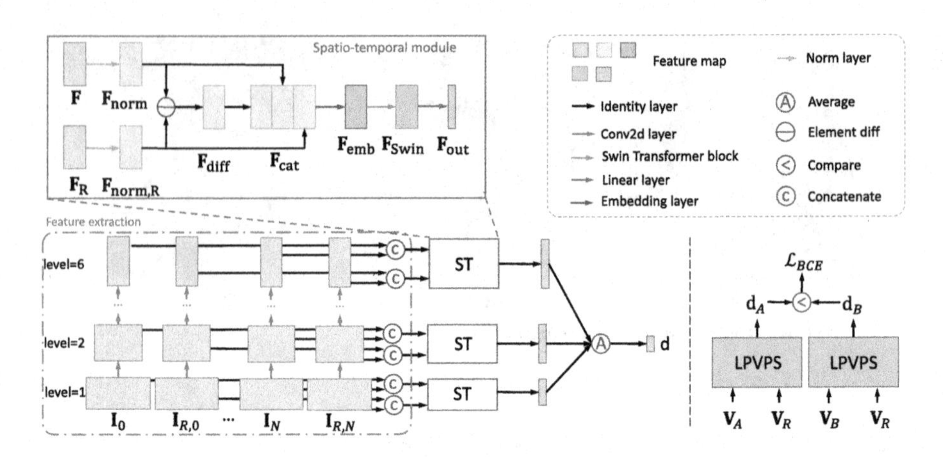

Fig. 4. Our network architecture. (Left) Learned Perceptual Video Patch Similarity (LPVPS) takes a video $\mathbf{V}$ and its reference $\mathbf{V}_R$ as input and predicts their perceptual similarity d. (Right) Our network is trained in a Siamese manner.

the fifth level, the numbers of channels are 16, 32, 64, 96, and 128. It can be represented as follows.

$$\{\mathbf{F}_{level,i}\}_{level=1}^{L} = f_{ext}(\mathbf{I}_i), \tag{1}$$

where $f_{ext}(\cdot)$ indicates the operation of the extraction network. $\mathbf{F}$ indicates the extracted features. L indicates the number of feature levels. The feature extraction network shares weights for all frames and videos. For each level, we concatenate the features from all frames as the final features,

$$\mathbf{F}_{level} = f_{cat}(\mathbf{F}_{level,0}, \cdots, \mathbf{F}_{level,N}), \tag{2}$$

where $f_{cat}(\cdot)$ is the concatenation operation. N is the number of frames.

Spatio-temporal (ST) module is designed to measure the perceptual distance between the feature maps $\mathbf{F}$ and its reference $\mathbf{F}_R$. It first estimates the difference between two normalized features as follows.

$$\mathbf{F}_{diff} = \|f_{norm}(\mathbf{F}) - f_{norm}(\mathbf{F}_R)\|_1, \tag{3}$$

where $f_{norm}(\cdot)$ indicates the operation of unit normalization in the channel dimension [97]. As reported in previous work on image perceptual metrics [26,51,52], the perceptual similarity is not only related to the difference between two images, but also the content of images. Therefore, our network leverages both the feature difference $\mathbf{F}_{diff}$ and the source features.

$$\mathbf{F}_{cat} = f_{cat}(\mathbf{F}_{diff}, f_{norm}(\mathbf{F}), f_{norm}(\mathbf{F}_R)). \tag{4}$$

As shown in Fig. 4, we first adopt a linear embedding layer to project the features $\mathbf{F}_{cat}$ to a fixed dimension 32.

$$\mathbf{F}_{emb} = f_{emb}(\mathbf{F}_{cat}), \tag{5}$$

Table 1. Comparison with state-of-the-art methods.

Method		VFIPS (val.)	BVI-VFI [19] (test)		
		2AFC	SROCC	PLCC	KROCC
Image	PSNR	0.763	0.742	0.722	0.656
	SSIM [79]	0.784	0.739	0.746	0.639
	MS-SSIM [80]	0.794	0.772	0.789	0.667
	LPIPS (VGG) [97]	0.808	0.628	0.796	0.517
	DISTS [21]	0.801	0.597	0.763	0.517
	PIM-1 [6]	0.787	0.492	0.668	0.428
	Watson-DFT [17]	0.800	0.628	0.706	0.538
Video	STRRED [7]	0.777	0.614	0.682	0.539
	VMAF [69]	0.805	0.583	0.614	0.483
	DeepVQA [77]	0.588	0.369	0.271	0.300
	VSFA [41]	0.660	0.108	0.486	0.050
	Ours	**0.830**	**0.794**	**0.870**	**0.700**

where $f_{emb}(\cdot)$ indicates the linear embedding layer that consists of a `Conv2D` layer with a 1×1 kernel.

We adopt a Swin Transformer block [47] to learn the temporal information across frames. Compared to the original design of Swin Transformer block [47], we do not use `Layer Norm (LN)` layers, which is critical to measure the difference between two videos. `LN` layers normalize the differences between two features and compromise the accuracy of the network. We empirically find that the dimension of the embedded features as 32, the head number as 2, and the window sizes of 4 can get good results and require relatively less computation and memory.

ST modules are applied to the features at different levels, which can capture the spatio-temporal information at multiple scales. It is similar to the pyramid architectures of optical flow networks [45,75]. The final distance is calculated by averaging the outputs at all the L levels from ST modules.

$$d = \frac{1}{LN} \sum_{l=0}^{L} \sum_{n=0}^{N} \mathbf{F}^{n}_{Swin,l} \tag{6}$$

where N indicates the number of elements in $\mathbf{F}_{Swin,l}$.

Loss Function. As shown in Fig. 4, our network is trained in a Siamese manner. Given two videos with different distortions $\mathbf{V}_A$, $\mathbf{V}_B$, its reference video $\mathbf{V}_R$, and its judgement $\hat{h}$, we first predict the perceptual similarity d_A and d_B from ($\mathbf{V}_A$, $\mathbf{V}_R$) and ($\mathbf{V}_B$, $\mathbf{V}_R$), respectively. We calculate the probability p with a sigmoid layer,

$$p = sigmoid(d_A - d_B). \tag{7}$$

Table 2. Comparison on the X-TEST(4K) dataset [73].

Method	PSNR	SSIM [79]	MSSSIM [80]	LPIPS [97]	STRRED [7]	VMAF [69]	Ours
2AFC	0.752	0.637	0.737	0.748	0.722	0.735	**0.789**

Then, we calculate the binary cross entropy loss $\mathcal{L}_{BCE}$ as:

$$\mathcal{L}_{BCE} = -(\hat{h}\log(p) + (1-\hat{h})\log(1-p)). \tag{8}$$

Training. We use PyTorch to train our neural network. Following [97], our learning rate is set to 0.0001. We use a mini-batch size of 8 and train the network for 20 epochs. We randomly resize the videos in the scale of [0.5, 1]. Our network is randomly initialized and uses AdamW [34] as the optimizer.

5 Experiments

We use the BVI-VFI [19] dataset as our test set. It contains 36 reference videos at 3 different frame rates: 30 fps, 60 fps, and 120 fps. All the videos last 5 s. Each reference video has 5 distorted videos generated from different video frame interpolation algorithms, including frame repeating, frame averaging, DVF [48], QVI [93] and ST-MFNet [18]. Please note that video frame interpolation methods used in the BVI-VFI dataset are different from the ones used in our VFIPS dataset. To test our metric on the BVI-VFI dataset, we evaluate our metric in a sliding-window manner and take the average score. We selected Spearman Rank Order Correlation Coefficient (SROCC), Pearson Linear Correlation Coefficient (PLCC), and Kendall Rank Order Correlation Coefficient (KROCC) scores as our evaluation metrics. We calculated the scores for each reference video and reported the mean score as the final result. We also report our results on our VFIPS validation set. Following LPIPS [97], we report the Two Alternative Forced Choice (2AFC) scores [97]. We compare our method to the representative state-of-the-art metrics and conduct ablation studies to evaluate our metric.

5.1 Comparisons to Existing Metrics

We compare our method to both the state-of-the-art image perceptual metrics, including PSNR, SSIM [79], MS-SSIM [80], LPIPS(VGG) [97], DISTS [21], PIM-1 [6], and Watson-DFT [17], and the state-of-the-art video perceptual metrics, including VMAF [69], DeepVQA [32], and STRRED [74]. We also compare with the Non-Reference metrics, VSFA [41]. For each method, we obtained the results by using the official codes/models provided by their authors, except for DeepVQA which we adopt its re-implementation from Tencent [77]. For the image perceptual similarity metrics, we evaluate the score per frame and take the average as the final score for that video sequence.

As shown in Table 1, our method outperforms the state-of-the-art methods by a considerable margin on both the VFIPS dataset and the BVI-VFI

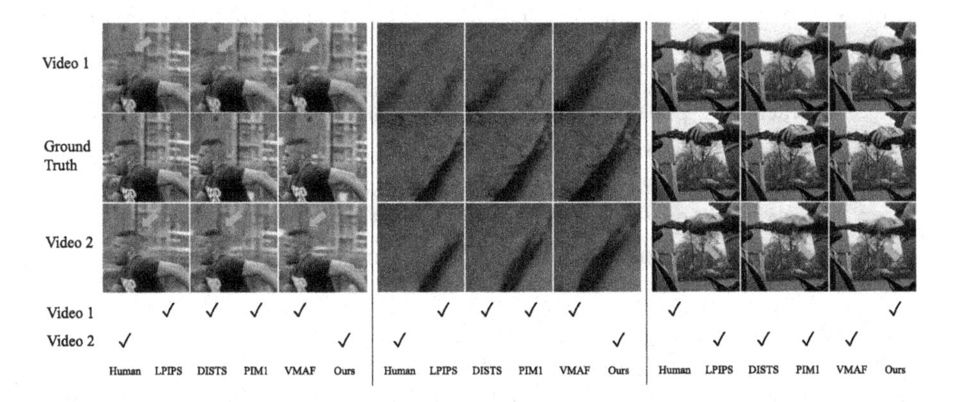

Fig. 5. Visual examples on the VFIPS dataset. Green arrows are used to label the area with noticeable difference. We mark the preference of each method using "✓". Compared to other methods, our method is consistent with humans. (Color figure online)

Ref						Ref					
Human	4th	3rd	5th	1st	2nd	Human	4th	1st	5th	3rd	2nd
VMAF	4th	3rd	5th	2nd	1st	VMAF	1st	2nd	3rd	4th	5th
LPIPS	3rd	4th	5th	1st	2nd	LPIPS	2nd	1st	5th	3rd	4th
Ours	4th	3rd	5th	1st	2nd	Ours	4th	1st	5th	3rd	2nd

Fig. 6. Visual examples on the BVI-VFI dataset [19]. Yellow rectangles are used to show the reference video. We report the rank for the distorted videos. Compared to other metrics, our metric is more consistent with humans. (Color figure online)

dataset. Especially when compared to the image-based similarity methods such as LPIPS, our method provides significant gains in terms of 2AFC on the VFIPS dataset and SROCC, PLCC, and KROCC on the BVI-VFI dataset. We attribute this improvement to the spatial-temporal module that captures spatio-temporal information. Compared to the video-based methods, our method again achieves significant performance gains on all metrics on both datasets, specifically, 0.025 in terms of 2AFC on the VFIPS validation set and 0.180 in SROCC, 0.188 in PLCC, 0.161 in KROCC on the BVI-VFI dataset. Figure 5 shows the visual examples on the VFIPS validation set. The predictions of our method are more consistent with humans. In the first example, Video 2 suffers less distortions to the person's head. In the second example, Video 2 is temporally consistent. In the third example, the hand in Video 2 is distorted more. As indicated, our predictions are consistent with humans in these examples. Figure 6 shows the visual examples from the BVI-VFI dataset [19]. For each reference video, it shows the rank for the distorted videos. Our predictions are more consistent with humans.

We also conduct our experiments on the X-TEST dataset [73], which has 15 testing videos with 4K resolution. We use SepConv, Super-Slomo, AdaCof, and XVFI to generate VFI results. Following Sim *et al.* [73], we set the temporal

Table 3. Effectiveness of different feature extraction network on the BVI-VFI datasets [19].

Extractor	SROCC	PLCC	KROCC	Param(M)	Runtime(ms)
AlexNet [37]	0.761	0.832	0.650	14.5	12.8
I3D [10]	0.659	0.758	0.550	20.3	33.2
Ours-3D	0.728	0.738	0.639	8.6	13.6
Ours-2D	**0.794**	**0.870**	**0.700**	**4.6**	**10.4**

distance of 32 frames for the large motion. For each video, we interpolate 32 frames and take the middle 12 frames for testing. There are 90 pairs in total. Each pair is annotated by 3 users. As reported in Table 2, our method outperforms other methods on such a large-motion dataset.

5.2 Ablation Studies

Feature Extraction Network. We investigate the impact of architectures of the feature extraction network in Table 3. Following LPIPS [97], we replace our extraction network with several classic architectures, including AlexNet [37], I3D [10], and ours-3D network. We use the official implementation of AlexNet and I3D [10]. For ours-3D, we replace the `Conv2D` layers in our feature extraction module with `Conv3D` layers with $3 \times 3 \times 3$ kernels. We measure the runtime for a 12-frame 256×256 video on a single Nvidia RTX A5000 GPU.

As shown in Table 3, our method not only achieves the best performance but also needs the least parameters. Compared to AlexNet, our feature extraction network achieves better performance while our method only needs 34.1% parameters and is 1.23× faster. Compared to I3D and ours-3D, our network achieves much better results and has less parameters. We attribute the improvement to that our feature extraction network keeps more temporal information. Specifically, I3D and ours-3D downsample the features temporally, while our network keeps features of all the frames. The temporal compression in the feature extraction network might hurt the performance of the whole metric.

Table 4. Effectiveness of the spatio-temporal module on the BVI-VFI dataset [19].

ST Module	SROCC	PLCC	KROCC
None	0.617	0.663	0.539
`Conv3D`	0.761	0.819	0.661
Original Swin	0.728	0.766	0.639
Ours-Swin w. `LN`	0.724	0.746	0.611
Ours-Swin	**0.794**	**0.870**	**0.700**

Spatio-Temporal Module. We study the spatio-temporal module in Table 4. We use different blocks for the spatio-temporal module. "`Conv3D`" indicates the `Conv3D` layers with $12 \times 3 \times 3$ kernels. "Original Swin" indicates the official Swin Transformer blocks. "Ours-Swin w. `LN`" indicates our Swin Transformer block with `LN` layers. Compared to the original Swin, it contains fewer parameters, including 32 vs. 96 (channels), 4 vs. 7 (window size). It has slightly worse performance. As shown in

Table 4, original Swin and ours-Swin w. `LN` do not perform as well as Conv3D. We attribute this performance loss to the `LN` normalizing the differences between features. Removing the `LN` layers improves the results by 0.070 on SROCC, 0.124 on PLCC, and 0.089 on KROCC on the BVI-VFI dataset.

Automatic Annotations by LPIPS. We study the impact of annotations in Table 5. As discussed in Sect. 3.2, our training set contains two parts: samples obtained through automatic annotation or human labeling. We train our network with different training subsets: human annotations, automatic annotations, and a combination of both. As shown in Table 5, the combined set achieves the best performance. The automatic training data can increase the diversity of the training set and improve the generalization capability of the metric.

Table 5. Effectiveness of annotations on the BVI-VFI dataset [19]

Annotation	SROCC	PLCC	KROCC
Human	0.719	0.753	0.611
Automatic	0.653	0.687	0.567
All	**0.794**	**0.870**	**0.700**

6 Limitations and Future Work

Many image-based quality assessment metrics can generate an error map between a distorted image and its reference to analyze altered pixels. However, our metric cannot produce such an error map for individual frames in the video since our spatio-temporal module contains convolutional layers that fuse features from these frames at an early stage in the inference. In the future, it will be interesting to design a network that can generate such error maps for each frame to provide more information on the distortion within and across the frames.

Recently, image perceptual quality metrics, such as LPIPS [97], are used as loss to create visually pleasant images or videos [63]. In theory, our network can also be used as a loss function to train a video frame interpolation network to produce visually pleasant videos. However, our metric has a high requirement for GPU memory. It takes about 2500 Mb memory to process one 256×256 12-frame image sequence, which might become a bottleneck in training a video frame interpolation network. In the future, we aim to reduce the network size and study how our video frame interpolation quality metric will help optimize video frame interpolation networks as a learned perceptual loss.

7 Conclusion

This paper presented a video perceptual quality metric for video frame interpolation. Our metric first extracts pyramid features for individual frames in the videos. Then it compares features at each level using a spatio-temporal module to capture the spatio-temporal information. The spatio-temporal module is composed of the Swin Transformer blocks to capture the spatio-temporal information. We also collected a dataset for video frame interpolation. Our annotations were from humans and a widely adopted image perceptual quality metric.

Our experiments showed that our dedicated video quality metric outperforms existing metrics for assessing video frame interpolation results.

Acknowledgment. This work was made possible in part thanks to Research Computing at Portland State University and its HPC resources acquired through NSF Grants 2019216 and 1624776. Source frames in Fig. 1, 3, and 5 are used under a Creative Commons license from Youtube users Ignacio, Scott, Animiles, H-Edits, 3 Playing Brothers, TristanBotteram, popconet, and billisa.

References

1. Antkowiak, J., et al.: Final report from the video quality experts group on the validation of objective models of video quality assessment (2000)
2. Baker, S., Scharstein, D., Lewis, J., Roth, S., Black, M.J., Szeliski, R.: A database and evaluation methodology for optical flow. Int. J. Comput. Vis. **92**(1), 1–31 (2011)
3. Bampis, C.G., Li, Z., Katsavounidis, I., Huang, T.Y., Ekanadham, C., Bovik, A.C.: Towards perceptually optimized end-to-end adaptive video streaming. arXiv preprint arXiv:1808.03898 (2018)
4. Bao, W., Lai, W.S., Ma, C., Zhang, X., Gao, Z., Yang, M.H.: Depth-aware video frame interpolation. In: Proceedings of the IEEE Conference on Computer Vision and Pattern Recognition, pp. 3703–3712 (2019)
5. Bao, W., Lai, W.S., Zhang, X., Gao, Z., Yang, M.H.: MEMC-Net: motion estimation and motion compensation driven neural network for video interpolation and enhancement. IEEE Trans. Pattern Anal. Mach. Intell. **43**, 933–948 (2019)
6. Bhardwaj, S., Fischer, I., Ballé, J., Chinen, T.: An unsupervised information-theoretic perceptual quality metric. In: Advances in Neural Information Processing Systems, vol. 33 (2020)
7. Bovik, A., Soundararajan, R., Bampis, C.: On the robust performance of the ST-RRED video quality predictor (2017)
8. van den Branden Lambrecht, C.J.: Color moving pictures quality metric. In: Proceedings of 3rd IEEE International Conference on Image Processing, vol. 1, pp. 885–888. IEEE (1996)
9. Carion, N., Massa, F., Synnaeve, G., Usunier, N., Kirillov, A., Zagoruyko, S.: End-to-end object detection with transformers. In: Vedaldi, A., Bischof, H., Brox, T., Frahm, J.-M. (eds.) ECCV 2020. LNCS, vol. 12346, pp. 213–229. Springer, Cham (2020). https://doi.org/10.1007/978-3-030-58452-8_13
10. Carreira, J., Zisserman, A.: Quo Vadis, action recognition? A new model and the kinetics dataset. In: Proceedings of the IEEE Conference on Computer Vision and Pattern Recognition, pp. 6299–6308 (2017)
11. Chen, H., et al.: Pre-trained image processing transformer. In: Proceedings of the IEEE/CVF Conference on Computer Vision and Pattern Recognition, pp. 12299–12310 (2021)
12. Chen, L.H., Bampis, C.G., Li, Z., Bovik, A.C.: Learning to distort images using generative adversarial networks. IEEE Signal Process. Lett. **27**, 2144–2148 (2020)
13. Cheng, X., Chen, Z.: Multiple video frame interpolation via enhanced deformable separable convolution. IEEE Trans. Pattern Anal. Mach. Intell. (2021)
14. Cheon, M., Yoon, S.J., Kang, B., Lee, J.: Perceptual image quality assessment with transformers. In; 2021 IEEE/CVF Conference on Computer Vision and Pattern Recognition Workshops (CVPRW), pp. 433–442 (2021)

15. Choi, M., Kim, H., Han, B., Xu, N., Lee, K.M.: Channel attention is all you need for video frame interpolation. In: Proceedings of the AAAI Conference on Artificial Intelligence, pp. 10663–10671 (2020)
16. Choi, M., Lee, S., Kim, H., Lee, K.M.: Motion-aware dynamic architecture for efficient frame interpolation. In: Proceedings of the IEEE/CVF International Conference on Computer Vision, pp. 13839–13848 (2021)
17. Czolbe, S., Krause, O., Cox, I., Igel, C.: A loss function for generative neural networks based on Watson's perceptual model. In: Advances in Neural Information Processing Systems, pp. 2051–2061 (2020)
18. Danier, D., Zhang, F., Bull, D.: ST-MFNet: a spatio-temporal multi-flow network for frame interpolation. In: Proceedings of the IEEE/CVF Conference on Computer Vision and Pattern Recognition, pp. 3521–3531 (2022)
19. Danier, D., Zhang, F., Bull, D.: A subjective quality study for video frame interpolation. arXiv preprint arXiv:2202.07727 (2022)
20. Dendi, S.V.R., Krishnappa, G., Channappayya, S.S.: Full-reference video quality assessment using deep 3D convolutional neural networks. In: 2019 National Conference on Communications (NCC), pp. 1–5. IEEE (2019)
21. Ding, K., Ma, K., Wang, S., Simoncelli, E.P.: Image quality assessment: unifying structure and texture similarity. IEEE Trans. Pattern Anal. Mach. Intell. 1 (2020)
22. Dosovitskiy, A., et al.: An image is worth 16 × 16 words: transformers for image recognition at scale. In: 9th International Conference on Learning Representations, ICLR 2021, Virtual Event, Austria, 3–7 May 2021. OpenReview.net (2021). https://openreview.net/forum?id=YicbFdNTTy
23. Gardiner, P., Ghanbari, M., Pearson, D., Tan, K.: Development of a perceptual distortion meter for digital video. In: 1997 International Broadcasting Convention, IBS 1997, pp. 493–497. IET (1997)
24. Ghadiyaram, D., Pan, J., Bovik, A., Moorthy, A., Panda, P., Yang, K.: LIVE-Qualcomm mobile in-capture video quality database (2017)
25. Ghadiyaram, D., Pan, J., Bovik, A.C.: A subjective and objective study of stalling events in mobile streaming videos. IEEE Trans. Circuits Syst. Video Technol. **29**(1), 183–197 (2017)
26. Goodman, N.: Seven strictures on similarity (1972)
27. Hamada, T., Miyaji, S., Matsumoto, S.: Picture quality assessment system by three-layered bottom-up noise weighting considering human visual perception. SMPTE J. **108**(1), 20–26 (1999)
28. Hu, P., Niklaus, S., Sclaroff, S., Saenko, K.: Many-to-many splatting for efficient video frame interpolation. In: Proceedings of the IEEE/CVF Conference on Computer Vision and Pattern Recognition, pp. 3553–3562 (2022)
29. Huang, Z., Zhang, T., Heng, W., Shi, B., Zhou, S.: RIFE: real-time intermediate flow estimation for video frame interpolation. arXiv preprint arXiv:2011.06294 (2020)
30. Jiang, H., Sun, D., Jampani, V., Yang, M.H., Learned-Miller, E., Kautz, J.: Super SloMo: high quality estimation of multiple intermediate frames for video interpolation. In: Proceedings of the IEEE Conference on Computer Vision and Pattern Recognition, pp. 9000–9008 (2018)
31. Kalluri, T., Pathak, D., Chandraker, M., Tran, D.: FLAVR: flow-agnostic video representations for fast frame interpolation. arXiv preprint arXiv:2012.08512 (2020)
32. Kim, J., Lee, S.: Deep learning of human visual sensitivity in image quality assessment framework. In: Proceedings of the IEEE Conference on Computer Vision and Pattern Recognition, pp. 1676–1684 (2017)

33. Kim, W., Kim, J., Ahn, S., Kim, J., Lee, S.: Deep video quality assessor: from spatio-temporal visual sensitivity to a convolutional neural aggregation network. In: Proceedings of the European Conference on Computer Vision (ECCV), pp. 219–234 (2018)
34. Kingma, D.P., Ba, J.: Adam: a method for stochastic optimization. arXiv preprint arXiv:1412.6980 (2014)
35. Korhonen, J.: Two-level approach for no-reference consumer video quality assessment. IEEE Trans. Image Process. **28**(12), 5923–5938 (2019)
36. Korhonen, J., Su, Y., You, J.: Blind natural video quality prediction via statistical temporal features and deep spatial features. In: Proceedings of the 28th ACM International Conference on Multimedia, pp. 3311–3319 (2020)
37. Krizhevsky, A., Sutskever, I., Hinton, G.E.: ImageNet classification with deep convolutional neural networks. In: Advances in Neural Information Processing Systems, vol. 25, pp. 1097–1105 (2012)
38. Kuroki, Y., Nishi, T., Kobayashi, S., Oyaizu, H., Yoshimura, S.: A psychophysical study of improvements in motion-image quality by using high frame rates. J. Soc. Inform. Display **15**(1), 61–68 (2007)
39. Kuroki, Y., Takahashi, H., Kusakabe, M., Yamakoshi, K.: Effects of motion image stimuli with normal and high frame rates on EEG power spectra: comparison with continuous motion image stimuli. J. Soc. Inf. Display **22**(4), 191–198 (2014)
40. Lee, H., Kim, T., Chung, T., Pak, D., Ban, Y., Lee, S.: AdaCoF: adaptive collaboration of flows for video frame interpolation. In: Proceedings of the IEEE Conference on Computer Vision and Pattern Recognition, pp. 5316–5325 (2020)
41. Li, D., Jiang, T., Jiang, M.: Quality assessment of in-the-wild videos. In: Proceedings of the 27th ACM International Conference on Multimedia, pp. 2351–2359 (2019)
42. Li, H., Yuan, Y., Wang, Q.: Video frame interpolation via residue refinement. In: IEEE International Conference on Acoustics, Speech and Signal Processing, pp. 2613–2617. IEEE (2020)
43. Li, Y., Zhang, K., Cao, J., Timofte, R., Van Gool, L.: LocalViT: bringing locality to vision transformers. arXiv preprint arXiv:2104.05707 (2021)
44. Li, Z., Aaron, A., Katsavounidis, I., Moorthy, A., Manohara, M.: Toward a practical perceptual video quality metric. Netflix Tech Blog **6**(2) (2016)
45. Liu, P., Lyu, M., King, I., Xu, J.: SelFlow: self-supervised learning of optical flow. In: Proceedings of the IEEE Conference on Computer Vision and Pattern Recognition, pp. 4571–4580 (2019)
46. Liu, Y.L., Liao, Y.T., Lin, Y.Y., Chuang, Y.Y.: Deep video frame interpolation using cyclic frame generation. In: Proceedings of the AAAI Conference on Artificial Intelligence, vol. 33, pp. 8794–8802 (2019)
47. Liu, Z., et al.: Swin transformer: hierarchical vision transformer using shifted windows. In: Proceedings of the IEEE/CVF International Conference on Computer Vision, pp. 10012–10022 (2021)
48. Liu, Z., Yeh, R.A., Tang, X., Liu, Y., Agarwala, A.: Video frame synthesis using deep voxel flow. In: Proceedings of the IEEE International Conference on Computer Vision, pp. 4463–4471 (2017)
49. Long, G., Kneip, L., Alvarez, J.M., Li, H., Zhang, X., Yu, Q.: Learning image matching by simply watching video. In: Leibe, B., Matas, J., Sebe, N., Welling, M. (eds.) ECCV 2016. LNCS, vol. 9910, pp. 434–450. Springer, Cham (2016). https://doi.org/10.1007/978-3-319-46466-4_26
50. Lubin, J.: A human vision system model for objective picture quality measurements. In: IEE Conference Publication, vol. 1, pp. 498–503. IET (1997)

51. Markman, A.B., Gentner, D.: Nonintentional similarity processing. The New Unconscious, pp. 107–137 (2005)
52. Medin, D.L., Goldstone, R.L., Gentner, D.: Respects for similarity. Psychol. Rev. **100**(2), 254 (1993)
53. Meyer, S., Cornillère, V., Djelouah, A., Schroers, C., Gross, M.: Deep video color propagation. In: Proceedings of the British Machine Vision Conference (BMVC) (2018)
54. Meyer, S., Djelouah, A., McWilliams, B., Sorkine-Hornung, A., Gross, M., Schroers, C.: PhaseNet for video frame interpolation. In: Proceedings of the IEEE Conference on Computer Vision and Pattern Recognition, pp. 498–507 (2018)
55. Meyer, S., Wang, O., Zimmer, H., Grosse, M., Sorkine-Hornung, A.: Phase-based frame interpolation for video. In: Proceedings of the IEEE Conference on Computer Vision and Pattern Recognition, pp. 1410–1418 (2015)
56. Min, X., Zhai, G., Zhou, J., Farias, M.C., Bovik, A.C.: Study of subjective and objective quality assessment of audio-visual signals. IEEE Trans. Image Process. **29**, 6054–6068 (2020)
57. Niklaus, S., Liu, F.: Context-aware synthesis for video frame interpolation. In: Proceedings of the IEEE Conference on Computer Vision and Pattern Recognition, pp. 1701–1710 (2018)
58. Niklaus, S., Liu, F.: Softmax splatting for video frame interpolation. In: Proceedings of the IEEE/CVF Conference on Computer Vision and Pattern Recognition, pp. 5437–5446 (2020)
59. Niklaus, S., Mai, L., Liu, F.: Video frame interpolation via adaptive convolution. In: Proceedings of the IEEE Conference on Computer Vision and Pattern Recognition, pp. 670–679 (2017)
60. Niklaus, S., Mai, L., Liu, F.: Video frame interpolation via adaptive separable convolution. In: Proceedings of the IEEE International Conference on Computer Vision, pp. 261–270 (2017)
61. Niklaus, S., Mai, L., Wang, O.: Revisiting adaptive convolutions for video frame interpolation. arXiv preprint arXiv:2011.01280 (2020)
62. Niklaus, S., Mai, L., Wang, O.: Revisiting adaptive convolutions for video frame interpolation. In: Proceedings of the IEEE/CVF Winter Conference on Applications of Computer Vision, pp. 1099–1109 (2021)
63. Niklaus, S., et al.: Learned dual-view reflection removal. In: Proceedings of the IEEE/CVF Winter Conference on Applications of Computer Vision, pp. 3713–3722 (2021)
64. Park, J., Ko, K., Lee, C., Kim, C.S.: BMBC: bilateral motion estimation with bilateral cost volume for video interpolation. arXiv preprint arXiv:2007.12622 (2020)
65. Park, J., Lee, C., Kim, C.S.: Asymmetric bilateral motion estimation for video frame interpolation. In: Proceedings of the IEEE/CVF International Conference on Computer Vision, pp. 14539–14548 (2021)
66. Pinson, M.H., Wolf, S.: A new standardized method for objectively measuring video quality. IEEE Trans. Broadcast. **50**(3), 312–322 (2004)
67. Prashnani, E., Cai, H., Mostofi, Y., Sen, P.: PieAPP: perceptual image-error assessment through pairwise preference. In: IEEE Conference on Computer Vision and Pattern Recognition, pp. 1808–1817 (2018)
68. Rakêt, L.L., Roholm, L., Bruhn, A., Weickert, J.: Motion compensated frame interpolation with a symmetric optical flow constraint. In: Bebis, G., et al. (eds.) ISVC 2012. LNCS, vol. 7431, pp. 447–457. Springer, Heidelberg (2012). https://doi.org/10.1007/978-3-642-33179-4_43

69. Rassool, R.: VMAF reproducibility: validating a perceptual practical video quality metric. In: 2017 IEEE International Symposium on Broadband Multimedia Systems and Broadcasting (BMSB), pp. 1–2 (2017). https://doi.org/10.1109/BMSB.2017.7986143
70. Reda, F.A., et al.: Unsupervised video interpolation using cycle consistency. In: Proceedings of the IEEE International Conference on Computer Vision, pp. 892–900 (2019)
71. Seshadrinathan, K., Soundararajan, R., Bovik, A.C., Cormack, L.K.: Study of subjective and objective quality assessment of video. IEEE Trans. Image Process. **19**(6), 1427–1441 (2010)
72. Shi, Z., Xu, X., Liu, X., Chen, J., Yang, M.H.: Video frame interpolation transformer. In: Proceedings of the IEEE/CVF Conference on Computer Vision and Pattern Recognition, pp. 17482–17491 (2022)
73. Sim, H., Oh, J., Kim, M.: XVFI: extreme video frame interpolation. In: Proceedings of the IEEE/CVF International Conference on Computer Vision, pp. 14489–14498 (2021)
74. Soundararajan, R., Bovik, A.C.: Video quality assessment by reduced reference spatio-temporal entropic differencing. IEEE Trans. Circuits Syst. Video Technol. **23**(4), 684–694 (2013)
75. Sun, D., Yang, X., Liu, M.Y., Kautz, J.: PWC-Net: CNNs for optical flow using pyramid, warping, and cost volume. In: Proceedings of the IEEE Conference on Computer Vision and Pattern Recognition, pp. 8934–8943 (2018)
76. Talebi, H., Milanfar, P.: NIMA: neural image assessment. IEEE Trans. Image Process. **27**(8), 3998–4011 (2018)
77. Tencent: Deep learning-based Video Quality Assessment (2022). https://github.com/Tencent/DVQA. Accessed 19 July 2022
78. Tu, Z., Wang, Y., Birkbeck, N., Adsumilli, B., Bovik, A.C.: UGC-VQA: benchmarking blind video quality assessment for user generated content. IEEE Trans. Image Process. **30**, 4449–4464 (2021)
79. Wang, Z., Bovik, A.C., Sheikh, H.R., Simoncelli, E.P.: Image quality assessment: from error visibility to structural similarity. IEEE Trans. Image Process. **13**(4), 600–612 (2004)
80. Wang, Z., Simoncelli, E.P., Bovik, A.C.: Multiscale structural similarity for image quality assessment. In: The Thirty-Seventh Asilomar Conference on Signals, Systems & Computers, vol. 2, pp. 1398–1402. IEEE (2003)
81. Watson, A.B.: Toward a perceptual video-quality metric. In: Human Vision and Electronic Imaging III, vol. 3299, pp. 139–147. International Society for Optics and Photonics (1998)
82. Watson, A.B., Hu, Q.J., McGowan, J.F., III.: Digital video quality metric based on human vision. J. Electron. Imaging **10**(1), 20–29 (2001)
83. Webster, A.A., Jones, C.T., Pinson, M.H., Voran, S.D., Wolf, S.: Objective video quality assessment system based on human perception. In: Human Vision, Visual Processing, and Digital Display IV, vol. 1913, pp. 15–26. International Society for Optics and Photonics (1993)
84. Winkler, S.: Digital Video Quality: Vision Models and Metrics. Wiley, Hoboken (2005)
85. Winkler, S., Mohandas, P.: The evolution of video quality measurement: from PSNR to hybrid metrics. IEEE Trans. Broadcast. **54**(3), 660–668 (2008)
86. Wolf, S., Pinson, M.H., Webster, A.A., Cermak, G.W., Tweedy, E.P.: Objective and subjective measures of mpeg video quality. Soc. Motion Picture Telev. Eng. **160**, 178 (1997)

87. Wu, C.Y., Singhal, N., Krahenbuhl, P.: Video compression through image interpolation. In: Proceedings of the European Conference on Computer Vision (ECCV), pp. 416–431 (2018)
88. Wu, H.R., Rao, K.R.: Digital Video Image Quality and Perceptual Coding. CRC Press (2017)
89. Wulff, J., Black, M.J.: Temporal interpolation as an unsupervised pretraining task for optical flow estimation. In: Brox, T., Bruhn, A., Fritz, M. (eds.) GCPR 2018. LNCS, vol. 11269, pp. 567–582. Springer, Cham (2019). https://doi.org/10.1007/978-3-030-12939-2_39
90. Xiao, F., et al.: DCT-based video quality evaluation. Final Project for EE392J 769 (2000)
91. Xu, M., Chen, J., Wang, H., Liu, S., Li, G., Bai, Z.: C3DVQA: full-reference video quality assessment with 3D convolutional neural network. In: 2020 IEEE International Conference on Acoustics, Speech and Signal Processing (ICASSP), ICASSP 2020, pp. 4447–4451. IEEE (2020)
92. Xu, X., Li, M., Sun, W.: Learning deformable kernels for image and video denoising. arXiv preprint arXiv:1904.06903 (2019)
93. Xu, X., Siyao, L., Sun, W., Yin, Q., Yang, M.H.: Quadratic video interpolation. In: Advances in Neural Information Processing Systems, vol. 32 (2019)
94. Xue, T., Chen, B., Wu, J., Wei, D., Freeman, W.T.: Video enhancement with task-oriented flow. Int. J. Comput. Vis. **127**(8), 1106–1125 (2019)
95. You, J., Korhonen, J.: Transformer for image quality assessment. In: 2021 IEEE International Conference on Image Processing (ICIP), pp. 1389–1393. IEEE (2021)
96. Yu, Z., et al.: Training weakly supervised video frame interpolation with events. In: Proceedings of the IEEE/CVF International Conference on Computer Vision, pp. 14589–14598 (2021)
97. Zhang, R., Isola, P., Efros, A.A., Shechtman, E., Wang, O.: The unreasonable effectiveness of deep features as a perceptual metric. In: Proceedings of the IEEE Conference on Computer Vision and Pattern Recognition, pp. 586–595 (2018)

Adaptive Feature Interpolation for Low-Shot Image Generation

Mengyu Dai[1(✉)], Haibin Hang[2], and Xiaoyang Guo[3]

[1] Salesforce, Cambridge, USA
mdai@salesforce.com
[2] Amazon, New York, USA
haibinh@amazon.com
[3] Meta, Menlo Park, USA
xiaoyangg@fb.com

Abstract. Training of generative models especially Generative Adversarial Networks can easily diverge in low-data setting. To mitigate this issue, we propose a novel implicit data augmentation approach which facilitates stable training and synthesize high-quality samples without need of label information. Specifically, we view the discriminator as a metric embedding of the real data manifold, which offers proper distances between real data points. We then utilize information in the feature space to develop a fully unsupervised and data-driven augmentation method. Experiments on few-shot generation tasks show the proposed method significantly improve results from strong baselines with hundreds of training samples.

1 Introduction

Majority of learning algorithms today favor the feed of large training data. However, it is often difficult to collect sufficient amount of high-quality data for usage. In addition, intelligent systems like human brains do not need millions of samples to learn useful patterns and are energy-efficient. On the premise of it, learning with small data has been an important research area in various tasks [5,13,16,24,27,32,34,38,40]. Among numerous promising works along the direction, a limited amount target on generative models. Training of generative models especially Generative Adversarial Networks (GANs) [11] can easily diverge in low-data setting. To overcome the issue, people come up with methods focusing on different aspects in GAN training, such as data augmentation [17,43], network architecture design [18,26] and applying regularization [35,41]. Data augmentation can substantially increase the size of usable samples and enable stable training [43].

Unlike above data augmentation approaches for generative models which target on image domain, we propose a simple yet effective method to implicitly augment training data without supervision. To our knowledge, it is the first attempt to interpolate the multidimensional output feature of the discriminator for data generation. This can possibly be due to the fact that applications

S. Avidan et al. (Eds.): ECCV 2022, LNCS 13675, pp. 254–270, 2022.
https://doi.org/10.1007/978-3-031-19784-0_15

using GAN frameworks usually adopt objectives with 1-dimensional discriminator output, such as vanilla GAN [11] and Wasserstein GAN [2]. Recently, Dai and Hang [8] introduce a metric learning perspective to the GAN discriminator with multidimensional output and reveal an interesting *flattening effect*: along the training process, the learned metric gradually becomes more uniform and flat. The observation inspire us to explore the possibility of implementing augmentation in feature space in an unsupervised fashion.

In this paper, we propose a way to implicitly augment training data by taking the advantage of the flattening effect of discriminators with multiple output neurons. An intuitive understanding is that, compared to highly sparse and nonlinear nature of real data manifold [3], the low-dimensional feature space is relatively dense and flat. Hence applying interpolation in feature space yields a feasible way for augmentation with higher fidelity. An example of the effect of feature interpolation during training with StyleGAN2 architectures is shown in Fig. 1. Both sessions utilized adaptive image augmentation [17], while feature interpolation significantly stabilized training. The novelty of this work is summarized as follows: 1. We propose an implicit data augmentation method in multidimensional feature space of discriminator output. To our knowledge, this is the first attempt in unsupervised image generation. 2. We develop a data-driven approach for augmentation, with criteria based on the underlying structure of feature space during training. 3. Results from few-shot image generation experiments show significant improvements on several benchmark datasets.

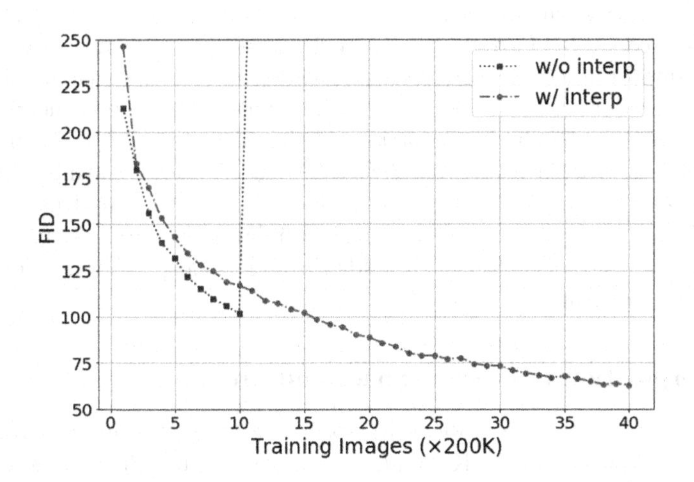

Fig. 1. FID (↓) during training on *Shells* dataset. Red: without feature interpolation; Green: with feature interpolation. Here "training images" refers to training iteration × batch size. (Color figure online)

2 Related Work

2.1 Low-Shot Data Generation

Recent works contribute to low-shot data generation from different perspectives. Karras *et al.* [17] implemented augmentation on images with adaptive probabilities by using a validation set. Zhao *et al.* [43] proposed DiffAugment which applies random differentiable augmentation on both real and generated images. The above methods significantly improve the amount of available training data in image domain thus remarkably prevent discriminator from overfitting. Tseng *et al.* [35] designed a regularization loss term on predictions of discriminator by tracking the moving averages of discriminator predictions during training. Zhang *et al.* [41] proposed consistency regularization for GANs, with the argument on the invariance of samples after transformation. Liu *et al.* [26] proposed an SLE module and an encoder-decoder reconstruction regularization on discriminator to improve training stability in low-data training settings. For other directions, one can refer to [14,15,23,28]. Different from above techniques, our proposed method is implemented in the multi-dimensional feature space of discriminator output. It does not conflict with data augmentation techniques in the image domain and is independent of network architectures.

2.2 Geometric Interpretations in GANs

The key idea of this paper comes from the interesting flattening effect of discriminator observed in [8]. Particularly, in their paper [8], Dai and Hang interpret the discriminator as a metric generator which learns some intrinsic metric of real data manifold such that the manifold is flat under the learned metric. In this paper, we observe the similar behaviors in the *geometric* GAN [25] framework with hinge loss. Specifically, *geometric* GAN use SVM separating hyper-plane to maximizes the margin between real/fake feature vectors and use hinge loss for discriminator which is simple and fast. Similar effects are also mentioned in [30] by Shao *et al.* which shows manifolds learned by deep generative models are close to zero curvature.

2.3 Interpolation in Feature Space and Mixup

Combining features in the embedding space are shown to be helpful in image retrieval [1,6,7,36]. Recently, Ko *et al.* [21] proposed embedding expansion which utilized a combination of embeddings and performs hard negative mining to learn informative feature representations. DeVries and Taylor [10] claimed that simple transformations to feature space results in plausible synthetic data due to manifold unfolding in feature space. Verma *et al.* [37] introduced Manifold Mixup, which implemented interpolation between hidden feature vectors to obtain smoother decision boundaries at multiple levels of representation. Furthermore, in [37] authors also indicated the flattening effect of Manifold Mixup in learning representations.

Another branch of data augmentation techniques takes advantage of the prior knowledge in learning tasks to interpolate both training images and the corresponding labels. Zhang *et al.* [42] suggested that linear interpolations of data samples should lead to linear interpolations of their associated labels. Kim *et al.* [20] applied batch mixup and formulated the optimal construction of a batch of mixup data. Other works along the track also show improvements on various discriminative learning tasks [19,33,37,39]. Despite the promising progress, applying these methods require augmentation in training data's associated labels, which does not suit the use case in this paper. Different from above augmentation methods which are mostly used in supervised discriminative tasks, in this work we develop an implicit augmentation method using data-driven feature interpolation, which is suitable for generative tasks in a unsupervised fashion.

3 Methodology

In this section we introduce the main idea of the paper and the proposed implicit augmentation algorithm for low-shot generation in detail.

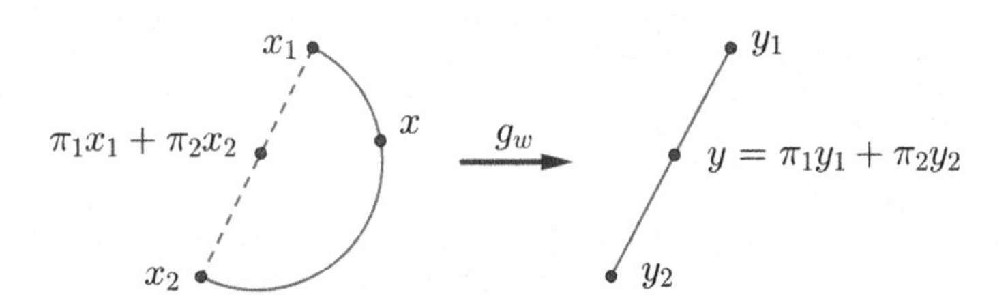

Fig. 2. Direct interpolation of real data likely returns points far away from real data manifold; With flattening effect, we propose interpolation in feature space which returns feature y as an approximation of using some imaginary "real" sample x.

3.1 The Flattening Effect of Discriminator

In this paper, we denote y as a *valid* feature vector iff. $y = g_w(x)$ for some data x from real data manifold given some deep metric learning network g_w. A simple illustration is shown in Fig. 2. One question we are interested in is: How far away is the interpolation of a group of valid feature vectors from some individual valid feature vector? In the following we will address this question in a GAN framework using observations from experimental results.

We adopt the default training setting in [26] and *Shells* dataset which contains 64 diverse images for experimentation. [26] uses hinge loss as GAN objective, thus the discriminator has metric learning effect and is equipped with multi-dimensional output. The hinge objective can be formulated as:

$$L_G = \mathbb{E}_{z \sim \mathcal{N}}[-D(G(z))] \tag{1}$$

$$L_D = \mathbb{E}_{x \sim P_{real}}[\max(0, 1 - D(x))] + \mathbb{E}_{x \sim P_{fake}}[\max(0, 1 + D(x))], \tag{2}$$

where $D(x)$ is also named $g_w(x)$ in this paper.

We utilize the following way introduced in [8,30] to detect the change of the learned metric along the training process: (i) For each iteration i, sample some real data points to form a finite metric space X_i; (ii) Construct the normalized distance matrix of X_i under the learned metric; (iii) Apply multidimensional scaling (MDS) to the normalized distance matrix to obtain the decreasing (finite) sequence of eigenvalues $C_i = \{\lambda_1^{(i)}, \lambda_2^{(i)}, \cdots, \lambda_b^{(i)}\}$, where b is the number of sample points.

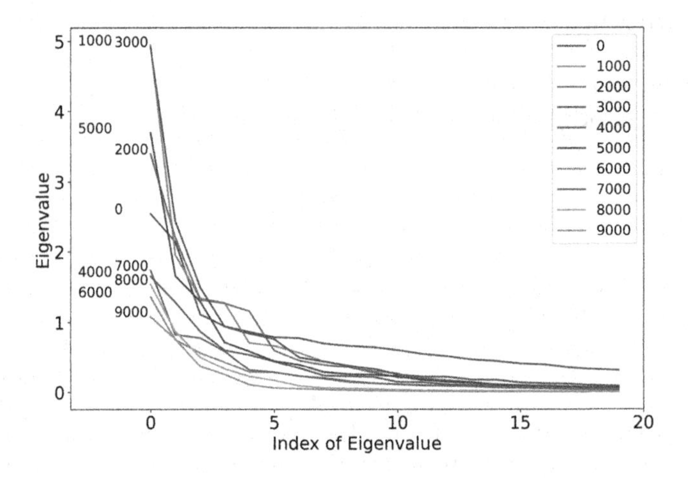

Fig. 3. Each curve represents the first 20 eigenvalues obtained using multidimensional scaling (MDS) of the 64×64 normalized distance matrix from 64 real images under learned metric during training on *Shells* dataset. We draw curves for iterations $0, 1000, \cdots, 9000$.

Now we observe the eigenvalues to see how the Euclidean distance among feature vectors evolve during the training process. As shown in Fig. 3, the curve of eigenvalues becomes closer and closer to x-axis with training going forward. At the iteration 9000, only the first few eigenvalues are non-trivial which implies that the valid feature vectors are compressed on a low-dimensional hyperplane. Compared to the input data dimension $m = 1024 \times 1024 \times 3$, the valid feature subspace is significantly flat and uniform. This experimental result is consistent with results in [8], even though the training settings being used are very different. An example of how to infer the shape of data set using eigenvalue curve is shown in Fig. 4.

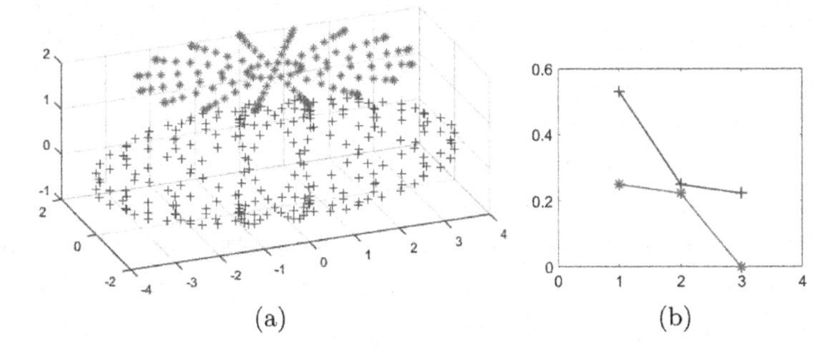

Fig. 4. (a) Blue points are located on an ellipsoid and red points are located on a flat disc; (b) Each curve represents the eigenvalue curve of corresponding point set with the same color. (Color figure online)

The above observation suggests interesting facts to the question at the beginning of this section: If a set of valid feature vectors $y_1, \cdots, y_k$ are close to each other, then any interpolation point $y = \sum_{i=1}^{k} \pi_i y_i$ with $\sum \pi_i = 1$ and $0 \leq \pi_i \leq 1$ is *very likely* a valid feature vector. Next, we explore how this flattening effect helps with data augmentation.

3.2 Implicit Data Augmentation

Given some neural network g_w, loss function L and training samples x_i, one direct way of augmenting data is to generate synthetic data sample x and use it as training data. All the efforts a synthetic data x could make end up with the calculation of the gradients:

$$\frac{\partial L(g_w(x))}{\partial w}.$$

Given valid feature vectors $y_1, y_2, \cdots, y_k$ which are extracted from some training samples $x_1, x_2, \cdots, x_k$. For any interpolation $y = \sum_{i=1}^{k} \pi_i y_i$ with $0 \leq \pi_i \leq 1$ and $\sum_{i=1}^{k} \pi_i = 1$, based on the flattening effect, *very likely* there exists a virtual real data point x such that $y = g_w(x)$. Even though it is not obvious to construct x explicitly, we are able to estimate its contribution to gradients *implicitly* by taking the average of the contributions of $x_1, \cdots, x_k$:

$$\frac{\partial L(g_w(x))}{\partial w} \approx \sum_{i=1}^{k} \pi_i \frac{\partial L(g_w(x_i))}{\partial w},$$

when $y_1, \cdots, y_k$ are close enough.

The above assertion is summarised in the following:

Lemma 1. *Given some neural network $g_w : \mathbb{R}^m \to \mathbb{R}^n$ and some differentiable loss function $L : \mathbb{R}^n \to \mathbb{R}$. Fix $y = g_w(x)$. Then for a set of nearby points*

$y_i = g_w(x_i)$, $i = 1, \cdots, k$ such that $y = \sum_{i=1}^{k} \pi_i y_i$, $\sum_{i=1}^{k} \pi_i = 1$, $0 \leq \pi_i \leq 1$, we have:

$$\left| \frac{\partial L(g_w(x))}{\partial w} - \sum_{i=1}^{k} \pi_i \frac{\partial L(g_w(x_i))}{\partial w} \right| = O(\max_i \|y - y_i\|).$$

Proof. In the following, we use ∂_j to represent the partial derivative of the j-th coordinate.

$$\begin{aligned}
&\frac{\partial L(g_w(x))}{\partial w} - \sum_{i=1}^{k} \pi_i \frac{\partial L(g_w(x_i))}{\partial w} \\
=&\sum_{j=1}^{n} \partial_j L(y) \frac{\partial g_w^{(j)}(x)}{\partial w} - \sum_{i=1}^{k} \pi_i \sum_{j=1}^{n} \partial_j L(y_i) \frac{\partial g_w^{(j)}(x_i)}{\partial w} \\
=&\sum_{i,j} \partial_j L(y) \pi_i \frac{\partial g_w^{(j)}(x_i)}{\partial w} - \sum_{i,j} \pi_i \partial_j L(y_i) \frac{\partial g_w^{(j)}(x_i)}{\partial w} \\
=&\sum_{i,j} \pi_i \frac{\partial g_w^{(j)}(x_i)}{\partial w} (\partial_j L(y) - \partial_j L(y_i)) \\
=&\frac{\partial g_w(x)}{\partial w} O(\max_i \|y - y_i\|) = O(\max_i \|y - y_i\|)
\end{aligned}$$

In summary, one can update the network parameters by using the average of the gradients raised by some set of training samples when performing gradient descent, if their embedded features are close to each other. In the following sections, we introduce the proposed data-driven augmentation algorithm in detail.

3.3 Nearest Neighbors Interpolation

Denote a data point (image) x_i, its feature vector $y_i = g_w(x_i)$, and the k nearest neighbours (including itself) as $y_{ij}, j = 1, \cdots, k$. We define an interpolated feature for y_i using its k nearest neighbours as

$$\tilde{y_i} = \sum_{j=1}^{k} \pi_{ij} y_{ij} \tag{3}$$

where $\sum_{j=1}^{k} \pi_{ij} = 1$ and $0 \leq \pi_{ij} \leq 1$.

For each y_i, π_{ij} in Eq. (3) follows Dirichlet distribution:

$$\pi_{ij} \sim \text{Dir}(\alpha_{ij}),\ i = 1, 2, \cdots, k \tag{4}$$

One can decide the concentration parameters α_{ij} to control the weights of the nearest neighbors. For example, when $\alpha_{ij} = 1$ for all js, the weights are uniformly distributed. Here we leverage distances between features and geometry of manifold to inform the parameters. The detailed procedure is described as follows.

For the i-th feature y_i and its nearest neighbor y_{ij}, $j = 1, 2, \cdots, k$:

$$\alpha_{ij} = T(M(y_i, y_{ij}))^t, \tag{5}$$

where $T(x) : \mathbb{R}^* \to \mathbb{R}^+$ is a monotonically decreasing function. ($\mathbb{R}^* = \{x \in \mathbb{R}, x >= 0\}, \mathbb{R}^+ = \{x \in \mathbb{R}, x > 0\}$). $M(y_i, y_{ij})$ is the distance between y_i and y_{ij} and $t > 0$ is used to control the skewness of the interpolation. There are lots of choices for $T(x)$, for example, $T(x) = \frac{1}{1+x}$. The intuition is to have larger weights for closer neighbors, as shown in Fig. 5. In terms of t, a smaller t gives more uniform/smooth interpolation while larger t prefers more weights on nearer neighbors. For simplicity we set $t = 1$ as the default choice.

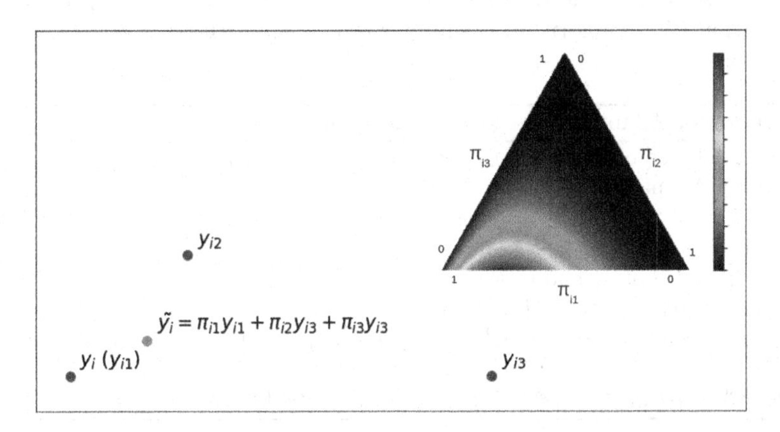

Fig. 5. Illustration of using Dirichlet distribution to interpolate features. $y_{ij}, j = 1, 2, 3$ are the first 3 nearest neighbors of y_i. $\tilde{y_i} = \pi_{i1} y_{i1} + \pi_{i3} y_{i2} + \pi_{i3} y_{i3}$ is the interpolation where π_{ij} are sampled from Dirichlet distribution with concentration parameter α_{ij} in proportion to the distance $M(y_i, y_{ij})$.

3.4 Data-Driven Adaptive Augmentation

To facilitate usage of feature interpolation, we consider taking advantage of the flattening effect during training to decide the aggressiveness of augmentation. Under the context, the aggressiveness can mainly be interpreted by (1) choices of nearest neighbour interpolation, (2) shape of Dirichlet distribution and (3) proportion of augmented features to use. To address (1), when the embedding space is more flat, one may use a larger k for sampling interpolated points. On the contrary, a small k leads to the interpolated features only cover a small portion of the feature space, thus may result in limited augmentation effect and bias in recovering real data manifold. One important question is how to reflect the degree of "flatness" of data manifold. As mentioned earlier, such information can be reached by the (multidimensional scaling) MDS of pairwise distances between features [8,30]. The flatness can be reflected by the number of large eigenvalues of MDS, where fewer number of large eigenvalues indicate approximately smaller dimensions of the space. Empirically we count number of eigenvalues $\{\lambda_i\}$ bigger than 10% of the largest eigenvalue λ_{max} in a batch b as the effective dimensions, and use $k = I(\lambda_i < 0.1\lambda_{max})$ as number of near neighbours.

The shape of Dirichlet distribution is controlled by M obtained from data itself as discussed in Sect. 3.3. We also involve augmentation probability p which decides the proportion of interpolated points used for training. $p = 0$ refers to no augmentation, and $p = 1$ refers to when all features are from interpolation. Similarly, we let $p = (k-1)/b$ which introduces more aggressive augmentation with fewer effective dimensions. In practice one can also find other ways to define p, or use fixed ps for simplicity. In experiments we observe using a reasonable choice of p (such as $p = 0.6$) is sufficient for stabilizing training. We will discuss the behaviors of these parameters later in Sect. 4.2. The whole Adaptive Feature Interpolation (AFI) algorithm is summarized in Algorithm 1.

Algorithm 1. Adaptive Feature Interpolation.

Input: A batch of features $\{y_i\}$ extracted from real data;
Output: Augmented batch of features $\{y_i^*\}$;

1: Calculate distance matrix M from $\{y_i\}$;
2: Solve for MDS of M and return its eigenvalues $\{\lambda_i\}$;
3: Calculate $\{\alpha\}, k, p$ from M and $\{\lambda_i\}$;
4: For each i, sample interpolated features $\tilde{y_i}$ using Eq. (4) with its k near neighbours;
5: For each i, set $y_i^* = \tilde{y_i}$ using Eq. (3) with probability p else $y_i^* = y_i$.

4 Experiments

In this section, we explore the behavior of the proposed method and provide evaluation results on multiple datasets.

4.1 Datasets and Implementation Details

We conducted unconditional generation experiments on several benchmark datasets or their subsets, including Shells, Art, Anime Face, Pokemon provided by [26], Cat, Dog [31], Obama, Grumpy cat [43] and CIFAR-10 [22]. We utilized various metrics for evaluations, including Frechet Inception Distance (FID) [12], Kernel Inception Distance (KID) [4] and Precision and Recall (PR) [29]. By default we generated 50K samples against real data for evaluation. Lower FID, KID scores and higher PR indicate better results.

We adopted StyleGAN2 [17] and FastGAN [26] network architectures, and used consistent parameter settings provided in their papers for experimentation.

To facilitate experiments with multidimensional discriminator output, the number of output neurons n of discriminator in [17] was set to 20, and the output logits of discriminator in [26] were reshaped to fit multidimensional setting. Feature interpolation is performed on features extracted from both real and fake images, which is empirically shown to be beneficial in experiments. Experiments were conducted using PyTorch framework on Tesla V100 GPUs.

4.2 Ablation Study

In this section we study the behaviors of proposed feature interpolation (FI) along with experiments using direct image interpolation (II) for comparison.

We first study the effect of p in simple cases using StyleGAN2 architectures [17,18]. In each setting we recorded FID during training with $p = 0.3, 0.6, 0.9$ as shown in Fig. 6. Here in II sessions we performed direct interpolation on images in a *mixup* style [42]. For FI sessions we did not utilize any image augmentation techniques. One can see that compared with FI experiments, the II sessions diverged earlier in all cases. The II session with $p = 0.9$ has less positive effect compared to when $p = 0.6$, which indicates image interpolation does not favor large ps in this case. In contrast, the FI experiments had more stable training sessions even with $p = 0.9$. Best results were obtained with dynamic p in this case. These results suggest that FI may better enjoy the "non-leaking" property [17] in training. In experiments on different datasets we find the use of small amount of original valid features is a necessary regularization for stable training. Note that for II it is not obvious to apply dynamic ks. One reason is that using Euclidean distances between pixels to find near neighbors seems worthless. In addition, with fixed real images one cannot implement dynamic augmentation based on the flattening effect.

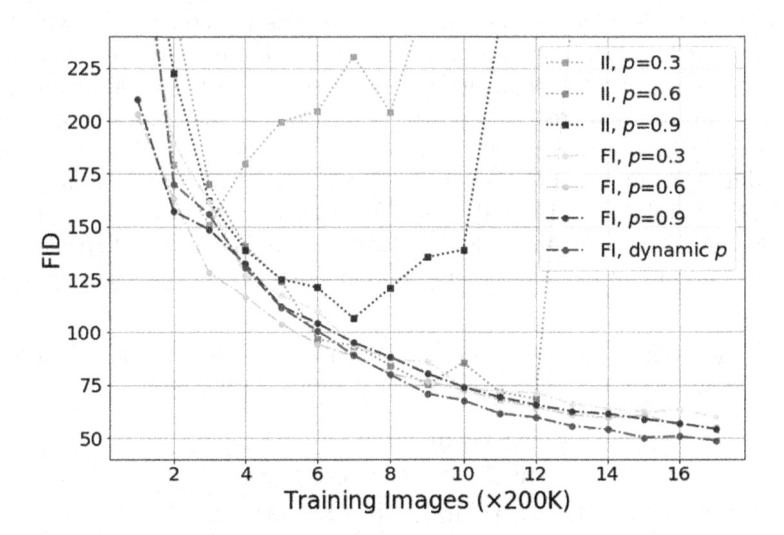

Fig. 6. FID evaluation during training on *Shells* dataset using image interpolation (II) and feature interpolation (FI) with StyleGAN2 architectures.

Next we experimented on settings both uniform ($t = 0$) and skewed ($t = 1$) distributions to study the behavior of Dirichlet distributions on feature interpolation. In each experiment we used batch size 8, and trained 50K iterations with fixed $p = 0.9$ using FastGAN architectures. In this case we employed DiffAugment for image augmentation and used 1K generated images for fast evaluation. In each setting we report the best FID during training along with its corresponding iteration in Table 1. Overall, we notice that FIDs with skewed distributions are in general better than ones with uniform distributions. Intuitively, with skewed distribution the interpolated features are likely closer to original features, thus may lead to smaller bias in training.

Table 1. FID evaluation of generated samples on *Shells* dataset using FastGAN architectures with fixed ks and $p = 0.9$.

	k	1	2	3	4	5	6	7	8
Uniform (t = 0)	FID	165.72	140.89	148.95	141.82	145.58	138.54	136.13	144.90
	Iter(K)	10	30	20	40	35	45	35	20
Skewed (t = 1)	FID	165.72	130.98	136.33	141.51	131.54	137.62	139.90	135.85
	Iter(K)	10	35	50	40	50	35	35	45

One interesting question is the relation between size of dataset N and value of k in training. We recorded ks up to 500K training images in CIFAR-10 experiments and 200K training images in *Art* experiments with effective batch size 8 on each GPU, and computed averaged k against different Ns. Table 2 also shows that under each setting, overall averaged k decreases with N increasing, which corresponds to less augmentation with more real training data. This dynamic mechanism reduces the risk of introducing more biased augmentation with larger N as mentioned in [17], where authors point out an interesting phenomenon that the positive effect of data augmentation decreases with size of real training data increasing. In addition, results in Table 2 also reveal that more training data enlarges the effective dimension of data manifold. Here we also present averaged k during training on CIFAR-10 in Fig. 7 which provides some insights on the behaviors of flatness reflected by k. We notice that smaller Ns suggest fewer effective dimensions of data manifold and stronger augmentation. Especially with small Ns ($N = 100$ and 500), the use of augmentation becomes more and more

Table 2. Averaged k during training under different amount of training data N with batch size 8 on each GPU.

Images	CIFAR-10				Art	
	100	500	5000	50000	100	1000
$p = 0$	4.39	4.02	2.34	2.14	3.41	2.43
AFI	4.62	3.86	2.18	2.26	3.53	2.59

aggressive during training, except for the beginning of training sessions. At the beginning stage, the averaged k fluctuates or even decreases before it reaches the point of inflection. The longer period of this training phase for larger Ns may suggest that the discriminator needs more training to learn meaningful feature representations.

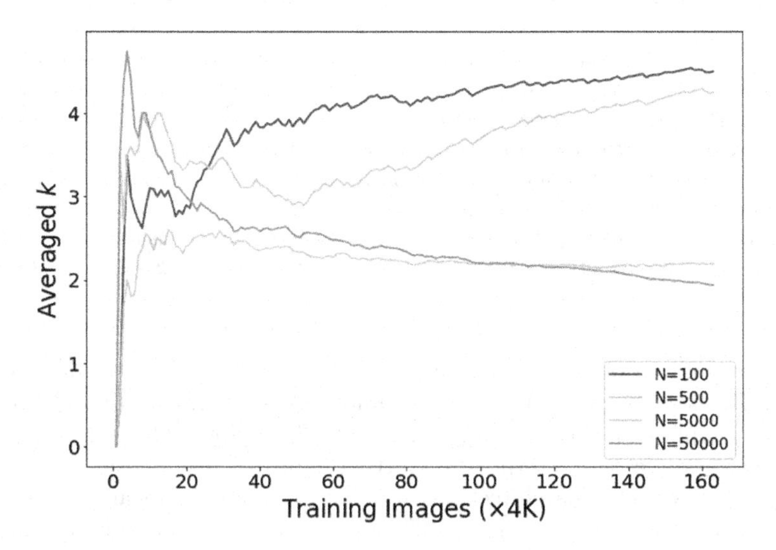

Fig. 7. Averaged k during training on CIFAR-10 with different Ns using effective batch size 8 on each GPU.

To study the effect of feature interpolation on other datasets, we performed experiments using AFI without applying transformation-based image augmentations such as [17,43]. We also present results using interpolation methods similar as input and manifold mixup [37,42], except that in the unsupervised task no labels are available for interpolation. In each setting we augmented both real and fake images (or features) with the corresponding method only. Interpolations using [37,42] were implemented on images and features respectively. Table 3 shows AFI significantly improves baseline results on all datasets. In contrast, directly applying feature interpolation in a *mixup* style led to the worst result.

Table 3. FID evaluation of experiments without transformation-based augmentations.

	Shells	Anime	Art	Pokemon	Cat	Dog	Grumpy cat	Obama
StyleGAN2	229.83	197.01	145.08	194.93	78.32	225.63	43.88	104.64
+ Input	135.66	91.99	86.51	174.91	60.35	118.79	26.98	36.65
+ Feature	319.80	308.16	117.35	376.83	236.99	270.66	118.80	160.57
+ AFI	48.71	85.97	81.08	131.86	57.47	143.60	37.46	31.88

4.3 Results

In the following we provide final evaluations results on various datasets.

Table 4. FID evaluations on 1024 × 1024 and 256 × 256 experiments using FastGAN and StyleGAN2 architectures.

Dataset	Shells	Anime	Art	Pokemon	Dog	Cat	Grumpy cat	Obama
Image size	1024	1024	1024	1024	256	256	256	256
Number of images	64	120	1000	800	389	160	100	100
FastGAN [26]	152.53	60.04	48.44	57.05	51.24	39.30	27.59	40.52
+ AFI	**124.80**	**55.35**	**43.09**	**50.47**	**50.89**	**35.18**	**25.02**	**36.43**
StyleGAN2 [17]	123.66	60.51	72.36	75.39	59.07	39.78	31.58	44.04
Hinge loss, $n = 20$	101.72	55.67	58.42	55.32	56.43	40.24	28.69	40.18
+ AFI	**62.99**	**33.48**	**43.94**	**44.79**	**49.14**	**35.26**	**22.03**	**31.99**

FID evaluation of 1024 × 1024 and 256 × 256 experiments are displayed in Table 4. In these experiments we incorporated image augmentation techniques, where [17] applies adaptive augmentation and [26] employs DiffAugment. Table 4 shows that using feature interpolation further improved results from strong baselines. With StyleGAN2 architectures we observe more significant gains across datasets. Note that simply using $n = 20$ without FI, one can already see improvements compared to results from [17]. This behavior is consistent with theoretical analysis and experimental results in [9] that multidimensional discriminator output has its advantage in GAN training. We further present evaluations of KID and precision-recall with StyleGAN2 architectures in Table 5 and Table 6 for reference. Examples of randomly generated 1024 × 1024 images are displayed in Fig. 8. As shown in the figure, samples from experiments with FI have consistently better qualities.

We also tested the effect of feature interpolation on CIFAR-10 with small amount of partial data using the default setting in [17] as baseline. Using feature interpolation improves FID from 42.80 to 27.62 with only 0.2% training data (100 images), and from 19.69 to 13.50 with 1% training data.

Table 5. KID($x10^3$)(↓) and Precision-Recall (PR)(↑) evaluations on 1024×1024 experiments.

	Shells		Anime		Art		Pokemon	
	KID	PR	KID	PR	KID	PR	KID	PR
[17]	20	(0.789, 0.085)	15	(0.966, 0.933)	26	(0.574, 0.823)	28	(0.621, 0.727)
+ AFI	**2**	**(0.852, 0.132)**	**4**	**(0.984, 0.974)**	**9**	**(0.887, 0.965)**	**12**	**(0.948, 0.922)**

Table 6. KID($\times 10^3$) and Precision-Recall (PR) evaluations on 256×256 experiments.

	Dog		Cat		Grumpy cat		Obama	
	KID	PR	KID	PR	KID	PR	KID	PR
[17]	18	(0.874, **0.948**)	6	(**0.974**, **0.951**)	5	(0.845, 0.794)	13	(0.930, 0.860)
+ AFI	**14**	(**0.922**, 0.932)	**4**	(**0.976**, **0.950**)	**3**	(**0.973**, **0.953**)	**10**	(**0.979**, **0.970**)

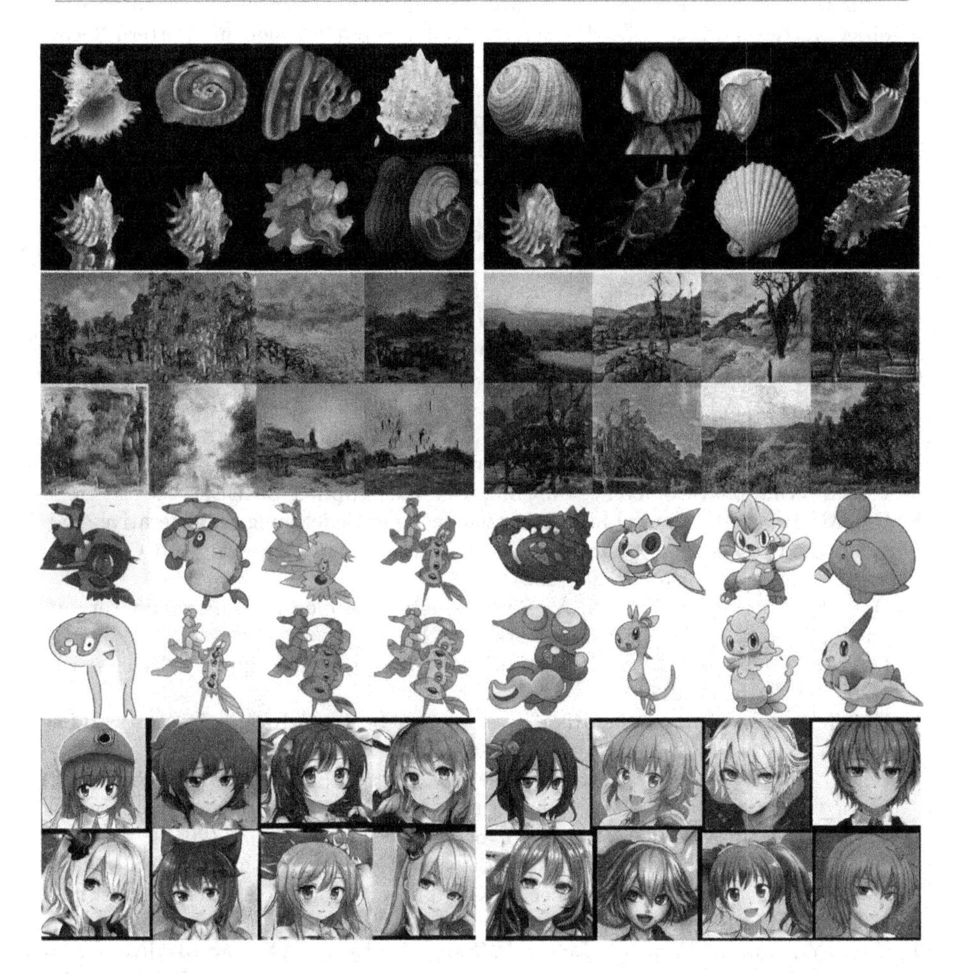

Fig. 8. Randomly generated 1024×1024 samples on different datasets. From top to bottom: Shells, Art, Pokemon and Anime. Left: StyleGAN2-ADA [17]; Right: with Adaptive Feature Interpolation.

5 Discussion

In this paper we have proposed an adaptive augmentation approach for low-shot data generation. Instead of producing new training images, the method functions in the multidimensional feature space of discriminator output by utilizing the flattening effect of feature space during training. Experiments show the proposed method improves results from strong baselines in low-data regime.

References

1. Arandjelović, R., Zisserman, A.: Three things everyone should know to improve object retrieval. In: 2012 IEEE Conference on Computer Vision and Pattern Recognition, pp. 2911–2918. IEEE (2012)
2. Arjovsky, M., Chintala, S., Bottou, L.: Wasserstein generative adversarial networks. In: Proceedings of the 34th International Conference on Machine Learning. Proceedings of Machine Learning Research, vol. 70, pp. 214–223. PMLR (2017)
3. Balestriero, R., Pesenti, J., LeCun, Y.: Learning in high dimension always amounts to extrapolation (2021)
4. Bińkowski, M., Sutherland, D.J., Arbel, M., Gretton, A.: Demystifying MMD GANs. In: International Conference on Learning Representations (2018)
5. Choe, J., Park, S., Kim, K., Hyun Park, J., Kim, D., Shim, H.: Face generation for low-shot learning using generative adversarial networks. In: Proceedings of the IEEE International Conference on Computer Vision (ICCV) Workshops (2017)
6. Chum, O., Mikulik, A., Perdoch, M., Matas, J.: Total recall II: query expansion revisited. In: CVPR 2011, pp. 889–896. IEEE (2011)
7. Chum, O., Philbin, J., Sivic, J., Isard, M., Zisserman, A.: Total recall: automatic query expansion with a generative feature model for object retrieval. In: 2007 IEEE 11th International Conference on Computer Vision, pp. 1–8. IEEE (2007)
8. Dai, M., Hang, H.: Manifold matching via deep metric learning for generative modeling. In: Proceedings of the IEEE/CVF International Conference on Computer Vision (ICCV), pp. 6587–6597 (2021)
9. Dai, M., Hang, H., Srivastava, A.: Rethinking multidimensional discriminator output for generative adversarial networks (2021)
10. DeVries, T., Taylor, G.W.: Dataset augmentation in feature space. arXiv preprint arXiv:1702.05538 (2017)
11. Goodfellow, I., et al.: Generative adversarial nets. In: Advances in Neural Information Processing Systems, vol. 27 (2014)
12. Heusel, M., Ramsauer, H., Unterthiner, T., Nessler, B., Hochreiter, S.: GANs trained by a two time-scale update rule converge to a local Nash equilibrium. In: Advances in Neural Information Processing Systems, vol. 30 (2017)
13. Hong, J., et al.: Reinforced attention for few-shot learning and beyond. In: Proceedings of the IEEE/CVF Conference on Computer Vision and Pattern Recognition (CVPR), pp. 913–923 (2021)
14. Hong, Y., Niu, L., Zhang, J., Liang, J., Zhang, L.: DeltaGAN: towards diverse few-shot image generation with sample-specific delta. arXiv preprint arXiv:2009.08753 (2020)
15. Hong, Y., Niu, L., Zhang, J., Zhao, W., Fu, C., Zhang, L.: F2GAN: fusing-and-filling GAN for few-shot image generation. In: Proceedings of the 28th ACM International Conference on Multimedia, pp. 2535–2543 (2020)
16. Huang, K., Geng, J., Jiang, W., Deng, X., Xu, Z.: Pseudo-loss confidence metric for semi-supervised few-shot learning. In: Proceedings of the IEEE/CVF International Conference on Computer Vision (ICCV), pp. 8671–8680 (2021)
17. Karras, T., Aittala, M., Hellsten, J., Laine, S., Lehtinen, J., Aila, T.: Training generative adversarial networks with limited data. In: Proceedings of the NeurIPS (2020)

18. Karras, T., Laine, S., Aittala, M., Hellsten, J., Lehtinen, J., Aila, T.: Analyzing and improving the image quality of StyleGAN. In: Proceedings of the CVPR (2020)
19. Kim, J.H., Choo, W., Song, H.O.: Puzzle mix: exploiting saliency and local statistics for optimal mixup. In: International Conference on Machine Learning (ICML) (2020)
20. Kim, J., Choo, W., Jeong, H., Song, H.O.: Co-mixup: saliency guided joint mixup with supermodular diversity. In: International Conference on Learning Representations (2021)
21. Ko, B., Gu, G.: Embedding expansion: augmentation in embedding space for deep metric learning. In: Proceedings of the IEEE Conference on Computer Vision and Pattern Recognition (2020)
22. Krizhevsky, A., Nair, V., Hinton, G.: CIFAR-10 (Canadian institute for advanced research)
23. Li, Y., Zhang, R., Lu, J., Shechtman, E.: Few-shot image generation with elastic weight consolidation. arXiv preprint arXiv:2012.02780 (2020)
24. Li, Y., et al.: Few-shot object detection via classification refinement and distractor retreatment. In: Proceedings of the IEEE/CVF Conference on Computer Vision and Pattern Recognition (CVPR), pp. 15395–15403 (2021)
25. Lim, J.H., Ye, J.C.: Geometric GAN (2017)
26. Liu, B., Zhu, Y., Song, K., Elgammal, A.: Towards faster and stabilized GAN training for high-fidelity few-shot image synthesis. In: International Conference on Learning Representations (2021)
27. Liu, S., Wang, Y.: Few-shot learning with online self-distillation. In: Proceedings of the IEEE/CVF International Conference on Computer Vision (ICCV) Workshops, pp. 1067–1070 (2021)
28. Ojha, U., et al.: Few-shot image generation via cross-domain correspondence. In: CVPR (2021)
29. Sajjadi, M.S.M., Bachem, O., Lučić, M., Bousquet, O., Gelly, S.: Assessing generative models via precision and recall. In: Advances in Neural Information Processing Systems (NeurIPS) (2018)
30. Shao, H., Kumar, A., Thomas Fletcher, P.: The Riemannian geometry of deep generative models. In: Proceedings of the IEEE Conference on Computer Vision and Pattern Recognition (CVPR) Workshops (2018)
31. Si, Z., Zhu, S.C.: Learning hybrid image templates (hit) by information projection. IEEE Trans. Pattern Anal. Mach. Intell. **34**, 1354–1367 (2012)
32. Stojanov, S., Thai, A., Rehg, J.M.: Using shape to categorize: low-shot learning with an explicit shape bias. In: Proceedings of the IEEE/CVF Conference on Computer Vision and Pattern Recognition (CVPR), pp. 1798–1808 (2021)
33. Thulasidasan, S., Chennupati, G., Bilmes, J.A., Bhattacharya, T., Michalak, S.: On mixup training: improved calibration and predictive uncertainty for deep neural networks. In: Advances in Neural Information Processing Systems, vol. 32. Curran Associates, Inc. (2019)
34. Tian, Y., Wang, Y., Krishnan, D., Tenenbaum, J.B., Isola, P.: Rethinking few-shot image classification: a good embedding is all you need? arXiv abs/2003.11539 (2020)
35. Tseng, H.Y., Jiang, L., Liu, C., Yang, M.H., Yang, W.: Regularing generative adversarial networks under limited data. In: CVPR (2021)
36. Turcot, P., Lowe, D.G.: Better matching with fewer features: the selection of useful features in large database recognition problems. In: 2009 IEEE 12th International Conference on Computer Vision Workshops, ICCV Workshops, pp. 2109–2116. IEEE (2009)

37. Verma, V., et al.: Manifold mixup: better representations by interpolating hidden states. In: International Conference on Machine Learning, pp. 6438–6447. PMLR (2019)
38. Yao, H., et al.: Automated relational meta-learning. In: International Conference on Learning Representations (2020)
39. Yun, S., Han, D., Oh, S.J., Chun, S., Choe, J., Yoo, Y.: CutMix: regularization strategy to train strong classifiers with localizable features. In: Proceedings of the IEEE/CVF International Conference on Computer Vision (ICCV) (2019)
40. Zhang, C., Song, N., Lin, G., Zheng, Y., Pan, P., Xu, Y.: Few-shot incremental learning with continually evolved classifiers. In: Proceedings of the IEEE/CVF Conference on Computer Vision and Pattern Recognition (CVPR), pp. 12455–12464 (2021)
41. Zhang, H., Zhang, Z., Odena, A., Lee, H.: Consistency regularization for generative adversarial networks. In: International Conference on Learning Representations (2020)
42. Zhang, H., Cisse, M., Dauphin, Y.N., Lopez-Paz, D.: Mixup: beyond empirical risk minimization. In: International Conference on Learning Representations (2018). https://openreview.net/forum?id=r1Ddp1-Rb
43. Zhao, S., Liu, Z., Lin, J., Zhu, J.Y., Han, S.: Differentiable augmentation for data-efficient GAN training. In: Conference on Neural Information Processing Systems (NeurIPS) (2020)

PalGAN: Image Colorization with Palette Generative Adversarial Networks

Yi Wang[1], Menghan Xia[2], Lu Qi[3], Jing Shao[4], and Yu Qiao[1(✉)]

[1] Shanghai AI Laboratory, Shanghai, China
{wangyi,qiaoyu}@pjlab.org.cn
[2] Tencent AI Lab, Shenzhen, China
[3] UC Merced, Merced, USA
[4] SenseTime Research, Science Park, Hong Kong
shaojing@senseauto.com

Abstract. Multimodal ambiguity and color bleeding remain challenging in colorization. To tackle these problems, we propose a new GAN-based colorization approach PalGAN, integrated with palette estimation and chromatic attention. To circumvent the multimodality issue, we present a new colorization formulation that estimates a probabilistic palette from the input gray image first, then conducts color assignment conditioned on the palette through a generative model. Further, we handle color bleeding with chromatic attention. It studies color affinities by considering both semantic and intensity correlation. In extensive experiments, PalGAN outperforms state-of-the-arts in quantitative evaluation and visual comparison, delivering notable diverse, contrastive, and edge-preserving appearances. With the palette design, our method enables color transfer between images even with irrelevant contexts.

Keywords: Image colorization · Generative adversarial networks · Attention · Color transfer

1 Introduction

Colorization means to predict the missing chrome information from the given gray image. It is an interesting and practical task in computer vision, widely used in legacy footage processing [27], color transfer [1,39], and other visual editing applications [3,52]. It is also exploited as a proxy task for self-supervised learning [25], since predicting perceptually natural colors from the given grayscale image heavily relies on scene understanding. However, even the ground-truth color is available for supervision, it is still very challenging to predict pixel colors from gray images, due to the ill-posed nature that one input grayscale could correspond to multiple possible color variants.

Most current methods [3,12,17,23,26,38,49,54,56] formulate colorization as a pixel-level regression task, suffering from multimodal representation more or less.

Supplementary Information The online version contains supplementary material available at https://doi.org/10.1007/978-3-031-19784-0_16.

S. Avidan et al. (Eds.): ECCV 2022, LNCS 13675, pp. 271–288, 2022.
https://doi.org/10.1007/978-3-031-19784-0_16

With the large-scale training data and end-to-end learning models, they can learn the color distribution prior conveniently, *e.g.* vegetation greenish tones, human skin colors, *etc.*. Anyhow, when it comes to objects with inherently color ambiguity (*e.g.* human clothes, cars, and other man-made stuff), these approaches tend to predict the brownish average colors. To tackle such multi-modality, researches [24,54,56] proposed to formulate the color prediction as pixel-level color classification, which allows multiple colors to be assigned to each pixel based on posterior probability. Unfortunately, these suffer from regional color inconsistency due to the independent pixel-wise sampling mechanism. In this regard, means of utilizing the sequential modeling [12,23] can only partially help the sampling issue, because the unidirectional sequential dependence of 2D flattened pixel primitives causes error accumulation and hinders the learning efficiency.

Apart from the multimodal issue, color bleeding is another common issue in colorization due to inaccurate identification of semantic boundaries. To suppress such visual artifacts, most works [3,17,26,38,49,54,56] resort to Generative adversarial networks (GAN) to encourage the generated chrome distribution to be indistinguishable from that of the real-life color images. Currently, no special algorithms or modules for deep models have been proposed to enhance the performance of this aspect, which matters the visual pleasantness considerably.

To avoid modeling the color multimodality pixel-wisely, we propose a new colorization framework PalGAN that predicts the pixel colors in a coarse-to-fine paradigm. The key idea is to first predict the global palette probability (*e.g.* palette histogram) from the grayscale. It does not collapse into a single specific colorization solution but represents a certain color distribution of the potential color variants. Then, the uncertainty about the per-pixel color assignment is modeled with a generative model in the GAN framework, conditioned on the grayscale and palette histogram. Therefore, multiple colorization results could be achieved by changing the palette histogram input.

To guarantee the color assignment with semantic correctness and regional consistency, we study color affinities by a proposed chromatic attention module. It explicitly aligns color affinity with both semantics and low-level characteristics. In structure, chromatic attention includes global interaction and local delineation. The former enables global context utilization for color inference by using semantic features in the attention mechanism. The latter preserves regional details by mapping the gray input to color through local affine transformation. The transformation is explicitly parameterized by the correlation between gray input and color feature. Experiments illustrate the effectiveness of our method. It achieves impressive visual results (Fig. 1) and quantitative superiority over state-of-the-art approaches over ImageNet [9] and COCO-Stuff [5]. Our method also works well with the user-specified palette histogram from a reference image, which could even have no content correlation with the input grayscale. So, by nature, our method supports diverse coloring results with certain controllability. Our code and pre-trained models are available at https://github.com/shepnerd/PalGAN.

Generally, our contributions are three-fold: i) We propose a new colorization framework PalGAN that decomposes colorization to palette estimation and pixel-wise assignment. It circumvents the challenges of color ambiguity and

Fig. 1. Our colorization results. 1_{st} row: inputs, and 2_{nd} row: our predictions.

regional homogeneity effectively, and supports diverse and controllable colorization by nature. ii) We explore the less-touched color affinities and propose an effective module named chromatic attention. It considers both semantic and local detail correspondence, applying such correlations to color generation. It alleviates notable color bleeding effects. iii) Our method surpasses state-of-the-arts in perceptual quality (FID [16] and LPIPS [55]) notably. It is known that there exists a trade-off between perceptual and fidelity results in multiple low-level tasks. We argue perceptual effects matter more than fidelity as colorization aims to produce realistic colorized results rather than restore identical pixel-wise colors as the ground truth. Regardless, our method can achieve best both fidelity (PSNR and SSIM) and perceptual performance with proper tuning.

2 Related Work

2.1 Colorization

User Guided Colorization. Some of early works [6–8,18,21,29,34,39,47] in colorization turn to a reference image for transferring its color statistics to the given gray one. With the prevalence of deep learning, such color transfer is characterized in neural feature space for introducing semantic consistency [15]. These works perform decently when the reference and input share similar semantics. Its applications are limited by the reference retrieval quality, especially when handling complicated scenes.

Besides of reference images, several systems require users to give sufficient local color hints (usually in scribble form) before colorizing inputs [21,27,37, 52]. Then approaches propagate the given colors based on their local affinities. Besides, some attempts are made [3] to explore other modalities like languages to instruct what colors are used and how they are distributed.

Learning-Based Colorization. This line of work [10,17,19,24,54,56] gives colorful images only from the gray inputs, learning a pixel-to-pixel mapping. Large-scale datasets are exploited in a self-supervised fashion, converting colorful pictures to gray ones for pair-wise training. Iizuka *et al.* [17] utilize image-level labels for associating predicted color with global semantics, using a global-and-local convolutional neural network. Larsson *et al.* [24] and Zhang *et al.* [54] introduce

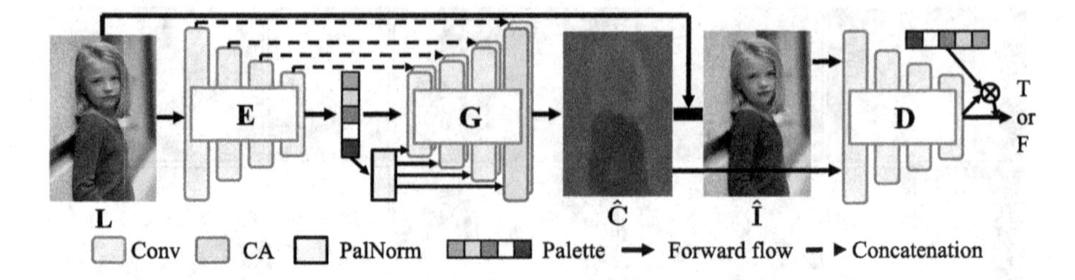

Fig. 2. Our colorization system framework.

pixel-level color distribution matching by classification, alleviating color unbalance and multi-modal outputs. Besides, extra input hints are integrated into learning systems by simulation in [56], providing automatic and semi-automatic ways to colorize images. Recently, transformer architectures are explored for this task considering their expressiveness on non-local modeling [23].

Some work explicitly exploits additional priors from pretrained models for colorization. Su *et al.* [38] study leveraging instance-level annotations (e.g., instance bounding boxes and classes) by using an off-the-shelf detector. It will make the colorization model focuses on color rendering without the need of recognizing high-level semantics. In addition to the mentioned pretrained discriminative models, pretrained generative ones are also exploitable in improving colorization performance in diversity. Wu *et al.* [49] explore to incorporate generative color prior from a pretrained BigGAN [4] to help a deep model produce colored results with diversities. They design an extra encoder to project the given gray image into latent code, then estimate colorful images from BigGAN. With such primary predictions, they further refine the color results by the intermediate features in BigGAN. Afifi *et al.* [1] propose employing a pretrained StyleGAN [20] for image recoloring, and color is controlled by histogram features.

2.2 GAN-Based Image-to-Image Translation

Image-to-image translation aims to learn the transformation between the input and output image. Colorization can be formulated to this task and handled by Generative Adversarial Networks [11] (GAN) based approaches [19,30,35,41,44]. They employ an adversarial loss that learns to discriminate between real and generated images, and then minimize this loss by updating the generator to make the produced results look realistic [28,31,36,42,45,46,50,51,57].

3 Method

PalGAN aims to colorize grayscale images. It formulates colorization as a palette prediction and assignment problem. Compared with directly learning the pixel-to-pixel mapping from gray to color as adopted by most learning-based methods, this disentanglement fashion not only brings empirical colorization improvements (Sect. 4), but also enables us to manipulate global color distributions by adjusting or regularizing palettes.

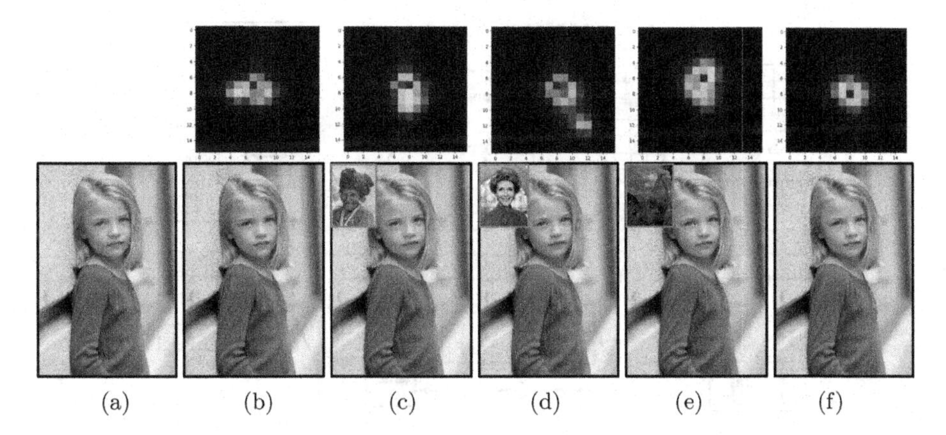

(a) (b) (c) (d) (e) (f)

Fig. 3. Visualizations of palettes (1_{st} row, shown in jet colormap) and how they work on colorization (2_{nd} row). (a) Input, (b) the ground truth, (c)–(e) reference-based colorization, (f) automatic colorization.

For PalGAN, its input is a grayscale image (*i.e.* the luminance channel of color images) $\mathbf{L} \in \mathcal{R}^{h \times w \times 1}$, and the output is the estimated chromatic map $\hat{\mathbf{C}} \in \mathcal{R}^{h \times w \times 2}$ that will be used as the complementary *ab* channels together with $\mathbf{L}$ in CIE *Lab* color space. PalGAN consists of palette generator $\mathcal{T}_{\mathbf{E}}$, palette assignment generator $\mathcal{T}_{\mathbf{G}}$, and a color discriminator $\mathbf{D}$. In inference, only $\mathcal{T}_{\mathbf{E}}$ and $\mathcal{T}_{\mathbf{G}}$ are employed. The whole framework is given in Fig. 2.

3.1 Palette Generator

$\mathcal{T}_{\mathbf{E}}$ estimates the global palette probabilities from the given gray image as $\hat{\mathbf{h}} = \mathcal{T}_{\mathbf{E}}(\mathbf{L})$. We employ a 2D chromatic histogram $\hat{\mathbf{h}} \in \mathcal{R}^{N_a \times N_b \times 1}$ to represent palette probabilities (N_a and N_b denotes bin numbers of a and b axes respectively), modeling the chromatic information statistics instead of learning a deterministic one. $\mathcal{T}_{\hat{P}}$ is an encoder network with several convolutional layers and a few multiple-layer perceptions (MLP), ended with a sigmoid function. The former is to extract features and the latter is to transform spatial features to a histogram (in vector form). With the explicit representation of the color palette in histogram form, we find it not only makes global color distribution more predictable, but also manipulative by introducing proper regularizations.

The user-guided colorization [6,34,56] has demonstrated the effectiveness of utilizing the color histogram of a reference image for colorizing images. Compared to the existing practice [6,34,56], we make one step further, *i.e.* synthesizing a palette histogram conditioned on the input grayscale instead of taking that from a user-specified reference image. This design brings two non-trivial advantages. First, it makes our method to be a self-contained fully automatic colorization system, instead of depending on any outside guidance (i.e. a reference image) to work. Second, in general cases, the palette histogram estimated each specific grayscale may offer more accurate and instructive information for the colorization process, than that from a reference image selected in the wild.

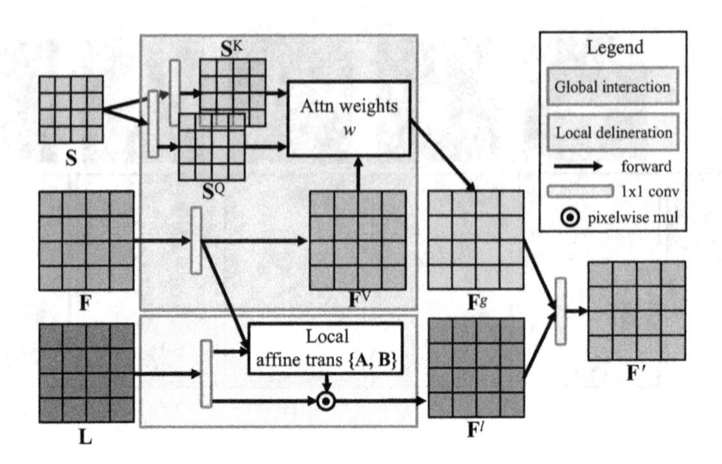

Fig. 4. The illustration of chromatic attention.

We empirically demonstrate this in Sect. 4.4. In Fig. 3, we visualize the predicted palette histogram (f), in comparison with the ground-truth (b) and those of reference images (c e).

3.2 Palette Assignment Generator

$\mathcal{T}_{\mathbf{G}}$ conducts color assignment task via conditional image generation. It produces the corresponding *ab* from the gray image conditioned on palette histogram $\hat{\mathbf{h}}$ and extra latent code z (sampled from a normal distribution), as $\hat{\mathbf{C}} = \mathcal{T}_{\mathbf{G}}(\mathbf{L}|\hat{\mathbf{h}}, z)$. It is a convolutional generator is composed of common residual blocks used in image translation [14,19], together with our customized Palette Normalization (PN) layer and Chromatic Attention (CA) module. The palette normalization is designed to promote the conformity of the generated chromatic channels to the palette guidance $\hat{\mathbf{h}}$, which is used along with each Batch Normalization layers. Specifically, the PN layer normalizes its input feature first and then performs an affine transformation parameterized by $g(\hat{\mathbf{h}})$ (where $g(\cdot)$ is a fully-connected layer). Besides, we propose a chromatic attention module (Fig. 4) to explicitly align color affinity to their corresponding semantic and low-level characteristics, which mitigates potential color bleeding or semantic misunderstanding effectively. We discuss the designs below in detail, along with a visualization of the effects of its components shown in Fig. 5.

Chromatic Attention. The proposed Chromatic Attention (CA) module incorporates both semantic and low-level affinities into constructing color relations. These two are realized by *global interaction* and *local delineation* submodules (Fig. 4). Specifically, inputs to CA are a high-resolution feature map $\mathbf{F}$ (of the size $\mathcal{R}^{128\times128\times64}$ from $\mathcal{T}_{\mathbf{G}}$), high-level feature map $\mathbf{S}$, and resized gray input $\mathbf{L}$. It outputs two feature maps $\mathbf{F}^g$ and $\mathbf{F}^l$ from global interaction and local delineation respectively, and fuses them into a feature map residual, adding back to the input feature map, as:

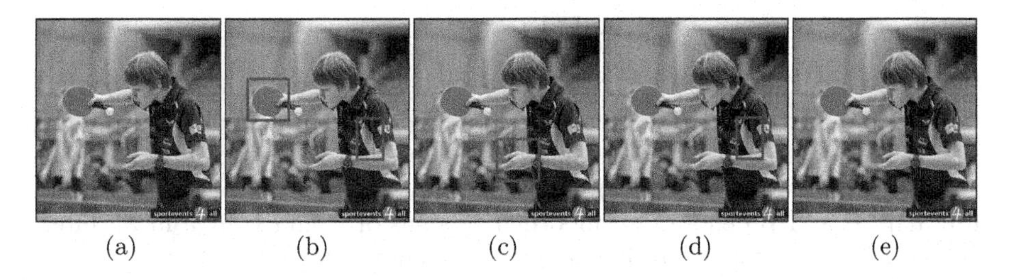

Fig. 5. Ablation studies of chromatic attention (CA). (a) Input, (b) wo CA, (c) w Global, (d) w Local, (e) full CA. Please zoom in.

$$\text{CA}(\mathbf{F}, \mathbf{S}, \mathbf{L}) = \mathbf{F} + \mathbf{F}' = \mathbf{F} + f(\mathbf{F}^g \oplus \mathbf{F}^l) = \mathbf{F} + f(\text{CA}_g(\mathbf{F}|\mathbf{S}) \oplus \text{CA}_l(\mathbf{L}|\mathbf{F})), \quad (1)$$

where $f(\cdot)$ is a nonlinear fusion operation formed by two consecutive convolutional layers, and $\oplus$ is channel-wise concatenation operation. $\text{CA}_g(\cdot)$ and $\text{CA}_l(\cdot)$ denote *global interaction* and *local delineation*, respectively. In this paper, we use $\mathbf{F}, \mathbf{F}' \in \mathcal{R}^{128\times 128\times 64}$.

Global Interaction. We reconstruct every regional feature point from the input feature map using a weighted sum of other ones, and such local weight is computed according to their semantic similarity. Formally, it is written as $\mathbf{F}^g_p = \sum_{q\in\mathbf{F}} w_{pq}\mathbf{F}^{\text{V}}_q$, where p and q denote a patch centering at pixel location p and q within $\mathbf{F}$, respectively. And w_{pq} is calculated from the region-wise interaction in the learned high-level feature maps from the input gray images. The region-wise feature interaction is measured by the cosine similarity between the normalized regional features, as:

$$w_{pq} = \frac{\exp(w'_{pq})}{\sum_{k\in\mathbf{S}}\exp(w'_{pk})} \quad \text{where} \quad w'_{pq} = \frac{\mathbf{S}^{\text{K}}_p \cdot \mathbf{S}^{\text{Q}}_q}{|\mathbf{S}^{\text{K}}_p||\mathbf{S}^{\text{Q}}_q|}, \quad (2)$$

where $\mathbf{S}$ denote high-level feature map, extracted from intermediate representation of the encoder $\mathcal{T}_{\hat{P}}$. $\mathbf{S}^{\text{K}}$ and $\mathbf{S}^{\text{Q}}$ denote two translated feature maps from $\mathbf{S}$ using convolution.

Local Delineation. Though color changes in texture and edges are delicate, overlooking these subtle variances leads to notable visual degradation. To preserve these details, we design a local delineation module to complement global interaction. We adopt the assumption that local color affinity is linearly correlated with its corresponding intensity [40,58]. We propose to learn such local relationship in the guided filter manner [13,48], which preserves edges from the guidance well. Our given local preserving module computes a learnable local affine transformation $\{\mathbf{A} \in \mathcal{R}^{128\times 128\times 64}, \mathbf{B} \in \mathcal{R}^{128\times 128\times 64}\}$ to map the gray image $\mathbf{L} \in \mathcal{R}^{256\times 256\times 1}$ to its corresponding *ab* feature map, as:

$$\mathbf{F}^l = \mathbf{A} \odot \mathbf{L}\downarrow + \mathbf{B}, \quad (3)$$

where $\odot$ is the element-wise multiplication operator and $\downarrow$ is downsampling one to ensure the spatial size of $\mathbf{L}$ is the same as $\mathbf{F}^l$. $\{\mathbf{A}, \mathbf{B}\}$ are parameterized by the a learnable local correlation between $\mathbf{L}$ and $\mathbf{F}$, as:

$$\mathbf{A} = \Psi(\frac{\text{cov}(\mathbf{F}, \mathbf{L})}{\text{var}(\mathbf{L}) + \epsilon}), \ \mathbf{B} = \overline{\mathbf{F}} - \mathbf{A} \odot \overline{\mathbf{L}} \tag{4}$$

where Ψ is a learnable transformation parameterized by a small convolutional net, $\text{cov}(\cdot, \cdot)$ computes the local covariance between two feature maps (within a fixed window size) while var() computes the local variance of the given feature map. $\overline{\mathbf{F}}$ and $\overline{\mathbf{L}}$ denote the smoothed versions of $\mathbf{F}$ and $\mathbf{L}$ by a mean filter, respectively. ϵ is a small positive number for computational stability.

Palette Optimization. To further ensure the proposed palette assignment generator is responsive to the given palette, we minimize the discrepancy between the palette extracted from the predicted chromatic channels and that from the corresponding ground truth. However, common histograms from images are non-differentiable due to the hard thresholds. Follow the practice of [1], we regard the palette histogram as a joint distribution over a and b, represented by a weighted sum of kernels. Formally, the color histogram is written as:

$$\mathbf{h}(a, b) = \frac{1}{Z} \sum_{x} k(\mathbf{C}_a(x), \mathbf{C}_b(x), a, b), \tag{5}$$

where $\mathbf{C}_a(x)$ and $\mathbf{C}_b(x)$ denote the values of pixel x in a and b channels, respectively. k is the used kernel function to measure the difference between $(\mathbf{C}_a(x), \mathbf{C}_b(x))$ and a given (a, b), Z is a normalization factor. In this paper, we adopt inverse-quadratic kernel [1], which is:

$$k(\mathbf{C}_a(x), \mathbf{C}_b(x), a, b) = \prod_{i \in \{a,b\}} (1 + (\frac{|\mathbf{C}_i(x) - i|}{\sigma})^2)^{-1}, \tag{6}$$

where σ controls the smoothness of adjacent bins. We find $\sigma = 0.1$ works best.

Regularization. To diversify the predicted colors, we introduce palette regularization, combating against the dull colors brought by imbalanced color distribution. On one hand, we employ *ab* histogram in probabilistic palette form to measure the color distribution in the predicted color map and ground truth. Minimizing their discrepancies explicitly considers different color ratios, avoiding converging to a few dominant ones. On the other hand, we diversify the produced colors by increasing the possibility of rare colors (statistically in training samples). We exploit the entropy of the probabilistic palette to control such diversity. Formally, the entropy of $\hat{\mathbf{h}}$ is $E(\hat{\mathbf{h}}) = -\sum_{i=1}^{|\hat{\mathbf{h}}|} \hat{\mathbf{h}}_i \log \hat{\mathbf{h}}_i$. To improve the color diversity in $\hat{\mathbf{h}}$, we can maximize $E(\hat{\mathbf{h}})$.

3.3 Color Discriminator

We give a color discriminator utilizing the palette, improving the result from the adversarial training. We incorporate the palette into the discriminator in a condition projection manner [33]. We employ convolutional discriminator $\mathbf{D}$,

Table 1. Quantitative results on the validation sets from different methods.

Method	ImageNet (ctest10k)				ImageNet (val50k)				COCO-Stuff			
	PSNR ↑	SSIM ↑	LPIPS ↓	FID ↓	PSNR ↑	SSIM ↑	LPIPS ↓	FID ↓	PSNR ↑	SSIM ↑	LPIPS ↓	FID ↓
CIColor [54]	22.30	0.902	0.221	12.20	22.26	0.902	0.221	9.39	21.84	0.895	0.234	22.32
UGColor [56]	24.26	0.918	0.174	7.49	24.26	0.919	0.173	4.60	24.34	0.924	0.165	14.74
Lei *et al.* [26]	24.52	0.917	0.202	12.60	24.03	0.918	0.189	6.35	24.59	0.922	0.191	23.10
Deoldify [2]	23.54	0.914	0.187	5.78	22.97	0.911	0.185	3.87	23.98	0.939	0.161	12.75
ColTrans [23]	21.81	0.892	0.218	6.37	22.12	0.894	0.216	3.81	22.11	0.898	0.210	11.65
Ours[1]	24.19	0.917	**0.161**	**4.60**	24.25	0.917	**0.161**	**2.78**	24.56	0.924	**0.148**	**7.70**
Ours[2]	**24.66**	**0.920**	0.170	5.24	**24.54**	**0.920**	0.168	3.62	**24.72**	**0.944**	0.156	8.93
InstColor* [38]	23.03	0.909	0.191	7.35	23.06	0.910	0.190	4.94	22.35	0.838	0.238	12.24
GPColor* [49]	21.66	0.871	0.230	5.46	21.81	0.880	0.230	3.62	N/A	N/A	N/A	N/A
Ours*	27.75	0.932	0.110	4.20	27.53	0.913	0.118	2.42	28.28	0.936	0.105	7.21

converting the input (the concatenation between the *ab* image and its converted RGB one) into a 1D feature embedding $\mathbf{g} \in \mathbb{R}^{\mathbf{256}\times\mathbf{1}}$. Then such feature is fused with the palette by the inner product. The likelihood of the realness of the input is given as:

$$p(\mathbf{C} \oplus \mathbf{I}) = (\mathbf{W}\mathbf{g})^{\mathrm{T}}\mathbf{h}, \tag{7}$$

where $\mathbf{W} \in \mathbb{R}^{n^2\times 256}$ is a learnable linear projection, and $\mathbf{I} \in \mathbb{R}^{h\times w\times 3}$ is the converted *rgb* version of $\mathbf{C}$ and $\mathbf{L}$.

3.4 Learning Objective

Palette estimation and assignment are trained with different optimization targets. For palette estimation, it is learned concerning palette reconstruction and regularization as:

$$\mathcal{L}_{\mathbf{E}} = \underbrace{\lambda_{\text{rec1}}|\mathbf{h}-\hat{\mathbf{h}}|_1}_{\text{reconstruction}} - \underbrace{\lambda_{\text{rg}}E(\hat{\mathbf{h}})}_{\text{regularization}}, \tag{8}$$

where λ_{rec1} and λ_{rg} balance the influences of different terms, set to 5.0 and 1.0, respectively.

The optimization target for palette assignment is formed by pixel-level regression, palette reconstruction, and adversarial training, as:

$$\mathcal{L}_{\mathbf{G}} = \underbrace{\lambda_{\text{reg}}|\mathbf{C}-\hat{\mathbf{C}}|_1}_{\text{regression}} + \underbrace{\lambda_{\text{rec2}}|\mathbf{h}-\tilde{\mathbf{h}}|_1}_{\text{reconstruction}} + \underbrace{\lambda_{\text{adv}}\mathcal{L}_{\text{adv}}}_{\text{adversarial}}, \tag{9}$$

where $\tilde{\mathbf{h}}$ are extracted from $\hat{\mathbf{C}}$ using Eq. 5. λ_{reg}, λ_{rec2}, and λ_{adv} are set to 5.0, 1.0, 1.0, respectively.

For the used adversarial loss, we employ hinge loss version. Its training target of generator is

$$\mathcal{L}_{adv} = -\mathrm{E}_{\mathbf{L}\sim\mathbb{P}_{\mathbf{L}}}\mathbf{D}(\hat{\mathbf{C}} \oplus \hat{\mathbf{I}}), \tag{10}$$

where $\hat{\mathbf{C}} = \mathcal{T}_{\mathbf{G}}(\mathbf{L}|\mathcal{T}_{\mathbf{E}}(\mathbf{L}))$, $\hat{\mathbf{I}}$ is a converted *rgb* version from $\hat{\mathbf{C}}$ and $\mathbf{L}$, and $\mathbb{P}_{\mathbf{L}}$ denotes the gray-scale image distribution. The optimization goal for the discriminator is

$$\mathcal{L}_{adv}^{\mathbf{D}} = \mathrm{E}_{\mathbf{I}\sim\mathbb{P}_{\mathbf{I}}}[\max(0, 1-\mathbf{D}(\mathbf{C}\oplus\mathbf{I}))] + \mathrm{E}_{\mathbf{L}\sim\mathbb{P}_{\mathbf{L}}}[\max(0, 1+\mathbf{D}(\hat{\mathbf{C}}\oplus\hat{\mathbf{I}})]. \tag{11}$$

where $\mathbb{P}_{\mathbf{I}}$ denotes the *rgb* image distribution, and $\mathbf{C}$ is converted from $\mathbf{I}$.

Training. We jointly train palette generator $\mathcal{T}_{\mathbf{E}}$ and palette assignment generator $\mathcal{T}_{\mathbf{G}}$ in an progressive fashion. Specifically, for the inputs $\{\mathbf{L}_i\}$ to $\mathcal{T}_{\mathbf{E}}$, the corresponding inputs to $\mathcal{T}_{\mathbf{G}}$ are $\{\mathbb{1}(p_{\mathbf{h}} > 0.8)\mathbf{h}_i + (1 - \mathbb{1}(p_{\mathbf{h}} > 0.8))\hat{\mathbf{h}}_i\}$, where $\mathbb{1}$ is an indicator function of value 1 if its condition holds true, and 0, otherwise. $p_{\mathbf{h}}$ is sampled from a uniform distribution $\mathcal{U}[\tau, 1]$. We start training with $\tau = 1$, then linearly decrease it to 0 when approaching the end of learning.

4 Experiments

We evaluate our method along with existing representative works on ImageNet [9] and COCO-Stuff [5]. On ImageNet we take two evaluation protocols. One is to evaluate all methods on a selective subset *ctest10k* (with 10K pictures) of its validate data (with 50K pictures) following the protocols in [24]. Another is to run on the full validation set, same as in [49]. For COCO-Stuff, we test all methods on its 5K validating images.

4.1 Implementation

We employ spectral normalization [32] on the whole model and a two time-scale update rule in training (lr for the generator and discriminator are $1e{-}4$ and $4e{-}4$, respectively) to stabilize learning. Adam [22] optimizer with $\beta_1 = 0$ and $\beta_2 = 0.9$ is used. For the applied batch normalization, we take the sync version. We train our method on the training set of ImageNet with 40 epochs with 8 TiTAN 2080ti using batch size 64. Images in training are randomly cropped in a fixed size (256×256) from the resized ones with aspect ratio unchanged. In testing, we resize images into 256×256 ones and do evaluations.

Baselines. We focus on the recent learning-based colorization methods for comparisons. Deoldify [2], CIColor [54], UGColor [56], Video Colorization [26], InstColor [38], ColTrans [23], and GPColor [49] are employed for comparisons. Note InstColor is learned with a pretrained object detection model (requiring both labels and bounding boxes), and GPColor exploits a pretrained (on ImageNet with labels) BigGAN. Other approaches including ours are only trained with paired gray-colorful images. For UGColor, we use its fully automatic version where no color hints are used. We use their released model for testing.

Metrics. We employ pixel-wise similarity measures PSNR, SSIM, image-level perceptual metric LPIPS [55], and Fréchet Inception Distance (FID) [16] to quantitatively evaluate colorization results. LPIPS and FID are more consistent with human evaluations compared with PSNR and SSIM.

4.2 Quantitative Evaluations

Compared with other methods, our proposed PalGAN (ours[1] in Table 1) gives the best perceptual scores (FID & LPIPS) both on ImageNet (FID: 4.60 and 2.78,

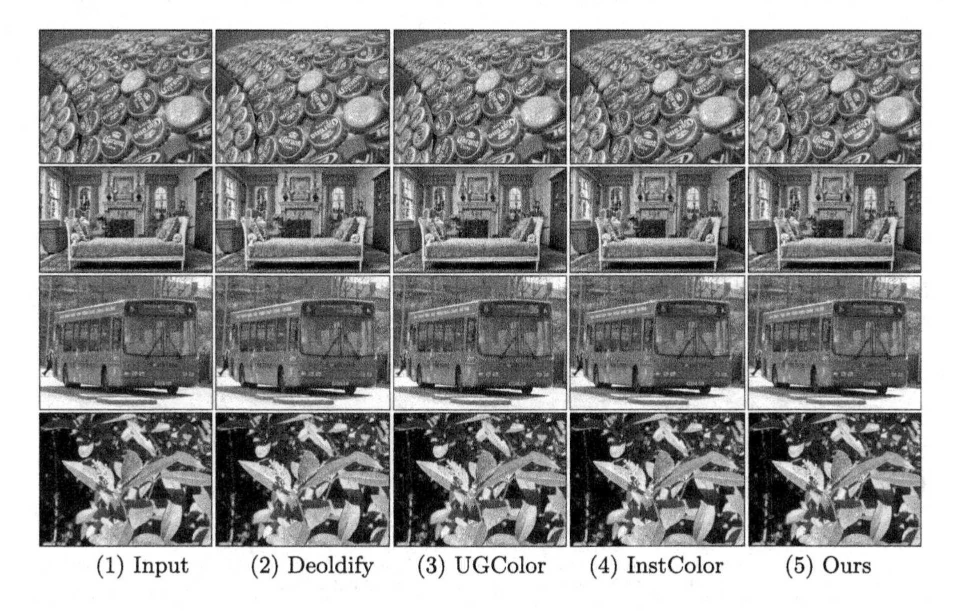

(1) Input (2) Deoldify (3) UGColor (4) InstColor (5) Ours

Fig. 6. Visual comparison on ImageNet and COCO-Stuff.

Table 2. User study. Each entry gives the percentage of cases where colorization results are favored compared with GT.

Method	Ours	Coltrans	GPCol	InstCol	Deoldify	UGCol
Rate	**47.20%**	41.50%	39.25%	37.50%	41.13%	42.50%

LPIPS: 0.161 and 0.161 from ctest10K and val50K, respectively) and COCO-Stuff (FID: 7.70, LPIPS: 0.148) without exploiting any annotations or hints, which outperforms other methods. It validates the superiority in realness and diversity of our results. We also achieve competitive fidelity scores (PSNR & SSIM) among all. It shows the well-behaved color restoration ability of PalGAN. If given the ground truth palette, our method (ours* in Table 1) can deliver impressive fidelity performance as well as a generative one. It shows the upper bound performance of our method for reference. For methods in Table 1 denoted with *, they employ external priors *e.g.* annotations.

Considering the trade-off between fidelity and perceptual results, we can get the best of both worlds on all benchmarks compared with others (ours^2 in Table 1) with proper training setting ($\lambda_{adv} = 0.1$ and other regularization coefficients remain still).

4.3 Qualitative Evaluations

As shown in Fig. 6, our colorization results give natural, diverse, and fine chrome predictions considering both semantic correspondence and local gradient change.

Fig. 7. Our method on legacy images. (1) Inputs, (2) our automatic results, (3) our reference-based results.

Fig. 8. Our reference-based colorization.

It suffers less from the common color bleeding compared with other methods, owing to chromatic attention. More results are given in Supp.

User Studies. Table 2 gives human evaluations on our methods with the compared ones. Following the protocol in [23, 54], we conduct a colorization Turning test. Specifically, the ground truth color image and its corresponding colorization result (from ours or other methods) are given to 20 participants in random order. These participants need to determine which one is more realistic than the other for no more than 2 s. There are 40 colorization predictions from each method, randomly chosen from ImageNet ctest10k. Table 2 presents that our method beats the competitors with a large margin.

Colorization of Legacy Photos. Though our model is trained in a self-supervised manner using synthetic data, it generalizes well on real-world black-and-white legacy pictures (from [15]), as given in Fig. 7. Color boundaries and consistency are well handled in these cases, working well on the object and portrait.

Reference-Based Colorization. With the intermediate palette, our approach can conduct reference-based (or example-based) colorization by feeding it with the palette from the reference color image, as given in Fig. 7 and 8. Note even using palettes from an image without semantic correlations with the input (Fig. 8), PalGAN still well tunes the given color distribution according to the semantics of the given image, keeping color regionally consistent. Note the car appearances in Fig. 8 present impressive diversity and realness.

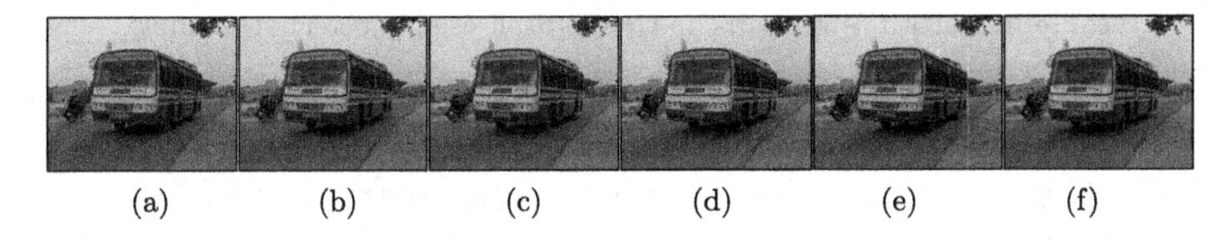

(a) (b) (c) (d) (e) (f)

Fig. 9. Ablation studies of model structures. (a) Input, (b) AE, (c) VAE, (d) PalGAN w PatchD, (e) PalGAN wo $E(\hat{\mathbf{h}})$, (f) full PalGAN.

4.4 Ablation Studies

Our key designs are ablated on COCO-Stuff as follows.

Palette Prediction and Assignment. We validate the effectiveness of our colorization formulation with the proposed model structure compared with a naive autoencoder (AE) and variational one (VAE). Specifically, AE shares the same computational units with PalGAN, except it generates feature maps instead of the palette from its encoder, and utilizes common BN instead of PalNorm in its decoder. VAE is almost the same as PalGAN but it changes the intermediate product palette into a latent vector constrained by Normal distribution. The optimization of AE and VAE is nearly the same as ours except they do not have the palette reconstruction and regularization term, and VAE employs one more term for regularizing the intermediate latent code.

In Table 3, we find PalGAN gets significant improvements on FID compared with AE and VAE, while its fidelity performance (PSNR and SSIM) is inferior to AE. It suggests intermediate latent code (in PalGAN and VAE) performs better at color generation than feature maps (in AE), and feature maps excel at fidelity restoration. It validates the effectiveness of our formulation and method on the usage of palette considering its generative performance. Moreover, Fig. 9 illustrates visual differences between varied structures in one example. A high fidelity score of AE does not guarantee the realism of its result.

The effectiveness of the predicted palette is studied. We use palettes from random reference images to simulate failed palette estimations in our method, and give the corresponding evaluation in Table 3 (PalGAN w rand ref). It shows the dramatic fidelity and generative performance drop, meaning our palette generator can learn effective chrome distribution for colorization. This is also supported by the visualizations of palettes and their corresponding images in Figure 3.

Chromatic Attention. We explore how the proposed chromatic attention affects colorization, given in Table 3 and 4. Compared with naive self-attention (PalGAN w SA in Table 3, and SA is applied on the high-level feature maps $\mathbf{S}$), our chromatic attention enhances both generative and fidelity performance notably. In Table 4, with global interaction in chromatic attention, the generative performance will be improved non-trivially on FID ($9.90 \rightarrow 8.34$). It is consistent with the observations in prior image generation works [4,43,44,53] that employing

Table 3. Quantitative results on COCO-Stuff using different structures.

Structure	PSNR ↑	SSIM ↑	LPIPS ↓	FID ↓
AE	**25.89**	**0.928**	**0.146**	14.15
VAE	23.21	0.905	0.179	11.76
UGC w CA	24.52	0.923	0.162	11.38
PalGAN w rand ref	20.88	0.883	0.240	9.64
PalGAN w SA	22.68	0.892	0.175	9.02
PalGAN w PatchD [19,41]	23.07	0.895	0.183	8.44
PalGAN w BN	22.36	0.895	0.209	9.97
PalGAN w SPADE [35]	24.06	0.916	0.167	7.90
PalGAN wo $E(\hat{\mathbf{h}})$	24.58	0.924	0.149	8.17
PalGAN	24.56	0.924	0.148	**7.70**

Table 4. Quantitative results on COCO-Stuff by ablating chromatic attention.

G	L	PSNR ↑	SSIM ↑	LPIPS ↓	FID ↓
✗	✗	21.93	0.902	0.203	9.90
✗	✓	24.52	**0.924**	**0.146**	9.97
✓	✗	23.32	0.907	0.174	8.34
✓	✓	**24.56**	**0.924**	0.148	**7.70**

Table 5. Quantitative results on COCO-Stuff about palette with different bins.

#Bins	16	64	256	576	1024
PSNR ↑	23.52	24.48	**24.56**	23.34	23.31
SSIM ↑	0.917	0.919	**0.924**	0.913	0.915
LPIPS ↓	0.172	0.153	**0.148**	0.152	0.159
FID ↓	8.24	7.92	**7.70**	8.04	8.16

attention will boost generation results. For the local delineation, it focuses on pixel-level restoration, giving notable fidelity increase on PSNR (21.93 → 24.52) and SSIM (0.902 → 0.924). Generally, CA achieves the best of both worlds as it enhances both pixel- and perceptual-level performance. Moreover, we give visual comparison of the ablation study on the chromatic attention in Fig. 5.

Note CA is a generic parametric module. It can be applied to previous methods *e.g.* UGC [56], and it can further improve the corresponding quantitative results (24.34 → 24.51, 0.924 → 0.925, 0.165 → 0.162, and 14.74 → 11.38 on PSNR, SSIM, LPIPS, and FID, respectively).

PalNorm and Color Discriminator. In Table 3, we find PalNorm yields better quantitative results than BN and SPADE [35] (we use gray-input as semantic layout to generate pixel-wise affine transformation). Besides, PalGAN (default with Color Discriminator) beats PalGAN with Patch Discriminator [41]. These show the effectiveness of our designed PalNorm and Color discriminator.

Palette Configuration. We systematically explore different factors of the employed palette. Table 5 shows how the number of bins of palette affects the colorization results. Generally, when #Bins is relatively small, increasing it (16 →256) will lead to a performance increase on almost all used metrics; while

#Bins is relatively large, increasing it (256 →1024) will lead to a performance drop. We conjecture this is caused by the tradeoff between the fineness and sparsity of the used palette. The rise of #Bins enhances both its fineness and sparsity. The former reduces ambiguities of palette and the latter makes the optimization of palette reconstruction harder. Empirically, #Bins is 256 is an acceptable setting, used in all experiments.

Also, as given in Table 3 (PalGAN wo $E(\hat{\mathbf{h}})$), applying diversity regularization on the estimated palette can improve our generative performance.

Limitation. In the user-guided colorization, current PalGAN lacks fine-grained control as we utilize a global palette. In addition, PalGAN cannot well address small-size regions with independent semantics (*e.g.* small objects), since the global interaction in CA cannot well represent these areas using semantic embeddings from small-scale feature maps. Failure cases are given in the supp.

5 Concluding Remarks

In this paper, we study multimodal challenges and color bleeding in colorization from a new perspective. We give a new formulation of colorization for multimodal representation. It introduces the palette as an intermediate variable. This leads to a new and workable colorization method by palette estimation and assignment, yielding diverse and controllable colorful outputs. Additionally, we address the color bleeding issue by explicitly studying color affinities using chromatic attention. It not only leverages semantic affinities to coordinate color, but also exploits the correlation between intensity and their corresponding chrome to delineate color details. Our method is experimentally proven effective and surpasses existing state-of-the-arts non-trivially.

Acknowledgment. This work is partially supported by the Shanghai Committee of Science and Technology (Grant No. 21DZ1100100).

References

1. Afifi, M., Brubaker, M.A., Brown, M.S.: HistoGAN: controlling colors of GAN-generated and real images via color histograms. In: CVPR, pp. 7941–7950 (2021)
2. Antic, J.: A deep learning based project for colorizing and restoring old images (and video!). https://github.com/jantic/DeOldify
3. Bahng, H., et al.: Coloring with words: guiding image colorization through text-based palette generation. In: Ferrari, V., Hebert, M., Sminchisescu, C., Weiss, Y. (eds.) ECCV 2018. LNCS, vol. 11216, pp. 443–459. Springer, Cham (2018). https://doi.org/10.1007/978-3-030-01258-8_27
4. Brock, A., Donahue, J., Simonyan, K.: Large scale GAN training for high fidelity natural image synthesis. arXiv preprint arXiv:1809.11096 (2018)
5. Caesar, H., Uijlings, J., Ferrari, V.: COCO-stuff: thing and stuff classes in context. In: CVPR, pp. 1209–1218 (2018)

6. Chang, H., Fried, O., Liu, Y., DiVerdi, S., Finkelstein, A.: Palette-based photo recoloring. TOG **34**(4), 139 (2015)
7. Charpiat, G., Hofmann, M., Schölkopf, B.: Automatic image colorization via multimodal predictions. In: Forsyth, D., Torr, P., Zisserman, A. (eds.) ECCV 2008. LNCS, vol. 5304, pp. 126–139. Springer, Heidelberg (2008). https://doi.org/10.1007/978-3-540-88690-7_10
8. Chia, A.Y.S., et al.: Semantic colorization with internet images. TOG **30**(6), 1–8 (2011)
9. Deng, J., Dong, W., Socher, R., Li, L.J., Li, K., Fei-Fei, L.: ImageNet: a large-scale hierarchical image database. In: CVPR, pp. 248–255 (2009)
10. Deshpande, A., Lu, J., Yeh, M.C., Jin Chong, M., Forsyth, D.: Learning diverse image colorization. In: CVPR, pp. 6837–6845 (2017)
11. Goodfellow, I., et al.: Generative adversarial nets. In: NeurIPS, pp. 2672–2680 (2014)
12. Guadarrama, S., Dahl, R., Bieber, D., Norouzi, M., Shlens, J., Murphy, K.: Pixcolor: pixel recursive colorization. arXiv preprint arXiv:1705.07208 (2017)
13. He, K., Sun, J., Tang, X.: Guided image filtering. TPAMI **35**(6), 1397–1409 (2012)
14. He, K., Zhang, X., Ren, S., Sun, J.: Deep residual learning for image recognition. In: CVPR, pp. 770–778 (2016)
15. He, M., Chen, D., Liao, J., Sander, P.V., Yuan, L.: Deep exemplar-based colorization. TOG **37**(4), 1–16 (2018)
16. Heusel, M., Ramsauer, H., Unterthiner, T., Nessler, B., Hochreiter, S.: GANs trained by a two time-scale update rule converge to a local nash equilibrium. In: NeurIPS, pp. 6626–6637 (2017)
17. Iizuka, S., Simo-Serra, E., Ishikawa, H.: Let there be color! joint end-to-end learning of global and local image priors for automatic image colorization with simultaneous classification. TOG **35**(4), 1–11 (2016)
18. Ironi, R., Cohen-Or, D., Lischinski, D.: Colorization by example. Render. Tech. **29**, 201–210 (2005)
19. Isola, P., Zhu, J.Y., Zhou, T., Efros, A.A.: Image-to-image translation with conditional adversarial networks. In: CVPR, pp. 1125–1134 (2017)
20. Karras, T., Laine, S., Aila, T.: A style-based generator architecture for generative adversarial networks. arXiv preprint arXiv:1812.04948 (2018)
21. Kim, E., Lee, S., Park, J., Choi, S., Seo, C., Choo, J.: Deep edge-aware interactive colorization against color-bleeding effects. In: ICCV, pp. 14667–14676 (2021)
22. Kingma, D.P., Ba, J.: Adam: a method for stochastic optimization. arXiv preprint arXiv:1412.6980 (2014)
23. Kumar, M., Weissenborn, D., Kalchbrenner, N.: Colorization transformer. arXiv preprint arXiv:2102.04432 (2021)
24. Larsson, G., Maire, M., Shakhnarovich, G.: Learning representations for automatic colorization. In: Leibe, B., Matas, J., Sebe, N., Welling, M. (eds.) ECCV 2016. LNCS, vol. 9908, pp. 577–593. Springer, Cham (2016). https://doi.org/10.1007/978-3-319-46493-0_35
25. Larsson, G., Maire, M., Shakhnarovich, G.: Colorization as a proxy task for visual understanding. In: CVPR (2017)
26. Lei, C., Chen, Q.: Fully automatic video colorization with self-regularization and diversity. In: CVPR, pp. 3753–3761 (2019)
27. Levin, A., Lischinski, D., Weiss, Y.: Colorization using optimization. In: SIGGRAPH, pp. 689–694 (2004)
28. Li, W., Lin, Z., Zhou, K., Qi, L., Wang, Y., Jia, J.: MAT: mask-aware transformer for large hole image inpainting. In: CVPR, pp. 10758–10768 (2022)

29. Liu, X., et al.: Intrinsic colorization. In: SIGGRAPH Asia, pp. 1–9 (2008)
30. Liu, X., Yin, G., Shao, J., Wang, X., et al.: Learning to predict layout-to-image conditional convolutions for semantic image synthesis. In: NeurIPS, pp. 570–580 (2019)
31. Liu, Z., Wang, Y., Qi, X., Fu, C.W.: Towards implicit text-guided 3d shape generation. In: CVPR, pp. 17896–17906, June 2022
32. Miyato, T., Kataoka, T., Koyama, M., Yoshida, Y.: Spectral normalization for generative adversarial networks. arXiv preprint arXiv:1802.05957 (2018)
33. Miyato, T., Koyama, M.: cGANs with projection discriminator. arXiv preprint arXiv:1802.05637 (2018)
34. Nguyen, R.M., Price, B., Cohen, S., Brown, M.S.: Group-theme recoloring for multi-image color consistency. In: Computer Graphics Forum, vol. 36, pp. 83–92. Wiley Online Library (2017)
35. Park, T., Liu, M.Y., Wang, T.C., Zhu, J.Y.: Semantic image synthesis with spatially-adaptive normalization. In: CVPR, pp. 2337–2346 (2019)
36. Qi, L., et al.: Open-world entity segmentation. arXiv preprint arXiv:2107.14228 (2021)
37. Qu, Y., Wong, T.T., Heng, P.A.: Manga colorization. TOG **25**(3), 1214–1220 (2006)
38. Su, J.W., Chu, H.K., Huang, J.B.: Instance-aware image colorization. In: CVPR, pp. 7968–7977 (2020)
39. Tai, Y.W., Jia, J., Tang, C.K.: Local color transfer via probabilistic segmentation by expectation-maximization. In: CVPR, vol. 1, pp. 747–754. IEEE (2005)
40. Torralba, A., Freeman, W.T.: Properties and applications of shape recipes (2002)
41. Wang, T.C., Liu, M.Y., Zhu, J.Y., Tao, A., Kautz, J., Catanzaro, B.: High-resolution image synthesis and semantic manipulation with conditional GANs. In: CVPR, pp. 8798–8807 (2018)
42. Wang, Y., Chen, Y.-C., Tao, X., Jia, J.: VCNet: a robust approach to blind image inpainting. In: Vedaldi, A., Bischof, H., Brox, T., Frahm, J.-M. (eds.) ECCV 2020. LNCS, vol. 12370, pp. 752–768. Springer, Cham (2020). https://doi.org/10.1007/978-3-030-58595-2_45
43. Wang, Y., Chen, Y.C., Zhang, X., Sun, J., Jia, J.: Attentive normalization for conditional image generation. In: CVPR, pp. 5094–5103 (2020)
44. Wang, Y., Qi, L., Chen, Y.C., Zhang, X., Jia, J.: Image synthesis via semantic composition. In: ICCV, pp. 13749–13758 (2021)
45. Wang, Y., Tao, X., Qi, X., Shen, X., Jia, J.: Image inpainting via generative multi-column convolutional neural networks. In: NeurIPS (2018)
46. Wang, Y., Tao, X., Shen, X., Jia, J.: Wide-context semantic image extrapolation. In: CVPR (2019)
47. Welsh, T., Ashikhmin, M., Mueller, K.: Transferring color to greyscale images. In: Proceedings of the 29th Annual Conference on Computer Graphics and Interactive Techniques, pp. 277–280 (2002)
48. Wu, H., Zheng, S., Zhang, J., Huang, K.: Fast end-to-end trainable guided filter. In: CVPR, pp. 1838–1847 (2018)
49. Wu, Y., Wang, X., Li, Y., Zhang, H., Zhao, X., Shan, Y.: Towards vivid and diverse image colorization with generative color prior. In: ICCV, pp. 14377–14386 (2021)
50. Xia, M., Wang, Y., Han, C., Wong, T.T.: Enhance convolutional neural networks with noise incentive block. arXiv preprint arXiv:2012.12109 (2020)
51. Xu, X., Wang, Y., Wang, L., Yu, B., Jia, J.: Conditional temporal variational autoencoder for action video prediction. arXiv preprint arXiv:2108.05658 (2021)
52. Yatziv, L., Sapiro, G.: Fast image and video colorization using chrominance blending. TIP **15**(5), 1120–1129 (2006)

53. Zhang, H., Goodfellow, I., Metaxas, D., Odena, A.: Self-attention generative adversarial networks. arXiv preprint arXiv:1805.08318 (2018)
54. Zhang, R., Isola, P., Efros, A.A.: Colorful image colorization. In: Leibe, B., Matas, J., Sebe, N., Welling, M. (eds.) ECCV 2016. LNCS, vol. 9907, pp. 649–666. Springer, Cham (2016). https://doi.org/10.1007/978-3-319-46487-9_40
55. Zhang, R., Isola, P., Efros, A.A., Shechtman, E., Wang, O.: The unreasonable effectiveness of deep features as a perceptual metric. In: CVPR (2018)
56. Zhang, R., et al.: Real-time user-guided image colorization with learned deep priors. arXiv preprint arXiv:1705.02999 (2017)
57. Zhu, J.Y., Park, T., Isola, P., Efros, A.A.: Unpaired image-to-image translation using cycle-consistent adversarial networks. In: ICCV, pp. 2223–2232 (2017)
58. Zomet, A., Peleg, S.: Multi-sensor super-resolution. In: Sixth IEEE Workshop on Applications of Computer Vision (WACV), pp. 27–31. IEEE (2002)

Fast-Vid2Vid: Spatial-Temporal Compression for Video-to-Video Synthesis

Long Zhuo[1], Guangcong Wang[2], Shikai Li[3], Wayne Wu[1,3], and Ziwei Liu[2](✉)

[1] Shanghai AI Laboratory, Shanghai, China
zhuolong@pjlab.org.cn
[2] S-Lab, Nanyang Technological University, Singapore, Singapore
{guangcong.wang,ziwei.liu}@ntu.edu.sg
[3] SenseTime Research, Science Park, Hong Kong
lishikai@sensetime.com

Abstract. Video-to-Video synthesis (Vid2Vid) has achieved remarkable results in generating a photo-realistic video from a sequence of semantic maps. However, this pipeline suffers from high computational cost and long inference latency, which largely depends on two essential factors: 1) network architecture parameters, 2) sequential data stream. Recently, the parameters of image-based generative models have been significantly compressed via more efficient network architectures. Existing methods mainly focus on slimming network architectures and ignore the size of the sequential data stream. Moreover, due to the lack of temporal coherence, image-based compression is not sufficient for the compression of the video task. In this paper, we present a spatial-temporal compression framework, **Fast-Vid2Vid**, which focuses on data aspects of generative models. It makes the first attempt at time dimension to reduce computational resources and accelerate inference. Specifically, we compress the input data stream spatially and reduce the temporal redundancy. After the proposed spatial-temporal knowledge distillation, our model can synthesize key-frames using the low-resolution data stream. Finally, Fast-Vid2Vid interpolates intermediate frames by motion compensation with slight latency. On standard benchmarks, Fast-Vid2Vid achieves around real-time performance as 20 FPS and saves around 8× computational cost on a single V100 GPU. Code and models are publicly available (Project page: https://fast-vid2vid.github.io/, Code and models: https://github.com/fast-vid2vid/fast-vid2vid).

Keywords: Video-to-video synthesis · GAN compression

1 Introduction

Video-to-video synthesis (Vid2Vid) [42] targets at synthesizing a photo-realistic video given a sequence of semantic maps as input. A wide range of applications are derived from this task, such as face-talking video generation

Supplementary Information The online version contains supplementary material available at https://doi.org/10.1007/978-3-031-19784-0_17.

S. Avidan et al. (Eds.): ECCV 2022, LNCS 13675, pp. 289–305, 2022.
https://doi.org/10.1007/978-3-031-19784-0_17

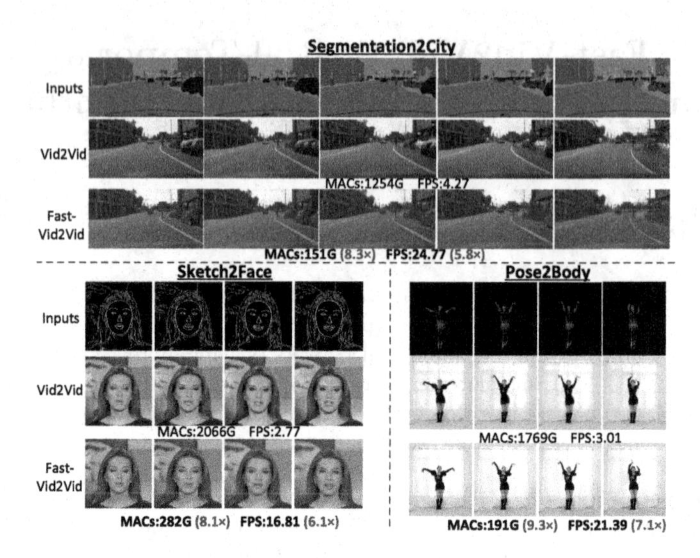

Fig. 1. Fast-Vid2Vid. Our proposed Fast-Vid2Vid accelerates the video-to-video synthesis and generates photo-realistic videos more efficiently compared to the original Vid2Vid model.

(Sketch2Face) [41,42], driving video generation (Segmentation2City) [41,42] and human pose transferring generation (Pose2Body) [4,25,51]. With the advance of Generative Adversarial Networks (GANs) [14], Vid2Vid models [41,42] have made significant progress in video quality. However, these approaches need large-scale computational resources to yield the results, and they are computationally prohibitive and environmentally unfriendly. For example, the standard Vid2Vid [42] consumes 2066 G MACs for generating each frame, which is 500× more than ResNet-50 [16]. Recent studies demonstrate that lots of recognition compression approaches have been successfully extended to image-based GAN compression methods [1,6,10,24,27,29]. Can we directly employ these existing image-based GAN compression methods to achieve promising Vid2Vid compression models? (Fig. 1).

In the literature, image-based GAN compression methods can be roughly categorized into three groups, including knowledge distillation [1,2,6,24,29], network pruning [29,40], and neural architecture search (NAS) [10,12,13,27,28]. They focus on obtaining a compact network by cutting the network architecture parameters of the original network. However, the input data, another factor that significantly affects the inference speed of a deep neural network, has been ignored by the existing GAN compression methods. Moreover, since they are image-based synthesis tasks, they do not consider redundant temporal information hidden in neighbor frames of a video. Therefore, directly applying image-based compression models to Vid2Vid synthesis is difficult to achieve the desired results. In this work, we aim to compress the input data stream while maintaining the well-designed network parameters, and generate the photo-realistic results for Vid2Vid synthesis. Furthermore, we make an initial attempt at removing temporal redundancy to accelerate Vid2Vid model.

There are three critical challenges for Vid2Vid compression. First, the typical Vid2Vid model [42] consists of several encoder-decoders to capture both spatial and temporal features. It is difficult to reduce the parameters from such a complicated structure due to the intricate connections between these encoders and decoders. Second, it is a challenge to compress the input data stream and achieve decent performance for GAN generation since the perceptual fields of GAN are much more erratic than image recognition. Third, transferring knowledge from a teacher model to a student model temporally is challenging to align with the spatial knowledge distillation, as the temporal knowledge is implicitly hidden within adjacent frames and is more difficult to capture than the spatial knowledge.

To address the above issues, in this paper, we propose a novel spatial-temporal compression framework for Vid2Vid synthesis, named **Fast-Vid2Vid**. As shown in Fig. 2, we reduce the computational resources by only compressing the input data stream through Motion-Aware Inference (MAI) without destroying the well-designed and complicated network parameters of the original Vid2Vid model, which addresses challenge 1. For challenges 2 and 3, we propose a Spatial-Temporal Knowledge Distillation method (STKD) that transfers spatial and temporal knowledge from the original model to the student network using compressed input data. Our goal is to transfer the knowledge from large-size synthesized videos to small-size synthesized ones to make GAN robust enough to gain promising visual performance when the input data is compressed.

We first train a spatially low-demand generator by taking low-resolution sequences as input but generating the full-resolution sequences. We perform Spatial Knowledge Distillation (Spatial KD) that transfers the spatial knowledge from the original generator to the spatially low-demand generator to obtain high-resolution information. Furthermore, we train a part-time generator by uniformly sampling video frames from sequences. We perform Temporal-aware Knowledge Distillation (Temporal KD) and distill the temporal knowledge of the original generator to the part-time student generator to obtain full-time motion information by introducing two losses, i.e., local temporal knowledge distillation loss and global temporal knowledge distillation loss. This design aims to capture the implicit knowledge in the time dimension.

To summarize, to the best of our knowledge, we make the first attempt to tackle the Vid2Vid compression problem at data aspects. On a single V100 GPU, Fast-Vid2Vid achieves 18.56 FPS (6.1× acceleration) with 8.1× less computational cost on Sketch2Face, 24.77 FPS (5.8× acceleration) with 8.3× less computational cost on Segmentation2City, and 21.39 FPS (7.1× acceleration) with 9.3× less computational cost on Pose2Body. The main contributions of this paper are concluded as three-fold:

- We present **Fast-Vid2Vid**, an sequential data stream compression method in spatial and temporal dimensions to greatly accelerate the Vid2Vid model.
- We introduce Spatial KD that transfers knowledge from a teacher model with high-resolution input data to a student model with low-resolution input data to learn high-resolution information.

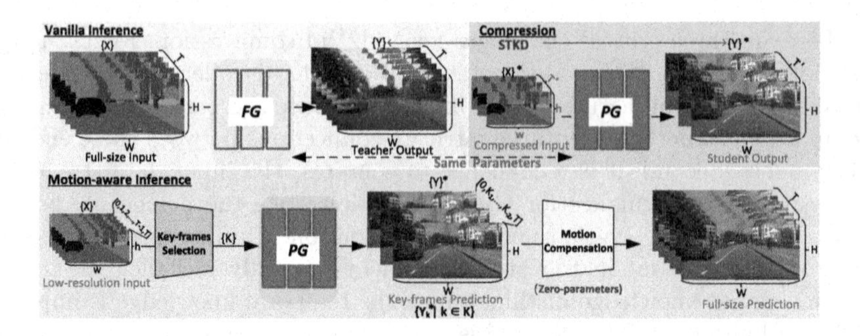

Fig. 2. The pipeline of our Fast-Vid2Vid. It maintains the same amount of parameters as the original generator but compresses the input data in space and time dimensions.

– We propose Temporal KD to distill knowledge from a full-time teacher model to a part-time student model. A new temporal knowledge distillation loss globally is further presented to capture the time-series correlation.

2 Related Work

Video-to-Video Synthesis. Video-to-video synthesis (Vid2Vid) is a computer vision task that generates a photo-realistic sequence using the corresponding semantic sequence. Based on high-resolution image-based synthesis [43], Wang *et al.* [42] developed a standard Vid2Vid synthesis model by introducing temporal coherence. Few-shot Vid2Vid model [41] further extended to a few-shot training version. Recently, Vid2Vid has been successfully extended to a wide range of video generation tasks, including video super-resolution [7,35,46], video inpainting [47,52], image-to-video synthesis [36,37] and human pose-to-body synthesis [4,11,25,51]. Most of these methods exploited temporal information to improve the performance of generated videos. However, they do not focus on Vid2Vid synthesis compression but on better visual performance.

Model Compression. Model compression aims at reducing superfluous parameters of deep neural networks to accelerate inference. In the computer vision task, lots of model pruning approaches [15,17,22,26,30,40,50] have greatly cut the weights of neural networks. Commonly, compression methods [17,18,23,26] reduced the unnecessary channels with low activations. GAN compression has been proved by [49] that it is far more difficult than normal CNN compression. A content-aware approach [29] was proposed to use salient regions to identify specific redundancy for GAN pruning. Wang *et al.* [40] reduced the redundant weights by NAS. Notably, the mentioned methods simplify the network structure but ignore the input information. Furthermore, these approaches do not consider the essential temporal coherence for video-based GAN, and thus yield sub-optimal results for compressing Vid2Vid models.

Knowledge Distillation. Knowledge Distillation aims to make a student network imitate its teacher. Hinton *et al.* [20] proposed an effective framework for model distillation in classification. Knowledge distillation has been widely used in recognition models [5,6,31,32,48]. Recently, lots of response-based knowledge distillation methods [1,2,6,10] were proposed for image-based GAN compression. For example, Jin *et al.* [24], developed distillation techniques from [27], used a global kernel alignment module to gain more potential information. Liu *et al.* [29] utilized a salient mask to guide the knowledge distillation process based on the norm. These methods only address image-based knowledge distillation, and thus only spatial knowledge is exploited, and they do not consider movements. Different from these spatial-aware knowledge distillation, we consider both spatial information and temporal information into knowledge distillation, which tailors for Vid2Vid model compression. Most recently, Feng *et al.* [9] have presented a resolution-aware knowledge distillation method that ignores the network parameters and compresses the input information for image recognition. In our work, it is the first time to introduce the input data compression for GAN synthesis.

3 Fast-Vid2Vid

3.1 A Revisit of GAN Compression

The function of a deep neural network (DNN) can be written as $f(X) = W * X$, where W denotes the parameters of the networks, $*$ represents the operation of DNN and X denotes the input data. Two essential factors accounting for computational cost are the parameters and the input data. Existing GAN compression methods [1,6,10,24,27,29] have intended to cut the computational cost by reducing the parameters of network structures. However, the network structures of GAN for specific tasks are carefully designed and it would cause poor visual results if the network parameters are cut arbitrarily. Another way to reduce computational cost is compressing the input data. In this work, we seek for compressing the input data instead of the parameters of well-designed networks. To the best of our knowledge, there is little literature working on compressing data for GAN compression.

3.2 Overview of Fast-Vid2Vid

The typical Vid2Vid framework [42] takes a sequence of semantic maps $\{X\}_0^T \in \mathbb{R}^{T\times H\times W}$ with T frames and the initial real frames as input and predicts a photo-realistic video sequence $\{Y\}_0^T \in \mathbb{R}^{T\times H\times W}$. H and W denotes the height and weight of each frame. The vanilla inference of the Vid2Vid model (the full-time teacher generator) that utilizes a full-size sequential input data stream is a consecutive process that synthesizes a video sequence frame by frame. Considering both image synthesis and temporal coherence, a Vid2Vid model often contains several encoder-decoders to capture spatial-temporal cues, which are computationally prohibitive. In this paper, we propose a Fast-Vid2Vid compression framework, an input data compression method, to reduce the computational resources of the Vid2Vid framework in both space and time dimensions.

Figure 2 illustrates the overview of the proposed method. Fast-Vid2Vid first replaces the resBlock of the original Vid2Vid generator [42] with decomposed

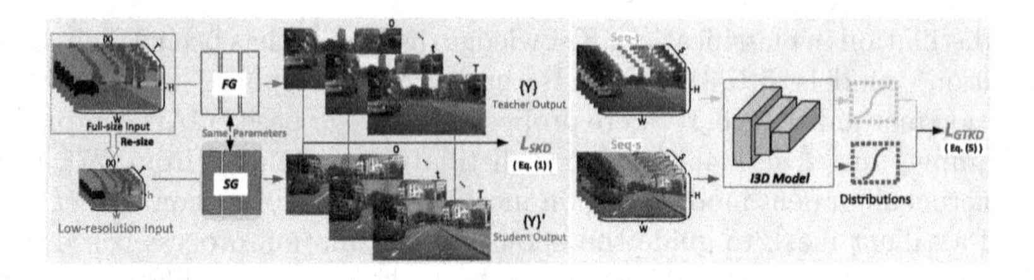

Fig. 3. Left: The proposed Spatial Knowledge Distillation (Spatial KD). **Right:** The proposed temporal loss for temporal global knowledge distillation.

convolutional block [21] to obtain a modern network architecture, which is similar to [27]. During knowledge distillation, we train a student generator using the compressed data and distill knowledge from the teacher generator by our proposed spatial-temporal knowledge distillation method (STKD). STKD, including spatial knowledge distillation (Spatial KD) and temporal knowledge distillation (Temporal KD), performs **spatial resolution compression** and **temporal sequential data compression**. After STKD, a part-time generator cooperating with motion compensation synthesizes a full-size sequence by **motion-aware inference (MAI)**.

3.3 Spatial Resolution Compression for Vid2Vid

To reduce the spatial input data, a straightforward way [9] is to predict the low-resolution results and re-size them into the full-resolution by a distortion algorithm. However, in our preliminary experiments, this method leads to severe artifacts since the distortion algorithm lacks high-frequency information and loses the important textures. Therefore, we make an adaptive change for Vid2Vid synthesis. We replace the downsampling layers with the normal convolution layers to generate the high-resolution results. The modified generator only takes the low-resolution semantic sequence $\{X\}_0^{'T} \in \mathbb{R}^{T \times h \times w}$ as the input, where $h \times w = \frac{1}{(2^d)^2} H \times W$, and d is the number of the modified downsampling layers. d is set to 1. In this way, we obtain a spatially low-demand generator.

Next, the spatially low-demand generator is required to learn high-frequency representation from the full-time teacher generator. We present a spatial knowledge distillation method (Spatial KD) to model high-frequency knowledge from the teacher net. Specifically, as shown in the left figure of Fig. 3, Spatial KD shrinks the margin between the low-resolution domain and the high-resolution domain to improve the performance of the student network. Spatial KD implicitly transfers spatial knowledge from the teacher net to the student net. Particularly, Spatial KD applies a knowledge distillation loss to mimic the visual features of the teacher net, and the spatial knowledge distillation loss L_{SKD} can be written as:

$$L_{SKD} = \frac{1}{T} \sum_{t=0}^{T} [MSE(Y_t, Y_t') + L_{per}(Y_t, {Y'}_t)], \tag{1}$$

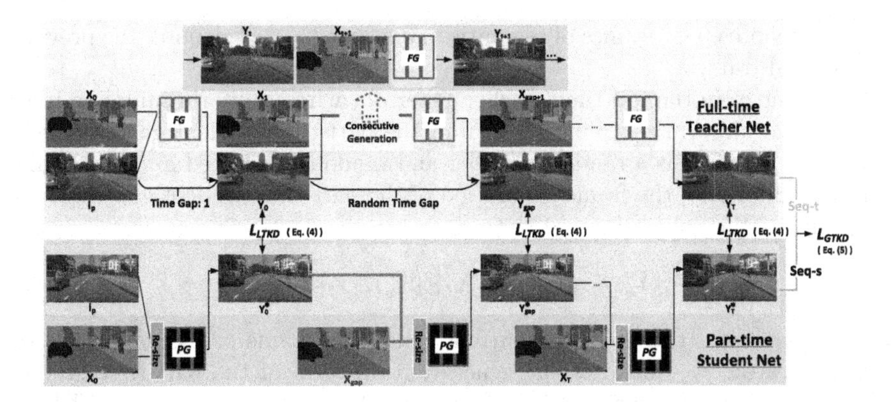

Fig. 4. The proposed temporal-aware knowledge distillation (Temporal KD).

where t means the current timestamp, T is the total timestamps of the sequences, Y is the output sequence of the teacher net and Y' is the predicted sequence of the spatially low-demand generator. MSE represents a mean squared error between two frames, and L_{per} denotes a perceptual loss [42].

3.4 Temporal Sequential Data Compression for Vid2Vid

Each video sequence consists of dense video frames, which brings an enormous burden on computational devices. How to efficiently synthesize dense frames is a difficult yet important issue for lightweight Vid2Vid models.

In Sect. 3.3, we obtain a spatially low-demand generator. To ease the burden of generating dense frames for each video, we re-train the spatially low-demand generator on sparse video sequences, which are uniformly sampled from dense video sequences. The sampling interval is randomly selected in each training iteration. The original Vid2Vid generator is regarded as a full-time teacher generator and the re-trained spatially low-demand generator is regarded as a part-time student generator. To distill the temporal knowledge from the full-time teacher generator to the part-time student generator, we propose a local temporal-aware knowledge distillation method and a global temporal-aware knowledge distillation method for temporal distillation, as shown in Fig. 4.

Both the full-time teacher generator and the part-time student generator take the previous $p-1$ synthesized frames $\{Y\}_1^{p-1}$ and p semantic maps $\{X\}_1^p$ as input and generate the next frame. The previous frames are used to capture the temporal coherence of the sequences and generate more coherent video frames. The generation process of the full-time teacher generator is the consecutive generation. More generally, each generation iteration of the full-time teacher generator can be formulated as follows:

$$Y_k = f_{FG}(\{X\}_{k-p}^{k}, \{Y\}_{k-p}^{k-1}), \tag{2}$$

where Y_k denotes the predicted current generative frame of the full-time teacher generator. f_{FG} denotes the generation function of the full-time teacher generator.

$\{X\}_{k-p}^{k}$ denotes $p+1$ frames of semantic maps, and $\{Y\}_{k-p}^{k-1}$ denotes the previous p generated frames.

Different from the full-time teacher generator whose uniform sampling interval is 1, the uniform sampling interval of the part-time student generator is g, where $1 < g < T$. g is a random number and randomly selected in each training iteration. Similarly, the frame generation of the part-time student generator can be formulated as follows:

$$Y_k^* = f_{PG}(f_R^d(\{X_t\}^{*k}_{k-p}), \{Y\}^{*k-1}_{k-p}), \tag{3}$$

where Y_k^* denotes the predicted current generative frame of the part-time student generator. f_{PG} denotes the generation function of the part-time student generator, f_R^d denotes the function of reducing the resolution into $\frac{1}{(2^d)^2}$, $\{X\}^{*k}_{k-p}$ includes p frames of the sparse semantic sequences and $\{Y\}^{*k-1}_{k-p}$ is the previous frames of the synthesized sparse sequences. To better illustrate Temporal KD, we set $p = 1$ in Fig. 4.

Because the full-time teacher generator is trained on sequences with dense frames and is learned to generate dense coherent frames, the full-time teacher generator cannot directly skip partial frames to generate sparse frames, leading to expensive computational cost. The part-time student generator generate sparse frames and interpolate intermediate frames with the slight computational cost. However, since the part-time student generator is trained on sequences with sampled sparse frames, the low sample rate could be two times less than temporal motion frequency and thus leads to aliasing according to Nyquist-Shannon sampling theorem. Our preliminary experiments also show that the large changes between two non-adjacent frames cause remarkable inter-frame incoherence and generate a bad result.

Local Temporal-Aware Knowledge Distillation. We first introduce the local temporal-aware knowledge distillation. Our goal is to distill the knowledge from the full-time generator to the part-time student generator to reduce aliasing. A straightforward idea is to align the outputs of the full-time generator and the outputs of the part-time student generator and reduce the distances between the corresponding synthesized frame pairs. The loss function of local temporal-aware knowledge distillation loss L_{LTKD} is given by

$$L_{LTKD} = \frac{1}{T}\sum_{t=0}^{T}[MSE(Y_t^*, Y_t) + L_{per}(Y_t^*, Y_t)], \tag{4}$$

where Y' denotes the resulting frame of the spatially low-demand generator and Y denotes the resulting frame of the teacher net. MSE represents a mean squared error between two frames. L_{per} denotes a perceptual loss [42].

Global Temporal-Aware Knowledge Distillation. Local temporal-aware knowledge distillation allows the part-time student generator to imitate the local motion of the full-time teacher generator. However, it does not consider the global semantic consistency. Therefore, we further propose a global temporal-aware knowledge distillation to distill global temporal coherence.

It is observed that the current frame synthesis deeply relies on the results of the previous synthesis. This indicates that the generated current frame implicitly contains information from the previous frames. The global temporal-aware knowledge distillation exploits the generated non-adjacent frames generated by the full-time teacher generator to distill the hidden information of temporal coherence. The frames generated by the full-time teacher generator at the same timestamps as the non-adjacent frames by the part-time student generator are extracted and concatenated into a predicted sequence $\{Y^*\}$ (Seq-s). Similarly, the results of the full-time teacher generator are concatenated into a sequence $\{Y^S\}$ (Seq-t) in time order, as shown in the right figure of Fig. 3. We introduce a global temporal loss to minimize the distance between $\{Y^*\}$ and $\{Y^S\}$, namely I3D-loss. The global temporal-aware knowledge distillation employs a well-trained I3D [3] model, a well-known video recognition model, to extract the time-series features of neighbor frames of $\{Y^S\}$ and $\{Y^S\}$. The global temporal-aware knowledge distillation loss is given by

$$L_{GTKD} = MSE(f_{I3D}(\{Y^*\}), f_{I3D}(\{Y^S\})), \tag{5}$$

where MSE calculates the mean squared error between two feature vectors, f_{I3D} is the function of the pre-trained I3D model. Finally, we obtain a temporal knowledge distillation loss, as written as,

$$L_{TKD} = \alpha L_{LTKD} + \beta L_{GTKD}, \tag{6}$$

where α and β control the importance of local and global losses. We set $\alpha = 2$ and $\beta = 15$ to enhance the global temporal coherence.

Full Objective Function. We integrate local temporal-aware and global temporal-aware knowledge distillation into a unified optimization framework, which enables the part-time student generator to imitate both global and local motions of the full-time teacher generator. The full objective function is given by

$$L_{KD} = \sigma L_{STD} + \gamma L_{TKD}, \tag{7}$$

where σ and γ control the weights of spatial and temporal KD losses, respectively. In particular, σ is set as 1 and γ is set as 2 to learn more knowledge of the temporal features from the teacher net.

3.5 Semantic-Driven Motion Compensation for Interpolation Synthesis

The temporal compression further greatly reduces the computational cost compared with the original Vid2Vid generator. However, the part-time student generator can only synthesize sparse frames $\{Y'\}$. To compensate for this problem, we use a fast motion compensation method [33], a zero-parameter algorithm, to complete the sequence. Motion compensation enables the synthesis of the inter frames between key-frames. As the adjacent frames are with slight changes, the final results remain a reliable visual performance by reducing the temporal redundancy. During inference, another question is which frames should be synthesized by the part-time student generator as sparse frames and how to determine these key-frames without sufficient photo-realistic frames. In this paper, we surprisingly find that we can distinguish key-frames $\{X'_k | k \in K\}$, where K is a set containing the numbers of key-frame, from semantic maps $\{X\}_0^{'T}$. With the key semantic maps, the part-time student generator generates sparse frames $\{Y'_k | k \in K\}$ and finally, we interpolate other inter frames to a full-size result sequence $\{Y\} \in \mathbb{R}^{T \times H \times W}$ by the fast motion compensation method.

4 Experiments

4.1 Experimental Setup

Models. We conduct our experiments using Vid2Vid [42] model. The original Vid2Vid model uses a coarse-to-fine generator for high-resolution output. To simplify the compression problem, we only retrain the first-scale Vid2Vid model based on the official repository[1]. We also evaluate three compression methods for image synthesis models, including NAS compression [27], CA compression [29] and CAT [24]. Since Vid2Vid is a 2D-based generation framework, these three methods can be easily transferred into this task with adjustments.

Datasets. We evaluate our proposed compression method on several datasets. We pre-process the following datasets as the same as the settings of Vid2Vid, including Face Video dataset [34], Cityscape Video dataset [8] and Youtube Dancing Video dataset [42]. Since we only apply the first scale of the Vid2Vid model, we re-size the datasets for convenience. Face Video dataset is re-sized into 512×512, Cityscape Video dataset is re-size into 256×512, and Youtube Dancing Video dataset is re-sized to 384×512.

Evaluation Metrics. We apply three metrics for quantitative evaluation, including FID [19], FVD [39] and pose error(PE) [41]. The lower score of the three metrics represents better performance.

Key-Frame Selection. We first calculate the residual maps between the adjacent frames, and sum up each map to draw smooth statistical curves using sliding windows. Thus, the peak ones of the curves represent the local maximum of the difference between two adjacent frames and are used as keyframes. Note that our keyframe selection only consumes about 0.5 ms for processing a video of 30 frames.

Motion Compensation. Motion compensation is to predict a video frame given the previous frames and future frames in video compression. We adopt the overlapped block motion compensation (OBMC) [33] and the enhanced predictive zonal search (EPZS) method [38] to generate the non-keyframes by "FFMPEG" toolbox. EPZS and OBMC consumes about 0.0008146G MACs for a 512×512 frame, which is much less than our generative model part.

[1] https://github.com/NVIDIA/vid2vid.

Table 1. Quantitative results of Fast-Vid2Vid. m.MACs represents the mean of the total MACs of the calculation resources for a sequence of video.

Task	Method	m.MACs(G)	FPS	Metric		
				FID (↓)	FVD (↓)	PE (↓)
Sketch2Face	Original	2066	2.77	34.17	6.74	—
	NAS	303	10.50	33.64	6.86	—
	CA	290	11.07	35.82	6.95	—
	CAT	294	10.89	33.90	6.43	—
	Ours	**282**	**18.56**	**29.02**	**5.79**	—
Segmentation2City	Original	1254	4.27	—	2.76	—
	NAS	277	12.44	—	2.99	—
	CA	187	15.48	—	3.98	—
	CAT	178	13.29	—	3.44	—
	Ours	**132**	**24.77**	—	**2.33**	—
Pose2Body	Original	1769	3.01	—	12.31	2.60
	NAS	280	12.57	—	12.53	3.28
	CA	253	13.92	—	15.89	4.85
	CAT	257	12.48	—	15.75	4.18
	Ours	**191**	**21.39**	—	**10.03**	**2.18**

4.2 Quantitative Results

We compare our method with the state-of-the-art GAN compression methods, NAS [27], CA [29] and CAT [24] on three benchmark datasets to validate the effectiveness of our approach. For a fair comparison, we perform a decent pruning rate by removing around 60% channels of the Vid2Vid model in CA and CAT, and use NAS to find out the best network configuration with similar mMACs.

The experimental results are shown in Table 1. We can see that given the lower computational budget, our method achieves the best FID and FVD on three datasets. Specifically, other GAN compression methods have lower performance than the full-size model while our method outperforms the original model. Other compression methods speed up the original model by simply cutting the network structures, but they ignore the temporal coherence. Meanwhile, the original Vid2Vid model significantly accumulates losses during inference. Our proposed motion-aware inference accumulates less losses since it only generates several frames of the sequence. Such results show the advantage of our spatial-temporal aware compression methods.

4.3 Ablation Study

We adopt Face Video as the benchmark dataset for our ablation study.

Effectiveness of Temporal KD Loss. Based on the spatially low-demand generator mentioned before, we analyze the temporal knowledge distillation for the

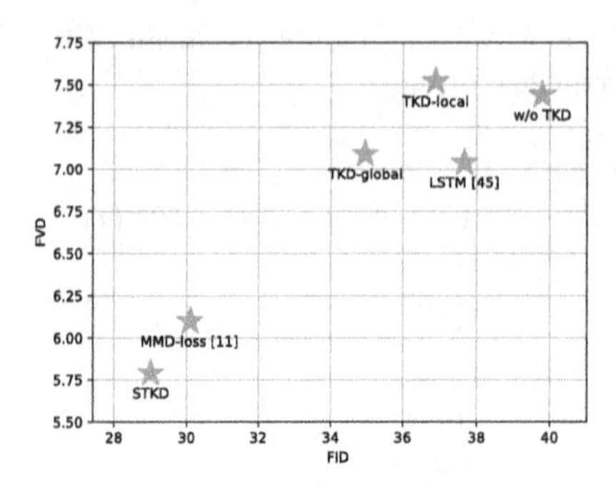

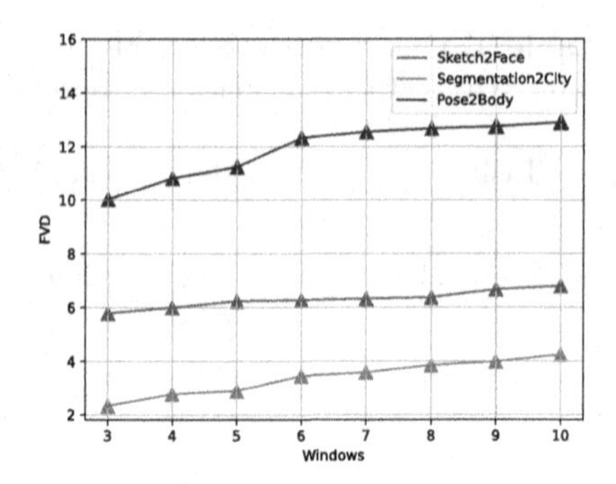

Fig. 5. Ablation study for fast Vid2Vid. **Left:** Temporal Loss ablation study for temporal loss. **Right:** The trade-off experiments for the windows of key-frames selector.

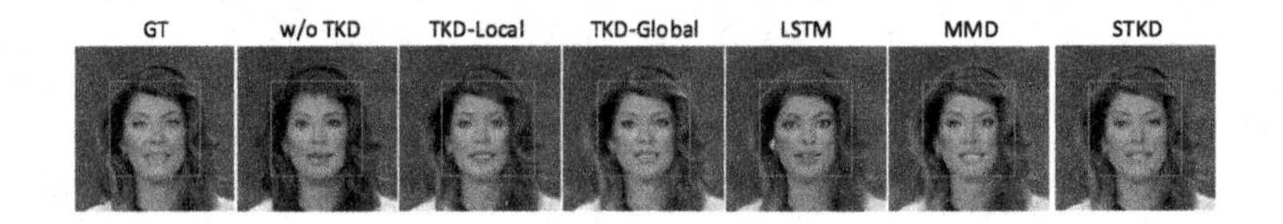

Fig. 6. The comparison of the results of different knowledge distillation.

Vid2Vid model. We set 6 different distillation loss schemes as: (1) w/o TKD: the spatially low-demand generator was retrained on the dataset; (2) TKD-local: the spatially low-demand generator is transferred the only local knowledge from the teacher net; (3) TKD-global: the spatially low-demand generator is transferred the only global knowledge from the teacher net; (4) MMD: the spatially low-demand generator is transferred the knowledge using MMD-loss [9]. (5) LSTM: the spatially low-demand generator is transferred the knowledge based on LSTM regulation [45]. (6) TKD: the spatially low-demand generator is transferred both local and global knowledge from the teacher net.

As shown in the left figure in Fig. 5, the local knowledge distillation loss improves the performance of the model without KD. The temporal KD loss globally further improves the performance of the common local KD loss, especially in FVD. Our proposed KD loss outperforms MMD-loss and LSTM-based KD loss. It indicates that the temporal KD loss effectively enhances the similarity of distribution on the video recognition network between the videos generated by the teacher network and the student network. We also provide the comparison results in Fig. 6 and our STKD generates more realistic frame than others.

Effectiveness of Spatial KD Loss. We conduct an ablative study for Spatial KD on the Sketch2Face benchmark. The spatial compression methods are used together with our proposed Temporal KD to perform Vid2Vid compression. Table 2 shows that our proposed Spatial KD performs better than other image compression methods. Our Spatial KD does not destroy network structures of the original GAN while other methods tweak the sophisticated parameters.

Table 2. Ablation study for spatial compression with the proposed Temporal KD.

Method	MACs(G)	FPS	FID	FVD
CA	331	17.00	36.65	6.76
CAT	310	18.02	35.64	6.85
NAS	344	16.78	32.41	6.71
Spatial KD	**282**	**18.56**	**29.02**	**5.79**

Table 3. Ablation study for the interpolation.

Interpolation	FID	FVD
Linear interpolation	31.55	6.45
Motion compensation	**29.02**	**5.79**

Table 4. Ablation study for the time gap.

Gap	FID	FVD
Fixed time gap	32.21	6.85
Random time gap	31.43	6.22
Key-frames	**29.02**	**5.79**

Effectiveness of Windows for Key-Frame Selection. We investigate the sliding windows to select the key-frames to find out the best trade-off between the sliding windows and the performance. The larger sliding windows mean that there are fewer key-frames selected and thus use less computational resources. Interestingly, as shown in the right figure of Fig. 5, it rises significantly in FVD when increasing the sliding windows, and achieves the best performance when the sliding windows are three in three tasks. It indicates that the part-time student generator needs enough independent motion to maintain decent performance.

Effectiveness of Interpolation. We apply two common interpolation methods for completing the video, namely linear interpolation and motion compensation. We also conduct ablative studies on interpolation methods. As shown in Table 3, motion compensation outperforms the simple linear interpolation. Therefore, we apply motion compensation as our zero-parameters interpolation method.

Effectiveness of Time Gap. We also study the ways of selected frames for the generation of the part-time student generator. We use the same numbers of the selected frames. Specifically, the fixed time gap strategy selects the frames in fixed time intervals, the random time gap strategy chooses the frames in random intervals, and the key-frame strategy selects the key-frames as the frames to be synthesized by the part-time student generator. As shown in Table 4, the key-frame strategy outperforms other strategies since the key-frames of a sequence consist of all essential motions and texture.

4.4 Qualitative Results

We illustrate the output sequences of the mentioned methods in Fig. 7. The generators in Vid2Vid synthesis would accumulate the visual losses sequentially since the current frame relies on the previously generated frames. Compared with other GAN compression methods, our proposed method generates more realistic results. We have provided more details in supplementary materials.

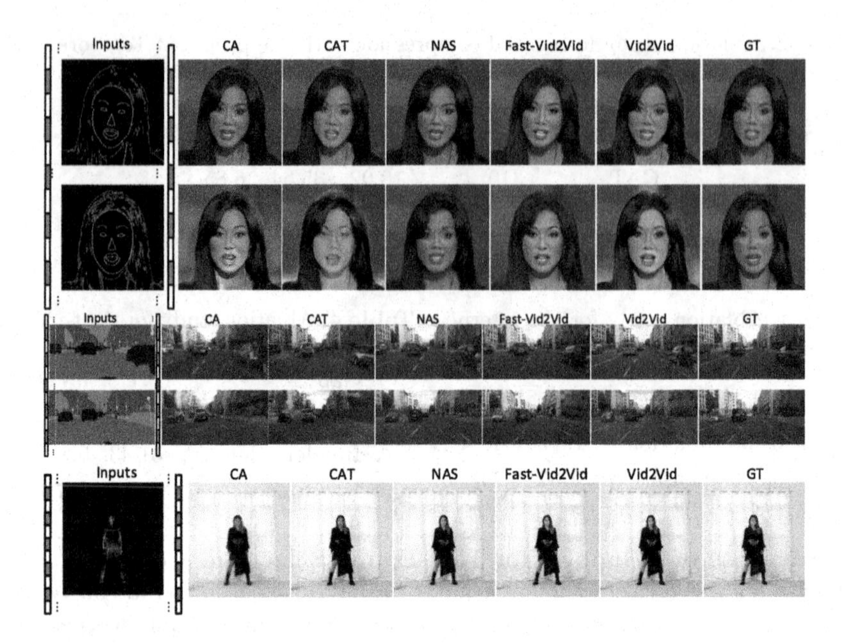

Fig. 7. Qualitative results compared with the advanced GAN compression methods in the task of Sketch2Face, Segmentation2City and Pose2Body.

5 Conclusion

In this paper, we present Fast-Vid2Vid to accelerate Vid2Vid synthesis. We propose a spatial-temporal compression framework to accelerate the inference by compressing the sequential input data stream but maintaining the parameters of the network. In the space dimension, we distill knowledge from the full-resolution domain to the low-resolution domain. In the time dimension, we use temporal-aware knowledge distillation for local and global knowledge to obtain a part-time generator. Finally, the part-time generator is used for motion-aware inference where it only generates the key-frames of the sequence and interpolates the middle frames by motion compensation. By reducing the resolution of the input data and extracting the key-frames of the data stream, Fast-Vid2Vid saves the computational resources significantly.

Discussion. We discuss some future directions for this work. Recently, sequence-in and sequence-out methods, like transformer, are challenging for model compression. On the contrary, our Fast-Vid2Vid accelerates Vid2Vid by optimizing a part-time student generator (via Temporal KD) and a lower-resolution spatial generator (via Spatial KD), which is versatile for various networks. When combined with seq-in and seq-out transformers like visTR [44], Fast-Vid2Vid first synthesizes a partial video by a part-time transformer-based student generator (via fully parallel computation) and then recovers the full video by motion compensation.

Acknowledgements. This work is supported by NTU NAP, MOE AcRF Tier 1 (2021-T1-001-088), and under the RIE2020 Industry Alignment Fund - Industry Collaboration Projects (IAF-ICP) Funding Initiative, as well as cash and in-kind contribution from the industry partner(s).

References

1. Aguinaldo, A., Chiang, P.Y., Gain, A., Patil, A., Pearson, K., Feizi, S.: Compressing GANs using knowledge distillation. arXiv preprint arXiv:1902.00159 (2019)
2. Belousov, S.: MobileStyleGAN: a lightweight convolutional neural network for high-fidelity image synthesis. arXiv preprint arXiv:2104.04767 (2021)
3. Carreira, J., Zisserman, A.: Quo vadis, action recognition? A new model and the kinetics dataset. In: Proceedings of the IEEE Conference on Computer Vision and Pattern Recognition, pp. 6299–6308 (2017)
4. Chan, C., Ginosar, S., Zhou, T., Efros, A.A.: Everybody dance now. In: Proceedings of the IEEE/CVF International Conference on Computer Vision, pp. 5933–5942 (2019)
5. Chen, G., Choi, W., Yu, X., Han, T., Chandraker, M.: Learning efficient object detection models with knowledge distillation. In: Advances in Neural Information Processing Systems 30 (2017)
6. Chen, H., et al.: Distilling portable generative adversarial networks for image translation. In: Proceedings of the AAAI Conference on Artificial Intelligence, vol. 34, pp. 3585–3592 (2020)
7. Chu, M., Xie, Y., Mayer, J., Leal-Taixé, L., Thuerey, N.: Learning temporal coherence via self-supervision for GAN-based video generation. ACM Trans. Graph. (TOG) **39**(4), 75:1 (2020)
8. Cordts, M., et al.: The cityscapes dataset for semantic urban scene understanding. In: Proceedings of the IEEE Conference on Computer Vision and Pattern Recognition, pp. 3213–3223 (2016)
9. Feng, Z., Lai, J., Xie, X.: Resolution-aware knowledge distillation for efficient inference. IEEE Trans. Image Process. **30**, 6985–6996 (2021)
10. Fu, Y., Chen, W., Wang, H., Li, H., Lin, Y., Wang, Z.: AutoGAN-distiller: searching to compress generative adversarial networks. arXiv preprint arXiv:2006.08198 (2020)
11. Gafni, O., Wolf, L., Taigman, Y.: Vid2game: Controllable characters extracted from real-world videos. arXiv preprint arXiv:1904.08379 (2019)
12. Gao, C., Chen, Y., Liu, S., Tan, Z., Yan, S.: AdversarialNAS: adversarial neural architecture search for GANs. In: Proceedings of the IEEE/CVF Conference on Computer Vision and Pattern Recognition, pp. 5680–5689 (2020)
13. Gong, X., Chang, S., Jiang, Y., Wang, Z.: AutoGAN: neural architecture search for generative adversarial networks. In: Proceedings of the IEEE/CVF International Conference on Computer Vision, pp. 3224–3234 (2019)
14. Goodfellow, I., et al.: Generative adversarial nets. In: Advances in Neural Information Processing Systems, vol. 27 (2014)
15. Han, S., Pool, J., Tran, J., Dally, W.J.: Learning both weights and connections for efficient neural networks. arXiv preprint arXiv:1506.02626 (2015)
16. He, K., Zhang, X., Ren, S., Sun, J.: Deep residual learning for image recognition. In: Proceedings of the IEEE Conference on Computer Vision and Pattern Recognition, pp. 770–778 (2016)

17. He, Y., Kang, G., Dong, X., Fu, Y., Yang, Y.: Soft filter pruning for accelerating deep convolutional neural networks. arXiv preprint arXiv:1808.06866 (2018)
18. He, Y., Zhang, X., Sun, J.: Channel pruning for accelerating very deep neural networks. In: Proceedings of the IEEE International Conference on Computer Vision, pp. 1389–1397 (2017)
19. Heusel, M., Ramsauer, H., Unterthiner, T., Nessler, B., Hochreiter, S.: GANs trained by a two time-scale update rule converge to a local nash equilibrium. In: Advances in Neural Information Processing Systems 30 (2017)
20. Hinton, G., Vinyals, O., Dean, J.: Distilling the knowledge in a neural network. arXiv preprint arXiv:1503.02531 (2015)
21. Howard, A., et al.: Searching for MobileNetV3. In: Proceedings of the IEEE/CVF International Conference on Computer Vision, pp. 1314–1324 (2019)
22. Hu, H., Peng, R., Tai, Y., Tang, C., Trimming, N.: A data-driven neuron pruning approach towards efficient deep architectures. arXiv preprint arXiv:1607.03250 46 (2016)
23. Hu, H., Peng, R., Tai, Y.W., Tang, C.K.: Network trimming: a data-driven neuron pruning approach towards efficient deep architectures. arXiv preprint arXiv:1607.03250 (2016)
24. Jin, Q., et al.: Teachers do more than teach: Compressing image-to-image models. In: Proceedings of the IEEE/CVF Conference on Computer Vision and Pattern Recognition, pp. 13600–13611 (2021)
25. Kappel, M., et al.: High-fidelity neural human motion transfer from monocular video. In: Proceedings of the IEEE/CVF Conference on Computer Vision and Pattern Recognition, pp. 1541–1550 (2021)
26. Li, H., Kadav, A., Durdanovic, I., Samet, H., Graf, H.P.: Pruning filters for efficient convnets. arXiv preprint arXiv:1608.08710 (2016)
27. Li, M., Lin, J., Ding, Y., Liu, Z., Zhu, J.Y., Han, S.: GAN compression: efficient architectures for interactive conditional GANs. In: Proceedings of the IEEE/CVF Conference on Computer Vision and Pattern Recognition, pp. 5284–5294 (2020)
28. Lin, J., Zhang, R., Ganz, F., Han, S., Zhu, J.Y.: Anycost GANs for interactive image synthesis and editing. In: Proceedings of the IEEE/CVF Conference on Computer Vision and Pattern Recognition, pp. 14986–14996 (2021)
29. Liu, Y., Shu, Z., Li, Y., Lin, Z., Perazzi, F., Kung, S.Y.: Content-aware GAN compression. In: Proceedings of the IEEE/CVF Conference on Computer Vision and Pattern Recognition, pp. 12156–12166 (2021)
30. Liu, Z., Li, J., Shen, Z., Huang, G., Yan, S., Zhang, C.: Learning efficient convolutional networks through network slimming. In: Proceedings of the IEEE International Conference on Computer Vision, pp. 2736–2744 (2017)
31. Lopez-Paz, D., Bottou, L., Schölkopf, B., Vapnik, V.: Unifying distillation and privileged information. arXiv preprint arXiv:1511.03643 (2015)
32. Luo, P., Zhu, Z., Liu, Z., Wang, X., Tang, X.: Face model compression by distilling knowledge from neurons. In: Thirtieth AAAI Conference on Artificial Intelligence (2016)
33. Orchard, M.T., Sullivan, G.J.: Overlapped block motion compensation: an estimation-theoretic approach. IEEE Trans. Image Process. **3**(5), 693–699 (1994)
34. Rössler, A., Cozzolino, D., Verdoliva, L., Riess, C., Thies, J., Nießner, M.: FaceForensics: a large-scale video dataset for forgery detection in human faces. arXiv preprint arXiv:1803.09179 (2018)
35. Sajjadi, M.S., Vemulapalli, R., Brown, M.: Frame-recurrent video super-resolution. In: Proceedings of the IEEE Conference on Computer Vision and Pattern Recognition, pp. 6626–6634 (2018)

36. Siarohin, A., Lathuilière, S., Tulyakov, S., Ricci, E., Sebe, N.: First order motion model for image animation. In: Advances in Neural Information Processing Systems 32, pp. 7137–7147 (2019)
37. Siarohin, A., Woodford, O.J., Ren, J., Chai, M., Tulyakov, S.: Motion representations for articulated animation. In: Proceedings of the IEEE/CVF Conference on Computer Vision and Pattern Recognition, pp. 13653–13662 (2021)
38. Tourapis, A.M.: Enhanced predictive zonal search for single and multiple frame motion estimation. In: Visual Communications and Image Processing, vol. 4671, pp. 1069–1079. SPIE (2002)
39. Unterthiner, T., van Steenkiste, S., Kurach, K., Marinier, R., Michalski, M., Gelly, S.: Towards accurate generative models of video: a new metric & challenges. arXiv preprint arXiv:1812.01717 (2018)
40. Wang, H., Gui, S., Yang, H., Liu, J., Wang, Z.: GAN slimming: all-in-one GAN compression by a unified optimization framework. In: Vedaldi, A., Bischof, H., Brox, T., Frahm, J.-M. (eds.) ECCV 2020. LNCS, vol. 12349, pp. 54–73. Springer, Cham (2020). https://doi.org/10.1007/978-3-030-58548-8_4
41. Wang, T.C., Liu, M.Y., Tao, A., Liu, G., Kautz, J., Catanzaro, B.: Few-shot video-to-video synthesis. arXiv preprint arXiv:1910.12713 (2019)
42. Wang, T.C., et al.: Video-to-video synthesis. arXiv preprint arXiv:1808.06601 (2018)
43. Wang, T.C., Liu, M.Y., Zhu, J.Y., Tao, A., Kautz, J., Catanzaro, B.: High-resolution image synthesis and semantic manipulation with conditional GANs. In: Proceedings of the IEEE Conference on Computer Vision and Pattern Recognition (2018)
44. Wang, Y., et al.: End-to-end video instance segmentation with transformers. In: Proceedings of the IEEE/CVF Conference on Computer Vision and Pattern Recognition, pp. 8741–8750 (2021)
45. Xiao, Z., Fu, X., Huang, J., Cheng, Z., Xiong, Z.: Space-time distillation for video super-resolution. In: Proceedings of the IEEE/CVF Conference on Computer Vision and Pattern Recognition, pp. 2113–2122 (2021)
46. Xu, G., Xu, J., Li, Z., Wang, L., Sun, X., Cheng, M.M.: Temporal modulation network for controllable space-time video super-resolution. In: Proceedings of the IEEE/CVF Conference on Computer Vision and Pattern Recognition, pp. 6388–6397 (2021)
47. Xu, R., Li, X., Zhou, B., Loy, C.C.: Deep flow-guided video inpainting. In: Proceedings of the IEEE/CVF Conference on Computer Vision and Pattern Recognition, pp. 3723–3732 (2019)
48. Yim, J., Joo, D., Bae, J., Kim, J.: A gift from knowledge distillation: fast optimization, network minimization and transfer learning. In: Proceedings of the IEEE Conference on Computer Vision and Pattern Recognition, pp. 4133–4141 (2017)
49. Yu, C., Pool, J.: Self-supervised GAN compression. arXiv preprint arXiv:2007.01491 (2020)
50. Zhang, T., et al.: A systematic DNN weight pruning framework using alternating direction method of multipliers. In: Ferrari, V., Hebert, M., Sminchisescu, C., Weiss, Y. (eds.) ECCV 2018. LNCS, vol. 11212, pp. 191–207. Springer, Cham (2018). https://doi.org/10.1007/978-3-030-01237-3_12
51. Zhou, Y., Wang, Z., Fang, C., Bui, T., Berg, T.: Dance dance generation: motion transfer for internet videos. In: Proceedings of the IEEE/CVF International Conference on Computer Vision Workshops (2019)
52. Zou, X., Yang, L., Liu, D., Lee, Y.J.: Progressive temporal feature alignment network for video inpainting. In: Proceedings of the IEEE/CVF Conference on Computer Vision and Pattern Recognition, pp. 16448–16457 (2021)

Learning Prior Feature and Attention Enhanced Image Inpainting

Chenjie Cao, Qiaole Dong, and Yanwei Fu(✉)

School of Data Science, Fudan University, Shanghai, China
{20110980001,qldong18,yanweifu}@fudan.edu.cn

Abstract. Many recent inpainting works have achieved impressive results by leveraging Deep Neural Networks (DNNs) to model various prior information for image restoration. Unfortunately, the performance of these methods is largely limited by the representation ability of vanilla Convolutional Neural Networks (CNNs) backbones. On the other hand, Vision Transformers (ViT) with self-supervised pre-training have shown great potential for many visual recognition and object detection tasks. A natural question is whether the inpainting task can be greatly benefited from the ViT backbone? However, it is nontrivial to directly replace the new backbones in inpainting networks, as the inpainting is an inverse problem fundamentally different from the recognition tasks. To this end, this paper incorporates the pre-training based Masked AutoEncoder (MAE) into the inpainting model, which enjoys richer informative priors to enhance the inpainting process. Moreover, we propose to use attention priors from MAE to make the inpainting model learn more long-distance dependencies between masked and unmasked regions. Sufficient ablations have been discussed about the inpainting and the self-supervised pre-training models in this paper. Besides, experiments on both Places2 and FFHQ demonstrate the effectiveness of our proposed model. Codes and pre-trained models are released in https://github.com/ewrfcas/MAE-FAR.

Keywords: Image inpainting · Attention · Vision transformer

1 Introduction

Image inpainting aims to fill missing regions of images with semantically consistent and harmoniously textured contents. It has a wide range of practical applications, including object removal [11], image editing [20], and so on. Conventional inpainting algorithms [3,7,15,25,32] rely on visual low-level assumptions and structures to search similar patches for the reconstruction. But they fail to tackle complicated inpainting situations with the limited feature representation.

C. Cao and Q. Dong—Have contributed equally.

Supplementary Information The online version contains supplementary material available at https://doi.org/10.1007/978-3-031-19784-0_18.

S. Avidan et al. (Eds.): ECCV 2022, LNCS 13675, pp. 306–322, 2022.
https://doi.org/10.1007/978-3-031-19784-0_18

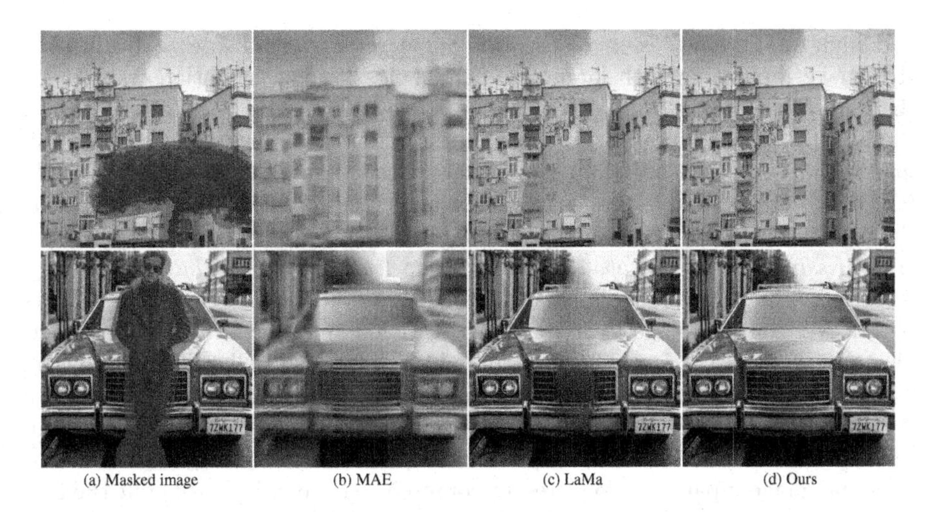

Fig. 1. The comparison of 1024 × 1024 high-resolution image inpainting. From left to right are masked images, results of pre-trained Masked AutoEncoder (MAE) [16], results of LaMa [34], and results from our method.

Deep Neural Networks (DNNs) have recently been exploited by researchers to achieve prominent improvements in image inpainting [5,26,28,34,36,49], which thanks to the great capabilities of Convolutional Neural Networks (CNNs) [23], Generative Adversarial Networks (GAN) [12], and attention-based Transformers [35]. However, repairing images corrupted by arbitrary masks with reasonable results is still challenging, especially in high-resolution cases. Because the inpainting model needs to understand the semantic information from masked images, which demands data-driven priors and sufficient model capacities. Furthermore, the dilemmas shown below should be solved (Fig. 1).

(i) *Limited capacities for modeling good priors.* Many pioneering works have tried to introduce prior information to inpainting models. Some works [5,28,33,36,44,45] propose multi-stage models, which repair various auxiliary information and corrupted images sequentially to enhance the image inpainting. These methods learn priors in specific fields, such as structures [5,28] or semantics [33] with good visual interpretability rather than features with more informative priors. Other methods [26,40] leverage auxiliary losses to introduce additional prior information without extending sufficient model capacities. Complex loss functions cause sophisticated hyper-parameter tuning and a more difficult inpainting training. Besides, some transformer based inpainting methods [36,45] heavily depend on low-resolution (LR) images generated by the time-consuming iterative sampling, and then upsample them with CNNs. Lahiri *et al.* [24] learn global latent priors with GAN, which can only solve simple scenes with single-object. Our method incorporates effective prior features from the transformer based representation learning [16] to enhance the inpainting, which make our method achieve superior results without overfitting the transformer results.

(ii) *Informative priors for high-resolution cases.* High-resolution (HR) image inpainting enjoys more practical implications with advanced electronic products

and high-quality images in real-world. Some researches devote to facilitating HR image inpainting with larger receptive fields [34,46], attention transfer for the high-frequency residual [41], and two-stage upsampling [47]. Unfortunately, these methods still tend to copy meaningless existing textures rather than really *understand* semantics of HR masked images without directly training with costly HR data. Our method leverages the continuous positional encoding to upsample the prior features for superior inpainting performance in HR images.
(iii) *Missing discussions about representation learning for inpainting.* Recently, self-supervised pre-training language models [4,9,30,31] have achieved great success in Natural Language Processing (NLP) fields. Such *masking and predicting* idea and transformer based architectures have been also well explored in vision tasks [2,16,39,51]. But these vision transformers only consider representation learning for classification tasks. To the best of our knowledge, no one has explored the application of self-supervised pre-training vision models to generative tasks, let alone image inpainting. We present comprehensive discussions about the pre-training based representation learning for image inpainting in this paper.

To address these dilemmas, we propose to guide the image inpainting with an efficient Masked AutoEncoder (MAE) pre-training model [16], which is called as prior Feature and Attention enhanced Restoration (FAR). Specifically, an MAE is firstly pre-trained with the masked visual prediction task. We replace some random masks with large and contiguous masks to make the MAE more suitable for the downstream task. Then, features from the MAE decoder are added to the inpainting CNN for the prior guided image inpainting. Moreover, we find that attention relations of MAE among masked and unmasked regions are compatible with CNN inpainting learning. So group convolutions are used to aggregate CNN features with attention scores from MAE, which can improve the inpainting performance a lot. Furthermore, our model can be effectively extended to HR inpainting with a little finetuning of bilinear resized MAE features and the Cartesian spatial grid. Besides, we discuss some pre-training and finetuning tricks to better utilize MAE for superior inpainting performance.

We highlight our contributions as follows. (1) We propose to learn image priors from pre-trained MAE features, which contain informative high-level knowledge and strengthen the inpainting model. (2) We propose to aggregate the inpainting CNN feature with attention scores from MAE to improve the performance. (3) Several pre-training and finetuning tricks are exploited to make our FAR learn better prior features and attentions from MAE. (4) Our method can be simply extended to HR cases and achieve state-of-the-art results. Extensive experiments on Places2 [50] and FFHQ [21] reveal that our proposed model performs better than other competitors.

2 Related Work

Image Inpainting with Auxiliaries. Inpainting with auxiliaries has demonstrated success in many previous works. Various priors have been leveraged to enhance the inpainting, such as edges [14,28], lines [5], gradients [40], segmentations [26,33], low-resolution images [36,45], and even latent priors [24]. Specifically, these methods can be categorized into two types. One is to employ the approach of first correcting auxiliary information and then guiding image inpainting with multi-stage models [5,24,28,33,36,44,45]. Since these methods enjoy superior performance and good interpretability, priors leveraged by them are not comprehensive enough to handle the image inpainting properly. Other methods [26,40] supervise the inpainting model with auxiliary information directly for introducing more positive priors. But capacities of these models are still stuck to tackle tough corrupted cases. Although work [14] combines both advantages mentioned above with the dual structure and texture learning, such low-level features are still insufficient to achieve results with rich semantics.

High-Resolution Image Inpainting. HR image inpainting with large mask areas is still challenging. In [47], Yu *et al.* propose an iterative inpainting method with a feedback mechanism to progressively fill holes. They further learn an upsampling network to handle HR inpainting results based on LR ones. Yi *et al.* [41] design a contextual residual aggregation mechanism for the restoration of high-frequency residuals, which are added to LR predictions. Besides, various dilated convolutions are used in [46] to enlarge receptive fields for HR inpainting. Furthermore, Suvorov *et al.* [34] leverage Fast Fourier Convolution (FFC) to learn a global receptive field in the frequency and achieve amazing HR inpainting results with periodic textures. However, these methods still suffer from copying meaningless textures without really 'understanding' semantics in HR images. In contrast to previous methods, we transfer prior features from masked autoencoders to HR cases with a continuous positional encoding, which greatly improves the quality of HR inpainting results with meaningful semantic priors.

Masked Visual Prediction. Masked Visual Prediction (MVP) is a self-supervised task for representation learning by masking and predicting image patches. This work is originated in the masked language model [9] of NLP. The Vision Transformer (ViT) [10] has studied self-supervised pre-training by masking patches. BEiT [2] and iBOT [51] learn MVP on high-level discrete tokens and self-distillation respectively. Moreover, MAE [16] proposes an efficient transformer-based masked autoencoder for visual representation learning. MaskFeat [39] further studies MVP, and proposes to use HOG features [8] to get excellent results efficiently. These MVP pre-training models can be finetuned to achieve excellent classification results. However, few discussions about MVP are explored for image generation. To the best of our knowledge, our work firstly studies MVP-based pre-training for image inpainting.

3 Method

Overview. The overall pipeline of our FAR is shown in Fig. 2. For the given masked image $\mathbf{I}_m$, we resize it into 256×256 and further enlarge the mask to patch-wise of 16×16. Thus we can get the masked image $\mathbf{I}'_m \in \mathbb{R}^{256 \times 256}$. Then the MAE is applied to encode the prior features as $\mathbf{F}_p = \mathrm{MAE}(\mathbf{I}_p^{(i)}), i \in \{1, 2, ..., N\}$, where $\mathbf{I}_p^{(i)}$ indicate total N unmasked patches from $\mathbf{I}'_m$ (Sect. 3.1).

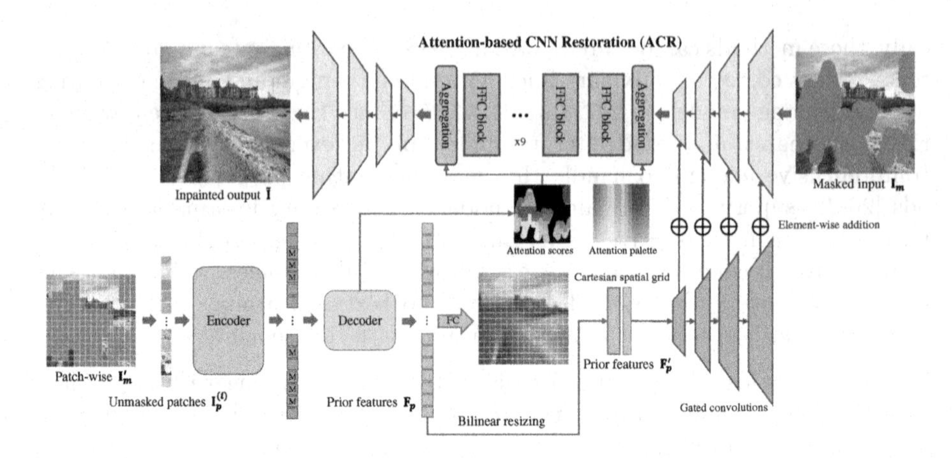

Fig. 2. The overview of our proposed FAR.

Prior features are resized to 1/8 of the original image size and concatenated with the Cartesian spatial grid as $\mathbf{F}'_p$. After that, $\mathbf{F}'_p$ is encoded by gated convolutions and added to the encoder of Attention-based CNN Restoration (ACR), which is used to restore the original masked image $\mathbf{I}_m$ (Sect. 3.2). Moreover, we leverage mean attention scores from the MAE decoder to aggregate unmasked features to masked regions in ACR and achieve the final inpainted image $\tilde{\mathbf{I}}$ (Sect. 3.2).

In this section, to better discuss the influence of pre-trained MAE on the inpainting model, we provide ablation studies on the subset of Places2 [50] with 5 scenes (about 25,000 training images, and 500 validation images, detailed in the Appendix). All methods are trained with 150k steps in 256×256. Although our MAE is pre-trained on the total Places2 training set, ablations among all MAE enhanced methods are still fair and meaningful. For the 512×512 ablations, we finetune the 256×256 model trained on the whole places2 for 150k steps with the dynamic resizing (Sect. 4) and test them on 1,000 validation images.

3.1 Masked Autoencoder for Inpainting

Training Settings. We use ViT-Base [10] as the backbone of MAE, which contains 12 encoder layers and 8 decoder layers. Although He *et al.* [16] have released pre-trained MAEs based on ImageNet-1K with random masks, there are still some domain gaps for the inpainting. Instead, we pre-train the MAE on the whole 1.8 million Places2 [50] and 68,000 FFHQ [21] for scene and face inpainting respectively. Validations of both datasets are excepted from the training set for a fair comparison. Moreover, the random mask used in the standard MAE is not amenable to inpainting. Although the masking ratio is high (75%), such noisy masks are easier to be restored by DNNs compared with continuous and large masks with even lower masking ratios [29]. Therefore, we blend continuous masks with 10% to 50% masking ratios and random masks, while the total masking rate remains at 75% as shown in Fig. 3. Specifically, both irregular and segmentation masks [5] are considered in continuous masks. Then we downsample continuous masks to 16×16 patch-wise forms with *max pooling* to ensure all masked patches are set to 1, while unmasked ones are 0, which enlarges masked regions. The mixed masking strategy can improve the learning of prior attention as shown in Table 4 of Sect. 3.2. Other training details are discussed in Sect. 4.

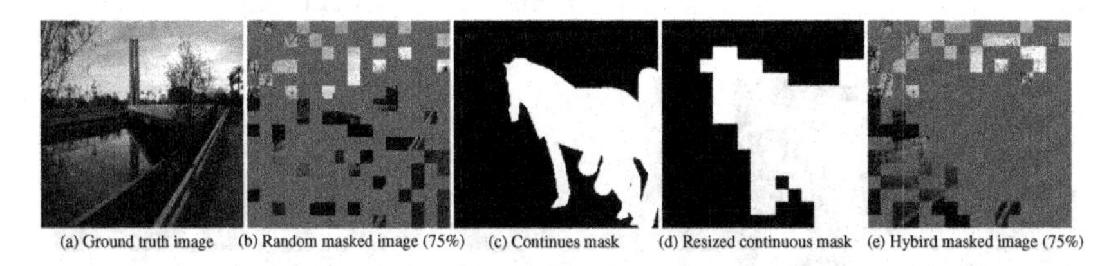

Fig. 3. The illustration for masking strategies. In (c), the continuous mask is combined with an irregular mask and a segmentation mask. We upsample the enlarged continues mask (16×16) in (d) for a better visualization.

Prior Features from MAE. In MAE [16], only 25% unmasked patches are applied into the encoder, while learnable masked tokens are shared in the decoder for the reconstruction. These masked tokens will be used to predict pixel values in masked regions. This trick makes the encoder enjoy much lower memory cost, and it can be pre-trained with more capacities for better performance in classification tasks. However, during the inpainting, features from the MAE encoder are insufficient. Because we should achieve masked features encoded by masked tokens with stacked attention modules. Therefore, features from decoder layers are more compatible with the inpainting task. To ensure good decoder learning, we chose to use balanced encoder-decoder ViT-Base architecture, which contains 12 encoder layers and 8 decoder layers rather than further enlarging the encoder capacity. In this work, we choose to use features from the last layer of the MAE decoder before the pixel linear projection as the prior features. Since the predicted images are blurry, which may contain limited information, especially for the HR inpainting. Thus an interesting future work would be exploring features from different transformer layers for inpainting.

Finetuning for Partially Masked Patches. Since the input of MAE is patch-wise tokens in 16×16, we have to enlarge some partially masked patches as shown in Fig. 3(c) (d), which may lose information. Intuitively, we further finetune MAE for those partially masked patches with 50 epochs in Places2. Specifically, masked embeddings of these partially masked patches are re-encoded by a new initialized linear layer of decoder. Inputs of these partially masked patches are composed of the concatenation of RGB pixel values (masked pixels are all 0 in 3 channels) and 0–1 masking maps. From the ablations in Table 1, the model trained with finetuned MAE performs slightly worse than the original one in FID. As shown in Fig. 4, we think that the finetuned MAE learns more explicit results, which makes the CNN restoration overfit MAE features. So these enlarged masked regions increase training difficulty and can be seen as noise regularization.

Table 1. Ablations of our full model enhanced by MAE with/without finetuning for partially masked patches (Partial F.T.).

Partial F.T	PSNR↑	SSIM↑	FID↓	LPIPS↓
×	24.51	**0.864**	**25.49**	0.113
✓	**24.67**	**0.864**	26.00	**0.111**

Fig. 4. Qualitative results of our full model enhanced by MAE with and without finetuning for partially masked patches.

3.2 Attention-Based CNN Restoration (ACR)

The design of CNN modules in ACR is referred to LaMa [34], which is an encoder-decoder model including 4 downsampling convolutions, 9 Fast Fourier Convolution (FFC) blocks, and 4 upsampling convolutions. FFC has been demonstrated that it can tackle some HR inpainting cases with strong periodic textures. We further enhance ACR with prior features and attentions from MAE as follows.

Prior Features Upsampling. To overcome inpainting tasks with arbitrary resolutions, the local feature ensemble [6] is leveraged to facilitate the feature warping from MAE to ACR in various resolutions. Given 16×16 patch-wise prior features $\mathbf{F}_p \in \mathbb{R}^{16\times16\times d}$ with dimension d from the MAE decoder, we resize them into 1/8 of the original image size with the bilinear interpolation. To indicate the continuous position in HR, the Cartesian spatial grid [19] *i.e.*, normalized 2d coordinates is concated to resized features as

$$\mathbf{F}'_p = \mathrm{Concat}(\mathrm{BilinearResize}(\mathbf{F}_p), \mathbf{C}) \in \mathbb{R}^{\frac{h}{8}\times\frac{w}{8}\times(d+2)}, \quad (1)$$

where h, w represent the original image height and width respectively; $\mathbf{C} \in \mathbb{R}^{\frac{h}{8}\times\frac{w}{8}\times 2}$ means normalized 2d coordinates grid valued from -1 to 1. As shown in the ablations of Table 2, such positional information can improve the HR inpainting results, and is complementary to learn smooth feature representations.

Prior Features Combination. Four gated deconvolutions [43] are leveraged to upsample $\mathbf{F}'_p$. Then these upsampled features are applied to the encoder of ACR. The gated mechanism works for making ACR filter corrupted features adaptively. To integrate prior features to ACR, the general solution is to element-wise add upsampled prior features to the downsampled ones of ACR. However,

Table 2. Ablations of models finetuned in 512 × 512 Places2 and tested in 1,000 images.

2D-coordinate	Norm-pixel	PSNR↑	SSIM↑	FID↓	LPIPS↓
×	✓	24.35	0.879	25.47	0.135
✓	✓	24.31	**0.880**	25.33	0.135
✓	×	**24.37**	**0.880**	**25.30**	**0.132**

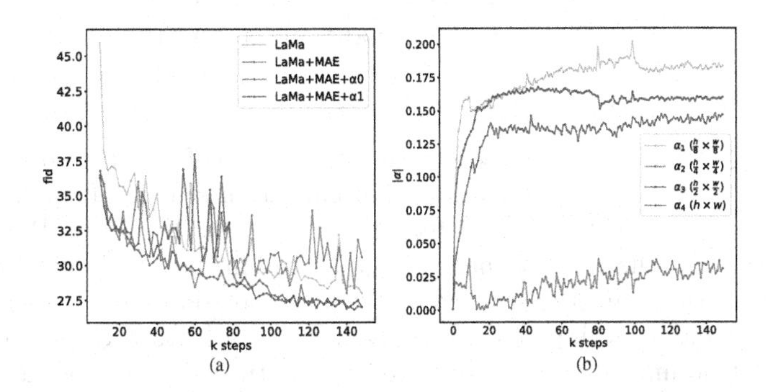

Fig. 5. (a) Ablations of various prior features combination methods. LaMa [34] is the baseline. 'LaMa+MAE' means directly adding MAE prior features to the ACR encoder. '+α0' and '+α1' indicate that prior features are multiplied with 0 and 1 initialized learnable-parameter α before the addition. (b) The line chart of $\alpha_j, j \in \{1,2,3,4\}$ shows absolute values for ACR features with different resolutions.

we observed an unstable training process with the vanilla element-wise addition as shown in the ablations of Fig. 5(a). We try to multiply trainable parameters $\alpha_j, j \in \{1,2,3,4\}$ to the prior features for the element-wise addition to ACR encoder features in $(\frac{h}{8}\times\frac{w}{8})$, $(\frac{h}{4}\times\frac{w}{4})$, $(\frac{h}{2}\times\frac{w}{2})$, and $(h\times w)$ respectively. Moreover, ablation studies of α_j initialized with 0 and 1 are shown in Fig. 5(a) and Table 3. The zero initialization enjoys a much more stable convergence. Therefore, the 0-initialized addition is adopted in our model, which is omitted in the follows for simplicity. We further analyze tendencies of different α_j trained with MAE layer 8 (*i.e.*, row 3 of Table 3) in Fig. 5(b). From Fig. 5(b), both high-level $(\frac{h}{8}\times\frac{w}{8})$ and low-level $(h\times w, \frac{h}{2}\times\frac{w}{2})$ features are important for the prior learning, since absolute values of $\alpha_1, \alpha_3, \alpha_4$ are large. And features in $\frac{h}{4}\times\frac{w}{4}$ seem have less effect during the inpainting.

Prior Attentions. Many inpainting researches show that the attention mechanism is useful for the image inpainting [41–43,52]. The classical contextual attention [42] used in inpainting is to aggregate masked features $\mathbf{F}_m$ with unmasked ones $\mathbf{F}_u$. The aggregation is based on the attention score $\mathbf{R}_{u,m}$ as

$$\begin{aligned} \cos_{u,m} &= \left\langle \frac{\mathbf{F}_u}{||\mathbf{F}_u||}, \frac{\mathbf{F}_m}{||\mathbf{F}_m||} \right\rangle \\ \mathbf{R}_{u,m} &= \mathrm{softmax}_u(\cos_{u,m}), \end{aligned} \tag{2}$$

Table 3. Ablations of models enhanced with different initialized parameters α. Column 'MAE' means that whether to use prior features from MAE. Column 'init-α' indicates initialized values of learnable parameter α for prior feature combination.

MAE	init-α	PSNR↑	SSIM↑	FID↓	LPIPS↓
×	No	24.12	0.859	28.01	0.124
✓	No	24.14	0.860	28.01	0.127
✓	0	**24.34**	0.860	**26.83**	0.117
✓	1	24.27	**0.861**	26.90	**0.115**

where $cos_{u,m}$ indicates normalized cosine similarities of masked and unmasked features; softmax$_u$ means the softmax normalization among all unmasked features. Then we can get the aggregated masked features $\mathbf{F}'_m = \sum_u \mathbf{R}_{u,m}\mathbf{F}_u$.

Unfortunately, the improvement of the attention module is not orthogonal to other effective inpainting strategies. We add two contextual attention (CA) modules [42] to ACR in the same positions as the prior feature aggregation shown in Fig. 2. But no improvement is achieved by the trainable contextual attention as shown in Table 4. We think that the restricted improvement is caused by the limited capacity of CNN models. Therefore, we try to use attention relations from the decoder of MAE to overcome the limited capacity. For the decoder layer l of MAE, we can get attention scores $\mathbf{R}^{(l)}_{u,m}$ as

$$\mathbf{R}^{(l)}_{u,m} = \text{softmax}(\frac{\mathbf{Q}^{(l)}\mathbf{K}^{(l)^T}}{\sqrt{d}} - inf \cdot \mathbf{M}), \tag{3}$$

where $\mathbf{Q}^{(l)}, \mathbf{K}^{(l)}$ mean query and key of the attention; d is the channels; and $\mathbf{M}$ indicates a 0–1 mask map, where 1 means masked regions. We make the scores of masked regions to 0 in Eq. (3) that means all masked regions should only pay attention to unmasked ones. Then we average values of total 8 decoder attention scores, and get the prior attention scores $\mathbf{R}_p$ as

$$\mathbf{R}_p = \frac{\sum_{l=1}^{L} \mathbf{R}^{(l)}_{u,m}}{L}, L = 8. \tag{4}$$

The aggregations for getting masked features $\mathbf{F}'_m$ in ACR are executed in the start and the end of FFC blocks as shown in Fig. 2. Then $\mathbf{F}'_m$ is added to the original features as the residual. Note that we also multiply a zero-initialized learnable parameter to the $\mathbf{F}'_m$ before the addition instead of using LayerNorm as discussed in [1]. Besides, from Table 4, random masking used in the vanilla MAE fails to learn proper attention relations compared with mixed one discussed in Sect. 3.1. The ablation study about different attention layers from MAE are discussed in the Appendix.

Table 4. Ablations of models with different attention and MAE masking strategies; 'mixed' mask type means MAE pre-trained with both random, enlarged irregular and segmentation masks; 'prior attention' means using prior attention aggregation from MAE; 'trainable CA' indicates contextual attention [42].

MAE mask type	Attention type	PSNR↑	SSIM↑	FID↓	LPIPS↓
Mixed	No attention	24.34	0.860	26.84	0.117
Mixed	Trainable CA	24.13	0.859	26.99	0.123
Random	Prior attention	24.39	0.861	26.25	0.117
Mixed	Prior attention	**24.51**	**0.864**	**25.49**	**0.113**

3.3 Loss Functions

Loss Functions of MAE. Our MAE loss is the mean squared error (MSE) between MAE predictions and ground truth pixel values for masked patches as in [16]. Besides, He *et al.* study to use normalized pixel values of each masked patch as the self-supervised target, which normalizes each masked patch with the mean and standard of this patch. Such a trick can improve the classification quality in their experiments. However, for the inpainting, the global relation among different patches is also important. The patch-wise normalization makes MAE learn more bias for each patch rather than the global information. Thus we study the MAE pre-training ablations with and without the patch-wise normalization as shown in Table 2. We find that non-normalized targets can achieve slightly better quality in HR inpainting.

Loss Functions of ACR. ACR is trained as a regular adversarial inpainting model. We adopt the same loss functions as LaMa [34], which include L1 loss, adversarial loss, feature match loss, and high receptive field (HRF) perceptual loss. Specifically, L1 loss is only used to constrain unmasked regions. The discriminator loss is based on PatchGAN [18]. And WGAN-GP [13] is used as the generator loss. The feature match loss [37] based on L1 loss between true and fake discriminator features is also used to stable the GAN training. Furthermore, we also leverage the segmentation pre-trained ResNet50 for the HRF loss as proposed in [34] to improve the inpainting quality. More details about the loss functions of ACR are in the Appendix.

4 Experiments

Datasets. Our model is trained on Places2 [50] and FFHQ [21]. For Places2, both the pre-training of MAE and training of our full model are based on the whole 1.8 million training images, and tested with 36,500 testing images. We prepare additional 1,000 testing images for 512×512 experiment. Detailed settings about main experiments and ablations are in the Appendix. We pre-train another MAE on 68,000 training set of FFHQ, which is also used to train the

Table 5. Left: Results on 256 × 256 FFHQ and Places2 with mixed masks compared among Co-Mod (Co.) [49], LaMa (La.) [34], EC [28], CTSDG (CT.) [14] and ours. **Right:** Results on 512 × 512 Places2 testset with mixed masks compared among HiFill (Hi.) [41], Co-Mod (Co.) [49], LaMa (La.) [34] and ours. Metrics are PSNR (P.), SSIM (S.), FID (F.) and LPIPS (L.).

	FFHQ			Places2				
	Co.	La.	Ours	EC	Co.	La.	CT.	Ours
P.↑	25.2	26.6	**26.8**	23.3	22.5	24.3	23.4	**24.5**
S.↑	.889	.903	**.906**	.839	.843	.869	.835	**.871**
F.↓	**5.85**	6.38	6.15	6.21	1.49	1.63	11.2	**1.31**
L.↓	.085	.078	**.074**	.149	.246	.155	.185	**.101**

	P.↑	S.↑	F.↓	L.↓
Hi.	20.1	.764	65.4	.291
Co.	22.0	.843	30.0	.166
La.	24.1	.877	27.8	.149
Ours	**24.3**	**.880**	**25.3**	**.119**

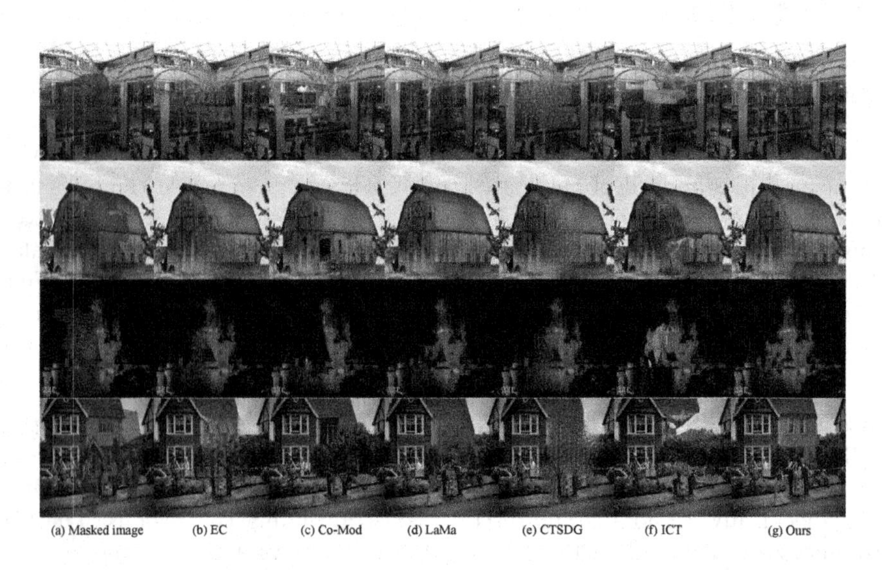

Fig. 6. Qualitative results of places2 256 × 256 images. From left to right are masked image, EC [28], Co-Mod [49], LaMa [34], CTSDG [14], ICT [36], and our results.

Fig. 7. Qualitative (A) 256 × 256 FFHQ, and (B) 1024 × 1024 results from network.

face inpainting model. And other 2,000 images work for testing. Our pre-trained MAE on Places2 and FFHQ can be well generalized to most real-world cases.

Implementation Detials. Our FAR is implemented with Pytorch based on two 48 GB NVIDIA RTX A6000 gpus. MAE is pre-trained on Places2 and FFHQ for 200 and 450 epochs respectively with batch size 512, while other settings follow the released codes[1] except the masking strategy and the partially masked finetuning. For ACR, we employ the Adam optimizer with $\beta_1 = 0.9$ and $\beta_2 = 0.999$, the learning rates are 1e−3 and 1e−4 for the generator and the discriminator. We train our models with 850k steps on Places2 and 150k steps on FFHQ. For every 200k steps on Places2 and 100k steps on FFHQ, the learning rate is reduced by half. To save the computation, we finetune our model on images with higher resolutions, which are dynamically resized from 256 to 512 for 150k steps on Places2.

Masks Settings. To solve real-world application problems, we adopt the masking strategy in [5]. The masks consist of irregular brush masks and COCO [27] segmentation masks ranged from 10% to 50%. During the training, we combine these two types masks in 20%.

Comparison Methods. We compared our model with other state-of-the-art models including Edge Connect (EC) [28], Co-Modulation GAN (Co-Mod) [49], Large Mask inpainting (LaMa) [34], Conditional Texture and Structure Dual Generation (CTSDG) [14] and Image Completion with Transformers (ICT) [36] using the official pre-trained models. We also retrain LaMa on Places2 and FFHQ, and further finetune it with the same settings as ours in the HR inpainting for a fair comparison.

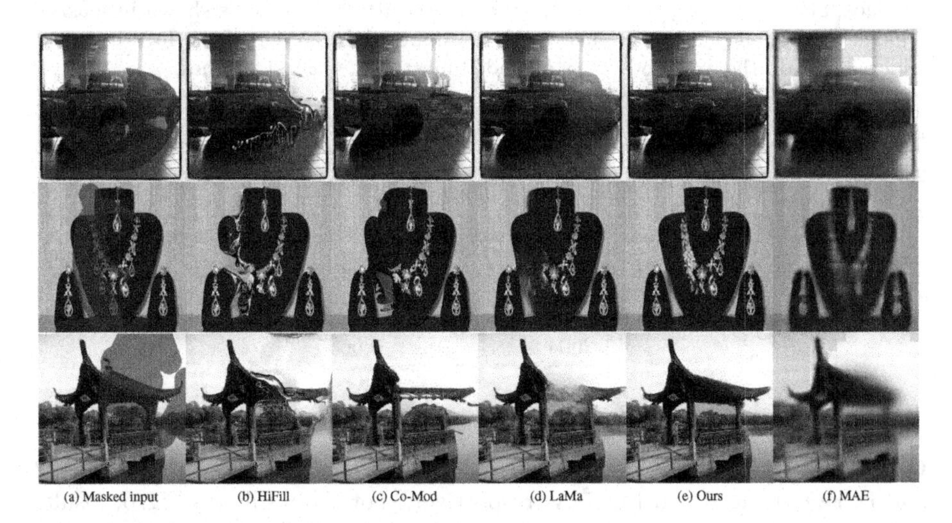

Fig. 8. Qualitative results of places2 512×512 images. From left to right are masked image, HiFill [41], Co-Mod [49], LaMa [34], our results, and MAE predictions.

[1] https://github.com/facebookresearch/mae.

Quantitative Comparisons. We evaluate PSNR, SSIM [38], FID [17], and LPIPS [48] for both 256×256 and 512×512 in Table 5. For 256×256 results shown in the left of Table 5, our method achieves significant improvements based on the LaMa baseline. Moreover, Co-Mod results are also competitive. For the FFHQ results, since our MAE enhanced method can achieve much more stable and faithful results for face images, PSNR and SSIM of our method is better than Co-Mod. The powerful stylegan [22] architecture helps Co-Mod learn better textures with good FID, but our results have superior human perception (lower LPIPS). For the Places2, our results achieve best results in all metrics.

For the HR 512×512 results listed in the right of Table 5, our method can outperform all other competitors. Note that most methods fail to tackle HR inpainting tasks in complex scenes. Results of baseline method LaMa are competitive, but our methods still achieve certain advantages compared with LaMa.

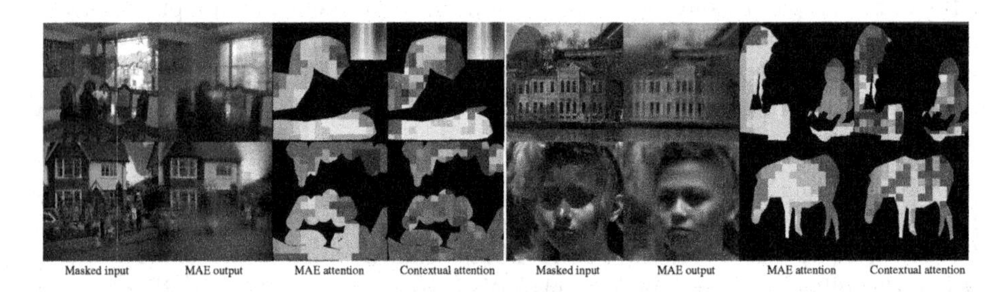

Fig. 9. Visualization of attentions from MAE and learned by contextual attention mechanism [42]. For the visualizations of attention maps, unmasked regions are ignored as black, and the patch attended with the highest attention score is shown in masked regions. Attention palettes are shown in the upper right corner of the first instance, which illustrate the exact positions attended mostly by masked patches.

Table 6. Efficiency of our model without pre-processing. Results are reported on 256×256 images with training and testing batch size 24 and 1 respectively.

MAE	Attention	Training (sec/batch)	Inference (sec/image)
×	×	1.0101	0.0425
✓	×	1.2510	0.0656
✓	✓	1.2604	0.0665

Qualitative Comparisons. We show 256×256 qualitative results of Places2 and FFHQ in Fig. 6 and Fig. 7(A) respectively. For the results in Fig. 6, results of EC and CTSDG are blurry and have obvious artifacts. The results of Co-Mod and LaMa have proper qualities. But these two methods generate some unreasonable building architectures, which also cause artifacts. For the results of ICT, they fail to get good quality due to the poor LR reconstruction. Moreover, the slow sampling strategy of ICT makes it hard to test large-scale image datasets for a quantitative comparison. Our method can achieve better inpainting results in both structures and textures. For face images, our method can achieve stable inpainted results with consistent eye gaze.

For qualitative results in 512×512, other compared algorithms fail to reconstruct reasonable semantics for certain objects in HR, such as car, necklace, and temple in Fig. 8, except our FAR. Such good results are benefited from informative prior features of MAE, which make the model understand the real categories of objects in HR cases. Besides, as shown in Fig. 8, our method avoids overfitting MAE outputs due to the enlarged masking strategy discussed in Sect. 3.1. We further provide some 1024 results in Fig. 7(B), which show that MAE can provide expressive priors for both structure and texture reconstructions.

Visualization of Attention. We show the visualization about the attention scores in Fig. 9. Patches attended with the highest attention score are shown in MAE and contextual attention maps, which can be located by the palettes in the first instance of Fig. 9. Attention scores from MAE are more consistent and reasonable compared with learning-based contextual attention [42]. Specifically, some irrelevant patches are attended by masked regions in the contextual attention, which leads to confused attention maps. Besides, although some MAE results are blurry, their attention relations are still effective.

Computation and Complexity. Benefited by the efficient design from [16] with 25% unmasked input tokens to the encoder, we should claim that the MAE pre-training is not heavier than CNNs. As discussed in implemented details, our MAE has been trained for 200 epochs in Places2 with about 10 days on two NVIDIA RTX A6000 gpus, which costs almost the same time for training a LaMa in Places2 for just 800k steps (only 28.4 epochs). Besides, we list times for both training and inference stages tested on A6000 in Table 6. It shows that the computation of the prior attention is negligible and can be ignored. MAE and gated convolutions increase about 0.024 s for predicting each image, which affects the training a little compared with the time-consuming GAN training. And the inference with 0.0665 s for each image is efficient enough for the user interaction, *e.g.*, object removal.

5 Conclusions

This paper proposes an MAE enhanced image inpainting model called FAR. We utilize a masked visual prediction based vision transformer – MAE to provide features for the CNN based inpainting model, which contain rich informative priors with meaningful semantics. Moreover, we apply the prior attention scores from the pre-trained MAE to aggregate masked features, which is proved to work better than learning contextual attention from scratch. Besides, many constructive issues about the pre-trained MAE and image inpainting are discussed in this paper. Our experiments show that our method can achieve good improvements with the prior features and attentions from MAE. Social impacts of our model, especially working on the face dataset are discussed in the Appendix.

Acknowledgements. This work was supported by SMSTC (2018SHZDZX01).

References

1. Bachlechner, T., Majumder, B.P., Mao, H.H., Cottrell, G.W., McAuley, J.: ReZero is all you need: fast convergence at large depth. arXiv preprint arXiv:2003.04887 (2020)
2. Bao, H., Dong, L., Wei, F.: BEiT: BERT pre-training of image transformers. arXiv preprint arXiv:2106.08254 (2021)
3. Bertalmío, M., Sapiro, G., Caselles, V., Ballester, C.: Image inpainting. Proceedings of the 27th Annual Conference on Computer Graphics and Interactive Techniques (2000)
4. Brown, T.B., et al.: Language models are few-shot learners. arXiv:2005.14165 (2020)
5. Cao, C., Fu, Y.: Learning a sketch tensor space for image inpainting of man-made scenes. arXiv preprint arXiv:2103.15087 (2021)
6. Chen, Y., Liu, S., Wang, X.: Learning continuous image representation with local implicit image function. In: Proceedings of the IEEE/CVF Conference on Computer Vision and Pattern Recognition, pp. 8628–8638 (2021)
7. Criminisi, A., Pérez, P., Toyama, K.: Object removal by exemplar-based inpainting. In: Proceedings of 2003 IEEE Computer Society Conference on Computer Vision and Pattern Recognition, vol. 2, p. II (2003)
8. Dalal, N., Triggs, B.: Histograms of oriented gradients for human detection. In: 2005 IEEE Computer Society Conference on Computer Vision and Pattern Recognition (CVPR 2005), vol. 1, pp. 886–893 (2005). https://doi.org/10.1109/CVPR.2005.177
9. Devlin, J., Chang, M.W., Lee, K., Toutanova, K.: BERT: pre-training of deep bidirectional transformers for language understanding. arXiv preprint arXiv:1810.04805 (2018)
10. Dosovitskiy, A., et al.: An image is worth 16x16 words: transformers for image recognition at scale. CoRR abs/2010.11929 (2020). https://arxiv.org/abs/2010.11929
11. Elharrouss, O., Almaadeed, N., Al-Maadeed, S., Akbari, Y.: Image inpainting: a review. Neural Process. Lett. **51**(2), 2007–2028 (2020)
12. Goodfellow, I., et al.: Generative adversarial nets. In: Advances in Neural Information Processing Systems 27 (2014)
13. Gulrajani, I., Ahmed, F., Arjovsky, M., Dumoulin, V., Courville, A.: Improved training of Wasserstein GANs. arXiv preprint arXiv:1704.00028 (2017)
14. Guo, X., Yang, H., Huang, D.: Image inpainting via conditional texture and structure dual generation. In: Proceedings of the IEEE/CVF International Conference on Computer Vision (ICCV), pp. 14134–14143 (2021)
15. Hays, J., Efros, A.A.: Scene completion using millions of photographs. ACM Trans. Graph. (SIGGRAPH 2007) **26**(3), 4-es (2007)
16. He, K., Chen, X., Xie, S., Li, Y., Doll'ar, P., Girshick, R.B.: Masked autoencoders are scalable vision learners. arXiv: 2111.06377 (2021)
17. Heusel, M., Ramsauer, H., Unterthiner, T., Nessler, B., Hochreiter, S.: GANs trained by a two time-scale update rule converge to a local nash equilibrium (2018)
18. Isola, P., Zhu, J.Y., Zhou, T., Efros, A.A.: Image-to-image translation with conditional adversarial networks. In: Proceedings of the IEEE Conference on Computer Vision and Pattern Recognition, pp. 1125–1134 (2017)

19. Jaderberg, M., Simonyan, K., Zisserman, A., Kavukcuoglu, K.: Spatial transformer networks. CoRR abs/1506.02025 (2015). http://arxiv.org/abs/1506.02025
20. Jo, Y., Park, J.: SC-FEGAN: face editing generative adversarial network with user's sketch and color. In: Proceedings of the IEEE/CVF International Conference on Computer Vision, pp. 1745–1753 (2019)
21. Karras, T., Laine, S., Aila, T.: A style-based generator architecture for generative adversarial networks. In: 2019 IEEE/CVF Conference on Computer Vision and Pattern Recognition (CVPR), pp. 4396–4405 (2019). https://doi.org/10.1109/CVPR.2019.00453
22. Karras, T., Laine, S., Aittala, M., Hellsten, J., Lehtinen, J., Aila, T.: Analyzing and improving the image quality of StyleGAN. In: Proceedings of the IEEE/CVF Conference on Computer Vision and Pattern Recognition, pp. 8110–8119 (2020)
23. Krizhevsky, A., Sutskever, I., Hinton, G.E.: ImageNet classification with deep convolutional neural networks. In: Advances in Neural Information Processing Systems 25, pp. 1097–1105 (2012)
24. Lahiri, A., Jain, A.K., Agrawal, S., Mitra, P., Biswas, P.K.: Prior guided GAN based semantic inpainting. In: Proceedings of the IEEE/CVF Conference on Computer Vision and Pattern Recognition, pp. 13696–13705 (2020)
25. Levin, A., Zomet, A., Weiss, Y.: Learning how to inpaint from global image statistics. In: Proceedings Ninth IEEE International Conference on Computer Vision, vol. 1, pp. 305–312 (2003)
26. Liao, L., Xiao, J., Wang, Z., Lin, C.-W., Satoh, S.: Guidance and evaluation: semantic-aware image inpainting for mixed scenes. In: Vedaldi, A., Bischof, H., Brox, T., Frahm, J.-M. (eds.) ECCV 2020. LNCS, vol. 12372, pp. 683–700. Springer, Cham (2020). https://doi.org/10.1007/978-3-030-58583-9_41
27. Lin, T.-Y., et al.: Microsoft COCO: common objects in context. In: Fleet, D., Pajdla, T., Schiele, B., Tuytelaars, T. (eds.) ECCV 2014. LNCS, vol. 8693, pp. 740–755. Springer, Cham (2014). https://doi.org/10.1007/978-3-319-10602-1_48
28. Nazeri, K., Ng, E., Joseph, T., Qureshi, F., Ebrahimi, M.: EdgeConnect: structure guided image inpainting using edge prediction. In: Proceedings of the IEEE/CVF International Conference on Computer Vision Workshops (2019)
29. Ntavelis, E., et al.: AIM 2020 challenge on image extreme inpainting. In: Bartoli, A., Fusiello, A. (eds.) ECCV 2020. LNCS, vol. 12537, pp. 716–741. Springer, Cham (2020). https://doi.org/10.1007/978-3-030-67070-2_43
30. Radford, A., Narasimhan, K.: Improving language understanding by generative pre-training (2018)
31. Radford, A., Wu, J., Child, R., Luan, D., Amodei, D., Sutskever, I.: Language models are unsupervised multitask learners (2019)
32. Roth, S., Black, M.J.: Fields of experts: a framework for learning image priors. In: 2005 IEEE Computer Society Conference on Computer Vision and Pattern Recognition (CVPR 2005), vol. 2, pp. 860–867 (2005)
33. Song, Y., Yang, C., Shen, Y., Wang, P., Huang, Q., Kuo, C.C.J.: SPG-Net: segmentation prediction and guidance network for image inpainting. arXiv preprint arXiv:1805.03356 (2018)
34. Suvorov, R., et al.: Resolution-robust large mask inpainting with Fourier convolutions. arXiv preprint arXiv:2109.07161 (2021)
35. Vaswani, A., et al.: Attention is all you need. In: Advances in Neural Information Processing Systems, pp. 5998–6008 (2017)
36. Wan, Z., Zhang, J., Chen, D., Liao, J.: High-fidelity pluralistic image completion with transformers. arXiv preprint arXiv:2103.14031 (2021)

37. Wang, T.C., Liu, M.Y., Zhu, J.Y., Tao, A., Kautz, J., Catanzaro, B.: High-resolution image synthesis and semantic manipulation with conditional GANs. In: Proceedings of the IEEE Conference on Computer Vision and Pattern Recognition, pp. 8798–8807 (2018)
38. Wang, Z., Bovik, A.C., Sheikh, H.R., Simoncelli, E.P.: Image quality assessment: from error visibility to structural similarity. IEEE Trans. Image Process. **13**(4), 600–612 (2004)
39. Wei, C., Fan, H., Xie, S., Wu, C., Yuille, A.L., Feichtenhofer, C.: Masked feature prediction for self-supervised visual pre-training. arXiv:2112.09133 (2021)
40. Yang, J., Qi, Z., Shi, Y.: Learning to incorporate structure knowledge for image inpainting. CoRR abs/2002.04170 (2020). https://arxiv.org/abs/2002.04170
41. Yi, Z., Tang, Q., Azizi, S., Jang, D., Xu, Z.: Contextual residual aggregation for ultra high-resolution image inpainting. In: Proceedings of the IEEE/CVF Conference on Computer Vision and Pattern Recognition, pp. 7508–7517 (2020)
42. Yu, J., Lin, Z., Yang, J., Shen, X., Lu, X., Huang, T.S.: Generative image inpainting with contextual attention. In: Proceedings of the IEEE Conference on Computer Vision and Pattern Recognition, pp. 5505–5514 (2018)
43. Yu, J., Lin, Z., Yang, J., Shen, X., Lu, X., Huang, T.S.: Free-form image inpainting with gated convolution. In: Proceedings of the IEEE/CVF International Conference on Computer Vision, pp. 4471–4480 (2019)
44. Yu, Y., et al.: WaveFill: a wavelet-based generation network for image inpainting. In: Proceedings of the IEEE/CVF International Conference on Computer Vision, pp. 14114–14123 (2021)
45. Yu, Y., et al.: Diverse image inpainting with bidirectional and autoregressive transformers. arXiv preprint arXiv:2104.12335 (2021)
46. Zeng, Y., Fu, J., Chao, H., Guo, B.: Aggregated contextual transformations for high-resolution image inpainting. arXiv preprint arXiv:2104.01431 (2021)
47. Zeng, Yu., Lin, Z., Yang, J., Zhang, J., Shechtman, E., Lu, H.: High-resolution image inpainting with iterative confidence feedback and guided upsampling. In: Vedaldi, A., Bischof, H., Brox, T., Frahm, J.-M. (eds.) ECCV 2020. LNCS, vol. 12364, pp. 1–17. Springer, Cham (2020). https://doi.org/10.1007/978-3-030-58529-7_1
48. Zhang, R., Isola, P., Efros, A.A., Shechtman, E., Wang, O.: The unreasonable effectiveness of deep features as a perceptual metric. In: Proceedings of the IEEE Conference on Computer Vision and Pattern Recognition, pp. 586–595 (2018)
49. Zhao, S., et al.: Large scale image completion via co-modulated generative adversarial networks. arXiv preprint arXiv:2103.10428 (2021)
50. Zhou, B., Lapedriza, A., Khosla, A., Oliva, A., Torralba, A.: Places: a 10 million image database for scene recognition. IEEE Trans. Pattern Anal. Mach. Intell. **40**(6), 1452–1464 (2017)
51. Zhou, J., et al.: iBOT: image BERT pre-training with online tokenizer. arXiv preprint arXiv:2111.07832 (2021)
52. Zhou, T., Ding, C., Lin, S., Wang, X., Tao, D.: Learning oracle attention for high-fidelity face completion. In: Proceedings of the IEEE/CVF Conference on Computer Vision and Pattern Recognition, pp. 7680–7689 (2020)

Temporal-MPI: Enabling Multi-plane Images for Dynamic Scene Modelling via Temporal Basis Learning

Wenpeng Xing and Jie Chen(✉)

Department of Computer Science, Hong Kong Baptist University, Kowloon Tong, Hong Kong
{cswpxing,chenjie}@comp.hkbu.edu.hk

Abstract. Novel view synthesis of static scenes has achieved remarkable advancements in producing photo-realistic results. However, key challenges remain for immersive rendering of dynamic scenes. One of the seminal image-based rendering method, the multi-plane image (MPI), produces high novel-view synthesis quality for static scenes. But modelling dynamic contents by MPI is not studied. In this paper, we propose a novel Temporal-MPI representation which is able to encode the rich 3D and dynamic variation information throughout the entire video as compact temporal basis and coefficients jointly learned. Time-instance MPI for rendering can be generated efficiently using mini-seconds by linear combinations of temporal basis and coefficients from Temporal-MPI. Thus novel-views at arbitrary time-instance will be able to be rendered via Temporal-MPI in real-time with high visual quality. Our method is trained and evaluated on Nvidia Dynamic Scene Dataset. We show that our proposed Temporal-MPI is much faster and more compact compared with other state-of-the-art dynamic scene modelling methods.

Keywords: Multi-plane image · Neural basis learning · Novel view synthesis

1 Introduction

Recent advancements on novel view synthesis have shown remarkable results on immersive rendering of static scenes using neural scene representations, such as Multi-plane Images (MPI) [17,35,40] and Neural Radiance Fields (NeRF) [4,18]. Neural basis expansion [35] and Plenoctree structures [38] have been recently proposed to further improve the rendering quality and efficiency. However, challenges still remain in modelling dynamic scenes, which require additional capacity to capture variations along time dimension.

Supplementary Information The online version contains supplementary material available at https://doi.org/10.1007/978-3-031-19784-0_19.

S. Avidan et al. (Eds.): ECCV 2022, LNCS 13675, pp. 323–338, 2022.
https://doi.org/10.1007/978-3-031-19784-0_19

Fig. 1. Reconstruction quality demonstration of the proposed Temporal-MPI for the testing sequences from the Nvidia Dynamic Scene dataset [37]. The dynamic visual effects can be viewed in Adobe PDF Reader. (See supplementary material for an animated version of the figure.)

To model dynamic contents, efforts have been made in training time-conditioned NeRF models [10,11,24]. Although photo-realistic view-synthesis results can be produced by these time-conditioned neural rendering methods, they normally require millions of ray-casting style queries during rendering, resulting in serious rendering delay and low frame rate. So, there is a popular branch of research on improving the rendering efficiency of neural scene representations by extracting the learned content into compact data structure, such as tree-based structure [38], or with occupancy priors [14] stored for efficient sampling. Another line of image-based rendering research, the MPI, focuses on rendering real-world forward-facing contents. MPI is highly efficient for real-time rendering due to its pre-computed 2.5D RGB-α volumes. In order to render dynamic scenes via MPI, pre-calculating and saving all time-instance MPIs is a straight-forward but engineering-oriented solution for time-space rendering. However, this method lacks temporal coherence and is expensive to save the bulky data incurred. *3DMaskVol21* [12] renders an image at a given timestamp by fusing a background MPI and instantaneous MPI using a 3D mask volume, which takes temporal-coherent information learned to be the background MPI. But generating these three volumes causes delay on rendering and heavy workload on caching. In comparison, our proposed method can generate arbitrary time-instance MPIs from one Temporal-MPI within mini seconds, which is much more efficient for real-time rendering and compact in storage.

In this paper, we propose a novel efficient representation for dynamic scenes, Temporal-MPI, for space-time immersive rendering. Different from previous methods [11,12,37] which rely on pre-trained optical flow model [6], ground-truth background images [12], pre-trained depth estimation model [25] or dynamic-static masks [37] as additional premise, we aim at creating a self-contained pipeline. In addition, our method does not need to explicitly store time-instance MPIs, which greatly decreases the requirement for storage space and being computationally efficient.

2 Related Work

Novel View Synthesis. Novel view synthesis is a long standing research issue that aims at synthesising novel views of a scene given arbitrary captured images, and has become one of the most popular classes of research topics in computer vision. Early researches on Light fields (LF) [21] represented the scene as a 4D Plenoptic Function [16] $L(x, y, s, t)$, where (x, y) represents spatial coordinates and (s, t) represents angular coordinates. The spatial-angular correlations embedded in LF images can be exploited for applications of depth estimation [7,20], super-resolution [8] and novel view rendering [9,30]. With recent advancements in deep learning, different learning-based scene representations were proposed. Novel views can be synthesized from monocular input, they are *SynSin20* [34], *3D Photo20* [28], *WorldSheet21* [5] and *MPIs20* [33]. They share a common rationale, which is integrating the learning of geometry and appearance from rendering loss. The second branch of research is using multi-view inputs that allow machine learning models to reason the scene's geometry using epipolar geometry and triangulation, the scene can be learned as a volume representation either explicitly or implicitly. Neural/implicit volume representations can encode the scene as a continuous volumetric function, they are Deep Voxels [29] and NeRF [18]. In addition to above continuous volumetric functions, a scene can be decomposed into a layered representation [27], they are MPI [40] and its followers [12,17,33,35]. Although these methods can produce photorealistic results, they can only model and render static scenes. The next key step of view rendering is rendering dynamic scenes.

Neural Spatial and Temporal Embedding for Novel-View Synthesis. A successful novel view synthesis requires accurate modeling of a scene's geometry. Modeling the geometry of non-rigid scenes with dynamic contents are ill-posed, and were tackled by reconstructing dynamic 3D meshes where priors like temporal information [1,32] or known template configurations [2,19]. Yet, these methods require 2D-to-3D matches or 3D point tracks. Thus, limiting their applicability to real world scenes or simulated scenes with complex textures.

Under the context of space-time view synthesis, adding time parameters to the input of static scene's representations is a straightforward implementation. There are time-conditioned warping fields in *D-NeRF21* [24], scene flow fields in *NeuralFlow21* [11] and radiance fields in *Neural3DVideo21* [10]. More specifically, *D-NeRF21* [24] added a time-conditioned deformation network to predict the time-dependent positional offsets to deform the canonical NeRF into a time-instance shape. *NeuralFlow21* [11] used temporal photometric consistency to encourage the time-conditioned NeRF to be learned from monocular videos. Neural3DVideo21 [10] also transformed NeRF into a space-time domain, and achieved frame-interpolation by interpolating time latent vectors. However, the time-consuming rendering process of above NeRF-style methods limit their capabilities to real-time applications. Directly warping images to novel views according to depth is an efficient view-synthesis pipeline. *DynSyn20* [37] combined multi-view and single-view depths to generate temporal consistent depths for dynamic views warping. However, their method has two drawbacks: first, it requires foreground masks that separate static and dynamic contents; second, their method can not handle occlusions well. *3DMaskVol21* [12] proposed a method of generating dynamic MPI with a 3D mask volume to alleviate artifacts around the integration boundary of background and instantaneous MPIs. However, their method

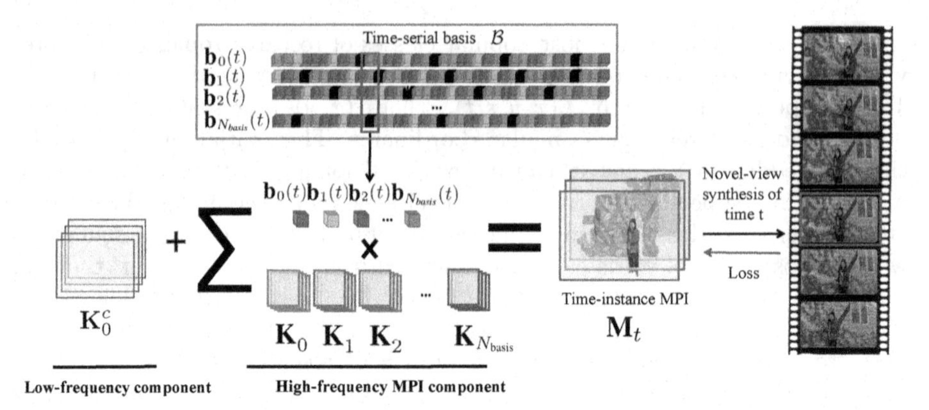

Fig. 2. Overall pipeline. The proposed Temporal-MPI contains three parts: low-frequency component $\mathbf{K}_0^c$, temporal basis $\mathcal{B}$ and high-frequency coefficients $\{\mathbf{K}_n\}_{n=1}^{N_{\text{basis}}}$. The alpha and color values in time-instantaneous MPI $\mathbf{M}_t$ are recovered as linear combinations of bases $\mathcal{B}$ and high-frequency coefficients $\{\mathbf{K}_n\}_{n=1}^{N_{\text{basis}}}$, and adding low-frequency component $\mathbf{K}_0^c$ from Temporal-MPI $\hat{\mathcal{M}}$. The color in the corresponding frame is rendered from time-instantaneous MPI $\mathbf{M}_t$ as MPI's alpha composition in Eq. (1). The overall pipeline is differentiable and optimized per scene by pixel rendering loss. (Color figure online)

requires two-step training and background images. Thus, limiting their general capabilities. Compared with NeRF-style methods, *DynSyn20* and *3DMaskVol21*, our method is efficient on rendering and compact on storage.

Neural Learnable Basis. Our method is closely related to basis learning [13]. In signal processing, data often contains underlying structure that can be processed intelligently by linear combinations of *subspaces*. Tang et al. [31] learned subspace minimization for low-level vision tasks, such as interactive segmentation, video segmentation, optical flow estimation and stereo matching. PCA-Flow [36] predicted video's optical flows as a weighted sum of the basis flow fields. We take inspiration from these works, and learn coefficients to combine globally shared time-wise subspace to draw instantaneous MPIs.

3 Approach

Given a set of synchronized multi-view videos $\{\mathbf{I}_t^k\}$ of a dynamic scene, where $t = 1, 2, \cdots T$ are the frame number, and $k = 1, 2, \cdots, K$ are camera indices, our goal is to construct a *compact* 3D representation which enables *real-time* and *novel-view* synthesis of the dynamic contents at a given time $t \in [1, T]$. To achieve the goal, one naive option is to calculate and save a set of separate MPI $\mathcal{M} = \{\mathbf{M}_t \in \mathbb{R}^{H\times W\times D\times 4}\}_{t=1}^T$ for every video frame. This, however, will be extremely memory- and computation-inefficient (generating $\mathcal{M}$ incurs more than $225\times T$ MB data and around 2 s delay when rendering at VGA resolution [12]). As such, rather than having to calculate and save MPIs for all video frames in advance or having to calculate an MPI on-the-run, we investigate a novel Temporal-MPI representation with learned temporal basis to compactly encode high-frequency variation throughout the entire video. An overall pipeline of our approach is shown in Fig. 2.

In the following, we will first briefly introduce the vanilla MPI representation in Sect. 3.1. Then, the temporal basis formulation will be elaborated in Sect. 3.2, and the novel-view temporal reconstruction in Sect. 3.3.

3.1 The Multi-plane Image Representation

Being one of the seminal representation frameworks for 3D content embedding and novel-view synthesis, Multi-plane Images (MPI) learn a layered depth decomposition of the scene from a set of multi-view references [17,40,40]. Following the MPI's illustration in Nex [35], let D denote the number of depth layers in a MPI, with the dimension of each layer being $H \times W \times 4$, where H and W denote the height and width of the MPI layers, 4 denotes 3-channel RGB and 1-channel alpha α. So we denote an MPI representation as $\mathbf{M} = \{\mathbf{C}_d, \mathbf{A}_d\}_{d=1}^{D}$, where $\mathbf{C}_d \in \mathbb{R}^{H\times W\times 3}$ are multiple layers of 3-channel RGB images and $\mathbf{A}_d \in \mathbb{R}^{H\times W\times 1}$ are one-channel alpha images, d denotes the depth plane index.

Synthesizing novel-views $\hat{\mathbf{I}}$ based on the MPI $\mathbf{M}$ involves two steps: first, warp all depth planes in the MPI homographically from a reference view to a source view; and second, render pixels using alpha-composition [23] over each layer's color:

$$\hat{\mathbf{I}} = \mathcal{O}(\mathcal{W}(\mathbf{A}), \mathcal{W}(\mathbf{C})), \tag{1}$$

here $\mathcal{W}$ denotes the warping operator, and $\mathcal{O}$ denotes the compositing operator. The compositing operator $\mathcal{O}$ is defined as:

$$\mathcal{O}(\mathbf{A}, \mathbf{C}) = \sum_{d=1}^{D} \mathbf{C}_d \mathcal{T}_d(\mathbf{A}), \tag{2}$$

$$\mathcal{T}_d(\mathbf{A}) = \mathbf{A}_d \prod_{i=d+1}^{D} (1 - \mathbf{A}_i). \tag{3}$$

where $\prod_{i=d+1}^{D}(1-\mathbf{A}_i)$ are accumulated transmittance, $\mathcal{T}_d$ are opacity. The output of $\mathcal{O}(\mathbf{A}, \mathbf{C})$ are final rendered colors. Both the composition $\mathcal{O}$ and the warping $\mathcal{W}$ operations are differentiable, thus allowing the representation $\mathbf{M}$ to learn the geometry and color information from final pixel rendering loss.

3.2 Temporal Basis Formulation

At a given time instance $t \in [1, T]$, we denote the time-instance MPI as $\mathbf{M}_t$. In order to render the entire novel view sequence at sequential timestamps, a set of time-instance MPIs $\mathcal{M} = \{\mathbf{M}_t \in \mathbb{R}^{H\times W\times D\times 4}\}_{t=1}^{T}$ are needed to generate. Based on the afore-analyzed reasons, we cannot exhaustively calculate and save $\mathcal{M}$. We propose a novel Temporal-MPI representation which is able to encode the rich 3D and high-frequency variation information throughout the entire video as compact temporal basis, and in the meantime, preserve high rendering efficiency for real-time novel-view synthesis. To achieve this, we divide the goal into two tasks, i.e., (i) learning the low-frequency color components as explicit parameters, and (ii) learning the high-frequency variation over a set of temporal basis.

Explicit Parameter Learning for Low-Frequency Component. Low-frequency contents in a video constitute the low-frequency part of the total energy along the time dimension, which can be well-captured and modeled explicitly by time-invariant parameters. By treating all the frames of the multi-view video $\{\mathbf{I}_t^k\}_{t,k}$ as source views equally and ignoring their respective frame indices, we can directly learn the multi-plane time-invariant RGB color parameters $\mathbf{K}_0^{\mathrm{c}} \in \mathbb{R}^{H\times W\times D/8\times 3}$ using the pixel rendering loss. $\mathbf{K}_0^{\mathrm{c}}$ models the low-frequency energy of the video, with possible blur over the dynamic area. Such an explicit modelling scheme for the low-frequency component proves to be important [35] and let the subsequent dynamic modelling to better focus on the temporal variation.

Temporal Basis Learning for High-Frequency Contents. Compared with low-frequency components, the high-frequency contents in $\mathcal{M}$ constitute the high-frequency energy along the time dimension. Being high-dimensional and with dynamic variations, the high-frequency contents still constitute a highly regularized manifold, considering the fact that (i) the video length is limited (we model video with 24 frames in length, although these frames could be extracted from longer video sequences), and (ii) time-variant pixels within a scene usually show consistent motion in clusters. This motivates us to compactly represent the high-frequency components based on a few learned time-variant temporal basis.

We denote the temporal basis as $\mathcal{B} \in \mathbb{R}^{4\times T\times N_{\mathrm{basis}}}$ which span the temporal variation space for $\mathcal{M}$. Here N_{basis} denotes the total number of basis. The first dimension of $\mathcal{B}$ is set to 4 which is reserved for modelling both the MPI color component (with 3 channels): $\mathcal{B}^c = \{\mathbf{b}_n^c\}_{n=1}^{N_{\mathrm{basis}}}$, and the alpha component $\mathcal{B}^\alpha = \{\mathbf{b}_n^\alpha\}_{n=1}^{N_{\mathrm{basis}}}$ (1 channel). Therefore $\mathcal{B} = [\mathcal{B}^c, \mathcal{B}^\alpha]$.

In our proposed framework, the temporal basis will be estimated by two time-dependent functions which are Multi-Layer Perceptron (MLP) networks $\mathcal{V}^c$ and $\mathcal{V}^\alpha$:

$$\{\mathbf{b}_n^c(t)\}_{n=1}^{N_{\mathrm{basis}}} = \mathcal{V}^c(\mathcal{E}(t)) : \mathbb{R} \mapsto \mathbb{R}^{3\times N_{\mathrm{basis}}}, \tag{4}$$

$$\{\mathbf{b}_n^\alpha(t)\}_{n=1}^{N_{\mathrm{basis}}} = \mathcal{V}^\alpha(\mathcal{E}(t)) : \mathbb{R} \mapsto \mathbb{R}^{1\times N_{\mathrm{basis}}}. \tag{5}$$

Here $\mathcal{E}(\cdot)$ is a time-encoding function which encodes time-sequential information into a high dimensional latent vector [10]. The temporal basis $\mathcal{B}$ learns a parsimonious frame that efficiently spans the temporal variation manifold. With a *pixel-specific* coding coefficient (to be elaborated in the next section), $\mathcal{B}$ can efficiently model the MPI pixel's temporal variation throughout the entire video.

3.3 Temporal Coding for Novel-View Synthesis

For an arbitrary frame index $t \in [1, T]$, a time-instance MPI $\mathbf{M}_t = [\mathbf{A}_t, \mathbf{C}_t]$ can be constructed based on the temporal basis $\mathcal{B}$ according to:

$$\mathbf{C}_t(\mathbf{x}) = \mathbf{K}_0^c(\mathbf{x}) + \sum_{n=1}^{N_{\text{basis}}} \mathbf{K}_n^c(\mathbf{x}) \times \mathbf{b}_n^c(t), \tag{6}$$

$$\mathbf{A}_t(\mathbf{x}) = \sum_{n=1}^{N_{\text{basis}}} \mathbf{K}_n^\alpha(\mathbf{x}) \times \mathbf{b}_n^\alpha(t). \tag{7}$$

Here $\mathbf{K}_n^\alpha(\mathbf{x})$ and $\mathbf{K}_n^c(\mathbf{x})$ are the coding coefficients for the respective temporal basis at a given MPI spatial location $\mathbf{x} \in \mathbb{R}^3$ (the 3 dimensions of $\mathbf{x}$ include its 2D coordinates and the depth plane index in $\mathcal{M}_t$). These coding coefficients are estimated by another set of MLPs $\mathcal{K}^c$ and $\mathcal{K}^\alpha$:

$$\{\mathbf{K}_n^c(\mathbf{x})\}_{n=1}^{N_{\text{basis}}} = \mathcal{K}^c(\mathcal{R}(\mathbf{x})) : \mathbb{R}^3 \mapsto \mathbb{R}^{3 \times N_{\text{basis}}}, \tag{8}$$

$$\{\mathbf{K}_n^\alpha(\mathbf{x})\}_{n=1}^{N_{\text{basis}}} = \mathcal{K}^\alpha(\mathcal{R}(\mathbf{x})) : \mathbb{R}^3 \mapsto \mathbb{R}^{1 \times N_{\text{basis}}}. \tag{9}$$

Similarly, here $\mathcal{R}(\cdot)$ is a position-encoding function which encodes the spatial information $\mathbf{x}$ into high-dimensional representations [18].

Based on Eq. (6) and (7), the time-instance MPI $\mathbf{M}_t$ can be warped and composited to any arbitrary viewing angles according to Eq. (2) and (1). In addition, by querying all elements $t = 1, \cdots, T$ along the temporal basis, we can construct the time-instance MPI for each video frame.

Remarks. (i) our proposed temporal MPI representation composes of an explicitly learned low-frequency multi-plane color component $\mathbf{K}_0^c \in \mathbb{R}^{H \times W \times D/8 \times 3}$, and a dynamically coded time-variant component via simultaneous basis and coefficient learning. We have achieved compression along the temporal dimension via the temporal basis, which compactly encodes time-variant color and geometry variation information throughout the entire video.

(ii) To maintain rendering efficiency and save storage-space, spatial-temporal information is efficiently encoded and propagated among different components in the Temporal-MPI. First, the low-frequency component $\mathbf{K}_0^c$ is temporally shared among all time frames, this ensures overall reconstruction quality and enables the high-frequency components to focus on time-dependent variations only; and second, the high-frequency coefficients, i.e., $\{\mathbf{K}_n^c(\mathbf{x})\}_{n=1}^{N_{\text{basis}}}$ and $\{\mathbf{K}_n^\alpha(\mathbf{x})\}_{n=1}^{N_{\text{basis}}}$, are point-wisely coded/learned, however, over a common set of temporal basis. This helps to remove the redundancy in modelling dynamic variation, and also helps to remove motion ambiguities for some pixels.

3.4 Training Loss Function

To let the Temporal-MPI focus on reconstruction quality, we ignore the sparsity of coding coefficients for this task. Coefficients and the temporal basis are jointly

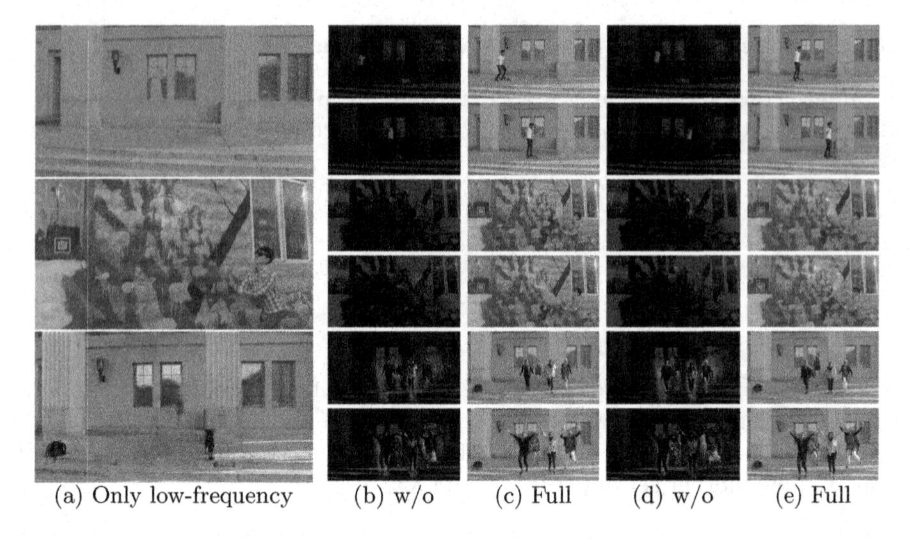

(a) Only low-frequency (b) w/o (c) Full (d) w/o (e) Full

Fig. 3. Low-frequency scene representation ablation study. The low-frequency components are rendered in (a); color output from the high-frequency components are rendered in (b) and (d); full rendering are visualised in (c) and (e).

learned and optimized. The whole system is optimized via the following loss function $\mathcal{L}$:

$$\mathcal{L} = ||\hat{\mathbf{I}}_t^k - \mathbf{I}_t^k||_2 + \lambda_1||\nabla\hat{\mathbf{I}}_t^k - \nabla\mathbf{I}_t^k||_1 + \lambda_2\text{TVC}(\mathbf{K}_0^c), \tag{10}$$

where $\hat{\mathbf{I}}_t^k$ is the rendered image at time t for the camera k, $\mathbf{I}_t^k$ is the ground truth image from the same view. The first term in $\mathcal{L}$ calculates the L_2 reconstruction loss. The second term penalises edge inconsistencies, with ∇ denoting the gradient operator. In the third term, TVC denotes total variation loss [3]. λ_1 and λ_2 are balancing weights for different loss terms.

4 Experiments

4.1 Implementation Details

Our model is implemented in PyTorch 1.10, using Adam as optimiser. The initial learning rate is set as 0.001, and decay by 0.1 every 2000 steps. The model takes 16 h to be trained on one Nvidia Geforce RTX 2070 Super GPU with a batch of 1500 rays, using 5.3 GB memory. The output resolution is 576 × 300. The position-encoding method in [18] is formulated as $\mathcal{R}(p) = [sin(2^0\frac{\pi}{2}p), sin(2^0\frac{\pi}{2}p), \ldots, sin(2^l\frac{\pi}{2}p), cos(2^l\frac{\pi}{2}p)]$ where the input location of scene point is normalised to $[-1, 1]$ and l is the index of encoding level set as 3. The index of time is embedded into a latent vector in size of 32 using dictionary learning as in [10]. For networks that parameterise $\mathcal{K}^c$ and $\mathcal{K}^\alpha$, we use MLP networks with 8 layers and 384 hidden nodes. Networks for $\mathcal{V}^c$ and $\mathcal{V}^\alpha$ are using MLP with 4 layers and

64 hidden nodes. The shape of high-frequency coefficients $\{\mathbf{K}_n^c(\mathbf{x})\}_{n=1}^{N_{\text{basis}}}$ and $\{\mathbf{K}_n^\alpha(\mathbf{x})\}_{n=1}^{N_{\text{basis}}}$ in Temporal-MPI is $320 \times 596 \times 32 \times 4 \times 5$ where 32 is the number of planes D, 596 and 320 are width W and height H including marginal offsets set as 10, 4 includes 3 channels of colors and 1 channel of alpha, and 5 is the number of basis N_{basis}. The shape of temporal basis $\mathcal{B}$ is $4 \times 5 \times 24$ where 5 is the number of basis N_{basis}, 24 is the total number of timestamps and 4 includes 3 channels for color and 1 channel for alpha. Low-frequency component $\mathbf{K}_0^c$ is in the shape of $320 \times 596 \times 4 \times 3$ before the repetition along depth dimension.

4.2 Dataset

Our model is trained and evaluated on the Nvidia Dynamic Scenes Dataset [37] that contains 8 scenes with motions recorded by 12 synchronized cameras. Nvidia Dynamic Scenes Dataset captures a dynamic scene with static background via stationary cameras which suit our goal of separately learning low- and high-frequency components well. We extract camera parameters for every camera using COLMAP [26]. We extracted 24 frames from the video sequence, and used multi-view images in selected frames for training. We select camera views 1–11 for training, and camera 12 for testing. So the total number of training images is 264. The camera location arrangement is shown in Fig. 4.

6	5	4	3	2	1
7	8	9	10	11	12

Fig. 4. Camera indexes in camera array.

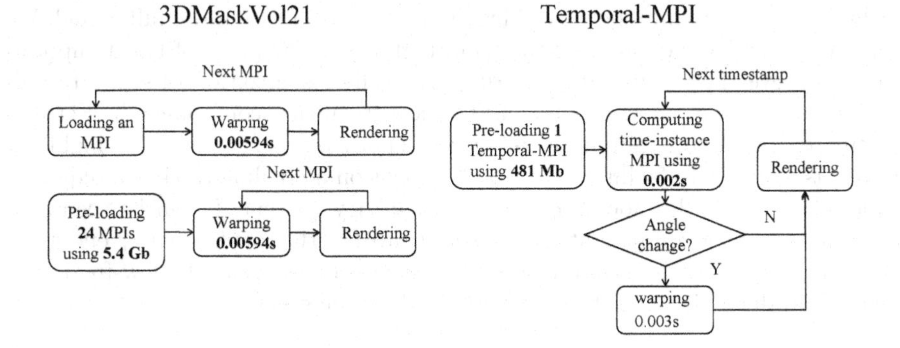

Fig. 5. Rendering pipeline comparison.

Table 1. Ablation study on PSNR vs. total number of timestamps. With D as 32 planes.

Scene/PSNR	Number of timestamps						
	8	16	24	32	40	48	60
Skating-2	30.323	28.813	28.575	28.612	27.324	25.431	25.012
Truck-2	28.441	28.174	28.056	27.951	27.591	25.332	24.963
Jumping	25.850	25.661	25.486	25.301	25.001	24.759	23.854
Balloon2-2	25.775	25.381	25.171	24.893	24.557	24.474	23.945

4.3 Ablation Study

In this section, we investigate the effectiveness of our main contributions in Temporal-MPI: high-frequency coefficients, temporal basis and low-frequency component.

Video Length. We evaluate the performance of our model trained with different length of videos. As shown in Table 1, We found performance degradation when the total number of timestamps T increased. This is due to reaching the representation threshold of temporal basis and high-frequency coefficients.

Low-Frequency Colors and High-Frequency Coefficients. To validate the contributions of low-frequency component $\mathbf{K}_0^c$, high-frequency components and time-serial basis ($\{\mathbf{K}_n^c(\mathbf{x})\}_{n=1}^{N_{\text{basis}}}$, $\{\mathbf{K}_n^\alpha(\mathbf{x})\}_{n=1}^{N_{\text{basis}}}$, $\mathcal{B}$), we performed experiments without these modules. As shown in Table 2, without low-frequency component or high-frequency components will lead to a worse result than the full model. We separately render the low-frequency and high-frequency parts of the Temporal-MPI to prove their individual contributions. The visualization of separate rendering settings are shown in Fig. 3. The low-frequency component in Fig. 3 is calculated by directly summing $\mathbf{K}_0^c$ across depth planes. We wish to highlight that it is designed to facilitate the MLP to focus on modelling the high-frequency residual by explicitly modelling the low frequency content. It can be seen that the low-frequency components successfully capture the low-frequency energy of the video, while the high-frequency components complement the low-frequency ones to produce high quality rendering for dynamic scenes.

Inference Speed. In this section, we investigate the relationships between inference speed and the size of Temporal-MPI. From Table 3, we can find that the computations of linear combinations of basis are efficient, and a big volume size of Temporal-MPI will not affect its real-time performance. Inference speed experiments are conducted on one Nvidia Tesla V100 GPU.

Table 2. Ablation study of low-frequency and high-frequency components.

Methods	No. planes	Metrics		
		SSIM (↑)	PSNR (↑)	LPIPS (↓)
w/o low-frequency	32	0.192	10.3	0.726
w/o high-frequency	32	0.611	22.6	0.213
Full	32	0.859	24.87	0.196

Table 3. Inference time vs. shape of temporal-MPI.

Resolution	No. basis	No. planes	Inference time (s, ↓)
596×320	5	32	0.002
596×320	13	32	0.003
1038×1940	5	32	0.025
1038×1940	13	32	0.029
1038×1940	5	192	0.030

4.4 Evaluation and Comparison

To prove the efficiency and compactness of our method, we first compare our method with state-of-the-art algorithms, *3DMaskVol21* [12] and *NeuralFlow21* [11], in terms of storage space and inference speed. Then, we evaluate the view synthesis quality with other methods.

Evaluation on Compactness of Representation. One of the main objectives of our approach is to learn a compact representation of a dynamic scene. So we evaluate the compactness of Temporal-MPI by comparing the number of network parameters and storage space with these two methods. As shown in Table 5, modeling a dynamic scene with 24 timestamps, Temporal-MPI only occupies 481 Mb for storage, which is 11 times smaller than *3DMaskVol21* on storage space. So our approach is extremely fast and compact for real-time rendering.

Evaluation on Efficiency. From Table 5, we can find that i) rendering NeuralFlow21 requires querying MLP exhaustively, so it is the slowest on rendering. ii) our rendering time is much faster than 3DMaskVol21: as shown in Fig. 5, 3DMaskVol21 requires per-frame warping, which is not a mandatory step of ours; it also requires per-frame MPI loading, but we only need to load once for the entire sequence, therefore longer sequence will bring more advantages to our efficiency. Considering both loading and rendering time, ours are much faster on rendering than both NeuralFlow21 and 3DMaskVol21 (on a $T = 24$ frame video). But 3DMaskVol21 is a generic method that generalizes to novel scenes, so it saves the cost of per-scene training. NeuralFlow21 has the highest rendering quality due to its advantages on dense sampling in depth dimensions.

Table 4. Quantitative evaluation of novel view synthesis on the Dynamic Scenes dataset. MV denotes whether the approach uses multi-view information or not, Ind. Src denotes the index of source views used to train the model. Ours (D = 32) denotes using 32 planes in MPI.

Methods	Ind. Src	MV	Metrics		
			SSIM (↑)	PSNR (↑)	LPIPS (↓)
SynSin20 [34]	3	No	0.488	16.21	0.295
MPIs20 [33]	3	No	0.629	19.46	0.367
3D Ken Burn19 [22]	3	No	0.630	19.25	0.185
3D Photo20 [28]	3	No	0.614	19.29	0.215
NeRF20 [18]	1–11	Yes	0.893	24.90	0.098
ConsisVideoDepth20 [15]	3	Yes	0.746	21.37	0.141
DynSyn20 [37]	1–11	Yes	0.761	21.78	0.127
NeuralFlow21 [11]	3	Yes	**0.928**	**28.19**	**0.045**
D-NeRF21 [24]	1–11	Yes	0.334	17.05	0.545
3DMaskVol21 [12]	3, 9	Yes	0.603	20.10	0.285
Ours (D = 32)	1–11	Yes	0.859	24.87	0.196

Evaluation on View-Synthesis Quality. We evaluate the effectiveness of our approach by comparing it to baseline methods quantitatively and qualitatively. We compare our approach with state-of-the-art single-view or multi-view novel view synthesis methods. For monocular methods, we compare with *SynSin20* [34] and *MPIs20* [33] trained on RealEstate 10K dataset [40]. *3D Photo20* [28] and *3D Ken Burns19* [22] were trained by wild images. For multi-view methods, we compare with *NeRF20* [18], *ConsisVideoDepth20* [15], *DynSyn20* [37], *NeuralFlow21* [11], *3DMaskVol21* [12] and *D-NeRF21* [24]. Results are referenced from recent publications [11,12]. We document the rendering quality in three error metrics: structural similarity index measure (SSIM), peak signal-to-noise ratio (PSNR), and perceptual similarity through LPIPS [39]. From Table 4, our algorithm has competitive average score across three metrics. Per-scene breakdown results are shown in Table 6.

Qualitative comparisons can be seen in Fig. 6, which show that our method achieves competitive rendering quality in both low- and high-frequency parts. The visual results of *3D Photo20* in Fig. 6 (a), *NeRF20* in Fig. 6 (b) and *DynSyn20* in Fig. 6 (c) are referenced from [11]. Observed from above images, *D-NeRF21* in Fig. 6 (e) produces blurry results, *DynSyn20* has great artifacts on thin structures, *3D Photo20* generates distortions, *3DMaskVol21* produces ghosting effects around the object's boundary given scenes with forward moving motions, such as Jumping and Umbrella.

4.5 Baseline for Brute-Force Scenario

To compare with brute-force scenario where an MPI is calculated for each time frame. We have tested LLFF19 [17] that includes all views 1–11 in a local fusion

Table 5. Comparison on real-time rendering/inference speed, output resolution and storage space. We assume the length of the modeled dynamic video is 24 frames. *3DMaskVol21* will require pre-loading 24 MPIs for real-time rendering of the whole sequence. *NeuralFlow21* is impossible for real-time tasks.

Time/methods	*NeuralFlow21*	*3DMaskVol21*	Ours
MPI generation time (sec, ↓)	–	2	**0.002**
MPI loading time (sec, ↓)	–	0.043	**0.083**/T
Warping time (sec, ↓)	–	0.00594	**0.003**
Rendering time (sec/frame, ↓)	6	0.049	**0.008**(T = 24)
Output resolution (pixel, ↑)	512 × 288	**640 × 360**	576× 300
Network parameters (million, ↓)	5.26	**1.17**	6.00
Storage space (Mb, ↓)	–	225×T (24) = 5400	**481** × 1(D = 32)
Training time (hour, ↓)	48†	120	16†

† denotes scene specific training

Table 6. Per-scene breakdown results from *DynSyn20*'s Dynamic Scenes dataset.

	Skating-2	Balloon1-2	Jumping	Playground	Balloon2-2	Truck-2	Average
PSNR(↑)	28.575	21.309	25.486	20.594	25.171	28.056	24.865
SSIM(↑)	0.925	0.802	0.886	0.7211	0.885	0.937	0.859
LPIPS(↓)	0.163	0.239	0.202	0.253	0.171	0.150	0.196

manner, and view 12 for testing. It takes 39.5649 s seconds to infer MPIs for a single frame, with average PSNR and SSIM 35.41 and 0.95, compared to 31.94 and 0.917 of the Temporal-MPI. Note that the baseline calculates and fuses several static MPIs for each frame, while we only calculate one neural MPI for the entire sequence.

5 Concluding Remarks

5.1 Limitations

Modeling dynamic scenes is challenging due to complex motions of dynamic objects over time, and specular surface and occlusions on angular domain. Our method makes the first attempt to use a compact temporal representation to reproduce dynamic scenes in time-sequences. Similar to *NeRF20*, our method requires optimization for each scene. Additionally, the output resolution is limited due to limited GPU memory. Furthermore, the rendering quality degrades when the length of sequence increases given default model parameters. Our approach is also only applicable to dynamic scenes without large camera motions that cause the change of background.

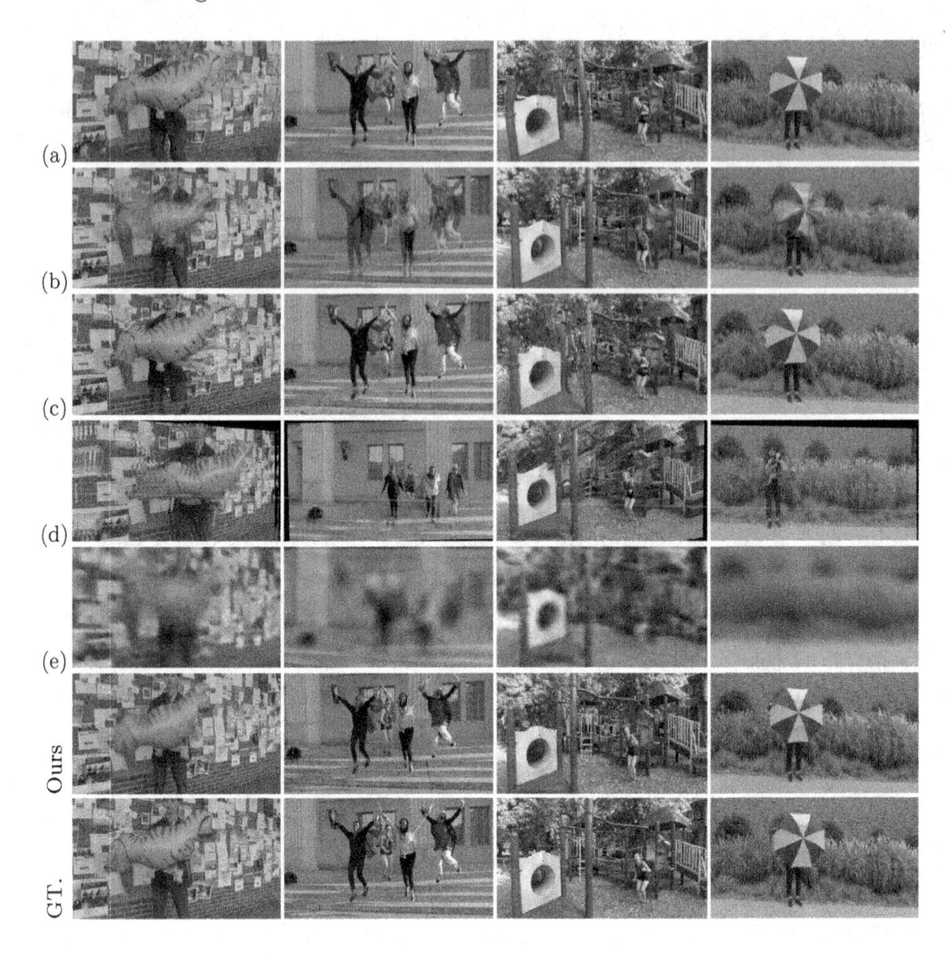

Fig. 6. Qualitative comparisons on the Dynamic Scenes dataset.

5.2 Conclusion

We have proposed a novel dynamic scene representation on top of Multi-plane Image (MPI) with basis learning. Our representation is efficient in computing, thus allowing real-time rendering of dynamics. Extensive studies on public dataset demonstrate the competitive rendering quality and efficiency of our approach. We believe using basis learning for temporal recovery and compression can be applied to the general problem of modeling dynamic contents and not limited to MPI. Using hierarchical encoding method to improve the learning power of MLP on modeling long-time-serial data could be a future extension of our work.

Acknowledgments. The research was supported by the Theme-based Research Scheme, Research Grants Council of Hong Kong (T45-205/21-N).

References

1. Agudo, A., Moreno-Noguer, F.: Simultaneous pose and non-rigid shape with particle dynamics. In: CVPR, pp. 2179–2187 (2015)
2. Bartoli, A., Gérard, Y., Chadebecq, F., Collins, T., Pizarro, D.: Shape-from-template. IEEE Trans. Pattern Anal. Mach. Intell. **37**(10), 2099–2118 (2015)
3. Chambolle, A., Lions, P.L.: Image recovery via total variation minimization and related problems. Numer. Math. **76**(2), 167–188 (1997)
4. Chen, A., et al.: MVSNeRF: fast generalizable radiance field reconstruction from multi-view stereo. In: CVPR, pp. 14124–14133 (2021)
5. Hu, R., Ravi, N., Berg, A.C., Pathak, D.: Worldsheet: wrapping the world in a 3D sheet for view synthesis from a single image. In: ICCV, pp. 12528–12537 (2021)
6. Ilg, E., Mayer, N., Saikia, T., Keuper, M., Dosovitskiy, A., Brox, T.: FlowNet 2.0: evolution of optical flow estimation with deep networks. In: CVPR, pp. 2462–2470 (2017)
7. Jeon, H.G., et al.: Accurate depth map estimation from a lenslet light field camera. In: CVPR, pp. 1547–1555 (2015)
8. Jin, J., Hou, J., Chen, J., Zeng, H., Kwong, S., Yu, J.: Deep coarse-to-fine dense light field reconstruction with flexible sampling and geometry-aware fusion. In: IEEE Transactions on Pattern Analysis and Machine Intelligence, p. 1 (2020)
9. Kalantari, N.K., Wang, T.C., Ramamoorthi, R.: Learning-based view synthesis for light field cameras. ACM Trans. Graph. **35**(6), 1–10 (2016)
10. Li, T., et al.: Neural 3D video synthesis from multi-view video. In: CVPR, pp. 5521–5531 (2022)
11. Li, Z., Niklaus, S., Snavely, N., Wang, O.: Neural scene flow fields for space-time view synthesis of dynamic scenes. In: CVPR, pp. 6498–6508 (2021)
12. Lin, K.E., Xiao, L., Liu, F., Yang, G., Ramamoorthi, R.: Deep 3D mask volume for view synthesis of dynamic scenes. In: ICCV, pp. 1749–1758 (2021)
13. Liu, G., Lin, Z., Yan, S., Sun, J., Yu, Y., Ma, Y.: Robust recovery of subspace structures by low-rank representation. IEEE Trans. Pattern Anal. Mach. Intell. **35**(1), 171–184 (2012)
14. Liu, L., Gu, J., Zaw Lin, K., Chua, T.S., Theobalt, C.: Neural sparse voxel fields. Adv. Neural. Inf. Process. Syst. **33**, 15651–15663 (2020)
15. Luo, X., Huang, J.B., Szeliski, R., Matzen, K., Kopf, J.: Consistent video depth estimation. ACM Trans. Graph. **39**(4), 71:1–71:13 (2020)
16. McMillan, L., Bishop, G.: Plenoptic modeling: an image-based rendering system. In: Proceedings of the 22nd Annual Conference on Computer Graphics and Interactive Techniques, pp. 39–46 (1995)
17. Mildenhall, B., et al.: Local light field fusion: practical view synthesis with prescriptive sampling guidelines. ACM Trans. Graph. **38**(4), 1–14 (2019)
18. Mildenhall, B., Srinivasan, P.P., Tancik, M., Barron, J.T., Ramamoorthi, R., Ng, R.: NERF: representing scenes as neural radiance fields for view synthesis. In: ECCV, pp. 405–421 (2020)
19. Moreno-Noguer, F., Fua, P.: Stochastic exploration of ambiguities for nonrigid shape recovery. IEEE Trans. Pattern Anal. Mach. Intell. **35**(2), 463–475 (2012)
20. Navarro, J., Buades, A.: Robust and dense depth estimation for light field images. IEEE Trans. Image Process. **26**(4), 1873–1886 (2017)
21. Ng, R., Levoy, M., Brédif, M., Duval, G., Horowitz, M., Hanrahan, P.: Light field photography with a hand-held plenoptic camera. Ph.D. thesis, Stanford University (2005)

22. Niklaus, S., Mai, L., Yang, J., Liu, F.: 3D Ken Burns effect from a single image. ACM Trans. Graph. **38**(6), 1–15 (2019)
23. Porter, T., Duff, T.: Compositing digital images. SIGGRAPH Comput. Graph. **18**(3), 253–259 (1984)
24. Pumarola, A., Corona, E., Pons-Moll, G., Moreno-Noguer, F.: D-NERF: neural radiance fields for dynamic scenes. In: CVPR, pp. 10318–10327 (2021)
25. Ranftl, R., Lasinger, K., Hafner, D., Schindler, K., Koltun, V.: Towards robust monocular depth estimation: mixing datasets for zero-shot cross-dataset transfer. In: IEEE Transactions on Pattern Analysis and Machine Intelligence, p. 1 (2020)
26. Schönberger, J.L., Frahm, J.M.: Structure-from-motion revisited. In: CVPR, pp. 4104–4113 (2016)
27. Shade, J., Gortler, S., He, L.W., Szeliski, R.: Layered depth images. In: ACM SIGGRAPH, pp. 231–242 (1998)
28. Shih, M.L., Su, S.Y., Kopf, J., Huang, J.B.: 3D photography using context-aware layered depth inpainting. In: CVPR, pp. 8025–8035 (2020)
29. Sitzmann, V., Thies, J., Heide, F., Nießner, M., Wetzstein, G., Zollhofer, M.: DeepVoxels: learning persistent 3D feature embeddings. In: ICCV, pp. 2437–2446 (2019)
30. Srinivasan, P.P., Wang, T., Sreelal, A., Ramamoorthi, R., Ng, R.: Learning to synthesize a 4D RGBD light field from a single image. In: ICCV, pp. 2243–2251 (2017)
31. Tang, C., Yuan, L., Tan, P.: LSM: learning subspace minimization for low-level vision. In: CVPR, pp. 6235–6246 (2020)
32. Tomasi, C., Kanade, T.: Shape and motion from image streams under orthography: a factorization method. Int. J. Comput. Vision **9**(2), 137–154 (1992)
33. Tucker, R., Snavely, N.: Single-view view synthesis with multiplane images. In: CVPR, pp. 551–560 (2020)
34. Wiles, O., Gkioxari, G., Szeliski, R., Johnson, J.: SynSin: end-to-end view synthesis from a single image. In: CVPR, pp. 7465–7475 (2020)
35. Wizadwongsa, S., Phongthawee, P., Yenphraphai, J., Suwajanakorn, S.: NeX: real-time view synthesis with neural basis expansion. In: CVPR, pp. 8534–8543 (2021)
36. Wulff, J., Black, M.J.: Efficient sparse-to-dense optical flow estimation using a learned basis and layers. In: CVPR, pp. 120–130 (2015)
37. Yoon, J.S., Kim, K., Gallo, O., Park, H.S., Kautz, J.: Novel view synthesis of dynamic scenes with globally coherent depths from a monocular camera. In: CVPR, pp. 5336–5345 (2020)
38. Yu, A., Li, R., Tancik, M., Li, H., Ng, R., Kanazawa, A.: PlenOctrees for real-time rendering of neural radiance fields. In: ICCV, pp. 5752–5761 (2021)
39. Zhang, R., Isola, P., Efros, A.A., Shechtman, E., Wang, O.: The unreasonable effectiveness of deep features as a perceptual metric. In: CVPR, pp. 586–595 (2018)
40. Zhou, T., Tucker, R., Flynn, J., Fyffe, G., Snavely, N.: Stereo magnification: learning view synthesis using multiplane images. ACM Trans. Graph. **37**(4), 1–12 (2018)

3D-Aware Semantic-Guided Generative Model for Human Synthesis

Jichao Zhang[1(✉)], Enver Sangineto[2], Hao Tang[3], Aliaksandr Siarohin[1,4], Zhun Zhong[1], Nicu Sebe[1], and Wei Wang[1]

[1] University of Trento, Trento, Italy
jichao.zhang@unitn.it
[2] University of Modena and Reggio Emilia, Modena, Italy
[3] ETH Zurich, Zurich, Switzerland
[4] Snap Research, Santa Monica, USA

Abstract. Generative Neural Radiance Field (GNeRF) models, which extract implicit 3D representations from 2D images, have recently been shown to produce realistic images representing rigid/semi-rigid objects, such as human faces or cars. However, they usually struggle to generate high-quality images representing non-rigid objects, such as the human body, which is of a great interest for many computer graphics applications. This paper proposes a 3D-aware Semantic-Guided Generative Model (3D-SGAN) for human image synthesis, which combines a GNeRF with a texture generator. The former learns an implicit 3D representation of the human body and outputs a set of 2D semantic segmentation masks. The latter transforms these semantic masks into a real image, adding a realistic texture to the human appearance. Without requiring additional 3D information, our model can learn 3D human representations with a photo-realistic, controllable generation. Our experiments on the DeepFashion dataset show that 3D-SGAN significantly outperforms the most recent baselines. The code is available at https://github.com/zhangqianhui/3DSGAN.

Keywords: Generative neural radiance fields · Human image generation

1 Introduction

Recent deep generative models can generate and manipulate high-quality images. For instance, Generative Adversarial Networks (GANs) [14], have been applied to different tasks, such as image-to-image translation [10,25,85], portrait editing [2,67,68,75], and semantic image synthesis [55], to mention a few. However, most state-of-the-art GAN models [19,20,28–32] are trained using 2D images only, operate in the 2D domain, and ignore the 3D nature of the world. Thus, they often struggle to disentangle the underlying 3D factors of the represented objects.

Supplementary Information The online version contains supplementary material available at https://doi.org/10.1007/978-3-031-19784-0_20.

S. Avidan et al. (Eds.): ECCV 2022, LNCS 13675, pp. 339–356, 2022.
https://doi.org/10.1007/978-3-031-19784-0_20

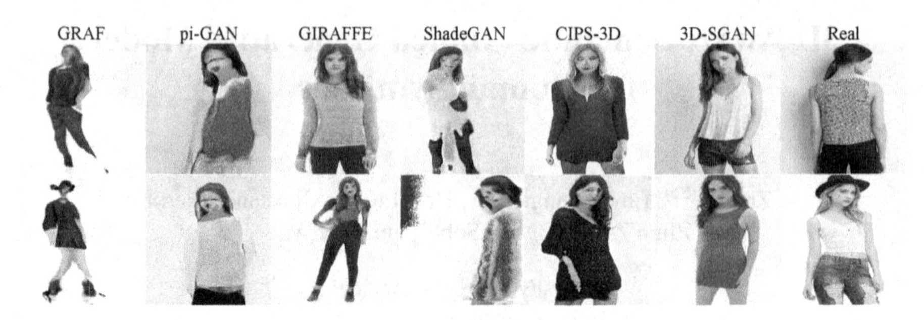

Fig. 1. A qualitative comparison between different generation methods: GRAF [66], pi-GAN [7], GIRAFFE [52], ShadeGAN [54] CIPS-3D [83], and 3D-SGAN (Ours).

Recently, different 3D-aware generative models [49,50,78] have been proposed to solve this problem. Since most of these methods do not need 3D annotations, they can create 3D content while reducing the hardware costs of common computer graphics alternatives. Differently from generating 3D untextured shapes [13,78], some of these methods [9,38,49,50,86] focus on 3D-aware realistic image generation and controllability. Generally speaking, these models mimic the traditional computer graphics rendering pipeline: they first model the 3D structure, then they use a (differentiable) projection module to project the 3D structure into 2D images. The latter may be a depth map [9], a sketch [86] or a feature map [49] which is finally mapped into the real image by a rendering module. During training, some methods require 3D data [9,86], and some [38,49,50] can learn a 3D representation directly from raw images.

An important class of *implicit* 3D representations are the Neural Radiance Fields (NeRFs), which can generate high-quality unseen views of complex scenes [7,12,26, 47,56–58]. Generative NeRFs (GNeRFs) combine NeRFs with GANs in order to condition the generation process with a latent code governing the object's appearance or shape [7,52,66]. However, these methods [7,52,54,66,83] focus on relatively simple and "rigid" objects, such as cars and faces, and they usually struggle to generate highly non-rigid objects such as the human body (e.g., see Fig. 1). This is likely due to the fact that the human body appearance is highly variable because of both its articulated poses and the variability of the clothes texture, being these two factors entangled with each other. Thus, adversarially learning the data distribution modeling all these factors, is a hard task, especially when the training set is relatively small.

To mitigate this problem, we propose to *split* the human generation process in two separate steps and use intermediate segmentation masks as the bridge of these two stages. Specifically, our 3D-aware Semantic-Guided Generative model (3D-SGAN) is composed of two generators: a GNeRF model and a texture generator. The GNeRF model learns the 3D structure of the human body and generates a semantic segmentation of the main body components, which is largely invariant to the surface texture. The texture generator translates the previous segmentation output into a photo-realistic image. To control the texture style, a Variational AutoEncoder (VAE [34]) approach with a StyleGAN-like [31] decoder is used to modulate the final decoding process. The similar idea has been used in [37], but their semantic generator is 2D, and it cannot perform 3D manipulations. We empirically show that splitting the human generation process into these two stages brings the following three advantages. First, the GNeRF model is able to learn the intrinsic 3D geometry of the human body, even when trained with a small dataset. Second, the texture generator can successfully translate semantic information into a textured object. Third, both generators can be controlled by explicitly varying their respective conditioning latent codes. Moreover, we propose two consistency losses to further disentangle the latent codes representing the garment type (which we call the "semantic" code) and the human pose. Finally, since there is no general metric which can be used to evaluate the 3D consistency of image generation with multiple viewpoints, we propose a point matching-based metric which we name *average Matched Points (aMP)*. Experiments conducted on the DeepFashion dataset [40] show that 3D-SGAN can generate high-quality person images significantly outperforming state-of-the-art approaches. In summary, the main contributions of this work are:

1) We propose 3D-SGAN, which combines a GNeRF with a VAE-conditioned texture generator for human-image synthesis.
2) We propose two consistency losses to increase the disentanglement between semantic information (e.g., garment type) and the human pose.
3) We show that 3D-SGAN generates high-quality human images, significantly outperforming the previous controllable state-of-the-art methods.
4) We propose a new metric (aMP) to evaluate the 3D-view consistency.

2 Related Work

3D-aware image synthesis is based on generative models which incorporate a 3D scene representation. This allows rendering photo-realistic images from different viewpoints. Early methods use GAN-based architectures for building 3D voxel [13,22,42,78] or mesh [21,60] representations. However, they mostly focus on learning untextured 3D structures. More recently, different methods learn textured representations directly from 2D images [12,49,51,52,66,77,86]. The resulting controllable 3D scene representation can be used for image synthesis. Some of these methods [9,86] require extra 3D data for disentangling shape from texture. The main idea is to generate an internal 3D shape and then project this shape into 2D sketches [86] or depth maps [9], which are finally rendered in a realistic image. Other methods are directly trained on 2D images without using 3D data [49,50,52,66,72,79]. For instance, inspired by StyleGANv2 [31], Thu et al. [49] propose HoloGAN, which predicts 3D abstract features using 3D convolutions, and then projects these features into a 2D representation which is finally decoded into an image. However, the learnable projection function, e.g., the decoder, results in an entangled representation, thus the view-consistency of the generated images is degraded. Katja et al. [66] use a NeRF to represent the 3D scene and a volume rendering technique to render the final image. However, this model works at relatively low image resolutions and it is restricted to single-object scenes. To tackle these issues, some works propose object-aware scene representations. For example, Liao et al. [38] combine a 2D generator and a projection module with a 3D generator which outputs multiple abstract 3D primitives. Every stage in this model outputs multiple terms to separately represent each object. Instead of abstract 3D primitives, Phuoc et al. [50] use

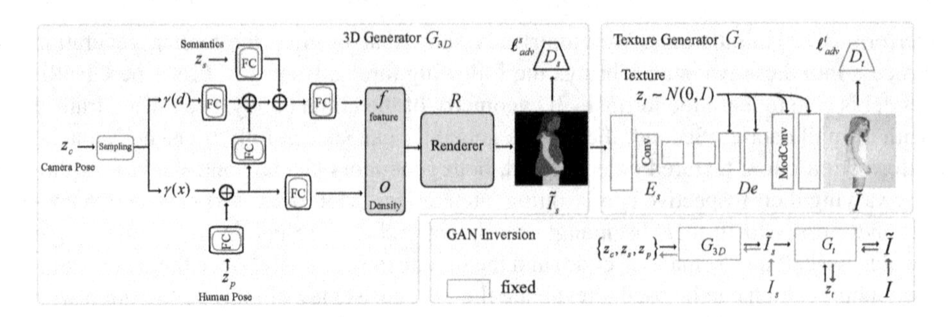

Fig. 2. An overview of the proposed 3D-SGAN architecture, composed of two main generators. G_{3D} (on the left) follows a GNeRF structure, with a NeRF kernel used to represent implicit 3D information, latent codes governing different appearance variations and a discriminator (D_s) which is used for adversarial training. The output of G_{3D} is the semantic masks $\tilde{I}_s$ (middle). The second generator (G_t, right) translates the semantic masks into a photo-realistic image $\tilde{I}$. Also G_t is trained adversarially (see top right, the second discriminator D_t). The human generation process can be controlled by interpolating different latent codes: the semantics $\boldsymbol{z}_s$, the pose $\boldsymbol{z}_p$, the camera $\boldsymbol{z}_c$, and the texture code $\boldsymbol{z}_t$. The bottom of the figure shows the GAN inversion scheme.

a voxel feature grid as the 3D representation, but their method fails to generate consistent images at high-resolution. Michael et al. [52] recently introduced GIRAFFE, a multiple-object scene representation based on NeRFs, jointly with an object composition operator. GIRAFFE is the state-of-the-art 3D-aware approach for both single and multiple object generation tasks. A few very recent papers [54,73] propose to learn an accurate object geometry by introducing a relighting module into the rendering process. Xu et al. [80], explicitly learn a structural and a textural representation (a feature volume), which is used jointly with the implicit NeRF mechanism. Chan et al. [8], propose to replace the 3D volume with three projection feature planes. Finally, 3D-consistency is addressed in [17,53,83], where, e.g., StyleGAN-based networks are used for neural rendering. However, most of these works fail to disentangle the semantic attributes.

While previous methods can achieve impressive rigid-object generation and manipulation results, they usually struggle to deal with non-rigid objects with complex pose and texture variations. For instance, the human body is a non-rigid object which is very important in many generative applications.

GANs for Human Generation. GANs [14] have been widely used for different object categories, such as, for instance, faces [6,28,31,32], cars [31,32,50,66], and churches [52]. However, GANs still struggle to produce high-quality full-human body images, because of the complex pose variations. Very recently, Sarkar et al. [64] proposed a VAE-GAN model for the pose transfer and the part sampling tasks. In more detail, this model extracts an UV texture map from the input image using DensePose [3], and then encodes the texture into a Gaussian distribution. Then, it samples from this distribution and warps the sample into the target pose space. Finally, the warped latent code is used as input to the decoder. Compared to Sarkar et al. [64], our method does not use an SMPL nor DensePose to extract the point correspondences as additional supervised information. Despite that, 3D-SGAN can learn 3D representations of the human body and control the generation process (e.g., by changing the input camera parameters).

StylePeople [16] is based on a full-body human avatar, which combines StyleGANv2 [31] with *neural dressing*. The StyleGANv2 module samples neural textures, and these textures are superimposed on the meshes of an SMPL. The textured meshes are finally rendered into an image. In contrast, our 3D-SGAN can perform semantic disentanglement and manipulation using semantic codes.

Pose transfer aims to synthesize person images in a novel view or in a new pose. This is a very challenging task, since it requires very complicated spatial transformations to account for different poses. Most works in this field can be categorized by the way in which the human pose is represented. Early works are based on keypoints [5,24, 27,43–46,59,61,62,69–71,74,81,84,87]. More recent methods [15,39,48,63,65] use correspondences between pixel location in 2D images and points in SMPL [41] (usually estimated using DensePose [18]). However, these approaches usually struggle to simultaneously provide a realistic and a 3D controllable person generation.

3 Preliminaries

NeRF [47] is an implicit model which represents a 3D scene using the weights of a multilayer perceptron (MLP). This MLP (h) takes as input a 3D coordinate $\boldsymbol{x} \in \mathbb{R}^3$ and a view direction $\boldsymbol{d} \in \mathbb{R}^2$, and outputs the density (or "opacity", o) and the view-dependent RGB color value $\boldsymbol{c}$:

$$(\boldsymbol{c}, o) = h(\gamma(\boldsymbol{x}), \gamma(\boldsymbol{d})), \tag{1}$$

where γ is a positional encoding function [76]. On the other hand, Generative NeRF (GNeRF) [66] is a NeRF conditioned on the latent codes $\boldsymbol{z}_g$ and $\boldsymbol{z}_a$, respectively representing the geometric shape and the object appearance, and drawn from a priori distributions. GNeRFs [52,66] are trained using an adversarial approach. In GIRAFFE [52], the color value ($\boldsymbol{c}$ in Eq. 1) is replaced by an intermediate feature vector $\boldsymbol{f}$:

$$(\boldsymbol{f}, o) = h(\gamma(\boldsymbol{x}), \gamma(\boldsymbol{d}), \boldsymbol{z}_g, \boldsymbol{z}_a). \tag{2}$$

$\boldsymbol{f}$ is mapped into a photo-realistic image using a volume and neural rendering module R and fed to a discriminator (more details in [52,66]).

Our 3D generator (see Sect. 4) is inspired by GIRAFFE [52]. However, it learns to produce a segmentation image, a simpler task with respect to directly generating a photo-realistic image (see Sect. 1).

4 The Proposed 3D-SGAN

Figure 2 shows the proposed 3D-SGAN architecture, composed of two main modules: a 3D-based segmentation mask generator and a texture generator. The former (G_{3D}) generates semantic segmentation masks of the human body which correspond to the main body parts and depend on the type of clothes, the camera viewpoint and the human pose. On the other hand, the texture generator (G_t) takes as input these segmentation masks and translates them into a photo-realistic image, adding a texture style randomly drawn from a pre-learned marginal distribution. The two modules are trained separately.

4.1 3D Generator for Semantic Mask Rendering

Given a set of 2D human image samples $\{I^i\}_{i=1}^N$, we first use an off-the-shelf human parsing tool [4] to obtain the corresponding ground-truth semantic segmentation masks $\{I_s^i\}_{i=1}^N$. Using $T = \{(I^i, I_s^i)\}_{i=1}^N$ as our training set, the goal is to train a two-step generative model:

$$\tilde{I} = G(\boldsymbol{z}_c, \boldsymbol{z}_s, \boldsymbol{z}_p, \boldsymbol{z}_t) = G_t(G_{3D}(\boldsymbol{z}_c, \boldsymbol{z}_s, \boldsymbol{z}_p), \boldsymbol{z}_t), \tag{3}$$

where $\tilde{I}$ is the final generated image. The latent codes $\boldsymbol{z}_c \sim P_c$ (see Sect. 5), $\boldsymbol{z}_s \sim \mathcal{N}(0, \boldsymbol{I})$, $\boldsymbol{z}_p \sim \mathcal{N}(0, \boldsymbol{I})$, and $\boldsymbol{z}_t \sim \mathcal{N}(0, \boldsymbol{I})$ represent, respectively: the camera viewpoint, the semantics (i.e., the garment type), the body pose and the human texture.

The structure of our 3D Generator G_{3D} is inspired by GIRAFFE [52] (Sect. 3). However, differently from [52,66], which learn to generate a textured object, in our case, h learns to generate a semantically segmented image. Specifically, we use a latent semantic code ($\boldsymbol{z}_s$) to condition the final segmentation output on the type of garment. As shown in Fig. 2, $\boldsymbol{z}_s$ does not influence the opacity generation branch, and it is injected into the direction-dependent branch, which finally outputs a feature vector $\boldsymbol{f}$, representing a point-wise semantic content. Formally, we have:

$$(\boldsymbol{f}, o) = h(\gamma(\boldsymbol{x}), \gamma(\boldsymbol{d}), \boldsymbol{z}_c, \boldsymbol{z}_s, \boldsymbol{z}_p). \tag{4}$$

Following [52,66], we generate a set of pairs $\{(\boldsymbol{f}, o)\}$ which are finally projected into the 2D plane using a rendering module R [47,52] (see Sect. 3), and represented by the segmentation masks $\tilde{I}_s$. Specifically, $\tilde{I}_s$ is a tensor composed of n_s channels, where each channel represents a segmentation mask of the same spatial resolution of the real images in T (Fig. 2).

G_{3D} is trained jointly with a discriminator D_s, which learns to discriminate between real (I_s) and fake ($\tilde{I}_s$) segmentation masks (more details in Sect. 4.4).

4.2 VAE-Conditioned Texture Generator

The goal of our texture generator G_t is twofold: (1) mapping the segmentation masks $\tilde{I}_s$ generated by G_{3D} into a textured human image and (2) learning a marginal distribution of the human texture using the dataset T. The latter is obtained using a Variational AutoEncoder (VAE [34]) framework, which we use to learn how to *modulate* the texture style of the decoder. Specifically, as shown in Fig. 3, G_t is composed of a semantic encoder E_s, a texture encoder E_t, and a decoder De. De is based on a StyleGANv2 architecture [31], in which a style code is used to "demodulate" the weights of each convolutional layer. We modify this architecture using a variational approach, in which the style code, *at inference time*, is extracted from a learned marginal distribution. In more detail, given a segmentation tensor I_s, we use E_s to extract the semantic content which is decoded into the final image using De and a texture code z_t. The latter is sampled using the VAE encoder E_t, which converts a real input image I into a latent-space normal distribution ($\mathcal{N}(\boldsymbol{\mu}, \boldsymbol{\sigma})$), from which $\boldsymbol{z}_t$ is randomly chosen:

$$(\boldsymbol{\mu}, \boldsymbol{\sigma}) = E_t(I), \boldsymbol{z}_t \sim \mathcal{N}(\boldsymbol{\mu}, \boldsymbol{\sigma}), \tilde{I} = De(E_s(I_s), \boldsymbol{z}_t). \tag{5}$$

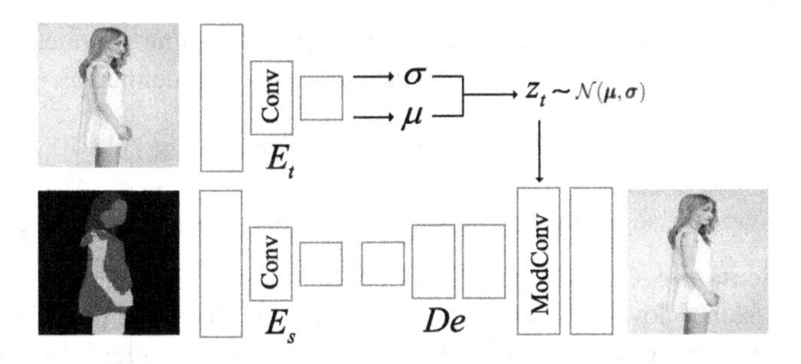

Fig. 3. The proposed VAE-conditioned texture generator. ModConv stands for "Modulated Convolution" [31].

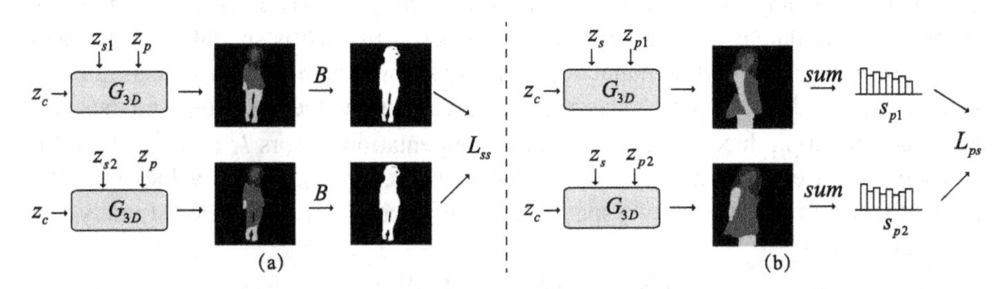

Fig. 4. A schematic representation of our consistency losses ℓ_{ss} (a) and ℓ_{ps} (b).

G_t is trained using the pairs in T. Specifically, given a pair of samples (I^i, I^i_s), we use an adversarial loss ℓ^t_{adv} (and a dedicated discriminator D_t), jointly with a reconstruction loss ℓ_r, and a standard Kullback-Leibler divergence ($\mathcal{D}_{kl}$) loss (ℓ_{kl}) [34]:

$$\ell_r = ||G_t(I^i_s, \boldsymbol{z}_t) - I^i||_1, \ell_{kl} = \mathcal{D}_{kl}(\mathcal{N}(\boldsymbol{\mu}, \boldsymbol{\sigma})||\mathcal{N}(\boldsymbol{0}, \boldsymbol{I})). \tag{6}$$

Note that, in the reconstruction process, G_t cannot ignore the segmentation tensor (I^i_s) and its corresponding encoder E_s. In fact, the information extracted from the real image I, and encoded using E_t, is not enough for the decoder to represent the image content, since $\boldsymbol{z}_t$ is used only as a style modulator in De.

4.3 Consistency Losses for Semantics and Pose Disentanglement

In G_{3D}, the opacity value (o), computed by h, does not depend on the latent code $\boldsymbol{z}_s$. Despite that, we have empirically observed that the semantics ($\boldsymbol{z}_s$) and the pose ($\boldsymbol{z}_p$) representations are highly entangled. We presume this is due to the convolutional filters in R (Sect. 3), where the two latent factors are implicitly merged. In order to increase the disentanglement of these factors, we propose two self-supervised consistency losses.

Silhouette-Based Geometric Consistency. This loss is based on the idea that two different body segmentations (e.g., long-sleeve vs. short- sleeve, etc.), produced using two different semantic codes $\boldsymbol{z}_{s1}$ and $\boldsymbol{z}_{s2}$, *but keeping fixed the pose and the camera codes*, once they are binarized, should correspond to roughly the same silhouette (see Fig. 4 (a)). Formally, the proposed geometric consistency loss ℓ_{ss} is defined as:

$$\ell_{ss} = \| B(G_{3D}(\boldsymbol{z}_c, \boldsymbol{z}_{s1}, \boldsymbol{z}_p)) - B(G_{3D}(\boldsymbol{z}_c, \boldsymbol{z}_{s2}, \boldsymbol{z}_p)) \|_1, \tag{7}$$

where $B(\tilde{I}_s)$ maps the segmentation masks $\tilde{I}_s$ into a binary silhouette image.

Pose-Based Semantic Consistency. Analogously to ℓ_{ss}, the proposed pose-based semantic consistency loss is based on the idea that two different pose codes should produce a similar body segmentation. However, as shown in Fig. 4 (b), despite the body being partitioned in similar semantic segments (e.g., because the clothes have not changed), when the human pose changes, the overall spatial layout of these segments can also change (e.g., see the two different arm positions in Fig. 4 (b)). For this reason, we formulate a semantic consistency loss (ℓ_{ps}) which is spatial-invariant, and it is based on the channel-by-channel comparison of two segmentation masks. In more detail, given two different pose codes $\boldsymbol{z}_{p1}$ and $\boldsymbol{z}_{p2}$, and fixing the semantics and the camera code, we first produce two corresponding segmentation tensors $\tilde{I}_s^1$ and $\tilde{I}_s^2$. Then, for each tensor and each channel, we sum all the channel-specific mask values over the spatial dimension and we get two spatial invariant vectors $\boldsymbol{s}_{p1}, \boldsymbol{s}_{p2} \in \mathbb{R}^{n_s}$. Finally, ℓ_{ps} is given by:

$$\ell_{ps} = \sum_{i=1}^{n_s} [\max(\frac{\mid \boldsymbol{s}_{p1}[i] - \boldsymbol{s}_{p2}[i] \mid}{\boldsymbol{s}_{p1}[i] + \epsilon}, \rho) - \rho], \tag{8}$$

where $\boldsymbol{s}[i]$ is the i-th channel value of vector $\boldsymbol{s}$, ϵ is a small value used for numerical stability, and ρ is a margin representing the tolerable channel-wise difference.

4.4 Training and Inference

G_{3D} is trained using an adversarial loss (ℓ^s_{adv}) jointly with ℓ_{ss} and ℓ_{ps} (Sect. 4.3):

$$\ell_{3D} = \ell^s_{adv} + \lambda_1 \ell_{ss} + \lambda_2 \ell_{ps}, \tag{9}$$

where λ_1 and λ_2 are hyper-parameters controlling the contribution of each loss term.

G_t is trained using a variational-adversarial approach (VAE-GAN [36]):

$$\ell_{tr} = \ell^t_{adv} + \lambda_3 \ell_r + \lambda_4 \ell_{kl}, \tag{10}$$

where λ_3 and λ_4 are hyper-parameters, and ℓ_{tr} is the overall objective function of G_t.

G_{3D} and G_t are trained separately. However, at inference time, the tensor $\tilde{I}_s$, generated by G_{3D}, is fed to G_t, along with a texture code $\boldsymbol{z}_t$, randomly drawn from a standard normal distribution:

$$\boldsymbol{z}_t \sim \mathcal{N}(\boldsymbol{0}, \boldsymbol{I}), \tilde{I} = De(E_s(\tilde{I}_s), \boldsymbol{z}_t). \tag{11}$$

4.5 Real Image Editing Using GAN Inversion

The variational method proposed in Sect. 4.2 cannot completely reconstruct the input image. For real image editing, we use a GAN inversion technique [1] to optimize the

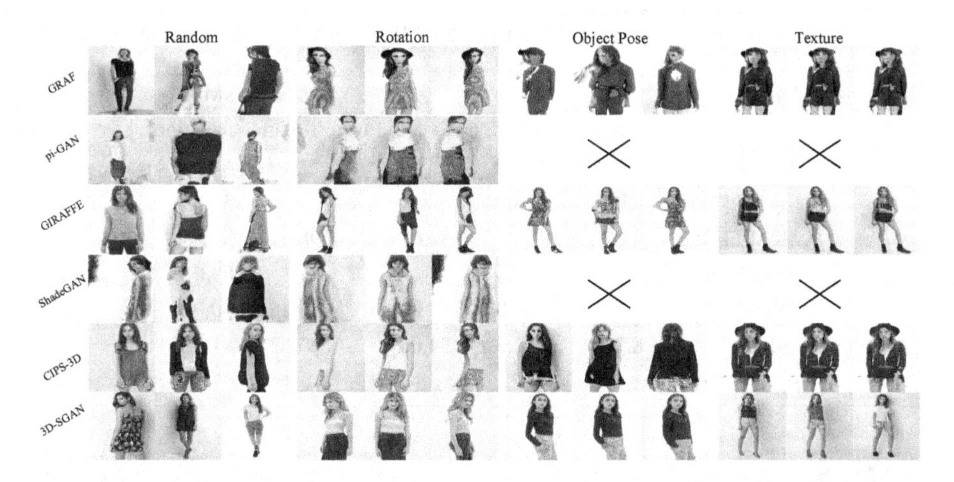

Fig. 5. A qualitative comparison. 'Random' means that the results are generated by randomly sampling the latent codes from the corresponding learned marginal distributions. The other 3 columns show controllable person generations with respect to the rotation, the human pose, and the texture attribute. The lack of the 'Object Pose' and the 'Texture' results for both pi-GAN and ShadeGAN is due to the fact that both methods use a single latent code to model both the texture and the geometry.

values of the latent codes corresponding to a real input image I. Since we have two separate generators (G_{3D} and G_t), the optimization process is based on two steps (see Fig. 2, bottom). Specifically, given a pair of real image and its corresponding segmentation masks (extracted using [4], see Sect. 4.1) (I, I_s), we first generate $\tilde{I}_s = G_{3D}(\boldsymbol{z}_c, \boldsymbol{z}_s, \boldsymbol{z}_p)$ and we optimize $||\tilde{I}_s - I_s||_1$ with respect to $\boldsymbol{z}_c, \boldsymbol{z}_s$ and $\boldsymbol{z}_p$. Let $\boldsymbol{z}_c^*, \boldsymbol{z}_s^*$ and $\boldsymbol{z}_p^*$ be the optimal values so found, and let $\tilde{I}_s^* = G_{3D}(\boldsymbol{z}_c^*, \boldsymbol{z}_s^*, \boldsymbol{z}_p^*)$. Then, we use $\tilde{I} = G_t(\tilde{I}_s^*, \boldsymbol{z}_t)$ and we optimize $||\tilde{I} - I||_1 + \tau LPIPS(\tilde{I}, I)$ with respect to $\boldsymbol{z}_t$, where $LPIPS(I^1, I^2)$ is the $LPIPS$ distance between two images [82] and we use $\tau = 10$.

Once obtained the latent codes $(\boldsymbol{z}_c^*, \boldsymbol{z}_s^*, \boldsymbol{z}_p^*, \boldsymbol{z}_t^*)$ corresponding to a real image, editing can be easily done by changing these codes.

5 Experiments

Datasets. We use the DeepFashion In-shop Clothes Retrieval benchmark [40], which consists of 52,712 high-resolution (1101×750 pixels) person images with various appearances and poses. This dataset has been widely used in pose transfer tasks. We use the following preprocessing. First, we remove overly cropped images, such as incomplete images of humans. Then, the remaining 42,977 images are resized into a 256×256 resolution, and are divided into 41,001 training and 1,976 testing images.

Training Details. Following GIRAFFE [52], the camera distribution P_c can be implemented by first sampling the camera code from a uniform distribution over the dataset-dependent camera elevation angles, and then applying an object affine transformation

Table 1. A quantitative comparison using the FID (↓) and the aMP (↑) scores.

Method	FID ↓				aMP ↑
	Random	Rotation	Object-Pose	Texture	Rotation
GRAF [66]	52.68	176.9	57.76	220.9	64.0
pi-GAN [7]	137.6	213.7	-	-	58.0
GIRAFFE [52]	42.73	123.4	82.61	98.41	51.0
ShadeGAN [54]	134.7	232.4	-	-	**89.0**
CIPS-3D [83]	69.45	156.9	233.6	**36.16**	60.0
3D-SGAN	**8.240**	**117.3**	**54.00**	60.63	81.0

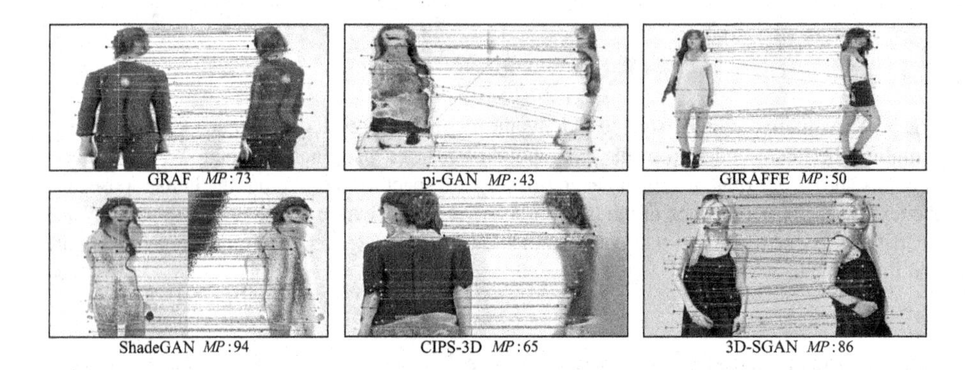

Fig. 6. Computing MP between pairs of generated images with 2 different viewpoints.

to sample 3D points and rays. Both G_t and G_{3D} are trained using the RMSprop optimizer [35]. The learning rate for both the discriminator and the generator is set to 10^{-4}. For the loss weights, we use: $\lambda_1 = 0.01$, $\lambda_2 = 0.01$, $\lambda_3 = 1$, and $\lambda_4 = 1$. For GAN inversion, we use the Adam optimizer [33] with a learning rate of 10^{-2}.

Baselines. We compare 3D-SGAN with five state-of-the-art 3D-aware generative approaches, i.e., GRAF [66], pi-GAN [7], GIRAFFE [52], ShadeGAN [54] and CIPS-3D [83]. For each baseline, we use the corresponding publicly available code with a few minor adaptations for the DeepFashion dataset. Note that some concurrent methods, such as StyleNeRF [17], GRAM [11], Tri-plane [8], StyleSDF [53] achieve a performance very similar to CIPS-3D. Moreover, for some of them there is no released code yet, thus, a direct comparison is not possible. The comparison with the 2D-GAN model HumanGAN [64] can be found in the supplementary material.

Metrics. We adopt the widely used FID [23] scores to evaluate the quality of the generated human images, following common protocols (e.g., using 5,000 fake samples, etc.). And we propose the average Matched Points (aMP) to evaluate the 3D-view consistency of the generated images. The supplementary document provides the introduction of this metric.

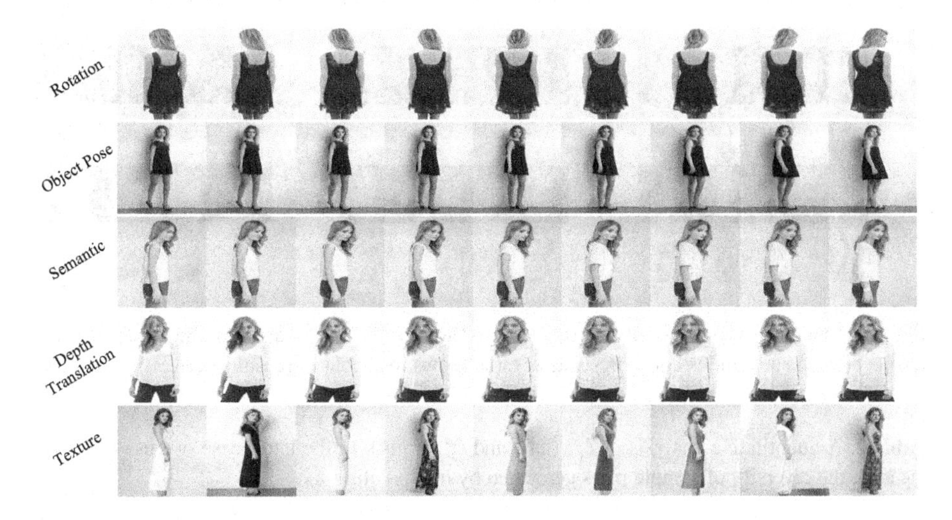

Fig. 7. Controllable person generation by interpolating latent codes (Rows 1–4). The fifth row shows texture generation results obtained randomly sampling $\boldsymbol{z}_t$.

5.1 Comparisons with State-of-the-Art Methods

Unconditioned Human Generation. Figure 5 ("Random" column) shows a qualitative comparison between image samples generated by all the models. GRAF [66], pi-GAN [7] and ShadeGAN [54] fail to generate realistic human images. GIRAFFE [52] and CIPS-3D [83] generate reasonable human images, but they suffer from visual artifacts and texture blurs. In contrast, 3D-SGAN synthesizes much better and more photo-realistic images. This qualitative analysis is confirmed in Table 1, where the corresponding FID scores show that 3D-SGAN significantly outperforms all the other baselines.

Controllable Human Generation. We analyse the representation controllability of all the models, which reflects the ability to disentangle different attributes from each other. We do this by manipulating a single latent code while fixing the others. Figure 5 (columns "Rotation", "Object Pose" and "Texture") shows a qualitative comparison by varying only a single latent code. We observe that all the models can rotate the camera viewpoint. However, GRAF and CIPS-3D fail to disentangle the object pose and the texture. Moreover, pi-GAN and ShadeGAN also suffer from the same problem, since they use one single latent code to model both texture and geometry. On the other hand, both GIRAFFE [52] and 3D-SGAN can effectively disentangle the different variation factors, but GIRAFFE [52] suffers from multi-view inconsistencies and mode collapse for the texture generation. In Table 1, we use FID scores to evaluate the realistic degree of each attribute (e.g., "Rotation", etc.). This is done computing FIDs using only the manipulated (e.g., rotated) fake images, which are compared with all the real images in the dataset. Note that this protocol cannot measure the attribute-based consistency. In most cases, 3D-SGAN has better FID scores than the other baselines.

In order to evaluate the 3D-view consistency, we use our proposed aMP metric (Sect. 5). Table 1 shows that 3D-SGAN gets the best aMP scores with respect to all the

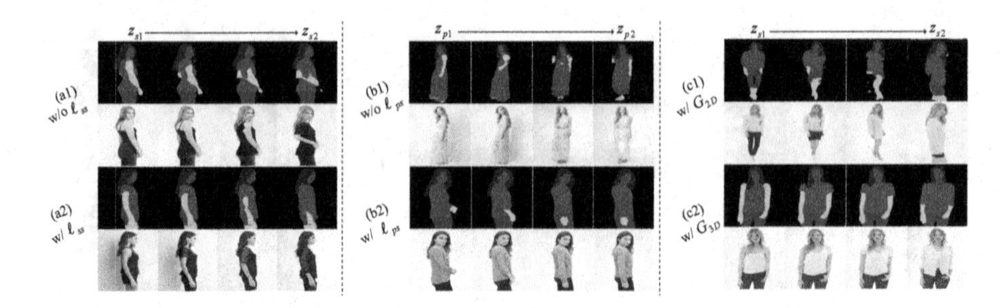

Fig. 8. A qualitative analysis of ℓ_{ss} (a), ℓ_{ps} (b), and G_{3D} (c). (a) and (c) show interpolation results between semantics codes $\boldsymbol{z}_{s1}$ and $\boldsymbol{z}_{s2}$. (b) shows interpolation results between pose codes $\boldsymbol{z}_{p1}$ and $\boldsymbol{z}_{p2}$.

Table 2. A quantitative analysis of ℓ_{ss} (left) and ℓ_{ps} (right). In the latter case, we use LPIPS to measure the diversity of sample pairs generated by interpolating $\boldsymbol{z}_p$.

Metrics	w/o ℓ_{ss}	w/ ℓ_{ss}	Metrics	w/o ℓ_{ps}	w/ ℓ_{ps}
L1 ↓	5.2489	**3.9614**	LPIPS ↓	0.1132	**0.0393**

other methods except from ShadeGAN, which however generates much less realistic images, as testified by the very high FIDs (134.7 vs. our 8.24, Table 1, first column) and qualitatively shown in Fig. 6.

Figure 7 shows additional controllable human image generation results obtained with 3D-SGAN. The generated images are realistic and, in most cases, the attributes are effectively disentangled. Specifically, Fig. 7 (1-st row) shows camera rotation results. The images generated by interpolating the camera pose parameter are consistent, and the transition from one image to the next is smooth, while simultaneously preserving the other attributes such as the texture and the pose. On the other hand, the second row shows images generated by interpolating the pose code. We again observe that human identity has been well preserved. Similarly, the other rows show that the non-target attributes have been well preserved. Finally, the third row shows that the head poses from left to right undergo only minor changes ("face frontalization"). This is likely due to both the limited training data and the data bias of the typical fashion images, where people have a frontal face. Additional results are shown in the Supplementary Material.

5.2 Ablation Study

The Consistency Losses. Figure 8 (a) shows a comparison between the results generated by 3D-SGAN with and without ℓ_{ss}. The effectiveness of ℓ_{ss} is shown by observing that, when removed, the generation process suffers from serious geometric inconsistencies. Specifically, the segmentation masks in Fig. 8 (a1) have undesirable pose variations, while Fig. 8 (a2) shows that ℓ_{ss} can largely alleviate this problem. To quantitatively evaluate this effect, we randomly sample two different semantic codes and we

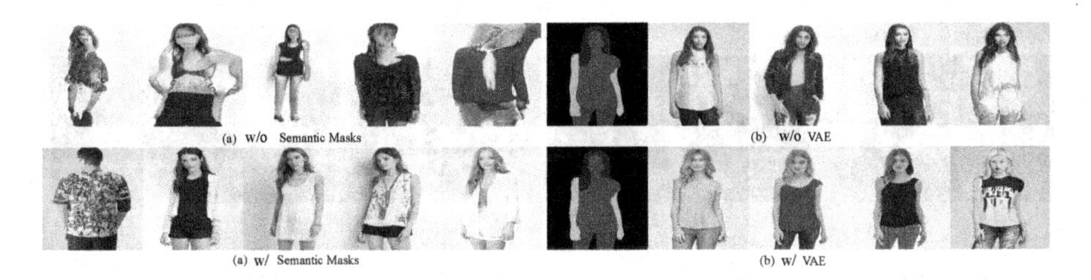

Fig. 9. An analysis of the impact of the semantic masks (a) and the VAE-conditioned texture generator (b).

Table 3. A quantitative analysis of the 3D generator G_{3D}, the semantic masks (SMs) and the VAE in our 3D-SGAN.

Metrics	w/ G_{2D}	w/o SMs	w/o VAE	full
FID ↓	13.24	66.79	14.35	**8.240**

compute the $L1$ distance between the silhouettes of the corresponding generated segmentations. We average the scores over 500 different samples. The results reported in Table 2 (left) validate the effectiveness of this loss for improving the geometric consistency.

Analogously, Fig. 8 (b) qualitatively evaluates the impact of ℓ_{ps} with respect to the semantic consistency over different pose codes. For instance, in Fig. 8 (b1) there is no "red" region in the segmentation masks in the first and in the second column. However, this region is present in columns 3 and 4. Conversely, Fig. 8 (a2) shows that ℓ_{ps} can alleviate this phenomenon. To quantitatively evaluate l_{ps}, we use LPIPS [82], and we measure the average *pairwise* diversity of the sample pairs generated by interpolating z_p (the lower the diversity, the higher the intra-pair consistency). Table 2 (right) shows that the full model achieves a lower diversity than the variant without ℓ_{ps}.

The 3D Generator. To evaluate the benefit of using a GNeRF based generator, we replace it with a vanilla GAN (G_{2D}), which takes the pose and the semantics code as inputs. In this experiment, we keep all the other modules fixed. Note that G_{2D} cannot manipulate the camera parameters and, thus, it cannot generate images from multiple viewpoints. Moreover, G_{3D} can better disentangle the semantics and the pose factors with respect to G_{2D}, as demonstrated by Fig. 8 (c), where we show interpolation results between two different semantic codes. Table 3 shows the G_{3D} (the full model) achieves significantly better FID scores than G_{2D}.

The Semantic Masks and the Texture Generator. Existing methods such as GRAF [66] and GIRAFFE [52] do not use an additional texture generator which translates semantic masks into textured images. In contrast, the effectiveness of our semantic-based approach is shown in Fig. 5 and Table 1. However, to provide an apple-to-apple comparison and further verify the effectiveness of the semantic masks, we use an additional baseline. Specifically, in this baseline, we render the 3D representation of G_{3D} into features rather than semantic masks and we use the texture generator to

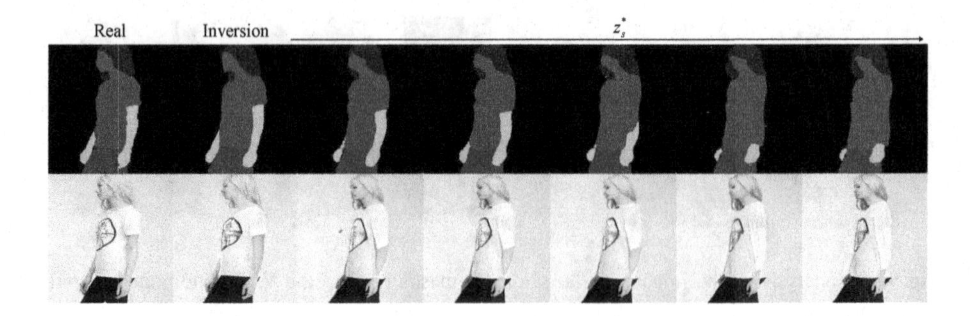

Fig. 10. Real data semantic editing results using GAN inversion.

map these features into the final image. Figure 9 (a) shows the comparison of our full model with this baseline. We observe that the baseline (w/o semantic masks) fails to generate high-quality human images. Table 3 shows that the full model quantitatively outperforms this baseline in terms of FID scores.

The Variational Autoencoder. We evaluate the effect of conditioning G_t using a VAE (Sect. 4.2). This is done by removing the texture encoder E_t jointly with ℓ_{kl} and ℓ_r from Eq. (10). Figure 9 (b) shows the comparison between the VAE-based approach and this variant. Both models generate human images with a high texture variability. However, the variant w/o VAE fails to preserve semantic information, i.e., the coherence between the semantic masks, describing the clothes layout, and the final generated clothes. This shows that our VAE-based G_t learns to effectively map the semantic tensors to human images while modeling the texture distribution with the latent code z_t. Table 3 shows that this variant is significantly outperformed by the proposed VAE-based encoder.

5.3 Real Human Image Editing

In this section, we use GAN inversion for real data editing tasks. The second column of Fig. 10 shows that the optimal code values $(z_c^*, z_s^*, z_p^*, z_t^*)$, obtained using the procedure described in Sect. 4.5, lead to an effective reconstruction of the real input data (first column). In the other columns, we linearly manipulate the semantic code z_s^* while keeping fixed the other codes. Specifically, the second row of Fig. 10 shows the generated images corresponding to the semantic masks in the first row. These results demonstrates the effectiveness of the GAN inversion mechanism and the possibility to apply our model to a wide range of human image editing tasks. We computed the average LPIPS and MS-SSIM scores between real and inversion images, respectively obtaining 0.0301 and 0.912, which confirms the high reconstruction quality of our inversion.

6 Conclusion

We proposed a 3D-aware Semantic-Guided Generative model (3D-SGAN) for human synthesis. We use a generative NeRF to implicitly represent the 3D human body and we render the 3D representation into 2D segmentation masks. Then, these masks are mapped into the final images using a VAE-conditioned texture generator. Moreover, we propose two consistency losses further disentangle the pose and the semantics factors. Our experiments show that the proposed approach generates human images which are significantly more realistic and more controllable than state-of-the-art methods.

Acknowledgements. This work was supported by the EU H2020 projects AI4Media (No. 951911) and SPRING (No. 871245).

References

1. Abdal, R., Qin, Y., Wonka, P.: Image2stylegan: how to embed images into the stylegan latent space? In: ICCV (2019)
2. Abdal, R., Zhu, P., Mitra, N., Wonka, P.: Styleflow: attribute-conditioned exploration of stylegan-generated images using conditional continuous normalizing flows. ACM TOG **40**(3), 1–21 (2020)
3. Alp Güler, R., Neverova, N., Kokkinos, I.: Densepose: dense human pose estimation in the wild. In: CVPR (2018)
4. Badrinarayanan, V., Kendall, A., Cipolla, R.: Segnet: a deep convolutional encoder-decoder architecture for image segmentation. IEEE TPAMI **39**(12), 2481–2495 (2017)
5. Balakrishnan, G., Zhao, A., Dalca, A.V., Durand, F., Guttag, J.: Synthesizing images of humans in unseen poses. In: CVPR (2018)
6. Brock, A., Donahue, J., Simonyan, K.: Large scale gan training for high fidelity natural image synthesis. In: ICLR (2019)
7. Chan, E., Monteiro, M., Kellnhofer, P., Wu, J., Wetzstein, G.: pi-gan: periodic implicit generative adversarial networks for 3d-aware image synthesis. In: CVPR (2021)
8. Chan, E.R., et al.: Efficient geometry-aware 3d generative adversarial networks. arXiv preprint. arXiv:2112.07945 (2021)
9. Chen, X., Cohen-Or, D., Chen, B., Mitra, N.J.: Towards a neural graphics pipeline for controllable image generation. CGF **40**(2), 127–140 (2021)
10. Choi, Y., Choi, M., Kim, M., Ha, J.W., Kim, S., Choo, J.: Stargan: unified generative adversarial networks for multi-domain image-to-image translation. In: CVPR (2018)
11. Deng, Y., Yang, J., Xiang, J., Tong, X.: Gram: generative radiance manifolds for 3d-aware image generation (2022)
12. DeVries, T., Bautista, M.A., Srivastava, N., Taylor, G.W., Susskind, J.M.: Unconstrained scene generation with locally conditioned radiance fields. In: ICCV (2021)
13. Gadelha, M., Maji, S., Wang, R.: 3d shape induction from 2d views of multiple objects. In: 3DV (2017)
14. Goodfellow, I., et al.: Generative adversarial nets. In: NeurIPS (2014)
15. Grigorev, A., Sevastopolsky, A., Vakhitov, A., Lempitsky, V.: Coordinate-based texture inpainting for pose-guided image generation. In: CVPR (2019)
16. Grigorev, A., et al.: Stylepeople: a generative model of fullbody human avatars. In: CVPR (2021)
17. Gu, J., Liu, L., Wang, P., Theobalt, C.: Stylenerf: a style-based 3d-aware generator for high-resolution image synthesis. ICLR (2022)
18. Güler, R.A., Neverova, N., Kokkinos, I.: Densepose: dense human pose estimation in the wild. In: CVPR (2018)
19. Gulrajani, I., Ahmed, F., Arjovsky, M., Dumoulin, V., Courville, A.: Improved training of wasserstein gans. In: NeurIPS (2017)
20. He, Z., Kan, M., Shan, S.: Eigengan: layer-wise eigen-learning for gans. In: ICCV (2021)

21. Henderson, P., Ferrari, V.: Learning single-image 3d reconstruction by generative modelling of shape, pose and shading. International Journal of Computer Vision **128**(4), 835–854 (2019). https://doi.org/10.1007/s11263-019-01219-8
22. Henzler, P., Mitra, N.J., Ritschel, T.: Escaping plato's cave: 3d shape from adversarial rendering. In: ICCV (2019)
23. Heusel, M., Ramsauer, H., Unterthiner, T., Nessler, B., Hochreiter, S.: Gans trained by a two time-scale update rule converge to a local nash equilibrium. In: NeurIPS (2017)
24. Huang, S., et al.: Generating person images with appearance-aware pose stylizer. In: IJCAI (2020)
25. Huang, X., Liu, M.Y., Belongie, S., Kautz, J.: Multimodal unsupervised image-to-image translation. In: ECCV (2018)
26. Jain, A., Tancik, M., Abbeel, P.: Putting nerf on a diet: semantically consistent few-shot view synthesis. In: ICCV (2021)
27. Jinsong, Z., Kun, L., Yu-Kun, L., Jingyu, Y.: PISE: person image synthesis and editing with decoupled gan. In: CVPR (2021)
28. Karras, T., Aila, T., Laine, S., Lehtinen, J.: Progressive growing of gans for improved quality, stability, and variation. In: ICLR (2018)
29. Karras, T., Aittala, M., Hellsten, J., Laine, S., Lehtinen, J., Aila, T.: Training generative adversarial networks with limited data. In: NeurIPS (2020)
30. Karras, T., et al.: Alias-free generative adversarial networks. In: NeurIPS (2021)
31. Karras, T., Laine, S., Aittala, M., Hellsten, J., Lehtinen, J., Aila, T.: Analyzing and improving the image quality of stylegan. In: CVPR (2020)
32. Karras, T., Laine, S., Aila, T.: A style-based generator architecture for generative adversarial networks. In: CVPR (2019)
33. Kingma, D.P., Ba, J.: Adam: a method for stochastic optimization. In: ICLR (2015)
34. Kingma, D.P., Welling, M.: Auto-encoding variational bayes. arXiv preprint. arXiv:1312.6114 (2013)
35. Kingma, D.P., Welling, M.: Auto-encoding variational bayes. In: ICLR (2013)
36. Larsen, A.B.L., Sønderby, S.K., Larochelle, H., Winther, O.: Autoencoding beyond pixels using a learned similarity metric. In: ICML (2016)
37. Lassner, C., Pons-Moll, G., Gehler, P.V.: A generative model of people in clothing. In: Proceedings of the IEEE International Conference on Computer Vision, pp. 853–862 (2017)
38. Liao, Y., Schwarz, K., Mescheder, L., Geiger, A.: Towards unsupervised learning of generative models for 3d controllable image synthesis. In: CVPR (2020)
39. Liu, W., Piao, Z., Tu, Z., Luo, W., Ma, L., Gao, S.: Liquid warping gan with attention: a unified framework for human image synthesis. IEEE TPAMI **44**(9), 5114–5132 (2021)
40. Liu, Z., Luo, P., Qiu, S., Wang, X., Tang, X.: Deepfashion: powering robust clothes recognition and retrieval with rich annotations. In: CVPR (2016)
41. Loper, M., Mahmood, N., Romero, J., Pons-Moll, G., Black, M.J.: SMPL: a skinned multi-person linear model. ACM TOG **34**(6), 1–16 (2015)
42. Lunz, S., Li, Y., Fitzgibbon, A.W., Kushman, N.: Inverse graphics GAN: learning to generate 3d shapes from unstructured 2d data. CoRR abs/2002.12674 (2020). https://arxiv.org/abs/2002.12674
43. Lv, Z., Li, X., Li, X., Li, F., Lin, T., He, D., Zuo, W.: Learning semantic person image generation by region-adaptive normalization. In: CVPR (2021)
44. Ma, L., Jia, X., Sun, Q., Schiele, B., Tuytelaars, T., Van Gool, L.: Pose guided person image generation. In: NeurIPS (2017)
45. Ma, L., Sun, Q., Georgoulis, S., Van Gool, L., Schiele, B., Fritz, M.: Disentangled person image generation. In: CVPR (2018)
46. Men, Y., Mao, Y., Jiang, Y., Ma, W.Y., Lian, Z.: Controllable person image synthesis with attribute-decomposed gan. In: CVPR (2020)

47. Mildenhall, B., Srinivasan, P.P., Tancik, M., Barron, J.T., Ramamoorthi, R., Ng, R.: Nerf: representing scenes as neural radiance fields for view synthesis. In: ECCV (2020)
48. Neverova, N., Alp Guler, R., Kokkinos, I.: Dense pose transfer. In: ECCV (2018)
49. Nguyen-Phuoc, T., Li, C., Theis, L., Richardt, C., Yang, Y.L.: Hologan: unsupervised learning of 3d representations from natural images. In: ICCV (2019)
50. Nguyen-Phuoc, T., Richardt, C., Mai, L., Yang, Y.L., Mitra, N.: Blockgan: learning 3d object-aware scene representations from unlabelled images. In: NeurIPS (2020)
51. Niemeyer, M., Geiger, A.: CAMPARI: camera-aware decomposed generative neural radiance fields. In: 3DV (2021)
52. Niemeyer, M., Geiger, A.: GIRAFFE: representing scenes as compositional generative neural feature fields. In: CVPR (2021)
53. Or-El, R., Luo, X., Shan, M., Shechtman, E., Park, J.J., Kemelmacher-Shlizerman, I.: StyleSDF: high-Resolution 3D-Consistent Image and Geometry Generation. In: CVPR (2022)
54. Pan, X., Xu, X., Loy, C.C., Theobalt, C., Dai, B.: A shading-guided generative implicit model for shape-accurate 3d-aware image synthesis. In: NeurIPS (2021)
55. Park, T., Liu, M.Y., Wang, T.C., Zhu, J.Y.: Semantic image synthesis with spatially-adaptive normalization. In: CVPR (2019)
56. Peng, S., et al.: Animatable neural radiance fields for human body modeling. In: ICCV (2021)
57. Peng, S.,et al.: Neural body: implicit neural representations with structured latent codes for novel view synthesis of dynamic humans. In: CVPR (2021)
58. Reiser, C., Peng, S., Liao, Y., Geiger, A.: Kilonerf: speeding up neural radiance fields with thousands of tiny mlps. In: ICCV (2021)
59. Ren, Y., Yu, X., Chen, J., Li, T.H., Li, G.: Deep image spatial transformation for person image generation. In: CVPR (2020)
60. Rezende, D.J., Eslami, S.M.A., Mohamed, S., Battaglia, P., Jaderberg, M., Heess, N.: Unsupervised learning of 3d structure from images. In: NeurIPS (2016)
61. Sanyal, S., et al.: Learning realistic human reposing using cyclic self-supervision with 3d shape, pose, and appearance consistency. In: ICCV (2021)
62. Sanyal, S., et al.: Learning realistic human reposing using cyclic self-supervision with 3d shape, pose, and appearance consistency. In: Proceedings of the IEEE/CVF International Conference on Computer Vision, pp. 11138–11147 (2021)
63. Sarkar, K., Golyanik, V., Liu, L., Theobalt, C.: Style and pose control for image synthesis of humans from a single monocular view. arXiv preprint. arXiv:2102.11263 (2021)
64. Sarkar, K., Liu, L., Golyanik, V., Theobalt, C.: Humangan: a generative model of humans images (2021)
65. Sarkar, K., Mehta, D., Xu, W., Golyanik, V., Theobalt, C.: Neural re-rendering of humans from a single image. In: Vedaldi, A., Bischof, H., Brox, T., Frahm, J.-M. (eds.) ECCV 2020. LNCS, vol. 12356, pp. 596–613. Springer, Cham (2020). https://doi.org/10.1007/978-3-030-58621-8_35
66. Schwarz, K., Liao, Y., Niemeyer, M., Geiger, A.: Graf: generative radiance fields for 3d-aware image synthesis. In: NeurIPS (2020)
67. Shen, Y., Gu, J., Tang, X., Zhou, B.: Interpreting the latent space of gans for semantic face editing. In: CVPR (2020)
68. Shen, Y., Zhou, B.: Closed-form factorization of latent semantics in gans. In: CVPR (2021)
69. Siarohin, A., Lathuilière, S., Sangineto, E., Sebe, N.: Appearance and pose-conditioned human image generation using deformable GANs. IEEE TPAMI **43**(4), 1156–1171 (2020)
70. Siarohin, A., Lathuilière, S., Tulyakov, S., Ricci, E., Sebe, N.: First order motion model for image animation. In: NeurIPS (2019)
71. Song, S., Zhang, W., Liu, J., Mei, T.: Unsupervised person image generation with semantic parsing transformation. In: CVPR (2019)

72. Sun, J., Wang, X., Zhang, Y., Li, X., Zhang, Q., Liu, Y., Wang, J.: Fenerf: face editing in neural radiance fields. arXiv preprint. arXiv:2111.15490 (2021)
73. Tan, F., et al.: Volux-gan: a generative model for 3d face synthesis with HDRI relighting. arXiv preprint. arXiv:2201.04873 (2022)
74. Tang, H., Bai, S., Zhang, L., Torr, P.H.S., Sebe, N.: XingGAN for person image generation. In: Vedaldi, A., Bischof, H., Brox, T., Frahm, J.-M. (eds.) ECCV 2020. LNCS, vol. 12370, pp. 717–734. Springer, Cham (2020). https://doi.org/10.1007/978-3-030-58595-2_43
75. Tov, O., Alaluf, Y., Nitzan, Y., Patashnik, O., Cohen-Or, D.: Designing an encoder for stylegan image manipulation. ACM TOG **40**(4), 1–14 (2021)
76. Vaswani, A., et al.: Attention is all you need. In: NeurIPS (2017)
77. Wang, X., Gupta, A.: Generative image modeling using style and structure adversarial networks. In: Leibe, B., Matas, J., Sebe, N., Welling, M. (eds.) ECCV 2016. LNCS, vol. 9908, pp. 318–335. Springer, Cham (2016). https://doi.org/10.1007/978-3-319-46493-0_20
78. Wu, J., Zhang, C., Xue, T., Freeman, W.T., Tenenbaum, J.B.: Learning a probabilistic latent space of object shapes via 3d generative-adversarial modeling. In: NeurIPS (2016)
79. Xu, X., Pan, X., Lin, D., Dai, B.: Generative occupancy fields for 3d surface-aware image synthesis. In: Advances in Neural Information Processing Systems (NeurIPS) (2021)
80. Xu, Y., Peng, S., Yang, C., Shen, Y., Zhou, B.: 3d-aware image synthesis via learning structural and textural representations. In: CVPR (2022)
81. Yildirim, G., Jetchev, N., Vollgraf, R., Bergmann, U.: Generating high-resolution fashion model images wearing custom outfits. In: Proceedings of the IEEE/CVF International Conference on Computer Vision Workshops (2019)
82. Zhang, R., Isola, P., Efros, A.A., Shechtman, E., Wang, O.: The unreasonable effectiveness of deep features as a perceptual metric. In: CVPR (2018)
83. Zhou, P., Xie, L., Ni, B., Tian, Q.: CIPS-3D: a 3D-aware generator of gans based on conditionally-independent pixel synthesis (2021)
84. Zhou, X., et al.: Cocosnet v2: full-resolution correspondence learning for image translation. In: CVPR (2021)
85. Zhu, J.Y., Park, T., Isola, P., Efros, A.A.: Unpaired image-to-image translation using cycle-consistent adversarial networks. In: ICCV (2017)
86. Zhu, J.Y., et al.: Visual object networks: image generation with disentangled 3D representations. In: NeurIPS (2018)
87. Zhu, Z., Huang, T., Shi, B., Yu, M., Wang, B., Bai, X.: Progressive pose attention transfer for person image generation. In: CVPR (2019)

Temporally Consistent Semantic Video Editing

Yiran Xu[1(✉)], Badour AlBahar[2], and Jia-Bin Huang[1]

[1] University of Maryland, College Park, USA
yiranx@umd.edu
[2] Virginia Tech, Blacksburg, USA

Abstract. Generative adversarial networks (GANs) have demonstrated impressive image generation quality and semantic editing capability of real images, e.g., changing object classes, modifying attributes, or transferring styles. However, applying these GAN-based editing to a video independently for each frame inevitably results in temporal flickering artifacts. We present a simple yet effective method to facilitate temporally coherent video editing. Our core idea is to minimize the temporal photometric inconsistency by optimizing both the latent code and the pre-trained generator. We evaluate the quality of our editing on different domains and GAN inversion techniques and show favorable results against the baselines.

Keywords: Video editing · GAN editing · Video consistency

1 Introduction

Generative adversarial models (GANs) [16] have shown remarkable ability to generate photorealistic images in various domains such as faces and common objects [10,28,29]. GANs take a latent code (usually sampled from a Gaussian distribution) as input and produce an image as the output. *GAN inversion* techniques allow us to project a *real image* onto the latent space of a pretrained GAN and retrieve its corresponding latent code. The pretrained GAN generator can then reconstruct that image using the estimated latent code. Modifying this estimated latent code opens up exciting new opportunities to perform a wide range of high-level editing tasks that are traditionally challenging, e.g., changing semantic object classes, modifying high-level attributes of the object/scene, or even applying 3D geometric transformations. We refer to the modification of the latent code with a semantic change in the image as *semantic editing*.

Supplementary Information The online version contains supplementary material available at https://doi.org/10.1007/978-3-031-19784-0_21.

S. Avidan et al. (Eds.): ECCV 2022, LNCS 13675, pp. 357–374, 2022.
https://doi.org/10.1007/978-3-031-19784-0_21

Fig. 1. Temporally consistent video semantic editing. We present a method for editing the semantic attributes of a video using a pre-trained StyleGAN model. Here we showcase free-form text based editing from SytleCLIP [44] to make the person appear "angry" (2nd row) or wear "eyeglasses" (3rd row).

Semantic Editing in Images. A recent line of research work [1,2,6,21,49,65,66] has shown promising results in reconstructing an input image by either optimizing the latent code (or latent variables) or directly predicting the latent code via an image encoder. These GAN inversion techniques enable interesting semantic photo editing applications. For image-level editing applications, several approaches [22,51,52] find specific semantic directions in the latent space, e.g., changing poses, colors, or age, while others [15] aim to change the global style, e.g., photo → sketch. We denote them as *In-domain* and *Out-of-domain* editing, respectively. With these *image* GAN inversion-based semantic editing approaches, how can we extend them to *videos*?

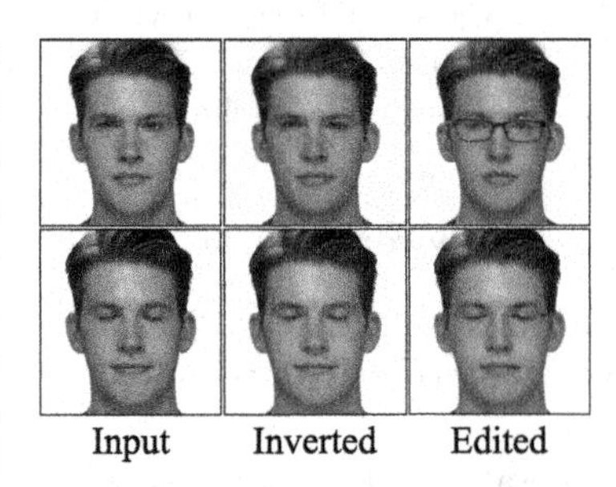

Fig. 2. Issues with per-frame editing. While current methods achieve faithful inversion and photorealistic editing, the results are inconsistent across frames (*eyeglasses*) and may fail to preserve details of the input video (*lips*).

Per-frame Editing. One straightforward way is to apply existing GAN inversion techniques [6,21,49,65] for each frame in a video *independently*. Figure 2 shows an example of applying a StyleCLIP mapper [44] on two frames. The input and the independently reconstructed frames look plausible when viewed individually, but two edited frames exhibit inconsistency (e.g., the frame of the eyeglasses). Recently, Yao et al. [62] learns to predict per-frame semantic editing directions for editing face videos. However, the edited videos suffer from apparent temporal flickering and fail to preserve facial identity.

Our Work. In this paper, we present a method for *temporally consistent* video semantic editing. We start from the existing GAN inversion approaches [6,49] to obtain the latent code for each frame. We first modify the latent code to achieve the initial per-frame editing results. However, such a direct editing approach results in temporal inconsistencies in the modified video's appearance or style. To deal with this challenge, we propose to compute bi-directional optical flow estimated from a frame pair sampled from the video. We can then adjust the latent code and the generator to minimize the photometric loss (along with valid flow vectors). We present a two-phase optimization strategy. In the first phase, we update only the latent codes via an MLP (with generator parameters frozen) to adjust the consistency of the detailed appearance. In the second phase, we finetune the generator with a local regularization to maintain the editability of the latent space. Our two-phase optimization approach helps achieve significantly improved temporal consistency while preserving the edited contents.

Concurrent Work. Two concurrent work [7,57] also apply StyleGAN for video editing. These methods either use per-frame pivot tuning [49] for maintaining the similarity between the edited and input frame [57] or apply latent vector smoothing [7] with StyleGAN3 [27]. Our method differs in 1) the use of explicit temporal consistency optimization and 2) the applicability of performing both in-domain and out-of-domain editing.

Our Contributions

- We tackle a task on GAN-based semantic editing in videos. We propose a simple yet effective flow-based approach to mitigate the temporal inconsistency of a directly (frame by frame) edited video.
- We present a two-phase optimization approach for updating the latent code *and* generator to preserve the video details.
- Our method is agnostic and can be applied to different GAN inversion and editing approaches.

2 Related Work

Generative Adversarial Networks. The quality and resolution of generated images have been achieved rapidly in recent years [10,25,27–29]. These GAN models can map a random latent code (a noise vector) to a photorealistic image. Many recent efforts have been devoted to improving the generator architectures [24,27–29], training strategies [10], loss function designs [18,41], and regularization [42]. Our work builds upon existing pretrained StyleGAN models as they demonstrate disentangled latent space for editing. Instead of *generating synthetic images*, our goal is to *edit real videos*.

GAN Inversion. GAN inversion [61,66] allows us to reconstruct real images by projecting them onto a pretrained GAN's latent space. These techniques facilitate interesting photo editing applications. They can be split into encoder-based [6,11,40,43,48,48,55,56,56,58], optimization-based [1,2,13,14,17,21,45,

54], and hybrid methods [8,49,65]. Our method is *agnostic* to different GAN inversion approaches for initializing the latent code, e.g., our experiments explore using PTI [49] for in-domain editing and Restyle encoder [6] for out-of-domain editing.

Semantic Image Editing in Latent Space. Semantic image manipulation and editing allow us to change the content and style of an image. It can be grouped into In-Domain editing and Out-of-Domain editing. Targeting at finding semantic directions in the latent space of a pretrained generator, in-domain editing [3–5,22,37,44,50–52,59,60,63] manipulates the attributes of the object, but keeps the same style. Out-of-domain [15,23,33], however, aims to change the style of the image. These techniques usually perform well on a single image but fail to maintain temporal consistency if applied to a video.

Semantic Video Editing. Recent and concurrent work [7,57,62] explore *video editing* with a pre-trained StyleGAN. The methods in [57,62] apply per-frame editing and show coherent editing without using any temporal information. However, these methods support only in-domain editing. For *localized editing* (e.g., adding eyeglasses), we find that the method in [62] produces inconsistency and fails to preserve identity. The work [7] applies temporal smoothing on the *inverted latent vectors* in StyleGAN3 [27]. Our approach, in contrast, directly minimizes the temporal photometric inconsistency at the *synthesized frames.*

Video Editing and Temporal Consistency. Temporal consistency is one critical criterion in video editing. Existing methods achieve temporal consistency often by enforcing the output videos to satisfy the constraints imposed by 2D optical flow [12,20]. Alternatively, several methods first estimate an unwarped 2D texture map (either explicitly [46] or implicitly [30]) and then perform editing. The editing can then be propagated to the original video via the estimated UV mapping. Several *blind* methods enhance the temporal consistency as a *post-processing* step [9,34,36]. However, they typically have difficulty in handling videos with significant appearance changes. Our work shares similar ideas with these methods to enforce temporal consistency, using the optical flow fields estimated from the initial edited video. Instead of directly optimizing the *pixel values*, our core idea is to leverage the pretrained generator, update the latent code and generator to achieve temporal consistent *and* photorealistic results.

3 Method

3.1 Overview

GAN Inversion. Given an input video $V_{input} = \{I_1, \cdots, I_T\}$ of T frames, our goal is to semantically edit all the video frames while preserving the temporal coherence of the edited video. To edit the input video V_{input}, we first align its frames by using a facial alignment method [19]. Then we use existing GAN inversion techniques (e.g., [6,49]) to invert the frames back to the latent code such that the inverted frame $I_t^{inv} = G(W_t^{inv}; \theta^{inv})$ is similar to the input frame:

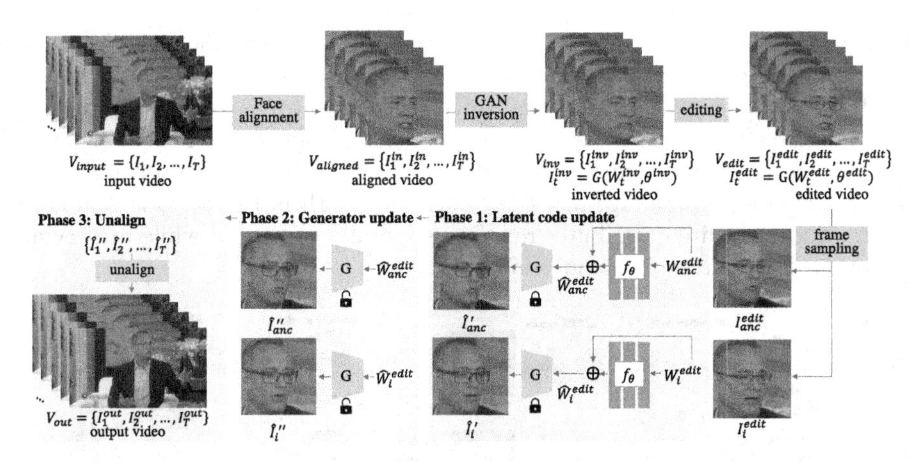

Fig. 3. Video editing with flow-based temporal consistency. Given an input video of T frames V_{input}, we first spatially align the video frames using an off-the-shelf face landmark detector. We then use existing GAN inversion techniques [6,49] to obtain the inverted frames $\{I_1^{inv}, I_2^{inv}, \cdots, I_T^{inv}\}$ and their corresponding latent code in the $\mathcal{W}^+$-space of StyleGAN $\{W_1^{inv}, W_2^{inv}, \cdots, W_T^{inv}\}$. We independently perform semantic editing on these inverted frames to obtain $\{I_1^{edit}, I_2^{edit}, \cdots, I_T^{edit}\}$ and their corresponding latent code $\{W_1^{edit}, W_2^{edit}, \cdots, W_T^{edit}\}$. To achieve temporal consistency, we choose an anchor frame I_{anc}^{edit} as the reference frame, and each time sample another frame I_i^{edit} from the edited video. To generate a temporally consistent edited video, we first refine the latent codes of the directly edited video W_{anc}^{edit} and $\{W_i^{edit}\}_{i \neq anc}$ to $\hat{W}_{anc}^{edit}$ and $\{\hat{W}_i^{edit}\}_{i \neq anc}$ by optimizing an MLP f_θ (phase 1). These refined latent codes result in the temporally consistent frames $\hat{I}_{anc}^{'}$ and $\hat{I}_i^{'}$. To further improve the temporal consistency, we keep the refined latent codes $\hat{W}_{anc}^{edit}$ and $\hat{W}_i^{edit}$ and only update the generator parameters (phase 2). This will generate $\hat{I}_{anc}^{''}$ and $\hat{I}_i^{''}$ with improved temporal consistency. After our two phase optimization, we finally unalign the frames to generate our final edited video V_{out} (phase 3).

$I_t^{inv} \approx I_t$. With the inverted frames, we can edit the inverted video $V_{inv} = \{I_1^{inv}, I_2^{inv}, \cdots, I_T^{inv}\}$ by *independently* editing its frames I_t^{inv}. We denote this frame-by-frame editing approach as "direct editing".

In-Domain and Out-of-Domain GAN-Based Editing. Commonly used *image-based* editing techniques via a GAN include (1) in-domain and (2) out-of-domain editing. We refer to an *in-domain editing* [22,51,52,60] as the editing that only manipulates the latent code, given a *fixed* pretrained generator. That is, the generator parameters θ^{inv} remain frozen ($\theta^{inv} = \theta^{edit}$), and only the latent code W_t^{edit} is updated. The in-domain editing usually changes semantic attributes such as color, age, or facial expressions. On the other hand, out-of-domain editing may involve updating the pretrained generator to produce an entirely new style (as shown in [15]). Here, the latent code remains the same $W_t^{edit} = W_t^{inv}$ and only the generator θ^{edit} changes.

Direct Editing on a Video. When applying both types of editing techniques to a video independently for each frame, we obtain an edited video

$V_{edit} = \{I_1^{edit}, I_2^{edit}, \cdots, I_T^{edit}\}$. For each directly edited frame I_t^{edit}, there is a corresponding latent code W_t^{edit} such that $I_t^{edit} = G(W_t^{edit}; \theta^{edit})$. Due to the per-frame, independent process, the edited video V_{edit} often suffers from temporal inconsistency. Moreover, due to the poor disentanglement of this per-frame editing, not only will the edited attributes differ among frames, but other existing facial attributes also change (see the change in mouth in Fig. 5). Our goal is to ensure that the edited attributes remain temporally consistent while preserving the other details from the input video.

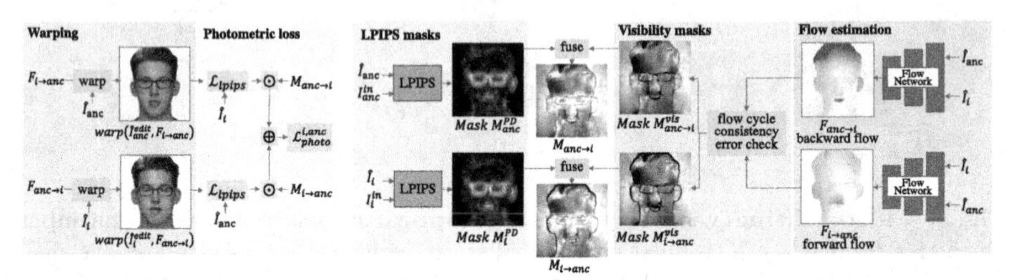

(a) $\mathcal{L}_{photo}$ computation (b) Flow and mask computation

Fig. 4. Photometric loss for temporal consistency. Given a frame pair $\hat{I}_i$ and $\hat{I}_{anc}$ (either from phase 1 or phase 2), we compute the forward and backward flows $F_{i\rightarrow anc}$ and $F_{anc\rightarrow i}$ using RAFT [53]. We then use these two flow fields to compute the visibility masks by performing a forward-backward and backward-forward flow consistency error check. For in-domain editing, we also use LPIPS to obtain a semantic mask that highlights the difference between the aligned input frames I_i^{in} and I_{anc}^{in} and our edited frames $\hat{I}_i$ and $\hat{I}_{anc}$. We then fuse both the LPIPS semantic masks and the visibility masks to get our final masks $M_{anc\rightarrow i}$ and $M_{i\rightarrow anc}$. To compute the photometric loss (Eq. 1), we use the flows to warp the directly edited frames and utilize the fuzed masks as shown in (a).

Overview of Our Approach. To achieve this goal, we propose a two-phase optimization approach: phase 1 updates the *latent code* via an MLP and phase 2 updates the *generator*. In both phases, we optimize the temporal photometric loss across frames. With the finetuned latent code and generator, we unalign the edited frames to produce an edited video. Figure 3 outlines our workflow. Below, we describe the details and the losses of our approach.

3.2 Flow-Based Temporal Consistency

We present a flow-based approach to explicitly encourage temporal consistency in the edited video V_{edit}.

Frame Sampling. As we cannot fit an entire video into the GPU memory, we choose to perform our optimization from a *pair of frames* at a time. We choose to use an anchor frame I_{anc}^{edit} as one of the pair, which we set as the middle frame of the video. This is inspired by recent video representation work [47], where a video is represented by a key frame and a flow network. At each iteration, we sample a latent code W_i^{edit}, corresponding to the frame I_i^{edit} and optimize the pair of frames $\{I_{anc}^{edit}, I_i^{edit}\}$. We perform our optimization in two phases (Sect. 3.3). In phase 1, we generate temporally consistent pairs $\{\hat{I}'_{anc}, \hat{I}'_i\}_{i \neq anc}$ as a result. In phase 2, we further improve the temporal consistency, recover other affected attributes brought by the per-frame editing due to the poor disentanglement, and generate the pairs $\{\hat{I}''_{anc}, \hat{I}''_i\}_{i \neq anc}$.

Flow Estimation and Warping. We use RAFT [53] to compute the forward and backward flows $F_{i \to anc}$ and $F_{anc \to i}$ of the pair $\{\hat{I}_{anc}, \hat{I}_i\}$. This pair is either the output of phase 1 $\{\hat{I}'_{anc}, \hat{I}'_i\}$ or phase 2 $\{\hat{I}''_{anc}, \hat{I}''_i\}$. We then use these two flows to warp the pair of frames $\{\hat{I}_{anc}, \hat{I}_i\}$.

Visibility Masks. To highlight the *non-occluded* regions, we compute the visibility masks $M_{anc \to i}^{vis}$ and $M_{i \to anc}^{vis} \in [0, 1]$. This mask shows lower weights for occluded pixels and higher weights for the non-occluded pixels (Fig. 4). To compute the visibility masks, we first compute forward-backward and backward-forward flow consistency error maps $\epsilon_{anc \to i}$ and $\epsilon_{i \to anc}$ and compute the error map by $\epsilon_{i \to anc}(p) = ||p - F_{anc \to i}(p + F_{anc \to j}(p))||_2$, where p is a pixel in the flow field. These resultant error maps are mapped to $[0, 1]$ using an exponential function such that $M_{anc \to i}^{vis} = \exp(-10\epsilon_{anc \to i})$ and $M_{i \to anc}^{vis} = \exp(-10\epsilon_{i \to anc})$.

Perceptual Difference Mask. For in-domain editing, because the introduced editing is temporally inconsistent, we observe that the visibility masks do *not* emphasize those edited parts (e.g., eyeglasses). To highlight those edited parts, we compute the soft semantic perceptual difference masks M_{anc}^{PD} and M_i^{PD} between the pair of frames and their corresponding aligned input frames using LPIPS [64] (Fig. 4). Due to the significant appearance differences, we cannot use these semantic perceptual difference masks for out-of-domain editing.

Fused Masks. For in-domain editing, we fuse the visibility masks and the semantic perceptual difference masks such that $M_{anc \to i} = (M_{anc \to i}^{vis} \oplus M_i^{PD})$ and $M_{i \to anc} = (M_{i \to anc}^{vis} \oplus M_{anc}^{PD})$. The masks will also be clamped to [0, 1]. This fusion is shown in Fig. 4. On the other hand, for out-of-domain editing, $M_{anc \to i} = M_{anc \to i}^{vis}$ and $M_{i \to anc} = M_{i \to anc}^{vis}$.

Bi-directional Photometric Loss. We use the warped frames and the final computed masks to compute the bi-directional photometric loss to achieve a temporally consistent video. This loss measures the difference between the two frames to calculate the deviation in the non-occluded parts.

$$\begin{aligned}\mathcal{L}_{photo} = \sum_{\hat{I}_i, \hat{I}_{anc} \in P} & M_{i \to anc} \mathcal{L}_{LPIPS}(\hat{I}_{anc}, warp(\hat{I}_i, F_{anc \to i})) \\ & + M_{anc \to i} \mathcal{L}_{LPIPS}(\hat{I}_i, warp(\hat{I}_{anc}, F_{i \to anc})),\end{aligned} \tag{1}$$

where $\hat{I}_t$ is either the output of phase 1 $\hat{I}'_t$ or phase 2 $\hat{I}''_t$. Intuitively, this bi-directional photometric loss ensures colors along the valid (forward-backward or backward-forward consistent) vectors across frames are as similar as possible.

3.3 Two-Phase Optimization Strategy

We split our optimization into two phases. In the first phase, we refine the latent codes $\{W_t^{edit}\}$ by only optimizing an MLP f_θ. While in the second phase, we only update the generator weights θ^{edit}.

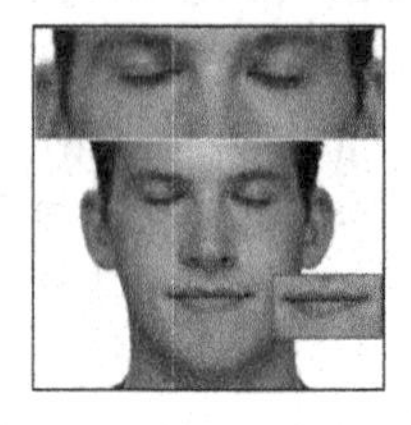
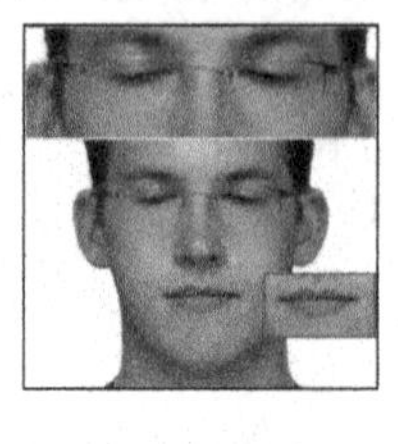
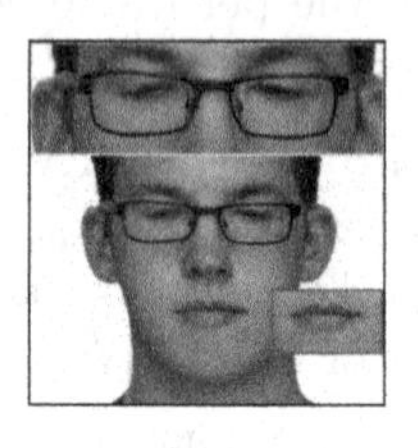
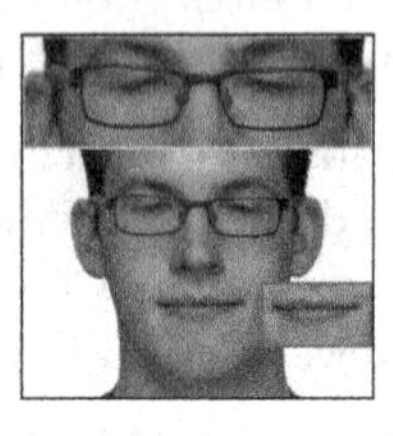

Input Direct Editing W updated W, G updated

Fig. 5. Motivation for two-phase optimization. Updating latent code W brings in the eyeglasses, and tuning G with the perceptual difference mask recovers the expression in the input.

Motivation. We use a two-phase optimization approach for in-domain editing because we observe that only refining the latent codes (phase 1) sometimes introduces undesired changes to *other* facial attributes. We show an example in Fig. 5. When we only update the latent codes, we achieve temporal consistency of the introduced glasses; however, the mouth expression of the person changes. To address this in the case of in-domain editing, we update the generator weights (phase 2) using the perceptual difference mask to enforce the pixels outside the mask to be the same as the input. This will maintain the facial expression of the aligned input frame. The primary source of inconsistency for out-of-domain editing is the global inconsistency (e.g., background). Hence, updating the generator (phase 2) introduces this desired global change.

Phase 1: Latent Code Update. In this phase, we update the latent code W_t^{edit} using a Multi-layer Perceptron (MLP) $f_\theta = (w; \theta_f)$ implicitly. We use the same architecture as StyleCLIP mapper [44]. We use this MLP to predict a residual for the latent codes and update the parameters of the MLP instead of directly optimizing the latent codes explicitly, such that:

$$\hat{W}_t^{edit} = W_t^{edit} + \alpha f_\theta(W_t^{edit}; \theta_f), \tag{2}$$

then for a pair of directly edited frames $\{I_{anc}^{edit}, I_i^{edit}\}$, we can get the updated frames $\hat{I}_i' = G(\hat{W}_i^{edit})$, $\hat{I}_{anc}' = G(\hat{W}_{anc}^{edit})$.

Our goal is to minimize:

$$\underset{\theta_f}{\operatorname{argmin}} \mathcal{L}_I = \underset{\theta_f}{\operatorname{argmin}} \sum_{t \neq anc} \mathcal{L}_{photo} + \lambda_{rf}\mathcal{L}_{rf} + \lambda_\epsilon \mathcal{L}_\epsilon, \tag{3}$$

where $\mathcal{L}_{photo}$ is the photometric loss, and

$$\mathcal{L}_{rf} = ||f_\theta(W_t^{edit}; \theta_f)||_1 + ||f_\theta(W_{anc}^{edit}; \theta_f)||_1 \tag{4}$$

is a regularization term to make sure we do not deviate too much from W_t^{edit}. We set $\lambda_{rf} = 0.1$ for the experiments. $\mathcal{L}_\epsilon = ||\epsilon_{anc\rightarrow i}||_1 + ||\epsilon_{i\rightarrow anc}||_1$ is the norm of error maps, and we set $\lambda_\epsilon = 10$.

The reason we use an MLP to update the latent code *implicitly* is that we observe that *explicitly* optimizing the latent codes results in an unstable optimization when using a large learning rate. However, the running time becomes too long when using a small learning rate. To address this, we introduce an MLP to predict the residual and update the latent codes *implicitly*. This leads to a more stable optimization. We show an example of x-t scanline in Fig. 6 to demonstrate the effectiveness of introducing the MLP.

Fig. 6. x-t slices between updating latent codes explicitly and implicitly with an MLP. We visualize the optimized frames and an x-t slice at $y = 500$. Explicitly updating latent code W gives us an unstable x-t scanline, while updating W implicitly with an MLP gives a smooth scanline.

Phase 2: Generator Update. For in-domain editing, in this phase, we use the updated latent codes $\{\hat{W}_t^{edit}\}_{t=1}^T$ from phase 1, and our goal is to finetune the generator only to minimize:

$$\hat{\theta}^{edit} = \underset{\hat{\theta}^{edit}}{\operatorname{argmin}} \mathcal{L}_{II} = \underset{\hat{\theta}^{edit}}{\operatorname{argmin}} \sum_{t\neq anc} \mathcal{L}_{photo} + \lambda_\epsilon \mathcal{L}_\epsilon + \lambda_r \mathcal{L}_r + \lambda_M \mathcal{L}_M , \quad (5)$$

$$\mathcal{L}_M = (1 - M_i^{PD})\mathcal{L}_{LPIPS}(\hat{I}_i^{''}, I_i^{in}) + (1 - M_{anc}^{PD})\mathcal{L}_{LPIPS}(\hat{I}_{anc}^{''}, I_{anc}^{in}) . \quad (6)$$

M_i^{PD} is the perceptual difference mask computed between $\hat{I}_i^{''} = G(\hat{W}_t^{edit}; \hat{\theta}^{edit})$ and aligned input I_i^{in}, and $\mathcal{L}_{LPIPS}(\cdot,\cdot)$ is the LPIPS distance loss [64]. We initialize $\hat{\theta}^{edit}$ as θ^{edit}. The LPIPS term also plays a role to maintain the sharpness of the edited frames. This is because the consistency can be achieved by pushing all the frames to become blurry.

Here, $\mathcal{L}_r$ is the regularization loss for the generator and λ_r is the strength of regularization. We introduce this loss to help prevent the generator G from losing its latent space editability as we do not wish to *ruin* its pretrained latent space. Therefore, similar to [49], we use this *local regularization* to preserve the editing ability of our generator. More specifically, we first obtain a latent code W_r by linearly interpolating between the current latent code $\hat{W}_t^{edit}$ and a randomly sampled code W_z with an interpolation parameter α_{interp}: $W_r =$

$\hat{W}_t^{edit} + \alpha_{interp} \frac{W_z - \hat{W}_t^{edit}}{||W_z - \hat{W}_t^{edit}||_2}$. This gives us a new latent code in a local region around $\hat{W}_t^{edit}$. To ensure that we do not lose the editing capability of the original generator, we add a penalty on the distance between the generated image from the new generator and the old one such that:

$$\mathcal{L}_r = \mathcal{L}_{LPIPS}(x_r, \hat{x}_r) + \lambda_{\ell_2}^r \mathcal{L}_{\ell_2}(x_r, \hat{x}_r)\,, \tag{7}$$

where $x_r = G(W_r; \theta^{edit}), \hat{x}_r = G(W_r; \hat{\theta}^{edit})$, $\lambda_{\ell_2}^r$ is the weight for ℓ_2 loss. This regularization can alleviate the side effects from updating G within a local area. This is desirable since for a video, the latent codes for the same identity tend to gather locally.

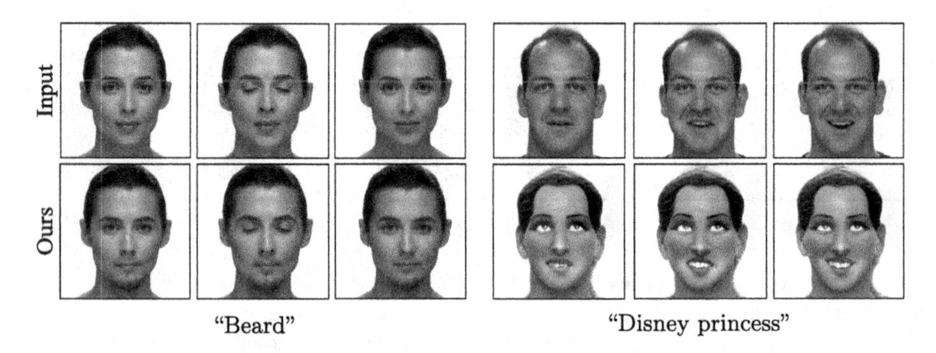

Fig. 7. Visual results on RAVDESS dataset [39]. We show both in-domain ("beard") and out-of-domain ("Disney princess") editing results. Our results maintain consistent changes with time preserving the temporal coherence.

For out-of-domain editing, unlike in-domain editing, we cannot rely on the perceptual difference mask, so the optimization goal reduces to:

$$\hat{\theta}^{edit} = \underset{\hat{\theta}^{edit}}{\operatorname{argmin}}\, \mathcal{L}_{II} = \underset{\hat{\theta}^{edit}}{\operatorname{argmin}} \sum_{t \neq anc} \mathcal{L}_{photo} + \lambda_r \mathcal{L}_r + \lambda_\epsilon \mathcal{L}_\epsilon\,. \tag{8}$$

To compensate for the regularization effect of the perceptual difference mask, we freeze the last eight layers of the synthesis network in G to avoid blurry results. As all the computations, including the GAN generator, flow estimation network, spatial warping, and photometric losses, are *differentiable*, we can backpropagate the errors all the way back. After phase 1 and 2, we will have $\{\hat{W}_t^{edit}\}_{t=1}^T$ and $G(\cdot; \hat{\theta}^{edit})$ as a result.

3.4 Phase 3: Unalign

After our two-phase optimization, we perform *stitch tuning* approach [57] as post-processing to put the aligned frames back to the original video to generate our final edited video. Note that this is only feasible for the in-domain editing because the out-of-domain editing has a global appearance compared to the input video.

4 Experimental Results

4.1 Experimental Setup

Implementation Details. We use StyleGAN-ADA [26] as our pre-trained generator. We experiment with in-domain and out-of-domain editing techniques to validate our approach for different GAN inversion methods. Specifically, for in-domain editing, we use the PTI inversion [49] (based on e4e [56]) and StyleCLIP mapper [44]. For out-of-domain editing, we use the Restyle encoder [6] and the StyleGAN-NADA [15]. We will release the source code and pretrained models. In the following, we show sample results from the video frames. We encourage the readers to view the videos in the supplementary material for video results.

Fig. 8. Visual comparison with DVP [36]. DVP achieves temporal consistency by severely smoothing the image and hence losing its sharpness. Our method, however, can achieve a balance between consistency and sharpness. In "eyeglasses" example (left), DVP shows a different pair of eyeglasses across the time (zoom-in for better visualization), while ours remain a good consistency for the eyeglasses; in "Disney princess" (right), DVP shows a blurry result with an unstable x-t scanline, while ours is sharper and shows a stable consistency in the scanline.

Fig. 9. Results on Internet videos. Results on the Internet videos. We change the first person to "surprised" expression, and change the second person to "angry".

Datasets. We conduct our metric evaluation using 20 videos randomly sampled from RAVDESS dataset [39]. We conduct 5 types of in-domain editing for each video and 5 types of out-of-domain editing. To further demonstrate the capabilities of our method to handle *real* videos, we also apply our approach to Internet videos and show the visual results.

Table 1. Out-of-domain editing comparison.

	$E_{warp}\downarrow$		LPIPS$\downarrow$	
Direct editing	0.0098		0.0000	
Editing categories	DVP [36]	Ours	DVP	Ours
Sketch	0.0036	0.0085	0.2404	0.1314
Pixar	0.0031	0.0025	0.1074	0.1178
Disney princess	0.0040	0.0078	0.2062	0.1204
Elf	0.0042	0.0108	0.2289	0.1310
Zombie	0.0040	0.0085	0.2033	0.1370
Average perfomance	0.0038	0.0076	0.1972	0.1275

Metrics. We aim to evaluate the method using two main aspects: 1) temporal consistency and 2) perceptual similarity with the semantically edited frames. To evaluate temporal consistency, we measure the *Warping Error* E_{warp}:

$$E_{warp}(I_t, I_{t+1}) = \frac{1}{\sum_{i=1}^{N} M_t(p_i)} \cdot \sum_{i=1}^{N} M_t(p_i)||I_t(p_i) - \hat{I}_{t+1}(p_i)||_2^2, \tag{9}$$

where $\hat{I}_{t+1} = warp(I_{t+1}, F_{t\rightarrow t+1})$, N is the number of pixels, p_i is the i-th pixel, M_t is a binary non-occlusion mask, which shows non-occluded pixels, we compute it using the forward-backward consistency error the threshold in [35,38].

We also measure the LPIPS perceptual similarity score [64] (with AlexNet [32]) between the directly edited video $V^{edit} = \{I_1^{edit}, I_2^{edit}, \cdots, I_T^{edit}\}$ and the output of our phase 2 $\{\hat{I}_1'', \hat{I}_2'', \cdots, \hat{I}_T''\}$ by measuring the averaged perceptual similarity between the corresponding frames. The purpose of these two metrics is to evaluate whether the method can achieve a balance between *temporal consistency* and *fidelity degradation*. This is an inherent trade-off. Preserving all the details of per-frame editing inevitably leads to temporal flickering artifacts. Focusing only on temporal consistency may easily lead to blurry videos. Our goal is that the final output video is visually similar to the directly (per-frame) edited video.

4.2 Out-of-domain Results

Setup. We first invert the videos frame by frame using the Restyle encoder [6] (psp-based [48]). We then directly apply five different out-of-domain editing effects produced by StyleGAN-NADA [15]. We perform our two-phase optimization approach on the directly edited video using Adam optimizer [31]. For phase

1, we set the learning rate to $\alpha_I = 0.005$, and update the latent codes for 5 epochs. In Eq. 2, we set $\alpha = 0.04$ for all the editing directions. For phase 2, we set the learning rate to $\alpha_{II} = 8 \times 10^{-4}$, and finetune G for 5 epochs. We set the regularization weight λ_r to 200.

Table 2. In-domain editing comparison.

	$E_{warp}\downarrow$			LPIPS↓	
Direct editing	0.0076			0.0000	
Editing categories	LT [62]	DVP [36]	Ours	DVP	Ours
Angry	–	0.0033	0.0032	0.2452	0.1100
Beard	0.0064	0.0038	0.0030	0.2444	0.1033
Eyeglasses	0.0066	0.0039	0.0034	0.1226	0.1097
Depp	–	0.0037	0.0031	0.2452	0.2024
Surprised	–	0.0035	0.0028	0.1415	0.1012
Average perfomance	0.0065	0.0036	0.0031	0.1760	0.1253

Evaluation. Table 1 shows that our method decreases the temporal error of the directly edited video. The primary sources of inconsistency in out-of-domain editing can be seen in the flickering background and the details of the hair. We show our visual results in Fig. 7. Our method preserves the temporal consistency and maintains the sharpness of the input video.

4.3 In-domain Editing Results

Setup. We first invert the videos frame by frame by using the PTI method [49]. We then directly apply five different semantic editing directions discovered by StyleCLIP mapper [44]. Next, we perform our two-phase optimization approach on the directly edited video using Adam optimizer [31]. For phase 1, we set the learning rate $\alpha_I = 0.05$, and update f_θ for 10 epochs. In Eq. 2, we set $\alpha = 0.12$ for the "eyeglasses", and $\alpha = 0.04$ for the rest of the semantic directions. For phase 2, we set the learning rate of G to $\alpha_{II} = 0.0001$, and finetune G for 5 epochs. We set the regularization weight λ_r to 200.

Evaluation. Table 2 shows that our approach improves the temporal consistency over the directly edited video baseline by a large margin. When dealing with in-domain editing, the primary source of inconsistency is the details of the newly added attributes, e.g., glasses or beard and some background flickering. We show sample visual results in Fig. 7, where the introduced changes are consistent among the different frames.

4.4 Two-phase Optimization Strategy Ablation Study

We demonstrate the effect of our two-phase optimization strategy of updating the latent codes first and following that with finetuning the generator G. We compare our two-phase approach to: (1) No optimization (i.e., direct editing), (2) update latent code only (phase 1), and (3) finetune generator G only.

Table 3. Two-stage optimization strategy ablation study.

Optimization stage		In-domain editing		Out-of-domain editing	
Update W_t^{edit}	Update G	$E_{warp} \downarrow$	LPIPS↓	$E_{warp} \downarrow$	LPIPS↓
–	–	0.0076	0.0000	0.0098	0.0000
✓	–	0.0064	0.2108	0.0094	0.1428
–	✓	**0.0027**	0.2463	**0.0057**	0.1375
✓	✓	0.0031	**0.1253**	0.0076	**0.1275**

We show the results in Table 3. When we only update generator G, we can achieve a low warping error E_{warp}. However, this is not desirable since finetuning G pushes the video to be consistent globally without modifying the local details. Therefore, the output video is different from the directly edited video (i.e., high LPIPS distance). Thus, we follow our two-phase optimization of a) updating the latent codes via an MLP f_θ (to improve local consistency), b) finetuning the generator G (to modify the global effect).

4.5 Comparison with Latent Transformer

We compare our method with Latent Transformer (LT) [62]. We show a quantitative comparison with two overlapped editing types, "beard" and "eyeglasses" in Table 2. LT edits video by updating the projected latent code *independently* for each frame without using temporal constraints. Our method, in contrast, uses flow-based loss to improve the temporal consistency, and our second phase uses a perceptual difference mask as a regularization to preserve the facial details other than the edited parts. As a result, our method can improve temporal consistency and preserve personal identity.

4.6 Comparison with Deep Video Prior (DVP)

We compare our method with DVP [36], a state-of-the-art blind video consistency approach, using their default setting. We show the in-domain editing comparison in Table 2 and the out-of-domain editing comparison in Table 1. For warping error E_{warp}, our method achieves improved results for in-domain editing and comparable results for out-of-domain editing. However, in terms of LPIPS distance, our visual results are more similar to the directly edited video for both in-domain and out-of-domain editing. We show visual comparison in Fig. 8. DVP can achieve temporally consistent results (i.e., low E_{warp}). However, this is at the cost of losing local details in the "eyeglasses" example or excessively smoothing the results to get a blurry video as in the "Disney Princess" example.

5 Conclusions

We have presented a novel method for video-based semantic editing by leveraging image-based GAN inversion and editing. Our approach starts from direct per-frame editing, and we refine the editing results by a flow-based method to minimize the bi-directional photometric loss. Our core approach is two-phase, by adjusting the latent codes via an MLP and tuning G to achieve temporal consistency. We show that our method can achieve temporal consistency and preserve its similarity to the direct editing results. Finally, our model-agnostic method is applicable to different GAN inversion and manipulation techniques.

Potential Negative Impacts. Malicious use of our technique may lead to video manipulation of public figures for spreading misinformation.

References

1. Abdal, R., Qin, Y., Wonka, P.: Image2stylegan: how to embed images into the stylegan latent space? In: ICCV (2019)
2. Abdal, R., Qin, Y., Wonka, P.: Image2stylegan++: how to edit the embedded images? In: CVPR (2020)
3. Abdal, R., Zhu, P., Mitra, N.J., Wonka, P.: Styleflow: attribute-conditioned exploration of stylegan-generated images using conditional continuous normalizing flows. ACM Trans. Graph. (TOG) **40**(3), 1–21 (2021)
4. Afifi, M., Brubaker, M.A., Brown, M.S.: Histogan: controlling colors of gan-generated and real images via color histograms. In: Proceedings of the IEEE Conference on Computer Vision and Pattern Recognition (2021)
5. Alaluf, Y., Patashnik, O., Cohen-Or, D.: Only a matter of style: age transformation using a style-based regression model. arXiv preprint arXiv:2102.02754 (2021)
6. Alaluf, Y., Patashnik, O., Cohen-Or, D.: Restyle: a residual-based stylegan encoder via iterative refinement. In: Proceedings of the IEEE/CVF International Conference on Computer Vision (ICCV) (2021)
7. Alaluf, Y., et al.: Third time's the charm? image and video editing with stylegan3. arXiv preprint arXiv:2201.13433 (2022)
8. Bau, D., Strobelt, H., Peebles, W., Zhou, B., Zhu, J.Y., Torralba, A., et al.: Semantic photo manipulation with a generative image prior. arXiv preprint arXiv:2005.07727 (2020)
9. Bonneel, N., Tompkin, J., Sunkavalli, K., Sun, D., Paris, S., Pfister, H.: Blind video temporal consistency. ACM TOG **34**(6), 1–9 (2015)
10. Brock, A., Donahue, J., Simonyan, K.: Large scale gan training for high fidelity natural image synthesis (2019)
11. Chai, L., Wulff, J., Isola, P.: Using latent space regression to analyze and leverage compositionality in gans. In: International Conference on Learning Representations (2021)
12. Chen, D., Liao, J., Yuan, L., Yu, N., Hua, G.: Coherent online video style transfer. In: ICCV (2017)
13. Collins, E., Bala, R., Price, B., Susstrunk, S.: Editing in style: uncovering the local semantics of gans. In: Proceedings of the IEEE/CVF Conference on Computer Vision and Pattern Recognition, pp. 5771–5780 (2020)

14. Daras, G., Odena, A., Zhang, H., Dimakis, A.G.: Your local gan: designing two dimensional local attention mechanisms for generative models. In: Proceedings of the IEEE/CVF Conference on Computer Vision and Pattern Recognition, pp. 14531–14539 (2020)
15. Gal, R., Patashnik, O., Maron, H., Chechik, G., Cohen-Or, D.: Stylegan-nada: clip-guided domain adaptation of image generators (2021)
16. Goodfellow, I., et al.: Generative adversarial nets. In: Advances in Neural Information Processing Systems, pp. 2672–2680 (2014)
17. Gu, J., Shen, Y., Zhou, B.: Image processing using multi-code gan prior. In: Proceedings of the IEEE/CVF Conference on Computer Vision and Pattern Recognition, pp. 3012–3021 (2020)
18. Gulrajani, I., Ahmed, F., Arjovsky, M., Dumoulin, V., Courville, A.C.: Improved training of wasserstein gans (2017)
19. Guo, J., Zhu, X., Yang, Y., Yang, F., Lei, Z., Li, S.Z.: Towards fast, accurate and stable 3D dense face alignment. In: Vedaldi, A., Bischof, H., Brox, T., Frahm, J.-M. (eds.) ECCV 2020. LNCS, vol. 12364, pp. 152–168. Springer, Cham (2020). https://doi.org/10.1007/978-3-030-58529-7_10
20. Huang, J.B., Kang, S.B., Ahuja, N., Kopf, J.: Temporally coherent completion of dynamic video. ACM TOG **35**(6), 1–11 (2016)
21. Huh, M., Zhang, R., Zhu, J.-Y., Paris, S., Hertzmann, A.: Transforming and projecting images into class-conditional generative networks. In: Vedaldi, A., Bischof, H., Brox, T., Frahm, J.-M. (eds.) ECCV 2020. LNCS, vol. 12347, pp. 17–34. Springer, Cham (2020). https://doi.org/10.1007/978-3-030-58536-5_2
22. Härkönen, E., Hertzmann, A., Lehtinen, J., Paris, S.: Ganspace: discovering interpretable gan controls. In: Proceeding NeurIPS (2020)
23. Jang, W., Ju, G., Jung, Y., Yang, J., Tong, X., Lee, S.: Stylecarigan: caricature generation via stylegan feature map modulation. ACM Trans. Graph. (TOG) **40**(4), 1–16 (2021)
24. Karras, T., Aila, T., Laine, S., Lehtinen, J.: Progressive growing of gans for improved quality, stability, and variation. In: International Conference on Learning Representations (2018)
25. Karras, T., Aittala, M., Hellsten, J., Laine, S., Lehtinen, J., Aila, T.: Training generative adversarial networks with limited data. arXiv preprint arXiv:2006.06676 (2020)
26. Karras, T., Aittala, M., Hellsten, J., Laine, S., Lehtinen, J., Aila, T.: Training generative adversarial networks with limited data. In: Proceeding NeurIPS (2020)
27. Karras, T., et al.: Alias-free generative adversarial networks. In: Proceeding NeurIPS (2021)
28. Karras, T., Laine, S., Aila, T.: A style-based generator architecture for generative adversarial networks. In: CVPR (2019)
29. Karras, T., Laine, S., Aittala, M., Hellsten, J., Lehtinen, J., Aila, T.: Analyzing and improving the image quality of stylegan. In: CVPR (2020)
30. Kasten, Y., Ofri, D., Wang, O., Dekel, T.: Layered neural atlases for consistent video editing. In: ACM TOG (2021)
31. Kingma, D.P., Ba, J.: Adam: a method for stochastic optimization. arXiv preprint arXiv:1412.6980 (2014)
32. Krizhevsky, A., Sutskever, I., Hinton, G.E.: Imagenet classification with deep convolutional neural networks. Adv. Neural. Inf. Process. Syst. **25**, 1097–1105 (2012)
33. Kwong, S., Huang, J., Liao, J.: Unsupervised image-to-image translation via pretrained stylegan2 network. IEEE Trans. Multimedia **24**, 1435–1448 (2021)

34. Lai, W.S., Huang, J.B., Wang, O., Shechtman, E., Yumer, E., Yang, M.H.: Learning blind video temporal consistency. In: ECCV (2018)
35. Lai, W.S., Huang, J.B., Wang, O., Shechtman, E., Yumer, E., Yang, M.H.: Learning blind video temporal consistency. In: European Conference on Computer Vision (2018)D
36. Lei, C., Xing, Y., Chen, Q.: Blind video temporal consistency via deep video prior. In: Advances in Neural Information Processing Systems (2020)
37. Li, B., et al.: Dystyle: dynamic neural network for multi-attribute-conditioned style editing. arXiv preprint arXiv:2109.10737 (2021)
38. Liu, Y.L., Lai, W.S., Yang, M.H., Chuang, Y.Y., Huang, J.B.: Learning to see through obstructions. In: CVPR (2020)
39. Livingstone, S.R., Russo, F.A.: The ryerson audio-visual database of emotional speech and song (ravdess): a dynamic, multimodal set of facial and vocal expressions in North American English. PLoS ONE, **13**(5), e0196391 (2018)
40. Luo, J., Xu, Y., Tang, C., Lv, J.: Learning inverse mapping by autoencoder based generative adversarial nets. In: International Conference on Neural Information Processing, pp. 207–216. Springer (2017). https://doi.org/10.1007/978-3-319-70096-0_22
41. Mao, X., Li, Q., Xie, H., Lau, R.Y., Wang, Z., Paul Smolley, S.: Least squares generative adversarial networks. In: Proceedings of the IEEE International Conference on Computer Vision, pp. 2794–2802 (2017)
42. Miyato, T., Kataoka, T., Koyama, M., Yoshida, Y.: Spectral normalization for generative adversarial networks (2018)
43. Nitzan, Y., Bermano, A., Li, Y., Cohen-Or, D.: Face identity disentanglement via latent space mapping. ACM Trans. Graph. (TOG) **39**, 1–14 (2020)
44. Patashnik, O., Wu, Z., Shechtman, E., Cohen-Or, D., Lischinski, D.: Styleclip: text-driven manipulation of stylegan imagery. In: Proceedings of the IEEE/CVF International Conference on Computer Vision (ICCV), pp. 2085–2094 (2021)
45. Raj, A., Li, Y., Bresler, Y.: Gan-based projector for faster recovery with convergence guarantees in linear inverse problems. In: Proceedings of the IEEE/CVF International Conference on Computer Vision, pp. 5602–5611 (2019)
46. Rav-Acha, A., Kohli, P., Rother, C., Fitzgibbon, A.: Unwrap mosaics: a new representation for video editing. In: ACM TOG (2008)
47. Rho, D., Cho, J., Ko, J.H., Park, E.: Neural residual flow fields for efficient video representations. arXiv preprint arXiv:2201.04329 (2022)
48. Richardson, E., et al.: Encoding in style: a stylegan encoder for image-to-image translation. In: IEEE/CVF Conference on Computer Vision and Pattern Recognition (CVPR) (2021)
49. Roich, D., Mokady, R., Bermano, A.H., Cohen-Or, D.: Pivotal tuning for latent-based editing of real images. arXiv preprint arXiv:2106.05744 (2021)
50. Saha, R., Duke, B., Shkurti, F., Taylor, G.W., Aarabi, P.: Loho: latent optimization of hairstyles via orthogonalization. In: Proceedings of the IEEE/CVF Conference on Computer Vision and Pattern Recognition, pp. 1984–1993 (2021)
51. Shen, Y., Yang, C., Tang, X., Zhou, B.: Interfacegan: interpreting the disentangled face representation learned by gans. In: TPAMI (2020)
52. Shen, Y., Zhou, B.: Closed-form factorization of latent semantics in gans. In: CVPR (2021)
53. Teed, Z., Deng, J.: Raft: recurrent all-pairs field transforms for optical flow. arXiv preprint arXiv:2003.12039 (2020)
54. Tewari, A., et al.: Pie: portrait image embedding for semantic control. arXiv preprint arXiv:2009.09485 (2020)

55. Tewari, A., et al.: Stylerig: rigging stylegan for 3d control over portrait images. In: Proceedings of the IEEE/CVF Conference on Computer Vision and Pattern Recognition, pp. 6142–6151 (2020)
56. Tov, O., Alaluf, Y., Nitzan, Y., Patashnik, O., Cohen-Or, D.: Designing an encoder for stylegan image manipulation. ACM Trans. Graph. (TOG) **40**(4), 1–14 (2021)
57. Tzaban, R., Mokady, R., Gal, R., Bermano, A.H., Cohen-Or, D.: Stitch it in time: gan-based facial editing of real videos. arXiv preprint arXiv:2201.08361 (2022)
58. Viazovetskyi, Y., Ivashkin, V., Kashin, E.: StyleGAN2 distillation for feed-forward image manipulation. In: Vedaldi, A., Bischof, H., Brox, T., Frahm, J.-M. (eds.) ECCV 2020. LNCS, vol. 12367, pp. 170–186. Springer, Cham (2020). https://doi.org/10.1007/978-3-030-58542-6_11
59. Wu, Y., Yang, Y.L., Xiao, Q., Jin, X.: Coarse-to-fine: facial structure editing of portrait images via latent space classifications. ACM Trans. Graph. (TOG) **40**(4), 1–13 (2021)
60. Wu, Z., Lischinski, D., Shechtman, E.: Stylespace analysis: disentangled controls for stylegan image generation. In: Proceedings of the IEEE/CVF Conference on Computer Vision and Pattern Recognition (CVPR), pp. 12863–12872 (2021)
61. Xia, W., Zhang, Y., Yang, Y., Xue, J.H., Zhou, B., Yang, M.H.: Gan inversion: a survey. arXiv preprint arXiv:2101.05278 (2021)
62. Yao, X., Newson, A., Gousseau, Y., Hellier, P.: A latent transformer for disentangled face editing in images and videos. In: Proceedings of the IEEE/CVF International Conference on Computer Vision, pp. 13789–13798 (2021)
63. Yüksel, O.K., Simsar, E., Er, E.G., Yanardag, P.: Latentclr: a contrastive learning approach for unsupervised discovery of interpretable directions. arXiv preprint arXiv:2104.00820 (2021)
64. Zhang, R., Isola, P., Efros, A.A., Shechtman, E., Wang, O.: The unreasonable effectiveness of deep features as a perceptual metric. In: Proceedings of the IEEE Conference on Computer Vision and Pattern Recognition, pp. 586–595 (2018)
65. Zhu, J., Shen, Y., Zhao, D., Zhou, B.: In-Domain GAN inversion for real image editing. In: Vedaldi, A., Bischof, H., Brox, T., Frahm, J.-M. (eds.) ECCV 2020. LNCS, vol. 12362, pp. 592–608. Springer, Cham (2020). https://doi.org/10.1007/978-3-030-58520-4_35
66. Zhu, J.-Y., Krähenbühl, P., Shechtman, E., Efros, A.A.: Generative visual manipulation on the natural image manifold. In: Leibe, B., Matas, J., Sebe, N., Welling, M. (eds.) ECCV 2016. LNCS, vol. 9909, pp. 597–613. Springer, Cham (2016). https://doi.org/10.1007/978-3-319-46454-1_36

Error Compensation Framework for Flow-Guided Video Inpainting

Jaeyeon Kang[1], Seoung Wug Oh[2], and Seon Joo Kim[1](✉)

[1] Yonsei University, Seoul, South Korea
seonjookim@yonsei.ac.kr
[2] Adobe, California, USA

Abstract. The key to video inpainting is to use correlation information from as many reference frames as possible. Existing flow-based propagation methods split the video synthesis process into multiple steps: flow completion → pixel propagation → synthesis. However, there is a significant drawback that the errors in each step continue to accumulate and amplify in the next step. To this end, we propose an Error Compensation Framework for Flow-guided Video Inpainting (ECFVI), which takes advantage of the flow-based method and offsets its weaknesses. We address the weakness with the newly designed flow completion module and the error compensation network that exploits the error guidance map. Our approach greatly improves the temporal consistency and the visual quality of the completed videos. Experimental results show the superior performance of our proposed method with the speed up of ×6, compared to the state-of-the-art methods. In addition, we present a new benchmark dataset for evaluation by supplementing the weaknesses of existing test datasets.

Keywords: Video inpainting · Object removal · Video restoration

1 Introduction

Video inpainting is the task of filling missing regions in frames with realistic content. Its applications cover object removal, watermark/logo/subtitle removal, and even corrupted video restoration. It is challenging as the hole regions should be filled with synthesized contents that are unnoticeable with the surrounding background and temporally consistent.

Existing video inpainting methods usually extract useful information from reference frames and merge them to fill the target frame. The key is to use as many frames as possible since most correlated pixels are somewhere in other frames. The common pipeline for video inpainting can be categorized into two groups: 1) direct synthesis, 2) flow-based propagation.

Supplementary Information The online version contains supplementary material available at https://doi.org/10.1007/978-3-031-19784-0_22.

S. Avidan et al. (Eds.): ECCV 2022, LNCS 13675, pp. 375–390, 2022.
https://doi.org/10.1007/978-3-031-19784-0_22

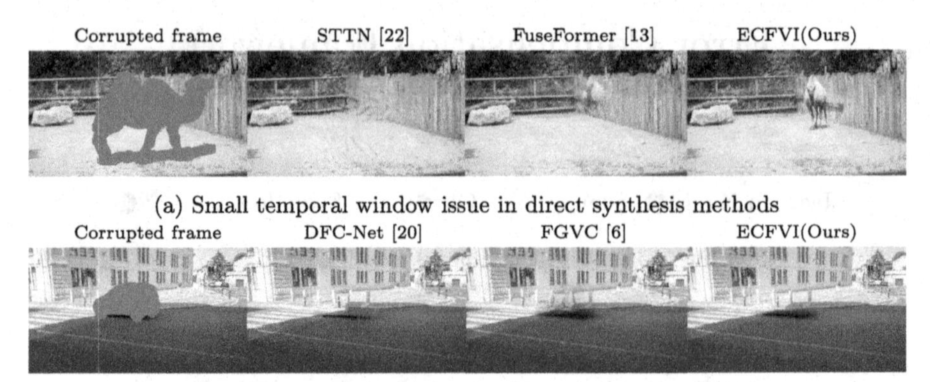

(a) Small temporal window issue in direct synthesis methods

(b) Pixel misalignment and brightness inconsistency issues in flow-based methods

Fig. 1. Qualitative comparison with other video inpainting methods on object removal scenarios. (a) The rear camel is originally shown in all frames, but it disappears in the result of [22,23] due to the small temporal window. (b) Wrong pixels are propagated, and the brightness inconsistency issue occurred in [6,20].

In general, direct synthesis methods adopt convolution-based [3,4,7,23] and attention-based [13,22] networks. They usually take corrupted frames and corresponding pixel-wise masks as input and directly output completed frames. However, due to the computational complexity, these methods have relatively small temporal windows resulting in only a few reference frames to be used (up to 10 frames). Serious temporal inconsistencies can occur if some key frames that contain unique pixel values and textures needed to fill the hole are not selected. Figure 1(a) illustrates failure cases of [13,22] when an occluded object is not properly included in the small reference frame set.

Flow-based methods [6,20] split the synthesis process as into three steps: flow completion, pixel propagation, and synthesis. During the flow completion, the corrupted optical flow maps are processed by inpainting the holes within the flow maps. In the pixel propagation step, pixel values from reference frames are propagated to the holes with the guidance of the inpainted flows. Finally, in the synthesis step, only the remaining holes are synthesized using an image inpainting method. Flow-based approaches show better long-term temporal consistency compared to direct synthesis methods as they explicitly propagate pixel values from other frames using flows.

Nevertheless, there are some drawbacks in flow-based methods. *(1) Flow completion.* In flow inpainting, it is crucial to infer spatio-temporal relationships from other corrupted flows. However, even adjacent flows can have different values and directions due to inaccurate flow estimation or non-linear motion between frames, resulting in temporally inconsistent flows. Also, without taking the underlying video contents, errors from the flow estimator like *From corrupted* in Fig. 2 cannot be handled. *(2) Pixel propagation.* Simply copy-and-pasting propagated pixels causes pixel misalignment and brightness inconsistency. As the pixel values/textures can be conveyed from distant frames, there are often perspective mismatch such as scale and shape. In addition, the brightness inconsistency may arise due to the changes in the lighting condition or camera exposure in the given video sequence. Models will output visual artifacts without the proper compensation for these errors. While existing methods attempt to address these issues with checking flow consistency [6,20] and poisson blending [6], visual artifacts still remain as shown in Fig. 1(b).

To this end, we propose a simple yet effective Error Completion Framework for Flow-guided Video Inpainting (ECFVI) that addresses both drawbacks. We follow the three steps in flow-based methods (i.e., flow completion, pixel propagation, synthesis) but significantly improve the first two steps by correcting errors, preventing the error accumulation. For the flow completion, we make the flow inpainting be aware of RGB values. Hallucinating missing flow values based on the underlying video contents can infer temporally coherent flows. Specifically, we first roughly complete RGB pixels on holes, considering the local temporal relationship. Then, the locally coherent RGB inpainting results guide the flow inpainting. Note that the local RGB inpainting here is only used for guiding the flow, and the actual inpainting results are obtained in later steps. For the pixel propagation step, we introduce additional compensation network to detect and correct the misalignment and brightness inconsistency errors from the pixel propagation. The network utilizes the error map outside the hole (called error guidance map) to predict errors inside the hole. Such carefully designed components can prevent errors from accumulating at the following stages, which significantly improves the visual quality and temporal consistency of completed videos.

For quantitative evaluation, many video inpainting methods construct their own datasets by randomly masking regions on a video. They usually set the unmasked video as the ground-truth and evaluate their models. However, when the randomly generated mask covers an entire object like in Fig. 6(a), a model will output object-removed results which are different from the unmasked video. Evaluation values from this case are not proportional to human perception. Therefore, we propose a new benchmark dataset by using an object segmentation algorithm [5] so that the mask only partially covers the object. The results on our dataset are identical to the unmasked video, which can provide comparative analysis into video inpainting methods.

In summary, the contributions of the paper are as follows:

- We propose a simple yet effective way to compensate for the limitations of the flow-based video inpainting. Specifically, we improve the flow completion with RGB-awareness and propose to compensate errors in pixel propagation.
- Our method outperforms previous state-of-the-art video inpainting methods in terms of PSNR/SSIM and visual quality. Compared with the state-of-the-art in flow-based method FGVC [6], we produce better results while reducing the computation time (×6 faster).
- We provide a new benchmark dataset for quantitative evaluation of video inpainting, which can be very useful for evaluating future work on this topic by providing a common ground.

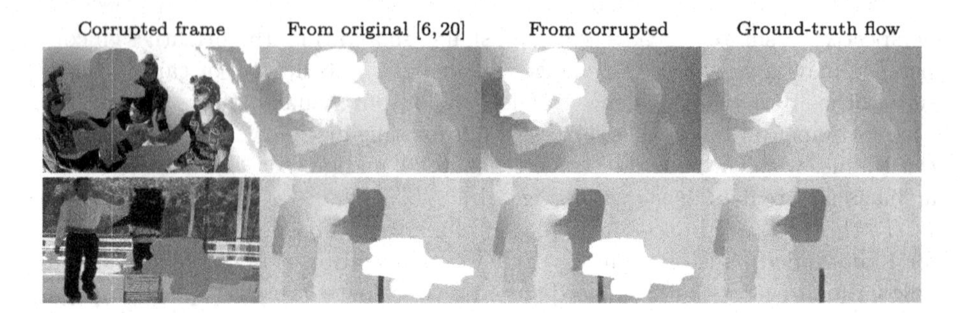

Fig. 2. This figure shows the corrupted flows with different settings. In [6,20], they take *original* frames as input to the flow estimator and remove the values in the hole. However, if the original frames are unavailable, errors can occur when the corrupted frames are used as input. The flow values near the hole are different from the ground-truth flow.

2 Related Work

2.1 Direct Synthesis Methods

From the success of deep learning, many deep-based video inpainting methods have emerged. [3,10,17] proposed to use 3D encoder-decoder networks for enhancing the efficiency and the temporal consistency. [13,22] exploited attention-based methods, which use transformer modules for matching similar patches on inter-intra frames. Zou et al. [23] used the optical flow to progressively merge target frames with reference frames to enrich feature representations on temporal relationships. Due to the memory constraints, above methods can only refer to a few frames. On the other hand, Oh et al. [14] exploited a non-local pixel matching to have a global temporal window by using the memory network. Although they refer to all frames, corruption of information occurs because the frames are continuously referenced in an implicit way. While Lee et al. [12] designed a network to take information from long-distance frames with global affine matrices, it cannot handle non-rigid complex motions.

2.2 Flow-Based Methods

Huang et al. [8] adopted spatial patch matching and flow-based method but suffered from mismatching issues caused by corrupted flows. Instead, [6,19] first inpainted the corrupted flows and propagated pixels along their flow trajectories. Despite the success, they still have some limitations to produce better results: First, to estimate the corrupted flows, they entered *original* frames to the flow estimator and removed the values with masks. This procedure works well in object removal scenarios where the original frames exist. The problem is when the original frames are not available as in video restoration scenarios. If the flow estimator takes the corrupted frames as input, the masks can interfere with estimating the motion of objects in the video (Fig. 2). The errors in corrupted flows will affect the flow completion stage, resulting in incorrect flow values. Second, it is difficult to use spatio-temporal information between the corrupted flows for flow completion as mentioned in previous section.

Wrong estimation from the flow completion would lead to the pixel misalignment issue. Both methods checked the flow consistency and only propagated trustworthy pixels to deal with the issue, but the consistency is not well preserved as the completed flow itself is incorrect. Our framework focuses on offsetting these limitations of flow-based methods.

3 Method

Let $X = [x_1, x_2, ..., x_T]$ be a set of corrupted video frames of length T. $M = [m_1, m_2, ..., m_T]$ denotes the corresponding frame-wise masks, where the spatial resolution is the same as X. For each mask, value 0 indicates valid regions, and 1 indicates hole (missing) regions. Video inpainting aims to predict the original or object removed video $\hat{Y} = [\hat{y}_1, \hat{y}_2, ..., \hat{y}_T]$ given X and M as inputs. We call the frame to be inpainted as the target frame and other frames as reference frames.

3.1 Overview

We illustrate our video inpainting framework in Fig. 3. The overall procedure can be summarized as follows. **(1) Flow completion with RGB guidance:** Given the corrupted frames, we first generate locally coherent frames. Then, we estimate complete flows between adjacent frames using our flow estimator (Sect. 3.2). Before the propagation, we estimate bi-directional completed flows between all frames. **(2) Pixel Propagation with Error Compensation:** With the guidance of completed flows, we propagate valid pixels from reference frames to target holes. In this process, we prevent the errors from propagating to the following stages using our error compensation network (Sect.3.3). In the order close to the target frame, we iterate propagation and compensation procedures until the hole regions are entirely removed or all frames are referenced. **(3) Synthesis:** If there are remaining holes to fill, we synthesize using an existing video inpainting method. This case usually occurs due to occlusion, where some pixels cannot be found in any other frames. For the synthesis, we used FuseFormer [13] with the weights from their website.

3.2 Flow Completion with RGB Guidance

Previous methods [6,20] first estimate corrupted flows with the flow estimator [9, 16] and complete flow values with their methods. However, without considering the original video contents, there is no ability to handle the errors from the corrupted flows. Therefore, we design our flow completion module to be aware of RGB values and directly output completed flows from the flow estimator.

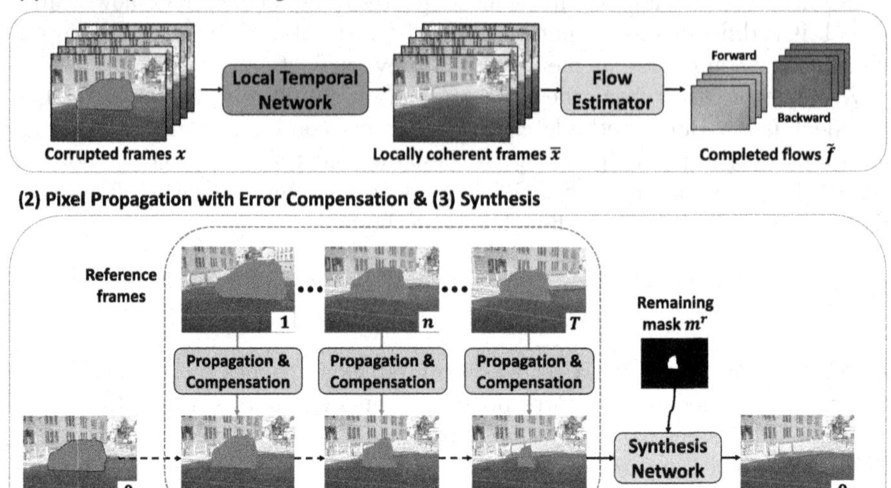

Fig. 3. Overview of our error compensation framework for flow-guided video inpainting (ECFVI). At first, we estimate all completed flows between adjacent frames using our flow completion module. With the guidance of the completed flows, we iteratively propagate and compensate pixels from reference frames to fill target holes. One iteration of the process is shown in Fig. 4. The remaining holes are further synthesized with the existing video inpainting method.

A simple approach would be to have the flow estimator take the corrupted frames and complete the flow values. However, since the flow estimator [9,16] iteratively takes two frames to estimate entire flows, completed flows are created without considering spatio-temporal information of other flows. Also, it is more challenging to run flow estimation and completion simultaneously. Therefore, we roughly complete RGB pixels through local neighboring frames (larger than two frames) and pass the results to our flow estimator. This can further exploit temporal information for estimating completed flows. Here, our flow estimator takes locally coherent frames $\bar{x}$ from a local temporal network LTN. Note that the intermediate frames $\bar{x}$ are only used for guiding the flow completion.

We design the local temporal network LTN to consist of an encoder, multiple spatio-temporal transformer layers [13,22] and a decoder. We first extract features of each input from the encoder and exploit transformer layers to merge the information from the reference in the deep encoding space. The decoder takes the output of the transformer layers to reconstruct locally coherent frames $\bar{x}$ as follows:

$$\bar{x}_i = LTN(x_{i-N:i+N}, m_{i-N:i+N}), \tag{1}$$

where N is the temporal radius and $N = 5$ is used in our experiment. Then, we estimate a completed flow between adjacent frames as follows:

$$\tilde{f}_{t \to t+1} = F(\bar{x}_t, \bar{x}_{t+1}, m_t, m_{t+1}), \tag{2}$$

where F is our flow estimator. Backward flow $\tilde{f}_{t \to t-1}$ can be estimated in the same manner. For classifying the regions where the original and coarsely completed content exist, we use masks m as additional input. The flow estimator is initialized with pretrained weights from RAFT [16] except for the first layer to take the masks as additional input. We jointly train our local temporal network and flow estimator as follows:

$$\mathcal{L} = \mathcal{L}_{rec} + \lambda_{flow} \mathcal{L}_{flow}, \tag{3}$$

where $\mathcal{L}_{rec}$ is the reconstruction L1 loss defined as $||\bar{x} - y||_1$. λ_{flow} is a coefficient for the loss terms and we set it to 2 in, our experiment. For the flow loss $\mathcal{L}_{flow}$, inspired by [15], we employ hard example mining mechanism to weigh more on the difficult areas as follows:

$$\mathcal{L}_{flow} = ||h_t \odot (\tilde{f}_{t \to t+1} - f_{t \to t+1})||_1, \tag{4}$$

where $\odot$ is element-wise multiplication and ground-truth $f_{t \to t+1}$ is calculated from original frames Y. We set the hard mining weight h_t with $(1 + |\bar{x}_t - y_t|)^2$. It encourages the model to implicitly deal with errors from the local temporal network. More details of our network are shown in the supplementary material.

This design choice has advantages in two aspects. First, it alleviates the problems of corrupted flows mentioned in Sect.2.2. Second, with the strong initialization of pretrained weights and no need to inpaint flow values from scratch, it can be trained much easier. Although there still could be errors from the two networks, we can further compensate it in the next stage.

3.3 Pixel Propagation with Error Compensation

The next step, pixel propagation with error compensation, is illustrated in Fig. 4. With the guidance of the completed flows, we iterate propagation and compensation steps so that the errors do not accumulate or amplify in the next iteration.

Propagation. Let x_i and x_j be the target and the reference frame respectively. With the completed flow $\tilde{f}_{i \to j}$, only the valid pixels in reference frame x_j are forwarded to target holes by using backward warping function w.

$$\begin{aligned} m_i^p &= m_i \odot (1 - w(m_j, \tilde{f}_{i \to j})), \\ \tilde{x}_i &= x_i + m_i^p \odot w(x_j, \tilde{f}_{i \to j}), \end{aligned} \tag{5}$$

where $\tilde{x}_i$ is the filled target frame. To find the regions where the matched and valid regions are in the reference frame, we warp the reference mask m_j and get valid regions $1 - w(m_j, \tilde{f}_{i \to j})$ in the warped reference frame. m_i^p denotes a

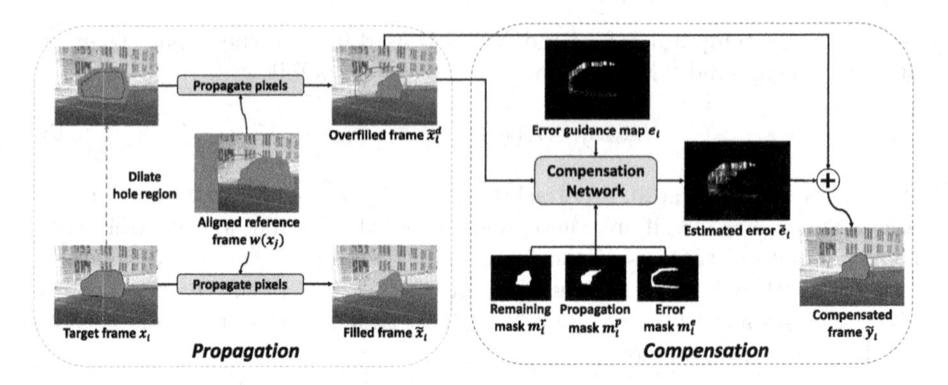

Fig. 4. For simplicity, we show the propagation and compensation stages from one reference frame x_j to the target frame x_i. We dilate the original hole region to get the overfilled frame $\tilde{x}_i^d$. Pixels corresponding to the blue region are further propagated from the aligned reference frame. The filled frame $\tilde{x}_i$ and overfilled frame $\tilde{x}_i^d$ are used to estimate the error guidance map e_i. Note that the green and blue regions correspond to propagation and error mask respectively.

propagation mask where the propagated pixels exist, shown by the green region in Fig. 4. This mask is used at the compensation stage to know where to compensate for errors. After the propagation, there may be a remaining mask m_i^r, where $m_i^r = m_i - m_i^p$. The remaining mask is further used for the next propagation or input for the synthesis network.

Compensation. As shown in Fig. 5, we observe that directly propagating pixels from the reference frame may cause misalignment or brightness inconsistency issues. To remedy these issues, one solution is taking filled frame $\tilde{x}_i$ and corresponding mask m_i^p as input and processing them through a GAN framework. However, it is difficult to train such a network because it cannot easily detect what types of problem occurred.

We approach this problem by introducing the error guidance map e_i as input to our network. To know what is wrong with the propagation, we need valid (ground-truth) values. Since only regions outside the holes in the given corrupted frame have valid values, we propagate more pixels by dilating the original hole regions like in Fig. 4. Then, we calculate the errors on enlarged parts, which is the error guidance map e_i. This process assumes that the propagated pixels on the enlarged and the propagated regions have similar error tendencies. We set the dilation factor as 17 pixels in our experiment.

We set dilated mask as m_i^d and overfilled frame as $\tilde{x}_i^d$. The error guidance map e_i and the corresponding error regions (mask) m_i^e are computed as follows:

$$\begin{aligned} m_i^e &= (m_i^d - m_i) \odot (1 - w(m_j, \tilde{f}_{i\to j})), \\ e_i &= m_i^e \odot (\tilde{x}_i^d - \tilde{x}_i). \end{aligned} \tag{6}$$

In Fig. 5, this error guidance map intuitively shows the problems of pixel propagation. If it has no issue, it will be black (zero). On the other hand, if it has a

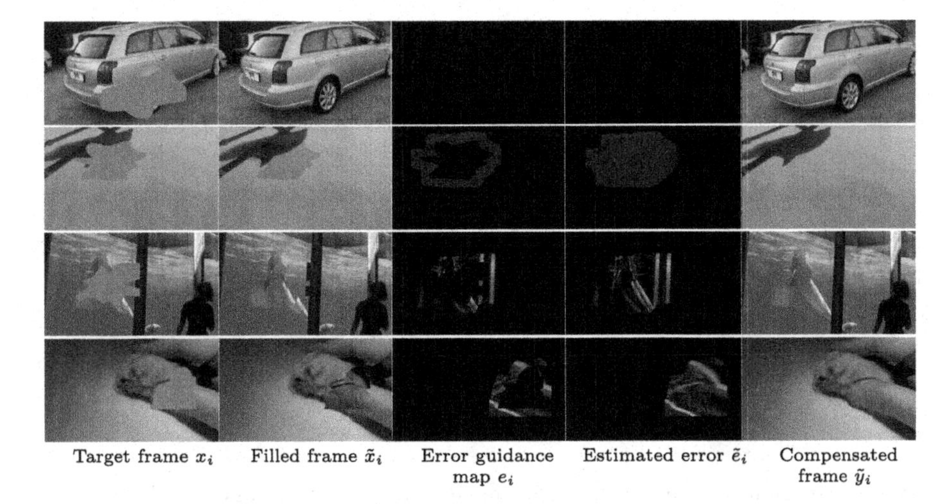

Fig. 5. Problems from the propagation stage and corresponding error maps. From top to bottom: no issue, brightness inconsistency, pixel misalignment, and combined issues. The estimated errors and compensated frames are computed using our method.

pixel misalignment issue, it will result in the edges being visible in the error map. This additional information helps our network be more aware of the problems from the propagation stage.

We design our error compensation network with a similar structure as our local temporal network LTN, but with different input and output settings:

$$\tilde{e}_i = ECN(\tilde{x}^d_{i-N:i+N}, e_{i-N:i+N}, m^{[e,p,r]}_{i-N:i+N}), \tag{7}$$

where ECN is our error compensation network and $m^{[e,p,r]}$ denotes error, propagation, remaining mask respectively. For simplicity, we will omit the index from now on. $\tilde{x}^d, e, m^e$ are concatenated in the channel dimension and forwarded to our encoder, and each of the two masks m^p, m^r are used at the transformer layers. Note that the overfilled frames $\tilde{x}^d$ are used instead of filled frames $\tilde{x}$ as input. It makes our network to learn the relationship between overfilled frame $\tilde{x}^d$ and the error guidance maps e on the error regions m^e. Since the propagated regions have some structure information, based on the relationship, our network can estimate error values on the regions m^p with accuracy. A final compensated frame is $\tilde{y} = \tilde{x}^d + \tilde{e}$. The results and corresponding errors are shown in Fig. 5.

The loss to train our error compensation network consists of reconstruction loss $\mathcal{L}_{rec}$ and adversarial loss $\mathcal{L}_{adv}$:

$$\mathcal{L} = \mathcal{L}_{rec} + \lambda_{adv}\mathcal{L}_{adv}, \tag{8}$$

where λ_{adv} is a coefficient for the loss terms and we set 0.01 in our experiment. $\mathcal{L}_{rec}$ is a L1 loss between the ground-truth frame y_i and compensated output $\tilde{y}_i$. For $\mathcal{L}_{adv}$, we use the model from Temporal PatchGAN (T-PatchGAN) [3]. Further details on our network are included in the supplementary material.

(a) Object removal scenario

(b) Video restoration scenario

Fig. 6. Two common scenarios for video inpainting: object removal and video restoration. The gray region in the corrupted frame denotes a mask. In the object removal scenario, the result and the original frame are different. Only the video restoration scenario should be used for qualitative evaluation. (Color figure online)

4 Experiments

4.1 Training Details

Our whole networks are trained on Youtube-VOS [19] dataset. It consists of 4,453 videos, which are split into 3,471/474/508 for training, validation and testing respectively. We randomly select 256×256 frames and free-form masks from [22] as input. To train our compensation network, we freeze the weights of flow completion. And for handling the brightness inconsistency issue, we further modify the brightness, hue and saturation of reference frames. Adam optimizer [11] is used where the learning rate starts with 1e−4 and is divided by 2 every 40,000 iterations. For each module, the total iteration is 120,000 with mini-batch size 4, and it takes about 2 d using two NVIDIA RTX 2080 TI GPUs.

4.2 Youtube-VI Dataset

Most previous works created their own random masks and measured the performance on Youtube-VOS [19] and DAVIS [2] datasets. However, when a randomly generated mask coincidentally covers an entire object, a model will create the object-removed result like in Fig. 6(a). The result is reasonable, but it is far from the original frame. Only the case in Fig. 6(b) should be used for quantitative evaluation. Also, if there is no motion in the mask and its vicinity, it is not suitable for video inpainting, which cannot trace any pixels from other frames.

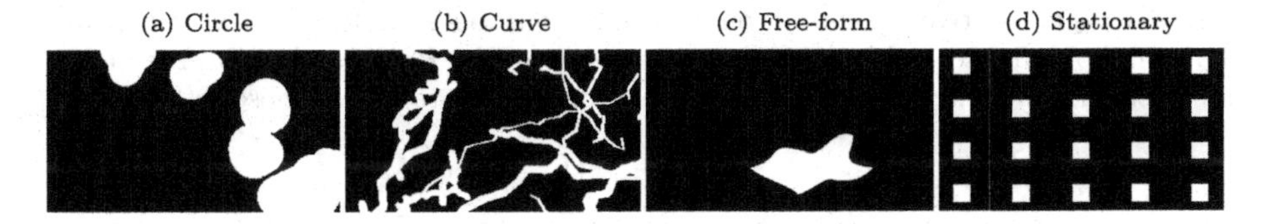

Fig. 7. Four different types of mask. In FVI [23] dataset, they use (a), (b) as a moving mask and (c) as a stationary mask. In our Youtube-VI dataset, (c) and (d) is used as a moving mask and stationary mask respectively.

TSAM [23] publicly provides frame-wise masks for the FVI [21] dataset. It consists of 100 scenes, each with 15 frames. For each scene, three types of masks are provided, some of which are shown in Fig. 7. While the masks cover diverse scenarios, 15 frames in a video are still limited compared to real-world videos. Also, the problems mentioned above on the mask still exist.

To this end, we design the new Youtube Video Inpainting (Youtube-VI) dataset, which will be publicly available. 100 scenes are selected in order of name from the beginning of Youtube-VOS dataset, and each scene has 50 frames. We generated two types of masks, which are shown in Fig. 7 and described as follows:
(1) Moving mask. We follow the same rules for generating free-form mask in STTN [22]. But we use a video segmentation network STCN [5] so that an object is not fully covered. With the help of STCN, we randomly generate one moving mask that partially overlaps with objects. Then we modify the mask again if the scene and mask do not look visually proper.
(2) Stationary mask. This case is usually used to cover the logo/subtitle removal scenarios. We find that using free-form masks is unsuitable since there may be no motion near the mask. Instead, we use 5×4 small square masks like FGVC [6].

4.3 Comparisons

We quantitatively compare with others on published FVI [23] and our new dataset for video restoration scenarios. Like [13,22,23], we use DAVIS [2] dataset for qualitative comparison in object removal scenarios. DAVIS dataset consists of 150 videos in total, in which 90 videos are annotated with the pixel-wise object masks. In addition, DAVIS-shadow dataset [8] where the mask covers both an object and its shadows are used for more visual comparison.

DFC-Net [20] and FGVC [6] only take original frames as input to the flow estimator from their published code. It is acceptable in object removal scenarios, but it is not fair to evaluate quantitative performance where the original frame is equal to ground-truth. Therefore, we also experiment with corrupted frames as input, and denote them as DFC-Net* and FGVC* respectively.

Quantitative Results. We evaluate video inpainting quality on PSNR and SSIM. In Table 1, we achieve the best performance on PSNR and SSIM except for the curve mask on FVI dataset. It is worth noting that the curve mask is a particular case in real-world scenarios. As we assume our error guidance map for common scenarios, the error values in dilated regions can be overlapped, which

Table 1. Quantitative evaluation on FVI and Youtube-VI datasets. Best values are shown in bold and second best values are underlined. For PSNR and SSIM, the higher is better. Missing entries indicate the method impossible to run at those settings.

	FVI [23]						Youtube-VI				
	Stationary mask		Circle mask		Curve mask		Stationary mask		Moving mask		
	PSNR	SSIM	PSNR	SSIM	PSNR	SSIM	PSNR	SSIM	PSNR	SSIM	Time(s)
DFC-Net [20]	29.20	0.9632	24.77	0.9219	28.03	0.9436	29.03	0.9609	32.94	0.9771	95
DFC-Net* [20]	28.51	0.9609	24.12	0.9141	26.60	0.9295	27.73	0.9536	32.14	0.9747	95
OPN [14]	31.59	0.9689	26.37	0.9264	29.79	0.9519	31.91	0.9721	33.56	0.9767	154
CPN [12]	31.76	0.9693	26.90	0.9315	31.09	<u>0.9664</u>	31.32	0.9687	34.15	0.9783	24
STTN [22]	32.18	0.9699	27.04	0.9288	30.01	0.9515	31.96	0.9716	34.66	0.9789	17
FGVC [6]	30.88	0.9690	26.14	0.9369	**31.50**	**0.9676**	32.25	0.9750	<u>35.68</u>	<u>0.9841</u>	734
FGVC* [6]	30.60	0.9682	25.55	0.9302	29.90	0.9565	31.21	0.9701	34.05	0.9795	734
TSAM [23]	31.74	0.9695	26.66	0.9250	<u>31.18</u>	0.9578	–	–	–	–	–
FuseFormer [13]	<u>33.19</u>	<u>0.9733</u>	<u>27.85</u>	<u>0.9368</u>	30.85	0.9577	<u>33.04</u>	<u>0.9755</u>	35.29	0.9806	18
ECFVI(Ours)	**33.27**	**0.9749**	**28.24**	**0.9469**	31.15	0.9638	**33.53**	**0.9795**	**36.80**	**0.9860**	121

worsens our results in such thin and overlapped masks. On the other hand, on the moving and stationary mask, our framework clearly outperforms by a large margin compared to DFC-Net [20] and FGVC [6], where the flow-based propagation is used. Such results show that our overall framework plays critical role in addressing the limitations of existing methods. More comparisons on other metrics, Video-based Fréchet Inception Distance (VFID) [18] and optical flow based warping error (EWarp) [18], are shown in the supplementary material.

We measure the execution time on 864×480 resolution for inpainting 50 frames to compare the efficiency. Since we design our entire components with deep frameworks, ours is $\times 6$ faster compared to FGVC [6] where the optimization-based methods are used.

Qualitative Results. In Fig. 8, we show our model's qualitative results compared with other methods, including STTN [22], FGVC [6] and FuseFormer [13]. To visualize the temporal consistency, we show the temporal profile [1] of the resulting videos below the completed frames. Sharp and smooth edges in the temporal profile indicate that the video has much less flickering. As can be seen, our method shows more visually pleasing results compared to others.

For a more comprehensive comparison, we conducted a user study to subjectively compare our method against others on the DAVIS dataset through Amazon Mechanical Turk. The experiments were performed on 30 videos that excludes easy (Tennis, Flamingo, etc.) and difficult (India, Drone, etc.) cases where every methods failed. The videos like Fig. 8 were shown to 20 participants. We unlabeled and shuffled the order in all videos, and asked to rank the results from 1 to 4 (1 is the highest preference). As shown in Fig. 9, the results of our model showed higher preference over others, supporting that our approach can produce more visually pleasing outputs.

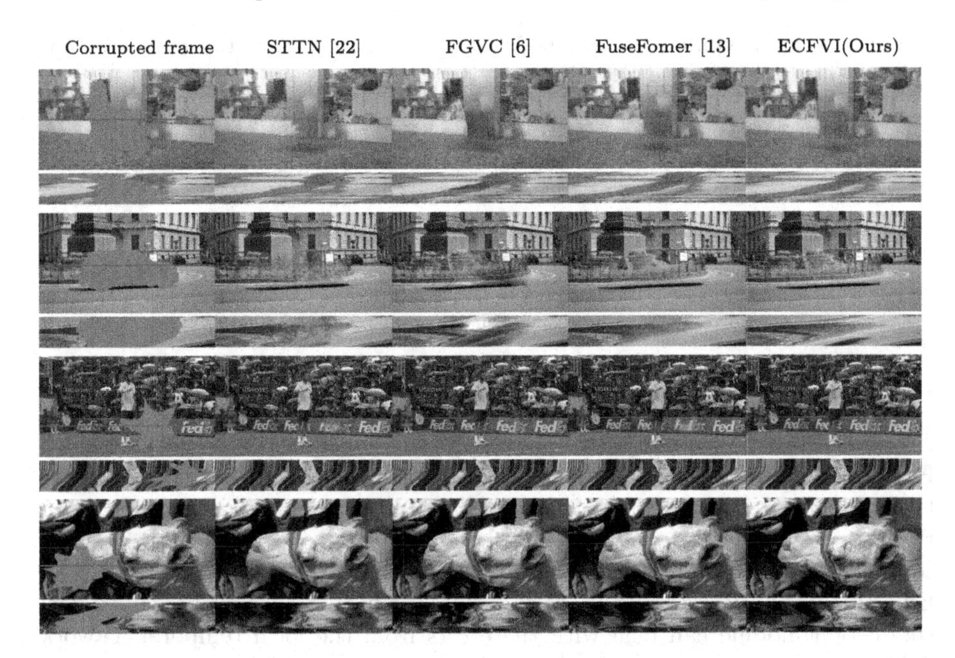

Fig. 8. Qualitative results compared with [6,13,22]. The temporal profiles of the red scan line are shown below the results. **Best viewed in zoom.** (Color figure online)

4.4 Ablation Study

Effectiveness of Flow Completion. In Table 2(a), we show the flow end-point-error (EPE) metric to compare our flow completion with others. We set the estimated flows from RAFT [16] as pseudo ground-truth flows. In the table, FGVC/* means FGVC and FGVC* respectively. In DFC-Net [20] and FGVC [6], the performance is significantly reduced when the corrupted frames are used as input to the flow estimator. It verifies that the errors in corrupted flows prevent estimating correct flows. In the same setting where the corrupted frames are used, ours achieves the best performance.

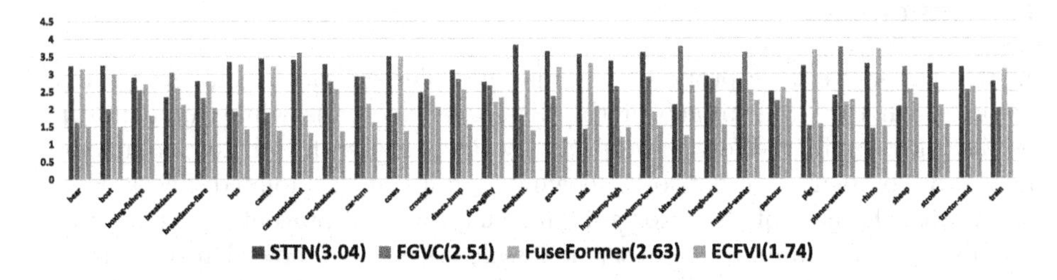

Fig. 9. A user study on DAVIS [2] object removal. The lower is better.

Table 2. Ablation studies on Youtube-VI dataset. For Flow-EPE, the lower is better. FGVC/* means FGVC and FGVC* respectively.

(a) Flow completion

	Stationary Mask	Moving Mask	
	Flow-EPE	Flow-EPE	Time(s)
DFC-Net/* [19]	0.32/0.89	0.38/0.86	43
FGVC/* [6]	0.12/0.57	0.28/0.80	412
FF [13] + RAFT [16]	0.49	0.80	52
w/o LTN	0.71	0.83	24
w/o Hard	0.40	0.58	52
Ours	0.33	0.49	52

(b) Video inpainting

	Stationary Mask		Moving Mask	
	PSNR	SSIM	PSNR	SSIM
FGVC* + Comp	32.87	0.9762	35.59	0.9823
GT flow + Comp	33.64	0.9796	37.17	0.9865
w/o Comp	30.05	0.9691	34.04	0.9810
w/o Guide	30.48	0.9692	33.81	0.9797
Ours	33.53	0.9795	36.80	0.9860

In addition, we conduct a series of ablation studies to analyze the effectiveness of our flow completion module. First, we sequentially run pretrained FuseFormer [13] followed by original RAFT [16] (FF + RAFT). To demonstrate the effectiveness of local temporal network LTN, we show the results only using the flow estimator (w/o LTN). The flow estimator is trained with the same setting as ours. In addition, we train our flow estimator with equal weights in Eq. (4) instead of h (w/o Hard). Such results verify that our jointly trained flow completion module can deal with the errors from the local temporal network LTN and produce more accurate completed flows.

Effectiveness of Compensation. In Table 2(b), we show the performance of the error compensation network by replacing our flow completion with FGVC's method (FGVC* + Comp). Although the flow completion from FGVC* is worse than ours, it outperforms the full procedure of FGVC*. We also show the pseudo ground-truth flow from RAFT [16] (GT flow + Comp). These two experiments demonstrate our error compensation network is robust to errors from the flow completion. Also, we test our model without compensation stage (w/o Comp) and without error guidance map (w/o Guide). In w/o Guide experiment, we use naive GAN where filled frames with propagation masks and remaining masks are only used as input. The results show severe performance drops compared to our method. These two experiments demonstrate that our compensation network using the error guidance map is necessary to handle the errors from the previous stages.

5 Conclusion

In this paper, we have proposed a simple yet effective video inpainting framework, which takes advantage of the flow-based methods while compensating for its shortcomings. We can propagate valid pixels with our flow completion and error compensation network, showing high-quality completed videos. Especially with the help of the error guidance map, we can prevent the errors from accumulating and amplifying at the following stages. We show that our method achieves state-of-the-art performance and demonstrate the benefits of our framework through extensive experiments. We also provide a new benchmark dataset, which provides comparative analysis into video inpainting methods.

Acknowledgement. This work was supported by Institute of Information & Communications Technology Planning & Evaluation (IITP) grant funded by the Korea government (MSIT) (No. 2014-3-00123, Development of High Performance Visual BigData Discovery Platform for LargeScale Realtime Data Analysis), and No. 2020-0-01361, Artificial Intelligence Graduate School Program (Yonsei University).

References

1. Caballero, J., et al.: Real-time video super-resolution with spatio-temporal networks and motion compensation. In: CVPR, pp. 4778–4787 (2017)
2. Caelles, S., et al.: The 2018 davis challenge on video object segmentation. arXiv preprint arXiv:1803.00557 (2018)
3. Chang, Y.L., Liu, Z.Y., Lee, K.Y., Hsu, W.: Free-form video inpainting with 3d gated convolution and temporal patchgan. In: ICCV, pp. 9066–9075 (2019)
4. Chang, Y.L., Liu, Z.Y., Lee, K.Y., Hsu, W.: Learnable gated temporal shift module for deep video inpainting. In: BMVC (2019)
5. Cheng, H.K., Tai, Y.W., Tang, C.K.: Rethinking space-time networks with improved memory coverage for efficient video object segmentation. In: NeurIPS (2021)
6. Gao, C., Saraf, A., Huang, J.-B., Kopf, J.: Flow-edge guided video completion. In: Vedaldi, A., Bischof, H., Brox, T., Frahm, J.-M. (eds.) ECCV 2020. LNCS, vol. 12357, pp. 713–729. Springer, Cham (2020). https://doi.org/10.1007/978-3-030-58610-2_42
7. Hu, Y.-T., Wang, H., Ballas, N., Grauman, K., Schwing, A.G.: Proposal-based video completion. In: Vedaldi, A., Bischof, H., Brox, T., Frahm, J.-M. (eds.) ECCV 2020. LNCS, vol. 12372, pp. 38–54. Springer, Cham (2020). https://doi.org/10.1007/978-3-030-58583-9_3
8. Huang, J.B., Kang, S.B., Ahuja, N., Kopf, J.: Temporally coherent completion of dynamic video. ACM Trans. Graph. (TOG) **35**(6), 1–11 (2016)
9. Ilg, E., Mayer, N., Saikia, T., Keuper, M., Dosovitskiy, A., Brox, T.: Flownet 2.0: evolution of optical flow estimation with deep networks. In: CVPR, pp. 2462–2470 (2017)
10. Kim, D., Woo, S., Lee, J.Y., Kweon, I.S.: Deep video inpainting. In: CVPR, pp. 5792–5801 (2019)
11. Kingma, D.P., Ba, J.: Adam: a method for stochastic optimization. arXiv preprint arXiv:1412.6980 (2014)
12. Lee, S., Oh, S.W., Won, D., Kim, S.J.: Copy-and-paste networks for deep video inpainting. In: ICCV, pp. 4413–4421 (2019)
13. Liu, R., et al.: Fuseformer: fusing fine-grained information in transformers for video inpainting. In: ICCV, pp. 14040–14049 (2021)
14. Oh, S.W., Lee, S., Lee, J.Y., Kim, S.J.: Onion-peel networks for deep video completion. In: ICCV, pp. 4403–4412 (2019)
15. Shrivastava, A., Gupta, A., Girshick, R.: Training region-based object detectors with online hard example mining. In: CVPR, pp. 761–769 (2016)
16. Teed, Z., Deng, J.: RAFT: recurrent all-pairs field transforms for optical flow. In: Vedaldi, A., Bischof, H., Brox, T., Frahm, J.-M. (eds.) ECCV 2020. LNCS, vol. 12347, pp. 402–419. Springer, Cham (2020). https://doi.org/10.1007/978-3-030-58536-5_24
17. Wang, C., Huang, H., Han, X., Wang, J.: Video inpainting by jointly learning temporal structure and spatial details. In: AAAI, vol. 33, pp. 5232–5239 (2019)

18. Wang, T.C., et al.: Video-to-video synthesis. arXiv preprint arXiv:1808.06601 (2018)
19. Xu, N., et al.: Youtube-vos: sequence-to-sequence video object segmentation. In: ECCV, pp. 585–601 (2018)
20. Xu, R., Li, X., Zhou, B., Loy, C.C.: Deep flow-guided video inpainting. In: CVPR, pp. 3723–3732 (2019)
21. Yu, J., Lin, Z., Yang, J., Shen, X., Lu, X., Huang, T.S.: Free-form image inpainting with gated convolution. In: ICCV, pp. 4471–4480 (2019)
22. Zeng, Y., Fu, J., Chao, H.: Learning joint spatial-temporal transformations for video inpainting. In: Vedaldi, A., Bischof, H., Brox, T., Frahm, J.-M. (eds.) ECCV 2020. LNCS, vol. 12361, pp. 528–543. Springer, Cham (2020). https://doi.org/10.1007/978-3-030-58517-4_31
23. Zou, X., Yang, L., Liu, D., Lee, Y.J.: Progressive temporal feature alignment network for video inpainting. In: CVPR, pp. 16448–16457 (2021)

Scraping Textures from Natural Images for Synthesis and Editing

Xueting Li[1(✉)], Xiaolong Wang[2], Ming-Hsuan Yang[1], Alexei A. Efros[3], and Sifei Liu[4]

[1] UC Merced, Merced, USA
xuetingl@nvidia.com
[2] UC San Diego, San Diego, USA
[3] UC Berkeley, Berkeley, USA
[4] NVIDIA, Santa Clara, USA

Abstract. Existing texture synthesis methods focus on generating large texture images given a small texture sample. But such samples are typically assumed to be highly curated: rectangular, clean, and stationary. This paper aims to scrape textures directly from natural images of everyday objects and scenes, build texture models, and employ them for texture synthesis, texture editing, etc. The key idea is to jointly learn image grouping and texture modeling. The image grouping module discovers clean texture segments, each of which is represented as a texture code and a parametric sine wave by the texture modeling module. By enforcing the model to reconstruct the input image from the texture codes and sine waves, our model can be learned via self-supervision on a set of cluttered natural images, without requiring any form of annotation or clean texture images. We show that the learned texture features capture many natural and man-made textures in real images, and can be applied to tasks like texture synthesis, texture editing and texture swapping.

Keywords: Texture synthesis · Segmentation and grouping

1 Introduction

Texture synthesis aims at generating a texture image of arbitrary size given a small texture example. Existing texture synthesis approaches [13,30,42,58,65] rely on carefully curated texture datasets [6,8,10], where each image is rectangular and only includes a single texture pattern. Collecting such datasets requires tedious human effort. At the same time, most natural images already contain abundant textures found on most objects and materials (e.g., the starfish, the corn, and the snake in Fig. 1). A question ensues – can we *scrape* textures directly from natural images, i.e., learn texture models in an unsupervised fashion?

Project page: https://sunshineatnoon.github.io/texture-from-image/.

Supplementary Information The online version contains supplementary material available at https://doi.org/10.1007/978-3-031-19784-0_23.

S. Avidan et al. (Eds.): ECCV 2022, LNCS 13675, pp. 391–408, 2022.
https://doi.org/10.1007/978-3-031-19784-0_23

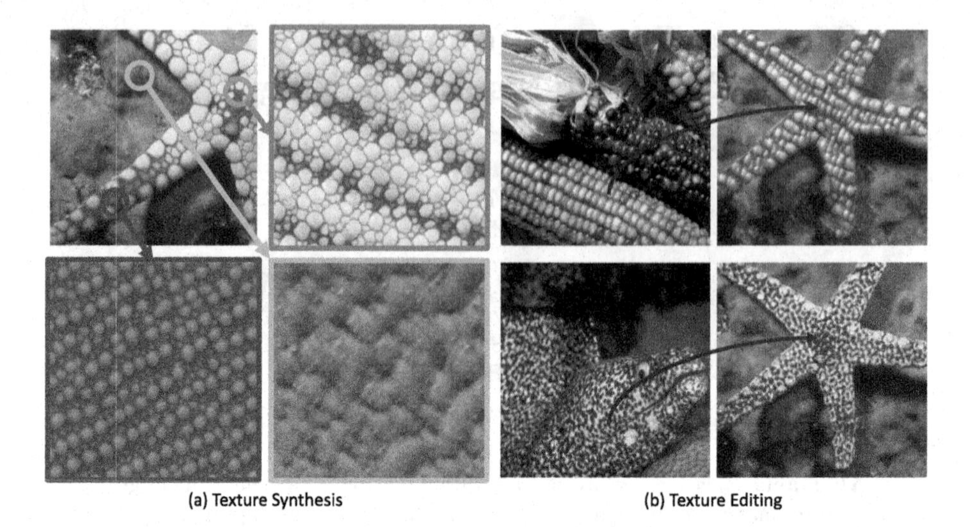

Fig. 1. Texture synthesis and editing. (a) Scraping textures from a natural image (upper left) to synthesize texture images. (b) Scraping textures from one image (left) to transfer onto another image (right).

This is a highly challenging task since: a) a natural image usually includes multiple different texture regions that cannot be readily used as supervision for texture feature learning, and b) the shapes of these texture regions are difficult to discover without segmentation annotation. This is why existing texture synthesis methods [12,13,16,17,33,37,58,65] require clean, rectangular texture patches as training data. Yet, even with clean texture patches, modeling and synthesizing diverse real world texture patterns is non-trivial. Contemporary methods mainly have two ways to synthesize texture given texture patches: a) encoding texture statistics in spatial features and resorting to transpose convolution for synthesis [42,45] or b) map a Gaussian distribution to a texture manifold [37,58,65]. However,the former method can only synthesize new texture image of limited size while the latter method fails to capture structural textures (e.g., the grid pattern in Fig. 3) due to a lack of structure in the Gaussian noise input. Furthermore, the limited representation capacity of neural networks may preclude the model to generalize to unseen images and produces inferior texture synthesis results.

To resolve these issues, we introduce a texture encoder that decomposes an image into a small number of clean texture segments and encode the texture pattern in each segment with a relatively low-dimensional vector texture code. We then use a decoder to reconstruct the input image given these texture codes. Our key insight is, by limiting the number of texture codes, we create an information bottleneck. In order to reconstruct the input image from this information bottleneck, our model is forced to group pixels of the same texture together without any segmentation annotation. Overview of our method is shown in Fig. 2.

Furthermore, to faithfully model the more structured textures in natural images, we introduce a parametric sine wave for each texture segment, besides the vector texture code. The periodicity of this sine wave allows our model

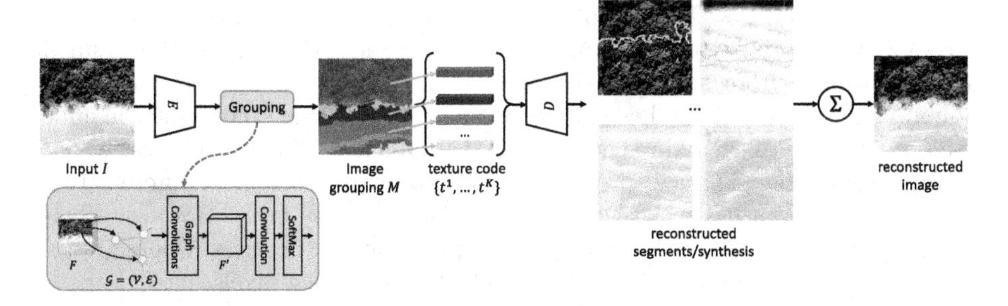

Fig. 2. Overview. Given a natural image I, the grouping module (blue box) produces a grouping mask M. For each group, we reconstruct the corresponding texture segment (shown inside the yellow contour) and extend it through texture synthesis. The reconstructed image is composited from all reconstructed texture segments and shown in the last column. Detailed illustration can be found in the supplementary. (Color figure online)

to capture both regular (e.g., the corn in Fig. 3) and irregular (e.g., the zebra in Fig. 3) texture patterns, as well as enables arbitrary-sized texture synthesis during inference. To efficiently generalize our model to unseen images, we first train the model on a natural image dataset such that the encoder can learn grouping prior from divergent samples. We then generalize the model to unseen images with test-time adaptation [35,38,57].

During inference, our model can be readily applied for arbitrary-sized texture synthesis, editing and swapping as shown in Fig. 1. Besides, it also produces texture segments that often adhere nicely to object boundaries as shown in Fig. 6. The key contribution of our work is a method for jointly learning image grouping and texture modeling from natural images using an autoencoder formulation. The rest of the paper will discuss prior work, describe our method in detail, and present results.

2 Related Work

Texture synthesis aims to generate a texture image conditioned on a given texture exemplar, while image grouping attempts to cluster pixels into segments based on their appearance. Since our work learns the two tasks jointly, we discuss the related work in both domains.

2.1 Texture Synthesis

Scraping Textures from Natural Images. Most existing texture synthesis methods require a rectangular and clean texture patch as exemplar. As a result, large amount of efforts have been made to collect and clean texture datasets [6,8,10]. A few methods tried to overcome this limitation and learn texture synthesis directly using natural images by making strong assumptions of the training images. For instance, Bergmann et al. [5] proposed PSGAN to encode few texture patterns in a natural image by learning a model with small patches cropped from an ultra-resolution image. However, the strong assumption in PSGAN – small patches cropped from ultra-resolution image only include one texture pattern easily breaks for normal sized natural images. Closest to our work, Rosenberger et al. [52] propose to decompose an image into different texture layers and synthesize larger texture images layer by layer. However, their model is only applicable to a very specific type of natural textures, where grouping process can simply be carried out by clustering on raw pixels.

Texture Synthesis from Well-Cropped Patches. Most existing texture synthesis methods learn from well-cropped texture patches. Existing methods can roughly be categorized into non-parametric [12,13,60] and optimization-based [5,17,29,42,58,63,65].

The non-parametric methods sequentially search and copy a pixel [13] or a patch [12] from the given texture exemplar that best fits for the current location. By bypassing the challenging task of analysing and modeling the complex structure and statistics of texture patterns, the algorithm produces realistic results that resemble the given exemplar. However, the sequential texture growing process can be unstable – it may start to produce "garbage" texture should it fail to find a suitable patch fitting the current location. It also takes a long time for high-resolution texture image synthesis without a specially designed mechanisms for efficient searching [4,60].

One line of optimization-based methods [17,22,23,30,33,50,54,62] try to map a prior distribution (e.g., a Gaussian noise) to a texture manifold by repeatedly modifying an initial noise image until it presents the same pattern as the given texture exemplar. For instance, the seminal work from Gatys et al. [17] starts with a random noise image and iteratively updates it by comparing the Gram matrices of the output and the exemplar. However, these methods suffer from the slow optimization process that has to be carried out for each given exemplar. They are also weaker at modeling structured textures than non-parametric methods due to a lack of structure in the noise image.

Another line of optimization-based works learn deep neural networks to speed up the synthesis process. They either encode the statistics of a single texture [29,55,58,68], or a limited number of textures [2,5,16,27,36,65] in a model while failing to generalize to unseen textures. Recently, two works [42,45] successfully produced a single model to synthesize unseen textures. They represent texture patterns by spatial feature maps and either pose texture synthesis as image upsampling in the Fourier domain or formulate texture expansion as transposed convolutions. However, all these texture synthesis methods require well-cropped texture patches as exemplar images and can only expand the texture to a limited size.

2.2 Perceptual Grouping and Image Segmentation

Perceptual grouping refers to the principles, first outlined by the Gestaltists [61] by which the visual system groups elements of the visual world into higher-order perceptual units (i.e. image regions). In computer vision, the area of image segmentation has been focused on operationalizing these principles, but despite some early successes, e.g. the classic Normalized Cuts algorithms [56], this problem still remains largely unsolved. Indeed, most subsequent methods gave up on partitioning an image into large, perceptual segments and instead focused on over-segmenting an image into superpixels [51], based on either hand-crafted low-level features such as color [1,14,15], texture and brightness [51,56] or deep features learned by neural networks [26,59,64]. The produced superpixels are compact and preserve sufficient local image structure that can be utilized for downstream tasks. Some later image segmentation approaches focus on producing larger segments by merging the pre-computed superpixels [25,39,40,44,49,51]. For example, the MCG method [49] produces region proposals by aligning segmentation from the normalized cuts algorithm [56] at different resolutions. The DeepGrouping method [39] learns a hierarchical graph neural network [53] and utilizes the features from the low- to high-level for texture, material, part, and object segmentation, respectively. In this work, we learn image grouping to produce texture segments that we then use as supervision for texture model learning. Our model learns to group pre-computed superpixels [1] into large texture segments without using any segmentation or grouping annotation as supervision.

2.3 Shape and Appearance Disentanglement

Disentangling an image into shape and appearance has been extensively studied in the past decades. Classical methods [19,20] such as the Primal Sketch [19] decomposes an image into "sketchable" and "non-sketchable" components, which essentially capture the structure and texture of the image respectively. More recent works [34,47,69] disentangle an image into structure and texture represented as deep features. The structure component is either represented as a spatial feature map [47] or approximated by a set of primitives [34,69]. The texture is encoded as a vector feature [47] or simply RGB values [34] or pre-defined stroke patterns [69]. Our model explicitly decomposes an image into structure represented as image grouping and texture captured by vector texture codes and sine waves.

3 Method

We propose a texture-based image auto-encoder that decomposes a natural image into a set of clean texture segments, each of which is represented by a 256-dimension texture code and a parametric sine wave. Our key idea is to create a information bottleneck with low-dimensional representations, so that when they are used to reconstruct the input images, the model is encouraged to:

(i) cluster pixels belonging to the same texture, and (ii) learn compact texture representation for each segment.

In the following, we first provide a high-level overview of the proposed framework in Sect. 3.1, then, we describe the encoder that is responsible for texture grouping and representation learning in Sect. 3.2. In Sect. 3.3, we introduce the decoder that reconstructs the input image from the compact texture representations. We describe how to employ our model for texture synthesis in Sect. 3.4. Finally, the self-supervised training objectives are discussed in Sect. 3.5.

3.1 Overview

The framework of the proposed algorithm is shown in Fig. 2. Given an input image $I \in \mathcal{R}^{3\times H\times W}$, the encoder E maps it to a per-pixel feature map $F \in \mathcal{R}^{C_F\times H\times W}$. The grouping module (blue box in Fig. 2) takes the feature map F as input and produces a grouping map $M \in \mathcal{R}^{K\times H\times W}$ that assigns each pixel to one of K groups. Each group represents a texture pattern in the image and is encoded by a vector texture code $t^k \in \mathcal{R}^{C_T}$ and a learnable sine wave $S^k, k = 1, \ldots, K$. For each texture segment, the decoder D is tasked to simultaneously learn to reconstruct it and synthesize a similar texture image.

3.2 Texture Encoder

The texture encoder aims to discover a set of clean texture segments from a highly cluttered natural image and encode their texture patterns for texture synthesis. To this end, we propose to build a graph neural network on top of superpixels [1,26,51,64] to group superpixels of the same texture together. The rationale of using superpixels instead of pixels is that superpixels are compact, consistent with boundaries between different texture regions while preserving sufficient receptive field for feature learning. Meanwhile, a graph neural network allows nearby superpixels to share or differentiate features depending on their similarities and yields features fit for the final grouping mask computation. We note that an alternative solution to obtain texture segments is by directly utilizing readily available semantic segmentation masks obtained by manual annotation [3,41,67] or pre-trained models [7,43]. However, these masks are defined based on semantic prior, where each region may contain several different texture patterns, e.g., a human face includes hair, skin, eyes, etc. Our encoder further encodes the texture pattern in each segment compactly using a texture code and a parametric sine wave. We introduce the details of the image grouping and texture modeling module in the following.

Image Grouping. The detailed structure of the grouping module is illustrated in the blue box in Fig. 2. First, we introduce a CNN backbone E to produce per-pixel feature vectors F, which is then input to the grouping module that contains the GNN. Each vertex $v \in \mathcal{V}$ in the graph $\mathcal{G} = (\mathcal{V}, \mathcal{E})$ is a superpixel, whose feature is computed as the averaged F of all pixels within the superpixel. The weight of each edge $e \in \mathcal{E}$ connecting two superpixels in the graph is computed as the cosine similarity of two adjacent superpixel features. Given such a graph, we first employ two graph convolution layers [32] to propagate information from each superpixel to its neighboring superpixels. Then a per-pixel feature map $F' \in \mathcal{R}^{C_F \times H \times W}$ is recovered by setting the feature of each pixel as the feature of the superpixel it belongs to. Finally, a convolution layer followed by a SoftMax layer is applied to group all pixels into K segments. The predicted grouping mask $M \in \mathcal{R}^{K \times H \times W}$ decomposes the input image into K different groups, each includes a unique texture pattern and is captured by a vector texture code and a parametric sine wave discussed next.

Texture Code Computation. Given the per-pixel feature $F' \in \mathcal{R}^{C_F \times H \times W}$ and the grouping mask $M \in \mathcal{R}^{K \times H \times W}$ from the grouping module discussed above, we first compute the features $F_K \in \mathcal{R}^{K \times C_F}$ for all groups as $F_K = flatten(M) \times flatten(F')^T$, where $flatten(M) \in \mathcal{R}^{K \times (HW)}$ and $flatten(F') \in \mathcal{R}^{C_F \times (HW)}$ are the flattened version of M and F' respectively. The kth row of F_K thus stores the feature of group k. The texture code t^k for each group is computed by two linear layers, taking ${F_K}^k$ as input.

Parametric Sine Wave Computation. Besides the texture code, we introduce a parametric sine wave to improve texture modeling. Our key insight is that the periodic sine wave well fits for the repetitive nature of textures, especially for regular patterns that resemble strong periodicity. Inspired by PSGAN [5], we formulate the parametric sine wave with learnable frequency P^k as:

$$S^k = \sin(P^k C + \Phi^k), k = 1, \ldots, K \quad (1)$$

where d is a hyperparameter indicating the channel number of the sine wave, $C \in \mathcal{R}^{2 \times (HW)}$ is a standard 2D coordinate grid, and Φ^k is a shift randomly sampled from $[0, 2\pi)$ to mimic random positional shift of texture segments. Intuitively, the frequency parameter P^k determines the repeating period of a texture and thus is predicted by applying two linear layers on the statistic texture code t^k.

3.3 Decoding Texture Codes and Sine Waves into RGB Images

Given the texture code and sine wave of each segment, we adopt a decoder to reconstruct the input image by taking the sine wave composition of different texture segments as input and modulating the feature at each layer by the texture code). Specifically, given the feature map F_l from layer l of the decoder, a pixel $F_l(i, j)$ assigned to group k is modulated by its corresponding texture code t^k as:

$$\gamma^k \frac{F_l(i,j) - \mu^k}{\sigma^k} + \beta^k \tag{2}$$

where the modulation parameters γ^k and β^k are projected from the texture code t^k by linear layers and μ^k, σ^k are the mean and standard deviation in each channel of F_l.

3.4 Texture Synthesis

Next, we task our decoder for texture synthesis to enhance texture feature learning. Specifically, we randomly choose a texture segment $q \in \{1, \ldots, K\}$ from the K groups and synthesize a corresponding texture image resembling the chosen texture segment. To this end, we feed the corresponding sine wave S^q into the decoder and modulate all pixels in the feature maps at each decoder layer using the texture code t^q. As a result, the decoder synthesizes a texture image that presents the same pattern resembling the chosen texture group q. More importantly, thanks to the inherent periodicity of the sine wave, our model can synthesize an arbitrary-sized texture image by simply extending the input sine wave during inference.

However, the segments produced by the grouping module may be noisy (e.g., a texture segment that includes more than one texture pattern), especially at the early training stage. To clean up the texture segments and provide better supervision for texture features learning, we split each group into spatially connected regions [26]. The texture segment used as the supervision for synthesis is then chosen from these spatially connected regions. This is based on the intuition that spatially connected regions are more likely to be clean texture segments with a single texture pattern, introducing less noise for texture features learning.

3.5 Objectives

Reconstruction Objectives. For each texture segment, we apply the L1, the VGG-based perceptual [29,66] loss and the focal frequency loss [28] between the input and the reconstructed segment.

Texture Synthesis Objectives. Since there is no direct supervision for the regions outside the given texture segment, we apply the Gram matrix matching objective [17,18,42] and a patch discriminator [47]. The former ensures the synthesized texture match the statistics of the texture segment while the latter encourages realistic texture synthesis.

As discussed in Sect. 1, texture segments in natural images often have irregular shapes. Thus we utilize the grouping mask produced by the grouping module to mask out irrelevant regions in both objectives. Specifically, when computing the gram matrix of a texture segment, we only take the pixels within the texture segment to calculate the correlation matrix. A detailed mathematical formulation can be found in the supplementary material.

For the patch discriminator, we randomly crop ten patches from the corresponding texture segment in the input image as real data and ten patches from the synthesized texture image as fake data. To ensure patches cropped from the synthesized texture images are comparable with those cropped from input images, we fill irrelevant areas in the former with corresponding pixels from the latter. The objective of the patch discriminator is:

$$L_{PGAN} = \mathbb{E}_{I,\hat{I}_t}[-\log(D_p(\text{crops}(\hat{I}_t) \cdot m + \text{crops}(I) \cdot (1-m), \text{crops}(I)))] \quad (3)$$

where D_p is the patch discriminator [47], crops$(\cdot)$ is the operation that randomly crops patches from an image and m is the mask patch from M_t corresponding to the cropped patches from the input image (i.e., crops(I)).

3.6 Generalization to Unseen Textures

To generalize our model to an unseen image, we first train our model on a small image dataset and then fine-tune it on a single image. Importantly, we make the following differences during test-time tuning compared to the main network training: a) we do not include any form of data augmentation, i.e., the training image is cropped once and fixed. b) we sample one texture code and synthesize two images from it. One image using a sine wave with zero offset and the other using a random offset. c) we apply reconstruction loss on the former synthesis and gram matrix matching and patch GAN to the later. d) we use Meanshift [9] to allow each image to have different group numbers. The rationale of these differences is that our model has to reconstruct the given texture segment well before extending it. By fixing the offset and the training image, we allow the model to achieve such a goal.

4 Experimental Results

4.1 Experimental Settings

Implementation Details. Our encoder is composed of multiple Residual blocks [21] and two graph convolution layers. The decoder is the same as the generator in [47]. We cluster pixels in each image into $K = 10$ groups for the main network training if not otherwise specified. Empirically we find that progressive training stabilizes the process and yields the best performance. Specifically, we train the model without the grouping module for 100 epochs and fine-tune with the grouping module for another 100 epochs. The texture synthesis objective is then added, and the entire pipeline is jointly trained until convergence. For each unseen image, we fine-tune the model for 5000 iterations with the same objectives introduced in Sect. 3.5. We use the Adam optimizer [31] with an initial learning rate of 5×10^{-5}. More details of the network architecture and optimization hyperparameters can be found in the supplementary material. The proposed model is implemented in PyTorch [48].

Datasets. We use the BSDS500 dataset [3,46] as our training dataset. Compared to other datasets such as COCO [41] or the ImageNet [11], the BSDS500 dataset provides larger and richer texture regions that are more fit for our texture synthesis task. For texture synthesis evaluation, we compose a dataset of 104 texture-abundant images not included in the training dataset from the ImageNet dataset [11]. Although our main focus is on texture synthesis, we validate the learned grouping masks by evaluating the image grouping module on the testing dataset from the BSDS500 dataset [3,46], which includes 200 testing images with grouping annotations. We note that these grouping annotations maybe semantic and include multiple textures in a single segment.

Baselines. We quantitatively evaluate our model on the task of texture synthesis and compare with the WCT [37], DeepTexture [17], PSGAN [5], and Image Quilting [12]. None of the baseline models can generalize to unseen texture images, nor can they synthesize texture directly from natural images or even irregular-shaped texture segments. Thus, to ensure best performance of the baselines, we provide texture patches that are as clean as possible for the baseline model training and train a separate baseline model for each testing image. As shown in Fig. 3 (e), for each target texture segment, we crop a 100×100 texture patch at the center of the texture segment to serve as the training exemplar for the baseline methods. The cropped texture segment maximally suppresses irrelevant textures. We emphasize that the proposed method does not require clean texture patches for training, thus avoids such a clean process.

4.2 Texture Synthesis, Swapping and Editing

Texture Synthesis. We present texture synthesis results as well as comparisons with baseline methods in Fig. 3. Compared to the WCT and DeepTexture method, our model is able to capture both regular (e.g., the grid in the first row of Fig. 3) and irregular texture patterns thanks to the parametric sine wave. The texture images produced by our method are also more realistic and uniform compared to the DeepTexture method, which naively repeats the texture region. Furthermore, by introducing the grouping module, our method is able to synthesize texture images from a texture segment. Instead, both the Image Quilting [13] and PSGAN [5] method require clean texture patches as exemplar and cannot automatically ignore other irrelevant textures in the exemplar. Arbitrary-sized texture synthesis by our method can be found in the supplementary material.

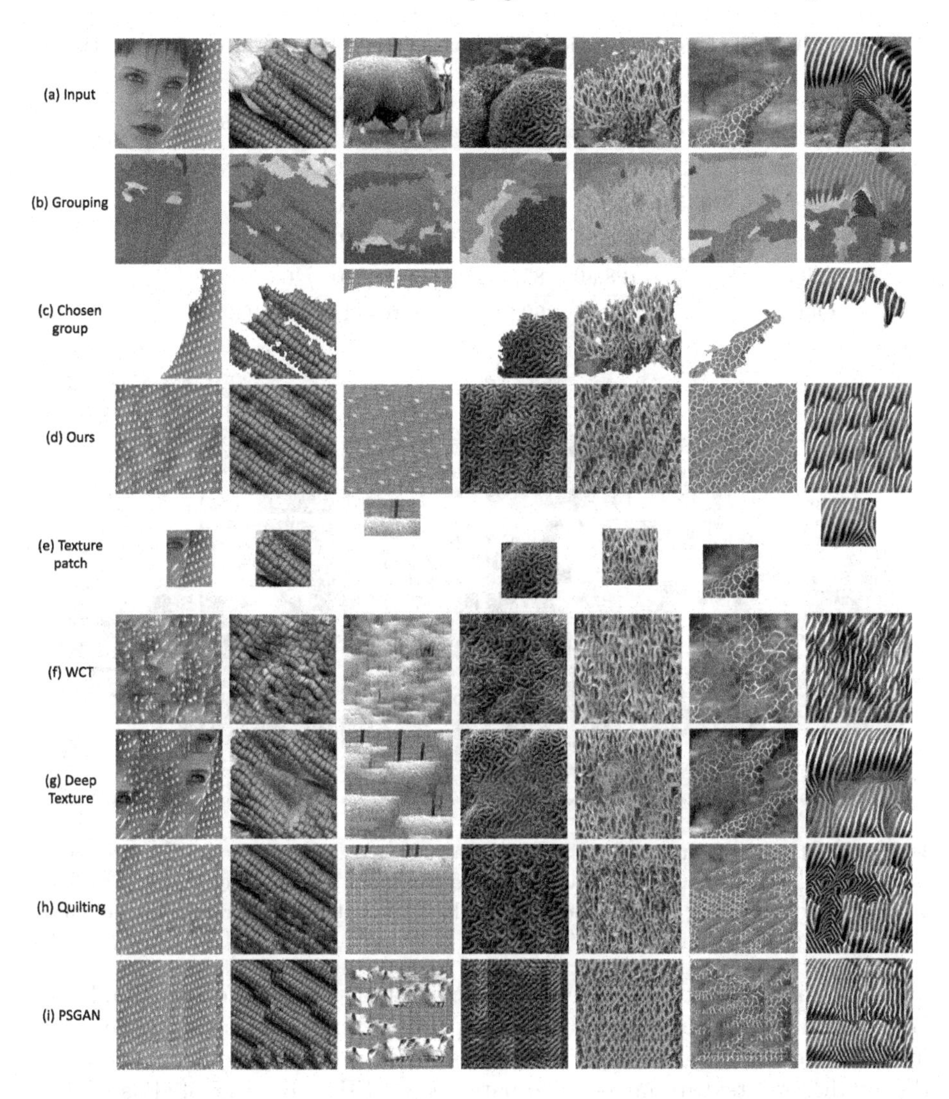

Fig. 3. Texture synthesis. (a) Input image. (b) Grouping mask predicted by our encoder. (c) A randomly chosen texture segment. (d) Texture synthesis by our method from the chosen texture segment in (c). (e) A patch cropped at the center of the chosen texture segment that serves as texture exemplar for the baselines models. (f)–(i) Texture synthesis results by WCT [37], DeepTexture [17], Image quilting [12] and PSGAN [5].

Quantitative Evaluation on Texture Synthesis. We use the Fréchet Inception Distance (FID) [24,47,55] and the Perceptual Similarity Distance (LPIPS) [66] to quantitatively evaluate how realistic the generated texture images are.

Since we only have access to texture segments of irregular shapes, we compare patches cropped from the texture segments in input images and synthesized texture images similarly as [42]. Specifically, we synthesize three different texture

Table 1. Quantitative evaluation of texture synthesis results. All metrics are the lower the better.

Metric	c-FID	c-FID (mask)	c-LPIPS	c-LPIPS (mask)
WCT [37]	149.47	82.65	0.3576	0.3547
DeepTexture [58]	151.91	107.34	0.3930	0.3825
Quilting [13]	76.80	63.64	0.3519	0.3568
PSGAN [5]	128.80	83.32	0.3810	0.3697
Ours	72.36	60.91	0.2838	0.3193

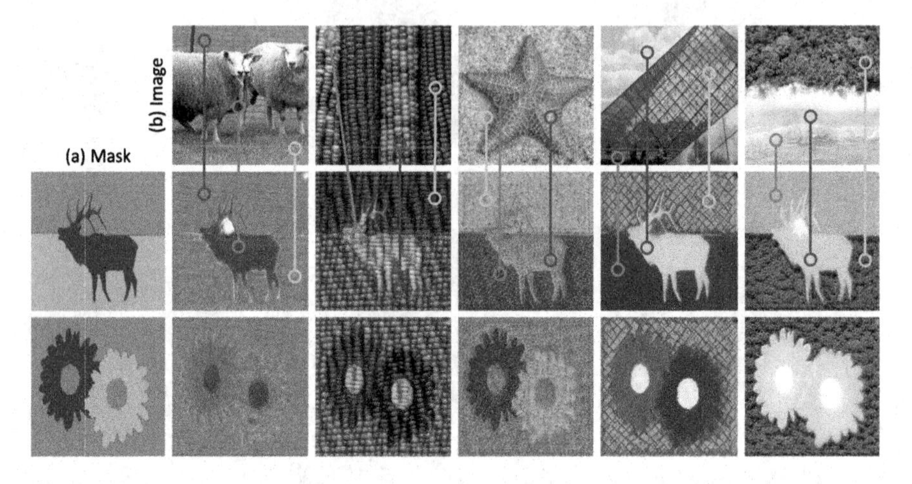

Fig. 4. Texture editing. Given the mask drawn by a user in (a) and a reference image in (b), we fill each region in the mask with textures from the reference image. The lines show the corresponding texture specified by the user that fills each region in the mask.

images from the texture segments of each testing image. We then randomly crop ten 64×64 patches from the corresponding texture segment in the input image defined by the grouping segment mask and ten corresponding patches from the synthesized texture image. The patch-based FID score (denoted as c-FID) as well as LPIPS score (denoted as c-LPIPS) are computed between patches cropped from the input image and synthesized texture images. Note that we use a dimension of 2048 for c-FID as opposite in [42] since we have sufficient number of cropped patches for evaluation. Furthermore, since the cropped patches in the input images may include irrelevant textures, we also mask the irrelevant textures in both the input image patches and the synthesized texture patches using our grouping masks. The FID and LPIPS metric of this setting are shown in the third and fifth column of Table 1. As shown in Table 1, the proposed method achieves lower c-FID and c-LPIPS score, indicating that it can synthesize more realistic textures from natural images.

Fig. 5. Texture swapping. Given a pair of images, our model predicts their groupings in (b)(e). We swap the texture codes of segments in each image chosen by a user. The yellow line indicates the swapped regions in the two images. (Color figure online)

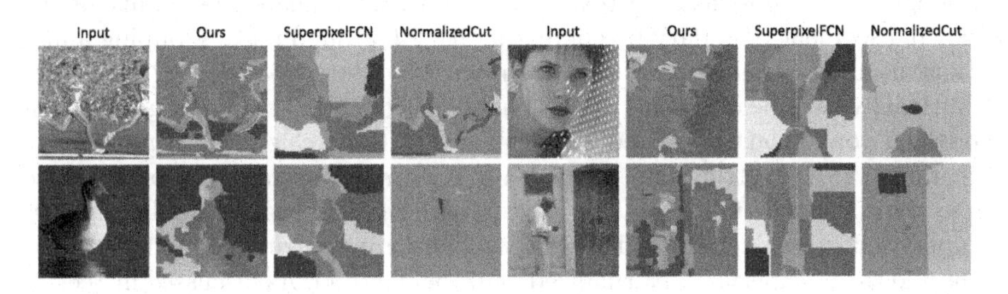

Fig. 6. Image Grouping. We show image grouping results by our method, the supervsied SuperpixelFCN [64] and the unsupervised Normalized Cuts [56] method visually.

Texture Editing. We further apply the learned model to texture editing and show results in Fig. 4. In this task, a user draws a mask (Fig. 4(a)) and chooses the texture segments in the input image (Fig. 4(b)) to provide texture codes. A texture feature map is composed by filling each region in the mask with the texture code of the chosen segment. The decoder combines the composed texture feature map with random noise to produce the final editing image. As shown in Fig. 4, each texture can be successfully transferred and realistically synthesized onto the corresponding region in the mask, demonstrating the flexibility of the learned texture codes in controlled texture editing tasks.

Texture Swapping. We also show the application of texture swapping in Fig. 5. Given a pair of images, we first predict their grouping masks as shown in Fig. 5(b)(e). A user defines which segments to be swapped out in each image (yellow lines in Fig. 5(b)(e)), the swapping is then carried out by swapping the texture codes of corresponding regions. The final swapped image (Fig. 5(c)(f)) is obtained by feeding the spatial code together with the swapped texture codes to the decoder.

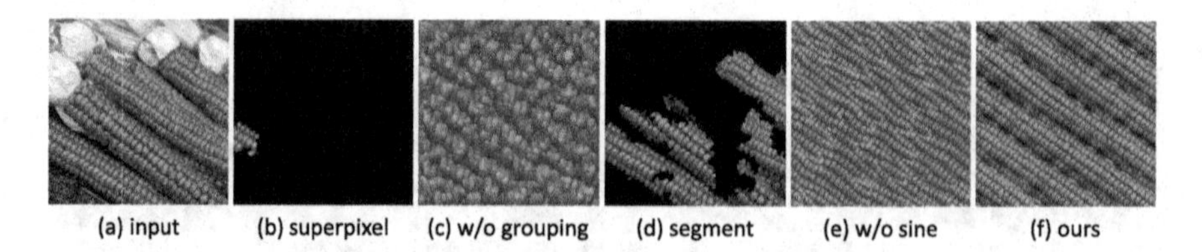

Fig. 7. Ablation study. (a) Input image. (b) A superpixel predicted by the SLIC method [1]. (c) Texture synthesis result by the baseline model using the superpixel in (b). (d) A texture segment predicted by our grouping model. (e) Texture synthesis by a baseline model learned with noise instead of sine waves. (f) Texture synthesis of our full model.

4.3 Image Grouping Evaluation

Although our work focuses on texture synthesis, we validate the quality of the predicted segment boundary by comparing it with ground truth grouping annotations using the Boundary Recall (BR) and Boundary Precision (BP) metric. Additionally, we measure the upper bound performance of any segmentation model on top of our predicted segments using the Achievable Segmentation Accuracy (ASA) metric. The definition and detailed explanation of all these three metrics can be found in [26]. We compare the proposed algorithm against a supervised superpixel method [64] trained with ground truth grouping annotations and the unsupervised Normalized Cuts [56] method. As discussed in Sect. 4, our model predicts $K = 10$ segments for each image. For fair comparisons, we adjust the hyper-parameters (i.e., the input image size in [64] and the merging threshold in [56]) such that these baselines produce a similar amount of groups (i.e., around ten groups). Table 2 shows the performance comparison and averaged group numbers by different methods. Figure 6 demonstrates the predicted grouping by each method visually. Both the proposed method and the Normalized Cuts [56] approach do not require ground truth grouping annotations. Yet, the proposed method achieves better boundary recall and serves better as a segmentation pre-processing step as it has a higher ASA score. On the other hand, the Normalized Cuts method aggressively merges superpixels when setting group number small (i.e., setting lower merging threshold), as shown in Fig. 6(d). This is also why the Normalized Cuts method has higher BP since it misses many boundaries in the image. Although self-supervisedly learned, our method achieves a comparable BP score with fewer groups than the supervised superpixel method [64]. The superpixel method achieves a higher ASA score since it has access to the ground truth grouping annotations during model training.

4.4 Ablation Studies

Image Grouping. To validate the contribution of image grouping in texture features learning, we introduce a baseline model that takes the SLIC superpixels [1] as texture segments instead of learning image grouping together with

Table 2. Image grouping evaluation on the BSDS500 [3,46] dataset

Metric	BR ↑	BP ↑	ASA ↑	Groups
SuperpixelFCN [64]	0.35	0.25	0.83	16
NormalizedCut [56]	0.3	0.38	0.58	14
Ours	0.67	0.27	0.68	10

texture modeling. We show qualitative comparison with the full model in Fig. 7 (c) and (f). The baseline model is limited by the small receptive field of superpixels and cannot fully capture the texture in each segment (e.g., the grid and corn kernel in Fig. 7). While our method is able to synthesize realistic textures that resemble the chosen texture segment.

Sine Wave. The inherent periodicity of the sine wave helps our model to synthesize regular pattern such as the grid shown in the first row of Fig. 7. On the contrary, if we feed a noise instead of a sine wave into our decoder, the baseline model fails to capture stationary pattern as shown in Fig. 7 (e).

5 Conclusion

In this work, we propose a framework to scrape textures from natural images by decomposing an image into a set of texture segments and represent each texture segment with a vector texture code. By reconstructing the input image from these texture codes, our model learns to group pixels of the same texture without any supervision. Through encoding each texture segment by the texture code, along with a parametric sine wave, we can learn texture representations directly from cluttered images without requiring any well-cropped texture patches. We demonstrate that these texture representations can be used for texture synthesis and editing, and the emergent image segments are often consistent with object boundaries. While our paper demonstrates some promising results, plenty of work remains in both texture modeling as well as image segmentation. The lesson of our work is that these two tasks would likely benefit from being considered jointly.

References

1. Achanta, R., Shaji, A., Smith, K., Lucchi, A., Fua, P., Süsstrunk, S.: SLIC superpixels compared to state-of-the-art superpixel methods. TPAMI **34**(11), 2274–2282 (2012)
2. Alanov, A., Kochurov, M., Volkhonskiy, D., Yashkov, D., Burnaev, E., Vetrov, D.: User-controllable multi-texture synthesis with generative adversarial networks. arXiv preprint arXiv:1904.04751 (2019)
3. Arbelaez, P., Maire, M., Fowlkes, C., Malik, J.: Contour detection and hierarchical image segmentation. TPAMI **33**(5), 898–916 (2011)

4. Barnes, C., Shechtman, E., Finkelstein, A., Goldman, D.B.: PatchMatch: a randomized correspondence algorithm for structural image editing. ACM Trans. Graph. (Proc. SIGGRAPH) **28**(3) (2009)
5. Bergmann, U., Jetchev, N., Vollgraf, R.: Learning texture manifolds with the periodic spatial GAN. arXiv preprint arXiv:1705.06566 (2017)
6. Brodatz, P.: Textures: A Photographic Album for Artists and Designers. Dover Pub, New York (1966)
7. Chen, L.C., Papandreou, G., Kokkinos, I., Murphy, K., Yuille, A.L.: DeepLab: semantic image segmentation with deep convolutional nets, atrous convolution, and fully connected CRFs. TPAMI **40**(4), 834–848 (2017)
8. Cimpoi, M., Maji, S., Kokkinos, I., Mohamed, S., Vedaldi, A.: Describing textures in the wild. In: CVPR (2014)
9. Comaniciu, D., Meer, P.: Mean shift: a robust approach toward feature space analysis. TPAMI **24**(5), 603–619 (2002)
10. Dai, D., Riemenschneider, H., Van Gool, L.: The synthesizability of texture examples. In: CVPR (2014)
11. Deng, J., Dong, W., Socher, R., Li, L.J., Li, K., Fei-Fei, L.: ImageNet: a large-scale hierarchical image database. In: CVPR (2009)
12. Efros, A.A., Freeman, W.T.: Image quilting for texture synthesis and transfer. SIGGRAPH (2001)
13. Efros, A.A., Leung, T.K.: Texture synthesis by non-parametric sampling. In: ICCV (1999)
14. Felzenszwalb, P.F., Huttenlocher, D.P.: Efficient graph-based image segmentation. IJCV **59**(2), 167–181 (2004). https://doi.org/10.1023/B:VISI.0000022288.19776.77
15. Fiedler, M., Alpers, A.: Power-SLIC: Diagram-based superpixel generation. arXiv preprint arXiv:2012.11772 (2020)
16. Frühstück, A., Alhashim, I., Wonka, P.: TileGAN: synthesis of large-scale non-homogeneous textures. TOG **38**(4), 1–11 (2019)
17. Gatys, L., Ecker, A.S., Bethge, M.: Texture synthesis using convolutional neural networks. In: NeurIPS (2015)
18. Gatys, L.A., Ecker, A.S., Bethge, M.: Image style transfer using convolutional neural networks. In: CVPR (2016)
19. Guo, C.e., Zhu, S.C., Wu, Y.N.: Primal sketch: integrating structure and texture. Comput. Vis. Image Underst. **106**(1), 5–19 (2007)
20. Han, F., Zhu, S.C.: Bottom-up/top-down image parsing with attribute grammar. TPAMI **31**(1), 59–73 (2008)
21. He, K., Zhang, X., Ren, S., Sun, J.: Deep residual learning for image recognition. In: CVPR (2016)
22. Heeger, D.J., Bergen, J.R.: Pyramid-based texture analysis/synthesis. In: Proceedings of the 22nd annual conference on Computer graphics and interactive techniques, pp. 229–238 (1995)
23. Heitz, E., Vanhoey, K., Chambon, T., Belcour, L.: A sliced Wasserstein loss for neural texture synthesis. In: CVPR (2021)
24. Heusel, M., Ramsauer, H., Unterthiner, T., Nessler, B., Hochreiter, S.: GANs trained by a two time-scale update rule converge to a local NASH equilibrium. In: NeurIPS (2017)
25. Hoiem, D., Efros, A.A., Hebert, M.: Geometric context from a single image. In: Tenth IEEE International Conference on Computer Vision (ICCV 2005) Volume 1, vol. 1, pp. 654–661. IEEE (2005)

26. Jampani, V., Sun, D., Liu, M.Y., Yang, M.H., Kautz, J.: Superpixel sampling networks. In: ECCV (2018)
27. Jetchev, N., Bergmann, U., Vollgraf, R.: Texture synthesis with spatial generative adversarial networks. arXiv preprint arXiv:1611.08207 (2016)
28. Jiang, L., Dai, B., Wu, W., Loy, C.C.: Focal frequency loss for image reconstruction and synthesis. In: ICCV (2021)
29. Johnson, J., Alahi, A., Fei-Fei, L.: Perceptual losses for real-time style transfer and super-resolution. In: Leibe, B., Matas, J., Sebe, N., Welling, M. (eds.) ECCV 2016. LNCS, vol. 9906, pp. 694–711. Springer, Cham (2016). https://doi.org/10.1007/978-3-319-46475-6_43
30. Kaspar, A., Neubert, B., Lischinski, D., Pauly, M., Kopf, J.: Self tuning texture optimization. In: Computer Graphics Forum (2015)
31. Kingma, D.P., Ba, J.: Adam: a method for stochastic optimization. arXiv preprint arXiv:1412.6980 (2014)
32. Kipf, T.N., Welling, M.: Semi-supervised classification with graph convolutional networks. arXiv preprint arXiv:1609.02907 (2016)
33. Kwatra, V., Essa, I., Bobick, A., Kwatra, N.: Texture optimization for example-based synthesis. In: SIGGRAPH (2005)
34. Li, T.M., Lukáč, M., Gharbi, M., Ragan-Kelley, J.: Differentiable vector graphics rasterization for editing and learning. TOG **39**(6), 1–15 (2020)
35. Li, X., Liu, S., De Mello, S., Kim, K., Wang, X., Yang, M.H., Kautz, J.: Online adaptation for consistent mesh reconstruction in the wild. In: NeurIPS (2020)
36. Li, Y., Fang, C., Yang, J., Wang, Z., Lu, X., Yang, M.H.: Diversified texture synthesis with feed-forward networks. In: CVPR (2017)
37. Li, Y., Fang, C., Yang, J., Wang, Z., Lu, X., Yang, M.H.: Universal style transfer via feature transforms. In: NeurIPS (2017)
38. Li, Y., Hao, M., Di, Z., Gundavarapu, N.B., Wang, X.: Test-time personalization with a transformer for human pose estimation. In: NeurIPS (2021)
39. Li, Z., Bao, W., Zheng, J., Xu, C.: Deep grouping model for unified perceptual parsing. In: CVPR (2020)
40. Lin, Q., Zhong, W., Lu, J.: Deep superpixel cut for unsupervised image segmentation. In: ICPR, pp. 8870–8876 (2021)
41. Lin, T.-Y., et al.: Microsoft COCO: common objects in context. In: Fleet, D., Pajdla, T., Schiele, B., Tuytelaars, T. (eds.) ECCV 2014. LNCS, vol. 8693, pp. 740–755. Springer, Cham (2014). https://doi.org/10.1007/978-3-319-10602-1_48
42. Liu, G., et al.: Transposer: universal texture synthesis using feature maps as transposed convolution filter. arXiv preprint arXiv:2007.07243 (2020)
43. Long, J., Shelhamer, E., Darrell, T.: Fully convolutional networks for semantic segmentation. In: CVPR (2015)
44. Malisiewicz, T., Efros, A.A.: Improving spatial support for objects via multiple segmentations. In: British Machine Vision Conference (BMVC) (2007)
45. Mardani, M., Liu, G., Dundar, A., Liu, S., Tao, A., Catanzaro, B.: Neural FFTs for universal texture image synthesis. In: NeurIPS (2020)
46. Martin, D., Fowlkes, C., Tal, D., Malik, J.: A database of human segmented natural images and its application to evaluating segmentation algorithms and measuring ecological statistics. In: ICCV (2001)
47. Park, T., et al.: Swapping autoencoder for deep image manipulation. In: NeurIPS (2020)
48. Paszke, A., et al.: Pytorch: an imperative style, high-performance deep learning library. In: NeurIPS (2019)

49. Pont-Tuset, J., Arbelaez, P., Barron, J.T., Marques, F., Malik, J.: Multiscale combinatorial grouping for image segmentation and object proposal generation. TPAMI **39**(1), 128–140 (2016)
50. Portilla, J., Simoncelli, E.P.: A parametric texture model based on joint statistics of complex wavelet coefficients. IJCV **40**(1), 49–70 (2000)
51. Ren, X., Malik, J.: Learning a classification model for segmentation. In: ICCV (2003)
52. Rosenberger, A., Cohen-Or, D., Lischinski, D.: Layered shape synthesis: automatic generation of control maps for non-stationary textures. ACM Trans. Graph. (TOG) **28**(5), 1–9 (2009)
53. Scarselli, F., Gori, M., Tsoi, A.C., Hagenbuchner, M., Monfardini, G.: The graph neural network model. IEEE Trans. Neural Netw. **20**(1), 61–80 (2008)
54. Sendik, O., Cohen-Or, D.: Deep correlations for texture synthesis. TOG **36**(5), 1–15 (2017)
55. Shaham, T.R., Dekel, T., Michaeli, T.: SinGAN: learning a generative model from a single natural image. In: ICCV (2019)
56. Shi, J., Malik, J.: Normalized cuts and image segmentation. TPAMI **22**(8), 888–905 (2000)
57. Sun, Y., Wang, X., Liu, Z., Miller, J., Efros, A., Hardt, M.: Test-time training with self-supervision for generalization under distribution shifts. In: ICML (2020)
58. Ulyanov, D., Lebedev, V., Vedaldi, A., Lempitsky, V.S.: Texture networks: feed-forward synthesis of textures and stylized images. In: ICML (2016)
59. Wang, Y., Wei, Y., Qian, X., Zhu, L., Yang, Y.: AINet: association implantation for superpixel segmentation. arXiv preprint arXiv:2101.10696 (2021)
60. Wei, L.Y., Levoy, M.: Fast texture synthesis using tree-structured vector quantization. In: SIGGRAPH (2000)
61. Wertheimer, M.: Laws of organization in perceptual forms. Psycologische Forschung 4 (1923)
62. Wu, Q., Yu, Y.: Feature matching and deformation for texture synthesis. TOG **23**(3), 364–367 (2004)
63. Wu, Z., Lin, D., Tang, X.: Deep Markov random field for image modeling. In: Leibe, B., Matas, J., Sebe, N., Welling, M. (eds.) ECCV 2016. LNCS, vol. 9912, pp. 295–312. Springer, Cham (2016). https://doi.org/10.1007/978-3-319-46484-8_18
64. Yang, F., Sun, Q., Jin, H., Zhou, Z.: Superpixel segmentation with fully convolutional networks. In: CVPR (2020)
65. Yu, N., Barnes, C., Shechtman, E., Amirghodsi, S., Lukac, M.: Texture mixer: a network for controllable synthesis and interpolation of texture. In: CVPR (2019)
66. Zhang, R., Isola, P., Efros, A.A., Shechtman, E., Wang, O.: The unreasonable effectiveness of deep features as a perceptual metric. In: CVPR (2018)
67. Zhou, B., Zhao, H., Puig, X., Fidler, S., Barriuso, A., Torralba, A.: Scene parsing through ade20k dataset. In: CVPR (2017)
68. Zhou, Y., Zhu, Z., Bai, X., Lischinski, D., Cohen-Or, D., Huang, H.: Non-stationary texture synthesis by adversarial expansion. arXiv preprint arXiv:1805.04487 (2018)
69. Zou, Z., Shi, T., Qiu, S., Yuan, Y., Shi, Z.: Stylized neural painting. In: CVPR (2021)

Single Stage Virtual Try-On Via Deformable Attention Flows

Shuai Bai, Huiling Zhou, Zhikang Li, Chang Zhou(✉), and Hongxia Yang

DAMO Academy, Alibaba Group, Hangzhou, China
{baishuai.bs,zhule.zhl,zhikang.lzk,ericzhou.zc,yang.yhx}@alibaba-inc.com

Abstract. Virtual try-on aims to generate a photo-realistic fitting result given an in-shop garment and a reference person image. Existing methods usually build up multi-stage frameworks to deal with clothes warping and body blending respectively, or rely heavily on intermediate parser-based labels which may be noisy or even inaccurate. To solve the above challenges, we propose a single-stage try-on framework by developing a novel Deformable Attention Flow (DAFlow), which applies the deformable attention scheme to multi-flow estimation. With pose keypoints as the guidance only, the self- and cross-deformable attention flows are estimated for the reference person and the garment images, respectively. By sampling multiple flow fields, the feature-level and pixel-level information from different semantic areas is simultaneously extracted and merged through the attention mechanism. It enables clothes warping and body synthesizing at the same time which leads to photo-realistic results in an end-to-end manner. Extensive experiments on two try-on datasets demonstrate that our proposed method achieves state-of-the-art performance both qualitatively and quantitatively. Furthermore, additional experiments on the other two image editing tasks illustrate the versatility of our method for multi-view synthesis and image animation. Code will be made available at https://github.com/OFA-Sys/DAFlow.

Keywords: Virtual try-on · Single stage · Deformable attention flows

1 Introduction

Virtual try-on aims to generate a photo-realistic and reasonable try-on result based on an in-shop garment and a reference person image. In recent years, due to its potential applications in the fashion and e-commerce industries, it has received more and more attention. Recent methods [4,5,9,10,43] have achieved considerable progress in generating realistic results and preserving details. However, this task is still challenging, especially under complex poses and large deformations, where most of existing methods are still suffering from misalignment or obvious artifacts.

Supplementary Information The online version contains supplementary material available at https://doi.org/10.1007/978-3-031-19784-0_24.

S. Avidan et al. (Eds.): ECCV 2022, LNCS 13675, pp. 409–425, 2022.
https://doi.org/10.1007/978-3-031-19784-0_24

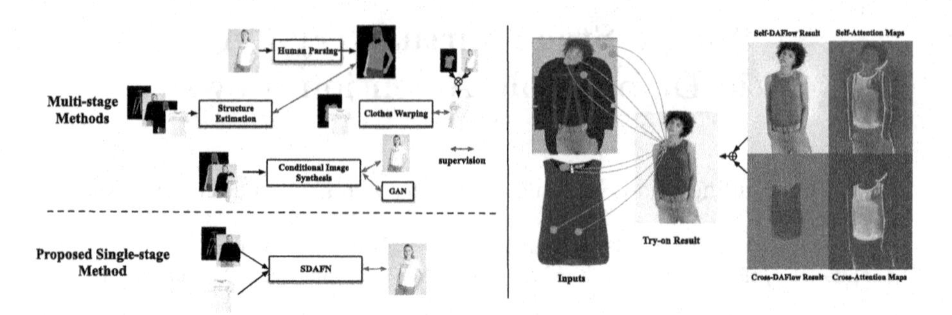

Fig. 1. The comparison between multi-stage methods and our single-stage approach. With multiple flow fields sampled, different regions are associated. Attention maps learn the 3D priors to make the try-on more realistic without 3D information input.

Most prior try-on systems adopt a multi-stage approach [9,15,41,43] shown in Fig. 1, including clothes warping, structure estimation, and image synthesis. Clothes warping is to align the garment to the target pose while preserving the texture details. Structure estimation predicts the segmentation map of the human body to guide the image synthesis. Given the warped clothing and the intermediate semantic labels, image synthesis is performed as a conditional generation task for pixel-level refinement.

Clothes warping in recent works [5,10,14,43] trains an extra flow network to warp clothes. Such flow operation retains the realistic texture but usually predicts the inaccurate structure. Some approaches [10,14] limit the unsmooth deformation of flow by introducing additional constraints but fail in dealing with complex poses. To further improve the quality of the generated results, some works [5,43] introduce more prior knowledge from some external pre-trained models, such as human parsing [11], densepose [13], and 3D depth. However, the inaccurate intermediate predictions may lead to results with noticeable artifacts. In addition, although the generative adversarial networks (GAN) [12] help preserve the sharp details of the generated images, they modify some attributes of clothes, such as color or style, which are not desirable for virtual try-on.

To address the above problems, we propose a Single-stage Deformable Attention Flow Network (SDAFN) to perform an end-to-end try-on task. we build a Deformable Attention Flow (DAFlow) module by applying a deformable attention scheme to multi-flow estimation. As shown in Fig. 1, it estimates multiple flow fields from different semantic areas and then synchronously merges the feature-level and pixel-level information with the attention mechanism in cascade. This allows the extract deformable attention flows to not only warp clothes but also synthesize photo-realistic human torsos and shadows at the same time. Then, we combine self- and cross-DAFlows to deal with the human body and the garment respectively, generating the fitting results in one single pass. In addition, our method only need pose keypoints as guidance. To our best knowledge, we are the first one-stage pure flow-based virtual try-on method.

The main contributions of this paper can be summarized as follows:

- We propose a single-stage virtual try-on framework, which applies self- and cross-DAFlows to deal with the reference person and garment images in parallel and generate realistic fitting results in an end-to-end manner.
- We propose a novel deformable attention flow module to estimate the reasonable structure while retaining the vivid texture in cascade. It works well even with a large misalignment between the clothes and the person and can be applied to a variety of resolutions.
- Extensive experiments show that our proposed method not only achieves superior performance on virtual try-on, but also can be extended to other image editing tasks, such as multi-view synthesis and image animation.

2 Related Work

2.1 Virtual Try-On

The study on the virtual try-on task mainly consists of 3D model-based approaches [1,2,23,30] and 2D image-based approaches [15,24,41,45]. Recently, 2D methods have attracted more and more attention because of highly accessible data and photo-realistic results. VITON [15] uses a two-stage coarse-to-fine strategy to generate a clothed person. It first estimates the rough human body shape, then warps clothes and refines the details of the clothed person according to the shape. To improve the accuracy, SwapNet [32] and VTNFP [45] adopt semantic segmentation as guidance. ACGPN [43] introduces an additional stage to predict the semantic layout of the reference image. DCTON [9] and ZFlow [5] add more accurate descriptor, like densepose [13] or UV [8] projection. VITON-HD [4] improves the performance of the conditional GAN structure on high-resolution images. Although the generated images get better quality, the pipeline is becoming more complex and relies on more external information. It results in a reduction in efficiency and obvious artifacts caused by the inaccurate intermediate labels. Recently WUTON [19] and PFAFN [10] adopt the "teacher-tutor-student" scheme to get rid of the extra information at inference time and achieve good performance. This proves that the network has the ability to perform try-on without heavy dependency. Our proposed method is able to obtain realistic fitting performance with only pose information as guidance.

2.2 Spatial Transform Module

The spatial transform module [20,34,38,44,47] is widely applied in optical flow estimation, image transformation and object recognition tasks. In virtual try-on, the spatial transform module is mainly used to wrap clothes to match the posture of the person. VITON [15] exploits a Thin-Plate Spline (TPS) [7] based warping method to deform the in-shop clothes to the refined result with a composition mask. CP-VTON [41] uses a neural network to learn the transformation parameters of TPS warping rather than using image descriptors. ClothFlow [14] introduces denser flow predictions through a cascade scheme, which makes deformation has a high dimension of freedom. However, dense flows usually present

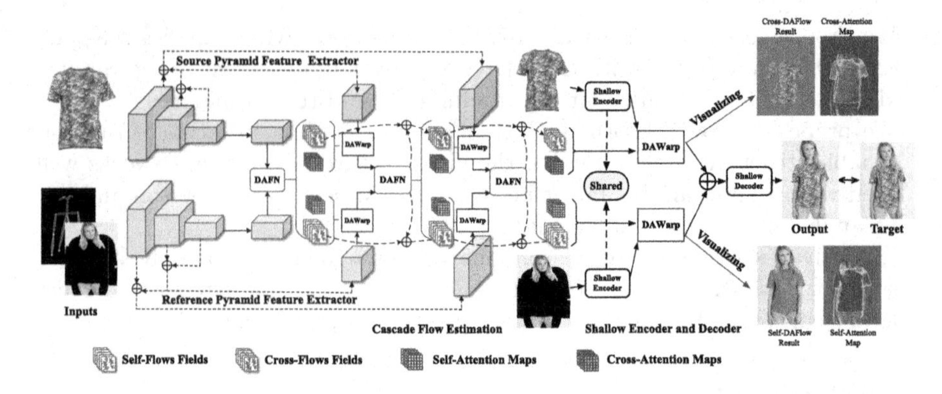

Fig. 2. The overall framework of our SDAFN. The garment, person and pose images are firstly fed into the unshared pyramid feature extractors. Then, both self- and cross-deformable attention flows are estimated in cascade. The final try-on result is obtained by applying shallow encoder and decoder together with the learned flows.

unappealing artifacts. To avoid this problem, ClothFlow uses flow variation regularization to enforce smoothness. PFAPN [10] adds a second-order smooth constraint to encourage the co-linearity of neighboring appearance flows. These constraints are proposed to make the garment smoother after deformation, but they still face the challenge when there exists a huge discrepancy between the target clothes and the original clothes on the model.

2.3 Efficient Attention Mechanism

The attention mechanism [40] is widely used in transformer models, benefiting from its long-distance association ability, and achieves good performance on image segmentation, object detection, and image generation. However, with the increase in feature resolution, dense attention will make a huge computational cost. This makes the attention mechanism usually used in low-resolution features [33]. Recently sparse attention has been introduced into detection and recognition tasks. Common methods include pre-defined local attention patterns [26,27,31] or learned sparse attention [22,39,42,48]. Combing the effectiveness of flow operation on preserving details and the capability of attention mechanism on estimating accurate structures, our proposed method naturally extend learned sparse attention to pixel-level image transformation.

3 Our Approach

In this section, we introduce the overall framework of our proposed single-stage virtual try-on model called SDAFN. Then, we describe the core DAFlow module as well as the training loss design in detail.

3.1 Architecture

Different from the previous methods who apply the clothes warping and the conditional generation in different stages, we use self- and cross-deformable attention flows to obtain the try-on results directly. In addition, our framework only relies on the pose keypoints as guidance. Specifically, as shown in Fig. 2, our model is composed of three parts: pyramid feature extraction, cascade flow estimation, and shallow encoder-decoder generation.

Pyramid Feature Extraction. Our model has two pyramid feature extractors, including a reference branch and a source branch. The reference branch takes the concatenation of the person and pose images as input, where the upper body of the person is masked. The garment image is fed into the source branch. The two feature extractors have the same Feature Pyramid Network (FPN) [25] structure with unshared parameters. The FPN network consists of N encoding layers where each layer has a downsampling convolution with a stride of 2 followed by two residual blocks [16].

Cascade Flow Estimation. It is difficult to directly and accurately predict large deformations, especially for the virtual try-on application, where the target and source images are not domain-unified. Hence, similar to the recent methods [10,14], we adopt the cascade flow estimation from coarse to fine.

Given the hierarchical reference and source features $\{\boldsymbol{x}_r^n, \boldsymbol{x}_s^n\}_{n=1}^N$ extracted by the pyramid feature extractors, as illustrated in Fig. 2, two types of flows and attention maps are estimated. The first is the self-flow fields and self-attention maps $(\boldsymbol{o}_{daf,r}^n, \boldsymbol{a}_r^n)$ from the reference branch. The second is the cross-flow fields and cross-attention maps $(\boldsymbol{o}_{daf,s}^n, \boldsymbol{a}_s^n)$ from the interaction of the both branches. The features $(\boldsymbol{x}_r^1, \boldsymbol{x}_s^1)$ at the lowest resolution are fed into the proposed Deformable Attention Flow Network (DAFN) to predict the initial flow fields $(\boldsymbol{o}_{daf,r}^1, \boldsymbol{o}_{daf,s}^1)$ and attention maps $(\boldsymbol{a}_r^1, \boldsymbol{a}_s^1)$. Then they will be refined and updated in cascade. Specifically, the reference and source feature maps $(\boldsymbol{x}_r^n, \boldsymbol{x}_s^n), n \in [2, N]$ are first transformed by the self- and cross-DAFlows $(\boldsymbol{o}_{daf,r}^{n-1}, \boldsymbol{o}_{daf,s}^{n-1}, \boldsymbol{a}_r^{n-1}, \boldsymbol{a}_s^{n-1})$ of the previous scale, called Deformable Attention Warping (DAWarp). Then, the transformed features $(\hat{\boldsymbol{x}}_r^n, \hat{\boldsymbol{x}}_s^n)$ are applied to predict the residual flows and the new attention maps in which finer flows are obtained. The above process continues until $n = N$. In our implementation, N is set to 5, and the process shown in Fig. 2 is illustrated with $N = 3$. Both DAFN and DAWarp modules are described in Sect. 3.2 in details.

Shallow Encoder-Decoder Generation. The shallow encoder projects the images from RGB to high dimensional space. With the final estimated flows and attention maps, the reference person and garment images in high dimensional feature space are transformed and merged with DAWarp. Then the merged result is fed into a shallow decoder to obtain the final try-on result. The shallow encoder and decoder both have two convolution layers without downsampling. In particular, both garment and reference person images share the same encoder and decoder. It is found in our experiment that the shallow encoder-decoder

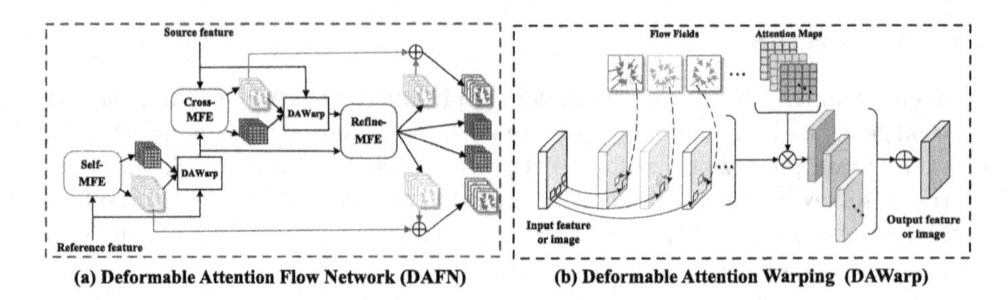

Fig. 3. Illustrations of DAFN network and DAWarp operation.

structure is a simple and effective way to enhance the representation capability at the pixel level and improve the flow generation quality.

3.2 Deformable Attention Flows

Revisiting the Conventional Flow Field. The flow field estimation [5,10,14] is extensively used in the cloth warping of virtual try-on. It takes the reference (e.g., semantic segmentation, pose keypoints or depth map) feature $\boldsymbol{x}_r \in \mathbb{R}^{C\times H\times W}$ and the source (e.g., clothes) feature $\boldsymbol{x}_s \in \mathbb{R}^{C\times H\times W}$ as inputs, and only estimates one 2D coordinate offset for each sampling position. C is the feature dimension, H and W denote the height and width of feature map. This process of predicting offset map $\boldsymbol{o} \in \mathbb{R}^{2\times H\times W}$ can be written as:

$$\boldsymbol{o} = F([\boldsymbol{x}_r, \boldsymbol{x}_s]), \tag{1}$$

where F indicates flow field estimator network. Accordingly, the warped feature at a 2D reference location p is calculated as :

$$\mathcal{W}(\boldsymbol{x}_s, \boldsymbol{o})_p = \boldsymbol{x}_s(p + \boldsymbol{o}_p), \tag{2}$$

where $\mathcal{W}(\cdot)$ denotes the warp operation, $p+\boldsymbol{o}_p$ is the estimated sampling position. The value $\boldsymbol{x}_s(p+\boldsymbol{o}_p)$ of each sampling point is computed by bilinear interpolation from the nearby grid points on the feature map.

The flow operation directly samples from images, retaining the realistic textures. However, structural information is often associated with multiple locations. For example, the shape of the clothes on the human body is determined by the posture, body shape, and the type of garment. Hence, only one flow field cannot estimate accurate structure, so most existing methods only apply the flow to warp the garment in virtual try-on. To mitigate this weakness, we propose a deformable attention flow (DAFlow) module, which preserves detailed textures and accurately estimates structures. The DAFlow module consists of the Multiple Flow field Estimator (MFE) and the DAWarp operation.

Multiple Flow Field Estimator. Inspired by deformable attention [48], our MFE predicts the fixed number K of sampling points, regardless of the spatial size of the feature maps. In contrast to the conventional single flow field

$\boldsymbol{o} \in \mathbb{R}^{2\times H\times W}$, our MFE forecasts multiple flow fields $\boldsymbol{o}_{daf} \in \mathbb{R}^{2K\times H\times W}$ and attention weight maps $\boldsymbol{a} \in \mathbb{R}^{K\times H\times W}$:

$$\boldsymbol{o}_{daf}, \boldsymbol{a} = F([\boldsymbol{x}_r, \boldsymbol{x}_s]), \tag{3}$$

where K is the sampled key number. In the implementation, both $\boldsymbol{o}_{daf}$ and $\boldsymbol{a}$ are obtained via the same ConvNet but different channels.

Deformable Attention Warping. As shown in the **(b)** of Fig. 3, each location is associated with features from multiple locations. Besides, DAFlows learn a variety of possible flows. The multiple flow fields and the attention operation are simultaneously applied to the features and images. In terms of the feature, the source or reference features $\boldsymbol{x} \in \mathbb{R}^{C\times H\times W}$ are effectively integrated into the desired target position as:

$$\mathcal{W}_{DAF}(\boldsymbol{x}, \boldsymbol{o}_{daf}, \boldsymbol{a})_p = \sum_{k=1}^{K} \frac{exp(\boldsymbol{a}_{pk})}{\sum_{i=1}^{K} exp(\boldsymbol{a}_{pi})} \boldsymbol{x}(p + \boldsymbol{o}_{daf,pk}), \tag{4}$$

where $(p + \boldsymbol{o}_{daf,pk})$ is the k_{th} sampling position for the reference location p, and $\mathcal{W}_{DAF}(\boldsymbol{x}_s, \boldsymbol{o}_{daf}, \boldsymbol{a})$ denotes the warped feature with DAFlows. As for the image level, merging multiple warped images has more generation possibilities and is recombined into new images with reasonable structures and realistic textures. As illustrated in Fig. 7, the neck and arm parts are realistically generated, which are not in the original image. Additionally, the warping operation applies bilinear interpolation, enabling the estimated offsets optimized with back-propagation.

3.3 Deformable Attention Flow Network

As illustrated in Sect. 3.1, we adopt cascade flow estimation to predict the deformation and attention from coarse to fine. Given the reference and source feature maps $(\boldsymbol{x}_r^n, \boldsymbol{x}_s^n), n \in [2, N]$, the feature maps are transformed to $(\hat{\boldsymbol{x}}_r^n, \hat{\boldsymbol{x}}_s^n)$ in accordance with the flows $(\boldsymbol{o}_{daf,r}^{n-1}, \boldsymbol{o}_{daf,s}^{n-1})$ and attention $(\boldsymbol{a}_r^{n-1}, \boldsymbol{a}_s^{n-1})$ from the previous scale:

$$\hat{\boldsymbol{x}}_{s/r}^n = \mathcal{W}_{DAF}(\boldsymbol{x}_{s/r}^n, \mathcal{U}(\boldsymbol{o}_{daf,s/r}^{n-1}), \mathcal{U}(\boldsymbol{a}_{s/r}^{n-1})), \tag{5}$$

where $\mathcal{U}(\cdot)$ denotes the bilinear samping with the scale factor of 2.

For each scale, DAFN is utilized to estimate the DAFlows. The **(a)** of Fig. 3 illustrates that the DAFN consists of three flow estimators (F_r^n, F_s^n, F_m^n), namely, Self-MFE, Cross-MFE and Refine-MFE. Firstly, the transformed reference feature $\hat{\boldsymbol{x}}_r^n$ is fed into F_r^n to estimate the initial residual self-flows and the self-attention maps $(\Delta\dot{\boldsymbol{o}}_{daf,r}^n, \dot{\boldsymbol{a}}_r^n)$:

$$\Delta\dot{\boldsymbol{o}}_r^n, \dot{\boldsymbol{a}}_r^n = F_r^n(\hat{\boldsymbol{x}}_r^n). \tag{6}$$

Secondly, the reference feature is transformed with $(\Delta\dot{\boldsymbol{o}}_r^n, \dot{\boldsymbol{a}}_r^n)$, and concatenated with source feature $\hat{\boldsymbol{x}}_s^n$ to forecast the residual cross-flow fields and cross-attention maps $(\Delta\dot{\boldsymbol{o}}_{daf,s}^n, \dot{\boldsymbol{a}}_s^n)$:

$$\Delta\dot{\boldsymbol{o}}_{daf,s}^n, \dot{\boldsymbol{a}}_s^n = F_s^n([\hat{\boldsymbol{x}}_s^n, \mathcal{W}_{DAF}(\hat{\boldsymbol{x}}_r^n, \Delta\dot{\boldsymbol{o}}_r^n, \dot{\boldsymbol{a}}_r^n)]). \tag{7}$$

Finally, the source feature is converted with $(\Delta\dot{\boldsymbol{o}}^n_{daf,s}, \dot{\boldsymbol{a}}^n_s)$, and combined with the newly transformed reference features to predict the refinements $(\Delta\ddot{\boldsymbol{o}}^n_{daf,r}, \Delta\ddot{\boldsymbol{o}}^n_{daf,s})$ and the final attention maps $(\boldsymbol{a}^n_r, \boldsymbol{a}^n_s)$:

$$\Delta\ddot{\boldsymbol{o}}^n_{daf,s}, \Delta\ddot{\boldsymbol{o}}^n_{daf,r}, \boldsymbol{a}^n_s, \boldsymbol{a}^n_r = F^n_m([\mathcal{W}_{DAF}(\hat{\boldsymbol{x}}^n_s, \Delta\dot{\boldsymbol{o}}^n_s, \dot{\boldsymbol{a}}^n_s), \mathcal{W}_{DAF}(\hat{\boldsymbol{x}}^n_r, \Delta\dot{\boldsymbol{o}}^n_r, \dot{\boldsymbol{a}}^n_r]). \tag{8}$$

The final self- and cross-flow fields at the n_{th} scale are:

$$\boldsymbol{o}^n_{daf,r} = \mathcal{U}(\boldsymbol{o}^{n-1}_{daf,r}) + \Delta\dot{\boldsymbol{o}}^n_{daf,r} + \Delta\ddot{\boldsymbol{o}}^n_{daf,r}, \tag{9}$$

$$\boldsymbol{o}^n_{daf,s} = \mathcal{U}(\boldsymbol{o}^{n-1}_{daf,s}) + \Delta\dot{\boldsymbol{o}}^n_{daf,s} + \Delta\ddot{\boldsymbol{o}}^n_{daf,s}, \tag{10}$$

and the outputs $\boldsymbol{a}^n_s, \boldsymbol{a}^n_r$ of Refine-MFE are applied as the final attention maps.

3.4 Train Losses

Discarding the dependence on intermediate parser-based labels, our model only uses the try-on result as supervision in an end-to-end manner. As for the loss functions, without any constraints or regularization on the flows, the model is directly optimized by comparing the similarity between generated results and ground truths in all scales. The similarity functions consist of the L1 loss, the perceptual loss [21], and the style loss. The L1 loss is formulated as:

$$\mathcal{L}_{L1} = ||\boldsymbol{I}_{out} - \boldsymbol{I}_{target}||_1, \tag{11}$$

where $||\cdot||_1$ indicates the $L1$ distance, and $\boldsymbol{I}_{out}$ and $\boldsymbol{I}_{target}$ represent the predicted image and target image. The perceptual loss proposed in [21] calculates the L1 distance of the feature maps extracted by the VGG-19 [37] network:

$$\mathcal{L}_{prec} = \sum_{i=1}^{5} ||\phi_i(\boldsymbol{I}_{out}) - \phi_i(\boldsymbol{I}_{taregt})||_1, \tag{12}$$

where $\phi_i(\boldsymbol{I}_{out})$ is the features of the i_{th} layer. In addition, the style loss optimizes the statistical error between the feature maps:

$$\mathcal{L}_{style} = \sum_{i} ||G^{\phi}_i(\boldsymbol{I}_{out}) - G^{\phi}_i \boldsymbol{I}_{taregt})||_1, \tag{13}$$

where G^{ϕ}_i denotes the Gram matrix of features. The overall loss is presented as:

$$\mathcal{L} = \sum_{n=1}^{N} (n+1) * (\lambda_{L1}\mathcal{L}^n_{L1} + \lambda_{prec}\mathcal{L}^n_{prec} + \lambda_{style}\mathcal{L}^n_{style}) \tag{14}$$

where $\mathcal{L}^n$ is the loss of the n_{th} scale, and the scale closer to the output is given a larger weight $n-1$.

4 Experiments

4.1 Datasets

We conduct experiments on VITON [15] and MPV [6] datasets. VITON dataset is commonly used in virtual try-on. it contains a training set of 14,221 image pairs and a testing set of 2,032 image pairs. Each pair has a front-view photo and an in-shop clothes image with the resolution 256 × 192. MPV dataset is a recently virtual try-on dataset with multiple views, containing 35,687/13,524 person/clothes images at 256 × 192 resolution where a test set of 4175 image pairs is split out. To fair comparison, following [10,19], the images tagged as back ones are removed since the target garment is only from the front.

4.2 Implementation Details

Network structure. The two encoders have the FPN [25] structure with five layers, each layer consists of a downsampling convolution with the stride of 2, followed by two residual blocks. The MFE flow estimator in each layer comprises ConvNets with four convolution layers, and the hidden dimensions are $[256, 128, 64, 32]$. To obtain a large receptive field, the kernel size of the last three convolution layers is set to 7. the shallow encoder and decoder are ConvNets with two convolution layers without downsampling, and the hidden dimensions are $[32, 64]$. The $K = 6$ in our experiment. Our approach only has a single stage, the computational efficiency is similar to the clothes warping stage in [10,14].

Training Details. We adopt the same training parameters for the two datasets. All our experiments are conducted using Pytorch on Tesla V100 GPUs. The AdamW [28] optimizer is adopted with a batch size of 8. We train the model for 200 epochs where the initial learning rate is 5×10^{-5} and is reduced to 0.1 times the original every 50 epochs. We set the weight of the loss function $\lambda_{L1} = 1, \lambda_{prec} = 1, \lambda_{style} = 100$.

4.3 Qualitative Results

Results on VITON. To qualitatively evaluate our method, we visually compare our method with three recently proposed virtual try-on works with available code implementations, including CP-VITON+ [29], ACGPN [43], PFAPN [10], as shown in Fig. 4. It shows that all comparing works are able to roughly align clothes with the target person pose. However, noticeable artifacts are observed from their results where there are complex poses or misalignment occurs between the target clothes and the person.

As shown in the first row of Fig. 4, baseline methods fail to preserve the striped pattern after warping, especially around the highly non-rigid body parts like the forearm and waist. The second row shows the results with the side view where baseline methods are not able to deal with large deformation in poses and lead to blur or incorrect fitting results. In comparison, our proposed method is capable of extracting accurate structural and textural information

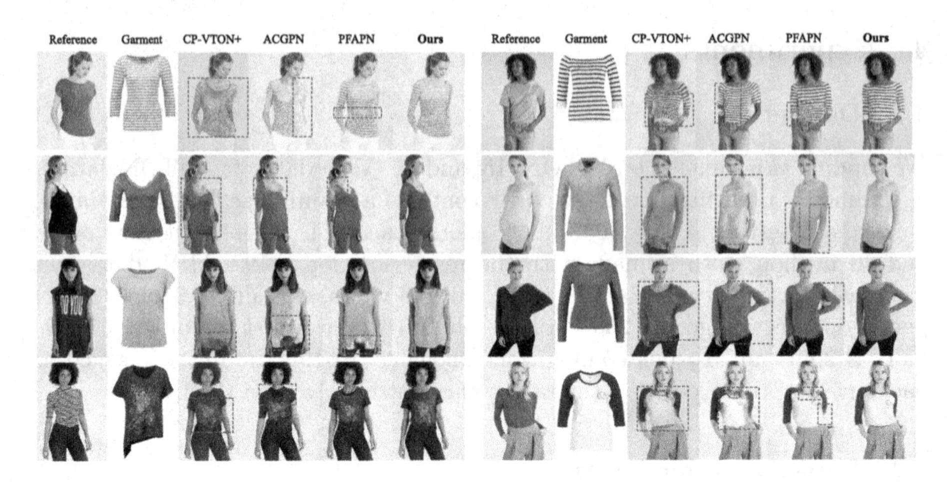

Fig. 4. Visual comparison on VITON dataset. Different regions are marked in red. (Color figure online)

Table 1. Comparisons with State-of-the-art methods on VITON under paired setting.

Methods	CP-VTON [41]	ClothFlow [14]	ACGPN [43]	ZFlow [5]	**SDAFN**(Ours)
FID (↓)	30.50	23.68	–	15.17	**10.97**
SSIM (↑)	0.784	0.843	0.845	0.885	**0.888**
IS (↑)	2.757	–	2.829	–	**2.859**
PSNR (↑)	21.01	23.60	–	25.46	**26.48**

and performing reasonable warping even when there exists huge discrepancy (e.g. large poses or long-sleeve in target clothes while short-sleeve in reference image). The last two columns that parser-based methods like CP-VITON+ [29] and ACGPN [43] are delicate to segmentation errors while learning warping flows. They do not always produce reasonable results for areas like necklines and lower-body parts. Even though PFAPN is a parser-free framework with less distortion in clothes warping, it cannot preserve or generate the body parts well which results in blurry arms and shoulders. Unlike them, our method clearly preserves the characteristics of both the target clothes as well as the body parts, benefiting from the proposed self- and cross-DAFlows.

Results on MPV. To verify the performance of our algorithm on multi-view data, we visualize the results on the MPV [6] dataset. As illustrated in Fig. 5, the reference pose images are shown in the first row while the target garments are shown in the first column. The manipulated results are presented in other columns. It can be seen that our method captures the texture and pose well even with large variations in clothes design and viewpoint change. Besides, it generates realistic body parts even if they are unseen from the reference images which demonstrates the robustness of our method.

Fig. 5. From diverse perspectives, our approach generates high-quality images and accurately preserves vivid attributes on MPV dataset.

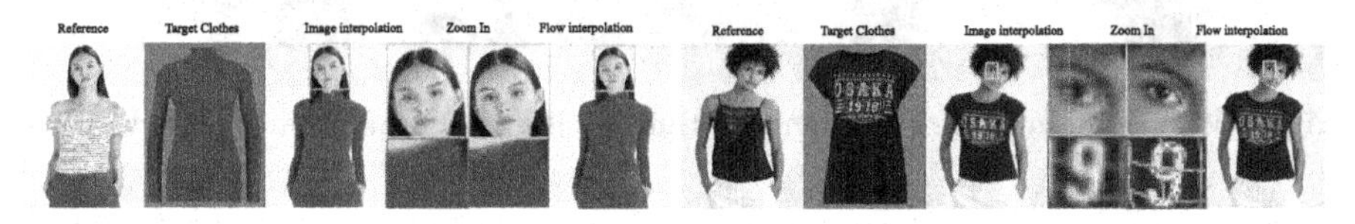

Fig. 6. Comparisons with image interpolation and flow interpolation. Flow interpolation has the advantage of scaling to the higher resolution.

Inference at Higher Resolutions. In contrast to previous methods that can only predict at the inference stage the fixed size of images which is the same as training, our method is able to perform try-on at a higher resolution without retraining. We apply the model trained at 256×192 resolution to test the images with the size of 512×384 from VITON-HD [15]. It is obvious to be seen in Fig. 6 that our pure flow-based approach can retain finer texture information from the original image by linear interpolating the flows, as compared with the image-level interpolation scheme. It helps preserve the clarity of the characters and the patterns of the clothes, which verifies the scalability of our model.

4.4 Quantitative Results

We conduct both paired and unpaired settings to quantitatively compare our work with baseline methods including CP-VTON [41], Clothflow [14], ACGPN [43], PFAFN [10], WUTON [19] and ZFlow [5].

Table 2. Comparisons with state-of-the-art methods on VITON and MPV datasets under unpaired setting. For FID, the lower is the better.

Methods	CP-VTON [41]	ClothFlow [14]	ACGPN [43]	SDAFN (Ours)	WUTON [19]	PF-APN [10]	**SDAFN+ (Ours)**
VITON (FID↓)	24.43	14.43	15.67	12.05	–	10.09	**9.46**
MPV (FID↓)	–	–	–	8.245	7.927	6.429	**5.805**

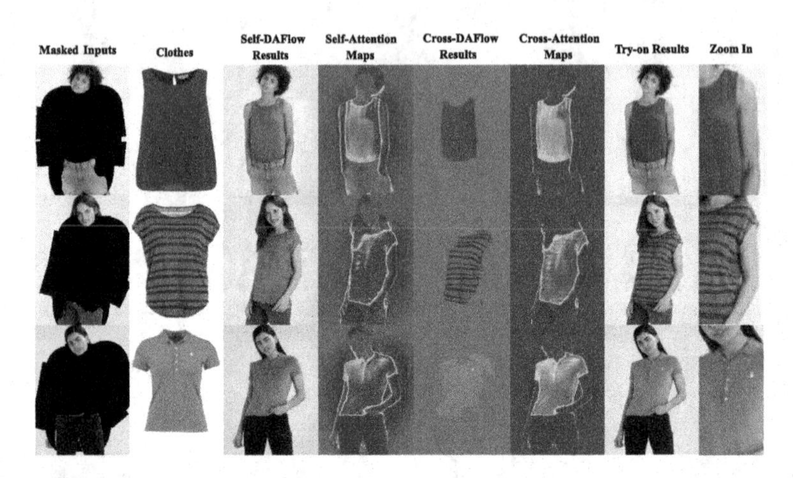

Fig. 7. Visualization of the attention maps and results of deformable attention warping.

Paired Setting. In the paired setting, we use the Structure Similarity Index Measure (SSIM) [36], the Peak Signal-to-Noise Ratio (PNSR) [18] and the Fréchet Inception Distance(FID) [17] to measure the similarity between the synthesized image and ground truth image. The Inception Score (IS) [35] is applied to evaluate the realism of the generated images. We take the target image (the same person wearing the same clothes) as the ground truth which is used to compare with the synthesized image for computing these metrics. It is noted that PFAFN and WUTON were removed from those measurements as they need to take the target image as input for inference.

As shown in Table 1, our approach consistently outperforms the baselines methods under all the metrics. Achieving the best FID and IS scores shows our ability to obtain better visual quality in results and higher SSIM and PSNR scores demonstrate our advantage of accuracy.

Unpaired Setting. Under the unpaired setting, there is no ground truth image for comparison. We directly adopt FID [17] to evaluate the similarity between the generated images and the real images. As reported in Table 2, we compare both parser-free and parser-based methods. In particular, for a fair comparison, we train SDAFN+ model like PFAPN [10] to compare, which allows us to directly input the original image without pose keypoints. In detail, SDAFN+ applies the prediction of SDAFN to replace the masked person and pose keypoints as

Table 3. Performances of models with different sampling number.

K	SSIM	PSNR
1	0.833	22.43
2	0.843	23.25
4	0.868	24.71
6	**0.888**	**26.48**
8	0.878	25.35

Table 4. The effect of different modules.

Config	SSIM	PSNR
Baseline	0.818	20.88
+Cascade	0.827	21.83
+Shallow En.	0.833	22.43
+DAFN	**0.888**	**26.48**

the input for training. Compared with parser-based methods, we achieve obvious improvements without segmentation guidance in FID. Compared with the parser-free method, the better FID on both VITON and MPV datasets indicates the generality of our method.

4.5 Ablation Study

In this section, we mainly evaluate the effectiveness of the proposed deformable attention flow module.

Deformable Attention Flow. The choice of the sampling key number K in deformable attention flow is studied. Setting K to 1 can be regarded as the traditional flow operation. With the increase of K, the performance gradually improves and drops slightly after it reaches 6, as shown in Table 3. It shows that sampling multiple attention regions can effectively improve the flow operation. We set it to be 6 as a balance between performance and computational cost. In addition, as K increases, more GPU memory is consumed for the sampling and warp operations but increases limited time consumption.

Visualization of Attention. In Fig. 7, we visualize self- and cross-DAFlow along with the intermediate attention maps. After the self-DAFlow process, the clear clothes and torso shape are generated along with shadows which renders the realistic human body. The cross-DAFlow not only predicts accurate warping results but also retrains the fine textural details of the clothes, even with the folds on the t-shirt as shown in the second row. The attention maps show strong evidence that our flow module is able to learn the 3D priors where the clothes fit the skin and make the final try-on more realistic. In general, our method effectively decouples the textural and structural information and learns to add 3D shadow effect without introducing auxiliary models or labels.

Modular Ablation Study. The ablation study on the effectiveness of each module in our framework is reported in Table.4. It is shown that both SSIM and PSNR metrics increase with each of the modules added, including cascaded flow estimation, shallow encoder and decoder, and deformable attention flow network (DAFN). Among them, DAFN contributes the most to the improvement. The best performance is reported when integrating all the modules together.

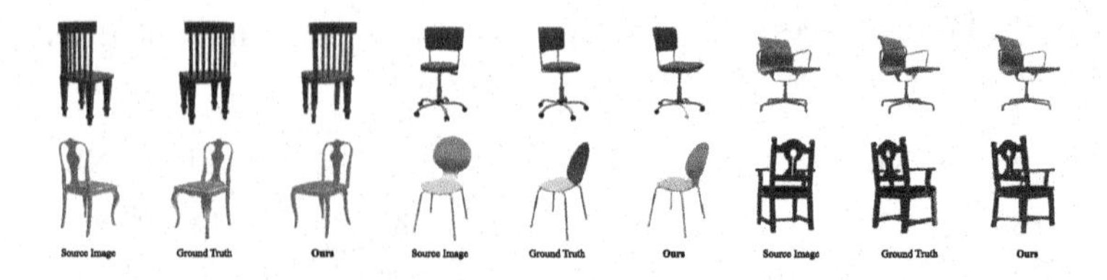

Fig. 8. Qualitative results of the multi-view synthesis on ShapeNet.

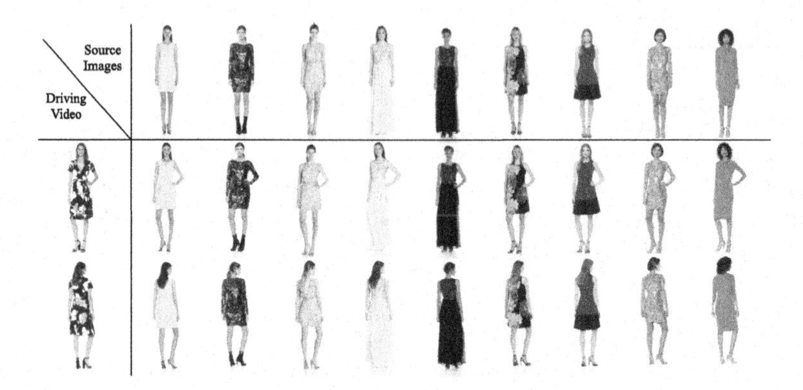

Fig. 9. Qualitative results of the image animation task on FashionVideo.

5 Applications on Other Tasks

In this section, we mainly verify the versatility of the proposed deformable attention flow on two other image editing tasks, namely multi-view synthesis and image animation. To deal with only one image transformation (compared to paired images given in try-on task), we remove the self-MFE in DAFN from our model. More details for implementation can be found in the supplementary.

Multi-view Synthesis. View synthesis aims to generate a novel view of an object given one single image as input. We apply our proposed SDAFN to learn the structural correlations of the same object under different viewing. We conduct experiments on the ShapeNet [3] chairs dataset. As demonstrated in Fig. 8, our method is able to predict the one-shot input chair under different views with accurate textures. It successfully reconstructs the unseen parts which are under occlusion from the input image.

Image Animation. Given an input image and a driving video, image animation is to generate a video sequence so that the object in the source image is animated according to the motion of the driving video. We experiment on the Fashion Video dataset [46], using 500 videos for training and 100 videos for testing. Example animations produced by our SDAFN on two actions are shown in Fig. 9, where accurate reconstruction of the input pose is generated even in the case of complex motion like the back of the body.

6 Conclusions

In this paper, we present a novel single-stage virtual try-on framework. With only pose map as guidance, our model generates photo-realistic fitting results in an end-to-end manner. The proposed deformable attention flow (DAFlow) module estimates the accurate structure while retaining the vivid texture. It synthesizes photo-realistic human torso and fitting clothes with 3D shadows. Extensive experiments and evaluations show that our proposed method not only achieves superior performance on virtual try-on, but also can be extended to other image editing tasks, such as multi-view synthesis and image animation.

References

1. Bertiche, H., Madadi, M., Escalera, S.: CLOTH3D: clothed 3D humans. In: Vedaldi, A., Bischof, H., Brox, T., Frahm, J.-M. (eds.) ECCV 2020. LNCS, vol. 12365, pp. 344–359. Springer, Cham (2020). https://doi.org/10.1007/978-3-030-58565-5_21
2. Bhatnagar, B.L., Tiwari, G., Theobalt, C., Pons-Moll, G.: Multi-garment net: Learning to dress 3d people from images. In: Proceedings of the IEEE/CVF International Conference on Computer Vision, pp. 5420–5430 (2019)
3. Chang, A.X., et al.: Shapenet: an information-rich 3d model repository. arXiv preprint arXiv:1512.03012 (2015)
4. Choi, S., Park, S., Lee, M., Choo, J.: Viton-hd: High-resolution virtual try-on via misalignment-aware normalization. In: Proceedings of the IEEE/CVF Conference on Computer Vision and Pattern Recognition, pp. 14131–14140 (2021)
5. Chopra, A., Jain, R., Hemani, M., Krishnamurthy, B.: Zflow: Gated appearance flow-based virtual try-on with 3d priors. In: Proceedings of the IEEE/CVF International Conference on Computer Vision, pp. 5433–5442 (2021)
6. Dong, H., et al.: Towards multi-pose guided virtual try-on network. In: Proceedings of the IEEE/CVF International Conference on Computer Vision, pp. 9026–9035 (2019)
7. Duchon, J.: Splines minimizing rotation-invariant semi-norms in sobolev spaces. In: Constructive Theory of Functions of Several Variables, pp. 85–100. Springer (1977). https://doi.org/10.1007/BFb0086566
8. Feng, Y., Wu, F., Shao, X., Wang, Y., Zhou, X.: Joint 3d face reconstruction and dense alignment with position map regression network. In: Proceedings of the European Conference on Computer Vision (ECCV), pp. 534–551 (2018)
9. Ge, C., Song, Y., Ge, Y., Yang, H., Liu, W., Luo, P.: Disentangled cycle consistency for highly-realistic virtual try-on. In: Proceedings of the IEEE/CVF Conference on Computer Vision and Pattern Recognition, pp. 16928–16937 (2021)
10. Ge, Y., Song, Y., Zhang, R., Ge, C., Liu, W., Luo, P.: Parser-free virtual try-on via distilling appearance flows. In: Proceedings of the IEEE/CVF Conference on Computer Vision and Pattern Recognition, pp. 8485–8493 (2021)
11. Gong, K., Liang, X., Zhang, D., Shen, X., Lin, L.: Look into person: Self-supervised structure-sensitive learning and a new benchmark for human parsing. In: Proceedings of the IEEE Conference on Computer Vision and Pattern Recognition, pp. 932–940 (2017)
12. Goodfellow, I., et al.: Generative adversarial nets. In: Advances in Neural Information Processing Systems 27 (2014)

13. Güler, R.A., Neverova, N., Kokkinos, I.: Densepose: dense human pose estimation in the wild. In: Proceedings of the IEEE Conference on Computer Vision and Pattern Recognition, pp. 7297–7306 (2018)
14. Han, X., Hu, X., Huang, W., Scott, M.R.: Clothflow: A flow-based model for clothed person generation. In: Proceedings of the IEEE/CVF International Conference on Computer Vision, pp. 10471–10480 (2019)
15. Han, X., Wu, Z., Wu, Z., Yu, R., Davis, L.S.: Viton: An image-based virtual try-on network. In: Proceedings of the IEEE Conference on Computer Vision and Pattern Recognition, pp. 7543–7552 (2018)
16. He, K., Zhang, X., Ren, S., Sun, J.: Deep residual learning for image recognition. In: Proceedings of the IEEE Conference on Computer Vision and Pattern Recognition (CVPR) (2016)
17. Heusel, M., Ramsauer, H., Unterthiner, T., Nessler, B., Hochreiter, S.: Gans trained by a two time-scale update rule converge to a local nash equilibrium. In: Advances in Neural Information Processing Systems 30 (2017)
18. Hore, A., Ziou, D.: Image quality metrics: Psnr vs. ssim. In: 2010 20th International Conference on Pattern Recognition, pp. 2366–2369. IEEE (2010)
19. Issenhuth, T., Mary, J., Calauzènes, C.: Do not mask what you do not need to mask: a parser-free virtual try-on. In: European Conference on Computer Vision, pp. 619–635. Springer (2020). https://doi.org/10.1007/978-3-030-58565-5_37
20. Jaderberg, M., Simonyan, K., Zisserman, A., et al.: Spatial transformer networks. In: Advances in Neural Information Processing Systems 28 (2015)
21. Johnson, J., Alahi, A., Fei-Fei, L.: Perceptual losses for real-time style transfer and super-resolution. In: Leibe, B., Matas, J., Sebe, N., Welling, M. (eds.) ECCV 2016. LNCS, vol. 9906, pp. 694–711. Springer, Cham (2016). https://doi.org/10.1007/978-3-319-46475-6_43
22. Kitaev, N., Kaiser, Ł., Levskaya, A.: Reformer: the efficient transformer. arXiv preprint arXiv:2001.04451 (2020)
23. Lahner, Z., Cremers, D., Tung, T.: Deepwrinkles: accurate and realistic clothing modeling. In: Proceedings of the European Conference on Computer Vision (ECCV), pp. 667–684 (2018)
24. Li, K., Chong, M.J., Zhang, J., Liu, J.: Toward accurate and realistic outfits visualization with attention to details. In: Proceedings of the IEEE/CVF Conference on Computer Vision and Pattern Recognition, pp. 15546–15555 (2021)
25. Lin, T.Y., Dollár, P., Girshick, R., He, K., Hariharan, B., Belongie, S.: Feature pyramid networks for object detection. In: Proceedings of the IEEE Conference on Computer Vision and Pattern Recognition, pp. 2117–2125 (2017)
26. Liu, P.J., et al.: Generating wikipedia by summarizing long sequences. arXiv preprint arXiv:1801.10198 (2018)
27. Liu, Z., et al.: Swin transformer: Hierarchical vision transformer using shifted windows. In: Proceedings of the IEEE/CVF International Conference on Computer Vision, pp. 10012–10022 (2021)
28. Loshchilov, I., Hutter, F.: Decoupled weight decay regularization. arXiv preprint arXiv:1711.05101 (2017)
29. Minar, M.R., Tuan, T.T., Ahn, H., Rosin, P., Lai, Y.K.: Cp-vton+: clothing shape and texture preserving image-based virtual try-on. In: CVPR Workshops (2020)
30. Mir, A., Alldieck, T., Pons-Moll, G.: Learning to transfer texture from clothing images to 3d humans. In: Proceedings of the IEEE/CVF Conference on Computer Vision and Pattern Recognition, pp. 7023–7034 (2020)
31. Qiu, J., Ma, H., Levy, O., Yih, S.W.t., Wang, S., Tang, J.: Blockwise self-attention for long document understanding. arXiv preprint arXiv:1911.02972 (2019)

32. Raj, A., Sangkloy, P., Chang, H., Lu, J., Ceylan, D., Hays, J.: Swapnet: garment transfer in single view images. In: Proceedings of the European Conference on Computer Vision (ECCV), pp. 666–682 (2018)
33. Ren, Y., Wu, Y., Li, T.H., Liu, S., Li, G.: Combining attention with flow for person image synthesis. In: Proceedings of the 29th ACM International Conference on Multimedia, pp. 3737–3745 (2021)
34. Ren, Y., Yu, X., Chen, J., Li, T.H., Li, G.: Deep image spatial transformation for person image generation. In: Proceedings of the IEEE/CVF Conference on Computer Vision and Pattern Recognition, pp. 7690–7699 (2020)
35. Salimans, T., Goodfellow, I., Zaremba, W., Cheung, V., Radford, A., Chen, X.: Improved techniques for training gans. In: Advances in Neural Information Processing Systems 29 (2016)
36. Seshadrinathan, K., Bovik, A.C.: Unifying analysis of full reference image quality assessment. In: 2008 15th IEEE International Conference on Image Processing, pp. 1200–1203. IEEE (2008)
37. Simonyan, K., Zisserman, A.: Very deep convolutional networks for large-scale image recognition. arXiv preprint arXiv:1409.1556 (2014)
38. Sun, D., Yang, X., Liu, M.Y., Kautz, J.: Pwc-net: Cnns for optical flow using pyramid, warping, and cost volume. In: Proceedings of the IEEE Conference on Computer Vision and Pattern Recognition, pp. 8934–8943 (2018)
39. Tay, Y., Bahri, D., Yang, L., Metzler, D., Juan, D.C.: Sparse sinkhorn attention. In: International Conference on Machine Learning, pp. 9438–9447. PMLR (2020)
40. Vaswani, A., et al.: Attention is all you need. In: Advances in Neural Information Processing Systems 30 (2017)
41. Wang, B., Zheng, H., Liang, X., Chen, Y., Lin, L., Yang, M.: Toward characteristic-preserving image-based virtual try-on network. In: Proceedings of the European Conference on Computer Vision (ECCV), pp. 589–604 (2018)
42. Wang, S., Li, B.Z., Khabsa, M., Fang, H., Ma, H.: Linformer: self-attention with linear complexity. arXiv preprint arXiv:2006.04768 (2020)
43. Yang, H., Zhang, R., Guo, X., Liu, W., Zuo, W., Luo, P.: Towards photo-realistic virtual try-on by adaptively generating-preserving image content. In: Proceedings of the IEEE/CVF Conference on Computer Vision and Pattern Recognition, pp. 7850–7859 (2020)
44. Yu, H., Chen, X., Shi, H., Chen, T., Huang, T.S., Sun, S.: Motion pyramid networks for accurate and efficient cardiac motion estimation. In: Martel, A.L., Abolmaesumi, P., Stoyanov, D., Mateus, D., Zuluaga, M.A., Zhou, S.K., Racoceanu, D., Joskowicz, L. (eds.) MICCAI 2020. LNCS, vol. 12266, pp. 436–446. Springer, Cham (2020). https://doi.org/10.1007/978-3-030-59725-2_42
45. Yu, R., Wang, X., Xie, X.: Vtnfp: an image-based virtual try-on network with body and clothing feature preservation. In: Proceedings of the IEEE/CVF International Conference on Computer Vision, pp. 10511–10520 (2019)
46. Zablotskaia, P., Siarohin, A., Zhao, B., Sigal, L.: Dwnet: dense warp-based network for pose-guided human video generation. arXiv preprint arXiv:1910.09139 (2019)
47. Zhou, T., Tulsiani, S., Sun, W., Malik, J., Efros, A.A.: View synthesis by appearance flow. In: Leibe, B., Matas, J., Sebe, N., Welling, M. (eds.) ECCV 2016. LNCS, vol. 9908, pp. 286–301. Springer, Cham (2016). https://doi.org/10.1007/978-3-319-46493-0_18
48. Zhu, X., Su, W., Lu, L., Li, B., Wang, X., Dai, J.: Deformable detr: deformable transformers for end-to-end object detection. arXiv preprint arXiv:2010.04159 (2020)

Improving GANs for Long-Tailed Data Through Group Spectral Regularization

Harsh Rangwani[1(✉)], Naman Jaswani[1], Tejan Karmali[1,2], Varun Jampani[2], and R. Venkatesh Babu[1]

[1] Indian Institute of Science, Bengaluru, India
harshr@iisc.ac.in
[2] Google Research, Mountain View, USA

Abstract. Deep long-tailed learning aims to train useful deep networks on practical, real-world imbalanced distributions, wherein most labels of the tail classes are associated with a few samples. There has been a large body of work to train discriminative models for visual recognition on long-tailed distribution. In contrast, we aim to train conditional Generative Adversarial Networks, a class of image generation models on long-tailed distributions. We find that similar to recognition, state-of-the-art methods for image generation also suffer from performance degradation on tail classes. The performance degradation is mainly due to class-specific mode collapse for tail classes, which we observe to be correlated with the spectral explosion of the conditioning parameter matrix. We propose a novel group Spectral Regularizer (gSR) that prevents the spectral explosion alleviating mode collapse, which results in diverse and plausible image generation even for tail classes. We find that gSR effectively combines with existing augmentation and regularization techniques, leading to state-of-the-art image generation performance on long-tailed data. Extensive experiments demonstrate the efficacy of our regularizer on long-tailed datasets with different degrees of imbalance. Project Page: https://sites.google.com/view/gsr-eccv22.

1 Introduction

Generative Adversarial Networks (GAN) [7] are consistently at the forefront of generative models for image distributions, also being used for diverse applications like image-to-image translation [23], super resolution [20] etc. One of the classic applications of GAN is class specific image generation, by conditioning on the class label y. The generated images in ideal case should associate to class label y, be of high quality and exhibit diversity. The conditioning is usually achieved with conditional Batch Normalization (cBN) [5] layers which induce class-specific (y) features at each layer of the generator. The additional class conditioning information enables GAN models like the state-of-the-art (SOTA) BigGAN [2] to generate diverse images, in comparison to unconditional models [13].

Supplementary Information The online version contains supplementary material available at https://doi.org/10.1007/978-3-031-19784-0_25.

S. Avidan et al. (Eds.): ECCV 2022, LNCS 13675, pp. 426–442, 2022.
https://doi.org/10.1007/978-3-031-19784-0_25

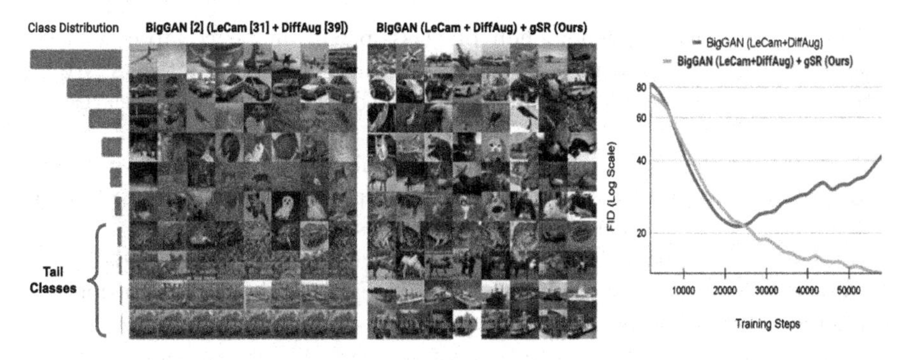

Fig. 1. Regularizing GANs on long-tailed training data. *(left)* Images generated from BigGAN trained on long-tailed CIFAR-10. *(right)* FID scores vs. Training steps. The proposed gSR regularizer prevents mode collapse, for the tail classes [2,36,45].

Recent works [45] demonstrate that performance of models like BigGAN deteriorates from mode collapse when limited training data is presented. The differentiable data augmentation approaches [14,35,45] attempt to mitigate this degradation by enriching the training data through augmentations. On the other hand, model based regularization techniques like LeCam [36] are also proposed to prevent the degradation of image quality in such cases.

Although these methods lead to effective increase in image generation quality, they are designed to increase performance when trained on balanced datasets (*i.e.* even distribution of samples across classes). We find that the SOTA methods like BigGAN (LeCam) with augmentation, which are designed for limited data, also suffer from the phenomenon of class-specific mode collapse when trained on long-tailed datasets. By *class-specific mode collapse*, we refer to deteriorating quality of generated images for tail classes, as shown in Fig. 1.

In this work we aim to investigate the cause of the class-specific mode collapse that is observed in the generated images of tail classes. We find that the class-specific mode collapse is correlated with spectral explosion (*i.e.* sudden increase in spectral norms, ref. Fig. 2) of the corresponding class-specific (cBN) parameters (when grouped into a matrix, described in Sect. 3.3). To prevent this spectral explosion, we introduce a novel class-specific *group Spectral Regularizer* (gSR), which constrains the spectral norm of class-specific cBN parameters. Although there are multiple spectral [37,42] regularization (and normalization [26]) techniques used in deep learning literature, all of them are specific to model weights W, whereas our regularizer focuses on cBN parameters. We, through our analysis, show that our proposed gSR leads to reduced correlation among tail classes' cBN parameters, effectively mitigating the issue of class-specific mode collapse.

We extensively test our regularizer's effectiveness by combining it with popular SNGAN [26] and BigGAN [2] architectures. It also complements discriminator specific SOTA regularizer (LeCam+DiffAug [36]), as it's combination with gSR ensures improved quality of image generation even across tail classes (Fig. 1). In summary, we make the following contributions:

- We first report the phenomenon of class-specific mode collapse, which is observed when cGANs are trained on long-tailed imbalance datasets. We find that spectral norm explosion of class-specific cBN parameters correlates with its mode collapse.
- We find that even existing techniques for limited data [15,22,36] are unable to prevent class-specific collapse. Hence, we propose a novel group Spectral Regularizer (gSR) which helps in alleviating such collapse.
- Combining gSR with existing SOTA GANs with regularization leads to large average relative improvement (of ∼25% in FID) for image generation on 5 different long-tailed dataset configurations.

2 Related Work

Generative Adversarial Networks: Generative Adversarial Networks [7] are a combination of Generator G and Discriminator D aimed at learning a generative model for a distribution. GANs have enabled learning of models for complex distributions like high resolution images etc. One of the inflection point for success was the invention of Spectral Normalization (SN) for GANs (SNGAN) which enabled GANs to scale to datasets like ImageNet [6] (1000 classes). The GAN training was further scaled by BigGAN [2] which demonstrated successful high resolution image generation, using a deep ResNet network.

Regularization: Several regularization techniques [16,21,23,44,46,48] are developed to alleviate the problem of mode collapse in GANs like Gradient Penalty [8], Spectral Normalization [26] etc. These include LeCam [36] and Differentiable Augmentations [15,45] which are the regularization techniques specifically designed to prevent mode collapse in limited data scenarios. Commonality among majority of these techniques are that they (a) designed for the data which is balanced across classes, and (b) focus on discriminator network D. In this work, we aim to regularize the cBN parameters in generator G, which makes our regularizer complementary to earlier works.

Long-Tailed Imbalanced Data: Long-tailed imbalance is a form of distribution in which majority of the data samples are present in head classes and the occurrence of per class data samples decreases exponentially as we move towards tail classes (Fig. 1(*left*)). This family of distribution represents natural distribution for species' population [10], objects [38] etc. As these distributions are natural, a lot of work has been done to learn discriminative models (*i.e.* classifiers) [3,4,12,25,41,47] which work across all classes, despite training data following a long-tailed distribution. However, even though there has been a lot of interest, there are still only a handful works which aim to learn generative models for long-tailed distribution. Mullick *et al.* [28] developed GAMO which aims to learn how to oversample using a generative model. Class Balancing GAN (CBGAN) with a Classifier in the Loop [31] is the only work which aims to learn a GAN to generate good quality images across classes (in a long-tailed distribution). However, their model is an unconditional GAN which requires a trained

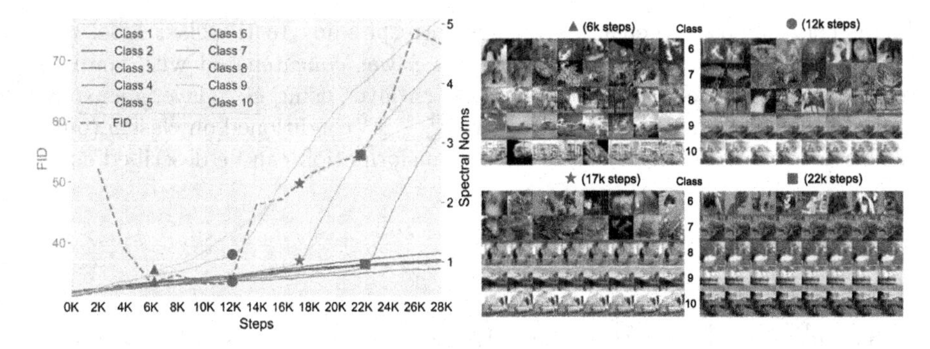

Fig. 2. Correlation between class-specific mode collapse and spectral explosion. *(left)* FID/Spectral Norms of class-specific gain parameter of conditional BatchNorm layer on CIFAR-10. Symbols on plot indicate that FID score's increase correlates with onset of spectral explosion on 4 tail classes respectively. *(right)* Images generated for tail classes at these train steps reveals corresponding class-specific mode collapse.

classifier to guide the GAN. The requirement of such a classifier can be restrictive. In this work we aim to develop conditional GANs which use class labels, does not require external classifier and generate good quality images (even for tail classes) when trained using long-tailed training data.

3 Approach

We start by describing conditional Generative Adversarial Networks (Sect. 3.1) and the associated class-specific mode collapse (Sect. 3.2). Following that we introduce our regularizer formulation, and explain the decorrelation among features caused by gSR that mitigates the mode collapse for tail classes (Sect. 3.3).

3.1 Conditional Generative Adversarial Networks

Generative Adversarial Networks (GAN) are a combination of two networks the generator G and discriminator D. The discriminator's goal is to classify images from training data distribution (P_r) as real and the images generated through G as fake. In a conditional GAN (cGAN), the generator and discriminator are also enriched with the information about the class label y associated with the image $\mathbf{x} \in \mathbb{R}^{3\times H\times W}$. The conditioning information y helps the cGAN in generating diverse images of superior quality, in comparison to unconditional GAN. The cGAN objectives can be described as:

$$\begin{aligned}
&\max_D V(D) = \mathop{\mathbb{E}}_{x\sim P_r}[f_{\mathcal{D}} D(x|y)] + \mathop{\mathbb{E}}_{z\sim P_z}[f_{\mathcal{G}}(1 - D(G(z|y)))] \\
&\min_G \mathcal{L}_G = \mathop{\mathbb{E}}_{z\sim P_z}[f_{\mathcal{G}}(1 - D(G(z|y)))]
\end{aligned} \tag{1}$$

where p_z is the prior distribution of latents, $f_{\mathcal{D}}, f_{\mathcal{G}}$, and $g_{\mathcal{G}}$ refer to the mapping functions from which different formulations of GAN can be derived (ref. [22]).

The generator $G(z|y)$ generates images corresponding to the class label y. In earlier works the conditioning information y was concatenated with the noise vector z, however recently conditioning each layer using cBN layer has shown improved performance [27]. For a feature $\mathbf{x}_\mathbf{y}^\mathbf{l} \in \mathbb{R}^d$ conditioned on class y (out of K classes) corresponding to layer l, the transformation can be described as:

$$\hat{\mathbf{x}}_\mathbf{y}^\mathbf{l} = \frac{\mathbf{x}_\mathbf{y}^\mathbf{l} - \mu_\mathbf{B}^\mathbf{l}}{\sqrt{{\sigma_\mathbf{B}^\mathbf{l}}^2 + \epsilon}} \rightarrow \gamma_\mathbf{y}^\mathbf{l}\hat{\mathbf{x}}_\mathbf{y}^\mathbf{l} + \beta_\mathbf{y}^\mathbf{l} \tag{2}$$

The $\mu_\mathbf{B}^\mathbf{l}$ and ${\sigma_\mathbf{B}^\mathbf{l}}^2$ are the mean and the variance of the batch respectively. The $\gamma_\mathbf{y}^\mathbf{l} \in \mathbb{R}^d$ and the $\beta_\mathbf{y}^\mathbf{l} \in \mathbb{R}^d$ are the cBN parameters which enable generation of the image for specific class y. We focus on the behaviour of these parameters in the subsequent sections.

3.2 Class-Specific Mode Collapse

Due to widespread use of conditional GANs [2,27], it is important that these models are able to learn across various kinds of training data distributions. However, while training a conditional GAN on long-tailed data distribution, we observe that GANs suffer from model collapse on tail classes (Fig. 1). This leads to only a single pattern being generated for a particular tail class. To investigate the cause of this phenomenon, we inspect the class-specific parameters of cBN, which are gain $\gamma_\mathbf{y}^\mathbf{l}$ and bias $\beta_\mathbf{y}^\mathbf{l}$. In existing works, characteristics of groups of features have been insightful for analysis of neural networks and have led to development of regularization techniques [9,40]. Hence for further analysis we also create n_g groups of the $\gamma_\mathbf{l}^\mathbf{y}$ parameters and stack them to obtain $\mathbf{\Gamma}_\mathbf{y}^\mathbf{l} \in \mathbb{R}^{n_g \times n_c}$, where n_c are the number of columns after grouping. It is observed that the value of spectral norm ($\sigma_{\max}(\mathbf{\Gamma}_\mathbf{y}^\mathbf{l}) \in \mathbb{R}$) explodes (*i.e.* increases abnormally) as mode collapse occurs for corresponding tail class y as shown in Fig. 2. We observe this phenomenon consistently across both the smaller SNGAN [26] (Fig. 2) and the larger BigGAN [2] (Fig. 1) model. We observe similar spectral explosion for BigGAN model as in Fig. 2 (empirically shown in Fig. 9). In the earlier works, mode collapse could be detected by anomalous behaviour of spectral norm of discriminator (refer to suppl. material for details). However in the class-specific mode collapse the discriminator's spectral norms show normal behavior and are unable to signal such collapse. Here, our analysis of $\sigma_{\max}(\mathbf{\Gamma}_\mathbf{y}^\mathbf{l})$ helps in detecting class-specific mode collapse.

3.3 Group Spectral Regularizer (gSR)

Our aim now is to prevent the class-specific mode collapse while training. To achieve this, we introduce a regularizer for the generator G which prevents spectral explosion. We would like to emphasize that earlier works including Augmentations [45], LeCam [36] regularizer etc. are applied on discriminator, hence our regularizer's focus on G is complementary to those of existing techniques. As we observe that spectral norm explodes for the $\gamma_\mathbf{y}^\mathbf{l}$ and $\beta_\mathbf{y}^\mathbf{l}$, we deploy a group

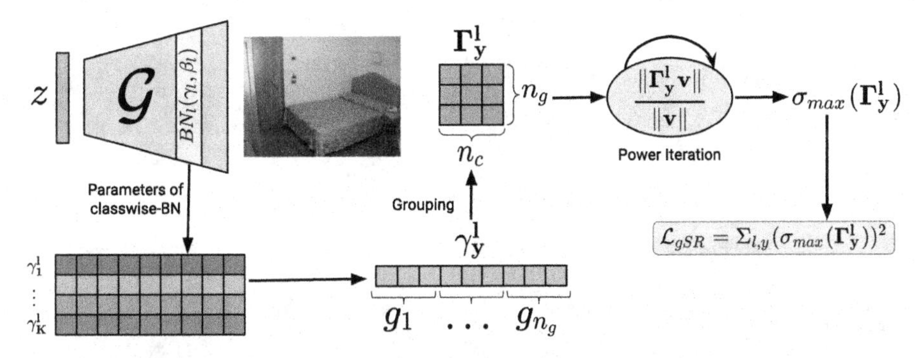

Fig. 3. Algorithmic overview. During each training step, 1) we extract the gain $\gamma_{\mathbf{y}}^{\mathbf{l}}$ for each cBN layers in G, 2) group them into matrix $\mathbf{\Gamma}_{\mathbf{y}}^{\mathbf{l}}$ and estimate $\sigma_{\max}(\Gamma_y^l)$. 3) We repeat the same procedure with bias $\beta_{\mathbf{y}}^{\mathbf{l}}$ to obtain $\sigma_{\max}(\mathbf{B}_{\mathbf{y}}^{\mathbf{l}})$. 4) Finally, we regularize both as described in $\mathcal{L}_{gSR}$ (Eq. 5).

Spectral Regularizer (gSR) to prevent mode collapse. Steps followed by gSR for estimation of $\sigma_{\max}$ of $\gamma_{\mathbf{y}}^{\mathbf{l}}(\in \mathbb{R}^d)$ are described below (also given in Fig. 3):

$$\textit{Grouping}: \ \mathbf{\Gamma}_{\mathbf{y}}^{\mathbf{l}} = \Pi(\gamma_{\mathbf{y}}^{\mathbf{l}}, n_g) \in \mathbb{R}^{n_g \times n_c} \tag{3}$$

$$\textit{Power Iteration}: \ \sigma_{\max}(\mathbf{\Gamma}_{\mathbf{y}}^{\mathbf{l}}) = \max_{\mathbf{v}} \|\mathbf{\Gamma}_{\mathbf{y}}^{\mathbf{l}}\mathbf{v}\| / \|\mathbf{v}\| \tag{4}$$

$\mathbf{v}(\in \mathbb{R}^d)$ is a random vector for power iterations. n_g and n_c are the number of groups and number of columns respectively, such that $n_g \times n_c = d$. After estimation of $\sigma_{\max}(\mathbf{\Gamma}_{\mathbf{y}}^{\mathbf{l}})$ and similarly $\sigma_{\max}(\mathbf{B}_{\mathbf{y}}^{\mathbf{l}})$, the regularized loss objective for generator can be written as:

$$\min_G \mathcal{L}_G + \lambda_{gSR}\mathcal{L}_{gSR}; \ \text{ where } \ \mathcal{L}_{gSR} = \sum_l \sum_y \lambda_y(\sigma^2_{\max}(\mathbf{\Gamma}_{\mathbf{y}}^{\mathbf{l}}) + \sigma^2_{\max}(\mathbf{B}_{\mathbf{y}}^{\mathbf{l}})) \tag{5}$$

As the spectral explosion is prominent for the tail classes, we weigh the spectral regularizer term with λ_y which has an inverse relation with number of samples n_y in class y. Prior work [3] shows that directly using $1/n_y$ can be over-aggressive hence, we use the effective number of samples (a soft version of inverse relation) formally given as (where $\alpha = 0.99$): $\lambda_y = (1-\alpha)/(1-\alpha^{n_y})$.

The regularized objective is used to update weights using backpropagation. Spectral regularizers are used in earlier works [37,42] but they are applied on network weights W, whereas to the best of our knowledge, ours is the first work that proposes the regularization of the batch normalization (BatchNorm) parameters. There exist other form of techniques like Spectral Normalization and Spectral Restricted Isometry Property (SRIP) [1] regularizer, which we empirically did not find to be effective for our work (comparison in Sect. 5.3).

Decorrelation Effect (Relation with Group Whitening): Group Whitening [9] is a recent work which whitens the activation map X by grouping, normalizing and whitening using Zero-phase Component Analysis (ZCA) to obtain

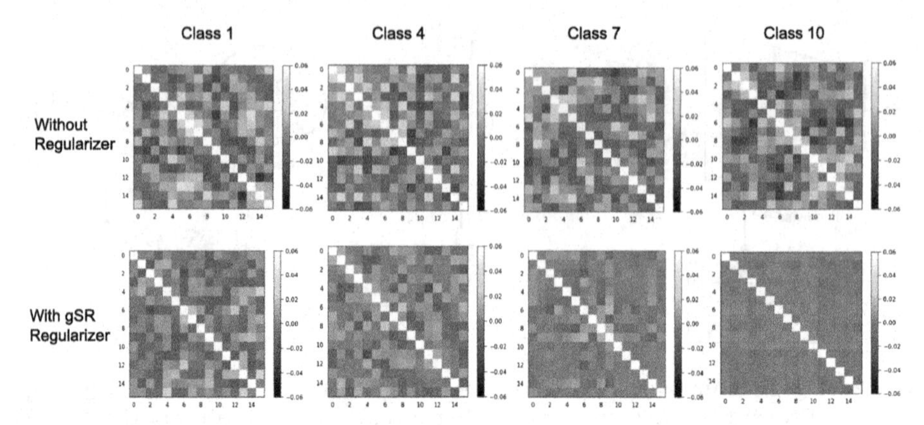

Fig. 4. Covariance matrices of $\mathbf{\Gamma}^l_{\mathbf{y}}$ for ($l = 1$) for SNGAN baseline. After using gSR (for tail classes with high λ) the covariance matrix converges to a diagonal matrix in comparison to without gSR (where large correlations exist). This demonstrates the decorrelation effect of gSR on $\gamma^l_{\mathbf{y}}$, which alleviates class-specific mode collapse.

$\hat{X}$. Due to whitening, the rows of $\hat{X}_g$ get decorrelated, which can be verified by finding the similarity of covariance matrix $\frac{1}{n_c}\hat{X}_g\hat{X}_g{}^{\mathsf{T}}$ to a diagonal matrix. The Group Whitening transformation significantly improves the generalization performance by learning diverse features. As our regularizer also operates on groups of features, we find that minimizing the $\mathcal{L}_{gSR}$ loss also leads to decorrelation of the rows of $\mathbf{\Gamma}^l_{\mathbf{y}}$. We verify this phenomenon by visualizing the covariance matrix $\frac{1}{n_c}[\mathbf{\Gamma}^l_{\mathbf{y}} - \mathbb{E}[\mathbf{\Gamma}^l_{\mathbf{y}}]]([\mathbf{\Gamma}^l_{\mathbf{y}} - \mathbb{E}[\mathbf{\Gamma}^l_{\mathbf{y}}]])^{\mathsf{T}}$.

In Fig. 4, we plot the covariance matrices for both the SNGAN and SNGAN with regularizer (gSR). We clearly observe that for tail classes with high λ_y the covariance matrix is more closer to a diagonal matrix which confirms the decorrelation of parameters caused by gSR. We find that decorrelation is required more in layers with more class-specific information (*i.e.* earlier layers of generator) rather than layers with generic features like edges. We provide the visualizations for more layers in the suppl. material.

Recent theoretical results [11,39] for supervised learning show that decorrelation of parameters mitigates overfitting, and leads to better generalization. This is analogous to our observation of decorrelation being able to prevent mode collapse and helpful in generating diverse images.

4 Experimental Evaluation

We perform extensive experiments on various long-tailed datasets with different resolution. For the controlled imbalance setting, we perform experiments on CIFAR-10 [18] and LSUN [43], which are commonly used in the literature [3,4,33] (Sect. 4.1). We also show results on challenging real-world datasets (with skewed data distribution) of iNaturalist2019 [10] and AnimalFaces [17] (Sect. 4.2).

Datasets: We use the CIFAR-10 [18] and a subset (5 classes) of LSUN dataset [43] (50k images balanced across classes) for our experiments. The choice

Table 1. Quantitative results on the CIFAR-10 and LSUN dataset. On an average, we observe a relative improvement in FID of 20.33% and 39.08% over SNGAN and BigGAN baselines respectively.

	CIFAR-10				LSUN			
Imb. Ratio (ρ)	100		1000		100		1000	
	FID (↓)	IS(↑)	FID (↓)	IS(↑)	FID (↓)	IS(↑)	FID (↓)	IS(↑)
CBGAN [31]	$33.01_{\pm 0.12}$	$6.58_{\pm 0.05}$	$44.82_{\pm 0.12}$	$5.92_{\pm 0.05}$	$37.41_{\pm 0.10}$	$2.82_{\pm 0.03}$	$44.70_{\pm 0.13}$	$2.77_{\pm 0.02}$
LSGAN [24]	$24.36_{\pm 0.01}$	$7.77_{\pm 0.07}$	$51.47_{\pm 0.21}$	$6.54_{\pm 0.05}$	$37.64_{\pm 0.05}$	$3.12_{\pm 0.01}$	$41.50_{\pm 0.04}$	$2.74_{\pm 0.02}$
SNGAN [26]	$30.62_{\pm 0.07}$	$6.80_{\pm 0.02}$	$54.58_{\pm 0.19}$	$6.19_{\pm 0.01}$	$38.17_{\pm 0.02}$	$3.02_{\pm 0.01}$	$38.36_{\pm 0.11}$	$2.99_{\pm 0.01}$
+ gSR (Ours)	$18.58_{\pm 0.10}$	$7.80_{\pm 0.09}$	$48.69_{\pm 0.04}$	$5.92_{\pm 0.01}$	$28.84_{\pm 0.09}$	$3.50_{\pm 0.01}$	$35.76_{\pm 0.05}$	$3.56_{\pm 0.01}$
BigGAN [2]	$19.55_{\pm 0.12}$	$8.80_{\pm 0.09}$	$50.78_{\pm 0.23}$	$6.50_{\pm 0.05}$	$38.65_{\pm 0.05}$	$\mathbf{4.02}_{\pm 0.01}$	$45.89_{\pm 0.30}$	$3.25_{\pm 0.01}$
+ gSR (Ours)	$\mathbf{12.03}_{\pm 0.08}$	$\mathbf{9.21}_{\pm 0.07}$	$\mathbf{38.38}_{\pm 0.01}$	$\mathbf{7.24}_{\pm 0.04}$	$\mathbf{20.18}_{\pm 0.07}$	$3.67_{\pm 0.01}$	$\mathbf{24.93}_{\pm 0.09}$	$\mathbf{3.68}_{\pm 0.01}$

of 5 class subset is for a direct comparison with related works [31,33] which identify this subset as challenging and use that for experiments. For converting the balanced subset to the long-tailed dataset with imbalance ratio (ρ) (*i.e.* ratio of highest frequency class to the lowest frequency class), we remove the additional samples from the training set. Prior works [3,4,25] follow this standard process to create benchmark long-tailed datasets. We keep the validation sets balanced and unchanged to evaluate the performance by treating all classes equally. We provide additional details about datasets in the suppl. material. We perform experiments on the imbalance ratio of 100 and 1000. In case of CIFAR-10 for $\rho = 1000$ the majority class contains 5000 samples whereas the minority class has only 5 samples. For performing well in this setup, the GAN has to robustly learn from many-shot (5000 sample class) as well as at few-shot (5 sample class) together, making this benchmark challenging.

Evaluation: We report the standard Inception Score (IS) [32] and Fréchet Inception Distance (FID) metrics for the generated datasets. We report the mean and standard deviation of 3 evaluation runs similar to the protocol followed by DiffAug [45] and LeCam [36]. We use a held out set of 10k images for calculation of FID for both the datasets. The held out sets are balanced across classes for fair evaluation of each class.

Configuration: We perform experiments by modifying the PyTorch-StudioGAN implemented by Kang *et al.* [13], which serves as the baseline for our framework. We generate 32×32 sized images for the CIFAR-10 dataset and 64×64 sized images for the LSUN dataset. For the CIFAR-10 experiments, we use $5D$ steps for each G step. Unless explicitly stated, we by default add the following two SOTA regularizers to obtain strong generic baselines for all the experiments (except CBGAN for which we follow exact setup described by Rangwani *et al.* [31]):

- DiffAugment [45]: We apply the differential augmentation technique with the colorization, translation, and cutout augmentations.
- LeCam [36]: LeCam regularizer prevents divergence of discriminator D by constraining its output through a regularizer term R_{LC} (ref. suppl. material).

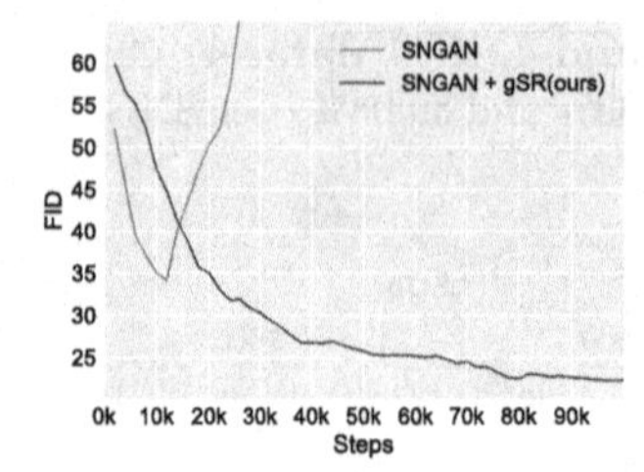

Fig. 5. Stability. Addition of gSR (to baseline) stabilizes the training to continually improve, as indicated by the FID scores.

Fig. 6. Qualitative evaluations of SNGAN baseline on LSUN dataset. Each row represents images from a class. Note the class-specific mode collapse observed in tail-classes in SNGAN (last two rows), which is alleviated after addition of gSR to generate diverse images.

Any improvement over these strong regularizers published recently is meaningful and shows the effectiveness of the proposed methods. We use a batch size of 64 for all our CIFAR-10 and LSUN experiments. For sanity check of the implementation we run the experiments for the balanced dataset (CIFAR-10) case where our FID is similar to the one obtained in LeCam [36], details are in the suppl. material.

Baselines: We compare our regularizer with the recent work of Class Balancing GAN (CBGAN) [31] which uses an auxiliary classifier for long-tailed image generation. We use the public codebase and configuration provided by the authors. The auxiliary classifiers are obtained using the LDAM-DRW as suggested by CBGAN authors. We use the SNGAN [26] (with projection discriminator [27]) as our base method on which we apply the Augmentation and LeCam regularizer for a strong baseline. We also compare our method with LSGAN [24], which is shown to be effective in preventing the mode-collapse (we use the same configuration as in SNGAN for fairness of comparison). To demonstrate improvement over large scale GANs we also use BigGAN [2] with LeCam and DiffAug regularizers as baseline. We then add our group Spectral Regularizer (gSR) in the loss terms for BigGAN and SNGAN, and report the performance in Table 1. We do not use ACGAN as our baseline as it leads to image generation which doesn't match the conditioned class label (*i.e.* class confusion) [31].

4.1 Results on CIFAR-10 and LSUN

Stability: Figure 5 shows the FID vs iteration steps for the SNGAN and SNGAN +gSR configuration. Using gSR regularizer with SNGAN is able to effectively prevent the class-specific mode collapse, which in turn helps the GAN to improve for a long duration of time. SNGAN without gSR starts degrading quickly and stops improving very soon, this shows the stabilizing effect imparted by gSR regularizer in the training process. The stabilizing effect is similarly observed even for the BigGAN (ref. FID plot in Fig. 1).

Comparison of Quality: We observe that application of regularizer effectively avoids mode collapse and leads to a large average improvement (of 7.46) in FID for the (SNGAN + gSR) case, in comparison to SNGAN baseline across the four datasets (Table 1). Our method is also effective on BigGAN where it is able to achieve SOTA FID and IS significant improvement in almost all cases. Although SNGAN and BigGAN baselines are already enriched with SOTA regularizers of (LeCam + DiffAug) to improve results, yet the addition of our gSR regularizer significantly boosts performance by harmonizing with other regularizers. It also shows that our regularizer complements the existing work and effectively reduces mode collapse. Figure 6 shows a comparison of the generated images for the different methods, where gSR is able to produce better images for the tail classes for LSUN dataset (refer Fig. 1 for qualitative comparison on CIFAR-10 ($\rho = 100$)). To quantify improvement over each class, we compute class-wise FID and mean FID (*i.e.* Intra FID) in Fig. 7. We find that gSR leads to very significant improvement in tail class FID as it prevents the collapse. Due to the stabilizing effect of gSR we find that head class FID are also improved, clearly demonstrating the benefit of gSR for all classes. We also provide additional metrics (precision [19], recall [19], density [29], coverage [29] and Intra-FID) in suppl. material. We find that almost all metrics show similar improvement as seen in FID (Table 1).

4.2 Results on Naturally Occurring Distributions

To show the effectiveness of our regularizer on natural distributions we perform experiments on two challenging datasets: iNaturalist-2019 [10] and AnimalFace [34]. The iNaturalist dataset is a real-world long-tailed dataset with 1010 classes of different species. There is high diversity among the images of each class, due to their distinct sources of origin. The dataset follows a long-tailed distribution with around 260k images. The second dataset we experiment with is the Animal Face Dataset [17] which contains 20 classes with around 110 samples per class. We generate 64×64 images for both datasets using the BigGAN with a batch size of 256 for iNaturalist and 64 for AnimalFaces. The BigGAN baseline is augmented with LeCam and DiffAug regularizers. We compare our method with the baselines described in [31]. We evaluate each model using the FID on a training subset which is balanced across classes. For baselines we directly report results from Rangwani *et al.*[31] (indicated by *) in Table 2.

The BigGAN baseline achieves an FID of 6.87 on iNaturalist 2019 dataset, which improves relatively by 7.42% (−0.51 FID) when proposed gSR is combined with BigGAN. Our approach is also able to achieve FID better than SOTA CBGAN on iNaturalist 2019 dataset. Table 2 shows the performance of the BigGAN baseline over the AnimalFace dataset, where after combining with our gSR regularizer we see FID improvement by 6.90% (−2.65 FID). The improvements on both the large long-tailed dataset and few shot dataset of AnimalFace shows that gSR is able to effectively improve performance on real-world datasets. We provide additional experimental details and results in the suppl. material.

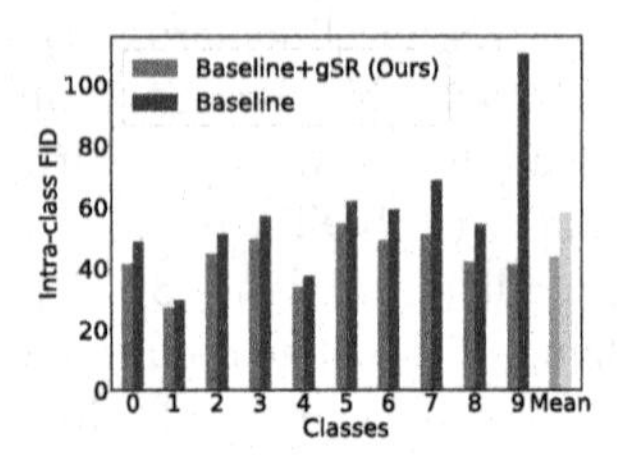

Fig. 7. Class-Wise FID and mean FID (Intra-FID) of BigGAN on CIFAR-10 over $5K$ generated images($\rho = 100$).

Table 2. Quantitative results on iNaturalist-2019 and AnimalFace Dataset. We compare mean FID (↓) with other existing approaches.

	iNaturalist 2019		AnimalFace
Method	cGAN	FID(↓)	FID(↓)
SNResGAN* [26]	✗	13.03$\pm$0.07	–
CBGAN* [31]	✗	9.01$\pm$0.08	–
ACGAN* [30]	✓	47.15$\pm$0.11	–
SNGAN* [27]	✓	21.53$\pm$0.03	–
BigGAN [2]	✓	6.87$\pm$0.04	38.41$\pm$0.04
+ gSR (Ours)	✓	**6.36**$\pm$0.04	**35.76**$\pm$0.04

5 Analysis

5.1 Ablation over Regularizers

We use the combination of existing regularizers (LeCam + DiffAug) with our regularizer (gSR) to obtain the best performing models. For further analysis of importance of each, we study their effect in comparison to gSR in this section. We perform experiments by independently applying each of them on vanilla SNGAN. Table 3 shows that existing regularizer in itself is not able to effectively reduce FID, whereas gSR is effectively able to reduce FID independently by 3.8 points. However, we find that existing regularizers along with proposed gSR, make an effective combination which further reduces FID significantly (by 9.27) on long-tailed CIFAR-10 ($\rho = 100$). This clearly shows that our regularizer effectively complements the existing regularizers.

5.2 High Resolution Image Generation

As the LSUN dataset is composed of high resolution scenes we also investigate if the class-specific mode collapse phenomenon when GANs are trained for high resolution image synthesis. For this we train SNGAN and BigGAN baselines for (128×128) using the DiffAugment and LeCAM regularizer (details in suppl. material). We find that similar phenomenon of spectral explosion leading to class-specific collapse occurs (as in 64×64 case), which is mitigated when the proposed gSR regularizer is combined with the baselines (Fig. 8). The gSR regularizer leads to significant improvement in FID (Table 4) also seen in qualitatively in Fig. 8.

5.3 Comparison with Related Spectral Regularization and Normalization Techniques

As gSR constrains the exploding spectral norms for the cBN parameters (Fig. 9) to evaluate its effectiveness, we test it against other variants of spectral normalization and regularization techniques on SGAN for CIFAR-10 ($\rho = 100$).

Fig. 8. Qualitative comparison of BigGAN variants on LSUN dataset (ρ=100) (128 × 128). Each row represents images from a distinct class.

Table 3. Ablation over regularizers on SNGAN. We report FID and IS on CIFAR-10 dataset (with $\rho = 100$).

LeCam+ DiffAug	gSR	FID(↓)	IS(↑)
✗	✗	$31.73_{\pm 0.08}$	$7.18_{\pm 0.02}$
✗	✓	$27.85_{\pm 0.05}$	$7.09_{\pm 0.02}$
✓	✗	$30.62_{\pm 0.07}$	$6.80_{\pm 0.02}$
✓	✓	$\mathbf{18.58}_{\pm 0.10}$	$\mathbf{7.80}_{\pm 0.09}$

Table 4. Image Generation (128 × 128). We report FID on both SNGAN and BigGAN on LSUN dataset (for ρ = 100 and ρ = 1000).

Imb. Ratio (ρ)	100	1000
	FID (↓)	FID (↓)
SNGAN [26]	$53.91_{\pm 0.02}$	$72.37_{\pm 0.08}$
+ gSR (Ours)	$\mathbf{25.31}_{\pm 0.03}$	$\mathbf{31.86}_{\pm 0.03}$
BigGAN [2]	$61.63_{\pm 0.11}$	$77.17_{\pm 0.18}$
+ gSR (Ours)	$\mathbf{16.56}_{\pm 0.02}$	$\mathbf{45.08}_{\pm 0.10}$

Group Spectral Normalization (gSN) of BatchNorm Parameters: In this setting, rather than using sum of spectral norms (Eq. 5) as regularizer for the class-specific parameters of cBN in gSR, we normalize them by dividing it by group spectral norms (*i.e.* $\frac{\gamma_{\mathbf{y}}^{\mathbf{l}}}{\sigma_{max}(\mathbf{\Gamma}_{\mathbf{y}}^{\mathbf{l}})}$) [26].

Group Spectral Restricted Isometry Property (gSRIP) Regularization: Extending SRIP [1], the class-specific parameters of cBN which are grouped to form a matrix $\mathbf{\Gamma}_{\mathbf{y}}^{\mathbf{l}}$, the regularizer is the sum of square of spectral norms of $({\mathbf{\Gamma}_{\mathbf{y}}^{\mathbf{l}}}^{T}\mathbf{\Gamma}_{\mathbf{y}}^{\mathbf{l}} - \mathbf{I})$, (instead that of $\mathbf{\Gamma}_{\mathbf{y}}^{\mathbf{l}}$ in gSR (Eq. 5)). We report our findings in Table 5. It can be inferred that all three techniques, namely gSN, gSRIP, and gSR, lead to significant improvements over the baseline. This also confirms our hypothesis that reducing (or constraining) spectral norm of cBN parameters alleviates class-specific mode collapse. However, it is noteworthy that gSR gives the highest boost over the baseline by a considerable margin in terms of FID.

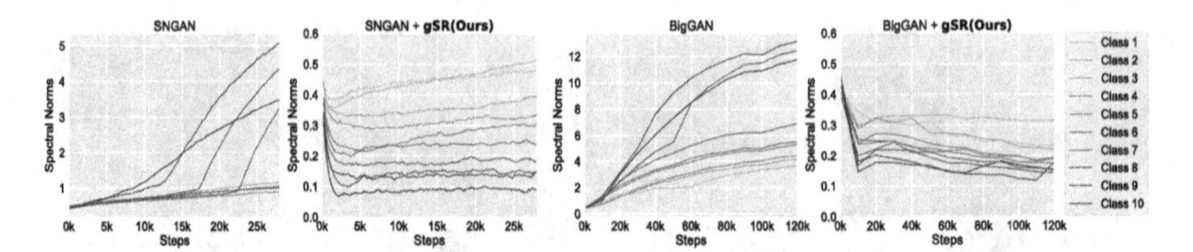

Fig. 9. Effect of gSR on spectral norms of $\mathbf{\Gamma^l_y}$ (CIFAR-10). We observe a spectral explosion both for SNGAN(*left*) and BigGAN(*right*) baselines of tail classes' cBN parameters. This is prevented by addition of gSR as shown on corresponding right.

5.4 Analysis of gSR

In this section (and suppl. material) we provide ablations of gSR using long-tailed CIFAR-10 (ρ=100).

Can gSR work with StyleGAN-2? We train and analyze the StyleGAN2-ADA implementation available [13] on long-tailed datasets, where we find it also suffers from class-specific mode collapse. We then implement gSR for StyleGAN2 by grouping 512 dimensional class conditional embeddings in mapping network to 16×32 and calculating their spectral norm which is added to loss (Eq. 5) as R_{gSR}. We find that gSR is able to effectively prevent the mode collapse (Fig. 10) and also results in significant improvement in FID in comparison to StyleGAN2-ADA baseline. Further analysis and results are present in suppl. material.

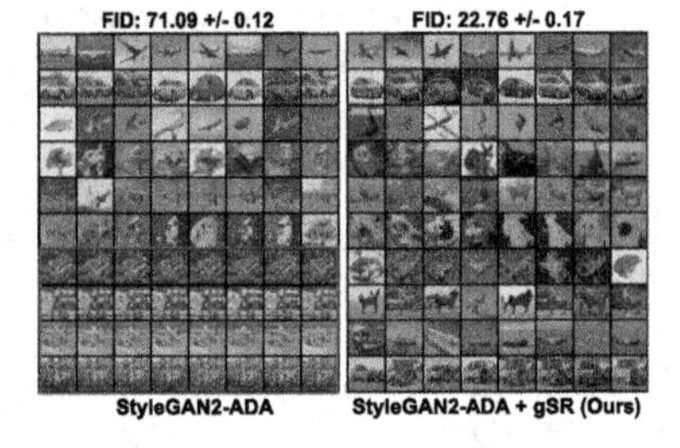

Fig. 10. StyleGAN2-ADA On CIFAR-10 (ρ = 100), comparison of gSR with the baseline.

What is gSR's Effect on Spectral Norms? We plot spectral norms of class-specific gain parameter of 1st layer of generator in SNGAN. Spectral norms explode for the tail classes without gSR, while they remain stable when gSR is used. Figure 5 for same experiment shows that while using gSR the FID keeps improving, whereas it collapses without using gSR. This confirms our hypothesis that constraining spectral norms stabilizes the training. We find that a similar phenomenon also occurs for BigGAN (Fig. 5) which uses SN in G, which shows that gSR is complementary to SN.

What Should be the Ideal Number of Groups? Grouping of the ($\gamma^l_\mathbf{y}$) into $\mathbf{\Gamma^l_y} \in \mathbb{R}^{n_g \times n_c}$ is a central operation in our regularizer formulation (Eq. 3). We group $\gamma^l_\mathbf{y}$ (and $\beta^l_\mathbf{y}$) $\in \mathbb{R}^{256}$ into a matrix $\mathbf{\Gamma^l_y}$ (and $\mathbf{B^l_y}$) ablate over different combinations of n_g and n_c. Table 6 shows that FID scores do not change much significantly with n_g. As we also use power iteration to estimate the spectral norm $\sigma_{\max}(\mathbf{\Gamma^l_y})$, we report iteration complexity (multiplications). Since grouping into square matrix(n_g = 16) gives slightly better FIDs while also being time

Table 5. Quantitative comparison of spectral regularizers. Comparison against Different Spectral Norm Regularizers on grouped cBN parameters.

	FID(↓)	IS(↑)
SNGAN [27]	30.62$\pm$0.07	6.80$\pm$0.07
+ gSN [26]	23.97$\pm$0.13	7.49$\pm$0.05
+ gSRIP [1]	23.67$\pm$0.02	7.79$\pm$0.06
+ gSR (Ours)	**18.58**$\pm$0.10	**7.80**$\pm$0.09

Table 6. Group size ablations. We report average FID and IS on CIFAR-10 dataset. $n_g = 16$ gives the best FID while also being computationally efficient, measured by per Iteration (Iter.) complexity. (Iter. complexity for power iteration method is calculated as $(n_g^2 + n_c^2)$ x number of power iterations (4 in our setting)).

n_g	n_c	FID(↓)	IS(↑)	Iter. Complexity(↓)
4	64	20.16$\pm$0.03	**7.96$\pm$0.01**	16448
8	32	18.69$\pm$0.06	7.80$\pm$0.01	4352
16	16	**18.58$\pm$0.10**	7.80$\pm$0.09	**2048**
32	8	20.19$\pm$0.06	7.85$\pm$0.01	4352

efficient we use it for our experiments. We also provide additional mathematical intuition for the optimality of choice of $n_c = n_g$ in suppl. material.

6 Conclusion and Future Work

In this work we identify a novel failure mode of *class-specific mode collapse*, which occurs when conditional GANs are trained on long-tailed data distribution. Through our analysis we find that the class-specific collapse for each class correlates closely with a sudden increase (explosion) in the spectral norm of its (grouped) conditional BatchNorm (cBN) parameters. To mitigate the spectral explosion we develop a novel group Spectral Regularizer (gSR), which constrains the spectral norms and alleviates mode collapse. The gSR reduces spectral norms (estimated through power iteration) of grouped parameters and leads to decorrelation of parameters, which enables GAN to effectively improve on long-tailed data distribution without collapse. Our empirical analysis shows that gSR: a) leads to improved image generation from conditional GANs (by alleviating class-specific collapse), and b) effectively complements exiting regularizers on discriminator to achieve state-of-the-art image generation performance on long-tailed datasets. One of the limitations present in our framework is that it introduces additional hyperparameter λ for the regularizer. Developing an hyperparameter free decorrelated parameterization for alleviating class-specific mode collapse is a good direction for future work. We hope that this work leads to further research on improving GANs for real-world long-tailed datasets.

Acknowledgements. This work was supported in part by SERB-STAR Project (Project:STR/2020/000128), Govt. of India and a Google Research Award. Harsh Rangwani is supported by Prime Minister's Research Fellowship (PMRF). We thank Lavish Bansal for help with StyleGAN experiments.

References

1. Bansal, N., Chen, X., Wang, Z.: Can we gain more from orthogonality regularizations in training deep networks? In: Advances in Neural Information Processing Systems, vol. 31 (2018)
2. Brock, A., Donahue, J., Simonyan, K.: Large scale GAN training for high fidelity natural image synthesis. arXiv preprint arXiv:1809.11096 (2018)
3. Cao, K., Wei, C., Gaidon, A., Arechiga, N., Ma, T.: Learning imbalanced datasets with label-distribution-aware margin loss. In: Advances in Neural Information Processing Systems (2019)
4. Cui, Y., Jia, M., Lin, T.Y., Song, Y., Belongie, S.: Class-balanced loss based on effective number of samples. In: CVPR (2019)
5. De Vries, H., Strub, F., Mary, J., Larochelle, H., Pietquin, O., Courville, A.C.: Modulating early visual processing by language. In: Advances in Neural Information Processing Systems (2017)
6. Deng, J., Dong, W., Socher, R., Li, L.J., Li, K., Fei-Fei, L.: ImageNet: a large-scale hierarchical image database. In: 2009 IEEE Conference on Computer Vision and Pattern Recognition (2009)
7. Goodfellow, I., et al.: Generative adversarial nets. In: Advances in neural information processing systems (2014)
8. Gulrajani, I., Ahmed, F., Arjovsky, M., Dumoulin, V., Courville, A.C.: Improved training of Wasserstein GANs. In: Advances in neural information processing systems (2017)
9. Huang, L., Zhou, Y., Liu, L., Zhu, F., Shao, L.: Group whitening: balancing learning efficiency and representational capacity. In: Proceedings of the IEEE/CVF Conference on Computer Vision and Pattern Recognition (2021)
10. iNaturalist: The inaturalist 2019 competition dataset. http://github.com/visipedia/inat_comp/tree/2019 (2019)
11. Jin, G., Yi, X., Zhang, L., Zhang, L., Schewe, S., Huang, X.: How does weight correlation affect generalisation ability of deep neural networks? Adv. Neural. Inf. Process. Syst. **33**, 21346–21356 (2020)
12. Kang, B., et al.: Decoupling representation and classifier for long-tailed recognition. In: International Conference on Learning Representations (2019)
13. Kang, M., Park, J.: Contrastive generative adversarial networks. arXiv preprint arXiv:2006.12681 (2020)
14. Karras, T., Aittala, M., Hellsten, J., Laine, S., Lehtinen, J., Aila, T.: Training generative adversarial networks with limited data. arXiv preprint arXiv:2006.06676 (2020)
15. Karras, T., Aittala, M., Hellsten, J., Laine, S., Lehtinen, J., Aila, T.: Training generative adversarial networks with limited data. In: Proceedings NeurIPS (2020)
16. Kavalerov, I., Czaja, W., Chellappa, R.: CGANs with multi-hinge loss. arXiv preprint arXiv:1912.04216 (2019)
17. Kolouri, S., Zou, Y., Rohde, G.K.: Sliced Wasserstein kernels for probability distributions. In: Proceedings of the IEEE Conference on Computer Vision and Pattern Recognition (2016)
18. Krizhevsky, A.: Learning multiple layers of features from tiny images. Technical Report (2009)
19. Kynkäänniemi, T., Karras, T., Laine, S., Lehtinen, J., Aila, T.: Improved precision and recall metric for assessing generative models. In: Advances in Neural Information Processing Systems, vol. 32 (2019)

20. Ledig, C., et al.: Photo-realistic single image super-resolution using a generative adversarial network. In: Proceedings of the IEEE Conference on Computer Vision and Pattern Recognition (2017)
21. Liu, K., Tang, W., Zhou, F., Qiu, G.: Spectral regularization for combating mode collapse in GANs. In: Proceedings of the IEEE/CVF International Conference on Computer Vision, pp. 6382–6390 (2019)
22. Liu, M.Y., Huang, X., Yu, J., Wang, T.C., Mallya, A.: Generative adversarial networks for image and video synthesis: algorithms and applications. Proc. IEEE **109**(5), 839–862 (2021)
23. Mao, Q., Lee, H.Y., Tseng, H.Y., Ma, S., Yang, M.H.: Mode seeking generative adversarial networks for diverse image synthesis. In: Proceedings of the IEEE Conference on Computer Vision and Pattern Recognition (2019)
24. Mao, X., Li, Q., Xie, H., Lau, R.Y., Wang, Z., Paul Smolley, S.: Least squares generative adversarial networks. In: Proceedings of the IEEE International Conference on Computer Vision (2017)
25. Menon, A.K., Jayasumana, S., Rawat, A.S., Jain, H., Veit, A., Kumar, S.: Long-tail learning via logit adjustment. In: International Conference on Learning Representations (2021)
26. Miyato, T., Kataoka, T., Koyama, M., Yoshida, Y.: Spectral normalization for generative adversarial networks. arXiv preprint arXiv:1802.05957 (2018)
27. Miyato, T., Koyama, M.: cGANs with projection discriminator. In: International Conference on Learning Representations (2018)
28. Mullick, S.S., Datta, S., Das, S.: Generative adversarial minority oversampling. In: The IEEE International Conference on Computer Vision (ICCV) (2019)
29. Naeem, M.F., Oh, S.J., Uh, Y., Choi, Y., Yoo, J.: Reliable fidelity and diversity metrics for generative models (2020)
30. Odena, A., Olah, C., Shlens, J.: Conditional image synthesis with auxiliary classifier GANs. In: Proceedings of the 34th International Conference on Machine Learning-Volume, vol. 70 (2017)
31. Rangwani, H., Mopuri, K.R., Babu, R.V.: Class balancing GAN with a classifier in the loop. arXiv preprint arXiv:2106.09402 (2021)
32. Salimans, T., Goodfellow, I., Zaremba, W., Cheung, V., Radford, A., Chen, X.: improved techniques for training GANs. In: Advances in neural information processing systems (2016)
33. Santurkar, S., Schmidt, L., Madry, A.: A classification-based study of covariate shift in GAN distributions. In: International Conference on Machine Learning, pp. 4480–4489 (2018)
34. Si, Z., Zhu, S.C.: Learning hybrid image templates (HIT) by information projection. IEEE Trans. Pattern Anal. Mach. Intell. **34**(7), 1354–1367 (2011)
35. Tran, N.T., Tran, V.H., Nguyen, N.B., Nguyen, T.K., Cheung, N.M.: On data augmentation for GAN training. IEEE Trans. Image Process. **30**, 1882–1897 (2021)
36. Tseng, H.Y., Jiang, L., Liu, C., Yang, M.H., Yang, W.: Regularizing generative adversarial networks under limited data. In: Proceedings of the IEEE/CVF Conference on Computer Vision and Pattern Recognition (2021)
37. Vahdat, A., Kautz, J.: NVAE: a deep hierarchical variational autoencoder. In: Neural Information Processing Systems (NeurIPS) (2020)
38. Wang, Y.X., Ramanan, D., Hebert, M.: Learning to model the tail. In: Advances in Neural Information Processing Systems (2017)
39. Wang, Z., Xiang, C., Zou, W., Xu, C.: MMA regularization: decorrelating weights of neural networks by maximizing the minimal angles. Adv. Neural. Inf. Process. Syst. **33**, 19099–19110 (2020)

40. Wu, Y., He, K.: Group normalization. In: Proceedings of the European conference on computer vision (ECCV) (2018)
41. Yang, Y., Xu, Z.: Rethinking the value of labels for improving class-imbalanced learning. In: NeurIPS (2020)
42. Yoshida, Y., Miyato, T.: Spectral norm regularization for improving the generalizability of deep learning. arXiv preprint arXiv:1705.10941 (2017)
43. Yu, F., Zhang, Y., Song, S., Seff, A., Xiao, J.: Lsun: Construction of a large-scale image dataset using deep learning with humans in the loop. CoRR abs/1506.03365 (2015)
44. Zhang, H., Zhang, Z., Odena, A., Lee, H.: Consistency regularization for generative adversarial networks. arXiv preprint arXiv:1910.12027 (2019)
45. Zhao, S., Liu, Z., Lin, J., Zhu, J.Y., Han, S.: Differentiable augmentation for data-efficient GAN training. In: Advances in Neural Information Processing Systems, vol. 33 (2020)
46. Zhao, Z., Singh, S., Lee, H., Zhang, Z., Odena, A., Zhang, H.: Improved consistency regularization for GANs. In: Proceedings of the AAAI Conference on Artificial Intelligence, vol. 35, pp. 11033–11041 (2021)
47. Zhou, B., Cui, Q., Wei, X.S., Chen, Z.M.: BBN: Bilateral-branch network with cumulative learning for long-tailed visual recognition, pp. 1–8 (2020)
48. Zhou, P., Xie, L., Ni, B., Geng, C., Tian, Q.: Omni-GAN: On the secrets of cGANs and beyond. In: Proceedings of the IEEE/CVF International Conference on Computer Vision, pp. 14061–14071 (2021)

Hierarchical Semantic Regularization of Latent Spaces in StyleGANs

Tejan Karmali[1,2](✉), Rishubh Parihar[1], Susmit Agrawal[1], Harsh Rangwani[1], Varun Jampani[4], Maneesh Singh[3], and R. Venkatesh Babu[1]

[1] Indian Institute of Science, Bengaluru, India
tejankarmali@iisc.ac.in
[2] Google Research, Bengaluru, India
[3] Motive Technologies, Inc, Oxford, UK
[4] Google Research, Cambridge, USA

Abstract. Progress in GANs has enabled the generation of high-resolution photorealistic images of astonishing quality. StyleGANs allow for compelling attribute modification on such images via mathematical operations on the latent style vectors in the $\mathcal{W}/\mathcal{W}+$ space that effectively modulate the rich hierarchical representations of the generator. Such operations have recently been generalized beyond mere attribute swapping in the original StyleGAN paper to include interpolations. In spite of many significant improvements in StyleGANs, they are still seen to generate unnatural images. The quality of the generated images is predicated on two assumptions; (a) The richness of the hierarchical representations learnt by the generator, and, (b) The linearity and smoothness of the style spaces. In this work, we propose a Hierarchical Semantic Regularizer (HSR) (Project Page and code: https://sites.google.com/view/hsr-eccv22) which aligns the hierarchical representations learnt by the generator to corresponding powerful features learnt by pretrained networks on large amounts of data. HSR is shown to not only improve generator representations but also the linearity and smoothness of the latent style spaces, leading to the generation of more natural-looking style-edited images. To demonstrate improved linearity, we propose a novel metric - Attribute Linearity Score (ALS). A significant reduction in the generation of unnatural images is corroborated by improvement in the Perceptual Path Length (PPL) metric by 16.19% averaged across different standard datasets while simultaneously improving the linearity of attribute-change in the attribute editing tasks.

1 Introduction

Recent years have seen tremendous advances in Generative Adversarial Network (GAN) [15] architectures and their training methods to produce highly

T. Karmali—Work done while at Indian Institute of Science.
M. Singh—Work done while at Verisk Analytics.

Supplementary Information The online version contains supplementary material available at https://doi.org/10.1007/978-3-031-19784-0_26.

S. Avidan et al. (Eds.): ECCV 2022, LNCS 13675, pp. 443–459, 2022.
https://doi.org/10.1007/978-3-031-19784-0_26

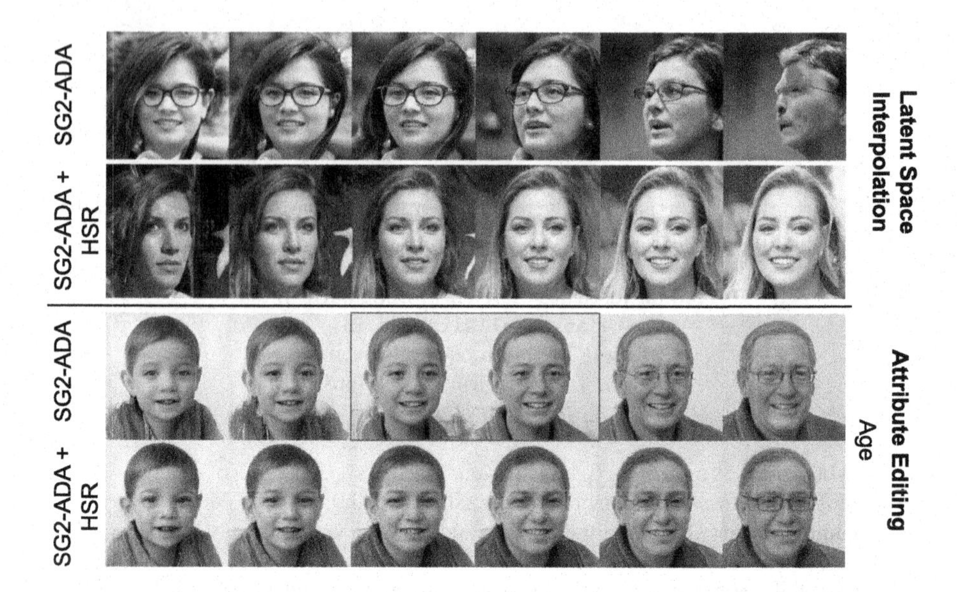

Fig. 1. Hierarchical Semantic Regularizer (HSR) improves the latent space to semantic image mapping to produce more natural-looking images. **Top:** We show latent interpolation for images from bottom 10%-ile image pairs ranked by PPL, a metric to measure smoothness of latent space. **Bottom:** Latent space using HSR mitigates artefacts in images during attribute edit transition and can transition smoothly (young to old (SG2-ADA) vs. young to middle-age to old (SG2-ADA+HSR), in continuous attributes like "Age"). Zoom in to observe the effects.

photorealistic images [7,28]. Progress in the StyleGAN family of GAN architectures has shown promise by improving both the image quality, as well as the quality of latent space representations which enables controlled image generation. This is achieved by transforming an input noise space $\mathcal{Z}$ to a latent style space $\mathcal{W}$ which modulates a synthesis network at various levels of representation hierarchies to generate an image with that style. This enables generation of compelling synthetic images with novel styles as well as practically useful applications such as GAN-based image attribute editing, style mixing, etc. [3,21,41,42,45,46]. Nonetheless, such networks still often produce unrealistic images (ref. Fig. 1).

These quality issues in StyleGANs can have the following sources: (a) the hierarchical representation spaces in the synthesis network, (b) the latent style space, in particular the linearity and smoothness of such spaces, and (c) the functions used to transform the representation spaces in (a) using the corresponding hierarchical style vectors in (b). Our work seeks to address some of these issues.

We take inspiration from the recent advances in self-supervised and supervised learning [8,10,19,49] which have allowed for the learning of semantically rich image representations translating into significant performance improves on image classification and other vision tasks [14,31,49]. Training on large datasets of natural images, like ImageNet [12], allows these techniques to learn hierarchically organized feature spaces capturing richer statistical patterns in natural images: shallower layer capturing low-level image features and the deeper layers abstract features highly correlated with visual semantics. Such pre-trained representations can be harnessed to enhance the representational power of StyleGANs.

In fact, we demonstrate that transferring rich pretrained representations mentioned above allow us to mitigate simultaneously the challenges associated with both the representation spaces in the synthesis network as well as the latent style spaces modulating these representations. To allow for such a transfer, we propose to use a regularization mechanism, called the Hierarchical Semantic Regularizer (HSR) which aligns the generator's features to those from an appropriate, state of the art, pretrained feature extractor at several corresponding scales (levels) of the generator network. The architecture is shown in Fig. 3.

Karras *et al.* [27] introduced the Perceptual Path Length (PPL) metric to measure the smoothness of mapping from a latent space to the output image and showed its correlation with the generated image quality. We demonstrate that HSR regularization in StyleGAN training leads to 16.19% relative improvement in PPL over StyleGAN2, leading to more realistic interpolations (refer Fig. 1).

A standard approach for controlled synthesis of novel images is via linear (convex) interpolation between attributes[1] corresponding to real images. Applications such as image editing utilize such capabilities under the presumption that style spaces are both linear as well as decorrelated allowing for desired controlled edits. Since, PPL does not measure linearity, we propose a novel metric, Attribute Linearity Score (ALS), to measure linearity in the attibute space. We demonstrate that HSR simultaneously improves linearity leading to smoother edits with significantly reduced editing artifacts (Fig. 1). A mean relative improvement of 15.5% over StyleGAN2-ADA is achieved on the ALS metric.

Our contributions are: (a) A novel Hierarchical Semantic Regularizer (HSR) improving the generation of natural-looking synthetic images from StyleGANs. HSR is presented in (Sect. 3 with an analysis of design choices 3.3). (b) Extensive bench-marking of improvements by HSR regularization on popular datasets, especially when utilizing linear interpolations for attribute editing (Sect. 4). (c) Since linearity of the latent attribute space is very important for performing controlled edits, we propose a new metric, Attribute Linearity Score (ALS), in (Sect. 4.3) and demonstrate improved linearity over the baselines.

2 Related Works

Generative Adversarial Networks. GAN proposed by Goodfellow *et al.* [15] a combination of two neural networks, *i.e.* generator G and discriminator D. For image synthesis the goal of D is to differentiate between real and generated images, whereas the G tries to fool the discriminator into classifying generated images as real. In the recent years several improvements in architecture [17,26,29,37,44], optimization objectives [4,6,34,35] and regularization [18,36] have made GANs an ubiquitous choice for image synthesis. It has

[1] We use style and attributes interchangeably.

been observed that GANs developed for large scale datasets, suffer mode collapse when trained on limited data. Augmentation methods like DiffAugment [61], ADA [25], ContraD [22] , APA [23] etc. mitigate the collapse by reducing the discriminator's overfitting.

Hierarchical Representations. In classical vision, methods which decompose image into a hierarchy have been exploited for the tasks of image stitching, manipulation and fusion [2,11]. Building on this motivation Shocher *et al.* [47] develop an image translation and manipulation method, which exploits hierarchical consistency of features of generator and a classifier. However this method is restricted to single image translation and manipulation. In contrast our work we aim to train a smooth and generalizable GAN which can simultaneously generate diverse images, by using semantic hierarchical consistency of features.

Knowledge Transfer Using Pre-Trained Features. Using pre-trained features trained on large scale datasets (*e.g.* ImageNet etc.) [19,50] have been useful for various downstream tasks across applications [13,20,52,56]. The recent development of the self-supervised approaches for representation learning [9,16,44] have further immensely improved the quality of features learnt. These features are being used in various applications like part segmentation, localization etc. without being explicitly trained on such tasks [8], which motivates our work which aims to transfer these semantic properties to G's feature space. Currently much work for transfer learning for GANs has focused on the fine-tuning large GANs using a few images for adapting it to a different domain [33,38–40]. Recently a concurrent work [32] also aims to use pre-trained features to improve GANs. However their goal is to improve discriminator. On contrary we aim to enrich GAN feature space by imparting it with semantic properties, leading to a disentangled and smooth latent space.

Image Editing Using Latent Space Interpolations. Latent space of pre-trained StyleGAN models is highly structured [45] and is popularly used to perform realistic image edits in the generated images [1,3,21,45,46,54,58]. The primary idea in most of these approaches is to find a direction in the the extended latent space $\mathcal{W}+$ for editing attributes and transforming a latent code by moving in that direction to perform edits. StyleCLIP [42] learns the directions for attribute editing by getting the guidance from pretrained CLIP [43]. On the contrary, our work imposes constraints so that latent space has more naturally interpretable directions when used by the GAN-based image editing methods.

3 Approach

In this section, we first describe the objective of GAN framework, properties of StyleGAN, and its evaluation in Sect. 3.1. Then, we describe Hierarchical Semantic Regularizer (HSR) (Sect. 3.2) and discuss its design in Sect. 3.3.

3.1 Preliminaries

Generative Adversarial Networks. GAN involves two competing networks, namely a Generator G and a Discriminator D. Taking a noise $\mathbf{z}$ sampled from a distribution P_z as input, G generates an image $G(\mathbf{z}) \in \mathbb{R}^{3\times H\times W}$. Whereas, D takes an input image $x \in \mathbb{R}^{3\times H\times W}$, and tries to classify it as real or generated. The objective of G is to fool D into making it classify the generated image as a real one. Formally, the learning objective can be written as:

$$\begin{aligned} \max_{D} \mathcal{L}_D &= \mathop{\mathbb{E}}_{x\sim P_r}[log(D(x))] + \mathop{\mathbb{E}}_{\mathbf{z}\sim P_z}[log(1 - D(G(\mathbf{z})))] \\ \min_{G} \mathcal{L}_G &= \mathop{\mathbb{E}}_{\mathbf{z}\sim P_z}[log(1 - D(G(\mathbf{z})))] \end{aligned} \tag{1}$$

StyleGAN. In StyleGAN, an architectural modification is introduced where $\mathbf{z}$ is transformed into a semantic latent space through a sequence of linear layers called Mapping Network G_m, before generating the image I through a Synthesis Network G_s as $I = G_s(\mathbf{w})$. Hence, $G = G_s \circ G_m$. The space learnt by G_m is known as $\mathcal{W}+$-space. It is observed that $\mathcal{W}+$ is more meaningful in terms of attributes learned from the training data as compared to noise space $\mathcal{Z}$. Several methods [21,45,46] propose ways to find attribute-specific directions in $\mathcal{W}+$ latent space.

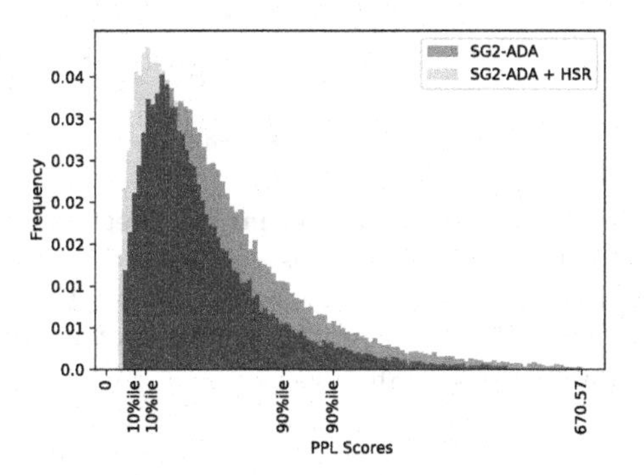

Fig. 2. Distribution of PPL over 50k images from SG2-ADA and SG2-ADA+HSR. HSR improves the perceptual quality of top and bottom 10%-ile images, thus leading to more natural-looking images.

Perceptual Path Length. To measure the smoothness of the mapping from a latent space to the output image, Karras *et al.* [28] proposed Perceptual Path Length (PPL). The requirement for this metric arises due to generation of unnatural images by GAN despite having low FID [28]. PPL aims to quantify the smoothness of latent space to output space mapping by measuring average of LPIPS [59] distances between two generated images under small perturbations in the latent space. A smoother latent space should have lesser PPL when compared to an uneven latent space. It is shown [28] that PPL correlates well with image quality, *i.e.* good quality images pairs will have less PPL, while if any one of the image is of bad quality, the PPL would be high. The images are sampled randomly without any truncation trick [7,30] to compute PPL. As observed in Fig. 6, the bottom 10%-ile by PPL (sorted in increasing order) among the generated images appear as out-of-distribution images. Hence, the mean PPL score can be used to quantify the extent of non-smooth regions of latent space which produce unnatural images. Hence, we will be using this metric as a primary metric for comparison of the smoothness of latent space learnt by the models.

3.2 Hierarchical Semantic Regularizer

Feature extractors of networks pretrained on large datasets (*e.g.* ImageNet etc.) of natural images using classification or self-supervised losses store strong priors

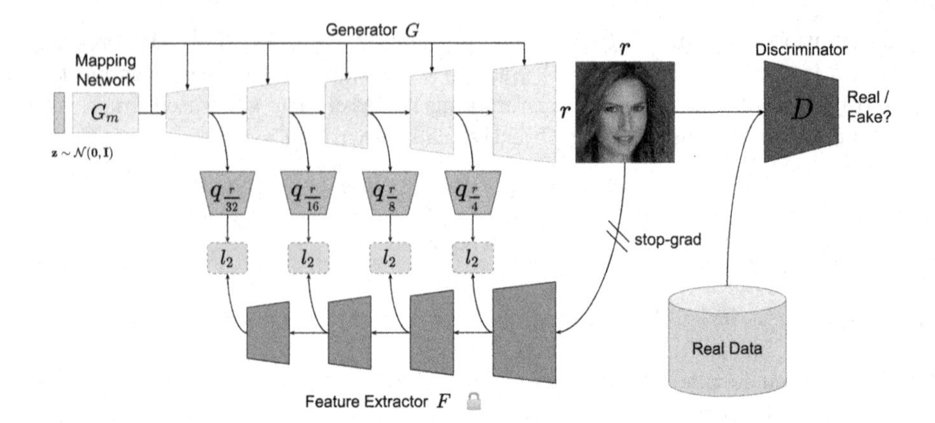

Fig. 3. Hierarchical Semantic Regularizer: We use a pre-trained network to extract features at various resolution hierarchically. We then train linear predictors over generator features to predict the pre-trained features hierarchically. This transfers the semantic knowledge to generator feature space, making it's latent space meaningful, disentangled and editable.

about the data, that are organized hierarchically. Each level of hierarchy captures a different semantic feature of data. The statistics of wide variety of natural images are captured by these networks [5,19,51]. Due to the inherent differences in the nature of tasks, discriminative models capture different kinds of features compared to generative ones. Therefore, we seek to enrich the G's intermediate feature space with guidance from a pretrained feature extractor.

We first give a general idea of the proposed regularizer and then dive into various design choices made in it's formulation. Given an image $\mathbf{x}$ as input, the feature extractor F returns semantically meaningful features from it. We attempt to make the generator aware of this explicit semantic feature space. To this end, we freeze the feature extractor and treat it as a fixed function that maps from image space to a semantically meaningful feature space.

Given such a mapping of the generated image, we try to align the Generator's features of this image through a set of feature predictors. This alignment is inspired by BYOL [16]. As illustrated in Fig. 3, we attach a predictor branch q to the Generator G. The objective of q is to learn a mapping from generator's intermediate feature map $G^{\pi_G^i}(\mathbf{z})$ to pretrained feature extractor's intermediate feature $F^{\pi_F^i}(G(\mathbf{z}))$, where π_G in π_F denote the ordered set of layer numbers in the G and F at which we attach the predictors (ref. Eq. 2). We attach multiple such predictor networks q_i at different scales of generator.

$$\mathcal{L}_G = \mathop{\mathbb{E}}_{\mathbf{z}\sim P_z}[log(1 - D(G(\mathbf{z})))] + \sum_{i=0}^{|\pi_G^i|} \mathop{\mathbb{E}}_{\mathbf{z}\sim P_z} \|q(G^{\pi_G^i}(\mathbf{z})) - F^{\pi_F^i}(G(\mathbf{z}))\|_2^2 \quad (2)$$

3.3 Design Choices

We analyse the effect of our Hierarchical Semantic Regularizer (HSR) against different design choices. For this purpose, we choose AnimalFace-Dog dataset which consists of 389 images. Since this is a low-shot dataset, we use StyleGAN2-ADA as our baseline. We perform all our experiments on 256×256 resolution.

What Should be the Choice of Feature Extractor? For this analysis, we choose 5 different feature extractors. We take combinations of CNN or transformer based networks trained using either self-supervised or supervised classification objective. We take ResNet-50 as the CNN backbone for both self-supervised (DINO) and supervised networks. For transformer-based networks, we use ViT-DINO and DeiT. Apart from trained networks, we also consider a randomly initialized ViT for baseline comparison.

We find that all pretrained feature extractors when used through HSR loss lead to introduction of meaningful semantic features in the intermediate latent spaces of the Generator. This is evidenced by reduction of PPL Score in Table 1, which signifies reduction in non-meaningful generations from the GAN. The reduction in PPL also implies improved disentanglement [28] and linearity in the $\mathcal{W}$ space of the Generator, which is a desired property for many applications. We get $\geq 6.2\%$ improvement in the PPL score when guided by these networks. ViT DINO's features stand apart, by improving the PPL score by 19% over the baseline. This is also supported by recent findings of Amir *et al.* [5], where they show several inherent properties of features from ViT-DINO, that are useful for computer vision tasks. With these results, we fix ViT DINO as the choice of the feature extractor for the rest of the experiments.

Which Layers of Generator are More Important? The StyleGAN generator G generates images using 7 synthesis blocks: starting from 4×4, up to full resolution of 256×256. Of these, we consider synthesis blocks having features of resolution 8, 16, 32, 64. This corresponds to scaling down of resolution r to $\frac{r}{32}$, $\frac{r}{16}$, $\frac{r}{8}$, and $\frac{r}{4}$. We choose these scales as it largely corresponds to the scales of downsampling by each block in SoTA CNN architectures [19,49]. The first block of G (which have low resolution, but are responsible for high-level semantics) are supervised by the last block of the feature extractor (as they also are responsible for high-level semantics). Similarly, other blocks of G are supervised by the blocks of the feature extractor that bring out similar level of semantics.

To decide which layers contribute the most to the improvement in PPL, we divide the 4 blocks into 3 groups. The 3 groups specialize in high ($\frac{r}{32}$, $\frac{r}{16}$), mid ($\frac{r}{16}$, $\frac{r}{8}$), and low ($\frac{r}{8}$, $\frac{r}{4}$) level of semantics. We observe, in Table 2, that it is the supervision at low-level semantics which is most useful for the G. We observe a gradation in the improvement over the baseline, as high-level semantic supervision is least useful, followed by middle, and low. Overall, supervision at all levels turns out to cause the highest improvement.

Does Path Length Regularizer (PLR) Complement HSR? Path Length Regularizer (PLR) was introduced in StyleGAN2 [28]. The intuition behind PLR is to promote fixed magnitude non-zero change in the resulting image when moving by a fixed step size in the $\mathcal{W}+$-space. As reported in Table 4, we find

Table 1. Feature space ablation: Ablating over different feature extractors for usage in HSR. HSR with ViT-DINO's features gives best results.

	FID ↓	PPL↓
StyleGAN2-ADA	53.28	59.27
+ ViT (RandInit)	53.65	56.97
+ ResNet50 DINO [8]	54.33	55.6
+ DeiT [51]	53.22	54.71
+ ResNet50 [19]	52.88	52.23
+ ViT DINO [8]	**51.58**	**48.02**

Table 2. Level of semantics: A gradation in the improvement over the baseline is observed as we supervise from high-level semantics to low-level semantics. Best results are obtained when all the levels are supervised.

	FID↓	PPL↓
StyleGAN2-ADA	53.28	59.27
+ High-level ($\frac{r}{32}$, $\frac{r}{16}$)	53.15	57.73
+ Mid-level ($\frac{r}{16}$, $\frac{r}{8}$)	52.91	54.18
+ Low-level ($\frac{r}{8}$, $\frac{r}{4}$)	53.66	51.77
+ All levels	**51.58**	**48.02**

Table 3. FFHQ-140k Results: We report FID, Precision, Recall and PPL for different methods. With large data our method (SG2+HSR) produces better results even compared to SG2-ADA.

FFHQ-140k	FID↓	Precision↑	Recall↑	PPL↓
SG2	3.92	**0.68**	0.45	175.09
+ HSR	**3.74**	**0.68**	**0.48**	**144.59**
SG2-ADA	**4.30**	0.69	**0.40**	163.11
+ HSR	5.26	**0.70**	0.38	**131.41**

Table 4. Performance wrt PLR. PLR and HSR complement each other, while being equally effective individually.

PLR	HSR	FID↓	PPL↓
✗	✗	57.97	75.63
✓	✗	53.28	59.27
✗	✓	52.98	58.60
✓	✓	**51.58**	**48.02**

that HSR itself gives slightly better improvement than the PLR over the baseline. While the best effect is noted when both, PLR and HSR, are applied together.

Insight. PLR's objective is to improve latent space smoothness, which leads to better PPL. Since PPL and image quality (natural-ness of image) are correlated, applying PLR improves the image quality. Whereas in HSR, we enforce the generator to predict in a feature space learnt from natural images using a pretrained feature extractor as prior. We observe that this objective, which targets bringing feature space of generator closer to a "natural" feature space also leads to improvement in the smoothness of latent space, as measured by PPL. This shows that image quality and latent space smoothness are complementary and related concepts. Therefore, optimizing for both gives better PPL score.

4 Experiments

In this section, we demonstrate the effectiveness of HSR experimentally. We first describe the experimental setup for all our experiments. Then, we evaluate the quantitative performance on several real-world datasets of varying sizes. Finally, we show improved linearity of latent space through attribute editing.

Table 5. Results on Limited Data. We present results on different limited data cases for FFHQ (left) dataset and on real-world datasets (right). We apply our regularizer on the strong baseline of StyleGAN2+ADA which is designed for limited data. We observe a significant decrease in PPL over baselines which implies a smooth, disentangled and meaningful latent space, while preserving photorealism (comparable FID).

Dataset	Method	FID↓	PPL↓	Dataset	Method	FID↓	PPL↓
FFHQ-1k	StyleGAN2-ADA	**19.14**	98.79	AnimalFace	StyleGAN2-ADA	53.28	59.27
	+ HSR	21.76	**90.39**	Dog	+ HSR	**51.58**	**48.02**
FFHQ-2k	StyleGAN2-ADA	**14.74**	136.14	AnimalFace	StyleGAN2-ADA	**39.50**	50.76
	+ HSR	15.53	**115.38**	Cat	+ HSR	40.25	**40.75**
FFHQ-10k	StyleGAN2-ADA	**7.16**	164.21	CUB	StyleGAN2-ADA	**5.78**	265.46
	+ HSR	8.08	**126.35**		+ HSR	6.15	**237.81**

4.1 Experimental Setup

Datasets. We run our experiments on FFHQ [24] (70k images), AnimalFace-Dog (389 images), AnimalFace-Cat (160 images) [48], CUB200 (12k images) [53], and LSUN-church [57] (126k images) (ref. supl. mat.) datasets. We augment the datasets by taking the horizontal flip of every image, doubling the number of images in the original dataset. **Implementation Details.** We use StyleGAN2-ADA (SG2-ADA) as the baseline GAN, with its architecture for 256×256 images, with batch size of 16. Predictors q contain `Conv1x1-LeakyReLU-Conv1x1`, with hidden dimension of 4096. We make use of 2 A6000 GPUs for training our models.

4.2 Results

On standard full dataset of FFHQ, we compare over both StyleGAN2 [28] (SG2) and StyleGAN2-ADA [25]. This is because StyleGAN2 shows slightly better performance against the ADA variant on large datasets. We also evaluate our method for limited data sizes. Traditionally, GANs have shown to perform poorly on smaller datasets, until recently several approaches [23,25,55] have been proposed which enables GANs to learn well on limited data. We observe that irrespective of dataset size, asking the generator to be predictive of semantic features of rich feature extractors via HSR improves the smoothness of the latent space, as it is evident by an average relative improvement in PPL scores of about 14.2% on average in Table 5, while that of 17.42% in case of full FFHQ (ref. Table 3). This is also evident qualitatively in Fig. 4 and 6, where we observe an improved latent-to-image mapping even in bottom 10%-ile images, when ranked by PPL scores. We also present images sampled randomly in Fig. 5, where we observe the mitigation of the unnatural faces and artefacts (highlighted images) upon application of HSR. Thus, HSR raises the lower bound for the natural-ness of

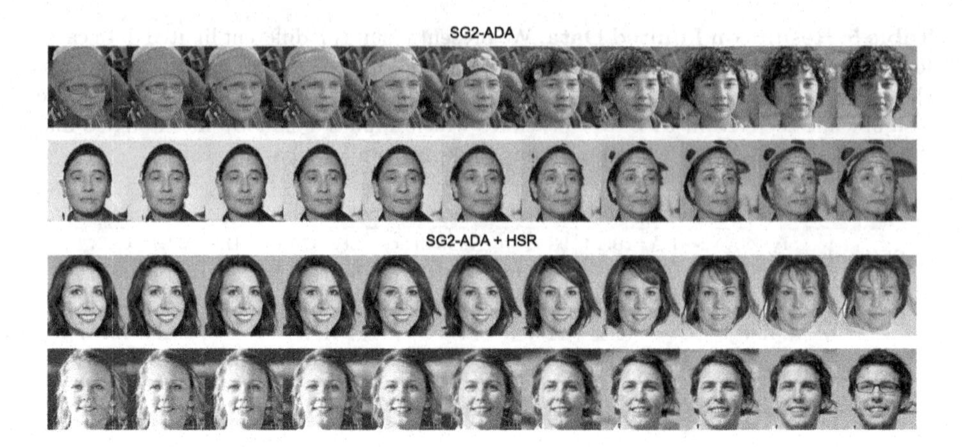

Fig. 4. Latent space interpolation of top 10-%ile images, ranked by PPL score. SG2-ADA images show traces of artifacts which are absent after applying HSR.

the images produced by a generator (also ref. Fig. 2).

4.3 Analysis of Linearity of Latent Space

Motivation. Latent space of a pre-trained StyleGAN has meaningful directions embedded in it. Shen *et al.* [45] shows that $\mathcal{W}+$ latent space is disentangled with respect to image semantics and there exist linear directions $\mathbf{d}$ in this space that control specific semantic attributes in the generated images. This is an important property of the latent space which is commonly used in controlled image synthesis [42] and image editing [1], as it leads to smooth interpolation between any two generated images. Furthermore, it is observed that the magnitude of latent transformations linearly correlates with the magnitude of the attribute changes in generated images [60]. Although, multiple works [1,21,42,45,54] are built upon this property to generate desired image transformations, there is no established metric to evaluate the extent of this linear correlation in the latent space. To this end, we propose a new metric called *Attribute Linearity Score (ALS)* for quantifying this linear correlation between the extent of latent transformations and the attribute changes.

Attribute Linearity Score (ALS). Let the attribute strength be given by attribute score (logit value) from a pretrained attribute classifier C [27]. Consider two latent codes $\mathbf{w_0}$ and $\mathbf{w_1} \in \mathcal{W}+$ and their corresponding generated images $G(\mathbf{w_0})$ and $G(\mathbf{w_1})$ (using the generator G). Convex combinations of $\mathbf{w_0}$ and $\mathbf{w_1}$ generate interpolated latent codes $\mathbf{w_t}$ (Eq. 3) on the line segment joining the two latent codes $\mathbf{w_0}$ and $\mathbf{w_1}$. Let the corresponding generated images be denoted by $G(\mathbf{w_t})$. Linearity of the latent space [60] with respect to the attribute strength C implies that the attribute score for the image $G(\mathbf{w_t})$ should be the same convex combination of the attribute strengths of $G(\mathbf{w_0})$ and $G(\mathbf{w_1})$ (Eq. 4).

Fig. 5. Comparison over uniformly random sampled images from StyeGAN2 and StyleGAN2+HSR. StyleGAN2 produces unnatural faces and artefacts over it such as peculiar eyeglasses (as shown in the highlighted images).

$$\mathbf{w_t} = \mathbf{w_0} + t * (\mathbf{w_1} - \mathbf{w_0}),\ t \in (0,1) \tag{3}$$

$$C(G(\mathbf{w_t})) \approx C(G(\mathbf{w_0})) + t * (C(G(\mathbf{w_1})) - C(G(\mathbf{w_0}))),\ t \in (0,1) \tag{4}$$

Consider the example shown in Fig. 7a, where we depict the transformation of the smile attribute. On the left, we show the plot of attribute scores with the interpolation parameter t using smile classifier C_s and on the right we show the image samples $G(\mathbf{w_t})$ for $t \in (0,1)$. A model with a linear latent space structure should have this plot close to the "ideal" (shown in dotted) straight line between the two end points. Similar plots are shown for the "smile" and "male" attributes in Fig. 7b. In both cases, we observe a significant departure from linearity.

The ALS score quantifies the deviation from the line segment defined in Eq. 4 using the mean squared error metric. To compute this, we first define a set of equally-spaced interpolation points $t \in \{0, \frac{1}{N}, \frac{2}{N}, \ldots, 1\}$. For each attribute $j \in \{1,\ldots,M\}$), the squared difference (Δ_{tj}) is computed using Eq. 5. The ALS score (Δ_T) is defined as the mean of Δ_{tj} over all M attributes and N interpolation points (*i.e.* $\Delta_T = \frac{1}{NM}\sum_{t=1}^{N}\sum_{j=1}^{M}\Delta_{tj}$).

$$\Delta_{tj} = ||C_j(G(\mathbf{w_t})) - C_j(G(\mathbf{w_0})) - t * (C_j(G(\mathbf{w_1})) - C_j(G(\mathbf{w_0})))||^2 \tag{5}$$

In the following sections, we first evaluate effect of linearity on applying HSR, by measuring ALS. Then we show it's application in measuring edits in images. We use StyleGAN2-ADA model as the baseline trained on FFHQ-10k for results in the rest of this section.

ALS Evaluation. Our proposed HSR is able to provide a smooth structure to the latent space which is evident by the lower ALS scores of our model.

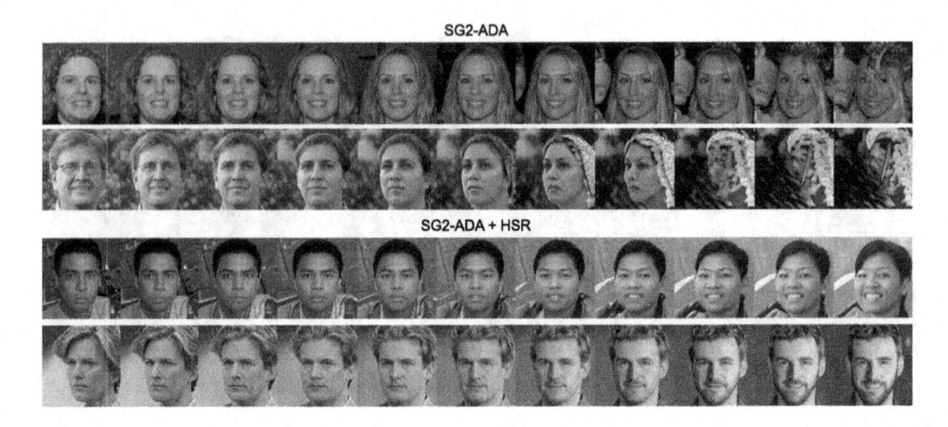

Fig. 6. Latent space interpolation of bottom 10-%ile images, ranked by PPL score. SG2-ADA latent space accommodates more unnatural images, leading to increase in PPL score. The latent space maps to more natural face-like images in SG2-ADA-HSR.

To further analyse the structure of the latent space we perform latent space interpolations and generate a sequence of images. To quantitatively evaluate the interpolation results, we used the proposed ALS scores for the interpolations. The lower ALS score represent the latent space is well structured and the magnitude of the attributes are linearly correlated with the latent transformation. The ALS scores for our model and baseline model in Table 8 for following set of popular attributes {gender, smile, age, hair, bangs, beard} [21,45,46]. Additionally, Fig. 8 (right) shows the variation of the mean attribute delta ($\Delta_{t,\cdot}$) with the interpolation parameter t. We can observe that in the middle region $t \in [0.4, 0.8]$ the baseline model has high deviation from linear behaviour, which is significantly less in our HSR regularized model. This is also seen quantitatively through proposed ALS-attribute score, in which our model outperforms baseline by **15%** of relative improvement. We can observe that the interpolations generated using the HSR results in smooth transitions and has high visual quality throughout the interpolation. The StyleGAN2+ADA model without HSR has sudden transitions in between and has some artifacts present (ref. Fig. 1).

Editability. The semantically rich structure of the latent space is widely used for performing semantic edits on the generated images [1,3,42,45,54,58]. For instance, if we have to add the attribute smile to the generated face image, one can edit the latent code as $\mathbf{w_{edit}} = \mathbf{w} + \alpha\mathbf{d}$ where α is edit strength and $\mathbf{d}$ is the direction for the smile attribute edit operation. However, often, the attribute scores of the edits performed by such methods does not change linearly with the edit strength parameter α as observed in Fig. 9. To this end, we perform the following experiment: Given an input source image I_0, we first perform attribute edit on it using latent space transformation to obtain I_1 using an existing approach [45]. Then, we use the latent code optimization to find the corresponding latent codes $\mathbf{w_0}$ and $\mathbf{w_1}$ in the latent space. Finally, we fol-

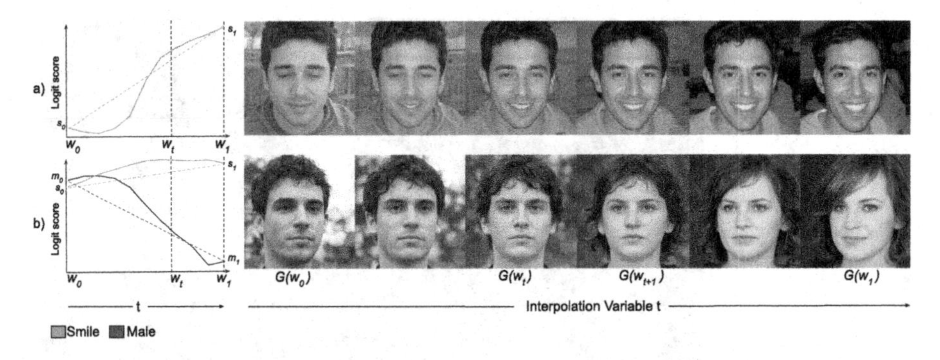

Fig. 7. Linearity of the latent space: Here we show the transition images generated by the intermediate latent code $\mathbf{w_t}$ in the right and the corresponding attribute scores s_t for for smile (row 1 and 2) and m_t for male attribute (row 2). For brevity we have written $s_t = C_s(G(\mathbf{w_t}))$ and $m_t = C_m(G(\mathbf{w_t}))$.

	Gender	Smile	Age	Hair	Bangs	Beard	Mean
SG2-ADA	1.38	1.48	1.18	1.96	1.95	1.60	1.59
+HSR	**1.12**	**0.99**	**1.15**	**1.87**	**1.62**	**1.16**	**1.32**

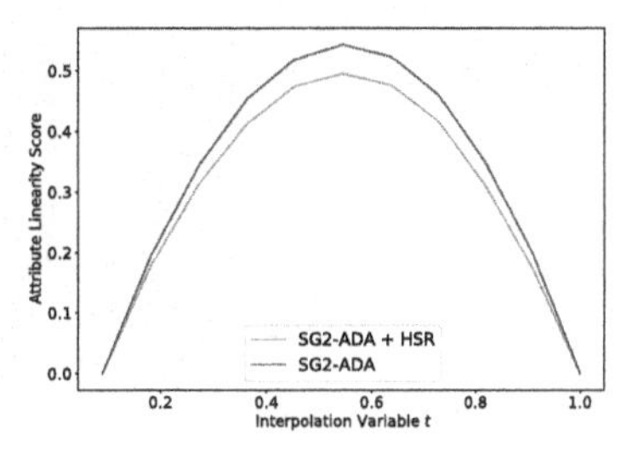

Fig. 8. ALS score comparison upon adding HSR. *(Right)*: Mean ALS computed for each value of the interpolation variable t. HSR is able to achieve a lower value of ALS supporting the linearity induced by ALS. *(Bottom)*: ALS score computed for all the face attributes separately.

lowed the same approach explained in Sect. 4.3 to generate intermediate images I_t using w_t for $t\{\in 0, \frac{1}{N}, \frac{2}{N}, ...1\}$. The results of the interpolation for edits are shown in Fig. 9. We compared StyleGAN2-ADA with and without HSR in this experiment. One can observe that in all the cases, adding HSR resulted in added linearity in the attribute scores plots. This property is highly desired in editing methods as it provides a fine-grained control over the attributes in the generated images. Also, observe that both the models evaluated are following the linear line closely in the first two examples. This suggests that the transitions along the age attribute is much more interpretable as it follows linearity. In all the three examples the model with HSR is able to approximate the linear line the better than the baseline without HSR. From the images, we can visually observe that the interpolations produced smooth transitions and their is no sudden jump in the attribute when using HSR. Also note that, the first and last images from both the models do not match "pixel perfectly", as they is generated by

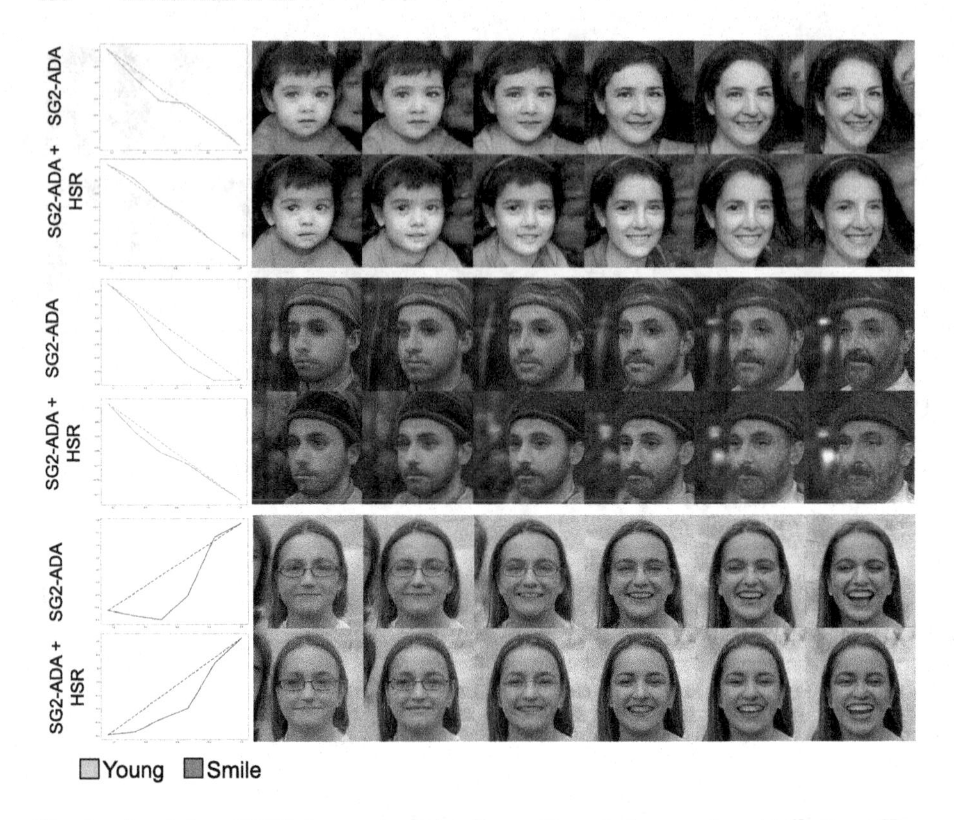

Fig. 9. Applying HSR improves the linearity of change in attributes. Here we show improved linearity for "Young" and "Smile" attributes. Plots show attribute score on Y-axis, interpolation variable t on X-axis.

optimization of latent code by different models (with and without HSR). Note that HSR improves the reconstruction quality of real images when embedded in the latent space using projection (ref. supl. mat.).

5 Conclusion

We proposed a novel, hierarchical semantic regularizer called HSR which allows us to regularize the latent representations in StyleGANs by aligning them to semantically rich ones learnt by state-of-the-art classifiers trained on large datasets. HSR is shown to significantly improve the quality of the generated images, especially those created via linear interpolation between attributes corresponding to real images. It further has a desirable property that the latent attribute space becomes more linear. To measure linearity, a novel metric *Attribute Linearity Score (ALS)* was introduced. Copious experiments on standard benchmarks validate the benefits of HSR and demonstrate statistically significant improvement in the quality of synthesized images. This leads us to interesting avenues for the future work: Enforcing structural priors (*e.g.* linear) in the latent space while training a GAN, which can lead to easier and fine-grained attribute editing.

Acknowledgements. This work was supported by MeitY (Ministry of Electronics and Information Technology) project (No. 4(16)2019-ITEA), Govt. of India and Uchhatar Avishkar Yojana (UAY) project (IISC_010), MHRD, India.

References

1. Abdal, R., Zhu, P., Mitra, N.J., Wonka, P.: StyleFlow: attribute-conditioned exploration of stylegan-generated images using conditional continuous normalizing flows. ACM Trans. Graph. (TOG) **40**(3), 1–21 (2021)
2. Adelson, E.H., Anderson, C.H., Bergen, J.R., Burt, P.J., Ogden, J.M.: Pyramid methods in image processing. RCA Engineer **29**(6), 33–41 (1984)
3. Alaluf, Y., Patashnik, O., Cohen-Or, D.: Only a matter of style: age transformation using a style-based regression model. ACM Trans. Graph. **40**(4), 1–12 (2021)
4. Albuquerque, I., Monteiro, J., Doan, T., Considine, B., Falk, T., Mitliagkas, I.: Multi-objective training of generative adversarial networks with multiple discriminators. In: ICML (2019)
5. Amir, S., Gandelsman, Y., Bagon, S., Dekel, T.: Deep ViT features as dense visual descriptors. CoRR abs/2112.05814 (2021), https://arxiv.org/abs/2112.05814
6. Arjovsky, M., Chintala, S., Bottou, L.: Wasserstein Generative Adversarial Networks. In: ICML (2017)
7. Brock, A., Donahue, J., Simonyan, K.: Large scale GAN training for high fidelity natural image synthesis. In: ICLR (2019)
8. Caron, M., Touvron, H., Misra, I., Jégou, H., Mairal, J., Bojanowski, P., Joulin, A.: Emerging properties in self-supervised vision transformers. In: Proceedings of the IEEE/CVF International Conference on Computer Vision (ICCV) (2021)
9. Chen, T., Kornblith, S., Norouzi, M., Hinton, G.: A simple framework for contrastive learning of visual representations. In: ICML (2020)
10. Chen, X., Fan, H., Girshick, R., He, K.: Improved baselines with momentum contrastive learning. arXiv preprint arXiv:2003.04297 (2020)
11. Choudhary, B.K., Sinha, N.K., Shanker, P.: Pyramid method in image processing. J. Inf. Syst. Commun. **3**(1), 269 (2012)
12. Deng, J., Dong, W., Socher, R., Li, L.J., Li, K., Fei-Fei, L.: ImageNet: a large-scale hierarchical image database. In: CVPR (2009)
13. Donahue, J., et al.: Decaf: a deep convolutional activation feature for generic visual recognition. In: ICML (2014)
14. Gatys, L.A., Ecker, A.S., Bethge, M.: Image style transfer using convolutional neural networks. In: 2016 IEEE Conference on Computer Vision and Pattern Recognition (CVPR) (2016)
15. Goodfellow, I., et al.: Generative adversarial nets. In: Ghahramani, Z., Welling, M., Cortes, C., Lawrence, N., Weinberger, K.Q. (eds.) Advances in Neural Information Processing Systems, vol. 27 (2014)
16. Grill, J.B., et al.: Bootstrap your own latent-a new approach to self-supervised learning. In: Advances in Neural Information Processing Systems (2020)
17. Gulrajani, I., Ahmed, F., Arjovsky, M., Dumoulin, V., Courville, A.C.: Improved training of Wasserstein GANs. In: Advances in neural information processing systems, vol. 30 (2017)

18. Gulrajani, I., Ahmed, F., Arjovsky, M., Dumoulin, V., Courville, A.C.: Improved training of Wasserstein GANs. In: NeurIPS (2017)
19. He, K., Zhang, X., Ren, S., Sun, J.: Deep residual learning for image recognition. In: CVPR (2016)
20. Huh, M., Agrawal, P., Efros, A.A.: What makes imagenet good for transfer learning? arXiv preprint arXiv:1608.08614 (2016)
21. Härkönen, E., Hertzmann, A., Lehtinen, J., Paris, S.: GANspace: discovering interpretable GAN controls. In: Proc. NeurIPS (2020)
22. Jeong, J., Shin, J.: Training GANs with stronger augmentations via contrastive discriminator. In: International Conference on Learning Representations (2021)
23. Jiang, L., Dai, B., Wu, W., Loy, C.C.: Deceive D: Adaptive Pseudo Augmentation for GAN training with limited data. In: NeurIPS (2021)
24. Karras, T., Aila, T., Laine, S., Lehtinen, J.: Progressive growing of GANs for improved quality, stability, and variation. In: ICLR (2018)
25. Karras, T., Aittala, M., Hellsten, J., Laine, S., Lehtinen, J., Aila, T.: Training generative adversarial networks with limited data. In: Proceedings NeurIPS (2020)
26. Karras, T., et al.: Alias-free generative adversarial networks. In: NeurIPS (2021)
27. Karras, T., Laine, S., Aila, T.: A style-based generator architecture for generative adversarial networks. In: CVPR (2019)
28. Karras, T., Laine, S., Aittala, M., Hellsten, J., Lehtinen, J., Aila, T.: Analyzing and improving the image quality of styleGAN. In: Proceedings of the IEEE/CVF Conference on Computer Vision and Pattern Recognition (CVPR) (2020)
29. Karras, T., Laine, S., Aittala, M., Hellsten, J., Lehtinen, J., Aila, T.: Analyzing and improving the image quality of styleGAN. In: CVPR (2020)
30. Kingma, D.P., Dhariwal, P.: Glow: generative flow with invertible 1x1 convolutions. In: Advances in Neural Information Processing Systems (2018)
31. Kornblith, S., Shlens, J., Le, Q.V.: Do better ImageNet models transfer better? In: CVPR (2019)
32. Kumari, N., Zhang, R., Shechtman, E., Zhu, J.Y.: Ensembling off-the-shelf models for gan training. arXiv preprint arXiv:2112.09130 (2021)
33. Liu, B., Zhu, Y., Song, K., Elgammal, A.: Towards faster and stabilized GAN training for high-fidelity few-shot image synthesis. In: International Conference on Learning Representations (2020)
34. Mao, X., Li, Q., Xie, H., Lau, R.Y., Wang, Z., Smolley, S.P.: Least squares generative adversarial networks. In: CVPR (2017)
35. Mescheder, L., Geiger, A., Nowozin, S.: Which training methods for GANs do actually converge? In: ICML (2018)
36. Miyato, T., Kataoka, T., Koyama, M., Yoshida, Y.: Spectral normalization for generative adversarial networks. arXiv preprint arXiv:1802.05957 (2018)
37. Miyato, T., Koyama, M.: cGANs with projection discriminator. In: ICLR (2018)
38. Mo, S., Cho, M., Shin, J.: Freeze the discriminator: a simple baseline for fine-tuning GANs. In: CVPRW (2020)
39. Noguchi, A., Harada, T.: Image generation from small datasets via batch statistics adaptation. In: ICCV (2019)
40. Ojha, U., et al.: Few-shot image generation via cross-domain correspondence. In: CVPR (2021)
41. Patashnik, O., Wu, Z., Shechtman, E., Cohen-Or, D., Lischinski, D.: StyleCLIP: text-driven manipulation of styleGAN imagery. In: Proceedings of the IEEE/CVF International Conference on Computer Vision (ICCV) (2021)
42. Patashnik, O., Wu, Z., Shechtman, E., Cohen-Or, D., Lischinski, D.: StyleCLIP: text-driven manipulation of styleGAN imagery. In: ICCV (2021)

43. Radford, A., et al.: Learning transferable visual models from natural language supervision. In: ICML (2021)
44. Radford, A., Metz, L., Chintala, S.: Unsupervised representation learning with deep convolutional generative adversarial networks. In: ICLR (2016)
45. Shen, Y., Yang, C., Tang, X., Zhou, B.: InterFaceGAN: interpreting the disentangled face representation learned by GANs. IEEE TPAMI (2020)
46. Shen, Y., Zhou, B.: Closed-form factorization of latent semantics in GANs. In: CVPR (2021)
47. Shocher, A., et al.: Semantic pyramid for image generation. In: CVPR (2020)
48. Si, Z., Zhu, S.C.: Learning hybrid image templates (HIT) by information projection. PAMI **34**(7), 1354–1367 (2011)
49. Simonyan, K., Zisserman, A.: Very deep convolutional networks for large-scale image recognition. In: ICLR (2015)
50. Tan, M., Le, Q.: EfficientNet: rethinking model scaling for convolutional neural networks. In: ICML (2019)
51. Touvron, H., Cord, M., Douze, M., Massa, F., Sablayrolles, A., Jegou, H.: Training data-efficient image transformers & distillation through attention. In: International Conference on Machine Learning (2021)
52. Tzeng, E., Hoffman, J., Saenko, K., Darrell, T.: Adversarial discriminative domain adaptation. In: CVPR (2017)
53. Wah, C., Branson, S., Welinder, P., Perona, P., Belongie, S.: Caltech-ucsd birds-200-2011 (cub-200-2011). Technical Report CNS-TR-2011-001, California Institute of Technology (2011)
54. Wu, Z., Lischinski, D., Shechtman, E.: StyleSpace analysis: disentangled controls for styleGAN image generation. In: Proceedings of the IEEE/CVF Conference on Computer Vision and Pattern Recognition (2021)
55. Yang, C., Shen, Y., Xu, Y., Zhou, B.: Data-efficient instance generation from instance discrimination. In: Advances in Neural Information Processing Systems (2021)
56. Yosinski, J., Clune, J., Bengio, Y., Lipson, H.: How transferable are features in deep neural networks? In: NeurIPS (2014)
57. Yu, F., Zhang, Y., Song, S., Seff, A., Xiao, J.: LSUN: construction of a large-scale image dataset using deep learning with humans in the loop. arXiv preprint arXiv:1506.03365 (2015)
58. Yüksel, O.K., Simsar, E., Er, E.G., Yanardag, P.: LatentCLR: a contrastive learning approach for unsupervised discovery of interpretable directions. In: Proceedings of the IEEE/CVF International Conference on Computer Vision (2021)
59. Zhang, R., Isola, P., Efros, A.A., Shechtman, E., Wang, O.: The unreasonable effectiveness of deep features as a perceptual metric. In: CVPR (2018)
60. Zhao, S., et al.: Large scale image completion via co-modulated generative adversarial networks. In: ICLR (2021)
61. Zhao, S., Liu, Z., Lin, J., Zhu, J.Y., Han, S.: Differentiable augmentation for data-efficient GAN training. In: NeurIPS (2020)

IntereStyle: Encoding an Interest Region for Robust StyleGAN Inversion

Seung-Jun Moon[1] and Gyeong-Moon Park[2(✉)]

[1] KLleon Tech., Seoul, South Korea
seungjun.moon@klleon.io
[2] Kyung Hee University, Seoul, South Korea
gmpark@khu.ac.kr

Abstract. Recently, manipulation of real-world images has been highly elaborated along with the development of Generative Adversarial Networks (GANs) and corresponding encoders, which embed real-world images into the latent space. However, designing encoders of GAN still remains a challenging task due to the trade-off between distortion and perception. In this paper, we point out that the existing encoders try to lower the distortion not only on the interest region, *e.g.*, human facial region but also on the uninterest region, *e.g.*, background patterns and obstacles. However, most uninterest regions in real-world images are located at out-of-distribution (OOD), which are infeasible to be ideally reconstructed by generative models. Moreover, we empirically find that the uninterest region overlapped with the interest region can mangle the original feature of the interest region, *e.g.*, a microphone overlapped with a facial region is inverted into the white beard. As a result, lowering the distortion of the whole image while maintaining the perceptual quality is very challenging. To overcome this trade-off, we propose a simple yet effective encoder training scheme, coined IntereStyle, which facilitates encoding by focusing on the interest region. IntereStyle steers the encoder to disentangle the encodings of the interest and uninterest regions. To this end, we filter the information of the uninterest region iteratively to regulate the negative impact of the uninterest region. We demonstrate that IntereStyle achieves both lower distortion and higher perceptual quality compared to the existing state-of-the-art encoders. Especially, our model robustly conserves features of the original images, which shows the robust image editing and style mixing results. We will release our code with the pre-trained model after the review.

Keywords: StyleGAN · Robust GAN inversion · Interest region · Interest disentanglement · Uninterest filter

1 Introduction

Recently, as Generative Adversarial Networks (GANs) [15] have been remarkably developed, real-world image editing through latent manipulation has been

Supplementary Information The online version contains supplementary material available at https://doi.org/10.1007/978-3-031-19784-0_27.

S. Avidan et al. (Eds.): ECCV 2022, LNCS 13675, pp. 460–476, 2022.
https://doi.org/10.1007/978-3-031-19784-0_27

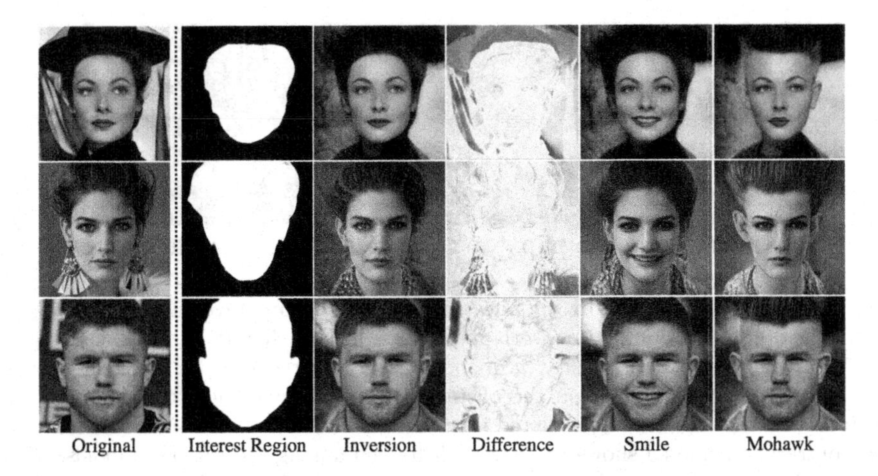

Fig. 1. Encoding of IntereStyle. Original images, interest region, inversion, the difference between the original images and their inversions, and the editing results (smile and Mohawk). The magnitude of the difference between the original and inversion images is colored in red. Our model successfully minimizes distortion on the interest region, even without the interest region mask for the inference. Moreover, even with low distortion, our model shows high editability. (Color figure online)

prevalent [25,27,28,30,32,33]. Especially, the strong disentangled property of StyleGAN [18,19] latent space, *i.e.*, W, enables scrupulous image editing [22,32], which can change only desirable features, *e.g.*, facial expression, while maintaining the others, *e.g.*, identity and hairstyle. For editing the image precisely with StyleGAN, it is required to get the suitable style latent, from which StyleGAN can reconstruct the image that has low distortion, high perceptual quality, and editability without deforming the feature of the original image.

Though StyleGAN is generally known to construct the image with high perceptual quality, the original style space W is not enough to represent every real-world image with low distortion. Consequently, a vast majority of recent StyleGAN encoders, including optimization-based methods, embed images into $W+$ space [1,3,24]. W uses the identical style vector for every layer in StyleGAN, obtained by the mapping function. On the other hand, $W+$ space provides a different style vector per layer and can even provide a random style vector in $\mathbb{R}^{512}$. However, as the distribution of style latent is far from W, reconstructed images show low perceptual quality and editability [25,30]. Consequently, lowering the distortion while keeping the other factors is still challenging.

In this paper, we claim that training to lower distortion on the entire region of the image directly is undesirable. In most cases, images contain regions that cannot be generated due to the inherent limitation of generators. Figure 1 shows clear examples of real-world images in the facial domain, which contain regions that are infeasible to be generated, *e.g.*, hats, accessories, and noisy backgrounds. Therefore, an encoder needs to concentrate on the generable region for inver-

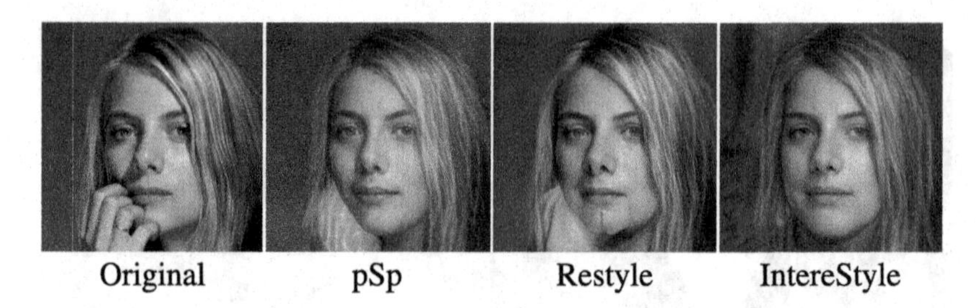

Fig. 2. Lowering distortion on the uninterest region. Inversion results of *pSp* [24], Restyle [3], and ours. An overlapped obstacle (*i.e.*, hand) on the facial region precludes clean inversion. Firstly, *pSp* shows high distortion on the eyes and generates unrealistic facial shapes on the obstacle region. Restyle tries to reconstruct the obstacle region, but the reconstructed image shows artifacts on the nose and chin. On the contrary, our model shows the lowest distortion among the existing models, while maintaining high perceptual quality as shown above.

sion while ignoring the un-generable region (*e.g.*, non-facial region for StyleGAN trained with FFHQ). This strategy helps the latents inverted from the generable region to be close to W, which show high perceptual quality and editability, as shown in Fig. 1.

Another observation is that an attempt to reconstruct the region which is not generable induces severe distortion even on the other generable regions. For example, in Fig. 2, a hand overlapped with the facial region is not generable by GAN encoders. Restyle [3], which shows the lowest distortion among all encoder-based inversion models until now, tries to lower distortion on the hand too, which rather causes catastrophic distortions on the nose and chin.

In the light of these observations, it is important to distinguish the region to be reconstructed elaborately from the rest. Here we define the term *interest region*, where the model focuses on the precise reconstruction with low distortion and high perceptual quality. Practically, in most cases, the interest region is aligned with the generable region of the image. For example, in facial image generation, the main interests are the face and hair parts of the output images, which are easier to generate than backgrounds. By focusing on the interest region, we can reduce distortion without any additional task, such as an attempt to encode latent excessively far from W [8].

Contributions. We propose a simple yet effective method for training a StyleGAN encoder, coined IntereStyle, which steers the encoder to invert given images by focusing on the interest region. In particular, we introduce two novel training schemes for the StyleGAN encoder: (a) Interest Disentanglement (InD) and (b) Uninterest Filter (UnF). First, InD precludes the style latent, which includes the information on the uninterest region, from distorting the inversion result of the interest region. Second, UnF filters the information of the uninterest region, which prevents our model from redundantly attending to the uninterest region. UnF boosts the effect of InD by forcing the model not to focus on the uninterest region overly. In addition, we propose a very simple yet effective scheme for determining the interest region, required only at the training stage.

We demonstrate that IntereStyle, combined with the iterative refinement [3], effectively reduces the distortion at the interest region of CelebA-HQ-test dataset. To the best of our knowledge, IntereStyle achieves the lowest distortion among the existing state-of-the-art encoder-based StyleGAN inversion models without generator tuning. Moreover, we qualitatively show that our model robustly preserves features of the original images even with overlapped obstacles, while other baselines fail to. Lastly, we show the experimental results for image editing via InterFaceGAN [28], StyleCLIP [22], and style mixing [18] results, where our model shows remarkably robust outputs when input images contain significant noises, *e.g.*, obstacles on the face.

2 Related Work

GAN Inversion. GAN inversion aims to transform given real-world images into latent vectors from which a pre-trained GAN model can reconstruct the original image. In the early stage of GAN inversion, the majority of models rely partially [4,5,35,36] or entirely [1,2,10–12,16,23,26] on the optimization steps per image. Though the optimization-based models show high inversion qualities, these models should perform numerous optimization steps per every input image [20], which are extremely time-consuming. Thus, training encoders that map images into the latent space has been prevalent to invert images in the real-time domain [3,24,29–31,35]. However, regardless of encoding methods, the existing state-of-the-art GAN inversion models focus on the whole region of images [2,3,24,30], including both interest and uninterest regions. We propose that focusing mainly on the interest region during GAN inversion improves the perceptual quality and editability of inverted images.

GAN Inversion Trade-Off. The desirable GAN inversion should consider both distortion and perceptual quality of inverted images [6,25,30]. However, due to the trade-off between two features, maintaining low distortion while enhancing the perceptual quality remains a challenging task [25,30]. Especially in StyleGAN, an inverted image from the latent far from W distribution achieves lower distortion [1,2,24] but shows lower perceptual quality [30] than the image from W distribution. Moreover, the latent far from W distribution shows lower editability [25,30], which makes editing the inverted image harder. Here the variance among latents for all layers can be used as an indicator of distance from W, where W shows a zero variance due to the identical latent per layer.[1] As shown

[1] Technically, we should identify whether the latents are from the mapping network of StyleGAN or not, but for simplicity, we only use the variance.

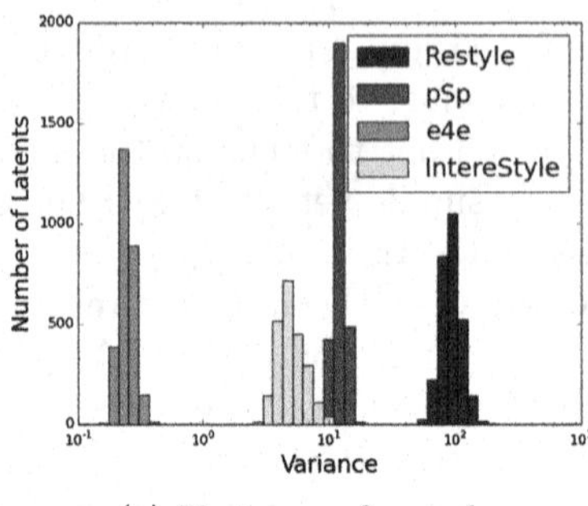

(a) Variance of encoders

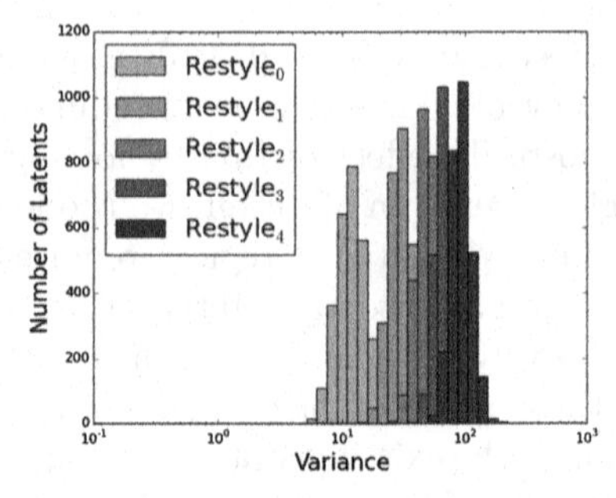

(b) Variance of Restyle per iteration

Fig. 3. Variance of latents from each encoder. We plot the variance of latents, derived from 2,800 CelebA-HQ test images with each encoder-based StyleGAN inversion model. The existing iterative refinement-based model, Restyle, shows relatively high variance among style latents per all layers. The variance of Restyle latents at iteration i, Restyle_i, increases along with i. *e4e*, which engages variation minimization loss term at the training scheme, shows relatively lower variance. We plot with a log-scale x-axis for better visualization.

in Fig. 3, the existing StyleGAN inversion models that show low distortion but suffer from low perceptual quality, *e.g.*, *pSp* [24] and Restyle [3], show relatively high variance among latents for all layers of StyleGAN. Especially, Fig. 3b shows that Restyle gradually increases the variance of latents as the iteration refinement progresses. In the case of *e4e* [30], it encodes images into latents close to W but with high distortion. In contrast to the existing methods, our model focuses on lowering distortion at the interest region, *i.e.*, hair and face. Since it is much easier than lowering at the uninterest region, *i.e.*, irregular backgrounds, hats, and accessories, our model successfully achieves lower distortion than the existing models while avoiding the drop of high perceptual quality.

3 Method

In this section, we propose a simple yet effective StyleGAN encoder training scheme named IntereStyle. We first introduce our notation and the model architecture. Then, we introduce how to determine the interest region in input images. Next, we propose two novel methods: interest disentanglement and uninterest filter in Sect. 3.3 and Sect. 3.4, respectively. Finally, we describe the whole training scheme of IntereStyle in Sect. 3.5.

3.1 Notation and Architecture

Our architecture is shown in Fig. 4, which is based on *pSp* [24] model, combined with the iterative refinement [3,31]. At i-th iteration of the iterative refinement, our encoder E receives a latent calculated at the previous step, $\mathbf{w}_{i-1}$[2], together

[2] When $i = 1$, we utilize $\mathbf{w}_0$ as an average latent of StyleGAN.

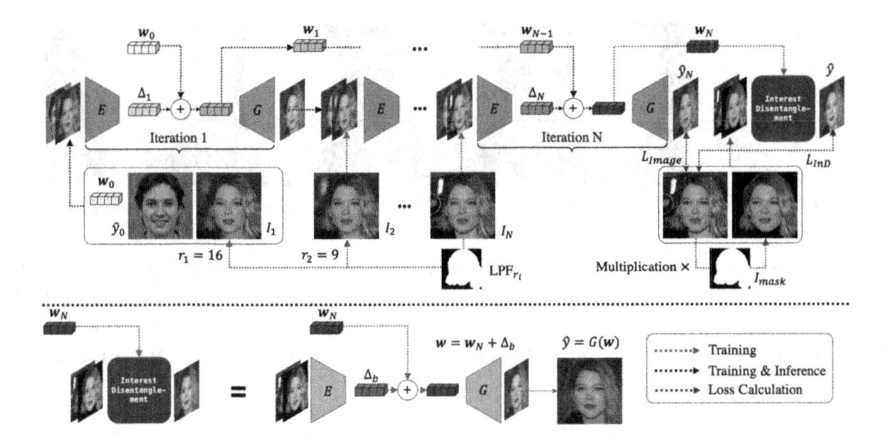

Fig. 4. Overall structure of IntereStyle. InterеStyle trained with total of N-th iteration, receives N images, *i.e.*, $I_1, I_2, ..., I_N$. I_i is an original image passed through a low-pass filter with radius r_i, which steers the model to focus on the coarse features at the early step of iterations, and I_i gets clearer as i grows. The Encoder targets to embed the difference of two images I_i and $\hat{y}_{i-1}$ into the latent space, Δ_{i-1}. After the N-th iteration is finished, We apply interest disentanglement. We multiply $\hat{y}_N$ with the I_{mask}, to wipe out the uninterest region. We yield Δ_b, by computing the difference between I and $I \cdot I_{mask}$. By adding Δ_b to the obtained latent w_N, we yield the output $\hat{y}$. At the inference stage, we yield the final output $\hat{y}_N$, without applying interest disentanglement.

with a pair of images. The pair consists of $(\hat{y}_{i-1}, I_i)$, where $\hat{y}_{i-1}$ is a decoded result of $\mathbf{w}_{i-1}$ via generator G, *i.e.*, $\hat{y}_{i-1} = G(\mathbf{w}_{i-1})$, and I_i is a preprocessed ground truth image, I, by our proposed method in Sect. 3.4. E targets to encode the difference between $\hat{y}_{i-1}$ and I_i into the latent, Δ_i. Consequently, G can yield an image $\hat{y}_i$, which is more similar to I_i than $\hat{y}_{i-1}$, by decoding the latent $\mathbf{w}_i = \mathbf{w}_{i-1} + \Delta_i$. Our model iteratively refines the latent with a total of N iterations. Finally, we utilize a loss function L_{image} for training, consisting of the weighted sum of L_2, $LPIPS$ [34], and L_{ID} [24]. We explain the details of each loss in Appendix A.

3.2 Interest Region

To guide the model to focus on the interest region, we need to label the interest region first. The interest region can be designated arbitrarily according to the usage of the inversion model. For instance, facial and hair regions for the facial domain, and the whole body for the animal domain can be set as the interest region. For labeling this interest region, the off-the-shelf segmentation networks are used, which is described in Sect. 4. However, directly using the segmentation masks from networks causes the distortion of facial boundaries in the generated

Fig. 5. Mask dilation. We compare the interest region obtained by the raw mask and the dilated mask. In the raw mask, the interest region excludes the information related to the facial boundary, which occurs distortion of a facial shape. Consequently, we dilate the mask to force the interest region to include the facial boundary information, as shown in $I \cdot$ Dilated Mask.

image. To accommodate the boundary information, we use the dilated segmentation mask containing the interest region, as shown in Fig. 5. Without dilation, the loss term on the interest region cannot penalize the inverted face on the uninterest region. Consequently, we dilate the mask to penalize the overflowed reconstruction of the interest region boundary. Though the small part of the uninterest region would be included in the interest region through the dilation, we empirically find that our model still precisely generates the interest region without any distortion of boundaries. We visually show the effect of mask dilation at the ablation study in Sect. 4.1.

3.3 Interest Disentanglement

To enforce the model to invert the interest region into the latent space precisely, we should train the model to concentrate on the interest region regardless of the uninterest region. However, due to the spatial-agnostic feature of Adaptive Instance Normalization (AdaIN) [21], inverted style latents considering the uninterest region may deteriorate the inversion quality of the interest region. To prevent the encoded style of the uninterest region from deforming the inverted result of the interest region, the inversion of each region should be *disentangled.*

As E encodes the difference of the input pair of images in ideal, the latent Δ obtained by encoding the pair of images that only differ on the uninterest region does not contain the information of the interest region. In other words, the decoding results from the latents $\mathbf{w}$ and $\mathbf{w} + \Delta$ should be the same on the interest region. Motivated by the above, we propose a simple method named *Interest Disentanglement* (InD) to distinguish the inversions of the interest region and uninterest region. We construct the pair of input images for InD as follows: the original image I, and the same I but multiplied with the interest region mask. Then, as shown in Fig. 4, we can yield the pair of images which only differs in the uninterest region, and the corresponding latent from E, Δ_b. Ideally, the information in Δ_b is solely related to the uninterest region, which implies $\mathbf{w}_N$

generates the interest region robustly even Δ_b is added. Consequently, we define InD loss, L_{InD} as follows;

$$L_{InD} := L_{image}(I \cdot I_{mask}, \hat{y} \cdot I_{mask}), \quad (1)$$

where $\hat{y}$ is the inversion result from the latent $\mathbf{w} = \mathbf{w}_N + \Delta_b$. We apply Interest Disentanglement only at the training stage, which enables the inference without the interest region mask. We empirically find that IntereStyle focuses on the interest region without any prior mask given, after the training.

3.4 Uninterest Filter

At the early steps of the iterative refinement, E focuses on reducing the distortion of the uninterest region [3]. Due to the spatial-agnostic feature of AdaIN, we claim that excessively focusing on the uninterest region hinders the inversion of the interest region. We propose a method named *Uninterest Filter* (UnF), to make E concentrate on the interest region at every iteration consistently. UnF eliminates details of the uninterest region, which is inherently infeasible to reconstruct. Thus, E can reduce the redundant attention on the uninterest region for the low distortion. In detail, UnF eases calculating Δ_i by blurring the uninterest region of I at each iteration, with a low-pass Gaussian Filter with radius r, LPF_r. As shown in Fig. 4, UnF *gradually* reduces the radius of Gaussian filter of LPF as iterations progress, with the following two reasons; First, the redundant attention on the uninterest region is considerably severe at the early stage of the iterations [3]. Consequently, we should blur the image at the early iterations heavily. Second, excessive blur results in the severe texture difference between the interest and uninterest region. We claim that E is implicitly trained to encode the difference of the whole region, which can be biased to produce the blurred region when the blurred images are consistently given. For the realistic generation, the input at the N-th iteration, I_N is deblurred, *i.e.*, identical to I. We calculate the input image at the i-th iteration as below:

$$I_i = \begin{cases} I \cdot I_{mask} + LPF_{r_i}(I \cdot (1 - I_{mask})), & 0 < i < N \\ I, & i = N. \end{cases} \quad (2)$$

3.5 Training IntereStyle

At the training stage, we jointly train the model with the off-the-shelf encoder training loss [24] L_{image} and L_{InD}. However, in contrast to Restyle [3], which back-propagates N times per batch, we back-propagate only once after the N-th iteration is over. Thus, ours show relatively faster training speed compared to Restyle. Our final training loss is defined as below:

$$L_{total} := L_{image}(I \cdot I_{mask}, \hat{y}_N \cdot I_{mask}) + \lambda L_{InD}. \quad (3)$$

Our proposed methods, InD and UnF are synergistic at the training; While InD disentangles the inversion of the uninterest region, UnF forces to look at the interest region. Though applying L_{image} to the images multiplied with I_{mask} inherently drives E to focus on the interest region, InD is essential for robust training. Without L_{InD}, we find E implicitly contains information of the uninterest region into Δ_i, which affects the inversion of the interest region by AdaIN.

4 Experiments

In this section, we briefly introduce our datasets and baselines first. The implementation details are described in Appendix A. Then, we compare the inversion results with baselines and ablation scenarios, both in qualitative and quantitative ways. Next, we compare the image manipulation of our model, together with baselines. Finally, we look into the iterative scheme of our model with Restyle. Though we mainly show the results on the facial domain, we note that our method shows remarkable results in various domains. We show the experimental results on the animal domain in Fig. 6 briefly and the plenty experimental results in Appendix D.

Datasets. For the facial domain, we trained the encoder using the FFHQ dataset, consisting of 70,000 human facial images. For the validation, we used the CelebA-HQ test set, consisting of 2,800 human facial images. We did not add or change any alignment or augmentation procedures compared to the existing encoder training methods [3,24,30] for a fair comparison. To generate the interest and uninterest region masks, we used the pre-trained Graphonomy [14] model. For the animal domain, we used AFHQ wild dataset [9] for training and validation, which consists of 4,730 and 500 images, respectively. We used the pre-trained DEtection TRansformer (DE-TR; [7]) for obtaining the interest region.

Baselines. We compared our model with the several well-known StyleGAN encoders: IDGI [35], *pSp* [24], *e4e* [30], and Restyle [3]. Moreover, in the case of the qualitative comparison of inversion, we additionally compared it with the optimization-based model [1,2], which is well-known for its outstanding performance. For the baseline models, we used the pre-trained weights that are publicly available for evaluation. Please refer to Appendix C for more detailed information of each baseline.

4.1 Inversion Evaluation

Qualitative Evaluation. Figure 6 shows the inverted images of IntereStyle and two StyleGAN inversion baselines, Image2StyleGAN [1] and Restyle [3]. In this figure, we show the entire inversion results along with the cropped images, which correspond to the areas of overlapping of the interest and uninterest regions. Without our robust inversion schemes, mere attempts to lower distortion over the entire region often occurred severe artifacts or feature deformations. Indeed,

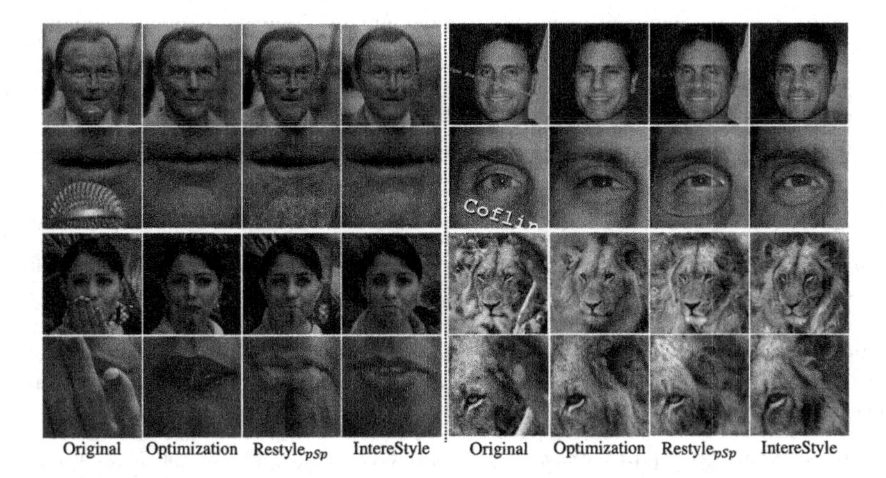

Fig. 6. Qualitative comparison. Comparison of various StyleGAN inversion methods. IntereStyle effectively disentangled the uninterest regions (*e.g.*, mic, letter, fingers and wood) from the interest region, which enabled robust handling of artifacts. However, the baseline models suffered from artifacts, which significantly deformed the feature of original images. Best viewed in zoom-in.

Table 1. Quantitative comparison. We calculated each result by multiplying the mask, *i.e.*, interest and facial masks, for the exact comparison of inversion quality of each region. IntereStyle showed the lowest L_2 and LPIPS on both the interest and facial regions among the state-of-the-art StyleGAN inversion models. Moreover, IntereStyle showed the best ID similarity among the baselines.

Model region	IDGI		pSp		e4e		Restyle$_{pSp}$		Restyle$_{e4e}$		IntereStyle	
	Interest	Face	Interest	Face	Interest	Face	Interest	Face	Interest	Face	Interest	Face
L_2	0.030	0.010	0.018	0.006	0.023	0.007	0.015	0.004	0.021	0.007	**0.013**	**0.003**
LPIPS	0.116	0.053	0.095	0.046	0.111	0.051	0.088	0.040	0.109	0.054	**0.075**	**0.036**
ID Similarity	0.18		0.56		0.47		0.65		0.51		**0.68**	

the baselines produced artifacts, which severely lower the perceptual quality. In addition, they mangled features of original images in some cases. For instance, Restyle$_{pSp}$ turned the microphone into a white beard, which does not exist in the original image. The optimization-based inversion relatively mitigated artifacts among the baselines, but still suffered from them. Moreover, it required more than 200 times longer inference time compared to the one of IntereStyle. In contrast, IntereStyle showed the most robust outputs compared to all baselines, which best preserved the features of original images without artifacts.

In Fig. 7, we qualitatively showed the effectiveness of the mask dilation. Without dilation, the model could not precisely reconstruct the original boundary of the interest region, which was denoted as the red region. In contrast, with the mask dilation, our model minimized the boundary deformation.

Fig. 7. Ablation study on dilated mask. Comparison of the results in the cases of using raw mask and dilated mask. For measuring the equivalence of the facial boundaries between the original and inverted images, we used Graphonomy to obtain the facial region of each image. We then calculated the difference of each facial region from the original one, denoted by red on the right side. While the model without the dilated mask could not reconstruct the exact boundary, our model minimized this error effectively.

Table 2. Ablation study. Performance comparison by adding each component of IntereStyle. InD and UnF contributed to the improvement of the model, which showed better results than naïvely applying L_{image} on the interest region. Especially, InD is advantageous for reducing L_2, and UnF mainly reduces LPIPS.

Method	L_2		LPIPS	
	Interest	Face	Interest	Face
Baseline Restyle [3]	0.015	0.004	0.088	0.040
+ L_{image} on the interest region	0.013	0.005	0.084	0.038
+ Interest Disentanglement (InD)	**0.012**	**0.003**	0.078	0.037
+ Uninterest Filter (UnF)	0.013	**0.003**	**0.075**	**0.036**

Quantitative Evaluation. We used L_2 and LPIPS [34] losses and measured ID similarity [24] by utilizing the pre-trained Curricularface [17], which shows the state-of-the-art performance on facial recognition. We measured the quality on the interest region, the facial and hair regions in this paper, which need to be inverted precisely. To this end, we multiplied the interest region mask at the calculation of ID similarity to preclude the facial recognition model from being affected by the non-facial region. Since the facial recognition performance is dependent on the features of the non-facial region [13], the inverted images are prone to be identified as similar faces with the original images due to the resemblance of non-facial regions. To compare the models solely on the facial region, we should wipe out the uninterest region.

As shown in Table 1, IntereStyle showed low distortion on both the interest and facial regions and preserved the identity well simultaneously. We conclude that focusing on the interest region is indeed helpful for robust inversion. Table 2 shows the ablation study by sequentially applying each component of our method to measure the effectiveness of the model performance. InD reduced the negative effect of the uninterest region, which indeed lowered distortion, compared to

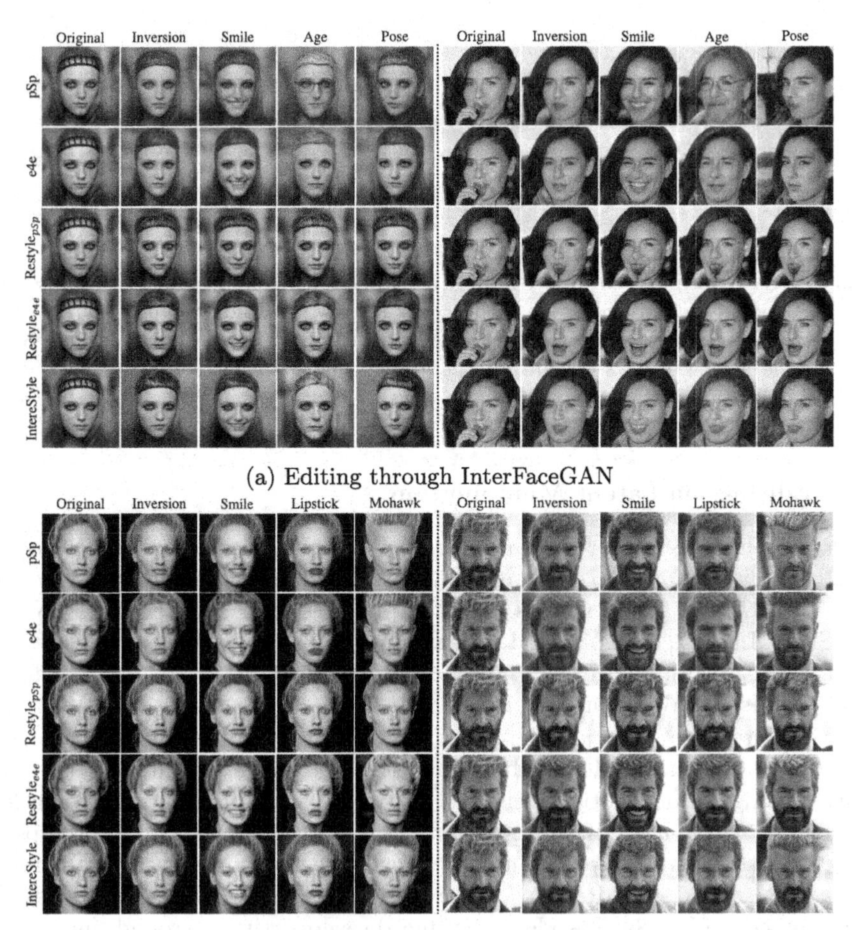

(a) Editing through InterFaceGAN

(b) Editing through StyleCLIP

Fig. 8. Editing comparison. Edited images via (a) InterFaceGAN and (b) StyleCLIP. Our model showed robust inversion results together with high editability consistently, while the baselines failed in various cases. First of all, *pSp* and *e4e* failed to invert robustly, which ignored makeups and the detailed appearance of each image. In the cases of Restyle$_{pSp}$ and Restyle$_{e4e}$, both were vulnerable with the overlapped obstacles. In the right image of (a), Restyle$_{pSp}$ generated severe artifacts, while Restyle$_{e4e}$ distorted the shape of mouth significantly. Moreover, the Restyle-based models showed poor editability. In (b), Restyle$_{pSp}$ and Restyle$_{e4e}$ failed to change the hairstyle in the Mohawk case.

naïvely applying L_{image} on the interest region. UnF lowered LPIPS by forcing the model to preserve features of the interest region. Please refer to Appendix D for more detailed results of each ablation model.

Fig. 9. Style mixing results. We interpolated latents from Source A to Source B. We took styles corresponding to either coarse (4^2–8^2), middle (16^2–32^2), or fine (64^2–1024^2) resolution from source B and took the rest from source A. IntereStyle showed the robust results on the interpolation, even with obstacles on the original source images, while Restyle$_{pSp}$ suffered from severe artifacts.

4.2 Editing via Latent Manipulation

Inversion of GAN is deeply related to the image manipulation on the latent space. In this section, we compare the quality of edited images produced by various StyleGAN inversion models [3,24,30], manipulated via InterFaceGAN [28] and StyleCLIP [22] methods, and style mixing [18,19]. Figure 8 shows the results of editing real-world images via InterFaceGAN and StyleCLIP, together with the inversion results. We changed three attributes for each method; smile, age, and pose for InterFaceGAN, and "smile", "lipstick", and "Mohawk hairstyle" for StyleCLIP. Our model showed high inversion and perceptual qualities consistently among various editing scenarios, even with strong makeups or obstacles. However, *pSp* and *e4e* missed important features of images, such as makeups or eye shapes. Moreover, *pSp* produced artifacts in several editing scenarios. In the cases of Restyle$_{pSp}$ and Restyle$_{e4e}$, they failed to robustly handle obstacles. In the right side of Fig. 8a, Restyle$_{pSp}$ produced severe artifacts around the mouth, while Restyle$_{e4e}$ totally changed the shape. Moreover, the Restyle-based models showed low editability in specific cases, such as "Mohawk".

To attribute to the superior disentanglement feature of StyleGAN latent space [28], we can separately manipulate the coarse and fine features of images. Following the settings from the StyleGAN [18] experiment, we took styles corresponding to either coarse, middle, or fine spatial resolution, respectively, from the latent of source B, and the others were taken from the latent of source A. Moreover, we mixed more than one hard case, *e.g.*, obstacles on faces and extreme poses, to evaluate the robustness of our model. As shown in Fig. 9, our model showed higher perceptual quality on the interpolated images compared to Restyle$_{pSp}$. Restyle$_{pSp}$ produced images with texture shift, *i.e.*, images with cartoon-like texture, distorted facial shape, and undesirable artifacts during the style mixing. In contrast, our model generated stable facial outputs. Additional qualitative results are shown in Appendix D.2.

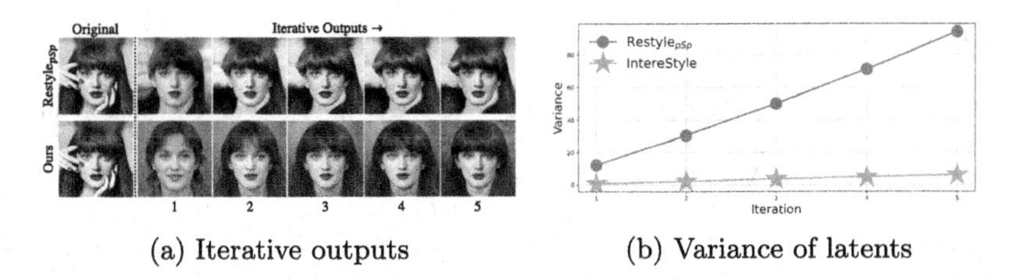

(a) Iterative outputs (b) Variance of latents

Fig. 10. Outputs and variance at each iteration. We compared intermediate outputs $\hat{y}_i$ of an example from CelebA-HQ test dataset, and the average of variances of latents among all test images at each iteration. Restyle showed severe artifacts on the interest region and a steep variance increment at each iteration, while IntereStyle showed robust inversion and maintained low variances.

4.3 Iterative Refinement of IntereStyle

We compared the progress of iterative refinement between Restyle [3] and IntereStyle in Fig. 10. Restyle reconstructed most of the coarse features within a few steps, while the variance of Restyle increased consistently as iteration progressed. In other words, the reduction of distortion is marginal, though Restyle excessively focuses on this. Consequently, the latent from Restyle was located far from W, which yields an image with low perceptual quality. In contrast, IntereStyle concentrated on the interest region that could be generated without a broad extension from W. Consequently, IntereStyle effectively reduced distortion on the interest region by iteration while maintaining high perceptual quality.

5 Conclusions

For StyleGAN inversion, focusing on the interest region is essential but under-explored yet. We found excessive attention on the uninterest region occurs the drop of perceptual quality and high distortion on the interest region. We proposed a simple yet effective StyleGAN encoder training scheme, coined InterStyle, composed of Interest Disentanglement and Uninterest Filter. We demonstrated that InterStyle showed both low distortion and high perceptual inversion quality, and enabled various latent manipulations robustly for image editing. We look forward to our work to be widely used in future research or the industry field, which needs a delicate inversion of the interest region for image editing.

Acknowledgments. This work was supported by the National Research Foundation of Korea (NRF) grant funded by the Korea government (MSIT) (No. 2021R1G1A1094379), and in part by the Institute of Information and Communications Technology Planning and Evaluation (IITP) grant funded by the Korea Government (MSIT) (Artificial Intelligence Innovation Hub) under Grant 2021-0-02068.

References

1. Abdal, R., Qin, Y., Wonka, P.: Image2StyleGAN: how to embed images into the StyleGAN latent space? In: Proceedings of the IEEE/CVF International Conference on Computer Vision, pp. 4432–4441 (2019)
2. Abdal, R., Qin, Y., Wonka, P.: Image2StyleGAN++: how to edit the embedded images? In: Proceedings of the IEEE/CVF Conference on Computer Vision and Pattern Recognition, pp. 8296–8305 (2020)
3. Alaluf, Y., Patashnik, O., Cohen-Or, D.: ReStyle: a residual-based StyleGAN encoder via iterative refinement. In: Proceedings of the IEEE/CVF International Conference on Computer Vision, pp. 6711–6720 (2021)
4. Bau, D., et al.: Semantic photo manipulation with a generative image prior. arXiv preprint arXiv:2005.07727 (2020)
5. Bau, D., et al.: Seeing what a GAN cannot generate. In: Proceedings of the IEEE/CVF International Conference on Computer Vision, pp. 4502–4511 (2019)
6. Blau, Y., Michaeli, T.: The perception-distortion tradeoff. In: Proceedings of the IEEE Conference on Computer Vision and Pattern Recognition, pp. 6228–6237 (2018)
7. Carion, N., Massa, F., Synnaeve, G., Usunier, N., Kirillov, A., Zagoruyko, S.: End-to-end object detection with transformers. In: Vedaldi, A., Bischof, H., Brox, T., Frahm, J.-M. (eds.) ECCV 2020. LNCS, vol. 12346, pp. 213–229. Springer, Cham (2020). https://doi.org/10.1007/978-3-030-58452-8_13
8. Chen, L.C., Zhu, Y., Papandreou, G., Schroff, F., Adam, H.: Encoder-decoder with atrous separable convolution for semantic image segmentation. In: Proceedings of the European conference on computer vision (ECCV), pp. 801–818 (2018)
9. Choi, Y., Uh, Y., Yoo, J., Ha, J.W.: StarGAN v2: diverse image synthesis for multiple domains. In: Proceedings of the IEEE/CVF Conference on Computer Vision and Pattern Recognition, pp. 8188–8197 (2020)
10. Collins, E., Bala, R., Price, B., Susstrunk, S.: Editing in style: uncovering the local semantics of GANs. In: Proceedings of the IEEE/CVF Conference on Computer Vision and Pattern Recognition, pp. 5771–5780 (2020)
11. Creswell, A., Bharath, A.A.: Inverting the generator of a generative adversarial network. IEEE Trans. Neural Netw. Learn. Syst. **30**(7), 1967–1974 (2018)
12. Daras, G., Odena, A., Zhang, H., Dimakis, A.G.: Your local GAN: designing two dimensional local attention mechanisms for generative models. In: Proceedings of the IEEE/CVF Conference on Computer Vision and Pattern Recognition, pp. 14531–14539 (2020)
13. Deng, J., Guo, J., Xue, N., Zafeiriou, S.: ArcFace: additive angular margin loss for deep face recognition. In: Proceedings of the IEEE/CVF Conference on Computer Vision and Pattern Recognition, pp. 4690–4699 (2019)
14. Gong, K., Gao, Y., Liang, X., Shen, X., Wang, M., Lin, L.: Graphonomy: universal human parsing via graph transfer learning. In: Proceedings of the IEEE/CVF Conference on Computer Vision and Pattern Recognition, pp. 7450–7459 (2019)
15. Goodfellow, I., et al.: Generative adversarial nets. Adv. Neural Inf. Process. Syst. **27** (2014)
16. Gu, J., Shen, Y., Zhou, B.: Image processing using multi-code GAN prior. In: Proceedings of the IEEE/CVF Conference on Computer Vision and Pattern Recognition, pp. 3012–3021 (2020)
17. Huang, Y., et al.: CurricularFace: adaptive curriculum learning loss for deep face recognition. In: proceedings of the IEEE/CVF Conference on Computer Vision and Pattern Recognition, pp. 5901–5910 (2020)

18. Karras, T., Laine, S., Aila, T.: A style-based generator architecture for generative adversarial networks. In: Proceedings of the IEEE/CVF Conference on Computer Vision and Pattern Recognition, pp. 4401–4410 (2019)
19. Karras, T., Laine, S., Aittala, M., Hellsten, J., Lehtinen, J., Aila, T.: Analyzing and improving the image quality of StyleGAN. In: Proceedings of the IEEE/CVF Conference on Computer Vision and Pattern Recognition, pp. 8110–8119 (2020)
20. Kim, H., Choi, Y., Kim, J., Yoo, S., Uh, Y.: StyleMapGAN: exploiting spatial dimensions of latent in GAN for real-time image editing. arXiv preprint arXiv:2104.14754 (2021)
21. Park, T., Liu, M.Y., Wang, T.C., Zhu, J.Y.: Semantic image synthesis with spatially-adaptive normalization. In: Proceedings of the IEEE/CVF Conference on Computer Vision and Pattern Recognition, pp. 2337–2346 (2019)
22. Patashnik, O., Wu, Z., Shechtman, E., Cohen-Or, D., Lischinski, D.: StyleCLIP: text-driven manipulation of StyleGAN imagery. In: Proceedings of the IEEE/CVF International Conference on Computer Vision, pp. 2085–2094 (2021)
23. Raj, A., Li, Y., Bresler, Y.: Gan-based projector for faster recovery with convergence guarantees in linear inverse problems. In: Proceedings of the IEEE/CVF International Conference on Computer Vision, pp. 5602–5611 (2019)
24. Richardson, E., et al.: Encoding in style: a StyleGAN encoder for image-to-image translation. In: Proceedings of the IEEE/CVF Conference on Computer Vision and Pattern Recognition, pp. 2287–2296 (2021)
25. Roich, D., Mokady, R., Bermano, A.H., Cohen-Or, D.: Pivotal tuning for latent-based editing of real images. arXiv preprint arXiv:2106.05744 (2021)
26. Saha, R., Duke, B., Shkurti, F., Taylor, G.W., Aarabi, P.: LOHO: latent optimization of hairstyles via orthogonalization. In: Proceedings of the IEEE/CVF Conference on Computer Vision and Pattern Recognition, pp. 1984–1993 (2021)
27. Shen, Y., Gu, J., Tang, X., Zhou, B.: Interpreting the latent space of GANs for semantic face editing. In: Proceedings of the IEEE/CVF Conference on Computer Vision and Pattern Recognition, pp. 9243–9252 (2020)
28. Shen, Y., Yang, C., Tang, X., Zhou, B.: Interfacegan: Interpreting the disentangled face representation learned by GANs. IEEE Trans. Pattern Anal. Mach. Intell. (2020)
29. Tewari, A., et al.: StyleRig: rigging StyleGAN for 3D control over portrait images. In: Proceedings of the IEEE/CVF Conference on Computer Vision and Pattern Recognition, pp. 6142–6151 (2020)
30. Tov, O., Alaluf, Y., Nitzan, Y., Patashnik, O., Cohen-Or, D.: Designing an encoder for StyleGAN image manipulation. ACM Trans. Graph. (TOG) **40**(4), 1–14 (2021)
31. Wei, T., et al.: A simple baseline for StyleGAN inversion. arXiv preprint arXiv:2104.07661 (2021)
32. Wu, Z., Lischinski, D., Shechtman, E.: StyleSpace analysis: disentangled controls for StyleGAN image generation. In: Proceedings of the IEEE/CVF Conference on Computer Vision and Pattern Recognition, pp. 12863–12872 (2021)
33. Yang, G., Fei, N., Ding, M., Liu, G., Lu, Z., Xiang, T.: L2M-GAN: learning to manipulate latent space semantics for facial attribute editing. In: Proceedings of the IEEE/CVF Conference on Computer Vision and Pattern Recognition, pp. 2951–2960 (2021)
34. Zhang, R., Isola, P., Efros, A.A., Shechtman, E., Wang, O.: The unreasonable effectiveness of deep features as a perceptual metric. In: Proceedings of the IEEE Conference on Computer Vision and Pattern Recognition, pp. 586–595 (2018)

35. Zhu, J., Shen, Y., Zhao, D., Zhou, B.: In-domain GAN inversion for real image editing. In: Vedaldi, A., Bischof, H., Brox, T., Frahm, J.-M. (eds.) ECCV 2020. LNCS, vol. 12362, pp. 592–608. Springer, Cham (2020). https://doi.org/10.1007/978-3-030-58520-4_35
36. Zhu, J.-Y., Krähenbühl, P., Shechtman, E., Efros, A.A.: Generative visual manipulation on the natural image manifold. In: Leibe, B., Matas, J., Sebe, N., Welling, M. (eds.) ECCV 2016. LNCS, vol. 9909, pp. 597–613. Springer, Cham (2016). https://doi.org/10.1007/978-3-319-46454-1_36

StyleLight: HDR Panorama Generation for Lighting Estimation and Editing

Guangcong Wang, Yinuo Yang, Chen Change Loy, and Ziwei Liu(✉)

S-Lab, Nanyang Technological University, Singapore, Singapore
{guangcong.wang,ccloy,ziwei.liu}@ntu.edu.sg, YANG0689@e.ntu.edu.sg

Abstract. We present a new lighting estimation and editing framework to generate high-dynamic-range (HDR) indoor panorama lighting from a single limited field-of-view (LFOV) image captured by low-dynamic-range (LDR) cameras. Existing lighting estimation methods either directly regress lighting representation parameters or decompose this problem into LFOV-to-panorama and LDR-to-HDR lighting generation sub-tasks. However, due to the partial observation, the high-dynamic-range lighting, and the intrinsic ambiguity of a scene, lighting estimation remains a challenging task. To tackle this problem, we propose a coupled dual-StyleGAN panorama synthesis network (**StyleLight**) that integrates LDR and HDR panorama synthesis into a unified framework. The LDR and HDR panorama synthesis share a similar generator but have separate discriminators. During inference, given an LDR LFOV image, we propose a focal-masked GAN inversion method to find its latent code by the LDR panorama synthesis branch and then synthesize the HDR panorama by the HDR panorama synthesis branch. StyleLight takes LFOV-to-panorama and LDR-to-HDR lighting generation into a unified framework and thus greatly improves lighting estimation. Extensive experiments demonstrate that our framework achieves superior performance over state-of-the-art methods on indoor lighting estimation. Notably, StyleLight also enables intuitive lighting editing on indoor HDR panoramas, which is suitable for real-world applications. Code is available at https://style-light.github.io/.

Keywords: Lighting estimation and editing · Panorama generation

1 Introduction

Environmental lighting models aim to approximate realistic lighting effects in a scene. Existing state-of-the-art methods [5,11,19,25] use HDR spherical panoramas for lighting. In many real-world applications, however, only LDR LFOV images are available. Therefore, it is important to estimate HDR panorama illumination from an LDR LFOV image, aided with the capability of lighting editing to flexibly control lighting conditions. Lighting estimation and editing have a

Supplementary Information The online version contains supplementary material available at https://doi.org/10.1007/978-3-031-19784-0_28.

S. Avidan et al. (Eds.): ECCV 2022, LNCS 13675, pp. 477–492, 2022.
https://doi.org/10.1007/978-3-031-19784-0_28

Fig. 1. StyleLight. The proposed StyleLight estimates HDR panorama lighting from an LDR LFOV image and performs lighting addition and removal.

wide range of applications in mixed reality such as object insertion and relighting in virtual meetings and games. For example, given an LDR LFOV background, a person can be inserted into the background and realistically relit by controllable panorama illumination (Fig. 1).

There are three main challenging problems in lighting estimation. First, only a partial scene is observed in a limited FOV image. The other unobserved part of a scene is ambiguous and could significantly affect the illumination distribution in a full scene due to ambiguous lighting sources with different numbers, shapes, and directions. Second, only a partial range of lighting is observed in typical LDR images. Physical lighting has a wide range of intensity levels in real scenes, ranging from shadows to bright light sources. However, in LDR images, underexposed areas are clipped to 0, and overexposed areas are clipped to 1. Out-of-range lighting cannot be observed in LDR images. Third, lighting estimation of unobserved views depends on intrinsic attributes of observed objects (e.g., material and 3D geometry), but these annotations are often unavailable in real-world applications.

Existing methods on illumination estimation provide partial solutions to illumination estimation, which could be roughly classified into two groups, i.e., regression-based estimation methods [4–6,9,14,31] and generation-based estimation methods [23,24,32]. In the first group, researchers propose to directly regress panorama lighting information from a given LDR LFOV image. The main difficulty of this pipeline is to parameterize lighting representations as target labels for regression, including light representation (e.g., number, shape, directions and intensity) [4], Spherical Harmonics [6], and wavelet transformation [31]. However, these lighting labels mainly focus on strong lighting sources and thus lose the details of a scene. When a scene contains reflective materials (e.g., glass, mirror, and metal), this pipeline will fail to provide a desired rendered visual effect. In image-based rendering, the HDR map can be also applied as background. Moreover, regression-based models often suffer from unstable training of neural networks due to partial observed view, a wide range of lighting, unknown geometry, and material of objects. For example, EMlight [32] claimed that it is necessary to train a regression model by gradually increasing training examples to avoid training collapse. Second, generation-based estimation methods aim at generating full HDR panoramas from an LDR LFOV image. They typically decompose illumination estimation as several sub-tasks, such as LFOV-to-panorama completion, LDR-to-HDR regression [32], and 3D geometry prediction [23]. These methods either require additional annotations or multiple optimization stages.

Our proposed method belongs to the second group. Unlike the existing generation-based estimation methods, we propose a coupled dual-StyleGAN panorama synthesis network (**StyleLight**) that solves LFOV-to-panorama and LDR-to-HDR problems in a unified framework and does not require any 3D geometry annotation. Specifically, StyleLight is a coupled dual-StylGAN that maps noises to both HDR and LDR panoramas. The LDR and HDR panorama synthesis share a similar generator. At the output of the generator, a tone mapping is adopted for the translation between HDR and LDR. Considering the fact that an HDR panorama consists of extremely different weak lighting distribution and strong lighting distribution, we use two discriminators to distinguish real/fake HDR and LDR panorama distributions, respectively. Given an LDR LFOV image, we propose a focal-masked GAN inversion method to find its latent code through the LDR panorama synthesis branch and then generate its corresponding HDR panorama as illumination maps through the HDR synthesis branch. Moreover, we propose a structure-preserved GAN inversion method for lighting editing with a trained StyleLight model to flexibly control the environment lighting of panoramas, such as turning on/off lights, opening/closing doors and windows, making our method well-suited for various real-world applications.

The main contributions of this paper can be summarized as follows: **1)** We propose a coupled dual-styleGAN synthesis network (StyleLight) that integrates HDR and LDR panorama synthesis in a unified framework. During inference, we propose a focal-masked GAN inversion method to solve both LFOV-to-panorama and LDR-to-HDR generation. **2)** We propose a structure-preserved GAN inversion method for lighting editing with the trained StyleLight model to flexibly control panorama lighting. **3)** Extensive experiments demonstrate the superiority and the effectiveness of our proposed lighting estimation method over state-of-the-art methods on indoor HDR panoramas and show promising applications with our proposed lighting editing.

2 Related Work

Lighting estimation is one of the long-standing problems in computer vision and graphics. Early work [3,27,29] typically simplifies the models by considering fixed number of lights [17], annotated light positions [13] and geometry [13,16,18]. Recent lighting estimation methods relax such assumptions and focus on more challenging lighting estimation. We briefly review the work most relevant to this paper.

Lighting Estimation. Existing lighting estimation methods can be roughly classified into two groups, i.e., regression-based methods and generation-based methods. Regression-based models [4,6,9,14,31,32] seek for effective lighting representations of HDR panoramas as labels for lighting regression. These methods take an LDR LFOV image as input and predict lighting labels, such as direct lighting representation (e.g., numbers, shapes, directions, and intensity)

[4], Spherical Harmonics (SH) [6], and wavelet transformation [31]. For example, Gardner *et al.* [5] trained a lighting classifier on an annotated LDR panorama dataset to predict the location of lighting sources of LDR LFOV images and fine-tuned the prediction of lighting intensity on a small HDR panorama dataset. Gardner *et al.* [4] regressed the illumination conditions such as the light source positions in 3D, areas, intensities, and colors. LeGendre *et al.* [14] designed a joint camera-sphere device to collect LDR LFOV images that include three reflective spheres, and regressed HDR lighting from an LDR LFOV image. Garon *et al.* [6] regressed the 5th-order SH coefficients for the lighting at a certain location given a single image and a 2D location in an image. Regression-based models ignore the low-intensity lighting details and thus fail to provide a desired rendered quality for mirror materials. Besides, they often suffer from unstable training of neural networks due to the high dynamic range of lighting.

Generation-based methods [5,23,24] focus on decomposing illumination estimation as several sub-tasks, such as LFOV-to-panorama completion and LDR-to-HDR regression. Song *et al.* [23] decomposed illuminations prediction into several simpler differentiable sub-tasks including geometry estimation, scene completion, and LDR-to-HDR estimation. Zhan *et al.* [32] decomposed the illumination map into spherical light distribution, light intensity and the ambient term for illumination regression. This aforementioned paradigm either decomposes illumination estimation as several sub-tasks or collects expensive annotations for training.

Image Editing via GAN Inversion. Image editing is a task that modifies a target attribution of a given image while preserving other details [7,15,20–22,26,28,33]. Recent state-of-the-art image editing methods focus on generative adversarial network (GAN) inversion techniques to control the latent space of GANs for image editing. For example, Ganspace [7] used Principal Component Analysis (PCA) to identify important latent directions and create interpretable controls for image attributes, such as viewpoint, aging, lighting, and time of day. InterfaceGAN [21] proposed to edit the latent code $\mathbf{z}$ with $\mathbf{z}_{edit} = \mathbf{z} + \alpha\mathbf{n}$ for each semantic attribute where $\mathbf{n}$ is a normal vector of a linear SVM's hyper-plane. SeFa [22] proposed a closed-form factorization algorithm for latent semantic discovery by directly decomposing the pre-trained weights. StyleFlow [1] formulated conditional exploration as an instance of conditional continuous normalizing flows in the GAN latent space conditioned by attribute features. These image editing methods focus on controlling the latent code to control attributes of objects. However, in our lighting editing setting, our goal is to edit a light source that is located at a small area of a panorama given a bounding box, which is different from the typical image editing.

3 Our Approach

3.1 Overview of the Proposed Framework

An overview of our proposed framework is depicted in Fig. 2. The framework consists of lighting estimation and lighting editing. The goal of lighting estimation is

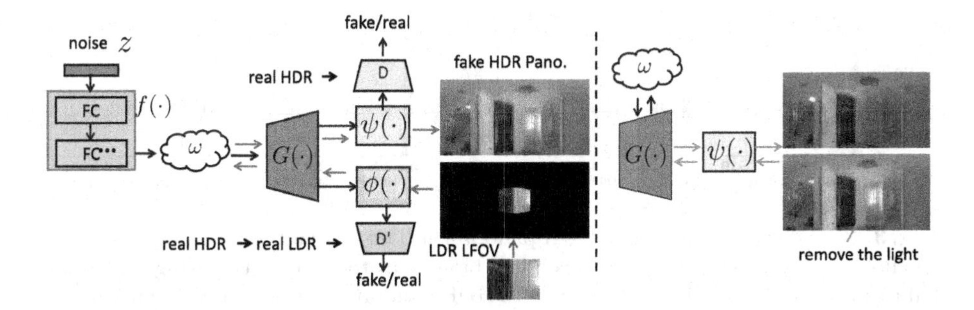

Fig. 2. Overview of the proposed framework. **Left:** Lighting estimation. We first propose a coupled dual-StyleGAN that generates HDR and LDR panoramas from noises. Given an LDR LFOV image, we map it into a panorama. We introduce a focal-masked GAN inversion method to find its latent code through the LDR branch and compute the HDR panorama through the HDR branch. **Right:** Lighting editing. With the trained dual-StyleGAN, we perform lighting editing to control panoramas lighting.

to predict HDR panorama from an LDR LFOV image. To achieve this goal, we introduce a novel coupled dual-StyleGAN synthesis network (StyleLight) that learns to synthesize both HDR and LDR panoramas from noises with a shared generator and two specific discriminators. With the trained StyleLight model, we propose a focal-masked GAN inversion method to find the latent code of an LDR LFOV image and predict the HDR panorama as an illumination map (Sect. 3.2). To control various lighting conditions of the target HDR panorama, we introduce a structure-preserved lighting editing method to solve this problem (Sect. 3.3). To the best of our knowledge, this is the first attempt to perform lighting editing on HDR panoramas for controlling the scene lighting.

3.2 Lighting Estimation

Generation-based lighting estimation methods typically decompose illumination estimation into several sub-tasks, such as LFOV-to-panorama completion and LDR-to-HDR regression. These methods optimize sub-tasks independently and could lead to sub-optimal estimation results. Moreover, these methods can be hardly scalable to lighting editing. Different from these methods, we propose a coupled dual-StyleGAN that integrates LFOV-to-panorama, LDR-to-HDR, and lighting editing into a unified framework.

Coupled Dual-StyleGAN for Joint LDR and HDR Panorama Synthesis. HDR panoramas are more difficult to synthesize than LDR panoramas due to the complex characteristics of distributions. Figure 3 shows the distributions of two HDR panorama examples (others are similar). An HDR panorama consists of two components, i.e., low-intensity pixels (blue, denoted as LDR, from 0 to 1) and high-intensity pixels that are out of range LDR pixels (red, denoted as HDR^{-}, larger than 1). The distributions suggest three key points. First, LDR distributions are similar to a Gaussian distribution. Most of the pixels are LDR

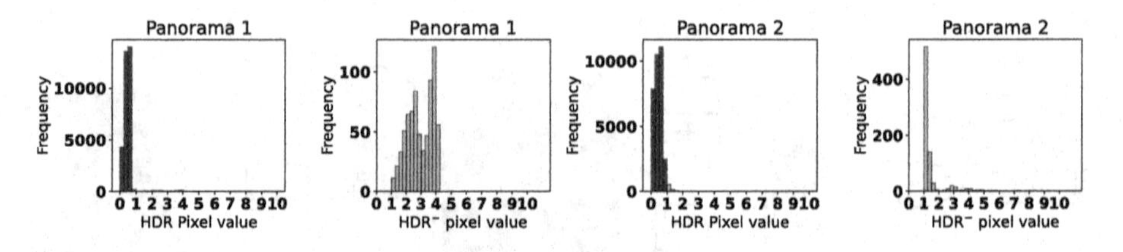

Fig. 3. Pixel distributions of two HDR panoramas. The first and third histograms cover the full range of pixel values. The second and fourth histograms visualize high-intensity values of the first and third histograms. An HDR panorama consists of two components, i.e., low-intensity pixels (blue, from 0 to 1, denoted as LDR) and high-intensity pixels that are out of range LDR pixels (red, larger than 1, denoted as HDR^-). (Color figure online)

pixels and they are limited in the narrow range. This suggests that LDR distributions are easy to generate. Second, HDR^- distributions are dissimilar and cover a few pixels in the high dynamic range (much larger than 1). This implies that HDR^- distributions are difficult to generate and could lead to unstable training of HDR synthesis. If we only use an HDR discriminator to distinguish entire HDR panoramas, the dynamic HDR^- pixels (e.g., >100) largely affect the discriminator and other regions (0–1) are often ignored through convolution. $D_{\prime}$ can be regarded as a regularization term to preserve image semantics in LDR regions. We analyze the distributions of 1401 training images. We fit a Gaussian distribution for both LDR and HDR^- of each image, respectively. We find that most of LDR distributions (99.86% LDR images) follow Gaussian distributions and most of HDR^- distributions (70.38% HDR^- images) fail to converge to Gaussian distributions. Note that LDR panoramas play a key role in the synthesis from LDR LFOV images to HDR panoramas in the next section. We expect HDR and LDR images to be well synthesized. It is noteworthy that although HDR^- distributions are difficult to generate, HDR^- and LDR have a strong semantic relation. LDR pixels provide the location and shape contextual information of lighting sources while HDR pixels provide intensity information.

Motivated by the fact that LDR panoramas are easy to generate and they provide location and shape contextual information of lighting for HDR synthesis, we introduce a coupled dual-StyleGAN for joint HDR and LDR panorama synthesis. With the help of LDR panorama synthesis, dual-StyleGAN is able to generate accurate HDR panoramas. Specifically, the dual-StyleGAN takes a noise vector $\boldsymbol{z} \in z$ as input and maps the noise $\boldsymbol{z}$ into a intermediate latent space ω that is much less entangled than the z space, as analyzed in StyleGAN2 [12]. The mapping function $\boldsymbol{w} = f(\boldsymbol{z})$ consists of eight fully-connected layers with activation layers. The latent code $\boldsymbol{w} \in \omega$ is then fed into the generator G. After that, two transformations $\psi(\cdot)$ and $\phi(\cdot)$ are used to transform the output of G into HDR and LDR panoramas, respectively. That is, $x_{\text{hdr}} = \psi(G(\boldsymbol{w}))$ and $x_{\text{ldr}} = \phi(G(\boldsymbol{w}))$, as illustrated in Fig. 2. The HDR output $\psi(\cdot)$ is an inverse tone-mapping plus clamping (positive pixels) and the LDR output $\phi(\cdot)$ is clampped from 0 to 1. To ease the stable training, we pre-process the HDR training panoramas by $x^{'} = \alpha x^{1/\gamma}$, where $\gamma = 2.4$ and α is set to 0.5 divided by 50-th percentile of all pixel values. We use $clamp(x^{'}, 0, +\infty))$ for HDR and $clamp(x^{'}, 0, 1)$ for LDR. During inference, we perform inverse tone-mapping $clamp(x^{'}, 0, +\infty))$ into x. Other tone-mappings are also suitable for our framework. The idea behind the shared generator network is that HDR and LDR panoramas share the same contextual semantics. Formally, the proposed coupled dual-StyleGAN is formulated as

$$\begin{aligned} \min_{G} \max_{D,D^{'}} \mathcal{J}(G, D, D^{'}) = & \ E_{x \backsim P_{data}(x)}[\log(D(x))] \\ & + E_{x \backsim P^{'}_{data}(x)}[\log(D^{'}(x))] \\ & + E_{w \backsim P_{w}}[\log(1 - D(\psi(G(\boldsymbol{w}))))] \\ & + E_{w \backsim P_{w}}[\log(1 - D^{'}(\phi(G(\boldsymbol{w}))))], \end{aligned} \tag{1}$$

where $D(\cdot)$ and $D^{'}(\cdot)$ denote two discriminators for HDR synthesis and LDR synthesis, respectively. In Eq. 1, we optimize G by minimizing the third and fourth terms. We optimize D and $D^{'}$ simultaneously by maximizing the probability of predicting real and fake panoramas. Note that we do not use the same discriminator because the distributions of HDR and LDR panoramas are different, as illustrated in Fig. 3.

Focal-Masked GAN Inversion for LDR-to-HDR Transformation and LFOV-to-Panorama Completion. After obtaining the coupled dual-StyleGAN, we can perform LDR-to-HDR transformation and LFOV-to-panorama completion. In the coupled dual-StyleGAN, HDR synthesis and LDR synthesis share the same generator G and thus guarantee pixel-level alignment between the LDR-to-HDR transformation. Thanks to the coupled dual-StyleGAN that integrates HDR synthesis and LDR synthesis into a unified framework, we can find a shared latent code of an HDR panorama and an LDR LFOV image in the ω space, as illustrated in Fig. 2(left). Given an LDR LFOV image, we map it into a spherical panorama as a masked panorama. We then extend a GAN inversion method [20] into a masked GAN inversion to tailor for our coupled dual-StyleGAN. We find the latent code of the masked panorama in the ω space. We forward the latent code through the HDR branch to obtain an HDR panorama. Let $x^{\text{ldr}}_{\text{lfov}}$ denote a masked panorama of the LDR LFOV image and $\boldsymbol{n}$ denote noise maps of all layers. We first optimize latent code $\boldsymbol{w} \in \omega$ such that synthesized LDR LFOV matches the target LDR LFOV as much as possible, which can be formulated as

$$\boldsymbol{w}_{*}, \boldsymbol{n}_{*} = \underset{\boldsymbol{w}, \boldsymbol{n}}{\text{argmin}}\, \mathcal{L}_{\text{LPIPS}}(M \odot \phi(G(\boldsymbol{w}, \boldsymbol{n}; \theta)), x^{\text{ldr}}_{\text{lfov}}) + \lambda_{n} \mathcal{L}_{n}(\boldsymbol{n}), \tag{2}$$

where $\mathcal{L}_{\text{LPIPS}}$ represents the VGGNet based Perceptual loss [10] for distance measure and $\mathcal{L}_{n}(\boldsymbol{n})$ represents a noise regularization term to constrain noise maps $\boldsymbol{n}$ and λ_{n} controls its importance. We re-use the symbol G with latent code $\boldsymbol{w} \in \omega$, noise maps $\boldsymbol{n}$ and weights θ as input. The variable M denotes a mask that preserves the pixels corresponding to the LFOV image $x^{\text{ldr}}_{\text{lfov}}$. Because it is hard to directly reconstruct the target LDR LFOV image by optimizing $\boldsymbol{w}$ and $\boldsymbol{n}$, we then fine-tune the weights θ of G by

$$\begin{aligned}\mathcal{L}_G(\theta) = &\mathcal{L}_{\text{LPIPS}}(M \odot \phi(G(\boldsymbol{w}_*, \boldsymbol{n}_*; \theta)), x^{\text{ldr}}_{\text{lfov}}) \\ &+ \lambda^R_{L2}\mathcal{L}_{L2}(M \odot \phi(G(\boldsymbol{w}_*, \boldsymbol{n}_*; \theta)), x^{\text{ldr}}_{\text{lfov}}),\end{aligned} \tag{3}$$

where $\mathcal{L}_G$ represents the similarity loss between a masked generated LDR image and the LFOV HDR image $x^{\text{ldr}}_{\text{lfov}}$. We use the Mean-Squared-Error loss and the VGGNet based Perceptual loss for distance measure. To further improve visual quality, we also use a regularization term, which is given by

$$\begin{aligned}\mathcal{L}_R(\theta) = &\mathcal{L}_{\text{LPIPS}}(\phi(G(\boldsymbol{w}_r, \boldsymbol{n}_*; \theta)), \phi(G(\boldsymbol{w}_r, \boldsymbol{n}_*; \theta_0))) \\ &+ \lambda^{R'}_{L2}\mathcal{L}^R_{L2}(\phi(G(\boldsymbol{w}_r, \boldsymbol{n}_*; \theta)), \phi(G(\boldsymbol{w}_r, \boldsymbol{n}_*; \theta_0))),\end{aligned} \tag{4}$$

where $\boldsymbol{w}_r = \boldsymbol{w}_* + \alpha(\boldsymbol{w}_z - \boldsymbol{w}_*)/||\boldsymbol{w}_z - \boldsymbol{w}_*||_2$ and $\boldsymbol{w}_z$ is a latent code that is generated by a random $\boldsymbol{z}$. The θ_0 is the old weights of G before finetuning. The α is a interpolation parameter. Finally, we combine Eqs. 3 and 4, and fine-tune G by

$$\theta_* = \underset{\theta}{\text{argmin}}\, \mathcal{L}_G(\theta) + \eta\mathcal{L}_R(\theta). \tag{5}$$

Simply extending the GAN inversion method [20] into a masked GAN inversion for the coupled dual-StyleGAN cannot guarantee that the pixels of lights in LDR panoramas will be transformed into high-intensity HDR$^-$. Strong lighting sources are the main factors for rendering environment, but high-intensity pixels are much less than low-intensity pixels, as shown in Fig. 3. Without any constraint, the model tends to fit the major LDR pixels and ignore the few HDR$^-$ pixels. Therefore, we propose a focal-masked to highlight the strong lighting sources. In our setting, we highlight top 10% strongest pixels. Given an LDR LFOV image $x^{\text{ldr}}_{\text{lfov}}$, we compute the top 10% strongest pixels and obtain a focal mask, which denoted as $FOCAL(x^{\text{ldr}}_{\text{lfov}})$. We re-write Eq. 2 as

$$\begin{aligned}\boldsymbol{w}_*, \boldsymbol{n}_* = &\underset{\boldsymbol{w},\boldsymbol{n}}{\text{argmin}}\, \mathcal{L}_{\text{LPIPS}}(M \odot \phi(G(\boldsymbol{w}, \boldsymbol{n}; \theta)), x^{\text{ldr}}_{\text{lfov}}) + \lambda_n\mathcal{L}_n(\boldsymbol{n}) \\ &+ \beta_{L2}\mathcal{L}_{L2}(FOCAL(\phi(G(\boldsymbol{w}, \boldsymbol{n}; \theta))), FOCAL(x^{\text{ldr}}_{\text{lfov}})).\end{aligned} \tag{6}$$

Note that $FOCAL(\phi(G(\boldsymbol{w}, \boldsymbol{n}; \theta)))$ is consistent with $FOCAL(x^{\text{ldr}}_{\text{lfov}})$ in Eq. 6. Similarly, we re-write Eq. 3 as

$$\begin{aligned}\mathcal{L}_G(\theta) = &\mathcal{L}_{\text{LPIPS}}(M \odot \phi(G(\boldsymbol{w}_*, \boldsymbol{n}_*; \theta)), x^{\text{ldr}}_{\text{lfov}}) \\ &+ \lambda^R_{L2}\mathcal{L}_{L2}(M \odot \phi(G(\boldsymbol{w}_*, \boldsymbol{n}_*; \theta)), x^{\text{ldr}}_{\text{lfov}}) \\ &+ \beta_{L2}\mathcal{L}_{L2}(FOCAL(\phi(G(\boldsymbol{w}_*, \boldsymbol{n}; \theta))), FOCAL(\boldsymbol{w}^{\text{ldr}}_{\text{lfov}})).\end{aligned} \tag{7}$$

3.3 Lighting Editing

Lighting estimation provides a way to replicate real-world lighting conditions of a scene from a single LDR LFOV image and can be used to re-render inserted objects. In some occasions, however, the estimated lighting conditions do not meet our design/modeling requirements. For example, we would like to turn on an indoor light and "turn off" the light from windows at night given daytime HDR panoramas. Although light sources are located in a local area, it affects the global panorama lighting. Existing approaches do not consider this problem. Here, we introduce a lighting editing method to control panorama lighting through exploiting the powerful StyleLight.

In general, lighting editing includes three operations, i.e., adding a new light, removing a light, and controlling lighting intensity. In essence, the first two operations can be accomplished by the third one. In light estimation, we obtain a latent code w_0 that recovers an HDR panorama from an LDR LFOV image. Here we assume that $\boldsymbol{w}_0$ is known. Our goal is to find a editable direction of $\boldsymbol{w}_0$ to adjust the lighting intensity such that the lighting intensity at a candidate location changes while the geometry structure of the scene is preserved. Given a bounding box of a candidate lighting location as a mask M, we introduce a structure-preserved lighting editing loss to find the editable direction of $\boldsymbol{w}$, which is formulated as

$$\begin{aligned}\boldsymbol{w}_*, \boldsymbol{n}_* = & \operatorname*{argmin}_{\boldsymbol{w}, \boldsymbol{n}} \mathcal{L}_{\text{LPIPS}}((1-M) \odot \phi(G(\boldsymbol{w}, \boldsymbol{n}; \theta)), (1-M) \odot \phi(G(\boldsymbol{w}_0, \boldsymbol{n}; \theta)) \\ & + \mathcal{L}_{L2}((1-M) \odot \phi(G(\boldsymbol{w}, \boldsymbol{n})), (1-M) \odot \phi(G(\boldsymbol{w}_0, \boldsymbol{n}; \theta)) \\ & + \frac{\delta}{N} \sum M \odot \phi(G(\boldsymbol{w}, \boldsymbol{n}; \theta)),\end{aligned} \tag{8}$$

where δ denotes a lighting adjusting factor. When δ is positive, the lighting intensity decreases; when δ is negative, the lighting intensity increases. N is the number of masked pixels. The third term aims to adjust lighting intensity given a bounding box, and the first and second terms preserve other regions.

4 Experiments

Datasets and Experimental Settings. We evaluate our StyleLight model on the Laval Indoor HDR dataset [5] that contains 2,100 HDR panoramas. Following the train/test split of [5], we use 1,719 panoramas for training and 289 panoramas for testing. For the test set, we crop one HDR LFOV image from each panorama and transform it into an LDR LFOV image by tone mapping [32]. Since lighting estimation is a challenging task, some compared methods use extra annotations. For example, Gardner2017 [5] used annotated lighting sources of SUN360 [30] to learn to predict the locations and shapes of lighting sources. Gardner2019 [4] manually annotated the Laval Indoor dataset using EnvyDepth [2] and used per-pixel depth maps to improve lighting estimation.

Implementation Details. We use StyleGAN2 [12] as the backbone. The output of the generator is a 256 × 512 image. We resize it into 128 × 256 for evaluation, which is consistent with other methods. The resolution of feature maps begins with a 4 × 8 constant map and increases by ×2 upsampling each block, yielding

256×512 patch-based feature maps. In the full training of StyleLight, 4000k real images (kimg) are shown to the discriminators. We train StyleLight with 8 v100 GPUs for 7 h. The batch size is set to 32. We use Adam optimizer for the generators and two discriminators with a learning rate of 0.0025 and set $\beta_1 = 0$, $\beta_2 = 0.99$.

Evaluation Metrics. Similar to Gardner2019 [4] and EMlighting [32], we use three spheres with different materials for evaluation, including mirror, matte silver, and gray diffuse, as illustrated in Fig. 4. We render the three spheres by ground-truth HDR panoramas and the estimated illumination maps. Similar to [32], we use Root Mean Square Error (RMSE), scale-invariant RMSE (si-RMSE), and Angular Error for evaluation. RMSE and si-RMSE measure pixel-level errors of lighting intensity. Angular Error measures the ratio of R, G, and B components. Different from [32], we compute the metrics on the three rendered spheres and ignore the area of the rendered black background, which is invariant to the area of the background. Because there are no public codes for evaluation, we design our own hyper-parameters for the three spheres. We also use Fréchet Inception Distance (FID) [8] to measure the distribution similarity between the ground-truth and the estimated illumination maps.

Fig. 4. Three spheres with different materials: mirrored silver, matte silver, and diffuse grey. (Color figure online)

4.1 Quantitative Evaluation

We compare StyleLight with three representative state-of-the-art illumination estimation methods, including a generation-based method [32] and two regression-based methods [4,5]. Note that Gardner2017 [5] used extra lighting annotations and Gardner2019 [4] used extra depth annotations for training. StyleLight and EMlight [32] models do not require any extra annotation. For each method, we render three typical spheres (Fig. 4) by the predicted illumination maps and ground-truth HDR panoramas on the test set and evaluate RMSE, si-RMSE, and Angular Error. We also evaluate FID to measure the distribution distance of ground-truth and predicted illumination maps.

As shown in Table 1, our StyleLight outperforms the compared methods on all metrics and materials. Among them, Gardner2017 and Gardner2019 focus on strong lighting sources and ignore low-intensity lighting, so they fail to achieve a desired RMSE and si-RMSE on the mirror sphere and do not preserve components of colors. EMlight first regresses a Gaussian map and fuses it with the LFOV image for illumination generation. The drawback is that the multi-step optimization could lead to sub-optimal performance. Since EMlight contains regression learning, it is difficult to train. EMlight[1] claimed that the model

[1] https://github.com/fnzhan/Illumination-Estimation.

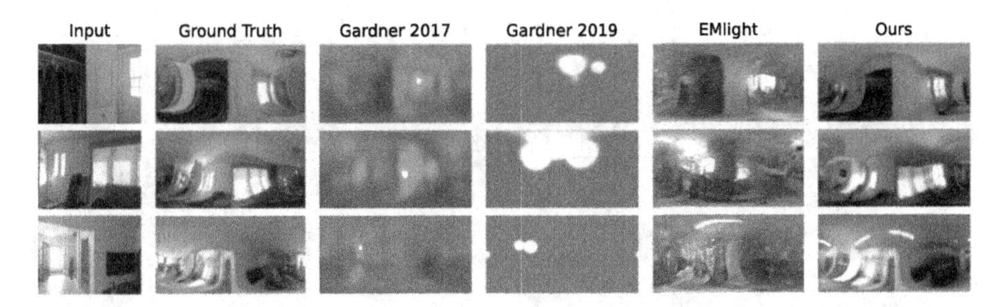

Fig. 5. Visual comparisons of our method with several SOTA methods.

has to be trained on subsets of 100, 1000, 2500, ... and the full set gradually and could collapse occasionally during training. We also see that our StyleLight largely outperforms other methods on FID, indicating that our predicted lighting distribution is closest to ground truth on a wide range of lighting. The results also suggest that StyleLight can synthesize plausible scenes and can achieve a promising result when rendering mirror materials. Our StyleLight also provides controllable lighting editing.

Table 1. Comparison with previous work. Gardner2017 and Gardner2019 are regression-based methods. EMlight is a generation-based method. Annotation denotes required manual annotations. The evaluation metrics include Angular Error, RMSE, si-RMSE and FID. M, S, and D denote mirror, matte silver, and diffuse material spheres, respectively. ↓ denotes that smaller is better.

Methods	Gardner2017 [5]			Gardner2019 [4]			EMlight [32]			StyleLight		
	M	S	D	M	S	D	M	S	D	M	S	D
Annotation	Lighting sources			Depth			No			No		
Angular Error↓	6.68	6.03	5.47	4.63	3.91	3.24	4.67	3.47	2.59	**4.30**	**2.96**	**2.41**
RMSE↓	0.70	0.38	0.19	0.85	0.49	–	0.59	0.33	0.17	**0.56**	**0.30**	**0.15**
si-RMSE↓	0.65	0.35	0.15	0.70	0.40	0.14	0.58	0.31	0.12	**0.55**	**0.29**	**0.11**
FID↓	307.5			344.1			263.8			**137.7**		

4.2 Qualitative Evaluation

In this section, we provide visual results to show the effectiveness of the proposed StyleLight model, including visual comparison with state-of-the-art lighting estimation methods and lighting editing.

Visual Comparison with the State-of-the-Art Lighting Estimation Methods. As depicted in Fig. 5, the predicted illumination maps by StyleLight are similar to the ground truth, and the proposed method obtains high-quality panoramas. Compared with the ground truth, Gardner2017 and Gardner2019 highlight strong lighting sources, which are not suitable to render mirror materials. EMlight considers both low-intensity and high-intensity lighting, but it

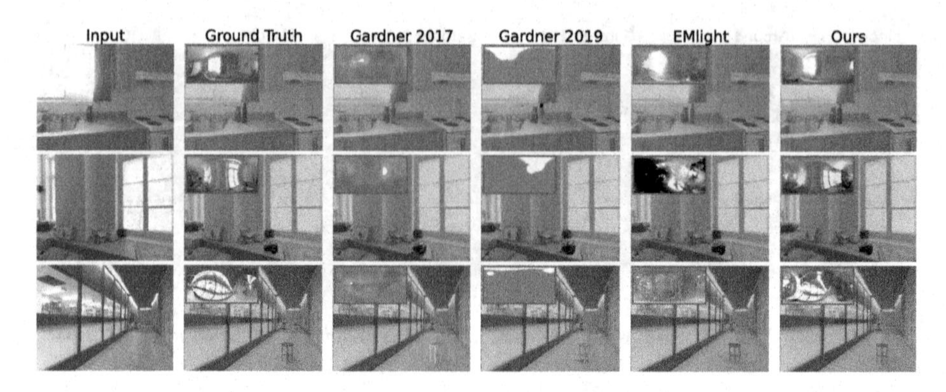

Fig. 6. Visual comparisons on object insertion and lighting. For each input LDR LFOV image, the four different methods estimate illumination maps (at the top-left of LFOV images) for rendering newly inserted objects.

obtains blurred images and it is difficult to train because of the unstable regression. Our StyleLight can synthesize panoramas of better quality, which can be further used for mirror rendering and lighting editing.

Visual Comparison with the State of the Art on Object Insertion and Lighting. We further compare StyleLight with state-of-the-art methods on object insertion and lighting. The predicted HDR panorama is a sphere. The object is placed at the center of the panorama and a warping operator [5] is used when inserted objects are not at the center of a panorama. Finally, image-based rendering is applied. The object insertion technique is similar to EMlight. The four different methods estimate illumination maps for rendering newly inserted objects. As shown in Fig. 6, compared with nearby objects in LFOV images, the inserted objects by Gardner2017 and Gardner2019 tend to lose low-intensity lighting information and generate inaccurate shadows. Thanks to the unique design of coupled dual-StyleGAN that integrates both HDR and LDR panoramas into a unified framework, StyleLight achieves a better visual outputs.

Lighting Editing with StyleLight. As shown in Fig. 7, given a panorama (the first column), we can remove lights in the second column and add new lights in the third column with the method presented in Sect. 3.3.

Effectiveness of the Coupled Dual-StyleGAN Structure. We conduct an ablation study to analyze the effectiveness of the coupled dual-StyleGAN framework. As shown in Fig. 8, the first row shows the quality of StyleLight that uses two discriminators to learn LDR and HDR distributions. The second row shows the LDR panorama quality by using one discriminator to train HDR synthesis (denoted as HDR-only) and the LDR panoramas are computed by tone mapping. It is observed that LDR panoramas from the 'HDR-only' suffer from severe distortion. This ablation demonstrates the effectiveness of the coupled dual-StyleGAN structure.

Fig. 7. Lighting editing with StyleLight. For each original panorama (Row 1), StyleLight removes a light (red box, Row 2) and then adds a new light (green box, Row 3). (Color figure online)

Fig. 8. Effectiveness of the coupled dual-StyleGAN structure. The dual-StyleGAN generates both HDR and LDR panoramas. HDR-only directly generates HDR panoramas.

Effectiveness of the Focal-Masked Loss Function. We conduct an ablation study for the new loss in Eq. 7. We analyze RMSE and si-RMSE as shown in Table 2. The existing GAN inversion is unavailable to LFOV images. We first develop a masked version and then extend it to a focal-masked version. The table shows the improvement of the proposed focal-masked loss function.

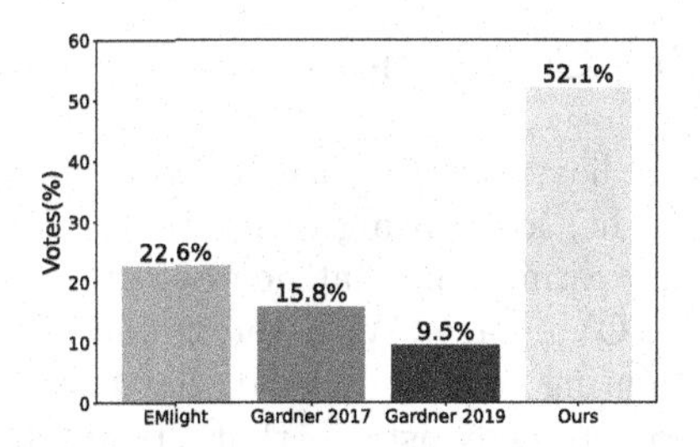

Fig. 9. User preference.

User Study. We conduct a user study to analyze predicted illumination maps. Figure 10 shows two examples of the user study. For each example, we insert an object into an image. We render the inserted object with illumination maps predicted by four methods. Users were asked to pick the images that were more closer to the images rendered by ground-truth HDR panorama. We rendered different objects under 10 scenes, and 100 users participated in the study. Results are shown in Fig. 9. We can observe that 52.1% prefer our rendered results, which significantly outperform other methods.

Table 2. Effectiveness of the focal-masked loss function. M, S, and D denote mirror, matte silver, and diffuse material spheres, respectively. ↓ denotes that smaller is better.

Metrics	Baseline [20]	Masked			Focal-masked		
	N/A	M	S	D	M	S	D
RMSE↓	N/A	0.57	0.31	0.16	**0.56**	**0.30**	**0.15**
si-RMSE↓	N/A	0.56	0.30	0.12	**0.55**	**0.29**	**0.11**

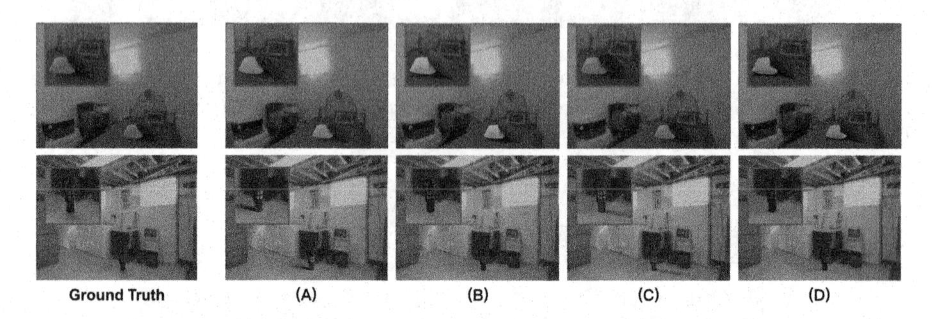

Fig. 10. For each row, the inserted objects are rendered by HDR illumination maps estimated by four methods. Which one do you think is more closer to the object rendered by the ground-truth HDR illumination map? First row: (A) EMlight, (B) Gardner 2017, (C) Ours, (D) Gardner 2019; Second row: (A) Gardner 2019, (B) Ours, (C) Gardner 2017, (D) EMlight

5 Conclusion

We have presented a coupled dual-StyleGAN panorama synthesis network (StyleLight) for lighting estimation and editing. StyleLight offers high-quality lighting estimation that generates the full indoor panorama lighting from a single LFOV image. With the trained StyleLight, we propose a structure-preserved lighting editing method that enables flexible lighting editing. Extensive experiments demonstrate that StyleLight achieves superior performance over state-of-the-art methods on indoor lighting estimation and also enables promising lighting editing on indoor HDR panoramas.

Limitation. StyleLight only considers indoor scenes in this paper. When applied to outdoor scenes, StyleLight needs a new set of hyper-parameters because lighting sources (e.g., area and intensity) are different between outdoor scenes and indoor scenes. An intuitive idea is to develop a generalizable method that adaptively tailors for various scenes, which is left to future research.

Acknowledgement. This work is supported by NTU NAP, MOE AcRF Tier 2 (T2EP20221-0033), and under the RIE2020 Industry Alignment Fund - Industry Collaboration Projects (IAF-ICP) Funding Initiative, as well as cash and in-kind contribution from the industry partner(s).

References

1. Abdal, R., Zhu, P., Mitra, N.J., Wonka, P.: StyleFlow: attribute-conditioned exploration of StyleGAN-generated images using conditional continuous normalizing flows. ACM Trans. Graph. (TOG) **40**(3), 1–21 (2021)
2. Banterle, F., Callieri, M., Dellepiane, M., Corsini, M., Pellacini, F., Scopigno, R.: EnvyDepth: an interface for recovering local natural illumination from environment maps. In: Computer Graphics Forum, vol. 32, pp. 411–420. Wiley Online Library (2013)
3. Barron, J.T., Malik, J.: Intrinsic scene properties from a single RGB-D image. In: Proceedings of the IEEE Conference on Computer Vision and Pattern Recognition, pp. 17–24 (2013)
4. Gardner, M.A., Hold-Geoffroy, Y., Sunkavalli, K., Gagné, C., Lalonde, J.F.: Deep parametric indoor lighting estimation. In: Proceedings of the IEEE/CVF International Conference on Computer Vision, pp. 7175–7183 (2019)
5. Gardner, M.A., et al.: Learning to predict indoor illumination from a single image. arXiv preprint arXiv:1704.00090 (2017)
6. Garon, M., Sunkavalli, K., Hadap, S., Carr, N., Lalonde, J.F.: Fast spatially-varying indoor lighting estimation. In: Proceedings of the IEEE/CVF Conference on Computer Vision and Pattern Recognition, pp. 6908–6917 (2019)
7. Härkönen, E., Hertzmann, A., Lehtinen, J., Paris, S.: GANSpace: discovering interpretable GAN controls. Adv. Neural. Inf. Process. Syst. **33**, 9841–9850 (2020)
8. Heusel, M., Ramsauer, H., Unterthiner, T., Nessler, B., Hochreiter, S.: GANs trained by a two time-scale update rule converge to a local nash equilibrium. Adv. Neural Inf. Process. Syst. **30** (2017)
9. Hold-Geoffroy, Y., Sunkavalli, K., Hadap, S., Gambaretto, E., Lalonde, J.F.: Deep outdoor illumination estimation. In: Proceedings of the IEEE Conference on Computer Vision and Pattern Recognition, pp. 7312–7321 (2017)
10. Johnson, J., Alahi, A., Fei-Fei, L.: Perceptual losses for real-time style transfer and super-resolution. In: Leibe, B., Matas, J., Sebe, N., Welling, M. (eds.) ECCV 2016. LNCS, vol. 9906, pp. 694–711. Springer, Cham (2016). https://doi.org/10.1007/978-3-319-46475-6_43
11. Kanamori, Y., Endo, Y.: Relighting humans: occlusion-aware inverse rendering for full-body human images. arXiv preprint arXiv:1908.02714 (2019)
12. Karras, T., Laine, S., Aittala, M., Hellsten, J., Lehtinen, J., Aila, T.: Analyzing and improving the image quality of StyleGAN. In: Proceedings of the IEEE/CVF Conference on Computer Vision and Pattern Recognition, pp. 8110–8119 (2020)
13. Karsch, K., Hedau, V., Forsyth, D., Hoiem, D.: Rendering synthetic objects into legacy photographs. ACM Trans. Graph. (TOG) **30**(6), 1–12 (2011)
14. LeGendre, C., et al.: DeepLight: learning illumination for unconstrained mobile mixed reality. In: Proceedings of the IEEE/CVF Conference on Computer Vision and Pattern Recognition, pp. 5918–5928 (2019)
15. Li, H., Liu, J., Bai, Y., Wang, H., Mueller, K.: Transforming the latent space of StyleGAN for real face editing. arXiv preprint arXiv:2105.14230 (2021)
16. Lombardi, S., Nishino, K.: Reflectance and illumination recovery in the wild. IEEE Trans. Pattern Anal. Mach. Intell. **38**(1), 129–141 (2015)
17. Lopez-Moreno, J., Hadap, S., Reinhard, E., Gutierrez, D.: Compositing images through light source detection. Comput. Graph. **34**(6), 698–707 (2010)

18. Maier, R., Kim, K., Cremers, D., Kautz, J., Nießner, M.: Intrinsic3D: High-quality 3D reconstruction by joint appearance and geometry optimization with spatially-varying lighting. In: Proceedings of the IEEE International Conference on Computer Vision, pp. 3114–3122 (2017)
19. Pandey, R., et al.: Total relighting: learning to relight portraits for background replacement. ACM Trans. Graph. (TOG) **40**(4), 1–21 (2021)
20. Roich, D., Mokady, R., Bermano, A.H., Cohen-Or, D.: Pivotal tuning for latent-based editing of real images. arXiv preprint arXiv:2106.05744 (2021)
21. Shen, Y., Gu, J., Tang, X., Zhou, B.: Interpreting the latent space of GANs for semantic face editing. In: Proceedings of the IEEE/CVF Conference on Computer Vision and Pattern Recognition, pp. 9243–9252 (2020)
22. Shen, Y., Zhou, B.: Closed-form factorization of latent semantics in GANs. In: Proceedings of the IEEE/CVF Conference on Computer Vision and Pattern Recognition, pp. 1532–1540 (2021)
23. Song, S., Funkhouser, T.: Neural illumination: lighting prediction for indoor environments. In: Proceedings of the IEEE/CVF Conference on Computer Vision and Pattern Recognition, pp. 6918–6926 (2019)
24. Srinivasan, P.P., Mildenhall, B., Tancik, M., Barron, J.T., Tucker, R., Snavely, N.: Lighthouse: predicting lighting volumes for spatially-coherent illumination. In: Proceedings of the IEEE/CVF Conference on Computer Vision and Pattern Recognition, pp. 8080–8089 (2020)
25. Sun, T., et al.: Single image portrait relighting. ACM Trans. Graph. **38**(4), 79–1 (2019)
26. Tewari, A., et al.: StyleRig: rigging StyleGAN for 3D control over portrait images. In: Proceedings of the IEEE/CVF Conference on Computer Vision and Pattern Recognition, pp. 6142–6151 (2020)
27. Valgaerts, L., Wu, C., Bruhn, A., Seidel, H.P., Theobalt, C.: Lightweight binocular facial performance capture under uncontrolled lighting. ACM Trans. Graph. **31**(6), 187–1 (2012)
28. Wang, T., Zhang, Y., Fan, Y., Wang, J., Chen, Q.: High-fidelity GAN inversion for image attribute editing. arXiv preprint arXiv:2109.06590 (2021)
29. Wu, C., Wilburn, B., Matsushita, Y., Theobalt, C.: High-quality shape from multi-view stereo and shading under general illumination. In: CVPR 2011, pp. 969–976. IEEE (2011)
30. Xiao, J., Ehinger, K.A., Oliva, A., Torralba, A.: Recognizing scene viewpoint using panoramic place representation. In: 2012 IEEE Conference on Computer Vision and Pattern Recognition, pp. 2695–2702. IEEE (2012)
31. Zhan, F., et al.: Sparse needlets for lighting estimation with spherical transport loss. In: Proceedings of the IEEE/CVF International Conference on Computer Vision, pp. 12830–12839 (2021)
32. Zhan, F., et al.: EMLight: lighting estimation via spherical distribution approximation. arXiv preprint arXiv:2012.11116 **2** (2020)
33. Zhuang, P., Koyejo, O., Schwing, A.G.: Enjoy your editing: controllable GANs for image editing via latent space navigation. arXiv preprint arXiv:2102.01187 (2021)

Contrastive Monotonic Pixel-Level Modulation

Kun Lu[✉], Rongpeng Li, and Honggang Zhang

Zhejiang University, Hangzhou, China
{lukun199,lirongpeng,honggangzhang}@zju.edu.cn

Abstract. Continuous one-to-many mapping is a less investigated yet important task in both low-level visions and neural image translation. In this paper, we present a new formulation called MonoPix, an unsupervised and contrastive continuous modulation model, and take a step further to enable a pixel-level spatial control which is critical but can not be properly handled previously. The key feature of this work is to model the monotonicity between controlling signals and the domain discriminator with a novel contrastive modulation framework and corresponding monotonicity constraints. We have also introduced a selective inference strategy with logarithmic approximation complexity and support fast domain adaptations. The state-of-the-art performance is validated on a variety of continuous mapping tasks, including AFHQ cat-dog and Yosemite summer-winter translation. The introduced approach also helps to provide a new solution for many low-level tasks like low-light enhancement and natural noise generation, which is beyond the long-established practice of one-to-one training and inference. Code is available at https://github.com/lukun199/MonoPix.

Keywords: Pixel-level modulation · Continuous monotonic translation · Contrastive training

1 Introduction

Deep learning has significantly revolutionized the way we process and represent images. Though in most high-level tasks, we may safely assume that the input-ground truth pairs are deterministic and perfectly aligned [17,23,45] (e.g., class labels, bounding boxes, and key points), it does not hold true in many low-level scenarios. In low-light image enhancement [6,46] and natural noise generation [9,32] for instance, the ever-changing environment makes it even impossible to provide firm ground truth pairs. Directly adopting a one-to-one (O2O) mapping on these pseudo labels can result in several problems: in the training stage, it ignores the one-to-many (O2M) mapping property, and inherently invites certain divergence and instability; for evaluation and deployment purposes, it incurs large variance in different environments [38] and lacks certain scalability.

Supplementary Information The online version contains supplementary material available at https://doi.org/10.1007/978-3-031-19784-0_29.

S. Avidan et al. (Eds.): ECCV 2022, LNCS 13675, pp. 493–510, 2022.
https://doi.org/10.1007/978-3-031-19784-0_29

To enable O2M mappings, a direct solution is to introduce conditional generative models [7,29], or carry out a multi-branch ensemble on several submodules [47,53]. The major obstacle comes from the lack of dense and varied discrete image pairs to support the subsequent training. Some other methods approach continuous generation by interpolating between two models, or tuning an extra adaptation module [21,22,37,62,68]. Nevertheless, methods in this category involve two correlated networks which are usually obtained from a fine-tuning or re-training. As a result, they not only prevent an end-to-end training, but also increase the storage burden and inference complexity. Moreover, all these methods can suffer from a training-inference gap that, a discrete training is not sufficient to provide a continuous modulation at inference stage.

On the other hand, through the content and style disentanglement [12,31,36], it is shown that a single generative network is capable of handling varied styles and attributes in an end-to-end manner, and provides a smooth generation by style interpolation [13,31]. To better secure the intermediate generation quality, works on continuous cross-domain translation further propose to constrain the in-between samples by coordinating two domain discriminators [18,51], or introducing some extra constraints [71]. However, to obtain a desired output, a user either has to change the random style noise at an unpredictable time complexity, or is required to provide a proper reference style which may not always be plausible. In addition, these methods are primarily focused on an image-level holistic style translation, but not typically designed for many tasks where a pixel-level modulation is preferred.

These limitations pique our interest; and we therefore devote to providing a new solution that simultaneously enables an unsupervised, end-to-end, continuous, and monotonic pixel-level image control. By injecting a spatial controlling signal into the feature space, and benefiting from the inductive bias of convolutions, we render it possible to directly manipulate from a fine-grained image-like interface. The monotonicity control is inspired by the recent progress on

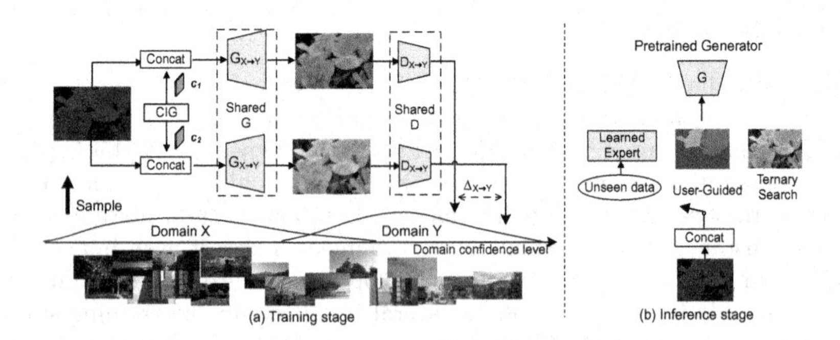

Fig. 1. Overview of MonoPix framework. Blue blocks are learnable parts. CIG denotes contrastive intensity generator (a procedure) that randomly produces two spatial control signals c_1 and c_2 for contrastive training. $\Delta_{X \mapsto Y}$ is the confidence difference. (a) During training, we model the monotonicity between control signal and domain confidence. (b) We enable multiple inference strategies or perform fast adaptation through another expert network (Color figure online)

contrastive learning [10,72] and image-to-image translation [18,51], where we further devise a contrastive modulation scheme that connects the control monotonicity with the confidence level of a domain discriminator. In this way, one image more "resembles" the target domain will get rewarded with a higher discriminator score; and the training could proceed without any other style level supervisions. A similar previous work CUT [54] introduces patchwise contrastive loss (on different patches) for image translation, while it does not support image editing and O2M mapping. A schematic illustration of our approach can be found in Fig. 1. Simple though these ideas are, they construct a competitive new solution for continuous domain translation, typically those in conjunction with conventional low-level tasks.

We summarize the contributions of this work:

- We investigate a new research area of continuous pixel-level image modulation which has not been particularly investigated before, and provide the corresponding solution. The surveyed topic enables an interesting unsupervised continuous translation and provides a fine-grained image editing tool.
- We propose a new contrastive learning-based solution for continuous image translation, which can be seamlessly integrated into existing GAN-based one-to-one mapping schemes, and is both training and inference-efficient. The proposed approach does not need any domain-specific labels and has no training and inference gap.
- We provide several basic evaluation metrics for continuous generation, and validate the state-of-the-art performance of the proposed model on both the translation continuity and quality. Extensive results on low-level tasks also demonstrate the effectiveness of our proposed approach.
- We are one of the first that revisit the long-established O2O mapping protocol in many low-level tasks. Our work suggests it is plausible and probably more flexible to formulate them in an O2M or M2M manner instead.

2 Related Work

One-to-Many Mappings in Low-Level Task. O2M mapping commonly occurs in low-level image processing. It is evidenced in recent literature that many tasks, such as deraining [53], noise modeling [7,21,22], and low-light image enhancement [29,34,64] can be categorized into this scope. Data-guided approaches model the diversity with synthesized one-to-many pairs (usually with physical models) and generate them with the corresponding control signals [5,7,22,29,57]. As an alternative to the dense training pairs, image modulation methods address this task in a two-stage training manner. Upchurch *et al.* [66] propose to train two networks and implement feature interpolations. In CFSNet [67] the extra network is further simplified into only a tuning branch. Inspired by the normalization operations, He *et al.* [21] propose to insert a per-layer modulation module and adjust them when trained on another enhancing level. Similarly, DynamicNet [62] performs adaptation with several extra tuning blocks. Besides feature interpolation, it is also shown that modulations can take place via interpolation on the network parameters [68] or image residuals [22].

There are also some efforts that attempt to perform consecutive enhancement [58,75], but the increased complexity and lack of control range also limit their applications.

In contrast, we present a continuous modulation in a totally unsupervised[1], end-to-end, and inference-efficient manner by training a single network that executes all these tasks. There is no training and inference gap in the proposed approach, which by design allows a continuous and realistic generation.

Image-to-Image Translation. CNN-based image translation mainly takes the style representation as an auxiliary loss [16,30,39,63]. Different from designing explicit style descriptors, it is observed by Dumoulin *et al.* [15] that incorporating an affine transformation after the instance normalization (IN) [65] layer can lead to a conditional style generation, which further inspires data-driven style manipulation methods like AdaIN [25] and WCT [41]. StyleBank [8], on the other hand models the style translation with a trainable bank that disentangles the style representation. It is also shown in [76] and [40] that by injecting the style information as part of the input, the model can generate varied samples with the same set of parameters.

As another line of work, GAN [19] provides an important tool for image translation. Pioneering works like Pix2Pix [27] and CycleGAN [80] introduce adversarial and unsupervised conditional generation, but are O2O mapping models. To enable a diverse translation, non-one-to-one mappings are thereafter investigated [3,12,26,48]. As one of the most popular and successful schemes, disentangling the content and style representation exhibits a great success on a wealth of attribute and style translation tasks [13,26,31,36,73], and enables a continuous translation by interpolating between two latent vectors [13,31]. Recently, works on continuous cross-domain translation further refine the quality of intermediate images by introducing an interpolation discriminator [43,71], constraining the intermediate results with discriminators from both sides [18,51], or by exploiting the path of interpolation and translation manifold [11,44,56].

Different from these existing works, the proposed approach eschews the need for either any extra discriminator or a weighted bi-lateral supervision that can hardly scale to model many-to-one mappings. Instead, we teach one discriminator by contrastively feeding two samples and modeling the consistency between control signals and translation monotonicity.

Image Editing. Our work is also related to image editing. This task can be achieved generally by learning a conditional image-level mapping [27,35,49,80] or by exploring the latent semantic trajectories [2,14,60,79,82]. Modern image editing methods prove to be effective in modeling multiple image-level features such as semantic contents, geometric shapes, and styles [42,61]; but there are a few that provide a spatial control [33,55,81], and typically in an unsupervised manner. The proposed method, further, provides a fine-grained, unsupervised, and spatial-level editing, whilst exhibiting both fidelity to inputs and a strong scalability for style and content manipulation.

[1] In this paper, we typically refer to methods that perform domain translation without the need for paired images and intermediate domain intensity labels.

3 Continuous Cross-Domain Pixel-Level Control

First we formalize a brief description of the newly examined task. Consider two domains X and Y each with corresponding image samples i.e. $\{\boldsymbol{x}_i\}_{i=1}^{M}$ and $\{\boldsymbol{y}_i\}_{i=1}^{N}$, together with a control signal $\boldsymbol{c}$ of the shape $W \times H$ representing the translation intensity of each pixel. We learn a mapping function $G_{X \mapsto Y} : X \times C \mapsto Y$ that translates from X to Y, and from Y to X if necessary. Moreover, the translation is required to be monotonic. Thus, for each image or patch $\boldsymbol{x}_i$ and two control signals $\boldsymbol{c_1}$ and $\boldsymbol{c_2}$, we require the translated result $G_{X \mapsto Y}(\boldsymbol{x}_i, \boldsymbol{c_2})$ better belongs to the target domain Y than its counterpart if $c_2^{i,j} > c_1^{i,j}$ for each pixel position. In the following, we will elaborate on the detailed framework and corresponding loss functions that lead to accomplishing this goal.

3.1 Pixel-Level Contrastive Generation

In existing works on unsupervised domain translation, the conditional style vector is usually constructed as a flat one [12,31] or with coarse spatial shapes [33]. These approaches are moderately satisfactory when controlling the local attributes, but can not enable a fine-grained control.

To tackle this problem, we first visit the pixel-level control signal. This idea is plain and easy to implement, where we can directly formulate the control signal as an extra input channel. Without loss of generality, we always assume that the pixel-level control signal is bounded in $[0, 1]$. Then, the translated result $\hat{\boldsymbol{y}}$ of an input $\boldsymbol{x}$ could be represented as (and vice versa from Y to X)

$$\hat{\boldsymbol{y}} = G_{X \mapsto Y}(\texttt{concat}(\boldsymbol{x}, \boldsymbol{c})) \quad \text{for each} \quad c^{i,j} \in [0, 1] \tag{1}$$

However, to enable such a pixel-level training is non-trivial. First, modern training schemes and datasets are only focused on holistic attributes, but there are no pixel-level labels. Second, we do not have a gallery of domain-specific intermediate labels that guide the modulation process, except for the coarse binary tags - 0 or 1, which can not model the domain distribution and precludes conventional absolute intensity-based training. To overcome these problems, we propose to formulate a contrastive learning-based approach instead, that trains the domain generator G by relative intensity. In the training stage, we first randomly generate two pixel-level control signals $\boldsymbol{c_1}$ and $\boldsymbol{c_2}$ for each input patch, then we encourage the discriminator to yield a higher domain-specific confidence level for that with a higher translation strength. Since the discriminator D is trained to examine both the quality and domain-specific features of generated images, a higher confidence level for D means a better likelihood that the output belongs to the target domain. This is different from DLOW [18], who needs to train two discriminators and overlooks the underlying data distributions in the source domain. Note that in the training stage, we empirically set a control signal $\boldsymbol{c}$ as one filled with the same value (i.e., $c^{1,1} = c^{1,2} = ... = c^{W,H}$), considering the randomness of control signal and the inductive bias of convolutional operation. On the contrary, assigning each $c^{i,j}$ with randomly generated numbers during training is tedious, and also violates the smoothness of neighboring pixels.

3.2 Normalization-Compatible Translator

Although concatenating the pixel-level control signal $\boldsymbol{c}$ can be direct and simple, it is incompatible with commonly used instance normalization modules [65]. Denoted by $\{\mathcal{K}^i\}_{i=1}^C$ a convolution kernel where i represents the i-th channel, $\mathtt{Idx}$ the channel index function and $\otimes$ the convolution operation, the convoluted result for $\mathtt{concat}(\boldsymbol{x}, \boldsymbol{c})$ can be calculated as $\boldsymbol{x} \otimes \mathcal{K}^{i \neq \mathtt{Idx}(\boldsymbol{c})} + \boldsymbol{c} \otimes \mathcal{K}^{i=\mathtt{Idx}(\boldsymbol{c})}$. As a result, the numerical difference between two contrastive samples is expected to be $(\boldsymbol{c_2} - \boldsymbol{c_1}) \otimes \mathcal{K}^{i=\mathtt{Idt}(\boldsymbol{c})}$, which becomes a constant when $\boldsymbol{c}$ is element-wise the same (in convolutions, a kernel is shared across the spatial dimension). It could also be concluded that any linear operation before instance normalization would cause the same problem (for example adding the broadcast control signal to the input) as the constant difference on two channels would be eliminated by IN. To prevent this, we introduce a simple non-linear transformation like leaky ReLU [50] when adopting instance normalization. Visualizations on the activated domain-specific features (by identifying the positions where a non-linear transformation takes place) can be found in the supplement. Batch normalization, on the other hand, introduces certain contrastivity inside the training samples but does not impede the overall monotonicity.

3.3 Objective Functions

Based on the proposed contrastive pixel-level modulation framework, we provide the corresponding constraints steering towards a monotonic and cross-domain control. Since the translation can be bi-directional, we will always take that from X to Y as an example.

Monotonicity Loss. We employ the output value of target discriminator to model the likelihood of belonging to this domain. Given an input image $\boldsymbol{x}$ and two pixel-level control signal $\boldsymbol{c_1}$ and $\boldsymbol{c_2}$ where we assume $c_2^{i,j} > c_1^{i,j}$, the confidence difference $\boldsymbol{\Delta}_{X \mapsto Y}^{tar}$ between translated results can be calculated as:

$$\boldsymbol{\Delta}_{X \mapsto Y}^{tar} = D_Y(G_{X \mapsto Y}(\boldsymbol{x}, \boldsymbol{c_2})) - D_Y(G_{X \mapsto Y}(\boldsymbol{x}, \boldsymbol{c_1})) \tag{2}$$

Under the monotonicity constraint $c_2^{i,j} > c_1^{i,j}$, we encourage the confidence difference $\boldsymbol{\Delta}_{X \mapsto Y}^{tar}$ to be positive for any $\boldsymbol{c_1}$ and $\boldsymbol{c_2}$:

$$\mathcal{L}_{mono-X2Y} = \mathbb{E}_{\boldsymbol{x} \sim X, \boldsymbol{c_1}, \boldsymbol{c_2} \sim C} \| \max\left(\epsilon - \boldsymbol{\Delta}_{X \mapsto Y}^{tar}, 0\right) \|_2^2 \tag{3}$$

where $\epsilon > 0$ is a hyperparameter controlling the strength of monotonicity, which we call a *margin* here. As illustrated in Fig. 2, the above loss exerts a quadratic punishment when $\boldsymbol{\Delta}_{X \mapsto Y}^{tar} < \epsilon$, and thus introduces not only a positive $\boldsymbol{\Delta}_{X \mapsto Y}^{tar}$, but also a positive margin. Omitting ϵ can cause a trivial solution where the confidence difference is only a little greater than zero and finally a weak control.

Domain Fidelity Loss. Although monotonicity loss enables a continuous cross-domain control, there exists an asymmetry issue. As we know that a well-trained discriminator can only produce a high confidence level for a *real* image, constraining the confidence level to be low, as in the case of $\boldsymbol{c_1}$, may not be promising.

The discrepancy can lead to a biased control that lacks in the diversity and control quality. Though this can be mitigated by applying a weak paired loss (e.g., *L1* or *MSE*) between the input image and the translated one, a rigid identity mapping also introduces certain instability for $\boldsymbol{c_2}$ and involves much parameter tuning. As a surrogate, we not only require a higher confidence level for $\boldsymbol{c_2}$ from the target domain discriminator, but also a higher confidence level for $\boldsymbol{c_1}$ from the source discriminator. We formulate the domain fidelity loss as:

$$\mathcal{L}_{df-X2Y} = \mathbb{E}_{\boldsymbol{x}\sim X, \boldsymbol{c_1}, \boldsymbol{c_2}\sim C} \| \max\left(\epsilon - \boldsymbol{\Delta}^{src}_{\boldsymbol{X}\mapsto\boldsymbol{Y}}, 0\right) \|_2^2 \tag{4}$$

where $\boldsymbol{\Delta}^{src}_{\boldsymbol{X}\mapsto\boldsymbol{Y}}$ is defined as:

$$\boldsymbol{\Delta}^{src}_{\boldsymbol{X}\mapsto\boldsymbol{Y}} = D_X(G_{X\mapsto Y}(\boldsymbol{x}, \boldsymbol{c_1})) - D_X(G_{X\mapsto Y}(\boldsymbol{x}, \boldsymbol{c_2})) \tag{5}$$

GAN Loss. MonoPix also utilizes the adversarial loss [19] and cycle consistency loss [80] to provide a proper gradient when training the bi-directional translation network. GAN loss is specialized in constraining the overall quality:

$$\min\max\ \mathcal{L}_{adv-X2Y} = \mathbb{E}_{\boldsymbol{y}\sim Y}[D_Y(\boldsymbol{y})] + \mathbb{E}_{\boldsymbol{x}\sim X, \boldsymbol{c}\sim C}[1 - D_Y(\hat{\boldsymbol{y}})] \tag{6}$$

We adopt cycle consistency loss since in many low-level tasks we are more concentrated on the intensity and stability of enhancement instead of the diversity of appearance. To this end, we expect that the input image could be recovered from a reverse translation with the same intensity $\boldsymbol{c}$:

$$\mathcal{L}_{cyc-X2Y} = \mathbb{E}_{\boldsymbol{x}\sim X}\left[\|G_{Y\mapsto X}(G_{X\mapsto Y}(\boldsymbol{x}, \boldsymbol{c}), \boldsymbol{c}) - \boldsymbol{x}\|_1\right] \tag{7}$$

Overall Objective. To keep the presentation succinct, we take the translation from X to Y as an example when constructing these loss functions. Similarly losses in the reverse direction could also be formulated. For the generator $G = \{G_{X\mapsto Y}, G_{Y\mapsto X}\}$ and discriminator $D = \{D_X, D_Y\}$, the overall training objective is expressed as:

$$\begin{aligned} \min G \quad & \mathcal{L}_G = \mathcal{L}_{adv} + \lambda_{cyc}\mathcal{L}_{cyc} + \lambda_{mn}\mathcal{L}_{mono} + \lambda_{df}\mathcal{L}_{df} \\ \min D \quad & \mathcal{L}_D = \mathcal{L}_{adv} + \lambda_{mn}\mathcal{L}_{mono} + \lambda_{df}\mathcal{L}_{df} \end{aligned} \tag{8}$$

where λ_{cyc}, λ_{mn} and λ_{df} are hyperparameters balancing different losses.

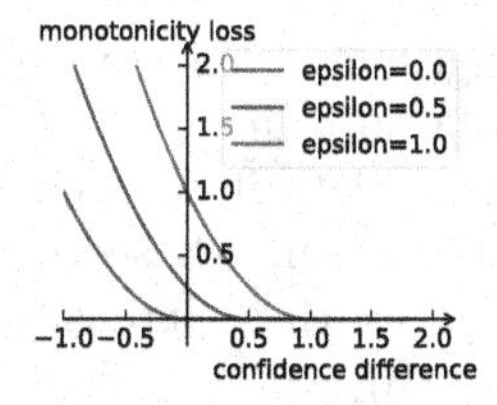

Fig. 2. Example curves of the monotonicity loss. A positive ϵ encourages a margin on confidence difference

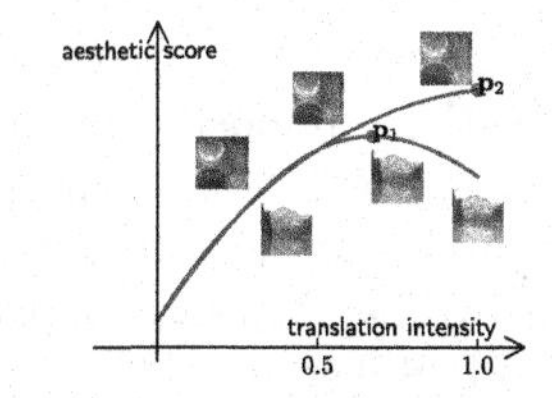

Fig. 3. Illustration of the aesthetic gap. $\boldsymbol{p}$ denotes the best balance point

3.4 Selective Inference

MonoPix naturally generates a gallery of candidate results for a given input by constraining the monotonicity. However, a high translation intensity, and hence a high domain discriminator's confidence level does not necessarily yield a satisfactory result. As shown in Fig. 3, an image can much "resemble" the target domain but is inconsistent with human perception (aesthetic score).

We assume there lies a perfect balance point $\boldsymbol{p}$ that yields the best aesthetic score, where this score can be any reasonable metric, such as human perception, and quantitative metrics like PSNR and SSIM [69] when a reference image is available. To approximate $\boldsymbol{p}$, MonoPix enables a variety of inference strategies with different execution complexity:

Exhaustive Inference. Benefiting from the monotonicity, we can always sample N control signals and choose the best result. The complexity is $O(N)$.

Ternary Search. Since the translation intensity - aesthetic score curve can be a monotonically increasing one or consists of two continuous monotonic parts, it is *unimodal-like*. Finding the local maxima can thus be efficiently solved by using *ternary search*. The details of this method are described in the supplementary material. It has the $O(\log N)$ complexity.

Guidance from Learned Expert. We can further decrease the complexity to $O(1)$ by introducing an expert network P that mimics target distribution. It can be trained in a few-shot manner and even on unseen datasets for fast adaptation.

4 Experiments and Results

We carry out experiments on both the commonly used domain translation datasets AFHQ cat-dog [13] and Yosemite summer-winter [80], and low-level tasks including prevalent LOL low-light enhancement [70] and SIDD noise generation [1]. For a fair comparison, we will always report the results when $\boldsymbol{c}$ is set to be element-wise the same when not specified.

4.1 Metrics

For continuous cross-domain image generation, we evaluate both the dynamic translation process and the quality of translated images:

AL/Rg: As we expect that an image should change monotonically from the input one, the foremost concern comes from the absolute linearity (AL) and control range (Rg). For an input image, we first modulate it by varying the intensity level $\boldsymbol{c}$ from 0 to 1 with an interval of 0.1. Then, AL score can be evaluated by the *Pearson correlation coefficient* between these intensity levels and the LPIPS distances [77] against the input image. Rg score is calculated by the *range* of these values, which reflects both the fidelity and difference.

RL/Sm: Besides absolute linearity, we follow [71] to measure the smoothness of adjacent intermediate pairs. To both overcome the drawback of *standard deviation* as used in [71] (which is sensitive to the absolute input value), and keep in comparison with AL, we calculate relative linearity (RL) from the linearity of cumulative adjacent LPIPS distances, where a high RL score indicates the translation is even between adjacent samples. The smoothness (Sm) can be evaluated by measuring the maximum value of these adjacent LPIPS distances.

ACC/FID: Following [51,71], we adopt a pre-trained ResNet [23] binary classifier to measure the success of translation. For each image, we report the highest confidence level of the classifier among the consecutive modulation process. The quality of translation can be evaluated by Fréchet inception distance (FID) [24] between the whole trajectory (with interval of 0.1) and real images, as in [44].

We use the corresponding paired evaluation metrics in low-level tasks. For low-light image enhancement, we adopt PSNR, SSIM, and LPIPS. For blind noise generation, we leverage average KL divergence (AKLD) [74].

4.2 Implementation Details

Following [28], we use the U-net [59] structure as the generator and discriminator, and Adam as the optimizer with β_1 and β_2 set to be 0.5 and 0.9 respectively. LSGAN [52] is adopted to provide the basic GAN loss. As to the hyperparameter settings, we always set λ_{cyc} as 10, as in CycleGAN, and λ_{mn} as 1, which is the same as the weight of $\mathcal{L}_{adv}$. λ_{df} and ϵ, along with training epochs vary in different datasets, which can be found in detail from the supplementary material. The learning rate is set to be 10^{-4} initially and linearly drops to zero.

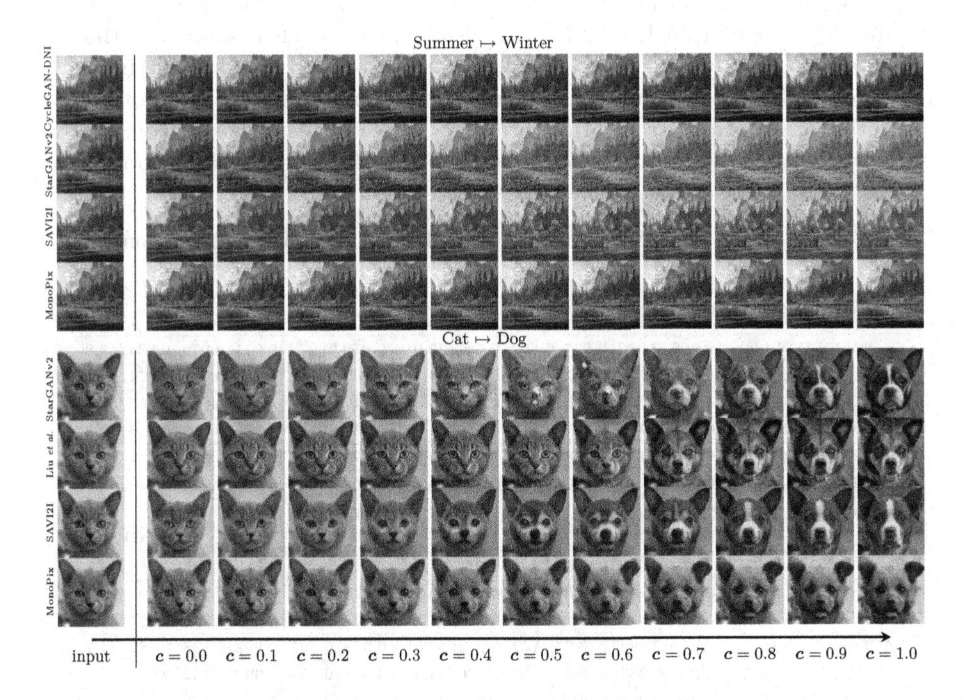

Fig. 4. Visual comparisons on continuous domain translation

4.3 Results on Continuous Cross-Domain Translation

We compare MonoPix with deep network interpolation method CycleGAN-DNI [68], style interpolation method StarGANv2 [13], and recently proposed continuous cross-domain translation model Liu *et al.* [44] and SAVI2I [51]. For style-guided methods, we randomly select their reference images from the target domain to provide the style information and report the average scores in 5 experiments (we call this variant "-Rdm"), and calculate the average target style information (style centroid) so that they can generate deterministic interpolations with $O(1)$ complexity (we call this variant "-Cent").

The visual and quantitative results can be found in Fig. 4 and Table 1 respectively. We first notice that two-stage interpolation method CycleGAN-DNI and our contrastive modulation approach MonoPix show a better fidelity on the input images, while style-guided solutions bear an obvious distortion on the original textures and attributes. Though CycleGAN-DNI exhibits the highest control range in Table 1, the intermediate smoothness is significantly worse than the rest candidates. We also find that MonoPix provides a competitive absolute linearity, but the modifications between adjacent translation intensities are not strongly guaranteed to be evenly distributed, as in Yosemite scenario. This is partially related to the non-linearity of domain discriminator, but is not a key practical problem since we can execute a finer exploration in "intensity of interests", which we illustrate more in the supplementary material. As to the realness and quality, MonoPix also performs favorably against other state of the arts, and strongly demonstrates the benefits of introducing a contrastive modulation scheme that helps to explore domain-specific features. Nevertheless, we note that evaluating FID on the whole trajectory, or on a specific intensity may produce different results for MonoPix, as we do not assume all input images share the same optimal translation intensity. More discussions on the qualitative evaluations can be found in the supplementary material. It is also noteworthy

Table 1. Quantitative results on cross-domain continuous translation

Methods	Summer ↦ winter			Winter ↦ summer		
	AL↑/Rg↑	RL↑/Sm↓	ACC↑/FID↓	AL↑/Rg↑	RL↑/Sm↓	ACC↑/FID↓
CycleGAN-DNI [68]	0.969/0.286	0.937/0.067	0.766/35.0	0.979/0.279	0.911/0.079	0.834/42.8
StarGANv2-Rdm [13]	0.805/0.134	0.992/0.021	0.616/46.1	0.741/0.107	0.996/0.018	0.771/50.1
StarGANv2-Cent [13]	0.719/0.100	0.985/0.016	0.650/49.3	0.417/0.055	0.996/0.010	0.720/52.8
SAVI2I-Rdm [51]	0.959/0.185	0.994/0.013	0.680/44.5	0.939/0.168	0.995/0.013	0.706/47.6
SAVI2I-Cent [51]	0.934/0.145	0.988/0.010	0.718/47.5	0.873/0.111	0.993/0.009	0.721/49.2
MonoPix (Ours)	0.939/0.154	0.940/0.015	0.950/38.1	0.960/0.145	0.964/0.019	0.949/47.3
Methods	Cat ↦ dog			Dog ↦ cat		
	AL↑/Rg↑	RL↑/Sm↓	ACC↑/FID↓	AL↑/Rg↑	RL↑/Sm↓	ACC↑/FID↓
StarGANv2-Rdm [13]	0.872/0.279	0.979/0.299	0.924/33.9	0.840/0.293	0.974/0.295	0.980/25.7
StarGANv2-Cent [13]	0.738/0.242	0.980/0.284	0.906/50.2	0.742/0.263	0.974/0.282	0.991/38.0
Liu *et al.* -Rdm [44]	0.930/0.281	0.967/0.174	0.956/37.1	0.933/0.313	0.956/0.216	0.997/27.5
SAVI2I-Rdm [51]	0.973/0.294	0.990/0.142	0.942/28.6	0.968/0.292	0.992/0.138	0.982/24.2
SAVI2I-Cent [51]	0.955/0.235	0.988/0.141	0.976/53.6	0.957/0.251	0.995/0.132	0.997/31.4
MonoPix (Ours)	0.986/0.398	0.990/0.050	0.970/41.1	0.983/0.407	0.984/0.076	0.997/14.4

to mention that interpolating towards the style centroid leads to a smoother and deterministic translation (variant "-Cent"), yet taking such a trajectory can cause duplicated or unrealistic patterns for style-guided approaches.

4.4 Applications in Low-Level Image Processing

In addition to cross-domain continuous translation, we show how MonoPix, along with the inference strategies can be used to facilitate low-level tasks.

Low-Light Image Enhancement. We integrate the proposed contrastive modulation approach with EnlightenGAN [28], a previous unsupervised one-to-one mapping (US-O2O) method in this area, and compare the performance on LOL test set with several state-of-the-art methods, including supervised one-to-one mapping (SL-O2O) method TBEFN [47], and reference-guided supervised (SL-Ref) approach KinD++ [78]. For fairness, we retrain EnlightenGAN on LOL training set, which yields an improved performance. We choose PSNR to guide the ternary search for this task unless specified.

To figure out the potential of our controllable generation scheme, we first implement exhaustive inference (EI), whose results can be viewed as the upper bound. As can be found in Table 2, MonoPix yields an average PSNR of 21.65, which is considerably higher than the one-to-one mapping baselines. This is because a rigid O2O mapping is not sufficient to model the diversity of data distribution. Interestingly, this result is even comparable to state-of-the-art reference-guided supervised approach KinD++, but note that MonoPix can produce competitive results without any expert knowledge or paired supervision, and streamlines the enhancement pipeline notably. The superiority of MonoPix is also supported from SSIM and LPIPS scores. When using a 7-times ternary search, we observe a tiny and acceptable performance drop. This evidence not only validates the monotonicity and continuity of the shaped generation manifold, but also demonstrates the effectiveness of our inference strategy. Next, we formulate a non-reference-guided variant by training an expert network with 10% of paired supervision (during training the backbone conditional generator is frozen). Shown in the experiment named "LP", the model then degrades to

Table 2. Quantitative results on LOL testset. ELGAN stands for EnlightenGAN

Method	Type	PSNR↑	SSIM↑	LPIPS↓
TBEFN [47]	SL-O2O	17.35	0.777	0.210
KinD++ [78]	SL-Ref	21.80	0.829	0.158
KinD++ [78]	SL-O2O	17.75	0.758	0.198
ELGAN [28]	US-O2O	19.01	0.709	0.274
Ours-EI	US-ref	21.65	0.731	0.273
Ours-TS	US-ref	21.55	0.731	0.273
Ours-LP	Few shot	19.75	0.725	0.272

Table 3. Quantitative results on SIDD natural noise generation

Method	Type	AKLD↓
CBDNet [20]	SL	0.728
ULRD [4]		0.545
GRDN [32]		0.443
DANet [74]		0.212
Ours-EI	US	0.137
Ours-TS		0.139

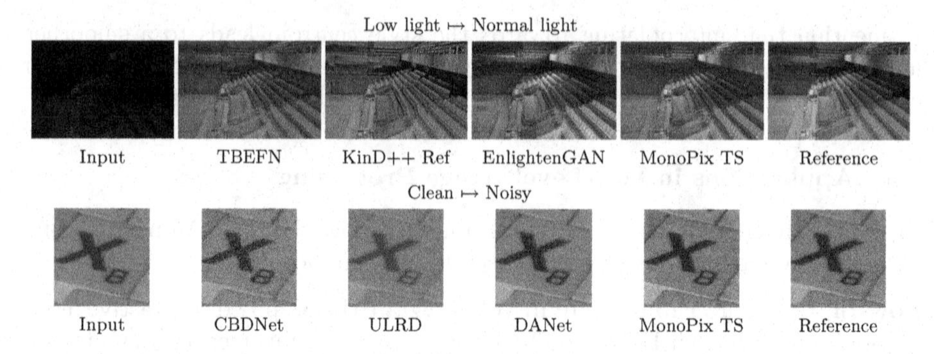

Fig. 5. Visual comparisons on the LOL low-light enhancement task and SIDD natural noise generation

Fig. 6. Examples on fine-grained pixel-level control. From left to right, we provide the input image, spatial control signal, and the modulation result respectively

an O2O mapping-based translator where the previous unsupervised contrastive training stage now serves as a pre-text task. Though not competitive as the ternary search variant, it still produces competitive results, yet with a complexity of only $O(1)$. Visual comparisons on this task can be found in Fig. 5.

Blind Image Noise Generation. Different from the commonly discussed task of image denoising, noise generation takes the reverse consideration, and typically is an O2M mapping task. We show in this part the potential and benefits of introducing the proposed continuous modulation approach into this field, and compare with state-of-the-art methods CBDNet [20], ULRD [4], GRDN [32], and DANet [74]. Quantitative and visual results can be found in Table 3 and Fig. 5 respectively, where our model ("TS" variant is guided by AKLD) again exhibits its superiority and generalization ability. To the best of our knowledge, MonoPix is the first work that proposes to continuously model natural noise in an unsupervised and spatially-controlled manner. More visual comparisons can be found in the supplementary material.

Pixel-Level Monotonic Control. In all the previous experiments, we change the overall translation strength by assigning each pixel with the same intensity. Recall that a fine-grained pixel-level control can be naturally obtained by customizing $c^{i,j}$, we here provide a few examples in Fig. 6. Note that the linearity of manifold further enables a slightly out-of-the-bound inference, especially for uni-directional tasks like illumination and noise control.

5 Discussion

5.1 Ablation Studies

To further examine the impact of the introduced method, we conduct ablation studies on Yosemite summer-winter translation with two key components that construct our approach: the monotonicity loss (carried along with the contrastive training scheme) and domain fidelity loss. Shown in Table 4 (a) and Fig. 7, the variant with none of these losses degrades to an O2O mapping, as we have expected. Note that the AL score is near zero since Pearson correlation coefficient can be negative. When we further add the monotonicity loss, the model succeeds in catching confidence differences and produces a coarse yet continuous translation. Our full model, as shown in variant (c), yields a better fidelity and wider control range. We find the translation quality is also improved under this setting, which is because fidelity loss also strengthens the translation monotonicity and encourages to explore more domain-specific features.

5.2 Sensitivity Analysis

We provide the sensitivity analysis on the hyperparameter ϵ in $\mathcal{L}_{mono}$ and $\mathcal{L}_{df}$, robustness of guiding criterion in ternary search, and the impact of inference complexity in this part. For better comparisons, we label the variants with varied ϵ consecutively following Table 4(c). The results in experiment (d) illustrate that merely asking the discriminator to produce a higher confidence level is insufficient and easily collapses to a trivial solution. In contrast, if we set ϵ to 1.0, as in experiment (e), the whole loss function is then largely dominated by $\mathcal{L}_{mono}$ and $\mathcal{L}_{df}$, which ultimately leads to an unpleasant translation. Here we note that variant (e) achieves a better FID score than (d), as the latter suffers from a severe mode collapse issue; however considering the unrealistic translation results, setting a high ϵ inevitably introduces a quality degradation for our full model (c). Concluded from these experiments, hyperparameter ϵ provides a balance for control intensity and visual quality, and is suggested to increase from a small value. Typically, setting $\epsilon = 0.5$ can yield acceptable translations

Table 4. Ablation studies on Yosemite dataset

Variant	$\mathcal{L}_{mono}$	$\mathcal{L}_{df}$	ϵ	Summer → Winter			Winter → Summer		
				AL↑/Rg↑	RL↑/Sm↓	ACC↑/FID↓	AL↑/Rg↑	RL↑/Sm↓	ACC↑/FID↓
(a)			0.5	0.143/0.000	0.999/0.000	0.733/49.9	0.105/0.000	1.000/0.000	0.702/60.0
(b)	√			0.927/0.091	0.914/0.008	0.866/41.3	0.963/0.097	0.964/0.010	0.847/50.3
(c)	√	√		0.939/0.154	0.940/0.015	0.950/38.1	0.960/0.145	0.964/0.019	0.949/47.3
(d)	√	√	0.0	0.415/0.000	1.000/0.000	0.735/50.0	0.250/0.000	1.000/0.000	0.712/59.5
(e)	√	√	1.0	0.771/0.569	0.861/0.210	0.996/39.7	0.774/0.571	0.873/0.204	0.994/48.7

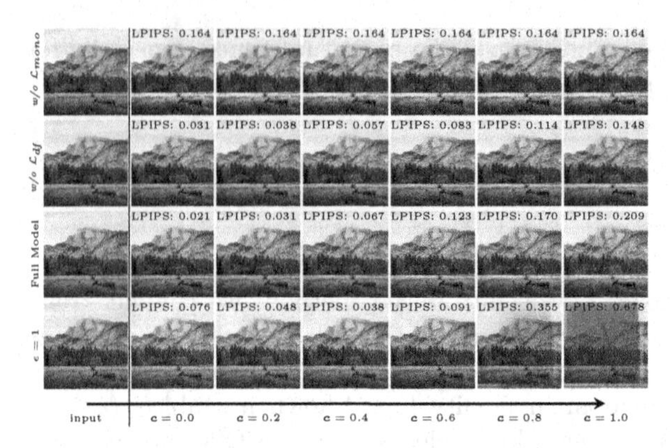

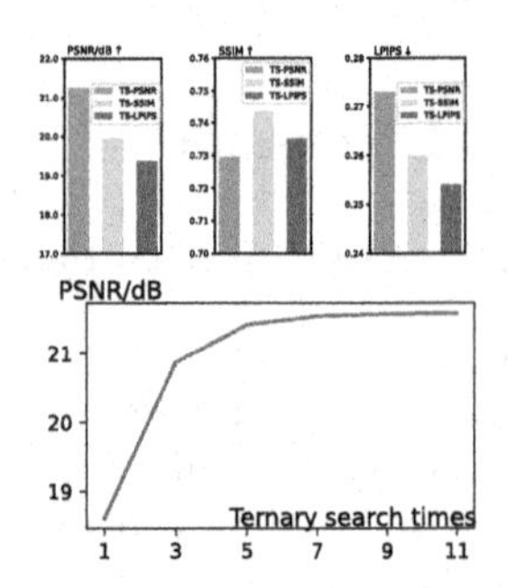

Fig. 7. Visual comparisons on the ablation study and sensitivity analysis. We label for each translated image the LPIPS score against the original input to provide a better comparison

Fig. 8. Sensitivity analysis on the choice of ternary search criterion (top) and impact of ternary search complexity (bottom)

for most of the examined tasks. Figure 8 shows that MonoPix can always produce pleasant results and correspondingly achieves the best score when guided with a specific criterion. Though not surprising, these results also provide a strong evidence on the feasibility and robustness of our proposals. We have also tested the impact of N which directly controls the inference complexity, where a 5 to 7 times of ternary search is found sufficient to provide competitive results.

6 Conclusions

We investigate in this paper an interesting yet important task - unsupervised continuous pixel-level modulation, and present MonoPix, the first successful solution by devising a novel contrastive modulation framework and the corresponding constraints. MonoPix can be deployed with selective inference complexity, and achieves a state-of-the-art performance on both the translation continuity and quality. The proposed contrastive modulation approach also provides a new alternative to paired or one-to-one mapping pipelines in low-level visions.

We also note that there are still several challenges under the current MonoPix framework, such as pixel-level multi-modal and multi-attribute control, absolute-relative linearity balance, better constraints and evaluation metrics etc. We hope the proposed method could attract more research interest in continuous pixel-level modulation, and help to shape our view on non-O2O mappings.

Acknowledgment. This work was supported in part by the National Natural Science Foundation of China under Grants 61731002 and 62071425, in part by the Zhejiang Key Research and Development Plan under Grants 2019C01002 and 2019C03131, in part by Huawei Cooperation Project, and in part by the Zhejiang Provincial Natural Science Foundation of China under Grant LY20F010016.

References

1. Abdelhamed, A., Lin, S., Brown, M.S.: A high-quality denoising dataset for smartphone cameras. In: CVPR, pp. 1692–1700 (2018)
2. Ali, J., Lucy, C., Phillip, I.: On the "steerability" of generative adversarial networks. In: ICLR (2020)
3. Almahairi, A., Rajeshwar, S., Sordoni, A., Bachman, P., Courville, A.: Augmented CycleGAN: learning many-to-many mappings from unpaired data. In: ICML, pp. 195–204 (2018)
4. Brooks, T., Mildenhall, B., Xue, T., Chen, J., Sharlet, D., Barron, J.T.: Unprocessing images for learned raw denoising. In: CVPR, pp. 11036–11045 (2019)
5. Cai, H., He, J., Qiao, Y., Dong, C.: Toward interactive modulation for photo-realistic image restoration. In: CVPR, pp. 294–303 (2021)
6. Chen, C., Chen, Q., Xu, J., Koltun, V.: Learning to see in the dark. In: CVPR, pp. 3291–3300 (2018)
7. Chen, D., Fan, Q., Liao, J., Aviles-Rivero, A., Yuan, L., Yu, N., Hua, G.: Controllable image processing via adaptive filterbank pyramid. TIP **29**, 8043–8054 (2020)
8. Chen, D., Yuan, L., Liao, J., Yu, N., Hua, G.: StyleBank: an explicit representation for neural image style transfer. In: CVPR, pp. 1897–1906 (2017)
9. Chen, J., Chen, J., Chao, H., Yang, M.: Image blind denoising with generative adversarial network based noise modeling. In: CVPR, pp. 3155–3164 (2018)
10. Chen, T., Kornblith, S., Norouzi, M., Hinton, G.: A simple framework for contrastive learning of visual representations. In: ICML, pp. 1597–1607 (2020)
11. Chen, Y.C., Xu, X., Tian, Z., Jia, J.: Homomorphic latent space interpolation for unpaired image-to-image translation. In: CVPR, pp. 2408–2416 (2019)
12. Choi, Y., Choi, M., Kim, M., Ha, J.W., Kim, S., Choo, J.: StarGAN: unified generative adversarial networks for multi-domain image-to-image translation. In: CVPR, pp. 8789–8797 (2018)
13. Choi, Y., Uh, Y., Yoo, J., Ha, J.W.: StarGAN v2: diverse image synthesis for multiple domains. In: CVPR, pp. 8188–8197 (2020)
14. Collins, E., Bala, R., Price, B., Susstrunk, S.: Editing in style: uncovering the local semantics of GANs. In: CVPR, pp. 5771–5780 (2020)
15. Dumoulin, V., Shlens, J., Kudlur, M.: A learned representation for artistic style. In: ICLR (2017)
16. Gatys, L., Ecker, A.S., Bethge, M.: Texture synthesis using convolutional neural networks. In: NeurIPS, vol. 28 (2015)
17. Girshick, R.: Fast R-CNN. In: ICCV (2015)
18. Gong, R., Li, W., Chen, Y., Gool, L.V.: DLOW: domain flow for adaptation and generalization. In: CVPR, pp. 2477–2486 (2019)
19. Goodfellow, I., et al.: Generative adversarial nets. In: NeurIPS, vol. 27 (2014)
20. Guo, S., Yan, Z., Zhang, K., Zuo, W., Zhang, L.: Toward convolutional blind denoising of real photographs. In: CVPR, pp. 1712–1722 (2019)
21. He, J., Dong, C., Qiao, Y.: Modulating image restoration with continual levels via adaptive feature modification layers. In: CVPR, pp. 11056–11064 (2019)
22. He, J., Dong, C., Qiao, Yu.: Interactive multi-dimension modulation with dynamic controllable residual learning for image restoration. In: Vedaldi, A., Bischof, H., Brox, T., Frahm, J.-M. (eds.) ECCV 2020. LNCS, vol. 12365, pp. 53–68. Springer, Cham (2020). https://doi.org/10.1007/978-3-030-58565-5_4

23. He, K., Zhang, X., Ren, S., Sun, J.: Deep residual learning for image recognition. In: CVPR (2016)
24. Heusel, M., Ramsauer, H., Unterthiner, T., Nessler, B., Hochreiter, S.: GANs trained by a two time-scale update rule converge to a local nash equilibrium. In: NeurIPS, vol. 30 (2017)
25. Huang, X., Belongie, S.: Arbitrary style transfer in real-time with adaptive instance normalization. In: ICCV, pp. 1501–1510 (2017)
26. Huang, X., Liu, M.Y., Belongie, S., Kautz, J.: Multimodal unsupervised image-to-image translation. In: ECCV, pp. 172–189 (2018)
27. Isola, P., Zhu, J.Y., Zhou, T., Efros, A.A.: Image-to-image translation with conditional adversarial networks. In: CVPR, pp. 1125–1134 (2017)
28. Jiang, Y., et al.: EnlightenGAN: deep light enhancement without paired supervision. TIP **30**, 2340–2349 (2021)
29. Jo, S.Y., Lee, S., Ahn, N., Kang, S.J.: Deep arbitrary HDRI: inverse tone mapping with controllable exposure changes. TMM **24**, 2713–2726 (2021)
30. Johnson, J., Alahi, A., Fei-Fei, L.: Perceptual losses for real-time style transfer and super-resolution. In: Leibe, B., Matas, J., Sebe, N., Welling, M. (eds.) ECCV 2016. LNCS, vol. 9906, pp. 694–711. Springer, Cham (2016). https://doi.org/10.1007/978-3-319-46475-6_43
31. Karras, T., Laine, S., Aila, T.: A style-based generator architecture for generative adversarial networks. In: CVPR, pp. 4401–4410 (2019)
32. Kim, D.W., Ryun Chung, J., Jung, S.W.: GRDN: grouped residual dense network for real image denoising and GAN-based real-world noise modeling. In: CVPRW (2019)
33. Kim, H., Choi, Y., Kim, J., Yoo, S., Uh, Y.: Exploiting spatial dimensions of latent in GAN for real-time image editing. In: CVPR, pp. 852–861 (2021)
34. Klyuchka, M., Neiterman, E.H., Ben-Artzi, G.: CEL-net: continuous exposure for extreme low-light imaging. arXiv:2012.04112 (2020)
35. Lee, C.H., Liu, Z., Wu, L., Luo, P.: MaskGAN: towards diverse and interactive facial image manipulation. In: CVPR, pp. 5549–5558 (2020)
36. Lee, H.-Y., Tseng, H.-Y., Huang, J.-B., Singh, M., Yang, M.-H.: Diverse image-to-image translation via disentangled representations. In: Ferrari, V., Hebert, M., Sminchisescu, C., Weiss, Y. (eds.) ECCV 2018. LNCS, vol. 11205, pp. 36–52. Springer, Cham (2018). https://doi.org/10.1007/978-3-030-01246-5_3
37. Lee, H., Kim, T., Son, H., Baek, S., Cheon, M., Lee, S.: Smoother network tuning and interpolation for continuous-level image processing. arXiv:2010.02270 (2020)
38. Li, C., Guo, C., Han, L.H., Jiang, J., Cheng, M.M., Gu, J., Loy, C.C.: Low-light image and video enhancement using deep learning: a survey. TPAMI (2021)
39. Li, C., Wand, M.: Combining Markov random fields and convolutional neural networks for image synthesis. In: CVPR, pp. 2479–2486 (2016)
40. Li, Y., Fang, C., Yang, J., Wang, Z., Lu, X., Yang, M.H.: Diversified texture synthesis with feed-forward networks. In: CVPR, pp. 3920–3928 (2017)
41. Li, Y., Fang, C., Yang, J., Wang, Z., Lu, X., Yang, M.H.: Universal style transfer via feature transforms. In: NeurIPS, vol. 30 (2017)
42. Ling, H., Kreis, K., Li, D., Kim, S.W., Torralba, A., Fidler, S.: EditGAN: high-precision semantic image editing. In: NeurIPS, vol. 34 (2021)
43. Lira, W., Merz, J., Ritchie, D., Cohen-Or, D., Zhang, H.: GANHopper: multi-hop GAN for unsupervised image-to-image translation. In: Vedaldi, A., Bischof, H., Brox, T., Frahm, J.-M. (eds.) ECCV 2020. LNCS, vol. 12371, pp. 363–379. Springer, Cham (2020). https://doi.org/10.1007/978-3-030-58574-7_22

44. Liu, Y., et al.: Smoothing the disentangled latent style space for unsupervised image-to-image translation. In: CVPR, pp. 10785–10794 (2021)
45. Long, J., Shelhamer, E., Darrell, T.: Fully convolutional networks for semantic segmentation. In: CVPR, pp. 3431–3440 (2015)
46. Lore, K.G., Akintayo, A., Sarkar, S.: LLNet: a deep autoencoder approach to natural low-light image enhancement. PR **61**, 650–662 (2017)
47. Lu, K., Zhang, L.: TBEFN: a two-branch exposure-fusion network for low-light image enhancement. TMM **23**, 4093–4105 (2020)
48. Lu, Y., Tai, Y.-W., Tang, C.-K.: Attribute-guided face generation using conditional CycleGAN. In: Ferrari, V., Hebert, M., Sminchisescu, C., Weiss, Y. (eds.) ECCV 2018. LNCS, vol. 11216, pp. 293–308. Springer, Cham (2018). https://doi.org/10.1007/978-3-030-01258-8_18
49. Lv, Z., et al.: Learning semantic person image generation by region-adaptive normalization. In: CVPR, pp. 10806–10815 (2021)
50. Maas, A.L., Hannun, A.Y., Ng, A.Y., et al.: Rectifier nonlinearities improve neural network acoustic models. In: ICML, vol. 30, p. 3 (2013)
51. Mao, Q., Tseng, H.Y., Lee, H.Y., Huang, J.B., Ma, S., Yang, M.H.: Continuous and diverse image-to-image translation via signed attribute vectors. IJCV **130**, 517–549 (2022)
52. Mao, X., Li, Q., Xie, H., Lau, R.Y., Wang, Z., Paul Smolley, S.: Least squares generative adversarial networks. In: ICCV, pp. 2794–2802 (2017)
53. Ni, S., Cao, X., Yue, T., Hu, X.: Controlling the rain: from removal to rendering. In: CVPR, pp. 6328–6337 (2021)
54. Park, T., Efros, A.A., Zhang, R., Zhu, J.-Y.: Contrastive learning for unpaired image-to-image translation. In: Vedaldi, A., Bischof, H., Brox, T., Frahm, J.-M. (eds.) ECCV 2020. LNCS, vol. 12354, pp. 319–345. Springer, Cham (2020). https://doi.org/10.1007/978-3-030-58545-7_19
55. Park, T., Liu, M.Y., Wang, T.C., Zhu, J.Y.: Semantic image synthesis with spatially-adaptive normalization. In: CVPR, pp. 2337–2346 (2019)
56. Pizzati, F., Cerri, P., de Charette, R.: CoMoGAN: continuous model-guided image-to-image translation. In: CVPR, pp. 14288–14298 (2021)
57. Pumarola, A., Agudo, A., Martinez, A.M., Sanfeliu, A., Moreno-Noguer, F.: GANimation: anatomically-aware facial animation from a single image. In: ECCV, pp. 818–833 (2018)
58. Ren, D., Zuo, W., Hu, Q., Zhu, P., Meng, D.: Progressive image deraining networks: a better and simpler baseline. In: CVPR, pp. 3937–3946 (2019)
59. Ronneberger, O., Fischer, P., Brox, T.: U-Net: convolutional networks for biomedical image segmentation. In: Navab, N., Hornegger, J., Wells, W.M., Frangi, A.F. (eds.) MICCAI 2015. LNCS, vol. 9351, pp. 234–241. Springer, Cham (2015). https://doi.org/10.1007/978-3-319-24574-4_28
60. Shen, Y., Zhou, B.: Closed-form factorization of latent semantics in GANs. In: CVPR, pp. 1532–1540 (2021)
61. Shi, J., Xu, N., Xu, Y., Bui, T., Dernoncourt, F., Xu, C.: Learning by planning: language-guided global image editing. In: CVPR, pp. 13590–13599 (2021)
62. Shoshan, A., Mechrez, R., Zelnik-Manor, L.: Dynamic-net: tuning the objective without re-training for synthesis tasks. In: ICCV, pp. 3215–3223 (2019)
63. Simonyan, K., Zisserman, A.: Very deep convolutional networks for large-scale image recognition. arXiv:1409.1556 (2014)
64. Sun, X., Li, M., He, T., Fan, L.: Enhance images as you like with unpaired learning. arXiv:2110.01161 (2021)

65. Ulyanov, D., Vedaldi, A., Lempitsky, V.: Instance normalization: the missing ingredient for fast stylization. arXiv:1607.08022 (2016)
66. Upchurch, P., et al.: Deep feature interpolation for image content changes. In: CVPR, pp. 7064–7073 (2017)
67. Wang, W., Guo, R., Tian, Y., Yang, W.: CFSNet: toward a controllable feature space for image restoration. In: ICCV, pp. 4140–4149 (2019)
68. Wang, X., Yu, K., Dong, C., Tang, X., Loy, C.C.: Deep network interpolation for continuous imagery effect transition. In: CVPR, pp. 1692–1701 (2019)
69. Wang, Z., Bovik, A.C., Sheikh, H.R., Simoncelli, E.P.: Image quality assessment: from error visibility to structural similarity. TIP **13**(4), 600–612 (2004)
70. Wei, C., Wang, W., Yang, W., Liu, J.: Deep Retinex decomposition for low-light enhancement. In: BMVC (2018)
71. Wu, P.W., Lin, Y.J., Chang, C.H., Chang, E.Y., Liao, S.W.: RelGAN: multi-domain image-to-image translation via relative attributes. In: ICCV, pp. 5914–5922 (2019)
72. Ye, M., Zhang, X., Yuen, P.C., Chang, S.F.: Unsupervised embedding learning via invariant and spreading instance feature. In: CVPR, pp. 6210–6219 (2019)
73. Yu, X., Chen, Y., Liu, S., Li, T., Li, G.: Multi-mapping image-to-image translation via learning disentanglement. In: NeurIPS, vol. 32 (2019)
74. Yue, Z., Zhao, Q., Zhang, L., Meng, D.: Dual adversarial network: toward real-world noise removal and noise generation. In: Vedaldi, A., Bischof, H., Brox, T., Frahm, J.-M. (eds.) ECCV 2020. LNCS, vol. 12355, pp. 41–58. Springer, Cham (2020). https://doi.org/10.1007/978-3-030-58607-2_3
75. Zamir, S.W., et al.: Multi-stage progressive image restoration. In: CVPR, pp. 14821–14831 (2021)
76. Zhang, H., Dana, K.: Multi-style generative network for real-time transfer. In: Leal-Taixé, L., Roth, S. (eds.) ECCV 2018. LNCS, vol. 11132, pp. 349–365. Springer, Cham (2019). https://doi.org/10.1007/978-3-030-11018-5_32
77. Zhang, R., Isola, P., Efros, A.A., Shechtman, E., Wang, O.: The unreasonable effectiveness of deep features as a perceptual metric. In: CVPR, pp. 586–595 (2018)
78. Zhang, Y., Guo, X., Ma, J., Liu, W., Zhang, J.: Beyond brightening low-light images. IJCV **129**(4), 1013–1037 (2021)
79. Zhu, J., Shen, Y., Zhao, D., Zhou, B.: In-domain GAN inversion for real image editing. In: Vedaldi, A., Bischof, H., Brox, T., Frahm, J.-M. (eds.) ECCV 2020. LNCS, vol. 12362, pp. 592–608. Springer, Cham (2020). https://doi.org/10.1007/978-3-030-58520-4_35
80. Zhu, J.Y., Park, T., Isola, P., Efros, A.A.: Unpaired image-to-image translation using cycle-consistent adversarial networks. In: ICCV, pp. 2223–2232 (2017)
81. Zhu, P., Abdal, R., Qin, Y., Wonka, P.: SEAN: image synthesis with semantic region-adaptive normalization. In: CVPR, pp. 5104–5113 (2020)
82. Zhuang, P., Koyejo, O., Schwing, A.G.: Enjoy your editing: controllable GANs for image editing via latent space navigation. In: ICLR (2021)

Learning Cross-Video Neural Representations for High-Quality Frame Interpolation

Wentao Shangguan, Yu Sun, Weijie Gan, and Ulugbek S. Kamilov(✉)

Washington University in St. Louis, St. Louis, USA
kamilov@wustl.edu
https://cigroup.wustl.edu/

Abstract. This paper considers the problem of temporal video interpolation, where the goal is to synthesize a new video frame given its two neighbors. We propose *Cross-Video Neural* ***Re****presentation (****CURE****)* as the first video interpolation method based on *neural fields (NF)*. NF refers to the recent class of methods for neural representation of complex 3D scenes that has seen widespread success and application across computer vision. CURE represents the video as a continuous function parameterized by a coordinate-based neural network, whose inputs are the spatiotemporal coordinates and outputs are the corresponding RGB values. CURE introduces a new architecture that conditions the neural network on the input frames for imposing space-time consistency in the synthesized video. This not only improves the final interpolation quality, but also enables CURE to learn a prior across multiple videos. Experimental evaluations show that CURE achieves the state-of-the-art performance on video interpolation on several benchmark datasets. (This work was supported by CCF-2043134.)

Keywords: Neural video representation · Neural fields · Video interpolation · Video enhancement

1 Introduction

The problem of temporal video interpolation seeks to synthesize new video frames given the observation of the neighboring frames. Video interpolation is crucial for many applications, including video compression [25], frame-rate conversion [3], novel view synthesis [9], and frame recovery [51]. Traditional video interpolation approaches have considered variational optimization [2], nearest neighbor search [5], or kernel filtering [49]. Recent work in the area has focused on deep learning (DL), where convolutional neural networks (CNNs) are trained to synthesize the desired frames. Two most widely-used

W. Shangguan and Y. Sun—These authors contributed equally and are ranked in the alphabetical order.

S. Avidan et al. (Eds.): ECCV 2022, LNCS 13675, pp. 511–528, 2022.
https://doi.org/10.1007/978-3-031-19784-0_30

Fig. 1. Visual illustration of a frame interpolated by CURE, NSFF [19], and AMBE [36]. NSFF and AMBE are the state-of-the-art methods for dynamic-scene view synthesis and video interpolation, respectively. Note how CURE substantially improves the visual quality and reduces the residual error in the second row.

classes of DL video interpolation methods are kernel-based [10,31–33] or flow-based [1,11,23,34,40,41,50,53] methods.

There has been considerable recent interest in the class of methods known as *neural fields (NF)* [37,46]. NF seeks to represent varying physical quantities of spatial and temporal coordinates using fully-connected coordinate-based neural networks. The NF neural networks take spatial or temporal coordinates at the input and produce the corresponding physical quantity at the output. This coordinate-based representation frees NF methods from pre-defined pixel/voxel grids, thus enabling the learning of *continuous* fields from discrete observations. For example, the popular *neural radiance fields (NeRF)* [28] and its various extensions [21,26] have achieved the state-of-the-art results in synthesizing novel views of complex 3D scenes from a set of 2D images. The NF approach has also been extended to the video setting by learning the representation of temporally-varying scenes from a set of videos [8,17,18,39,52]. Despite the recent activity, to the best of our knowledge, there is no NF method for temporal video interpolation that jointly addresses the lack of the camera-pose information, the need for test-time optimization, and the use of priors on the unknown video frames.

In this paper, we propose ***C****ross-Video Ne****u****ral* ***Re****presentation (****CURE****)* as a novel NF method for high-quality temporal video interpolation. Unlike traditional NF methods for representing dynamic scenes, CURE uses a deep neural network to directly represent a video as a continuous function of time and space. This conceptual change significantly simplifies the setup by circumventing the reliance on the camera pose information, which is usually difficult to estimate in certain applications such as handheld filming. The direct application of NF to video interpolation leads to poor performance due to the challenge of representing complex space-time variations from limited data. CURE addresses this issue by introducing a novel strategy based on conditioning the neural field on the input frames using our *spatiotemporal encoding module (STEM)*. For each queried pixel, STEM generates a local feature code that contains the spatial information around the pixel as well as the temporal information from the neighboring frames. STEM thus incorporates a space-time consistency prior into the representation network for mitigating visual artifacts. The architecture of conditioning also allows CURE to learn a video prior across different videos, enabling fast frame interpolation with a simple feed-forward pass. This is fundamentally different from the traditional NF approach of optimizing the weights of neural network for every new test video, thus making the inference computationally expensive.

Contributions: The technical contributions of this work are as follows:

1. We propose CURE as a novel NF approach for temporal video interpolation without any information on the camera pose. CURE is fundamentally different from the existing flow-based and kernel-based DL approaches.
2. We develop a new architecture for CURE that leverages prior information over video frames, thus enforcing plausibility of the synthesized frames. The prior is obtained by conditioning the neural network on local feature codes encoding both temporal and spatial information extracted from the neighboring frames.
3. We perform extensive validation of CURE on several benchmark datasets, including UCF101 [47], Vimeo90K [53], SNU Films [7], Nvidia Dynamic Scene Dataset [54], and Xiph4K. The empirical results show the state-of-the-art performance of CURE for temporal video interpolation as well as highlight its advantage over the current NF methods (see Fig. 1).

2 Related Work

In this section, we briefly review the DL-based methods for video interpolation and discuss several recent results in the area of neural fields.

2.1 Deep Learning for Video Interpolation

A number of DL methods have been developed for video frame interpolation. Early work has proposed direct [24] and phase-base [27] methods that pioneered the usage of deep learning for the task. Recently, the research effort has shifted to flow-based and kernel-based methods due to their improved performance. In the following, we highlight some recent methods from these two classes.

Flow-based methods [13,23] rely on the optical flow to synthesize new temporal frames by warping the existing frames. A common scheme consists of two modules, where the first module predicts the optical flow and the second module performs frame synthesis. A recent line of work has considered improvements due to the estimation of task-oriented flow in a self-supervised manner [53], leveraging of multiple optical flow maps for characterizing bilateral motion [35,36], and adoption of recursive multi-scale flow learning [44]. Another line of work has investigated different warping strategies, such as warping both the original frames and their feature representations for extra information [29] and softmax splatting for differentiable forward warping [30].

Kernel-based methods [33] view frame interpolation as convolution operations over local patches. The idea was first introduced in AdaConv [31], which uses spatial-adaptive kernels to capture both the local motion between the input frames and the coefficients for pixel synthesis in the interpolated frame. The empirical success of AdaConv has spurred various follow-up work, including on the use of separable convolutional kernels [32], jointly leveraging optical flow and adaptive convolutions [1], introducing spatiotemporal deformable convolutions [15,43], and combining channel attention modules [7].

The proposed CURE method represents a new approach based on continuous cross-video neural representation of videos using conditional NF.

2.2 Neural Fields

NF has emerged as a popular vision paradigm with applications in 3D image and shape synthesis [6,37,45], tomography [20,22,42,48], audio processing [46], and view rendering [26,28,55]. A line of work related to this paper has considered applying NF for view synthesis of dynamic scenes, where the goal is to render a novel image of a temporally-varying scene from an arbitrary viewpoint. Examples include learning a moving human body from sparse multi-view videos [39], jointly learning deformable fields and scenes [8,19,38], and 3D video synthesis from multi-view videos [17]. Although these method can be applied for video interpolation (e.g. by fixing the viewpoint in space), they are not designed for this task and usually lead to poor interpolation quality. Another related line of work has applied NF to low-level video processing, including video denoising [4,8], video super-resolution [8], and video compression [4]. To the best of our knowledge, CURE is the first NF model designed for temporal frame interpolation.

3 Cross-Video Neural Representation

We now present the technical details of CURE. Figure 3 illustrates the overall structure of the proposed method. We begin by introducing the idea of representing a video segment as a function and then explain the architecture of STEM that infuses the consistency prior into CURE.

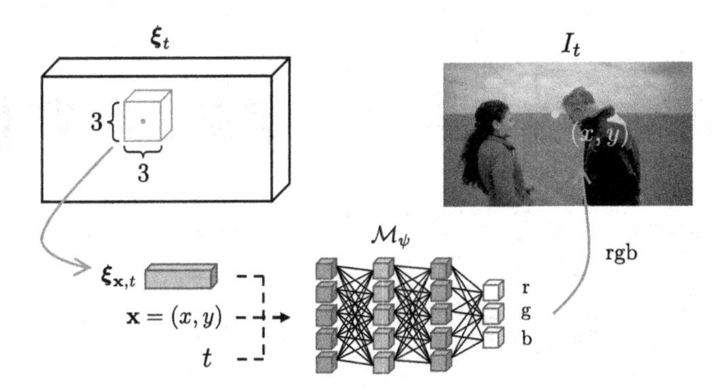

Fig. 2. CURE represents a video segment as a continuous vector-valued function, which maps $(\mathbf{x}, t, \boldsymbol{\xi}_{\mathbf{x},t})$ to a single RGB color. The vector $\boldsymbol{\xi}_{\mathbf{x},t}$ is the local feature code associated with the 3×3 region centered at $\mathbf{x}$ for the desired frame t.

3.1 Representing Video as a Conditional Neural Field

CURE represents a video segment between two observed frames as a continuous vector-valued function (see the visual illustration in Fig. 2). The input to the function is the spatial pixel location $\mathbf{x} = (x, y) \in \mathbb{R}^2$, the relative time of the desired frame $0 \leq t \leq 1$, and an associated local feature code $\boldsymbol{\xi}_{\mathbf{x},t} = \{\xi_{(x-1,y-1,t)}, \cdots, \xi_{(x+1,y+1,t)}\}$ of the 3×3 region centered at $\mathbf{x}$. The output of the function is the single RGB value $\mathbf{c} = (r, g, b) \in \mathbb{R}^3$ at $\mathbf{x}$ in the predicted frame at time t. Here, $\xi_{(u,w,\tau)}$ denotes the latent code associated with location (u, w) and time τ, and we selected a 3×3 sub-region in order to impose pixel smoothness. The formulation in CURE does not need any information on the camera pose, simplifying its use. Additionally, the continuous formulation enables CURE to synthesize any frame given by the relative time t (see Fig. 7).

CURE uses a neural network $\mathcal{M}_\psi : (\mathbf{x}, t, \boldsymbol{\xi}_{\mathbf{x},t}) \rightarrow \mathbf{c}$ to approximate the coordinate-to-function mapping, where the ψ denotes the network weights. The network contains 7 fully-connected layers with 256 neurons. Two skip connections are included in the second and fourth layers to concatenate the input vectors with the intermediate results. Figure 4(c) shows the architectural details of CURE. It is worth mentioning that as discussed in Sect. 4.3 the exclusion of the spatiotemporal coordinates from CURE leads to suboptimal empirical performance. CURE is trained over multiple videos for learning a general representation of videos. This is achieved by using the neighboring frames to generate the local feature codes. In the experiments, we trained CURE over a standard video interpolation dataset containing frame triplets.

3.2 Spatio-Temporal Encoding

Directly applying NF to video interpolation leads to poor results due to the lack of spatial and temporal consistency. Figure 8 illustrates the results obtained by the vanilla NF, which simply maps (x, y, t) to the corresponding RGB value. The

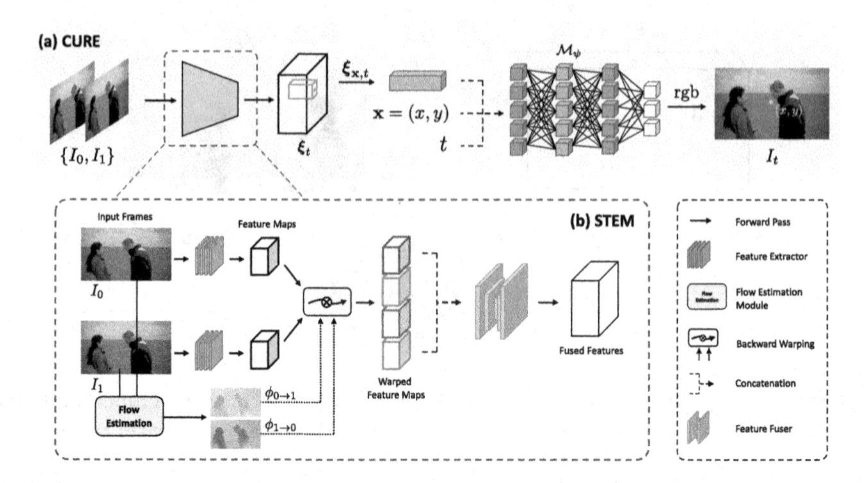

Fig. 3. (a) The overall architecture of CURE. (b) The spatiotemporal encoding module (STEM) that generates $\boldsymbol{\xi}_t$ for the predicted frame I_t.

motivation for spatio-temporal encoding is to provide additional prior information in both space and time to improve the neural representation performance.

We design STEM to generate the local feature codes $\boldsymbol{\xi}_t = \{\xi_{(1,1,t)}, ..., \xi_{(H,W,t)}\}$ from the two neighboring frames $\{I_0, I_1\}$ of the desired frame I_t. Here, H and W denote pixel width and height of the video. Figure 3(b) illustrates the detailed architecture of STEM, which we denote as $\mathcal{S}_\theta : \{I_0, I_1\} \rightarrow \boldsymbol{\xi}_t$ for convenience.

Feature Extraction. STEM first extracts two individual $H \times W \times 64$ feature maps $\{F_0, F_1\}$ from the input frames by using a customized residual network (ResNet) $\mathcal{R} : I \rightarrow F$,

$$F_0 = \mathcal{R}(I_0), \quad F_1 = \mathcal{R}(I_1). \tag{1}$$

The architecture of $\mathcal{R}$ is shown in Fig. 4(a). Note that these feature maps are later used to form the latent feature map for frame I_t via a two-step procedure: feature warping and fusion.

Feature Warping. Feature warping aims to interpolate the latent feature maps associated with time t by warping $\{F_0, F_1\}$. This step ensures the temporal consistency between existing and predicted frames in the latent space. To improve the interpolation accuracy, STEM adopts the bilateral motion approximation technique [35] to generate *four* bilateral motion maps (fw: forward & bw: backward) from the bidirectional optical flows $\phi_{0\to1}$ and $\phi_{1\to0}$,

$$\phi^{\text{fw}}_{t\to0} = (-t) \times \phi_{0\to1}, \tag{2a}$$

$$\phi^{\text{fw}}_{t\to1} = (1-t) \times \phi_{0\to1}, \tag{2b}$$

$$\phi^{\text{bw}}_{t\to0} = t \times \phi_{1\to0}, \tag{2c}$$

$$\phi^{\text{bw}}_{t\to1} = -(1-t) \times \phi_{1\to0}. \tag{2d}$$

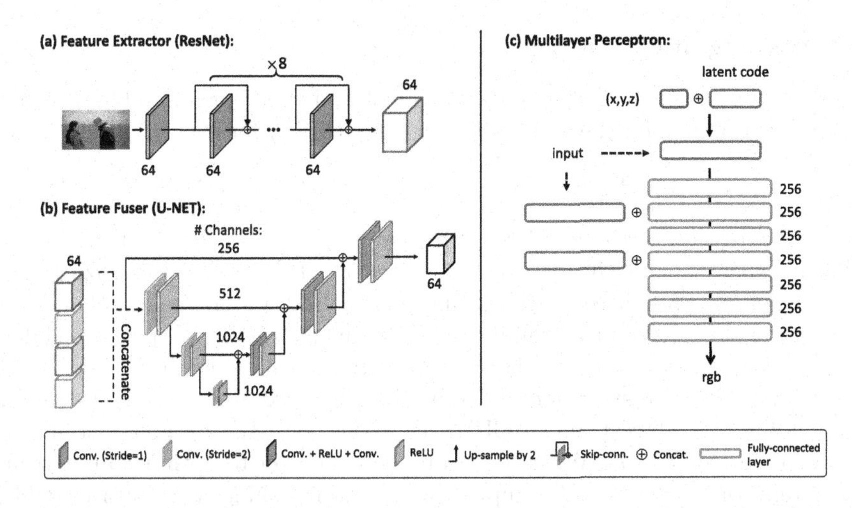

Fig. 4. Visual illustration of the network architectures used in CURE.

Here, $\phi_{0\to1}$ and $\phi_{1\to0}$ are estimated by an off-the-shelf differentiable optical flow estimator. In our case, RAFT [50] is used for flow estimation. STEM used the spatial transform network [12] to warp the feature maps. In particular, we warp a target feature map F_{tar} into a reference feature F_{ref} by using the motion map from the reference to target

$$F_{\text{ref}}(\mathbf{x}) = F_{\text{tar}}(\mathbf{x} + \phi_{\text{ref}\to\text{tar}}(\mathbf{x})) \tag{3}$$

By applying (3) and (2) to $\{F_0, F_1\}$, we can obtain four warped feature maps via cross combination

$$F^{\text{bw}}_{0\to t}(\mathbf{x}) = F_0(\mathbf{x} + \phi^{\text{bw}}_{t\to0}(\mathbf{x})), \tag{4a}$$
$$F^{\text{bw}}_{1\to t}(\mathbf{x}) = F_1(\mathbf{x} + \phi^{\text{bw}}_{t\to1}(\mathbf{x})), \tag{4b}$$
$$F^{\text{fw}}_{0\to t}(\mathbf{x}) = F_0(\mathbf{x} + \phi^{\text{fw}}_{t\to0}(\mathbf{x})), \tag{4c}$$
$$F^{\text{fw}}_{1\to t}(\mathbf{x}) = F_1(\mathbf{x} + \phi^{\text{fw}}_{t\to1}(\mathbf{x})), \tag{4d}$$

where a single warped feature map has the spatial dimension of $H \times W \times 64$. In feature warping, STEM adopted bilinear interpolation to handle off-the-grid pixels.

Feature Fusion. To output the final local feature codes for I_t, a feature fusion step is then employed to fuse the warped feature maps into a single $H \times W \times 64$ feature map,

$$\boldsymbol{\xi}_t = \mathcal{U}(F_{\text{concat}}) \tag{5}$$
$$\text{where} \quad F_{\text{concat}} := F^{\text{fw}}_{0\to t} \oplus F^{\text{fw}}_{1\to t} \oplus F^{\text{bw}}_{0\to t} \oplus F^{\text{bw}}_{1\to t}.$$

Here, $\oplus$ denotes feature concatenation along the channel dimension, and $\mathcal{U}$ represents the fusion network. In STEM, we implemented a customized U-Net as $\mathcal{U}$, which is shown in Fig. 4(b).

3.3 Training of CURE

We jointly optimize the parameters of STEM and the coordinate-based neural network by minimizing the mean squared error (MSE) for each pixel location

$$\mathcal{L}(\psi,\theta) = \|\mathcal{M}_\psi(\mathbf{x}, t, \boldsymbol{\xi}_{\mathbf{x},t}) - I_t(\mathbf{x})\|_2^2 \quad \text{where} \quad \boldsymbol{\xi}_t = \mathcal{S}_\theta(I_0, I_1). \tag{6}$$

The vector θ includes the weights of the feature extraction network $\mathcal{R}$, feature fusion network $\mathcal{U}$, and the optical flow estimator RAFT. We initialize RAFT with the default pre-trained weights, and set the RAFT iteration number to 50.

We trained CURE on the training dataset Vimeo90K [53]. We performed data augmentation by vertically and horizontally flipping the video frames. CURE is trained by taking the start and end frames as input and predicting the middle frame at $t = 0.5$. Once trained, CURE can continuously interpolate frames at any $t \in (0, 1)$ (see Fig. 7). We used Adam [14] optimizer to optimize all network parameters for 70 epochs. We implemented a diminishing learning rate which decreases by 0.95 at every epoch. The training was performed on a machine equipped with an AMD Ryzen Threadripper 3960X 24-Core Processor and 2 NVIDIA GeForce RTX 3090 GPUs. It took roughly 3 weeks to train the model.

4 Experiments

We numerically validated CURE on several standard video interpolation datasets. Our experiments include comparisons with the state-of-the-art (SOTA) methods and ablation studies highlighting the key components of the proposed architecture.

4.1 Datasets

We consider the following four datasets for numerical evaluation:

Vimeo90K: The Vimeo90K training dataset contains 51,312 frame triplets with 448×256 pixel resolution, and the validation dataset contains 3,782 triplets that we use for testing. We additionally increased the training data by extracting 91,701 triplets (i.e. $\{I_0, I_2, I_4\}$) from the septulets dataset.

UCF101: The UCF101 [47] dataset contains realistic action videos collected from YouTube. We used a subset collected by [23] for evaluation. The subset contains 379 triplets with 256×256 resolution.

SNU-FILM: The dataset [7] contains 1,240 triplets extracted from different videos with up to 1280×720 resolution. The triplets are categorized to *easy*, *medium*, *hard*, and *extreme*, based on the motion strength between frames.

Xiph4K: The dataset contains eight randomly-selected 4K videos downloaded from the Xiph website[1]. For each video, we extracted the first fifty odd-numbered frames to form the testing triplets, all resized to 2048×1080 pixels.

Nvidia Dynamic Scene (ND Scene): ND Scene [54] is a video dataset that is commonly used for novel view synthesis. The dataset contains videos of

[1] https://media.xiph.org/video/derf/.

Table 1. Average PSNR/SSIM obtained by CURE and the current state-of-the-art (SOTA) methods on UCF101, Vimeo90K, and SNU-FILM, NDScene, and Xiph4K datasets. We use bold and underline to highlight the highest and second highest values, respectively. Note the superior results of CURE over SOTA methods.

	UCF 101	Vimeo 90K	SNU-FILM				ND Scene	Xiph4K	**Average**
			Easy	Medium	Hard	Extreme			
SepConv [32] (ICCV17)	27.97 0.9401	27.32 0.9481	30.33 0.9661	27.33 0.9451	22.80 0.8794	18.06 0.7747	24.38 0.9294	22.83 0.9748	25.13 0.9072
ToFlow [53] (IJCV19)	34.58 0.9667	33.73 0.9682	39.08 0.9890	34.39 0.9740	28.44 0.9180	23.39 0.8310	30.05 0.9466	27.53 0.9850	31.40 0.9361
BMBC [35] (ECCV20)	35.16 0.9689	35.01 0.9764	39.90 0.9902	35.31 0.9774	29.33 0.9270	23.92 0.8432	36.01 0.9827	28.03 0.9131	32.83 0.9474
CAIN [7] (AAAI20)	34.91 0.9690	34.65 0.9730	39.89 0.9900	35.61 0.9776	29.90 0.9292	24.78 0.8507	34.81 0.9789	29.94 0.9184	33.06 0.9484
RRIN [16] (ICASSP20)	34.93 0.9496	35.22 0.9643	39.31 0.9897	35.23 0.9776	29.79 0.9301	24.59 0.8511	33.89 0.9750	29.30 0.9232	32.78 0.9451
AdaCoF [15] (CVPR20)	34.90 0.9680	34.47 0.9730	39.80 0.9900	35.05 0.9754	29.46 0.9244	24.31 0.8439	35.75 0.9822	28.57 0.9093	32.79 0.9458
XVFI [44] (ICCV21)	35.09 0.9685	35.07 0.9760	39.76 0.9991	35.12 0.9769	29.30 0.9245	23.98 0.8417	35.76 0.9819	28.46 0.9121	32.82 0.9476
ABME [36] (ICCV21)	35.38 0.9698	36.18 0.9805	39.59 0.9901	35.77 0.9789	30.58 0.9364	25.42 0.8639	33.17 0.9736	30.74 0.9366	33.35 0.9537
CURE	35.36 0.9705	35.73 0.9789	39.90 0.9910	35.94 0.9797	30.66 0.9373	25.44 0.8638	36.24 0.9839	30.94 0.9389	33.78 0.9555

eight scenes, recorded by a static camera rig with twelve GoPro cameras. We separate each video into triplets with a temporal sliding window of three frames.

X4K: The X4K dataset [44] contains eight 4K videos with 33 frames. We used this dataset for testing continuous multi-frame interpolation, with all videos resized to 2048 × 1080 pixels. In the test, we used the first and last frames as input frames to predict seven middle frames, that is, $\{I_4, I_8, I_{12}, I_{15}, I_{20}, I_{24}, I_{28}\}$.

Quantitative Metrics: We use *peak-signal-to-noise ratio (PSNR)* and *structural similarity index (SSIM)* to quantify the performance of all methods.

4.2 Performance Evaluation

Comparison with Video Interpolation Methods. Table 1 summarizes the results of comparing CURE against several SOTA algorithms. The results are obtained by running the published code with the pre-trained weights. The average PSNR and SSIM values obtained by each algorithm on the considered datasets are presented. Note how CURE outperforms all SOTA methods in terms of the PSNR/SSIM values averaged over all datasets. For example, CURE outperforms ABME, which is the latest SOTA method, on most datasets and tied its performance (i.e. −0.1 dB in PSNR or −0.001 in SSIM) on the rest.

Figure 5 visually compares CURE to several representative SOTA methods on an example from Vimeo90K. We highlight the visual differences by zooming in the yellow box and plotting the corresponding residual error map. In this comparison, SloMO, ToFlow, and BMBC reconstruct video frames with significant

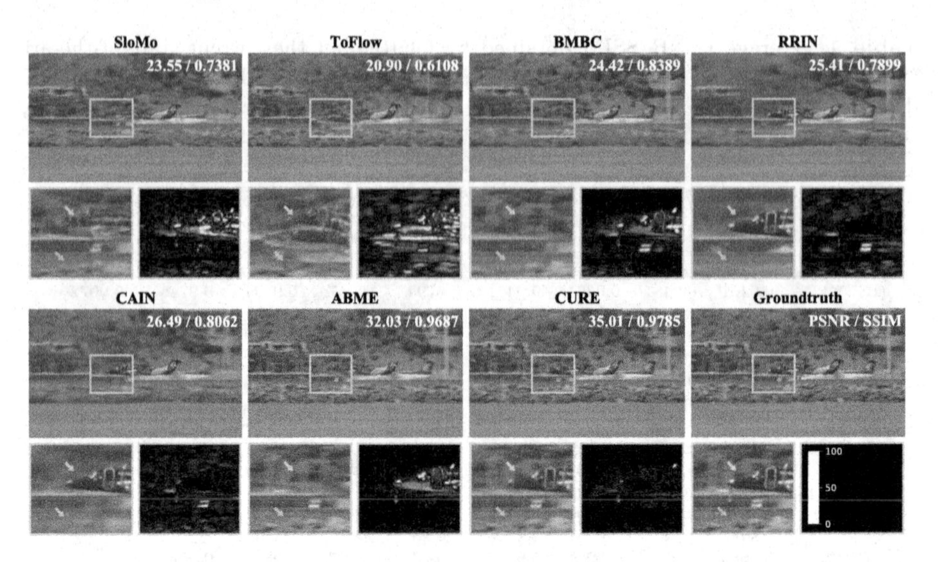

Fig. 5. Visual comparison between CURE and current SOTA methods on an example from Vimeo90K. The visual differences are highlight by the yellow boxes paired with the residual error maps. Note how CURE outperforms the SOTA methods in term of sharpness (see green arrows) and accuracy (yellow arrows). (Color figure online)

blurry and ghosting artifacts, failing to capture the shape of the airplane front (see yellow arrows). While RRIN and CAIN successfully recover the airplane front, they completely blur the mountain in the background (see green arrows). Note how CURE clearly reconstructs both the moving airplane as well as the background mountain with fine details. The qualitative visual improvements are also corroborated by the higher quantitative values and lower residual errors.

Table 2. Avg. PSNR/SSIM values obtained on X4K.

Methods	X4K
SuperSloMo	19.58 / 0.7099
ToFlow	22.33/0.7259
BMBC	24.28 / 0.7777
XVFI	24.12/0.7566
CURE	30.05 / 0.8998

Figure 6 provides a visual results on two videos from SNU-FILM (hard). These two examples were selected due to the fast motion of the recorded objects, which makes the frame interpolation challenging. ToFlow completely fails to reconstruct the bird's wingtip in the second video. Other algorithms, such as BMBC, RRIN, CAIN, and AMBE, produce blurry frames that smooth out the details on the feathers. CURE successfully handles the motion by reconstructing a sharp frame. Note the visual similarity between the frame synthesized by CURE and the groundtruth. Similar results can be observed in the comparison on the first video of a man skateboarding, where CURE significantly outperforms other algorithms.

Figure 7 highlights CURE's ability to perform continuous interpolation on a challenging video from X4K. We include the result by XVFI as the baseline. Each figure plots both image and the corresponding residual error map with respect to

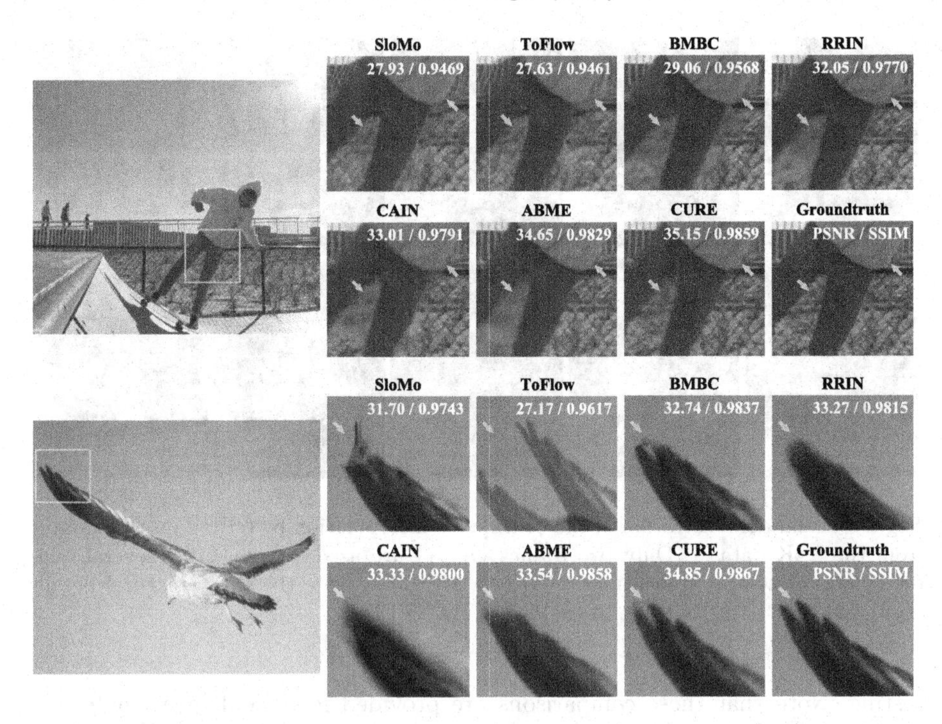

Fig. 6. Visual comparison between CURE and current SOTA methods on two examples from SNU-FILM (hard). We highlight the differences by zooming in the regions with the presence of strong motion (the man's waist and the bird's wingtip). Note how CURE outperforms the SOTA methods by removing distortions and enhancing sharpness.

the groundtruth. The visual differences on the textural details are highlighted in the green boxes. Note that both CURE and XVFI were only trained to predict the frame at $t = 0.5$ on the Vimeo90K dataset. CURE produces high-quality and accurate frames for every temporal location, significantly outperforming XVFI in terms of both visual quality and precision. Note how CURE recovers the vehicle license plate that is completely distorted by XVFI. Corresponding quantitative results are summarized in Table 2, corroborating the excellent performance of CURE.

Comparison with NF Methods. We compared CURE with two different NF algorithms: (1) vanilla NF (V-NF) that directly learns a coordinate-based neural network to map $(\mathbf{x}, t)$ to the corresponding color pixel, and (2) NSFF [19] that learns the dynamic scene in the video and applies view synthesis to generate new frames. Both algorithms need to re-optimize their weights for every test video/scene, while CURE learns a general video prior. We implemented V-NF by adopting the neural network architecture of NeRF and setting the positional encoding L to 10 and 2 for (x, y) and t, respectively. We trained V-NF by using the even-number frames. For NSFF, we downloaded the pre-trained models for

Fig. 7. Visual demonstration of the continuous interpolation by CURE on an example from the X4K dataset. Only two frames $\{I_0, I_1\}$ are used for predicting several middle frames at temporal locations $t \in \{0.25, 0.5, 0.75\}$. Note the sharp textural details preserved by CURE, but lost in the results by XVFI.

testing. Note that these comparisons are provided for completeness only; the performance of these NF algorithms on video interpolation should not be interpreted as an indication of their performance on their original tasks.

We evaluated the comparison on ND Scene. For each scene, we randomly selected one camera video from the videos captured by twelve cameras. We resized each video to accommodate the setup used by the pre-trained models of NSFF. Table 3 summarizes the details and quantitative results on each test video. As illustrated, CURE significantly outperforms V-NF and NSFF in every test scenario, and the PSNR improvement is usually over 5 dB. Figure 8 presents a visual comparison of the *truck* video to better highlight the difference. Without any consistency constraint, V-NF faithfully recovers the general content in the frame, but loses the feature details and suffers strong gridding artifacts. For example, the residual map highlights the missing edges in the interpolated frame in V-NF. On the other hand, NSFF generates a visually pleasing frame with faithful details. However, the problem of NSFF is that it predicts the wrong position of the moving truck, which significantly reduces the PSNR values. CURE avoids all these problems and interpolates a high-quality frame with both fine details and correct position of the truck. These results highlight the contribution of CURE to the research on NF-based video frame interpolation.

4.3 Ablation Studies

We highlighted the key components of the proposed model by conducting several ablation experiments. We focus on showing the effectiveness of the coordinate-based neural representation by considering two baseline variants:

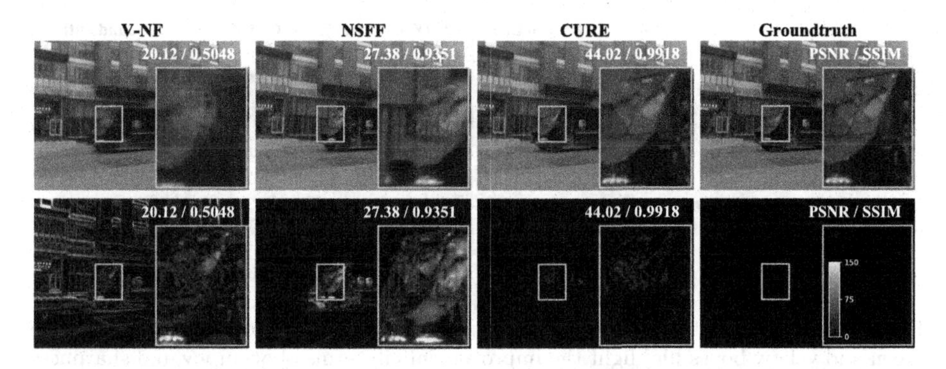

Fig. 8. Visual comparison between CURE, vanilla-NeRF, and NSFF methods on ND Scene. The first row shows the images, and the second row shows the corresponding residual error map. We highlighted the visual difference in the yellow box. CURE achieves substantially higher PSNR (about 17 dB) over NSFF. (Color figure online)

Table 3. Quantitative (PSNR/SSIM) comparisons of different video representation algorithms on the ND Scene dataset. This table highlights that CURE provides better video representation by learning a general prior across videos.

Dataset	Camera Post	Resolution	Algorithms		
			V-NF	NSFF [19]	CURE
Ballon1-2	1	540 × 288	10.75/0.4582	24.58/0.8980	**31.05/0.9742**
Dynamic Face-2	3	547 × 288	23.09/0.8317	27.13/0.9608	**40.81/0.9930**
Jumping	5	540 × 288	26.07/0.8099	27.19/0.9281	**31.23/0.9624**
Playground	2	573 × 288	22.45/0.7222	24.64/0.8912	**36.15/0.9938**
Skating-2	4	541 × 288	28.44/0.8648	34.05/0.9782	**39.67/0.9899**
Truck-2	9	542 × 288	28.37/0.8075	34.74/0.9751	**39.78/0.9887**
Umbrella	11	543 × 288	23.91/0.5876	23.91/0.8441	**39.66/0.9880**

CURE-w/o-Rep: This method directly trains the STEM network as a frame interpolator by adding an additional head to predict the RGB images. We design CURE-w/o-Rep to show the importance of the NF module.

CURE-w/o-Crd: This method simplifies CURE by removing the input of the spatiotemporal coordinate $(\mathbf{x}, t)$. We design CURE-w/o-Crd to show the importance of the coordinate indexing for synthesizing the desired RGB values.

Figure 9 visually compares CURE with its two baseline variants on an example from SNU (hard). Without neural representation, CURE-w/o-Rep predicts the wrong location of the bird's wingtip shown in the green box, losing enough representation power to properly represent the video. On the other hand, CURE-w/o-Crd produces a blurry image that loses the texture details by missing the exact spatiotemporal coordinate. Table 4 summarizes the numerical values on all

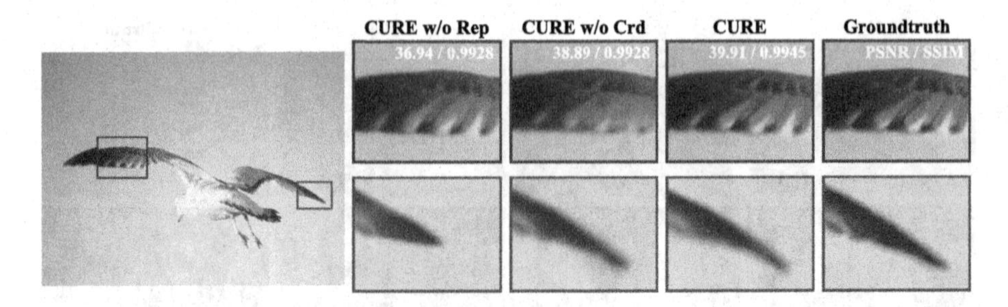

Fig. 9. Visual illustration of the importance of the various components of CURE. The green and yellow boxes highlight the improvements in terms of accuracy and sharpness due to the full CURE compared to its ablated variants. (Color figure online)

Table 4. Average PSNR/SSIM obtained by the full CURE and its ablated variants.

	UCF 101	Vimeo 90K	SNU-FILM				ND Scene	Xiph4K	**Average**
			Easy	Medium	Hard	Extreme			
CURE w/o Rep	34.80 0.9682	35.15 0.9769	39.82 0.9903	35.71 0.9787	30.36 0.9350	25.14 0.8601	35.13 0.9802	30.45 0.9359	33.32 0.9530
CURE w/o Crd	35.11 0.9689	35.25 0.9772	39.76 0.9900	35.63 0.9784	30.39 0.9350	25.26 0.8609	35.68 0.9816	30.52 0.9341	33.45 0.9531
CURE	**35.36** **0.9705**	**35.73** **0.9789**	**39.90** **0.9910**	**35.94** **0.9797**	**30.66** **0.9373**	**25.44** **0.8638**	**36.24** **0.9839**	**30.94** **0.9389**	**33.78** **0.9555**

the considered datasets. The results clearly show that CURE achieves the best results by jointly leveraging STEM and coordinate-based representation.

5 Conclusion

This work proposes a novel temporal video frame interpolation algorithm, CURE, which is based on the recent class of techniques known as neural fields (NF). The key idea of CURE is to represent video segment between two observed frames as a continuous vector-valued function of the spatiotemporal coordinates, which is parameterized by a fully-connected coordinate-based neural network. Unlike the traditional NF variants, CURE enforces space-time consistency in the synthesized frames by conditioning its NF on the information from the neighboring frames. This conditioning enables CURE to learn a prior across different videos. Experimental evaluations clearly show that the CURE outperforms the state-of-the-art methods on five benchmark datasets. The comparisons with the traditional NF methods and with the ablated variants of CURE highlight the synergistic effect of various components of the full CURE.

References

1. Bao, W., Lai, W.S., Zhang, X., Gao, Z., Yang, M.H.: MEMC-Net: motion estimation and motion compensation driven neural network for video interpolation and enhancement. IEEE Trans. Pattern Anal. Mach. Intell. **43**(3), 933–948 (2021)
2. Brox, T., Bruhn, A., Papenberg, N., Weickert, J.: High accuracy optical flow estimation based on a theory for warping. In: Pajdla, T., Matas, J. (eds.) ECCV 2004. LNCS, vol. 3024, pp. 25–36. Springer, Heidelberg (2004). https://doi.org/10.1007/978-3-540-24673-2_3
3. Castagno, R., Haavisto, P., Ramponi, G.: A method for motion adaptive frame rate up-conversion. IEEE Trans. Circuits Syst. Video Technol. **6**(5), 436–446 (1996)
4. Chen, H., He, B., Wang, H., Ren, Y., Lim, S.N., Shrivastava, A.: NeRV: neural representations for videos. Adv. Neural. Inf. Process. Syst. **34**, 21557–21568 (2021)
5. Chen, Z., Jin, H., Lin, Z., Cohen, S., Wu, Y.: Large displacement optical flow from nearest neighbor fields. In: Proceedings of the IEEE Conference on Computer Vision and Pattern Recognition, pp. 2443–2450 (2013)
6. Chen, Z., Zhang, H.: Learning implicit fields for generative shape modeling. In: Proceedings of the IEEE Conference on Computer Vision and Pattern Recognition, pp. 5939–5948 (2019)
7. Choi, M., Kim, H., Han, B., Xu, N., Lee, K.M.: Channel attention is all you need for video frame interpolation. In: Proceedings of the AAAI Conference on Artificial Intelligence, vol. 34, pp. 10663–10671 (2020)
8. Du, Y., Zhang, Y., Yu, H.X., Tenenbaum, J.B., Wu, J.: Neural radiance flow for 4D view synthesis and video processing. In: Proceedings of the IEEE/CVF International Conference on Computer Vision, pp. 14324–14334 (2021)
9. Flynn, J., Neulander, I., Philbin, J., Snavely, N.: DeepStereo: learning to predict new views from the world's imagery. In: Proceedings of the IEEE Conference on Computer Vision and Pattern Recognition, pp. 5515–5524 (2016)
10. Gupta, A., Aich, A., Roy-Chowdhury, A.K.: ALANET: adaptive latent attention network forjoint video deblurring and interpolation. arXiv:2009.01005 [cs.CV] (2020)
11. Hui, T.W., Tang, X., Loy, C.C.: LiteFlowNet: a lightweight convolutional neural network for optical flow estimation. In: Proceedings of the IEEE Conference on Computer Vision and Pattern Recognition, pp. 8981–8989 (2018)
12. Jaderberg, M., Simonyan, K., Zisserman, A., et al.: Spatial transformer networks. Adv. Neural Inf. Process. Syst. **28** (2015)
13. Jiang, H., Sun, D., Jampani, V., Yang, M.H., Learned-Miller, E., Kautz, J.: Super-SloMo: high quality estimation of multiple intermediate frames for video interpolation. In: Proceedings of the IEEE Conference on Computer Vision and Pattern Recognition, pp. 9000–9008 (2018)
14. Kingma, D., Ba, J.: Adam: a method for stochastic optimization. In: International Conference on Learning Representations (ICLR) (2015)
15. Lee, H., Kim, T., Chung, T.Y., Pak, D., Ban, Y., Lee, S.: AdaCof: adaptive collaboration of flows for video frame interpolation. In: Proceedings of the IEEE/CVF Conference on Computer Vision and Pattern Recognition, pp. 5316–5325 (2020)
16. Li, H., Yuan, Y., Wang, Q.: Video frame interpolation via residue refinement. In: IEEE International Conference on Acoustics, Speech and Signal Processing (ICASSP), pp. 2613–2617 (2020)
17. Li, T., et al.: Neural 3D video synthesis. arXiv:2103.02597 (2021)

18. Li, Z., Niklaus, S., Snavely, N., Wang, O.: Neural scene flow fields for space-time view synthesis of dynamic scenes. arXiv:2011.13084 (2020)
19. Li, Z., Niklaus, S., Snavely, N., Wang, O.: Neural scene flow fields for space-time view synthesis of dynamic scenes. In: Proceedings of the IEEE/CVF Conference on Computer Vision and Pattern Recognition, pp. 6498–6508 (2021)
20. Lindell, D.B., Martel, J.N.P., Wetzstein, G.: AutoInt: automatic integration for fast neural volume rendering. In: Proceedings of the IEEE/CVF Conference on Computer Vision and Pattern Recognition (2021)
21. Liu, L., Gu, J., Lin, K.Z., Chua, T.S., Theobalt, C.: Neural sparse voxel fields. Adv. Neural. Inf. Process. Syst. **33**, 15651–15663 (2020)
22. Liu, R., Sun, Y., Zhu, J., Tian, L., Kamilov, U.S.: Zero-shot learning of continuous 3D refractive index maps from discrete intensity-only measurements. arXiv:2112.00002 (2021)
23. Liu, Z., Yeh, R.A., Tang, X., Liu, Y., Agarwala, A.: Video frame synthesis using deep voxel flow. In: Proceedings of the IEEE International Conference on Computer Vision, pp. 4463–4471 (2017)
24. Long, G., Kneip, L., Alvarez, J.M., Li, H., Zhang, X., Yu, Q.: Learning image matching by simply watching video. In: European Conference on Computer Vision, pp. 434–450 (2016)
25. Lu, G., Zhang, X., Chen, L., Gao, Z.: Novel integration of frame rate up conversion and HEVC coding based on rate-distortion optimization. IEEE Trans. Image Process. **27**(2), 678–691 (2017)
26. Martin-Brualla, R., Radwan, N., Sajjadi, M.S., Barron, J.T., Dosovitskiy, A., Duckworth, D.: NeRF in the wild: neural radiance fields for unconstrained photo collections. In: Proceedings of the IEEE Conference on Computer Vision and Pattern Recognition, pp. 7210–7219 (2021)
27. Meyer, S., Djelouah, A., McWilliams, B., Sorkine-Hornung, A., Gross, M., Schroers, C.: PhaseNet for video frame interpolation. In: Proceedings of the IEEE Conference on Computer Vision and Pattern Recognition, pp. 498–507 (2018)
28. Mildenhall, B., Srinivasan, P.P., Tancik, M., Barron, J.T., Ramamoorthi, R., Ng, R.: NeRF: representing scenes as neural radiance fields for view synthesis. In: European Conference on Computer Vision, pp. 405–421 (2020)
29. Niklaus, S., Liu, F.: Context-aware synthesis for video frame interpolation. In: Proceedings of the IEEE Conference on Computer Vision and Pattern Recognition, pp. 1701–1710 (2018)
30. Niklaus, S., Liu, F.: Softmax splatting for video frame interpolation. In: Proceedings of the IEEE/CVF Conference on Computer Vision and Pattern Recognition, pp. 5437–5446 (2020)
31. Niklaus, S., Mai, L., Liu, F.: Video frame interpolation via adaptive convolution. In: Proceedings of the IEEE Conference on Computer Vision and Pattern Recognition, pp. 670–679 (2017)
32. Niklaus, S., Mai, L., Liu, F.: Video frame interpolation via adaptive separable convolution. In: Proceedings of the IEEE International Conference on Computer Vision, pp. 261–270 (2017)
33. Niklaus, S., Mai, L., Wang, O.: Revisiting adaptive convolutions for video frame interpolation. In: Proceedings of the IEEE/CVF Winter Conference on Applications of Computer Vision, pp. 1099–1109 (2021)
34. Oh, J., Kim, M.: DeMFI: deep joint deblurring and multi-frame interpolation with flow-guided attentive correlation and recursive boosting. arXiv:2111.09985 [cs.CV] (2021)

35. Park, J., Ko, K., Lee, C., Kim, C.S.: BMBC: bilateral motion estimation with bilateral cost volume for video interpolation. In: European Conference on Computer Vision, pp. 109–125 (2020)
36. Park, J., Lee, C., Kim, C.S.: Asymmetric bilateral motion estimation for video frame interpolation. In: Proceedings of the IEEE/CVF International Conference on Computer Vision, pp. 14539–14548 (2021)
37. Park, J.J., Florence, P., Straub, J., Newcombe, R., Lovegrove, S.: DeepSDF: learning continuous signed distance functions for shape representation. In: Proceedings of the IEEE Conference on Computer Vision and Pattern Recognition, pp. 165–174 (2019)
38. Park, K., et al.: Nerfies: deformable neural radiance fields. In: Proceedings of the IEEE/CVF International Conference on Computer Vision, pp. 5865–5874 (2021)
39. Peng, S., et al.: Neural body: implicit neural representations with structured latent codes for novel view synthesis of dynamic humans. In: Proceedings of IEEE Conference Computer Vision and Pattern Recognition (2021)
40. Ranjan, A., Black, M.J.: Optical flow estimation using a spatial pyramid network. In: Proceedings of the IEEE Conference on Computer Vision and Pattern Recognition, pp. 4161–4170 (2017)
41. Ren, Z., Yan, J., Ni, B., Liu, B., Yang, X., Zha, H.: Unsupervised deep learning for optical flow estimation. In: Proceedings of the AAAI Conference on Artificial Intelligence, vol. 31 (2017)
42. Shen, L., Pauly, J., Xing, L.: NeRP: implicit neural representation learning with prior embedding for sparsely sampled image reconstruction. arXiv:2108.10991 [eess.IV] (2021)
43. Shi, Z., Liu, X., Shi, K., Dai, L., Chen, J.: Video frame interpolation via generalized deformable convolution. IEEE Trans. Multimedia **24**, 426–439 (2021)
44. Sim, H., Oh, J., Kim, M.: XVFI: extreme video frame interpolation. In: Proceedings of the IEEE/CVF International Conference on Computer Vision, pp. 14489–14498 (2021)
45. Sitzmann, V., Martel, J., Bergman, A., Lindell, D., Wetzstein, G.: Implicit neural representations with periodic activation functions. In: Advances in Neural Information Processing Systems, vol. 33 (2020)
46. Sitzmann, V., Zollhoefer, M., Wetzstein, G.: Scene representation networks: continuous 3D-structure-aware neural scene representations. In: Advances in Neural Information Processing Systems, vol. 32 (2019)
47. Soomro, K., Zamir, A.R., Shah, M.: UCF101: a dataset of 101 human actions classes from videos in the wild. arXiv preprint arXiv:1212.0402 (2012)
48. Sun, Y., Liu, J., Xie, M., Wohlberg, B., Kamilov, U.S.: CoIL: coordinate-based internal learning for tomographic imaging. IEEE Trans. Comp. Imag. **7**, 1400–1412 (2021)
49. Takeda, H., Van Beek, P., Milanfar, P.: Spatio-temporal video interpolation and denoising using motion-assisted steering kernel (MASK) regression. In: Proceedings of the IEEE International Conference on Image Processing, pp. 637–640 (2008)
50. Teed, Z., Deng, J.: RAFT: recurrent all-pairs field transforms for optical flow. In: Vedaldi, A., Bischof, H., Brox, T., Frahm, J.-M. (eds.) ECCV 2020. LNCS, vol. 12347, pp. 402–419. Springer, Cham (2020). https://doi.org/10.1007/978-3-030-58536-5_24
51. Wu, J., Yuen, C., Cheung, N.M., Chen, J., Chen, C.W.: Modeling and optimization of high frame rate video transmission over wireless networks. IEEE Trans. Wireless Commun. **15**(4), 2713–2726 (2015)

52. Xian, W., Huang, J.B., Kopf, J., Kim, C.: Space-time neural irradiance fields for free-viewpoint video. In: Proceedings of the IEEE/CVF Conference on Computer Vision and Pattern Recognition, pp. 9421–9431 (2021)
53. Xue, T., Chen, B., Wu, J., Wei, D., Freeman, W.T.: Video enhancement with task-oriented flow. Int. J. Comput. Vision **127**(8), 1106–1125 (2019)
54. Yoon, J.S., Kim, K., Gallo, O., Park, H.S., Kautz, J.: Novel view synthesis of dynamic scenes with globally coherent depths from a monocular camera. In: Proceedings of the IEEE/CVF Conference on Computer Vision and Pattern Recognition, pp. 5336–5345 (2020)
55. Zhang, K., Riegler, G., Snavely, N., Koltun, V.: NeRF++: analyzing and improving neural radiance fields. arXiv:2010.07492 [cs.CV] (2020)

Learning Continuous Implicit Representation for Near-Periodic Patterns

Bowei Chen[(✉)], Tiancheng Zhi, Martial Hebert, and Srinivasa G. Narasimhan

Carnegie Mellon University, Pittsburgh, PA 15213, USA
{boweiche,tzhi,mhebert,srinivas}@andrew.cmu.edu

Abstract. Near-Periodic Patterns (NPP) are ubiquitous in man-made scenes and are composed of tiled motifs with appearance differences caused by lighting, defects, or design elements. A good NPP representation is useful for many applications including image completion, segmentation, and geometric remapping. But representing NPP is challenging because it needs to maintain global consistency (tiled motifs layout) while preserving local variations (appearance differences). Methods trained on general scenes using a large dataset or single-image optimization struggle to satisfy these constraints, while methods that explicitly model periodicity are not robust to periodicity detection errors. To address these challenges, we learn a neural implicit representation using a coordinate-based MLP with single image optimization. We design an input feature warping module and a periodicity-guided patch loss to handle both global consistency and local variations. To further improve the robustness, we introduce a periodicity proposal module to search and use multiple candidate periodicities in our pipeline. We demonstrate the effectiveness of our method on more than 500 images of building facades, friezes, wallpapers, ground, and Mondrian patterns in single and multi-planar scenes.

Keywords: Near-periodic patterns · Neural implicit representation · Single image optimization

1 Introduction

Patterns are all around us and help us understand our visual world. In the 1990s, a human preattentive vision experiment [43] showed that periodicity is a crucial factor in high-level pattern perception. But most real patterns are not composed of perfectly periodic (tiled) motifs. Consider the commonly occurring real-world building facade scene in Fig. 1(a). While the windows are laid out periodically, they vary in their individual appearances. There are several design elements (borders, texture), shading variations, or obstructions (tree, car, street lamp) that are not periodic. These factors make it challenging to create a good computational representation for such "Near-Periodic Patterns" (NPP).

Supplementary Information The online version contains supplementary material available at https://doi.org/10.1007/978-3-031-19784-0_31.

S. Avidan et al. (Eds.): ECCV 2022, LNCS 13675, pp. 529–546, 2022.
https://doi.org/10.1007/978-3-031-19784-0_31

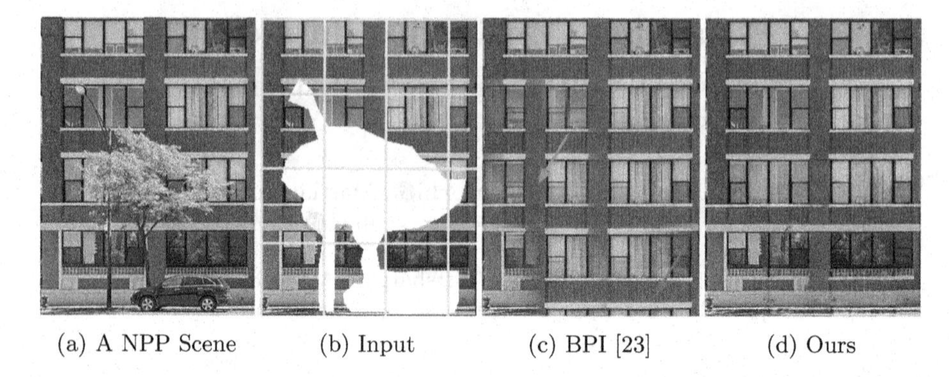

(a) A NPP Scene (b) Input (c) BPI [23] (d) Ours

Fig. 1. Inpainting to remove the tree, street lamp, and car from a near-periodic patterned (NPP) scene in (a). Input image (b) visualizes the mask (white unknown region) and detected (but inaccurate) NPP representation (yellow lattice). Guided by this inaccurate representation, the state of the art method BPI [23] (c) fails to generate windows occluded by the tree (orange arrow) and the white strip across the bottom (green arrow). Our NPP-Net (d) maintains global consistency and local variations, while preserving the known regions. (Color figure online)

A good NPP representation must preserve both *global consistency* (similar motifs layout) and *local variations* (different appearances). For global consistency, the distance (periods) and orientations between adjacent motifs should be accurate (*e.g.*, window layout). At the same time, the local details in the scene should be fully encoded (*e.g.*, appearance variations in windows or the horizontal design strips). In this paper, we present a novel method to learn such a representation that can be used for applications such as image completion (our main focus), segmentation of periodic parts, and resolution enhanced scene remapping, *e.g.*, transforming to a fronto-parallel view (see supplementary).

Existing image completion works applicable for NPP can be classified into two categories. The first category does not explicitly consider knowledge of periodicity. They complete images by training on large datasets [22,36,46,57,58] or by exploring single image statistics [2,44,50]. However, these methods fail to generate good global consistency, especially with a large unknown mask inside (interpolation) or outside (extrapolation) the image border, or severe perspective effect. The other category [16,23,24,32] models periodicity as prior for image completion. These works extract explicit NPP representations (*e.g.*, displacement vectors) and use them to guide image completion. These methods can generate good global periodic structure *if* the estimated periodicity is accurate. However, this is hard to achieve in the presence of strong local variations.

Our work is inspired by the recent progress on implicit neural representations [35] that map image coordinates to RGB values using coordinate-based multi-layer perceptrons (MLP). But, naively using this method fails on our task due to the lack of a good periodicity prior. Thus, we present a periodicity-aware coordinate-based MLP to learn a continuous implicit neural representation, which we call NPP-Net for short. The key idea is to extract periodicity information from a partially observed NPP scene and inject it into both the MLP input and the loss function to help optimize NPP representation.

Three novel steps are proposed for the above idea: (1) The *Periodicity Proposal* step extracts periodicity in the form of a set of candidate periods and orientations that are used together to handle inaccurate detections; (2) The *Periodicity-Aware Input Warping* injects periodicity into the MLP input by warping input coordinates according to the proposed periodicities. This step preserves global consistency and the MLP converges to a good periodic pattern easily; (3) Finally, the *Periodicity-Guided Patch Loss* samples observable patches according to periodicity to optimize the representation. This step preserves local variations, improves extrapolation ability, and removes high-frequency artifacts.

Our approach only requires a single image for optimization. This is important since there are no large dedicated NPP datasets. Thus, we evaluate our approach on a total of 532 NPP subclasses chosen from three datasets [1,10,48]. The scenes include building facades, friezes, ground patterns, wallpapers, and Mondrian patterns that are tiled on one or more geometric planes and perspectively warped. Our dataset is larger than those (157 at most) used in previous works [16,23,24,32] that are designed for NPP. We mainly apply NPP-Net for the image completion task, but extend it to resolution enhanced NPP remapping and NPP segmentation in the supplementary. We compare NPP-Net with four traditional [2,11,16,23] and five deep learning-based methods [44,46,50,57,58], and eight variants of NPP-Net. Experiments show that NPP-Net can interpolate and extrapolate images, in-paint large and arbitrarily shaped regions, recover blurry regions when images are remapped, segment periodic and non-periodic regions, in planar and multi-planar scenes. Figure 1 shows the effectiveness of NPP-Net, inpainting a complex NPP scene, compared to the state of the art BPI [23]. While our method is not designed for general scenes, it is a useful tool to understand a large class of man-made scenes with near-periodic patterns.

2 Related Work

Near-Periodic Patterns Completion: There are two types of image completion methods that can be applied to NPP. The first type of methods do not explicitly consider periodicity as prior for completion [6,22,36,46,51,57,58,63]. The second stream of methods takes advantage of periodicity to guide the completion. We focus on reviewing the second type of methods.

The first stage for these methods is to obtain an NPP representation to guide image completion. Existing methods aim to represent NPP by detecting the global periodicity despite local variations. The types of NPP arrangements vary [16,25,38,41,42,54] but commonly, periodic patterns are assumed to form a 2D lattice [14,21,24,26,27,37,39]. The first lattice-based work [14] for periodicity detection without human interaction finds correspondences using visual similarity and geometric consistency. Liu *et al.* [26] improve this process by incorporating generalized PatchMatch [3] and Markov Random Field. Furthermore, Lettry *et al.* [21] detect a repeated pattern model by searching in the feature space of a pre-trained CNN. Recently, Li *et al.* [24] design a compact strategy by searching on deep feature space without any implicit models. But it requires hyperparameter tuning to achieve competitive results. All existing methods describe periodicity using an explicit representation such as keypoints [14,26,42,49], feature-based motifs [41] or displacement vectors [21,24]. But they do not preserve both global consistency and local variations well.

The second stage is to generate or inpaint an NPP image guided by the NPP representation [13,16,23,24,28–30,32]. One common assumption is that the NPP lie on a single plane. Liu *et al.* [29] synthesize an NPP image through multi-model deformation fields given an input NPP patch and its representation. Mao *et al.* [32] propose GAN-based NPP generation. Huang *et al.* [16] and BPI [23] extend image completion to the multi-plane case. They detect periodicities in these planes [15,24,31] and use them to guide image completion, based on [53] and [2]. Unlike earlier work Huang *et al.*, BPI uses the periodicity detection method based on feature maps extracted from a pre-trained network. Also, the state of the art BPI's image completion step does not use their prior GAN-based method [32].

In summary, the above methods assume that NPP representation is good enough for guidance, which is not guaranteed. By contrast, we merge the two stages by optimizing the implicit representation using image reconstruction error.

Implicit Neural Representations: Recently, coordinate-based multi-layer perceptron (MLP) has been used to obtain implicit neural representation (INR). It maps coordinates to various signals such as shapes [9,12,44], scenes [33,35] and images [5,8,44]. Mildenhall *et al.* [35] represent a 3D scene from a sparse set of views for novel view synthesis. Siren [44] replaces ReLU by a periodic activation function and designs an initialization scheme for modeling finer details. Chen *et al.* [8] present a Local Implicit Image Function for the generation of arbitrary resolutions. Skorokhodov *et al.* [45] design a decoder based on INR with GAN training, for high-quality image generation.

NPP-Net differs from previous methods in two ways: (1) Directly using MLP [35,44] fails to learn accurate NPP representation without high-level structural understanding. We propose a periodicity-aware MLP. (2) Many works require a large dataset for training, while we optimize on a single image.

3 NPP-Net

We aim to build an MLP that maps image coordinates to pixel values, given a partial observation of an NPP image. We will describe NPP-Net using the image completion task. The unknown (masked) region is completed (or inpainted) by training on the remainder of the NPP image. For clarity, we first describe the method for single planar NPP scene and pre-warp the image to be fronto-parallel [60]. Then we will extend NPP-Net to handle multi-planar scenes.

Our key idea is to extract periodicity information from the known NPP region and inject it into the MLP input and the loss function. The initial pipeline of NPP-Net (Fig. 2) consists of three modules: (1) Periodicity-Aware Input Warping transforms image coordinates using the detected periodicity. (2) Coordinate-based MLP maps the transformed coordinates to the corresponding RGB value.

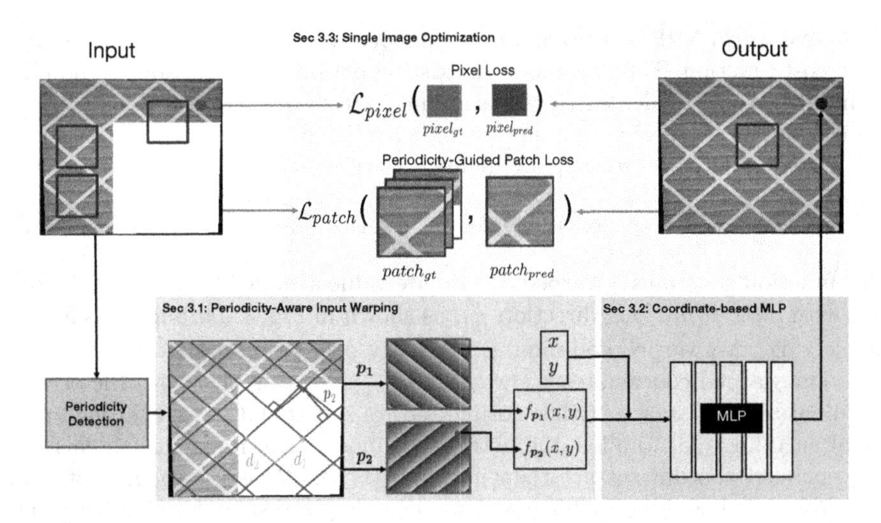

Fig. 2. Initial pipeline of NPP-Net consists of three modules. (1) *Periodicity-Aware Input Warping* (pink) warps input coordinates using detected periodicity. (2) *Coordinate-based MLP* (blue) maps warped and input coordinate features to an RGB value. (3) *Single Image Optimization* (yellow) uses pixel loss and periodicity-guided patch loss on a single NPP image. Final pipeline in Sect. 3.4 shows how multiple periodicities are automatically detected and utilized. (Color figure online)

(3) Single Image Optimization provides a periodicity-guided loss function for optimizing the MLP on a single image.

3.1 Periodicity-Aware Input Warping

A traditional MLP is not good at capturing global periodic structure without additional priors. In fact, previous works [55,64] have shown that a traditional MLP is unable to extrapolate a 1D periodic signal even with many training samples. The Periodicity-Aware Input Warping module thus explicitly injects periodicity information into the MLP by warping image coordinates (x, y).

Assuming a 2D lattice arrangement, the periodicity is represented as two displacement vectors $\boldsymbol{d}_1$ and $\boldsymbol{d}_2$ (orange arrows in Fig. 2). A perfect infinite periodic pattern is invariant if shifted by $\alpha \boldsymbol{d}_1 + \beta \boldsymbol{d}_2 (\alpha, \beta \in \mathbb{Z})$. This representation can be transformed into periods and orientations, visualized as the magnitudes and orientations of the red arrows ($\boldsymbol{p}_1$ and $\boldsymbol{p}_2$, called **periodicity vectors**). Mathematically, the transformation is obtained by solving $\boldsymbol{p}_1 \cdot \boldsymbol{d}_2 = \boldsymbol{d}_1 \cdot \boldsymbol{p}_2 = \boldsymbol{0}$ and $\boldsymbol{p}_1 \times \boldsymbol{d}_2 = \boldsymbol{d}_1 \times \boldsymbol{p}_2 = \boldsymbol{d}_1 \times \boldsymbol{d}_2$, where the cross product is defined using the corresponding 3D vectors on the plane $z = 0$. A **periodicity** is then defined as a vector pair $(\boldsymbol{p}_1, \boldsymbol{p}_2)$. Extension to circular patterns is available in supplementary.

One way to obtain the periodicity for an NPP image is to treat it as a learnable parameter and jointly optimize it with NPP-Net [7,17]. However, this is hard for two reasons. (1) Good periodicity is not unique (any multiple works). (2)

Many real-world NPP scenes contain strong local variations, leading to a complicated cost function. Thus we adopt an existing periodicity detection method [24] for input warping, which extracts feature maps from a pretrained CNN and performs brute force search to obtain a periodicity. Then, for each periodicity vector $\boldsymbol{p} = (p\cos\theta, p\sin\theta) \in \{\boldsymbol{p}_1, \boldsymbol{p}_2\}$, we define a warp as a bivariate function:

$$f_{\boldsymbol{p}}(x, y) = (x\cos\theta + y\sin\theta) \bmod p. \tag{1}$$

This function generates a warped coordinate value sampled from a periodic pattern with period $|\boldsymbol{p}|$ along direction $\frac{\boldsymbol{p}}{|\boldsymbol{p}|}$, as shown in Fig. 2. Through this feature engineering, the warped coordinates explicitly encode the periodicity information. The warped coordinates $f_{\boldsymbol{p}_1}(x, y)$ and $f_{\boldsymbol{p}_2}(x, y)$, together with the original coordinates x and y, are further normalized to $[-1, 1]$ and passed through positional encoding [35] to allow the network to model high frequency signals [47]. The encoded coordinates are then input to the MLP. We keep the notations of coordinates before and after positional encoding the same for simplicity. The dimension of the features is $4d$, including d frequencies in the positional encoding, and a set of four values for each frequency: $x, y, f_{\boldsymbol{p}_1}(x, y)$, and $f_{\boldsymbol{p}_2}(x, y)$.

3.2 Coordinate-Based MLP

We adopt coordinate-based MLP to represent NPP images. It is more effective and compact than a CNN to model periodic signals since coordinates are naturally suited for encoding positional (periodic) information. Specifically, we input the warped coordinate features to enforce global consistency, and also input the original coordinate features without warping to help preserve local variations. The output of the MLP is an RGB value corresponding to the input image coordinate. Since ReLU activation function has been proven to be ineffective to extrapolate periodic signals [55], we use the more suitable SNAKE function [64].

3.3 Single Image Optimization

Pixel Loss: Pixel loss is the most intuitive way to optimize coordinate-based MLP [35], which compares predicted and ground truth pixel values. For image coordinate $\boldsymbol{x} = (x, y)$, we adopt the robust loss function $\mathcal{L}_{rob}$ [4], given by:

$$\mathcal{L}_{pixel}(\boldsymbol{x}) = \mathcal{L}_{rob}(\hat{C}(\boldsymbol{x}), C(\boldsymbol{x})), \tag{2}$$

where $\hat{C}(\boldsymbol{x})$ and $C(\boldsymbol{x})$ are the output RGB values of the MLP and the ground truth RGB values from the input image at position $\boldsymbol{x}$, respectively. This loss is applied only to the known regions.

But merely adopting pixel loss like NeRF [35] fails to generate a good NPP for two reasons: (1) The high-dimensional input features result in the generation of high-frequency artifacts (Fig. 3(b)). See [47,61] for details about this problem. (2) Pixel loss does not enforce explicit constraints to model the correlation between a coordinate's features and its neighbors. This constraint is critical for preserving local variations since it helps capture local patch statistics. Thus, pixel loss fails to preserve local variations. For example, in Fig. 3(b), pixel loss generates some periodic artifacts in the top non-periodic region[1].

Periodicity-Guided Patch Loss: To address the limitations of pixel loss, we force the network to learn patch internal statistics by incorporating patch loss, which compares predicted and ground truth patches. The ground truth patches can be sampled at the same position as the predicted patch (for known regions), or sampled according to periodicity (for any region).

GT Patch at the Same Position: For a predicted patch in the known region, the ground truth patch at the same position is available. Specifically, for a square patch with size s centered at position $\boldsymbol{x}$, we input all the pixel coordinates in the patch into MLP to obtain a predicted RGB patch $\hat{I}_s(\boldsymbol{x})$. Let $I_s(\boldsymbol{x})$ be the corresponding ground truth at the same position and $M_s(\boldsymbol{x})$ be the mask of known pixels. We apply perceptual loss [59] on masked patches:

$$\mathcal{L}_p(\boldsymbol{x}) = \mathcal{L}_{pct}(\hat{I}_s(\boldsymbol{x}) \odot M_s(\boldsymbol{x}), I_s(\boldsymbol{x}) \odot M_s(\boldsymbol{x})), \tag{3}$$

where $\odot$ is the element-wise product.

GT Patches Sampled Based on Periodicity: To train on unknown regions, we propose to sample ground truth patches based on periodicity. This is an effective way to handle the MLP extrapolation problem, which cannot be solved by merely using input warping and SNAKE activation function [64]. The input and output images in Fig. 2 illustrate this sampling strategy.

Specifically, we sample multiple nearby ground truth patches for supervision by shifting position $\boldsymbol{x}$ based on the estimated periodicity. The shifted patch center is defined as $\boldsymbol{x}_{\alpha\beta} = \boldsymbol{x} + \alpha \boldsymbol{d}_1 + \beta \boldsymbol{d}_2 (\alpha, \beta \in \mathbb{Z})$, where $\boldsymbol{d}_1$ and $\boldsymbol{d}_2$ are the displacement vectors. Because the predicted and ground truth patches are not necessarily aligned, we adopt contextual loss $\mathcal{L}_{ctx}$ [34]:

$$\mathcal{L}_c(\boldsymbol{x}) = \frac{1}{N} \sum_{(\alpha,\beta)\in T_N} \mathcal{L}_{ctx}(\hat{I}_s(\boldsymbol{x}) \odot M_s(\boldsymbol{x}_{\alpha\beta}), I_s(\boldsymbol{x}_{\alpha\beta}) \odot M_s(\boldsymbol{x}_{\alpha\beta})), \tag{4}$$

where T_N is a set of (α, β) pairs corresponding to the N nearest ground truth patches, since local variations are preserved using nearby patches for supervision.

Patch Loss: Our patch loss combines the two sampling strategies: $\mathcal{L}_{patch}(\boldsymbol{x}) = \lambda_p \gamma(\boldsymbol{x}) \mathcal{L}_p(\boldsymbol{x}) + \lambda_c \mathcal{L}_c(\boldsymbol{x})$, where λ_p and λ_c are constant weights. $\gamma(\boldsymbol{x})$ is a binary function: 1 for $\boldsymbol{x}$ in known regions and 0 for $\boldsymbol{x}$ in unknown regions.

Total Loss: Our final loss is the combination of patch loss and pixel loss:

$$\mathcal{L} = \frac{\lambda_1}{|B_1|} \sum_{\boldsymbol{x}\in B_1} \mathcal{L}_{pixel}(\boldsymbol{x}) + \frac{\lambda_2}{|B_2|} \sum_{\boldsymbol{x}\in B_2} \mathcal{L}_{patch}(\boldsymbol{x}), \tag{5}$$

[1] Figure 3 is generated based on the final pipeline explained in Sect. 3.4.

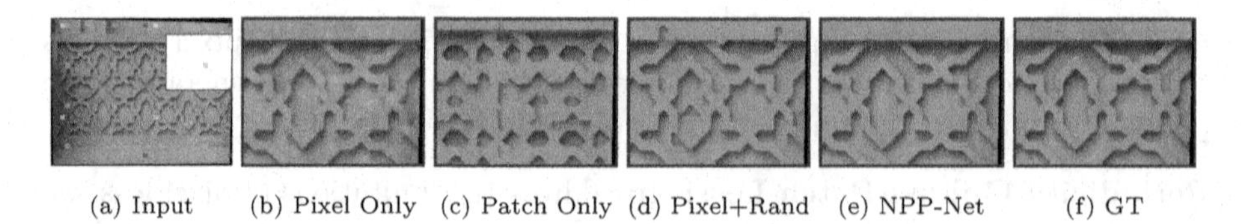

Fig. 3. Comparing different losses based on the final pipeline. The red, yellow and cyan dots in (a) visualize the Top-3 periodicities. Zoom-ins of the unknown area (white rectangle) are in (b)–(f). Merely using pixel loss (b) generates high-frequency artifacts across the image and periodic artifacts in the top part. Adopting only patch loss (c) removes the artifacts but has poor global structure. Using pixel loss and patch loss with random sampling (d) cannot preserve global consistency and local variations well since the ground truth patches are not sampled according to periodicity and might be far from the predicted patch. With pixel loss and periodicity-guided patch loss, NPP-Net (e) solves these issues. (Color figure online)

where λ_1 and λ_2 are constant weights. B_1 contains pixel coordinates that are randomly sampled in known areas, and B_2 contains the center coordinates sampled in both known and unknown areas in proportion.

Training with this loss preserves both global consistency and local variations, as shown in Fig. 3(e). In fact, only using patch loss cannot ensure global consistency if the detected periodicity is not accurate enough. In Fig. 3(c), the pattern structure is poorly reconstructed because it only focuses on the local structure. We also show the result for the combination of pixel loss and patch loss with patches that are randomly sampled in the known regions (we call it random sampling strategy) in Fig. 3(d). This fails to generate correct periodic patterns and good local details because the output and sampled patches have a large misalignment and are far away from each other.

3.4 Periodicity Proposal

Although the above initial pipeline shows good performance, it still fails to handle very inaccurate periodicity detection. To improve the robustness of NPP-Net, we design a Periodicity Proposal module to provide additional periodicity information. As shown in Fig. 4, we first search multiple candidate periodicities and then augment the input to MLP to handle inaccurate periodicity detection.

Periodicity Searching: Our searching strategy is based on the same periodicity detection method [24] we adopt in the initial pipeline. But the authors' original implementation requires manual hyperparameter tuning. Instead, we design an automatic tuning method, which evaluates each candidate periodicity (obtained from various hyperparameters) in the context of image completion. Specifically, we first generate M pseudo masks in the known regions and treat them as unknown masks for image completion. Then we execute the initial pipeline for each candidate periodicity, and compute its reconstruction error in pseudo mask regions for periodicity ranking. Since we focus on reconstructing a coarse global structure, we use a lightweight initial NPP-Net without patch loss for efficiency, which takes around 10 s for each periodicity in a Titan Xp GPU.

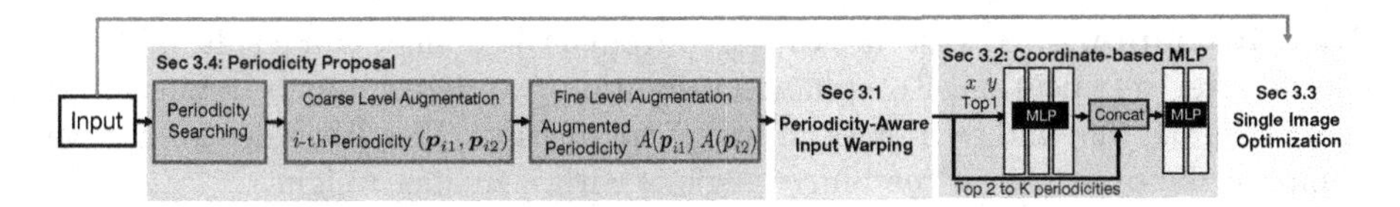

Fig. 4. Final pipeline of NPP-Net modifies two modules of the initial pipeline. (1) *Periodicity Proposal* (green) automatically searches and augments the input periodicity to handle inaccurate periodicity detection and encourage the global consistency. (2) *Coordinate-based MLP* (lavender blue) has two branches: (a) for Top-1 periodicity and original coordinates, and (b) for the rest. (Color figure online)

Periodicity Augmentation: Prior methods [16,23,24,32] also use one periodicity to guide completion as in our initial pipeline. This cannot guarantee global consistency if the periodicity is inaccurate, especially when the unknown mask is large (see experiments). So, we augment the pattern periodicity at two levels to improve robustness. At the coarse level, instead of searching the best periodicity, we keep Top-K candidates $\{(\boldsymbol{p}_{i1}, \boldsymbol{p}_{i2}) \mid i \in \mathbb{Z}^+, i \leq K\}$, to cover multiple possible solutions. This coarse-level augmentation encourages NPP-Net to move towards the most reasonable candidate periodicity. At the fine level, we augment periodicities with small offsets to better handle smaller errors. Specifically, a periodicity vector $\boldsymbol{p}$ is augmented to be $A(\boldsymbol{p}) = \{\boldsymbol{p} + \delta\frac{\boldsymbol{p}}{|\boldsymbol{p}|} \mid \delta \in \Delta\}$. We empirically define $\Delta = \{0, \pm 0.5, \pm 1\}$ (in pixels). Finally, we merge all the augmented periodicity vectors as $P = \bigcup_{i \in \mathbb{Z}^+, i \leq K} A(\boldsymbol{p}_{i1}) \cup A(\boldsymbol{p}_{i2})$. Note that $|P| = 2K|\Delta|$.

In our final pipeline, P contains $K|\Delta|$ periodicities. We perform input warping (Sect. 3.1) for each periodicity and input the transformed coordinate features into the MLP. We add an additional branch to the MLP, as shown in Fig. 4. Since the Top-1 periodicity is likely the most accurate (see experiments), we input the coordinate features warped using the Top-1 periodicity (including fine-level augmentation) and original coordinate features, to the first branch. The coordinate features warped using the Top-2 to K periodicities are sent to the second branch. For optimization, we sample patches according to the Top-1 periodicity. All other parts remain the same in our final pipeline. We evaluate these changes in our ablation study. See supplementary for implementation details including hyperparameters, network architecture, and runtime.

3.5 Extensions

Non-NPP Region Segmentation: Parts of a scene may not be near-periodic (*e.g.*, trees in front of a building facade). We thus segment the non-periodic regions in an NPP image in an unsupervised manner. We use a traditional segmentation method [18] to provide an initial guess for the non-periodic regions, treated as the unknown mask in image completion. After training NPP-Net, we relabel the initial non-periodic regions with low reconstruction error as periodic regions. *Similar strategy is adopted to serve as a pre-filtering step before applying our method to any arbitrary scene (see supplementary).*

NPP Remapping: NPP scenes captured from a tilted angle can result in blurry motifs after rectification. To enhance resolution, we detect blurry regions and treat them as the unknown mask in image completion. The difference is that we compute the pixel loss in the blurry regions with a smaller weight.

Multi-plane NPP Completion: Given an image with different NPPs on different planes, we first adopt a pre-trained plane segmentation network [56] to obtain a coarse plane segmentation. Then we select a bounding box in each plane as a reference to rectify the plane using TILT [60]. Note that, we do not require accurate segmentation since it is only used for bounding box selection. For each rectified plane, we first use our NPP segmentation method to segment the non-periodic regions (mainly from other planes) and treat them as invalid pixels. Then we perform NPP completion on each plane, transform it back to the original image coordinate system, and recompose the image. Figure 6 shows qualitative results for this extension. *Detailed implementation and experiments for these extensions are in supplementary.*

4 Experiments

Dataset: We evaluate NPP-Net on 532 images selected from three relevant datasets for NPP completion: PSU Near-Regular Texture Database (NRTDB) [1], Describable Textures Dataset (DTD) [10], and Facade Dataset [48]. There are 165 NPP images in the NRTDB dataset including facades, friezes, bricks, fences, grounds, Mondrian images, wallpapers, and carpets. Similarly, there are 258 NPP images in the DTD Dataset including honeycombs, grids, meshes and dots. The Facade Dataset has 109 rectified images of facades. Some of these facades are strictly *not* NPP because often the windows are not arranged periodically. But nonetheless we include these to evaluate our approach when the NPP assumption is not strictly satisfied. Finally, we also collect a small dataset with 11 NPP images for real-world applications (*e.g.*, removing trees in the scene). In general, scenes in NRTDB are more challenging than DTD since they contain more non-periodic regions (boundaries, trees, sky, etc.), complex illuminations and backgrounds, and multiple periodicities across an image (Fig. 5 row 3). We use TILT [60] to rectify all the images to be fronto-parallel if needed. See supplementary for more details including sampled images and mask generation.

Metrics: No single metric can evaluate NPP image completion comprehensively. So we adopt three metrics to cover different scales, including LPIPS (perceptual distance) [59], SSIM [52], and PSNR. Lower LPIPS, higher SSIM, and higher PSNR mean better performance. A known limitation for SSIM and PSNR is that blurry images also tend to receive high scores in these metrics [20], while LPIPS handles this issue better. See supplementary for FID [40] and RMSE metrics.

Ablation Study: We perform three studies. First, we compare to a "No Periodicity" variant, which uses a standard coordinated-based MLP without a periodicity prior. Results in Table 1 show that it fails to understand the arrangement of tiled motifs. Note that the Facade dataset may have different performances

Table 1. Comparison with baselines and NPP-Net variants for NPP completion and the metrics are evaluated only in unknown regions. The best and second-best results (excluding variants) are highlighted in bold and underline respectively. NPP-Net outperforms all the baselines on NRTDB and DTD. While Facade has some non-NPP images, NPP-Net can still outperform all other baselines except for Lama. See the supplementary for the results tested on the full images.

Category	Method	NRTDB [1]			DTD [10]			Facade [48]		
		LPIPS ↓	SSIM ↑	PSNR ↑	LPIPS ↓	SSIM ↑	PSNR ↑	LPIPS ↓	SSIM ↑	PSNR ↑
Large datasets	PEN-Net [57]	0.497	0.452	17.97	0.473	0.365	15.81	0.426	0.444	<u>15.78</u>
	ProFill [58]	0.401	0.300	16.35	0.443	0.249	14.30	0.374	0.391	14.73
	Lama [46]	<u>0.196</u>	<u>0.551</u>	<u>18.64</u>	<u>0.274</u>	<u>0.479</u>	<u>16.39</u>	**0.207**	<u>0.468</u>	15.24
Single image	Image Quilting [11]	0.428	0.074	13.25	0.415	0.077	12.18	0.550	0.002	10.28
	PatchMatch [2]	0.263	0.542	18.14	0.361	0.383	15.47	0.369	0.341	14.22
	DIP [50]	0.554	0.292	16.46	0.659	0.181	13.15	0.582	0.258	15.22
	Siren [44]	0.636	0.084	14.38	0.762	0.080	13.11	0.780	0.052	12.00
	Huang *et al.* [16]	0.287	0.410	16.99	0.302	0.320	14.88	0.387	0.279	13.75
	BPI [23]	0.254	0.442	16.86	0.303	0.305	14.82	0.458	0.173	12.20
NPP-Net variants	No periodicity	0.429	0.449	18.22	0.468	0.350	16.38	0.379	0.443	15.54
	Pixel only	0.308	0.618	20.20	0.397	0.473	17.86	0.427	0.458	15.54
	Patch only	0.322	0.340	15.83	0.395	0.275	13.22	0.412	0.164	11.99
	Pixel + Random	0.216	0.670	20.74	0.264	0.501	18.03	0.316	0.426	15.43
	Initial pipeline	0.213	0.647	20.38	0.293	0.462	17.26	0.277	0.438	15.11
	Top1 + Offsets	0.205	0.656	20.60	0.285	0.477	17.70	0.289	0.412	14.80
	Top5 + Offsets	0.197	0.661	20.73	0.259	0.492	18.03	0.265	0.483	15.51
	Top3 w/o Offsets	0.210	0.648	20.44	0.275	0.474	17.36	0.269	0.460	15.33
NPP-Net	Top3 + Offsets	**0.188**	**0.679**	**21.01**	**0.249**	**0.504**	**18.32**	<u>0.263</u>	**0.485**	**15.93**

because it has some non-NPP images. Second, to study the loss functions, we design three variants: (1) Only pixel loss, (2) Only patch loss, (3) Pixel loss and patch loss with random patch sampling. As discussed in Sect. 3.3, the results in Table 1 and Fig. 3 show that NPP-Net outperforms other variants.

Third, we study the effect of the periodicity augmentation (coarse level Top-K candidate periodicities and fine level offsets) by testing four variants: (1) Initial pipeline (No augmentation), (2) Top-1 with offsets, (3) Top-5 with offsets, (4) Top-3 without offsets. Table 1 shows that the initial pipeline performs the worst. Larger K (Top-5) hurts the performance as more inaccurate periodicities may be included. But, smaller K (Top-1) also performs badly because the correct periodicity may not be included. A suitable K (Top-3) without offsets performs worse since offsets can help better handle smaller errors. With the appropriate K and offsets, NPP-Net generates the best results. See more studies in supplementary.

Baselines: We compare against non-periodicity-guided and periodicity-guided baselines. For the former, we select two traditional methods, Image Quilting [11] and PatchMatch [2], that can handle pattern structure locally for some scenes with properly selected patch size. We then consider two learning-based methods DIP [50] (CNN-based) and Siren [44] (MLP-based) trained on a single image for inpainting. We also choose several learning-based methods: PEN-Net [57], ProFill [58] and Lama [46] that are trained on large real-world datasets [10,18, 19,62] since they show competitive NPP completion examples in their work.

For periodicity-guided methods, we choose two baselines - Huang *et al.* [16] and BPI [23]. Both works were designed for multi-planar scenes, but can be used for single-plane completion as well. BPI first segments and rectifies planes, then performs periodicity detection [24] on each plane, and inpaints each plane independently. For fair comparison in single-plane NPP image, we only compare with BPI's completion step to remove potential inconsistency introduced from other steps (*e.g.*, plane rectification). For Huang *et al.*, we use their pipeline without modification as their method works directly for a single plane and the completion step cannot be easily separated out. We will also compare with these two methods in multi-plane NPP images.

Comparison with Baselines: Table 1 shows the quantitative results for all methods. For NRTDB and DTD datasets, among the single-image baselines, BPI obtains the almost best LPIPS because it generates a more reasonable global structure guided by periodicity. PatchMatch obtains better SSIM and PSNR even if it generates blurred results for large masks. Lama achieves the best results among the baselines since it adopts fast Fourier convolution for a larger image receptive field, which allows it to implicitly learn the underlying periodicity from large datasets. Our NPP-Net outperforms all baselines on these two datasets by optimizing only on a single image. For the Facade dataset, even if some non-NPP images are included, NPP-Net performs better than other baselines (except for Lama). Lama effectively learns scene prior from large datasets and thus works well for non-NPP images, leading to the best performance in this dataset.

Qualitative results are shown in Fig. 5. Large rectangle masks are challenging since there is less information from which to estimate the representation. Perceptually, PatchMatch works well when the motifs are small (row 3) but results in blur with large masks (row 2 and 6). Although BPI and Huang *et al.* perform better than non-periodicity-guided baselines, they generate artifacts since the NPP representation (periodicity) has poor global consistency (row 1–6) or lacks local variations (row 6–9). Note that the Top-1 periodicity in row 1 (red dots) is inaccurate - the actual periods are half of the one shown. We show that NPP-Net can extrapolate NPP images well (row 5), generalize to irregular masks (row 6), and work for scenes that contain non-periodic regions (row 4). We show that NPP-Net can be extended to handle multi-planar scenes in Fig. 6. Among the baselines, Lama (trained on large datasets) can better handle local variations (row 2 purple box). Although it captures some global structure when the mask is not large, Lama performs worse than BPI when the mask is outside the image border for extrapolation (all rows) and perspective effect is severe (row 2 cyan box). Learning from the Top-K periodicities, NPP-Net produces the best images, maintaining global consistency and local variations. Finally, the output of NPP-Net can in turn improve periodicity detection, leading to better image completion for both BPI and NPP-Net (see supplementary).

Influence of Mask Size: We study the influence of different mask sizes for image completion. All figures are shown in the supplementary due to the space limitation. First, for each image, if the K-th periodicity has the smallest error among Top-3 periodicities, we assign the image to the K-th periodicity. In the

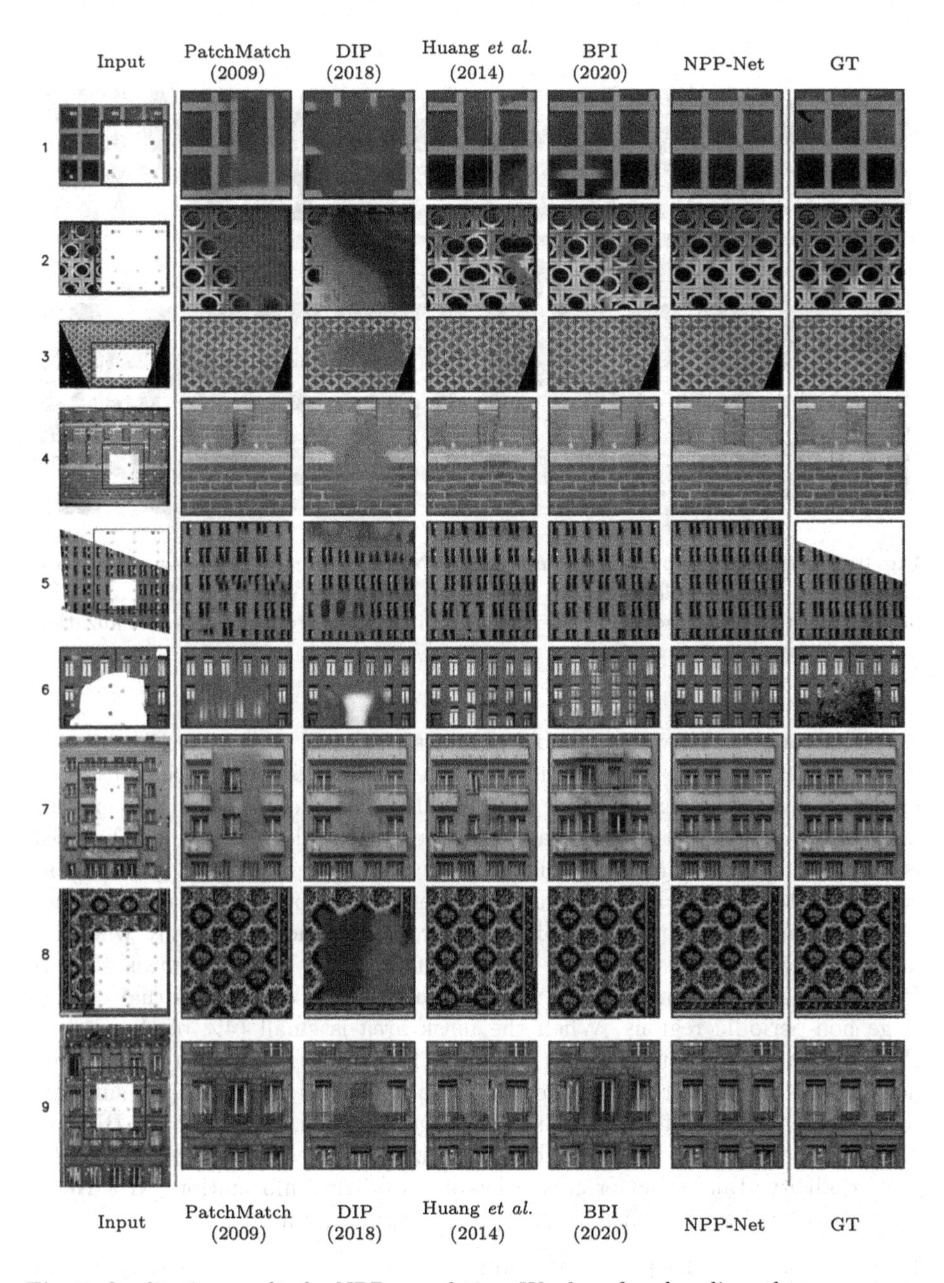

Fig. 5. Qualitative results for NPP completion. We show four baselines that operate on a single image. The red, yellow, and cyan dots in input images show the first, second, and third periodicity from periodicity searching module, respectively. For visualization, all periods are scaled by 2. Our NPP-Net outperforms all baselines for global consistency (rows 1–6) and local variations (rows 6–9). (Color figure online)

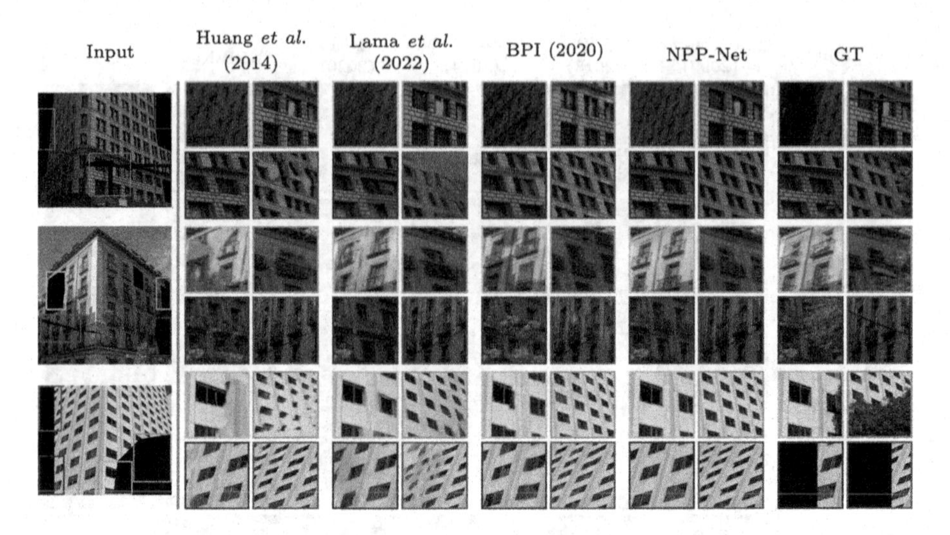

Fig. 6. Qualitative comparison for multi-plane NPP completion. We show three baselines, which are either designed for multi-plane NPP scenes (Huang *et al.* and BPI) or trained on large datasets (Lama). Some zoom-in boxes are resized for visualization. Full results are in supplementary.

supplementary, we show the number of images assigned to each periodicity with different mask sizes in Fig. 19, and illustrate how to compute periodicity error in detail. From the figure, while the Top 1st periodicity is the best one for most of the images with small mask size, the number decreases when the mask is large (64% of the image). This illustrates that the other two periodicities contain better periodicity and leveraging them by our periodicity augmentation strategy can be helpful for learning NPP representation, especially when the mask is large. Second, we show the performances for different mask sizes in Fig. 20 and 21 (supplementary). In particular, we filter out images that contain large non-periodic regions. When the mask area is small (4% of the image), PatchMatch slightly outperforms NPP-Net because the unknown regions may not contain pattern structure, and simply sampling nearby patches is sufficient to produce good results. Among single-image methods, Huang *et al.*, BPI and NPP-Net perform better when the mask size increases since they are guided by periodicity. Taking better advantage of periodicity information, NPP-Net is more robust to various mask sizes, especially for larger masks.

Limitations: (1) The periodicity proposal cannot be too erroneous, allowing tolerance of about 10%. (2) It assumes a multi-planar scene with translated, circular, and potentially other types of symmetrical NPP that can be modeled.

Conclusion: We show how to learn an effective neural representation of Near-Periodic Patterns. We design the periodicity proposal, periodicity-aware input warping, periodicity-guided loss to maintain global consistency and local variations. NPP-Net provides a strong tool to study a large class of man-made scenes.

Acknowledgement. This work was supported by a gift from Zillow Group, USA, and NSF Grants #CNS-2038612, #IIS-1900821.

References

1. PSU Near-Regular Texture Database. http://vivid.cse.psu.edu/
2. Barnes, C., Shechtman, E., Finkelstein, A., Goldman, D.B.: PatchMatch: a randomized correspondence algorithm for structural image editing. ACM Trans. Graph. **28**, 24 (2009)
3. Barnes, C., Shechtman, E., Goldman, D.B., Finkelstein, A.: The generalized PatchMatch correspondence algorithm. In: Daniilidis, K., Maragos, P., Paragios, N. (eds.) ECCV 2010. LNCS, vol. 6313, pp. 29–43. Springer, Heidelberg (2010). https://doi.org/10.1007/978-3-642-15558-1_3
4. Barron, J.T.: A general and adaptive robust loss function. In: IEEE Conference on Computer Vision and Pattern Recognition (CVPR), pp. 4331–4339 (2019)
5. Bemana, M., Myszkowski, K., Seidel, H.P., Ritschel, T.: X-fields: implicit neural view-, light-and time-image interpolation. ACM Trans. Graph. (TOG) **39**, 1–15 (2020)
6. Cao, C., Fu, Y.: Learning a sketch tensor space for image inpainting of man-made scenes. In: Proceedings of the IEEE/CVF International Conference on Computer Vision, pp. 14509–14518 (2021)
7. Chen, H., Liu, J., Chen, W., Liu, S., Zhao, Y.: Exemplar-based pattern synthesis with implicit periodic field network. In: Proceedings of the IEEE/CVF Conference on Computer Vision and Pattern Recognition, pp. 3708–3717 (2022)
8. Chen, Y., Liu, S., Wang, X.: Learning continuous image representation with local implicit image function. In: IEEE Conference on Computer Vision and Pattern Recognition (CVPR), pp. 8628–8638. IEEE (2021)
9. Chen, Z., Zhang, H.: Learning implicit fields for generative shape modeling. In: IEEE Conference on Computer Vision and Pattern Recognition (CVPR), pp. 5939–5948. IEEE (2019)
10. Cimpoi, M., Maji, S., Kokkinos, I., Mohamed, S., Vedaldi, A.: Describing textures in the wild. In: Proceedings of the IEEE Conference on Computer Vision and Pattern Recognition (CVPR) (2014)
11. Efros, A.A., Freeman, W.T.: Image quilting for texture synthesis and transfer. In: International Conference on Computer Graphics and Interactive Techniques (SIGGRAPH), pp. 341–346. Association for Computing Machinery (2001)
12. Genova, K., Cole, F., Vlasic, D., Sarna, A., Freeman, W.T., Funkhouser, T.: Learning shape templates with structured implicit functions. In: IEEE Conference on Computer Vision and Pattern Recognition (CVPR), pp. 7154–7164. IEEE (2019)
13. Halperin, T., et al.: Endless loops: detecting and animating periodic patterns in still images. ACM Trans. Grap. (TOG) **40**, 1–12 (2021)
14. Hays, J., Leordeanu, M., Efros, A.A., Liu, Y.: Discovering texture regularity as a higher-order correspondence problem. In: Leonardis, A., Bischof, H., Pinz, A. (eds.) ECCV 2006. LNCS, vol. 3952, pp. 522–535. Springer, Heidelberg (2006). https://doi.org/10.1007/11744047_40
15. He, K., Sun, J.: Statistics of patch offsets for image completion. In: Fitzgibbon, A., Lazebnik, S., Perona, P., Sato, Y., Schmid, C. (eds.) ECCV 2012. LNCS, vol. 7573, pp. 16–29. Springer, Heidelberg (2012). https://doi.org/10.1007/978-3-642-33709-3_2

16. Huang, J.B., Kang, S.B., Ahuja, N., Kopf, J.: Image completion using planar structure guidance. ACM Trans. Graph. (TOG) **33**, 1–10 (2014)
17. Jetchev, N., Bergmann, U., Vollgraf, R.: Texture synthesis with spatial generative adversarial networks. arXiv preprint arXiv:1611.08207 (2016)
18. Jiri, B., Jan, S., Jan, K., Habart, D.: Supervised and unsupervised segmentation using superpixels, model estimation, and graph cut. J. Electron. Imaging **26**, 061610 (2017)
19. Karras, T., Aila, T., Laine, S., Lehtinen, J.: Progressive growing of GANs for improved quality, stability, and variation. arXiv preprint arXiv:1710.10196 (2017)
20. Ledig, C., et al.: Photo-realistic single image super-resolution using a generative adversarial network. In: IEEE Conference on Computer Vision and Pattern Recognition (CVPR), pp. 4681–4690 (2017)
21. Lettry, L., Perdoch, M., Vanhoey, K., Van Gool, L.: Repeated pattern detection using CNN activations. In: IEEE Winter Conference on Applications of Computer Vision (WACV), pp. 47–55. IEEE (2017)
22. Li, J., Wang, N., Zhang, L., Du, B., Tao, D.: Recurrent feature reasoning for image inpainting. In: IEEE Conference on Computer Vision and Pattern Recognition (CVPR). IEEE (2020)
23. Li, Y., et al.: Multi-plane program induction with 3D box priors. In: Neural Information Processing Systems (NeurIPS) (2020)
24. Li, Y., Mao, J., Zhang, X., Freeman, W.T., Tenenbaum, J.B., Wu, J.: Perspective plane program induction from a single image. In: IEEE Conference on Computer Vision and Pattern Recognition (CVPR), pp. 4434–4443. IEEE (2020)
25. Liu, J., Liu, Y.: Grasp recurring patterns from a single view. In: IEEE Conference on Computer Vision and Pattern Recognition (CVPR), pp. 2003–2010. IEEE (2013)
26. Liu, S., Ng, T.T., Sunkavalli, K., Do, M.N., Shechtman, E., Carr, N.: PatchMatch-based automatic lattice detection for near-regular textures. In: IEEE Conference on Computer Vision and Pattern Recognition (CVPR), pp. 181–189. IEEE (2015)
27. Liu, Y., Collins, R.T., Tsin, Y.: A computational model for periodic pattern perception based on frieze and wallpaper groups. IEEE Trans. Pattern Anal. Mach. Intell. (TPAMI) **26**, 354–371 (2004)
28. Liu, Y., Lin, W.C.: Deformable texture: the irregular-regular-irregular cycle. Carnegie Mellon University, the Robotics Institute (2003)
29. Liu, Y., Lin, W.C., Hays, J.: Near-regular texture analysis and manipulation. ACM Trans. Graph. (TOG) **23**, 368–376 (2004)
30. Liu, Y., Tsin, Y., Lin, W.C.: The promise and perils of near-regular texture. Int. J. Comput. Vis. (IJCV) **62**, 145–159 (2005)
31. Lowe, D.G.: Object recognition from local scale-invariant features. In: Proceedings of the IEEE International Conference on Computer Vision (ICCV), pp. 1150–1157. IEEE (1999)
32. Mao, J., Zhang, X., Li, Y., Freeman, W.T., Tenenbaum, J.B., Wu, J.: Program-guided image manipulators. In: IEEE Conference on Computer Vision and Pattern Recognition (CVPR), pp. 4030–4039. IEEE (2019)
33. Martin-Brualla, R., Radwan, N., Sajjadi, M.S., Barron, J.T., Dosovitskiy, A., Duckworth, D.: NeRF in the wild: neural radiance fields for unconstrained photo collections. In: IEEE Conference on Computer Vision and Pattern Recognition (CVPR), pp. 7210–7219. IEEE (2021)
34. Mechrez, R., Talmi, I., Zelnik-Manor, L.: The contextual loss for image transformation with non-aligned data. In: Proceedings of the European Conference on Computer Vision (ECCV), pp. 768–783 (2018)

35. Mildenhall, B., Srinivasan, P.P., Tancik, M., Barron, J.T., Ramamoorthi, R., Ng, R.: NeRF: representing scenes as neural radiance fields for view synthesis. In: Vedaldi, A., Bischof, H., Brox, T., Frahm, J.-M. (eds.) ECCV 2020. LNCS, vol. 12346, pp. 405–421. Springer, Cham (2020). https://doi.org/10.1007/978-3-030-58452-8_24
36. Nazeri, K., Ng, E., Joseph, T., Qureshi, F., Ebrahimi, M.: EdgeConnect: structure guided image inpainting using edge prediction. In: The IEEE International Conference on Computer Vision (ICCV) Workshops. IEEE (2019)
37. Park, M., Brocklehurst, K., Collins, R.T., Liu, Y.: Deformed lattice detection in real-world images using mean-shift belief propagation. IEEE Trans. Pattern Anal. Mach. Intell. (TPAMI) **31**, 1804–1816 (2009)
38. Park, M., Brocklehurst, K., Collins, R.T., Liu, Y.: Translation-symmetry-based perceptual grouping with applications to urban scenes. In: Kimmel, R., Klette, R., Sugimoto, A. (eds.) ACCV 2010. LNCS, vol. 6494, pp. 329–342. Springer, Heidelberg (2011). https://doi.org/10.1007/978-3-642-19318-7_26
39. Park, M., Collins, R.T., Liu, Y.: Deformed lattice discovery via efficient mean-shift belief propagation. In: Forsyth, D., Torr, P., Zisserman, A. (eds.) ECCV 2008. LNCS, vol. 5303, pp. 474–485. Springer, Heidelberg (2008). https://doi.org/10.1007/978-3-540-88688-4_35
40. Parmar, G., Zhang, R., Zhu, J.Y.: On aliased resizing and surprising subtleties in GAN evaluation. In: IEEE Conference on Computer Vision and Pattern Recognition (CVPR) (2022)
41. Pritts, J., Chum, O., Matas, J.: Detection, rectification and segmentation of coplanar repeated patterns. In: IEEE Conference on Computer Vision and Pattern Recognition (CVPR), pp. 2973–2980. IEEE (2014)
42. Pritts, J., Rozumnyi, D., Kumar, M.P., Chum, O.: Coplanar repeats by energy minimization. In: Proceedings of the British Machine Vision Conference (BMVC), pp. 107.1–107.12. IEEE (2016)
43. Rao, A., Lohse, G.: Identifying high level features of texture perception. CVGIP: Graph. Models Image Process. **55**, 218–233 (1993)
44. Sitzmann, V., Martel, J., Bergman, A., Lindell, D., Wetzstein, G.: Implicit neural representations with periodic activation functions. Adv. Neural Inf. Process. Syst. **33**, 7462–7473 (2020)
45. Skorokhodov, I., Ignatyev, S., Elhoseiny, M.: Adversarial generation of continuous images. In: IEEE Conference on Computer Vision and Pattern Recognition (CVPR), pp. 10753–10764. IEEE (2021)
46. Suvorov, R., et al.: Resolution-robust large mask inpainting with Fourier convolutions. In: Proceedings of the IEEE/CVF Winter Conference on Applications of Computer Vision, pp. 2149–2159 (2022)
47. Tancik, M., et al.: Fourier features let networks learn high frequency functions in low dimensional domains. In: Neural Information Processing Systems (NeurIPS) (2020)
48. Teboul, O., Simon, L., Koutsourakis, P., Paragios, N.: Segmentation of building facades using procedural shape priors. In: 2010 IEEE Computer Society Conference on Computer Vision and Pattern Recognition, pp. 3105–3112. IEEE (2010)
49. Torii, A., Sivic, J., Pajdla, T., Okutomi, M.: Visual place recognition with repetitive structures. In: IEEE Conference on Computer Vision and Pattern Recognition (CVPR), pp. 883–890. IEEE (2013)
50. Ulyanov, D., Vedaldi, A., Lempitsky, V.: Deep image prior. In: IEEE Conference on Computer Vision and Pattern Recognition (CVPR), pp. 9446–9454. IEEE (2018)

51. Wang, T., Ouyang, H., Chen, Q.: Image inpainting with external-internal learning and monochromic bottleneck. In: Proceedings of the IEEE/CVF Conference on Computer Vision and Pattern Recognition, pp. 5120–5129 (2021)
52. Wang, Z., Bovik, A.C., Sheikh, H.R., Simoncelli, E.P.: Image quality assessment: from error visibility to structural similarity. IEEE Trans. Image Process. **13**, 600–612 (2004)
53. Wexler, Y., Shechtman, E., Irani, M.: Space-time completion of video. IEEE Trans. Pattern Anal. Mach. Intell. **29**, 463–476 (2007)
54. Wu, C., Frahm, J.-M., Pollefeys, M.: Detecting large repetitive structures with salient boundaries. In: Daniilidis, K., Maragos, P., Paragios, N. (eds.) ECCV 2010. LNCS, vol. 6312, pp. 142–155. Springer, Heidelberg (2010). https://doi.org/10.1007/978-3-642-15552-9_11
55. Xu, K., Zhang, M., Li, J., Du, S.S., Kawarabayashi, K.I., Jegelka, S.: How neural networks extrapolate: from feedforward to graph neural networks. In: International Conference on Learning Representations (ICLR) (2021)
56. Yu, Z., Zheng, J., Lian, D., Zhou, Z., Gao, S.: Single-image piece-wise planar 3D reconstruction via associative embedding. In: CVPR, pp. 1029–1037 (2019)
57. Zeng, Y., Fu, J., Chao, H., Guo, B.: Learning pyramid-context encoder network for high-quality image inpainting. In: The IEEE Conference on Computer Vision and Pattern Recognition (CVPR), pp. 1486–1494 (2019)
58. Zeng, Yu., Lin, Z., Yang, J., Zhang, J., Shechtman, E., Lu, H.: High-resolution image inpainting with iterative confidence feedback and guided upsampling. In: Vedaldi, A., Bischof, H., Brox, T., Frahm, J.-M. (eds.) ECCV 2020. LNCS, vol. 12364, pp. 1–17. Springer, Cham (2020). https://doi.org/10.1007/978-3-030-58529-7_1
59. Zhang, R., Isola, P., Efros, A.A., Shechtman, E., Wang, O.: The unreasonable effectiveness of deep features as a perceptual metric. In: IEEE Conference on Computer Vision and Pattern Recognition (CVPR) (2018)
60. Zhang, Z., Ganesh, A., Liang, X., Ma, Y.: TILT: transform invariant low-rank textures. Int. J. Comput. Vis. (IJCV) **99**, 1–24 (2012)
61. Zheng, J., Ramasinghe, S., Lucey, S.: Rethinking positional encoding. arXiv preprint arXiv:2107.02561 (2021)
62. Zhou, B., Lapedriza, A., Khosla, A., Oliva, A., Torralba, A.: Places: a 10 million image database for scene recognition. IEEE Trans. Pattern Anal. Mach. Intell. **40**, 1452–1464 (2017)
63. Zhou, Y., Barnes, C., Shechtman, E., Amirghodsi, S.: TransFill: reference-guided image inpainting by merging multiple color and spatial transformations. In: Proceedings of the IEEE/CVF Conference on Computer Vision and Pattern Recognition, pp. 2266–2276 (2021)
64. Ziyin, L., Hartwig, T., Ueda, M.: Neural networks fail to learn periodic functions and how to fix it. In: Neural Information Processing Systems (NeurIPS), pp. 1583–1594 (2020)

End-to-End Graph-Constrained Vectorized Floorplan Generation with Panoptic Refinement

Jiachen Liu[1], Yuan Xue[2], Jose Duarte[3], Krishnendra Shekhawat[4], Zihan Zhou[5], and Xiaolei Huang[1](✉)

[1] College of Information Sciences and Technology, The Pennsylvania State University, State College, USA
suh972@psu.edu
[2] Department of Electrical and Computer Engineering, Johns Hopkins University, Baltimore, USA
[3] College of Arts and Architecture, The Pennsylvania State University, State College, USA
[4] Department of Mathematics, BITS Pilani, Pilani, India
[5] Manycore Tech Inc., Hangzhou, China

Abstract. The automatic generation of floorplans given user inputs has great potential in architectural design and has recently been explored in the computer vision community. However, the majority of existing methods synthesize floorplans in the format of rasterized images, which are difficult to edit or customize. In this paper, we aim to synthesize floorplans as sequences of 1-D vectors, which eases user interaction and design customization. To generate high fidelity vectorized floorplans, we propose a novel two-stage framework, including a *draft stage* and a multi-round *refining stage*. In the first stage, we encode the room connectivity graph input by users with a graph convolutional network (GCN), then apply an autoregressive transformer network to generate an initial floorplan sequence. To polish the initial design and generate more visually appealing floorplans, we further propose a novel panoptic refinement network (PRN) composed of a GCN and a transformer network. The PRN takes the initial generated sequence as input and refines the floorplan design while encouraging the correct room connectivity with our proposed geometric loss. We have conducted extensive experiments on a real-world floorplan dataset, and the results show that our method achieves state-of-the-art performance under different settings and evaluation metrics.

1 Introduction

House design is essential yet challenging work for professional architects, usually requiring extensive collaboration and multi-round refinement. Floorplan design

J. Liu and Y. Xue—These authors contributed equally to this work.

Supplementary Information The online version contains supplementary material available at https://doi.org/10.1007/978-3-031-19784-0_32.

S. Avidan et al. (Eds.): ECCV 2022, LNCS 13675, pp. 547–562, 2022.
https://doi.org/10.1007/978-3-031-19784-0_32

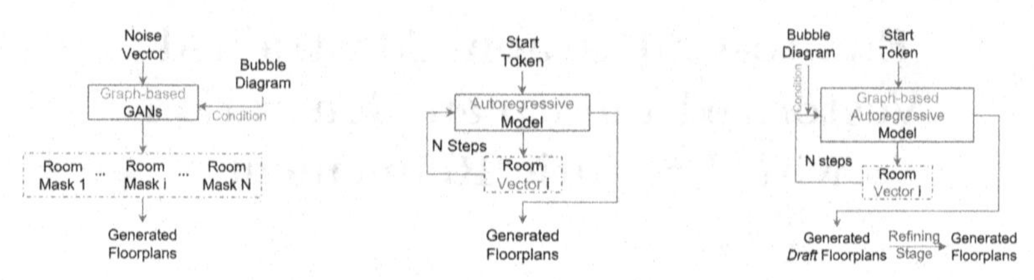

(a) Previous rasterized methods [17,18]. **(b)** Previous vectorized methods [20]. **(c)** Our proposed vectorized method.

Fig. 1. Comparisons between different floorplan generation pipelines.

is a crucial part of house design, which involves designing the room layouts and their connectivities such as walls and doors. With the availability of several large-scale floorplan benchmarks [4,5,24] and advance in generative models such as generative adversarial networks (GANs) [6], generating floorplans automatically has recently attracted the attention and interest of architects as well as computer vision researchers [10,19].

Motivated by the interactive design process, previous works generated floorplans with different input constraints, such as building boundaries [9,24], room dimensions [22], or bubble diagrams that describe room numbers, room types, and adjacency or connectivity information [17,18]. There is also work which explores floorplan generation from scratch in an unconditional way (*e.g.*, [20]). Since architects usually preset room categories and their connectivities before designing the floorplans, we find the setting by HouseGAN [17] to be the closest to the design practice. We thus follow their setup throughout this paper for interactive floorplan generation.

Current state-of-the-art methods with conditional inputs [17,18] use graph-based GANs to generate room layouts as rasterized images. Although they have shown superior generation realism, we argue that GAN-generated rasterized images have several limitations in representing floorplan layouts. First, it is challenging for GANs to learn geometric layout properties such as axis-aligned walls. Second, some input conditions such as room numbers cannot be explicitly expressed in the rasterized floorplans. Last but not least, rasterized masks are irregular representations, which have restricted the options for users or architects to refine or customize the generated floorplans. In this paper, we propose to represent floorplan layouts as sequences of 1-D vectors describing the room bounding box coordinates, and generate vectorized floorplans in an end-to-end fashion. With a concise and user-friendly representation, generating vectorized floorplans enables efficient inference, more accessible user interactions, and connected space representation.

Mimicking the floorplan design process, we propose a novel two-stage framework conditioned on input bubble diagrams and gradually refine the generated floorplan through multiple steps. Figure 1 illustrates the difference between our proposed framework and previous methods. In the first stage (*i.e.*, *draft stage*), we apply graph convolutional networks (GCNs) to encode the room connectivity from the bubble diagram, then an initial room sequence is generated with an autoregressive self-attention transformer network following previous layout generation works [7,20]. However, the generated draft result may be unsatisfactory since the autoregressive models cannot access the global information from the entire sequence at each generation timestamp. Thus, in the second stage (*i.e.*, refining stage in Fig. 2), we propose a panoptic refinement network (PRN) to refine the draft floorplan design for a more visually appealing appearance while keeping the correct room connectivity. Taking the entire floorplan sequence from the draft stage as input, a GCN in the panoptic refinement network first processes the generated sequence to refine the connectivity relationship. Another transformer is further applied to exploit the contextual information of the whole sequence. In addition, a novel geometric loss based on layout overlapping is proposed to encourage the correct layout of the generated floorplan.

The main contributions of our paper can be summarized as follows:

- We propose a novel two-stage framework to generate vectorized floorplans from user inputs. To the best of our knowledge, our work is the first to allow the end-to-end conditional generation of vectorized floorplans.
- In the draft stage, we integrate GCNs into a transformer-based autoregressive model and show how to effectively capture the room connectivity relationships provided in the user inputs.
- In the refining stage, we propose a panoptic refinement network for multi-round refinements of the draft floorplans, while encouraging the correct room connectivity with our proposed geometric loss.
- Extensive experiments on a real-world floorplan dataset show that our method outperforms previous state-of-the-art approaches under most settings.

2 Related Work

Data-Driven Floorplan Generation. Various data-driven approaches have been proposed in recent years for automatic floorplan generation. Mainstream methods primarily focused on generating floorplans as rasterized images. RPLAN [24] created a large-scale annotated dataset from residential buildings and proposed a two-stage approach to generate rooms from a given boundary. Graph2Plan [9] proposed to first generate layout-constrained floorplans as rasterized images via convolutional neural networks (CNNs) and graph neural networks (GNNs), then applied an offline optimization algorithm to align the rooms and transform them into a vectorized floorplan, given the boundary and graph-constrained layout as input. HouseGAN [17] took the bubble diagram of the rooms as user input, and introduced a GAN-based architecture to generate each room. HouseGAN++ [18] was the extension of HouseGAN and the authors proposed an improved GAN-based network, using a GT-conditioning training strategy as well as a list of test-time meta-optimization algorithms.

While previous rasterized floorplan generation methods have achieved promising results, their applicability is often limited by the rasterized representations. In practice, architects mostly work with vectorized representation [25] of floorplans due to its flexibility and geometric compatibility. To leverage vectorized representations, existing methods [18,24] often converted rasterized floorplans into vectorized representations via postprocessing steps [3]. However, the representation power is yet limited by the originally generated rasterized images. To this end, we directly represent the floorplan rooms as vectorized bounding boxes in our method. More recently, Para *et al.* [20] also utilized vectorized representation to generate floorplans using an autoregressive model and focused more on unconditional generation. More specifically, a transformer was used to predict elements including room coordinates and sizes in an autoregressive manner, followed by another transformer to generate edges such as walls and doors. A post-processing optimization step was adopted to refine the output from the networks to obtain final layouts. Despite using the same floorplan representations, our work has several essential differences from [20]. First, our model enables interaction with graph-based user input, which satisfies the room connectivity constraints. Second, we encode constraints such as doors into the learning objectives of the same network, skipping the need to use a separate model to learn constraints as in [20]. Moreover, unlike [20] which applied linear programming regularization to optimize network output, we propose a panoptic refinement network that refines the draft floorplan while encouraging the correct room connectivity in an end-to-end fashion.

Autoregressive Models for Layout Generation. Floorplan generation can be interpreted as a special case of layout generation. In the literature, the autoregressive model has been widely used in layout or shape generation tasks [2,7,11,15,16] due to its superior representation ability to model sequential relationships. LayoutVAE [11] proposed a variational auto-encoder (VAE) [13] architecture conditioned on priors such as element numbers and types. LayoutTransformer [7] applied a transformer-based architecture to learn layout attributes autoregressively in an unconditional manner. VTN [2] instantiated the VAE framework with a transformer network and generated layouts in a self-supervised manner. However, there are two potential limitations for transformer-based autoregressive models in layout generation [15]. First, the information flow is unidirectional and immutable in the autoregressive model since the model only has access to previously generated results during the iterative generation. Second, most existing transformers only considered unconditional layout generation without user inputs and thus are not suitable for an interactive generation. To mitigate such limitations, we enable interactive generation by integrating GCNs into the transformer for draft floorplan generation. Unlike existing refinement method for scene layout generation which only refines the output once [26], we resolve the unidirectional generation issue by applying another panoptic refinement network that utilizes the whole vectorized floorplan sequence and perform multi-round refinement for the draft floorplan.

3 Methodology

Given a bubble diagram input from the user which encodes the number of rooms, room types, as well as their connections through doors, our proposed framework generates vectorized floorplans in two stages. In the draft stage, we employ a

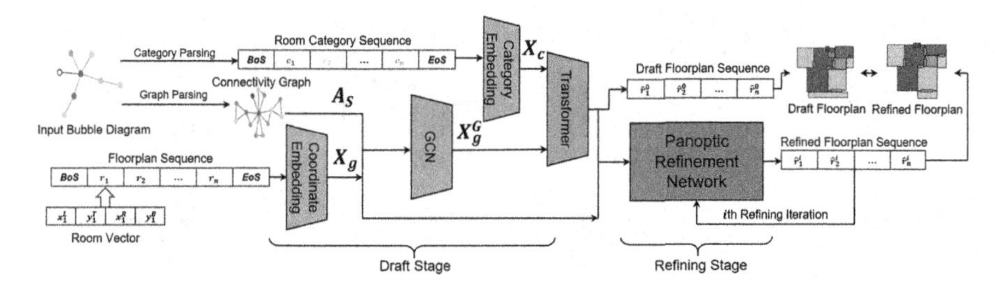

Fig. 2. Overview of our method. User inputs are first processed by a GCN and room elements are generated by a transformer in the *draft stage*. The layouts are further refined by a panoptic refinement network in the *refining stage*.

GCN for the layout coordinate embeddings to encode the connectivity relationship between rooms and their connecting doors. The GCN-processed embeddings as well as the category embeddings are then passed into an autoregressive transformer network to generate room layout elements. In the generated floorplan, each room or door is represented by a set of bounding box coordinates. In the refining stage, we apply another transformer network to integrate the global information from the entire floorplan sequence to refine the layouts of the initial floorplan generated in the draft stage. An overview of the proposed framework can be found in Fig. 2. More details are introduced in the remaining part of this section and implementation details can be found in the supplementary materials.

3.1 Draft Stage with GCN-Constrained Autoregressive Transformer

Layout Representation. Previous layout generation works [7,15] represented each object with 5 attributes (c, x, y, w, h), where $c \in C$ is the category, (x, y) is the center location and (w, h) is the object size. We adopt a similar setting but with the exception of c since in our work, c is given as part of the input conditions encoded in the input bubble diagram therefore there is no need to generate c. c can be directly obtained for each room from the input, where doors are treated as special cases of rooms. As illustrated in Fig. 2, we use the geometric representation (x^L, y^T, x^R, y^B) for an object, where (x^L, y^T) is the top-left corner and (x^R, y^B) is the bottom-right corner. We then discretize the continuous x- and y-coordinates with 8-bit uniform quantization (from 0 to 255) [7,16] for the ease of representation. Following the convention of autoregressive generation [7], we include an additional starting token and an ending token for the layout sequence. We flatten the whole layout with a sequence of 1-D vectors as:

$$S = [\langle BoS\rangle, \boldsymbol{r}_1, \boldsymbol{r}_2, ..., \boldsymbol{r}_N, \langle EoS\rangle], \quad (1)$$

where N is the total number of rooms and doors, $\langle BoS\rangle$ and $\langle EoS\rangle$ are the starting and ending token of the sequence, respectively. Room or door vector is represented by $\boldsymbol{r}_i = [x_i^L, y_i^T, x_i^R, y_i^B]$. Geometric coordinate embedding X_g is learned through a learnable embedding layer which projects the original quantized coordinates to an embedding space.

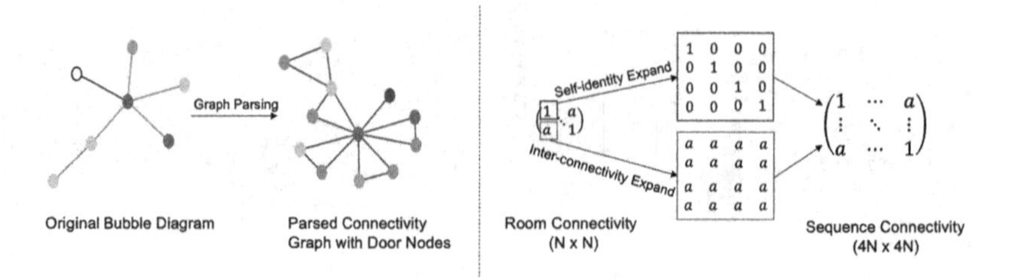

Fig. 3. Left: the process of parsing an original bubble diagram to a connectivity graph adapted for GCN. **Right:** the process of transforming the room connectivity matrix A into the sequence connectivity matrix A_S.

Also, the room categories corresponding to the objects in the layout sequence are encoded in a separate room category sequence. Following [7], we encode the categories c as discrete values in the range of $[256, 256+N_c]$, where N_c is the number of room categories. Then, a category embedding X_c is also learned through a learnable embedding layer sharing weights with the geometric embedding layer to convert quantized category values to the embedding space.

Encoding Connectivity Graph with GCN. To encode the connectivity relationships, we build a connectivity graph A_S which considers the connectivity between rooms via doors from the input bubble diagram. Throughout the paper, we treat all doors as special cases of rooms, as the nodes of the input graph. As illustrated in the left side of Fig. 3, two nodes are parsed as connected by an edge if there are two rooms connected via a shared door or if a room and a door are connected. Inspired by the success of GCN [14] in various tasks, we propose to encode the connectivity graph from the bubble diagram with a GCN. The GCN takes the input geometric embedding X_g as well as a pre-computed sequence connectivity matrix A_S as input. Let $\tilde{A}_S = A_S + I$, where I is the identity matrix indicating that each room is connected to itself, we adopt a learnable layer W to get the updated feature representation X_g^G using the row-normalized connectivity matrix $\hat{A}_S$ as:

$$X_g^G = f(X_g, A_S) = \hat{A}_S X_g W, \quad \hat{A}_S = \tilde{D}^{-\frac{1}{2}} \tilde{A}_S \tilde{D}^{-\frac{1}{2}}, \quad \tilde{D}_{ii} = \sum_j \tilde{A}_{S_{ij}}. \tag{2}$$

Now we present how to generate the sequence connectivity matrix A_S as in the right side of Fig. 3. For efficient processing by GCN, we adapt the original room connectivity matrix $A \in N \times N$, where N is the room number, to a sequence connectivity matrix A_S for the vectorized floorplan sequence. We then flatten the quantified $N \times 4$ bounding boxes to a sequence of 1-D vectors with length $4N + 2$, including special tokens $\langle BoS \rangle$ and $\langle EoS \rangle$. For each element $\boldsymbol{r}_i$, we get its connectivity value with another element $\boldsymbol{r}_j$ by querying $A_{i'j'}$ as $A_{S_{ij}} = A_{i'j'}$, where i' and j' are the room indexes of $\boldsymbol{r}_i$ and $\boldsymbol{r}_j$. For $\langle BoS \rangle$ and $\langle EoS \rangle$, we fox their connectivities to other elements as 0. A_S is used as the input of GCN. Note that during an intermediate step t of the autoregressive generation process, only the generated sequence embeddings $X_{g_{1:t}}$ and a partial graph $A_{S_{1:t}}$ are used as the input of GCN. Thus, the initial floorplan generated in the draft stage may not fully utilize the information of the entire sequence and neglect certain connectivities between rooms. We discuss how to resolve this issue in Sect. 3.2.

Autoregressive Generation with Self-attention Transformer. Following previous autoregressive generation works [7,16,20], we generate the floorplan sequence through an iterative process. The joint distribution of autoregressive generation can be represented by the chain rule:

$$p(s_{1:4N+2}) = \prod_{i=1}^{4N+2} p(s_i|s_{1:i-1}). \quad (3)$$

A transformer network is applied for autoregressive generation, with a teacher-forcing strategy during training. For the input, we have obtained the GCN-processed geometric embedding X_g^G and the category embedding X_c for each element. To help encode the geometric position of each room inside the floorplan sequence, we also build another embedding layer to learn the positional embeddings X_p [23] for each room element. The input of the transformer is the summation of the three types of embeddings: $X = X_g^G + X_c + X_p$. At the test time, student-forcing is used, and only the starting token is used as input to the transformer network. For the transformer architecture, we apply multi-head self-attention as in LayoutTransformer [7] and use GPT-1 [21] as our backbone transformer network. Note that our model is flexible and also compatible with other transformer models.

3.2 Refining Stage with Panoptic Refinement Network

A major limitation of transformer-based autoregressive generation is that, for each generation step, only previous states can be taken as input. Thus, the model cannot leverage information from the future states, and may neglect the connectivity between certain elements. For example, in vectorized floorplan generation, it is challenging to properly place a room or door even for human experts, if only limited context information is provided. However, if the initial states for all rooms are given, a transformer model can access the panoptic features and has a better sense of their relative positions with corresponding categories. This motivates us to design a panoptic refining network to improve the floorplan generated in the draft stage by encouraging correct global connectivities.

In the refining stage, we introduce a novel panoptic refinement network (PRN) to refine the initial generations from the autoregressive transformer. The network is also implemented as a GCN and a transformer network with multi-head self-attention modules. Although the network architectures of the two stages are similar, their learning targets are fundamentally different. In the draft stage, our model aims to predict one element at a time given all previously generated rooms; while in the multi-round refining stage, the PRN aims to learn the refined elements based on the entire sequence.

During training, we use greedy decoding to convert the learned probabilities into the coordinate sequence with discrete values. Note that we do not include $\langle BoS \rangle$ and $\langle EoS \rangle$ in the input of PRN. For the input of the refinement network, since the category embeddings and positional embeddings remain unchanged, we directly get them from the input of the draft stage. Only the geometric embeddings will be updated, which is denoted as X_{gr}. The refinement GCN processes X_{gr} and obtain the GCN-updated embeddings X_{gr}^{G}. Then the input embeddings X_r for PRN transformer can be represented in a similar manner as the draft stage: $X_r = X_{gr}^{G} + X_c + X_p$.

Finally, the embedding sequence is passed into the PRN transformer to get the refined floorplan sequences. The end-to-end refinement process of PRN can be performed iteratively by taking the refined output from iteration i as the input of the next iteration $i+1$. More analysis is given in Sect. 4.4.

3.3 Training Strategies and Losses

Hybrid Sorting Strategy. As shown in previous works [16], ordering is important for training autoregressive models. Layout generation methods [7,16] often sorted the elements by the relative spatial positions. However, this strategy used in unconditional generation cannot be directly applied to category-conditioned generation. Since conditional categories are provided as constraints, a category needs to be strictly associated with a generated element. Thus, we propose a hybrid sorting strategy for a more effective conditional autoregressive generation in both training and testing. To leverage categorical spatial information, we first compute the average spatial position for each room category throughout the whole training set, then sort room categories by their average positions and place two types of doors (*i.e.*, interior and front doors) at the end.

Reconstruction Loss. We apply the standard cross-entropy loss on the generations from both stages to learn a categorical distribution for the quantized coordinates. The loss is averaged by the sequence length T and the refinement iterations N_r, which can be represented as:

$$L_{recon} = \frac{1}{T}\sum_{t=1}^{T} L_{ar} + \frac{1}{TN_r}\sum_{t=1}^{T}\sum_{i=1}^{N_r} L_{ref}, \tag{4}$$

where L_{ar} denotes for the loss from autoregressive model and L_{ref} denotes for the loss from panoptic refinement model.

Geometric Loss. Since the reconstruction loss focuses more on single elements, the connectivity between rooms is not explicitly measured by the reconstruction loss. To further encourage correct room layout and connectivity, especially for the refining stage, we introduce a global geometric constraint by calculating the intersection-over-union (IoU) values between the generated and real layouts. To calculate the IoU values, we need to first transform the predicted probabilities on the quantized coordinates to continuous values. However, direct conversion is not differentiable and thus cannot be trained via back-propagation. To enable end-to-end training, we borrow ideas from the differentiable soft-argmax strategy originally proposed in the stereo matching task [12]. The ith element's prediction x_i is then represented as $x_i = \sum_{j=1}^{N} x_j \cdot \sigma(x_j)$, where $\sigma(x_j)$ is the probability of quantized coordinate x_j implemented with the softmax function. Now we can compute a geometric IoU matrix $\mathcal{G}_{gen}$ for the generated sequence and build a geometric loss L_{geo} measuring the L_1 distance between $\mathcal{G}_{gen}$ and its groundtruth geometric IoU matrix $\mathcal{G}_{gt}$:

$$L_{geo} = \frac{1}{N_l} \sum_{i_l=1}^{N_l} ||\mathcal{G}_{gen}^{i_l} - \mathcal{G}_{gt}^{i_l}||_1, \tag{5}$$

where i_l and N_l are the index and the total number of pair-wise objects in a floorplan, respectively. L_{geo} is applied in both draft stage and refining stage. Our final loss objective is the combination of two losses without scaling factors.

4 Experiments

4.1 Dataset and Pre-processing

Similar to [18,20], we conducted experiments on RPLAN dataset [24] which provides vector-graphic floorplans from real-world residential buildings. We followed the pre-processing pipeline used in HouseGAN++ with a modification for bounding-box (bbox) vectorized floorplan generation. When we transformed the non-standard polygons into rectangular bounding boxes, we found some cases where frontal doors or interior doors were incorrectly kept inside a room (see Fig. 1 in supplementary materials as an example). To filter out such noisy cases, we computed the overlapping between doors and rooms, then discarded a sample if its frontal doors overlapped with any rooms or any of its interior doors overlapped with any non-living rooms over a threshold $\tau = 50\%$.

4.2 Baselines and Evaluation Metrics

We compared our method with two state-of-the-art floorplan generation methods using the same input setting [17,18] as ours. We also compared with LayoutTransformer [7] which we adopted as the backbone generation network. For fair comparisons, we retrained all baseline methods with their official implementations and applied the same data pre-processing as described in Sect. 4.1.

Following [17,18] which also took bubble diagram as input constraints, we evaluated our method in terms of realism, diversity, and compatibility. For realism, we invited 10 respondents with architecture design expertise (professors and graduate students from the architecture department) and 7 amateurs to

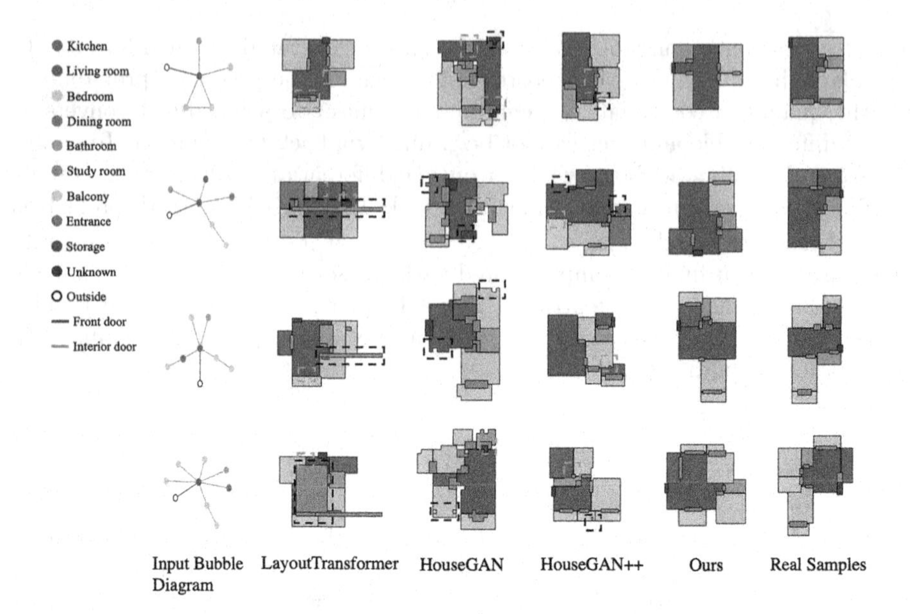

Fig. 4. Sampled floorplan generations from different methods, including real floorplans and their corresponding input bubble diagrams. We use the same color legend as in [18]. Areas with salient connectivity issues and visual artifacts are highlighted with green and **black** dashed boxes, respectively. (Color figure online)

give rankings on 100 generation examples based on visual quality and functionality. Our method and two other competing methods [7,18] each generated 100 floorplans and respondents ranked the methods after reviewing each set of 3 generated floorplans that correspond to the same input bubble diagram. A method obtained a score +1 for being ranked as $1st$, 0 for $2nd$, and -1 for $3rd$. Diversity was computed as the FID score [8] between the generated floorplans from a method and the real floorplans, to assess the diversity of the generated floorplans given the same input. The compatibility [18] validates the connectivity correctness of a generated floorplan by computing the graph edit distance (GED) [1] between the input graph with the reconstructed graph from the generated floorplan. For a generated plan to achieve a high compatibility score, a door must be correctly placed between two rooms (or between a room and the outside area). We generated 1,000 testing samples on each data split and ran each model for 5 rounds to get the mean and standard deviation of FID and GED scores.

For a comprehensive comparison, we evaluated our methods under both the *separate training* and *mixed training* settings. Under the separate training setting [18], floorplans were split into training and testing sets based on different numbers of rooms. Training was done using any number of rooms except the number of rooms reserved for testing. Comparisons using the separate training are provided in Table 2. In addition to the separate training, we also reported metrics calculated using the *mixed training*. Specifically, we mixed floorplans

Table 1. Quantitative comparison between methods using the mixed training. Diversity and compatibility metrics are reported as "mean/std.".

Model	Visual realism ↑	Functional realism ↑	Diversity ↓	Compatibility ↓
LayoutTransformer [7]	−0.58	−0.57	**12.3/0.2**	4.6/0.0
HouseGAN++ [18]	0.12	0.09	15.7/0.2	3.5/0.0
Ours	**0.46**	**0.48**	13.8/0.4	**2.8/0.1**

Table 2. Quantitative comparison between different methods using the separate training. Integers refer to train on floorplans with other number of rooms and test with the designated number of room.

Model	Diversity ↓				Compatibility ↓			
	5	6	7	8	5	6	7	8
LayoutTransformer [7]	21.5/0.3	27.8/0.5	25.6/0.6	28.4/0.8	2.4/0.1	4.0/0.1	5.6/0.1	6.4/0.1
HouseGAN [17]	57.6/1.3	70.1/1.2	61.7/0.9	57.7/0.6	3.5/1.6	4.3/1.7	5.2/1.9	6.1/1.9
HouseGAN++ [18]	24.1/0.6	17.7/0.7	15.0/0.7	22.2/0.4	1.6/0.0	2.5/0.1	**2.8/0.0**	**3.8/0.0**
Ours	**15.4/0.5**	**16.4/0.5**	**14.6/0.4**	**14.1/0.4**	**1.3/0.0**	**1.9/0.0**	2.9/0.1	4.6/0.0

with different numbers of rooms together and split them into training and testing without overlapping in input bubble diagrams. We believe this is a more general setting since floorplans with different number of rooms should all be available for training. Comparisons using mixed training are illustrated in Table 1.

4.3 Result Analysis

Table 2 shows the quantitative comparisons among different methods under the separate training setting. Our method outperforms all the previous methods on diversity by a large margin. For compatibility, our model is significantly better than existing approaches on the split containing 5 or 6 rooms, and is comparable with HouseGAN++ [18] on the 7 room split. Under our proposed mixed training setting in Table 1, which we believe is more suitable for realistic scenarios, our method clearly outperforms HouseGAN++ on all metrics. Note that evaluation of visual and functional realism by ten architecture experts and seven amateurs was conducted using floorplans generated in the mixed training setting, and the result on realism comparison is reported in Table 1. One can see that, for realism, our method outperforms [7,18] on both visual quality and functionality by a large margin. Qualitatively, from the samples shown in Fig. 4, one can see that LayoutTransformer [7] often fails to generate doors correctly, especially when the input bubble diagrams are complex. The rooms generated by HouseGAN [17] have salient artifacts due to the limitations of rasterized representation. There are fewer artifacts in floorplans generated by HouseGAN++ [18], but non-straight room boundaries still exist. Another non-negligible drawback of rasterized representation is that several disconnected regions could be generated for one specific room (*e.g.*, the living room of the *3rd* case by HouseGAN++, Fig. 4). In comparison, our method only generates objects as connected components due to

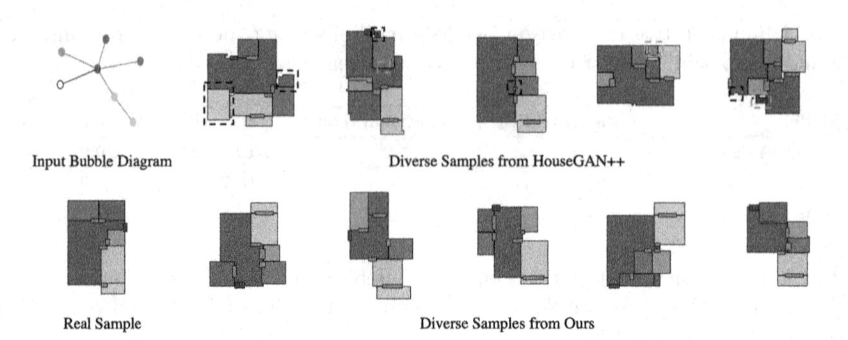

Fig. 5. Diversity comparison between HouseGAN++ and ours. Salient connectivity issues and visual artifacts are highlighted with green and **black** boxes. (Color figure online)

Table 3. Ablation study of our proposed method. Base refers to the backbone GPT-1 [21] transformer network. HS, PR, GL represent our proposed hybrid sorting, panoptic refinement, geometric loss, respectively.

(a) Our proposed components.

Components					Diversity (↓)	Compatibility (↓)
Base	HS	GCN	PR	GL	6	6
✓					27.8/0.5	4.0/0.1
✓	✓				29.8/0.6	3.6/0.1
✓	✓	✓			14.5/0.2	3.7/0.0
✓	✓		✓		**14.0/0.3**	3.4/0.1
✓	✓	✓	✓		18.3/0.2	2.1/0.1
✓	✓	✓	✓	✓	15.8/0.5	**1.9/0.0**

(b) Refinement iterations of PRN.

# of iteration	Diversity (↓)	Compatibility (↓)
	6	6
1-iteration	15.1/0.8	2.7/0.0
3-iterations	**14.4/0.3**	2.2/0.0
5-iterations	15.8/0.5	**1.9/0.0**
7-iterations	16.9/0.4	2.1/0.1

the advantage of vectorized representation. Furthermore, as shown in Fig. 5, our model better keeps the fidelity and connectivity while generating diverse samples than the state-of-the-art HouseGAN++ [18].

4.4 Ablation Study

Proposed Components. We conducted an ablation study on each of our proposed components, including the hybrid sorting (HS) strategy, the use of GCN to encode connectivity graph, the proposed panoptic refinement (PR) network, and the geometric loss (GL). Specifically, we conducted ablation experiments on the 6 room generation split. The base model corresponds to the result of LayoutTransformer [7], which we used as the unconditional backbone network. As shown in Table 3a, applying the hybrid sorting strategy improved the compatibility compared with the base model. Adding GCN or refinement model independently upon the autoregressive model significantly improved the diversity, but had little benefits for compatibility. A possible reason is that, passing a partial graph containing previously generated elements to GCN can smooth the feature for past elements, but is not beneficial to the generation for future elements. Applying the two modules simultaneously greatly improved compatibility since

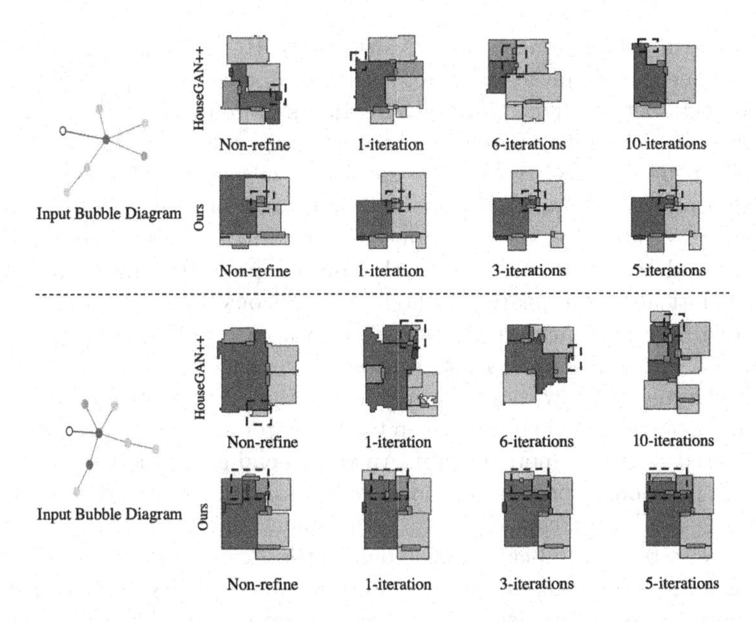

Fig. 6. Floorplan refinement between ours and [18]. Salient differences across different iterations are highlighted with **black** dashed boxes.

a complete connectivity graph drives GCN to capture the connectivity relationship of the whole sequence, and the refinement transformer can further exploit global contexts. With the proposed geometric loss on the generated vectors, both diversity and compatibility got further improved as the correct layouts and room connectivities were encouraged during the training.

Refinement Iterations. We also demonstrate the effect of applying different numbers of refinement iterations. As shown in Table 3b, there were no significant differences in diversity among various iterations. For compatibility, the best result was achieved when we refined the initial generated sequence for 5 iterations. We have tried to refine more iterations, such as $N_r = 7$, but the experiments showed no further improvement, which may indicate that PRN has adequately refined the draft floorplan. We also show some qualitative visualizations in Fig. 6 to help examine the effect of our multi-round refinements.

5 Discussion

HouseGAN++ [18] also proposed to refine floorplans iteratively using heuristics and optimizations. Our proposed PRN refinement network is very different, however, since it is an end-to-end model which refines the draft floorplan while keeping correct layouts. As illustrated in Fig. 6, applying one iteration of PRN refinement (*i.e.*, 1-iteration) significantly improved the generation quality compared with no refinement (*i.e.*, 'Non-refine'). Refining more iterations consistently made details more accurate, such as reducing overlaps, making neighboring objects better aligned, and placing doors more properly. Compared with the floorplans after multiple refinement iterations by HouseGAN++, the floorplans refined by our PRN were more consistent in layout and inherited correct room connectivities acquired in previous iterations of refinement. Meanwhile, refinements by HouseGAN++ introduced more drastic changes across different iterations, and occasionally regions with desirable properties from previous iterations got lost and disappeared in the later iterations.

Although we have achieved superior performance under most settings, our method still has limitations. First, our model represents rooms as rectangular boxes, which may not always be optimal in practice. We try to mitigate the issue by removing overlapping parts between the living room and other rooms, making the generated floorplan more general. An end-to-end generation with vectorized multi-edge polygons representing rooms could be a future direction. Second, we did not achieve better compatibility performance than HouseGAN++ on the split of 7 or 8 rooms, indicating that our method can be further improved for modeling longer floorplan sequences. More advanced GNNs and transformers may perform better in modeling more complex graphs and longer sequences.

During the user study, besides rankings of methods which we used to calculate realism scores, we also collected feedback in the form of free-form comments from architects. Most of the architects agreed that our method has made substantial progress in automated floorplan generation, and can potentially change the way in which initial floorplan designs are created and visualized. They especially liked that methods like ours can easily generate multiple candidate initial designs. However, they also mentioned that certain domain knowledge should be embedded in the generation process, for example, circulation spaces must be smaller, kitchen and dining must be connected, service blocks like washroom should be at the edge of the building. We plan to explore how to incorporate such domain knowledge into the generation process in our future work.

6 Conclusion

In this paper, we propose a novel two-stage vectorized floorplan generation framework. Floorplans can be generated and refined in an end-to-end fashion given an input bubble diagram encoding room number, type and connectivity. Compared to previous state-of-the-art floorplan generation methods which generate rasterized floorplan images, our proposed vectorized floorplan generation method generates more realistic and usable floorplan designs. Endorsed by the architect experts in our user study, we believe our method has great potential in computer-aided floorplan design.

Acknowledgements. This work is supported in part by NSF Award #1815491. We appreciate the help from professors and graduate students from College of Arts and Architecture at Penn State with the user study. We also would like to thank Enyan Dai for meaningful discussions on GNN.

References

1. Abu-Aisheh, Z., Raveaux, R., Ramel, J.Y., Martineau, P.: An exact graph edit distance algorithm for solving pattern recognition problems. In: 2015 4th International Conference on Pattern Recognition Applications and Methods (2015)
2. Arroyo, D.M., Postels, J., Tombari, F.: Variational transformer networks for layout generation. In: Proceedings of the IEEE/CVF Conference on Computer Vision and Pattern Recognition, pp. 13642–13652 (2021)
3. Chen, J., Liu, C., Wu, J., Furukawa, Y.: Floor-SP: inverse cad for floorplans by sequential room-wise shortest path. In: Proceedings of the IEEE/CVF International Conference on Computer Vision, pp. 2661–2670 (2019)
4. Cruz, S., Hutchcroft, W., Li, Y., Khosravan, N., Boyadzhiev, I., Kang, S.B.: Zillow indoor dataset: annotated floor plans with 360deg panoramas and 3D room layouts. In: Proceedings of the IEEE/CVF Conference on Computer Vision and Pattern Recognition, pp. 2133–2143 (2021)
5. Fan, Z., Zhu, L., Li, H., Chen, X., Zhu, S., Tan, P.: FloorPlanCAD: a large-scale cad drawing dataset for panoptic symbol spotting. In: Proceedings of the IEEE/CVF International Conference on Computer Vision, pp. 10128–10137 (2021)
6. Goodfellow, I., et al.: Generative adversarial nets. Adv. Neural Inf. Process. Syst. **27** (2014)
7. Gupta, K., Lazarow, J., Achille, A., Davis, L.S., Mahadevan, V., Shrivastava, A.: LayoutTransformer: layout generation and completion with self-attention. In: Proceedings of the IEEE/CVF International Conference on Computer Vision, pp. 1004–1014 (2021)
8. Heusel, M., Ramsauer, H., Unterthiner, T., Nessler, B., Hochreiter, S.: GANs trained by a two time-scale update rule converge to a local nash equilibrium. Adv. Neural Inf. Process. Syst. **30** (2017)
9. Hu, R., Huang, Z., Tang, Y., Van Kaick, O., Zhang, H., Huang, H.: Graph2plan: learning floorplan generation from layout graphs. ACM Trans. Graph. (TOG) **39**(4), 118–1 (2020)
10. Huang, W., Zheng, H.: Architectural drawings recognition and generation through machine learning (2018)
11. Jyothi, A.A., Durand, T., He, J., Sigal, L., Mori, G.: LayoutVAE: stochastic scene layout generation from a label set. In: Proceedings of the IEEE/CVF International Conference on Computer Vision, pp. 9895–9904 (2019)
12. Kendall, A., et al.: End-to-end learning of geometry and context for deep stereo regression. In: Proceedings of the IEEE International Conference on Computer Vision, pp. 66–75 (2017)
13. Kingma, D.P., Welling, M.: Auto-encoding variational bayes. arXiv preprint arXiv:1312.6114 (2013)
14. Kipf, T.N., Welling, M.: Semi-supervised classification with graph convolutional networks. arXiv preprint arXiv:1609.02907 (2016)
15. Kong, X., Jiang, L., Chang, H., Zhang, H., Hao, Y., Gong, H., Essa, I.: BLT: bidirectional layout transformer for controllable layout generation. arXiv preprint arXiv:2112.05112 (2021)
16. Nash, C., Ganin, Y., Eslami, S.A., Battaglia, P.: PolyGen: an autoregressive generative model of 3D meshes. In: International Conference on Machine Learning, pp. 7220–7229. PMLR (2020)
17. Nauata, N., Chang, K.-H., Cheng, C.-Y., Mori, G., Furukawa, Y.: House-GAN: relational generative adversarial networks for graph-constrained house layout generation. In: Vedaldi, A., Bischof, H., Brox, T., Frahm, J.-M. (eds.) ECCV 2020.

LNCS, vol. 12346, pp. 162–177. Springer, Cham (2020). https://doi.org/10.1007/978-3-030-58452-8_10
18. Nauata, N., Hosseini, S., Chang, K.H., Chu, H., Cheng, C.Y., Furukawa, Y.: House-GAN++: generative adversarial layout refinement networks. arXiv preprint arXiv:2103.02574 (2021)
19. Newton, D.: Generative deep learning in architectural design. Technol.—Archit.+ Design **3**(2), 176–189 (2019)
20. Para, W., Guerrero, P., Kelly, T., Guibas, L.J., Wonka, P.: Generative layout modeling using constraint graphs. In: Proceedings of the IEEE/CVF International Conference on Computer Vision, pp. 6690–6700 (2021)
21. Radford, A., Narasimhan, K., Salimans, T., Sutskever, I.: Improving language understanding by generative pre-training (2018)
22. Shekhawat, K., Upasani, N., Bisht, S., Jain, R.N.: A tool for computer-generated dimensioned floorplans based on given adjacencies. Autom. Constr. **127**, 103718 (2021)
23. Vaswani, A., et al.: Attention is all you need. Adv. Neural Inf. Process. Syst. **30** (2017)
24. Wu, W., Fu, X.M., Tang, R., Wang, Y., Qi, Y.H., Liu, L.: Data-driven interior plan generation for residential buildings. ACM Trans. Graph. (TOG) **38**(6), 1–12 (2019)
25. Xu, L., et al.: BlockPlanner: city block generation with vectorized graph representation. In: Proceedings of the IEEE/CVF International Conference on Computer Vision, pp. 5077–5086 (2021)
26. Yang, C.F., Fan, W.C., Yang, F.E., Wang, Y.C.F.: LayoutTransformer: scene layout generation with conceptual and spatial diversity. In: Proceedings of the IEEE/CVF Conference on Computer Vision and Pattern Recognition, pp. 3732–3741 (2021)

Few-Shot Image Generation with Mixup-Based Distance Learning

Chaerin Kong[1], Jeesoo Kim[2], Donghoon Han[1], and Nojun Kwak[1](✉)

[1] Seoul National University, Seoul, South Korea
{veztylord,dhk1349,nojunk}@snu.ac.kr
[2] NAVER WEBTOON AI, Seongnam-si, South Korea
jeesookim@webtoonscorp.com

Abstract. Producing diverse and realistic images with generative models such as GANs typically requires large scale training with vast amount of images. GANs trained with limited data can easily memorize few training samples and display undesirable properties like "stairlike" latent space where interpolation in the latent space yields discontinuous transitions in the output space. In this work, we consider a challenging task of pretraining-free few-shot image synthesis, and seek to train existing generative models with minimal overfitting and mode collapse. We propose mixup-based distance regularization on the feature space of both a generator and the counterpart discriminator that encourages the two players to reason not only about the scarce observed data points but the relative distances in the feature space they reside. Qualitative and quantitative evaluation on diverse datasets demonstrates that our method is generally applicable to existing models to enhance both fidelity and diversity under few-shot setting. Codes are available (https://github.com/reyllama/mixdl).

Keywords: Generative Adversarial Networks (GANs) · Few-shot image generation · Latent mixup

1 Introduction

Remarkable features of Generative Adversarial Networks (GANs) such as impressive sample quality and smooth latent interpolation have drawn enormous attention from the community, but what we have enjoyed with little gratitude claim their worth in a data-limited regime. As naive training of GANs with small datasets often fails both in terms of fidelity and diversity, many have proposed novel approaches specifically designed for few-shot image synthesis. Among the most successful are those adapting a pretrained source generator to the target domain [26,31,34] and those seeking generalization to unseen categories through feature fusion [16,19]. Despite their impressive synthesis quality, these approaches

Supplementary Information The online version contains supplementary material available at https://doi.org/10.1007/978-3-031-19784-0_33.

S. Avidan et al. (Eds.): ECCV 2022, LNCS 13675, pp. 563–580, 2022.
https://doi.org/10.1007/978-3-031-19784-0_33

Fig. 1. Cross Domain Correspondence [34] adaptation of FFHQ source generator on various target domains (10-shot). Finding a semantically similar source domain is crucial for CDC as large domain gap greatly harms the transfer performance. We later show that our method outperforms CDC without any source domain pretraining even on the semantically related domains.

are often critically constrained in practice as they all require semantically related large source domain dataset to pretrain on [34], as illustrated in Fig. 1. For some domains like abstract art paintings, medical images and cartoon illustrations, it is very difficult to collect thousands of samples, while at the same time, finding an adequate source domain to transfer from is not straightforward either. To train GANs from scratch with limited data, several augmentation techniques [22,55] and model architecture [27] have been proposed. Although these methods have presented promising results on low-shot benchmarks consisting of hundreds to thousands of training images, they fall short for few-shot generation where the dataset is even more constrained (*e.g.*, $n = 10$).

GANs trained with small dataset typically display one of the two behaviors: severe quality degradation [22,55] or near-perfect memorization [13], as visible from Fig. 2 (*left*). Hence producing *novel* samples of *reasonable* quality is the ultimate goal of few-shot generative models. We note that memorization differs from the classic mode collapse problem, as the former is not just lack of diversity, but the *fundamental inability to generate unseen samples.*

As directly combatting memorization with as little as 10 training samples is extremely difficult if not impossible, we choose to tackle a surrogate problem instead. Our key observation is that strongly overfitted generators are only capable of producing a limited set of samples, resulting in discontinuous transitions in the image space under latent interpolation. We call this *stairlike latent space phenomenon*, which has been pointed out by previous works [8,36] as an indicator for memorization. Figure 2 (*right*) demonstrates that previous methods designed for diversity preservation [4] or low-shot synthesis [27] all display such behavior under few-shot setting ($n = 10$). Therefore, instead of pursuing the seemingly insurmountable task of suppressing memorization, we directly target *stairlike latent space problem* and propose effective distance regularizations to explicitly *smooth* the latent space of the generator (G) and the discriminator (D), which we empirically show is equivalent to fighting memorization in effect.

Our high level idea is to maximally exploit the scarce data points by continuously exploring their semantic mixups [52]. The discriminator overfitted to few real samples, however, shows overly confident and abrupt decision boundaries, leaving the generator with no choice but to faithfully replicate them in order

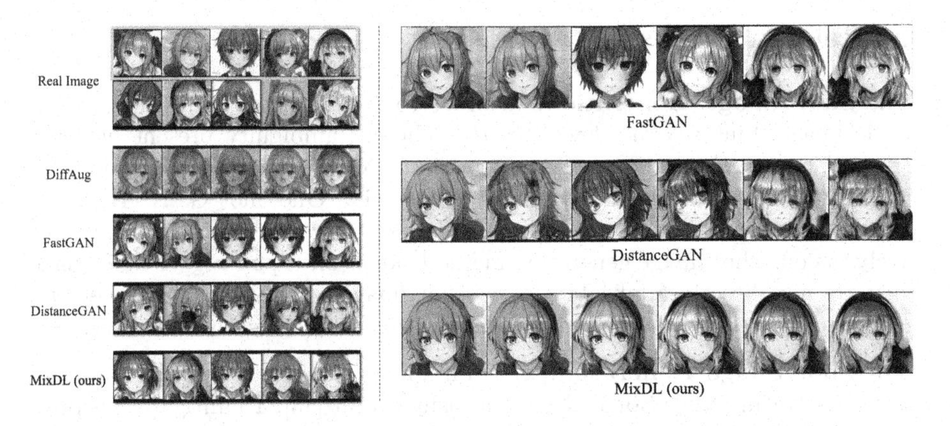

Fig. 2. Training GANs with as little as 10 real samples typically results in either complete collapse or severe memorization *(left)*. Strongly overfitted generators can only generate a limited set of images, hence displaying *stairlike* latent interpolation *(right)*.

to convince the opponent. This results in aforementioned *stairlike latent space* for both G and D, rendering smooth semantic mixups impossible. To tackle this problem, we explore G's latent space with a randomly sampled interpolation coefficient $\mathbf{c}$, enforcing relative semantic distances between samples to follow the mixup ratio. By simultaneously imposing similar regularization on D's feature space, we prohibit the discriminator from embedding images to arbitrary locations for its convenience of memorizing, and guide its feature space to be aligned by semantic distances. Our objective is inspired by the formulation of [34] that aims to transfer diversity information from source domain to target domain. We tailor it for our single domain setting, where no source domain is available to import diversity from, and show that our method is able to produce diverse novel samples with convincing quality even with as little as 10 training images.

We further observe that models trained with our regularizations resist mode collapse surprisingly well even with no special augmentation. We believe that our distance regularizations encourage the model to preserve inherent diversity present in early stages throughout the course of training. Resistance to overfitting and mode collapse combined opens up doors for sample diversity under rigorous data constraint, which we demonstrate later with experimental results.

In sum, our contributions can be summarized as:

- We propose a two-sided distance regularization that encourages learning of smooth and mode-preserved latent space through controlled latent interpolation.
- We introduce a simple framework for few-shot image generation without a large source domain dataset that is compatible with existing architectures and augmentation techniques.
- We evaluate our approach on a wide range of datasets and demonstrate its effectiveness in generating diverse samples with convincing quality.

2 Related Works

One-Shot Image Generation. In order to create diverse outcomes from a single image, SinGAN [39] leverages the inherent ambiguity present in downsampled image. Based on SinGAN, ConSinGAN [18] proposes a technique to control the trade-off between fidelity and diversity. One-Shot GAN [41] uses a dual-branch discriminator where each head identifies context and layout, respectively. As one-shot image generation methods focus on exploiting a single image, they are not directly applicable to few-shot image generation tasks where the generator must learn the underlying distribution of a collection of images.

Low-Shot Image Generation. Given a limited amount of training data, the discriminator in conventional GAN can easily overfit [43]. To mitigate this problem, DiffAugment [55] imposes differentiable data augmentation to both real and fake samples while ADA [22] devises non-leaking adaptive discriminator augmentation. FastGAN [27] suggests a skip-layer excitation module and a self-supervised discriminator, which saves computational cost and stabilizes low-shot training. GenCo [11] shows impressive results on low-shot image generation task by using multiple discriminators to alleviate overfitting. Despite their promising performances on low-shot benchmarks, these methods often show significant instability under stricter data constraint, namely in *few-shot* setting.

Few-Shot Generation with Auxiliary Dataset. Thus far, the *few-shot* image generation task ($n \approx 10$) mostly required pretraining on larger dataset with similar semantics [37,47,48,54] mainly due to its inherent difficulty. A group of works [3,16,19,20] learns transferable generation ability on *seen categories* and seek generalization into *unseen categories* through fusion-based methods. FreezeD [31] and EWC [26] further improves transfer learning framework for GANs. Meanwhile, CDC [34] computes the similarities between samples within each domain and encourages the corresponding similarity distributions to resemble each other. It aims to directly transfer the structural diversity of the source domain to the target, yielding impressive performance. In this paper, we modify the formulation of CDC and propose a novel few-shot generation framework that does not require any auxiliary data or separate pretraining step.

Generative Diversity. Mode collapse has been a long standing obstacle in GAN training. [2,30] introduce divergence metrics that are effective at stabilizing GAN training while [12,14] tackle this problem by training multiple networks. Another group of works [4,28,29,42,50] proposes regularization methods to preserve distances in the generated output space. Unlike these works, we consider the few-shot setting where the diversity is restricted mainly due to memorization, and introduce an interpolation-based distance regularization as an effective remedy.

Latent Mixup. Since [52], mixup methods have been actively explored to enforce smooth behaviors in between training samples [5,6,44]. In generative models, [36] emphasizes the importance of smooth latent transition as a counterevidence for memorization, but as state-of-the-art GAN models trained with sufficient data naturally possess such property [8,24], it has been mainly studied

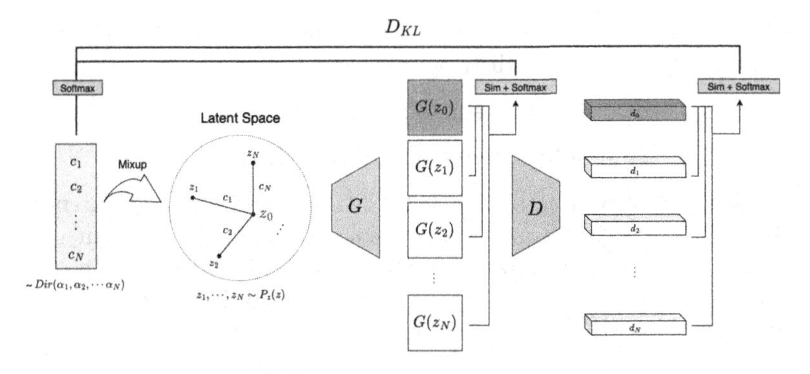

Fig. 3. Overview of our **Mixup-based Distance Learning (MixDL)**. We sample mixup coefficients from a Dirichlet distribution and generate an anchor point z_0 through interpolation. Then we enforce pairwise similarities between intermediate generator activations to follow the interpolation coefficients. Similar regularization is imposed on discriminator's penultimate activation, which is linearly projected before similarity calculation. The proposed regularization terms can be added on top of any traditional adversarial framework.

with autoencoders. [7,35] regularize autoencoders to learn smooth latent space while [38,49] explore their potential as generative models through interpolation.

3 Approach

We consider the situation where only few train examples (e.g., $n = 10$) are available with no semantically similar source domain. Hence, we would like to train a generative model from scratch, *i.e.*, with no auxiliary dataset or separate pretraining step, using only a handful of images. Under such challenging constraints, overfitting greatly restricts a model's ability to learn data distribution and produce diverse samples. We identify its byproduct *stairlike latent space* as the core obstacle, as it not only indicates memorizing but also prohibits hallucination through semantic mixup. We observe that both the generator and the discriminator suffer from the problem with insufficient data, evidenced by discontinuous latent interpolation and overly confident decision boundary, respectively.

To this end, we propose mixup-based distance learning (MixDL) framework that guides the two players to form soft latent space and leverage it to generate diverse samples. We further discover that our proposed regularizers effectively combat mode collapse, a problem particularly more devastating with a small dataset, by preserving diversity present in early training stages. As our formulation is inspired by [34], we first introduce their approach in Sect. 3.1, and formally state our methods in Sect. 3.2 and Sect. 3.3. Our final learning framework and the corresponding details can be found in Sect. 3.4.

3.1 Cross-Domain Correspondence

In CDC [34], the authors propose to transfer the relationship learned in a source domain to a target domain. They define a probability distribution from pairwise similarities of generated samples in both domains and bind the latter to the former. Formally, they define distributions as

$$p^l = \text{softmax}(\{\text{sim}(G_s^l(z_0), G_s^l(z_i))\}_{i=1}^N) \quad (1)$$

$$q^l = \text{softmax}(\{\text{sim}(G_{s\rightarrow t}^l(z_0), G_{s\rightarrow t}^l(z_i))\}_{i=1}^N) \quad (2)$$

where G^l is the generator activation at the lth layer and $\{z_i\}_0^N$ are latent vectors. Note that G_s and $G_{s\rightarrow t}$ correspond to source and target domain generator, respectively, and p^l, q^l are N-way discrete probability distributions consisting of N pairwise similarities. Then, along with adversarial objective $\mathcal{L}_{adv}$, they impose a KL-divergence-based regularization of the following form:

$$\mathcal{L}_{dist} = \mathbb{E}_{z\sim p_z(z)}[D_{KL}(q^l||p^l)]. \quad (3)$$

The benefits of this auxiliary objective are twofold: it prevents distance collapse in the target domain and transfers diversity from the source to target via one-to-one correspondence. However, as visible from Fig. 1, the synthesis quality is greatly affected by the semantic distance between source and target. Hence, we propose MixDL, which modifies CDC for pretraining-free few-shot image synthesis and provides consistent performance gains across different benchmarks.

3.2 Generator Latent Mixup

In [34], the anchor point z_0 could be chosen arbitrarily from the prior distribution $p_z(z)$ since they were transferring the rich structural diversity of the source domain to the target latent space. As this is no longer applicable in our setting, we propose to resort to diverse *combinations* of given samples. Hence, preserving the modes and learning interpolable latent space are our two main desiderata. To this end, we define our anchor point using Dirichlet distribution as follows:

$$z_0 = \sum_{i=1}^N c_i z_i, \quad \mathbf{c} \sim Dir(\alpha_1, \cdots, \alpha_N) \quad (4)$$

where $\mathbf{c} \triangleq [c_1, \cdots, c_N]^T$. Using Eq. 4, the latent space can be navigated in a quantitatively controlled manner. Defining probability distribution of pairwise similarities as in [34], we bind it to the interpolation coefficients $\mathbf{c}$ instead. The proposed distance loss is defined as follows:

$$\mathcal{L}_{dist}^G = \mathbb{E}_{z\sim p_z(z),\mathbf{c}\sim Dir(\alpha)}[D_{KL}(q^l||p)], \quad (5)$$

$$q^l = \text{softmax}(\{\text{sim}(G^l(z_0), G^l(z_i))\}_{i=1}^N), \quad (6)$$

$$p = \text{softmax}(\{c_i\}_{i=1}^N), \quad (7)$$

where $Dir(\alpha)$ denotes the Dirichlet distribution with parameters $\alpha = (\alpha_1, \cdots, \alpha_N)$. This efficiently accomplishes our desiderata. Intuitively, unlike naive generators that gradually converge to few modes, our regularization forces the generated samples to differ from each other by a controlled amount, making mode collapse very difficult. At the same time, we constantly explore our latent space with continuous coefficient vector $\mathbf{c}$, explicitly enforcing smooth latent interpolation. An anchor point similar to [34] can be obtained with one-hot coefficients $\mathbf{c}$.

3.3 Discriminator Feature Space Alignment

While the generator distance regularization can alleviate mode collapse and stairlike latent space problem surprisingly well, the root cause of constrained diversity still remains unresolved, *i.e.*, discriminator overfitting. As long as the discriminator delivers overconfident gradient signals to the generator based on few examples it observes, generator outputs will be strongly pulled towards the small set of observed data. To encourage the discriminator to provide smooth signals to the generator based on reasoning about continuous semantic distances rather than simply memorizing the data points, we impose similar regularization on its feature space. Formally, we define our discriminator $D(x) = (d^{(2)} \circ d^{(1)})(x)$ where $d^{(2)}(x)$ refers to the final FC layer that outputs {real, fake}. When a set of generated samples $\{G(z_i)\}_{i=1}^N$ and the interpolated sample $G(z_0)$ is provided to D, we construct an N-way distribution similar to Eq. 6 as

$$r = \text{softmax}(\{\text{sim}(proj(d_0^{(1)}), proj(d_i^{(1)}))\}_{i=1}^N) \tag{8}$$

where $proj$ refers to a linear projection layer widely used in self-supervised learning literature [9,10,15] and $d_j^{(1)} \triangleq d^{(1)}(G(z_j))$. Without the linear projector, we found the constraint too rigid that it harms overall output quality. We define our distance regularization for the discriminator as

$$\mathcal{L}_{dist}^D = \mathbb{E}_{z\sim p_z(z),\mathbf{c}\sim Dir(\alpha)}[D_{KL}(r||p)]. \tag{9}$$

This regularization penalizes the discriminator for storing memorized real samples in arbitrary locations in the feature space and encourages the space to be aligned with relative semantic distances. Thus it makes memorization harder while guiding the discriminator to provide smoother and more semantically meaningful signals to the generator.

3.4 Final Objective

Figure 3 shows an overall concept of our method. Our final objective takes the form:

$$\mathcal{L}^G = \mathcal{L}_{adv}^G + \lambda_G \mathcal{L}_{dist}^G \tag{10}$$

$$\mathcal{L}^D = \mathcal{L}_{adv}^D + \lambda_D \mathcal{L}_{dist}^D \tag{11}$$

where we generally set $\lambda_G = 1000$ and $\lambda_D = 1$.

As our method is largely independent of model architectures, we apply our method to two existing models, StyleGAN2[1] [24] and FastGAN [27]. We keep their objective functions as they are and simply add our regularization terms. For StyleGAN2, we interpolate in $\mathcal{W}$ rather than $\mathcal{Z}$, which has been shown to have better properties such as disentanglement [1,45,56]. Mixup coefficients $\mathbf{c}$ is sampled from a Dirichlet distribution of parameters all equal to one. Patch-level discrimination [21,34] is applied for mixup images to encourage our generator to be *creative* while exploring the latent space.

[1] https://github.com/rosinality/stylegan2-pytorch.

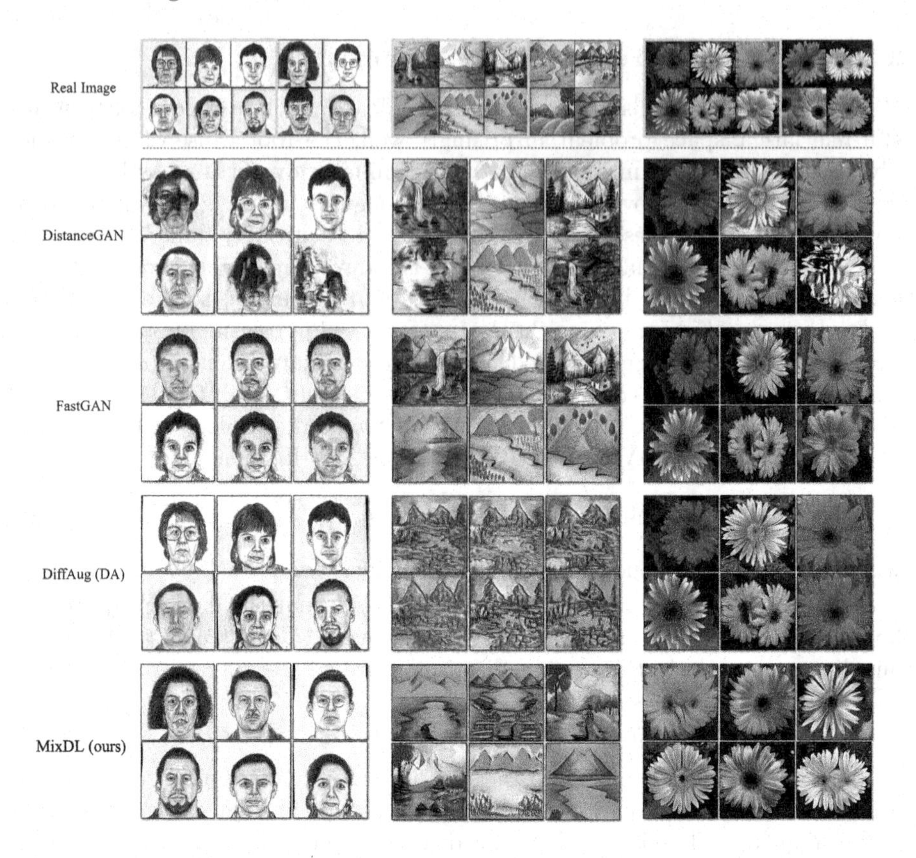

Fig. 4. 10-shot image generation results. While baseline methods either collapse or **simply replicate the training samples (yellow box)**, our method actively encourages the generator to explore semantic mixups of given samples, which enables synthesis of various unseen samples. (Color figure online)

4 Experiments

Baselines. We mainly apply our method to the state-of-the-art unconditional GAN model, StyleGAN2 [24]. Data augmentation techniques introduced by [55] and [22] show promising performance on low-shot image generation task, so we evaluate them along with ours and refer to them as *DA* and *ADA* respectively. We additionally apply our method to FastGAN [27], which is a light-weight GAN architecture that allows faster convergence with limited data. Although methods designed for alleviating mode collapse [4,28,29] are not directly targeted for data-limited setting, we further adopt these as baselines considering the similarity in objective formulation. We implement them on StyleGAN2 for better synthesis quality and fair comparison. Transfer based methods such as EWC [26] and CDC [34] fundamentally differ from ours as they require a large scale pretraining and thus are not directly comparable. However, we include CDC [34] since our method adjusts it for a more general single domain setting.

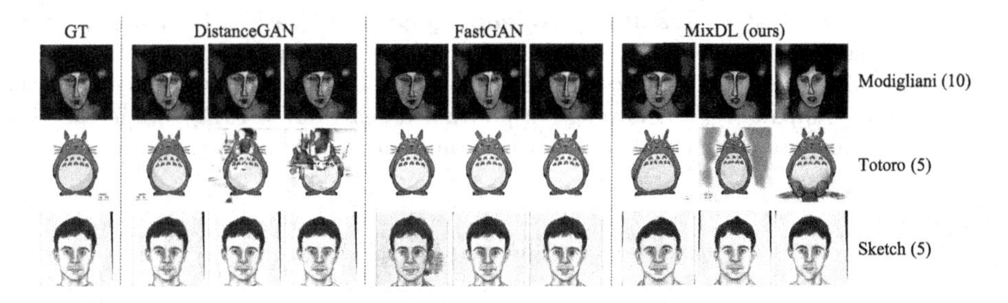

Fig. 5. Uncurated collection of samples sharing the same training image as nearest neighbor. Images from baselines are largely identical, but those produced by ours are all different. Numbers in parentheses indicate the dataset size.

Datasets. For quantitative evaluation, we use Animal-Face Dog [40], Oxford-flowers [33], FFHQ-babies [23], face sketches [46], Obama and Grumpy Cat [55], anime face [27] and Pokemon (pokemon.com, [27]). Aforementioned datasets contain 100 to 8189 samples, so we simulate few-shot setting by randomly sampling 10 images, if not stated otherwise. For qualitative evaluation, we further experiment on paintings of Amedeo Modigliani [51], landscape drawings [34] and web-crawled images of Totoro. All the images are 256×256. Additional synthesis results and information about datasets can be found in the supplementary.

Evaluation Metrics. We measure *Fréchet Inception Distance* (FID) [17], sFID [32] and precision/recall [56] for datasets containing a sufficient number ($\geq$100) of samples along with pairwise *Learned Perceptual Image Patch Similarity* (LPIPS) [53]. For simulated few-shot tasks, the FID and sFID are computed against the full dataset as in [26,34]. We further use LPIPS as a distance metric for demonstrating interpolation smoothness and mode preservation.

4.1 Qualitative Result

Figure 4 shows generated samples from 10-shot training. We observe that baseline methods either collapse to few modes or severely overfit to the training data, resulting in inability to generate novel samples. Ours is the only method that produces a variety of convincing samples that are not present in the training set. Our method combines visual attributes such as hairstyle, beard and glasses in a natural way, producing distinctive samples under harsh data constraint.

The difference is more distinguished when we take a closer look. In Fig. 5 we display uncurated sets of generated images along with their nearest neighbor real images. Samples from *DistanceGAN* [4] and *FastGAN* [27] are either defective or largely identical to the corresponding GT, but our method generates unique samples with recognizable visual features. We believe this is because our distance regularization enforces outputs from different latent vectors to differ from each other, proportionally to the relative distances in the latent space.

Table 1. Quantitative results on 10-shot generation task. FID and sFID are computed against the full dataset and LPIPS is calculated between generated samples. The best and the second best scores are in bold and underlined. Although CDC† is not directly comparable as it leverages a pretrained generator (FFHQ), we include it for the relevancy to our method. Clear performance drops are observed with increased domain gap (*e.g.*, FFHQ → Dogs).

Method	Anime Face			Animal-Face Dog			Oxford Flowers			Face Sketches			Pokemon		
	FID ↓	sFID ↓	LPIPS ↑	FID ↓	sFID ↓	LPIPS ↑	FID ↓	sFID ↓	LPIPS ↑	FID ↓	sFID ↓	LPIPS ↑	FID ↓	sFID ↓	LPIPS ↑
FastGAN [27]	123.7	127.9	0.341	103.0	117.4	0.633	182.7	111.2	0.667	76.3	81.8	0.148	123.5	105.7	0.578
StyleGAN2 [23]	166.0	111.4	0.363	177.5	127.7	0.569	177.3	143.0	0.537	94.2	84.4	0.435	257.6	136.5	0.439
StyleGAN2 + DA [55]	162.0	96.8	0.204	136.1	123.5	0.559	187.0	154.4	0.687	43.1	59.9	0.438	280.1	148.9	0.179
StyleGAN2 + ADA [22]	130.2	108.0	0.288	236.5	126.2	0.636	167.8	83.5	0.719	62.8	67.3	0.399	214.3	95.5	0.496
FastGAN + Ours	107.6	98.5	0.478	99.8	111.7	0.625	180.5	75.5	0.657	45.0	58.0	0.416	144.0	118.3	0.584
StyleGAN2 + Ours	73.1	**92.8**	0.548	96.0	99.9	0.682	136.6	67.6	0.734	39.4	**43.3**	0.479	117.0	**57.7**	0.539
StyleGAN2 + DA + Ours	**70.2**	94.1	0.551	96.4	107.6	0.682	129.9	66.9	0.705	**35.6**	50.1	0.471	**114.3**	79.0	**0.607**
StyleGAN2 + ADA + Ours	75.0	96.5	**0.571**	**94.1**	**96.6**	**0.684**	**127.7**	**52.5**	**0.763**	39.2	45.7	**0.482**	155.5	65.7	0.544
StyleGAN2 + CDC† [34]	93.4	107.4	0.469	206.7	110.1	0.545	107.5	99.9	0.518	45.7	46.1	0.428	126.6	79.1	0.342

Table 2. Quantitative comparison with diversity preservation methods on 10-shot image generation task. MixDL is equivalent to *StyleGAN2+Ours*.

Method	Anime Face			Animal-Face Dog			FFHQ-babies		
	FID ↓	sFID ↓	LPIPS ↑	FID ↓	sFID ↓	LPIPS ↑	FID ↓	sFID ↓	LPIPS ↑
N-Div [28]	175.4	176.4	0.425	150.4	153.6	0.632	177.1	177.1	0.510
MSGAN [29]	138.6	100.5	0.536	165.7	123.0	0.630	165.4	120.1	0.569
DistanceGAN [4]	84.1	93.0	0.543	102.6	114.2	0.678	105.7	102.9	0.640
MixDL (ours)	**73.1**	**92.8**	**0.548**	**96.0**	**99.9**	**0.682**	**83.4**	**73.9**	**0.643**

4.2 Quantitative Evaluation

Table 1 shows FID, sFID and LPIPS scores for several low-shot generation methods [22,27,55] on 10-shot image generation task. We can see that our method consistently outperforms the baselines, often with significant margins. Moreover, our regularizations can be applied concurrently to data augmentations to obtain further performance gains. Note that while StyleGAN2 armed with advanced data augmentations fails to converge from time to time, our method guarantees stable convergence to a better optimum across all datasets. Surprisingly, ours outperforms CDC [34] on all metrics even when the two domains are closely related, *e.g. anime-face* and *face sketches*. For dissimilar domains like *pokemon*, CDC tends to sacrifice diversity (*i.e.*, LPIPS) for better fidelity, which nevertheless falls short overall. We present training snapshots in the supplementary.

Additional quantitative comparison with diversity preserving methods is displayed in Table 2. Although these methods have some similarities with ours, especially MixDL-G, we can observe steady improvements with MixDL. As the baselines are simply designed to minimize mode collapse, we believe they are relatively prone to memorization, which is a far devastating issue in few-shot setting.

While pretraining-free 10-shot image synthesis task has not been studied much, several works [27,55] have previously explored generative modeling with as little as 100 samples. We present quantitative evaluations on popular low-shot

Table 3. FID comparison on low-shot benchmarks. LPIPS measures in-domain diversity.

Dataset	Obama	Cat	Flowers	Obama	Cat
Shot	100	100	100	10	10
LPIPS	0.615	0.613	0.795	0.598	0.598
StyleGAN2	63.1	43.3	192.2	174.7	76.4
+ DA	46.9	27.1	91.6	66.8	45.6
+ Ours	58.4	26.6	82.0	62.7	41.1
+ DA + Ours	**45.4**	**26.5**	**64.0**	**57.9**	**39.3**

Table 4. Precision and recall metrics on 100-shot benchmarks.

Method	Obama		Cat	
	Prec.	Rec.	Prec.	Rec.
StyleGAN2	0.47	0.07	0.15	0.12
+MixDL	**0.52**	**0.32**	**0.86**	**0.50**
FastGAN	0.90	0.36	0.90	0.43
+MixDL	**0.91**	**0.47**	**0.91**	**0.50**

Table 5. Ablation on MixDL-G and MixDL-D. Two regularizations combined generally yields the best performances.

MixDL		Dog (10-shot)		Babies (100-shot)		Flowers (100-shot)	
G	*D*	FID ↓	LPIPS ↑	FID ↓	LPIPS ↑	FID ↓	LPIPS ↑
		177.5	0.569	131.0	0.574	192.2	0.747
	✓	118.4	0.649	83.4	0.638	94.1	0.775
✓		**95.4**	0.673	71.7	0.638	84.0	0.780
✓	✓	96.0	**0.682**	**63.4**	**0.647**	**82.0**	**0.782**

Table 6. Mixup coefficient sampling distribution ablation. We adopt $\alpha = 1$ for simplicity.

Distribution	Dirichlet			Gaussian	Uniform
Parameter	$\alpha = 0.1$	$\alpha = 1$	$\alpha = 10$	standard	-
FID (↓)	76.4	**73.1**	80.8	76.0	74.8
LPIPS (↑)	0.536	**0.548**	0.532	**0.548**	0.546

benchmarks in Table 3. We observe that our method consistently improves the baseline, and the margin is larger for more challenging tasks, *i.e.*, dataset with greater diversity or fewer training samples. We discuss experiments on these benchmarks in depth in Sect. 5. Table 4 shows precision and recall [25] for these benchmarks, where MixDL boosts scores especially in terms of diversity.

4.3 Ablation Study

We further evaluate the effects of the proposed regularizations, MixDL-G (generator) and MixDL-D (discriminator), through ablation under different settings. In Table 5, we observe that in general, our regularizations both contribute to better quality and diversity, while in some special cases, only adding MixDL-G leads to better FID score. We conjecture that aligning discriminator's feature vectors with the interpolation coefficients can impose overly strict constraint for some datasets. We nonetheless observe consistent improvements on diversity.

Figure 6 shows the evaluation across different subset sizes. Since FFHQ-babies and Oxford-flowers contain more than 2,000 and 8,000 images respectively, we randomly sample subsets of size 10, 100 and 1,000. We can see that the performance of StyleGAN2 steadily improves with more training samples, but it consistently benefits from MixDL. Hence, we believe that with limited data in general, our method can be broadly used to improve model performance. Lastly in Table 6, the effect of using different Dirichlet concentration parameters and sampling distribution for mixup is illustrated. We find that setting $\alpha = 1$ yields the best performance, so we uniformly use this throughout the experiments.

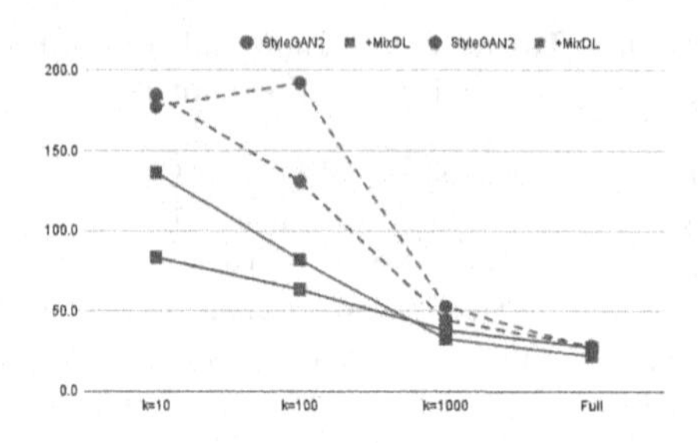

(a) FID scores for different dataset sizes.

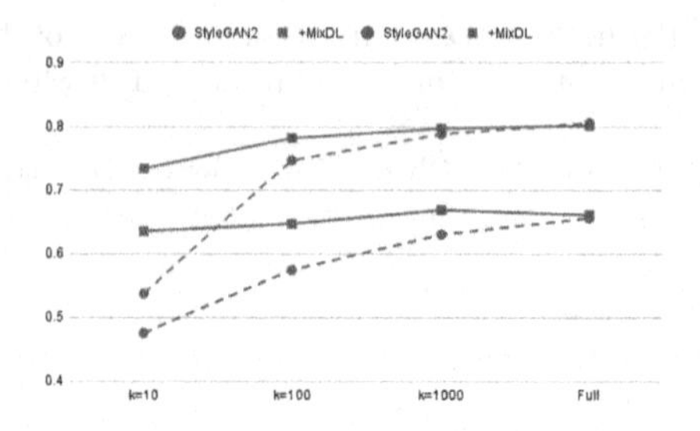

(b) LPIPS for different dataset sizes.

Fig. 6. Shot ablation results. Red indicates FFHQ-babies and blue represents flowers. Our method consistently improves both metrics with limited data. (Color figure online)

4.4 Latent Space Smoothness

Smooth latent space interpolation is an important property of generative models that disproves overfitting and allows synthesis of novel data samples. As our proposed method focuses on diversity through latent smoothing, we quantitatively evaluate this using a variant of Perceptual Path Length (PPL) proposed by [23].

PPL was originally introduced as a measure of latent space disentanglement under the assumption that a more disentangled latent space would show smoother interpolation behavior [23]. As we wish to directly quantify latent space smoothness, we slightly modify the metric by taking 10 subintervals between any two latent vectors and measure their perceptual distances. Table 7 reports the subinterval mean, standard deviation, and the mean for the full path (*End*). Note that as PPL is a quadratic measure, the sum of subinterval means can be smaller than the endpoint mean. All four models show similar endpoint mean, suggesting that the overall total perceptual distance is consistent, while ours displays the lowest PPL standard deviation. As low PPL variance across subintervals is a direct sign of perceptually uniform latent transitions, we can verify the effectiveness of our method in smoothing the latent space. Similar insight can be found from Fig. 7 where the baselines display *stairlike* latent transition while ours shows smooth semantic interpolation. More details on PPL computation can be found in the supplementary materials.

4.5 Preserving Diversity

As opposed to [34] that preserves diversity in the source domain, our method can be interpreted as preserving the diversity inherently present in the early stages throughout the course of training, by constantly exploring the latent space and enforcing relative similarity between samples. To validate our hypothesis, we keep track of pairwise LPIPS of generated samples and the number of *modes* in the early iterations. Figure 8 shows the result, where the number of *modes* is represented by the number of unique training samples (real images) that are the nearest neighbor to any of the generated images. In Fig. 8a, we can see that

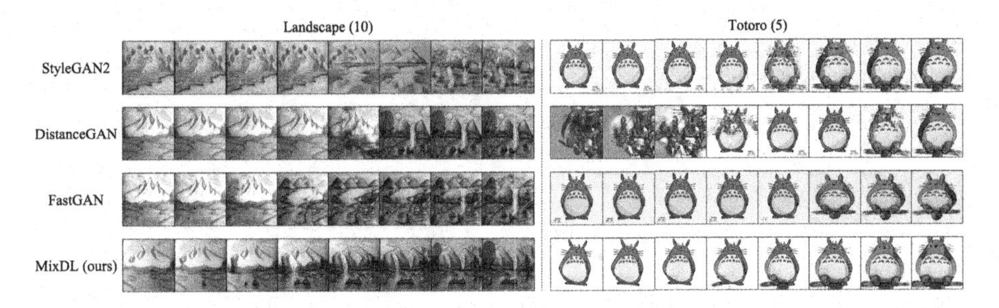

Fig. 7. Latent space interpolation result. Ours shows smooth transitions with high quality while others show defective or abrupt transitions.

Table 7. Perceptual Path Length uniformity. We generate 5000 latent interpolation paths and subdivide each into 10 subintervals to compute perceptual distances. Standard deviation (std) is computed across the subintervals, indicating perceptual uniformity of latent transition.

Dataset	Landscape			Totoro		
Metric	Mean	Std.	End	Mean	Std.	End
StyleGAN2	21.91	12.66	60.90	16.43	15.39	56.53
DistanceGAN	23.07	21.53	70.71	16.76	14.82	61.50
FastGAN	15.49	15.00	67.75	10.03	12.14	54.16
MixDL	12.82	**4.19**	64.28	11.75	**6.44**	56.83

vanilla StyleGAN2 and our method show similar LPIPS in the beginning, but the baseline quickly loses diversity while ours maintains relatively high level of diversity throughout the training. Figure 8b delivers similar implication that FastGAN trained with MixDL better preserves modes compared to the baseline.

Combined with latent space smoothness explained in Sect. 4.4, generators equipped with MixDL learn rich mode-preserving latent space with smooth interpolable landscape. This naturally allows generative diversity particularly appreciated under the constraint of extremely limited data.

5 Discussion

The trade-off between fidelity and diversity in GANs has been noted by many [8, 23]. Truncation trick, a technique widely used in generative models, essentially denotes that diversity can be traded for fidelity. In few-shot generation task, it is very straightforward to obtain near-perfect fidelity at the expense of diversity as one can simply overfit the model, while generating diverse *unseen* data points is very challenging. This implies that with only a handful of data, the diversity should be credited no less than the fidelity.

However, we believe that the widely used low-shot benchmarks, e.g., 100-shot Obama and Grumpy Cat, inherently favor faithful reconstruction over audacious

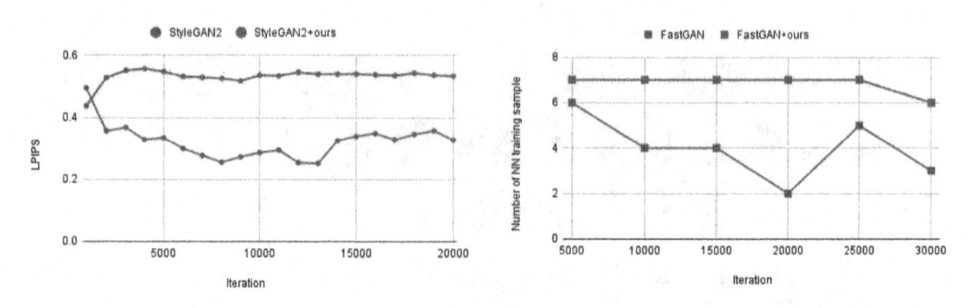

(a) LPIPS in early iterations. (b) Number of unique NN training samples.

Fig. 8. (a) shows that our method produces samples with greater diversity. (b) indicates the number of unique training samples that are nearest neighbor to the generated samples. We generate 500 samples for the analysis. Note that 10 is the upper bound for 10-shot setting. Snapshots are available in the supplementary.

exploration. The main limitations we find in these datasets are twofold: (i) the intra-diversity is too limited as they contain photos of a single person or object, evidenced by low LPIPS in Table 3 and (ii) FID is computed based on the 100 samples that were used for training. We acknowledge that (ii) is a common practice in generative models, but the problem with these benchmarks is that the number of samples is too limited, making it possible for some models to simply *memorize* a large portion of them. These two combined results in benchmarks that allow relatively easy replication and reward it generously at the same time. In other words, we believe that a model's capacity to explore continuous image manifold and *be creative* can potentially backfire in these benchmarks.

To address these limitations, in Table 3 we extend the benchmark with three additional datasets: 100-shot flowers, 10-shot Obama and Grumpy Cat. The first one challenges the model with greater diversity while the last two evaluate its capacity to learn distribution in a generalizable manner, as the FID is still computed against the full 100 images. As our method mainly aims for modeling diversity, we observe marginal performance gains in the traditional benchmarks. However on the extended benchmarks, it shows significant contributions, confirming its excellence at learning diversity even under challenging situations.

6 Conclusion

We propose MixDL, a set of distance regularizations that can be directly added to existing models for few-shot image generation. Unlike previous works, MixDL enables high-quality synthesis of novel images with as few as 5 to 10 training samples, even without any source domain pretraining. Thorough evaluations on diverse benchmarks demonstrate the effectiveness of our framework. We hope our work facilitates future research on data efficient generative modeling.

Acknowledgments. This work was supported by the National Research Foundation of Korea (NRF) grant (2021R1A2C3006659) and IITP grants (NO. 2021-0-01343, No. 2021-0-00537), both funded by the Korea government (MSIT). It was also supported by Samsung Electronics Co., Ltd (IO201223-08260-01).

References

1. Alaluf, Y., Patashnik, O., Cohen-Or, D.: ReStyle: a residual-based StyleGAN encoder via iterative refinement. In: Proceedings of the IEEE/CVF International Conference on Computer Vision, pp. 6711–6720 (2021)
2. Arjovsky, M., Chintala, S., Bottou, L.: Wasserstein generative adversarial networks. In: International Conference on Machine Learning, pp. 214–223. PMLR (2017)
3. Bartunov, S., Vetrov, D.: Few-shot generative modelling with generative matching networks. In: International Conference on Artificial Intelligence and Statistics, pp. 670–678. PMLR (2018)
4. Benaim, S., Wolf, L.: One-sided unsupervised domain mapping. In: Advances in Neural Information Processing Systems, vol. 30 (2017)
5. Berthelot, D., et al.: ReMixMatch: semi-supervised learning with distribution alignment and augmentation anchoring. arXiv preprint arXiv:1911.09785 (2019)
6. Berthelot, D., Carlini, N., Goodfellow, I., Papernot, N., Oliver, A., Raffel, C.: MixMatch: a holistic approach to semi-supervised learning. arXiv preprint arXiv:1905.02249 (2019)
7. Berthelot, D., Raffel, C., Roy, A., Goodfellow, I.: Understanding and improving interpolation in autoencoders via an adversarial regularizer. arXiv preprint arXiv:1807.07543 (2018)
8. Brock, A., Donahue, J., Simonyan, K.: Large scale GAN training for high fidelity natural image synthesis. arXiv preprint arXiv:1809.11096 (2018)
9. Chen, T., Kornblith, S., Norouzi, M., Hinton, G.: A simple framework for contrastive learning of visual representations. In: International Conference on Machine Learning, pp. 1597–1607. PMLR (2020)
10. Chen, X., He, K.: Exploring simple Siamese representation learning. In: Proceedings of the IEEE/CVF Conference on Computer Vision and Pattern Recognition, pp. 15750–15758 (2021)
11. Cui, K., Huang, J., Luo, Z., Zhang, G., Zhan, F., Lu, S.: Genco: generative co-training on data-limited image generation. arXiv preprint arXiv:2110.01254 (2021)
12. Durugkar, I., Gemp, I., Mahadevan, S.: Generative multi-adversarial networks. arXiv preprint arXiv:1611.01673 (2016)
13. Feng, Q., Guo, C., Benitez-Quiroz, F., Martinez, A.M.: When do GANs replicate? On the choice of dataset size. In: Proceedings of the IEEE/CVF International Conference on Computer Vision, pp. 6701–6710 (2021)
14. Ghosh, A., Kulharia, V., Namboodiri, V.P., Torr, P.H., Dokania, P.K.: Multi-agent diverse generative adversarial networks. In: Proceedings of the IEEE Conference on Computer Vision and Pattern Recognition, pp. 8513–8521 (2018)
15. Grill, J.B., et al.: Bootstrap your own latent: a new approach to self-supervised learning. arXiv preprint arXiv:2006.07733 (2020)
16. Gu, Z., Li, W., Huo, J., Wang, L., Gao, Y.: LoFGAN: fusing local representations for few-shot image generation. In: Proceedings of the IEEE/CVF International Conference on Computer Vision, pp. 8463–8471 (2021)

17. Heusel, M., Ramsauer, H., Unterthiner, T., Nessler, B., Hochreiter, S.: GANs trained by a two time-scale update rule converge to a local Nash equilibrium. In: Advances in Neural Information Processing Systems, vol. 30 (2017)
18. Hinz, T., Fisher, M., Wang, O., Wermter, S.: Improved techniques for training single-image GANs. In: Proceedings of the IEEE/CVF Winter Conference on Applications of Computer Vision, pp. 1300–1309 (2021)
19. Hong, Y., Niu, L., Zhang, J., Zhang, L.: MatchingGAN: matching-based few-shot image generation. In: 2020 IEEE International Conference on Multimedia and Expo (ICME), pp. 1–6. IEEE (2020)
20. Hong, Y., Niu, L., Zhang, J., Zhao, W., Fu, C., Zhang, L.: F2GAN: fusing-and-filling GAN for few-shot image generation. In: Proceedings of the 28th ACM International Conference on Multimedia, pp. 2535–2543 (2020)
21. Isola, P., Zhu, J.Y., Zhou, T., Efros, A.A.: Image-to-image translation with conditional adversarial networks. In: Proceedings of the IEEE Conference on Computer Vision and Pattern Recognition, pp. 1125–1134 (2017)
22. Karras, T., Aittala, M., Hellsten, J., Laine, S., Lehtinen, J., Aila, T.: Training generative adversarial networks with limited data. arXiv preprint arXiv:2006.06676 (2020)
23. Karras, T., Laine, S., Aila, T.: A style-based generator architecture for generative adversarial networks. In: Proceedings of the IEEE/CVF Conference on Computer Vision and Pattern Recognition, pp. 4401–4410 (2019)
24. Karras, T., Laine, S., Aittala, M., Hellsten, J., Lehtinen, J., Aila, T.: Analyzing and improving the image quality of StyleGAN. In: Proceedings of the IEEE/CVF Conference on Computer Vision and Pattern Recognition, pp. 8110–8119 (2020)
25. Kynkäänniemi, T., Karras, T., Laine, S., Lehtinen, J., Aila, T.: Improved precision and recall metric for assessing generative models. In: Advances in Neural Information Processing Systems, vol. 32 (2019)
26. Li, Y., Zhang, R., Lu, J., Shechtman, E.: Few-shot image generation with elastic weight consolidation. arXiv preprint arXiv:2012.02780 (2020)
27. Liu, B., Zhu, Y., Song, K., Elgammal, A.: Towards faster and stabilized GAN training for high-fidelity few-shot image synthesis. In: International Conference on Learning Representations (2020)
28. Liu, S., Zhang, X., Wangni, J., Shi, J.: Normalized diversification. In: Proceedings of the IEEE/CVF Conference on Computer Vision and Pattern Recognition, pp. 10306–10315 (2019)
29. Mao, Q., Lee, H.Y., Tseng, H.Y., Ma, S., Yang, M.H.: Mode seeking generative adversarial networks for diverse image synthesis. In: Proceedings of the IEEE/CVF Conference on Computer Vision and Pattern Recognition, pp. 1429–1437 (2019)
30. Mao, X., Li, Q., Xie, H., Lau, R.Y., Wang, Z., Paul Smolley, S.: Least squares generative adversarial networks. In: Proceedings of the IEEE International Conference on Computer Vision, pp. 2794–2802 (2017)
31. Mo, S., Cho, M., Shin, J.: Freeze the discriminator: a simple baseline for fine-tuning GANs. arXiv preprint arXiv:2002.10964 (2020)
32. Nash, C., Menick, J., Dieleman, S., Battaglia, P.W.: Generating images with sparse representations. arXiv preprint arXiv:2103.03841 (2021)
33. Nilsback, M.E., Zisserman, A.: A visual vocabulary for flower classification. In: 2006 IEEE Computer Society Conference on Computer Vision and Pattern Recognition (CVPR 2006), vol. 2, pp. 1447–1454. IEEE (2006)
34. Ojha, U., et al.: Few-shot image generation via cross-domain correspondence. In: Proceedings of the IEEE/CVF Conference on Computer Vision and Pattern Recognition, pp. 10743–10752 (2021)

35. Oring, A., Yakhini, Z., Hel-Or, Y.: Autoencoder image interpolation by shaping the latent space. arXiv preprint arXiv:2008.01487 (2020)
36. Radford, A., Metz, L., Chintala, S.: Unsupervised representation learning with deep convolutional generative adversarial networks. arXiv preprint arXiv:1511.06434 (2015)
37. Robb, E., Chu, W.S., Kumar, A., Huang, J.B.: Few-shot adaptation of generative adversarial networks. arXiv preprint arXiv:2010.11943 (2020)
38. Sainburg, T., Thielk, M., Theilman, B., Migliori, B., Gentner, T.: Generative adversarial interpolative autoencoding: adversarial training on latent space interpolations encourage convex latent distributions. arXiv preprint arXiv:1807.06650 (2018)
39. Shaham, T.R., Dekel, T., Michaeli, T.: SinGAN: learning a generative model from a single natural image. In: Proceedings of the IEEE/CVF International Conference on Computer Vision, pp. 4570–4580 (2019)
40. Si, Z., Zhu, S.C.: Learning hybrid image templates (HIT) by information projection. IEEE Trans. Pattern Anal. Mach. Intell. **34**(7), 1354–1367 (2011)
41. Sushko, V., Gall, J., Khoreva, A.: One-shot GAN: learning to generate samples from single images and videos. In: Proceedings of the IEEE/CVF Conference on Computer Vision and Pattern Recognition, pp. 2596–2600 (2021)
42. Tran, N.T., Bui, T.A., Cheung, N.M.: DIST-GAN: an improved GAN using distance constraints. In: Proceedings of the European Conference on Computer Vision (ECCV), pp. 370–385 (2018)
43. Tseng, H.Y., Jiang, L., Liu, C., Yang, M.H., Yang, W.: Regularizing generative adversarial networks under limited data. In: Proceedings of the IEEE/CVF Conference on Computer Vision and Pattern Recognition, pp. 7921–7931 (2021)
44. Verma, V., et al.: Interpolation consistency training for semi-supervised learning. Neural Networks (2021)
45. Wang, T., Zhang, Y., Fan, Y., Wang, J., Chen, Q.: High-fidelity GAN inversion for image attribute editing. arXiv preprint arXiv:2109.06590 (2021)
46. Wang, X., Tang, X.: Face photo-sketch synthesis and recognition. IEEE Trans. Pattern Anal. Mach. Intell. **31**(11), 1955–1967 (2008)
47. Wang, Y., Gonzalez-Garcia, A., Berga, D., Herranz, L., Khan, F.S., van de Weijer, J.: MineGAN: effective knowledge transfer from GANs to target domains with few images. In: Proceedings of the IEEE/CVF Conference on Computer Vision and Pattern Recognition, pp. 9332–9341 (2020)
48. Wang, Y., Wu, C., Herranz, L., van de Weijer, J., Gonzalez-Garcia, A., Raducanu, B.: Transferring GANs: generating images from limited data. In: Proceedings of the European Conference on Computer Vision (ECCV), pp. 218–234 (2018)
49. Wertheimer, D., Poursaeed, O., Hariharan, B.: Augmentation-interpolative autoencoders for unsupervised few-shot image generation. arXiv preprint arXiv:2011.13026 (2020)
50. Yang, D., Hong, S., Jang, Y., Zhao, T., Lee, H.: Diversity-sensitive conditional generative adversarial networks. arXiv preprint arXiv:1901.09024 (2019)
51. Yaniv, J., Newman, Y., Shamir, A.: The face of art: landmark detection and geometric style in portraits. ACM Trans. Graph. (TOG) **38**(4), 1–15 (2019)
52. Zhang, H., Cisse, M., Dauphin, Y.N., Lopez-Paz, D.: Mixup: beyond empirical risk minimization. arXiv preprint arXiv:1710.09412 (2017)
53. Zhang, R., Isola, P., Efros, A.A., Shechtman, E., Wang, O.: The unreasonable effectiveness of deep features as a perceptual metric. In: Proceedings of the IEEE Conference on Computer Vision and Pattern Recognition, pp. 586–595 (2018)

54. Zhao, M., Cong, Y., Carin, L.: On leveraging pretrained GANs for generation with limited data. In: International Conference on Machine Learning, pp. 11340–11351. PMLR (2020)
55. Zhao, S., Liu, Z., Lin, J., Zhu, J.Y., Han, S.: Differentiable augmentation for data-efficient GAN training. arXiv preprint arXiv:2006.10738 (2020)
56. Zhu, P., Abdal, R., Qin, Y., Femiani, J., Wonka, P.: Improved StyleGAN embedding: where are the good latents? arXiv preprint arXiv:2012.09036 (2020)

A Style-Based GAN Encoder for High Fidelity Reconstruction of Images and Videos

Xu Yao[1(✉)], Alasdair Newson[1], Yann Gousseau[1], and Pierre Hellier[2]

[1] LTCI, Télécom Paris, Institut Polytechnique de Paris, Paris, France
[2] InterDigital R&I, Rennes, France
yaoxu.fdu@gmail.com

Abstract. We present a new encoder architecture for the inversion of Generative Adversarial Networks (GAN). The task is to reconstruct a real image from the latent space of a pre-trained GAN. Unlike previous encoder-based methods which predict only a latent code from a real image, the proposed encoder maps the given image to both a latent code and a feature tensor, simultaneously. The feature tensor is key for accurate inversion, which helps to obtain better perceptual quality and lower reconstruction error. We conduct extensive experiments for several style-based generators pre-trained on different data domains. Our method is the first feed-forward encoder to include the feature tensor in the inversion, outperforming the state-of-the-art encoder-based methods for GAN inversion. Additionally, experiments on video inversion show that our method yields a more accurate and stable inversion for videos. This offers the possibility to perform real-time editing in videos. *Code is available at* https://github.com/InterDigitalInc/FeatureStyleEncoder.

Keywords: GAN inversion · StyleGAN encoder · Latent space

1 Introduction

The image synthesis power of Generative Adversarial Networks [12] (GAN) has been amply demonstrated by a great quantity of work on such architectures. However, since a GAN only decodes an image from a probabilistic latent space, a significant research problem is how to *encode* images into the latent space of a pretrained GAN, especially in the case of real (photographic) images, as opposed to *synthetic* images, which are generated by sampling in the latent space. Recent studies [17,34,35,44] have shown that it is possible to control semantic attributes of synthetic images by manipulating the latent space of a pre-trained GAN. However, an efficient encoding method, necessary for real images, still remains an open problem especially in the case of these editing tasks.

Among the many studies on GAN inversion, recent works have been primarily focused on style-based generators [21–23], because of their excellent performance

Supplementary Information The online version contains supplementary material available at https://doi.org/10.1007/978-3-031-19784-0_34.

S. Avidan et al. (Eds.): ECCV 2022, LNCS 13675, pp. 581–597, 2022.
https://doi.org/10.1007/978-3-031-19784-0_34

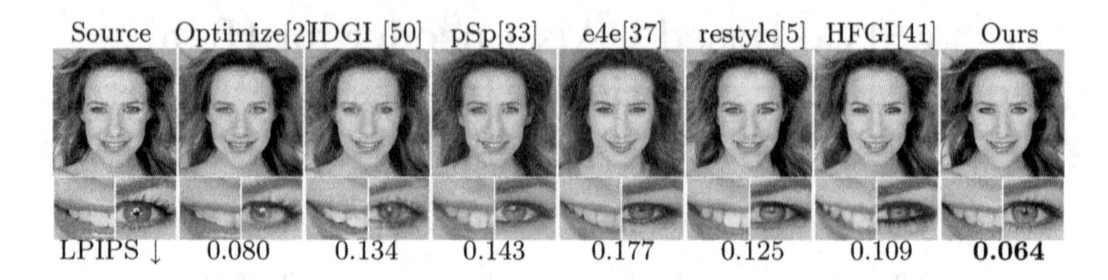

Fig. 1. Inversion of a real image in the latent space of StyleGAN2. We compare our model against state-of-the-art for the inversion of StyleGAN2 [23] pre-trained on face domain. Our method outperforms the state-of-the-art by up to 20%–50% in LPIPS distance [49].

in high quality image synthesis. Unlike traditional generative models which feed the latent code though the input layer only, style-based generators feed latent code to each scale of the generator to control the style of the generated image. This multi-scale generation is one of the main strengths of style-based generators. It is clear that the architecture's success is based on the separation of the latent code, which is a vector that acts *globally* at each scale, from the feature tensor, which has a *spatial* organisation. Exploiting this separation is a key component of the encoder we propose here (Fig. 1).

To invert a pretrained GAN, the current solutions can be divided into two types: optimization-based methods and encoder-based methods. Optimization-based methods use some form of gradient descent in the latent space to find the code that minimizes a reconstruction error. Encoder-based methods, on the other hand, train a neural network that projects from the image space to the latent space. While the optimization is straight-forward to perform, it has significant shortcomings. Firstly, the inverted latent codes do not necessarily lie on the original latent space, since the optimization is carried out locally with respect to one image, making them difficult to use for editing tasks, as shown by [5,37]. Secondly, the approach is computationally expensive, requiring an optimization for each image. The encoder-based approach is much faster and the inverted latent codes are more regularized and *better suited for editing.*

Currently, all the encoder-based GAN inversion methods use the latent code only, ignoring the above-mentioned feature tensors. Unfortunately, while the inversion results are globally perceptually similar to the input, they lack crisp finer spatial details and appear over-smoothed. This is coherent, since the latent codes act *globally*, thus spatially localized details are difficult to preserve. Several optimization-based approaches have identified this problem, and show that considering the feature tensors leads to results of better perceptual quality. However, they present all the drawbacks of optimization-based approaches mentioned above (slow, unconstrained latent code).

To have the best of both worlds, we consider an encoder-based approach which modifies both the latent code and the feature tensors simultaneously, an approach that currently does not exist in the literature. We learn an encoder which maps an image to a feature tensor and a latent code, simultaneously. This design significantly improves the perceptual quality of the inversion and achieves a balanced trade-off between reconstruction quality and editing capacity. The main contributions of our paper can be summarized as follows:

- We propose a new GAN encoder architecture, which is the first feed-forward encoder to include the feature tensor in the inversion. To train the encoder, we present a new training process, which learns two inversions simultaneously, on both real and synthetic images, which significantly improves the perceptual quality;
- We present a novel latent space editing approach, which allows us to leverage existing editing methods for style-based generators. This way, we achieve editing results that are comparable to state-of-the-art methods while preserving the high fidelity inversion;
- We conduct extensive experiments to show that our model greatly outperforms state-of-the-art methods on inversion and editing tasks on images and videos. In particular, we improve the perceptual metrics by a very large margin (50%). In addition, we show that the video inversion results of our method is more consistent and stable, which favors further editing on videos.

2 Related Works

The goal of our work is to learn an encoder for projecting real images to the latent space of a pre-trained GAN. Much of the recent literature on GAN inversion pays particular attention to style-based generators [20–23], as their latent spaces are better disentangled and have improved editing properties.

Style-Based Generator. Karras *et al.* proposed the first style-based generator, named StyleGAN [22]. Unlike traditional generative models which feed the latent code though the input layer only, a style-based generator feeds latent code through adaptive instance normalization at each convolution layer to control the style of the generated image. The perceptual quality and variety of the StyleGAN synthetic images surpassed previous image generative models [7,19]. In StyleGAN2 [23], the image quality was improved further by introducing weight demodulation and path length regularization and redesigning the generator normalization. The StyleGAN2-Ada [20] explored the possibility to train a GAN model with limited data regimes, by using an adaptive discriminator augmentation mechanism that significantly stabilizes training. The third generation, alias-free GAN [21], addressed the aliasing artifacts in the generator, by employing small architectural changes to discard unwanted information and boost the generator to be fully equivariant to translation and rotation.

Latent Space Editing. The motivation of GAN inversion is to achieve real image editing using the latent space of a pretrained GAN model. Various studies show it possible to edit synthetic images by manipulating the corresponding latent code. Local semantic editing can be achieved by optimizing the latent code directly [2,27]. To explore high level semantic information in the latent space, learning based techniques have been proposed. These techniques include unsupervised exploration [39], learning linear SVM models [34], principle component analysis on the latent codes [17], and k-means clustering of the activation features [10]. To achieve better disentangled editing, [3,14,30,36,47] proposed to learn neural networks in the latent space. The recent works [35,40] discovered interpretable transforms by directly decomposing the weights or feature maps of pre-trained GANs. Additionally, [6,26] modify the style-based GAN architecture and retrain it for better disentanglement in image generation. [24,29,31] train jointly an encoder and a style-based decoder architecture for image manipulations.

GAN Inversion. The goal of GAN inversion is to encode a real image to the latent space of a pretrained GAN, so that the image generated from the inverted latent code is the reconstruction of the input image. Among the rich literature on GAN inversion [45], the approaches addressing style-based generators can be classified into two main types: optimization-based methods [1,16,18,46,51] and encoder-based models [5,33,37,41,43]. There are also hybrid methods [8,48,50] which mix the two previous ones. The optimization-based methods produce the inverted latent code by minimizing the reconstruction error on the input image. For StyleGAN inversion, Abdal *et al.* [1] proposed to embed the input image in an extended latent space $\mathcal{W}^+$, which offers greater flexibility and improves the reconstruction quality. Recent works show that including a feature tensor in the optimization helps preserve more spatial details and improves the perceptual quality [18,51]. Despite the satisfying reconstruction quality, optimization-based methods usually present lower editing quality due to the random element of the optimization process. To better regularize/stabilize the inversion, encoder-based methods train an encoder to map real images to the latent space of the pretrained generator. Richardson *et al.* [33] proposed the first baseline to learn an encoder for StyleGAN inversion. To improve editing capacity, Tov *et al.* [37] proposed a regularization term which forces the inverted latent code in $\mathcal{W}^+$ to lie closer to the original latent space. A recent concurrent work of Wang *et al.* [41] formulated the inversion task to a data compression problem and proposed an adaptive distortion alignment module to improve the reconstruction quality. On the other hand, hybrid methods take the inverted latent code from a pretrained encoder as initialization and perform optimization on it. Zhu *et al.* [50] proposed to learn a domain-guided encoder and use it as a regularizer for domain-regularized optimization. However, despite the gain in the reconstruction quality, the optimization step makes hybrid methods less suited for video inversion and editing.

3 Method

3.1 Overview

A style-based generator, such as StyleGAN [21–23], consists of a mapping network and a generator $\mathbf{G}$. The mapping network first maps a random latent code $\mathbf{z} \in \mathcal{Z} \subset \mathbb{R}^{512}$ to an intermediate latent code $\mathbf{w} \in \mathcal{W} \subset \mathbb{R}^{512}$, which is further used to scale and bias the feature tensors. We denote a feature tensor (also called

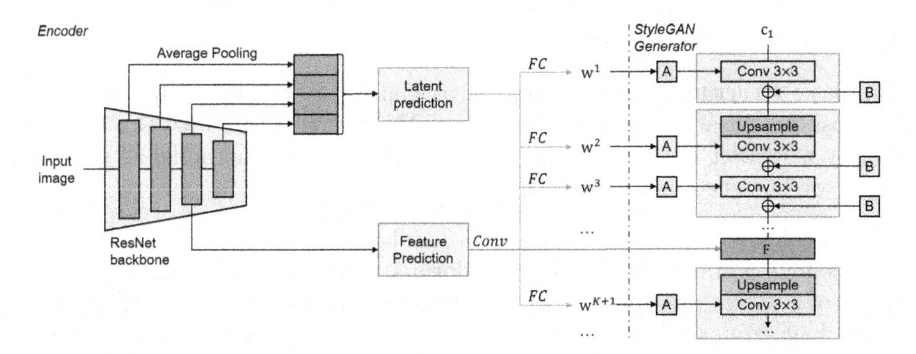

Fig. 2. Encoder Architecture. Our model consists of a ResNet backbone and two output branches: one for latent code prediction, the other for feature tensor prediction. The inverted latent code $\mathbf{w}$ is a concatenation of N latent blocks $\{\mathbf{w}^1, \mathbf{w}^2, ..., \mathbf{w}^N\}$, each controlling a separate convolution layer in the generator. During generation, we replace the feature tensors at the Kth convolution layer of the generator with the encoded feature tensor $\mathbf{F}$, and synthesize the inversion with the latent blocks $\{\mathbf{w}^{K+1}, ..., \mathbf{w}^N\}$. K is a fixed parameter, chosen so that reconstruction is accurate and editing can be performed efficiently

feature map) with $\mathbf{F} \in \mathcal{F} \subset \mathbb{R}^{h \times w \times c}$. The parameters (h, w, c) correspond to the spatial size and the number of channels of the tensor. Thus, contrary to the latent codes, the feature tensors have a 2D spatial organisation. Each feature tensor is the output of an upsampling from a lower resolution, followed by an AdaIn layer controlled by the latent codes, and finally a convolution. At the coarsest layer, the input is a constant feature tensor, which is learned during training. See Fig. 2 for an illustration. In this Figure, we can see the clear separation between latent codes and feature tensors, which is so important to StyleGAN's success.

To project a *synthetic* image $\mathbf{G}(\mathbf{w})$ to the latent space, it is possible to compute the latent code in the original latent space $\mathcal{W}$ and achieve a satisfying inversion. However, it is much more difficult to project a real image to the original latent space [23], due to the gap between the real data distribution and the synthetic one. An alternative is to project real images to an extended latent space $\mathcal{W}^+$ [1], where $\mathbf{w} \in \mathcal{W}^+$ is a concatenation of N latent blocks $\{\mathbf{w}^1, \mathbf{w}^2, ..., \mathbf{w}^N\}$, each controlling a feature tensor in the generator.

Current encoder based methods for GAN inversion learn only the latent codes. Their inversion results are globally perceptually similar to the input but fail to capture finer details. Therefore, it is preferable to include a learned feature tensor to preserve these spatial details. The optimization-based methods [18,51] show that including these feature tensors in the optimization process help to preserve spatial details. Performing optimization on both the latent code and feature tensors yields near perfect reconstruction on real images.

In our work, we aim to have the best of both worlds: we wish to achieve this high reconstruction fidelity, while maintaining the speed and editing capacity of an encoder. Thus, we propose an encoder architecture which projects an image to a latent code $\mathbf{w} \in \mathcal{W}^+$, and a feature tensor $\mathbf{F} \in \mathcal{F}$. This feature tensor is chosen at a fixed layer K of the generator.

3.2 Encoder Architecture

The basic structure of our encoder is modelled on the classic approach used by most previous works on style-based GAN inversion [5,33,37,41,43], which employ a ResNet backbone. Different from the existing methods, as shown in Fig. 2, we have two output branches:

- A latent prediction branch to encode the latent code $\mathbf{w} \in \mathcal{W}^+$. The ResNet backbone contains four blocks, each down-sampling the input tensors by a factor of 2. Given an input image, we extract the tensors after each block. Then the four groups of tensors are passed through an average pooling layer, concatenated and flattened to produce the latent prediction. This is then mapped to the latent code $\mathbf{w} = \{\mathbf{w}^1, \mathbf{w}^2, ..., \mathbf{w}^N\}$. Each latent block $\mathbf{w}^i$ is generated from a different mapping network, expressed by a single fully connected layer.
- A feature prediction branch, where the ResNet tensors, extracted after the penultimate block of the ResNet backbone, are passed through a convolutional network to encode the feature tensor $\mathbf{F} \in \mathcal{F}$ (see Fig. 2). This network is implemented with two convolutional layers, with batch normalization in between. Please note that the ResNet tensors are not the same as those of StyleGAN. We refer to the StyleGAN tensors (which control the generation) as *feature tensors*. Let $\mathbf{G}(\mathbf{w}^{1:K})$ denote the feature tensors at the Kth convolution layer of the generator. We replace $\mathbf{G}(\mathbf{w}^{1:K})$ with the encoded feature tensor $\mathbf{F}$, and use the rest of the latent codes $\{\mathbf{w}^{K+1}, ..., \mathbf{w}^N\}$ to generate the inversion $\mathbf{G}(\mathbf{F}, \mathbf{w}^{K+1:N})$. We choose $K = 5$ for a balance between the inversion quality and editing capacity, leading to $\mathcal{F} \subset \mathbb{R}^{16\times16\times512}$.

To summarize the entire process, our encoder produces the Kth feature tensor of the StyleGAN generator, simultaneously with all the latent codes. Due to the sequential nature of StyleGAN, $\mathbf{G}(\mathbf{w}^{1:K})$ and $\{\mathbf{w}^{K+1}, ..., \mathbf{w}^N\}$ completely determine the output image. Another way of seeing this is that we have "started" the generation from layer K, ignoring the previous layers. Note that even if the latent codes of previous layers $\{\mathbf{w}^1, ..., \mathbf{w}^K\}$ are not used for the inversion, they will be used later for *editing*. The choice of K is crucial to achieve a balance between good reconstruction and style editing. We have studied this choice carefully, and show results for different values in the Supplementary Material.

3.3 Editing

In a style-based generator, the styles corresponding to coarse layers control high-level semantic attributes, the styles of the middle layers control smaller scale features, and the last layers control micro structures. Given a latent code $\mathbf{w}$, let us consider that we have a modified latent code $\tilde{\mathbf{w}} = \mathbf{w} + \Delta\mathbf{w}$ corresponding to a desired editing, obtained from a latent space editing method [17,34,35].

Contrary to the case of inversion, we now wish to modify the latent codes of *all* layers, to achieve editing. For this, we start by obtaining the initial inversion of the input image, which gives us the latent code $\mathbf{w}$ and the feature tensor $\mathbf{F}$. Recall that the inversion is determined by the feature tensor $\mathbf{F}$ and the latent codes

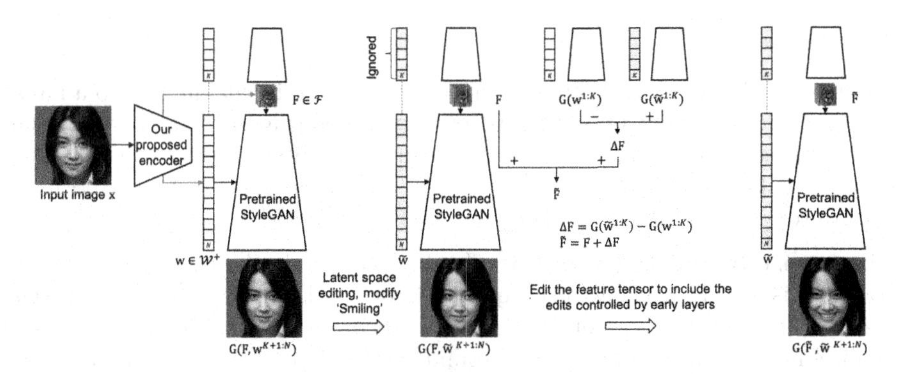

Fig. 3. Editing process. To edit an input image $\mathbf{x}$, we first encode it to the latent code $\mathbf{w}$ and the feature tensor $\mathbf{F}$. We then use a latent space editing method to transform $\mathbf{w}$ into $\tilde{\mathbf{w}}$, which corresponding to a desired manipulation. If we use $\tilde{\mathbf{w}}$ and $\mathbf{F}$ to generate the output image $\mathbf{G}(\mathbf{F}, \tilde{\mathbf{w}}^{K+1:N})$, the changes corresponding to early layers will be ignored. Therefore, we need to modify $\mathbf{F}$ to include these edits. We do this by extracting the feature tensors at the Kth layer $\mathbf{G}(\mathbf{w}^{1:K})$ and $\mathbf{G}(\tilde{\mathbf{w}}^{1:K})$, computing the difference and adding it to $\mathbf{F}$ to obtain the modified feature tensor $\tilde{\mathbf{F}}$. Finally, we use this new feature tensor $\tilde{\mathbf{F}}$ and $\tilde{\mathbf{w}}$ to generate the edited image $\mathbf{G}(\tilde{\mathbf{F}}, \tilde{\mathbf{w}}^{K+1:N})$

$\{\mathbf{w}^{K+1}, ..., \mathbf{w}^{N}\}$. Thus, the changes corresponding to early layers $\{\mathbf{w}^{1}, ..., \mathbf{w}^{K}\}$ will be ignored. To preserve these editing effects, we need to modify $\mathbf{F}$. As shown in Fig. 3, we do this by extracting the feature tensors at the Kth layer $\mathbf{G}(\mathbf{w}^{1:K})$ and $\mathbf{G}(\tilde{\mathbf{w}}^{1:K})$. We suppose that the difference between them should be close to that between $\mathbf{F}$ and the modified feature tensor $\tilde{\mathbf{F}}$. Therefore, we can find $\tilde{\mathbf{F}}$ as follows:

$$\tilde{\mathbf{F}} = \mathbf{F} + \mathbf{G}(\tilde{\mathbf{w}}^{1:K}) - \mathbf{G}(\mathbf{w}^{1:K}). \tag{1}$$

Finally, we use this new feature tensor $\tilde{\mathbf{F}}$ and the rest of the modified latent codes $\{\tilde{\mathbf{w}}^{K+1}, ..., \tilde{\mathbf{w}}^{N}\}$ to generate the edited image $\mathbf{G}(\tilde{\mathbf{F}}, \tilde{\mathbf{w}}^{K+1:N})$.

4 Training

Previous methods on GAN inversion [5,33,37,41,43] use only real images as training data. However, the perceptual quality of their inversion results is not as good as the synthetic images generated by StyleGAN. An intuitive explanation is that there is a difference between the data distributions of the real and synthetic images. Recall that the encoder project a given image to the extended latent space $\mathcal{W}^+$ while synthetic images can be reconstructed from the original latent space $\mathcal{W}$. If synthetic images are not viewed by the encoder, the resulting latent codes may not perform in the same way as those of the original latent space. Therefore, we use both synthetic and real images as training data.

4.1 Losses

As mentioned above, the proposed encoder inverts an input image $\mathbf{x}$ to a latent code $\mathbf{w}$, and a feature tensor $\mathbf{F}$. To ensure the editing capacity of the latent code, the encoder is trained on two inversions simultaneously - one generated with only the latent code $\tilde{\mathbf{x}}_1 = \mathbf{G}(\mathbf{w})$ and the other generated with both the feature tensor and the latent code $\tilde{\mathbf{x}}_2 = \mathbf{G}(\mathbf{F}, \mathbf{w}^{K+1:N})$.

Pixel-Wise Reconstruction Loss. In the case of a synthetic image, the reconstruction is measured using mean squared error (MSE) on $\tilde{\mathbf{x}}_1$ only. In this special case, the ground-truth latent code exists so theoretically a perfect inversion can be obtained. For a real image, the ground-truth latent code may not exist, so a per-pixel constraint may be too restrictive. The loss is expressed as:

$$\mathcal{L}_{mse} = ||\mathbf{G}(\mathbf{w}) - \mathbf{x}||_2. \tag{2}$$

Multi-scale Perceptual Loss. A common problem of the previous methods is the lack of sharpness of the inversion results, despite using the per-pixel MSE. To tackle this, we propose a multi-scale loss design which enables the reconstruction of finer details. Given an input image $\mathbf{x}$ and its inversion $\tilde{\mathbf{x}}$, the multi-scale perceptual loss is defined as:

$$\mathcal{L}_{m\_lpips}(\tilde{\mathbf{x}}) = \sum_{i=0}^{2} ||\mathbf{V}(\lfloor \tilde{\mathbf{x}} \rfloor_i) - \mathbf{V}(\lfloor \mathbf{x} \rfloor_i)||_2, \tag{3}$$

where $\lfloor . \rfloor_i$ refers to downsampling by a scale factor 2^i and $\mathbf{V}$ denotes the feature extractor. This design allows the encoder to capture the perceptual similarities at different scales, which favors the perceptual quality of the inversion. This loss is applied on both inversions.

Feature Reconstruction. To ensure the possibility of using Eq. (1) to edit the encoded feature tensor $\mathbf{F}$, $\mathbf{F}$ should be similar to the feature tensors at the Kth convolution layer of the generator, denoted by $\mathbf{G}(\mathbf{w}^{1:K})$. Therefore, we propose a feature reconstruction loss, which favors the encoded feature tensor to stay close to the original feature space. This term is defined as:

$$\mathcal{L}_{f\_recon} = ||\mathbf{F} - \mathbf{G}(\mathbf{w}^{1:K})||_2. \tag{4}$$

The total loss is defined as:

$$\mathcal{L}_{total} = \mathcal{L}_{mse} + \lambda_1 \mathcal{L}_{m\_lpips} + \lambda_2 \mathcal{L}_{f\_recon}, \tag{5}$$

where $\lambda_1 = 0.2$ and $\lambda_2 = 0.01$ are weights balancing each loss.

Face Inversion. For the inversion of a styleGAN model pre-trained on face domain, we adopt the multi-layer identity loss and the face parsing loss introduced by [43]. Given an input image $\mathbf{x}$ and its inversion $\tilde{\mathbf{x}}$, the multi-layer identity loss is defined as:

$$\mathcal{L}_{id}(\tilde{\mathbf{x}}) = \sum_{i=1}^{5}(1 - \langle \mathbf{R}_i(\tilde{\mathbf{x}}), \mathbf{R}_i(\mathbf{x}) \rangle), \tag{6}$$

where $\mathbf{R}$ is the pre-trained ArcFace network [11]. The face parsing loss is defined as:

$$\mathcal{L}_{parse}(\tilde{\mathbf{x}}) = \sum_{i=1}^{5}(1 - \langle \mathbf{P}_i(\tilde{\mathbf{x}}), \mathbf{P}_i(\mathbf{x}) \rangle), \quad (7)$$

where $\mathbf{P}$ is a pre-trained face parsing model [52]. These two above-mentioned losses are applied on both inversions. Hence the total loss for face inversion is:

$$\mathcal{L}_{face} = \mathcal{L}_{total} + \lambda_3 \mathcal{L}_{id} + \lambda_4 \mathcal{L}_{parsing}, \quad (8)$$

where $\lambda_3 = 0.1$ and $\lambda_4 = 0.1$ are weights balancing the identity preserving and face parsing terms.

4.2 Implementation Details

We train the proposed encoder for the inversion of several style-based generators pre-trained on various domains, specifically, for StyleGAN2 [23] on faces and cars, and StyleGAN2-Ada [20] on cats and dogs. In addition, we show preliminary results for the inversion of StyleGAN3 [21] on faces. For each generator pre-trained on a specific domain, a separate encoder is trained. During the training, we use a batch size of 4, each batch containing two real images and two synthetic images. The model is trained for 12 epochs, using $10K$ iterations per epoch. The learning rate is 10^{-4} for the first 10 epochs and is divided by ten for the last 2 epochs. For the face domain, we minimize Eq. (8), using FFHQ [22] for training, and CelebA-HQ [19] for evaluation. For the car domain, we minimize Eq. (5), using Stanford Cars [25] training set for training, and the corresponding test set for evaluation. For the cat/dog domain, we minimize Eq. (5), using AFHQ Cats/Dogs [9] train set for training, and the corresponding test set for evaluation.

5 Experiments

In this section, we compare our method with the current state-of-the-art GAN inversion methods. We conduct the evaluation from two aspects: inversion quality and editing capacity. We also show results on videos as well as ablation studies.

5.1 Inversion

We evaluate our model against the current state-of-the-art encoder-based GAN inversion methods: pSp [33], e4e [37], restyle [5] and HFGI [41]. We first perform comparisons for the inversion of the StyleGAN2 model pre-trained on the FFHQ dataset. For each method we use the official implementation [4,32,38,42] to generate the results.

Qualitative Results. Figure 4 shows the inversion results of the different methods. Overall, visual inspection shows that our method outperforms other models. Firstly, faces are more faithfully reconstructed globally. Secondly, zoom-in

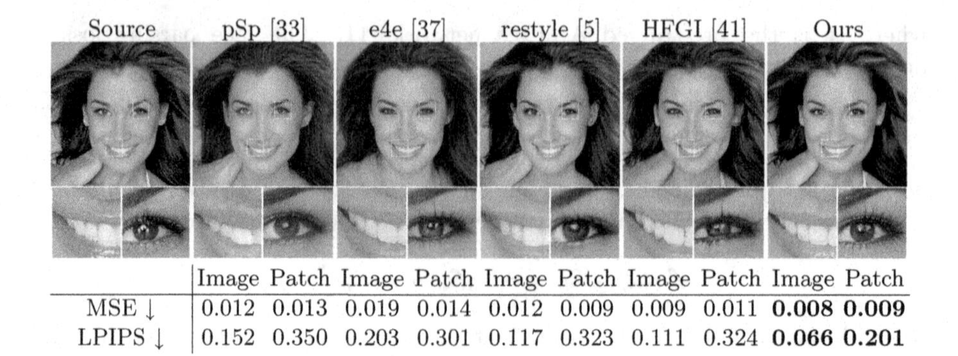

	Image	Patch	Image	Patch	Image	Patch	Image	Patch	Image	Patch
MSE ↓	0.012	0.013	0.019	0.014	0.012	0.009	0.009	0.011	**0.008**	**0.009**
LPIPS ↓	0.152	0.350	0.203	0.301	0.117	0.323	0.111	0.324	**0.066**	**0.201**

Fig. 4. Inversion of StyleGAN2 pretrained on face domain. We compare our model against state-of-the-art encoder-based methods [5,33,37,41] for the inversion of StyleGAN2 pre-trained on face domain. Our inversion results are visually more faithful and zoom-in patches show that they exhibit much more details and sharpness. Pixel-wise reconstruction errors (MSE error, lower is better) and perceptual quality (LPIPS distance, lower is better) confirm this visual conclusion on these examples

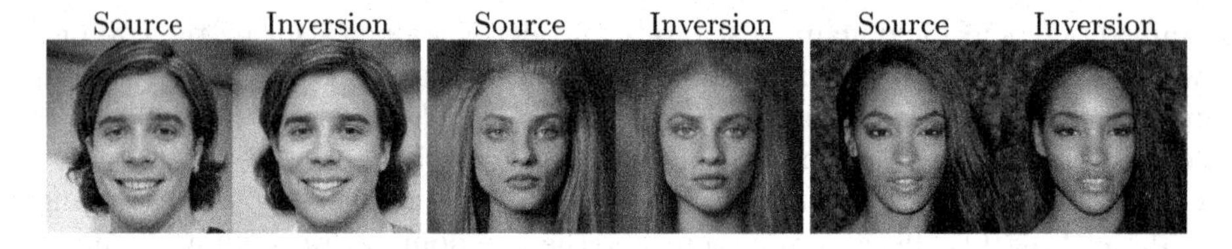

Fig. 5. Inversion of StyleGAN3 pretrained on face domain. We show preliminary inversion results of the 3rd generation of StyleGAN [21] on face domain. Compared with StyleGAN2, the architecture of StyleGAN3 has several important changes. Despite the architectural changes, our method still yields satisfying inversion results

patches show that more details are preserved and that the images produced by our framework are significantly sharper than those of the concurrent methods.

Inversion of StyleGAN3. We show preliminary inversion results of the third generation of StyleGAN [21] pretrained on FFHQ. Compared with StyleGAN2, the architecture of StyleGAN3 has several important changes. The input tensor passed into the generator is no longer constant, but synthesized from the latent code. The spatial size of the input tensor is increased from 4×4 to 36×36. As shown in Fig. 5, despite the architectural changes, our proposed encoder still yields satisfying inversion results.

Quantitative Comparison. We evaluate our approach quantitatively against the aforementioned encoder based methods [5,33,37,41] and a hybrid method (in-domain GAN) [50]. We compare each method on the inversion of StyleGAN2 pretrained on FFHQ, using the first $1K$ images of CelebA-HQ as evaluation data. To measure the reconstruction error, we compute *SSIM*, *PSNR* and *MSE*.

Table 1. Quantitative evaluation. We use *SSIM*, *PSNR* and *MSE* to measure the reconstruction error, and *LPIPS* [49] for the perceptual quality. We also measure the *identity similarity* (ID) between the inversion and the source image, which refers to the cosine similarity between the features in CurricularFace [15] of both images. To measure the discrepancy between the real data distribution and the inversion one, we use *FID* [13]. Overall, our method outperforms the state-of-the-art baselines by up to 10%–20%. In terms of perceptual quality (LPIPS), we improve the result by 50%

Method	SSIM ↑	PSNR ↑	MSE ↓	LPIPS ↓	ID ↑	FID ↓
IDGI [50]	0.554	22.06	0.034	0.136	0.18	36.83
pSp [33]	0.509	20.37	0.040	0.159	0.56	34.68
e4e [37]	0.479	19.17	0.052	0.196	0.51	36.72
restyle [5]	0.537	21.14	0.034	0.130	0.66	32.56
HFGI [41]	0.595	22.07	0.027	0.117	0.68	26.53
Ours	**0.641**	**23.65**	**0.019**	**0.066**	**0.80**	**19.03**

(a) Inversion on car domain. (b) Inversion on cat/dog domain.

Fig. 6. Inversion on other domains. In (a), we show the inversion results of StyleGAN2 pre-trained on car domain. Our method captures better the details than e4e [37] and restyle-e4e [5]. In (b), we show the inversion results of StyleGAN2-Ada pre-trained on the cat and dog domains, respectively

To measure the perceptual quality, we measure the *LPIPS* [49] distance. Additionally, we measure the *identity similarity* (ID) between the inversion and the source image, which refers to the cosine similarity between the features in CurricularFace [15] of the two images. To measure the discrepancy between the real data distribution and the inversion one, we use the Frechet Inception Distance [13] (*FID*). Table 1 presents the quantitative evaluation of all the methods. Our method significantly outperforms the state-of-the-art methods on all the metrics. In terms of perceptual quality (LPIPS), improvement can attain 50%.

Inversion for Other Domains. Figure 6(a) shows the inversion for StyleGAN2 pretrained on the car domain. We train the encoder with Stanford Car dataset [25]. Compared with e4e [37] and restyle-e4e [5], our inversion achieves a better reconstruction quality, preserving better the details. Figure 6(b) shows the inversion for StyleGAN2-Ada pretrained on AFHQ Cat/Dog dataset [9]. Our encoder achieves nearly perfect inversions. Here we did not compare with [5,37], as the official pre-trained model is unavailable.

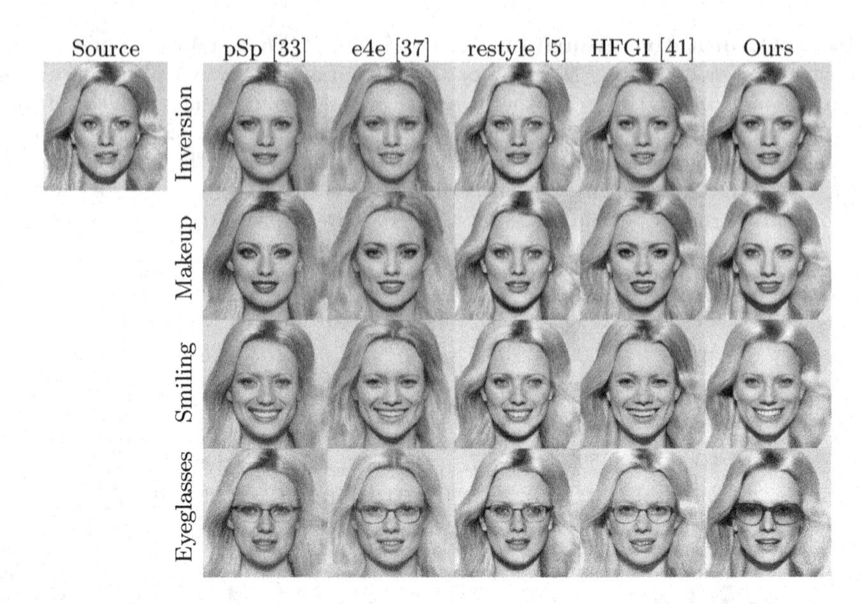

Fig. 7. Latent space editing. For each method, we apply InterFaceGAN [34] to perform latent editing for facial attribute manipulation. Our method yields plausible editing results, while at the same time preserving better the identity and the sharpness. More editing results are presented in the Supplementary Material

5.2 Editing

In this experiment, we consider the task of real image editing via latent space manipulation. We compare our approach with the state-of-the-art encoder-based GAN inversion methods [5,33,37,41] on the facial image editing via the latent space of StyleGAN2 pretrained on the FFHQ dataset. As such, for each inversion model, we generate the inverted latent codes for the first $10K$ images of CelebA-HQ, and use InterFaceGAN [34] to find the editing directions in the latent space. Figure 7 shows facial attribute editing results for all methods. Compared with the state-of-the-art, our method yields visually plausible editing results, while preserving better the identity and sharpness.

Additionally, we show style mixing results in Fig. 8, generated from the latent code of one image with the feature tensor of another image. From this experiment we observe that the geometric structures such as pose and facial shape are encoded by the feature tensor, while the appearance styles like eye color and makeup are encoded by the latent code.

5.3 Video Inversion

In this section, we discuss the possibility of integrating our proposed encoder into a video editing pipeline [47]. We compare the inversion quality and stability of different encoders on videos. Figure 9 shows the inversion results on several images extracted from the same video sequence. The last two frames are extreme

Fig. 8. Style mixing. The 2nd and 5th column show the inversions of two images $\mathbf{x}_A$ and $\mathbf{x}_B$, denoted by $\mathbf{G}(\mathbf{F}_A, \mathbf{w}_A)$ and $\mathbf{G}(\mathbf{F}_B, \mathbf{w}_B)$, respectively. The 3rd column is generated from the feature tensor of $\mathbf{x}_A$ and the latent code of $\mathbf{x}_B$, denoted by $\mathbf{G}(\mathbf{F}_A, \mathbf{w}_B)$, and vice versa for the 4th column, denoted by $\mathbf{G}(\mathbf{F}_B, \mathbf{w}_A)$. This shows that the feature tensor encodes the geometric structures such as pose and facial shape, whereas the latent code controls the appearance styles like eye color and makeup

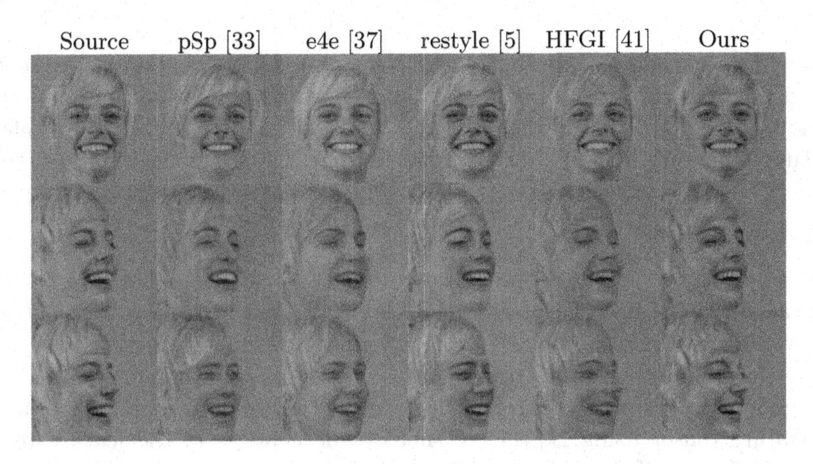

Fig. 9. Video inversion. For each method, we show the inversion results of several frames extracted from a video sequence. Our inversion method preserves better the identity along the video and yields a better reconstruction for the extreme poses

poses. As can be observed, other methods [5,33,37,41] fail to invert non-frontal poses, thus damaging the consistent reconstruction along the sequence. Our approach yields consistent inversion of high fidelity, which favors further editing on videos. Please refer to the supplementary material for the *video editing results.*

We evaluate our encoder quantitatively against the state-of-the-art for video inversion on RAVDESS [28], a dataset of talking face videos. From which we sample randomly 120 videos as evaluation data. For each method, we perform the inversion on each video and compute the quantitative metrics on the inversion results. As shown in Table 2, our approach outperforms the competing approaches on both the reconstruction error and the perceptual quality.

5.4 Ablation Study

We conduct an ablation study on the training setup to analyze how each part of the losses contributes to the inversion quality. We compare the inversion quality in the case of removing the per-pixel loss in Eq. 2, the identity loss in Eq. (6)

Table 2. Quantitative evaluation on video inversion. We sample randomly 120 videos from RAVDESS dataset [28], perform the inversion using each method and compute the quantitative metrics. Our method outperforms the competing approaches on both the reconstruction error and the perceptual quality

Method	SSIM ↑	PSNR ↑	MSE ↓	LPIPS ↓	ID ↑
pSp [33]	0.736	22.30	0.025	0.196	0.687
e4e [37]	0.713	20.57	0.037	0.220	0.620
restyle [5]	0.761	23.17	0.021	0.189	0.781
HFGI [41]	0.783	24.04	0.017	0.182	0.810
Ours	**0.818**	**26.64**	**0.009**	**0.122**	**0.895**

and the face parsing loss in Eq. (7). We also compare the multi-scale perceptual loss in Eq. (3) to a normal LPIPS loss [49]. We further analyze the importance of the feature prediction branch and the choice of training data. Please refer to the supplementary material for the quantitative analysis of the ablation study.

5.5 Limitations

The main limitation of the proposed encoder lies in its global manipulation capacity. In the architecture of StyleGAN, the global attributes are controlled by lower layers, while the smaller local styles are controlled by higher layers. Our method yields better editing results on the attributes controlled by layers $>K$. To handle the attributes controlled by lower layers, we have proposed to modify the feature tensor using Eq. (1) to include the changes in the feature tensor. However, if the difference between the original feature tensor and the inverted one is important, this simple subtraction may generate artifacts. Moreover, the details reconstructed solely by the feature tensor are hard to change. In the future, it could be helpful to study further improvements for the feature tensor editing, *e.g.*., by including masks for the area of interest, or by training another network to generate the corresponding editing for the feature tensor.

6 Conclusion

In this paper, we propose a new encoder architecture for style-based GAN inversion and explore its editing capacity on images and videos. To the best of our knowledge, this is the first *feed-forward encoder* to include the feature tensor in the inversion, which significantly improves the perceptual quality of the inversion results, outperforming competitive state-of-the-art methods. Next, we introduce a *novel editing approach*, which makes the proposed encoder amenable to existing latent space editing methods. Experiments show that the editing capacity of our encoder is comparable to state-of-the-art methods while the editing results are of higher perceptual quality. Moreover, the experiments on video inversion show that our method yields a more accurate and stable inversion for videos. This could significantly facilitate *real-time editing in videos.*

References

1. Abdal, R., Qin, Y., Wonka, P.: Image2StyleGAN: how to embed images into the StyleGAN latent space? In: Proceedings of the IEEE/CVF International Conference on Computer Vision, pp. 4432–4441 (2019)
2. Abdal, R., Qin, Y., Wonka, P.: Image2StyleGAN++: how to edit the embedded images? In: Proceedings of the IEEE/CVF Conference on Computer Vision and Pattern Recognition, pp. 8296–8305 (2020)
3. Abdal, R., Zhu, P., Mitra, N., Wonka, P.: StyleFlow: attribute-conditioned exploration of StyleGAN-generated images using conditional continuous normalizing flows. arXiv e-prints, pp. arXiv-2008 (2020)
4. Alaluf, Y., Patashnik, O., Cohen-Or, D.: Official implementation of ReStyle: a residual-based StyleGAN encoder via iterative refinement (2021). https://github.com/yuval-alaluf/restyle-encoder
5. Alaluf, Y., Patashnik, O., Cohen-Or, D.: ReStyle: a residual-based StyleGAN encoder via iterative refinement. In: Proceedings of the IEEE/CVF International Conference on Computer Vision, pp. 6711–6720 (2021)
6. Alharbi, Y., Wonka, P.: Disentangled image generation through structured noise injection. In: Proceedings of the IEEE/CVF Conference on Computer Vision and Pattern Recognition, pp. 5134–5142 (2020)
7. Brock, A., Donahue, J., Simonyan, K.: Large scale GAN training for high fidelity natural image synthesis. In: International Conference on Learning Representations (2019). https://openreview.net/forum?id=B1xsqj09Fm
8. Chai, L., Zhu, J.Y., Shechtman, E., Isola, P., Zhang, R.: Ensembling with deep generative views. In: Proceedings of the IEEE/CVF Conference on Computer Vision and Pattern Recognition, pp. 14997–15007 (2021)
9. Choi, Y., Uh, Y., Yoo, J., Ha, J.W.: StarGAN v2: diverse image synthesis for multiple domains. In: Proceedings of the IEEE/CVF Conference on Computer Vision and Pattern Recognition, pp. 8188–8197 (2020)
10. Collins, E., Bala, R., Price, B., Susstrunk, S.: Editing in style: uncovering the local semantics of GANs. In: Proceedings of the IEEE/CVF Conference on Computer Vision and Pattern Recognition, pp. 5771–5780 (2020)
11. Deng, J., Guo, J., Xue, N., Zafeiriou, S.: ArcFace: additive angular margin loss for deep face recognition. In: Proceedings of the IEEE/CVF Conference on Computer Vision and Pattern Recognition, pp. 4690–4699 (2019)
12. Goodfellow, I., et al.: Generative adversarial nets. In: Advances in Neural Information Processing Systems, pp. 2672–2680 (2014)
13. Heusel, M., Ramsauer, H., Unterthiner, T., Nessler, B., Hochreiter, S.: GANs trained by a two time-scale update rule converge to a local Nash equilibrium. In: Advances in Neural Information Processing Systems, vol. 30 (2017)
14. Hou, X., Zhang, X., Liang, H., Shen, L., Lai, Z., Wan, J.: GuidedStyle: attribute knowledge guided style manipulation for semantic face editing. Neural Networks (2021)
15. Huang, Y., et al.: CurricularFace: adaptive curriculum learning loss for deep face recognition. In: Proceedings of the IEEE/CVF Conference on Computer Vision and Pattern Recognition, pp. 5901–5910 (2020)
16. Huh, M., Zhang, R., Zhu, J.-Y., Paris, S., Hertzmann, A.: Transforming and projecting images into class-conditional generative networks. In: Vedaldi, A., Bischof, H., Brox, T., Frahm, J.-M. (eds.) ECCV 2020. LNCS, vol. 12347, pp. 17–34. Springer, Cham (2020). https://doi.org/10.1007/978-3-030-58536-5_2

17. Härkönen, E., Hertzmann, A., Lehtinen, J., Paris, S.: GANSpace: discovering interpretable GAN controls. In: Proceedings of the NeurIPS (2020)
18. Kang, K., Kim, S., Cho, S.: GAN inversion for out-of-range images with geometric transformations. In: Proceedings of the IEEE/CVF International Conference on Computer Vision, pp. 13941–13949 (2021)
19. Karras, T., Aila, T., Laine, S., Lehtinen, J.: Progressive growing of GANs for improved quality, stability, and variation. In: International Conference on Learning Representations (2018)
20. Karras, T., Aittala, M., Hellsten, J., Laine, S., Lehtinen, J., Aila, T.: Training generative adversarial networks with limited data. In: Proceedings of the NeurIPS (2020)
21. Karras, T., et al.: Alias-free generative adversarial networks. In: Proceedings of the NeurIPS (2021)
22. Karras, T., Laine, S., Aila, T.: A style-based generator architecture for generative adversarial networks. In: Proceedings of the IEEE Conference on Computer Vision and Pattern Recognition, pp. 4401–4410 (2019)
23. Karras, T., Laine, S., Aittala, M., Hellsten, J., Lehtinen, J., Aila, T.: Analyzing and improving the image quality of StyleGAN. In: Proceedings of the IEEE/CVF Conference on Computer Vision and Pattern Recognition, pp. 8110–8119 (2020)
24. Kim, H., Choi, Y., Kim, J., Yoo, S., Uh, Y.: Exploiting spatial dimensions of latent in GAN for real-time image editing. In: Proceedings of the IEEE/CVF Conference on Computer Vision and Pattern Recognition, pp. 852–861 (2021)
25. Krause, J., Stark, M., Deng, J., Fei-Fei, L.: 3D object representations for fine-grained categorization. In: Proceedings of the IEEE International Conference on Computer Vision Workshops, pp. 554–561 (2013)
26. Kwon, G., Ye, J.C.: Diagonal attention and style-based GAN for content-style disentanglement in image generation and translation. In: Proceedings of the IEEE/CVF International Conference on Computer Vision, pp. 13980–13989 (2021)
27. Ling, H., Kreis, K., Li, D., Kim, S.W., Torralba, A., Fidler, S.: EditGAN: high-precision semantic image editing. arXiv preprint arXiv:2111.03186 (2021)
28. Livingstone, S.R., Russo, F.A.: The Ryerson audio-visual database of emotional speech and song (RAVDESS): a dynamic, multimodal set of facial and vocal expressions in North American English. PLoS ONE **13**(5), e0196391 (2018)
29. Park, T., et al.: Swapping autoencoder for deep image manipulation. In: Larochelle, H., Ranzato, M., Hadsell, R., Balcan, M.F., Lin, H. (eds.) Advances in Neural Information Processing Systems, vol. 33, pp. 7198–7211. Curran Associates, Inc. (2020). https://proceedings.neurips.cc/paper/2020/file/50905d7b2216bfeccb5b41016357176b-Paper.pdf
30. Patashnik, O., Wu, Z., Shechtman, E., Cohen-Or, D., Lischinski, D.: StyleCLIP: text-driven manipulation of StyleGAN imagery. In: Proceedings of the IEEE/CVF International Conference on Computer Vision, pp. 2085–2094 (2021)
31. Pidhorskyi, S., Adjeroh, D.A., Doretto, G.: Adversarial latent autoencoders. In: Proceedings of the IEEE/CVF Conference on Computer Vision and Pattern Recognition, pp. 14104–14113 (2020)
32. Richardson, E., et al.: Official implementation of encoding in style: a StyleGAN encoder for image-to-image translation (2020). https://github.com/eladrich/pixel2style2pixel
33. Richardson, E., et al.: Encoding in style: a StyleGAN encoder for image-to-image translation. In: Proceedings of the IEEE/CVF Conference on Computer Vision and Pattern Recognition, pp. 2287–2296 (2021)

34. Shen, Y., Gu, J., Tang, X., Zhou, B.: Interpreting the latent space of GANs for semantic face editing. In: Proceedings of the IEEE/CVF Conference on Computer Vision and Pattern Recognition, pp. 9243–9252 (2020)
35. Shen, Y., Zhou, B.: Closed-form factorization of latent semantics in GANs. In: Proceedings of the IEEE/CVF Conference on Computer Vision and Pattern Recognition, pp. 1532–1540 (2021)
36. Tewari, A., et al.: StyleRig: rigging StyleGAN for 3D control over portrait images. In: Proceedings of the IEEE/CVF Conference on Computer Vision and Pattern Recognition, pp. 6142–6151 (2020)
37. Tov, O., Alaluf, Y., Nitzan, Y., Patashnik, O., Cohen-Or, D.: Designing an encoder for StyleGAN image manipulation. ACM Trans. Graph. (TOG) **40**(4), 1–14 (2021)
38. Tov, O., Alaluf, Y., Nitzan, Y., Patashnik, O., Cohen-Or, D.: Official implementation of designing an encoder for StyleGAN image manipulation (2021). https://github.com/omertov/encoder4editing
39. Voynov, A., Babenko, A.: Unsupervised discovery of interpretable directions in the GAN latent space. In: International Conference on Machine Learning, pp. 9786–9796. PMLR (2020)
40. Wang, B., Ponce, C.R.: The geometry of deep generative image models and its applications. arXiv preprint arXiv:2101.06006 (2021)
41. Wang, T., Zhang, Y., Fan, Y., Wang, J., Chen, Q.: High-fidelity GAN inversion for image attribute editing. arXiv preprint arXiv:2109.06590 (2021)
42. Wang, T., Zhang, Y., Fan, Y., Wang, J., Chen, Q.: Official implementation of high-fidelity GAN inversion for image attribute editing (2021). https://github.com/Tengfei-Wang/HFGI
43. Wei, T., et al.: A simple baseline for StyleGAN inversion. arXiv preprint arXiv:2104.07661 (2021)
44. Wu, Z., Lischinski, D., Shechtman, E.: StyleSpace analysis: disentangled controls for StyleGAN image generation. In: Proceedings of the IEEE/CVF Conference on Computer Vision and Pattern Recognition, pp. 12863–12872 (2021)
45. Xia, W., Zhang, Y., Yang, Y., Xue, J.H., Zhou, B., Yang, M.H.: GAN inversion: a survey. arXiv preprint arXiv:2101.05278 (2021)
46. Xu, Y., Du, Y., Xiao, W., Xu, X., He, S.: From continuity to editability: inverting GANs with consecutive images. In: Proceedings of the IEEE/CVF International Conference on Computer Vision, pp. 13910–13918 (2021)
47. Yao, X., Newson, A., Gousseau, Y., Hellier, P.: A latent transformer for disentangled face editing in images and videos. In: Proceedings of the IEEE/CVF International Conference on Computer Vision, pp. 13789–13798 (2021)
48. Yu, C., Wang, W.: Adaptable GAN encoders for image reconstruction via multi-type latent vectors with two-scale attentions. arXiv preprint arXiv:2108.10201 (2021)
49. Zhang, R., Isola, P., Efros, A.A., Shechtman, E., Wang, O.: The unreasonable effectiveness of deep features as a perceptual metric. In: Proceedings of the IEEE Conference on Computer Vision and Pattern Recognition, pp. 586–595 (2018)
50. Zhu, J., Shen, Y., Zhao, D., Zhou, B.: In-domain GAN inversion for real image editing. In: Vedaldi, A., Bischof, H., Brox, T., Frahm, J.-M. (eds.) ECCV 2020. LNCS, vol. 12362, pp. 592–608. Springer, Cham (2020). https://doi.org/10.1007/978-3-030-58520-4_35
51. Zhu, P., Abdal, R., Femiani, J., Wonka, P.: Barbershop: GAN-based image compositing using segmentation masks. arXiv preprint arXiv:2106.01505 (2021)
52. zllrunning: Face parsing network pre-trained on CelebAMask-HQ dataset (2019). https://github.com/zllrunning/face-parsing.PyTorch

FakeCLR: Exploring Contrastive Learning for Solving Latent Discontinuity in Data-Efficient GANs

Ziqiang Li[1], Chaoyue Wang[2(✉)], Heliang Zheng[2], Jing Zhang[3], and Bin Li[1(✉)]

[1] University of Science and Technology of China, Hefei, China
iceli@mail.ustc.edu.cn, binli@ustc.edu.cn
[2] JD Explore Academy, Beijing, China
chaoyue.wang@outlook.com, zhenghl@mail.ustc.edu.cn
[3] The University of Sydney, Camperdown, Australia
jing.zhang1@sydney.edu.au

Abstract. Data-Efficient GANs (DE-GANs), which aim to learn generative models with a limited amount of training data, encounter several challenges for generating high-quality samples. Since data augmentation strategies have largely alleviated the training instability, how to further improve the generative performance of DE-GANs becomes a hotspot. Recently, contrastive learning has shown the great potential of increasing the synthesis quality of DE-GANs, yet related principles are not well explored. In this paper, we revisit and compare different contrastive learning strategies in DE-GANs, and identify (i) the current bottleneck of generative performance is the discontinuity of latent space; (ii) compared to other contrastive learning strategies, Instance-perturbation works towards latent space continuity, which brings the major improvement to DE-GANs. Based on these observations, we propose FakeCLR, which only applies contrastive learning on perturbed fake samples, and devises three related training techniques: Noise-related Latent Augmentation, Diversity-aware Queue, and Forgetting Factor of Queue. Our experimental results manifest the new state of the arts on both few-shot generation and limited-data generation. On multiple datasets, FakeCLR acquires more than 15% FID improvement compared to existing DE-GANs. Code is available at https://github.com/iceli1007/FakeCLR.

Keywords: Data-efficient training · GANs · Contrastive learning

This work was performed when Ziqiang Li was visiting JD Explore Academy as a research intern.

Supplementary Information The online version contains supplementary material available at https://doi.org/10.1007/978-3-031-19784-0_35.

S. Avidan et al. (Eds.): ECCV 2022, LNCS 13675, pp. 598–615, 2022.
https://doi.org/10.1007/978-3-031-19784-0_35

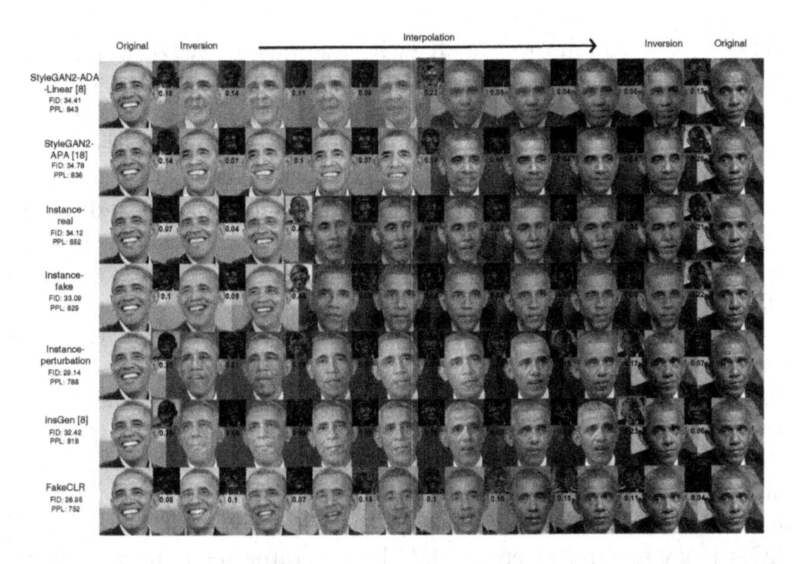

Fig. 1. Training GANs with as little as 100 training samples (Obama dataset) typically results in severe discontinuity in latent space, *i.e.* under-diversity interpolation on the top four rows. StyleGAN2-ADA-Linear [13] is the baseline model with data augmentation. Generator trained with Instance-perturbation and ours show more accurate inversion, smoother latent space, more diverse interpolation, and better FID and PPL. The small grey images visualize the difference between the two face images. The red numbers are mean of pixel-wise difference, we highlighted the difference score > 0.2 with red border. (Color figure online)

1 Introduction

Generative Adversarial Networks [1] are capable of generating realistic images [2] indistinguishable from real ones, which acquire remarkable achievements in image processing [3–6] and computer vision [7–10]. The success of GANs relies on large-scale training datasets, for example, CelebA consists of around 200K images, and LSUN has more than 1M images. However, many scenarios may have a limited amount of training data, *e.g.* there are just 10 examples per artist in the Artistic-Faces dataset [11]. Therefore, Data-Efficient Generative Adversarial Networks (DE-GANs [12]) are important to many real-world practices and attract more and more attention. More analyses can be found in the survey [12].

Previous studies [14,15] demonstrate that severe discriminator overfitting leads to the training failure of DE-GANs. The less training data there is, the earlier the discriminator overfitting happens. Specifically, the discriminator is overly confiding in a few real samples and shows abnormal decision boundaries. Then, the gradients passing to generator are inaccurate, and the training instability problems occur [16]. To mitigate it, many technologies [17–21] have been proposed. Among them, data augmentation strategies [14,15,22,23] largely alleviate the training instability issue.

Currently, how to improve the diversity and quality of DE-GANs generated images becomes the major objective of the community. Contrastive learning (CL) [24,25], as a representative self-supervised learning algorithm, has been applied to different components of GANs, *e.g.* real or fake samples. Although existing methods demonstrate promising performance on DE-GANs, the motivation and principle behind different CL strategies are not well explored and justified. In this paper, we revisit the principle of different contrastive learning strategies and explore a novel perspective of understanding the connections between contrastive learning and DE-GANs' training.

In particular, we experimentally analyze the results of thee different contrastive learning strategies (Instance-real: Eq. (6), Instance-fake: Eq. (7), and Instance-perturbation: Eq. (8)) on DE-GANs. Among them, Instance-perturbation contributes most and achieves significant gains in terms of FID, while others bring slight or even negative improvement (Table 1). To dig out how Instance-perturbation works, we analyze the latent space by GAN inversion and interpolation. As shown in Fig. 1 (top four rows), the training data of DE-GANs is sparse and discrete, and the generator tends to remember these samples, which leads to the discontinuity of the latent space. It can be observed that Instance-perturbation (the fifth row) can alleviate this problem, which achieves better inversion results and more continuous latent interpolation. Such results indicate that current DE-GANs, *e.g.* the most representative method StyleGAN2-ADA, are troubled by the discontinuities on latent space, and reasonable contrastive learning methods may help to solve this problem. Based on these findings, we propose a novel contrastive learning method termed FakeCLR for DE-GANs, which only applies Instance-perturbation and additional three exquisite strategies inspired by continuously changing fake images. These three strategies are Noise-related Latent Augmentation, Diversity-aware Queue, and Forgetting Factor of Queue. Finally, the proposed FakeCLR achieves state-of-the-art performance and contributes to more continuous latent space. Overall, our contributions include:

- *Illustrating the major bottleneck of DE-GANs.* We explore and emphasize the bottleneck, discontinuous of the latent space, of existing DE-GAN methods.
- *Exploring the connections between contrastive learning and DE-GANs.* Through comprehensive experiments, we revisit three popular contrastive learning strategies that can be applied to DE-GANs, and identify that only Instance-perturbation brings the major improvement on generative performance.
- *A new contrastive learning method for DE-GANs.* We propose new contrastive learning with noise perturbation on fake images for DE-GANs called FakeCLR. Meanwhile, three technical innovations have been proposed for improving generative performance.
- *Comprehensive experimental validations on various training settings.* Besides data augmentation, the proposed FakeCLR achieves significant improvements on both limited-data generation and few-shot generation tasks.

2 Background

2.1 Data-Efficient Generative Adversarial Networks

Generative Adversarial Networks (GANs) [1] act a two-player adversarial game, where the generator $G(z)$ is a distribution mapping function that transforms low-dimensional latent distribution p_z to target distribution p_g. And the discriminator $D(x)$ evaluates the divergence between the generated distribution p_g and real distribution p_r. The generator and discriminator minimize and maximize the adversarial loss function. This min-max game can be expressed as:

$$\min_{\phi} \max_{\theta} f(\phi, \theta) = \mathbb{E}_{\mathbf{x}\sim p_r}\left[g_1\left(D_\theta(\mathbf{x})\right)\right] + \mathbb{E}_{\mathbf{z}\sim p_z}\left[g_2\left(D_\theta\left(G_\phi(\mathbf{z})\right)\right)\right], \quad (1)$$

where ϕ and θ are parameters of the generator G and discriminator D, respectively. g_1 and g_2 are different functions for various GANs. Data-Efficient GANs is a special case of GANs, which targets at obtaining a generator that can fit the real data distribution with limited, few-shot training samples. Formally, given the limited data distribution p_l (Generally, samples of p_l is the subset of p_r), we expect the generated distribution p_g obtained by training with p_l to be as close as possible to p_r. Concretely, loss functions of each network are formalized as:

$$\begin{aligned} \mathcal{L}_D &= -\mathbb{E}_{\mathbf{x}\sim p_l}\left[g_1\left(D_\theta(\mathbf{x})\right)\right] - \mathbb{E}_{\mathbf{z}\sim p_z}\left[g_2\left(D_\theta\left(G_\phi(\mathbf{z})\right)\right)\right], \\ \mathcal{L}_G &= \mathbb{E}_{\mathbf{z}\sim p_z}\left[g_2\left(D_\theta\left(G_\phi(\mathbf{z})\right)\right)\right]. \end{aligned} \quad (2)$$

Previous studies attribute the degradation of DE-GANs to the overfitting of the discriminator. Many data augmentation, regularization, architectures, and pre-training techniques [12] have been proposed to mitigate this issue. (i) Data Augmentation [14,15,26–28] is a striking method for mitigating overfitting of the discriminator and orthogonal to other ongoing researches on training, architecture, and regularization. Popular augmentation strategies, such as Adaptive Data Augmentation (ADA), are employed as the essential complements in most DE-GANs. (ii) Regularization [13,29,30] is also a kind of popular technology for enhancing generalization in discriminator by introducing priors or extra supervision tasks. For instance, Tseng *et al.* [21] proposed an anchors-based regularization term to slow down the training of discriminator under limited data. (iii) Architecture is also the key point to improve the performance and stability of GANs. Some light-weight networks [31] have been employed in DE-GANs. Additionally, some model compression methods [32] have also been introduced in DE-GANs [20]. (iv) Pre-training [17–19,33–36] provides another solution to decrease the demand for data. Some studies argued that generator pre-training introduces some image-diversity [37] and latent-smooth [38] priors, and discriminator pre-training [37] provides accurate gradient for DE-GANs.

According to the scale of training data, data-efficient generation can be roughly divided into two tasks: few-shot generation and limited-data generation. Few-shot generation is a challenging task due to the employed tiny amount of data, for instance, 10-shot and 100-shot. Compared to few-shot generation, limited-data generation employs more data, such as 1K or 2K. Generally, the employed technologies, apart from data augmentation, are notably different between few-shot and limited-data generation. Our proposed FakeCLR is a universal technology achieving remarkable performance on both tasks.

2.2 Contrastive Learning

Contrastive learning [24,25,39] is a kind of self-supervised method to extract representations. Generally, given an image $\mathbf{x}$, two random views (query: $\mathbf{x}_q$ and key: $\mathbf{x}_q^+$) are created by different data augmentations (T_1 and T_2), defined as:

$$\mathbf{x}_q = T_1(x), \quad \mathbf{x}_q^+ = T_2(x). \tag{3}$$

Positive pair is defined as such query-key pair, between $\mathbf{x}_q$ and $\mathbf{x}_q^+$, from the same image. Negative pairs are defined as pairs from different images, *i.e.* between $\mathbf{x}_q$ and $\left\{\mathbf{x}_{k_i}^-\right\}_{i=1}^N$, where $\mathbf{x}_{k_i}^- = T_2(\mathbf{x}_i)$ and $\mathbf{x} \neq \mathbf{x}_i$, $i = 1, \cdots, N$. All views are passed through the feature extractor $F(\cdot)$ to acquire representation $\mathbf{v}$:

$$\mathbf{v}_q = F\left(\mathbf{x}_q\right), \quad \mathbf{v}_q^+ = F\left(\mathbf{x}_q^+\right), \quad \mathbf{v}_{k_i}^- = F\left(\mathbf{x}_{k_i}^-\right), \quad i = 1 \ldots N. \tag{4}$$

Contrastive learning aims to maximize the similarity of positive pairs and makes negative pairs dissimilar. Therefore, the InfoNCE loss [39] is designed as:

$$\begin{aligned} &\mathcal{C}_{F(\cdot),\phi(\cdot)}\left(\mathbf{x}_q, \mathbf{x}_q^+, \left\{\mathbf{x}_{k_i}^-\right\}_{i=1}^N\right) = \\ &\quad -\log \frac{\exp\left(\phi\left(\mathbf{v}_q\right)^T \phi\left(\mathbf{v}_q^+\right)/\tau\right)}{\exp\left(\phi\left(\mathbf{v}_q\right)^T \phi\left(\mathbf{v}_q^+\right)/\tau\right) + \sum_{i=1}^N \exp\left(\phi\left(\mathbf{v}_q\right)^T \phi(\mathbf{v}_{k_i}^-)/\tau\right)}, \end{aligned} \tag{5}$$

where $\phi(\cdot)$ is the head network to project the representation to different spaces, and τ is the temperature coefficient. Generally, contrastive learning strategies are employed in discriminative models and have achieved convincing performance on representation learning [24,25,40,41]. There are still a few works [13,42] to apply contrastive learning into unconditional generative tasks. Specifically, Jeong *et al.* [42] proposed Contrastive Discriminator (ContraD), a way of training discriminators of GANs using improved SimCLR [25]. Yang *et al.* [13] proposed InsGen, a instance discrimination realized by MoCo-v2 [43], achieving the state-of-the-art performance on limited-data generation. Different from them, our FakeCLR only applies contrastive learning to the fake images and adopts three advanced strategies designed specifically for generative tasks, acquiring remarkable performance on both few-shot and limited-data generation.

3 How Contrastive Learning Benefits DE-GANs?

Previous works tend to attribute the success of contrastive learning to extra discrimination tasks. For example, in [13], contrastive learning is considered as an instance discrimination task applying to both real and fake images by two additional heads $\phi^r(\cdot)$ and $\phi^f(\cdot)$ besides the original bi-classification head $\phi^d(\cdot)$. In this section, we explore the question of *How contrastive learning benefits Data-Efficient GANs?* by conducting comprehensive comparisons among different contrastive learning strategies.

3.1 Comparisons and Analyses on Contrastive Learning Strategies

Following existing study [13], three main contrastive learning strategies can be applied in Data-Efficient GANs, *i.e.* Instance-real, Instance-fake, and Instance-perturbation. Specifically, *Instance-real* introduces an independent real head $\phi^r(\cdot)$ to conduct additional instance discrimination on real images, *i.e.*,

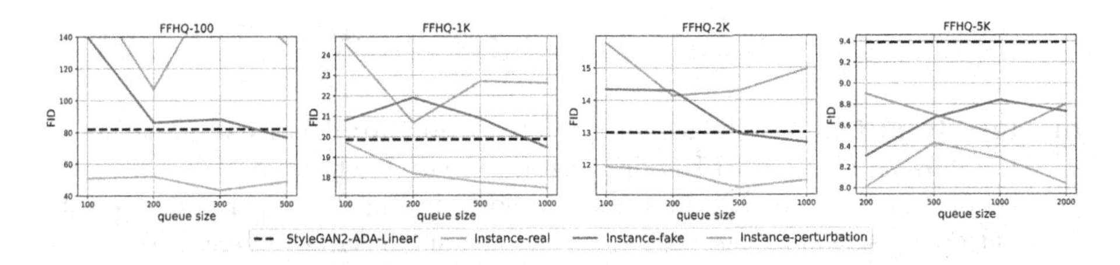

Fig. 2. Part of the ablation experiments of different contrastive learning strategies with various queue sizes and datasets.

$$\mathcal{C}^r = \mathcal{C}_{d(\cdot),\phi^r(\cdot)}\left(T_q(\mathbf{x}_q), T_{k^+}(\mathbf{x}_q), \{T_{k_i}(\mathbf{x}_{k_i})\}_{i=1}^N\right), \tag{6}$$

where T_q, T_{k^+}, and T_{k_i} are different augmentations, $\mathbf{x}_q$ and $\mathbf{x}_{k_i}$ are sampled from the real distribution. Similarly, *Instance-fake* introduces an independent fake head $\phi^f(\cdot)$ to conduct instance discrimination on fake images through contrastive learning and whose objective is formalized as:

$$\mathcal{C}^f = \mathcal{C}_{d(\cdot),\phi^f(\cdot)}\left(T_q(G(\mathbf{z}_q)), T_{k^+}(G(\mathbf{z}_q)), \{T_{k_i}(G(\mathbf{z}_{k_i}))\}_{i=1}^N\right), \tag{7}$$

where $\mathbf{z}_q$ and $\mathbf{z}_{k_i}$ are sampled from the latent distribution. Furthermore, *Instance-perturbation* introduces a noise perturbation strategy [44] to further improve the power of Instance-fake. The objective is formalized as:

$$\mathcal{C}_p^f = \mathcal{C}_{d(\cdot),\phi^f(\cdot)}\left(T_q(G(\mathbf{z}_q)), T_{k^+}(G(\mathbf{z}_q + \epsilon_q)), \{T_{k_i}(G(\mathbf{z}_{k_i}))\}_{i=1}^N\right), \tag{8}$$

where ϵ_q stands for the perturbation of latent augmentation, which is sampled from a Gaussian distribution with small variance.

In the following, we adopt StyleGAN2-ADA [15] and InsGen [13] as reference and conduct ablation studies[1] to explore *"How Contrastive Learning Benefits Data-Efficient GANs?"*. Related results are reported in Table 1 and Fig. 2.

[1] In InsGen [13], ADA-Linear rather than ADA is used as augmentation T when the number of images in the dataset is less than 5K. For fairness, ADA-Linear, rather than ADA and InsGen, should be regraded as the baseline for comparison.

Table 1. The whole ablation experiments of different contrastive learning strategies. We demonstrate the best FID metric for different queue sizes on various datasets. Results with * are best on four different queue sizes, which are better than those reported in [13].

256×256 resolution	FFHQ-100	FFHQ-1K	FFHQ-2K	FFHQ-5K
StyleGAN2-ADA	85.8	21.29	15.39	10.96
StyleGAN2-ADA-Linear	82	19.86	13.01	9.39
Instance-real	107	20.68	14.15	8.50
Instance-fake	76.6	19.45	12.69	8.31
Instance-perturbation	**43.35**	17.46	**11.29**	8.01
InsGen [13]	53.93	19.58	11.92	-
InsGen*	45.75*	18.21*	11.47*	**7.83***

Instance-Real Causes Performance Drop on Small Datasets. Although on large dataset such as FFHQ-5K, Instance-real[2] is effective, it would cause performance drop on smaller datasets. As shown in Fig. 2, the performance with different queue size is consistently decreased on small datasets, *e.g.* FFHQ-100, FFHQ-1K, and FFHQ-2K. Such a degradation is much severer when the queue size is 100 or 300 on FFHQ-100 dataset.

Instance-Fake Improves Performance Marginally on Small Datasets. When adding Instance-fake[2] strategy, compared to the baseline StyleGAN2-ADA-Linear, the best synthesis quality for different queue sizes is consistently improved in Table 1, regardless of the number of training data. However, on small datasets such as FFHQ-100, FFHQ-1K, and FFHQ-2K, the increase is marginal, and the performance is even decreased when queue size is small (as shown in Fig. 2).

Instance-Perturbation is the Key Point for DE-GANs. After using Instance-perturbation[2], FID obtains significant improvement regardless of queue size and the number of training images. Specifically, compared to Instance-fake, the best FID for different queue size as shown in Table 1 acquires consistent improvement. Furthermore, Instance-perturbation also obtains FID improvements of -2.4, -0.75, and -0.18 relative to InsGen (*i.e.* employs instance-real at the same time) on FFHQ-100, FFHQ-1K, and FFHQ-2K datasets, respectively.

The Optimal Queue Size is Related to the Diversity of Negative Samples. Following MoCo-v2 [24], negative samples of InsGen are stored in a queue to reduce the computational complexity. Two queues, fake instance queue and real instance queue, are adopted in InsGen. The size of them are defined as *fqs* and *rqs* respectively. Empirically, length of the queue tends to be the 5% number of the dataset in MoCo-v2, indicating *rqs* is related to the number of real data. According to our experiments (Fig. 2), we find that the optimal *fqs* in Instance-fake are larger than the optimal *rqs* in Instance-real under the same setting. We

[2] The objectives of discriminator are: i) Instance-real: $\mathcal{L}_D^{'} = \mathcal{L}_D + \lambda^r \mathcal{C}^r$; ii) Instance-fake:$\mathcal{L}_D^{'} = \mathcal{L}_D + \lambda^f \mathcal{C}^f$; iii) Instance-perturbation: $\mathcal{L}_D^{'} = \mathcal{L}_D + \lambda^f \mathcal{C}_p^f$, Respectively..

Table 2. Perceptual path length (PPL) of z and w spaces on FFHQ-2K (256×256)

Method	PPL (z)		PPL (w)	
	Mean	Std. dev	Mean	Std. dev
StyleGAN2-ADA-Linear	2163	1819	232	100.3
Instance-fake	2399	2082	217	95.6
Instance-perturbation	1729	1626	167	90.5
FakeCLR	**1411**	**1490**	**136**	**65.4**

argue that the optimal queue size is related to the diversity of negative samples in queues. In Instance-real, the diversity of negative samples is low with limited training data. Thus long *rqs* leads to the degradation of contrastive learning.

3.2 Instance-Perturbation and Latent Space Continuity

The above analyses demonstrate that Instance-perturbation is the key point of contrastive learning in DE-GANs and better performance than InsGen can be obtained by using Instance-perturbation alone. Moreover, we argue that Instance-perturbation mitigates the discontinuity of latent space effectively. Intuitively, Instance-perturbation introduces latent similarity prior that images synthesized from the latent codes within a neighbourhood are close to each other, which makes the latent space continuous. Experimentally, Fig. 1 illustrates that Instance-perturbation owns more continuous interpolation, more accurate inversion, better FID, and better PPL compared to StyleGAN2-LDA, StyleGAN2-APA [23], and other contrastive learning strategies on Obama dataset. Furthermore, Table 2 shows that PPL [45] of Instance-perturbation outperform other methods in Mean and Std. We also show some similar results of FFHQ-100 dataset in Sec. A.1 of the supplementary materials.

4 Methodology

As discussed above, real instance discrimination is not conducive to the DE-GANs' training. Hence, we only adopt Instance-perturbation, which is illustrated on the top part of Fig. 3. Furthermore, bottom part of Fig. 3 introduces three innovative strategies, that are Noise-related Latent Augmentation, Diversity-aware Queue, and Forgetting Factor of Queue, for improving the efficiency of contrastive learning on fake images. In summary, the proposed complete objective function[3] of FakeCLR is optimized using:

$$\hat{\mathcal{L}}_D = \mathcal{L}_D + \lambda^f \hat{\mathcal{C}}_p^f, \quad \hat{\mathcal{L}}_G = \mathcal{L}_G + \lambda_G \mathcal{C}^f,$$
$$\text{where} \quad \hat{\mathcal{C}}_p^f = \hat{\mathcal{C}}_{d(\cdot),\phi^f(\cdot)}\left(T_q(G(\mathbf{z}_q)), T_{k+}(G(\mathbf{z}_q + \hat{\epsilon}_q)), \{T_{k_i}(G(\mathbf{z}_{k_i})), \mathbf{m}_i\}_{i=1}^{\hat{N}}\right). \tag{9}$$

[3] Followed InsGen [13], Instance-fake ($\mathcal{C}^f$) is added into the generator loss to improve the generated diversity.

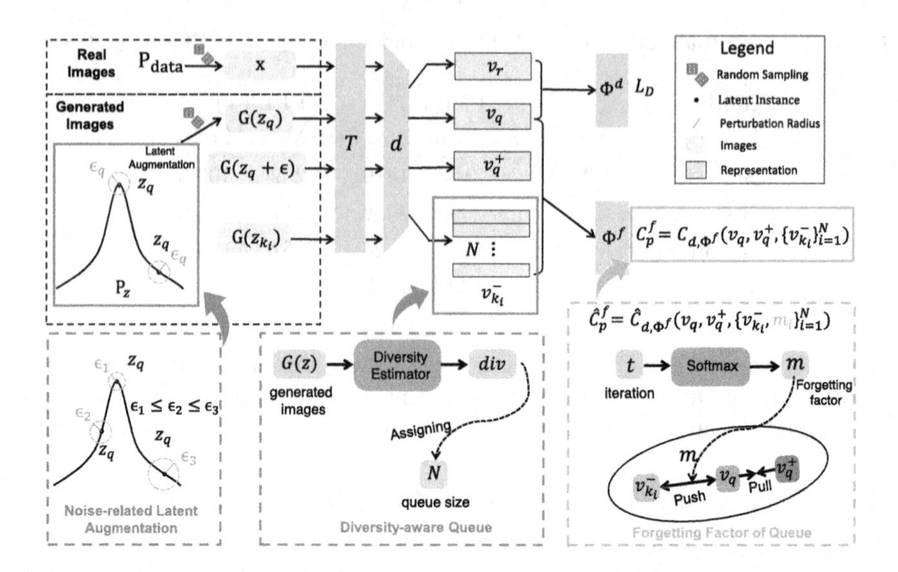

Fig. 3. Overview of FakeCLR. Overall, the pipeline only adopting fake instance discrimination with latent augmentation is illustrated on the top of the figure, where the representation is learned from bi-classification loss ($\mathcal{L}_D$) and Instance-perturbation loss ($\mathcal{C}_p^f$). Here, N is the number of negative samples (queue size), $T(\cdot)$ is the data augmentation, d is the backbone of discriminator, ϕ^d and ϕ^f are two independent heads. Furthermore, three strategies proposed in this paper are illustrated at the bottom of the figure.

Noise-Related Latent Augmentation. From Sect. 3.2, noise perturbation is the key point for contrastive learning in DE-GANs, which alleviates the discontinuity of latent space effectively. From this perspective, noise perturbation introduces latent similarity prior to the discriminator, which provides latent-continuity guidance for the generator. However, the original strategy holds the same perturbation radius (ϵ_q), which is not desirable due to the nonuniform latent distribution. More sensibly, a noise-related perturbation with different latent sampling z_q should be adopted. Generally, $\epsilon_q \propto |z_q|$ when $z_q \sim \mathcal{N}(0,1)$. It can be interpreted as stronger similarity prior should be required in low-density regions of the latent distribution. In this paper, we adopt a simple implementation that is $\hat{\epsilon}_q = l_1 \cdot |z_q|$, where l_1 is a coefficient.

Diversity-Aware Queue. The number of negative samples is critical in contrastive learning. Short queue leads to inhibiting convergence [25], while long queue suffers from redundancy and inappropriate negative samples. Empirically, length of the queue is related to the size of dataset, such as 5% number of the dataset in MoCo-v2 [24]. However, this empirical hypothesis does not seem to hold up in GANs due to the continuously changing generated images over training iteration. As described in Sect. 3.1, the optimal queue size may be related to the diversity of samples in queue. Furthermore, the diversity of fake queues is indeed higher than that of real queues and decreases gradually accompanying with training due to the convergence and stability of generator. Therefore, a diversity-aware queue is required for contrastive learning in fake images. Generally, $N \propto DE(G(z))$, where $DE(\cdot)$ is a diversity estimator. In this paper, we also adopt a simple implementation that is $\hat{N} = l_2 \cdot t$ to reduce the computational complexity, where l_2 is a negative coefficient and t is the iteration of training. Therefore, queue size in FakeCLR decreases linearly with respect to the training.

Forgetting Factor of Queue. The choice of negative pairs has drawn much attention in contrastive learning. [46,47] demonstrate that hard negative samples capture desirable generalization properties and improve the performance of downstream task. Different from real images, fake images are updated with training iteration. As a result, varying levels of generated samples exist in the negative queue. However, vanilla contrastive learning considers them to have the same importance. Some iteration-based prior, *i.e.*, the negative samples generated by the current iteration should be given more attention, should be introduced. In this paper, we apply a forgetting factor to negative samples of the queue. Specifically, the contribution of previous samples should be smaller than current samples to contrastive loss. This makes the contrastive term focus more on recently generated samples. The iteration-based InfoNCE is formulated as:

$$\begin{aligned} &\hat{\mathcal{C}}_{F(\cdot),\phi(\cdot)}\left(\mathbf{x}_q, \mathbf{x}_k^+, \{\mathbf{x}_{k_i}^-, \mathrm{m}_i\}_{i=1}^N\right) \\ &= -\log \frac{\exp\left(\phi\left(\mathbf{v}_q\right)^T \phi\left(\mathbf{v}_k^+\right)/\tau\right)}{\exp\left(\phi\left(\mathbf{v}_q\right)^T \phi\left(\mathbf{v}_k^+\right)/\tau\right) + \sum_{i=1}^N \exp\left(\left(\phi\left(\mathbf{v}_q\right)^T \phi\left(\mathbf{v}_{k_i}^-\right) + \mathrm{m}_i\right)/\tau\right)}, \end{aligned} \quad (10)$$

where m_i is the introduced forgetting factor to adjust the importance of different negative samples, as defined by:

$$\begin{aligned} \mathbf{m}_i &= \frac{\exp\left(\hat{\mathbf{t}}_i/\tau_{\mathbf{m}}\right)}{\sum_{j=1}^N \exp\left(\hat{\mathbf{t}}_j/\tau_{\mathbf{m}}\right)}, \quad \text{where} \quad \hat{\mathbf{t}}_i = \frac{\mathbf{t}_i - \min(\mathbf{t})}{\max(\mathbf{t}) - \min(\mathbf{t})}, \\ \mathbf{t} &= \{\mathbf{t_1}, \mathbf{t_2}, \cdots, \mathbf{t}_N\}, \quad i = 1, 2, \cdots, N, \end{aligned} \quad (11)$$

$\tau_{\mathbf{m}}$ is a temperature coefficient that controls the distribution $\mathbf{m} := \{\mathbf{m_1}, \cdots, \mathbf{m}_N\}$, and t_i means that the negative sample $\mathbf{v}_{k_i}^-$ is generated by the generator of the t_i-th iteration. The distribution of $\mathbf{m}$ under different temperature $\tau_{\mathbf{m}}$ and queue size N has be shown in Sec. A.2 of supplementary materials. The proposed forgetting factor ($\mathbf{m}$) indeed improves the importance of current negative samples to contrastive term. The proof and discussion are in Sec. A.3 of supplementary materials. Furthermore, a PyTorch-like pseudocode of iteration-based contrastive learning has also be shown in Sec. A.4 of supplementary materials.

Table 3. Comparison with the state of the arts over FFHQ (256 × 256): Training with 100, 1K, 2K, and 5K samples. FID (lower is better) is reported as the evaluation metric. Our method performs the best consistently. All methods adopt the StyleGAN2 architecture. Results with * are best on four different queue sizes, which are better than those reported in [13].

Method	Augmentation	FFHQ-100	FFHQ-1K	FFHQ-2k	FFHQ-5K
StyleGAN2 [2]	No	179	100.16	54.3	49.68
GenCo [48]	No	148	65.31	47.32	27.96
DISP [49]	Yes	-	-	21.06	-
LeCam-GAN [21]	Yes	-	21.70	-	-
ADA [15]	Yes	85.8	21.29	15.39	10.96
ADA-Linear [13]	Yes	82	19.86	13.01	9.39
APA [23]	Yes	65	18.89	16.90	8.38
InsGen [13]	Yes	53.93	19.58	11.92	-
InsGen*	Yes	45.75*	18.21*	11.47*	7.83*
FakeCLR (ours)	Yes	**42.56**	**15.92**	**9.90**	**7.25**

5 Experiments

We evaluate the advancement of our proposed FakeCLR on multiple datasets. After introducing implementation details, we present the comparison on FFHQ [45] dataset. Section 5.3 presents the comparison on few-shot datasets (Obama [14], Grumby Cat [14], Panda [14], and AnimalFace [50]). Furthermore, ablation studies showing the importance of different components are demonstrated in Sect. 5.4.

5.1 Implementation Details

Similar to InsGen [13], our FakeCLR also can be easily implemented on any GAN framework. In this paper, we take the state-of-the-art GANs model, StyleGAN2 [2], as an example to demonstrate how FakeCLR is implemented.

InsGen reuses the network structure and learning hyper-parameters of StyleGAN2 [2], the augmentation pipeline of ADA [15], and contrastive learning settings of MoCo-v2 [24] achieving state-of-the-art results with limited training data. Additionally, a linear hyper-parameter that controls the strength of augmentation (ADA-Linear) has also been proposed to replace the adaptive hyper-parameter in ADA [15]. Therefore, we exactly reuse the above settings in InsGen for a fair comparison. Besides, some new hyper-parameter in FakeCLR are setted as $l_1 = 0.1$ and $\tau_{\mathrm{m}} = 0.01$. All experiments are conducted on two NVIDIA Tesla A100 GPUs with 64 batch size.

5.2 Results on FFHQ

FFHQ contains 70K high-resolution images of human faces. To reduce the computational complexity, we resize images to 256 × 256. For the experiments of

Fig. 4. Generated images and the corresponding FID under various datasets. All images are synthesized randomly without truncation.

limited data, we collect a subset of training data by randomly sampling with 100, 1K, 2K, and 5K images. Regardless of the number of the training data, the Fréchet Inception Distance (FID) [51] metric is calculated between 50K fake images and all 70K images, evaluating the performance in terms of both reality and diversity. Moreover, the implementation of our FakeCLR is based on the official implementation of StyleGAN2-ADA and InsGen.

Table 3 shows the comparison with previous studies on FFHQ-100, FFHQ-1K, FFHQ-2K, and FFHQ-5k. We compare to most recent works, and our method achieves the new state-of-the-art consistently. Specifically, compared to the baseline method InsGen, our method improves the FID on FFHQ by 3.19, 2.29, 1.57, and 0.58 with 100, 1K, 2K, and 5K training images, respectively. Such results show the effectiveness of our proposed FakeCLR, especially for data-efficient scenarios. Some generated examples and the corresponding FID on FFHQ are presented on the top part of Fig. 4. More examples generated by InsGen [13] and our FakeCLR are presented in Sec. A.5 of the supplementary materials. Furthermore, we also analyze the computation cost in Sec. A.6 of the supplementary materials. Note that the three strategies adopted in FakeCLR only introduce a fraction of the cost. More importantly, our FakeCLR achieves better results even with 8.8% less computational cost than InsGen [13]. Some qualitative analysis of nearest neighbor test as done in DiffAugment [14] in the LPIPS and pixel space, and comparisons on PPL and KID metrics on the FFHQ-100 dataset have also been presented in Sec. A.7 and Sec. A.8 of the supplementary materials, respectively.

5.3 Results on Few-Shot Generation

We also evaluate the advancement of FakeCLR on few-shot generation that contains five datasets (*i.e.*, Obama, Grumpy Cat, Panada, AnimalFace Cat, and

Table 4. Comparison with previous works over 100-shot and AFHQ: Training with 100 (Obama, Grumpy Cat, and Panda), 160 (AFHQ Cat), and 389 (AFHQ Dog) samples. FID (lower is better) is reported as the evaluation metric. Our method performs the best in most datasets. All methods adopt the StyleGAN2 architecture except for FastGAN+DA.

Method	Massive augmentation	Pre-training w/ 70K images	100-shot			AnimalFace	
			Obama	Grumpy cat	Panda	Cat	Dog
Scale/shift [33]	No	Yes	50.72	34.20	21.38	54.83	83.04
MineGAN [34]	No	Yes	50.63	34.54	14.84	54.45	93.03
TransferGAN [35]	No	Yes	48.73	34.06	23.20	52.61	82.38
TransferGAN + DA [14]	Yes	Yes	39.85	29.77	17.12	49.10	65.57
FreezeD [36]	No	Yes	41.87	31.22	17.95	47.70	70.46
StyleGAN2 [2]	No	No	80.20	48.90	34.27	71.71	130.19
DA [14]	Yes	No	46.87	27.08	12.06	42.44	58.85
FastGAN+DA [31]	Yes	No	41.05	26.65	10.03	35.11	50.66
ADA [15]	Yes	No	45.69	26.62	12.90	40.77	56.83
ADA-Linear [13]	Yes	No	34.41	25.67	10.23	36.30	54.39
SGLP [30]	Yes	No	45.37	26.52	-	-	-
LeCam-GAN [21]	Yes	No	33.16	24.93	10.16	34.18	54.88
GenCo [48]	Yes	No	32.21	**17.79**	9.49	30.89	49.63
InsGen [13]	Yes	No	32.42	22.01	9.85	33.01	44.93
FakeCLR (Ours)	Yes	No	**26.95**	19.56	**8.42**	**26.34**	**42.02**

AnimalFace Dog) with 100, 100, 100, 160, and 389 training images, respectively. All models in this section are trained with the resolution of 256×256, and the datasets can be found at the link. FID [51] is calculated between 50K fake images and the training images. Other settings are the same as experiments on FFHQ.

Table 4 demonstrates the quantitative comparison on few-shot generation. Top five baselines in Table 4 pre-train the model with large FFHQ-70K. It can be observed that FakeCLR can even achieve better FID by using only 100–400 training samples. The synthesis quality is substantially improved by our method on most datasets except for Grumpy Cat datasets. Specifically, we obtain an FID improvements over GenCo [48] by 16.3%, −9.9%, 11.3%, 14.7%, and 15.3% on five datasets, respectively. Compared to InsGen [13], the FID improvements are 16.9%, 11.1%, 14.5%, 20.2%, and 15.3%, respectively. Moreover, some qualitative results are shown on the second row of Fig. 4. More examples generated by InsGen [13] and FakeCLR can be found in Sec. A.9 of the supplementary. Furthermore, nearest neighbor test in the LPIPS and pixel space, and comparisons on PPL and KID metrics on the Obama dataset have also been presented in Sec. A.7 and Sec. A.8 of the supplementary, respectively.

5.4 Ablation Study

Component-Level Investigation. We investigate the effectiveness of each component on the FFHQ dataset. As shown in Table 5, all components are effective and can improve the performance Among them, $\hat{C}_{d(\cdot),\phi^f(\cdot)}$ achieves the most

Table 5. Ablation study of FakeCLR. The proposed strategies in FakeCLR all improve the generation over the baseline. Here, the first line is baseline StyleGAN2-ADA-Linear, and the other lines mean adding the proposed strategies.

C_p^f	$\hat{\epsilon}_q$	$\hat{C}_{d(\cdot),\phi^f(\cdot)}$	$\hat{N}$	FFHQ-100	FFHQ-1K	FFHQ-2K	FFHQ-5K
				82.00	19.86	13.01	9.39
✓				43.35	17.46	11.29	8.01
✓	✓			44.10	17.09	10.65	7.89
✓	✓	✓		**41.62**	16.05	10.40	7.48
✓	✓	✓	✓	42.56	**15.92**	**9.90**	**7.25**

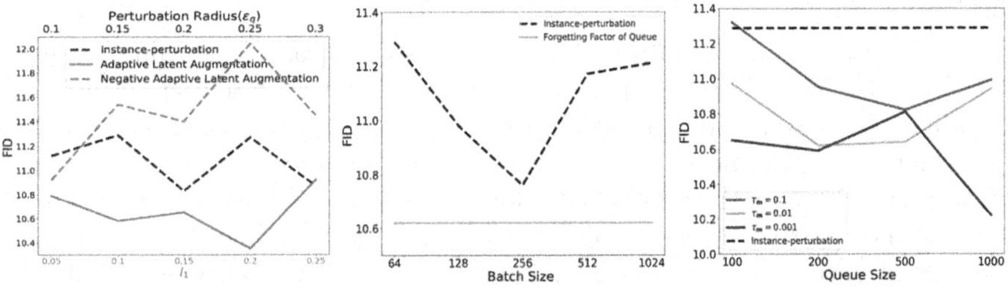

(a) Ablation study of perturbation radius

(b) Ablation study of batch size

(c) Ablation study of temperature(τ_m)

Fig. 5. Ablation studies of different components in strategies of FakeCLR. (a): Ablation study of perturbation radius (ϵ_q) in Noise-related Latent Augmentation. (b): Ablation study of batch size in Forgetting Factor of Queue. (c): Ablation study of temperature (τ_{m}) in Forgetting Factor of Queue.

significant improvement. In addition, combining all components performs the best on FFHQ-1K, FFHQ-2K, and FFHQ-5K. On FFHQ-100, Diversity-aware Queue does not further improve the performance because the training data size is extremely small. Detailed ablations on FFHQ-2K are introduced as follows and more results can be found in Sec. A.10, A.11, and A.12 of the supplementary.

Noise-Related Latent Augmentation. Perturbation Radius (ϵ_q) is an important hyper-parameter, and we compare different ϵ_q to our Noise-related latent augmentation ($\hat{\epsilon}_q$) in Fig. 5(a). The result shows that our strategy outperforms the best fixed strategy with a clear margin. Furthermore, negative prior that is $\epsilon_q \propto 1/|z_q|$ has been introduced in Negative Noise-related Latent Augmentation. The results demonstrate that negative prior severely compromise performance, which further supports our motivation of Noise-related Latent Augmentation.

Forgetting Factor of Queue. The queue of negative samples is always a key factor in contrastive learning. An effective negative queue is supposed to be with diverse and up-to-date generated samples. Simply tuning batch size would cause a trade-off, *i.e.*, small batch size involves old samples, and large batch size reduces data diversity. As shown in Fig. 5(b), 256 is the optimal batch size. And proposed Forgetting Factor of Queue provides a solution to break such trade-off, which suppresses old samples involved by small batch size and can achieve better performance. Note that for fair comparisons, we keep the training batch size constant and only adopt gradient accumulation to change the number of negative samples entering the queue for each iteration in contrastive learning (*i.e.*, batch size in Forgetting Factor of Queue). Figure 5(c) further shows the impact of Temperature ($\tau_{\mathbf{m}}$) in our method. As shown in Sec. A.2 of the supplementary materials, $\tau_{\mathbf{m}}$ controls the distribution of $\mathbf{m}$ and smaller $\tau_{\mathbf{m}}$ indicate higher weights for samples in current iteration. The results show that proposed method outperform the original method, and $\tau_{\mathbf{m}} = 0.001$ achieves the best performance.

6 Conclusions

In this paper, we revisit the principle of different contrastive learning strategies in DE-GANs. According to experiments, we identify that latent similarity prior introduced by Instance-perturbation makes the main contribution to DE-GANs' performance. Furthermore, we also explore and emphasize a major bottleneck of existing DE-GAN methods, discontinuous of the latent space and demonstrate that Instance-perturbation can mitigate it effectively. Based on these, we propose a new contrastive learning method, FakeCLR, in DE-GANs, which acquires a new state of the art on both few-shot generation and limited-data generation.

Acknowledgements. The work is partially supported by the National Natural Science Foundation of China under Grand No. U19B2044 and No. 61836011.

References

1. Goodfellow, I., et al.: Generative adversarial nets. In: Advances in Neural Information Processing Systems, pp. 2672–2680 (2014)
2. Karras, T., Laine, S., Aittala, M., Hellsten, J., Lehtinen, J., Aila, T.: Analyzing and improving the image quality of StyleGAN. In: Proceedings of the IEEE/CVF Conference on Computer Vision and Pattern Recognition, pp. 8110–8119 (2020)
3. Wang, C., Xu, C., Wang, C., Tao, D.: Perceptual adversarial networks for image-to-image transformation. IEEE Trans. Image Process. **27**(8), 4066–4079 (2018)
4. Wenlong, Z., Yihao, L., Dong, C., Qiao, Y.: RankSRGAN: generative adversarial networks with ranker for image super-resolution. IEEE Trans. Pattern Anal. Mach. Intell. (2021)
5. Yang, Y., Wang, C., Liu, R., Zhang, L., Guo, X., Tao, D.: Self-augmented unpaired image dehazing via density and depth decomposition. In: Proceedings of the IEEE/CVF Conference on Computer Vision and Pattern Recognition (CVPR), pp. 2037–2046, June 2022
6. Qiu, J., Wang, X., Maybank, S.J., Tao, D.: World from blur. In: Proceedings of the IEEE/CVF Conference on Computer Vision and Pattern Recognition, pp. 8493–8504 (2019)
7. Yang, C., Wang, Z., Zhu, X., Huang, C., Shi, J., Lin, D.: Pose guided human video generation. In: Ferrari, V., Hebert, M., Sminchisescu, C., Weiss, Y. (eds.) ECCV 2018. LNCS, vol. 11214, pp. 204–219. Springer, Cham (2018). https://doi.org/10.1007/978-3-030-01249-6_13

8. Tao, R., Li, Z., Tao, R., Li, B.: ResAttr-GAN: unpaired deep residual attributes learning for multi-domain face image translation. IEEE Access **7**, 132594–132608 (2019)
9. Jiang, L., Zhang, C., Huang, M., Liu, C., Shi, J., Loy, C.C.: TSIT: a simple and versatile framework for image-to-image translation. In: Vedaldi, A., Bischof, H., Brox, T., Frahm, J.-M. (eds.) ECCV 2020. LNCS, vol. 12348, pp. 206–222. Springer, Cham (2020). https://doi.org/10.1007/978-3-030-58580-8_13
10. Qiu, J., Yang, Y., Wang, X., Tao, D.: Scene essence. In: Proceedings of the IEEE/CVF Conference on Computer Vision and Pattern Recognition, pp. 8322–8333 (2021)
11. Yaniv, J., Newman, Y., Shamir, A.: The face of art: landmark detection and geometric style in portraits. ACM Trans. Graph. (TOG) **38**(4), 1–15 (2019)
12. Li, Z., Wu, X., Xia, B., Zhang, J., Wang, C., Li, B.: A comprehensive survey on data-efficient GANs in image generation. arXiv preprint arXiv:2204.08329 (2022)
13. Yang, C., Shen, Y., Xu, Y., Zhou, B.: Data-efficient instance generation from instance discrimination. arXiv preprint arXiv:2106.04566 (2021)
14. Zhao, S., Liu, Z., Lin, J., Zhu, J.Y., Han, S.: Differentiable augmentation for data-efficient GAN training. In: Advances in Neural Information Processing Systems 33 (2020)
15. Karras, T., Aittala, M., Hellsten, J., Laine, S., Lehtinen, J., Aila, T.: Training generative adversarial networks with limited data. In: Advances in Neural Information Processing Systems 33 (2020)
16. Arjovsky, M., Bottou, L.: Towards principled methods for training generative adversarial networks. arXiv preprint arXiv:1701.04862 (2017)
17. Yang, C., et al.: One-shot generative domain adaptation. arXiv preprint arXiv:2111.09876 (2021)
18. Sauer, A., Chitta, K., Müller, J., Geiger, A.: Projected GANs converge faster. In: Advances in Neural Information Processing Systems 34 (2021)
19. Kumari, N., Zhang, R., Shechtman, E., Zhu, J.Y.: Ensembling off-the-shelf models for GAN training. arXiv preprint arXiv:2112.09130 (2021)
20. Chen, T., Cheng, Y., Gan, Z., Liu, J., Wang, Z.: Data-efficient GAN training beyond (just) augmentations: a lottery ticket perspective. In: Advances in Neural Information Processing Systems 34 (2021)
21. Tseng, H.Y., Jiang, L., Liu, C., Yang, M.H., Yang, W.: Regularizing generative adversarial networks under limited data. In: Proceedings of the IEEE/CVF Conference on Computer Vision and Pattern Recognition, pp. 7921–7931 (2021)
22. Tran, N.T., Tran, V.H., Nguyen, N.B., Nguyen, T.K., Cheung, N.M.: On data augmentation for GAN training. IEEE Trans. Image Process. **30**, 1882–1897 (2021)
23. Jiang, L., Dai, B., Wu, W., Loy, C.C.: Deceive D: adaptive pseudo augmentation for GAN training with limited data. In: Thirty-Fifth Conference on Neural Information Processing Systems (2021)
24. He, K., Fan, H., Wu, Y., Xie, S., Girshick, R.: Momentum contrast for unsupervised visual representation learning. In: Proceedings of the IEEE/CVF Conference on Computer Vision and Pattern Recognition, pp. 9729–9738 (2020)
25. Chen, T., Kornblith, S., Norouzi, M., Hinton, G.: A simple framework for contrastive learning of visual representations. In: International Conference on Machine Learning, pp. 1597–1607. PMLR (2020)
26. Zhao, Z., Zhang, Z., Chen, T., Singh, S., Zhang, H.: Image augmentations for GAN training. arXiv preprint arXiv:2006.02595 (2020)
27. Wang, C., Xu, C., Yao, X., Tao, D.: Evolutionary generative adversarial networks. IEEE Trans. Evol. Comput. **23**(6), 921–934 (2019)

28. Tran, N.T., Tran, V.H., Nguyen, N.B., Nguyen, T.K., Cheung, N.M.: Towards good practices for data augmentation in GAN training. arXiv preprint arXiv:2006.05338 (2020)
29. Li, Z., et al.: A systematic survey of regularization and normalization in GANs. arXiv preprint arXiv:2008.08930 (2020)
30. Kong, C., Kim, J., Han, D., Kwak, N.: Smoothing the generative latent space with mixup-based distance learning. arXiv preprint arXiv:2111.11672 (2021)
31. Liu, B., Zhu, Y., Song, K., Elgammal, A.: Towards faster and stabilized GAN training for high-fidelity few-shot image synthesis. In: International Conference on Learning Representations (2020)
32. Zhang, Z., Chen, X., Chen, T., Wang, Z.: Efficient lottery ticket finding: less data is more. In: International Conference on Machine Learning, pp. 12380–12390. PMLR (2021)
33. Noguchi, A., Harada, T.: Image generation from small datasets via batch statistics adaptation. In: Proceedings of the IEEE/CVF International Conference on Computer Vision, pp. 2750–2758 (2019)
34. Wang, Y., Gonzalez-Garcia, A., Berga, D., Herranz, L., Khan, F.S., Weijer, J.v.d.: MineGAN: effective knowledge transfer from GANs to target domains with few images. In: Proceedings of the IEEE/CVF Conference on Computer Vision and Pattern Recognition, pp. 9332–9341 (2020)
35. Wang, Y., Wu, C., Herranz, L., van de Weijer, J., Gonzalez-Garcia, A., Raducanu, B.: Transferring GANs: generating images from limited data. In: Ferrari, V., Hebert, M., Sminchisescu, C., Weiss, Y. (eds.) ECCV 2018. LNCS, vol. 11210, pp. 220–236. Springer, Cham (2018). https://doi.org/10.1007/978-3-030-01231-1_14
36. Mo, S., Cho, M., Shin, J.: Freeze the discriminator: a simple baseline for fine-tuning GANs. arXiv preprint arXiv:2002.10964 (2020)
37. Grigoryev, T., Voynov, A., Babenko, A.: When, why, and which pretrained GANs are useful? In: International Conference on Learning Representations (2022)
38. Ojha, U., et al.: Few-shot image generation via cross-domain correspondence. In: Proceedings of the IEEE/CVF Conference on Computer Vision and Pattern Recognition, pp. 10743–10752 (2021)
39. Oord, A.v.d., Li, Y., Vinyals, O.: Representation learning with contrastive predictive coding. arXiv preprint arXiv:1807.03748 (2018)
40. Chen, X., He, K.: Exploring simple Siamese representation learning. In: Proceedings of the IEEE/CVF Conference on Computer Vision and Pattern Recognition, pp. 15750–15758 (2021)
41. Xiao, T., Wang, X., Efros, A.A., Darrell, T.: What should not be contrastive in contrastive learning. arXiv preprint arXiv:2008.05659 (2020)
42. Jeong, J., Shin, J.: Training GANs with stronger augmentations via contrastive discriminator (2021)
43. Chen, X., Fan, H., Girshick, R., He, K.: Improved baselines with momentum contrastive learning. arXiv preprint arXiv:2003.04297 (2020)
44. Cheung, T.H., Yeung, D.Y.: Modals: modality-agnostic automated data augmentation in the latent space. In: International Conference on Learning Representations (2020)
45. Karras, T., Laine, S., Aila, T.: A style-based generator architecture for generative adversarial networks. In: Proceedings of the IEEE/CVF Conference on Computer Vision and Pattern Recognition, pp. 4401–4410 (2019)
46. Robinson, J., Chuang, C.Y., Sra, S., Jegelka, S.: Contrastive learning with hard negative samples. arXiv preprint arXiv:2010.04592 (2020)

47. Kalantidis, Y., Sariyildiz, M.B., Pion, N., Weinzaepfel, P., Larlus, D.: Hard negative mixing for contrastive learning. In: Advances in Neural Information Processing Systems 33, pp. 21798–21809 (2020)
48. Cui, K., Huang, J., Luo, Z., Zhang, G., Zhan, F., Lu, S.: GenCo: generative co-training for generative adversarial networks with limited data. arXiv preprint arXiv:2110.01254 (2021)
49. Mangla, P., Kumari, N., Singh, M., Krishnamurthy, B., Balasubramanian, V.N.: Data InStance Prior (DISP) in generative adversarial networks. In: Proceedings of the IEEE/CVF Winter Conference on Applications of Computer Vision, pp. 451–461 (2022)
50. Si, Z., Zhu, S.C.: Learning hybrid image templates (HIT) by information projection. IEEE Trans. Pattern Anal. Mach. Intell. **34**(7), 1354–1367 (2011)
51. Heusel, M., Ramsauer, H., Unterthiner, T., Nessler, B., Hochreiter, S.: GANs trained by a two time-scale update rule converge to a local nash equilibrium. In: Advances in Neural Information Processing Systems 30 (2017)

BlobGAN: Spatially Disentangled Scene Representations

Dave Epstein[1(✉)], Taesung Park[2], Richard Zhang[2], Eli Shechtman[2], and Alexei A. Efros[1]

[1] UC Berkeley, Berkeley, USA
dave@eecs.berkeley.edu
[2] Adobe Research, San Jose, USA

Abstract. We propose an unsupervised, mid-level representation for a generative model of scenes. The representation is mid-level in that it is neither per-pixel nor per-image; rather, scenes are modeled as a collection of spatial, depth-ordered "blobs" of features. Blobs are differentiably placed onto a feature grid that is decoded into an image by a generative adversarial network. Due to the spatial uniformity of blobs and the locality inherent to convolution, our network learns to associate different blobs with different entities in a scene and to arrange these blobs to capture scene layout. We demonstrate this emergent behavior by showing that, despite training without any supervision, our method enables applications such as easy manipulation of objects within a scene (*e.g.* moving, removing, and restyling furniture), creation of feasible scenes given constraints (*e.g.* plausible rooms with drawers at a particular location), and parsing of real-world images into constituent parts. On a challenging multi-category dataset of indoor scenes, BlobGAN outperforms StyleGAN2 in image quality as measured by FID. See our project page for video results and interactive demo: http://www.dave.ml/blobgan.

Keywords: Scenes · Generative models · Mid-level representations

1 Introduction

The visual world is incredibly rich. It is so much more than the typical ImageNet-style photos of solitary, centered objects (cars, cats, birds, faces, *etc.*), which are the mainstays of most current paper result sections. Indeed, it was long clear, both in human vision [9,27] and in computer vision [19,28,56,58,91], that understanding and modeling objects within the context of a *scene* is of the utmost importance. Visual artists have understood this for centuries, first by discovering and following the rules of scene formation during the Renaissance, and then by expertly breaking such rules in the 20th century (cf. the surrealists including Magritte, Ernst, and Dalí).

Supplementary Information The online version contains supplementary material available at https://doi.org/10.1007/978-3-031-19784-0_36.

S. Avidan et al. (Eds.): ECCV 2022, LNCS 13675, pp. 616–635, 2022.
https://doi.org/10.1007/978-3-031-19784-0_36

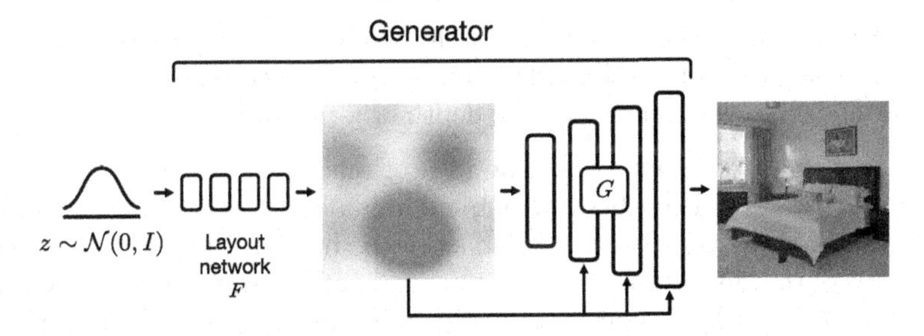

Fig. 1. In our generator, random noise is mapped by the layout network F to blob parameters. Blobs output by F are *splatted* spatially onto corresponding locations in the feature grid, used both as initial input and as spatially-adaptive modulation for the convolutional decoder G. Our blob representation automatically serves as a strong mid-level generative representation for scenes, discovering objects and their layouts.

However, in the current deep learning era, scene modeling for both analysis and synthesis tasks has largely taken a back seat. Images of scenes are either represented in a top-down fashion, no different from objects – *i.e.* for GANs or image classifiers, scene classes such as "bedrooms" or "kitchens" are represented the same way as object classes, such as "beds" or "chairs". Or, scenes are modeled in a bottom-up way by semantic labeling of each image pixel, *e.g.*, semantic segmentation, pix2pix [31], SPADE [59], *etc.* Both paths seem unsatisfactory because neither can provide easy ways of reasoning about parts of the scene as *entities*. The scene parts are either baked into a single entangled latent vector (top-down), or need to be grouped together from individual pixel labels (bottom-up).

In this paper, we propose an *unsupervised mid-level representation* for a generative model of scenes. The representation is mid-level in that it is neither per-pixel nor per-image; rather, scenes are modeled as a collection of spatial, depth-ordered Gaussian "blobs". This collection of blobs provides a bottleneck in the generative architecture, as shown in Fig. 1, forcing each blob to correspond to a specific object in the scene and thus causing a spatially disentangled representation to emerge. This representation allows us to perform a number of scene editing tasks (see Fig. 3) previously only achievable with extensive semantic supervision, if at all.

2 Related Work

Mid-level Scene Representations. Work on mid-level scene representations can be traced back to the 1970s, to the seminal papers of Yakimovsky and Feldman [91] and Ohta et al. [56], which already contained many key ideas including joint bottom-up segmentation and top-down reasoning. Other important developments were the line of work on normalized-cuts segmentation [20,78,96] and qualitative 3D scene interpretation [19,24,28,81] in the early 2000s. But most relevant to the present manuscript is the classic *Blobworld* work of Carson et al. [12], a region-based image retrieval system, with each image represented by a mixture-of-Gaussian blobs. Our model could be considered a generative version of this representation, except we also encode the depth ordering of the blobs.

Scene Analysis by Synthesis. The idea of modeling a complex visual scene by trying to generate it has been attempted a number of times in the past. Early methods, such as [84,85,87], introduced key ideas but were limited by the generative models of the time. To address this, several approaches tried non-parametric generation [30,47,73], with Scene Collaging [30] the most valiant attempt, showing layered scene representations despite very heavy computational burden. With the advancement of deep generative models, parametric analysis-by-synthesis techniques are having a renaissance, with some top-down [57,93,95] as well as bottom-up [22,62] techniques.

Conditional Image Generation. Conditional GANs [29,89,103], such as image-to-image translation setups [31], predict an image from a predefined representation, *e.g.* semantic segmentation maps [45,59], object-attribute graphs [8,33], text [53,66–68,71,99], pose [3,43,75], and keypoints [11]. Other setups include using perceptual losses [14], implicit likelihood estimation [44], and more recently, diffusion models [50,74]. [23,48,49,79,80,88] explore related intermediate representations to help generation (mostly of humans or objects) but none provide the ability to generate and manipulate high-quality scene images of our method.

Unconditional Generation and Disentanglement. Rather than use explicit conditioning, it is possible to learn an image "manifold" with a generative model such as a VAE [26,41] or GAN [18] and explore emergent capabilities. GANs have improved in image quality [10,16,35–38,65,98] and are our focus. Directions of variation naturally emerge in the latent space and can be discovered when guided by geometry/color changes [32], language or attributes [2,61,65,76,90], cognitive signals [17], or in an unsupervised manner [21,63,77]. Discovering disentangled representations remains a challenging open problem [46]. To date, most successful applications have been on data of objects, *e.g.* faces and cars, or changing textures for scenes [60]. Similar to us, an active line of work explores adding 3D inductive biases [51,52,54], but individual object manipulation has largely focused on simple diagnostic scenes [34]. Alternatively, the internal units of a pretrained GAN offer finer spatial control, with certain units naturally correlating with object classes [5,6,92]. The internal compositionality of GANs can be leveraged to harmonize images [13,15] or perform a limited set of edits on objects in a scene [6,97,101]. Crucially, while these works require semantic supervision to identify units and regions, our work uses a representation where these factors naturally emerge.

3 Method

Our method aims to learn a representation of scenes as spatial maps of blobs through the generative process. As shown in Fig. 1, a layout network maps from random noise to a set of blob parameters. Then, blobs are differentiably splatted onto a spatial grid – a "blob map" – which a StyleGAN2-like decoder [39] converts into an image. Finally, the blob map is used to modulate the decoder We train our model in an adversarial framework with an unmodified discriminator [38]. Interestingly, even without explicit labels, our model learns to decompose scenes into entities and their layouts.

Our generator model is largely divided into two parts. First, we apply an 8-layer MLP F to map random noise $z \in \mathbb{R}^{d_{\text{noise}}} \sim \mathcal{N}(0, I_d)$ to a collection of blobs parameterized by $\boldsymbol{\beta} = \{\beta_i\}_{i=1}^k$ which are splatted onto a spatial $H \times W \times d$ feature grid in a differentiable manner. This process is visualized in Fig. 2. The feature grid is then passed to a convolutional decoder G to produce final output images. In the remainder of this section, we describe the design of our representation as well as its implementation in detail.

3.1 From Noise to Blobs as Layout

We map from random Gaussian noise to distributions of blobs with an MLP F with dimension d_{hidden}. The last layer of F is decoded into a sequence of blob properties $\boldsymbol{\beta}$. We opt for a simple yet effective parametrization of blobs, representing them as ellipses by their center coordinates $x \in [0,1]^2$, scale $s \in \mathbb{R}$, aspect ratio $a \in \mathbb{R}$, and rotation angle $\theta \in [-\pi, \pi]$. Each blob is also associated with a structure feature $\phi \in \mathbb{R}^{d_{\text{in}}}$ and a style feature $\psi \in \mathbb{R}^{d_{\text{style}}}$. Altogether, our blob representation is:

$$\beta \in \mathbb{R}^{2+1+1+1+d_{\text{in}}+d_{\text{style}}} \triangleq (x, s, a, \theta, \phi, \psi)$$

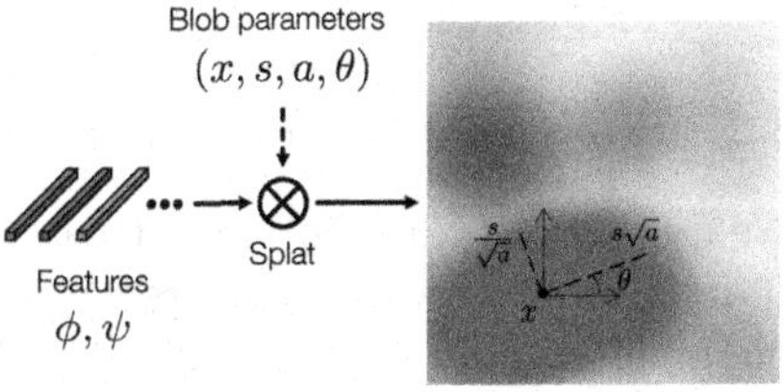

Blob parameters
(x, s, a, θ)
Splat
Features
ϕ, ψ

Fig. 2. Our elliptical blobs β are parametrized by centroid x, scale s, aspect ratio a, and angle θ. We composite multiple blobs with alpha values that smoothly decay from blob centers. The features ϕ or ψ are splatted on their corresponding blobs and passed to the decoder.

Next, we transform the blob parameters to a 2D feature grid by populating the ellipse specified by β with the feature vectors ϕ and ψ. We do this differentiably by assigning an opacity and spatial falloff to each blob. Specifically, we calculate a grid $\boldsymbol{\alpha} \in [0,1]^{H \times W \times k}$ which indicates each blob's opacity value at each location. We then use these opacity maps to *splat* the features ϕ, ψ at various resolutions, using a single broadcasted matrix multiplication operation.

In more detail, we begin by computing per-blob opacity maps $o \in [0,1]^{H \times W}$. For each grid location $x_{\text{grid}} \in \left\{\left(\frac{w}{W}, \frac{h}{H}\right)\right\}_{w=1,h=1}^{W,H}$ we find the squared Mahalanobis distance to the blob center x:

$$d(x_{\text{grid}}, x) = (x_{\text{grid}} - x)^T (R \Sigma R^T)^{-1} (x_{\text{grid}} - x), \tag{1}$$

where $\Sigma = c \begin{bmatrix} a & 0 \\ 0 & \frac{1}{a} \end{bmatrix}$, R is a 2D rotation matrix by angle θ, and $c = 0.02$ controls blob edge sharpness. The opacity of a blob at a given grid location is then:

$$o(x_{\text{grid}}) = \sigma\left(s - d(x_{\text{grid}}, x)\right), \tag{2}$$

where s acts as a control of blob size by shifting inputs to the sigmoid. Intuitively, this can be thought of as taking a soft thresholding operation on a Gaussian to define an in-region and an out-region. For example, our model can output a large negative $s < 0$ to effectively "turn off" a blob. Rather than taking the softmax to normalize values at each location, we use the alpha-compositing operation [64], which allows us to model occlusions and object relationships more naturally by imposing a 2.1-D z-ordering [55]:

$$\alpha_i(x_{\text{grid}}) = o_i(x_{\text{grid}}) \prod_{j=i+1}^{k} (1 - o_j(x_{\text{grid}})). \tag{3}$$

Lastly, our blob mapping network also outputs background vectors ϕ_0, ψ_0, with a fixed opacity $o_0 = 1$. Final features at each grid location are the convex combination of blob feature vectors, given by the (k+1) α_i scores.

3.2 From Blob Layouts to Scene Images

We now describe a function G that converts the representation of scenes as blobs $\boldsymbol{\beta}$ described in Sect. 3.1 into realistic, harmonized images. To do so, we build on the architecture of StyleGAN2 [38]. We modify it to take in a spatially-varying input tensor based on blob structure features rather than a single, global vector, and perform spatially-varying style modulation.

As opposed to standard StyleGAN, where the single style vector w must capture information about all aspects of the scene, our representation separates layout (blob locations and sizes) and appearance (per-blob feature vectors) by construction, naturally providing a foundation for a disentangled representation.

Concretely, we compute a feature grid Φ at 16×16 resolution using blob structure vectors ϕ_i and use Φ as input to G, removing the first two convolutional blocks of the base architecture to accommodate the increased resolution. We also apply spatial style-based modulation [59] at each convolution using feature grids $\Psi_{l\times l}$ for $l \in \{16, 32, \ldots, 256\}$ computed from blob style vectors ψ_i.

3.3 Encouraging Disentanglement

Intuitively, all activations within a blob are governed by the same feature vector, encouraging blobs to yield image regions of self-similar properties, *i.e.* entities in a scene. Further, due to the locality of convolution, the layout of blobs in the input must strongly inform the final arrangement of image regions. Finally, our latent space separates layout (blob location, shape, and size) from appearance (blob features) by construction. All the above help our model learn to bind individual blobs to different objects and arrange these blobs into a coherent layout, disentangling scenes spatially into their component parts.

To further nudge our network in this direction, we stochastically perturb blob representations $\boldsymbol{\beta}$ before inputting them to G, enforcing our model to be robust under these perturbations. We implement this by corrupting blob parameters with uniform noise δx, δs, and $\delta\theta$. This requires that blobs be independent of

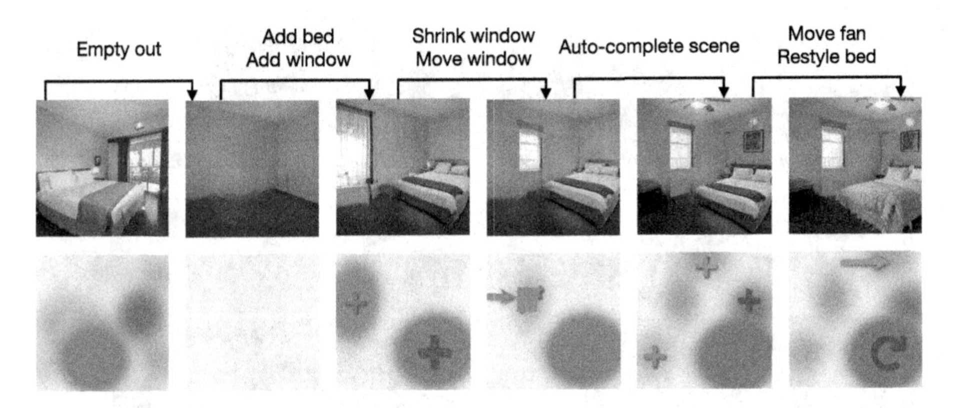

Fig. 3. Blobs allow extensive image manipulation: We apply a sequence of modifications to the blob map of an image generated by our model and show the resulting images outputs at each stage of the editing process, demonstrating the strength of our learned representation.

each other, promoting object discovery and discouraging degenerate solutions which rely on precise blob placement or shape.

We also experiment with style mixing, where with probability 0.2 we uniformly sample between 0 and k blobs to swap, and permute style vectors for these blobs among different batch samples. We find that this intervention harms our training process since it requires that all styles match all layouts, an assumption we show does not hold in Sect. 4.3. We also try randomly removing blobs from the forward process with some probability, but found this hurts training, since certain objects must always be present in certain kinds of scenes (*e.g.* kitchens are unlikely to have no refrigerator). This constraint led to a more distributed, and therefore less controllable, representation of scene entities.

4 Experiments

We evaluate our learned representation quantitatively and qualitatively and demonstrate that a spatially disentangled representation of scenes emerges. We begin by showing that our model learns to associate individual blobs with objects in scenes, and then show that our representation captures the distribution of scene layouts. We highlight some applications of our model in Fig. 3. Finally, we use our model to parse the layouts of real scene images via inversion. For more results, including on additional datasets and ablations, please see Appendix.

4.1 Training and Implementation

We largely follow the training procedure set forth in StyleGAN2 [38], with non-saturating loss [18], R1 regularization every 16 steps with $\gamma = 100$ but no path length regularization, and exponential moving average of model weights [35]. We use the Adam optimizer [40] with learning rate 0.002 and implement equalized learning rate for stability purposes as recommended by [35,38].

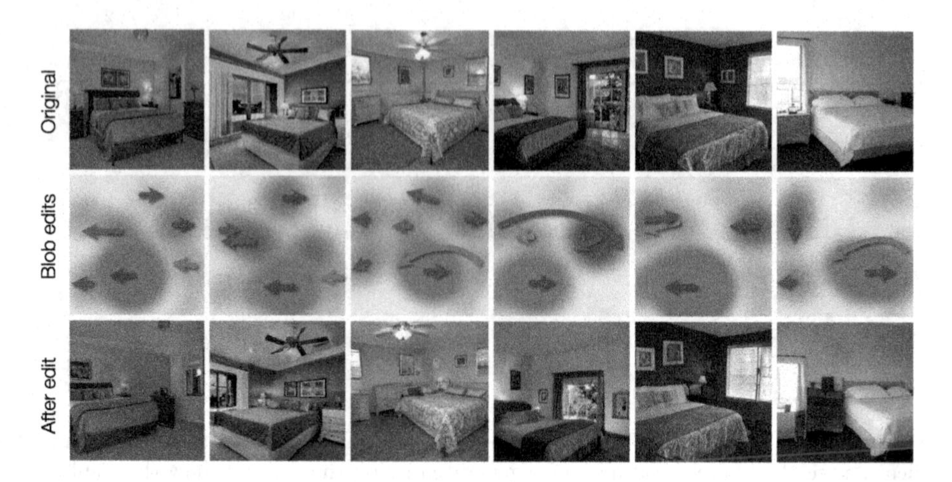

Fig. 4. Moving blobs to rearrange objects: By modifying coordinates x_i of certain blobs as shown in the middle row, we can perform operations such as rearranging furniture in bedrooms. Note that since our representation is layered, we can model occlusions, such as the bed and the dresser in the leftmost and rightmost images.

We set $d_{\text{noise}} = 512$. Our layout generator F is an 8-layer MLP with $d_{\text{hidden}} = 1024$ and leaky ReLU activations. We L2-normalize ϕ and ψ vectors output by the layout generator before splatting. Altogether, the dimension of the last layer is $d_{\text{out}} = k(d_{\text{in}} + d_{\text{style}} + 5) + d_{\text{in}} + d_{\text{style}}$. To compensate for the removal of the first two convolutional blocks in the generator G, we increase channel widths at all remaining layers by 50%. We set d_{in} and d_{style} depending on the number of blobs k, and values range between 256 and 768. We experiment with $k \in [5, 50]$ depending on the data considered. We set the blob sigmoid temperature $c = 0.02$ by visual inspection of blob edge hardness. Model performance is relatively insensitive to jittering parameters. We perturb blob parameters with uniform noise as $\delta x \in [-0.04, 0.04]$ (around 10px at 256px resolution), $\delta\theta \in [-0.1, 0.1]$ rad (around 6°), and $\delta s \in [-0.5, 0.5]$ (varying radii of blobs by around 5px).

We train our model on categories from the LSUN scenes dataset [94]. In particular, we train models on bedrooms; conference rooms; and the union of kitchens, living rooms, and dining rooms. In the following section, we show results of models trained on bedroom data with $k = 10$ blobs. Please see Appendix for results on more data (§A), further details on our blob parametrization and its implementation (§C), and ablations (§E).

4.2 Discovering Entities with Blobs

The ideal representation is able to disentangle complex images of scenes into the objects that comprise them. Here, we demonstrate through various image manipulation applications that this ability emerges in our model. Our unsupervised representation allows effortless rearrangement, removal, cloning, and restyling of

Fig. 5. Removing blobs: Despite the extreme rarity of bedless bedrooms in the training data, the ability to remove beds from scenes by removing corresponding blobs emerges. We can also remove windows, lamps and fans, paintings, dressers, and nightstands in the same manner.

objects in scenes. We also measure correlation between blob presence and semantic categories as predicted by an off-the-shelf network and thus empirically verify the associations discovered by our model.

Figure 4 shows the result of intervening to manipulate the center coordinates x_i of blobs output by our model, and thus **rearranging furniture configurations**. We are able to arbitrarily alter the position of objects in the scene by shifting their corresponding blobs without affecting their appearance. This interaction is related to traditional "image reshuffling" where rearrangement of image content is done in pixel space [4,72,82]. Our model's notion of depth ordering also allows us to easily de-occlude objects – *e.g.* curtains, dressers, or nightstands – that were hidden in the original images, while also enabling the introduction of new occlusions by moving one blob behind another.

In Fig. 5, we show the effect of **removing entirely certain blobs** from the representation. Specifically, we remove all blobs but the one responsible for beds, and show that our model is able to clear out the room accordingly. We also remove the bed blob but leave the rest of the room intact, showing a remarkable ability to create bedless bedrooms, despite training on a dataset of rooms with beds. Figure 3 shows the effect of resizing blobs to change window size; see Appendix A for further results on changing blob size and shape. In Fig. 6, we remove a blob that our model – trained on a challenging multi-category union of scene datasets – has learned to associate with tables across scene categories.

Our edits are not constrained to the set of blobs present in a layout generated by our model; we can also introduce new blobs. Figure 8 demonstrates the impact of **copying and pasting the same blob** in a new location. Our model is able to faithfully duplicate objects in scenes even when the duplication yields an image that is out of distribution, such as a room with two ceiling fans.

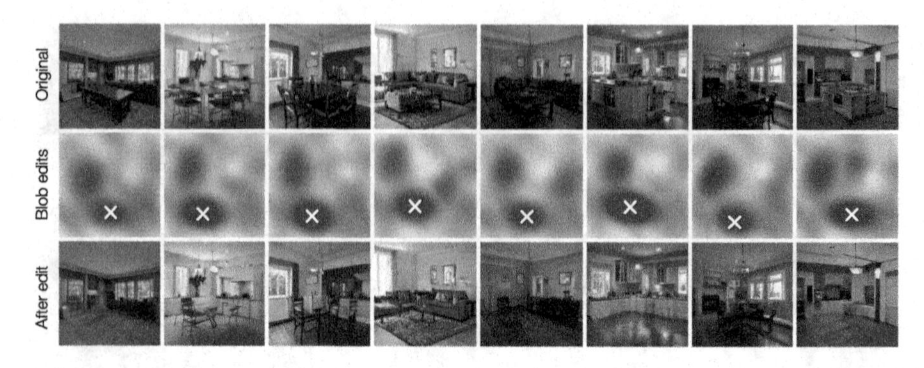

Fig. 6. Removing all sorts of tables: We train BlobGAN on a multi-category dataset of kitchens, living rooms, and dining rooms. We find that a particular blob specializes in generating tables across scene types, and feature vectors dictate whether it becomes a coffee table, kitchen island, or dining table. For many more editing operations on this dataset and others, please see Appendix.

Fig. 7. Swapping blob styles: Interchanging ψ_i vectors without modifying layout leads to localized edits which change the appearance of individual objects in the scene.

Our representation also allows performing edits across images. Figure 7 shows the **highly granular redecorating** enabled by swapping blob style vectors; we are able to copy objects such as bedsheets, windows, and artwork from one room to another without otherwise affecting the rendered scene.

Quantitative Blob Analysis. Next, we quantitatively study the strong associations between blobs and semantic object classes. We do so by randomly setting the size parameter s of a blob to a large negative number to effectively remove it. We then use an off-the-shelf segmentation model to measure which semantic class has disappeared. We visualize the correlation between classes and blobs in Fig. 9 *(left)*; the sparsity of this matrix shows that blobs learn to specialize into distinct scene entities. We also visualize the distribution of blob centroids in Fig. 9 *(right)*, computed by sampling many different random vectors z. The resultant heatmaps provide a glimpse into the distribution of objects in training data – our model learns to locate blobs at specific image regions and control the objects they represent by varying feature vectors.

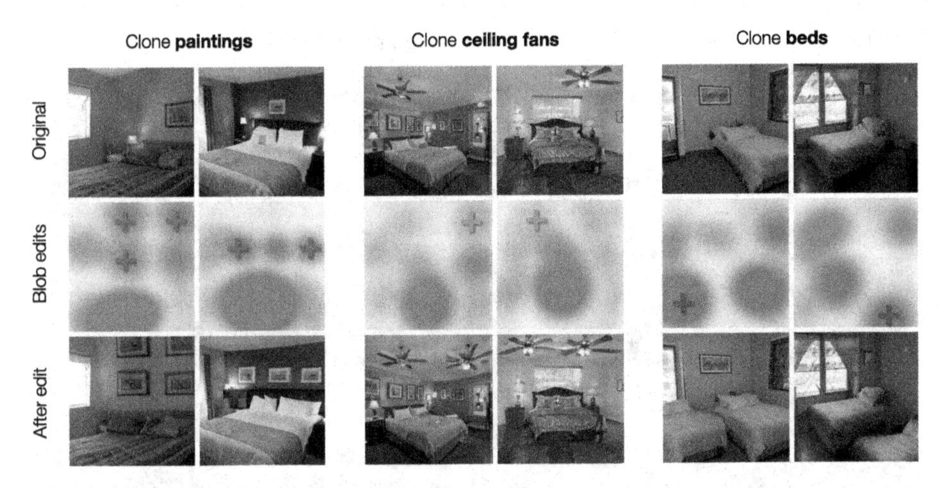

Fig. 8. Cloning blobs: We clone blobs in scenes, arrange them to form a new layout, and show corresponding model outputs. Added blobs are marked with a +.

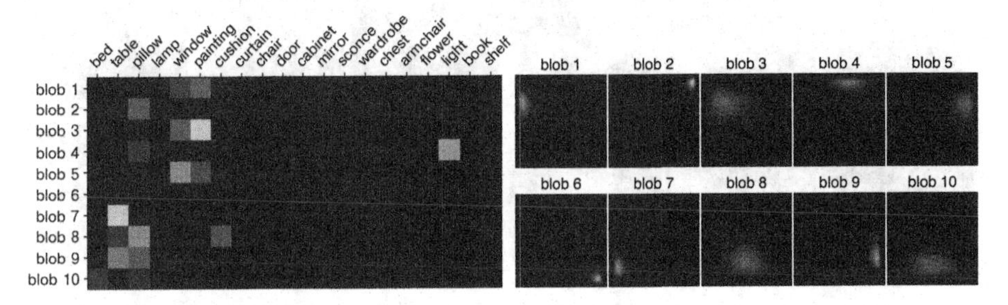

Fig. 9. Blob spatial preferences: Our model allocates each blob to a certain region of the image canvas, revealing patterns in the training distribution of objects. We visualize each blob's correlation with classes predicted by a segmentation model [83] *(left)* as well as the spatial distribution of blob centroids *(right)*.

4.3 Composing Blobs into Layouts

The ideal representation of scenes must go beyond simply disentangling images into their component parts, and capture the rich contextual relationships between these parts that dictate the process of scene formation [9,27]. In contrast to previous work in generative modeling of realistic images, our representation explicitly discovers the layout (*i.e.*, the joint distribution) of objects in scenes.

By solving a simple constrained optimization problem at test-time, we are able to sample realistic images that satisfy constraints about the underlying scene, a functionality we call "scene auto-complete". This auto-complete allows applications such as filling empty rooms with items, plausibly populating rooms given a bed or window at a certain location, and finding layouts that are compatible with certain sets of furniture.

We ground this ability quantitatively by demonstrating that "not everything goes with everything" [11] in real-world scenes – for example, not every room's

Fig. 10. Generating and populating empty rooms: We show different empty rooms, each with their own background vector ψ_0, as well as furnished rooms given by latents z optimized to match these background vectors. This simple sampling procedure yields a diverse range of layouts to fill the scenes. Note that while empty rooms do not appear in training data, our model is reasonably capable of generating them.

style can be combined with any room's layout. We show that our scene autocomplete yields images that are significantly more photorealistic than naïvely combining scene properties at random, and outperforms regular StyleGAN in image quality, diversity as well as in fidelity of edits.

Conditionally Sampling Scenes: We can construct an ad-hoc conditional distribution by optimizing random inputs to match a set of constraints in the form of properties c of a source image's blob map $\boldsymbol{\beta}$:

$$c \subset \bigcup_{i=0}^{k} \{x_i^{\text{src}}, s_i^{\text{src}}, a_i^{\text{src}}, \theta_i^{\text{src}}, \phi_i^{\text{src}}, \psi_i^{\text{src}}\} \tag{4}$$

For example, $c = \{\phi_0^{\text{src}}, \psi_0^{\text{src}}\}$ constrains the background of an output image to match that of a source image, and $c = \{x_i^{\text{src}}, s_i^{\text{src}}, a_i^{\text{src}}\}$ constrains the shape (but not the appearance) of the i-th blob to match the source.

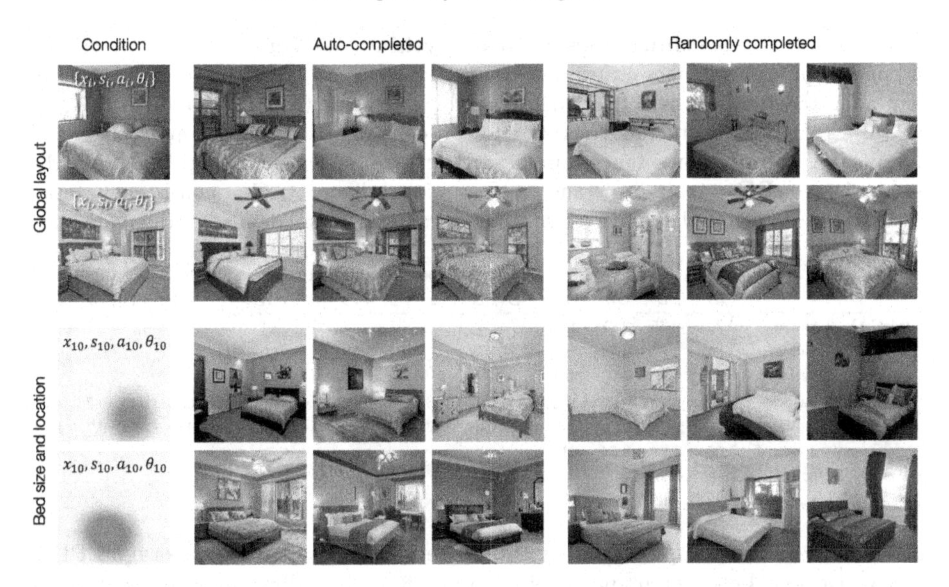

Fig. 11. Scene auto-complete: Various conditional generation problems fall under the umbrella of "scene auto-complete", *i.e.* using our layout network F to sample different scenes satisfying constraints on a subset of blob parameters. We show layout-conditioned style generation as well as prediction of plausible scenes given the location and size (but not style) of beds. Rather than using F to plausibly auto-complete scenes, we can also generate a random scene and simply override parameters of interest to match desired values. As shown on the right, such scenes have objects inserted, removed, reoriented, or otherwise disfigured due to incompatibility.

We obtain conditional samples by drawing initial noise vectors $z^{\text{init}} \sim \mathcal{N}(0, I_d)$ and optimizing $F(z^{\text{init}})$ to match the constraint set c with an L2 loss, leaving other parameters free. We use the Adam optimizer with learning rate 0.01 and find that between 50 and 300 iterations, which complete in around a second on an NVIDIA RTX 3090, give z^{optim} vectors that sufficiently match constraints. We then set the final layout to be $c \cup \{\boldsymbol{\beta}^{\text{optim}} \backslash c\}$, *i.e.* the initial constraints combined with the free parameters given by the optimized noise vectors, and decode layouts into images as described in Sect. 3.2. In effect, this process finds new scenes known by our model to be compatible with the specified constraints, as opposed to randomly drawn from an unconditional distribution.

We examine applications of our scene auto-complete and compare it to scenes generated by baseline approaches in Figs. 10 and 11. Scene auto-complete yields images that are both more realistic and more faithful to the desired image operations. We quantitatively demonstrate this in Table 1, where we show that using auto-complete to find target images whose properties to apply for conducting edits significantly outperforms the use of randomly sampled targets and/or models such as StyleGAN not trained with compositionality in mind.

To evaluate image photorealism after an edit, we calculate FID [25] on automatically edited images. We must also ensure that image quality does not come

Table 1. Not everything goes with everything: We edit images by overriding properties in target images either generated at random or conditionally sampled using our model. By varying the network depth at which we begin to swap styles in StyleGAN, we tune a knob between image quality and edit consistency. To further preserve global layout and improve consistency, our model can also use structure grids Φ from the source image. **PD** = paired distance, **GD** = global diversity, **C** = consistency. In all cases, scene auto-complete outperforms baselines. Metrics are defined in the main text.

		Layout → Styles				Window → Room				Bed → Room				Painting → Room			
		FID ↓	PD ↑	GD ↑	C ↑	FID	PD	GD	C	FID	PD	GD	C	FID	PD	GD	C
StyleGAN	2 coarse	4.23	0.75	0.77	46.5	–	–	–	–	–	–	–	–	–	–	–	–
	3 coarse	5.04	0.73	0.76	55.3	–	–	–	–	–	–	–	–	–	–	–	–
	4 coarse	5.58	0.71	0.76	62.9	–	–	–	–	–	–	–	–	–	–	–	–
BlobGAN	Random	8.10	0.72	0.74	47.9	6.41	0.72	0.73	17.4	10.88	0.67	0.73	52.6	6.31	0.72	0.73	7.7
	Conditional	4.59	0.70	0.74	55.2	4.75	0.67	0.72	27.2	7.12	0.64	0.72	60.0	4.58	0.69	0.73	13.0
	+ source Φ	5.06	0.68	0.74	63.6	–	–	–	–	–	–	–	–	–	–	–	–

at the expense of sample diversity; to this end, we measure the average LPIPS [100] distance between images before and after the edit and refer to this as Paired Distance (**PD**). We also measure the expected distance between pairs of edited images to gauge whether edits cause perceptual mode collapse, and call this Global Diversity (**GD**). Finally, we confirm that our editing operations stay faithful to the conditioning provided. For predicting style from layout, we simply report the fraction of image pixels whose predicted class label as output by a segmentation network [83] remains the same. For localized object edits, we report the intersection-over-union of the set of pixels whose prediction was the target class before and after edit. We refer to this metric as Consistency (**C**).

Our results verify the intuition that, *e.g.*, not every configuration of furniture can fit a bed at a given location. Please see Appendix D for more results.

4.4 Evaluating Visual Quality and Diversity

Our model achieves perceptual realism competitive with previous work. In Table 2, we report FID [25] as well as improved precision and recall [42], which capture realism and diversity of samples. Bedroom images generated by our model appear more realistic than StyleGAN's [37], but less diverse. We hypothesize this is due to the design of our representation, which rejects strange scene configurations that cannot be modeled by blobs. When trained on the challenging union of multiple LSUN indoor scene categories, *BlobGAN outperforms StyleGAN2*, indicating an ability to scale to harder data. See Appendix A for details.

Table 2. BlobGAN achieves visual quality competitive with StyleGAN2 [39] on LSUN Bedrooms. Our samples are more realistic but capture less of the data distribution [42], perhaps by rejecting unconventional or malformed scenes in the training data.

	FID ↓	Precision ↑	Recall ↑
StyleGAN2	3.85	0.5932	0.4492
BlobGAN	3.43	0.5974	0.4463

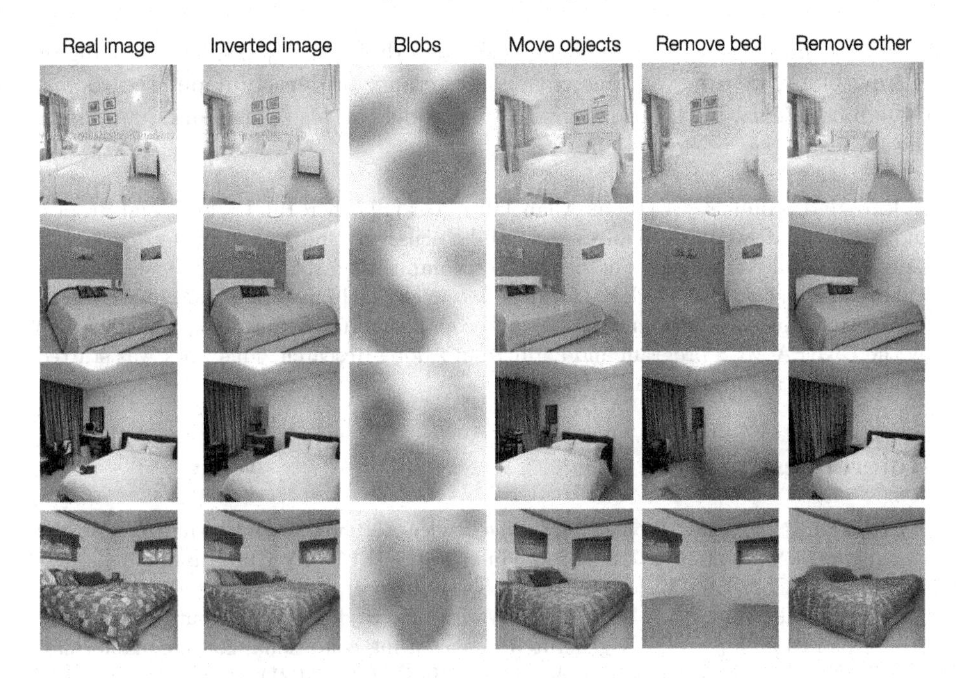

Fig. 12. Parsing real images via inversion: Our representation can also parse real images by inverting them into blob space. We can remove and reposition objects in real images – spot the differences from the original!

4.5 Parsing Images into Regions

Though our representation is learned on generated (*i.e.* fake) images, in Fig. 12 we show that it can represent real images via inversion. We follow best practices [1,7,69,86,102] for inversion: We train an encoder to predict blob parameters, reconstructing both real and fake images, and then optimize encoder predictions to better reconstruct specific inputs. While this method leads to uneditable, off-manifold latents in previous work [70], we find our blob representation to be more robust in this sense and amenable to naïve optimization. Importantly, we find that the same manipulations described above can be readily applied to real images after inversion. See Appendix B for more information.

5 Conclusion

We present BlobGAN, a mid-level representation for generative modeling and parsing of scenes. Taking random noise as input, our model first outputs a set of spatial, depth-ordered blobs, and then splats these blobs onto a feature grid. This feature grid is used as input to a convolutional decoder which outputs images. While conceptually simple, this approach leads to the emergence of a disentangled representation that discovers entities in scenes and their layout. We demonstrate a set of edits enabled by our approach, including rearranging layouts by moving blobs and editing styles of individual objects. By removing or cloning blobs, we are even able to generate empty or densely populated rooms, though none exist in the training set. Our model can also parse and manipulate the layout of real images via inversion.

Acknowledgements. We thank Allan Jabri, Assaf Shocher, Bill Peebles, Tim Brooks, and Yossi Gandelsman for endless insightful discussions and important feedback, and especially thank Vickie Ye for advice on blob compositing, splatting, and visualization. Thanks also to Georgios Pavlakos for deadline-week pixel inpsection and Shiry Ginosar for post-deadline-week guidance and helpful comments. Research was supported in part by the DARPA MCS program and a gift from Adobe Research. This work was started while DE was an intern at Adobe.

References

1. Abdal, R., Qin, Y., Wonka, P.: Image2StyleGAN: how to embed images into the StyleGAN latent space? In: Proceedings of the IEEE/CVF International Conference on Computer Vision, pp. 4432–4441 (2019)
2. Abdal, R., Zhu, P., Mitra, N.J., Wonka, P.: StyleFlow: attribute-conditioned exploration of StyleGAN-generated images using conditional continuous normalizing flows. ACM Trans. Graph. (TOG) **40**(3), 1–21 (2021)
3. AlBahar, B., Lu, J., Yang, J., Shu, Z., Shechtman, E., Huang, J.B.: Pose with style: detail-preserving pose-guided image synthesis with conditional StyleGAN. ACM Trans. Graph. **40**, 1–11 (2021)
4. Barnes, C., Shechtman, E., Finkelstein, A., Goldman, D.B.: PatchMatch: a randomized correspondence algorithm for structural image editing. ACM Trans. Graph. **28**(3), 24 (2009)
5. Bau, D., Liu, S., Wang, T., Zhu, J.-Y., Torralba, A.: Rewriting a deep generative model. In: Vedaldi, A., Bischof, H., Brox, T., Frahm, J.-M. (eds.) ECCV 2020. LNCS, vol. 12346, pp. 351–369. Springer, Cham (2020). https://doi.org/10.1007/978-3-030-58452-8_21
6. Bau, D., et al.: Gan dissection: visualizing and understanding generative adversarial networks. arXiv preprint arXiv:1811.10597 (2018)
7. Bau, D., et al.: Seeing what a GAN cannot generate. In: Proceedings of the IEEE/CVF International Conference on Computer Vision, pp. 4502–4511 (2019)
8. Bear, D., et al.: Learning physical graph representations from visual scenes. Adv. Neural. Inf. Process. Syst. **33**, 6027–6039 (2020)
9. Biederman, I.: On the semantics of a glance at a scene (1981)
10. Brock, A., Donahue, J., Simonyan, K.: Large scale GAN training for high fidelity natural image synthesis (2018)
11. Brooks, T., Efros, A.A.: Hallucinating pose-compatible scenes. arXiv preprint arXiv:2112.06909 (2021)
12. Carson, C., Thomas, M., Belongie, S., Hellerstein, J.M., Malik, J.: Blobworld: a system for region-based image indexing and retrieval. In: Huijsmans, D.P., Smeulders, A.W.M. (eds.) VISUAL 1999. LNCS, vol. 1614, pp. 509–517. Springer, Heidelberg (1999). https://doi.org/10.1007/3-540-48762-X_63
13. Chai, L., Wulff, J., Isola, P.: Using latent space regression to analyze and leverage compositionality in GANs. arXiv preprint arXiv:2103.10426 (2021)

14. Chen, Q., Koltun, V.: Photographic image synthesis with cascaded refinement networks. In: Proceedings of the IEEE International Conference on Computer Vision, pp. 1511–1520 (2017)
15. Collins, E., Bala, R., Price, B., Susstrunk, S.: Editing in style: uncovering the local semantics of GANs. In: Proceedings of the IEEE/CVF Conference on Computer Vision and Pattern Recognition, pp. 5771–5780 (2020)
16. Denton, E.L., Chintala, S., Fergus, R., et al.: Deep generative image models using laplacian pyramid of adversarial networks. Adv. Neural Inf. Process. Syst. **28** (2015)
17. Goetschalckx, L., Andonian, A., Oliva, A., Isola, P.: GANalyze: toward visual definitions of cognitive image properties. In: Proceedings of the IEEE/CVF International Conference on Computer Vision, pp. 5744–5753 (2019)
18. Goodfellow, I., et al.: Generative adversarial nets. Adv. Neural Inf. Process. Syst. **27** (2014)
19. Gupta, A., Efros, A.A., Hebert, M.: Blocks world revisited: image understanding using qualitative geometry and mechanics. In: Daniilidis, K., Maragos, P., Paragios, N. (eds.) ECCV 2010. LNCS, vol. 6314, pp. 482–496. Springer, Heidelberg (2010). https://doi.org/10.1007/978-3-642-15561-1_35
20. Hariharan, B., Arbeláez, P., Girshick, R., Malik, J.: Simultaneous detection and segmentation. In: Fleet, D., Pajdla, T., Schiele, B., Tuytelaars, T. (eds.) ECCV 2014. LNCS, vol. 8695, pp. 297–312. Springer, Cham (2014). https://doi.org/10.1007/978-3-319-10584-0_20
21. Härkönen, E., Hertzmann, A., Lehtinen, J., Paris, S.: GANSpace: discovering interpretable GAN controls. Adv. Neural. Inf. Process. Syst. **33**, 9841–9850 (2020)
22. He, K., Chen, X., Xie, S., Li, Y., Dollár, P., Girshick, R.: Masked autoencoders are scalable vision learners. arXiv preprint arXiv:2111.06377 (2021)
23. He, X., Wandt, B., Rhodin, H.: Latentkeypointgan: controlling GANs via latent keypoints. arXiv preprint arXiv:2103.15812 (2021)
24. Hedau, V., Hoiem, D., Forsyth, D.: Recovering the spatial layout of cluttered rooms. In: 2009 IEEE 12th International Conference on Computer Vision, pp. 1849–1856. IEEE (2009)
25. Heusel, M., Ramsauer, H., Unterthiner, T., Nessler, B., Hochreiter, S.: GANs trained by a two time-scale update rule converge to a local nash equilibrium. Adv. Neural Inf. Process. Syst. **30** (2017)
26. Higgins, I., et al.: Beta-VAE: learning basic visual concepts with a constrained variational framework (2016)
27. Hock, H.S., Romanski, L., Galie, A., Williams, C.S.: Real-world schemata and scene recognition in adults and children. Mem. Cogn. **6**(4), 423–431 (1978)
28. Hoiem, D., Efros, A.A., Hebert, M.: Recovering surface layout from an image. IJCV **75**(1), 151–172 (2007)
29. Huang, X., Liu, M.-Y., Belongie, S., Kautz, J.: Multimodal unsupervised image-to-image translation. In: Ferrari, V., Hebert, M., Sminchisescu, C., Weiss, Y. (eds.) ECCV 2018. LNCS, vol. 11207, pp. 179–196. Springer, Cham (2018). https://doi.org/10.1007/978-3-030-01219-9_11
30. Isola, P., Liu, C.: Scene collaging: analysis and synthesis of natural images with semantic layers. In: ICCV (2013)
31. Isola, P., Zhu, J.Y., Zhou, T., Efros, A.A.: Image-to-image translation with conditional adversarial networks. In: Proceedings of the IEEE Conference on Computer Vision and Pattern Recognition, pp. 1125–1134 (2017)
32. Jahanian, A., Chai, L., Isola, P.: On the "steerability" of generative adversarial networks. arXiv preprint arXiv:1907.07171 (2019)

33. Johnson, J., Gupta, A., Fei-Fei, L.: Image generation from scene graphs. In: Proceedings of the IEEE Conference on Computer Vision and Pattern Recognition, pp. 1219–1228 (2018)
34. Johnson, J., Hariharan, B., Van Der Maaten, L., Fei-Fei, L., Lawrence Zitnick, C., Girshick, R.: CLEVR: a diagnostic dataset for compositional language and elementary visual reasoning. In: Proceedings of the IEEE Conference on Computer Vision and Pattern Recognition, pp. 2901–2910 (2017)
35. Karras, T., Aila, T., Laine, S., Lehtinen, J.: Progressive growing of GANs for improved quality, stability, and variation (2018)
36. Karras, T., et al.: Alias-free generative adversarial networks. Adv. Neural Inf. Process. Syst. **34** (2021)
37. Karras, T., Laine, S., Aila, T.: A style-based generator architecture for generative adversarial networks. In: Proceedings of the IEEE/CVF Conference on Computer Vision and Pattern Recognition, pp. 4401–4410 (2019)
38. Karras, T., Laine, S., Aittala, M., Hellsten, J., Lehtinen, J., Aila, T.: Analyzing and improving the image quality of StyleGAN. In: Proceedings of the IEEE/CVF Conference on Computer Vision and Pattern Recognition, pp. 8110–8119 (2020)
39. Karras, T., Laine, S., Aittala, M., Hellsten, J., Lehtinen, J., Aila, T.: Analyzing and improving the image quality of StyleGAN (2020)
40. Kingma, D.P., Ba, J.: Adam: a method for stochastic optimization. arXiv preprint arXiv:1412.6980 (2014)
41. Kingma, D.P., Welling, M.: Auto-encoding variational bayes. arXiv preprint arXiv:1312.6114 (2013)
42. Kynkäänniemi, T., Karras, T., Laine, S., Lehtinen, J., Aila, T.: Improved precision and recall metric for assessing generative models. Adv. Neural Inf. Process. Syst. **32** (2019)
43. Lewis, K.M., Varadharajan, S., Kemelmacher-Shlizerman, I.: TryonGAN: body-aware try-on via layered interpolation. ACM Trans. Graph. (Proc. ACM SIGGRAPH 2021) **40**(4), 1–10 (2021)
44. Li, K., Zhang, T., Malik, J.: Diverse image synthesis from semantic layouts via conditional IMLE. In: Proceedings of the IEEE/CVF International Conference on Computer Vision, pp. 4220–4229 (2019)
45. Li, Y., Li, Y., Lu, J., Shechtman, E., Lee, Y.J., Singh, K.K.: Collaging class-specific GANs for semantic image synthesis. In: Proceedings of the IEEE/CVF International Conference on Computer Vision, pp. 14418–14427 (2021)
46. Locatello, F., et al.: Challenging common assumptions in the unsupervised learning of disentangled representations. In: International Conference on Machine Learning, pp. 4114–4124. PMLR (2019)
47. Malisiewicz, T., Efros, A.: Beyond categories: the visual memex model for reasoning about object relationships. Adv. Neural Inf. Process. Syst. **22** (2009)
48. Mejjati, Y.A., Milefchik, I., Gokaslan, A., Wang, O., Kim, K.I., Tompkin, J.: GaussiGAN: controllable image synthesis with 3D gaussians from unposed silhouettes. arXiv preprint arXiv:2106.13215 (2021)
49. Mejjati, Y.A., et al.: Generating object stamps. In: Computer Vision and Pattern Recognition Workshop on AI for Content Creation (CVPRW) (2020)
50. Meng, C., Song, Y., Song, J., Wu, J., Zhu, J.Y., Ermon, S.: SDEdit: image synthesis and editing with stochastic differential equations. arXiv preprint arXiv:2108.01073 (2021)
51. Nguyen-Phuoc, T., Li, C., Theis, L., Richardt, C., Yang, Y.L.: HoloGAN: unsupervised learning of 3D representations from natural images. In: Proceedings of

the IEEE/CVF International Conference on Computer Vision, pp. 7588–7597 (2019)
52. Nguyen-Phuoc, T.H., Richardt, C., Mai, L., Yang, Y., Mitra, N.: BlockGAN: learning 3D object-aware scene representations from unlabelled images. Adv. Neural. Inf. Process. Syst. **33**, 6767–6778 (2020)
53. Nichol, A., et al.: GLIDE: towards photorealistic image generation and editing with text-guided diffusion models. arXiv preprint arXiv:2112.10741 (2021)
54. Niemeyer, M., Geiger, A.: GIRAFFE: representing scenes as compositional generative neural feature fields. In: Proceedings of IEEE Conference on Computer Vision and Pattern Recognition (CVPR) (2021)
55. Nitzberg, M., Mumford, D.B.: The 2.1-D Sketch. IEEE Computer Society Press (1990)
56. Ohta, Y., Kanade, T., Sakai, T.: An analysis system for scenes containing objects with substructures. In: Proceedings of 4th International Joint Conference on Pattern Recognition (IJCPR 1978), pp. 752–754 (1978)
57. Oktay, D., Vondrick, C., Torralba, A.: Counterfactual image networks (2018). https://openreview.net/forum?id=SyYYPdg0-
58. Oliva, A., Torralba, A.: The role of context in object recognition. Trends Cogn. Sci. **11**(12), 520–527 (2007)
59. Park, T., Liu, M.Y., Wang, T.C., Zhu, J.Y.: Semantic image synthesis with spatially-adaptive normalization. In: Proceedings of the IEEE/CVF Conference on Computer Vision and Pattern Recognition, pp. 2337–2346 (2019)
60. Park, T., et al.: Swapping autoencoder for deep image manipulation. Adv. Neural. Inf. Process. Syst. **33**, 7198–7211 (2020)
61. Patashnik, O., Wu, Z., Shechtman, E., Cohen-Or, D., Lischinski, D.: StyleCLIP: text-driven manipulation of StyleGAN imagery. In: Proceedings of the IEEE/CVF International Conference on Computer Vision, pp. 2085–2094 (2021)
62. Pathak, D., Krähenbühl, P., Donahue, J., Darrell, T., Efros, A.: Context encoders: feature learning by inpainting (2016)
63. Peebles, W., Peebles, J., Zhu, J.-Y., Efros, A., Torralba, A.: The hessian penalty: a weak prior for unsupervised disentanglement. In: Vedaldi, A., Bischof, H., Brox, T., Frahm, J.-M. (eds.) ECCV 2020. LNCS, vol. 12351, pp. 581–597. Springer, Cham (2020). https://doi.org/10.1007/978-3-030-58539-6_35
64. Porter, T., Duff, T.: Compositing digital images. In: Proceedings of the 11th annual Conference on Computer Graphics and Interactive Techniques, pp. 253–259 (1984)
65. Radford, A., Metz, L., Chintala, S.: Unsupervised representation learning with deep convolutional generative adversarial networks. arXiv preprint arXiv:1511.06434 (2015)
66. Ramesh, A., Dhariwal, P., Nichol, A., Chu, C., Chen, M.: Hierarchical text-conditional image generation with clip latents. arXiv preprint arXiv:2204.06125 (2022)
67. Ramesh, A., et al.: Zero-shot text-to-image generation. In: International Conference on Machine Learning, pp. 8821–8831. PMLR (2021)
68. Reed, S., Akata, Z., Yan, X., Logeswaran, L., Schiele, B., Lee, H.: Generative adversarial text to image synthesis. In: International Conference on Machine Learning, pp. 1060–1069. PMLR (2016)
69. Richardson, E., et al.: Encoding in style: a StyleGAN encoder for image-to-image translation. In: Proceedings of the IEEE/CVF Conference on Computer Vision and Pattern Recognition, pp. 2287–2296 (2021)

70. Roich, D., Mokady, R., Bermano, A.H., Cohen-Or, D.: Pivotal tuning for latent-based editing of real images. arXiv preprint arXiv:2106.05744 (2021)
71. Rombach, R., Blattmann, A., Lorenz, D., Esser, P., Ommer, B.: High-resolution image synthesis with latent diffusion models. arXiv preprint arXiv:2112.10752 (2021)
72. Rott Shaham, T., Dekel, T., Michaeli, T.: SinGAN: learning a generative model from a single natural image. In: IEEE International Conference on Computer Vision (ICCV) (2019)
73. Russell, B., Efros, A., Sivic, J., Freeman, B., Zisserman, A.: Segmenting scenes by matching image composites. Adv. Neural Inf. Process. Syst. **22** (2009)
74. Saharia, C., et al.: Palette: Image-to-image diffusion models. arXiv preprint arXiv:2111.05826 (2021)
75. Sarkar, K., Golyanik, V., Liu, L., Theobalt, C.: Style and pose control for image synthesis of humans from a single monocular view (2021)
76. Shen, Y., Yang, C., Tang, X., Zhou, B.: InterFaceGAN: interpreting the disentangled face representation learned by GANs. IEEE Trans. Pattern Anal. Mach. Intell. (2020)
77. Shen, Y., Zhou, B.: Closed-form factorization of latent semantics in GANs. In: Proceedings of the IEEE/CVF Conference on Computer Vision and Pattern Recognition, pp. 1532–1540 (2021)
78. Shi, J., Malik, J.: Normalized cuts and image segmentation. IEEE Trans. Pattern Anal. Mach. Intell. **22**(8), 888–905 (2000)
79. Siarohin, A., Lathuilière, S., Tulyakov, S., Ricci, E., Sebe, N.: First order motion model for image animation. In: Conference on Neural Information Processing Systems (NeurIPS) (2019)
80. Siarohin, A., Woodford, O., Ren, J., Chai, M., Tulyakov, S.: Motion representations for articulated animation. In: CVPR (2021)
81. Silberman, N., Hoiem, D., Kohli, P., Fergus, R.: Indoor segmentation and support inference from RGBD images. In: Fitzgibbon, A., Lazebnik, S., Perona, P., Sato, Y., Schmid, C. (eds.) ECCV 2012. LNCS, vol. 7576, pp. 746–760. Springer, Heidelberg (2012). https://doi.org/10.1007/978-3-642-33715-4_54
82. Simakov, D., Caspi, Y., Shechtman, E., Irani, M.: Summarizing visual data using bidirectional similarity. In: CVPR. IEEE Computer Society (2008)
83. Strudel, R., Garcia, R., Laptev, I., Schmid, C.: Segmenter: transformer for semantic segmentation. In: Proceedings of the IEEE/CVF International Conference on Computer Vision, pp. 7262–7272 (2021)
84. Sudderth, E.B., Torralba, A., Freeman, W.T., Willsky, A.S.: Learning hierarchical models of scenes, objects, and parts. In: Tenth IEEE International Conference on Computer Vision (ICCV 2005) Volume 1, vol. 2, pp. 1331–1338. IEEE (2005)
85. Torralba, A., Willsky, A., Sudderth, E., Freeman, W.: Describing visual scenes using transformed Dirichlet processes. Adv. Neural Inf. Process. Syst. **18** (2005)
86. Tov, O., Alaluf, Y., Nitzan, Y., Patashnik, O., Cohen-Or, D.: Designing an encoder for StyleGAN image manipulation. ACM Trans. Graph. (TOG) **40**(4), 1–14 (2021)
87. Tu, Z., Chen, X., Yuille, A.L., Zhu, S.C.: Image parsing: unifying segmentation, detection, and recognition. Int. J. Comput. Vis. **63**(2), 113–140 (2005)
88. Wang, J., Yang, C., Xu, Y., Shen, Y., Li, H., Zhou, B.: Improving GAN equilibrium by raising spatial awareness. In: Proceedings of the IEEE/CVF Conference on Computer Vision and Pattern Recognition, pp. 11285–11293 (2022)

89. Wang, T.C., Liu, M.Y., Zhu, J.Y., Tao, A., Kautz, J., Catanzaro, B.: High-resolution image synthesis and semantic manipulation with conditional GANs. In: Proceedings of the IEEE Conference on Computer Vision and Pattern Recognition, pp. 8798–8807 (2018)
90. Wu, Z., Lischinski, D., Shechtman, E.: StyleSpace analysis: disentangled controls for StyleGAN image generation. In: Proceedings of the IEEE/CVF Conference on Computer Vision and Pattern Recognition, pp. 12863–12872 (2021)
91. Yakimovsky, Y., Feldman, J.A.: A semantics-based decision theory region analyser. In: IJCAI, pp. 580–588. William Kaufmann (1973)
92. Yang, C., Shen, Y., Zhou, B.: Semantic hierarchy emerges in deep generative representations for scene synthesis. Int. J. Comput. Vis. **129**(5), 1451–1466 (2021)
93. Yao, S., et al.: 3D-aware scene manipulation via inverse graphics. Adv. Neural Inf. Process. Syst. **31** (2018)
94. Yu, F., Seff, A., Zhang, Y., Song, S., Funkhouser, T., Xiao, J.: LSUN: construction of a large-scale image dataset using deep learning with humans in the loop. arXiv preprint arXiv:1506.03365 (2015)
95. Yu, H.X., Guibas, L.J., Wu, J.: Unsupervised discovery of object radiance fields. arXiv preprint arXiv:2107.07905 (2021)
96. Yu, S.X., Gross, R., Shi, J.: Concurrent object recognition and segmentation by graph partitioning. Adv. Neural Inf. Process. Syst. **15** (2002)
97. Zhang, C., Xu, Y., Shen, Y.: Decorating your own bedroom: locally controlling image generation with generative adversarial networks. arXiv preprint arXiv:2105.08222 (2021)
98. Zhang, H., Goodfellow, I., Metaxas, D., Odena, A.: Self-attention generative adversarial networks. In: International conference on machine learning, pp. 7354–7363. PMLR (2019)
99. Zhang, H., et al.: StackGAN: text to photo-realistic image synthesis with stacked generative adversarial networks. In: Proceedings of the IEEE International Conference on Computer Vision, pp. 5907–5915 (2017)
100. Zhang, R., Isola, P., Efros, A.A., Shechtman, E., Wang, O.: The unreasonable effectiveness of deep features as a perceptual metric. In: Proceedings of the IEEE Conference on Computer Vision and Pattern Recognition, pp. 586–595 (2018)
101. Zhu, J., Shen, Y., Xu, Y., Zhao, D., Chen, Q.: Region-based semantic factorization in GANs. arXiv preprint arXiv:2202.09649 (2022)
102. Zhu, J.-Y., Krähenbühl, P., Shechtman, E., Efros, A.A.: Generative visual manipulation on the natural image manifold. In: Leibe, B., Matas, J., Sebe, N., Welling, M. (eds.) ECCV 2016. LNCS, vol. 9909, pp. 597–613. Springer, Cham (2016). https://doi.org/10.1007/978-3-319-46454-1_36
103. Zhu, J.Y., et al.: Toward multimodal image-to-image translation. Adv. Neural Inf. Process. Syst. **30** (2017)

Unified Implicit Neural Stylization

Zhiwen Fan(✉), Yifan Jiang, Peihao Wang, Xinyu Gong, Dejia Xu, and Zhangyang Wang

The University of Texas at Austin, Austin, USA
{zhiwenfan,yifanjiang97,peihaowang,xinyu.gong,dejia,atlaswang}@utexas.edu

Abstract. Representing visual signals by implicit neural representation (INR) has prevailed among many vision tasks. Its potential for editing/processing given signals remains less explored. This work explores a new intriguing direction: training a **stylized** implicit representation, using a generalized approach that can apply to various 2D and 3D scenarios. We conduct a pilot study on a variety of INRs, including 2D coordinate-based representation, signed distance function, and neural radiance field. Our solution is a Unified **I**mplicit **N**eural **S**tylization framework, dubbed **INS**. In contrary to vanilla INR, INS decouples the ordinary implicit function into a style implicit module and a content implicit module, in order to separately encode the representations from the style image and input scenes. An amalgamation module is then applied to aggregate these information and synthesize the stylized output. To regularize the geometry in 3D scenes, we propose a novel self-distillation geometry consistency loss which preserves the geometry fidelity of the stylized scenes. Comprehensive experiments are conducted on multiple task settings, including fitting images using MLPs, stylization for implicit surfaces and sylized novel view synthesis using neural radiance. We further demonstrate that the learned representation is continuous not only spatially but also style-wise, leading to effortlessly interpolating between different styles and generating images with new mixed styles. Please refer to the video on our project page for more view synthesis results: https://zhiwenfan.github.io/INS.

1 Introduction

Implicit Neural Representation (INR) has gained remarkable popularity in representing concise signal representation in computer vision and computer graphics [48,52,57,68,79]. As an alternative to discrete grid-based signal representation, implicit representation is able to parameterize modern signals as samples of a continuous manifold, using multi-layer perceptions (MLP) to map between coordinates and signal values. Several seminal works [52,68,79] have verified the effectiveness of INR in representing image, video, and audio. Followups further

Z. Fan, Y. Jiang, P. Wang—Equal contribution.

Supplementary Information The online version contains supplementary material available at https://doi.org/10.1007/978-3-031-19784-0_37.

S. Avidan et al. (Eds.): ECCV 2022, LNCS 13675, pp. 636–654, 2022.
https://doi.org/10.1007/978-3-031-19784-0_37

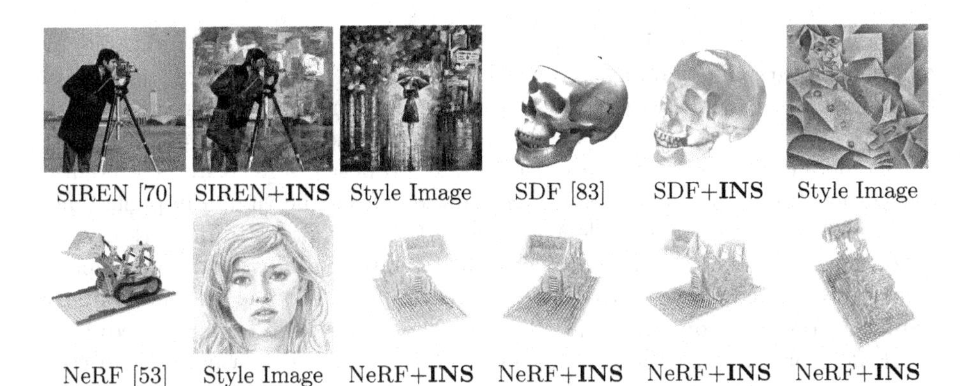

Fig. 1. Representative visual examples generated by the proposed method. We show stylized results on three different types of implicit representation, including 2D coordinate-based mapping function (SIREN [68]), Signed Distance Function (SDF [57]), and Neural Radiance Field (NeRF [52]).

apply INR to more challenging tasks including novel-view synthesis [6,7,52,82], 3D-aware generative model [9,10,86], and inverse problem [16,71]. While implicit neural representation reveals multiple advantages compared to conventional discrete signals, a general question of curiosity might be: *which and how modern visual signal processing approaches/tasks designed for discrete signals can also be applied to continuous representations?* Research pursuing this answer has been conducted on implicit neural representation since its origin. Chen *et al.* [16] apply a local implicit function to image super-resolution and they observe that INR can surpass bilinear and nearest upsampling. Sun *et al.* [71] demonstrate the effectiveness of INR in the context of sparse-view X-ray CT. Dupont *et al.* [22] propose to store the weights of a neural implicit function instead of pixel values, which surprisingly outperforms JPEG compression format. [83] further demonstrate superior video compression using similar ideas.

We investigate a novel setting: to yield visually pleasing stylized examples under various 2D and 3D scenarios, using a generalized approach leveraging implicit neural representations. Note that, training a stylized implicit neural representation still faces many hurdles. On one hand, the aforementioned works mostly have the access to dense measurements or at least sparse clean data, which enables training an implicit neural network **under the supervision of target signal**. In contrast to those tasks/approaches, current image stylization mechanisms are mostly conducted in an unsupervised manner, due to the absence of stylized ground truth data. Consequently, it is still unknown whether coordinate-based MLP can be optimized without accessing corresponding ground truth signals. On the other hand, existing style images are mostly based on 2D scenes, which raises obstacles when being considered as the appearance of 3D implicit representation. Prior art [18] attempted on marrying stylization with one specific type of Implicit Neural Representation, the neural radiance field (NeRF) [52]. Nevertheless, it still captures the statistics of style information by a series of pre-trained convolution-based hypernetwork to generate model weights, rather than a direct implicitly encoding stylization. As indicated by recent literature [45], training a robust hypernetwork requires a large amount of training samples, while novel-view synthesis tasks commonly hold no more than hundreds of views, potentially jeopardizing the synthesized visual quality.

To conquer the aforementioned fragility, we propose a Unified **I**mplicit **N**eural **S**tylization framework, coined as **INS**. Different from the vanilla implicit function which is built upon a single MLP network, the proposed framework divides an ordinary implicit neural representation to multiple individual components. Concretely speaking, we introduce a Style Implicit Module to the ordinary implicit representation, and coin the later one as Content Implicit Module in our framework. During the training process, the stylized information and content scene are encoded as one continuous representation, and then fused by another Amalgamation Module. To further regularize the geometry of given scenes, we utilize an additional self-distilled geometry consistency loss on top of the rendered density, for the stylization of NeRF. Eventually, INS is able to render view-consistent stylized scenes from novel views, with visually impressive texture details: a few examples are shown in Fig. 1.

Our contributions are outlined below:

- We propose INS, a unified implicit neural stylization framework, consists of a style implicit module, a content implicit module, and an amalgamation module, which enables us to synthesize promising stylized scenes under multiple 2D and 3D implicit representations.
- We conduct comprehensive experiments on several popular implicit representation frameworks in this novel stylization setting, including 2D coordinate-based framework (SIREN [68]), Neural Radiance Field (NeRF [52]), and Signed Distance Functions (SDF [57]). The rendering results are found to be more consistent, in both shape and style details, from different views.
- We further demonstrate that INS is able to learn representations that are continuous not only with regard to spatial placements (including views), but also in the style space. This leads to effortlessly interpolating between different styles and generating images rendered by the new mixed styles.

2 Related Works

2.1 Implicit Function

Recent research has exhibited the potential of Implicit Neural Representation (INR) to replace traditional discrete signals with continuous functions parameterized by multilayer perceptrons (MLP), in computer vision and graphics [68,72]. The coordinate-based neural representations [17,48,49] have become a popular representation for various tasks such as representing image/video [22, 68,83], 3D reconstruction [4,17,26,28,48,55–57,60,62], and 3D-aware generative modelling [10,20,29,32,47,54,63,86]. Analogously, as this representation is differentiable, prior works apply coordinate-based MLPs to many inverse problems in computational photography [3,12,16,64,69,74] and scientific computing [31,44,85].

2.2 Implicit 3D Scene Representation

Traditional 3D reconstruction methods utilizes discrete representations such as point cloud [2], meshes [42,65], multi-plane images [51], depth maps [30,78] and voxel grids [43]. Recently, INR has also prevailed among 3D scene representation tasks, which simply adopt an MLP that maps from any continuous input 3D coordinate to the geometry of the scene, including signed distance function (SDF) [5,28,39,40,49,57,70], 3D occupancy network [17,48], and so on. In addition to representing shape, INR has also been extended to encode object appearance. Among them, Neural Radiance Field (NeRF [52]) is one of the most effective coordinate-based neural representations for photo-realistic view synthesis that represents a scene as a field of particles. Draw inspiration from the preliminary success made by NeRF, a lot of following works further improve and extend it to wider application [6,6,7,7,8,13,23,33,46,50,58,59,59,75,80]. Different from the grid-based approaches, training a stylized implicit representation can not access ground truth signals, which further amplifies the difficulty of optimizing the implicit neural representation.

2.3 Stylization

Traditionally, image stylization is formulated as a painterly rendering process through stroke prediction [81,84]. The first neural style transfer method, proposed by Gatys *et al.* [25], builds an iterative framework to optimize the input image in order to minimize the content and style loss defined by a pre-trained VGG network. Due to the frustratingly large cost of training time, a number of follow-ups further explore how to design a feed-forward deep neural networks [41,73], which obtain real-time performance without sacrificing too much style information. Recently, several works extend it to video stylization [11,14,36,61] and 3D environment [18,27,34,35,53].

Ha-NeRF [15] is proposed for recovering a realistic NeRF at a different time of day from a group of tourism images, with a CNN to encode the appearance latent code. The most related NeRF-based stylization works: Style3D [18] and StylizedNeRF [37] still require a CNN-based hypernetwork or decoder to generate the stylized parameters for neural radiance field. In comparison, our proposed INS framework can work for more general implicit representations beyond neural radiance field, and can also be extended to encoding multiple styles. Experiments demonstrate that INS generate more faithful stylization on NeRF compared with Style3D [18].

3 Preliminary

This section introduces the relevant background on several implicit representations and volumetric radiance representations, including image fitting [68], neural radiance field [52] and signed distance function [57].

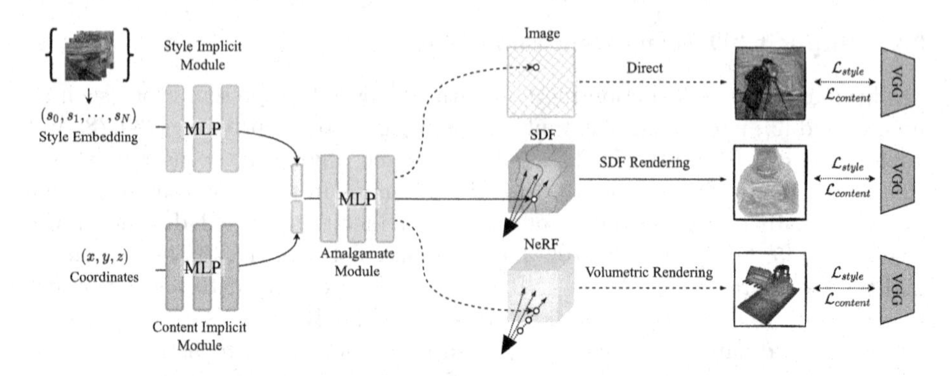

Fig. 2. The main pipeline of unified implicit neural stylization (INS) framework and its components. We took SDF with the proposed INS, for example, it inputs with implicit coordinates along with ray directions and style embeddings. Style Implicit Module (SIM) and Content Implicit Module (CIM) are used to extract conditional implicit style features and implicit scene features. Amalgamate Module (AM) is applied to fuse features in the two spaces, generating styliezed density and color intensity of each 3D point. An implicit rendering step is applied on the top of AM (i.e., Volume rendering for NeRF, surface rendering for SDF) to render the pixel intensity.

Implicit Image Fitting: The most prototypical example of neural implicit representation is image regression [68,72]. Consider fitting a function $f : \mathbb{R}^2 \rightarrow \mathbb{R}^3$ that encodes the pixel array of a given image $\mathcal{I}$ into a continuous representation. Function $f(\boldsymbol{x})$ takes pixel coordinates $\boldsymbol{x} \in \mathbb{R}^2$ as the inputs, and outputs the corresponding RGB colors $\boldsymbol{c} \in \mathbb{R}^3$. Parameterizing f with a multi-layer perception networks (MLPs), it can be optimized by the mean-squared error (MSE) loss function $\mathcal{L} = \mathbb{E}_{\boldsymbol{x} \sim \mathbb{P}(\mathcal{I})} \| f(\boldsymbol{x}) - \boldsymbol{c} \|_2^2$, where $\mathbb{P}(\mathcal{I})$ is a probability measure support in image lattice $\mathcal{I}$.

Neural Radiance Field: In contrast to point-wisely regression of implicit fields, NeRF [52] proposes to reconstruct a radiance field by inversing a differentiable rendering equation from captured images. Specifically, NeRF learns an MLP $f : (\boldsymbol{x}, \boldsymbol{\theta}) \mapsto (\boldsymbol{c}, \sigma)$ with parameters $\boldsymbol{\Theta}$, where $\boldsymbol{x}$ is the spatial coordinate in 3D space and $\boldsymbol{\theta}$ represents the view directions $\in [-\pi, \pi]^2$. The output $\boldsymbol{c} \in \mathbb{R}^3$ indicates the predicted color of the sampled point, $\sigma \in \mathbb{R}_+$ signifies its density value. The pixel color intensity can be obtained using volume rendering [21] by ray tracing, integrating the predicted color and density along the ray. To render a pixel on the image plane, NeRF casts a ray $\boldsymbol{r} = (\boldsymbol{o}, \boldsymbol{d}, \boldsymbol{\theta})$ through the pixel and accumulate the color and density of K point samples along the view direction in the 3D space. The pixel color intensity can be estimated:

$$\boldsymbol{C}(\boldsymbol{r}|\boldsymbol{\Theta}) = \sum_{k=1}^{K} T_k (1 - \exp(-\sigma_k \Delta t_k)) \boldsymbol{c}_k, \tag{1}$$

where $(\boldsymbol{c}_k, \sigma_k) = f(\boldsymbol{x}_k, \boldsymbol{\theta})$, $\boldsymbol{x}_k = \boldsymbol{o} + t_k \boldsymbol{d}$, t_k are the marching distance of sampled points, and $T_k = \exp(-\sum_{l=1}^{k-1} \sigma_l \Delta t_l)$ is known as the transmittance to model occlusion. $\Delta t_k = t_{k+1} - t_k$ indicates the distance of sampled point in 3D space. With this approximated rendering pipeline, the model weights are optimized by minimizing the L_2 distance between rendered ray colors $\boldsymbol{C}(\boldsymbol{r})$ and captured pixel colors $\widehat{\boldsymbol{C}}(\boldsymbol{r})$ as follows:

$$\boldsymbol{\Theta}^* = \arg\min_{\boldsymbol{\Theta}} \mathbb{E}_{\boldsymbol{r}\sim\mathbb{P}(\mathcal{R})} \left\| \boldsymbol{C}(\boldsymbol{r}|\boldsymbol{\Theta}) - \widehat{\boldsymbol{C}}(\boldsymbol{r}) \right\|_2^2, \tag{2}$$

where $\mathcal{R}$ is a collection of rays cast from all pixels in the training set, and $\mathbb{P}(\mathcal{R})$ defines a distribution over it.

Implicit Surface Representation: Signed Distance Function (SDF) [19] $f: \mathbb{R}^3 \to \mathbb{R}$ is an implicit representation of 3D geometries. SDF specifies each spatial point with the signed distance to the implicit iso-surface, where the sign indicates whether the point is inside or outside the object. Recent works of [57,67] propose to employ MLPs to represent this continuous field via direct supervision using point clouds. To optimize a textured SDF from multi-view images like NeRF [50], Yariv *et al.*[79] proposes a neural rendering pipeline, named IDR, which enables rendering images from an SDF. With this framework, one can indirectly supervise SDF using its multi-view projections. Suppose given a camera pose, we can cast rays $\boldsymbol{r} = (\boldsymbol{o}, \boldsymbol{v})$ through each pixel to trace an intersected point with the surface:

$$\hat{\boldsymbol{x}} = \boldsymbol{o} + t_0\boldsymbol{v} - \frac{\boldsymbol{v}}{\nabla f_0 \cdot \boldsymbol{v}_0} f(\boldsymbol{o} + t_0\boldsymbol{v}), \tag{3}$$

where t_0, $\boldsymbol{v}_0$ and f_0 are initial states when performing ray tracing (see [79]). After obtaining the intersection $\hat{\boldsymbol{x}}$ of ray and surface, IDR also lets the SDF network f output an appearance embedding $\hat{\boldsymbol{\gamma}}$, and computes the normal $\hat{\boldsymbol{n}} = \nabla f(\hat{\boldsymbol{x}})$. Then the ray color can be rendered by another rendering MLP conditioned on both point coordinate $\hat{\boldsymbol{x}}$ and normal $\hat{\boldsymbol{n}}$:

$$\boldsymbol{C}(\boldsymbol{r}|\boldsymbol{\Theta}) = r(\hat{\boldsymbol{x}}, \hat{\boldsymbol{n}}, \boldsymbol{d}, \hat{\boldsymbol{\gamma}}). \tag{4}$$

Similar to NeRF [50], f and r are simultaneously optimized by photometric loss between captured image pixels and rendered rays (see Eq. 2).

4 Method

We next illustrate the main pipeline of Unified **I**mplicit **N**eural **S**tylization (INS). INS consists of a Style Implicit Module (SIM) to transform the input style embedding into implicit style representations, a Content Implicit Module (CIM) to map the input coordinates into implicit scene representations, and an Amalgamate Module (AM) which amalgamates the two representation to predict RGB intensity. To preserve the geometry fidelity while generating the stylized texture of rendered views, a self-distilled geometry consistency regularization is applied upon the INS framework.

4.1 Implicit Style and Content Representation

Generating the stylized images $\boldsymbol{Y}$ can be formulated as an energy minimization problem [25]. It consists of a *content loss* and a *style loss*, defined under a pre-trained VGG network [66]. We build upon the prior work and thus propose our implicit stylization framework for SIREN, SDF and NeRF.

Content Representation. The content loss in 2D images stylization pipeline [25] $\mathcal{L}_{\text{content}}$ is defined as:

$$\mathcal{L}_{\text{content}}(\boldsymbol{C}, \boldsymbol{Y}) = \frac{1}{C_{i,j}W_{i,j}H_{i,j}} \sum_{(i,j)\in\mathcal{J}} \|F_{i,j}(\boldsymbol{C}) - F_{i,j}(\boldsymbol{Y})\|_F^2, \tag{5}$$

where $\boldsymbol{C}$ denotes the content ground truth image and $\boldsymbol{Y}$ denotes synthesized output, $F_{i,j}$ denotes the feature map extracted from a VGG-16 model pre-trained on ImageNet, i represents its i-th max pooling, and j represents its j-th convolutional layer after i-th max pooling layer. $C_{i,j}$, $W_{i,j}$ and $H_{i,j}$ are the dimensions of the extracted feature maps. We adapt the content loss to the intermediate layer of INS pipeline to preserve the content of the predicted color image patch, we choose $i = 2$, $j = 2$ by default.

Style Representation. To extract representation of the stylized information, [66] introduces a different feature space to capture texture information [24]. Similar to the content loss, the feature space is built upon the filter response in multiple layers of a pre-trained VGG network. By capturing the correlations of the filter responses expressed by the Gram matrix $\boldsymbol{G}_{i,j} \in \mathbb{R}^{C_{i,j}\times C_{i,j}}$ between the style image $\boldsymbol{S}$ and the synthesized image $\boldsymbol{Y}$, multi-scale representations can be obtained to capture the texture information from the style image and endow such texture on the stylized image. Here, we define our style loss $\mathcal{L}_{\text{style}}$ using the same layers of VGG-16 with [25] on the top of the prediction of implicit neural representations and the given style image:

$$\mathcal{L}_{\text{style}}(\boldsymbol{S}, \boldsymbol{Y}) = \sum_{(i,j)\in\mathcal{J}} \|\boldsymbol{G}_{i,j}(\boldsymbol{S}) - \boldsymbol{G}_{i,j}(\boldsymbol{Y})\|_F^2, \tag{6}$$

$$\text{where } [\boldsymbol{G}_{i,j}]_{c,c'}(\boldsymbol{Y}) = \frac{1}{C_{i,j}W_{i,j}H_{i,j}} \sum_{k=1}^{H\times W} F_{i,j}(\boldsymbol{Y})_{c,k} F_{i,j}(\boldsymbol{Y})_{c',k}, \tag{7}$$

where $\mathcal{J}$ are the indices of selected feature maps. In practice, we choose $\mathcal{J} = \{(1,2),(2,2),(3,3),(4,3)\}$ in our experiments.

Conditional INS Stylization. Conditional encoding has been widely applied in convolutional networks [38,77]. Similarly, we propose the conditional implicit representation by input with style conditioned embeddings using and extracting style-dependent features to render stylized color and density, which is shown in Fig. 2. In training, we prepare n style images with an n dimensional one-hot style-condition vector. A mini-batch is constructed with the combinations of one content training patch and all candidate style images. The one-hot vector is fed into SIM to extract the w-dimensional style features, they are then concatenated with the implicit representations output from CIM. The following layers of AM take the two features, aggregate them to render the pixel intensity and scene geometry along the rays. A pre-trained VGG [66] is appended on the top of the INS pipeline to apply implicit style and content constraints during training. During the inference stage, we discard the VGG network, the INS framework becomes a pure MLPs-based network.

4.2 Geometry Consistency for Neural Radiance Field

NeRF [52] learns reasonable 3D geometry inherently due to the particular design of implicit radiance field and the supervision from multiple views. However, the INS framework is expected to integrate style statistics from 2D image into 3D radiance field, where no multi-view style images accessible during training process. To specialize INS for neural radiance fields, we propose to regularize INS with proper geometry constraint to produce faithful shape and appearance. As the ground truth of target geometry is unavailable in most novel view synthesis benchmarks, we seek help from the self-distillation framework [76]. Concretely speaking, we first train the content implicit module (CIM) only to obtain a clean geometry σ_1, following the vanilla NeRF training pipeline. After that, we copy the trained weight of CIM and keep that fixed (as shown in the grey block in Fig. 3). In the next, we turn to optimize the whole INS framework (SIM, CIM and AM) with implicit stylization constrains. Meanwhile, a self-distilled geometry constraint between the original geometry σ_1 produced by the fixed CIM weight and the final stylized geometry σ_2 reconstructed by the implicit neural stylization framework. The objective of self-distilled geometry consistency loss is formulated as $\mathcal{L}_{\text{geo}} = |\sigma_1 - \sigma_2|$, where we adopt the mean-absolute error for the densities of each sampled point.

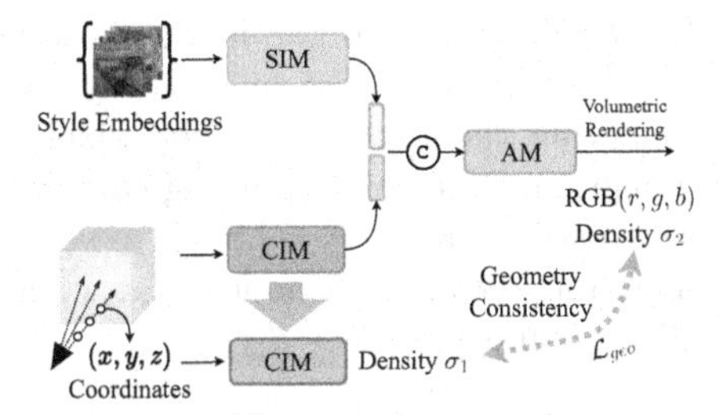

Fig. 3. The proposed self-distilled geometry consistency. The CIM weights in the grey box are from a pre-trained NeRF and are kept fixed since then. During fine-tuning, the output density of the fixed CIM serves as geometry constrains for the stylized density from the output of AM.

Sampling-Stride Ray Sampling. Neural Radiance Field casts a number of rays (typically not adjacent) from camera origin, intersecting the pixel, into the volume and accumulating the color based on density along the ray.

While our model input with rays intersected with an image patch of size $\mathcal{P} \in K \times K$, predicting the stylized patch $\mathcal{P}' \in K \times K$ with its texture closed to the given style images. 2D style transfer methods [11,36] typically crop the patch larger than 256×256. However, it is too expensive for the neural radiance field as it queries the MLPs more than $256 \times 256 \times N$ times of the MLP for each step [52], where N indicates sampled points number along each ray.

Similar to [47,63], we adopt a *Sampling-Stride Ray Sampling* strategy to enlarge the receptive field of the sampled patch to capture a more global context. The illustration of the ray sampling can be found in our supplementary materials, where a sampling stride larger than 1 result in a large receptive field while keeping computational cost fixed.

4.3 Optimization

Let $\boldsymbol{F}(\boldsymbol{T})$ denote an INS model parameterized by $\boldsymbol{\Theta}$, which synthesizes an image (patch) from the view specified by the camera pose $\boldsymbol{T} \in \mathbb{R}^{4\times 4}$ by marching and rendering the rays for all the pixels on the image (patch). Given multi-view images and the corresponding camera parameters $\mathcal{T} = \{\boldsymbol{C}_i, \boldsymbol{T}_i\}_{i=1}^{N}$, as well as a set of style images $\mathcal{S} = \{\boldsymbol{S}_i\}_{i=1}^{M}$, we train the INS using a combination of losses including reconstruction loss $\mathcal{L}_{\text{recon}}$, geometry consistency loss $\mathcal{L}_{\text{geo}}$, content loss $\mathcal{L}_{\text{content}}$ and style loss $\mathcal{L}_{\text{style}}$:

$$\begin{aligned}\mathcal{L}_{\text{total}}(\boldsymbol{\Theta}|\mathcal{T},\mathcal{S},\boldsymbol{\Theta}_{\text{vgg}}) &= \mathbb{E}_{(\boldsymbol{C},\boldsymbol{T})\sim\mathbb{P}(\mathcal{T})}\left[\lambda_0\mathcal{L}_{\text{recon}}(\boldsymbol{F}(\boldsymbol{T}),\boldsymbol{C}) + \lambda_1\mathcal{L}_{\text{geo}}(\boldsymbol{F}(\boldsymbol{T}))\right] \\ &+ \mathbb{E}_{\boldsymbol{S}\sim\mathbb{P}(\mathcal{S})}\,\mathbb{E}_{(\boldsymbol{C},\boldsymbol{T})\sim\mathbb{P}(\mathcal{T})}\left[\lambda_2\mathcal{L}_{\text{content}}(\boldsymbol{F}(\boldsymbol{T}),\boldsymbol{C}|\boldsymbol{\Theta}_{\text{vgg}}) + \lambda_3\mathcal{L}_{\text{style}}(\boldsymbol{F}(\boldsymbol{T}),\boldsymbol{S}|\boldsymbol{\Theta}_{\text{vgg}})\right],\end{aligned} \tag{8}$$

where λ_0, λ_1, λ_2, λ_3 control the strength of each loss term, $\boldsymbol{\Theta}_{\text{vgg}}$ denotes the parameters of the VGG network, and $\mathbb{P}(\cdot)$ defines a distribution over a support.

5 Experiments

5.1 Stylization on SIREN MLPs

As one representative example, we apply INS on fitting an image via SIREN [68] MLPs. We reuse the original SIREN framework [68] as CIM and follow its training recipes to fit images of 512×512 pixels. Besides that, we also incorporate the SIM and AM on SIREN. A pre-trained VGG-16 network is appended on the output to provide style and content supervisions during training. As seen in Fig. 4, the proposed framework successfully in representing the images with the given style statistics in an implicit way.

5.2 Novel View Synthesis with NeRF

Experimental Settings. We train our INS framework on NeRF-Synthetic [52] dataset and Local Light Field Fusion(LLFF) [51] dataset. NeRF-Synthetic consists of complex scenes with 360-degree views, where each scene has a central object with 100 inward-facing cameras distributed randomly on the upper hemisphere. Both rendered images and ground truth meshes are provided in NeRF-Synthetic dataset. LLFF dataset consists of forward-facing scenes, with fewer images. We implement INS on the same architecture and training strategy with

Fig. 4. Visual examples generated by applying INS to SIREN [68]. Given an input image used for fitting and style images, SIREN+INS can express the style statistics via an implicit manner(i.e., MLPs).

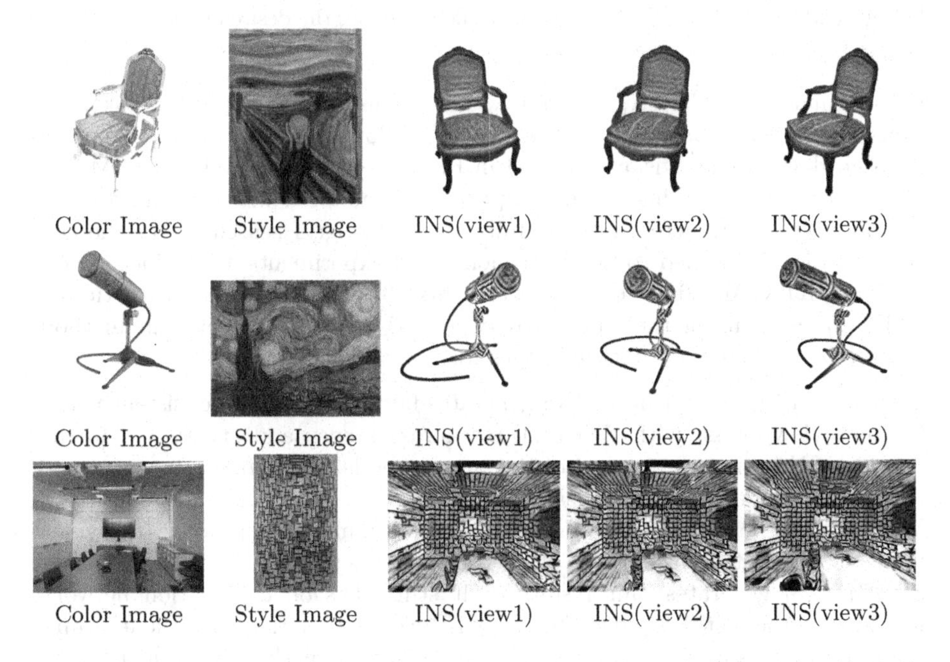

Fig. 5. Multi-view examples rendered by applying INS to the neural radiance field. The rendered scenes and objects show consistent stylized texture under different views.

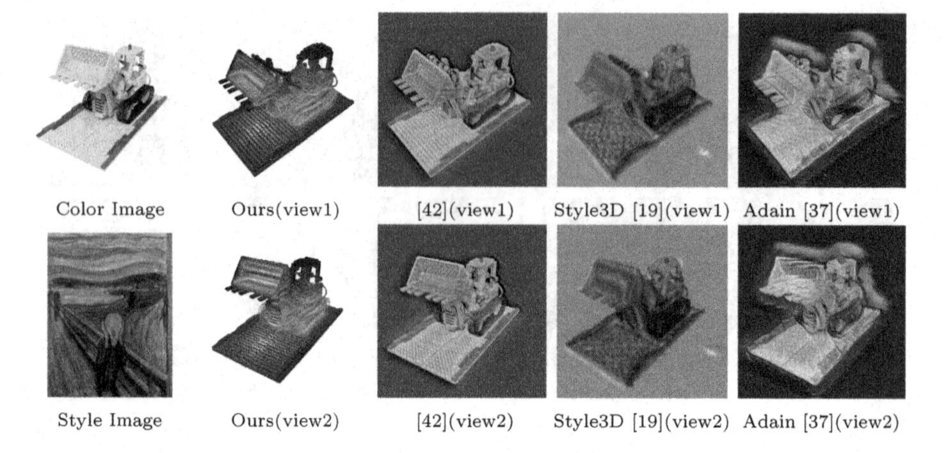

Fig. 6. Qualitative comparisons. We compare INS with several methods on novel view synthesis datasets. Three scenes with different style images are demonstrated, we can see our proposed INS method preserves the learned geometry better (e.g. with clean background) and capture global context while achieving the desired styles.

the original NeRF [52]. λ_1, λ_2 and λ_3 are set as zero in the first 150k iterations and then set as 1e6, 1 and 1e8 in the following 50k iterations. The self-distilled density supervision depicted in Fig. 3 is generated from the CIM with 150k iterations pre-training. Adam optimizer is adopted with learning rates of 0.0005. Hyper-parameters are carefully tuned via grid searches and the best configuration is applied to all experiments. All experiments are trained on one NVIDIA RTX A6000 GPU. We retrain Style3D [18] in NeRF-Synthetic and LLFF datasets using their provided code and setting. We train all methods using the same number of style images for fair comparisons.

Results. In Fig. 5, we can see INS generates faithful and view-consistent results for new viewpoints, with rich textures across scenes and styles. We further compare INS with three state-of-the art methods, including 3D neural stylization [18], image-based stylization methods [36,41]. As is shown in Fig. 6, We can see that stylizations from image-based methods produce noisy and view-inconsistent stylization as they transfer styles based on a single image. Style3D [18] generates blur results as it still relies on convolution networks (a.k.a. hypernetwork) to generate the MLP weights for the subsequent volume rendering. Our proposed implicit neural stylization method is trained to preserve correct scene geometry as well as capture global context, generating better view-consistent stylizations.

5.3 Stylization on Signed Distance Function

Experimental Settings. DeepSDF [57] only learns the 3D geometry from given inputs. Later work IDR [79] extend it to reconstruct both 3D surface and appearance. We follow IDR [79] to implement the implicit neural stylization framework.

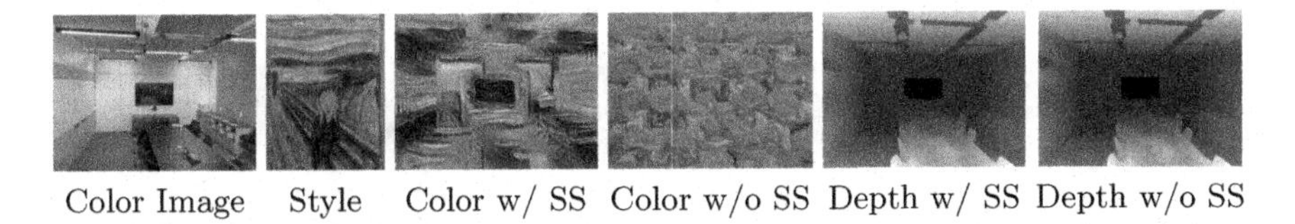

Fig. 7. Qualitative evaluation on the effectiveness of Sampling Stride technique (SS). We can see INS with a larger receptive field generate more satisfying stylization.

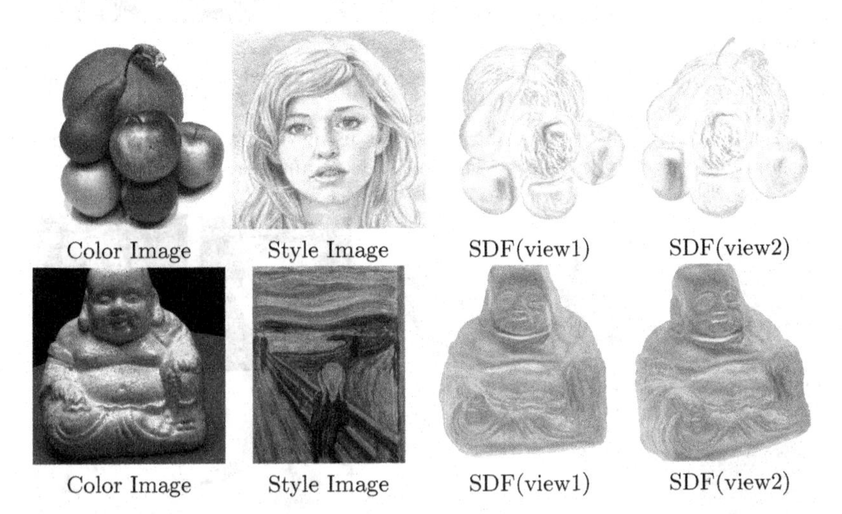

Fig. 8. Visualization results of applying INS to the Signed Distance Function. Given multi-view color images and style images, the proposed INS can learn the style statistics for the disentangled geometry and appearance.

To encode style statistic onto IDR, we project the learned textured SDF into multi-view images and implement our style loss on the rendered results. In the experiments, we picked 2 scenes from the DTU dataset [1], where each scene consists of 50 to 100 images and object masks captured from different angles. Similar to NeRF, we pre-train the IDR model for chosen scenes by minimizing the loss between the ground truth image and the rendered result. Then both the SDF network and rendering network are jointly optimized the proposed framework from projected views. Note that due to IDR's architectural design, we are no longer able to impose self-distilled geometry consistency loss. Instead, we employ content/style loss in the masked region, which is similar to [79]. Besides, we observe that the SDF representation is more sensitive to parameter variations. To maintain intact geometries, we adjust the learning rate for the SDF network to 10^{-11} times smaller than the rendering network.

Results. As are shown in Fig. 8, the visualizations of two-view SDF representation demonstrate that both the learned appearance and geometry have deformed to fit the given style statistics.

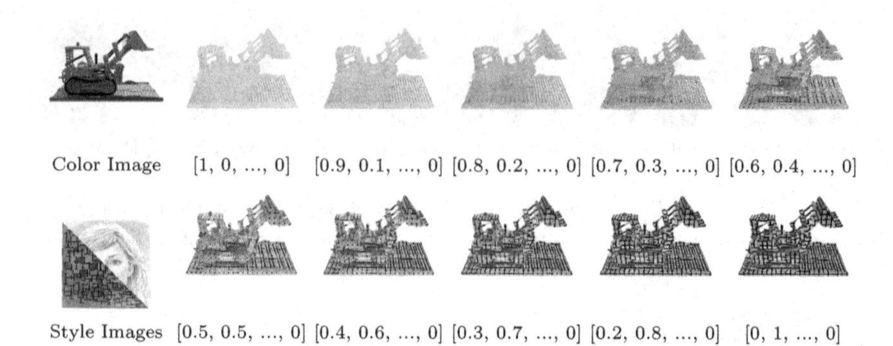

Fig. 9. Visualization of the mixture of two styles using different mixture weights. We can see the style smoothly transfers from one style to another.

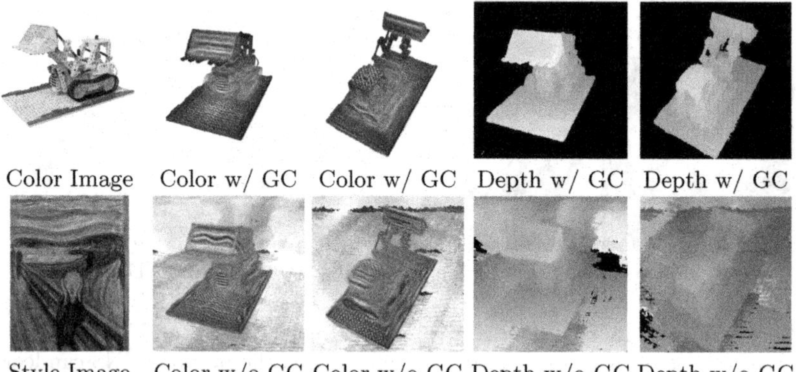

Fig. 10. Qualitative results of the self-distilled geometry consistency("GC"). Both the rendered images and depth maps are shown for validation.

5.4 Conditional Style Interpolation

Training with style-conditioned one-hot embedding, we can interpolate between style images to mix multiple styles with arbitrary weights. Specifically, we train INS on NeRF with two style images along with a two-dimensional one-hot vector as conditional code. After training, we mix the two style statistics by using a weighted two-dimensional vector. As is shown in Fig. 9, the synthesized results can smoothly transfer from the first style to the second style when we linearly mix the two style embeddings at inference time.

5.5 Ablation Study

Effect of Updating NeRF's Geometry Field. The geometry modification in the density branch enables a more flexible stylization, by stylizing shape tweaks on the object surface. As shown in Fig. 11, only updating the color branch easily

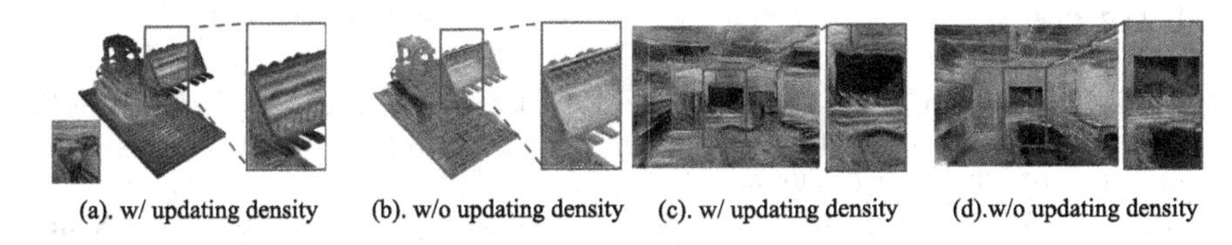

Fig. 11. Ablation study of optimizing style-conditioned density and color. By updating the density, it can render richer textures than only updating the color branch.

collapses to the low-level color transformation instead of the painterly texture transformation, violating our original goal. This phenomenon is also addressed in the previous 3D mesh stylization [42], where they explicitly model the 3D shape by deforming the mesh vertexes.

Effect of Self-distilled Geometry Consistency. To evaluate the effectiveness of the proposed geometry consistency regularizer, we visualize the front and back viewpoints of the synthesized color images and depth maps. As is shown in Fig. 10, the proposed self-distilled geometry consistency learns a good trade-off between stylization and clean geometry.

Should INS Learned with Larger Receptive Field? To investigate the effect by using the Sampling Stride (SS) Ray Sampling strategy, we conduct a comparison of ray sampling with and without sampling stride for NeRF stylization. For a fair comparison, we set the ray number as 64×64 in both settings. INS with sampling stride covers the content resolution of $(64 \times s) \times (64 \times s)$ where s indicates sampling strides depicted in Sect. 4.2 and here we set s=4. Figure 7 shows that INS with SS achieves significantly better visual results, as it results in a higher receptive field in perceiving content statistics.

6 Conclusions

In this work, we present a Unified Implicit Neural Stylization framework (INS) to stylize complex 2D/3D scenes using implicit function. We conduct a pilot study on different types of implicit representations, including 2D coordinate-based mapping function, Neural Radiance Field, and Signed Distance Function. Comprehensive experiments demonstrate that the proposed method yields photo-realistic images/videos with visually consistent stylized textures. One limitation of our work lies in the training efficiency issue, similar to most implicit representation, rendering a style scene requires several hours of training, precluding on-device training. Addressing this issue could become a future direction.

References

1. Aanæs, H., Jensen, R.R., Vogiatzis, G., Tola, E., Dahl, A.B.: Large-scale data for multiple-view stereopsis. Int. J. Comput. Vis. **120**(2), 153–168 (2016). https://doi.org/10.1007/s11263-016-0902-9
2. Aliev, K.-A., Sevastopolsky, A., Kolos, M., Ulyanov, D., Lempitsky, V.: Neural point-based graphics. In: Vedaldi, A., Bischof, H., Brox, T., Frahm, J.-M. (eds.) ECCV 2020. LNCS, vol. 12367, pp. 696–712. Springer, Cham (2020). https://doi.org/10.1007/978-3-030-58542-6_42
3. Attal, B., et al.: Time-of-flight radiance fields for dynamic scene view synthesis. In: Advances in Neural Information Processing Systems, vol. 34 (2021)
4. Atzmon, M., Haim, N., Yariv, L., Israelov, O., Maron, H., Lipman, Y.: Controlling neural level sets. In: Advances in Neural Information Processing Systems, vol. 32 (2019)
5. Atzmon, M., Lipman, Y.: Sal: sign agnostic learning of shapes from raw data. In: Proceedings of the IEEE/CVF Conference on Computer Vision and Pattern Recognition, pp. 2565–2574 (2020)
6. Barron, J.T., Mildenhall, B., Tancik, M., Hedman, P., Martin-Brualla, R., Srinivasan, P.P.: Mip-nerf: a multiscale representation for anti-aliasing neural radiance fields. In: Proceedings of the IEEE/CVF International Conference on Computer Vision, pp. 5855–5864 (2021)
7. Barron, J.T., Mildenhall, B., Verbin, D., Srinivasan, P.P., Hedman, P.: Mip-nerf 360: unbounded anti-aliased neural radiance fields. arXiv preprint. arXiv:2111.12077 (2021)
8. Bergman, A., Kellnhofer, P., Wetzstein, G.: Fast training of neural lumigraph representations using meta learning. In: Advances in Neural Information Processing Systems, vol. 34 (2021)
9. Chan, E.R., et al.: Efficient geometry-aware 3d generative adversarial networks. arXiv preprint. arXiv:2112.07945 (2021)
10. Chan, E.R., Monteiro, M., Kellnhofer, P., Wu, J., Wetzstein, G.: pi-gan: periodic implicit generative adversarial networks for 3d-aware image synthesis. In: Proceedings of the IEEE/CVF Conference on Computer Vision and Pattern Recognition, pp. 5799–5809 (2021)
11. Chen, D., Liao, J., Yuan, L., Yu, N., Hua, G.: Coherent online video style transfer. In: Proceedings of the IEEE International Conference on Computer Vision, pp. 1105–1114 (2017)
12. Chen, H., He, B., Wang, H., Ren, Y., Lim, S.N., Shrivastava, A.: Nerv: neural representations for videos. In: Advances in Neural Information Processing Systems, vol. 34 (2021)
13. Chen, T., Wang, P., Fan, Z., Wang, Z.: Aug-nerf: training stronger neural radiance fields with triple-level physically-grounded augmentations. In: Proceedings of the IEEE/CVF Conference on Computer Vision and Pattern Recognition, pp. 15191–15202 (2022)
14. Chen, X., Zhang, Y., Wang, Y., Shu, H., Xu, C., Xu, C.: Optical flow distillation: towards efficient and stable video style transfer. In: Vedaldi, A., Bischof, H., Brox, T., Frahm, J.-M. (eds.) ECCV 2020. LNCS, vol. 12351, pp. 614–630. Springer, Cham (2020). https://doi.org/10.1007/978-3-030-58539-6_37
15. Chen, X., et al.: Hallucinated neural radiance fields in the wild. In: Proceedings of the IEEE/CVF Conference on Computer Vision and Pattern Recognition, pp. 12943–12952 (2022)

16. Chen, Y., Liu, S., Wang, X.: Learning continuous image representation with local implicit image function. In: Proceedings of the IEEE/CVF Conference on Computer Vision and Pattern Recognition, pp. 8628–8638 (2021)
17. Chen, Z., Zhang, H.: Learning implicit fields for generative shape modeling. In: Proceedings of the IEEE/CVF Conference on Computer Vision and Pattern Recognition, pp. 5939–5948 (2019)
18. Chiang, P.Z., Tsai, M.S., Tseng, H.Y., Lai, W.S., Chiu, W.C.: Stylizing 3d scene via implicit representation and hypernetwork. In: Proceedings of the IEEE/CVF Winter Conference on Applications of Computer Vision, pp. 1475–1484 (2022)
19. Curless, B., Levoy, M.: A volumetric method for building complex models from range images. In: Proceedings of the 23rd Annual Conference on Computer Graphics and Interactive Techniques, pp. 303–312 (1996)
20. DeVries, T., Bautista, M.A., Srivastava, N., Taylor, G.W., Susskind, J.M.: Unconstrained scene generation with locally conditioned radiance fields. In: Proceedings of the IEEE/CVF International Conference on Computer Vision, pp. 14304–14313 (2021)
21. Drebin, R.A., Carpenter, L., Hanrahan, P.: Volume rendering. ACM Siggraph Comput. Graph. **22**(4), 65–74 (1988)
22. Dupont, E., Goliński, A., Alizadeh, M., Teh, Y.W., Doucet, A.: Coin: compression with implicit neural representations. arXiv preprint. arXiv:2103.03123 (2021)
23. Gao, C., Shih, Y., Lai, W.S., Liang, C.K., Huang, J.B.: Portrait neural radiance fields from a single image. arXiv preprint. arXiv:2012.05903 (2020)
24. Gatys, L., Ecker, A.S., Bethge, M.: Texture synthesis using convolutional neural networks. In: Advances in Neural Information Processing Systems, vol. 28 (2015)
25. Gatys, L.A., Ecker, A.S., Bethge, M.: A neural algorithm of artistic style. arXiv preprint. arXiv:1508.06576 (2015)
26. Genova, K., Cole, F., Vlasic, D., Sarna, A., Freeman, W.T., Funkhouser, T.: Learning shape templates with structured implicit functions. In: Proceedings of the IEEE/CVF International Conference on Computer Vision, pp. 7154–7164 (2019)
27. Gong, X., Huang, H., Ma, L., Shen, F., Liu, W., Zhang, T.: Neural stereoscopic image style transfer. In: Proceedings of the European Conference on Computer Vision (ECCV) (2018)
28. Gropp, A., Yariv, L., Haim, N., Atzmon, M., Lipman, Y.: Implicit geometric regularization for learning shapes. arXiv preprint. arXiv:2002.10099 (2020)
29. Gu, J., Liu, L., Wang, P., Theobalt, C.: Stylenerf: a style-based 3d-aware generator for high-resolution image synthesis. arXiv preprint. arXiv:2110.08985 (2021)
30. Gu, X., Fan, Z., Zhu, S., Dai, Z., Tan, F., Tan, P.: Cascade cost volume for high-resolution multi-view stereo and stereo matching. In: Proceedings of the IEEE/CVF Conference on Computer Vision and Pattern Recognition, pp. 2495–2504 (2020)
31. Han, J., Jentzen, A., Weinan, E.: Solving high-dimensional partial differential equations using deep learning. In: Proceedings of the National Academy of Sciences, vol. 115, no. 34, pp. 8505–8510 (2018)
32. Hao, Z., Mallya, A., Belongie, S., Liu, M.Y.: Gancraft: Unsupervised 3d neural rendering of minecraft worlds. In: Proceedings of the IEEE/CVF International Conference on Computer Vision, pp. 14072–14082 (2021)
33. Hedman, P., Srinivasan, P.P., Mildenhall, B., Barron, J.T., Debevec, P.: Baking neural radiance fields for real-time view synthesis. In: Proceedings of the IEEE/CVF International Conference on Computer Vision, pp. 5875–5884 (2021)
34. Höllein, L., Johnson, J., Nießner, M.: Stylemesh: Style transfer for indoor 3d scene reconstructions. arXiv preprint. arXiv:2112.01530 (2021)

35. Huang, H.P., Tseng, H.Y., Saini, S., Singh, M., Yang, M.H.: Learning to stylize novel views. In: Proceedings of the IEEE/CVF International Conference on Computer Vision, pp. 13869–13878 (2021)
36. Huang, X., Belongie, S.: Arbitrary style transfer in real-time with adaptive instance normalization. In: Proceedings of the IEEE International Conference on Computer Vision, pp. 1501–1510 (2017)
37. Huang, Y.H., He, Y., Yuan, Y.J., Lai, Y.K., Gao, L.: Stylizednerf: consistent 3d scene stylization as stylized nerf via 2d–3d mutual learning. In: Proceedings of the IEEE/CVF Conference on Computer Vision and Pattern Recognition, pp. 18342–18352 (2022)
38. Iizuka, S., Simo-Serra, E., Ishikawa, H.: Let there be color! joint end-to-end learning of global and local image priors for automatic image colorization with simultaneous classification. ACM Trans. Graph. (ToG) **35**(4), 1–11 (2016)
39. Jiang, C., Sud, A., Makadia, A., Huang, J., Nießner, M., Funkhouser, T., et al.: Local implicit grid representations for 3d scenes. In: Proceedings of the IEEE/CVF Conference on Computer Vision and Pattern Recognition, pp. 6001–6010 (2020)
40. Jiang, Y., Ji, D., Han, Z., Zwicker, M.: Sdfdiff: differentiable rendering of signed distance fields for 3d shape optimization. In: Proceedings of the IEEE/CVF Conference on Computer Vision and Pattern Recognition, pp. 1251–1261 (2020)
41. Johnson, J., Alahi, A., Fei-Fei, L.: Perceptual losses for real-time style transfer and super-resolution. In: Leibe, B., Matas, J., Sebe, N., Welling, M. (eds.) ECCV 2016. LNCS, vol. 9906, pp. 694–711. Springer, Cham (2016). https://doi.org/10.1007/978-3-319-46475-6_43
42. Kato, H., Ushiku, Y., Harada, T.: Neural 3d mesh renderer. In: Proceedings of the IEEE Conference on Computer Vision and Pattern Recognition, pp. 3907–3916 (2018)
43. Kutulakos, K.N., Seitz, S.M.: A theory of shape by space carving. Int. J. Comput. Vis. **38**(3), 199–218 (2000). https://doi.org/10.1023/A:1008191222954
44. Li, Z., et al.: Fourier neural operator for parametric partial differential equations. arXiv preprint. arXiv:2010.08895 (2020)
45. Lorraine, J., Duvenaud, D.: Stochastic hyperparameter optimization through hypernetworks. arXiv preprint. arXiv:1802.09419 (2018)
46. Martin-Brualla, R., Radwan, N., Sajjadi, M.S., Barron, J.T., Dosovitskiy, A., Duckworth, D.: Nerf in the wild: neural radiance fields for unconstrained photo collections. In: Proceedings of the IEEE/CVF Conference on Computer Vision and Pattern Recognition, pp. 7210–7219 (2021)
47. Meng, Q., et al.: Gnerf: gan-based neural radiance field without posed camera. In: Proceedings of the IEEE/CVF International Conference on Computer Vision, pp. 6351–6361 (2021)
48. Mescheder, L., Oechsle, M., Niemeyer, M., Nowozin, S., Geiger, A.: Occupancy networks: learning 3d reconstruction in function space. In: Proceedings of the IEEE/CVF Conference on Computer Vision and Pattern Recognition, pp. 4460–4470 (2019)
49. Michalkiewicz, M., Pontes, J.K., Jack, D., Baktashmotlagh, M., Eriksson, A.: Implicit surface representations as layers in neural networks. In: Proceedings of the IEEE/CVF International Conference on Computer Vision, pp. 4743–4752 (2019)
50. Mildenhall, B., Hedman, P., Martin-Brualla, R., Srinivasan, P., Barron, J.T.: Nerf in the dark: high dynamic range view synthesis from noisy raw images. arXiv preprint. arXiv:2111.13679 (2021)

51. Mildenhall, B., Srinivasan, P.P., Ortiz-Cayon, R., Kalantari, N.K., Ramamoorthi, R., Ng, R., Kar, A.: Local light field fusion: Practical view synthesis with prescriptive sampling guidelines. ACM Trans. Graph. (TOG) **38**(4), 1–14 (2019)
52. Mildenhall, B., Srinivasan, P.P., Tancik, M., Barron, J.T., Ramamoorthi, R., Ng, R.: NeRF: representing scenes as neural radiance fields for view synthesis. In: Vedaldi, A., Bischof, H., Brox, T., Frahm, J.-M. (eds.) ECCV 2020. LNCS, vol. 12346, pp. 405–421. Springer, Cham (2020). https://doi.org/10.1007/978-3-030-58452-8_24
53. Mu, F., Wang, J., Wu, Y., Li, Y.: 3d photo stylization: learning to generate stylized novel views from a single image. arXiv preprint. arXiv:2112.00169 (2021)
54. Niemeyer, M., Geiger, A.: Giraffe: Representing scenes as compositional generative neural feature fields. In: Proceedings of the IEEE/CVF Conference on Computer Vision and Pattern Recognition, pp. 11453–11464 (2021)
55. Niemeyer, M., Mescheder, L., Oechsle, M., Geiger, A.: Occupancy flow: 4d reconstruction by learning particle dynamics. In: Proceedings of the IEEE/CVF international conference on computer vision, pp. 5379–5389 (2019)
56. Oechsle, M., Mescheder, L., Niemeyer, M., Strauss, T., Geiger, A.: Texture fields: learning texture representations in function space. In: Proceedings of the IEEE/CVF International Conference on Computer Vision, pp. 4531–4540 (2019)
57. Park, J.J., Florence, P., Straub, J., Newcombe, R., Lovegrove, S.: Deepsdf: learning continuous signed distance functions for shape representation. In: Proceedings of the IEEE/CVF Conference on Computer Vision and Pattern Recognition, pp. 165–174 (2019)
58. Park, K., et al.: Nerfies: deformable neural radiance fields. In: Proceedings of the IEEE/CVF International Conference on Computer Vision, pp. 5865–5874 (2021)
59. Park, K., et al.: Hypernerf: a higher-dimensional representation for topologically varying neural radiance fields. arXiv preprint. arXiv:2106.13228 (2021)
60. Peng, S., Niemeyer, M., Mescheder, L., Pollefeys, M., Geiger, A.: Convolutional occupancy networks. In: Vedaldi, A., Bischof, H., Brox, T., Frahm, J.-M. (eds.) ECCV 2020. LNCS, vol. 12348, pp. 523–540. Springer, Cham (2020). https://doi.org/10.1007/978-3-030-58580-8_31
61. Ruder, M., Dosovitskiy, A., Brox, T.: Artistic style transfer for videos. In: Rosenhahn, B., Andres, B. (eds.) GCPR 2016. LNCS, vol. 9796, pp. 26–36. Springer, Cham (2016). https://doi.org/10.1007/978-3-319-45886-1_3
62. Saito, S., Huang, Z., Natsume, R., Morishima, S., Kanazawa, A., Li, H.: Pifu: pixel-aligned implicit function for high-resolution clothed human digitization. In: Proceedings of the IEEE/CVF International Conference on Computer Vision, pp. 2304–2314 (2019)
63. Schwarz, K., Liao, Y., Niemeyer, M., Geiger, A.: Graf: generative radiance fields for 3d-aware image synthesis. Adv. Neural. Inf. Process. Syst. **33**, 20154–20166 (2020)
64. Shen, S., et al.: Non-line-of-sight imaging via neural transient fields. IEEE Trans. Pattern Anal. Mach. Intell. **43**(7), 2257–2268 (2021)
65. Shrestha, R., Fan, Z., Su, Q., Dai, Z., Zhu, S., Tan, P.: Meshmvs: multi-view stereo guided mesh reconstruction. In: 2021 International Conference on 3D Vision (3DV), pp. 1290–1300. IEEE (2021)
66. Simonyan, K., Zisserman, A.: Very deep convolutional networks for large-scale image recognition. arXiv preprint. arXiv:1409.1556 (2014)
67. Sitzmann, V., Chan, E., Tucker, R., Snavely, N., Wetzstein, G.: Metasdf: meta-learning signed distance functions. Adv. Neural. Inf. Process. Syst. **33**, 10136–10147 (2020)

68. Sitzmann, V., Martel, J., Bergman, A., Lindell, D., Wetzstein, G.: Implicit neural representations with periodic activation functions. Adv. Neural. Inf. Process. Syst. **33**, 7462–7473 (2020)
69. Sitzmann, V., Rezchikov, S., Freeman, W.T., Tenenbaum, J.B., Durand, F.: Light field networks: neural scene representations with single-evaluation rendering. arXiv preprint. arXiv:2106.02634 (2021)
70. Sitzmann, V., Zollhöfer, M., Wetzstein, G.: Scene representation networks: continuous 3d-structure-aware neural scene representations. In: Advances in Neural Information Processing Systems, vol. 32 (2019)
71. Sun, Y., Liu, J., Xie, M., Wohlberg, B., Kamilov, U.S.: Coil: coordinate-based internal learning for imaging inverse problems. arXiv preprint. arXiv:2102.05181 (2021)
72. Tancik, M., et al.: Fourier features let networks learn high frequency functions in low dimensional domains. Adv. Neural. Inf. Process. Syst. **33**, 7537–7547 (2020)
73. Ulyanov, D., Lebedev, V., Vedaldi, A., Lempitsky, V.S.: Texture networks: feed-forward synthesis of textures and stylized images. In: ICML, vol. 1, p. 4 (2016)
74. Wang, P., Fan, Z., Chen, T., Wang, Z.: Neural implicit dictionary learning via mixture-of-expert training. In: International Conference on Machine Learning, pp. 22613–22624. PMLR (2022)
75. Xu, D., Jiang, Y., Wang, P., Fan, Z., Shi, H., Wang, Z.: Sinnerf: training neural radiance fields on complex scenes from a single image. arXiv preprint. arXiv:2204.00928 (2022)
76. Xu, Y., Qiu, X., Zhou, L., Huang, X.: Improving bert fine-tuning via self-ensemble and self-distillation. arXiv preprint. arXiv:2002.10345 (2020)
77. Yanai, K., Tanno, R.: Conditional fast style transfer network. In: Proceedings of the 2017 ACM on International Conference on Multimedia Retrieval, pp. 434–437 (2017)
78. Yao, Y., Luo, Z., Li, S., Fang, T., Quan, L.: Mvsnet: depth inference for unstructured multi-view stereo. In: Proceedings of the European Conference on Computer Vision (ECCV), pp. 767–783 (2018)
79. Yariv, L., et al.: Multiview neural surface reconstruction by disentangling geometry and appearance. Adv. Neural. Inf. Process. Syst. **33**, 2492–2502 (2020)
80. Yu, A., Ye, V., Tancik, M., Kanazawa, A.: pixelnerf: neural radiance fields from one or few images. In: Proceedings of the IEEE/CVF Conference on Computer Vision and Pattern Recognition, pp. 4578–4587 (2021)
81. Zeng, K., Zhao, M., Xiong, C., Zhu, S.C.: From image parsing to painterly rendering. ACM Trans. Graph. **29**(1), 2–1 (2009)
82. Zhang, K., Riegler, G., Snavely, N., Koltun, V.: Nerf++: analyzing and improving neural radiance fields. arXiv preprint. arXiv:2010.07492 (2020)
83. Zhang, Y., van Rozendaal, T., Brehmer, J., Nagel, M., Cohen, T.: Implicit neural video compression. arXiv preprint. arXiv:2112.11312(2021)
84. Zhao, M., Zhu, S.C.: Customizing painterly rendering styles using stroke processes. In: Proceedings of the ACM SIGGRAPH/Eurographics Symposium on Non-photorealistic Animation and Rendering, pp. 137–146 (2011)
85. Zhong, E.D., Bepler, T., Berger, B., Davis, J.H.: Cryodrgn: reconstruction of heterogeneous cryo-em structures using neural networks. Nat. Methods **18**(2), 176–185 (2021)
86. Zhou, P., Xie, L., Ni, B., Tian, Q.: Cips-3d: a 3d-aware generator of gans based on conditionally-independent pixel synthesis. arXiv preprint. arXiv:2110.09788 (2021)

GAN with Multivariate Disentangling for Controllable Hair Editing

Xuyang Guo[1,2], Meina Kan[1,2], Tianle Chen[1,2], and Shiguang Shan[1,2,3](✉)

[1] Key Lab of Intelligent Information Processing of Chinese Academy of Sciences (CAS), Institute of Computing Technology, Beijing 100190, China
{xuyang.guo,tianle.chen}@vipl.ict.ac.cn, {kanmeina,sgshan}@ict.ac.cn
[2] University of Chinese Academy of Sciences, Beiijng 100049, China
[3] Peng Cheng Laboratory, Shenzhen 518055, China

Abstract. Hair editing is an essential but challenging task in portrait editing considering the complex geometry and material of hair. Existing methods have achieved promising results by editing through a reference photo, user-painted mask, or guiding strokes. However, when a user provides no reference photo or hardly paints a desirable mask, these works fail to edit. Going a further step, we propose an efficiently controllable method that can provide a set of sliding bars to do continuous and fine hair editing. Meanwhile, it also naturally supports discrete editing through a reference photo and user-painted mask. Specifically, we propose a generative adversarial network with a multivariate Gaussian disentangling module. Firstly, an encoder disentangles the hair's major attributes, including color, texture, and shape, to separate latent representations. The latent representation of each attribute is modeled as a standard multivariate Gaussian distribution, to make each dimension of an attribute be changed continuously and finely. Benefiting from the Gaussian distribution, any manual editing including sliding a bar, providing a reference photo, and painting a mask can be easily made, which is flexible and friendly for users to interact with. Finally, with changed latent representations, the decoder outputs a portrait with the edited hair. Experiments show that our method can edit each attribute's dimension continuously and separately. Besides, when editing through reference images and painted masks like existing methods, our method achieves comparable results in terms of FID and visualization. Codes can be found at https://github.com/XuyangGuo/CtrlHair.

Keywords: Controllable editing · Hair editing · Style transfer · Style manipulation

1 Introduction

Hair consists of numerous, dedicated, and small strands, whose complex geometry and material lead to various hair appearances in color, structure, shape,

Supplementary Information The online version contains supplementary material available at https://doi.org/10.1007/978-3-031-19784-0_38.

S. Avidan et al. (Eds.): ECCV 2022, LNCS 13675, pp. 655–670, 2022.
https://doi.org/10.1007/978-3-031-19784-0_38

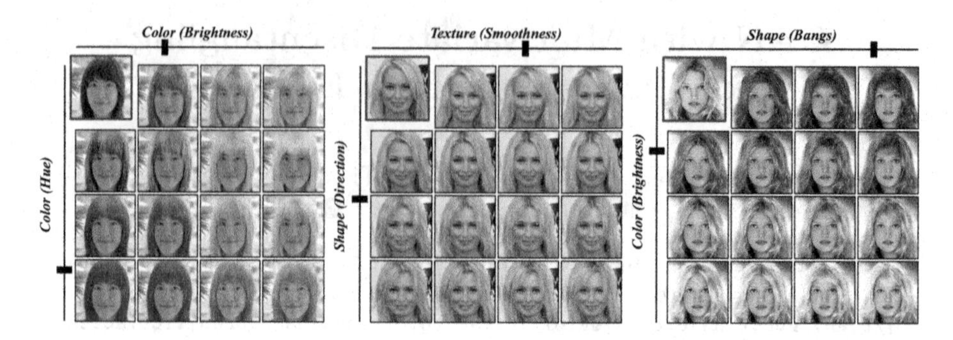

Fig. 1. Controllable hair editing by using CtrlHair. Given a portrait in the upper left corner of each subfigure, CtrlHair can edit the hair by sliding a set of bars. CtrlHair supports editing of fine-grained factors of an attribute simultaneously and separately

style, etc. The complex nature of hair makes it difficult to model, depict and generate. Hair editing, as a kind of image manipulation, is one of the most challenging and important components of portrait editing. The task of hair editing has valuable application scenarios, such as animation, games design, and virtual reality, and can also help to better understand the generation of face images.

The development of Generative Adversarial Nets [6] has greatly boosted the hairstyle transfer [2,21,24,28], i.e. transferring the hairstyle of a reference image to a target image including shape, appearance, etc. These methods especially MichiGAN [24] achieve impressive results, which divides the hair into attributes, including appearance, structure, and shape, and then edits hair according to painted mask, guiding strokes, or reference photos for better user controllability. However, when a user has no reference photo or hardly paints a desirable mask, these existing works fail to edit. Go a further step, our work endeavors to provide an efficiently controllable hair editing method, named as *CtrlHair*, with which a user can do arbitrary hair editing by simply sliding a set of control bars, providing reference photos, or painting a mask. Besides, rather than transferring or editing an entire attribute such as color, shape, texture, we want to explore manipulating each dimension of an attribute for finer control, e.g., change the direction of hair shape and the brightness of hair color, as shown in Fig. 1.

Specifically, we propose a GAN with a multivariate Gaussian disentangling module. Firstly, an encoder disentangles the hair's major attributes, including color, texture, and shape, to separate latent representations, one for each attribute. The latent representation of each attribute is modeled as a standard multivariate Gaussian distribution, to make each dimension can be changed *continuously* and *finely*. Benefiting from the Gaussian distribution, any manual editing including sliding a bar, providing a reference photo, painting a mask can be easily made, which is flexible for users to interact with. Finally, with changed latent representations, a decoder outputs an image with edited hair.

Among all attributes, shape editing not only change the hair region but also affects other parts of the portrait. To avoid the problem of tricky misalignment between hair and face region and inpainting caused by shape change, we especially propose a novel Shape Adaptor module. The shape adaptor takes the changed hair shape, face mask, and background mask as input, and automatically does alignment and inpainting to output an appropriate portrait mask including hair region, face region, and background region.

Briefly, our contributions are summarized in three-folds:

1. We propose a GAN with multivariate Gaussian disentangling, with which a user can do continuous and fine-grained hair editing by simply sliding a set of control bars, providing reference photos, or painting a mask.
2. Especially, for a realistic portrait with edited hair, a learning-based shape adaptor is proposed to automatically do alignment between hair and face region, as well as inpaint those uncovered regions caused by shape change.
3. Experiments show that our method can edit each dimension of each attribute continuously and separately. It also achieves comparable results as existing methods on task of hairstyle transfer.

2 Related Works

Image Generation. In recent years, attributing to the development of Generative Adversarial Networks [6], the realism and resolution of image generation have been significantly improved. The early method DCGAN [20] introduces a deconvolutional structure, which greatly improves realism. ProgressiveGAN [10] and StyleGANs [11,12] further propose a progressive generative network to achieve photo-realistic face images in high resolution. In addition, another kind of work focuses on conditional generation with supervised or unsupervised condition. The supervised conditional GAN methods, such as CGAN [15] and ACGAN [16], are trained with a label as a conditional signal to generate an image belonging to the condition-indicated class. Moreover, InfoGAN [3] learns disentangled representations in an unsupervised manner, by maximizing mutual information between conditional signal and generated images. Furthermore, conditional image generation is widely explored for different types of tasks, e.g., style transferring [9], generating images from segmentation masks [18], etc.

Face Editing. The above methods can generate images, but cannot edit a given image. So, there appear a few works using GAN to do image editing, especially face editing. Given an input image, these methods aim to generate an image where only one or several attributes are edited while the rest contents keep unchanged. Most face editing methods [4,5,7,14,17,19,22,23,27] manipulate specific face attributes by using binary attribute labels (e.g., with or without glasses/smiling). However, editing with binary attribute labels can only do coarsely editing. So, some works such as MaskGAN [13] are proposed to use facial segmentation masks to edit the face more flexibly and meticulously. This method can edit the shape or transfer the appearance style of the entire image. Different from encoding the appearance style of an image into one feature representation in MaskGAN [13], SEAN [29] further disentangles the appearance style of the entire image to multiple separate feature representations according to the segmentation masks, allowing more flexible editing.

Hair Editing. Among all face parts, hair has complex structures, shapes, appearances, etc., leading to many semantics and making fine annotations lacking. Hence, the general face editing methods mentioned above are hardly used for hair editing with meaningful semantic variation. So there are some works [2,21,24,28] that are specially dedicated to editing hair especially. Since hair is hardly labeled by discrete category for training like face editing, existing methods directly transfer the hair of a reference image to another. Chai et al. [2] and LOHO [21] disentangle the attributes of hair into two parts, including shape-relevant attributes and shape-irrelevant attributes, which can be separately transferred from a reference image to another image. Moreover, MichiGAN [24] and Barbershop [28] further disentangle hair into three orthogonal attributes, including shape, structure, and appearance, and users can edit these attributes by providing reference images, painted masks, or guiding strokes. Overall, these methods can edit or transfer the hairstyle by reference photos or painted masks as conditions. However, when a user has no reference photos or hardly paints a desirable mask, these works fail to edit. That is what our work attempt to deal with.

3 CtrlHair

Given an input portrait I, our CtrlHair attempts to edit the attributes of hair region continuously and finely. The overall editing process is as follows:

$$(X, S) = \Phi(I), \quad (\hat{X}, \hat{S}) = \text{CtrlHair}(X, S), \quad \hat{I} = \Psi(\hat{X}, \hat{S}, I), \tag{1}$$

where Φ parses the input portrait I to get a segmentation mask S, including hair mask S_H, face mask S_F and background mask S_B, i.e. $S = (S_H, S_F, S_B)$. X can be just the hair region of the original I or can be a basic feature of hair for a better feature presentation. Ψ blends the edited hair region and other regions including face and background to a portrait $\hat{I}$. Since our work mainly focuses on the hair editing, Φ and Ψ are directly obtained by using a pre-trained model, i.e. pre-trained BiSeNet [25,30] is used to obtain portrait segmentation mask S and SEAN [29] is used to obtain basic feature representation X and generate portrait with edited hair $\hat{X}$ and $\hat{S}$.

As stated above, the main goal of our method is to edit hair continuously and finely. Continuous editing means that users can edit a given hair to any other rational look, or a look specified by a reference photo and user-painted mask. Fine editing means that users can separately edit fine-grained variation factor of a hair attribute, such as length of shape, hue of color, etc. Besides, the edited hair region should be harmoniously blended with other regions including face and background to generate a realistic portrait.

For this goal, we propose a generative adversarial network with a multivariate Gaussian disentangling module. The proposed method consists of three successive modules, i.e. encoding module, interactive editing module, and decoding module as shown in Fig. 2. We first introduce the three modules in the editing procedure of CtrlHair, and then presents the training objectives in detail.

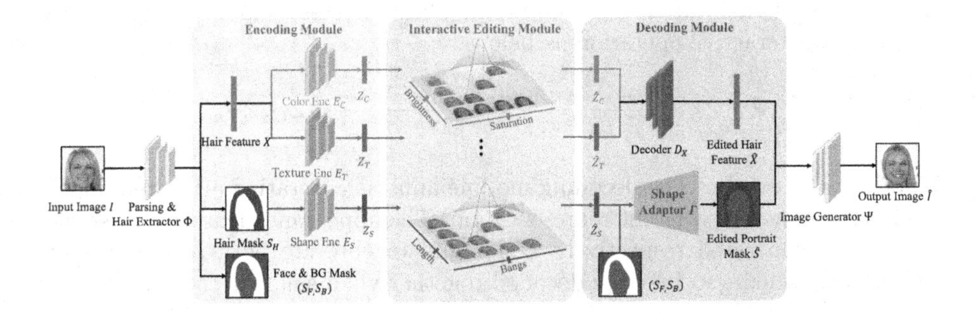

Fig. 2. An overview of CtrlHair, which consists of three modules: encoding, interactive editing, and decoding. The encoding module separates hair into distinct latent representations, of which each is formulated as a standard multivariate Gaussian distribution. In the interactive editing module, any manual editing including a set of bars, reference images, painted masks, can be made. The decoding module outputs an edited portrait with these edited latent representations

3.1 The Editing Procedure

Encoding Module. The encoding module aims at disentangling the hair into three latent representations, corresponding to color, texture, and shape. These latent representations are respectively obtained by three encoders as below:

$$Z_C = E_C(X), \quad Z_T = E_T(X), \quad Z_S = E_S(S_H), \tag{2}$$

where $Z_C, Z_T, Z_S \in \mathcal{N}(0, I)$ represent the disentangled representation of color, texture, and shape. E_C, E_T and E_S are corresponding encoders. For each attribute, expectedly it can be edited to any other look continuously and finely. Therefore, the representation of Z_C, Z_T and Z_S are all restricted to follow standard multivariate Gaussian distribution during training. Benefiting from the continuity of Gaussian distribution, any editing can be made by directly changing the value of Z_C, Z_T, and Z_S. Besides, all dimensions of standard multivariate Gaussian distribution are mutually independent, which makes that each dimension of an attribute can be separately edited for fine editing, such as length of shape, hue of color, etc. The meaning of each dimension of an attribute can be learned without supervision to implicitly mine variational factors (e.g. smoothness of texture), and can be also supervised by labels to explicitly learn meaning factors (e.g. brightness of color).

Interactive Editing Module. Given the disentangled latent representation Z_C, Z_T and Z_S, users can easily edit each dimension of each attribute, simply by sliding a set of bars, and get edited latent representation $\hat{Z}_C$, $\hat{Z}_T$ and $\hat{Z}_S$. Besides, it also naturally supports users to edit according to reference photos or painted masks like existing methods [21,24,28] by inputting them to the encoders to get their latent representation. These processes can be illustrated as one unified interactive operation as below:

$$(Z_C, Z_T, Z_S) \xrightarrow{f_{\text{sliding bars}};\ f_{\text{references}};\ f_{\text{painted mask}}} (\hat{Z}_C, \hat{Z}_T, \hat{Z}_S). \tag{3}$$

Decoding Module. The decoding module aims at generating an edited image based on the modified latent representation. The appearance feature and shape of hair are generated respectively. Color and texture are both related to the appearance of hair, so they are decoded together via a single decoder as:

$$\hat{X} = D_X(\hat{Z}_C, \hat{Z}_T), \tag{4}$$

where D_X is the decoder for color and texture.

Similarly, the shape of hair can be also decoded from Z_S like appearance in Eq. (4). However, different from color and texture, shape editing not only changes hair region but also affects the face and background region. On one hand, the edited shape should be well aligned with the face region in appropriate scale and width, e.g. a thin face with a large-shaped hair looks unrealistic. On the other hand, shape change would uncover some regions that are originally covered by hair. These regions should be inpainted as face or background. Existing methods ignore these two problems, or just throw them to GAN, leading to awkward portrait or inaccurate hair shape compared with the reference image. Therefore, we specifically propose a learning-based shape adaptor Γ to obtain a well-aligned hair shape and well-inpainted portrait mask as below:

$$\hat{S} = \Gamma(\hat{Z}_S, S_F, S_B). \tag{5}$$

In the portrait mask $\hat{S}$, the hair shape is slightly adjusted, and the background mask and face mask are inpainted if they are uncovered after changing hair shape, to ensure a harmonious and realistic portrait image generated by Eq. (1).

3.2 The Training Objectives

Expectedly, the generated image from the whole editing process in Eq. (1) should satisfy three requirements, i.e. the edited image $\hat{I}$ should look realistic; the hair of $\hat{I}$ is correctly edited; continuous and fine editing is supported. These three characteristics are formulated as three types of objectives including realism loss $\mathcal{L}^{real}$, reconstruction loss $\mathcal{L}^{rec}$ and distribution loss $\mathcal{L}^{dist}$ as follows:

$$\mathcal{L} = \lambda^{real}\mathcal{L}^{real} + \sum\nolimits_{k\in\{C,T,S\}}(\lambda_k^{rec}\mathcal{L}_k^{rec} + \lambda_k^{dist}\mathcal{L}_k^{dist}). \tag{6}$$

Firstly, the generated image should be realistic. Since the Φ and Ψ are pre-trained, we only need to ensure the generated $\hat{X}$ and $\hat{S}$ same as those from the real images. So, a realism adversarial loss $\mathcal{L}^{real}$, i.e., the first term in Eq. (6), is optimized by adversarial training between generated $(\hat{X}, \hat{S})$ and (X, S) from real images, same as the adversarial training in conventional GANs. *Secondly*, the editing should be correct, i.e. the generated image $\hat{I}$ should be with the edited attributes while other attributes remain unchanged. These are achieved by the reconstruction loss $\mathcal{L}^{rec}$, i.e. the second term in Eq. (6). *Thirdly*, the representation Z_C, Z_T, Z_S are expected to follow a standard multivariate Gaussian distribution for continuous and fine editing, which is formulated by distribution loss $\mathcal{L}^{dist}$, i.e., the third term in Eq. (6).

The reconstruction loss and distribution loss of each attribute are different respecting to their distinct natures, which are introduced respectively as follows.

Color. For color, the latent representation $Z_C \in \mathbb{R}^4$ is designed as a 4-dimensional vector. The first 3 dimensions represent the HSV color values, which capture the mean color of the hair region. The 4th dimension represents the variance considering the color of all pixels in a hair region is not identical but usually different. The reconstruction loss and distribution loss of color are formulated as below:

$$\min_{E_C} \mathcal{L}_C^{dist} = \mathbb{E}_{(X,\tilde{Z}_C)\sim P_d} \left\| E_C(X) - \tilde{Z}_C \right\|_2^2, \quad \tilde{Z}_C \sim \mathcal{N}(0, I) \tag{7}$$

$$\min_{D_X} \mathcal{L}_C^{rec} = \mathbb{E}_{\hat{Z}_C,\hat{Z}_T\sim\mathcal{N}(0,I)} \left\| E_C(\hat{X}) - \hat{Z}_C \right\|_2^2. \tag{8}$$

where P_d is the training set. Equation (7) ensures that the latent representation generated from real images follows gaussian distribution through a supervised manner. Since H, S, V, and variance of the hair region can be easily calculated without manual labeling, they are used as the supervised color label. Specifically, the distribution of (H, S, V, variance) of hair region from training images is transformed to standard Gaussian distribution dimension-wisely, achieved by analytically mapping the quantiles of the cumulative distribution of training samples to standard Gaussian distribution, denoted as $\tilde{Z}_C$. Then Eq. (7) enforces the encoded latent representation to be the same as its supervised label $\tilde{Z}_C$, i.e. enforce $Z_C = E_C(X)$ follow standard multivariate Gaussian distribution.

Equation (8) ensures that each $\hat{Z}_C$ sampled from $\mathcal{N}(0, I)$ can correctly manipulate the generated hair feature $\hat{X}$ via reconstruction between the sampled $\hat{Z}_C$ and encoded $E_C(\hat{X})$ from generated feature with it.

Besides color, any other supervised attributes can be easily added similarly.

Texture. The texture attribute captures the pattern and regularity of the hair, such as hair-strand thickness, smoothness, curliness, etc. These semantics are difficult to label manually, so we use unsupervised methods to discover them.

On one hand, the low dimensional latent representation Z_T should be capable of manipulating the image, formulated as two reconstruction loss as follows:

$$\min_{E_T,D_X} \mathcal{L}_T^{rec} = \mathbb{E}_{X\sim P_d} \|X - D_X(E_C(X), E_T(X))\|_2^2 \tag{9}$$

$$+\mathbb{E}_{\hat{Z}_C,\hat{Z}_T\sim\mathcal{N}(0,I)} \left\| E_T(\hat{X}) - \hat{Z}_T \right\|_2^2. \tag{10}$$

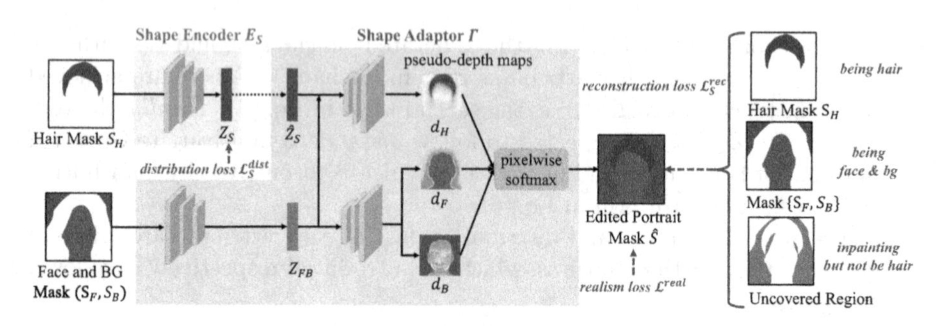

Fig. 3. The design of shape editing. The shape adaptor takes latent representation $\hat{Z}_S$ of edited hair shape, original face and background mask S_F, S_B as input; then adjusts them to obtain an edited portrait mask, whose hair region, inpainting region, and the rest face and background region are constrained accordingly

The reconstruction in Eq. (10) maximizes the mutual information [3] between $\hat{Z}_T$ and $\hat{X}$, which ensures each dimension of $\hat{Z}_T$ can manipulate some kind of variation in $\hat{X}$. The reconstruction in Eq. (9) enforces $E_T(X)$ to try to capture a meaningful variation of texture rather than other attributes or a trivial solution.

On the other hand, Z_T should follow standard multivariate Gaussian distribution. This is partly achieved by reconstructing sampled $\hat{Z}_T$ in above Eq. (10). Besides, an auxiliary constraint is used to directly enforce the low-order moments of $\hat{Z}_T$ to be same as that of standard Gaussian distribution. Here, one- and second-order moments (i.e. $\mu = \mathbf{0}$, and $\Sigma = \mathbf{I}$) are adopted, formulated as:

$$\min_{E_T, D_X} \mathcal{L}_T^{dist} = \|\mathbb{E}_{X\sim P_d} E_T(X) - \mu\|_2^2 + \|\mathbb{V}_{X\sim P_d} E_T(X) - \Sigma\|_F^2. \tag{11}$$

Shape. Shape is a complex attribute since shape not only changes the hair region but also affects the face and background region. Therefore, the shape editing should be not only correct but also reasonable, i.e. hair shape should be well aligned with the face region and the uncovered region should be inptained reasonably. Hence, a shape adaptor Γ is proposed to simultaneously adjust the shape to align with the face and also inpaint those uncovered regions.

To achieve it, during training, we conduct constraints by sampling a person's hair mask $S_H \in \{0,1\}^{H\times W\times 1}$, and adapt it to another person with his face mask and background mask $(S_F, S_B) \in \{0,1\}^{H\times W\times 2}$. The objective loss is calculated on the entire generated portrait mask $\hat{S}$ instead of only on the hair region. As shown in Fig. 3, the shape adaptor takes edited shape representation $\hat{Z}_S$, face and background mask (S_F, S_B) as input, and then generates the pseudo depth-map of hair, face, and background, denoted as $d_H, d_F, d_B \in \mathbb{R}^{H\times W\times 1}$. The pseudo depth-maps allow easy and soft composition [1] of them to get the adjusted portrait mask $\hat{S}_k = \frac{\exp(d_k)}{\sum_{m\in\{H,F,B\}} \exp(d_m)}$ $(k \in \{H, F, B\})$.

Table 1. Functionality comparison of different methods

Functionality	MichiGAN [24]	LOHO [21]	BarberShop [28]	CtrlHair (ours)
Interaction mode	References painted mask sketch	References	References painted mask	References painted mask sliding bars
Editing flexiblity	Coarse, discrete			Fine-grained, continuous
Shape editing	Replace directly			Shape adaptor

The shape reconstruction loss is calculated on the adjusted portrait mask $\hat{S}$ from the shape adaptor as below:

$$\min_{E_S,\Gamma} \mathcal{L}_S^{rec} = -\mathbb{E}_{S_H,S_F,S_B\sim P_d}\Bigg[S_H \log(\hat{S}_H) + (1-S_H)\sum_{k\in\{F,B\}} S_k \log(\hat{S}_k) \quad (12)$$

$$+ (1-S_H)\log(1-\hat{S}_H)\Bigg], \quad (13)$$

where the first term in Eq. (12) enforces that the generated hair shape should be almost the same as the input while allowing slightly adjustment for different (S_F, S_B) guided by adversarial realism loss $\mathcal{L}^{real}$. The second term in Eq. (12) ensures that the unaffected face and background region remain unchanged. The unconvered region is determined automatically via adversarial realism loss $\mathcal{L}^{real}$, but with an extra constraint in Eq. (13) to prohibit being inpainted as hair again.

Besides, the distribution loss $\mathcal{L}_S^{dist}$ is designed as adversarial loss between the distribution of generated latent representation and Gaussian distribution:

$$\min_{E_S}\max_{\delta} \mathcal{L}_S^{dist} = \mathbb{E}_{S_H\sim P_d}[\log(1-\delta(E_S(S_H))] + \mathbb{E}_{Z_S^r\sim\mathcal{N}(0,I)}[\log\delta(Z_S^r)], \quad (14)$$

where δ is the corresponding discriminator for adversarial training. The distribution loss can also be designed similarly to the texture. They are enforced differently according to each attribute's characteristics for better results.

Summary. The parameters of the whole network including E_C, E_T, E_S, D_F, and Γ are optimized by minimizing the above three types of objectives, i.e., realism loss, reconstruction loss, and distribution loss. More details can be found in the supplementary material.

3.3 Discussion

Overall, CtrlHair has three differences with existing methods as shown in Table 1. Firstly, CtrlHair supports more interaction modes including reference photos, painted masks and sliding a set of bars. Secondly, CtrlHair can continuously and finely manipulate hair attributes such as hue of color, length of shape, while existing methods can only transfer the entire attribute. Thirdly, we elaborately consider the alignment between hair and face, and inpainting of those uncovered regions, while existing methods directly ignore or throw them to GAN.

4 Experiments

In this section, we present the experimental setting, then compare our CtrlHair method with existing methods in terms of hairstyle transfer, and investigate the ability of continuously and finely editing followed by an ablation study.

4.1 Datasets and Implementation Details

For experimental evaluation, the high-quality CelebAMask-HQ [13] dataset and FFHQ [11] dataset are used. From the two datasets, we select 10413 images without severe hat-occlusion and large pose orientation for experiments. Among them, 1000 are randomly sampled for testing and the rest 9413 images are used for training. The resolution of each image is resized to 256×256.

The encoder of color E_C, encoder of texture E_T and decoder for both of them D_X are all designed as fully connection architecture while shape encoder E_S and shape adaptor Γ are designed as convolution architecture considering that shape is position sensitive. The dimensionality of latent representations Z_C, Z_T, Z_S are set as 4, 8 and 16 respectively. For texture attribute, besides unsupervised mining, we also manually annotate a small number of images (1180 for wavy, and 727 for straightness) with a binary label for curliness factor to do weakly supervised training. In total, we edit 3 attributes with 28 controllable dimensions. Among them, 11 fine-grained dimensions are manually found to be with obvious semantics, which are shown in the supplementary materials.

4.2 Comparison with Existing Methods on Hairstyle Transfer

Since existing methods can not do continuous hair editing, we firstly compare with them on the task of discretely hair editing, i.e. hairstyle transfer. Our CtrlHair method is compared to MichiGAN [24], LOHO [21], and Barbershop [28], in terms of editing correctness, realism, and computational efficiency. Among them, MichiGAN needs an additional inpainting module for uncovered region, but this part is not publicly available, so we use GatedConv [26] for its inpainting.

Editing Correctness. The editing results of the compared methods are shown in Fig. 4, where all methods transfer the appearance of a reference photo and its corresponding shape to a given input image. Firstly, we compare the results of hair-appearance transferring as shown in Fig. 4(a), (b), and (c). As for the hair region, all methods transfer the hair appearance favorably, i.e. look very similar to the reference image. As for the appearance harmony of hair and face region, MichiGAN looks slightly stiff, LOHO, BarberShop and our CtrlHar look naturally. Secondly, we compare the results of hair-shape transferring. As can be seen in Fig. 4(d), when the reference's face shape is larger, it is inappropriate to directly replace the shape mask of the input image with the reference mask, e.g. artifact appears in MichiGAN and shadow appears on the left side of the face in LOHO, due to a lack of considering alignment. Both BarberShop and our CtrlHair achieve better results. In Fig. 4(e) and (f), MichiGAN and LOHO incorrectly inpaint the uncovered region, i.e. the uncovered region is inpainted

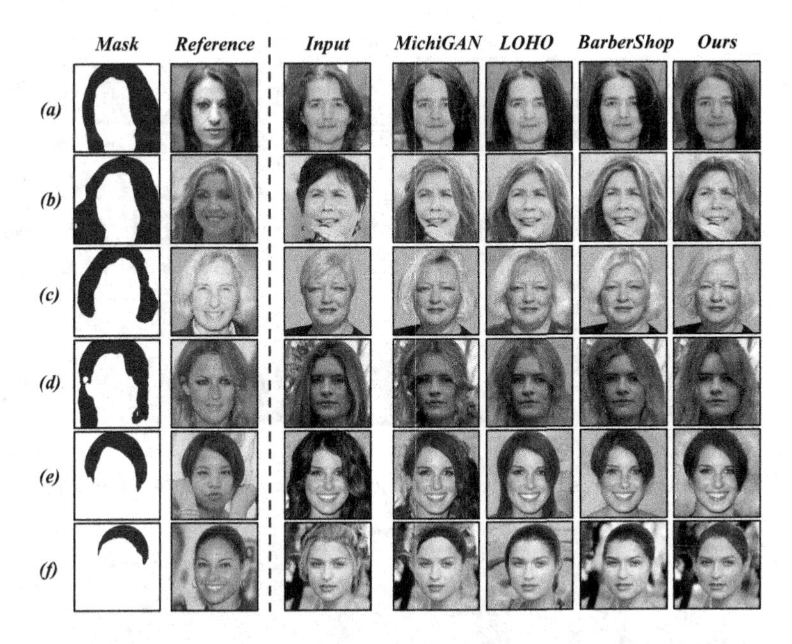

Fig. 4. Comparison with existing methods on hairstyle transfer with a reference image

Table 2. Comparison with the existing method in terms of realism and time efficiency

	MichiGAN [24]	LOHO [21]	BarberShop [28]	CtrlHair (ours)
FID↓	23.80	31.03	24.93	**21.39**
Time consuming↓	**1.6 s+3.1 s**[1]	960.1 s	395.8 s	**7.7 s**

Since face parsing and inpainting processes are not publicly available for MichiGAN, BiSeNet [25] and GatedConv [26] are alternatively used, which cost 3.1 s

with hair again while it should be inpainted with face or background. Our method produces more correct shape, benefiting from our elaborately designed shape adaptor. Please refer to the supplementary material for more comparison results.

Realism Comparison. To evaluate the generation realism, we compare the Fréchet Inception Distance (FID) [8] from different methods according to the same 3000 transferred images selected randomly. FID judges the similarity between edited images and real images. The results in Table 2 show that our method produces a lower FID than others, meaning a slightly better realism.

Time Consuming Comparison. The time consuming of inference on a single NVIDIA RTX is evaluated as shown in Table 2. Both LOHO and BarberShop need several minutes since they are optimization-based methods. As learning-based methods, both MichiGAN and our method only need several seconds, making them friendly for real-time interaction with users' manipulation.

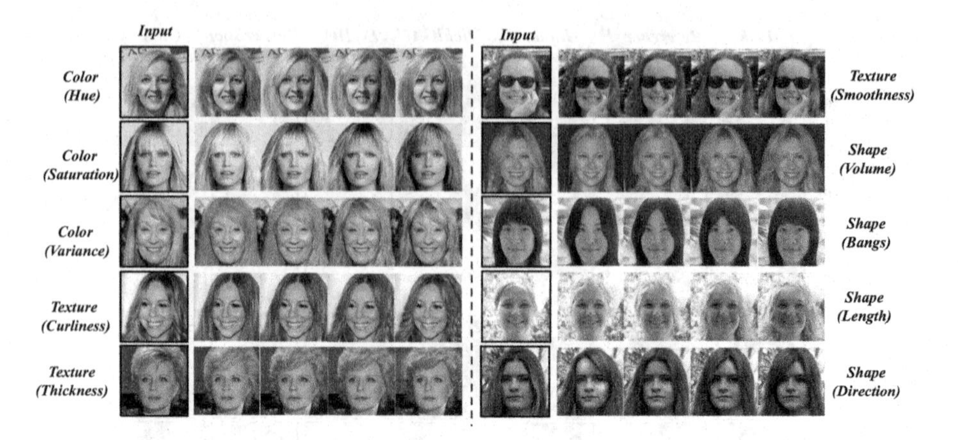

Fig. 5. Continuous hair editing at each dimension of an attribute via sliding a bar

Fig. 6. A wide variety of hair is obtained continuously for the person in the red box (Color figure online)

4.3 Ability of Continuous Editing

The major contribution of our method is continuous and fine editing of hair. Figure 1 shows some results of editing multiple attributes via sliding bars, and Fig. 5 further shows results of editing fine-grained variational factors of each attribute with no need of any reference image. These visual results show that our CtrlHair can correctly and separately edit the hair, i.e. correctly edit the desirable attribute while not affect other hair attributes or face region, in a continuous manner. Based on this capability, a wide variety of hair can be obtained continuously in Fig. 6, and continuous editing towards a target style indicated by a reference image can be naturally obtained as shown in Fig. 7.

Moreover, we quantitatively verify the ability of continuous editing of our method with results shown in Fig. 8. We take three attributes for examplar, including saturation of color, volume of shape, and length of shape since the real

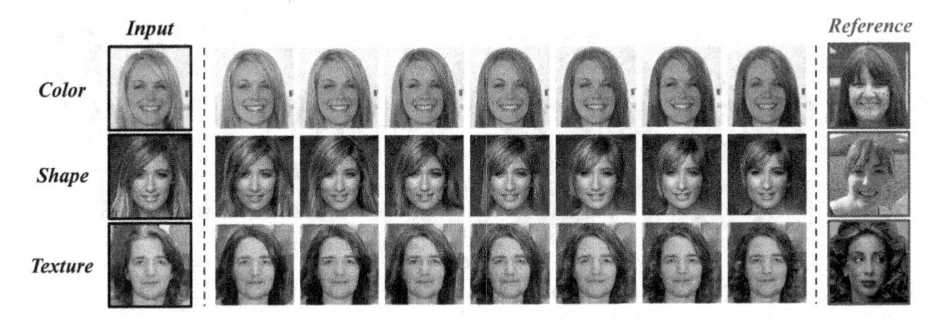

Fig. 7. Continuous editing towards a target style indicated by a reference image

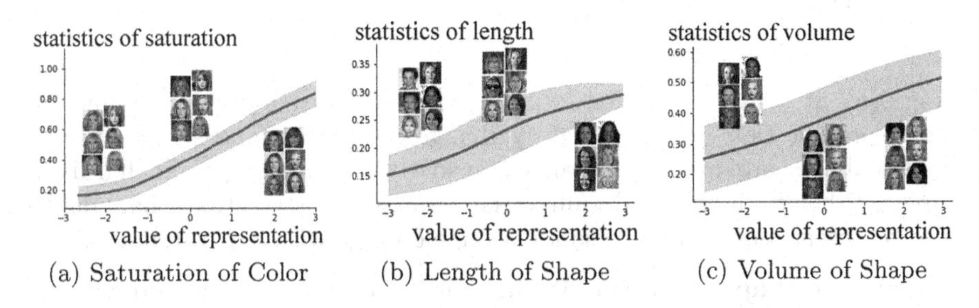

(a) Saturation of Color (b) Length of Shape (c) Volume of Shape

Fig. 8. Illustration of continuous hair editing via quantitative curve for three attributes. Each curve shows attribute value of edited image w.r.t. the latent representation value

value of these attributes are easily calculated as ground-truth for analysis. We continuously change the value of the latent representation (i.e. horizontal axis), such as saturation of color. Then, portrait images are generated with these latent representations. After that, the values corresponding to the manipulated factor in the generated image are calculated as values in the vertical axis, such as the real saturation of hair region of generated image. Each curve in Fig. 8 is plotted according to 100 sampling points during the interval $[-3\sigma, 3\sigma]$. Each point is calculated as an average 900 images with standard variance shown as blue region around the curve. As can be seen, the latent representation affects the attributes continuously and smoothly, validating the ability of our CtrlHair.

Please refer to the supplementary material for more results, more description of the values in Fig. 8, and editing demos with a user interface.

4.4 Ablation Study

In this section, we verify the effectiveness of standard multivariate Gaussian distribution, and necessity of shape adaptor.

Effectiveness of Gaussian Distribution. To evaluate how the standard multivariate Gaussian distribution affects the control of editing, we compare the hair editing results with and without it by randomly sampling texture and shape latent representations from Gaussian distribution. The results are shown

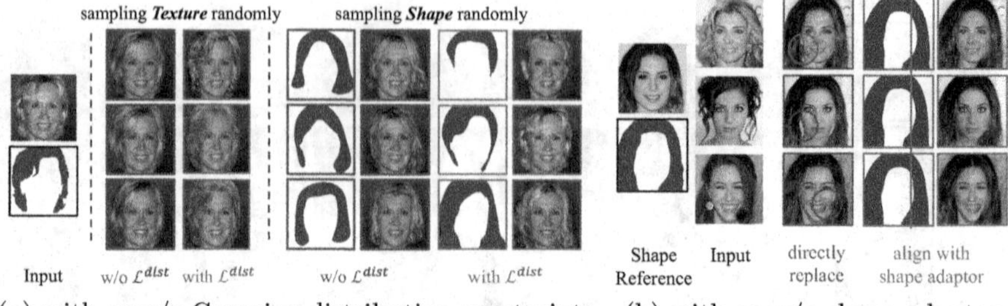

(a) with or w/o Gaussian distribution constraints (b) with or w/o shape adaptor

Fig. 9. Ablation study of the constraints of standard multivariate Gaussian distribution and shape adaptor.

in Fig. 9(a). It can be seen that, without Gaussian distribution, changing the latent representation can only slightly change the texture and shape since the hairs attributes are mapped to an expansive space, making them hardly be sampled. When with standard multivariate Gaussian distribution, changing the latent representation can traverse a reasonable range with rich variations.

Necessity of Shape Adaptor. We investigate our CtrlHair method with and without shape adaptor, i.e. directly replace the hair mask of an input image with that of the reference image. The results are shown in Fig. 9(b). It can be clearly seen that, in the edited images without using the shape adaptor, a large-shaped hair and a thin face does not match well, leading to awkward results. This comparison demonstrates the necessity and effectiveness of our shape adaptor.

5 Conclusion and Future Work

In this work, we propose an efficiently controllable editing method for hair, which can provide a set of bars for users to simply slide it to do continuous and fine hair editing. The proposed method also naturally supports editing with reference photos and user-painted mask. The method is designed as an encoding-editing-decoding framework, where the latent representations for each attribute of hair are formulated as standard multivariate Gaussian distribution supporting for continuous and fine editing. In this work, most variational factors of attributes are mined unsupervised, which does not necessarily correspond to a meaning semantic. In the future, we will explore disentangling them according to the intrinsic semantic of hair attributes.

Acknowledgements. This work is partially supported by the National Key Research and Development Program of China (No. 2017YFA0700800), the Natural Science Foundation of China (No. 62122074), and the Beijing Nova Program (Z191100001119123).

References

1. Burgess, C.P., et al.: MONet: unsupervised scene decomposition and representation. arXiv preprint arXiv:1901.11390 (2019)
2. Chai, M., Ren, J., Tulyakov, S.: Neural hair rendering. In: Vedaldi, A., Bischof, H., Brox, T., Frahm, J.-M. (eds.) ECCV 2020. LNCS, vol. 12363, pp. 371–388. Springer, Cham (2020). https://doi.org/10.1007/978-3-030-58523-5_22
3. Chen, X., Duan, Y., Houthooft, R., Schulman, J., Sutskever, I., Abbeel, P.: InfoGAN: interpretable representation learning by information maximizing generative adversarial nets. In: Neural Information Processing Systems (NIPS) (2016)
4. Choi, Y., Choi, M., Kim, M., Ha, J.W., Kim, S., Choo, J.: StarGAN: unified generative adversarial networks for multi-domain image-to-image translation. In: IEEE Conference on Computer Vision and Pattern Recognition (CVPR), pp. 8789–8797 (2018)
5. Choi, Y., Uh, Y., Yoo, J., Ha, J.W.: StarGAN v2: diverse image synthesis for multiple domains. In: IEEE/CVF Conference on Computer Vision and Pattern Recognition (CVPR), pp. 8188–8197 (2020)
6. Goodfellow, I., et al.: Generative adversarial nets. In: Neural Information Processing Systems (NIPS) (2014)
7. He, Z., Zuo, W., Kan, M., Shan, S., Chen, X.: AttGAN: facial attribute editing by only changing what you want. IEEE Trans. Image Process. **28**(11), 5464–5478 (2019)
8. Heusel, M., Ramsauer, H., Unterthiner, T., Nessler, B., Hochreiter, S.: GANs trained by a two time-scale update rule converge to a local Nash equilibrium. Adv. Neural Inf. Process. Syst. (2017)
9. Johnson, J., Alahi, A., Fei-Fei, L.: Perceptual losses for real-time style transfer and super-resolution. In: Leibe, B., Matas, J., Sebe, N., Welling, M. (eds.) ECCV 2016. LNCS, vol. 9906, pp. 694–711. Springer, Cham (2016). https://doi.org/10.1007/978-3-319-46475-6_43
10. Karras, T., Aila, T., Laine, S., Lehtinen, J.: Progressive growing of GANs for improved quality, stability, and variation. In: International Conference on Learning Representations (ICLR) (2018)
11. Karras, T., Laine, S., Aila, T.: A style-based generator architecture for generative adversarial networks. In: IEEE/CVF Conference on Computer Vision and Pattern Recognition (CVPR) (2019)
12. Karras, T., Laine, S., Aittala, M., Hellsten, J., Lehtinen, J., Aila, T.: Analyzing and improving the image quality of StyleGAN. In: IEEE/CVF Conference on Computer Vision and Pattern Recognition (CVPR), pp. 8110–8119 (2020)
13. Lee, C.H., Liu, Z., Wu, L., Luo, P.: MaskGAN: towards diverse and interactive facial image manipulation. In: IEEE/CVF Conference on Computer Vision and Pattern Recognition (CVPR), pp. 5549–5558 (2020)
14. Liu, M., et al.: STGAN: a unified selective transfer network for arbitrary image attribute editing. In: IEEE/CVF Conference on Computer Vision and Pattern Recognition (CVPR), pp. 3673–3682 (2019)
15. Mirza, M., Osindero, S.: Conditional generative adversarial nets. arXiv preprint arXiv:1411.1784 (2014)
16. Odena, A., Olah, C., Shlens, J.: Conditional image synthesis with auxiliary classifier GANs. In: International Conference on Machine Learning (ICML), pp. 2642–2651 (2017)

17. Or-El, R., Sengupta, S., Fried, O., Shechtman, E., Kemelmacher-Shlizerman, I.: Lifespan age transformation synthesis. In: Vedaldi, A., Bischof, H., Brox, T., Frahm, J.-M. (eds.) ECCV 2020. LNCS, vol. 12351, pp. 739–755. Springer, Cham (2020). https://doi.org/10.1007/978-3-030-58539-6_44
18. Park, T., Liu, M.Y., Wang, T.C., Zhu, J.Y.: Semantic image synthesis with spatially-adaptive normalization. In: IEEE/CVF Conference on Computer Vision and Pattern Recognition (CVPR), pp. 2337–2346 (2019)
19. Perarnau, G., Van De Weijer, J., Raducanu, B., Álvarez, J.M.: Invertible conditional GANs for image editing. arXiv preprint arXiv:1611.06355 (2016)
20. Radford, A., Metz, L., Chintala, S.: Unsupervised representation learning with deep convolutional generative adversarial networks. In: International Conference on Learning Representations (ICLR) (2015)
21. Saha, R., Duke, B., Shkurti, F., Taylor, G.W., Aarabi, P.: LOHO: latent optimization of hairstyles via orthogonalization. In: IEEE/CVF Conference on Computer Vision and Pattern Recognition (CVPR), pp. 1984–1993 (2021)
22. Shen, W., Liu, R.: Learning residual images for face attribute manipulation. In: IEEE Conference on Computer Vision and Pattern Recognition (CVPR), pp. 4030–4038 (2017)
23. Shoshan, A., Bhonker, N., Kviatkovsky, I., Medioni, G.: GAN-control: explicitly controllable GANs. In: IEEE/CVF International Conference on Computer Vision (ICCV), pp. 14083–14093 (2021)
24. Tan, Z., et al.: MichiGAN: multi-input-conditioned hair image generation for portrait editing. ACM Trans. Graph. **39**(4), 95 (2020)
25. Yu, C., Wang, J., Peng, C., Gao, C., Yu, G., Sang, N.: BiSeNet: bilateral segmentation network for real-time semantic segmentation. In: European Conference on Computer Vision (ECCV), pp. 325–341 (2018)
26. Yu, J., Lin, Z., Yang, J., Shen, X., Lu, X., Huang, T.S.: Free-form image inpainting with gated convolution. In: IEEE/CVF International Conference on Computer Vision (ICCV), pp. 4471–4480 (2019)
27. Zhang, G., Kan, M., Shan, S., Chen, X.: Generative adversarial network with spatial attention for face attribute editing. In: European Conference on Computer Vision (ECCV), pp. 417–432 (2018)
28. Zhu, P., Abdal, R., Femiani, J., Wonka, P.: Barbershop: GAN-based image compositing using segmentation masks. ACM Trans. Graph. **40**(6), 215:1–215:13 (2021)
29. Zhu, P., Abdal, R., Qin, Y., Wonka, P.: SEAN: image synthesis with semantic region-adaptive normalization. In: IEEE/CVF Conference on Computer Vision and Pattern Recognition (CVPR), pp. 5104–5113 (2020)
30. zllrunning: face-parsing.PyTorch. https://github.com/zllrunning/face-parsing.PyTorch

Discovering Transferable Forensic Features for CNN-Generated Images Detection

Keshigeyan Chandrasegaran[1(✉)], Ngoc-Trung Tran[1], Alexander Binder[2,3], and Ngai-Man Cheung[1]

[1] Singapore University of Technology and Design (SUTD), Singapore, Singapore
{keshigeyan,ngoctrung_tran,ngaiman_cheung}@sutd.edu.sg
[2] Singapore Institute of Technology (SIT), Singapore, Singapore
alexander.binder@singaporetech.edu.sg
[3] University of Oslo (UIO), Oslo, Norway
alexabin@uio.no

Abstract. Visual counterfeits (We refer to CNN-generated images as counterfeits throughout this paper.) are increasingly causing an existential conundrum in mainstream media with rapid evolution in neural image synthesis methods. Though detection of such counterfeits has been a taxing problem in the image forensics community, a recent class of forensic detectors – *universal detectors* – are able to surprisingly spot counterfeit images regardless of generator architectures, loss functions, training datasets, and resolutions [61]. This intriguing property suggests the possible existence of *transferable forensic features (T-FF)* in *universal detectors*. In this work, we conduct the first analytical study to discover and understand *T-FF* in *universal detectors*. Our contributions are 2-fold: 1) We propose a novel *forensic feature relevance statistic (FF-RS)* to quantify and discover *T-FF* in *universal detectors* and, 2) Our qualitative and quantitative investigations uncover an unexpected finding: *color* is a critical *T-FF* in *universal detectors*. Code and models are available at https://keshik6.github.io/transferable-forensic-features/.

1 Introduction

Visual counterfeits are increasingly causing an existential conundrum in mainstream media [22,23,37,43,51]. With rapid improvements in CNN-based generative modelling [1,6,12,21,26–29,31,32,45,47,56–58,64–66], detection of such counterfeits is increasingly becoming challenging and critical. Nevertheless, a recent class of forensic detectors known as *universal detectors* are able to surprisingly spot counterfeits regardless of generator architectures, loss functions, datasets and resolutions [61]. i.e.: Publicly released ResNet-50 [24] universal detector by Wang *et al.* [61] trained only on ProGAN [26] counterfeits, surprisingly generalizes well to detect counterfeits from unseen GANs including StyleGAN2 [29], StyleGAN [28], BigGAN [6], CycleGAN [66], StarGAN [11] and

Supplementary Information The online version contains supplementary material available at https://doi.org/10.1007/978-3-031-19784-0_39.

S. Avidan et al. (Eds.): ECCV 2022, LNCS 13675, pp. 671–689, 2022.
https://doi.org/10.1007/978-3-031-19784-0_39

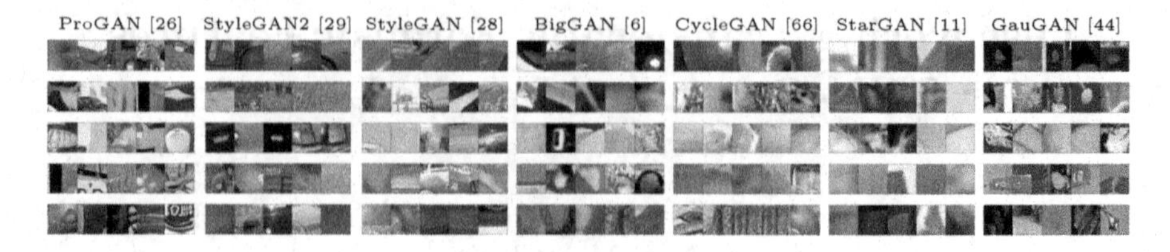

Fig. 1. Color is a critical *transferable forensic feature (T-FF)* in universal detectors: Large-scale study on visual interpretability of *T-FF* discovered through our proposed *forensic feature relevance statistic (FF-RS)*, reveal that color information is critical for cross-model forensic transfer. Each row represents a color-conditional *T-FF* and we show the LRP-max response regions for different GANs counterfeits for the publicly released ResNet-50 universal detector by Wang *et al.* [61]. This detector is trained with ProGAN [26] counterfeits [61] and cross-model forensic transfer is evaluated on unseen GANs. All counterfeits are obtained from the ForenSynths dataset [61]. The consistent color-conditional LRP-max response across all GANs for these *T-FF* clearly indicate that *color* is critical for cross-model forensic transfer in universal detectors. We further observe similar results using an EfficientNet-B0-based [55] universal detector following the exact training/test strategy proposed by Wang *et al.* [61] in Fig. 3. More visualizations are included in Supplementary.

GauGAN [44]. This intriguing cross-model forensic transfer property suggests the existence of *transferable forensic features (T-FF)* in universal detectors.

1.1 Transferable Forensic Features (T-FF) in Universal Detectors

This work is motivated by a profound and challenging thesis statement: *What are the transferable forensic features (T-FF) in universal detectors for counterfeit detection?* A more elemental representation of this thesis statement would be: given an image of a real car and a high fidelity synthetic car generated from an unseen GAN (i.e.: StyleGAN2 [29]), what *T-FF* are used by the universal detector, such that it detects the synthetic car as counterfeit accurately? Though Wang *et al.* [61] hypothesize that universal detectors may learn low-level CNN artifacts for detection, no qualitative or quantitative evidence is available in contemporary literature to understand *T-FF* in universal detectors. *Our work takes the first step towards discovering and understanding T-FF in universal detectors for counterfeit detection.* A foundational understanding on *T-FF* and their properties are of paramount importance to both image forensics research and image synthesis research. Understanding *T-FF* will allow to build robust forensic detectors and to devise techniques to improve image synthesis methods to avoid generation of forensic footprints.

1.2 Our Contributions

Our work conduct the *first analytical study to discover and understand T-FF in universal detectors for counterfeit detection.* We begin our study by comprehensively demonstrating that input-space attribution – using 2 popular algorithms namely Guided-GradCAM [50] and LRP [2] – of universal detector decisions are not informative to discover *T-FF*. Next, we study the forensic feature space of universal detectors to discover *T-FF*. But investigating the feature space is an extremely daunting task due to the sheer amount of feature maps present. i.e.: ResNet-50 [24] architecture contains approximately 27K feature maps. To tackle this challenging task, *we propose a novel forensic feature relevance statistic (FF-RS), to quantify and discover T-FF in universal detectors.* Our proposed FF-RS (ω) is a scalar which quantifies the ratio between positive forensic relevance of the feature map and the total unsigned relevance of the entire layer that contains the particular feature map. Using our proposed FF-RS (ω), we successfully discover *T-FF* in the publicly released ResNet-50 universal detector [61].

Next, to understand the discovered *T-FF*, *we introduce a novel pixel-wise explanation method based on maximum spatial Layer-wise Relevance Propagation response (LRP-max).* Particularly we visualize the pixel-wise explanations of each discovered *T-FF* in universal detectors independently using LRP-max visualization method. Large-scale study on visual interpretability of *T-FF* reveal that color information is critical for cross-model forensic transfer. Further large-scale quantitative investigations using median counterfeits probability analysis and statistical tests on maximum spatial activation distributions based on color ablation show that *color* is a critical *T-FF* in universal detectors. Our findings are intriguing and new to the research community, as many contemporary image forensics works focus on frequency discrepancies between real and counterfeit images [8,16,17,30,49,63]. In summary, our contributions are as follows:

- We propose a novel *forensic feature relevance statistic (FF-RS)* to quantify and discover *transferable forensic features (T-FF)* in universal detectors for counterfeit detection.
- We qualitatively – using our proposed LRP-max visualization for feature map activations – and quantitatively – using median counterfeits probability analysis and statistical tests on maximum spatial activation distributions based on color ablation – show that *color* is a critical *transferable forensic feature (T-FF)* in universal detectors for counterfeit detection.

2 Related Work

Counterfeit Detection: Recent works have studied counterfeit detection both in the RGB domain [13,38,42,48,60,61,63] and frequency domain [8,16–18,35]. Particularly, notable number of works have proposed to use hand-crafted features for counterfeit detection [8,16,17,42]. Some recent works have also proposed methods to detect and attribute counterfeits to the generating architectures [39, 62]. Anomaly detection techniques leveraging on pre-trained face recognition models have also been proposed [60].

Cross-Model Forensic Transfer: Most counterfeit detection works do not focus on cross-model forensic transfer. Among the works that study forensic transfer, Cozzolino *et al.* [13] and Zhang *et al.* [63] observed that counterfeit detectors generalized poorly during cross-model forensic transfer. In order to solve poor forensic transfer performance, Cozzolino *et al.* [13] proposed an autoencoder based adaptation framework to improve cross-model forensic transfer. Using simple experiments, Mccloskey *et al.* [40] showed that detection based on the frequency of over-exposed pixels can provide good discrimination between real images and counterfeits. The work by Wang *et al.* [61] was the first work to show that counterfeit detectors – universal detectors – can generalize well during cross-model forensic transfer without any re-training/fine-tuning/adaptation on the target samples suggesting the possible existence of *transferable forensic features*. Furthermore, Chai *et al.* [7] showed that patch-based forensic detectors with limited receptive fields often perform better at detecting unseen counterfeits compared to full-image based detectors.

Interpretability Methods: A number of interpretability methods in machine learning aim to summarize the relations which a model has learnt as a whole, such as PCA and t-SNE [36,46], or to explain single decisions of a neural network. The latter may follow very different lines of questioning, such as identifying similar training samples in k-NN and prototype CNNs [9,33], finding modified samples such as pertinent negatives [15], or model-based uncertainty estimates [20]. One class of algorithms aims at computing input space attributions. This includes Shapley values [10,34,54] suitable for tabular data types, and methods for data types for which dropping a feature is not well defined, relying on modified gradients such as Guided Backprop [52], Layer-wise Relevance Propagation (LRP) [2], Guided-GradCAM [50], Full-Grad [53], and class-attention-mapping inspired research [14,19,25,41,59]. Bau *et al.* proposed frameworks for interpreting representations at the feature map level used for GANs [3,4].

3 Dataset/Metrics

We use the ForenSynths dataset proposed by Wang *et al.* [61]. ForenSynths is the largest counterfeit benchmark dataset containing CNN-generated images from multiple generator architectures, datasets, loss functions and resolutions. In addition to ProGAN [26], we select 6 candidate GANs to comprehensively study cross-model forensic transfer in this work namely, StyleGAN2 [29], StyleGAN [28], BigGAN [6], CycleGAN [66], StarGAN [11] and GauGAN [44]. Following Wang *et al.* [61], we use AP (Average Precision) to measure cross-model forensic transfer of universal detectors. Particularly, we also show the accuracies for real and counterfeit images as we intend to understand counterfeit detection.

4 Discovering Transferable Forensic Features (T-FF)

4.1 Input-Space Attribution Methods

Interpretable machine learning algorithms are useful exploratory tools to visualize neural networks' decisions by input-space attribution [5,14,19,25,41,50,53,59]. We start from the following question: *Are interpretability methods suitable to discover T-FF in universal detectors?*

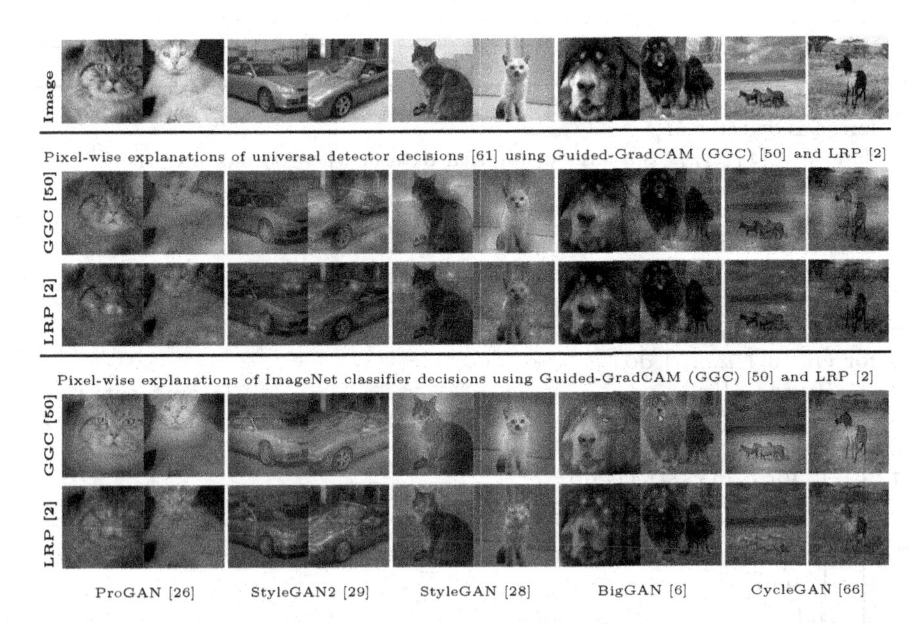

Fig. 2. Pixel-wise explanations of universal detector decisions are not informative to discover *T-FF*: We show pixel-wise explanations using Guided-GradCAM (GGC) (row 2) [50] and LRP (row 3) [2] for the ResNet-50 universal detector [61] for ProGAN [26], CycleGAN [66], StarGAN [11], BigGAN [6] and StyleGAN2 [29]. The universal detector predicts probability $p \geq 95\%$ for all counterfeit images shown above. All these counterfeits are obtained from the ForenSynths dataset [61]. For LRP [2], we only show the positive relevances. We also show the pixel-wise explanations of ImageNet classifier decisions for the exact counterfeits using GGC (row 4) and LRP (row 5). This is shown as a control experiment to emphasize the significance of our observations. As one can clearly observe, pixel-wise explanations of universal detector decisions are not informative to discover *T-FF* (rows 2, 3) as the explanations appear to be random and not reveal any meaningful visual features used for counterfeit detection. Particularly, it remains unknown as to why the universal detector outputs high detection probability ($p \geq 95\%$) for these counterfeits. On the other hand, pixel-wise explanations of ImageNet classifier decisions produce meaningful results. i.e.: GGC (row 4) and LRP (row 5) explanation results for cat samples (columns 1, 2, 5, 6) show that ImageNet uses features such as eyes and whiskers to classify cats. This shows that interpretability techniques such as GGC and LRP are not informative to discover *T-FF* in universal detectors. In other words, we are unable to discover any forensic footprints based on pixel-wise explanations of universal detectors. More examples shown in Supplementary.

We use 2 popular interpretation methods namely Guided-GradCAM [50] and LRP [2] to analyse the pixel-wise explanations of universal detector decisions. These methods were chosen due to their relatively low amount of gradient shattering noise. We show the pixel-wise explanation results of ResNet-50 universal detector [61] decisions for ProGAN [26] and 4 GANs not used for training – CycleGAN [66], StarGAN [11], BigGAN [6] and StyleGAN2 [29] – in Fig. 2. As one can observe in Fig. 2, pixel-wise explanations of universal detector decisions

Algorithm 1: Calculate FF-RS (ω) (Non-vectorized)

Input:
forensics detector M,
data $D = \{x\}_{i=1}^{n}$, D is a large counterfeit dataset where x_i indicates the ith counterfeit image.
Output:
$\omega(l_c)$ where l, c indicates the layer and channel index of forensic feature maps. Every forensic feature map can be characterized by a unique set of l, c.

1 $R \leftarrow [\,]$; /*List to store feature map relevances*/
2 Set M to evaluation mode
3 **for** i *in* $\{0, 1, , ..., n\}$ **do**
4     $f(x_i) \leftarrow M(x_i)$; /*logit output*/
5     $r_i \leftarrow LRP(M, x_i, f(x_i))$; /*calculate LRP scores for counterfeits*/
6   **for** l' *in* $r_i.size(0)$ **do**
7   **for** c' *in* $r_i.size(1)$ **do**
8             $r_i(l', c', h, w) \leftarrow \frac{max(0, r_i(l', c', h, w))}{\sum_{c,h,w} ||r_i(l', c, h, w)||}$
9             $R.append(r_i)$; /*r_i.size():(layer, channel, height, width)*/
10   **end**
11   **end**
12 **end**
13 $\omega(l_c) \leftarrow \sum_{h,w} \frac{1}{N} \sum_{i}^{n} R_i(l, c, h, w)$; /*forensic feature relevance*/
14 **return** $\omega(l_c)$

are not informative to discover *T-FF* due to their focus on spatial localization. Particularly, we are unable to discover any forensic footprints based on pixel-wise explanations of universal detector decisions. This is consistently seen across both Guided-GradCAM [50] and LRP [2] methods. We remark that these observations do not indicate failure modes of Guided-GradCAM [50] or LRP [2] methods, but rather suggest that universal detectors are learning more complex *T-FF* that are not easily human-parsable.

4.2 Forensic Feature Space

Given that input-space attribution methods are not informative to discover *T-FF*, we study the feature space to discover *T-FF* in universal detectors for counterfeit detection. Particularly, we ask the question: which feature maps in universal detectors are responsible for cross-model forensic transfer? This is a very challenging problem as it requires quantifying the importance of every feature map in universal detectors for counterfeit detection. The ResNet-50 universal detector [61] consists of approximately 27K intermediate feature maps.

Forensic Feature Relevance Statistic (FF-RS): We propose a novel *FF-RS (ω)* to quantify the relevance of every feature map in universal detectors for counterfeit detection. Specifically, for feature map at layer l and channel c, $\omega(l_c)$ computes the forensic relevance of this feature map for counterfeit detection. We

Table 1. Sensitivity assessments using feature map dropout showing that our proposed *FF-RS* (ω) successfully quantifies and discovers *T-FF*: We show the results for the publicly released ResNet-50 universal detector [61] (top) and our own version of EfficientNet-B0 [55] universal detector (bottom) following the exact training and test strategy proposed in [61]. We show the AP, real and GAN image detection accuracies for baseline [61], top-k, random-k and low-k forensic feature dropout. The random-k experiments are repeated 5 times and average results are reported. Feature map dropout is performed by suppressing (zeroing out) the resulting activations of target feature maps (i.e.: top-k). We can clearly observe that feature map dropout of top-k corresponding to *T-FF* results in substantial drop in AP and GAN detection accuracies across ProGAN and all 6 unseen GANs [6,11,28,29,44,66] compared to baseline, random-k and low-k results. This is consistently seen in both ResNet-50 and EfficientNet-B0 universal detectors. This shows that our proposed *FF-RS* (ω) can successfully quantify and discover the *T-FF* in universal detectors. $k \approx 0.5\%$ of total feature maps. More details included in Supplementary.

ResNet-50 $k = 114$	ProGAN [26]			StyleGAN2 [29]			StyleGAN [28]			BigGAN [6]			CycleGAN [66]			StarGAN [11]			GauGAN [44]		
	AP	Real	GAN	AP	Real	GAN	AP	Real	GAN	AP	Real	GAN	AP	Real	GAN	AP	Real	GAN	AP	Real	GAN
baseline [61]	100	100.0	100	99.1	95.5	95.0	99.3	96.0	95.6	90.4	83.9	85.1	97.9	93.4	92.6	97.5	94.0	89.3	98.8	93.9	96.4
top-k	**69.8**	**99.4**	**3.2**	**55.3**	**89.4**	**11.3**	**56.6**	**90.6**	**13.7**	**55.4**	**86.3**	**18.3**	**61.2**	**91.4**	**17.4**	**72.6**	**89.4**	**35.9**	**71.0**	**95.0**	**18.8**
random-k	100	99.9	96.1	98.6	89.4	96.9	98.7	91.4	96.1	88.0	79.4	85.0	96.6	81.0	96.2	97.0	88.0	91.7	98.7	91.9	97.1
low-k	100	100	100	99.1	95.6	95.0	99.3	96.0	95.6	90.4	83.9	85.1	97.9	93.4	92.6	97.5	94.0	89.3	98.8	93.9	96.4
EfficientNet-B0 $k = 27$	ProGAN [26]			StyleGAN2 [29]			StyleGAN [28]			BigGAN [6]			CycleGAN [66]			StarGAN [11]			GauGAN [44]		
	AP	Real	GAN	AP	Real	GAN	AP	Real	GAN	AP	Real	GAN	AP	Real	GAN	AP	Real	GAN	AP	Real	GAN
baseline	100	100	100	95.9	95.2	85.4	99.0	96.1	94.3	84.4	79.7	75.9	97.3	89.6	93.0	96.0	92.8	85.5	98.3	94.1	94.4
top-k	**50.0**	**100.**	**0.0**	**54.5**	**94.3**	**7.0**	**52.1**	**97.3**	**2.6**	**53.5**	**97.4**	**3.8**	**47.5**	**100.**	**0.0**	**50.0**	**100.**	**0.0**	**46.2**	**100.**	**0.0**
random-k	100	99.9	100	96.5	91.9	89.8	99.2	91.2	97.5	84.5	59.4	89.1	96.9	82.6	95.8	96.7	82.5	93.3	98.1	87.8	96.2
low-k	100	100	100	95.3	88.7	88.3	98.9	90.8	96.1	83.5	70.8	80.8	96.6	85.2	94.1	95.4	91.0	85.4	98.1	91.2	96.4

describe the important design considerations and intuitions behind our proposed *FF-RS* (ω) below and include the pseudocode in Algorithm 1:

- We postulate the existence of a set of feature maps in universal detectors that are responsible for cross-model forensic transfer. In particular, we hypothesize that there is a set of *common transferable forensic feature maps* that mostly gets activated when passing counterfeits from ProGAN [26] and unseen GANs.
- Our proposed *FF-RS* (ω) is a scalar that quantifies the forensic relevance of every feature map. In particular, ω for a feature map quantifies the ratio between positive forensic relevance of the feature map and the total unsigned forensic relevance of the entire layer that contains the particular feature map. This is shown in Line 8 in Algorithm 1. For the numerator we are only interested in positive relevance, therefore use a max operation to select only positive relevance (identical to the ReLU operation).
- The relevance scores are calculated using LRP [2] (More details on LRP [2] in Supplementary). This is shown in Line 5 in Algorithm 1 where $r_i(l, c, h, w)$ is the estimated relevance of the feature map at layer l, channel c at the spatial location h, w

- ω is calculated over large number of counterfeit images and is bounded between $[0, 1]$. i.e.: $\omega = 1$ indicates that the particular feature map is the most relevant forensic feature and $\omega = 0$ indicates vice versa.
- Finally we use ω to rank all the feature maps and identify the set of *T-FF*. We refer to this set as top-k in our experiments.

Experiments: Sensitivity Assessments of Discovered T-FF Using Algorithm 1. We perform rigorous sensitivity assessments using feature map dropout experiments to demonstrate that our proposed *FF-RS (ω)* is able to quantify and discover *T-FF*. Feature map dropout suppresses (zeroing out) the resulting activations of the target feature maps. Particularly, feature map dropout of *T-FF* should satisfy the following sensitivity conditions:

1. Significant reduction in overall AP across ProGAN [26] and all unseen GANs [6,11,28,29,44,66] indicating poor cross-model forensic transfer.
2. Significant reduction in GAN/counterfeit detection accuracies across ProGAN [26] and all unseen GANs [6,11,28,29,44,66] compared to real image detection accuracies as ω is calculated for counterfeits.

Test Bed Details: We use the ForenSynths test dataset proposed in [61]. ω is calculated using 1000 counterfeits from ProGAN [26] validation set in ForenSynths. We use the following experiment codes:

- top-k: Set of T-FF discovered using *FF-RS* (ω)
- random-k: Set of random feature maps used as a control experiment.
- low-k: Set of low-ranked feature maps corresponding to extremely small values of ω, i.e.: $\omega \approx 0$.

Results: We show the results in Table 1 for ResNet-50 and EfficientNet-B0 universal detectors. We clearly observe that feature map dropout of top-k features corresponding to *T-FF* satisfies both sensitivity conditions above indicating that our proposed *FF-RS (ω)* is able to quantify and discover *transferable forensic features*. We also observe that feature map dropout of low-k (low-ranked) forensic features has little/no effect on cross-model forensic transfer which further adds merit to our proposed *FF-RS (ω)*.

5 Understanding Transferable Forensic Features (T-FF)

Given the successful discovery of *T-FF* using our proposed *FF-RS (ω)*, in this section, we ask the following question: what counterfeit properties are detected by this set of *T-FF*? Though Wang *et al.* [61] hypothesize that universal detectors may learn low-level CNN artifacts for cross-model forensic transfer, no qualitative/quantitative evidence is available to understand as to what features in counterfeits are being detected during cross-model forensic transfer.

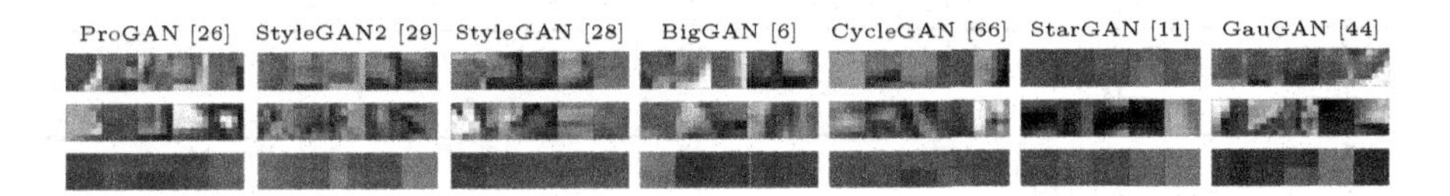

Fig. 3. Color is a critical *T-FF* in universal detectors: Large-scale study on visual interpretability of *T-FF* discovered through our proposed *FF-RS (ω)* reveal that color information is critical for cross-model forensic transfer. Each row represents a color-based *T-FF* and we show the LRP-max response regions for ProGAN and all 6 unseen GANs [6,11,28,29,44,66] counterfeits for our own version of EfficientNet-B0 [55] universal detector following the exact training/test strategy proposed by Wang *et al.* [61]. This detector is trained with ProGAN [26] counterfeits [61] and cross-model forensic transfer is evaluated on other unseen GANs. All counterfeits are obtained from the ForenSynths dataset [61]. The consistent color-conditional LRP-max response across all GANs for these *T-FF* clearly indicate that *color* is critical for cross-model forensic transfer in universal detectors. More visualizations are included in Supplementary.

5.1 LRP-Max Explanations of T-FF

We approach this problem from a visual interpretability perspective. In this section, we introduce a novel pixel-wise explanation method for feature map activations based on maximum spatial Layer-wise Relevance Propagation response (LRP-max). The idea behind LRP-max is to independently visualize which pixels in the input space correspond to maximum spatial relevance scores for each *T-FF*. Particularly, instead of back-propagating using the detector logits, we back-propagate from the maximum spatial relevance neuron of each *T-FF* independently. We remark that LRP-max does not depend on external modules such as segmentation used in Network Dissection [3] and GAN Dissection [4] methods. The pseudocode is included in Supplementary.

Color is a Critical T-FF in Universal Detectors: LRP-max visualizations of *T-FF* uncover the unexpected observation that a substantial amount of *T-FF* exhibits color-conditional activations. We show the LRP-max regions for ProGAN [26] and all unseen GANs [6,11,28,29,44,66] for ResNet-50 and EfficientNet-B0 universal detectors in Fig. 1 and 3 respectively. As one can observe, the consistent color-conditional LRP-max response across all GANs for these T-FF clearly indicate that color is critical for cross-model forensic transfer in universal detectors. This is very surprising and observed for the first time in contemporary image forensics research, yet shown qualitatively. In the next section, we conduct quantitative studies to rigorously verify that color is a critical *T-FF* in universal detectors.

5.2 Color Ablation Studies

In this section, we conduct 2 quantitative studies to show that *color* is a critical *transferable forensic feature* in universal detectors. Our studies measure the sensitivity of universal detectors before and after color ablation.

Algorithm 2: Statistical test over maximum spatial activations for *T-FF* (Non-vectorized)

Input:
forensics detector M,
data $D = \{x\}_{i=1}^{n}$, D is a large counterfeit dataset where x_i indicates the ith counterfeit image.
T-FF set S
Output:
$p(l_c)$ where l, c indicates the layer and channel index of forensic feature maps. p indicates p-value of the statistical test.
Every forensic feature map can be characterized by a unique set of l, c.

```
1  Set M to evaluation mode
2  for l', c' in S do
3      A_b ← [] ;                          /*store baseline counterfeits activations*/
4      A_g ← [] ;                          /*store grayscale counterfeits activations*/
5      for i in {0, 1, , ...., n} do
6          a_b ← GLOBAL_MAXPOOL(M_{l_c}(x_i)) ;                       /*baseline*/
7          a_g ← GLOBAL_MAXPOOL(M_{l_c}(grayscale(x_i))) ;            /*grayscale*/
8          A_b.append(a_b)
9          A_g.append(a_g)
10     end
11     p(l'_{c'}) ← MEDIAN − TEST(A_b, A_g) ;                         /*median test*/
12 end
13 return p(l_c)
```

Study 1: We investigate the change in probability distribution of universal detectors when removing color information in counterfeits during cross-model forensic transfer. We specifically study the change in median counterfeit probability when removing color information (median is not sensitive to outliers). The results for both ResNet-50 and EfficientNet-B0 universal detectors are shown in Fig. 4. As one can clearly observe, color ablation causes the median probability predicted by the universal detector to drop by more than 89% across all unseen GANs showing that *color* is a critical *T-FF* in universal detectors. This is observed in both ResNet-50 and EfficientNet-B0 universal detectors.

Study 2: In this study, we measure the percentage of *T-FF* that are color-conditional. Particularly, we conduct a statistical test to compare the maximum globally pooled spatial activation distributions of each *T-FF* before and after color ablation. The intuition is that with color ablation, color-conditional *T-FF* will produce lower amount of activations for the same sample and we perform a hypothesis test to measure whether the maximum spatial activation distributions are statistically different before (Baseline) and after color ablation (Grayscale). Particularly, we use Mood's median test (non-parametric) with a significance level of $\alpha = 0.05$ in our study. The pseudocode is shown in Algorithm 2. The results for ResNet-50 and EfficientNet-B0 universal detectors are shown in Table 2. Our results show that substantial amount of *T-FF* in universal detectors are color-

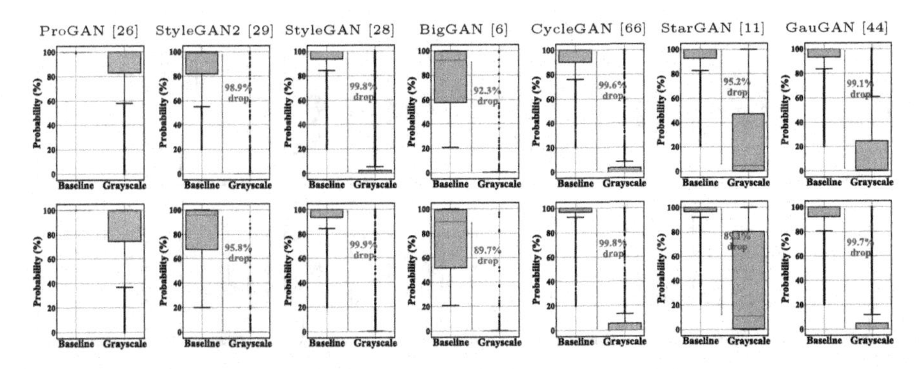

Fig. 4. *Color* is a critical *T-FF* in *universal detectors*: We show the box-whisker plots of probability (%) predicted by the universal detector for counterfeits before (Baseline) and after *color ablation* (Grayscale) for 7 GAN models. The red line in each box-plot shows the median probability. We show the results for the ResNet-50 universal detector [61] (top row) and our version of EfficientNet-B0 [55] universal detector following the exact training/test strategy proposed in [61] (bottom row). These detectors are trained with ProGAN [26] counterfeits [61] and cross-model forensic transfer is evaluated on other unseen GANs. All counterfeits are obtained from the ForenSynths dataset [61]. We clearly show that *color ablation* causes the median probability for counterfeits to drop by more than 89% across all unseen GANs. This is consistently seen across both universal detectors. These observations quantitatively show that *color* is a critical *T-FF* in universal detectors. AP and accuracies shown in Supplementary. (Color figure online)

conditional indicating that color is a critical *T-FF*. We also show the maximum spatial activation distributions for some color-conditional *T-FF* for ResNet-50 and EfficientNet-B0 universal detectors in Fig. 6 and 7 respectively. As one can observe maximum spatial activations are suppressed for these *T-FF* across ProGAN [26] and all other unseen GANs [6,11,28,29,44,66] when removing color information. This clearly suggests that these *T-FF* are color-conditional.

6 Applications: Color-Robust (CR) Universal Detectors

Reliance on substantial amount of color information for cross-model forensic transfer exposes universal detectors to attacks via color-ablated counterfeits. This is particularly unfavourable. In this section, we propose a data augmentation scheme to build Color-Robust (CR) universal detectors that do not substantially rely on color information for cross-model forensic transfer. The crux of the idea is to randomly remove color information from samples during training (both for real and counterfeit images). Particularly, we perform random Grayscaling during training with 50% probability to maneuver universal detectors to learn *T-FF* that do not substantially rely on color information.

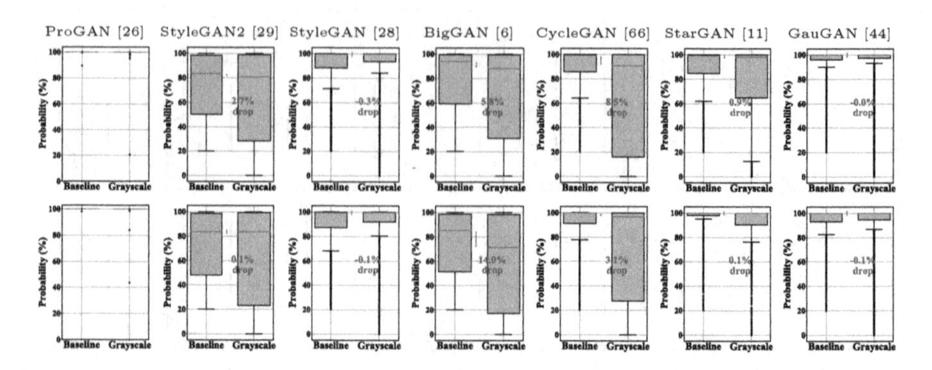

Fig. 5. *CR-universal detectors trained using our proposed data augmentation scheme (Sect. 6) are more robust to color ablation during cross-model forensic transfer:* These universal detectors are trained with data augmentation where color is ablated 50% of the time during training. This ensures that *T-FF* do not substantially rely on color information. We show the box-whisker plots of probability (%) predicted by the CR-universal detectors for counterfeits before (Baseline) and after *color ablation* (Grayscale) for 7 GAN models. The red line in each box-plot shows the median probability. We show the results for the ResNet-50 CR-universal detector [61] (top row) and EfficientNet-B0 [55] CR-universal detector (bottom row). We clearly observe that the median probability for counterfeits have similar values (compared to Fig. 4) before and after color ablation indicating CR-universal detectors are more robust to color-ablated counterfeit attacks. AP and accuracies shown in Supplementary. (Color figure online)

Results: Median probability analysis results for ResNet-50 and EfficientNet-B0 CR-universal detectors are shown in Fig. 4. We clearly observe that with our proposed data augmentation scheme, CR-universal detectors are more robust to color ablation during cross-model forensic transfer indicating that they learn *T-FF* that do not substantially rely on color information. We further show the percentage of color-conditional *T-FF* in Table 3. With our proposed data augmentation scheme, we quantitatively show that CR-universal detectors contain substantially lower amount of color-conditional *T-FF* (Fig. 5).

Table 2. *Significant amount of T-FF are color-conditional:* We show the percentage (%) of color-conditional T-FF in ResNet-50 and EfficientNet-B0 universal detectors measured using Mood's median test. We show the results for ProGAN [26] and all 6 unseen GANs [6,11,28,29,44,66]. Particularly, we consider a *T-FF* to be color conditional if the p-value of the median test is less than the significance level of $\alpha = 0.05$. As one can clearly observe, significant amount of *T-FF* are color-conditional. This quantitatively shows that color is a critical *T-FF* in universal detectors.

% Color-conditional	ProGAN [26]	StyleGAN2 [29]	StyleGAN [28]	BigGAN [6]	CycleGAN [66]	StarGAN [11]	GauGAN [44]
ResNet-50	85.1	83.3	84.2	86.8	86.8	86.0	94.7
EfficientNet-B0	51.9	55.6	55.6	48.1	44.4	55.6	63.0

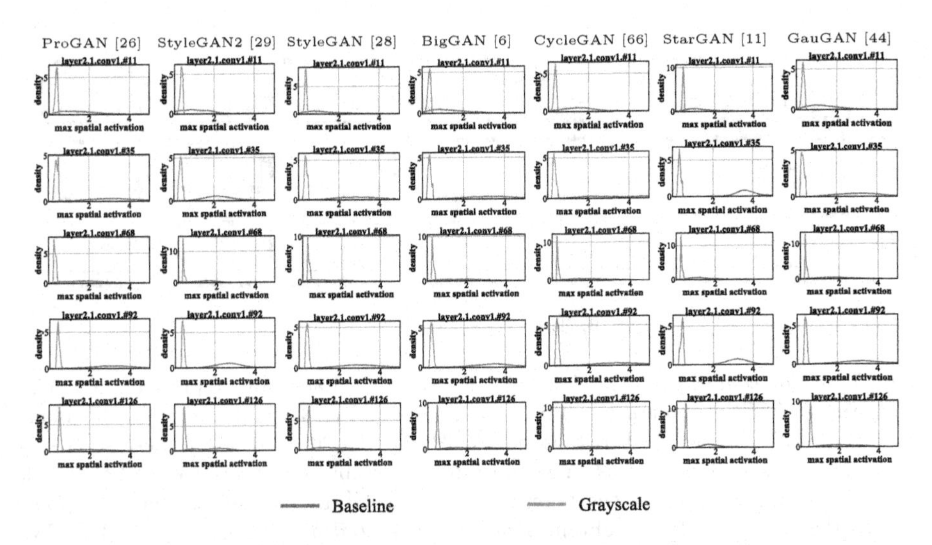

Fig. 6. *Color-conditional T-FF in ResNet-50:* Each row represents a color-conditional *T-FF* (exact same T-FF as shown in Fig. 1), and we show the maximum spatial activation distributions for 7 GAN models before (Baseline) and after color ablation (Grayscale). We remark that for each counterfeit in the ForenSynths dataset [61], we apply global max pooling to the specific T-FF to obtain a *maximum spatial activation* value (scalar). We can clearly observe that these *T-FF* are producing noticeably lower spatial activations (max) for the same set of counterfeits after removing color information. This clearly indicates that these *T-FF* are color-conditional.

Table 3. *CR-universal detectors have noticeably lower amount of color-conditional T-FF:* We show the percentage(%) of color-conditional T-FF in ResNet-50 and EfficientNet-B0 CR-universal detectors measured using Mood's median test. We show the results for ProGAN [26] and all 6 unseen GANs [6,11,28,29,44,66]. Particularly, we consider a *T-FF* to be color conditional if the p-value of the median test is less than the significance level of $\alpha = 0.05$. We clearly observe that training universal detectors using our proposed data augmentation scheme results in CR-universal detectors that contain noticeably lower amount of color-conditional *T-FF*.

% Color-conditional	ProGAN [26]	StyleGAN2 [29]	StyleGAN [28]	BigGAN [6]	CycleGAN [66]	StarGAN [11]	GauGAN [44]
ResNet-50	35.1	37.7	39.5	37.7	36.8	35.9	38.1
EfficientNet-B0	29.1	27.7	27.9	34.5	30.1	29.4	28.6

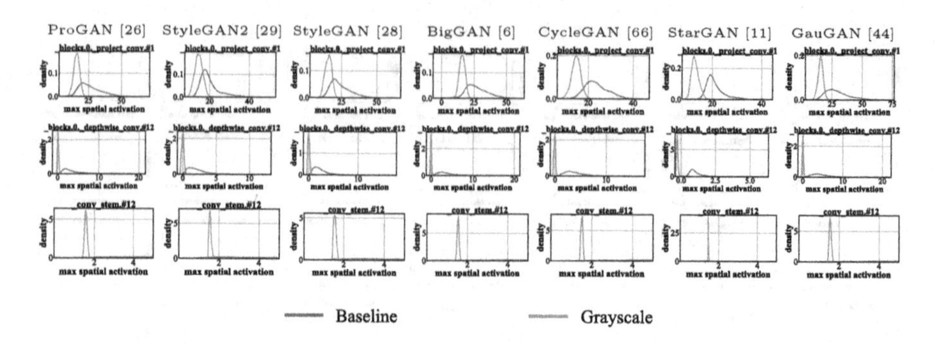

Fig. 7. *Color-conditional T-FF in EfficientNet-B0:* Each row represents a color-conditional *T-FF* (exact same T-FF as shown in Fig. 3), and we show the maximum spatial activation distributions for ProGAN [26] and all 6 unseen GANs [6,11,28,29,44,66] before (Baseline) and after color ablation (Grayscale). We remark that for each counterfeit in the ForenSynths dataset [61], we apply global max pooling to the specific T-FF to obtain a *maximum spatial activation* value (scalar). We can clearly observe that these *T-FF* are producing noticeably lower spatial activations (max) for the same set of counterfeits after removing color information. This clearly indicates that these *T-FF* are color-conditional.

7 Discussion and Conclusion

We conducted the *first analytical study to discover and understand transferable forensic features (T-FF) in universal detectors.* Our first set of investigations demonstrated that input-space attribution methods such as Guided-GradCAM [50] and LRP [2] are not informative to discover *T-FF*. In light of these observations, we study the forensic feature space of universal detectors. Particularly, we propose a novel *forensic feature relevance statistic (FF-RS)* to quantify and discover *T-FF* in universal detectors. Rigorous sensitivity assessments using feature map dropout convincingly show that our proposed FF-RS (ω) is able to successfully quantify and discover *T-FF*.

Further investigations on the *T-FF* uncover an unexpected finding: *color* is a critical *T-FF* in universal detectors. We show this critical finding qualitatively using our proposed LRP-max visualization of discovered *T-FF*, and quantitatively using median counterfeit probability analysis and statistical tests on maximum spatial activation distributions of *T-FF* based on color ablation. i.e.: We showed that ≈85% of *T-FF* are color-conditional in the publicly released ResNet-50 universal detector [61]. Finally, we propose a simple data augmentation scheme to train Color-Robust (CR) universal detectors. We remark that color is not the only *T-FF*, but it is a critical *T-FF* in universal detectors. A natural question would be why is color a critical *T-FF*. Though this is not a straight-forward question to answer, we provide our perspective: Color distribution of real images is non-uniform, and we hypothesize that GANs struggle to capture the diverse, multi-modal color distribution of real images. i.e.: low-density color regions. This may result in noticeable discrepancies between real and GAN images (counterfeits) in the color space, and such discrepancies can be used as forensic features to discriminate between real and GAN images. We include additional experiments/analysis in Supplementary.

Acknowledgements. This research is supported by the National Research Foundation, Singapore under its AI Singapore Programmes (AISG Award No.: AISG2-RP-2021-021; AISG Award No.: AISG-100E2018-005). This project is also supported by SUTD project PIE-SGP-AI-2018-01. Alexander Binder was supported by the SFI Visual Intelligence, project no. 309439 of the Research Council of Norway.

References

1. Arjovsky, M., Chintala, S., Bottou, L.: Wasserstein GAN (2017)
2. Bach, S., Binder, A., Montavon, G., Klauschen, F., Müller, K.R., Samek, W.: On pixel-wise explanations for non-linear classifier decisions by layer-wise relevance propagation. PLoS ONE **10**(7), e0130140 (2015)
3. Bau, D., Zhou, B., Khosla, A., Oliva, A., Torralba, A.: Network dissection: quantifying interpretability of deep visual representations. In: Proceedings of the IEEE Conference on Computer Vision and Pattern Recognition, pp. 6541–6549 (2017)
4. Bau, D., et al.: GAN dissection: visualizing and understanding generative adversarial networks. In: International Conference on Learning Representations (2018)
5. Binder, A., Montavon, G., Lapuschkin, S., Müller, K.-R., Samek, W.: Layer-wise relevance propagation for neural networks with local renormalization layers. In: Villa, A.E.P., Masulli, P., Pons Rivero, A.J. (eds.) ICANN 2016. LNCS, vol. 9887, pp. 63–71. Springer, Cham (2016). https://doi.org/10.1007/978-3-319-44781-0_8
6. Brock, A., Donahue, J., Simonyan, K.: Large scale GAN training for high fidelity natural image synthesis. In: International Conference on Learning Representations (2019). https://openreview.net/forum?id=B1xsqj09Fm
7. Chai, L., Bau, D., Lim, S.-N., Isola, P.: What makes fake images detectable? Understanding properties that generalize. In: Vedaldi, A., Bischof, H., Brox, T., Frahm, J.-M. (eds.) ECCV 2020. LNCS, vol. 12371, pp. 103–120. Springer, Cham (2020). https://doi.org/10.1007/978-3-030-58574-7_7
8. Chandrasegaran, K., Tran, N.T., Cheung, N.M.: A closer look at Fourier spectrum discrepancies for CNN-generated images detection. In: Proceedings of the IEEE/CVF Conference on Computer Vision and Pattern Recognition (CVPR), pp. 7200–7209, June 2021
9. Chen, C., Li, O., Tao, C., Barnett, A.J., Su, J., Rudin, C.: This looks like that: deep learning for interpretable image recognition. Curran Associates Inc., Red Hook (2019)
10. Chen, J., Song, L., Wainwright, M.J., Jordan, M.I.: L-shapley and C-shapley: efficient model interpretation for structured data. In: International Conference on Learning Representations (2019). https://openreview.net/forum?id=S1E3Ko09F7
11. Choi, Y., Choi, M., Kim, M., Ha, J.W., Kim, S., Choo, J.: StarGAN: unified generative adversarial networks for multi-domain image-to-image translation. In: Proceedings of the IEEE Conference on Computer Vision and Pattern Recognition (2018)
12. Choi, Y., Uh, Y., Yoo, J., Ha, J.W.: StarGAN v2: diverse image synthesis for multiple domains. In: Proceedings of the IEEE/CVF Conference on Computer Vision and Pattern Recognition (CVPR), June 2020

13. Cozzolino, D., Thies, J., Rössler, A., Riess, C., Nießner, M., Verdoliva, L.: ForensicTransfer: weakly-supervised domain adaptation for forgery detection. arXiv preprint arXiv:1812.02510 (2018)
14. Desai, S.S., Ramaswamy, H.G.: Ablation-CAM: visual explanations for deep convolutional network via gradient-free localization. In: 2020 IEEE Winter Conference on Applications of Computer Vision (WACV), pp. 972–980 (2020)
15. Dhurandhar, A., et al.: Explanations based on the missing: towards contrastive explanations with pertinent negatives. In: Proceedings of the 32nd International Conference on Neural Information Processing Systems, NIPS 2018, pp. 590–601. Curran Associates Inc., Red Hook (2018)
16. Durall, R., Keuper, M., Keuper, J.: Watch your up-convolution: CNN based generative deep neural networks are failing to reproduce spectral distributions. In: IEEE/CVF Conference on Computer Vision and Pattern Recognition (CVPR), June 2020
17. Dzanic, T., Shah, K., Witherden, F.: Fourier spectrum discrepancies in deep network generated images. In: Thirty-Fourth Annual Conference on Neural Information Processing Systems (NeurIPS), December 2020
18. Frank, J., Eisenhofer, T., Schönherr, L., Fischer, A., Kolossa, D., Holz, T.: Leveraging frequency analysis for deep fake image recognition. In: International Conference on Machine Learning, pp. 3247–3258. PMLR (2020)
19. Fu, R., Hu, Q., Dong, X., Guo, Y., Gao, Y., Li, B.: Axiom-based grad-CAM: towards accurate visualization and explanation of CNNs. In: BMVC (2020)
20. Gal, Y., Ghahramani, Z.: Dropout as a Bayesian approximation: representing model uncertainty in deep learning. In: Balcan, M.F., Weinberger, K.Q. (eds.) Proceedings of the 33rd International Conference on Machine Learning. Proceedings of Machine Learning Research, New York, USA, 20–22 June 2016, vol. 48, pp. 1050–1059. PMLR (2016). https://proceedings.mlr.press/v48/gal16.html
21. Goodfellow, I., et al.: Generative adversarial nets. In: Ghahramani, Z., Welling, M., Cortes, C., Lawrence, N., Weinberger, K.Q. (eds.) Advances in Neural Information Processing Systems, vol. 27, pp. 2672–2680. Curran Associates, Inc. (2014). https://proceedings.neurips.cc/paper/2014/file/5ca3e9b122f61f8f06494c97b1afccf3-Paper.pdf
22. Hao, K., Heaven, W.D.: The year deepfakes went mainstream, December 2020. https://www.technologyreview.com/2020/12/24/1015380/best-ai-deepfakes-of-2020/
23. Harrison, E.: Shockingly realistic Tom Cruise deepfakes go viral on TikTok, February 2021. https://www.independent.co.uk/arts-entertainment/films/news/tom-cruise-deepfake-tiktok-video-b1808000.html
24. He, K., Zhang, X., Ren, S., Sun, J.: Deep residual learning for image recognition. In: Proceedings of the IEEE Conference on Computer Vision and Pattern Recognition (CVPR), June 2016
25. Jiang, P.T., Zhang, C.B., Hou, Q., Cheng, M.M., Wei, Y.: LayerCAM: exploring hierarchical class activation maps for localization. IEEE Trans. Image Process. **30**, 5875–5888 (2021)
26. Karras, T., Aila, T., Laine, S., Lehtinen, J.: Progressive growing of GANs for improved quality, stability, and variation. In: International Conference on Learning Representations (2018). https://openreview.net/forum?id=Hk99zCeAb
27. Karras, T., Aittala, M., Hellsten, J., Laine, S., Lehtinen, J., Aila, T.: Training generative adversarial networks with limited data. In: Larochelle, H., Ranzato, M., Hadsell, R., Balcan, M.F., Lin, H. (eds.) Advances in

Neural Information Processing Systems, vol. 33, pp. 12104–12114. Curran Associates, Inc. (2020). https://proceedings.neurips.cc/paper/2020/file/8d30aa96e72440759f74bd2306c1fa3d-Paper.pdf
28. Karras, T., Laine, S., Aila, T.: A style-based generator architecture for generative adversarial networks. In: Proceedings of the IEEE/CVF Conference on Computer Vision and Pattern Recognition (CVPR), June 2019
29. Karras, T., Laine, S., Aittala, M., Hellsten, J., Lehtinen, J., Aila, T.: Analyzing and improving the image quality of StyleGAN. In: Proceedings of the IEEE/CVF Conference on Computer Vision and Pattern Recognition (CVPR), June 2020
30. Khayatkhoei, M., Elgammal, A.: Spatial frequency bias in convolutional generative adversarial networks, October 2020
31. Ledig, C., et al.: Photo-realistic single image super-resolution using a generative adversarial network. In: 2017 IEEE Conference on Computer Vision and Pattern Recognition (CVPR), pp. 105–114 (2017). https://doi.org/10.1109/CVPR.2017.19
32. Lim, S.K., Loo, Y., Tran, N.T., Cheung, N.M., Roig, G., Elovici, Y.: DOPING: generative data augmentation for unsupervised anomaly detection with GAN. In: 18th IEEE International Conference on Data Mining, ICDM 2018, pp. 1122–1127. Institute of Electrical and Electronics Engineers Inc. (2018)
33. Lloyd, S.: Least squares quantization in PCM. IEEE Trans. Inf. Theory **28**(2), 129–137 (1982). https://doi.org/10.1109/TIT.1982.1056489
34. Lundberg, S.M., Lee, S.I.: A unified approach to interpreting model predictions. In: Guyon, I., et al. (eds.) Advances in Neural Information Processing Systems, vol. 30. Curran Associates, Inc. (2017). https://proceedings.neurips.cc/paper/2017/file/8a20a8621978632d76c43dfd28b67767-Paper.pdf
35. Luo, Y., Zhang, Y., Yan, J., Liu, W.: Generalizing face forgery detection with high-frequency features. In: Proceedings of the IEEE/CVF Conference on Computer Vision and Pattern Recognition, pp. 16317–16326 (2021)
36. van der Maaten, L., Hinton, G.: Visualizing data using t-SNE. J. Mach. Learn. Res. **9**(86), 2579–2605 (2008). http://jmlr.org/papers/v9/vandermaaten08a.html
37. Mahmud, A.H.: Deep dive into deepfakes: frighteningly real and sometimes used for the wrong things, October 2021. https://www.channelnewsasia.com/singapore/deepfakes-ai-security-threat-face-swapping-2252161
38. Marra, F., Gragnaniello, D., Cozzolino, D., Verdoliva, L.: Detection of GAN-generated fake images over social networks. In: 2018 IEEE Conference on Multimedia Information Processing and Retrieval (MIPR), pp. 384–389 (2018). https://doi.org/10.1109/MIPR.2018.00084
39. Marra, F., Gragnaniello, D., Verdoliva, L., Poggi, G.: Do GANs leave artificial fingerprints? In: 2019 IEEE Conference on Multimedia Information Processing and Retrieval (MIPR), pp. 506–511. IEEE (2019)
40. McCloskey, S., Albright, M.: Detecting GAN-generated imagery using saturation cues. In: 2019 IEEE International Conference on Image Processing (ICIP), pp. 4584–4588. IEEE (2019)
41. Muhammad, M.B., Yeasin, M.: Eigen-CAM: class activation map using principal components. In: 2020 International Joint Conference on Neural Networks (IJCNN), pp. 1–7. IEEE (2020)
42. Nataraj, L., et al.: Detecting GAN generated fake images using co-occurrence matrices. Electron. Imaging **2019**(5), 532-1 (2019)
43. News, C.: Synthetic media: how deepfakes could soon change our world, October 2021. https://www.cbsnews.com/news/deepfake-artificial-intelligence-60-minutes-2021-10-10/

44. Park, T., Liu, M.Y., Wang, T.C., Zhu, J.Y.: Semantic image synthesis with spatially-adaptive normalization. In: Proceedings of the IEEE/CVF Conference on Computer Vision and Pattern Recognition, pp. 2337–2346 (2019)
45. Park, T., Liu, M.Y., Wang, T.C., Zhu, J.Y.: Semantic image synthesis with spatially-adaptive normalization. In: Proceedings of the IEEE/CVF Conference on Computer Vision and Pattern Recognition (CVPR), June 2019
46. Pearson, K.: LIII. on lines and planes of closest fit to systems of points in space. Lond. Edinb. Dublin Philos. Mag. J. Sci. **2**(11), 559–572 (1901). https://doi.org/10.1080/14786440109462720
47. Razavi, A., van den Oord, A., Vinyals, O.: Generating diverse high-fidelity images with VQ-VAE-2. In: Wallach, H., Larochelle, H., Beygelzimer, A., d' Alché-Buc, F., Fox, E., Garnett, R. (eds.) Advances in Neural Information Processing Systems, vol. 32, pp. 14866–14876. Curran Associates, Inc. (2019). https://proceedings.neurips.cc/paper/2019/file/5f8e2fa1718d1bbcadf1cd9c7a54fb8c-Paper.pdf
48. Rossler, A., Cozzolino, D., Verdoliva, L., Riess, C., Thies, J., Nießner, M.: FaceForensics++: learning to detect manipulated facial images. In: Proceedings of the IEEE/CVF International Conference on Computer Vision, pp. 1–11 (2019)
49. Schwarz, K., Liao, Y., Geiger, A.: On the frequency bias of generative models. In: Advances in Neural Information Processing Systems, vol. 34 (2021)
50. Selvaraju, R.R., Cogswell, M., Das, A., Vedantam, R., Parikh, D., Batra, D.: Grad-CAM: visual explanations from deep networks via gradient-based localization. In: Proceedings of the IEEE International Conference on Computer Vision, pp. 618–626 (2017)
51. Simonite, T.: What happened to the deepfake threat to the election? November 2020. https://www.wired.com/story/what-happened-deepfake-threat-election/
52. Springenberg, J.T., Dosovitskiy, A., Brox, T., Riedmiller, M.A.: Striving for simplicity: the all convolutional net. In: Bengio, Y., LeCun, Y. (eds.) 3rd International Conference on Learning Representations, ICLR 2015, San Diego, CA, USA, 7–9 May 2015, Workshop Track Proceedings (2015). arxiv.org/abs/1412.6806
53. Srinivas, S., Fleuret, F.: Full-gradient representation for neural network visualization. In: Advances in Neural Information Processing Systems, vol. 32 (2019)
54. Strumbelj, E., Kononenko, I.: An efficient explanation of individual classifications using game theory. J. Mach. Learn. Res. **11**, 1–18 (2010)
55. Tan, M., Le, Q.: EfficientNet: rethinking model scaling for convolutional neural networks. In: International Conference on Machine Learning, pp. 6105–6114. PMLR (2019)
56. Tran, N.-T., Bui, T.-A., Cheung, N.-M.: Dist-GAN: an improved GAN using distance constraints. In: Ferrari, V., Hebert, M., Sminchisescu, C., Weiss, Y. (eds.) Computer Vision – ECCV 2018. LNCS, vol. 11218, pp. 387–401. Springer, Cham (2018). https://doi.org/10.1007/978-3-030-01264-9_23
57. Tran, N.T., Tran, V.H., Nguyen, B.N., Yang, L., Cheung, N.M.M.: Self-supervised GAN: analysis and improvement with multi-class minimax game. In: Wallach, H., Larochelle, H., Beygelzimer, A., d' Alché-Buc, F., Fox, E., Garnett, R. (eds.) Advances in Neural Information Processing Systems, vol. 32. Curran Associates, Inc. (2019). https://proceedings.neurips.cc/paper/2019/file/d04cb95ba2bea9fd2f0daa8945d70f11-Paper.pdf
58. Tran, N.T., Tran, V.H., Nguyen, N.B., Nguyen, T.K., Cheung, N.M.: On data augmentation for GAN training. IEEE Trans. Image Process. **30**, 1882–1897 (2021)
59. Wang, H., et al.: Score-CAM: score-weighted visual explanations for convolutional neural networks. In: Proceedings of the IEEE/CVF Conference on Computer Vision and Pattern Recognition Workshops, pp. 24–25 (2020)

60. Wang, R., et al.: FakeSpotter: a simple yet robust baseline for spotting AI-synthesized fake faces. In: Proceedings of the Twenty-Ninth International Conference on International Joint Conferences on Artificial Intelligence, pp. 3444–3451 (2021)
61. Wang, S.Y., Wang, O., Zhang, R., Owens, A., Efros, A.A.: CNN-generated images are surprisingly easy to spot... for now. In: IEEE/CVF Conference on Computer Vision and Pattern Recognition (CVPR), June 2020
62. Yu, N., Davis, L.S., Fritz, M.: Attributing fake images to GANs: learning and analyzing GAN fingerprints. In: Proceedings of the IEEE/CVF International Conference on Computer Vision, pp. 7556–7566 (2019)
63. Zhang, X., Karaman, S., Chang, S.: Detecting and simulating artifacts in GAN fake images. In: 2019 IEEE International Workshop on Information Forensics and Security (WIFS), pp. 1–6 (2019). https://doi.org/10.1109/WIFS47025.2019.9035107
64. Zhao, S., Liu, Z., Lin, J., Zhu, J.Y., Han, S.: Differentiable augmentation for data-efficient GAN training. In: Conference on Neural Information Processing Systems (NeurIPS) (2020)
65. Zhao, Y., Ding, H., Huang, H., Cheung, N.M.: A closer look at few-shot image generation. In: Proceedings of the IEEE/CVF Conference on Computer Vision and Pattern Recognition, pp. 9140–9150 (2022)
66. Zhu, J.Y., Park, T., Isola, P., Efros, A.A.: Unpaired image-to-image translation using cycle-consistent adversarial networks. In: Proceedings of the IEEE International Conference on Computer Vision, pp. 2223–2232 (2017)

Harmonizer: Learning to Perform White-Box Image and Video Harmonization

Zhanghan Ke[1(✉)], Chunyi Sun[2], Lei Zhu[1], Ke Xu[1], and Rynson W. H. Lau[1]

[1] Department of Computer Science, City University of Hong Kong, Hong Kong, China
kezhanghan@outlook.com
[2] Australian National University, Canberra, Australia

Abstract. Recent works on image harmonization solve the problem as a pixel-wise image translation task via large autoencoders. They have unsatisfactory performances and slow inference speeds when dealing with high-resolution images. In this work, we observe that adjusting the input arguments of basic image filters, *e.g.*, brightness and contrast, is sufficient for humans to produce realistic images from the composite ones. Hence, we frame image harmonization as an image-level regression problem to learn the arguments of the filters that humans use for the task. We present a *Harmonizer* framework for image harmonization. Unlike prior methods that are based on black-box autoencoders, Harmonizer contains a neural network for filter argument prediction and several white-box filters (based on the predicted arguments) for image harmonization. We also introduce a cascade regressor and a dynamic loss strategy for Harmonizer to learn filter arguments more stably and precisely. Since our network only outputs image-level arguments and the filters we used are efficient, Harmonizer is much lighter and faster than existing methods. Comprehensive experiments demonstrate that Harmonizer surpasses existing methods notably, especially with high-resolution inputs. Finally, we apply Harmonizer to video harmonization, which achieves consistent results across frames and 56 *fps* at 1080P resolution.

1 Introduction

Extracting the foreground from one image and compositing it onto a background image is a popular operation in vision applications, *e.g.*, image editing [2,29] and stitching [44,45]. In order for the composite image to be more realistic, *i.e.*, cannot be easily distinguished by humans, the image harmonization task is introduced to remove the inconsistent appearances between the foreground and background. This task is challenging because many conditions, such as lighting and imaging device being used, can affect object visual appearances [3,46], and humans are sensitive to even fine inharmony in appearances [23,42].

Traditional methods [7,18,23,28,31,36,47] focus on matching the hand-crafted statistics between foreground and background regions, disregarding the

S. Avidan et al. (Eds.): ECCV 2022, LNCS 13675, pp. 690–706, 2022.
https://doi.org/10.1007/978-3-031-19784-0_40

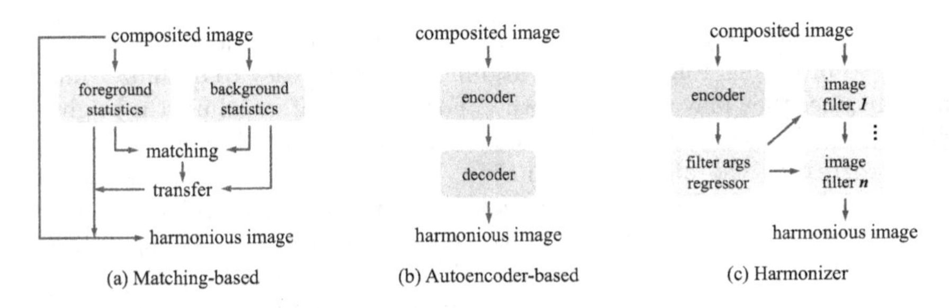

Fig. 1. Comparison of Harmonization Frameworks. (a) Traditional matching-based methods transfer the background appearance to foreground regions based on hand-crafted statistics. (b) Autoencoder-based methods use black-box models to obtain harmonious images via pixel-wise image translation. (c) Our Harmonizer regresses image-level filter arguments to perform image harmonization in a white-box manner.

semantic information which is vital for eliminating the large appearance gap. Recent deep-learning based methods [9,12,14,16,19,35,39] leverage the strong semantic representation capability of autoencoders [34] to lower the appearance gap. They regard image harmonization as a pixel-wise image translation task [17, 40] from a composite image to a harmonious version. Although they have achieved notable success, they also suffer from three key problems. First, their performances are unsatisfactory at high resolutions. Since using high-resolution images for training requires a huge amount of resources, these methods usually train and evaluate at low resolution. Second, these methods are not suitable for mobile devices or real-time applications, due to their high computational overheads. The size of recent autoencoder-based models [8,9,13,27] is larger than 100 MB, and their inference speed at 1080P (Full HD) resolution is only ~10 *fps* on a RTX3090 GPU. Third, the images generated by these methods may not be consistent with the input images in terms of textures/details, *i.e.*, the original image contents may be changed, because neural networks are still essentially black-box models.

To design an efficient strategy for resolution-independent image harmonization in a white-box manner, we conduct a user study to explore how humans perform this task. We observe that humans are able to produce realistic images by adjusting the input arguments of some basic image filters, such as brightness and contrast. These filters also do not suffer from the three aforementioned problems, *i.e.*, resolution-dependence, inefficiency, and black-box manner. Motivated by our observation, we formulate the image harmonization task as an image-level regression problem to learn the arguments of the filters used by humans, and present a *Harmonizer* framework for the task. The key idea of our design is to combine a neural network and white-box filters for image harmonization, rather than just using black-box autoencoders. Specifically, in Harmonizer, the network contains a backbone encoder and a regressor for filter argument prediction, while the white-box filters use the predicted arguments to harmonize the input composite images. Figure 1 summarizes the main differences between Harmonizer and existing frameworks.

To learn filter arguments more stably and precisely, we need to further consider two problems. First, the filter arguments are not easy to optimize simultaneously since they may affect each other. For example, if we adjust the brightness first before adjusting the highlight, we should consider the brightness argument when regressing the highlight argument. We note that utilizing a straightforward multiple-head regressor to predict each filter argument independently has unsatisfactory performances. We solve this problem by introducing a cascade regressor to predict each filter argument based on the features of the preceding filter arguments. Second, the loss of each filter output would accumulate all errors from the preceding filters, causing the regressor to bias towards some filters. We address this problem by introducing a dynamic loss strategy, which can balance the losses and helps Harmonizer pay more attention to the filters that are more difficult to learn. Besides, we design a simple but effective exponential moving average (EMA) based strategy to adapt Harmonizer to video harmonization.

We conduct extensive experiments to evaluate Harmonizer. The results on the iHarmony4 benchmark [9] demonstrate that Harmonizer outperforms prior state-of-the-art by a large margin. Harmonizer also has clear advantages in terms of model size and inference speed. Our ablation study verifies the effectiveness of each component of Harmonizer. For video harmonization, Harmonizer obtains consistent results across frames and an inference speed of 56 *fps* at 1080P resolution on a RTX3090 GPU.

2 Related Works

2.1 Image Harmonization

For an image composited of foreground image F with foreground mask M and background image B, the image harmonization task optimizes a harmonization function $\mathcal{H}$ that processes the foreground region MF in order to match with the visual appearance of B, *i.e.*, creating a natural image I, as:

$$I = \mathcal{H}(MF) + (1 - M)B. \tag{1}$$

Most traditional algorithms proposed $\mathcal{H}$ functions that concentrated on matching low-level appearance statistics, including color distributions [30,31,33,42], color templates [7], and gradient domain [18,29,38]. Some works further combined multi-scale statistics [36] or considered the visual realism of images [20,23].

In recent years, many methods based on CNNs have been proposed with notable successes. These works regarded image harmonization as a pixel-wise image translation task, and their $\mathcal{H}$ functions are implemented based on autoencoders. For example, Tsai *et al.* [39] trained an end-to-end autoencoder to explore high-level semantics. Cun *et al.* [10] introduced a spatial-separated attention module to leverage low-level appearances. Cong et al. [4,5] focused on finding more effective methods to guide the processing of the foreground using the information from the background. Ling *et al.* [27] related image harmonization with background-to-foreground style transfer. Guo *et al.* [13] considered the intrinsic image characteristics to handle reflectance and illumination. Guo *et al.* [12] replaced the CNN encoder with a Transformer to capture global background context.

In spite of the success, all the aforementioned methods suffer from poor performances and slow inference speeds at high resolution, due to the low-resolution images used in the training process and the high computational overheads of autoencoders. Instead, in this work, we formulate the image harmonization task as an image-level regression problem, and our proposed Harmonizer can solve the task with a consistent inference speed at high resolutions with negligible performance degradation.

2.2 White-Box Image Editing

Recently, some works combined neural networks with human understandable (*i.e.*, white-box) algorithms for image editing. These methods usually have more stable performance than using only black-box neural networks. In addition, while the results from black-box neural networks may not be invertible, white-box algorithms allow users to further edit the images or undo any unwanted operations. For example, Yan *et al.* [43] used model predictions to adjust pixel values. Zou *et al.* [48] proposed a generative framework with a renderer/blender to simulate the human painting process. Hu *et al.* [15] applied differentiable image operators for photo post-processing based on reinforcement learning. Wang *et al.* [41] finished cartoon stylization by tuning the representations decomposed from the images.

In the image harmonization task, existing deep learning based methods are all based on black-box autoencoders [8,9,12,13,27,39], except for a concurrent work that attempts to support high-resolution inputs [25]. In contrast, our proposed Harmonizer combines a neural network with image filters to perform image harmonization in a white-box manner.

The work most relevant to ours is probably Hu *et al.* [15]. However, they used reinforcement learning to predict both types and arguments of filters. Besides, their method regresses only one filter in each step and may perform the same filter multiple times, resulting in a slow inference speed. Instead, we regress the arguments of a set of filters simultaneously and performs each filter only once, avoiding redundant filter operations and greatly improving efficiency.

3 Harmonizer

3.1 Design Motivation

Harmonizer aims to address image harmonization from a new perspective - combining neural networks with a white-box strategy. Since the white-box strategy that we select should be understood by humans, we first conduct a two-stage user study to analyze how humans perform image harmonization.

In the first stage, we investigate the white-box strategy humans use for image harmonization. We ask 5 experts who work in the image editing field (2 photographers, 2 designer, and 1 painter) to process composite images with Photoshop.

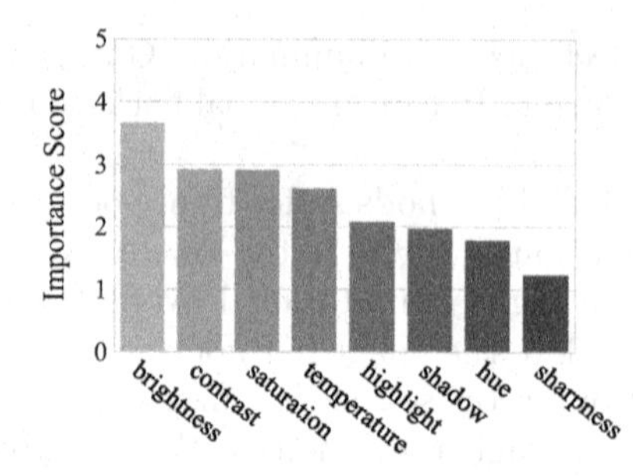

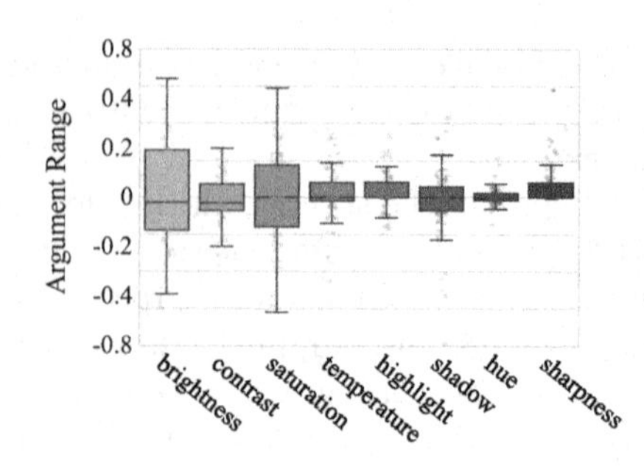

Fig. 2. Statistics of Our User Study. *Left:* We sort image filters by their average importance scores. *Right:* For each filter, we use a Gaussian to fit its input arguments from the participants and visualize the argument distribution via Boxplots.

We note that they accomplish this task mainly by modifying some image properties through tools that can be split into a set of image filters. For example, the "Levels" tool in Photoshop combines the highlight filter, the shadow filter, and the contrast filter. So, the 5 experts are essentially using image filters for image harmonization. In Harmonizer, we select the white-box strategy the same as the 5 experts: adjusting the arguments of appropriate filters to edit the foreground to match the background.

In the second stage, we study the importance of different filters in humans' perception, and the value ranges that humans tune the filter arguments. We build an image harmonization system based on the filters used in the first stage. We invite 26 participants. For each of them, our system will display 10 composite images, including 5 images that are identical among all participants and 5 images randomly selected for each participant. For each composite image, the participants are required to adjust the given filters to make it looks natural. Meanwhile, they should give an importance score for each filter, indicating its role in processing the composite images. The score values are between 1 and 5. The higher the score, the more important the filter is. We record the importance scores and the filter arguments input by the participants for statistics. As shown in Fig. 2, the average importance scores of filters (Fig. 2 *Left*) guide us to choose the filters with high scores, *i.e.*, the filters that are more important in humans' perception, for Harmonizer. The distributions of filter arguments (Fig. 2 *Right*) guide us to set appropriate value ranges for filter arguments, *i.e.*, the value ranges that humans use.

Based on the user study above, we determine the white-box strategy used in Harmonizer (the first stage). We also understand which filters are important and the appropriate argument ranges for the filters (the second stage).

3.2 Architecture

As shown in Fig. 3, the framework of Harmonizer contains a backbone encoder $\mathcal{E}$, a regressor $\mathcal{R}$, and a set of image filters $\mathcal{F} = \{\mathcal{F}_1, \ldots, \mathcal{F}_k\}$, where k indicates the number of filters. The backbone $\mathcal{E}$ in Harmonizer is EfficientNet-B0 [37]. Given

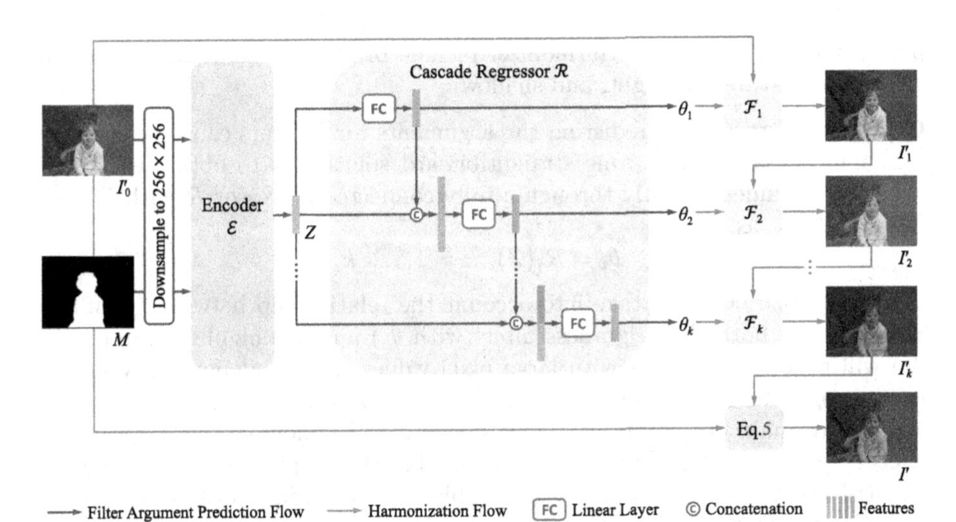

Fig. 3. The Harmonizer Framework. For an input image I'_0 with its foreground mask M, Harmonizer uses a neural network ($\mathcal{E}$+$\mathcal{R}$) to regress k image filter arguments $\theta = \{\theta_1, \ldots, \theta_k\}$ (*i.e.*, the Filter Argument Prediction Flow). The image filters $\mathcal{F} = \{\mathcal{F}_1, \ldots, \mathcal{F}_k\}$ in Harmonizer are then executed in sequence with the predicted arguments θ to obtain the output image I' (*i.e.*, the Harmonization Flow).

a composite image I'_0 and its corresponding foreground mask M, Harmonizer first downsamples them to the resolution of 256 × 256 and inputs them to $\mathcal{E}$ to extract image features Z (with 160 channels), as:

$$Z = \mathcal{E}(I'_0,\ M). \tag{2}$$

Then, Harmonizer processes Z by global pooling and uses $\mathcal{R}$ to regress filter arguments θ from it, as:

$$\theta = \{\theta_1, \ldots, \theta_k\} = \mathcal{R}(Z), \quad \text{where} \quad \theta_i \in [-1, 1],\ \ i = 1, \ldots, k. \tag{3}$$

With θ, Harmonizer executes the k filters in sequence on I'_0, as:

$$I'_i = \mathcal{F}_i(I'_{i-1},\ \theta_i),\ \ i = 1, \ldots, k. \tag{4}$$

Finally, the harmonious image I' is created by:

$$I' = MI'_k + (1 - M)I'_0. \tag{5}$$

Eq. 5 ensures that the background regions in I' are the same as I'_0, *i.e.*, the background pixels are not changed.

To balance the performance and the speed, we have also identified the preferred number of filters k and which k filters to use in Harmonizer. Our evaluations show that setting $k = 6$ is able to satisfy the real-time requirement (Table 4 *Right*). The six most important filters that we have selected based on human perception (Fig. 2 *Left*) for Harmonizer include brightness, contrast, saturation, color temperature, highlight, and shadow.

Cascade Regressor. Predicting the arguments for k filters can be considered as a multi-task problem. One straightforward solution is to obtain each filter argument θ_i independently through a fully connected regressor $\mathcal{R}$ with k heads $\{\mathcal{R}_1, \ldots, \mathcal{R}_k\}$, as:

$$\theta_i = \mathcal{R}_i(Z), \quad i = 1, \ldots, k. \tag{6}$$

However, Eq. 6 does not take into account the relationship between the filters. For example, both the brightness filter (with θ_b) and the highlight filter (with θ_h) will process the pixels with large pixel values. If we independently predict θ_b and θ_h from Z and constrain them with ground truth, both of them will attempt to make the composite input image look harmonious. As a result, the effects of these two filters will be accumulated in the output image, leading to an unsatisfactory result. To address this problem, we introduce a cascade regressor that uses the feature vector of the preceding filter arguments as conditions when regressing an argument θ_i, as:

$$\begin{aligned} &\theta_1 = \mathcal{R}_1(Z), \\ &\theta_i = \mathcal{R}_i(Z \,|\, \theta_{i-1}) = \mathcal{R}_i(Z \,|\, \theta_{i-1}, \ldots, \theta_1), \quad i = 2, \ldots, k. \end{aligned} \tag{7}$$

In practice, after predicting a filter argument, we concatenate its feature vector with Z to regress the next argument.

3.3 Training Strategy

We generate the composite input images from natural images for training. Since Harmonizer executes the k filters in a specific order $\mathcal{F}_1 \rightarrow, \ldots, \rightarrow \mathcal{F}_k$ on the composite image I'_0, we reverse this filter order to $\mathcal{F}_k \rightarrow, \ldots, \rightarrow \mathcal{F}_1$ to create I'_0 from a natural image I, as:

$$\begin{aligned} &I_k = I, \\ &I_{k-i} = F_{k-i+1}(I_{k-i+1}, \, \xi_{k-i+1}), \quad i = 1, \ldots, k, \\ &I'_0 = I_0, \end{aligned} \tag{8}$$

where ξ_{k-i+1} is the input arguments inside the range of $[-1, 1]$. However, some filters (*e.g.*, the color temperature filter) are sensitive to the input arguments and may drastically change the image appearance with even a small change in argument value, resulting in an irreversible I'_0, *i.e.*, we may not be able to recover I. To alleviate this problem, we propose to reduce the range of the argument values when generating I'_0. As shown in Fig. 2 *Right*, our user study provides a rough argument range for each filter, which can guarantee the reversibility of the composite images in most cases. Therefore, we sample the input argument ξ_i for the filter $\mathcal{F}_i$ from a Gaussian distribution $\mathcal{G}_i$, as:

$$\xi_i = \mathcal{G}_i(m_i, \, v_i), \quad i = 1, \ldots, k, \tag{9}$$

where m_i and v_i are the mean and variance from Fig. 2 *Right*, respectively.

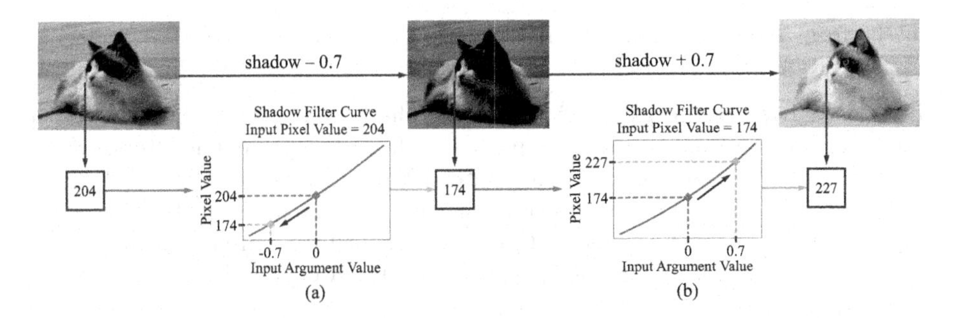

Fig. 4. Asymmetric Filter Operations. For the shadow filter that we use, if we (a) adjust the shadow with an argument of −0.7, the pixel value of 204 will drop to 174. After that, if we (b) adjust the shadow with an argument of 0.7, the pixel value of 174 will increase to 227, which is not equal to the original pixel value of 204.

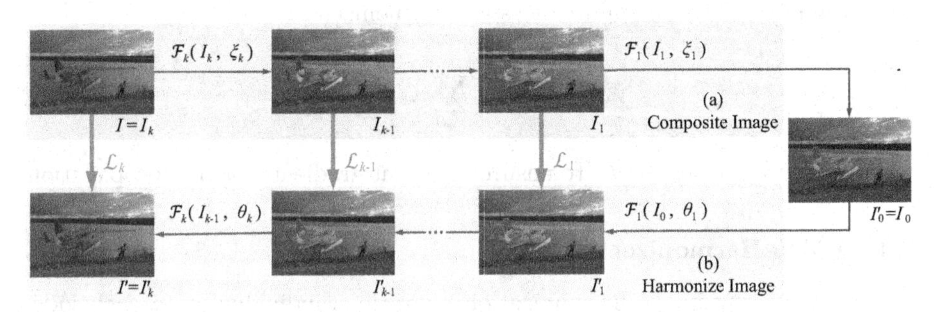

Fig. 5. Optimizing Filter Arguments θ. For (a) a composite image I'_0 generated from a natural image I using the filter arguments $\xi = \{\xi_1, \ldots, \xi_k\}$, Harmonizer tends to (b) predict a set of arguments θ to recover a harmonious image I' from I'_0. For some filters, the ground truth of θ_i is unknown. Hence, we optimize θ through the loss $\mathcal{L}_i$ between each filter output I'_i and its corresponding composite image I_i.

Note that some of the filters used are non-linear, and their operations are asymmetric. Here, we take the shadow filter $\mathcal{F}_s$ as an example. As illustrated in Fig. 4, if we use the shadow filter $\mathcal{F}_s$ with an input argument ξ_s to adjust an image I, we may not recover I using $\mathcal{F}_s$ with argument $-\xi_s$. Therefore:

$$I \neq \mathcal{F}_s(\mathcal{F}_s(I,\, \xi_s),\, -\xi_s). \tag{10}$$

Hence, $-\xi_s$ cannot be used as the ground truth of the shadow filter argument θ_s predicted by Harmonizer. As shown in Fig. 5, instead of regressing θ directly, we optimize each θ_i through the L2 loss between the filter output I'_i and its corresponding composite image I_i (calculated in Eq. 8), as:

$$\mathcal{L}_i = M \parallel I'_i - I_i \parallel_2 = M \parallel \mathcal{F}_i(I'_{i-1},\, \theta_i) - I_i \parallel_2, \;\; i = 1, \ldots, k. \tag{11}$$

Here the foreground mask M constrains the loss only on the foreground regions. We apply a loss on each output to ensure gradient propagation for the filters in the front. For the composite input image generated by GAN [11] (following [8,9,12] *etc.*) rather than Eq. 8, we only apply the loss $\mathcal{L}_k$ on the final output.

Dynamic Loss Strategy. During training, the Loss $\mathcal{L}_i$ usually increases with the filter index i due to the inaccurate I_{i-1} from the preceding filters, which may bias the regressor towards later filters. We introduce a dynamic strategy to balance the losses. We first subtract the errors accumulated by the preceding filters from $\mathcal{L}_i$. We then normalize $\mathcal{L}_i$ to enhance the loss of filters that introduce larger errors. Formally, the loss $\mathcal{L}_i$ is dynamically reweighted by:

$$\tilde{\mathcal{L}}_i = max\Big(\frac{\mathcal{L}_i - \mathcal{L}_{i-1}}{\mathcal{L}_k}, 0\Big), \quad i = 1, \ldots, k, \tag{12}$$

Note that we detach the gradients at the denominator $\mathcal{L}_k$. If $\mathcal{L}_i < \mathcal{L}_{i-1}$, we reweight $\mathcal{L}_i$ to 0 to focus on optimizing $\mathcal{L}_{i-1}$ as we consider $\mathcal{F}_i$ work well for the current input. The final training loss for Harmonizer is:

$$\mathcal{L} = \mu \sum_{i=1}^{k} \tilde{\mathcal{L}}_i. \tag{13}$$

where μ is used to rescale $\mathcal{L}_i$ to ensure sufficient gradients for backpropagation.

3.4 Video Harmonizer

Applying existing image harmonization algorithms individually on each video frame often leads to flickering of the foreground in the output sequence. Although some video processing methods [4,5,22,24] have been proposed to encourage the prediction consistency across video frames, they require a long processing time or additional modules for training. Therefore, obtaining stable results in real-time video harmonization is an unexplored problem.

We introduce here a simple but effective strategy for adapting Harmonizer to video harmonization. The idea behind our strategy is to ensure that the predicted filter arguments change smoothly across frames. We achieve this by smoothing the predicted arguments θ with exponential moving average (EMA), as:

$$\bar{\theta}^t = (1 - \alpha)\,\bar{\theta}^{t-1} + \alpha\,\theta^t, \tag{14}$$

where t is the frame index, and $\alpha = 0.9$ is an EMA coefficient.

4 Experiments

In this section, we first introduce the datasets, metrics, and training details for our experiments. We then compare Harmonizer with existing image harmonization methods (Sect. 4.1). We also show the effectiveness of adapting Harmonizer to video harmonization (Sect. 4.2). We further conduct ablation experiments to evaluate the effectiveness of individual components in Harmonizer (Sect. 4.3). Finally, we demonstrate the advantages of Harmonizer in real-world image/video harmonization applications through user studies (Sect. 4.4).

Table 1. Quantitative Comparison on iHarmony4 at 256×256 Resolution. All metrics are computed following the previous works. ↑ indicates the higher the better, while ↓ indicates the lower the better.

Dataset	Metric	DIH [39]	S2AM [10]	DOVE [9]	BARG [8]	IntrIH [13]	IHT [12]	Our
HAdobe5k	MSE↓	92.65	63.40	52.32	39.94	43.02	47.96	**21.89**
	fMSE↓	593.03	404.62	380.39	359.49	284.21	321.14	**170.05**
	PSNR↑	32.28	33.77	34.34	35.34	35.20	36.10	**37.64**
HFlickr	MSE↓	163.38	143.45	145.21	97.32	105.13	88.41	**64.81**
	fMSE↓	1099.13	785.65	985.79	769.02	716.60	617.26	**434.06**
	PSNR↑	29.55	30.03	29.75	31.34	31.34	32.37	**33.63**
HCOCO	MSE↓	51.85	41.07	36.72	24.84	24.92	20.99	**17.34**
	fMSE↓	798.99	542.06	551.01	489.94	416.38	377.11	**298.42**
	PSNR↑	34.69	35.47	35.83	37.03	37.16	37.87	**38.77**
Hday2night	MSE↓	82.34	76.61	56.92	50.98	55.53	58.14	**33.14**
	fMSE↓	1129.40	989.07	1067.19	853.61	797.04	823.68	**542.07**
	PSNR↑	34.62	34.50	35.53	35.88	35.96	36.38	**37.56**
All	MSE↓	76.77	59.67	52.36	37.82	38.71	37.07	**24.26**
	fMSE↓	778.41	537.23	541.53	513.16	400.29	395.66	**280.51**
	PSNR↑	33.41	34.35	34.75	35.88	35.90	36.71	**37.84**

Datasets. Following the recent works, we conduct our experiments on the iHarmony4 benchmark [9], which contains four subsets: HCOCO, HAdobe5k, HFlickr, and Hday2night. Each sample in iHarmony4 consists of a natural image, a foreground mask, and a composite image (with the foreground generated by GAN [11]). During training, we also create the composite images via Eq. 8. Note that this is a data augmentation method, without using any extra data.

Metrics. We evaluate the image harmonization performance by Mean Square Error (MSE), foreground MSE (fMSE), and Peak Signal-to-Noise Ratio (PSNR). fMSE calculates MSE only on the foreground regions rather than the whole image, as image harmonization does not change the background appearance.

Training. We train Harmonizer by Adam for 60 epochs. With a batch size of 16, the initial learning rate is set to $3e^{-4}$ and is multiplied by 0.1 after every 25 epochs. We set μ (in Eq. 12) to 10. In all experiments, except the ablation on the number of filters, we use Harmonizer with the 6 filters stated in Sect. 3.2.

4.1 Comparison with State-of-the-arts

We compare Harmonizer with recently proposed methods, including DIH [39], S2AM [10], DOVE [9], BARG [8], IntrIH [13], and IHT [12]. We use the pre-trained models released by their authors for evaluation. We first follow the prior works to evaluate all methods at low resolution, *i.e.*, the output harmonious images and the ground truths will be resized to 256×256 for metric calculation. As shown in Table 1, Harmonizer outperforms the existing methods on all four subsets of iHarmony4. Notably, compared to the state-of-the-art method, Harmonizer reduces the average MSE across all subsets by 35%.

Table 2. Quantitative Comparison on iHarmony4 at High Resolutions. All metrics are calculated at the original image resolution of the samples in iHarmony4. The inputs to the existing methods are in low-resolution. Their outputs are then bilinearly upsampled to high resolutions for metric calculation. We also apply Polynomial Color Mapping for upsampling (with subscript "+PCM").

Dataset	Metric	DOVE [9]	DOVE [9] +PCM	BARG [8]	BARG [8] +PCM	IHT [12]	IHT [12] +PCM	Our
HAdobe5k	MSE↓	68.16	72.08	77.96	88.20	56.90	63.28	**24.37**
	fMSE↓	511.02	579.21	560.49	689.58	465.72	547.61	**196.12**
	PSNR↑	33.30	32.82	32.65	32.17	33.63	33.04	**37.80**
HFlickr	MSE↓	172.80	159.46	159.34	150.67	135.49	127.10	**69.19**
	fMSE↓	1192.88	1110.22	1114.29	1096.91	994.23	976.08	**479.26**
	PSNR↑	28.81	29.71	29.01	29.88	29.59	30.44	**33.37**
HCOCO	MSE↓	56.49	47.13	52.84	46.62	44.95	40.16	**20.93**
	fMSE↓	1000.14	844.84	940.79	844.21	838.86	785.03	**374.96**
	PSNR↑	33.35	34.5	33.54	34.51	34.19	34.85	**37.69**
Hday2night	MSE↓	58.23	67.81	53.99	66.91	63.26	72.94	**37.28**
	fMSE↓	1125.46	1007.99	958.02	1109.15	988.56	1054.97	**640.74**
	PSNR↑	35.44	35.17	35.65	35.07	35.71	35.06	**37.15**
All	MSE↓	72.98	67.35	71.93	70.76	58.89	57.22	**27.62**
	fMSE↓	882.26	800.48	851.83	832.62	750.06	741.98	**339.23**
	PSNR↑	32.86	33.49	32.82	33.32	33.54	33.83	**37.23**

Table 3. Comparison on Inference Speed, Model Size, and GPU Memory. The speed evaluation is conducted at 1080P resolution on a RTX3090 GPU.

Metric	S2AM [10]	DOVE [9]	BARG [8]	IntrIH [13]	IHT [12]	Our
Inference Speed (*fps*) ↑	6.76	13.8	11.6	1.2	5.1	**56.3**
Model Size (MB) ↓	268.1	219.1	234.9	163.5	25.8	**21.7**
GPU Memory (GB) ↓	6.3	6.5	3.7	16.47	18.5	**2.3**

For practical applications, which typically use higher image resolutions, the quantitative results at 256×256 resolution as shown above may not reflect the actual image harmonization performance. To study this issue, we further evaluate Harmonizer with the strong baseline DOVE and the state-of-the-art BARG/IHT at high resolutions. We compute the metrics at the original resolutions of the images in iHarmony4. Note that the subsets in iHarmony4 have different resolutions, *e.g.*, the average image size for HCOCO is about 500×500 and for HAdobe5k is about 3000×3000. Since high-resolution inputs would significantly degrade the performances of the existing methods that are trained on 512×512 resolution, we still input low-resolution images to them and then bilinearly upsample their results to high resolutions for metric calculation. To avoid blurry outputs caused by bilinear upsampling, we also upsample their results using Polynomial Color Mapping (PCM) [1], which can transfer the foreground appearances in the low-resolution outputs to the high-resolution composite images without loss of details (has been used for visualization in

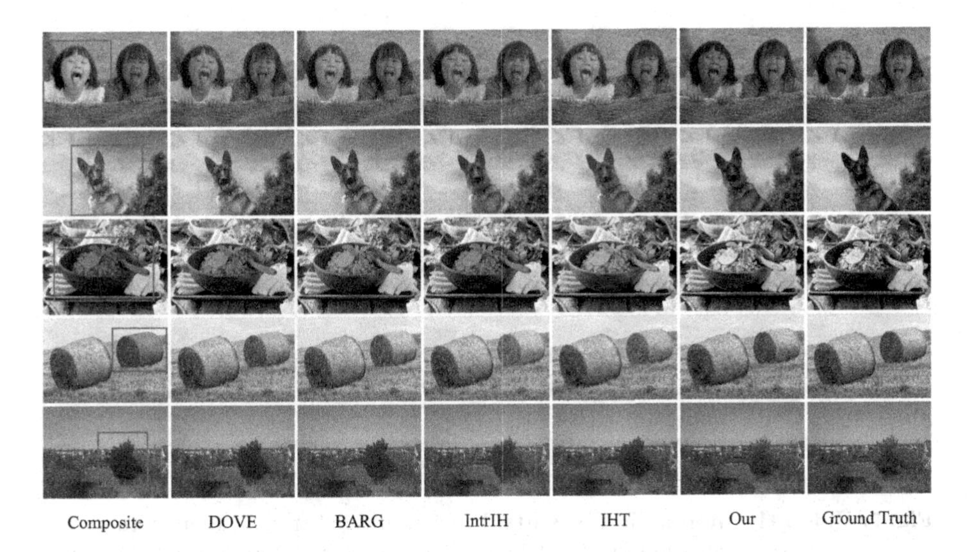

Fig. 6. Visual Comparison on iHarmony4. The red boxes in the composite images indicate the foreground regions. Zoom in for better visualization. (Color figure online)

previous works). In contrast, our Harmonizer can process the high-resolution inputs directly as its filters are resolution-independent. Table 2 shows the metrics computed at the original image resolutions. The performances of the existing methods are significantly lowered, *e.g.*, MSE/fMSE of IHT are increased from 37.07/395.66 to 57.22/741.98. In contrast, Harmonizer only has a small performance drop at high resolutions, and its MSE is now 50% lower than the state-of-the-art method. We provide some visual comparisons in Fig. 6.

Table 3 compares different methods in terms of inference speed, model size, and memory requirement. A fast inference speed is necessary for real-time applications, while a small model size and a low memory requirement facilitate deployment on mobile devices. Our results demonstrate that Harmonizer is faster, lighter, and more memory efficient than other methods. Remarkably, on a RTX3090 GPU, Harmonizer can process a 1080P (Full HD) video at 56 *fps*, about 4× faster than the recent fastest method DOVE [9]. Moreover, we observe that Harmonizer can be further accelerated by a fusion implementation of filters or using techniques like Halide Auto-Scheduler.

4.2 Video Harmonization Results

As shown in Fig. 7, by applying Eq. 14, Harmonizer obtains stable harmonization results across video frames. On the contrary, the results of the prior methods suffer from severe flickers. Unfortunately, our strategy proposed in Sect. 3.4 is not suitable for use in prior existing methods since they solve harmonization as a pixel-wise image translation problem in a black-box manner.

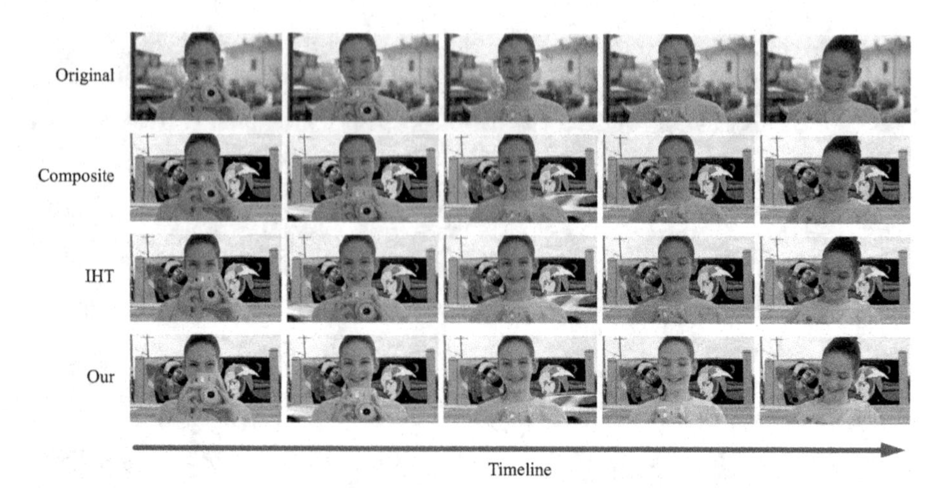

Fig. 7. Video Harmonization Results. Previous image harmonization methods (we only visualize the results from IHT [12] here due to space limitation) output the frames with inconsistent foreground appearances (*e.g.*, the cloth regions). Instead, our Harmonizer obtains consistent harmonization results across frames.

Table 4. Ablation of Harmonizer on iHarmony4. *Left*: Evaluating the effectiveness of different Harmonizer components at 256 × 256 resolution. *Right*: Evaluating the performance of Harmonizer with different numbers of filters. MSE and PSNR are calculated at 256 × 256 resolution, while *fps* is measured at 1080P resolution.

Cascade regressor	Dynamic loss strategy	Metrics	
		MSE ↓	PSNR ↑
		28.47	36.96
✓		26.85	37.23
✓	✓	**24.26**	**37.84**

Metrics	Number of filters			
	2	4	6	8
MSE ↓	74.16	29.60	24.26	**23.51**
PSNR ↑	33.49	36.75	37.84	**38.06**
fps ↑	**86.2**	63.5	56.3	51.9

4.3 Ablation Study

In Table 4 *Left*, we evaluate the cascade regressor and dynamic loss strategy proposed in Harmonizer. The results show that both techniques can improve the image harmonization performance. We also observe that even without these two techniques, the results of Harmonizer (MSE of 33.28) still surpass the previous state-of-the-art (MSE of 37.07), which demonstrates that the overall design of Harmonizer has advantages over the black-box autoencoders. In Table 4 *Right*, we analyze the impact of different numbers of filters on the performance and speed. Specifically, we validate Harmonizer with 2, 4, 6, and 8 filters. For the experiments with 6 or 8 filters, we choose the most important filters based on Fig. 2. For the experiments with 2 or 4 filters, we select the filters randomly and report the metrics averaged over 3 runs. The results show that the performance of Harmonizer increases as the number of filters increases. Besides, Harmonizer needs at least 4 filters to avoid large performance degradation.

Fig. 8. Visual Comparison on Real Composite Images. "Human" means that the results are produced by humans.

Table 5. User Study Results. We calculate B-T scores to quantify our user study results. For image harmonization, the results from humans are compared.

Metrics	Composite	DOVE [9]	BARG [8]	IntrIH [13]	IHT [12]	Our	Human
B-T Score (Image) ↑	0.412	0.639	0.618	0.663	0.724	<u>1.028</u>	**1.393**
B-T Score (Video) ↑	<u>1.173</u>	0.587	0.497	0.530	0.601	**2.042**	–

4.4 User Studies

We show the advantage of Harmonizer in real-world image/video harmonization through user studies. For image harmonization, we use the real composite images released in [39], which includes 99 images. Since these images have no labels, we ask a skilled human to process them for reference. For video harmonization, we composite the 20 foreground videos generated by the video matting methods [21, 26, 32] with 20 new background videos to create composite videos (one of them is shown in Fig. 7). We invite 12 participants to rank the results from different methods and the human. In Table 5, we follow prior works to use the Bradley-Terry model (B-T model) [6] for ranking. Harmonizer achieves the highest B-T scores. The B-T scores of applying prior methods to video harmonization are even lower than the original composite inputs due to severe flickering. Figure 8 visualizes two samples used in our image harmonization user study.

Fig. 9. A Failure Case of Harmonizer. We show a composite image with a strong hue difference (within the blue color range) between foreground and background. (Color figure online)

5 Conclusion

In this paper, we have studied the image harmonization process carried out by humans, which has inspired us to design Harmonizer. Benefited by our novel architecture and two proposed techniques, *i.e.*, a cascade regressor and a dynamic loss strategy, Harmonizer is lighter and faster than prior methods while achieving new state-of-the-art performances. Nevertheless, our method does have limitations. It may fail to handle color-specific appearance inconsistency or the different lighting conditions between the foreground and background. Figure 9 shows one case. As a future work, we would like to develop more complex image filters, *e.g.*, color-separated filters, to address the problem.

References

1. Afifi, M., Brown, M.S.: Deep white-balance editing. In: CVPR (2020)
2. Barnes, C., Shechtman, E., Finkelstein, A., Goldman, D.B.: Patchmatch: a randomized correspondence algorithm for structural image editing. ACM Trans. Graph. **28**(3), 24 (2009)
3. Barron, J.T., Malik, J.: Shape, illumination, and reflectance from shading. IEEE TPAMI **37**(8), 1670–1687 (2014)
4. Bonneel, N., Sunkavalli, K., Paris, S., Pfister, H.: Example-based video color grading. ACM Trans. Graph. **32**(4), 1–39 (2013)
5. Bonneel, N., Tompkin, J., Sunkavalli, K., Sun, D., Paris, S., Pfister, H.: Blind video temporal consistency. ACM Transa. Graph. **34**(6), 1–9 (2015)
6. Bradley, R.A., Terry, M.E.: The rank analysis of incomplete block designs - I. The method of paired comparisons. Biometrika **39**(3/4), 324–345 (1952)
7. Cohen-Or, D., Sorkine, O., Gal, R., Leyvand, T., Xu, Y.Q.: Color harmonization. ACM Trans. Graph. **25**(3), 624–630 (2006)
8. Cong, W., Niu, L., Zhang, J., Liang, J., Zhang, L.: Bargainnet: background-guided domain translation for image harmonization. In: ICME (2021)
9. Cong, W., et al.: Dovenet: deep image harmonization via domain verification. In: CVPR (2020)
10. Cun, X., Pun, C.M.: Improving the harmony of the composite image by spatial-separated attention module. IEEE Trans. Image Process **29**, 4759–4771 (2020)
11. Goodfellow, I.J., et al.: Generative adversarial nets. In: NeurIPS (2014)
12. Guo, Z., Guo, D., Zheng, H., Gu, Z., Zheng, B., Dong, J.: Image harmonization with transformer. In: ICCV (2021)
13. Guo, Z., Zheng, H., Jiang, Y., Gu, Z., Zheng, B.: Intrinsic image harmonization. In: CVPR (2021)
14. Hao, G., Iizuka, S., Fukui, K.: Image harmonization with attention-based deep feature modulation. In: BMVC (2020)
15. Hu, Y., He, H., Xu, C., Wang, B., Lin, S.: Exposure: a white-box photo post-processing framework. ACM Trans. Graph. **37**(2), 1–17 (2018)
16. Huang, H., Xu, S., Cai, J., Liu, W., Hu, S.: Temporally coherent video harmonization using adversarial networks. IEEE Trans. Image Process **29**, 214–224 (2020)
17. Isola, P., Zhu, J.Y., Zhou, T., Efros, A.A.: Image-to-image translation with conditional adversarial networks. In: CVPR (2017)
18. Jia, J., Sun, J., Tang, C.K., Shum, H.Y.: Drag-and-drop pasting. ACM Trans. Graph. **25**(3), 631–637 (2006)
19. Jiang, Y., et al.: A self-supervised framework for image harmonization. In: ICCV (2021)

20. Johnson, M.K., Dale, K., Avidan, S., Pfister, H., Freeman, W.T., Matusik., W.: Cg2real: improving the realism of computer generated images using a large collection of photographs. IEEE Trans. Vis. Comput. Graph. **17**(9), 1273-1285 (2010)
21. Ke, Z., Sun, J., Li, K., Yan, Q., Lau, R.W.: Modnet: real-time trimap-free portrait matting via objective decomposition. In: AAAI (2022)
22. Lai, W.S., Huang, J.B., Wang, O., Shechtman, E., Yumer, E., Yang, M.H.: Learning blind video temporal consistency. In: ECCV (2018)
23. Lalonde, J.F., Efros, A.A.: Using color compatibility for assessing image realism. In: ICCV (2007)
24. Lei, C., Xing, Y., Chen, Q.: Blind video temporal consistency via deep video prior. In: Neurips (2020)
25. Liang, J., Cun, X., Pun, C.: Spatial-separated curve rendering network for efficient and high-resolution image harmonization. arXiv abs/2109.05750 (2021)
26. Lin, S., Yang, L., Saleemi, I., Sengupta, S.: Robust high-resolution video matting with temporal guidance. In: WACV (2022)
27. Ling, J., Xue, H., Song, L., Xie, R., Gu, X.: Region-aware adaptive instance normalization for image harmonization. In: CVPR (2021)
28. Luan, F., Paris, S., Shechtman, E., Bala, K.: Deep painterly harmonization. In: EGSR (2018)
29. Pérez, P., Gangnet, M., Blake, A.: Poisson image editing. ACM Trans. Graph. **22**(3), 313–318 (2003)
30. Pitie, F., Kokaram, A.: The linear monge-kantorovitch linear colour mapping for example-based colour transfer. In: European Conference on Visual Media Production (2007)
31. Pitie, F., Kokaram, A., Dahyot, R.: N-dimensional probability density function transfer and its application to color. In: ICCV (2015)
32. Qin, X., Zhang, Z., Huang, C., Dehghan, M., Zaiane, O., Jagersand, M.: U2-net: going deeper with nested u-structure for salient object detection. Pattern Recogn. **106**, 107404 (2020)
33. Reinhard, E., Adhikhmin, M., Gooch, B., Shirley, P.: Color transfer between images. IEEE Comput. Graph. Appl. **21**(5), 34–41 (2001)
34. Ronneberger, O., Fischer, P., Brox, T.: U-Net: convolutional Networks for Biomedical Image Segmentation. In: Navab, N., Hornegger, J., Wells, W.M., Frangi, A.F. (eds.) MICCAI 2015. LNCS, vol. 9351, pp. 234–241. Springer, Cham (2015). https://doi.org/10.1007/978-3-319-24574-4_28
35. Sofiiuk, K., Popenova, P., Konushin, A.: Foreground-aware semantic representations for image harmonization. In: WACV (2021)
36. Sunkavalli, K., Johnson, M.K., Matusik, W., Pfister, H.: Multi-scale image harmonization. ACM Trans. Graph. **29**(4), 1–10 (2010)
37. Tan, M., Le, Q.: Efficientnet: rethinking model scaling for convolutional neural networks. In: ICML (2019)
38. Tao, M.W., Johnson, M.K., Paris, S.: Error-tolerant image compositing. In: Daniilidis, K., Maragos, P., Paragios, N. (eds.) ECCV 2010. LNCS, vol. 6311, pp. 31–44. Springer, Heidelberg (2010). https://doi.org/10.1007/978-3-642-15549-9_3
39. Tsai, Y.H., Shen, X., Lin, Z., Sunkavalli, K., Lu, X., Yang, M.H.: Deep image harmonization. In: CVPR (2017)
40. Wang, T.C., Liu, M.Y., Zhu, J.Y., Tao, A., Kautz, J., Catanzaro, B.: High-resolution image synthesis and semantic manipulation with conditional gans. In: CVPR (2018)
41. Wang, X., Yu, J.: Learning to cartoonize using white-box cartoon representations. In: CVPR (2020)

42. Xue, S., Agarwala, A., Dorsey, J., Rushmeier, H.: Understanding and improving the realism of image composites. ACM Trans. Graph. **31**(4), 1–10 (2012)
43. Yan, Z., Zhang, H., Wang, B., Paris, S., Yu, Y.: Automatic photo adjustment using deep neural networks. ACM Trans. Graph. **35**(2), 1–15 (2016)
44. Zaragoza, J., Chin, T.J., Brown, M.S., Suter, D.: As-projective-as-possible image stitching with moving DLT. In: CVPR (2013)
45. Zhang, F., Liu, F.: Parallax-tolerant image stitching. In: CVPR (2014)
46. Zhang, R., Tsai, P.S., Cryer, J.E., Shah, M.: Shape-from-shading: a survey. IEEE TPAMI **21**(8), 690–706 (1999)
47. Zhu, J.Y., Krahenbuhl, P., Shechtman, E., Efros, A.A.: Learning a discriminative model for the perception of realism in composite images. In: ICCV (2015)
48. Zou, Z., Shi, T., Qiu, S., Yuan, Y., Shi, Z.: Stylized neural painting. In: CVPR (2021)

Text2LIVE: Text-Driven Layered Image and Video Editing

Omer Bar-Tal[1(✉)], Dolev Ofri-Amar[1], Rafail Fridman[1], Yoni Kasten[2], and Tali Dekel[1]

[1] Weizmann Institute of Science, Rehovot, Israel
omer234@gmail.com
[2] NVIDIA Research, Tel Aviv, Israel

Abstract. We present a method for zero-shot, text-driven editing of natural images and videos. Given an image or a video and a text prompt, our goal is to edit the appearance of existing objects (e.g., texture) or augment the scene with visual effects (e.g., smoke, fire) in a semantic manner. We train a generator on an *internal dataset*, extracted from a single input, while leveraging an *external* pretrained CLIP model to impose our losses. Rather than directly generating the edited output, our key idea is to generate an *edit layer* (color+opacity) that is composited over the input. This allows us to control the generation and maintain high fidelity to the input via novel text-driven losses applied directly to the edit layer. Our method neither relies on a pretrained generator nor requires user-provided masks. We demonstrate localized, semantic edits on high-resolution images and videos across a variety of objects and scenes. Webpage: http://www.text2live.github.io.

Keywords: Text-guided image and video editing · CLIP

1 Introduction

Computational methods for manipulating the appearance and style of objects in natural images and videos have seen tremendous progress, facilitating a variety of editing effects to be achieved by novice users. Nevertheless, research in this area has been mostly focused in the Style-Transfer setting where the target appearance is given by a reference image (or domain of images), and the original image is edited in a global manner [14]. Controlling the localization of the edits typically involves additional input guidance such as segmentation masks. Thus, appearance transfer has been mostly restricted to global artistic stylization or to specific image domains or styles (e.g., faces, day-to-night, summer-to-winter). In this work, we seek to eliminate these requirements and enable more flexible and creative semantic appearance manipulation of real-world images and videos.

Inspired by the unprecedented power of recent Vision-Language models, we use simple text prompts to express the target edit. This allows the user to easily

O. Bar-Tal, D. Ofri-Amar and R. Fridman—Have contributed equally.

S. Avidan et al. (Eds.): ECCV 2022, LNCS 13675, pp. 707–723, 2022.
https://doi.org/10.1007/978-3-031-19784-0_41

Fig. 1. Text2LIVE performs *semantic*, *localized* edits to real-world images (b), or videos (c). Our key idea is to generate an *edit layer*–RGBA image representing the target edit when composited over the input (a). This allows us to use text to guide not only the final composite, but also the edit layer itself (target text prompts are shown above each image). Our edit layers are synthesized by training a generator on a *single* input, without relying on user-provided masks or a pretrained generator.

Fig. 2. Text2LIVE generates an edit layer, which is composited over the input. The text prompts expressing the target layer and the final composite are shown above each image. Our layered editing facilitates various effects including changing objects' texture or augmenting the scene with complex semi-transparent effects.

and intuitively specify the target appearance and the object/region to be edited. Specifically, our method enables *local*, *semantic* editing that satisfies a given target text prompt (e.g., Fig. 1 and Fig. 2). For example, given the cake image in Fig. 1(b), and the target text "oreo cake", our method automatically locates the cake region and synthesizes realistic, high-quality texture that combines naturally with the original image – the cream filling and the cookie crumbs "paint" the cake in a *semantically-aware* manner.

Our framework leverages the representation learned by a Contrastive Language-Image Pretraining (CLIP) model, which has been pretrained on 400 million text-image examples [32]. Various recent image editing methods [2,3,9,10,30] have demonstrated the richness of the visual and textual space spanned by CLIP. However, editing *existing* objects in *arbitrary, real-world* images remains challenging. Most existing methods combine a pre-trained generator (e.g., GAN/Diffusion model) in conjunction with CLIP. With GANs, the domain of images is restricted and requires inverting the input image to the GAN's latent space – a challenging task by itself [46]. Diffusion models [11,42] overcome these barriers but face an inherent trade-off between satisfying the target edit and maintaining high fidelity to the original content [2]. Furthermore, it is not straightforward to extend these methods to videos. In this work, we take a different route and propose to *learn a generator from a single input* – image or video and text prompts.

If no external generative prior is used, how can we steer the generation towards meaningful, high-quality edits? We achieve this via two key components: (i) we propose a novel text-guided *layered editing*, i.e., rather than directly generating the edited image, we represent the edit via an RGBA layer (color and opacity) that is composited over the input. This allows us to guide the content and localization of the generated edit via a novel objective function, including text-driven losses applied directly to the edit layer. As seen in Fig. 2, we use text prompts to express both the final edited image and the target effect (e.g., fire) represented by the edit layer. (ii) We train our generator on an *internal dataset* of diverse image-text training examples by applying various augmentations to the input image and text. We demonstrate that our internal learning approach serves as a strong regularization, enabling high quality generation of complex textures and semi-transparent effects.

We further take our framework to the realm of *text-guided video editing*. Real-world videos consist of complex objects and camera motion, providing abundant information about the scene. Yet, consistent video editing is difficult and cannot be achieved naïvely. We thus propose to decompose the video into a set of 2D *atlases* using [16]. Each atlas is a unified 2D image representing either a foreground object or the background throughout the video. This representation simplifies video editing: edits applied to a 2D atlas are consistently mapped back to the video automatically. Here, we extend our framework to perform edits in the atlas space while harnessing the rich information readily available in videos.

In summary, we present the following contributions:

- An end-to-end text-guided framework for performing localized, semantic edits of existing objects in real-world images.
- A novel layered editing approach and objective function that automatically guides the content and localization of the generated edit.
- We demonstrate the effectiveness of internal learning for training a generator on a single input in a zero-shot manner.
- An extension to video which harnesses the richness of information across time, and can perform consistent text-guided editing.

- We demonstrate various edits, ranging from changing objects' texture to generating complex semi-transparent effects, all achieved fully automatically across a wide-range of objects and scenes.

2 Related Work

Text-Guided Image Manipulation and Synthesis. There has been remarkable progress since the use of conditional GANs in both text-guided image generation [35,47–49], and editing [7,20,27]. ManiGAN [20] proposed a text-conditioned GAN for editing an object's appearance while preserving the image content. However, such multi-modal GAN-based methods are restricted to specific image domains and limited in the expressiveness of the text (e.g., trained on COCO [22]). DALL-E [33] addresses this by learning a joint image-text distribution over a massive dataset. While achieving remarkable text-to-image generation, DALL-E is not designed for editing existing images. GLIDE [28] takes this approach further, supporting both text-to-image generation and inpainting.

Instead of directly training a text-to-image model, a recent surge of methods leverage a pretrained generator, and use CLIP [32] for text-guided generation [3,10,23,30]. StyleCLIP [30] and StyleGAN-NADA [10] use a pretrained StyleGAN2 [15] for image manipulation, by either controlling the GAN's latent code [30], or by fine-tuning the StyleGAN's output domain [10]. However, editing a real input image using these methods requires first tackling the GAN-inversion challenge [36,44]. Furthermore, these methods can edit images from a few specific domains, and edit images in a *global* fashion. In contrast, we consider a different problem setting – *localized* edits that can be applied to real-world images spanning a variety of object and scene categories. A recent exploratory and artistic trend in the online AI community has demonstrated impressive text-guided image generation by steering the generation process of a pretrained VQ-GAN [8], or diffusion models [11,42] using CLIP. [17] takes this approach a step forward by optimizing the diffusion process itself. However, since the generation is *globally* controlled by the diffusion process, this method is not designed to support localized edits that are applied only to selected objects.

To enable region-based editing, user-provided masks are used to control the diffusion process for image inpainting [2]. In contrast, our goal is not to generate new objects but rather to manipulate the appearance of existing ones, while preserving the original content. Furthermore, our method is fully automatic and performs the edits directly from the text, without user edit masks.

Several works [9,12,19,26] take a *test-time optimization* approach and leverage CLIP without using a pre-trained generator. For example, CLIPDraw [9] renders a drawing that matches a target text by directly optimizing a set of vector strokes. To prevent adversarial solutions, various augmentations are applied to the output image, all of which are required to align with the target text in CLIP embedding space. CLIPStyler [19] takes a similar approach for *global* stylization. Our goal is to perform *localized* edits, which are applied only to specific objects. Furthermore, CLIPStyler optimizes a CNN that observes *only* the source image. In contrast, our generator is trained on an *internal dataset*, extracted from the

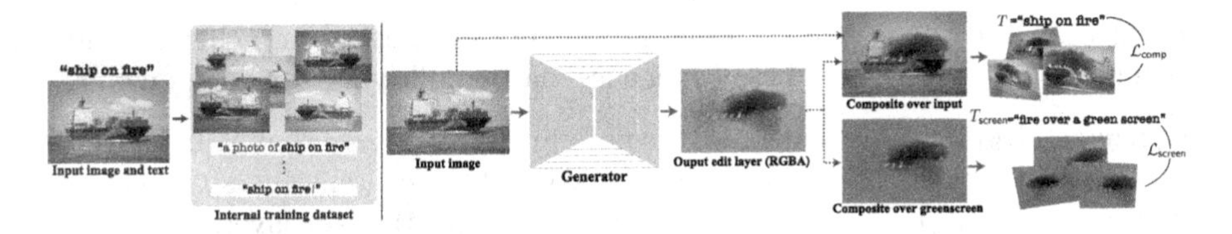

Fig. 3. *Image pipeline.* Our method consists of a generator trained on a single input image and target text prompts. *Left:* an *internal image-text dataset* of diverse training examples is created by augmenting both image and text (see Sect. 3.1). *Right:* Our generator takes as input an image and outputs an edit RGBA layer (color+opacity), which is composited over the input to form the final edited image. The generator is trained by minimizing several loss terms that are defined in CLIP space, and include: $\mathcal{L}_{\text{comp}}$, applied to the composite, and $\mathcal{L}_{\text{screen}}$, applied to the edit layer (when composited over a green background). We apply additional augmentations before CLIP (Sect. 3.1) (Color figure online)

input image and text. We draw inspiration from previous works that show the effectiveness of internal learning in the context of generation [38,40,45].

Other works use CLIP to synthesize [12] or edit [26] a single 3D representation (NeRF or mesh). The unified 3D representation is optimized through a differentiable renderer: CLIP loss is applied across different 2D rendered viewpoints. Inspired by this approach, we use a similar concept to edit videos. In our case, the "renderer" is a layered neural atlas representation of the video [16].

Consistent Video Editing. Existing approaches for consistent video editing can be roughly divided into (i) propagation-based, which use keyframes [13,43] or optical flow [37] to propagate edits throughout the video, and (ii) video layering-based, in which a video is decomposed into layers that are then edited [16,21, 24,25,34]. Lu et al. [24,25] estimate *omnimattes* – RGBA layers depicting a target object and its associated scene effects. Omnimattes facilitate various video effects (e.g., object removal). However, the layers are computed independently for each frame, hence cannot support consistent edit propagation across time. Kasten et al. [16] address this challenge by decomposing a video into unified 2D atlas layers (foreground/background). Edits applied to the 2D atlases are automatically mapped back to the video, achieving temporal consistency with minimal effort. Here, we treat the neural layered atlas model as a *video renderer*, leveraging it for text-guided video editing.

3 Text-Guided Layered Image and Video Editing

We focus on semantic, localized edits expressed by simple text prompts. Such edits include changing objects' texture or semantically augmenting the scene with complex semi-transparent effects (e.g., smoke, fire). To this end, we harness the potential of learning a generator from a *single input* image or video while leveraging CLIP, which is kept fixed and used to establish our losses [32]. Our task is ill-posed – numerous possible edits can satisfy the target text according to CLIP, some of which are noisy or undesired [9,23]. Thus, controlling edits' localization and preserving the original content are essential for achieving high-quality editing. We tackle these challenges via the following key components:

1. *Layered editing.* Our generator outputs an RGBA layer that is composited over the input image. This allows us to control the content and spatial extent of the edit via dedicated losses applied directly to the edit layer.
2. *Explicit content preservation and localization losses.* We devise new losses using the internal spatial features in CLIP space to preserve the original content, and to guide the localization of the edits.
3. *Internal generative prior.* We construct an internal dataset of examples by applying augmentations to the input image/video and text. These augmented examples are used to train our generator, whose task is to perform text-guided editing on a larger and more diverse set of examples.

3.1 Text to Image Edit Layer

As illustrated in Fig. 3, our framework consists of a generator G_θ that takes as input a source image I_s and synthesizes an *edit layer*, $\mathcal{E} = \{C, \alpha\}$, which consists of a color image C and an opacity map α. The final edited image I_o is given by compositing the edit layer over I_s:

$$I_o = \alpha \cdot C + (1 - \alpha) \cdot I_s \tag{1}$$

Our main goal is to generate $\mathcal{E}$ such that the final composite I_o would comply with a target text prompt T. In addition, generating an RGBA layer allows us to use text to further guide the generated content and its localization. To this end, we consider a couple of auxiliary text prompts: T_{screen} which expresses the target *edit layer*, when composited over a green background, and T_{ROI} which specifies a region-of-interest in the source image, and is used to initialize the localization of the edit. For example, in the *Bear* edit in Fig. 2, $T=$ *"fire out of the bear's mouth"*, $T_{\text{screen}} =$ *"fire over a green screen"*, and $T_{\text{ROI}} =$ *"mouth"*. We next describe in detail how these are used in our objective function.

Objective Function. Our novel objective function incorporates three main loss terms, all defined in CLIP's feature space: (i) $\mathcal{L}_{\text{comp}}$, which is the driving loss and encourages I_o to conform with T, (ii) $\mathcal{L}_{\text{screen}}$, which serves as a direct supervision on the edit layer, and (iii) $\mathcal{L}_{\text{structure}}$, a structure preservation loss w.r.t. I_s. Additionally, a regularization term $\mathcal{L}_{\text{reg}}$ is used for controlling the extent of the edit by encouraging sparse alpha matte α. Formally,

$$\mathcal{L}_{\text{Text2LIVE}} = \mathcal{L}_{\text{comp}} + \lambda_g \mathcal{L}_{\text{screen}} + \lambda_s \mathcal{L}_{\text{structure}} + \lambda_r \mathcal{L}_{\text{reg}}, \tag{2}$$

where λ_g, λ_s, and λ_r control the relative weights between the terms, and are fixed throughout all our experiments (see Supplementary Materials on our website - SM).

Composition Loss. $\mathcal{L}_{\text{comp}}$ reflects our primary objective of generating an image that matches the target text prompt and is given by a combination of a *cosine distance* loss and a *directional* loss [30]:

$$\mathcal{L}_{\mathsf{comp}} = \mathcal{L}_{\mathsf{cos}}\left(I_o, T\right) + \mathcal{L}_{\mathsf{dir}}(I_s, I_o, T_{\mathsf{ROI}}, T), \tag{3}$$

where $\mathcal{L}_{\mathsf{cos}} = \mathcal{D}_{\mathrm{cos}}\left(E_{\mathrm{im}}(I_o), E_{\mathrm{txt}}(T)\right)$ is the cosine distance between the CLIP embeddings for I_o and T. Here, E_{im}, E_{txt} denote CLIP's image and text encoders, respectively. The second term controls the direction of edit in CLIP space [10,30] and is given by: $\mathcal{L}_{\mathsf{dir}} = \mathcal{D}_{\mathrm{cos}}(E_{\mathrm{im}}(I_o) - E_{\mathrm{im}}(I_s), E_{\mathrm{txt}}(T) - E_{\mathrm{txt}}(T_{\mathsf{ROI}}))$.

Similar to most CLIP-based editing methods, we first augment each image to get several different views and calculate the CLIP losses w.r.t. each of them separately, as in [2]. This holds for all our CLIP-based losses. See SM for details.

Screen Loss. The term $\mathcal{L}_{\mathsf{screen}}$ serves as a direct text supervision on the edit layer $\mathcal{E}$. We draw inspiration from chroma keying [4]–a well-known technique by which a solid background (often green) is replaced by an image in a post-process. Chroma keying is extensively used in image and video post-production, and there is high prevalence of online images depicting various visual elements over a green background. We thus composite the edit layer over a green background I_{green} and encourage it to match the template $T_{\mathsf{screen}} :=$ "{ }*over a green screen*", (Fig. 3):

$$\mathcal{L}_{\mathsf{screen}} = \mathcal{L}_{\mathsf{cos}}\left(I_{\mathsf{screen}}, T_{\mathsf{screen}}\right) \tag{4}$$

where $I_{\mathsf{screen}} = \alpha \cdot C + (1 - \alpha) \cdot I_{\mathsf{green}}$.

A nice property of this loss is that it allows intuitive supervision on a desired *effect*. For example, when generating semi-transparent effects, e.g., *Bear* in Fig. 2, we can use this loss to focus on the fire regardless of the image content by using $T_{\mathsf{screen}} =$ "fire over a green screen". Unless specified otherwise, we plug in T to our screen text template in all our experiments. Similar to the composition loss, we first apply augmentations on the images before feeding to CLIP.

Structure Loss. We want to allow substantial texture and appearance changes while preserving the objects' original spatial layout, shape, and perceived semantics. While various perceptual content losses have been proposed in the context of style transfer, most of them use features extracted from a pretrained VGG [41]. Instead, we define our loss in CLIP feature space. This allows us to impose additional constraints to the resulting internal CLIP representation of I_o. Inspired by classical and recent works [18,39,45], we adopt the *self-similarity* measure. Specifically, we feed an image into CLIP's ViT and extract its K spatial tokens from the deepest layer. The self-similarity matrix, denoted by $S(I) \in \mathbb{R}^{K \times K}$, is used as structure representation. Each matrix element $S(I)_{ij}$ is defined by:

$$S(I)_{ij} = 1 - \mathcal{D}_{\mathrm{cos}}\left(\mathbf{t}^i(I), \mathbf{t}^j(I)\right) \tag{5}$$

where $\mathbf{t}_i(I) \in \mathbb{R}^{768}$ is the ith token of image I.

The term $\mathcal{L}_{\mathsf{structure}}$ is defined as the Frobenius norm distance between the self-similarity matrices of I_s, and I_o:

$$\mathcal{L}_{\mathsf{structure}} = \|S(I_s) - S(I_o)\|_F \tag{6}$$

Sparsity Regularization. To control the spatial extent of the edit, we encourage the output opacity map to be sparse. Following [24,25], we define the sparsity loss as a combination of L_1- and L_0-approximation regularization terms:

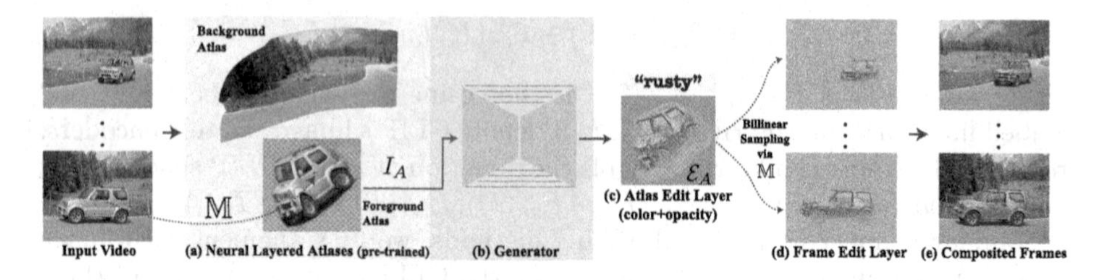

Fig. 4. *Video pipeline.* (a) a pretrained and fixed *neural layered atlas* model [16] is used as a "video renderer", which consists of: a set of 2D atlases, mapping functions from pixels to the atlases (and per-pixel fg/bg opacity values). Our framework takes in an atlas I_A and a target text prompt (e.g., "rusty car"), and trains a generator (b) to output an atlas edit layer $\mathcal{E}_A$ (c). The edited atlas (d) is rendered to frames using the pre-trained mapping network $\mathbb{M}$, and then (e) composited over the original video.

$$\mathcal{L}_{\mathsf{reg}} = \gamma||\alpha||_1 + \Psi_0(\alpha) \tag{7}$$

where $\Psi_0(x) \equiv 2\text{Sigmoid}(5x) - 1$ is a smooth L_0 approximation that penalizes non zero elements. We fix γ in all our experiments.

Bootstrapping. To achieve accurate localized effects without user-provided edit mask, we apply a text-driven *relevancy* loss to initialize our opacity map. Specifically, we use Chefer et al. [5] to automatically estimate a *relevancy map*[1] $R(I_s) \in [0,1]^{224\times224}$ which roughly highlights the image regions that are most relevant to a given text T_{ROI}. We use $R(I_s)$ to initialize α by minimizing:

$$\mathcal{L}_{\mathsf{init}} = \text{MSE}\left(R(I_s), \alpha\right) \tag{8}$$

Note that the relevancy maps are noisy, and only provide a rough estimation for the region of interest (Fig. 8(c)). Thus, we anneal this loss during training (see implementation details in SM). By training on diverse internal examples along with the rest of our losses, our framework dramatically refines this rough initialization, and produces accurate and clean opacity (Fig. 8(d)).

Training Data. Our generator is trained from scratch for each input (I_s, T) using an *internal dataset* of diverse image-text training examples $\{(I_s^i, T^i)\}_{i=1}^N$ that are derived from the input (Fig. 3 left). Specifically, each training example (I_s^i, T^i) is created by randomly applying a set of augmentations to I_s and T. Image augmentations include global crops, color jittering, and flip; text augmentations are sampled from predefined text templates (e.g., *"a photo of* { }*"*); see details in SM. The vast space of all augmentation combinations provides a rich and diverse dataset for training. The task is now to learn *one* mapping function G_θ for the *entire dataset*, posing a strong regularization. Specifically, for each individual example, G_θ has to generate a plausible edit layer $\mathcal{E}^i$ from I_s^i such that the composited image is well described by T^i. We demonstrate the effectiveness of our *internal learning* approach compared to test-time optimization in Sect. 4.

[1] [5] works with 224×224 images, so we resize I_s and α before applying loss (8).

3.2 Text to Video Edit Layer

A natural question is whether our image framework can be applied to videos. The key additional challenge is achieving a temporally consistent result. Naïvely applying our image framework on each frame independently yields unsatisfactory jittery results (see Sect. 4). To enforce temporal consistency, we utilize the Neural Layered Atlases (NLA) method [16], as illustrated in Fig. 4(a). We next provide a brief review of NLA and discuss in detail our extension to videos.

Preliminary: Neural Layered Atlases. NLA provides a unified 2D parameterization of a video: the video is decomposed into a set of 2D atlases, each can be treated as a 2D image, representing either one foreground object or the background throughout the entire video. An example of foreground and background atlases are shown in Fig. 4. For each video location $p = (x, y, t)$, NLA computes a corresponding 2D location (UV) in each atlas, and a foreground opacity value. This allows to reconstruct the original video from the set atlases. NLA comprises of several Multi-Layered Perceptrons (MLPs), representing the atlases, the mappings from pixels to atlases and their opacity. More specifically, each video location p is first fed into two mapping networks, $\mathbb{M}_b$ and $\mathbb{M}_f$:

$$\mathbb{M}_b(p) = (u_b^p, v_b^p), \qquad \mathbb{M}_f(p) = (u_f^p, v_f^p) \tag{9}$$

where (u_*^p, v_*^p) are the 2D coordinates in the background/foreground atlas space. Each video location is also fed to an MLP that predicts its foreground opacity. The predicted UV coordinates are then fed into an atlas network $\mathbb{A}$ that outputs RGB colors in each location. Thus, the original RGB value of p can be reconstructed by mapping p to the atlases, extracting the corresponding atlas colors, and blending them according to the predicted opacity. See [16] for full details.

Importantly, NLA enables consistent video editing: the continuous atlas (foreground/background) is first discretized to a fixed resolution image (e.g., 1000×1000 px). The user can edit the discretized atlas using image editing tools (e.g., Photoshop). The atlas edit is then mapped back to the video, and blended with the frames using the predicted UV mappings and foreground opacity. Here, we are interested in generating atlas edits in a *fully automatic* manner, solely via text.

Text to Atlas Edit Layer. Our video framework leverages NLA as a "video renderer", as illustrated in Fig. 4. Specifically, given a pretrained and fixed NLA model for a video, our goal is to generate a 2D *atlas edit layer*, either for the background or foreground, such that when mapped back to the video, each of the rendered frames would comply with the target text.

Similar to the image framework, we train a generator G_θ that takes a 2D discretized atlas I_A and generates an atlas edit layer $\mathcal{E}_A = \{C_A, \alpha_A\}$. The pretrained UV mapping $\mathbb{M}$ is used to bilinearly sample $\mathcal{E}_A$ to map it to each frame:

$$\mathcal{E}_{\mathrm{t}} = \mathrm{Sampler}(\mathcal{E}_A, \mathcal{S}) \tag{10}$$

$\mathcal{S} = \{\mathbb{M}(p) \mid p = (\cdot, \cdot, t)\}$ is the set of UV coordinates corresponding to frame t. The edited video is obtained by blending $\mathcal{E}_t$ with the frames, following [16].

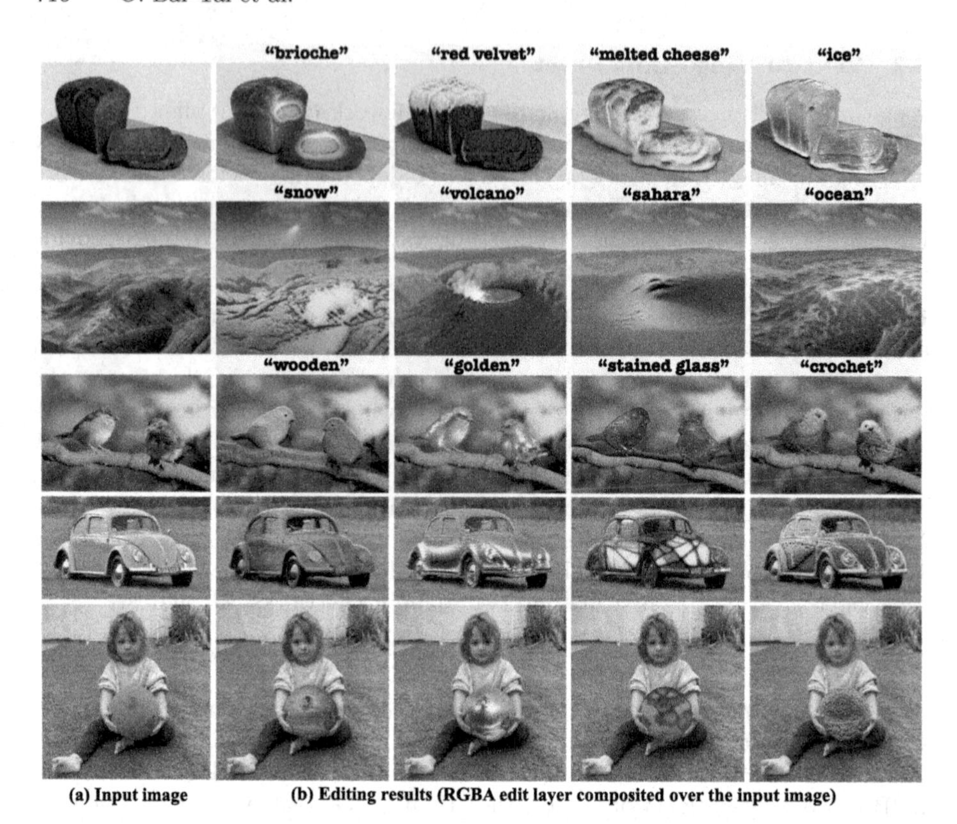

Fig. 5. *Text2LIVE image results.* Across rows: different images, across columns: different target edits. All results are produced fully automatically w/o any input masks.

Training. A straightforward approach to train G_θ is to treat I_A as an image and plug it into our image framework (Sect. 3.1). This will ensure temporal consistency, yet it has two main drawbacks: (i) the atlas often non-uniformly distorts the original structures (see Fig. 4), which may lead to low-quality edits, (ii) solely using the atlas while ignoring the video frames, disregards the abundant, diverse information available in the video such as different viewpoints or non-rigid object deformations, which can serve as "natural augmentations" to our generator. We overcome these drawbacks by mapping the atlas edit back to the video and applying our losses on the resulting edited *frames* with the same objective as in Eq. 2; we construct an internal dataset from the atlas.

More specifically, a training example is constructed by first extracting a crop from I_A. To ensure we sample informative atlas regions, we randomly crop a video segment in both space and time, and then map it to a corresponding atlas crop I_{Ac} using $\mathbb{M}$ (see SM for technical details). We then apply additional augmentations to I_{Ac} and feed it into the generator, resulting in an edit layer $\mathcal{E}_{Ac} = G_\theta(I_{Ac})$. We then map $\mathcal{E}_{Ac}$ and I_{Ac} back to the video, resulting in frame edit layer $\mathcal{E}_t$, and a reconstructed foreground/background crop I_t. This is done by bilinearly sampling $\mathcal{E}_{Ac}$ and I_{Ac} using Eq. (10), with $\mathcal{S}$ as the set of UV

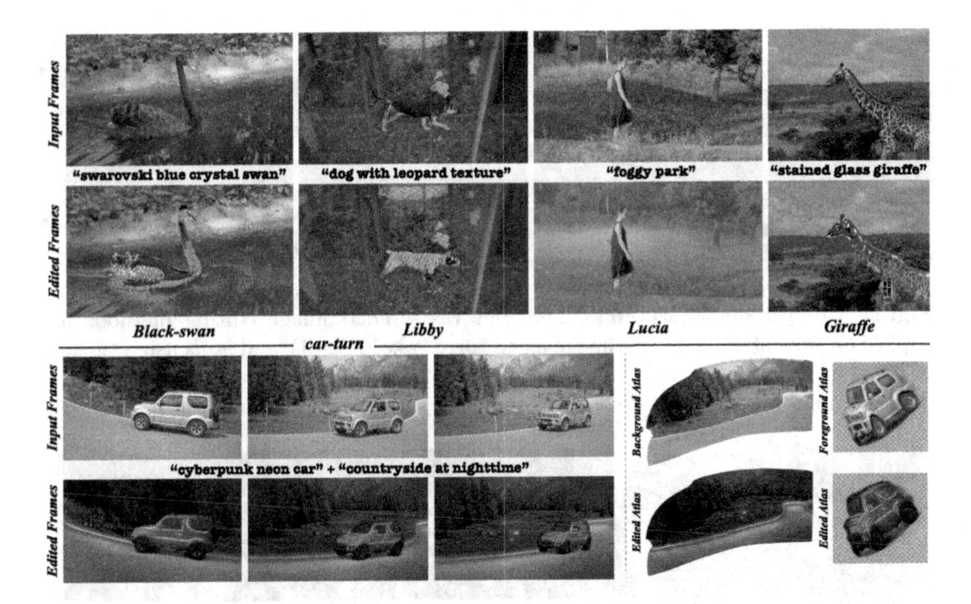

Fig. 6. *Text2LIVE video results.* A representative frame from the original and edited videos are shown for each example, along with the target text prompt. In *car-turn*, both foreground and background atlases are edited sequentially (see Sect. 4). The original and edited atlases are shown on the right. Full video results are included in the SM.

coordinates corresponding to the frame crop. Finally, we apply $\mathcal{L}_{\text{Text2LIVE}}$ from Eq. 2, where $I_s = I_t$ and $\mathcal{E} = \mathcal{E}_t$.

4 Results

We tested our method across various real-world, high-resolution images and videos. The image set contains 35 images collected from the web, spanning various object categories, including animals, food, landscapes, and more. The video set contains 7 videos from DAVIS [31]. We applied our method for various edits, ranging from text prompts that describe the texture/materials of specific objects to edits that express complex scene effects (e.g., smoke, fire, clouds). Sample results can be seen in Fig. 1, Fig. 2, and Fig. 5 for images, and Fig. 6 for videos (see SM for more). As seen, our method successfully generates photorealistic textures that are "painted" over the target objects in a semantically aware manner. Edits are accurately localized, even under partial occlusions, multiple objects (last and third row of Fig. 5) and complex scene composition (the dog in Fig. 2). Our method successfully augments the scene with complex semi-transparent effects without changing irrelevant content (see Fig. 1).

4.1 Comparison to Prior Work

To the best of our knowledge, there is no existing method tailored for our task: text-driven *semantic*, *localized* editing of *existing* objects in *real-world* images

Fig. 7. *Comparison to baselines.* Different text-guided image editing methods are applied on several images: *cake* image using "oreo cake" (Fig. 1); and *birds* using "golden birds" (Fig. 5). Manually created masks (a) are provided to the inpainting methods (b-c); the other methods are mask-free. Our results are shown in Fig. 1, and Fig. 5.

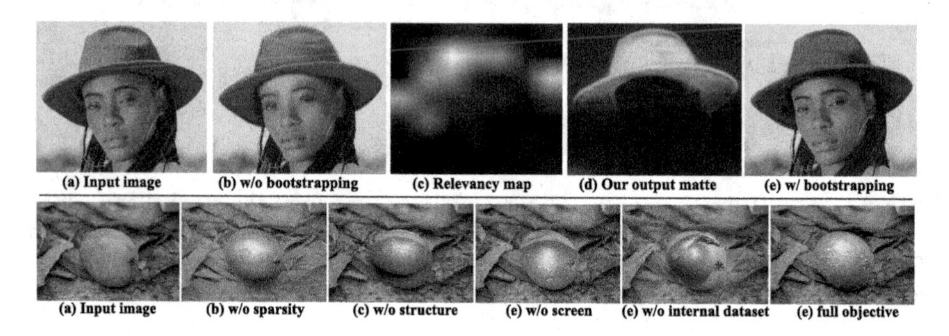

Fig. 8. *Top:* We illustrate the effect of our relevancy-based bootstrapping for (a) using "red hat" as the target edit. (b) w/o bootstrapping our edited image suffers from color bleeding. Our alpha-matte is initialized to capture the hat (T_{ROI} = "hat") using the raw relevancy map, which only provides a very rough supervision (c); during training, our method dramatically refines the matting (d). *Bottom:* We ablate each of our loss terms and the effect of internal learning ("mango" to "golden mango"). See Sect. 4.3.

and videos. We illustrate the key differences between our method and several prominent text-driven image editing methods. We consider those that can be applied to a similar setting to ours: editing real-world images without restriction to specific domains. Inpainting methods: Blended-Diffusion [2] and GLIDE [28], require user-provided editing mask. CLIPStyler, which performs image stylization; Diffusion+CLIP [1], and VQ-GAN+CLIP [6]: both combine CLIP with either a pretrained VQ-GAN or a Diffusion model. See SM for qualitative comparison to StyleGAN text-guided editing methods [10,30].

Figure 7 shows representative results (more in SM). As seen, none of these methods are designed for our task. The inpainting methods (b-c), even when supplied with tight edit masks, generate new content in the masked region rather than changing the texture of the existing one. CLIPStyler (d) modifies the image in a *global* artistic manner rather than performing *local* semantic editing (e.g., the background in both examples changed entirely, regardless of the image content). Diffusion+CLIP [1] (e) can often synthesize high-quality images, but with either low fidelity to the target text or the input image content (see more in SM).

Table 1. *AMT surveys evaluation (see Sect.* 4). We report the percentage of judgments in our favor (mean, std). Our method outperforms all baselines.

Image baselines			Video baselines	
CLIPStyler	VQ-GAN+CLIP	Diffusion+CLIP	Atlas baseline	Frames baseline
0.85 ± 0.12	0.86 ± 0.14	0.82 ± 0.11	0.73 ± 0.14	0.74 ± 0.15

VQ-GAN+CLIP [6] (f) fails to maintain fidelity to the input image and produces non-realistic images. Our method automatically locates the cake region and generates high-quality texture that naturally combines with the original content.

4.2 Quantitative Evaluation

Comparison to Image Baselines. We conduct an extensive human perceptual evaluation on Amazon Mechanical Turk (AMT). We adopt the Two-alternative Forced Choice (2AFC) protocol suggested in [18,29]. Participants are shown a reference image and a target editing prompt, along with two alternatives: ours and a baseline. We consider from the above baselines those not requiring user masks. Participants were asked: *"Which image better shows objects in the reference image edited according to the text"*. We collected 12,450 user judgments using 82 image-text pairs across several prominent text-guided image editing methods. Table 1 reports the percentage of votes in our favor. Our method outperforms all baselines by a large margin, including those using a strong generative prior.

Comparison to Video Baselines. We quantify the effectiveness of our key design choices for our video framework by comparing our video method against (i) *Atlas Baseline:* feeding the discretized 2D Atlas to our single-image method (Sect. 3.1), and using the same inference pipeline illustrated in Fig. 4 to map the edited atlas back to frames. (ii) *Frames Baseline:* treating all video frames as part of a single *internal dataset*, used to train our generator; at inference, we apply the trained generator independently to each frame.

We conduct a human perceptual evaluation in which we provide participants a target editing prompt and two video alternatives: ours and a baseline. Participants were asked *"Choose the video that has better quality and better represents the text"*. We collected 2,400 user judgments over 19 video-text combinations. Table 1 reports the percentage of votes in our favor. We first note that the *Frames baseline* produces temporally inconsistent edits. The *Atlas baseline* produces temporally consistent results yet struggles to generate high-quality textures and often produces blurry results. These observations support our hypotheses mentioned in Sect. 3.2 (see SM for visual comparisons).

4.3 Ablation Study

Figure 8 (top) illustrates the effect of our relevancy-based bootstrapping (Sect. 3.1). As seen, it allows us to achieve accurate object mattes, which significantly improves the rough, inaccurate relevancy maps.

Fig. 9. *Limitations.* CLIP often exhibits biases towards certain visual elements. Our method is designed to edit existing objects, thus new objects may not be visually pleasing. Yet, the desired edit can often be achieved by using more specific text (left).

We ablate the loss terms in our objective by qualitatively comparing our results obtained with our full objective (Eq. 2) and with a specific loss removed (Fig. 8). Without $\mathcal{L}_{\text{reg}}$, the output matte does not accurately capture the mango, resulting in a global color shift around it. Without $\mathcal{L}_{\text{structure}}$, the model outputs an image with the desired appearance but fails to preserve the mango shape. Without $\mathcal{L}_{\text{screen}}$, the object segmentation is noisy (color bleeding from the mango), and the texture quality is degraded. Lastly, we consider a test-time optimization baseline by not using our internal dataset but rather feeding G_θ the same input at each training step. As seen, this baseline results in lower-quality edits.

4.4 Limitations

We noticed that for some edits, CLIP exhibits a very strong bias towards a specific solution. For example, as seen in Fig. 9, given an image of a cake, the text "birthday cake" is highly associated with candles. Our method is not designed to significantly deviate from the original layout and to create new objects, thus generates unrealistic candles. Nevertheless, in many cases the desired edit can be achieved by using more specific text. For example, the text "moon" guides the generation toward a crescent. Instead, "a bright full moon" steers the generation toward a full moon (Fig. 9 left). Finally, as noted by prior works (e.g., [26]), we also noticed that slightly different text prompts describing similar concepts may lead to slightly different flavors of edits.

Our video framework relies on the pretrained NLA model. Thus, we are restricted to examples where NLA works well, as artifacts in the atlas representation can propagate to our edited video. An exciting avenue of future research may include fine-tuning the NLA representation jointly with our model.

5 Conclusion

We considered a new problem setting in the context of zero-shot text-guided editing: semantic, localized editing of real-world images and videos. Addressing this task requires careful control of several aspects of the editing: localization, preservation of the original content, and visual quality. We proposed to generate text-driven edit layers that allow us to tackle these challenges, without using a pretrained generator in the loop. We further demonstrated how to adopt our framework, with minimal changes, for consistent text-guided video editing. We believe that the key principles exhibited in the paper hold promise for leveraging large-scale multi-modal networks in tandem with an internal learning approach.

Acknowledgements. We thank Kfir Aberman, Lior Yariv, Shai Bagon for reviewing early drafts; Narek Tumanyan for assisting with the user evaluation. This project received funding from the Israeli Science Foundation (grant 2303/20).

References

1. Disco Diffusion. https://colab.research.google.com/github/alembics/disco-diffusion/blob/main/Disco_Diffusion.ipynb
2. Avrahami, O., Lischinski, D., Fried, O.: Blended diffusion for text-driven editing of natural images. In: Proceedings of the Conference on Computer Vision and Pattern Recognition (CVPR) (2022)
3. Bau, D., et al.: Paint by word. arXiv preprint arXiv:2103.10951 (2021)
4. Brinkmann, R.: The Art and Science of Digital Compositing: Techniques for Visual Effects, Animation and Motion Graphics. Morgan Kaufmann, Burlington (2008)
5. Chefer, H., Gur, S., Wolf, L.: Generic attention-model explainability for interpreting bi-modal and encoder-decoder transformers. In: Proceedings of the IEEE/CVF International Conference on Computer Vision (ICCV) (2021)
6. Crowson, K.: VQGAN+CLIP. https://colab.research.google.com/github/justinjohn0306/VQGAN-CLIP/blob/main/VQGAN%2BCLIP(Updated).ipynb
7. Dong, H., Yu, S., Wu, C., Guo, Y.: Semantic image synthesis via adversarial learning. In: Proceedings of the IEEE International Conference on Computer Vision, ICCV (2017)
8. Esser, P., Rombach, R., Ommer, B.: Taming transformers for high-resolution image synthesis. In: IEEE Conference on Computer Vision and Pattern Recognition (CVPR) (2021)
9. Frans, K., Soros, L., Witkowski, O.: CLIPDraw: exploring text-to-drawing synthesis through language-image encoders. arXiv preprint arXiv:2106.14843 (2021)
10. Gal, R., Patashnik, O., Maron, H., Chechik, G., Cohen-Or, D.: StyleGAN-NADA: CLIP-guided domain adaptation of image generators. arXiv preprint arXiv:2108.00946 (2021)
11. Ho, J., Jain, A., Abbeel, P.: Denoising diffusion probabilistic models. In: Advances in Neural Information Processing Systems (NeurIPS) (2020)
12. Jain, A., Mildenhall, B., Barron, J.T., Abbeel, P., Poole, B.: Zero-shot text-guided object generation with dream fields. In: Proceedings of the Conference on Computer Vision and Pattern Recognition (CVPR) (2022)
13. Jamriška, O., et al.: Stylizing video by example. ACM Trans. Graph. **38**, 1–11 (2019)
14. Jing, Y., Yang, Y., Feng, Z., Ye, J., Yu, Y., Song, M.: Neural style transfer: a review. IEEE Trans. Visual Comput. Graphics **26**(11), 3365–3385 (2019)
15. Karras, T., Laine, S., Aittala, M., Hellsten, J., Lehtinen, J., Aila, T.: Analyzing and improving the image quality of StyleGAN. In: IEEE/CVF Conference on Computer Vision and Pattern Recognition (CVPR) (2020)
16. Kasten, Y., Ofri, D., Wang, O., Dekel, T.: Layered neural atlases for consistent video editing. ACM Trans. Graph. (TOG) **40**(6), 1–12 (2021)
17. Kim, G., Ye, J.C.: DiffusionCLIP: text-guided image manipulation using diffusion models. arXiv preprint arXiv:2110.02711 (2021)

18. Kolkin, N.I., Salavon, J., Shakhnarovich, G.: Style transfer by relaxed optimal transport and self-similarity. In: IEEE Conference on Computer Vision and Pattern Recognition (CVPR) (2019)
19. Kwon, G., Ye, J.C.: CLIPstyler: image style transfer with a single text condition. arXiv preprint arXiv:2112.00374 (2021)
20. Li, B., Qi, X., Lukasiewicz, T., Torr, P.H.: ManiGAN: text-guided image manipulation. In: Proceedings of the IEEE/CVF Conference on Computer Vision and Pattern Recognition (CVPR) (2020)
21. Lin, S., Fisher, M., Dai, A., Hanrahan, P.: LayerBuilder: layer decomposition for interactive image and video color editing. arXiv preprint arXiv:1701.03754 (2017)
22. Lin, T.-Y., et al.: Microsoft COCO: common objects in context. In: Fleet, D., Pajdla, T., Schiele, B., Tuytelaars, T. (eds.) ECCV 2014. LNCS, vol. 8693, pp. 740–755. Springer, Cham (2014). https://doi.org/10.1007/978-3-319-10602-1_48
23. Liu, X., Gong, C., Wu, L., Zhang, S., Su, H., Liu, Q.: FuseDream: training-free text-to-image generation with improved CLIP+GAN space optimization. arXiv preprint arXiv:2112.01573 (2021)
24. Lu, E., et al.: Layered neural rendering for retiming people in video. ACM Trans. Graph. (2020)
25. Lu, E., Cole, F., Dekel, T., Zisserman, A., Freeman, W.T., Rubinstein, M.: Omnimatte: associating objects and their effects in video. In: Proceedings of the IEEE/CVF Conference on Computer Vision and Pattern Recognition (CVPR) (2021)
26. Michel, O., Bar-On, R., Liu, R., Benaim, S., Hanocka, R.: Text2Mesh: text-driven neural stylization for meshes. arXiv preprint arXiv:2112.03221 (2021)
27. Nam, S., Kim, Y., Kim, S.J.: Text-adaptive generative adversarial networks: manipulating images with natural language. In: Advances in Neural Information Processing Systems (NeurIPS) (2018)
28. Nichol, A., et al.: GLIDE: towards photorealistic image generation and editing with text-guided diffusion models. arXiv preprint arXiv:2112.10741 (2021)
29. Park, T., et al.: Swapping autoencoder for deep image manipulation. In: Advances in Neural Information Processing Systems (NeurIPS) (2020)
30. Patashnik, O., Wu, Z., Shechtman, E., Cohen-Or, D., Lischinski, D.: StyleCLIP: text-driven manipulation of StyleGAN imagery. In: Proceedings of the IEEE/CVF International Conference on Computer Vision (ICCV) (2021)
31. Pont-Tuset, J., Perazzi, F., Caelles, S., Arbeláez, P., Sorkine-Hornung, A., Van Gool, L.: The 2017 DAVIS challenge on video object segmentation. arXiv preprint arXiv:1704.00675 (2017)
32. Radford, A., et al.: Learning transferable visual models from natural language supervision. In: Proceedings of the 38th International Conference on Machine Learning (ICML) (2021)
33. Ramesh, A., et al.: Zero-shot text-to-image generation. In: Proceedings of the 38th International Conference on Machine Learning (ICML) (2021)
34. Rav-Acha, A., Kohli, P., Rother, C., Fitzgibbon, A.W.: Unwrap mosaics: a new representation for video editing. ACM Trans. Graph. (2008)
35. Reed, S.E., Akata, Z., Yan, X., Logeswaran, L., Schiele, B., Lee, H.: Generative adversarial text to image synthesis. In: Proceedings of the 33rd International Conference on Machine Learning (ICML) (2016)
36. Richardson, E., et al.: Encoding in style: a StyleGAN encoder for image-to-image translation. In: Proceedings of the IEEE/CVF Conference on Computer Vision and Pattern Recognition (CVPR) (2021)

37. Ruder, M., Dosovitskiy, A., Brox, T.: Artistic style transfer for videos. In: Rosenhahn, B., Andres, B. (eds.) GCPR 2016. LNCS, vol. 9796, pp. 26–36. Springer, Cham (2016). https://doi.org/10.1007/978-3-319-45886-1_3
38. Shaham, T.R., Dekel, T., Michaeli, T.: SinGAN: learning a generative model from a single natural image. In: Proceedings of the IEEE/CVF International Conference on Computer Vision (ICCV) (2019)
39. Shechtman, E., Irani, M.: Matching local self-similarities across images and videos. In: 2007 IEEE Computer Society Conference on Computer Vision and Pattern Recognition (CVPR) (2007)
40. Shocher, A., Bagon, S., Isola, P., Irani, M.: InGAN: capturing and retargeting the "DNA" of a natural image. In: 2019 IEEE/CVF International Conference on Computer Vision (ICCV) (2019)
41. Simonyan, K., Zisserman, A.: Very deep convolutional networks for large-scale image recognition. arXiv preprint arXiv:1409.1556 (2014)
42. Song, J., Meng, C., Ermon, S.: Denoising diffusion implicit models. In: 9th International Conference on Learning Representations (ICLR) (2021)
43. Texler, O., et al.: Interactive video stylization using few-shot patch-based training. ACM Trans. Graph. **39**(4), 73:1 (2020)
44. Tov, O., Alaluf, Y., Nitzan, Y., Patashnik, O., Cohen-Or, D.: Designing an encoder for StyleGAN image manipulation. ACM Trans. Graph. (TOG) **40**(4), 1–14 (2021)
45. Tumanyan, N., Bar-Tal, O., Bagon, S., Dekel, T.: Splicing ViT features for semantic appearance transfer. In: Proceedings of the IEEE/CVF Conference on Computer Vision and Pattern Recognition (2022)
46. Xia, W., Zhang, Y., Yang, Y., Xue, J.H., Zhou, B., Yang, M.H.: GAN inversion: a survey. arXiv preprint arXiv:2101.05278 (2021)
47. Xu, T., et al.: AttnGAN: fine-grained text to image generation with attentional generative adversarial networks. In: Proceedings of the IEEE Conference on Computer Vision and Pattern Recognition (CVPR) (2018)
48. Zhang, H., et al.: StackGAN: text to photo-realistic image synthesis with stacked generative adversarial networks. In: Proceedings of the IEEE International Conference on Computer Vision (ICCV) (2017)
49. Zhang, H., et al.: StackGAN++: realistic image synthesis with stacked generative adversarial networks. IEEE Trans. Pattern Anal. Mach. Intell. **41**(8), 1947–1962 (2019)

Digging into Radiance Grid for Real-Time View Synthesis with Detail Preservation

Jian Zhang[1], Jinchi Huang[1], Bowen Cai[1], Huan Fu[1(✉)], Mingming Gong[2], Chaohui Wang[3], Jiaming Wang[1], Hongchen Luo[1], Rongfei Jia[1], Binqiang Zhao[1], and Xing Tang[1]

[1] Tao Technology Department, Alibaba Group, Hangzhou, China
fuhuan.fh@alibaba-inc.com

[2] School of Mathematics and Statistics Melbourne Centre, University of Melbourne, Melbourne, Australia

[3] Univ Gustave Eiffel, CNRS, ENPC, LIGM, 77454 Marne-la-Vallée, France

Abstract. Neural Radiance Fields (NeRF) [31] series are impressive in representing scenes and synthesizing high-quality novel views. However, most previous works fail to preserve texture details and suffer from slow training speed. A recent method SNeRG [11] demonstrates that baking a trained NeRF as a Sparse Neural Radiance Grid enables real-time view synthesis with slight scarification of rendering quality. In this paper, we dig into the Radiance Grid representation and present a set of improvements, which together result in boosted performance in terms of both speed and quality. First, we propose an HieRarchical Sparse Radiance Grid (HrSRG) representation that has higher voxel resolution for informative spaces and fewer voxels for other spaces. HrSRG leverages a hierarchical voxel grid building process inspired by [30,55], and can describe a scene at high resolution without excessive memory footprint. Furthermore, we show that directly optimizing the voxel grid leads to surprisingly good texture details in rendered images. This direct optimization is memory-friendly and requires multiple orders of magnitude less time than conventional NeRFs as it only involves a tiny MLP. Finally, we find that a critical factor that prevents fine details restoration is the misaligned 2D pixels among images caused by camera pose errors. We propose to use the perceptual loss to add tolerance to misalignments, leading to the improved visual quality of rendered images.

Keywords: 3D representation · View synthesis · Real-time rendering

1 Introduction

Neural rendering has emerged as a promising new avenue towards controllable image and video generation of photo-realistic virtual worlds. In the past two years, there has been an explosive interest in neural volumetric representations,

J. Zhang, J. Huang and B. Cai—These authors contribute equally to this work.

S. Avidan et al. (Eds.): ECCV 2022, LNCS 13675, pp. 724–740, 2022.
https://doi.org/10.1007/978-3-031-19784-0_42

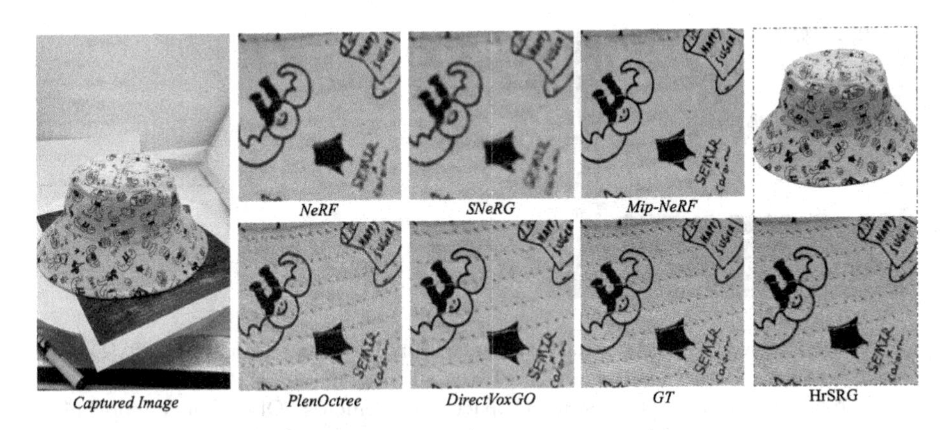

Fig. 1. The rendered image by our approach (HrSRG) is perceptually better than others while contains fine details. Achieving such a quality, HrSRG brings the real-time rendering practice and requires less storage space compared to SNeRG [11] and PlenOctree [55] (See Fig. 2). Zoom in for more details.

such as Neural Radiance Fields (NeRF) [31], due to their superiority in synthesizing photo-realistic novel views of scenes. NeRF takes a deep multilayer perceptron (MLP) that can map from 3D points to their volume densities and view-dependent emitted colors to represent a scene.

The main obstacle to the practical deployment of NeRF is the slow inference and training speed. It takes roughly 2 min to render a 1920×1080 (1080p) image on a high-performing GPU. Reconstructing a scene from high-definition (HD) images would take more than 48 h. One of the main reasons is that NeRF would query the MLP network hundreds of times even for rendering a single ray. Recent advances show that caching pre-computed values would bring real-time rendering experiences [11,55]; however, their compacted representations require a large memory footprint and often impose a quality loss of 1–2dB compared to NeRF. Towards the slow training speed issue, efforts have been made to learn image-based rendering by leveraging neural radiance fields [4,50]. Still, their formulations are not compatible with published real-time rendering approaches yet since they rely on nearby source images when performing rendering. In addition, they are overly sensitive to the pose distances between source and target views. Several concurrent works [32,44,54] study direct voxel grid optimization for super-fast training.

Another relatively understudied open problem is that NeRF fails to recover texture characteristics or fine details. To the best of our knowledge, Mip-NeRF [2] have discussed and partially remedied blurring and aliasing issues. Still, we find it unable to preserve fine details of real scenes. In fact, NeRF and its variants commonly generate excessively blurred in close-up views when representing real scenes instead of synthetic scenes. One possible reason is there are inevitable misaligned 2D pixels among images caused by camera pose errors. Unfortunately, neither NeRF nor its variants can handle this issue well.

In the paper, we dig into a Radiance Grid representation raised by SNeRG [11], which enjoys real-time synthesis at the cost of rendering quality. We present

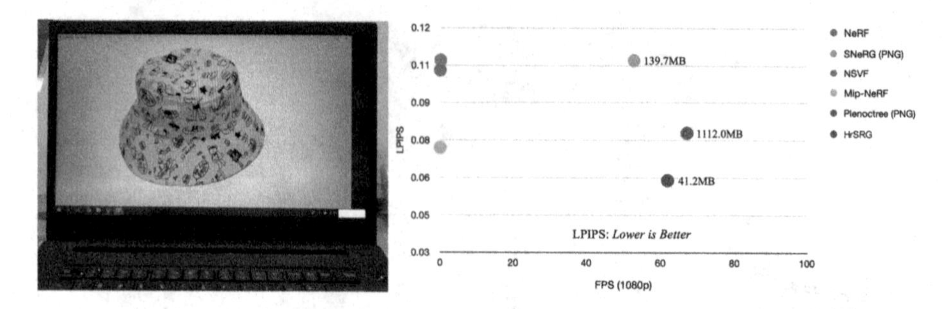

Fig. 2. LPIPS *vs.* **FPS.** We study the speed-quality tradeoff on Real 360° Objects (1080p images). HrSRG surpasses the two real-time methods [11,55] on rendering speed, rendering quality, and storage space requirements. Obtaining an optimized HrSRG requires about 5 h (vs. 1–2 days for NeRF) on a V100 GPU. We measure the rendering speed on a Lenovo ThinkPad P1 Gen 2 notebook. It has a 4GB NVIDIA Quadro T1000 mobile graphics card. Zoom in for the runtime texture details.

a series of improvements that lead to significantly boosted performance in terms of both speed and quality. First, we argue that baking a uniformly high-resolution voxel grid to represent a scene is memory intensive and not flexible. We notice that, for a ray, only the points sampled around the surfaces contributes to its color rendering. Thus, we introduce an *HieRarchical Sparse Radiance Grid* (HrSRG) representation that has higher voxel resolution around informative spaces and shows lower voxel resolution for other 3D spaces. We produce HrSRG for a scene by exploiting an hierarchical voxel grid building process leveraging [30]. A voxel contains volume density, radiance, and a three-dimensional feature vector. Second, in light of 2D texture atlas [12], we treat voxel values as parameters and directly optimize them, leading to surprisingly good texture characteristics in the rendered images. Finally, NeRF often cannot produce images with excellent quality for real scenes since per-pixel losses such as ℓ_p-norm neither address the aforementioned misalignment issue nor consider the perceptual quality. Luckily, training HrSRG is memory-friendly, which allows us to learn more than 200K rays simultaneously in one single iteration. We thus incorporate a perceptual loss [15] to improve the clarity of rendered images further and preserve more fine details. Our experiments show fitting a scene through HrSRG requires takes about six hours, requiring multiple orders of magnitude less time than previous high-performing NeRFs.

In summary, our main contributions are as follows:

- We present an HieRarchical Sparse Radiance Grid (HrSRG) representation leveraging a hierarchical voxel grid building process. It enables representing a scene at high resolution without excessive memory footprint.
- We show that it is possible to recover a scene's texture characteristics by directly optimizing its radiance grid.
- We introduce a perceptual loss to address the misalignment issue caused by camera pose errors. It can further improves the clarity of rendered images.

- Experiments demonstrate our HrSRG supports real-time rendering and accelerated training while surpasses other methods on rendering quality and required memory footprint for real scenes. See Fig. 1 and Fig. 2.

2 Related Work

The computer vision and graphics communities have put tremendous efforts into studying scene representation and novel view synthesis [5,9,10,16,19,22,27,40,43,46,48]. Recently, Neural Radiance Fields (NeRF) [31] have emerged as a promising technique for learning to represent 3D scenes and synthesizing novel views of the scene in impressive quality. Due to its superior performance, it has been extended to model deformable objects [7,35], dynamic scenes [23,34], and transparent objects [13]. Nevertheless, the vanilla NeRF fails to reconstruct fine details of scenes and suffers from slow training and inference speed.

Accelerated NeRF Training. Optimizing a NeRF to represent a scene typically takes around 1–2 days. To accelerate training, major efforts have been made to incorporate multi-view geometry into NeRF and utilize depths and pixel-wise correspondences as priors to guide the optimization process [4,24,42,45,47,53,56]. For example, GRF [47] and PixelNeRF [56] learn dense local features and retrieve feature vectors from each input image for an interested 3d point. NerfingMVS [53] predicts dense depth as priors to guide volume sampling in NeRF. It also exploits uncertainty based on multi-view consistency to determine the depth ranges of each ray. These image-based rendering approaches are not compatible with NeRF's real-time rendering techniques yet. Some other approaches [42,45] study weights initialization by exploring meta-learning for neural representations, which result in faster convergence during optimization.

Real-Time Volume Rendering. Due to the expensive sampling mechanism and costly neural network computations, rendering a NeRF is extremely slow. To speed up the rendering process, Decomposed Radiance Fields [36] and Kilo-NeRF [37] spatially decompose a scene into multiple cells to reduce the inference complexity. AutoInt [25] and DoNeRF [33] learns to reduce the samples for each queried ray. Other efforts have shown faster rendering even real-time rendering experiences can be reached by catching pre-computed values [8,11,55]. Especially, SNeRG [11] presents a deferred NeRF architecture so that it can store as many as NeRF values into a radiance grid data structure. PlenOctree [55] explores SH values to encode view-dependent effects and study octree representations for real-time view synthesis.

Preserving Details. A few works observe that NeRF, in its vanilla formulation, is limited in recovering fine details. The previous best practice is from Mip-NeRF [2]. It rendered anti-aliased conical frustums instead of rays to provide a partial remedy to the aliasing and blurring issues. Other works such as NerfMM [52] and BARF [24] learn to optimize camera poses to reduce the adverse impacts of imperfect camera poses. NeRF-ID [1] focuses on improving the hierarchical volume sampling technique of NeRF. However, none of these methods can preserve fine details for real scenes.

Relation to Plenoxels [54], **DirectVoxGO** [44], **and Instant-NGP** [32]. These are some great concurrent works that study direct voxel grid optimization for super-fast NeRF training [44,54]. Still, there are some difference betweens HrSRG and these works. Higher voxel resolution would better describe local details of scenes, but training a high-resolution voxel grid is memory intensive. For example, optimizing a 1024^3 voxel grid often leads to OMM errors on modern GPUs. Besides, high-resolution voxel grid requires large hardware storage spaces. [44,54] provide a solution by utilizing the "Trilinear Interpolation (Tri. Interp.)" operation for low-resolution voxel grids. However, incorporating "Tri. Interp." would degrade the rendering speed (about 3x). Our HrSRG presents to consider voxel resolutions for different 3D spaces of a scene. It can describe a scene at high resolution without excessive memory footprint. Instant-NGP [32] introduces a great hash encoding approach that can train NeRF in 5 min while preserving the texture details. In contrast, HrSRG directly optimize the color values (or 3D texture atlas) inspired by the 2D texture atlas optimization success [12].

3 Method

Beyond NeRF, our primary goal is to design a representation or approach that (1) can recover scenes' texture characteristics, (2) enables real-time rendering of 1080p images on conventional devices, (3) accelerates the training procedure, and (4) reduces the hardware storage space requirement. This section presents our solution HrSRG and explains how it can open a potent avenue towards these goals.

3.1 Review of NeRF

NeRF [31] takes a continuous volumetric function parameterized by a deep MLP to represent a scene. The MLP inputs a single 5D coordinates (x, y, z, θ, ϕ), *i.e.*, a 3D location $\mathrm{x} = (x, y, z)$ and 2D viewing direction $\mathrm{d} = (\theta, \phi)$, and outputs a radiance $\mathrm{c} = (r, g, b)$ and volume density σ at this location. The formulation is expressed as:

$$\mathrm{c}, \sigma = \mathrm{MLP}_{\Theta}(\mathrm{x}, \mathrm{d}), \tag{1}$$

where Θ denotes the MLP's weights that to be learned.

To render a 2D pixel, NeRF casts a ray r from its corresponding camera center o along the direction d passing through the pixel's center. Then, it samples N points along the ray, and approximate a volume rendering integral [28] to obtain the pixel's color $\hat{C}(\mathrm{r})$:

$$\hat{C}(\mathrm{r}) = \sum_{i=1}^{N} w_i \mathrm{c}_i, \tag{2}$$

$$w_i = T_i(1 - \exp(-\sigma_i \delta_i)), \tag{3}$$

where $T_i = \exp(-\sum_{j=1}^{i-1} \sigma_j \delta_j)$ is the accumulated transmittance along r, and δ_i denotes the distance of two consecutive samples.

NeRF simply minimizes the mean squared errors (MSE) between the rendered and true pixel colors ($\hat{C}(\mathrm{r})$ and $C(\mathrm{r})$) to learn its MLP:

$$\mathcal{L}_r = \sum_{\mathrm{r} \in \mathcal{R}} \|\hat{C}(\mathrm{r}) - C(\mathrm{r})\|, \tag{4}$$

where $\mathcal{R}$ is the sampled rays in each batch. In practice, NeRF simultaneously optimizes a "coarse" network and a "fine" network that have the same architectures. The two networks (or MLPs) allow NeRF to perform hierarchical volume sampling to secure better rendering quality.

3.2 Review of Deferred NeRF (Def-NeRF)

As mentioned in Sect. 1, the pre-caching techniques allow real-time rendering of NeRF. The volume density can be easily stored, but it is non-trivial to catch the emitted colors because NeRF relies on (θ, ϕ) and the geometry feature mapping from (x, y, z) to encode view-dependent effects. A subsequent work SNeRG [11] provides a remedy by introducing a deferred NeRF architecture. It contains a deep MLP that maps from a 3D location to the diffuse color c, 4-dimension feature vector v, volume density σ at this location.

$$\mathrm{c}, \sigma, \mathrm{v} = \mathrm{MLP}_{\Theta}(\mathrm{x}, \mathrm{d}). \tag{5}$$

Then, for a ray r, it passes the accumulated feature vector $V(\mathrm{r})$ and viewing direction d to a tiny MLP with parameters Φ to produce a specular color (or view-dependent residual). The final color $\hat{C}(\mathrm{r})$ is obtained by an addition of $\mathrm{MLP}_{\Phi}(V(\mathrm{r}), \mathrm{d})$ and the accumulated diffuse color $\hat{C}_d(\mathrm{r})$:

$$\begin{aligned} \hat{C}(\mathrm{r}) &= \mathrm{MLP}_{\Phi}(V(\mathrm{r}), \mathrm{d}) + \hat{C}_d(\mathrm{r}), \\ \hat{C}_d(\mathrm{r}) &= \sum_{i=1}^{N} T_i(1 - \exp(-\sigma_i \delta_i))\mathrm{c}_i. \end{aligned} \tag{6}$$

After training, this formulation enables baking diffuse colors, 4-dimension feature vectors, and opacities within a 3D voxel grid data structure.

We find the vanilla deferred NeRF [10] would produce noisier volume density (or geometry) for real-world objects than the vanilla NeRF. Thus, we initialize the opacity (or volume density) prediction MLP of the deferred NeRF from a non-converged vanilla NeRF. We also incorporate a background regularization term [27] to the vanilla NeRF pre-training process to ensure that it reconstructs the object of interest.

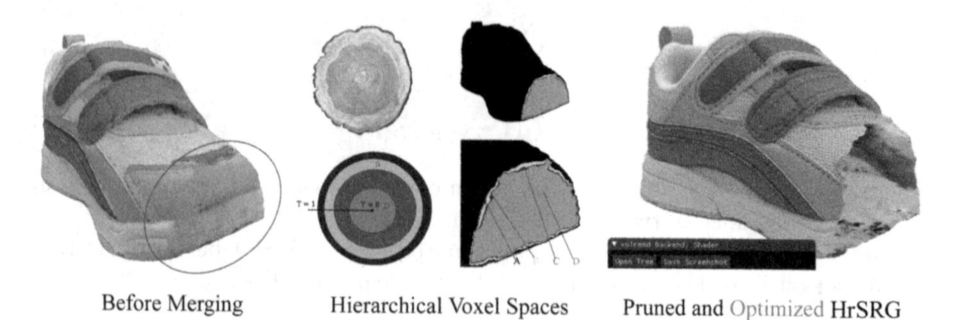

Fig. 3. Constructing HrSRG. We bake a non-converged Deferred NeRF Variant (Sect. 3.2) into a Hierarchical Sparse Radiance Grid structure. We refer to Sect. 3.3 for how we produce the HrSRG representation by exploiting the hierarchical voxel spaces. A voxel in HrSRG contains volume density σ, diffuse color c, and a 3-dimension feature v. Optimizing HrSRG means we treat these values as parameters to be learned (See Sect. 3.4). Zoom in for a better view.

Our experiments show training Def-NeRF several epochs is sufficient as the following HrSRG optimization stage is robust to an imperfect initialization. It is worth mentioning that we surprisingly find, for some cases, the specular feature v and the emitted color c can be learned from scratch in Sect. 3.4. The more important thing is to give a good geometry (or volume density σ) initialization.

3.3 Hierarchical Sparse Radiance Grid (HrSRG)

We first bake Def-NeRF into a N^3 voxle grid to represent a scene. A voxel contains opacity (or volume density), diffuse color, and a 3-dimensional specular feature. For a specific voxel, we randomly cast K rays from K origins passing through the voxel, and compute their accumulated transmittance values (See T_i in Eq. 2). With this operation, we can assign each voxel a maximum transmittance value. Then, we define four transmittance intervals (or 3D spaces) A, B, C, and D as shown in Fig. 3 (*Middle*) based on the transmittance thresholds τ_1, τ_2, and τ_3, respectively. We find from Fig. 3 that space A contains most of the high-frequency texture details while other spaces contribute slightly to the base color in the rendering process. This observation motives us to utilize higher voxel resolution for space A, and lower voxel resolution for spaces B, C, and D. For the purpose, we introduce the merging and splitting operations leveraging the octree data structure [29,30,55].

Merging. By converting the N^3 voxle grid into a tree structure with maximum depth L, we have a transmittance value for each node and leaf. For each node that is with depth L_i and has eight leaves, if the maximum transmittance value among itself and its leaves is below τ, we delete the eight leaves. We recursively consider depth L_1, L_2, and L_3 with thresholds τ_3, τ_2, and τ_1, respectively. As a results, we capture the approximate voxel resolutions $(N/2)^3$, $(N/4)^3$, $(N/8)^3$ for spaces B, C, and D, respectively.

Splitting. After the merging operation, if a leaf is with depth L, we take it as a parent of eight new leaves. A newly created leaf has the same voxel values (*i.e.*, $(\mathrm{c}, \sigma, \mathrm{v})$) as its parent. With this splitting operation, the space A would have a voxel resolution of $(2N)^3$.

In this paper, the hyperparameters are set to 512, 100, 0.3, 0.01, 0.001 for N, K, τ_1, τ_2, and τ_3, respectively. The HrSRG construction process takes about 30 min on average. One may capture more 3D space levels and tune τ for better performance in term of the memory footprint.

3.4 Optimizing HrSRG

We now have the HrSRG representation and a tiny MLP with weights Φ. We can treat $\{(\mathrm{c}, \sigma, \mathrm{v})\}$ as parameters and directly optimize them like Φ since the volume rendering process is differentiable with respect to these values. Note that, SNeRG [11] only fine-tunes its MLP_Φ after obtaining its sparse radiance grid.

Perceptual Loss. NeRF often yields blurry renderings for real scenes. A possible reason is that there are misalignment errors in estimated camera poses as discussed in [12]. Per-pixel losses such as ℓ_p-norm overlook this issue. Besides, ℓ_p-norm fails to encourage high-frequency crispness, thus in many cases would result in perceptually unsatisfying solutions with overly smooth textures, as analyzed in [14,15,20,21]. [12] has demonstrated a PatchGAN loss [14] can well tackle these two problems. Here, we simply introduce a perceptual loss [15]. Experiments show it tolerates the misalignment issue and can largely improve the perceptual quality of rendered images. It is worth mentioning that it is non-trivial for conventional NeRFs to incorporate a perceptual loss since training them is memory-intensive. Fortunately, optimizing HrSRG is memory-friendly as it does not involve a large MLP. It allows us simultaneously to learn more than 200K rays (*vs.* 6K for NeRF) in one single iteration. The perceptual loss encourages the rendered image $\hat{I}$ to be perceptually similar to its ground-truth image I by matching their semantic features:

$$\mathcal{L}_p = \frac{\lambda_p}{H_j * W_j} \sum_{h,w} \|\phi_j(\hat{I}) - \phi_j(I)\|, \tag{7}$$

where ϕ_j is the output of the jth convolution block of a pre-trained VGG-19 network [41], H_j and W_j are the spatial dimensions of the feature maps. In our experiments, we render a 384×384 image patch per-iteration during training. We consider the first three convolution blocks and set the hyperparameter λ_p to 0.01. It takes about 80 s to optimize HrSRG for 100 iterations (or to optimize 2.0×10^7 rays) on a single V100 GPU.

Table 1. Comparisons on Real Scenes. HrSRG achieves competitive or best performance. Especially on the Real 360° Objects dataset, it outperforms the compared methods by a large margin over the image quality metric LPIPS. It is worth mentioning that though HrSRG captures more details and produces perceptually better visual quality as shown in Fig. 1 and Fig. 4, it cannot yield an improvement on PSNR. As stated in [15,57], PSNR might fail to account for many nuances of human perception. We find that LPIPS might measure the perceptual quality and texture details better. SNeRG [11] has the voxel resolution of 1000^3 (the default hyperparameter).

	Tanks and Temples [18]			Real 360° Objects		
Method	PSNR ↑	SSIM ↑	LPIPS ↓	PSNR ↑	SSIM ↑	LPIPS ↓
NeRF [6,31]	27.94	0.904	0.168	26.42	0.911	0.103
NSVF [26]	28.40	0.901	0.155	28.48	0.908	0.107
SNeRG [11]	25.39	0.904	0.186	28.37	0.910	0.106
PlenOctree [55]	27.99	0.917	0.131	28.44	0.925	0.077
Mip-NeRF [2]	27.68	0.918	0.113	**29.44**	0.920	0.072
DirectVoxGo [44]	**28.41**	0.911	0.155	28.75	0.917	0.079
Plenoxels [54]	27.41	0.906	0.162	28.06	0.905	0.109
HrSRG	27.33	**0.919**	**0.099**	29.05	**0.927**	**0.059**

3.5 Further Pruning and Real-Time Rendering

On the Synthetic-NeRF dataset [31], our uncompressed HrSRG require an average storage space of 210 M (See Table 2). Note that, it has an approximate voxel resolution of 1,024 at 3D space A in Fig. 3. To minimize the file size, SNeRG [11] applies the PNG compression technique by flattening and reshaping the features and emitted colors into the image planes. The voxel values here have been prequantified to 8 bits. There is a uncompressing step that we need to reshape the PNG image to its original data structure when performing rendering. It is OK for computers but is not friendly to mobile devices as the uncompressing step might require many computation resources. We utilize the median cut algorithm to further prune our tree structure [3,55].

Finally, we obtain a compacted HrSRG of about 40 M, an acceptable size for transmission over the internet. The compacted HrSRG can be directly loaded without any uncompressing computations. HrSRG enables rendering 1080p images in unprecedented quality at 61 FPS (*vs.* 67 FPS of PlenOctree) on a laptop GPU. Our main bottleneck is the tiny MLP which targets encoding view-dependent effects. Removing it can bring an improved real-time rendering experience (up to ∼75 FPS) with a slight quality loss.

Table 2. Comparisons with Baselines. Opt. is the training time. HrSRG yields slightly lower PSNR and SSIM compared to PlenOctree [55] and other direct voxel grid optimization approaches [44,54] because the Def-NeRF structure [11] shows some limitations in modeling strong specular effects. The performance gain on the LPIPS metric is due to the perceptual loss. Our unpruned HrSRG already requires small hardware storage space. Applying the median cut (Med. Cut) algorithm [3,55] to HrSRG results in approximately 6x compression rates, *i.e.*. The PNG compression technique can reduce the file size by more than 10 times, but suffers from a decompressing operation [11,55]. Our rendering speed is slightly slower than PlenOctree because HrSRG contains a tiny specular MLP. The LPIPS score for SNeRG is the reproduced one as the official LPIPS code is unavailable now.

Method	PSNR ↑	SSIM ↑	LPIPS ↓	MB ↓	FPS ↑	Opt. ↓
Synthetic-NeRF@800^2 [31]						
NeRF [6,31]	31.69	0.953	0.068	-	0.03	> 1 day
JaxNeRF+ [6]	33.00	**0.962**	0.038	-	-	> 1 day
KiloNeRF [37]	31.00	0.950	**0.030**	-	-	-
NSVF [26]	31.75	0.954	0.048^{alex}	-	-	-
Mip-NeRF [2]	33.09	0.961	0.043	-	-	-
DirectVoxGo [44]	31.95	0.957	0.053	-	-	15 mins
Plenoxels [54]	31.71	0.958	0.049	-	-	11 mins
Instant-NGP [32]	**33.17**	-	-	-	-	5 mins
SNeRG (PNG) [11]	30.38	0.950	0.071*	86.7	87.1	> 1 day
PlenOctree-512 (PNG) [55]	31.67	0.957	0.053	340.2	**129.8**	> 1 day
HrSRG (w/o Med. Cut)	31.06	0.954	0.051	210.4	110.2	6 hrs
HrSRG	31.04	0.954	0.051	**35.8**	113.4	6 hrs
Real 360° Objects@1080p						
NeRF [31]	26.42	0.911	0.103	-	0.01	> 1 day
SNeRG (PNG) [11]	28.14	0.902	0.107	139.7	52.8	> 1 day
PlenOctree-1024 (PNG) [55]	28.39	0.923	0.078	1112.0	**67.1**	> 1 day
HrSRG (w/o Med. Cut)	**29.07**	**0.927**	**0.059**	250.7	60.3	5 hrs
HrSRG	29.05	0.927	0.059	**41.2**	61.9	5 hrs

4 Experiments

Implementation Details. As discussed before, our approach contains a Def-NeRF pre-training stage and a HrSRG optimization stage. For Def-NeRF, we pre-cache all the training rays. Then, we train the vanilla NeRF for 5 epochs with a batch size of 8,192. As stated in Sect. 3.2, we take the pre-trained MLP of NeRF as a backbone, and fixed its parameters to learn the tiny specular MLP for another 5 epochs. The learning rate begins at 5×10^{-3} and linearly decays to 5×10^{-5} over the training epochs. For real scenes, we resize the 1080p images to 960×540 and train Def-NeRF with mixed precision. When training HrSRG, we randomly render a 384×384 image patch per iteration. We train HrSRG in

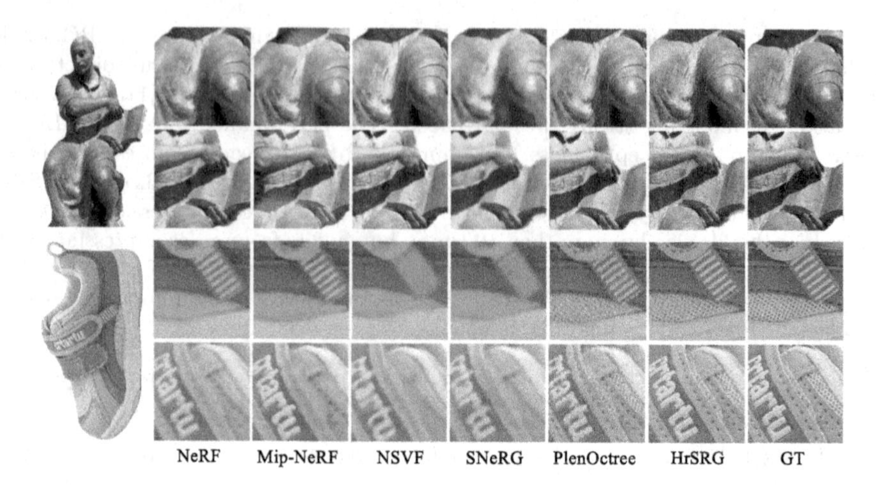

Fig. 4. Qualitative Comparisons. HrSRG can recover more fine details of scenes while improve the perceptual quality of the rendered images. See Table 1 and Table 2 for quantitative comparisons. Zoom in for a better view.

45 epochs and decay the learning rate from 1×10^{-2} to 1×10^{-5} over the course of optimization. We have not adopted the DDA (digital differential analyzer) algorithm [55] when optimizing HrSRG. We also incorporate a TV (total variation) loss as a regularization term. We use the Adam optimizer [17] for both the two training stages with default hyperparameters. It takes about 6 h (*vs.* 1–2 days for NeRF) on average to fitting a Synthetic-NeRF [31] scene. We leverage PlenOctree's codes [55] to implement the merging and splitting operations and the web renderer.

Datasets. We conduct experiments on three datasets, including Synthetic-NeRF [31], a subset of Tanks and Temples [18], and our Real 360° Objects dataset. Synthetic-NeRF contains eight objects with rendered 800×800 images, and the ground truth camera poses. There are 300 images per object, 100 for training and 200 for testing. We use the same Tanks and Temples subset as NSVF [26]. It provides five real scenes with 1080p images, their manually labeled masks, and known camera poses. To better study fine details of real scenes, we capture eight object-centric videos via a mobile phone to construct the Real 360° Objects dataset. Each object is with rich details in some regions. We select 300 1080p images per object, 100 for training and 200 for testing. We simply estimate the camera poses via COLMAP [38,39] without any post-processing. As discussed in [26] and [12], the camera pose errors are often unavoidable. Real 360° Objects may also help to investigate weather our approach is tolerant to the misalignment issue caused by camera pose errors or not.

Evaluation. We measure the render-time performance, training speed, storage cost, and rendering quality in our experiments. The rendering speed is evaluated

Table 3. Ablation Studies on the "Hat" case of Real 360° Objects. Def-NeRF-Plus is the converged Def-NeRF. HrSRG-minus is the diffuse version of HrSRG. HrSRG-N means we capture the HrSRG representation from a N^3 voxel grid (See Sect. 3.3).

Method	PSNR ↑	SSIM ↑	LPIPS ↓	MB ↓	FPS ↑
Def-NeRF-Plus	25.02	0.838	0.128	-	0.01
HrSRG (w/o $\mathcal{L}_p$)	26.43	0.894	0.061	48.0	59.7
HrSRG	26.64	0.895	0.056	48.0	59.7
HrSRG-minus	25.65	0.876	0.069	27.4	68.5
HrSRG-256	24.72	0.847	0.107	8.7	132.4
HrSRG-512	26.64	0.895	0.056	48.0	59.7

via a Lenovo ThinkPad P1 Gen 2 notebook with a 4GB NVIDIA Quadro T1000 mobile graphics card. We compute PSNR, SSIM [51], and LPIPS [57] as previously. The scores of the compared approaches are source (or computed) from their papers if provided. We make qualitative comparisons to showcase the texture details. We argue the LPIPS metric might measure the perceptual quality and fine details better.

4.1 Benchmark Comparisons

Compared Methods. We take SNeRG [11] and PlenOctree [55] as the baseline methods. Specifically, we adopt the "PlenOctree after fine-tuning" setting, which has been labeled as the complete model by the PlenOctree paper. On the Real 360° Objects dataset, we train PlenOctree at the voxel resolution of 1024^3. For SNeRG, we examine the further fine-tuned version. We also make comparisons with other represented NeRF variants such as NSVF [26], Mip-NeRF [2], DirectVoxGo [44], and [54]. For all the compared methods, we use their released codes (if not specified) and take care to follow their best configurations. On the synthetic and Tank and Temples datasets, the scores for NSVF are computed using the pre-trained models shared by KiloNeRF [37].

Performance. The scores are reported in Table 1 and Table 2. HrSRG achieves promising performance with respect to rendering quality and rendering speed, moreover requires less storage space than real-time NeRF baselines. Especially on the Real 360° Objects dataset, HrSRG outperforms the compared methods by a large margin over the image quality metric LPIPS. Some qualitative comparisons are presented in Fig. 1 and Fig. 4. HrSRG can produce high-quality rendering with rich fine details, while images rendered by other methods contain blurring artifacts. We find HrSRG yields slightly slower PSNR and SSIM on Synthetic-NeRF [31]. A possible reason is that the deferred NeRF structure [11] shows limitation to preserve strong specular effects. Furthermore, learning to represent a single scene takes around6 h and 5 h on Synthetic-NeRF and Real 360° Objects,

Fig. 5. A qualitative study of the impact of the key components. HrSRG-minus cannot produce reasonable renderings for specular materials. Zoom in to see the fine details.

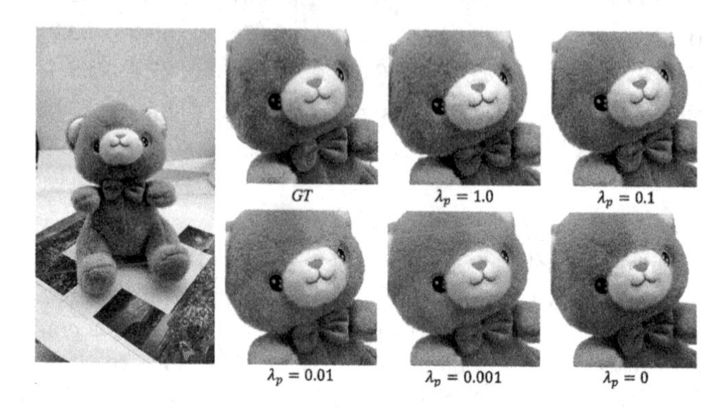

Fig. 6. Perceptual Loss. A very large λ_p yields incorrect colors (See nose and cravat of the "Bear") as the model would focus more on optimizing the perceptual quality. Zoom in to see the fine details.

respectively. Other methods take about 1 day to obtain a converged model. It is worth mentioning that our experiments show that fine-tuning a PlenOctree from a converged NeRF-SH cannot recover more fine details compared to NeRF-SH. We optimize HrSRG from a non-converged Def-NeRF. Overall, for real scenes, HrSRG surpasses the compared baselines [11,55] on the texture characteristic preservation ability and model size.

4.2 Ablation Studies

In Table 3 (*Top*) and Fig. 5, we discuss the core techniques of the HrSRG approach. We take Def-NeRF as the baseline. "Plus" means we train Def-NeRF until it has converged. Ablation "HrSRG (w/o $\mathcal{L}_p$)" shows (1) removing L_p has significant effect on the perceptual quality of rendered images; and (2) directly optimizing color values (*vs.* Def-NeRF-Plus) can impressively recover the fine details. We also study HrSRG-minus which removes the specular effects. We find it cannot handle specular materials. As analyzed in [11], a diffuse model may reasonably fake view-dependent effects by hiding mirrored versions of reflected content inside the objects' surfaces. Table 3 (*Bottom*) studies the HrSRG construction approach presented in Sect. 3.3. HrSRG-512 is the default HrSRG version in this paper. Moreover, we ablate τ_1 on the "Shoe" case. By setting τ_1 to 0.01, 0.3, 0.8, and 0.9, we capture models with sizes of 169.4M, 59.9M, 40.6M, and 35.5M, respectively. We find the threshold τ_1 does not have a significant impact on the rendering quality.

Figure 6 explores the impact on rendering quality of different λ_p for the perceptual loss (See Eq. 7). Our main target is to minimize Eq. 4, thus a very large λ_p yields inaccurate colors. A small λ_p could also help to recover more fine details (*vs.* HrSRG (w/o λ_p)), but has a minimal effect in improving the clarity of rendered images. We find that the perceptual loss does not has a significant affect on PSNR and SSIM, but does improving LPIPS.

5 Discussion and Limitation

SNeRG [11] prunes invalid voxels via a sparsity regularization term $\mathcal{L}_s$ and an opacity pruning (OP) policy. The OP approach culls macroblocks that the voxel visibilities (or maximum transmittance) are low. PlenOctree [55] also applies OP to reduce the memory footprint. However, our experiments show that enlarging the loss weight of $\mathcal{L}_s$ would easily filter some valid voxels out. As a result, a rendered image would show different degrees of transparency in some regions. Besides, OP with a very low threshold τ could also delete many voxels by mistakes. Examples are shown in the supplementary. Luckily, our 6-DoF viewing applications show HrSRG is robust to these issues.

One of the limitations of HrSRG is the slow training speed. It takes about 6 h to obtain a complete HrSRG model of a real scene on a V100 GPU. Though requiring several multiple orders of magnitude less time than conventional NeRFs, it is still not friendly for practical requirements. Taking E-commerce as an example, HrSRG can bring 6-DOF immersive viewing experiences of objects, but there are billions of goods to be reconstructed. We notice that several great concurrent works [32,44,54] have developed techniques leveraging the voxel grid representations that allows super-fast scene reconstruction. Another limitation is that HrSRG cannot well reconstruct the strong specular materials like other NeRFs. A recent formulation Ref-NeRF [49] may bring some inspirations.

6 Conclusion

We notice that previous NeRF works fail to preserve texture characteristics for real scenes. We dig into the radiance grid representation [11], a real-time paradigm of NeRF, and propose a set of improvements, which together produce surprising rendering quality and speed. In particular, we present a hierarchical voxel grid building approach that capture an Hierarchical Sparse Radiance Grid (HrSRG) to represent a scene. It has higher voxel resolution for informative spaces and fewer voxels for other spaces. Furthermore, taking inspiration from the 2D texture atlas success [12], we show that directly optimizing the color and specular feature of HrSRG could surprisingly reconstruct the fine details of scenes. Moreover, benefiting from our formulation, we introduce a perceptual loss to tackle the misalignment issues caused by camera pose errors. It can bring more details to rendered images while largely improving the perceptual quality. HrSRG enables rendering 1080p images at about 62 FPS in unprecedented quality on a notebook. A pruned HrSRG requires about 41M storage space to represent a scene.

References

1. Arandjelović, R., Zisserman, A.: Nerf in detail: learning to sample for view synthesis. arXiv preprint arXiv:2106.05264 (2021)
2. Barron, J.T., Mildenhall, B., Tancik, M., Hedman, P., Martin-Brualla, R., Srinivasan, P.P.: Mip-nerf: a multiscale representation for anti-aliasing neural radiance fields. In: ICCV (2021)
3. Bentley, J.L.: K-d trees for semidynamic point sets. In: Proceedings of the Sixth Annual Symposium on Computational Geometry, pp. 187–197 (1990)
4. Chen, A., et al.: Mvsnerf: fast generalizable radiance field reconstruction from multi-view stereo. In: ICCV (2021)
5. Davis, A., Levoy, M., Durand, F.: Unstructured light fields. In: Computer Graphics Forum, vol. 31, pp. 305–314. Wiley Online Library (2012)
6. Deng, B., Barron, J.T., Srinivasan, P.P.: JaxNeRF: an efficient JAX implementation of NeRF (2020). https://github.com/google-research/google-research/tree/master/jaxnerf
7. Gafni, G., Thies, J., Zollhofer, M., Nießner, M.: Dynamic neural radiance fields for monocular 4d facial avatar reconstruction. In: Proceedings of the IEEE/CVF Conference on Computer Vision and Pattern Recognition, pp. 8649–8658 (2021)
8. Garbin, S.J., Kowalski, M., Johnson, M., Shotton, J., Valentin, J.: Fastnerf: high-fidelity neural rendering at 200fps. arXiv preprint arXiv:2103.10380 (2021)
9. Gortler, S.J., Grzeszczuk, R., Szeliski, R., Cohen, M.F.: The lumigraph. In: Proceedings of the 23rd Annual Conference on Computer Graphics and Interactive Techniques, pp. 43–54 (1996)
10. Hedman, P., Philip, J., Price, T., Frahm, J.M., Drettakis, G., Brostow, G.: Deep blending for free-viewpoint image-based rendering. ACM Trans. Graph. (TOG) **37**(6), 1–15 (2018)
11. Hedman, P., Srinivasan, P.P., Mildenhall, B., Barron, J.T., Debevec, P.: Baking neural radiance fields for real-time view synthesis. In: ICCV (2021)
12. Huang, J., et al.: Adversarial texture optimization from rgb-d scans. In: CVPR, pp. 1559–1568 (2020)
13. Ichnowski, J., Avigal, Y., Kerr, J., Goldberg, K.: Dex-nerf: using a neural radiance field to grasp transparent objects. arXiv preprint arXiv:2110.14217 (2021)
14. Isola, P., Zhu, J.Y., Zhou, T., Efros, A.A.: Image-to-image translation with conditional adversarial networks. In: CVPR, pp. 1125–1134 (2017)
15. Johnson, J., Alahi, A., Fei-Fei, L.: Perceptual losses for real-time style transfer and super-resolution. In: Leibe, B., Matas, J., Sebe, N., Welling, M. (eds.) ECCV 2016. LNCS, vol. 9906, pp. 694–711. Springer, Cham (2016). https://doi.org/10.1007/978-3-319-46475-6_43
16. Kar, A., Häne, C., Malik, J.: Learning a multi-view stereo machine. arXiv preprint arXiv:1708.05375 (2017)
17. Kingma, D.P., Ba, J.: Adam: a method for stochastic optimization. arXiv preprint arXiv:1412.6980 (2014)
18. Knapitsch, A., Park, J., Zhou, Q.Y., Koltun, V.: Tanks and temples: benchmarking large-scale scene reconstruction. ACM Trans. Graph. (ToG) **36**(4), 1–13 (2017)

19. Kutulakos, K.N., Seitz, S.M.: A theory of shape by space carving. Int. J. Comput. Vis. **38**(3), 199–218 (2000). https://doi.org/10.1023/A:1008191222954
20. Larsen, A.B.L., Sønderby, S.K., Larochelle, H., Winther, O.: Autoencoding beyond pixels using a learned similarity metric. In: ICML, pp. 1558–1566. PMLR (2016)
21. Ledig, C., et al.: Photo-realistic single image super-resolution using a generative adversarial network. In: CVPR, pp. 4681–4690 (2017)
22. Levoy, M., Hanrahan, P.: Light field rendering. In: Proceedings of the 23rd Annual Conference on Computer Graphics and Interactive Techniques, pp. 31–42 (1996)
23. Li, Z., Niklaus, S., Snavely, N., Wang, O.: Neural scene flow fields for space-time view synthesis of dynamic scenes. In: Proceedings of the IEEE/CVF Conference on Computer Vision and Pattern Recognition, pp. 6498–6508 (2021)
24. Lin, C.H., Ma, W.C., Torralba, A., Lucey, S.: Barf: bundle-adjusting neural radiance fields. In: IEEE International Conference on Computer Vision (ICCV) (2021)
25. Lindell, D.B., Martel, J.N., Wetzstein, G.: Autoint: automatic integration for fast neural volume rendering. In: Proceedings of the IEEE/CVF Conference on Computer Vision and Pattern Recognition, pp. 14556–14565 (2021)
26. Liu, L., Gu, J., Lin, K.Z., Chua, T.S., Theobalt, C.: Neural sparse voxel fields. In: NeurIPS (2020)
27. Lombardi, S., Simon, T., Saragih, J., Schwartz, G., Lehrmann, A., Sheikh, Y.: Neural volumes: learning dynamic renderable volumes from images. arXiv preprint arXiv:1906.07751 (2019)
28. Max, N.: Optical models for direct volume rendering. IEEE TVCG **1**(2), 99–108 (1995)
29. Meagher, D.: Geometric modeling using octree encoding. Comput. Graphics Image Process. **19**(2), 129–147 (1982)
30. Mescheder, L., Oechsle, M., Niemeyer, M., Nowozin, S., Geiger, A.: Occupancy networks: Learning 3d reconstruction in function space. In: Proceedings of the IEEE/CVF Conference on Computer Vision and Pattern Recognition, pp. 4460–4470 (2019)
31. Mildenhall, B., Srinivasan, P.P., Tancik, M., Barron, J.T., Ramamoorthi, R., Ng, R.: Nerf: representing scenes as neural radiance fields for view synthesis. In: ECCV, pp. 405–421. Springer (2020). https://doi.org/10.1145/3503250
32. Müller, T., Evans, A., Schied, C., Keller, A.: Instant neural graphics primitives with a multiresolution hash encoding. arXiv preprint arXiv:2201.05989 (2022)
33. Neff, T., et al.: Donerf: towards real-time rendering of compact neural radiance fields using depth oracle networks. arXiv preprint arXiv:2103.03231 (2021)
34. Ost, J., Mannan, F., Thuerey, N., Knodt, J., Heide, F.: Neural scene graphs for dynamic scenes. In: Proceedings of the IEEE/CVF Conference on Computer Vision and Pattern Recognition, pp. 2856–2865 (2021)
35. Park, K., et al.: Deformable neural radiance fields. arXiv preprint arXiv:2011.12948 (2020)
36. Rebain, D., Jiang, W., Yazdani, S., Li, K., Yi, K.M., Tagliasacchi, A.: Derf: decomposed radiance fields. In: Proceedings of the IEEE/CVF Conference on Computer Vision and Pattern Recognition, pp. 14153–14161 (2021)
37. Reiser, C., Peng, S., Liao, Y., Geiger, A.: Kilonerf: speeding up neural radiance fields with thousands of tiny mlps. arXiv preprint arXiv:2103.13744 (2021)
38. Schönberger, J.L., Frahm, J.M.: Structure-from-motion revisited. In: Conference on Computer Vision and Pattern Recognition (CVPR) (2016)
39. Schönberger, J.L., Zheng, E., Frahm, J.-M., Pollefeys, M.: Pixelwise view selection for unstructured multi-view stereo. In: Leibe, B., Matas, J., Sebe, N., Welling, M.

(eds.) ECCV 2016. LNCS, vol. 9907, pp. 501–518. Springer, Cham (2016). https://doi.org/10.1007/978-3-319-46487-9_31
40. Seitz, S.M., Dyer, C.R.: Photorealistic scene reconstruction by voxel coloring. Int. J. Comput. Vis. **35**(2), 151–173 (1999). https://doi.org/10.1023/A:1008176507526
41. Simonyan, K., Zisserman, A.: Very deep convolutional networks for large-scale image recognition. In: ICLR (2015)
42. Sitzmann, V., Chan, E.R., Tucker, R., Snavely, N., Wetzstein, G.: Metasdf: meta-learning signed distance functions. arXiv preprint arXiv:2006.09662 (2020)
43. Sitzmann, V., Thies, J., Heide, F., Nießner, M., Wetzstein, G., Zollhofer, M.: Deepvoxels: learning persistent 3d feature embeddings. In: Proceedings of the IEEE/CVF Conference on Computer Vision and Pattern Recognition, pp. 2437–2446 (2019)
44. Sun, C., Sun, M., Chen, H.T.: Direct voxel grid optimization: super-fast convergence for radiance fields reconstruction. arXiv preprint arXiv:2111.11215 (2021)
45. Tancik, M., et al.: Learned initializations for optimizing coordinate-based neural representations. In: Proceedings of the IEEE/CVF Conference on Computer Vision and Pattern Recognition, pp. 2846–2855 (2021)
46. Thies, J., Zollhöfer, M., Nießner, M.: Deferred neural rendering: image synthesis using neural textures. ACM Trans. Graph. (TOG) **38**(4), 1–12 (2019)
47. Trevithick, A., Yang, B.: Grf: learning a general radiance field for 3d scene representation and rendering. arXiv:2010.04595 (2020)
48. Tulsiani, S., Zhou, T., Efros, A.A., Malik, J.: Multi-view supervision for single-view reconstruction via differentiable ray consistency. In: Proceedings of the IEEE Conference on Computer Vision and Pattern Recognition, pp. 2626–2634 (2017)
49. Verbin, D., Hedman, P., Mildenhall, B., Zickler, T., Barron, J.T., Srinivasan, P.P.: Ref-nerf: Structured view-dependent appearance for neural radiance fields. arXiv preprint arXiv:2112.03907 (2021)
50. Wang, Q., et al.: Ibrnet: learning multi-view image-based rendering. In: CVPR, pp. 4690–4699 (2021)
51. Wang, Z., Bovik, A.C., Sheikh, H.R., Simoncelli, E.P.: Image quality assessment: from error visibility to structural similarity. IEEE Trans. Image Process. **13**(4), 600–612 (2004)
52. Wang, Z., Wu, S., Xie, W., Chen, M., Prisacariu, V.A.: NeRF−: neural radiance fields without known camera parameters. arXiv preprint arXiv:2102.07064 (2021)
53. Wei, Y., Liu, S., Rao, Y., Zhao, W., Lu, J., Zhou, J.: Nerfingmvs: guided optimization of neural radiance fields for indoor multi-view stereo. In: Proceedings of the IEEE/CVF International Conference on Computer Vision, pp. 5610–5619 (2021)
54. Yu, A., Fridovich-Keil, S., Tancik, M., Chen, Q., Recht, B., Kanazawa, A.: Plenoxels: radiance fields without neural networks. arXiv preprint arXiv:2112.05131 (2021)
55. Yu, A., Li, R., Tancik, M., Li, H., Ng, R., Kanazawa, A.: PlenOctrees for real-time rendering of neural radiance fields. In: ICCV (2021)
56. Yu, A., Ye, V., Tancik, M., Kanazawa, A.: pixelnerf: neural radiance fields from one or few images (2020)
57. Zhang, R., Isola, P., Efros, A.A., Shechtman, E., Wang, O.: The unreasonable effectiveness of deep features as a perceptual metric. In: Proceedings of the IEEE Conference on Computer Vision and Pattern Recognition, pp. 586–595 (2018)

Author Index

Printed in the United States
by Baker & Taylor Publisher Services